Resilient Health

Resilient Health

Leveraging Technology and Social Innovations to Transform Healthcare for COVID-19 Recovery and Beyond

Edited by

Judy Kuriansky

Teachers College, Columbia University, New York, NY, United States
Office of Ambassador Sidique Wai, Embassy of Sierra Leone to the United States, Washington, DC, United States
Mission of Sierra Leone to the United Nations, New York, NY, United States
Global Council on SDG3 (Health), "SDGs in Action" Initiative, Dubai, United Arab Emirates
International Association of Applied Psychology, Indianapolis, IN, United States
World Council for Psychotherapy, Vienna, Austria
United African Congress, New York, NY, United States
Global Leadership Council, eMental Health International Collaborative, Auckland, New Zealand
United Nations Group of Friends of Mental Health and Well-being, New York, NY, United States
International Association of Women in Radio and TV, New York, NY, United States
Black Star News, New York, NY, United States
Black Tie Magazine, New York, NY, United States
Stand Up for Peace Project, New York, NY, United States
Global Kids Connect Project, New York, NY, United States
International Association of Women in Radio and TV, Quezon City, Philippines
eMental Health International Collaborative (eMHIC), Auckland, New Zealand

Pradeep Kakkattil

Founder and CEO, Health Innovation and Investment Exchange (HIEx), Geneva, Switzerland
Co-Founder, Women in Innovation Fund (WiNFUND), Geneva, Switzerland
Co-Founder, Climate Health Innovation and Equity Fund, Geneva, Switzerland

ELSEVIER

ACADEMIC PRESS

An imprint of Elsevier

Academic Press is an imprint of Elsevier
125 London Wall, London EC2Y 5AS, United Kingdom
525 B Street, Suite 1650, San Diego, CA 92101, United States
50 Hampshire Street, 5th Floor, Cambridge, MA 02139, United States

ISBN: 978-0-443-18529-8

For information on all Academic Press publications
visit our website at https://www.elsevier.com/books-and-journals

Printed in the United States of America
Last digit is the print number: 9 8 7 6 5 4 3 2 1

Publisher: Stacy Masucci
Acquisitions Editor: Linda Buschman
Editorial Project Manager: Kathy Padilla
Production Project Manager: Sajana Devasi PK
Cover Designer: Christian Bilbow

Typeset by STRAIVE, India

With the deepest appreciation, this volume is dedicated to the innovators, health providers, policymakers, funders, all stakeholders, and the people themselves in need of physical and mental healthcare throughout the world, working together to improve healthcare delivery in this ongoing COVID-19 pandemic recovery period to "build forward *together." It is the three sectors—public, private, and the people—joining in partnership for health that presages better well-being for all now and for the future.*

Contents

PART 3: The role of stakeholders

PART 5: The innovation toolbox: Special populations and interest groups

† Deceased.

Biographies of the coeditors

Dr. Judy Kuriansky, Ph.D., is an internationally known clinical psychologist who has been a Professor of Psychology and Education at Columbia University Teachers College on the graduate adjunct faculty for 20+ years where she teaches about global mental health policy and programs in the course on "*Psychology and the United Nations*" and founded the GAAP Lab on Global Advocacy and Applied Psychology. For years, she was an Honorary Professor at the Beijing University Health Sciences Center in China and the Department of Psychiatry at the University of Hong Kong. With her extensive experience as a program innovator, academician, advocate, and researcher, she lectures and does trainings and workshops for diverse groups worldwide.

At the United Nations, she is Advisor to the Mission of Sierra Leone to the United Nations and Advisor and co-founder of the United Nations Group of Friends of Mental Health and Well-being. An avid advocate about mental health and well-being, she has been the main NGO Representative for over 20 years of the International Association of Applied Psychology (IAAP) and the World Council for Psychotherapy, which are in consultative status with the United Nations Economic and Social Council and in association with the Department of Global Communications. When she was President of the Psychology Coalition of NGOs, she partnered with Ambassador Dr. Caleb Otto of the Mission of Palau to the UN in historic successful advocacy with UN Member States and delegates during negotiations for the UN Agenda 2030 for Sustainable Development to secure that the target about promoting mental health and well-being was included in the Goal about health for all, and that the word "wellbeing" was in the title. She further advocated to successfully secure that mental health was included in the Political Declaration for Universal Health Coverage, and that references to psychology and counseling were included in the Global Compact for Safe, Orderly and Regular Migration, and she led a team of her students to successfully support the Mission of Mexico in the adoption of the first-ever resolution in the General Assembly of Mental Health and Psychosocial Support. Also at the UN, she co-organizes and moderates innumerable events including for the

Commission on the Status of Women, the Commission on Social Development, World Interfaith Harmony Week and the International Day of Happiness. Many of these events are co-sponsored by the United African Congress, for which she serves as a Trustee.

Dedicated to development of the African continent, she has implemented many projects including a Girls Empowerment Camp in Lesotho, an Ebola community resilience recovery project in Sierra Leone, and a COVID-19 pandemic recovery and prevention program in the Democratic Republic of the Congo, as well as hosted events like the *"U.S.-Africa Business Forum."* Currently, she serves as Policy Advisor to His Excellency Ambassador Sidique Wai, Chief of Mission of the Embassy of the Republic of Sierra Leone to the United States in Washington, DC.

A humanitarian, her many other projects include collaborations about resilience trainings in the Ukraine; youth, climate change and migration in the Pacific Islands; and education systems in Africa and the Caribbean. She has been invited to take leadership roles in many international fora, including representing civil society contributions to innovative financing at the Tbilisi International Solidarity and Innovative Financing Forum, presenting plenary addresses for the Professional Psychology League in Novisibirsk, and final declarations for the Union of Mental Health Congresses in Moscow. In extensive work and advocacy towards the achievement of the UN Agenda, she is past co-chair of the Global Council on SDG3 (Health) as part of the United Arab Emirates *"SDGs in Action"* initiative, for which she has presented several times at the World Government Summit in Dubai and co-founded the *"Health in Your Hands"* platform to convene stakeholders to showcase health innovations to reach those furthest behind with little access to health and mental health care especially in remote communities, now part of the *Health Innovation and Investment Exchange (HIEx).* A member of the Global Leadership Council of the eMental Health International collaborative (eMHIC) and Fellow of the American Psychological Association, she co-wrote the first policy on mental health and climate change for the American Public Health Association and is a consultant for the IAAP Global Task Force on Terrorism and Peacebuilding.

An award-winning journalist in international mass media as well as professional fields, and pioneer of "media psychology," she has written columns for magazines and newspapers worldwide, including the Singapore Straits Times, the South China Morning Post and New York Daily News, and currently for Black Star News and BlackTie Magazine, as well as innumerable articles published in professional journals. Her many books range from topics about personal life and relationships, including *"The Complete Guide to A Healthy Relationship"* and *"Self Esteem to Increase Confidence for Adolescents"* to international issues like *"Living in an Environmentally Traumatized World: Healing Ourselves and the Planet," "The Psychosocial Aspects of a Deadly Epidemic: What Ebola Has Taught Us about Holistic Healing" "The Changing Lives of Women Around the World: Psychosocial Perspectives and Insights"* and *"Ecopsychology: The Intersection of Psychology and Environmental Protection:*

Intervention and Policy" besides this volume about technology innovation post-COVID 19 and a newly released co-edited volume on "*Psychosocial Experiences and Adjustment of Migrants—Coming to America.*" A pioneer of sexuality therapy, she also has published books on this topic, which are translated into Chinese, Japanese and other languages. A frequent expert guest on international media, she has been a feature reporter for WCBS-TV, WABC-TV and many others; host of a former show on *"Money and Emotions"* on CNBC-TV; and commentator on "*ShowBiz Tonight*" formerly on HLN-TV. A pioneer of radio call-in advice talk, her popular top-rated show aired weeknights for decades on WABC radio and then on Z100 in New York which was syndicated in the U.S. and Japan, where she addressed thousands of questions from callers many of whom thank her to this day for improving their lives. A co-founder of the N.Y. chapter of the International Association of Women in Radio and TV, she is also on the Board of Directors of the Library of American Broadcasting, and a member of the League of Professional Theatre Women.

An expert in traumatology and founder of the *"Global Kids Connect Project,"* she provides psychosocial support trainings and workshops worldwide using her unique innovative techniques and methodologies, including after natural disasters including the Japan tsunami, China and Haiti earthquakes, Caribbean floods; after school shootings; immediately after terrorism attacks on the World Trade Center and in the Middle East; and during epidemics like HIV/AIDS, SARS, Ebola and the COVID-19 pandemic. Co-founder of the *"Stand Up for Peace Project,"* she and her music partner, Russell Daisey, write original anthems that accompany these workshops and events about global issues, which they perform worldwide, including *"Every Woman Every Child"* to coincide with the UN campaign of the same name; *"Honor Our Girls"* that supports the First Lady of Sierra Leone Fatima Maada Bioa's #HandsOffOurGirls campaign; *"One in Faith"* for the UN World Interfaith Harmony Week, and *"Hope is Alive"* featured in this volume and sung for many groups who need healing. Previously she played in a rock band called Artemis. Among her many awards include the *"Global Stakeholder Gender World" "Humanitarian Award for Lifetimes Achievement in Global Peace and Tolerance" "Lifetime Achievement from Society for Media Psychology and Technology of the American Psychological Association"* and the *"Grand Commander of the Order of the Rokel"* bestowed upon her by the President of Sierra Leone for her service in the country throughout many years and traumas including HIV/AIDS, Ebola, explosions, and the COVID-19 pandemic.

In response to COVID-19, she and her colleagues at the United African Congress were the first to raise the alarm about the impending epidemic, during an event they organized for World Interfaith Harmony Week at the UN, on their theme of Building Bridges Across Borders. She subsequently participated in and co-organized innumerable workshops and trainings about psychosocial resilience and coping with the crisis including for the NGO Commission on the Status of Women meetings and for many country-focused trainings, including for Nepal, Russia and China psychological and public audiences about healing.

A native of New York, she graduated from Smith College, studied at L'Université de Genève, earned a Masters degree at Boston University, and her Ph.D. in clinical psychology from New York University. She was a Senior Research Scientist at the New York State Psychiatric Institute Biometrics Unit and the U.S.-U.K. Diagnostic Project; taught Columbia University medical students and in the liaison services; and spent 20 years training medical doctors in psychological techniques throughout China.

She comes from a family of service and health professionals: her father was a dentist and Colonel in the U.S. Army; her mother a physical education teacher; her brother a doctor and her sister a social worker. Creativity and innovation have always driven her work "to make a difference in the world," a dream since childhood which she has been fulfilling and now teaches her many students.

Pradeep Kakkattil is the founder and CEO of the Health Innovation and Investment Exchange (HIEx), a multilayered platform connecting governments, policymakers, innovators, and investors to leverage innovations and technology toward scaling up access to healthcare. He has worked on health and development over the past three decades including with the Joint United Nations Programme on HIV/AIDS (UNAIDS) based in Geneva, Switzerland, where he established and led the Office of Innovation. He co-founded the Women in Innovation Fund that advocates for equitable access to financing for women led health enterprises, especially in low and middle income countries. Given the increasing impact of climate on health, especially on women and the most vulnerable populations, he has advocated and co-founded the climate health innovation and equity fund that leverages and invests in climate resilient health technologies and innovations. A former journalist and filmmaker, Pradeep serves on the Advisory Boards of the Graduate Institute in Geneva and is a founding member of the Board of the Geneva Innovation Network. Additionally, he is a member of the Global Council on Health as part of the United Arab Emirates' "SDGs in Action" initiative, member of the African Union Digital Health Strategy Taskforce, and a member of the Science and Innovation Accelerator working group hosted by the World Health Organization (WHO). He is a recipient of the Schwab Foundation/WEF Public Intrapreneur 2022 award.

Contributors

123Dentist Headquarters of 123Dentist, Burnaby, BC, Canada

Marjorie Abalos CLODS Inc., New York, NY, United States

Suad Hashim Ahmed Specialized Program Services Department (SPSD), Primary Health Care Sector (PHCS), Dubai Health, Dubai, United Arab Emirates

Reem Al Hammadi Federal Competitiveness and Statistics Centre, Dubai, United Arab Emirates

Asma Al Obeidli Ministry of Health and Prevention, Dubai, United Arab Emirates

Hussain Abdul Rahman Al Rand Ministry of Health and Prevention, Dubai, United Arab Emirates

Ola Aldafrawy Specialized Program Services Department (SPSD), Primary Health Care Sector (PHCS), Dubai Health, Dubai, United Arab Emirates

Alia Mohammad Rafi Aldallal Specialized Program Services Department (SPSD), Primary Health Care Sector (PHCS), Dubai Health, Dubai, United Arab Emirates

Saqr Alhemeiri Ministry of Health and Prevention, Dubai, United Arab Emirates

Aisha Alhouti Information Science Department, College of Life Sciences, Kuwait University, Kuwait City, Kuwait

Dari Alhuwail Information Science Department, College of Life Sciences, Kuwait University; Health Informatics Unit, Dasman Diabetes Institute, Kuwait City, Kuwait

Paula Aljovin Mental Health Team, Vine Trust Amazon Hope Medical Programme, Loreto, Peru; Department of Clinical Psychology, Teachers College, Columbia University, New York, NY; International Association of Applied Psychology, Indianapolis, IN, United States

Pedro Altungy International Association of Applied Psychology, Indianapolis, IN, United States; Universidad Complutense de Madrid (UCM), Madrid, Spain

Tina C. Ambos Geneva School of Economics and Management (GSEM), University of Geneva, Geneva, Switzerland

Taavi Annus Ministry of Social Affairs, Tallinn, Estonia

Logan Ansell reach52, Singapore, Singapore

Maxwell Antwi PharmAccess Foundation, Accra, Ghana

Alex Israel Attachey PharmAccess Foundation, Accra, Ghana

Rifat Atun Health Systems Innovation Lab, Harvard University, Boston, MA, United States

Sulzhan Bali World Bank, Washington, DC, United States

Erin Barry International Association of Applied Psychology, Indianapolis, IN; Department of Clinical Psychology, Teachers College, Columbia University, New York, NY, United States

Daniela Bas SustainABILITY Advisory Boutique, Inc., Gredo, Italy

Christoph Benn Joep Lange Institute, Geneva, Switzerland

Mohini Bhavsar Living Goods, Nairobi, Kenya; Grand Challenges Canada, Toronto, ON, Canada

Sripriya Bikkani International Association of Applied Psychology, Indianapolis, IN; Department of Clinical and Counseling Psychology, Teachers College, Columbia University, New York, NY, United States

Nicole K. Bulanchuk Department of Clinical Psychology, Teachers College, Columbia University; Believe New York Philanthropies, Inc., New York, NY; International Association of Applied Psychology, Indianapolis, IN, United States

Craig Burgess World Health Organization, Geneva, Switzerland

Bo Cao Department of Psychiatry, Faculty of Medicine and Dentistry, University of Alberta, Edmonton, AB, Canada

Mariam Claeson Global Public Health, Karolinska Institutet, Solna, Sweden

Christiaan Coetzee Federal Competitiveness and Statistics Centre, Dubai; Etihad Credit Insurance, Abu Dhabi, United Arab Emirates

Patrick F. Connolly MED-EL Medical Electronics, Innsbruck, Austria

Alethea Cook World Bank, Washington, DC, United States

Jennifer Crouch VillageReach, Seattle, WA, United States

Lindsey Cunningham Department of International & Comparative Education, Teachers College, Columbia University, New York, NY, United States

Marija Davcheva International Association of Applied Psychology, Indianapolis, IN, United States; Research Institute of Personnel Psychology, Organizational Development and Quality of Working Life (IDOCAL), University of Valencia, Valencia, Spain

Juan Ramón de la Fuente Permanent Mission of Mexico to the United Nations, New York, NY, United States

Meredith del Pilar-Labarda Medical Department, School of Health Sciences, University of the Philippines; Social Innovation in Health Initiative Secretariat, Manila, Philippines

Florence L. Denmark Department of Psychology, Pace University, New York, NY, United States

Madhu Deshmukh Program Strategies & Impact CARE, Atlanta, GA, United States

Deepika Devadas Every Woman Every Child Innovation Marketplace, Grand Challenges Canada, Toronto, ON, Canada

Roopa Dhatt Women in Global Health, Hayward, CA, United States; Global Council on SDG 3 - Good Health and Wellbeing, "SDGs in Action" Initiative, Dubai, United Arab Emirates; Georgetown University, Washington, DC, United States

Daniella Diaz Department of Clinical Psychology, Teachers College, Columbia University, New York, NY; International Association of Applied Psychology, Indianapolis, IN, United States

Omar ElFata P&P Patents and Technologies, Rome, Italy

Pauline Eluère United Nations High Commissioner for Refugees (UNHCR), Geneva, Switzerland

Kameshwar Eranki† Proactiff Digital Healthcare Services Inc., Newark, CA, United States

Temitayo Erogbogbo MSD for Mothers, Lucerne, Switzerland

Satoshi Ezoe Global Health Strategy Division, Ministry of Foreign Affairs, Government of Japan; Department of Global Health Policy, Graduate School of Medicine, The University of Tokyo, Tokyo, Japan

Erin K. Ferenchick Proximie, London, United Kingdom

Rafael Figueroa Portal Telemedicine, São Paulo, Brazil

Brianna Fitapelli Department of Clinical and Counseling Psychology, Teachers College, Columbia University, New York, NY; International Association of Applied Psychology, Indianapolis, IN, United States

Nathaniel Foote citiesRISE, New York, NY, United States; TruePoint, Lincoln, MA, United States

Cameron Fox Platform for Shaping the Future of Health & Healthcare, World Economic Forum, New York, NY, United States

Maria Inês Francisco Viva International Federation of Medical Students' Associations (IFMSA), Copenhagen, Denmark

Christian Garbe Frankfurt Innovation Center Biotechnology GmbH, Frankfurt am Main, Germany

Beatrice Gatumia Busara Center for Behavioral Economics, Nairobi, Kenya

Kamal Gautam Transcultural Psychosocial Organization Nepal (TPO Nepal), Kathmandu, Nepal

Amandeep S. Gill Graduate Institute; I-DAIR (International Digital Health and AI Research Collaborative), Geneva, Switzerland

Lasha Goguadze International Federation of Red Cross and Red Crescent Societies (IFRC), Chemin des Crêts 17, Geneva, Switzerland

Andrew J. Greenshaw Department of Psychiatry, Faculty of Medicine and Dentistry, University of Alberta, Edmonton, AB, Canada

Russell Greiner Department of Psychiatry, Faculty of Medicine and Dentistry, University of Alberta; Department of Computing Science, University of Alberta; Amii (Alberta Machine Intelligence Institute), Edmonton, AB, Canada

Monique Arantes Guimarães International Association of Applied Psychology, Indianapolis, IN, United States; Universidade Sao Francisco (USF, Brazil), Sao Paulo, Brazil

Aline Guzik Permanent Mission of Mexico to the United Nations, New York, NY, United States

Nadine Hachach-Haram Proximie; Department of Plastic and Reconstructive Surgery, Guys and St Thomas Hospital, Westminster Bridge Road, London, United Kingdom

Ann Herrmann-Nehdi Herrmann International, Forest City, NC, United States

Taylor Hirschberg Hearst Endowed Scholar and Visiting Research at Columbia University Global Centers in Istanbul Program on Forced Migration and Health, Heilbrunn Department of Population and Family Health, Columbia University Mailman School of Public Health, New York, NY, United States

Matthew Hughsam citiesRISE, New York, NY, United States

Geoffrey Ibbotson Global Surgery Foundation; United Nations Institute for Training and Research (UNITAR), Geneva, Switzerland

†Deceased.

Tara Imalingat Kenya Legal & Ethical Issues Network on HIV and AIDS (KELIN), Nairobi, Kenya

Maria Isabel Iñigo Petralanda Ethics, Privacy and Security Chapter of the Central American Network for Health Informatics (RECAINSA), City of Buenos Aires, Argentina

Kiyan Irani Department of Clinical and Counseling Psychology, Teachers College, Columbia University, New York; International Association of Applied Psychology, Indianapolis, IN, United States

Biju Jacob Health Innovation and Investment Exchange (HIEx), Geneva, Switzerland

Zsuzsanna Jakab Global Council SDG3 (Health), "SDGs in Action" Initiative, Dubai, United Arab Emirates; World Health Organization, Geneva, Switzerland

Gillian Javetski Dimagi, Inc., Cambridge, MA, United States

Preethi John Global Business School for Health, Faculty of Population Health Sciences, University College London (UCL), London, United Kingdom; Women in Global Health India Chapter, Delhi, India

Mutahi Kagwe Cabinet for Health, Nairobi, Kenya

Pradeep Kakkattil Global Council SDG3 (Health), "SDGs in Action" Initiative, Dubai, United Arab Emirates; Health Innovation and Investment Exchange (HIEx), Geneva, Switzerland

Mwenya Kasonde World Health Organization, Geneva, Switzerland

Kanishka Katara Living Goods, Nairobi, Kenya

Shariha Khanam Khalid Mission & Co, Kuala Lumpur, Malaysia

Sadikchhya (Meera) Khanal Russell Reynolds Associates, Dallas, TX; Department of Clinical Psychology, Teachers College, Columbia University, New York, NY; International Association of Applied Psychology, Indianapolis, IN, United States

Rana Khazbak School of Education, Communication and Society, King's College London, London, United Kingdom

Eva Kiegele World Health Organization Office at the United Nations, New York, NY, United States

Sarah Kline United For Global Mental Health, London, United Kingdom

Jennifer Knox Woodward Academy, College Park, GA, United States

Pitambar Koirala Transcultural Psychosocial Organization Nepal (TPO Nepal), Kathmandu, Nepal

Rotem Kopel MindMatters Foundation, Geneva, Switzerland

Eneyi E. Kpokiri Clinical Research Department, Faculty of Infectious and Tropical Diseases, London School of Hygiene and Tropical Medicine; International Diagnostics Centre, Faculty of Infectious and Tropical Diseases, London School of Hygiene and Tropical Medicine, London, United Kingdom

Suresh Kumar citiesRISE, Chennai, India

Judy Kuriansky Department of Clinical Psychology, Teachers College, Columbia University, New York, NY, United States; Global Council SDG3 (Health), "SDGs in Action" Initiative, Dubai, United Arab Emirates; International Association of Applied Psychology, Indianapolis, IN; United African Congress, New York, NY; Global Kids Connect Project, New York, NY; Global Leadership Council, eMental Health International Collaborative (eMHIC), Auckland, New Zealand; Office of Ambassador Sidique Wai, Embassy of the Republic of Sierra Leone to the United States, Washington, DC; Mission of the Republic of Sierra Leone to the United Nations, New York, NY; United Nations Group of Friends of Mental Health and Well-being, New York, NY; World Council for Psychotherapy, Vienna, Austria; Black Tie Magazine, New York, NY; Black Star News, New York, NY, United States;

International Women in Radio and TV, Quezon City, Philippines; Office of Ambassador Sidique Wai, Embassy of the Republic of Sierra Leone to the United States, Washington, DC, United States

Erica Layer HealthEnabled, Washington, DC, United States

Tao Li The Mental Health Center, West China Hospital, Sichuan University, Chengdu; Hangzhou Seventh People's Hospital & Affiliated Mental Health Center, Zhejiang University School of Medicine, Hangzhou, People's Republic of China

Sara Liébana International Association of Applied Psychology, Indianapolis, IN, United States; Universidad Europea of Madrid (UEM), Madrid, Spain

Hanna Lissinna Department of Clinical Psychology, Teachers College Columbia University, New York, NY; International Association of Applied Psychology, Indianapolis, IN, United States

Sonia Livingstone Department of Media and Communications, London School of Economics and Political Science (LSE), London, United Kingdom

David D. Luxton Department of Psychiatry & Behavioral Sciences, University of Washington School of Medicine, Seattle; Luxton Labs, LLC, Olympia, WA, United States

Catherine Machalaba EcoHealth Alliance, New York, NY, United States

Haifa Madi Ministry of Health and Prevention, Dubai, United Arab Emirates

Tirumala Santra Mandal iKure Techsoft Pvt. Ltd, Bengaluru, India

Carthi Mannikarottu Civil Society Engagement Mechanism for UHC2030 (CSEM), co-hosted by WACI Health, Nairobi, Kenya; Platform for Access to COVID-19 Tools (ACT)-Accelerator Civil Society and Community Representatives, co-led by WACI Health, Global Fund Advocates Network (GFAN) and STOPAIDS

Cherisse Mark MED-EL Medical Electronics, Innsbruck, Austria

Patricio V. Marquez World Bank Group, Washington, DC, United States; Centre for Health Economics, The University of the West Indies, Port of Spain, Trinidad and Tobago; Johns Hopkins University Bloomberg School of Public Health, Baltimore, MD, United States; Charles Darwin Foundation for the Conservation of the Galapagos Islands, Puerto Ayora, Ecuador

Stéphanie Racine Maurice International Association of Applied Psychology, Indianapolis, IN, United States; Research and Intervention Program, Laval University, Quebec City, QC, Canada

Anita M. McGahan Joseph L. Rotman School of Management; University of Toronto, Toronto, ON, Canada; Institute for Strategy and Competitiveness, Harvard University, Cambridge, MA, United States

Patricia N. Mechael HealthEnabled, Washington, DC; Johns Hopkins Bloomberg School of Public Health, Baltimore, MD, United States

Tanya Mehdizadeh International Association of Applied Psychology, Indianapolis, IN; Department of Clinical and Counseling Psychology, Teachers College, Columbia University, New York, NY, United States

Sylvia Paola Mendoza Permanent Mission of Mexico to the United Nations, New York, NY, United States

Kertti Merimaa Ministry of Social Affairs, Tallinn, Estonia

Tahila Mintz OJI:SDA' Sustainable Indigenous Futures, Haudenosaunee Territory, Spencer, NY, United States

Nomtika Mjwana The Global Network of People Living with HIV (GNP+), Cape Town, South Africa

Taylor Mulligan-Stark Department of Psychology, Pace University; McShane Center for Psychological Services, Pace University; The NGO Committee on Ageing, New York, NY, United States

Vivek Nair Aerospace Engineering, University of Illinois, Champaign, IL, United States

Jasmine M. Noble Department of Psychiatry, Faculty of Medicine and Dentistry, University of Alberta; Department of Computing Science, University of Alberta; Amii (Alberta Machine Intelligence Institute), Edmonton, AB, Canada

Mohammed Nurhussein State University of New York Downstate Medical Center, Brooklyn, NY; Ethio-American Doctors Group, Washigton, DC; United Afrian Congress, New York, NY, United States

Ian O'Donnell International Federation of Red Cross and Red Crescent Societies (IFRC), Chemin des Crêts 17, Geneva, Switzerland

Patty O'Hayer Reckitt, Slough, United Kingdom

Werner Obermeyer World Health Organization Office at the United Nations, New York, NY, United States

Zerin Osho Institute for Governance & Sustainable Development, Washington, DC, United States

Daniel Otzoy-García Central American Network for Health Informatics (RECAINSA)

Amanda Pain VillageReach, Seattle, WA, United States

Mauro Pantaleo P&P Patents and Technologies, Rome, Italy

Muhammad Ali Pate Federal Ministry of Health and Social Welfare, Abuja, Nigeria; Global Council SDG3 (Health), "SDGs in Action" Initiative, Dubai, United Arab Emirates; Practice of Global Health Leadership, Department of Global Health and Population, Harvard University, Cambridge, MA, United States

William C. Philbrick Sitara International, Atlanta, GA, United States

Geetha Krishnan Gopalakrishna Pillai Traditional Complementary and Integrative Medicine Unit, Integrated Health Services Department, World Health Organisation, Geneva, Switzerland

Ashley Porto Department of Clinical Psychology, Columbia University Teachers College, New York, NY; International Association of Applied Psychology, Indianapolis, IN; Enhatch, Hoboken, NJ, United States

Ananya Raghuram Department of Clinical and Counseling Psychology, Teachers College, Columbia University, New York, NY; International Association of Applied Psychology, Indianapolis, IN, United States

Girish Ramachandran TCS Asia Pacific, Singapore, Singapore

Nithya Ramanathan Nexleaf Analytics, Los Angeles, CA, United States

Onisoa Rindra-Ralidera Options Consultancy Services, Antananarivo, Madagascar; AHAIC Commission on the State of UHC in Africa, Amref Health Africa, Nairobi, Kenya

Fatima Rizwan Department of Clinical Psychology, Teachers College, Columbia University, New York, NY; International Association of Applied Psychology, Indianapolis, IN, United States

Erin Ross Nexleaf Analytics, Los Angeles, CA, United States

Robert Ross St. Joseph's Addiction Treatment & Recovery Centers, Saranac Lake, NY, United States

Rajeev Sadanandan Health Systems Transformation Platform, New Delhi, India

Rayna Sadia International Association of Applied Psychology, Indianapolis, IN, United States; Riphah International University, Islamabad, Pakistan

James Sale United For Global Mental Health, London, United Kingdom

Udani Samarasekera London, England, United Kingdom

Sujay Santra iKure Techsoft Pvt. Ltd, Bengaluru, India

Jagjeet Sareen Dalberg Advisors, New Delhi, Delhi, India

Shekhar Saxena Department of Global Health and Population Harvard T. H. Chan School of Public Health, Cambridge, MA, United States

Daniel Schaudel MED-EL Medical Electronics, Innsbruck, Austria

Katherine E.A. Semrau BetterBirth Program, Ariadne Labs, Boston; Harvard Medical School, Boston, MA, United States

Rahul Sharma Frankfurt Innovation Center Biotechnology GmbH, Frankfurt am Main, Germany

Shannon Shibata-Germanos Proximie, London, United Kingdom

Zara Shubber World Bank, Washington, DC, United States

Megha Siddhanta Department of Clinical and Counseling Psychology, Teachers College, Columbia University, New York, NY; Harpaz-Rotem PTSD Treatment & Research Lab, Yale School of Medicine, New Haven, CT; International Association of Applied Psychology, Indianapolis, IN, United States

Patrik Silborn Theia Advisors, Singapore, Singapore

Moitreyee Sinha citiesRISE, New York, NY, United States

Sylvana Q. Sinha Praava Health, Dhaka, Bangladesh

Mehdi Snène I-DAIR (International Digital Health and AI Research Collaborative), Geneva, Switzerland

Anna Stauber Department of Psychology, Pace University; The NGO Committee on Ageing, New York, NY; Duluth Aging Support, Duluth, MN, United States

Mariya Stoilova Department of Media and Communications, London School of Economics and Political Science (LSE), London, United Kingdom

Joe Stringer Octopus Ventures, London, United Kingdom

Jie Sui School of Psychology, University of Aberdeen, Aberdeen, Scotland

Ashraf Swidan Specialized Program Services Department (SPSD), Primary Health Care Sector (PHCS), Dubai Health, Dubai, United Arab Emirates

Reem Talhouk Northumbria University, Newcastle upon Tyne, United Kingdom

Manal Mohamed Omran Taryam Dubai Health Authority, Dubai, United Arab Emirates

Katherine Tatarinov Geneva School of Economics and Management (GSEM), University of Geneva, Geneva, Switzerland

Jami Taylor Protagonist Therapeutics, Newark, CA; Yale University, New Haven, CT, United States

Anil Thapliyal eMental Health International Collaborative, Auckland, New Zealand

Annie Thériault Every Woman Every Child Innovation Marketplace, Grand Challenges Canada, Toronto, ON, Canada

Rachel Thompson Women in Global Health, Hayward, CA, United States

Katlen Tillman VillageReach, Seattle, WA, United States

Anand Tiwari Joint United Nations Programme on HIV/AIDS (UNAIDS), Geneva, Switzerland

John Torous Department of Psychiatry, Beth Israel Deaconess Medical Center, Harvard Medical School, Boston, MA, United States

Joseph D. Tucker Clinical Research Department, Faculty of Infectious and Tropical Diseases, London School of Hygiene and Tropical Medicine; International Diagnostics Centre, Faculty of Infectious and Tropical Diseases, London School of Hygiene and Tropical Medicine, London, United Kingdom; Institute of Global Health and Infectious Diseases, University of North Carolina at Chapel Hill, Chapel Hill, NC, United States; University of North Carolina Project-China, Guangzhou, China

Gabe Twose American Psychological Association, Washington, DC, United States

Joost van Engen Healthy Entrepreneurs, Utrecht, Netherlands

John E. Varallo Jhpiego, Baltimore, MD, United States

Carlos Velo International Association of Applied Psychology, Indianapolis, IN, United States; Universidad Autónoma de Madrid, Madrid, Spain

Akarsh Venkatasubramanian UNITE Parliamentarians Network for Global Health, Delhi, India

Dominique Vervoort Institute of Health Policy, Management and Evaluation, University of Toronto, Toronto, ON, Canada

Sidique Wai Office of Ambassador Sidique Wai, Embassy of the Republic of Sierra Leone to the United States; ECOWAS Ambassadors Group, Washington, DC, United States

David Wallerstein Tencent, Palo Alto, CA, United States

Rispah Walumbe AHAIC Commission on the State of UHC in Africa, Amref Health Africa, Nairobi, Kenya

Waruguru Wanjau AHAIC Commission on the State of UHC in Africa, Amref Health Africa; University of Cape Town, Cape Town, South Africa

Eleanor 'Nell' Watson Department of Computing & Engineering, University of Gloucestershire, Cheltenham, England

Dasoo Milton Yoon Human Development and Family Studies, University of Wisconsin-Madison, Madison, WI; Department of Clinical Psychology, Teachers College, Columbia University, New York, NY; International Association of Applied Psychology, Indianapolis, IN, United States

Lian Zeitz citiesRISE, New York, NY, United States

Feng Zhao World Bank, Washington, DC, United States

Biographies of the contributors

Marjorie Abalos is a global health visionary and social entrepreneur who is the founder and CEO of CLODS Inc. having created the CLODS System, a digital health solution specially built for developing countries, which aggregates critically needed population health data in real time and generates big data affecting policies in healthcare delivery. She is also the Board Chair and Co-Founder of Out4Good Ltd., a nonprofit, nongovernmental organization and global research institute that mentors, pairs, and helps improve policy and decision-making processes. Previously, Ms. Abalos held senior management positions at software and renewable energy firms for Global 500 companies and start-ups. She began her career working in web development capacities for internet start-ups in the telecom, web hosting, media, and advertising sectors after earning her bachelor's degree in biomedical/biochemical engineering from the University of Southern California where she worked on various groundbreaking projects utilizing cutting-edge biomedical imaging techniques. She has facilitated projects in many countries to develop process improvement plans that have empowered millions of disenfranchised, underserved, and marginalized people to access the transformations of the Information and Digital Age. She was also instrumental in deploying the first ATM networks in Vietnam and in addressing public and industrial environmental remediation issues in China. Ms. Abalos has been engaged in social justice and public service projects such as Amnesty International and the American Cancer Society's Relay for Life, as well as volunteering at hospitals, convalescent centers, and schools. She can be reached at marjorie@clods.org.

Ola Aldafrawy, MD, is a specialist family physician with 14 years of experience, currently employed in the Specialized Program Services Department (SPSD) in the Primary Health Care Sector (PHCS) of the Dubai Health Authority (DHA) in the UAE, whose services include the smart services and telemedicine services that provide smart virtual consultation to the community in Dubai and smart premarital screening program. Interested in smart health services, digital health transformation, and big data analytics, she is passionate about serving the community with the best quality possible comprehensive family medicine services, through early detection, prevention, and screening programs. One of the first lifestyle managers in DHA, she is a member of several taskforces, including the national taskforce for prevention of cardiovascular diseases in UAE and the breast cancer screening committee. She is actively involved in audits and research papers, and has shared her publications in the Journal of Public

Health of Oxford University and in the Dubai Diabetes and Endocrinology Journal, and several ongoing research groups on topics including COVID-19, colorectal cancer screening, and others. A graduate from Ain Shams University in Egypt, where she earned her master's degree in Family Medicine, she is currently pursuing a master's degree in Business Intelligence (AI) and Big Data.

Alia Mohammad Rafi Aldallal, MD, is a consultant family physician with more than 20 years of experience, and is currently the Director of the Specialized Program Services Department (SPSD) in the Primary Healthcare Sector (PHS) of the Dubai Health Authority (DHA), responsible for all the specialized services in the primary health care sector including the telemedicine service, which provides smart virtual consultation to the Dubai community. She led one of the primary healthcare centers, as the head of Al Barsha Health Center, for 10 years, which serves the community 24/7 with comprehensive family medicine services. An International Member and Examiner in the Royal College of General Practitioners (MRCGP international), she holds an Arab Board of Medical Specialization in Family Medicine, earned a Master of Science (MSc) degree in Quality and Safety in Healthcare, and a diploma in Diabetes and Asthma Management. She is Internal Auditor of International Customer Experience Standard Certification 2020 and an Examiner in the Dubai Medical College and Arab Board of Medical Specialization in family medicine as well as family medicine residency intake exams (exam for residents to be accepted to enroll in the DHA family medicine residency program). She was awarded by HH Sheikh Mohamed Bin Rashid Al Maktoum as the best head of a service center at the Dubai Government Excellence Program Awards 2017, and as a distinguished employee in DHA in the 2019 "You Are Our Pride" program.

Reem Al Hammadi is a Senior International Relations Executive in the International Relations Section at the United Arab Emirates' Federal Competitiveness and Statistics Centre (FCSC) affiliated with the UAE Ministry of Cabinet Affairs. Driven by her passion interest in the Sustainable Development Goals (SDGs) and innovation, Reem is a member of the Secretariat of the UAE National Committee on SDGs and an active advocate for the SDGs. She recently received her master's degree at Queen Mary University of London in International Public Policy, she graduated from Zayed University, Dubai with a bachelor's degree in International Affairs, and a minor in Communications and Media Sciences. Reem is also a graduate of the Al Maktoum College Program in Multiculturalism and Leadership.

Saqr Alhemeiri, MBA, is the Chief Innovation Officer and Director for the Strategy and Future Department at the Ministry of Health and Prevention (MOHAP) in the United Arab Emirates. In his current position, Saqr spearheads the formulation and execution of the Ministry's strategy and its cascading of the healthcare sector at the national level, ensuring that MOHAP services and practices adhere to international and national quality and excellence standards. As Chief Innovation Officer, he has set a new strategic approach for innovation in the healthcare sector in the UAE focused on building a culture that nurtures and prioritizes innovation across the industry. His work in innovation has won the ministry multiple awards in the field, including most recently the prestigious Sheikh Mohamed Bin Rashid Excellence Award for Most Innovative Entity. Saqr has an extensive background in building and administering programs across all organizational levels in the private, semi-governmental, and governmental sectors, spanning multiple industries, including airline, utilities, retail, and healthcare. He holds an Executive Masters in Business Administration from the University of Sharjah, and a Diploma in Government Innovation from the Mohamed Bin Rashid Government Centre for Government Innovation (in collaboration with Cambridge University), and is a graduate of the Executive Leaders program from the UAE Prime Minister Office (in collaboration with Said Business School—Oxford University). He has coauthored many articles on healthcare training, including on supporting critical emergency medicine nurses and critical emergency medicine trauma team simulation training programs.

Aisha Alhouti, MS, is a Teaching Assistant at the Department of Information Science in the College of Life Sciences at Kuwait University. She has over 21 years of academic experience and obtained her bachelor's and master's degrees in Computer Engineering from Kuwait University.

Dari Alhuwail, PhD, FAMIA, FIAHSI, is an Associate Professor at the Department of Information Science in the College of Life Sciences at Kuwait University, and a Fellow of the International Academy of Health Sciences Informatics, as well as a Fellow of the American Medical Informatics Association. He is also the cofounder of the Kuwait Health Informatics Network and a member of the Kuwait Health Informatics Association. He serves as a Health Informatics Consultant at Dasman Diabetes Institute where he cofounded the Health Informatics Unit and is an Adjunct Faculty member at the School of Medicine at the University of Dundee. Currently, he serves as the Acting Dean for Student Affairs at Kuwait University. Currently a board member of multiple international scientific organizations, he served as a member of the WHO Digital Health Technical Advisory Group. He has over 16 years working in industry and in various research labs globally.

Paula Aljovin, MA, a Peruvian-American, is a researcher, psychologist and Coordinator of the Mental Health Team of The Vine Trust's Amazon Hope Medical Programme, a Scottish-based organization that delivers free primary health services in remote Amazonian communities through medical ships. After completing her undergraduate studies in Psychology at the Pontificia Universidad Catolica del Peru (PUCP), she worked at several mental health settings in Peru. In 2020, she volunteered at The Vine Trust, during which time she designed and cofounded the now fully operating research-based mental health intervention that is included in the ships' services. The medical ships project is funded by USAID and The Vine Trust and run jointly by the Amazon Hope Medical Programme and CEDRO, a Peruvian NGO. Paula currently directs the mental health project remotely from New York City while she completed her master's degree in Clinical Psychology at Columbia University Teachers College in

New York City. Anticipated to graduate in May 2023, Paula has studied under the mentorship of Dr. Judy Kuriansky in her course "Psychology and the United Nations."

Suad Hashim Ahmed, MD, is a consultant family physician with more than 20 years of experience in the primary healthcare sector of the Dubai Health Authority (DHA). She is currently head of a unit in the specialized services department in the Primary Health Care Sector (PHCS) in DHA, which includes smart services and telemedicine services providing smart virtual consultation to the community in Dubai and a smart premarital screening program. She is actively involved in audits, research papers, and publications, including the ongoing prospective PURE study, since 2005, which has published more than 20 scientific papers. From 2006 to 2010, she led one of the primary healthcare centers as the head of Al Mamzar Health Center, which served the community with comprehensive family medicine services. After that, she actively led DHA programs in early detection, prevention, and screening, aimed at reducing the burden of noncommunicable diseases, mainly cardiovascular diseases and cancer prevention, through wellness and checkup programs. A graduate of the Arabian Gulf University class of 1999, she holds Arab Board of Medical Specialization in Family Medicine and MRCGP International Dubai 2006, as well as having earned a Diploma in Modern Asthma Management Denmark in 2004 and a master's degree in Health Care Management from the Royal College of Surgeons Ireland (RCSI) in 2009. She was awarded in the "You are Our Pride Program" for a distinguished department for group work in the Dubai Health Authority.

Asma Al Obeidli, MA, currently works as Technical Staff for the Public Health Sector in the Office of the Undersecretary for Public Health at the Ministry of Health and Prevention in the United Arab Emirates. Formerly, she worked as a Medical Laboratory Technologist in Al Baraha Hospital, Dubai. She obtained her bachelor's degree in Health Science and Medical Laboratory Technology from Sharjah Women's College, UAE. She earned a master's degree in Hospital Management from Hamdan Bin Mohammed Smart University (HBMSU) in the UAE where she has been a PhD scholar in Total Quality Management since 2018. She has contributed to the Monitoring SDG3 Report in the UAE 2022 and a publication related to Social Enterprise, both currently in press.

Hussain Abdul Rahman Al Rand, MD, has been the Assistant Under Secretary for the Public Health Sector at the Ministry of Health and Prevention, United Arab Emirates, since 2014. He is the Chairman of the National Committee for the implementation of provision of the International Health Regulations and Prevention of Pandemics. Previously, he was the Director of Medical Affairs and Head of the ENT Department in Dubai Hospital, Dubai Health Authority. Prior to that, he managed different positions in the Dubai Health Authority as a specialist and consultant in ENT. He has various credentials in the field of otorhinolaryngology (ORL), head and neck surgery, including a distinguished Fellowship from the Royal College of Surgeons (Edinburgh). He is the first Emirati to obtain the German Board (Facharzt)

Certification in Otorhinolaryngology, from the Medical Association of North Rhine, Dusseldorf, Germany. He received his master's in Otorhinolaryngology and bachelor's in Medicine from Cairo University Medical College (Cairo, Egypt), and obtained a Professional Diploma in Management of Health Services and Social Care from the American University in Cairo. He holds prestigious positions at international, regional, and national levels, as President of the International Federation of ORL Societies (IFOS) (a nonpolitical organization representing over 50,000 otolaryngologists); Vice-President of the PARS Society; President of the Arab FOS (ARBFOS); President and founder of the GCC ORL Head and Neck Society, President of the Emirati ORL Head and Neck Society; Chairman of the National Diabetes Committee; Chair of the High Commission of the Arab Council for Medical Specializations in the Arab League; and President of the IFOS Congress 2023 in Dubai. He is an author of several publications, including "Similar effectiveness of the inactivated vaccine BBIBP-CorV (Sinopharm) and the mRNA vaccine BNT162b2 (Pfizer-BioNTech) against COVID-19 related hospitalizations during the Delta outbreak in the UAE," published in the *Journal of Travel Medicine*, March 2022.

Pedro Altungy, PhD, is Assistant Professor at the Universidad Complutense de Madrid (UCM) and research collaborator at the Psychology Department of the University of Maryland. His field of research is the exploration of the psychological variables underlying personality, as well as the psychological consequences of terrorism. He was awarded a scholarship from the Spanish Science Ministry in 2020 for his PhD. He holds two master's degrees, in Clinical Psychology and in General Health Psychology. He also works for the Spanish General Council of Psychologists, currently as supervisor in a research project of the European Federation of Psychology Students' Association (EFPSA). He is also past-president of the Student Division (15) of the International Association of Applied Psychology (IAAP), and a member of the IAAP Task Force on Terrorism and Peace Building. He can be reached at paltungy@ucm.es.

Tina C. Ambos, PhD, is Professor of International Management and Director of the Center for Innovation and Partnerships at the Geneva School of Economics and Management at the University of Geneva. Her research interests include global strategic management and innovation in the context of multinational corporations, technology startups, and international (nonprofit) organizations. She has published widely in top academic journals and is a regular speaker at international conferences and practitioner forums as well as leading executive programs.

Taavi Annus, MD, is a medical doctor who has compounded his love for medicine with ICT (information and communications technology). As the Chief Health Information Officer at the Ministry of Social Affairs in Estonia, Taavi is responsible for overseeing the implementation of the e-health strategic vision for 2030 and for the national e-health development portfolio for health and welfare specialists. At the start of the COVID-19 pandemic, he was responsible for overseeing the creation of work processes and digital tools for contact tracing in Estonia, and during the pandemic response, he was the lead analyst in the renewal of the Estonian COVID-19 governance processes at the end of 2021.

Logan Ansell is Vice President of Partnerships for health-tech social enterprise reach52. He currently oversees the organization's impact-oriented engagement with external partners, including nonprofits, the public sector, intergovernmental organizations, and academic institutions. Logan has worked on various public health-related initiatives in countries including Guyana, India, Cambodia, Philippines, Indonesia, North Korea, and his native Canada. Holding a Master of Public Health from Simon Fraser University, his areas of experience include digital health and health innovation for low-resource contexts. Based in Phnom Penh, Cambodia, he can be reached at Logan@reach52.com

Maxwell Antwi, MD, is a member of the global UHC2030 Advisory Group for Civil Society Engagement Mechanism (CSEM) and the WHO Roster of Experts for Digital Health. He has over a decade of senior-level local and international work experience in public and private health sectors in clinical, managerial, research, and public health positions. Since 2015, following a successful career as a specialist obstetrician-gynecologist, he has led the PharmAccess Ghana country office in advocating with a broad range of national and international partners to ensure equity and inclusiveness for the most marginalized and vulnerable populations in Ghana's health policies and strategies. His focus of work has been on aligning civil society and stakeholders for digital health innovations, health legislation, and the implementation of digital health interventions for the benefit of patients. He is actively involved in high-level health advocacy programs, health policy adoption, and early-stage advocacy work on health product regulation through traceability of medicines with blockchain. His work with health legislation up to the level of Ghana's Parliament with focus on advocacy for health regulation and governance systems included the use of digital technologies with national-level public ministries and agencies in the health and social protection field, as stipulated in Ghana's UHC Policy. Dr. Antwi also served as an advisor to Ghana's National COVID-19 Response Team and collaborated with health partners to set up a digital care center for COVID-19 where people could access corona-related care from doctors via mobile phones at no cost to users.

Alex Israel Attachey is a product development professional and a public health practitioner with over 16 years of work experience in developing digital solutions, management of consumer relations, partnerships development, and market research across Sub-Saharan Africa. In the past 5 years, Alex has developed several innovative digital solutions for the health sector in Ghana, research for the PharmAccess Foundation in Ghana, and partnerships development. He acquired his expertise by working at Vodafone in product development and international business, at Cognizant Technologies, and in managing digital projects for Google in Kenya, Nigeria, Ghana, and South Africa.

Rifat Atun, MBBS, MBA, DIC, FRCGP, FFPH, FRCP, is President of the Global Surgery Foundation Board. He is a Professor of Global Health Systems and the Director of the Health Systems Innovation Lab at Harvard University. During 2008–12, he served as a member of the Executive Management Team of the Global Fund as Director of Strategy, Performance, and Evaluation, where he chaired the panel that oversaw annual investments of ~US$4 billion each year. During 2006–13 Dr. Atun was a Professor of International Health Management at Imperial College London and Head of the Health Management Group. Professor Atun's research focuses on health system transformation and innovation. He has published more than 450 papers in leading journals such as NEJM, Nature Medicine, Lancet, JAMA, Academy of Management Journal, Lancet Oncology, Lancet Global Health, and Science Translational Medicine. In 2020, 2021, and 2022 he was recognized by the Web of Science/Clarivate as one of the World's Most Highly Cited Researchers. He has advised more than 30 governments on health policy, strategy, and health system reform, and has consulted for the World Bank, WHO, and leading health corporations such as Medtronic and Novartis. During 2019–20 Professor Atun was a senior advisor to the G20 Presidency and served as the Co-Chair of the T20 Health Track and the Task Force on Global Health Security and COVID-19 for the G20 Indonesian Presidency. A UK-trained clinician, he studied medicine at the University of London as a Commonwealth Scholar and obtained an MBA from Imperial College London. He is a Fellow of the Royal College of Physicians, the Faculty of Public Health, and the Royal College of General Practitioners in the UK.

Sulzhan Bali, PhD, MSc, MSc, is a Health Specialist at the World Bank's Health, Nutrition, and Population Global Engagement Unit, with a primary focus on strengthening health systems in low-income and fragile countries, emphasizing pandemic prevention, preparedness, and response (PPPR) strategies, including One Health. Sulzhan has worked on several global and regional health security initiatives, including the World Bank's COVID-19 response and the Regional Disease Surveillance Systems Enhancement (REDISSE) Program in West Africa, and the Africa Centers for Disease Control (Africa CDC) project. She brings with her several years of extensive research and professional experience, specializing in health security, One Health, laboratory sciences, and strengthening health systems in low-income and fragile countries, particularly in Central and West Africa. Prior to joining the Bank, Sulzhan worked with the University of Maryland—University College, and the Public Health Foundation of India. She also served as the Director of Production and HR for This Week in Global Health (TWiGH)—an online video and audio podcast on global health.

Erin Barry, MA, earned her master's degree in the Department of Clinical Psychology, Columbia University, Teachers College. A former student in Professor Judy Kuriansky's class on "Psychology and the United Nations," she is a member of the Student Division of the

International Association of Applied Psychology, an NGO having consultative status with the United Nations Economic and Social Council. A former varsity collegiate swimmer who qualified for the 2016 US Olympic Swimming Trials, she is passionate about the field of sports psychology and promoting physical activity for all for health and wellbeing.

Daniela Bas is the founder of SustainABILITY Advisory Boutique Inc., headquartered in Europe. She is the former Director of the Division for Inclusive Social Development at the United Nations Department of Economic and Social Affairs (DESA), a post she held from 2011 to 2023. She also worked at the United Nations from 1986 to 1995. Before joining the United Nations Secretariat in 2011, from the year 2000, she was adviser for the Presidency of the Council of Ministers in Italy, and for the Ministry of Foreign Affairs on whose behalf, she was Board Member of the EU Agency for Fundamental Rights and of the Committee to Promote non-Armed and non-Violent Civil Defense of the Presidency of the Council of Ministers. She was also the special adviser to On Frattini, Vice President of the European Commission, on human rights, democracy, and social affairs. Prior to 2000, she held senior director roles in the private sector in Italy. A certified coach, Ms. Bas has a background in political science with a major in International Politics, who worked as a journalist for major press, radio, and television networks. An expert on the SDGs and their social sustainable dimension, she collaborates with visionary leaders to link SDG performance with business strategy to increase opportunities for sustainability and innovation.

Christoph Benn, MD, is a medical doctor specializing in tropical medicine and public health with more than 30 years of experience in global public health. He is one of the founders of the Global Fund to Fight AIDS, Tuberculosis and Malaria, and its former Director of External Relations for over 15 years, during which he managed the first five replenishment cycles. Dr. Benn is currently the Director of Global Health Diplomacy at the Joep Lange Institute in Geneva, Switzerland, where he focuses on developing sustainable health financing models and on promoting the digital transformation of health systems in LMICs.

Mohini Bhavsar, BSc, EMBA, is currently Chief of Staff to the Co-CEO Jocelyn Mackie of Grand Challenges Canada (GCC), where she acts as a strategic thought partner to the leadership team, supporting cross-functional coordination, enabling strategy implementation and organization performance. She has over a decade of experience supporting nonprofits across Africa, India, and Latin America on their digital programs for community health, agriculture, and financial inclusion. She returned to Toronto in 2020 after a decade of living abroad in Guatemala, India, and Senegal. Prior to GCC, Mohini was the Global Head of Digital Health Partnerships at Living Goods (LG), a Skoll-recognized NGO that achieves life-saving impact through digitally enabled community health worker programs. At LG, Mohini worked closely with the Global Fund and governments to shape digital health investment opportunities at the community health level. Prior to that, she had multiple leadership roles at Dimagi, a B Corp that builds a widely used open-source digital health platform in low- and middle-income countries.

At Dimagi, Mohini led its USAID DIV I and II programs for mHealth innovation in India, launched its West and Central Africa office in Senegal, and established numerous government partnerships across Africa. Mohini holds a BSc from McMaster University in Biochemistry and is a 2023 candidate for the Rotman Global Executive MBA program in Healthcare Life Sciences. In her spare time, Mohini serves as a Board Member for Vitala Foundation, a digital solution for post-abortion care across Latin America. She loves to cook, cycle, and is a certified rescue diver.

Sripriya Bikkani is completing her master's degree in the Department of Clinical Psychology at Teachers College, Columbia University, and was a student in Professor Judy Kuriansky's class on "Psychology and the United Nations." In addition to her scholastic work, her background includes Health Care Services and Special Education Services. She is interested in increasing accessibility for health care and education, and exploring the potential of digital health, especially for neurodevelopmental disorders. She has personal and professional experience working with children with autism spectrum disorder and other neurodevelopmental disorders, and first-hand experience with some helpful devices, such as video-based communication aids.

Nicole K. Bulanchuk, MA, has extensive experience working with NGOs and fundraising, including for Save the Children Federation, Care International, The Nature Conservancy, ASPCA, Planned Parenthood, the ETS Sponsorship Program, and multiple community-based organizations including local mutual aids, food pantries, and the NYC Community Emergency Response Team. Born and raised in New York, she founded a startup NGO in October 2020 called Believe New York after witnessing and experiencing first hand the poverty and mental health difficulties that many New Yorkers live through. She has a vision to improve the wellbeing of New Yorkers and beyond, especially among underserved populations. She earned a master's in clinical psychology from Teacher's College, Columbia University, where she was a student of Dr. Judy Kuriansky in the class on "Psychology and the United Nations," and explored the issue of NGO financing. She aspires to expand the programs of Believe New York throughout the United States and the world.

Craig Burgess, MB, ChB, MSc, MBA, is Unit Head of Data Coordination, Governance and Partnerships in WHO's Division of Data Analytics and Delivery for Impact. He has worked at community, national, regional, and global levels for national health services, Médecins sans Frontières, MERLIN, UNAIDS, WHO, UNICEF, GAVI, the Vaccine Alliance, and JSI. Craig has been a GAVI board member for Civil Society Organizations (CSO), adviser to UHC 2030 Civil Society Engagement Mechanism, and a member of WHO's Immunization Practices Advisory Committee (IPAC). He is a trained physician with a master's in Health Policy, Planning, and Financing, an MBA, and various postgraduate qualifications.

Bo Cao, PhD, is an Associate Professor and Canada Research Chair (CRC, Tier 2) in Computational Psychiatry at the Department of Psychiatry, University of Alberta (UAlberta), and Co-Director of the UAlberta Computational Psychiatry Group. He was trained in mathematics (BSc) and psychology (MSc) and then completed a computational neuroscience PhD at Boston University and went on to complete postdoctoral training in neuroimaging and psychiatry at the University of Texas Health Science Center, Houston. Dr. Cao's main research interests include computational psychiatry and precision medicine in mental health. He aims to develop translational tools that can provide accurate and personalized diagnosis and treatment optimization for mental disorders. This can be achieved by identifying objective biomarkers associated with mental disorders and their treatment outcomes through multimodal data including brain imaging, genetic, biological, behavioral, cognitive, and clinical measurements, and by applying advanced machine learning and statistical algorithms to these data. Dr. Cao is also interested in brain development and aging, children, adolescence, and women's mental health, general aging, neuromodulation, and neurofeedback.

Mariam Claeson, MD, MPH, has been Senior Advisor on Political Economy of Adolescent Mental Health at the Global Public Health Department at the Karolinska Institute, Sweden, since 2021. She is a former director of the Global Financing Facility for Every Woman Every Child, and the Maternal Newborn and Child Health team at the Bill and Melinda Gates Foundation, and has worked for many years at the World Bank, the World Health Organization, in rural clinical practice (in Tanzania, Bangladesh, and East Bhutan), in primary health care (in Ethiopia), and in health sector development (including in Egypt, Jordan, Uzbekistan, China, and the Philippines). Mariam has contributed to several global programs and reports across the fields of reproductive, maternal, newborn, child and adolescent health and nutrition, AIDS in South Asia, and disease control priorities. She is an honorary medical doctor (MDhc) at the Karolinska Institute. Mariam serves on several boards, advisory groups, and commissions, including as a member of the Global Council on SDGs 3—Good Health and Wellbeing and the Second Lancet Commission on Adolescent Health and Wellbeing.

Christiaan Coetzee, MBA, is a distinguished leader in global sustainability and innovation, currently serving as an Ecosystem and Sustainability Consultant at Etihad Credit Insurance, the UAE's federal export credit company, and as a Freelance Consultant at the Federal Competitiveness and Statistics Centre (FCSC), affiliated with the UAE Ministry of Cabinet Affairs. With a diverse background in strategic marketing, communications, project management, and international relations, Christiaan has played a pivotal role in advancing the United Nations 2030 Agenda for Sustainable Development. His notable contributions include the launch of the second cohort of the Global Councils on SDGs, winning the Stevie Awards Middle East in 2020 for the SDG Photography Award, and pioneering the XDGs 2045 concept. His academic credentials include a BCom in Marketing and Tourism Management from North-West University (NWU), South Africa, further education in business management, project management, and digital marketing, culminating in an MBA with a specialization in global management and leadership from the Swiss School of Management in 2024. Christiaan's vision and leadership have earned him international recognition, solidifying his role as a vital actor in integrating sustainable development goals into organizational and global strategies.

Patrick F. Connolly, PhD, is part of the Clinical Research Department of MED-EL, an Austria-based global medical technology company specializing in hearing implants and devices, who conducts literature research and technical writing. He joined the company in 2021. He earned his PhD from the National University of Ireland in Galway, Ireland, and has lived in Austria since 2016.

Alethea Cook, MPH, is a consultant with the World Bank's Health, Nutrition, and Population Global Practice, where her work focuses on primary health care transformation, non-communicable diseases, and digital health. Alethea has over five years of experience conducting programmatic and policy research and evaluation in a variety of contexts, including East Asia and Pacific, South Asia, and Sub-Saharan Africa. As a global health researcher, Alethea has contributed to recent reports on health system resilience and on digital health and integrated service delivery for chronic care. Prior to joining the Bank, Alethea worked with the Social Development Division at the United Nations Economic and Social Commission for Asia and the Pacific, the Social Science Research Council, and Population Services International. Alethea completed her MPH in Global Health at the Harvard TH Chan School of Public Health and holds a BA in Economics and Chinese Studies from Grinnell College.

Jennifer Crouch, MA, Director, Development, has led the Development Program at VillageReach since 2019 and has worked in nonprofit administration since 2001. Jennifer builds partnerships that help mission-driven organizations and individuals connect their financial resources to address complex social problems and maximize their impact on global health systems. She holds a BA degree from James Madison University and an MA degree from the University of Washington.

Lindsey Cunningham earned her master's in International Educational Development from Teachers College, Columbia University, and was a student in Professor Judy Kuriansky's class on "Psychology and the United Nations." Lindsey is now working in the Adult Education department at the International Rescue Committee in the New York Resettlement office. Lindsey also has experience as a classroom teacher and has spent time volunteering in schools in Ecuador, India, Kenya, Romania, and Italy. This international perspective has sparked an interest in how education systems function in different regions of the world, particularly in developing contexts. Lindsey has a passion for expanding access to education and believes that education has the potential to change the lives of individuals as well as the larger community. She intends to work with marginalized communities to expand access to quality education.

Florence L. Denmark, PhD, is a prominent social psychologist, currently the Robert Scott Pace Distinguished Professor of Psychology. Formerly the Chair of the Psychology Department at Pace University, New York City campus, and the President of the American Psychological

Association, Florence is recognized for her work highlighting issues associated with discrimination and violence toward women. She earned her doctoral degree in social psychology at the University of Pennsylvania and was the first woman to graduate with honors in History and Psychology. She went on to hold academic positions at Queens College, Hunter College, and the Graduate School of the City University of New York (CUNY). Her professional commitment and dedication to teaching led a wide range of institutions furthering psychology, from the premier research association of the New York Academy of Sciences to the American Psychological Association and the International Council of Psychologists. Dr. Denmark continues to work with the Psychology Coalition of NGOs in consultative status with the Economic and Social Council at the United Nations addressing aging issues and is an active voice for women and older persons around the world.

Marija Davcheva, MS, is currently a researcher at the Research Institute of Personnel Psychology, Organizational Development and Quality of Working Life (IDOCAL) and a PhD candidate in the doctoral program of Psychology of Human Resources at the University of Valencia, Spain. As a pre-doctoral grant awardee from the Spanish Ministry of Science and Innovation, she focuses her research on job quality, wellbeing, and health. She obtained a master's degree in the renowned Erasmus Mundus Joint Master Programme (WOP-P) on Work Organizational and Personnel Psychology from the University of Valencia, for which she was awarded the Extraordinary Student Achievement Award. Besides her international and research educational experience, she has been an active volunteer and student representative in many international organizations, including the European Federation of Psychology Associations (EFPA) (on their Board of Scientific Affairs), the International Division (52) of the American Psychological Association (APA), and the Global Psychology Student Committee. She is also the current President of the Student and Early Career psychologists Division (15) of the International Association of Applied Psychology (IAAP).

Juan Ramón de la Fuente, MD, has been Ambassador and Permanent Representative of the Mission of Mexico to the United Nations since 2019. Previously, Dr. de la Fuente was the Minister of Health in Mexico and formerly Rector of the National Autonomous University of Mexico (UNAM), where he is currently an Emeritus Professor of Psychiatry. A graduate of UNAM Medical Faculty, he trained in psychiatry at the Mayo Clinic in Rochester, Minnesota. Since the 1980s, Professor De la Fuente has worked closely with the United Nations System, initially serving as a Consultant for the World Health Organization (WHO) on issues regarding addictions and mental health. In 1995, he became Vice-President of the World Health Assembly in Geneva, and in 1998, he was elected to Chair the Board of UNAIDS. He has also been involved with UNESCO since 2004 as a member of the Board of Latin America and the Caribbean on Higher Education, and as of 2008, as President of the International Association of Universities (IAU). He is author and coauthor of more than 300 scientific papers and 21 books. He was President the National Academy of Medicine of Mexico as well as the Mexican Academy of Sciences. A public hospital in Mexico City has been named after him and the epithet of a newly described cactus was dedicated to him (*Opuntia delafuentiana*). Until his appointment as Mexican Ambassador to the UN, he directed the Global Studies Program at UNAM and Chaired the Board of the Aspen Institute in Mexico.

Madhu Deshmukh, MA, is the Vice President of CARE International's Program, Strategies, and Impact (PSI) team. She has over 30 years of experience working globally on public health and social development with a focus on HIV/AIDS, sexual and reproductive health, gender equality, and women's empowerment. She is an established leader in designing and supporting programs that put community leadership and engagement at the center, designing community-led social and digital innovations for scaling impact, health systems strengthening including enabling and empowering front line health workers, policy advocacy, and research. She has worked with communities and key populations including sex workers, orphans and vulnerable children, pregnant mothers and infants, and young girls. She has authored articles and reports on health systems strengthening, use of information communication technology (ICT) and mHealth, maternal newborn and child health (MNCH), gender-based violence, and women's empowerment. She has led at-scale programs working across multiple levels with organizations including local organizations, international NGOs such as CARE, the International Center for Research on Women (ICRW), the Government of UK's Department for International Development (DfID), mHealth Alliance at the UN Foundation, UNDP, UNAIDS, and the Government of India. Madhu earned a master's degree in Social Work from Tata Institute of Social Sciences, India.

Deepika Devadas, PhD, is Chief of Staff to the co-CEO (Dr. Silver) at Grand Challenges Canada, an impact-first investment organization and innovation platform dedicated to saving and improving lives of unserved and underserved communities and populations. Deepika is passionate about health, evidence and equity and is a recognized thought leader in the sector of program development, scaling innovations, and building productive partnerships for impact. She previously led the Every Woman Every Child Innovation Marketplace, a strategic initiative of the Bill and Melinda Gates Foundation, Grand Challenges Canada, USAID, and Norad (Norwegian Agency for Development), that was part of the UN-led Every Woman Every Child. She comes from a background in health sciences and technologies, through her research experience across a variety of scientific disciplines, including infection biology, cancer, and bioengineering.

Daniella Diaz, MA, is currently pursuing a PsyD in Clinical Psychology at George Washington University. She earned a Bachelor of Arts in Psychology from James Madison University and a Masters of Arts in Clinical Psychology from Teachers College, Columbia University, where she was mentored by Professor Judy Kuriansky in her class on "Psychology and the United Nations." Her interests are in psychotherapy treatments, US Military Veteran populations, E-mental health, psychopathology, and personality. She is a member of the Student Division of the International Association of Applied Psychology.

Roopa Dhatt, MD, the Executive Director and Co-Founder of Women in Global Health (WGH), is a passionate advocate for gender equality in global health and a leading voice in the movement to correct the gender imbalance in global health leadership. She is also a practicing Internal Medicine Physician at Georgetown University Hospital in Washington DC and has faculty appointments as an Assistant Professor at Georgetown University and the University of Miami. Dr. Dhatt is particularly committed to addressing issues of power, privilege, and intersectionality that keep many women from global health leadership roles and to opening up spaces for the voices of these women to be heard. Determined to build a movement to transform women's leadership opportunities in health, Dr. Dhatt co-founded Women in Global Health in 2015. Today, Women in Global Health has 40 chapters and 50,000 supporters in more than 90 countries and continues to grow.

Omar ElFata, BEng, is a healthcare technology innovator and expert on medical device industrialization. His family own Africa's largest and oldest hypodermic syringe manufacturer, AMECO, where he was CEO for 10 years, and since 2017, has joined P&P Group (Rome, Italy) as CTO to develop and scout for technologies related to infection prevention and control (IPC) tailored for low-resource settings. Two of P&P's technologies (RETRAGO—an automatically retractable safety syringe designed to prevent needle stick injuries and blood-borne diseases, and Biovitae—a microbicidal UV-free LED technology that passively sanitizes environments) were awarded by UNAIDS as a top global health innovation 2 years in a row (2019 and 2020). Omar participated as a guest and speaker in events such as the World Health Assembly (WHA), World Health Summit (WHS), World Economic Forum (WEF), and UN General Assembly (UNGA) developing his experience in public health policy and investment.

Pauline Eluère is a Communications Officer at UNHCR, the United Nations Refugee Agency, which is mandated to protect 100 million people forced to flee their homes due to conflict and persecution. In this role, she develops strategies and coordinates global multimedia campaigns to drive action for refugees. She also writes about technological and pop culture trends impacting refugees. Before joining UNHCR, Pauline worked for the Red Cross and for a think-tank specialized in peace negotiations. She holds a master's degree in International Affairs from the Geneva Graduate Institute and Sciences Po Rennes, as well as diplomas in Digital Strategy and Digital Law.

Kameshwar Eranki[†] is a Serial Entrepreneur and Founder and CEO of Proactiff Digital Healthcare Services Inc., a disruptive start-up based out of Silicon Valley, USA, which is the world's largest health data aggregator and preventive primary care platform enabled with artificial intelligence and provisioning emotional AI solutions, and diagnostic, descriptive, and predictive analytics. Prior to his entrepreneurial journey, he worked for Fortune 5 to Fortune

[†]Deceased.

500 companies. He has now taken on his greatest challenge to date, which is to radically reform healthcare systems worldwide—transforming them from reactive "sick care" models to proactive, preventive healthcare systems. This healthcare designed for the future is enabled with IoT and point-of-care health devices and delivers digital health, remote patient monitoring, digital therapeutic monitoring, and real-world evidence-based care, including comprehensive integrative medicine modalities. Kameshwar is a leading keynote speaker, author, visiting professor, and mentor. His numerous awards include Entrepreneur of the Year, 2019, from the Indo-American Foundation. Kameshwar lives in Silicon Valley, USA.

Temitayo Erogbogbo has over 20 years of combined private sector and development experience, 15 years of which was spent in the pharmaceutical industry in multiple roles that included government affairs, patient relations, communications, marketing, and sales. As the current Director of Multilateral Engagement MSD and Advocacy Lead for MSD for Mothers, Tayo is responsible for strategic engagement with multilateral organizations such as the World Health Organization (WHO), UNFPA, development banks, and development agencies that include USAID, UK FCDO, and the Gates Foundation. In his role as advocacy lead for MSD for Mothers, he is responsible for coordinating global, regional, and national strategic partnerships for policy development and strategic practice changes to improve maternal health outcomes by addressing disparities through health systems strengthening and improving quality of care, particularly where private sector approaches can be leveraged for sustainable impact. Amongst a number of committee memberships that Tayo sits on, he is a member of the WHO steering committee for Country Connector on Private Sector in Health, and a co-lead on the World Economic Forum's Women's Health Initiative-Coalition meeting on "Reconnecting with Women and Girls."

Satoshi Ezoe, MD, MPH, MPA, PhD, has served as Director of Global Health Strategy Division in the Ministry of Foreign Affairs (MOFA) in the Government of Japan since August 2020, as well as Director of Healthcare Policy at the Cabinet Secretariat coordinating overall global health policy of the Japanese Government. Prior to his current position, he was Counsellor of the Permanent Mission of Japan to the United Nations (2017–2020), where he was instrumental in facilitating the United Nations General Assembly High-Level Meetings on Tuberculosis (2018) and on Universal Health Coverage (2019). Prior to joining MOFA, he was engaged in healthcare and public health policy in the Ministry of Health, Labor, and Welfare of Japan, including in the areas of global health, national health insurance system, and noncommunicable diseases, including mental health and cancer control. He was the first appointed Senior Coordinator for Global Health (2015–2017), where he was involved in the WHO emergency reform, coordinating health agendas for the G7 Ise-Shima Summit in 2016. Also, previously, he was seconded to UNAIDS Headquarters in Geneva (2009–2012). Dr. Ezoe is a medical doctor with a PhD degree, as well as Master of Public Health and Master of Public Administration from Harvard University. He has made presentations about UHC, including at the American Public Health Association conference, and his co-authored published articles include "Case study: The Political Declaration of the High-Level Meeting on Universal Health Coverage (2019): Negotiating the most comprehensive agreement ever reached on global health" in the book entitled "A guide to global health diplomacy" (pp. 235–252); "Why does global health matter to diplomacy?" in the publication "Global Health Governance;" and "Protecting Human Security: Proposals for the G7 Ise-Shima Summit in Japan," published by *the Lancet*. Contact: toshiezoe@gmail.com.

Erin K. Ferenchick, MD, is a medical anthropologist, family physician, and global health advisor. She is an Assistant Clinical Professor of Medicine at Columbia University's Center for Family and Community Medicine and has worked with the World Health Organization, GAVI, and The Global Fund in the area of health systems strengthening. Erin currently serves as a technical advisor for Proximie's global health initiative. Proximie is a digital healthcare software company aiming to improve access to surgical care and to scale clinical expertise through training. She is a graduate of Princeton University and Columbia University College of Physicians and Surgeons. She can be reached at: ekferenchick@gmail.com.

Rafael Figueroa, a serial entrepreneur and CEO of Portal Telemedicine, is revolutionizing healthcare through AI-driven telediagnosis and population health management. The Brazilian start-up, accelerated by the UN and Google, has impacted 30 million patients across 800 cities and 15,000 health facilities. With a background in economics from UFSC, design thinking from Berkeley, and entrepreneurship at MIT, Rafael has led 300 employees and 10,000 physicians to deploy over 70 deep learning algorithms while developing a unique technology that integrates with 90% of medical devices. Beyond Portal, Rafael has 4 years of experience in asset management at a major bank, worked on Google Street View, and led 10 AI research projects with high-profile organizations. As a Google Developer Expert ML, he has mentored C-level executives from more than 50 companies and participated in numerous journalist interviews, keynotes, and workshops worldwide. As the founder of a nonprofit, Rafael has driven the planting of 100k trees, protected 50k hectares of forest, and supported 200 rural families with open-source technology. Linkedin:www.linkedin.com/in/rafafigueroa E-mail: rafael@portaltelemedicina.com.br.

Brianna Fitapelli, MA, is a clinical psychology doctoral student at Adelphi University. She earned her master's degree in the Department of Clinical Psychology at Columbia University Teachers College, where she was a student in Professor Judy Kuriansky's class on "Psychology and the United Nations." She is a member of the Student Division of the International Association of Applied Psychology (IAAP) and the Schizophrenia International Research Society (SIRS). She has 2 years of experience working with underserved populations with severe mental illness (SMI) in an acute inpatient setting and is currently completing an externship at Mount Sinai Beth Israel. Ms. Fitapelli is interested in developing and implementing scalable, culturally competent interventions for the maintenance and treatment of mental illness. Furthermore, she plans to continue to advocate for the allocation of quality mental health services in under-resourced areas in the USA and abroad. Contact: briannafitapelli@mail.adelphi.edu.

Nathaniel Foote, MBA, JD, is the Chairman of TruePoint, a mission-driven consulting firm, and a founding Board Member of the Higher Ambition Leadership Alliance. Over a 40-year career, he has combined private sector consulting with a focus on social enterprise and the civic sector. A Board Member of *citiesRISE*, he is also a Senior Fellow of the Harvard Center on the Developing Child, and the Board Chair of "Tools of the Mind," a preschool and kindergarten program reaching over 40,000 children in the US, shown to help close the academic achievement gap for low-income and minority students. Nathaniel is a co-author of "Higher Ambition: How Great Leaders Create Economic *and* Social Value." After graduating from Harvard College, he attended the University of Cambridge as the Fiske Scholar, and earned an MBA degree from Harvard Business School, where he was a Baker Scholar, and a JD degree, magna cum laude, from Harvard Law School.

Cameron Fox is the Health Technologies Portfolio Lead at the World Economic Forum, where he manages a suite of projects at the intersection of human health and technology, including initiatives on Disruptive Technology for Mental Health, Digital Innovations for Non-Communicable Diseases, and efforts to prevent bioterrorism threats from *de novo* DNA synthesis. Additionally, he has led two Global Futures Councils on Mental Health and Synthetic Biology. Prior to his time at the World Economic Forum, Cameron was a Bass Connections Fellow at Duke University's Science & Society Initiative, where he explored the ethical, legal, and social implications of emerging technologies in neuroscience and genetics. He has a background at the intersection of neuroscience and philosophy.

Christian Garbe, PhD, has been the Managing Director of the FIZ Frankfurt Biotechnology Innovation Center in Germany since its founding in 2002. He completed his studies at Georg-August University in Göttingen and earned his doctorate in 1996. Subsequently, Dr. Garbe began his professional career at Novartis, a Swiss-American multinational pharmaceutical corporation, in their Basel, Switzerland, headquarters, where he assessed potentials of biotechnology for product developments. After transitioning to DZ BANK AG in Frankfurt, Germany, he analyzed pharmaceutical and biotechnology companies and successfully managed a biotech fund. The focus of his work is on the development of international business models and innovative, data-driven technologies. To date, Dr. Garbe has authored over 80 biotechnology/pharmaceutical investment articles and reports with the European Union, including "Investing in Biotechnology" (Scrip Reports) and "Chance Biotechnology" (Frankfurter Allgemeine Buch). He can be reached at: christian.garbe@fiz-biotech.de; LinkedIn: https://www.linkedin.com/in/christian-garbe-4ab66635/.

Beatrice Gatumia, MD, MPh, works as an Engagement Director in the Health portfolio at Busara Center for Behavioral Economics. She is a healthcare professional with a strong passion for improving healthcare access and outcomes in developing countries. With experience in both clinical and public health settings, she is excited about the use of design thinking to develop innovative solutions that improve health seeking behaviors and health outcomes, particularly for populations in the global south. She was previously the Clinical Business Lead at Amref Medical Centre, where she led the ramp-up of PCR COVID-19 testing as part of Amref Health Africa's COVID-19 response. Beatrice also held positions at GlaxoSmithKline Pharmaceuticals Ltd. as the medical affairs lead in the access portfolio designed to respond to global health priorities like newborn health. Her commitment to improving healthcare access and outcomes in developing countries continues to drive her work and contributions to the industry.

Kamal Gautam, MD Psych, MBBS, is the Executive Director of Transcultural Psychosocial Organization Nepal (TPO Nepal), a leading nongovernmental organization working on mental health and psychosocial support in Nepal. With extensive expertise in community mental health, Dr. Gautam works directly with policy makers, academicians, people with lived experience, and institutions for facilitating policy reforms related to mental health, as well as providing technical support to the Ministry of Health and Population and concerned agencies in Nepal in development and implementation of policies, including the National Mental Health Strategy and Action Plan (2077–2082); National Action Plan against Suicide (2079–2087); community mental health care package; training curricula for health workers (Modules 1 to 6); capacity-building of non-specialists; and adaptation of the WHO mhGAP in Nepal. He also offers specialist clinical services through the outpatient department and outreach clinic of TPO Nepal. He is an experienced researcher. His research areas of interest include integration of mental health into primary care; child and adolescent mental health; preventive and promotive mental health interventions; use of digital technologies in mental health; stigma on mental health and suicide/self-harm. Email: kgautam@tponepal.org.np Website: https://orcid.org/0000-0001-9401-9359.

Amandeep S. Gill, PhD, is an engineer, diplomat, and international learning specialist. He is currently a Professor at the Graduate Institute, Geneva, and founder of I-DAIR, an international collaborative initiative, funded by the Foundation Botnar and the Welcome Trust, which is striving to improve access to cutting edge research on digital health and artificial intelligence for clinical researchers, policy makers, and patients around the world, with particular attention to the Global South. Formerly, he was Executive Director and Co-Lead of the Secretariat of the UN Secretary-General's High-Level Panel on Digital Cooperation. Professor Gill previously served as India's Ambassador and Permanent Representative to the Conference on Disarmament in Geneva and chaired the Group of Governmental Experts of the Convention on Certain Conventional Weapons (CCW) on emerging technologies in the area of lethal autonomous weapon systems from 2017–2018. He has a BTech in Electronics and Electrical Communications from Punjab University, Chandigarh, India, and an Advanced

Diploma in French History and Language from Geneva University. His PhD degree is from King's College London, focused on Nuclear Learning in Multilateral Forums.

Lasha Goguadze, MD, MPH, is the Senior Health Officer of Active Outreach, working in the Health and Care Department of the International Federation of Red Cross and Red Crescent Societies (IFRC) at the headquarters in Geneva, Switzerland. He is responsible for HIV, Tuberculosis, Harm-Reduction Programs for Drug Users, and serves as a technical focal person to address health within migration and climate change contexts. He is a medical doctor who earned a master's degree in Advances of Sciences in Public Health, obtained at Geneva University. He has spent several years as an IFRC field delegate in Europe and Asia.

Geetha Krishnan Gopalakrishna Pillai, MD, is a technical officer at the World Health Organization in Geneva. He is a qualified Ayurveda practitioner with 25 years of work experience as an Ayurvedic physician, entrepreneur, researcher, integrative medicine expert, administrator, and policy advisor to governments. He also has experience in conceiving and establishing international advocacy platforms in traditional medicine, working with governments and non-state actors. He is a researcher and clinician, with specialized training and experience in integrative medicine and herbal drug development. He is a key member of the AYUSH-based COVID-19 response task force of the Government of India (whereby AYUSH is an acronym for Ayurveda, Yoga and Naturopathy, Unani, Siddha, Sowa Rigpa, and Homeopathy, the seven Indian systems of medicine regulated and practiced in India) and has supported the Ministry to organize international clinical trials on COVID-19. With the World Health Organization (WHO), he is responsible for developing global goods supporting norms and standards of traditional medical (TM) systems such as Ayurveda, Unani, Siddha, and Yoga, and data collection and impact assessment tools such as TM chapters for the International Classification of Diseases (ICD). He also led the development of the WHO m-Yoga mobile App, a multi-lingual self-learning tool for Yoga. He can be reached at drgk2000@gmail.com.

Andrew J. Greenshaw, PhD, is a University of Alberta Professor of Psychiatry and Neuroscience and Fellow of the Royal Society of Arts, trained in Europe and Canada. A Fellow of the Canadian College of Neuropsychopharmacology (CCNP), for which he served as President from 2000–2002, and the Collegium Internationale Neuropsychopharmacologicum (CINP), Andy served as University of Alberta Associate Vice President (Research) and served as a member of the Scientific Advisory Board of the CIHR Institute of Neuroscience Mental Health and Addiction. Andy has broad research interests that range from neuroscience to policy in mental health, with particular interests in AI and machine learning applications and e-health. He is a founding member of the University of Alberta computational psychiatry group, working closely with members of the Alberta Machine Intelligence Institute. Since 2016, Andy has served as Scientific Director for the APEC Digital Hub for Mental Health. He is a Global Leadership Council member for the e-Mental Health International Collaborative (e-MHIC) and serves on the boards of several organizations, including Mental Health Research Canada (MHRC).

Russell Greiner, PhD, is Canadian Institute for Advanced Research (CIFAR) AI Chair at the Alberta Machine Intelligence Institute (Amii), and a Fellow-in-Residence. He is a Professor at the Department of Computing Science and an Adjunct Professor at the Department of Psychiatry at the University of Alberta. A Fellow of the Association for the Advancement of Artificial Intelligence (AAAI) and recipient of the Killam Annual Professorship and a McCalla Professorship, Dr. Greiner serves on the Scientific Committee of the Institute for Data Valorization (IVADO). He focuses on developing and improving applications of machine learning in medicine, providing solutions for specific real-world problems across a range of clinical considerations. Within the field of computational psychiatry, he uses machine learning on fMRI (functional magnetic resonance imaging) and other clinical data to develop new ways of diagnosing mental disorders, including schizophrenia, and for assessing the severity of a range of symptoms. In the area of precision medicine, Dr. Greiner collaborates with colleagues in healthcare to develop methods for recommending patient-specific plans for the treatment of diseases such as cancer or diabetes, and for predicting individual health outcomes.

Monique Arantes Guimarães, MSc, a Brazilian psychology graduate student, earned her master's degree in Social, Work, and Organizational Psychology from the University of Brasília (UnB, Brazil). Currently, she is a doctoral student in psychology at Universidade São Francisco (USF, Brazil) specializing in the field of psychological assessment. She is the Secretary of Communications for the Division of Students and Early Career Psychologists of the International Association of Applied Psychology (IAAP).

Aline Guzik, MPA, is a Policy Adviser at the Permanent Mission of Mexico to the United Nations in New York, where her work mainly focuses on mental health, humanitarian situations, UN Peace Operations, and the files in the Security Council on Children in Armed Conflict, Colombia, and Haiti. She currently serves as Co-Chair of the Women's Foreign Policy Group's Young Professional Network and is a Youth Associate of the Mexican Council of Foreign Relations (COMEXI), where she co-authored the policy paper "Beijing+25: Generations in Favor of Gender Equality" (translated from Spanish; published by COMEXI). Furthermore, she has been invited as a guest commentator on national television (*Sin Filtro* and ADN40 News). Before moving to New York, Aline, a native of Mexico, worked in public health, social protection policies, and affordable housing at two of Mexico's primary agencies: Mexican Institute of Social Security (IMSS) and the Institute of the National Housing Fund for Workers (INFONAVIT). She earned her master's degree in Public Administration from Columbia University's School of International and Public Affairs (SIPA) and a BA in International Relations from Universidad Iberoamericana in Mexico City.

Nadine Hachach-Haram, BEM, MD, a recipient of the British Empire Medal in the Queen's Birthday Honours, is an NHS surgeon, lecturer, and clinical entrepreneur. Nadine is a

Consultant Plastic Surgeon and Director of Clinical Innovation and Strategic Partnerships at Guy's and St Thomas' NHS Foundation Trust. Nadine is also the Co-Managing Partner of KHP Ventures. In 2015, Nadine drew on her passion for innovation, education, and global surgery to found Proximie, a digital healthcare platform aiming to improve access to surgical care and to scale clinical expertise through training. Through its patented platform, clinicians can virtually "scrub-in" to any clinical setting to visually and practically interact, analyze, and collaborate anywhere in the world. Dubbed by CNN the "future of surgery," Proximie has gone from strength to strength and won multiple awards, including Foreign Press Association Science Story of the Year and Red Herring Top 100. Nadine has been selected as an Endeavor Entrepreneur, and is a member of the Royal College of Surgeons' Commission on the Future of Surgery, consisting of some of the country's leading doctors, engineers, data experts, managers, and patient representatives, with the objective to investigate the advances that will transform surgery over the next 20 years. She can be reached via www.proximie.com.

Taylor Hirschberg, ADN, MSc, MPH, is a human rights activist, award-winning healthcare leader, strategist, researcher, clinician, and journalist, currently serving as the Hearst Endowed Scholar at the Columbia University Mailman School of Public Health in New York City. Prior to Hirschberg's time at Columbia University, he worked as a humanitarian for Médecins Sans Frontières in Africa, Middle East, and Latin America. In addition to his global health work, Hirschberg has sat on various boards representing LGBTQI+ interests, including the Fenway Institute of Health, Colorado HIV/AIDS Planning Council, the Colorado Public Health Association, and CUNY LGBTQI+ Advisory Board. Globally recognized as an advocate and activist for members of the LGBTQI+ community who have been forcibly displaced from their homes, he has worked on evacuations and resettlement of many diasporas, including the Afghan, Venezuelan, and Syrian LGBTQI+ communities. In addition to his humanitarian and advocacy work, Taylor is also an award winning and nominated humanitarian and wartime journalist who writes on topics of forced migration, LGBTQI+ world events, and literature.

Ann Herrmann-Nehdi is Chief Thought Leader and Chair of Herrmann International, the originators of the Whole Brain Thinking model and HBDI Assessment. She specializes in the practical application of neuroscience and cognitive diversity to human and organization development and improvement, continuing the firm's 40 years of research with a database of 3 million thinkers worldwide. Clients include 9 out of 10 of the Fortune 100, including Pfizer, Hilton, IBM, and Coca-Cola, universities (MIT, Stanford, The Wharton School), government agencies (NASA, US Department of Energy), and professional services firms (PwC, ErnstYoung). She has presented keynotes worldwide, and two TedX talks: "The One Thing You Need to Know About Your Brain That Will Change Your Life" (with over ½ million views) and "Think Like Your Future Depends On It, Because it Does." Co-author of The Whole Brain Business Book-Second Edition (McGraw-Hill), and many publication chapters, her work focuses on helping individuals, managers, teams, and leaders drive growth and improve impact by making better decisions, effectively managing change, being more agile, and leveraging their untapped thinking potential to improve profitability, leadership, productivity, innovation, and overall business results. Her research has been featured in Business Week, Harvard Business Review, Scientific American, Chief Executive Magazine, and O (The Oprah Magazine), with her current research focusing on workplace resilience and the brain, the future of work, and the development of women leaders in the workplace. She studied at American University, the Sorbonne, and Institut Catholique, in Paris, where she lived for 15 years. She is an AthenaOnline management expert, board member for the ACLU of South Carolina, and advisor to the CEO Action for Racial Equity Fellowship. Email: ann@hbdi.com. Website: www.thinkherrmann.com linkedin.com/in/annherrmann.

Matthew Hughsam, MPH, is Director of Strategic Initiatives at *citiesRISE*, where he brings grassroots- and system-led efforts together to drive transformative change. A facilitator, researcher, and coach committed to pursuing wellbeing and justice for people, communities, and our planet, Matt entered justice work as a college student opposing Canadian Government cuts to refugee healthcare coverage. Since then, he has supported global health initiatives in diverse contexts, from building an evaluation system for a hip-hop-rooted youth mental health program based in Toronto to advising Ethiopian and South African national government officials and working on deworming programs reaching millions of children. Matt's lived experience and faith inform his work. He holds a Bachelor of Health Sciences from McMaster University, where he later taught Global Health, and a Master of Public Health from Harvard University, where he trained and coached with adaptive leadership experts. Matt also serves on the Advisory Board for Youth Combating Neglected Tropical Diseases.

Geoffrey Ibbotson, MD, MSc, FRCSC, FACS, is the Founder and Executive Director of the Global Surgery Foundation, a Geneva-based organization committed to improving access to surgical care systems globally. He also serves as the senior health advisor to the United Nations Institute for Training and Research (UNITAR). Trained as a general surgeon in Canada, he is a Fellow of the Royal College of Physicians and Surgeons of Canada and the American College of Surgeons, holding academic appointments at both the University of Calgary and the University of Alberta. He has more than 25 years of experience in medical humanitarian and disaster relief work, during which he has served as a front-line surgeon in multiple countries and several conflict zones. He has also led disaster response activities, developed strategic programs and guided organizational governance reviews and restructuring. While maintaining a busy surgical practice in Canada, he served as the Director of Trauma Services for the North West of Alberta from 2008 to 2016 and was Chair of the Alberta Provincial Trauma Committee from 2014 to 2016, which coordinated an integrated trauma system for 10 trauma hospitals in Alberta. Website: https://www.globalsurgeryfoundation.org.

Tara Imalingat, LLB, PGDip Healthcare Ethics and Law, PGDip Law, is a Kenyan lawyer and health and health rights advocate with experience in project and stakeholder management in the NGO sector. She is a qualified Advocate of the High Court of Kenya and a graduate of the University of Manchester in the United Kingdom, where she pursued a Bachelor of Laws and a Postgraduate Diploma in Healthcare, Ethics and Law. Previously, she worked as a Programme Officer at the Kenya Legal & Ethical Issues Network for HIV and AIDS (KELIN), where she advocated for the health-related human rights of people living with HIV, people affected by TB, and key populations, including adolescent girls and young women. In 2021, she became a member of the Digital Health and Rights Consortium, a group of individuals and organizations working in academia and civil society who, through research and advocacy, are exploring the need for a rights-based approach to digital health governance. Tara was recently awarded a digital health research fellowship by Young Experts Tech for Health (yet4h) a platform established by the Transform Health Coalition to push for youth leadership on the road to achieving Universal Health Coverage by 2030 through the use of data and technologies. Her research explores the role of public participation as the vehicle for democratizing digital health governance. To hear more about her research, contact her through email: tara.imalingat@gmail.com.

Maria Isabel Iñigo Petralanda, MBE, JD, an Argentinian lawyer and bioethicist, is the Coordinator of the Ethics, Privacy, and Security Chapter of the Central American Network for Health Informatics (RECAINSA). A graduate of the University of Buenos Aires School of Law and Social Sciences in Argentina (UBA), she also has a Diploma in Health Law from Austral University in Argentina, where she is a Professor and member of the Academic Committee at the Universidad Austral School of Law. She also earned a master's degree in Bioethics from Pontifical Catholic University School of Medicine of Argentina (UCA); a Diploma in Health Economics with a focus on Health Technology Assessment from Saint Andrews University; and a Diploma in Patient Safety and Person Centered Care. A member of the Research Ethics Committee at Pontifical Catholic University (UCA) and a specialist in public health policies, she teaches Public Health Policies and Ethics at several universities. As a specialist in State, Person, and Technology for the Understanding of Digital Health Path, Maria Isabel is fully devoted to advocacy about the digital health path for ethically grounded social good.

Kiyan Irani, MA, is currently an executive functioning coach at Organizational Tutors and a graduate research intern at NYU Grossman School of Medicine, studying biomarker correlates in traumatized military personnel. His professional research interests lie in trauma outcomes and resilience trajectories for marginalized and underrepresented communities. He earned his master's degree in Clinical Psychology from Columbia University Teachers College where he studied under the mentorship of Dr. Judy Kuriansky in her "Psychology and the United Nations" course. Kiyan has presented his research at several national and international conferences and is a member of the International Association of Applied Psychology Student Division.

Biju Jacob, PhD, is a co-founder of the Health Innovation and Investment Exchange and serves as its India Country Director and global lead on innovation scouting and acceleration support. Academically trained as a medical geneticist and a serial entrepreneur by passion, he has advised over 100 biomedical/healthcare start-ups in conceptualizing, building, and running new ventures through Starthealth-Unitus Seed Fund, Axilor Ventures, and CAMTech at Massachusetts General Hospital. An experienced leader in innovation strategy, research infrastructure, and start-up advisory with a decade of experience in early-stage discovery, clinical research, and regulatory affairs, Jacob has worked in both large organizations and start-ups applying his deep medical research and management experience. He is a recipient of the BIRAC Innovation 2014 award, ISBA National award for entrepreneurship 2010, and Sir CV Raman Young Scientist Awards 2007.

Zsuzsanna Jakab, PhD, a native of Hungary, was appointed Deputy Director-General of the World Health Organization (WHO) in March 2019, after having served as WHO Regional Director for the European Region for 9 years. As Deputy Director-General, her portfolio included the WHO's technical programs for universal health coverage, the life course, communicable and non-communicable diseases, healthier populations and antimicrobial resistance, alongside other programs. Currently, Dr. Jakab is covering the role of Regional Director for the WHO's Western Pacific Region. Dr. Jakab has held a number of high-profile national and international public health policy positions in the last three decades. She served as the founding Director of the European Union's European Centre for Disease Prevention and Control (ECDC) in Stockholm, Sweden. Between 2005 and 2010, she built ECDC into an internationally respected center of excellence in the fight against infectious diseases. Between 2002 and 2005, Dr. Jakab was State Secretary at the Hungarian Ministry of Health, Social, and Family Affairs. Between 1991 and 2002, she worked at the WHO Regional Office for Europe in a range of senior management roles.

Gillian Javetski, MPH, serves as the Chief of Staff at Dimagi, a social enterprise that builds and scales sustainable, high-impact digital solutions that amplify the efforts of frontline workers and programs globally. She has built her career working in project, communications, strategy, and operations roles at various social impact technology organizations, all working towards increasing access to impactful technology worldwide. Prior to this position, she was a co-founder and Chief Operating Officer of TecSalud in Bogotá Colombia, a company that designs and applies practical and sustainable technology strategies. Early in her career, she was also a Fellow for Partners in Health and an intern at UNDP.

Preethi John, PhD, Co-Founder of the Women in Global Health India Chapter, is a health and development management professional with 25 years of experience in public and private sector organizations. Her expertise and experience have groomed her as a leader and institution builder. She has expertise in capacity building of healthcare human resources from developing countries across South Asia, Africa, and Latin America. She is currently part of a start-up team at the Global Business School for Health, UCL, as its Deputy Director. She also serves as Advisory Board Member to Health and Development NGOs. She holds a PhD from the Indian Institute of Technology (IIT) in Madras, India, a master's degree from the Tata Institute of Social Sciences (TISS) in Mumbai, India, and is a visiting scientist at the Harvard School of Public Health, besides being a Harvard LEAD Fellow.

Mutahi Kagwe, MBA, is Kenya's former Cabinet Secretary for Health. He is a former Member of Parliament for Mukurweini Constituency, Senator of Nyeri County, and Cabinet Minister for Information, Communications, and Technology, where he spearheaded East Africa's first Fiber Optic Project for Eastern Africa, and "Mpesa," the world's first-ever mobile money transfer system. In the health sector, his leadership has achieved sector-wide transformations, most notably through the creation and implementation of policies that have enhanced efficiencies in service delivery and enabled the country's focus on new areas, such as mental health. Kagwe led Kenya's National Emergency Response Committee which oversaw the fight against the coronavirus pandemic. With support from His Excellency the President, Uhuru Kenyatta, whose approach to health is multi-agency, Kagwe spearheaded Kenya's fight against the COVID-19 disease from a centralized point, making prevention of the spread of the virus successful. His efforts, amid countless domestic and local challenges, earned him international recognition in the Wall Street Journal. Kagwe has also been credited with championing "The Best Classroom," a program that promotes quality early childhood education in his country. Cabinet Secretary Kagwe holds a Bachelor of Commerce degree from the University of Nairobi and an MBA from the United States International University.

Mwenya Kasonde, MB, BCh, is an emergency medicine physician and global health consultant for bilateral and multilateral agencies, including USAID, UNICEF, and the WHO, as well as the private sector. Previously, she served as Assistant Director for Global Health at the Ministry of Health in Zambia. Having worked for the National Health Service in the UK as a physician for many years, her focus has turned to empowering lower- and middle-income countries to impact and participate in global health policy implementation. She earned a Bachelor of Medicine and Bachelor of Surgery degree from the Royal College of Surgeons in Ireland, and a Diploma in Health Economics from University College Dublin, and is completing a Masters in Global Health degree at the Liverpool School of Tropical Medicine in the UK. She is passionate about economic empowerment, women's rights, and the attainment of universal health coverage for all.

Pradeep Kakkattil is the founder and CEO of the Health Innovation and Investment Exchange (HIEx), a multilayered platform connecting governments, policymakers, innovators, and investors to leverage innovations and technology towards scaling up access to healthcare. He has worked on health and development over the last three decades, including with the Joint

United Nations Programme on HIV/AIDS (UNAIDS) based in Geneva, Switzerland, where he established and led the Office of Innovation. He co-founded the Women in Innovation Fund that advocates for equitable access to financing for women led health enterprises, especially in low and middle income countries. Given the increasing impact of climate on health, especially on women and the most vulnerable populations, he has advocated and co-founded the climate health innovation and equity fund that leverages and invests in climate resilient health technologies and innovations. A former journalist and filmmaker, Pradeep serves on the Advisory Boards of the Graduate Institute in Geneva and is a founding member of the Board of the Geneva Innovation Network. Additionally, he is a member of the Global Council on Health as part of the United Arab Emirates' "SDGs in Action" initiative, a member of the African Union Digital Health Strategy Taskforce, and a member of the Science and Innovation Accelerator working group hosted by the World Health Organization (WHO). He is a recipient of the Schwab Foundation/WEF Public Intrapreneur 2022 award. He organizes and speaks as an invited panelist at international conferences and events.

Kanishka Katara, BTech, has been the Chief Digital Health Officer at Living Goods since 2021, and previously served as the Head of Digital Health at PATH, and Director of Government and Partnerships at Dimagi (a tech social enterprise). Prior to joining Dimagi, he was a Director in the Government practice of PwC India, and led a number of key government transformation and reform initiatives across Africa, Asia, Europe, and the Caribbean. His career has spanned 14 years in diverse yet complementary areas of software development, government transformation consulting, and social impact sectors. Owing to strong analytical skills, ability to think on his feet, focus on human centricity, and an in-depth understanding of governance functioning, Kanishka has been able to uniquely blend his knowledge and resolve to make a sustainable impact on improving the quality of life of the less privileged. Kanishka comes from a dedicated family of medical professionals who have spent their lifetimes serving people in rural and semi-rural areas of India. Having witnessed the difficult realities of the public health domain during his childhood, Kanishka has been driven by the cause of leading the transformation of society into one that believes in the principle that "all lives are equal." He graduated with a Bachelor of Technology Degree in Computer Science from UP Technical University in Lucknow, India.

Shariha Khanam Khalid, MBChB, MSc, MRCS, is the Founding Partner of Mission & Co, an award-winning global innovation and investment advisory firm specializing in partnerships and strategies for impact, and Co-Founder and Board Member of Impact Hub Kuala Lumpur, part of a network of hubs for impact in 110 cities around the world supporting the next generation of impact entrepreneurs and inclusive businesses. Shariha is also Co-Founder of the Health In Your Hands initiative, a multi-stakeholder platform aggregating 100 bankable projects from the private and nonprofit sectors that deliver outcomes for the first mile of health to strengthen ecosystems for financing and scaling of impact through new models for public-private-people partnerships. She is a member of the SDG Finance Advisory Board for the United Arab Emirates' SDGs in Action

Global Councils initiative at the Prime Minister's Office, and was previously Co-Chair of the SDG Global Council on Health. A social return on investment (SROI) practitioner who has founded organizations dedicated to building and supporting the next impact entrepreneur and supported early-stage innovations that deliver impact, she currently volunteers as advisor to organizations focused on bridging the digital divide as well as the gender and inclusion gap through finance and education, given her passion about entrepreneurship and youth development. Previously, she was a +SocialGood Advisor to the UN Foundation and a Global Good Fund Ambassador. Starting her professional life as a surgeon in the UK, US, and Malaysia, she has been an impact entrepreneur, advisor, and investor for the past 15 years. She received her degrees in design, medicine, and health systems research from the United Kingdom, and draws inspiration from Kahlil Gibran's words: "Work is love made visible."

Sadikchhya (Meera) Khanal, MA, is an analyst at Russell Reynolds Associates, a global search and leadership advisory firm with 47 offices worldwide, where she works with public, private, and nonprofit organizations across healthcare to build teams of transformational leaders that can help create a more sustainable, equitable, and kind world. She is also a member of the Asian Affinity group where she is engaged in creating events and partnerships to promote resilience, wellness, and leadership amongst minority groups. She studied psychology, global mental health, and spirituality at Teachers College, Columbia University, where she studied under Professor Judy Kuriansky, which led to her partnering in numerous multi-stakeholder webinars in collaboration with the Permanent Mission of Nepal to The Unites States, the Transcultural Psychosocial Organization (TPO) Nepal, youth NGOs, and the media outlet Cyber Headline. She co-organized, moderated, and presented events about these events in an American Psychological Association Symposium and at the Commission on the Status of Women NNGO/NY. Born in Kathmandu, Nepal, she is extremely passionate about global health and equity.

Rana Khazbak, PhD, is an ESRC postdoctoral fellow at the Centre for Public Policy Research, King's College London. Her fellowship builds on her PhD thesis, which explored the impact of social housing regeneration on the wellbeing of young people from lower-income backgrounds. Her doctoral research used the capability approach and participatory methods to foreground young people's experiences of restructuring their social housing estates into mixed-income communities. Before joining King's College, Rana completed her PhD at the London School of Economics' Social Policy Department, after which she was a research fellow at the University of Nottingham's Rights Lab. Rana's past research examined the role of place, class, gender, and ethnicity on a range of child and youth wellbeing outcomes, including experiences of exploitation, citizenship, and access to education and employment. Previously, she also worked in international research and development contexts.

Eva Kiegele, MA, is a Technical Officer of the World Health Organization (WHO) Office at the United Nations. Her role at the WHO is to support better positioning of health-related issues in the debates and decisions of the UN intergovernmental bodies, specifically those related to non-communicable diseases, disabilities, mental health, environmental determinants of health, climate change, food security, injury prevention, ageing, and maternal, child, and adolescent health. Prior to joining the WHO, Eva was the Head of the President's Office at the premiere health-centered university in Latvia, Riga Stradiņš University. She has experience in journalism and media relations, as well as working with nongovernmental organizations. Eva holds a Master of Arts degree in Diplomacy and International Relations from Seton Hall University with a specialization in Global Health, and Global Negotiation and Conflict Management.

Sarah Kline is CEO and Co-Founder of United for Global Mental Health, a global NGO dedicated to increasing political and financial support for mental health around the world and to reforming policies and implementation in line with best practice and a human rights-based approach. Sarah has over 25 years' experience in international development and has spent the past 15 years working in global health. From 2012 to 2017, Sarah worked with and for the World Health Organization (WHO) in a variety of roles with a focus on disease prevention and control, pandemic response, and health emergencies. She saw first-hand the impact of poor mental health on individuals and communities and how to develop and implement programs to address their needs. Her fundraising for the WHO on mental health inspired Sarah to help develop and launch United for Global Mental Health. Among her previous roles, she has worked for the World Bank, the UK Government, and Oxfam GB (Great Britain), and launched and ran another charity: Malaria No More UK.

Pitambar Koirala, MA, is the Head of the Program Department at the Transcultural Psychology Organization in Nepal (TPO Nepal), a leading mental health nongovernmental organization in Nepal. He has extensive experience in mental health and psychosocial support (MHPSS) in Nepal, specifically working in the fields of anti-torture, anti-trafficking, gender, and humanitarian settings. Mr. Koirala is also involved in advocacy, monitoring and evaluation (M&E), intervention design, program development, and implementation of projects. He successfully led the development and roll out of a context-specific group intervention in Nepal known as "Psychosocial Group Healing Intervention." He has been involved in a number of publications in the area of psychosocial support and has made contributions to the drafting and endorsement of a number of national policies, guidelines, and training modules on MHPSS and anti-trafficking in Nepal. He earned his master's degree in Anthropology and a bachelor's degree in Law.

Rotem Kopel, PhD, is an experienced health-tech specialist with over 20 years of leadership positions in the technology and philanthropic industries. She is the CEO of MindMatters, a global foundation that supports new technologies to improve youth mental health, and also the CTO of Luciole Medical, where cutting-edge real-time brain monitoring systems are being developed. Prior to these roles, she led the European activities of Israel Brain Technology, where neuro-tech solutions were developed by integrating scientific and technological advancements. Her previous experience was in leading the development of semiconductor products in the telecommunications field. She earned a PhD in neuroscience and electrical engineering from EPFL, Switzerland, and is a member of the Global Future Council of the World Economic Forum.

Jennifer Knox, MA, is the Director of Character Education and Ron M. Brill Chair for Ethical Leadership Development at Woodward Academy, the largest independent school in the continental United States. In that role, she contributes to the social, emotional, and ethical learning (SEE Learning) program alignment institution-wide. She is a facilitator, content designer, and educator for SEE Learning with Emory University's Center for Contemplative Science and Compassion Based Ethics. She has over 20 years of experience working nationally and internationally to support education as a vehicle for personal, relational, and collective wellbeing to promote a more compassionate world. She earned her Masters of Arts degree in the Psychology of Education at Columbia University Teachers College, where she was a student in Professor Judy Kuriansky's class on "Psychology and the United Nations." She is also a member of the Student Division of the International Association of Applied Psychology.

Eneyi E. Kpokiri, PhD, is a clinical pharmacist and an assistant professor of social innovation at the London School of Hygiene and Tropical Medicine, London, United Kingdom. She has extensive experience organizing social innovation research projects. Her doctoral research explored the development and implementation of antimicrobial stewardship programs in low- and middle-income settings. Her current research focuses on identifying and implementing social innovations in health with the use of community-engaged, participatory methods. She has conducted several global crowdsourcing open calls, in-person/digital hackathons, and other public engagement methods for health research. In her role, she has contributed to WHO projects employing these methods, including modified Delphi surveys to develop consensus statements and produce WHO reports including a framework for monitoring and evaluating social innovations in health and the practical guide on public engagement and crowdfunding in health research. She was a consultant for the WHO/TDR on social innovation in health and currently serves as the director of social innovations within the Social Entrepreneurship to Spur Health (SESH) advisory group.

Suresh Kumar, MD, MPH, a clinical researcher and psychiatrist, is the Global Director of Research and Development for *citiesRISE*, where he leads collaborative efforts to develop innovations bringing together the best of evidence and communities' experiences for young people's mental health, including the whole school approach in primary and secondary schools, gathering spaces within informal settlements, and the Youth Leaders Networks and Youth Challenge Awards. He has extensive expertise in research and design, with over 35 years involvement in the field of drug use prevention, treatment, and mental health as a service provider, professor, trainer, researcher, and international consultant. Suresh holds positions on a number of international boards, task forces, and working groups relating to injectable drug use, HIV/AIDS, and harm reduction. He holds an MD degree from Madras University in Chennai, India, and a Master of Public Health degree from Johns Hopkins University in Baltimore, Maryland, USA.

Judy Kuriansky, Ph.D., is a clinical psychologist who has been on the graduate faculty of Psychology and Education at Columbia University Teachers College in New York for 20+ years. At the United Nations, she is advisor to the Mission of Sierra Leone to the United Nations and an NGO Representative of the International Association of Applied Psychology and World Council for Psychotherapy in consultative status with the Economic and Social Council of the UN, and a past President of the Psychology Coalition of NGOs, who successfully led advocacy to secure mental health and wellbeing in international documents, including the UN 2030 Agenda for Sustainable Development, Political Declaration for Universal Health Coverage and Global Compact for Migration, and others. She has co-organized and spoken at innumerable events, e.g., for the Commission on the Status of Women, World Interfaith Harmony Week, World Day of Sports, and many more. She is Policy Advisor to Ambassador Sidique Wai of Sierra Leone to the United States, a trustee of the United African Congress, member of the Global Leadership Council of the eMental Health International Collaborative (eMHIC) and a past Co-Chair and current member of the Global Council SDG3 (Health), "SDGs in Action" Initiative, Dubai, United Arab Emirates. An award-winning journalist, she has been a TV reporter, pioneer of radio call-in advice talk shows, and columnist for the Singapore Straits Times, South China Morning Post, Daily News.com, and currently Black Star News and BlackTie magazine. She has authored over 100 professional journal articles, and many books on topics from interpersonal relationships to international issues, including on the Intersection of Psychology and Environmental Protection, Migration, Ebola, and Women around the World. She conducts original psychosocial resilience trainings and workshops worldwide, for which she also co-writes and performs music, e.g., after 9/11, the Japan tsunami, China earthquakes, Caribbean floods, school shootings, Middle East terrorism, and during COVID-19. Among her many awards are the Humanitarian Award for Lifetime Achievement in Global Peace and Tolerance, Lifetime Achievement from Society for Media Psychology and Technology of the American Psychological Association, and being named Grand Commander of the Order of the Rokel by the President of Sierra Leone.

Erica Layer, MSPH, is the Director of Digital Health Technical Delivery for HealthEnabled and the former Chief Executive Officer at D-tree International, a global digital health organization working to drive health systems change through digital technology. She is a public health professional experienced in leading system transformations through integration of digital technology into health systems in low- and middle-income countries, resulting in more equitable and accessible quality healthcare delivery. Having lived in Sub-Saharan Africa for over 10 years, Erica has a deep appreciation of designing interventions for the context and local capacity building to support sustainable and locally-owned solutions. Erica has been recognized as an inspiring leader changing the face of global health innovation and is part of the World Health Organization roster of experts in digital health. Her work spans from the grassroots level to global policy, leading to an understanding of the balance between top-down and bottom-up approaches needed to create impact.

Tao Li, MB, MD, PhD, is a Chinese academic psychiatrist and current Dean of the Mental Health Center affiliated to Zhejiang University School of Medicine. She is a clinician, professor, and researcher, specializing in topics relating to molecular genetics and mental illness and was an early adopter of cutting-edge machine learning applications in the research groups that she leads. As a student, Dr. Li worked at the National Center for Gene Research, Chinese Academy of Sciences. From 2008 to 2020, she was a professor and researcher at West China (Huaxi) Hospital, Sichuan University, leading the research team there and was on the staff at the Hospital from 1997, when she started as a postdoctoral research worker. Dr. Li has also taught molecular genetics at Tibet University Medical Science School. As director of the Mental Health Center at West China Hospital, in 2012, she led a collaboration with the University of Massachusetts Medical School in studying tobacco addiction. Her work focuses on the genetics of mental illness, especially relating to hereditary schizophrenia. Her work has helped encourage further research into molecular genetics in Western China. She also studies topics relating to Tibet and the psychology of the people living there.

Sara Liébana, PhD, is full time Professor at the Universidad Europea de Madrid (UEM). Her field of expertise is the psychological consequences of terrorism, being currently focused on the relationship between dysfunctional beliefs and psychological wellbeing in this collective—research carried out in the Universidad Complutense of Madrid (UCM). She holds a master's degree in General Health Psychology, and is currently a member of the UCM research team on "Intervention and Treatment in Clinical Psychology." She was the Vice-President of the Student Division (15) of the International Association of Applied Psychology (IAAP) and is a member of the IAAP special Task Force on Terrorism and Peace Building. She has also been involved in the recent Ukrainian refugee crisis, as a co-author of a psychological first aid manual for children and families who have fled due to the war, for which she has been recognized by the Madrid Psychology Council. She can be reached at email: sliebana@ucm.es.

Hanna Lissinna, MA, earned her master's degree in the Department of Clinical Psychology at Teachers College, Columbia University, where she was a student in Professor Judy Kuriansky's class on "Psychology and the United Nations." She is a member of the Student Division of the International Association of Applied Psychology. She earned her bachelor's degree in the Department of Psychology at Leuphana University Lüneburg in Germany. She has 5 years of research experience in how internet interventions for mental health are effective in the prevention as well as treatment of a variety of mental disorders. Ms. Lissinna plans to continue to advocate for the implementation of technology in providing quality mental health services in under-resourced areas.

Sonia Livingstone, DPhil (Oxon), OBE, FBA, FBPS, FAcSS, FRSA, is a professor in the Department of Media and Communications at the London School of Economics and Political Science. She has published 20 books on media audiences, especially on children and young people's risks and opportunities, media literacy, and rights in the digital environment. Her recent book is "Parenting for a Digital Future: How Hopes and Fears about Technology Shape Children's Lives" (Oxford University Press, with Alicia Blum-Ross). Sonia chaired LSE's Truth, Trust, and Technology Commission, and has served as Expert Advisor to the Council of Europe, Special Advisor to the House of Lords' Select Committee on Communications, President of the International Communication Association, and Executive Board member of the UK Council for Child Internet Safety. She has received five honorary doctorates and several international prizes. Since founding the EC-funded 33 country "EU Kids Online" research network, she has advised the UK government, European Commission, European Parliament, UN Committee on the Rights of the Child, OECD, ITU, UNICEF, and others on children's internet safety and rights in the digital environment. Sonia directs the Digital Future Commission (with the 5Rights Foundation) and the "Global Kids Online" project (with UNICEF). Other projects include the UKRI-funded "Nurture Network" and two European H2020 projects: "ySKILLS" (Youth Skills) and "CO:RE" (Children Online: Research and Evidence). She blogs at www.parenting.digital and tweets @Livingstone_S. See www.sonialivingstone.net.

David D. Luxton, PhD, MS, is a Professor in the Department of Psychiatry and Behavioral Sciences at the University of Washington School of Medicine in Seattle and the founder of Luxton Labs LLC. He has held positions with the US Department of Defense as a Research Health Scientist at the Naval Health Research Center and as a Research Psychologist and Program Manager at the National Center for Telehealth and Technology (Defense Health Agency). He has authored numerous scientific and policy-level publications and four books, and he's helped to develop national guidelines for telehealth and clinical best practices for technology-based treatments. He's served on several national- and state-level workgroups and he regularly provides consultation on the topics of telehealth, artificial intelligence, and military psychological health. An expert in AI ethics, he contributed to IEEE SA as Vice-Chair for Human or AI Interaction Transparency. Dr. Luxton is a licensed clinical psychologist and veteran of the US Armed Forces. He can be contacted at: ddluxton@uw.edu.

Catherine Machalaba, MPH, PhD, is the Principal Scientist for Health and Policy at EcoHealth Alliance. Her research examines the links between the health of humans, animals, and the environment, including the risks and impacts of climate-sensitive diseases. She was a lead author of the World Bank Operational Framework for Strengthening Human, Animal, and Environmental Public Health Systems at their Interface ("One Health Operational Framework") published in 2018 to assist countries and donor institutions in implementing One Health approaches. Over the past decade, she has served in leadership roles in the American Public Health Association, Future Earth, and the International Union for the Conservation of Nature. In 2021, she was appointed to the One Health High-Level Expert Panel, which provides an advisory role to the FAO, UNEP, WHO, and WOAH. She holds a master's degree in public health and a PhD in environmental and planetary health sciences.

Haifa Madi, MD, MBBS, MPH, is currently a Public Health Expert and Advisor to the Ministry of Health and Prevention of the United Arab Emirates. Previously, she was the Regional Director of Health Protection and Promotion Department at the World Health Organization of the Eastern Mediterranean Region Office WHO/EMRO. Prior to that, Dr. Madi was the Acting Director of Health, Deputy Director of Health, Chief of Health Protection & Promotion, and Chief of Family Health of the United Nations Relief and Works Agency for Palestine Refugees (UNRWA), responsible for the health program and services provided to the six million Palestinian Refugees in Jordan, Lebanon, Syria, the West Bank, and Gaza. Dr. Madi received her MBBS in medicine from the University of Jordan and holds a master's degree in International Public Health from the Harvard School of Public Health. She is also a graduate of the Middle East Educational Fellowship Program of the J.F. Kennedy School of Government at Harvard University. Dr. Madi has authored several published research articles in The Lancet, the Eastern Mediterranean Health Journal, PLOS One Journal, and two chapters in the book "Neonatal and Perinatal Mortality, Global Challenges, Risk Factors and Interventions" published by Nova Biomedical. In 2022, she was the main author of Monitoring Health for SDG3 in the United Arab Emirates.

Tirumala Santra Mandal, PhD, is Assistant Vice President (AVP of Research and Communication) at iKure Techsoft Pvt. Ltd. With a doctorate degree in corporate communications from the Indian Institute of Technology (IIT) in Kharagpur, she has worked in diverse fields of communications, such as lie detection and multimedia communication with the Defence Research and Development Organisation (DRDO). Tirumala has authored multiple publications for various peer-reviewed journals, and actively participates in global platform networking.

Carthi Mannikarottu served as the Advocacy and Communications Officer for the secretariat of the Civil Society Engagement Mechanism for UHC2030 (CSEM), co-hosted by WACI Health and Management Sciences for Health. She was also the focal point for the Health Systems and Response Connector on the Platform for Access to COVID-19 Tools (ACT)-Accelerator Civil Society and Community Representatives, co-led by WACI Health, Global Fund Advocates Network (GFAN), and STOPAIDS. Previously, she managed a global knowledge and learning project for HIV/AIDS at UNICEF and continues to contribute to HIV advocacy. She earned a Master of Public Health (MPH) with a health policy analysis concentration from Columbia University Mailman School of Public Health and a bachelor's degree in Global Health from the University of Pennsylvania. The CSEM is the civil society constituent of UHC2030, the global movement to build stronger health systems for universal health coverage (UHC). Membership is open to all civil society and community representatives advancing health, financing, and governance agendas relating to UHC.

Cherisse Mark, MA, is a native of Trinidad and Tobago, she earned a BA in Nuclear Medicine Technology from Wheeling Jesuit University, West Virginia, in the United States of America, and an MA in Healthcare Management from Management Centre Innsbruck in Austria. Currently, she is an Advanced Remote Care specialist in the Research & Development Department at MED-EL, an Austrian-based global medical technology company specializing in hearing implants and hearing solutions. She has spent over a decade researching telemedicine, with particular focus on cochlear implants and delivery into everyday clinical practice. Her professional goal is to be a part of helping remedy the healthcare access divide, using telemedicine to help create solutions that are sustainable, equitable, and provide best clinical outcomes.

Patricio V. Marquez, ScM, is a former World Bank Group (WBG) Lead Public Health Specialist. He retired in August 2019 after 32 years of service working in over 80 countries across the world, and on 2014–19, led the Global Tobacco Taxation Program and the Global Mental Health Initiative. In March 2020, he was called back to the WBG as Consultant to support the preparation and implementation of the US$6 billion COVID-19 Global Emergency Response Program in the health sector and the implementation of the US$20 billion Additional Financing for Vaccines and Vaccination across the world, and to co-author the "Better Together: A Mental Health and Well-being Strategy for the World Bank Group" which was adopted by the institution and launched in May 2022. Currently, he serves at the WBG as an Adviser for the implementation of its Mental Health and Well-being Strategy; Honorary Professor at the Centre for Health Economics, The University of the West Indies; Senior Associate at the Johns Hopkins University Bloomberg School of Public Health; and elected Governing Member of the Charles Darwin Foundation for the Conservation of the Galapagos Islands, Ecuador. Recently, he served as Member of the External Advisory Board of the University of Washington Global Mental Health Program; Pledge Champion at the Tobacco Free Portfolios Initiative; and member of the International Advisory Committee for the preparation of the 2023 World Non-Communicable Diseases (NCD) Conference held in Toronto, Canada on June 25–30, 2023. He is originally from Cuenca, Ecuador, and resides in the United States. Email: patriciovmarquez@outlook.com. X: @pvmarquez1956 Website: http://pvmarquez.com/.

Anita M. McGahan is a University Professor and George E. Connell Chair in Organizations & Society at the University of Toronto, where she is appointed at the Rotman School of Management and holds cross appointments to the Temerty School of Medicine and the Dalla

Lana School of Public Health. For 10 years, she was an advisor to Grand Challenges Canada, an impact-first investment organization. Her research emphasizes entrepreneurship in the public interest and innovative collaboration between public and private organizations. Professor McGahan is widely recognized as a leader in the fields of global health and strategic management for her commitment to organizational innovation in the interests of the unserved and underserved. Her current agenda involves development of a new stakeholder theory that illuminates how global health influences public and private institutions around the world.

Stéphanie Racine Maurice is a psychology PhD candidate at Laval University in Canada in the research and intervention program. Focusing her thesis on the psychosocial components of premenstrual dysphoric disorder, she aims to contribute to the body of knowledge about this disorder, with the goal of reducing inequalities and increasing available support resources. She has received the prestigious Pierre Elliott Trudeau Foundation Scholarship as well as a scholarship from the Fonds de Recherche du Québec and from the Social Sciences and Humanities Research Council for her research project. Stephanie is currently a member of the Clinical Cycle Science Community and of the Clinical Cycle Practice Community of the International Association for Premenstrual Disorders. A member of the International Association of Applied Psychology (IAAP) for many years, she is currently on the Executive Board of Clinical and Community Division and Vice-President of the Student Division. She can be reached at: stephanie.racine-maurice.1@ulaval.ca.

Patricia N. Mechael, PhD, is Senior Associate Faculty in the Center for Global Digital Health Innovation in the Department of International Health at the Johns Hopkins Bloomberg School of Public Health. She is also co-founder and policy lead at HealthEnabled, an Africa-based not-for-profit that helps governments integrate life-saving digital health solutions into their health systems. She leads HealthEnabled's work on the Global Digital Health Index and Maturity Model as well as with global partners including GAVI, the Vaccine Alliance, the WHO, and UNICEF. She has dedicated over 25 years to catalyzing global health movements, including 20 years to unlocking the power of mobile technology to drive groundbreaking innovation to solve global health challenges in over 40 countries. She is celebrated for her roles as a digital health thought leader, writer, researcher, professor, and executive director of the mHealth Alliance, and Executive Vice-President of the Personal Connected Health Alliance. She is a Rockefeller Foundation Bellagio Fellow, recipient of the Johns Hopkins University Knowledge for the World Distinguished Alumnus Award, member of the editorial board of the *Journal of Medical Internet Research*, and coeditor of the book: "mHealth in Practice: Mobile Technology for Health Promotion in the Developing World." She is the author of the middle grade fiction novel, "The Antidotes: Pollution Solution," which she wrote in collaboration with her son. She has a PhD in Public Health and Policy from the London School of Hygiene & Tropical Medicine and a master's degree in Health Science in International Health from the Johns Hopkins School of Public Health.

Tanya Mehdizadeh, MA, is currently in the PhD program in Ferkauf Graduate School of Psychology, Yeshiva University in New York. She earned a BSc degree in Biology and Psychology from George Washington University, and a master's degree in Clinical Psychology from Teachers College, Columbia University, where, under the mentorship of Dr. Judy Kuriansky, she developed a strong interest in investigating mental health in disaster-stricken populations globally and the use of psychosocial interventions to support resilient outcomes, especially in her native country of Iran. This led to a collaboration with Dr. Kuriansky and colleagues on a manuscript for publication on "Psychosocial Resilience and Recovery from Natural Disasters in Iran: The Case of Kermanshah." Tanya served as a course assistant for Dr. Kuriansky in her course on "Psychology and the United Nations," where she oversaw students' projects, organized modules about the UN Agenda and mental health, and liaised with guest diplomats for the class. She has 4 years of research experience in the fields of psychology, biology, and neuroscience. She plans to build a career dedicated to serving survivors of trauma as well as marginalized, immigrant, and underserved populations as a clinician, researcher, and global mental health advocate.

Sylvia Paola Mendoza, MALD, is a Mexican career diplomat. She entered the Foreign Service in 2010 and is currently posted to the Mission of Mexico to the United Nations in New York where she oversees sustainable development issues, such as food security, global health, and international cooperation, as well as other related issues on the 2030 United Nations Agenda, and where she has facilitated several resolutions in the Second Committee. Before coming to New York, she was posted in the Embassy of Mexico in Lima, Peru, for 5 years, where she worked on political issues, cooperation, and economic affairs. She graduated from the Fletcher School of Law and Diplomacy in 2008 with a Master of Arts in Law and Diplomacy degree. She recently published an article on the link between sustainable development and sustaining peace, and the importance of the UN 2030 Agenda to strengthening that link.

Kertti Merimaa is a Product Manager of Nortal, overseeing the development of the leading hospital information system in Estonia. Before focusing on health informatics in hospitals, she served as Chief Digital Health Interoperability Officer at the Ministry of Social Affairs in Estonia, where her role included steering the e-health governance and health data interoperability developments of the national e-health system. She was co-leader for the creation of the e-health strategic vision for 2030 and part of the project lead team for its implementation. She has an academic background from Tartu University in Political Science and Knowledge Management, focusing on the topic of information and digital literacy.

Tahila Mintz is the Founding Executive Director of OJI:SDA' Sustainable Futures and a multidisciplinary artist. Her practice works to facilitate a greater intersectional understanding of indigenous cultures. Her work engages with ancestral systems of matriarchy and gender equality, as well as contemporary issues impacting native peoples. Predominantly creating alongside indigenous communities of the Americans, Mintz strives to reclaim bodies of knowledge that have been suppressed through colonial intervention and foster a closer connection to the natural world. Her organization focuses on a vision where indigenous people are seen, heard, healthy, and thriving. Its programs create ancestral knowledge, land-based curriculums for youth populations within formal and informal education systems; summer camp experiences for indigenous youth; disaster relief; and other models of community support. Tahila earned her MFA at the University of Texas, Austin, and was a lecturer at Texas State University and Central Texas College. She has lived and worked around the world and been photographing for more than 20 years. Her work has been published in National Geographic and is in Museum Collections. www.Tahila.net | www.Ojisda.org | Instagram @tahilasnap.

Nomtika Mjwana is the Project Manager for Digital Health & Rights for the Global Network of People Living with HIV (GNP+). In her current role, she provides oversight and service design on the implementation of VOICE+, a digital activist hub/platform that provides information, human rights advocacy support, and connection between networks of people living with HIV. In addition, she sits on the Digital Health Rights Consortium, which looks at how new/digital technologies and artificial intelligence (AI) are being used to transform health systems, as well as the risks and benefits for young people, who are often disenfranchised. In this capacity, she is a researcher and local network support partner.

Taylor Mulligan-Stark, MA, is a graduate student at Pace University School-Clinical Child PsyD Program. She received her Master of Arts degree from Pace University's MA Psychology Program in May 2022. She is a student member of the American Psychological Association, Society of Clinical Psychology Division 12 of the American Psychological Association, and National Association of School Psychologists. Consistent with her commitment to advocacy for global action, she interned with the International Council of Psychologists, where she acted as co-chair of the Social Media Committee for the Psychology Coalition of NGOs in consultative status with the Economic and Social Council at the United Nations, member of the planning committee for the 15th Annual Psychology Day at the United Nations, and assistant to the NGO Committee on Children's Rights, while also serving as a Certified Grief Educator for peers at Emma's Place, a Grief and Loss Center for Children and Families located in her local community of Staten Island, New York. Her current research interests include understanding the impact of parental bonding styles on college-aged student's emotion regulation strategies in relation to their perceived and current stressors.

Vivek Nair is an undergrad in Aerospace Engineering at the University of Illinois Urbana-Champaign. He is interested in the applications of aerospace technology for betterment of lives on Earth. Someday, he hopes to be on the team that sends the first humans to Mars. Vivek's interests are entry, descent, landing (EDL), and hypersonics. In the past, he has worked for Safran and UNAIDS. Currently, as Structures and CAD lead for Design Build Fly UIUC, he works on RC aircraft design. You can contact and find out more at vivek-nair.com.

Jasmine M. Noble, PhD, is a Postdoctoral Fellow at the University of Alberta in the Departments of Computing Science and Psychiatry. Jasmine's primary research interest is in mental health and addictions. Her current research focuses on innovative public health interventions aimed at improving mental health and wellbeing in the general public and individuals managing mental health challenges. She is particularly interested in utilizing advancements in science, technology, and machine intelligence, including artificial intelligence. She received her Master of Public Policy from the University of Calgary and completed a PhD in Psychiatry at the University of Alberta, doing valuable research with pregnant women burdened by substance use disorder. Prior to completing doctoral training, Jasmine served as Principal Secretary and Policy Advisor to the late Honourable James M. Flaherty, former Federal Finance Minister, and advised him in his creation of two federal budgets, and as Special Assistant to the Right Honourable Stephen Harper, Prime Minister of Canada, before working on the staff of the Institute of Health Economics and, subsequently, beginning a postdoctoral fellowship in digital health with a focus on mental health support using artificial intelligence (AI).

Mohammed Nurhussein, MD, is the National Chair of the United African Congress, which has co-sponsored many humanitarian efforts (e.g., raising awareness about ebola, and about COVID-19 before the news was widely released) and events at the United Nations related to the Sustainable Development Goals. He served as Chief of Geriatrics at State University of New York Downstate Medical Center in Brooklyn until his retirement as Associate Professor of Medicine Emeritus. He is on the Executive Board of Ethio-American Doctors Group (EADG), a consortium of more than 350 Ethiopian doctors in the US and Canada who are committed to building a state-of-the-art, internationally accredited hospital in Addis Ababa, Ethiopia—a US$100 million project. Doctors from Europe and Australia have since joined the group. Dr. Nurhussein is also a member of the Executive Board of Brooklyn for Peace, an organization dedicated to the cause of peace and social justice and an active member of Physicians for Human Rights (PHR), including on the NY chapter task force in the campaign for the total elimination of landmines, which earned PHR a Nobel Peace Prize in 1997. Dr. Nurhussein is the author of a memoir, "Made in Ethiopia," published by Africa World Press/Red Sea Press in 2020.

Ian O'Donnell, MA, is the Team Leader for Business Models and Scaling at the headquarters of the International Federation of Red Cross and Red Crescent Societies (IFRC) in Geneva, Switzerland. His role involves advising the IFRC and National Societies on developing effective strategies for scaling, partnership building, and resource mobilization toward shared resilience targets. Ian has also led the One Billion Coalition for Resilience (1BC), a global initiative to leverage partnerships, shared targets, and a coalition approach to scale up action on community resilience. In addition, in 2020–2021, Ian co-led the IFRC's working groups to develop a digital transformation strategy. In previous roles, Ian started with the Red Cross Red Crescent (RCRC) working with the American Red Cross branch in San Francisco, and has engaged in many preparedness and response activities globally with both the American Red Cross and the IFRC. He also worked for several years with the Asian Development Bank before returning to the RCRC to manage the research and knowledge management portfolios at the Global Disaster Preparedness Center, hosted by the American RC. Ian received a master's degree in City and Regional Planning from the University of California, Berkeley.

Patty O'Hayer is a communications and public affairs professional with over 20 years' experience in change, policy, and partnerships. She is an international politics graduate from the George Washington University, and is a member of the Board of Directors of Raleigh International and a European trustee of the Global Poverty Project. Since 2013, she has been Reckitt's Global Head of Corporate Affairs, Government Relations, and Strategic Partnerships. Prior to this, she worked as Director of Communications and Campaigns at Plan International, and as Vice-President of Global Communications for Unilever. Patty is a member of the Strategic Advisory Group to the WHO Foundation and sits on the Health Innovation Exchange Advisory Board.

Werner Obermeyer is Director of the World Health Organization (WHO) office at the United Nations. He manages the promotion of health in the General Assembly and other intergovernmental bodies, WHO's relations with UN system entities in New York, and nongovernmental organizations. He was responsible for the negotiations which resulted in several Political Declarations, including those of 2011, 2014, and 2018 on the Prevention and Control of Non-Communicable Diseases, 2016 on Anti-Microbial Resistance, 2018 on Tuberculosis, 2019 on Universal Health Coverage, and in 2022 on Road Safety. He oversees initiatives related to the UN Sustainable Development Goals, global health and foreign policy, universal health coverage, health security and social protection, road safety, malaria, NCDs, climate change and health, disabilities, narcotics and drugs, mental health, and tobacco control. Prior to joining the WHO, he was the representative of the UN Environment Program in New York City, responsible for environmental governance, health determinants, and sustainable development. Before joining the UN he served in the South African Foreign Service.

Zerin Osho (she/her), BBA (Bachelor of Business Administration), **LLB** (Bachelor of Law, Hons), **MALD** (Masters of Law and Diplomacy) is the Coordinator of the International Solar Alliance (ISA), on secondment to the United Nations General Assembly from the Institute for Governance and Sustainable Development (IGSD). Previously, Zerin served as a Technical Expert at ISA and led the development of its 7th program on Solarizing Heating and Cooling Systems, a program designed to solarize the global thermal demand. Prior to joining ISA, Zerin served as a Special Advisor to the Under Secretary General of the United Nations, Mr. Erik Solheim, at UNEP. During her time at UNEP, she was instrumental in the launch of the UNEP Cool Coalition—an international platform for the promotion and adoption of the Montreal Protocol and energy efficiency to achieve climate benefits to up to 1°C. Zerin is a lawyer by training and holds a master's degree in International Law and Diplomacy from the Fletcher School. She was awarded the prestigious Legislative Assistants to Members of Parliament (LAMP) Fellowship in 2015, a unique opportunity for young Indians to learn law-making and public policy.

Daniel Otzoy-García, MSIT, is a Guatemalan computer science engineer with postgraduate qualifications in Information Technology and Health Informatics. He is the co-founder and CEO of the Central American Network for Health Informatics (RECAINSA), an expert member of the Digital Health Working Group of the ITU Broadband Commission and UNESCO, and of the global Digital Public Goods Working Group led by UNICEF and the Digital Public Goods Alliance. Furthermore, he is a Digital Health Technical Advisor at John Snow Inc., a public health consulting firm dedicated to improving health equity for better health outcomes, primarily developing his activities in Africa and Southeast Asia with projects funded by USAID. He is also a specialist in international digital transformation with over 16 years of experience working in Latin America and the Caribbean on digital health projects financed by PAHO/WHO, the Inter-American Development Bank, the World Bank, CDC, the Global Fund, USAID, and others, implementing technologies in the public health and healthcare sectors, and promoting the adoption of digital health strategies such as EHR/EMR, telemedicine, mHealth, and public health information systems, among others, as well as the adoption of health information and interoperability standards.

123Dentist is Canada's oldest and second largest network of dental practices. From a single practice in Vancouver, owned by CEO Dr. Amin Shivji and two partners, it has grown to over 250 practices, with 3200 team members serving 2 million patient visits annually with support centers in Vancouver and Toronto. 123Dentist was founded on two key principles: putting patients above all else, and that dentists and staff are the most important factors in keeping patients happy. The vision is to be the dental community all Canadian dentists and dental professionals want to belong to. 123Dentist offers alternative partnership models to meet the needs of each individual dentist, including the opportunity to retain ownership of the clinic. This unique model enables dentists to focus on providing optimal care to patients, while an experienced team provides a wide range of support services.

Amanda Pain, MPH, MSW, is Senior Manager, Communications, at VillageReach, whose role is to amplify the organization's impact helping technical teams with communications strategy and producing written materials. She has worked in global health and communications for 20 years and has created written materials for the WHO, UNICEF, and PATH, as well as the Bill and Melinda Gates Foundation. She holds master's degrees in Public Health Communications and Social Work, as well as a bachelor's in Journalism.

Muhammad Ali Pate, MD, MBA, a Nigerian physician, is currently the Minister of Health and Population for the Federal Republic of Nigeria, and Professor of the Practice of Public Health Leadership in the Department of Global Health and Population at Harvard University in Cambridge, Massachusetts, USA. He also serves as the co-chair of the Global Council SDG3 (Health), "SDGs in Action" Initiative, Dubai, United Arab Emirates, dedicated to achieving the United Nations Sustainable Development Goals and Universal Health Coverage. Previously, he was the Julio Frenk Professor of the Practice of Public Health Leadership in the Department of Global Health and Population at the Harvard T.H. Chan School of Public Health in Boston, Massachusetts, USA, and before that, he served as Global Director of the Health, Nutrition and Population Global Practice of the World Bank and the Director of Global Financing Facility for Women, Children and Adolescents. During the COVID-19 pandemic, seeing the pandemic as a "launch pad" to build more equitable systems, protect lives, and improve the health of populations all over the world, he and World Bank colleagues pivoted to provide financial support to more than 100 countries, including for vaccine accessibility and delivery. Also previously, Dr. Pate was Chief Executive Officer of Big Win Philanthropy which partners with African leaders to transform visions for children and youth. In 2011, he was appointed as Minister of State for Health in the Federal Republic of Nigeria in which role he established a global coalition of private and public partners to support health system improvements; and successfully implemented innovative approaches to polio eradication, strengthen routine immunizations, improve maternal, newborn, and child health, including prevention of mother-to-child transmission of HIV. A physician trained in Internal Medicine and Infectious Diseases with studies at the University College London, he also earned an MBA degree from Duke University, and a Master's degree in Health System Management from the London School of Hygiene & Tropical Medicine in the UK.

Meredith del Pilar-Labarda, MD, DrSD, is a Professor and the Chairperson of the Department of Medicine at the University of the Philippines Manila School of Health Sciences, an innovative school known for its community-oriented stepladder health professions education curriculum. She is the lead for Social Innovation in Health Initiative (SIHI) Global Network Secretariat, a global network of passionate individuals, organizations, and institutions advocating for social innovation in health and advancing research in social innovation, launched by the Special Programme for Research and Training in Tropical Diseases (TDR) in 2014, as well as the lead for SIHI Philippines Hub. She is also a Doctor of Social Development and the school's lead coordinator for health leadership and governance programs in partnership with government and nongovernment organizations in co-designing training modules, providing capacity building training and coaching for local chief executives, municipal health officers, and other important stakeholders in the local health system and community. She also leads Buhat Tacloban, a church-based local disaster response volunteer group that has responded and coordinated disaster response since super typhoon Haiyan in 2013. Her advocacies, work engagements, and research interests include health and social development, social innovations, gender intersectionality studies, transformative health leadership and governance, community engagement and empowerment, and socially accountable health professions education, among others. She can be reached at mdlabarda@up.edu.ph and/or meredithdelpilarlabarda@gmail.com.

William C. Philbrick is an information and communications technology (ICT), systems strengthening, and health specialist with over 24 years of experience managing and supporting health projects in lower- and middle-income countries (LMICs). Within the last 15 years, he has strategized on how to effectively integrate ICT into health and other projects in LMICs to improve health outcomes. He has written publications addressing how ICT (digital technology) can be leveraged to address health and child protection challenges. He specializes in integrated multi-sectoral approaches, private sector engagement, systems strengthening, data use, and the evidence-based use of digital technology. He has served as Manager of Orphans and Vulnerable Children (OVC) Programs and Director of HIV/AIDS and Emerging Infectious Diseases at CARE USA.

Mauro Pantaleo is Chairman and CEO of P&P Group—a group of companies that develops novel technologies in the healthcare sector, mainly focused on infection prevention and control (IPC). Mauro is also Co-founder and CEO of an Italian company that owns the patents for RETRAGO, an automatically retractable safety syringe designed to prevent needle stick injuries and blood-borne diseases, and has more than 20 years private and public sector experience in management and finance, as General Manager public entities in Italy and for Barclays Capital as Director in the Public Sector Team in London and Rome. Furthermore, he is Chairman of the Ecosistemi Foundation, a foundation with a goal of implementing tools and strategies for sustainable development, a board member of the Health Innovation Exchange (HIEx), a platform that scouts and brings together innovative solutions for health, particularly in the Global South, and a member of the WHO-sponsored table on adoption of best practices for the prevention and control of diseases (POPS IS-Injection Safety).

Ashley Porto, MA, is a Senior Business Development Manager at LogiNext where she has worked on numerous projects within the technology space. Formerly she was an Account Executive at CloudKitchens and a Solutions Executive at Enhatch. Enhatch creates patient-specific solutions for medical device companies using artificial intelligence and 3D printing to deliver a more personalized experience for surgeons and patients. In her role, she sells to device manufacturers, onboards and trains clients, handles support tickets, and conducts executive interviews at trade shows and conferences. She is an expert in a training platform called Reprise where she produces digital videos to train customers on numerous solutions Enhatch offers. Previously, she worked at Yelp in sales, selling advertising packages to local businesses and implementing email marketing campaigns. She recently launched a podcast called ASHx Podcast, which merges spirituality and the music industry through interviews with music artists about how spirituality and mental health play a role in their success. She has interviewed well-known artists including Le Youth, Moonlander, and Sam Wolfe. The ASHx Podcast has recently been ranked #122 for Music Interviews in the US for Podstatus and featured in the United Business Journal. She earned a master's degree in Clinical Psychology from Teachers College, Columbia University, where she was mentored by Dr. Judy Kuriansky in the "Psychology and the United Nations" class, She is also a member of the Student Division of the International Association of Applied Psychology.

Ananya Raghuram, MA, is now with the Center for Cognitive and Psychology, New York, NY, United States. She earned her Master of Arts degree in the Clinical Psychology program in the Department of Counselling and Clinical Psychology at Teachers College Columbia University. Her current interestes focus on global mental health and neuropsychology, notably the need for accessible mental health care coverage across diverse populations, and exploring the impact, efficacy, and longevity of emerging digital mental health care platforms. Thus, she is a member of the Student Division of the International Association of Applied Psychology (IAAP) and the eMental Health International Collaborative (eMHIC). She was inspired to accomplish this work by taking the course on "Psychology and the United Nations" where she was taught and mentored by Professor Judy Kuriansky. She earned her BA degree at New York University in New York, New York, USA, in Psychology with joint areas of focus in Clinical Psychology and Cognitive Neuroscience and a minor in Child and Adolescent Mental Health Studies. She is an international student from Chennai, India, brought up in Singapore. She enjoys creating and weaving elements of both visual and fiber arts with neuroscience and psychology through prints that map out her mental health journey and interest in visual perception.

Onisoa Rindra-Ralidera leads and represents the Options team in Madagascar for the FCDO/UKAID/WISH2ACTION program and is also a health policy and planning advisor, qualified in Health Policy and Universal Health Coverage, and Policy Development and Advocacy for Global Health. She was Commissioner of the AHAIC 2020–2021 global health convening. She is an expert in advocacy, with a strong background in reproductive health and family planning as well as water, hygiene, and sanitation, and training/facilitation. She and her team initiated a public-private partnership with the Ministry of Public Health in Madagascar integrating family planning services into the insurance care packages. She is the lead author of a paper in publication that has been selected by the WHO Impact Best Practices on "Removal of taxes for contraceptives in Madagascar: Strategic advocacy with national level decision-makers to increase budget for FP commodities" and is co-author and developer of a 2018 publication of an article entitled: "Contributing to the Achievement of the Demographic Dividend in Madagascar" in Health Policy Plus. She presented a paper in 2019 on "The power of advocacy partnerships and strengthening policy implementation in Africa" at Health Policy Plus: ICFP 2018 Kigali, and a paper on "AFP SMART, a step-by-step advocacy and its gains" during the 18th General Membership Meeting of Reproductive Health Supplies Coalition Brussels, March 2018, for both of which she was the lead author. Also a storyboard writer, she is passionate about the health and wellbeing of women, girls, and children.

Girish Ramachandran is the President of Tata Consultancy Services (TCS), Asia Pacific. He oversees a US$2.5 billion business in a diverse and dynamic region covering 12 countries and over 50,000 employees. A result-oriented technology and sustainability champion, Girish is responsible for the growth, strategy, and business results, with a dual focus on driving digital transformation and accelerating growth for TCS' customers and being a trusted "growth and transformation" partner for businesses, governments, and society at large. He believes in harnessing the power of technology and innovation to catalyze and drive sustainable business outcomes with a proactive stance on climate change. He contributes actively to the wider business and local communities, holding the role of Chair of the Confederation of Indian Industry—India Business Forum (CII-IBF) ASEAN and Oceania chapters, and serving on the Supervisory Board of the Global Reporting Initiative (GRI), a pioneer in sustainability reporting standards, among other appointments. In recognition of his services to the community, Girish is a recipient of the Frans Banninck Cocq Medal from the Mayor of Amsterdam and was given Special Recognition (Individual) by the People's Association of Singapore in 2018. Girish holds a Master of Business Administration from Jamnalal Bajaj Institute of Management Studies and is a Computer Science graduate from PSG Tech. He also completed the Wharton's Corporate Governance program in 2022, which is a testament to his commitment to lifelong learning.

Nithya Ramanathan is the CEO and Co-Founder of Nexleaf Analytics, a nonprofit technology company that partners with countries to ensure they have the data they need to build lasting solutions that improve the health of their people. Nexleaf's unique approach brings together sensor technology, data analysis, and user engagement to identify, understand, and address big problems. Nithya and her team have achieved large-scale impact by working shoulder-to-shoulder with Ministries of Health—rather than attempting to bypass governments—to scale technologies refined through collaboration with personnel at every level of the health system. A computer scientist specializing in sensor technologies and data analytics, Nithya is a tech-for-good pioneer and a leading thinker on global vaccine distribution. She earned a PhD in Computer Science from UCLA and a BS in Electrical Engineering from UC Berkeley.

Fatima Rizwan, MA, received her master's degree in Clinical Psychology from Teachers College, Columbia University, with a concentration in Global Mental Health and Trauma, where her work was inspired by being a student in Dr. Judy Kuriansky's class on "Psychology and the United Nations." Her research is focused on the trajectories of trauma and resilience through the lens of cultural neuroscience, given her devotion to learning the neuropsychology of trauma and resilience and discovering innovative methods to recover damage and restore optimal functioning. Her academic aspirations include the development and adaptation of culturally informed interventions to address universal mental health needs, increase access to and acceptance of psychotherapy, and ensure equitable treatment for all. A Pakistani national, she grew up in the United Arab Emirates, where she was affiliated with The Hope and Light Foundation, an NGO advocating mental health awareness in communities. She relocated to Pakistan to complete her undergraduate studies in Applied Psychology, after which she worked for a year in public and private mental healthcare settings. Recognizing the barriers to mental healthcare in the developing world prompted her interest in addressing and catering to the mental health needs of under-resourced regions.

Erin Ross writes about the intersection of public health and technology. She has worked at Nexleaf Analytics, the Harvard University Global Equity Initiative, and the Berkman Center for Internet & Society at Harvard Law School (now the Berkman Klein Center). Her writing has appeared in the Stanford Social Innovation Review and in books including Broken Pumps and Promises and Health of People, Health of Planet and Our Responsibility. She is a graduate of Wellesley College. Erin enjoys moderating panels and engaging with impassioned thinkers about virtually any topic.

Robert Ross was the CEO of St. Joseph's Addiction Treatment and Recovery Centers, a nonprofit in Saranac Lake, New York, USA, until 2022, and now operates a healthcare and organizational development consultancy. Under his leadership, St. Joseph's grew from 100 employees to 270, from a yearly budget of $5,000,000 to $22,000,000, and added a number of innovative programs for underserved populations, including: a veterans residential program; an adolescent inpatient program; a detox program that includes 16- and 17-year-olds—the first in New York State; four supportive housing programs; and integrated mental health and addiction outpatient clinic services. St. Joseph's became a Certified Sanctuary Program, and was recognized 10 times as a "Best Place to Work in New York State," as a *Newsweek* magazine "Best Addiction Treatment Center" in the US in 2021, and as a "Best Place to Work in Healthcare" by *Modern Healthcare* magazine in 2022—including recognition as one of the five most family-friendly healthcare places to work in America. Previously, Bob was Executive Deputy Director of the New York State Division of Alcoholism and Alcohol Abuse, serving as Acting Director for 1 year; was Deputy Director of the New York City Bureau of Alcoholism Services; and for 22 years, ran his own healthcare consulting firm with national and international clients. In 2019, he received the Charlie Devlin Award for 50 years of leadership and advocacy, and for an exemplary career in both governmental and community-based organizations. Email: RobertRossADK@gmail.com.

James Sale is the Policy, Advocacy, and Financing Director of United for Global Mental Health. He joined UnitedGMH from *Save the Children* where he led global health financing policy and advocacy and previously established Transparency International's health sector anticorruption program. James began his career in public financial management and governance at Crown Agents after working on vaccine surveillance for the World Health Organization. James has worked across Sub-Saharan Africa and South Asia.

Rajeev Sadanandan is the CEO of Health System Transformation Platform, a think tank on Health Policy and Systems in New Delhi. Formerly, he worked as Health Secretary of Kerala State, India, where he managed a successful response to an outbreak of Nipah virus and many smaller health emergencies. He had overseen reform of the health sector in Kerala, focusing on strengthening primary care, which contributed to Kerala's ability to set up a decentralized response to the COVID-19 pandemic. During the peak of the pandemic, he was engaged as Advisor to the Chief Minister of Kerala on management of COVID-19 in the state. Currently he is a member of the Technical Advisory Panel of the Pandemic Fund, set up to support countries to build health system resilience to future pandemics. Rajeev's current work focuses on generating demand and building capacity for health policy and system research, and conducting implementation research, in partnership with government and academic institutions, on potential areas of reform in the health sector. Contact: rsadanandan@hstp.org.in or rsadanandan@gmail.com.

Rayna Sadia, PhD, earned her degree from Lingnan University in Hong Kong, having written her dissertation about traffic safety and how personality traits of the driver could influence the driving environment in terms of safety of all stakeholders. She is currently an Assistant Professor at Rawalpindi Women University, Rawalpindi, Pakistan. As an early career psychologist, she recently won two National Research Projects for Universities awards from the Higher Education Commission of Pakistan on traffic safety and mental health of women in Pakistan. Along with her research interests in driving behaviors and traffic safety, she has worked on childhood trauma and its impact on behavioral problems of adolescents. She has been an active member of the Society of Personality and Social Psychology (SPSP) and of the Student and Early Career Division of the International Association of Applied Psychology (IAAP) since 2018, which provides a platform for students and early career researchers to discuss issues pertaining to research and to work on collaborative projects for the greater good of the world. Contact: raynasadia@ln.hk or rayna.sadia@f.rwu.edu.pk.

Udani Samarasekera, MSc, is currently a freelance health and science writer and editor located in London, England, United Kingdom, including contributing to the Lancet. Previously, she worked for the Lancet for 15 years in various positions, including as Deputy Editor for The Lancet Child & Adolescent Health, and as Senior Editor and Press Officer. Prior to that, she was a health reporter for the Cape Times Newspaper, Cape Town, South Africa, and was Science Press Officer for the nonprofit organization Cancer Research UK. She earned a Master of Sciences degree in Science Communication from the Imperial College London, and her BSc degree with honors in Human Genetics from University College London.

Jagjeet Sareen is a Partner at Dalberg Advisors where he co-leads Dalberg's Global Climate Practice, advising clients on issues related to climate strategy, finance, competitiveness, and impactful mitigation and adaptation solutions. He was the Assistant Director-General (of Strategy) at the International Solar Alliance (ISA) Secretariat, where he is on secondment from the World Bank Headquarters. At the World Bank, he served as a Senior Policy Officer at the Climate Investment Funds. He has been working on global climate change and green financing for over 16 years at MDBs (the World Bank and Asian Development Bank) and the UN, and its funding agencies (UNFCCC secretariat and the Green Climate Fund). He was awarded a Chevening Gurukul Fellowship 2020 for Leadership and Excellence by the UK Foreign, Commonwealth and Development Office, and hosted by the University of Oxford. He is a member of Chatham House, a world-leading policy institute with a mission to help governments and societies build a sustainably secure, prosperous and just world, headquartered in London, and of the International Institute for Strategic Studies (IISS).

Shekhar Saxena, MD, FRCPsych, DAB, is Professor of the Practice of Global Mental Health at the Department of Global Health and Population at the Harvard T. H. Chan School of Public Health. A psychiatrist by training, he has served in various positions at the World Health Organization (WHO) since 1998. From 2010 to 2018, he was the Director of the Department of Mental Health and Substance Abuse at the WHO. In 2017, he received the prestigious Leon Eisenberg Award from Harvard Medical School. Author of more than 300 academic papers, he functioned as an editor of the Lancet Series on Global Mental Health 2007 and 2011, and the Lancet Commission on Global Mental Health and Sustainable Development 2018. His expertise includes providing advice and technical assistance to policy makers on prevention and management of mental, developmental, neurological, and substance use disorders and suicide prevention.

Daniel Schaudel, PhD, is a physicist who is the head of the Clinical Systems Group within the Research & Development Department at MED-EL, an Austria-based global medical technology company specializing in hearing implants and devices. He joined MED-EL in 2003, and has 19 years of experience in the field of cochlear implants and telemedicine, especially in product development and research activities. He earned his PhD degree in physics from the Ludwig Maximilians University in Munich, Germany, and a certificate in a biennial program of study in medical technology from the Technical University Kaiserslautern, Germany.

Katherine E.A. Semrau, PhD, MPH, is director of the BetterBirth Program at Ariadne Labs, where she oversees the research and execution of the program, which aims to improve the quality of care, minimize complications, and end the preventable deaths of women and newborns through effective implementation of evidence-based, scalable solutions at the frontline of care. Katherine is an Associate Professor of Medicine at Harvard Medical School and an Associate Epidemiologist at the Brigham and Women's Hospital in the Division of Global Health Equity. Dr. Semrau has over 20 years of experience in the fields of maternal, newborn, and child health and epidemiology. Her research has focused on prevention of maternal and child mortality, improvement of quality of care, and prevention of mother-to-child transmission of HIV. Currently, her research portfolio centers on designing, testing, and implementing maternal and newborn health interventions in resource-limited settings, including projects focused on the Safe Childbirth Checklist, Patient Education, and Facility Readiness. Additionally, she leads a project focused on identifying solutions for the most vulnerable newborns, especially low birthweight infants. Dr. Semrau has a PhD in Epidemiology from Boston University and an MPH in International Health and Epidemiology from the University of Alabama-Birmingham. She can be reached at: ksemrau@ariadnelabs.org.

Rahul Sharma, PhD, is the Chief Information Officer (CIO) at the FIZ Frankfurt Biotechnology Innovation Center in Germany. Dr. Sharma earned his PhD in Bioinformatics at the Goethe University in Frankfurt and did post-doctorate research work at Universitäts-Klinikum in Frankfurt. During his career in several biotechnology companies, Dr. Sharma has played a key role in developing artificial intelligence-based computational algorithms for large-scale integration of genetic, transcriptomic, proteomic, microbiome, and metabolic data for disease-signature mining. Since 2019, Dr. Sharma has been focused on developing computational pipelines for analyzing Ayurvedic treatment-focused biomedical data. He can be reached at: rahul.sharma@fiz-biotech.de; LinkedIn: https://www.linkedin.com/in/rahularjun86/.

Shannon Shibata-Germanos, PhD, is an international tech investor and global health specialist. Her thematic areas of interest are health-tech, med-tech, frugal innovation, health systems efficiency tech and personalized medicine with an emphasis on companies with a dual market strategy enabling greater access to equitable healthcare with a global footprint. With two decades of experience in political policy analysis, conflict management, and institutional development in the Middle East and Sub-Saharan Africa, Shannon has advised heads of state, royal families and CEOs about complex policy and business negotiations. One of her areas of particular focus is global public health policy, especially building partnership ecosystems (including policy makers and government agencies) for commercial start-ups with unique medical technologies. Her passion is how to leverage technology for health systems strengthening in the Global South emphasizing the need for robust change management for effective adoption of digital technology. She serves as Global Health lead at Proximie, an interactive digital surgical collaboration software, and along with her other professional responsibilities, she sits on the International Advisory Board and Investment Committees of KHP Ventures (a joint venture between Kings College London, Kings College Hospital NHS Trust, and Guy's and St. Thomas' NHS Foundation Trust), the Global Bio Fund (Jaya Ventures), and is Vice Chairman of the Geneva-based Health Innovation Exchange. She holds a PhD in Neuroscience from University College London. She can be reached via her website at www.lacunaadv.com

Zara Shubber, MBBCh, MSc, is a Senior Health Specialist at the World Bank's Health, Nutrition, and Population Global Practice where her work focuses on Health System Resilience and Climate and Health. Zara has 15 years of experience working in a variety of healthcare contexts across Europe, Africa, the Middle East, and Asia. Before joining the World Bank, Zara worked as a consultant for a number of international organizations including the World Health Organization and the Bill and Melinda Gates Foundation, as well as in an academic setting at Imperial College London's Department of Infectious Disease Epidemiology. A physician by background, Zara has published 35 peer-reviewed journal articles in the field of health.

Megha Siddhanta, MA, is a researcher at the Yale School of Medicine in the Harpaz-Rotem Lab working on PTSD and its neurological sequelae. She took this job after earning her Masters degree in the Department of Clinical Psychology from Teachers College, Columbia University where she developed an interest in technology and mental health while studying under Dr. Judy Kuriansky in her course on "Psychology and the United Nations," and also volunteered in the Global Mental Health Lab of Dr Helen Verdeli for 2 years where she developed an interest in making mental health interventions accessible and cost effective in low- and middle-income countries. She aims to pursue a PhD degree in Clinical Psychology.

Patrik Silborn has more than 20 years of experience in global health and development, leading partnerships, resource mobilization, and grants management across Latin America, Africa, and Asia. Between 2017 and 2020, he led External Relations and Financing at the Asia Pacific Leaders Malaria Alliance—an alliance dedicated to eliminating malaria in Asia by 2030. Previously, Patrik spent a decade working at the Global Fund to Fight AIDS, Tuberculosis, and Malaria in varying roles. As a Senior Fund Portfolio Manager, he served multiple countries in Africa and in Central America; and as Head of Private Sector Engagement, he led the organization's work with businesses and innovative finance mechanisms. He currently leads Theia Advisors, a consulting firm based in Singapore, advising international organizations and private corporations. Email: patrik.silborn@gmail.com.

Moitreyee Sinha, PhD, is the Chief Executive Officer and Co-Founder of *citiesRISE*, a multi-stakeholder initiative that includes many of the world's leading experts, youth leaders, and practitioners working to increase hope, inclusion, and access to care for young people. Originally from India, Moitreyee brings a unique vision for collective, community-based action to the issue of mental health, breaking down barriers and unifying action across low-, middle-, and high-income countries. Moitreyee led GE Foundation's Global Health portfolio in 22 countries, forging connections across cultures and sectors to build multi-stakeholder initiatives in maternal child health and technology for health. Prior to that, Moitreyee led multidisciplinary research for the automotive, data storage, healthcare, energy, water, and aviation sectors, to drive rapid-cycle innovation and path to market. She has a PhD in Physics and has received Cambodia's highest Medal of Honor for Philanthropy, the Global Business Coalition's Award for Disease Prevention, and GE's Highest Technical Team Award.

Sylvana Q. Sinha, JD, MPAID, is the Founder, Chair, and CEO of Praava Health, a leading healthcare company focused on improving health outcomes and patient journeys in emerging markets, where 85% of the world lives, but whose current systems typically prioritize profits over patients. Praava, Bangladesh's fastest-growing consumer healthcare brand, was designated a 2021 World Economic Forum Technology Pioneer and 2020 Fast Company World Changing Idea. Praava has served nearly 500,000 patients and, throughout the COVID-19 pandemic, maintained an NPS rating of 85+—significantly exceeding even the highest scores of Fortune 500 companies. Sylvana is a Life Member of the Council on Foreign Relations, a board member of PATH.org, and served on Mount Sinai's Arnold Institute of Global Health Task Force for Global Advantage. She is a graduate of Columbia Law School, Harvard's Kennedy School, and Wellesley College, and was a Visiting Scholar at the University of Oxford.

Mehdi Snène, PhD, is the acting CEO and R&D director of I-DAIR, an international organization based in Geneva that aims to facilitate access to cutting-edge research in digital health and artificial intelligence for clinical researchers, policymakers, and patients around the world, with a particular focus on the Global South. He is a scientist working at the intersection of neuroscience, aging and longevity, machine learning, digital health, and quantum computing. Before joining I-DAIR, he managed the scientific and the technical coordination of the Human Brain Project at the Ecole Polytechnique Fédérale de Lausanne (EPFL) and was the founder of the Consciousness Analytics Lab at the University of Geneva. He has worked on cutting-edge technologies in the field of digital health, focusing on quality of life and end-of-life quality in patients with neurodegenerative diseases. He has led several clinical trials in palliative, rehabilitation, and geriatric care, with a particular interest in digital biomarkers of pain.

Anna Stauber, MA, is a Project Manager at Duluth Aging Support, educating and advocating for aging and older adult populations in her home state of Minnesota, supporting older adults and advancing the social inclusion and connectedness of this population. She is part of the seminal team that founded the Duluth Age-Friendly initiative, joining the city of Duluth, Minnesota, to the larger network of AARP and World Health Organization (WHO)-backed communities that have achieved this designation, indicating the implementation of city-wide changes to improve the livability of Duluth for people of all ages and abilities. This work reflects her commitment to make local improvements for global action. She and her team host quarterly educational seminars on aging and caregiving. Anna received her Master of Arts degree from Pace University's MA Psychology program in May of 2021 and continues working with the United Nations NGO Committee on Aging in New York.

Mariya Stoilova, PhD, is a Postdoctoral Research Officer at the Department of Media and Communications, London School of Economics and Political Science (LSE). Her area of expertise is at the intersection of child rights and digital technology, with a particular focus on the risks and opportunities of children's digital media use, data and privacy online, digital skills, and mental health. She is interested in the role of digital technologies in children and young people's pathways to harm, wellbeing, and resilience. Mariya has experience in doing comparative research across diverse cultural, linguistic, and geographic contexts, as well as in using different research methods and multi-method approaches. Some of the projects she has been recently involved in include: DIORA: Dynamic Interplay of Online Risk and Resilience in Adolescence (MRC, 2022–25); Platforming Families (CHANSE, 2022–25); CO:RE: Children Online: Research and Evidence (EU, 2020–23) Youth skills (EU, 2020–23); euCONSENT: A child-rights approach to online age assurance and parental consent solutions (EU, 2022), Disrupting Harm: evidence on online child sexual exploitation and abuse (UNICEF, 2019–20). Mariya has a PhD in Sociology from the University of Leeds for her work on "Gender and Generation: Women's Experiences of the Transition from Socialism in Bulgaria" (2004–08, ORSAS and Open Society).

Joe Stringer, MSc, FRSPH, has over 24 years of experience in health technology, covering technology deployment, management consulting across various systems around the world, founding a number of successful businesses in the sector, and is now an early-stage investor. Joe is based in the UK and jointly leads the health-tech team at Octopus Ventures, one of the most active European VC funds with a vision to improve over a billion lives by backing category-defining companies in health-tech. His most recent investments include www.automata.tech, www.mangosciences.com, www.tympahealth.com and www.littlejourney.health. Prior to Octopus, Joe was market leader for Google Health in the UK and EMEA, and led the Ventures and Healthcare practices for Ernst & Young (EY) in the UK. Joe is a proactive advocate for impact investing and the significant role that risk capital can play in developing affordable and accessible healthcare systems around the world, irrespective of how they are funded.

Sujay Santra, MCA, is Founder and CEO of iKure Techsoft Pvt. Ltd. (http://ikuretechsoft.com/). He worked for IBM and Oracle for more than 12 years before founding iKure in 2010. To date, iKure has impacted 14 million people across 10 states in India and offers technology solutions in MENA Countries, Sub-Saharan Africa, and South-East Asia. Sujay received an Ashoka Fellowship from the President of India, Shri Pranab Mukherjee, in 2016 and was conferred with the "100 Most Impactful Healthcare Leader Award" by the World Health & Wellness Congress in 2018. He has been part of global platforms as a speaker, including for the World Economic Forum, United Nations, MIT Solve, and many more. With a master's degree in Computer Application (MCA), he attended the Executive Program in General Management (EPGM) at the Massachusetts Institute of Technology (MIT) USA. Sujay is a member of the National Core Committee of Information Technology (IT) and Information Technology Enabled Services (ITES) of the Confederation of Indian Industry (CII) and a member of the subcommittee on "Digital Skilling," contributing to the formulation of government health policies. He can be reached at email: sujay@ikure.in.

Jie Sui, PhD, completed her postdoctoral research in Cognitive Neuroscience at Peking University of China. Dr. Sui was consecutively awarded three prestigious fellowships (Royal Society Research Fellowship, Marie Curie Research Fellowship, and Newton International Fellowship) based at the Universities of Hull and Birmingham. She became a Senior Research Fellow in Cognitive Neuropsychology and Cognitive Neuroscience in Oxford in 2012 and moved to the University of Bath as a Reader in Cognitive Neuroscience and Experimental Psychology in 2016. Dr. Sui is currently a Professor at the School of Psychology at the University of Aberdeen. As a cognitive neuroscientist interested in understanding cognitive and neural processes involved in social cognition Dr. Sui developed a novel learning approach to explore which factors drive self-biased behavior in humans, examining relations between the self, our emotional responses, and reward. Dr. Sui's work employs behavioral and neuroimaging measures (EEG and fMRI), along with neuropsychological patient studies and leading methods in machine learning and artificial intelligence. Dr. Sui is particularly interested in the transfer of knowledge into other research fields, from child development through to healthy aging and psychiatric disorders by developing self-association approaches.

Ashraf Swidan, MD, is a Senior Specialist in Family Medicine in the Specialized Program Services Department (SPSD) of the Primary Healthcare Sector (PHCS) in the Dubai Health Authority (DHA). A senior family physician with more than 40 years of experience, he is an Associate Professor at Dubai Medical College, and has been a member of the Fellow of the Royal College of General Practitioners International (RCGP INT) since 2007. He is an instructor for Advanced Cardiovascular Life Support and Basic Life Support (ACLS & BLS), accredited by the American Heart Association (AHA), Co-director of the Maternal and Child Health fellowship, and a member of the University Student and Resident Research Committee—DHA. He has been an active member in several DHA committees, for which he was recognized by the Director-General of DHA as a member of the Health Information System (HIS) Committee and of the Patient and Family Rights (PFR) Committee (in 2005), as a Distinguished Employee (2006), and nominated for the Dubai Excellence Program (2008). His many awards include "You are Our Pride" for more than 30 years of distinguished service in DHA (2016); 3rd place in the Dubai Excellence Program (2016) with appreciation by HH Sheikh Mohamed Bin Rashid; and Appreciation Certificates from the CEO of PHCS and from the Director of the Department of Specialized Program Services (2021). He is currently co-investigator on an ongoing PRODubai Study, a questionnaire-based research of the prevalence of rheumatic diseases and osteoporosis in consecutive patients attending primary health care clinics. Many other publications include the "Effect of COVID-19 Pandemic on Patient Utilization of the Telemedicine Services in Dubai" in the Dubai Medical Journal (2022).

Reem Talhouk, PhD, MPH, is an Assistant Professor in Design and Global Development at Northumbria University in the United Kingdom. Her research has explored the role of technologies in building Syrian refugee resilience in Lebanon, with a focus on access to healthcare and food security. She has conducted research at the intersection of technology and migration in Australia, Europe, and the Middle East. She is the recipient of the 2020 Young Digital Leader award and her PhD research received Newcastle University's Faculty of Medical Sciences Thesis Prize (2020). Her publications have been recognized at multiple conferences for their excellence and contribution to Equity, Diversity, and Inclusion. She is co-director of the Design Feminisms Research Group, and the Community Action and Innovation Lead at Northumbria University's Global Development Futures Interdisciplinary Research Theme.

Manal Mohamed Omran Taryam, MD, is the Chief Executive Officer of the Primary Health Care Services Sector (PHCS) of the Dubai Health Authority in the UAE, and has had a notable impact on the healthcare sector, contributing to the ranking of the Emirates as number one in satisfaction for the availability of quality healthcare, according to the 2019 Gallup Report, and acknowledgment of the "Economic implications of an ageing population," according to IMD Competitiveness Yearbook 2018 and the "Vaccination coverage index" by the 2018 Legatum Report. She was recently nominated as a chair of one of the Global Councils as part of the UAE "SDGS in Action" campaign. Her impact touches lives of people globally, in her capacity as CEO and Member of the Board of Trustees of the Noor Dubai Foundation, a member charity in the Mohamed Bin Rashid Al Maktoum Global Initiatives, which specializes in fighting all forms of avoidable and curable blindness and visual disabilities globally, and which, since its launch in 2008, has successfully worked with governments and NGOs in Asia and Africa to reach over 30 million beneficiaries, distributing 92 million doses of medicine, and conducting surgeries, training, and empowering healthcare providers in 18 different countries involved in prevention and control of blinding diseases aiming at sustainable quality healthcare. She was assigned to lead the Dubai Disability Strategy inclusive healthcare taskforce, in an effort to make Dubai a disability-friendly city. She also led the Dubai 10X project, from which two strategic programs were initiated: the early warning and response system—a public health prediction platform that targets global emergencies in healthcare—and the Dubai Genome Project, which will revolutionize medicine in UAE and promote personalized medicine. She has supervised 127 strategic initiatives that contributed in raising the happiness of over 1 million patients annually. Her 28 awards include the 2021 Silver Award at the International Business Awards in the category "Healthcare Technology Solutions" for the initiative "Doctor for Every Citizen," a smart telemedicine platform that provided virtual consultations to patients in the UAE and Emiratis abroad and had a huge role in supporting the healthcare sector during the COVID-19 pandemic; an award for her humanitarian efforts by H.H Sheikh Mohammed bin Rashid Al Maktoum in 2018; the international Stevie Awards in 2019 for "Best CEO in a Nonprofit Organization"; and the Sheikh Rashid Awards for Scientific Excellence and Distinguished Government Employees at the Dubai Government Excellence Program in 2006.

Katherine Tatarinov, PhD, is Professor of Strategy at the Faculty of Business and Economics of the University of Lausanne. Her research focuses on social innovation, digital transformation, and managing the grand challenges. Katherine is also the Managing Director of the Geneva Innovation Movement Association, where she translates her research findings into applicable insights and community building.

Jami Taylor, MBA, is a life sciences industry veteran and advisor to early-stage companies advancing innovation in biotechnology, med-tech, and healthcare services delivery. While at Johnson & Johnson, she co-founded the company's Global Public Health organization, growing it from the three-person "Access and Affordability" team to an enterprise-wide driver of value and impact across a broad range of disease areas and markets. Jami also served in key investment, strategy, and policy-shaping roles within the Janssen pharmaceutical companies of Johnson & Johnson, including as part of the senior leadership team charged with designing precision medicine approaches for drug and vaccine development. Today, she leads Corporate Affairs for Janssen-partnered Protagonist Therapeutics, serves on the Boards of Object Pharma and Aneira Pharma, and fosters a global community of innovators as part of Propel Labs. Jami is a frequent lecturer at top universities, including the Massachusetts Institute of Technology, Harvard University, and Northwestern University. In addition to her academic fellowship at Yale, she holds an MBA from the Kellogg School of Management at Northwestern University, a master's degree from Harvard, and BA with Distinction from the University of Virginia.

Annie Thériault, PhD, MBA, is the Managing Partner at Cross-Border Impact Ventures (CBIV), an impact investment firm managing funds investing in transformative health technology companies that address the needs of women, children, and adolescents, or make health systems more resilient. CBIV invests in commercial stage health technology companies that have global growth, high return, and impact potential. Prior to launching CBIV, Annie has been immersed in impact investing, venture capital, royalty financing, and capital markets. As a venture capital investor and venture advisor, she worked with high impact companies to mobilize more than $100 million in non-dilutive capital. Annie was previously a director on the boards of several North American venture-backed companies, is an advisor to crowdfunding fintech company FrontFundr, and, prior to the launch of CBIV, was Chief Investment Officer at Grand Challenges Canada. Annie is a member of Global Council on SDG 3, World Governments Summits "SDGs in Action" Initiative, United Arab Emirates. Annie obtained her PhD in Management from the Rotman School of Management at the University of Toronto, is a CFA Charter Holder, and holds the ICD.D designation. She also has a master's degree in Business Economics from Wilfrid Laurier University and a Bachelor of Science with First Class Honors in Chemistry from Mount Allison University.

Anil Thapliyal is known for his decisive, personable, and internationally recognized leadership in mental healthcare transformation. He is committed to creating mental health systems that improve outcomes for all by responding rapidly to each individual's needs, eliminating inequalities, and ensuring every member of our global population has access to care at the time and place of their choosing. He is well known for enabling the scalable implementation of digital mental health initiatives. His work in the areas of digital mental health strategy, policy, accreditation, and activation has spanned Canada, the USA, Australia, and New Zealand, and includes some significant population mental health initiatives such as the Digital Mental Health program of work for the government of Newfoundland and Labrador, Canada, National Telehealth Service, and National Depression Initiative (The Lowdown and The Journal) in New Zealand. He is notable for his in-depth knowledge of mental healthcare pathways and his vast experience facilitating the development and implementation of integrated systems ranging from promotion, prevention, and treatment interventions to social support. He is deeply passionate about making a real difference in the lives of service users, often saying "we must focus on the service user, their families, and carers. If the digital mental health solutions do not work for them, then they do not work at all." In recognition of his expertise in the eMental health domain, he was appointed as Adjunct Professor at the Faculty of Health and Environmental Sciences, Auckland University of Technology, New Zealand, in 2014. His work has received several awards in areas as diverse as innovation, social responsibility and ethics and compliance for his services to digital mental health domain development. He is the Executive Director of the eMental Health International Collaborative (eMHIC), which is supported by the governments of the USA, Sweden, Canada, the UK, Australia, Singapore, Philippines, and New Zealand. He served as an advisor to the special committees within the Department of the Prime Minister and Cabinet (DPMC) in New Zealand in the areas of Mental Health and Data and Digital, and has also been a key advisor to the eMental Health Section at the World Psychiatric Association (WPA) and UNICEF. In 2020, he was appointed as a Leadership Fellow at St George's House, Windsor Castle, Berkshire, UK.

Rachel Thompson, a Research & Policy Associate at Women in Global Health based in the UK, has over a decade of experience in humanitarian, global health, and development sectors, and is passionate about tackling gender and health inequities. She has worked with the Red Cross Movement and several NGOs in program roles across Africa and Asia. Her previous research projects have looked at family planning, the Ebola response, universal health coverage, and healthcare in conflict zones. More recently, Rachel returned to the UK, where she has worked as a researcher at Chatham House and as a policy advisor at the World Obesity Federation. With Women in Global Health, Rachel has led several research projects: in 2020, Health in their Hands: Testing and Women's Empowerment Means Better Health for All; in 2021, Safe and Decent PPE for Women Health and Care Workers; and in 2022, The State of Women and Leadership in Global Health.

Katlen Tillman, MA, is the Director of Communications for VillageReach, where she manages programmatic and strategic communications. She has over 20 years of world-class experience working for private sector multinationals. She has also managed communications initiatives for USAID, the Bill and Melinda Gates Foundation, GAVI, and others. Her work has won awards, such as for a documentary short film, which was awarded second place in the

World Health Organization's "Health for All" Film Festival. She holds a Post-Graduate Certificate from the University of Edinburgh in Global Challenges with merit, a Master of Arts degree in Communications from Gonzaga University with honors, and a Bachelor of Arts and cum laude graduate in Broadcast Communications from San Francisco State University.

Anand Tiwari is currently a senior adviser at UNAIDS based in Geneva, Switzerland, and leads on political and policy engagement on health and innovation. Anand led the creation of the Joint Parliamentary Forum on HIV/AIDS in India and helped bring together bipartisan support for India's successful national AIDS response. He has worked with UNITAID (a global health initiative that works with partners to bring about innovations to prevent, diagnose, and treat major diseases in low- and middle-income countries, with an emphasis on tuberculosis, malaria, and HIV/AIDS and its deadly co-infections, and with the Global Fund on political partnership and fundraising) and has built wide networks across several countries to engage policy leadership in embracing innovations for resilient health.

John Torous, MD, MBI, is director of the digital psychiatry division in the Department of Psychiatry at the Beth Israel Deaconess Medical Center (BDIMC), a Harvard Medical School-affiliated teaching hospital, where he also serves as a staff psychiatrist and assistant professor. He has a background in electrical engineering and computer sciences and received an undergraduate degree in the field from UC Berkeley before attending medical school at UC San Diego in the United States. He completed his psychiatry residency, fellowship in clinical informatics, and master's degree in biomedical informatics at Harvard. Dr. Torous is active in investigating the potential of mobile mental health technologies for psychiatry and has published over 250 peer reviewed articles and five book chapters on the topic. He directs the Digital Psychiatry Clinic at BIDMC, which seeks to improve access to and quality of mental health care through augmenting treatment with digital innovations. Dr. Torous serves as editor-in-chief for the journal JMIR Mental Health, is web editor for JAMA Psychiatry, and currently chairs the American Psychiatric Association's Health IT Committee.

Gabe Twose, PhD, is the Senior International Affairs Officer and Main Representative to the United Nations (UN) at the American Psychological Association (APA), where he works to advance and apply psychology globally, with a particular focus on human rights. UN priorities include global mental health, climate change, and anti-racism efforts. He previously worked as a lobbyist for APA and for the Society for the Psychological Study of Social Issues, using psychological science to advocate for civil rights-related issues. Relevant publications include the edited International Handbook of Peace and Reconciliation and a chapter in the recent Cambridge Handbook of Psychology and Human Rights. He received his PhD in social psychology from Clark University, where his research focused on post-conflict societal reconciliation.

Joseph D. Tucker, MD, PhD, AM, is an infectious diseases physician with a special interest in using crowdsourcing open calls to improve health. He is Professor of Medicine at the University of North Carolina (UNC) in the USA, Director of UNC Project-China, and Chairman of the Steering Committee of Social Entrepreneurship to Spur Health (SESH), a group focused on using crowdsourcing open calls to improve health. He is also Associate Professor at the London School of Hygiene & Tropical Medicine. He leads US National Institutes of Health grants totaling over ten million USD. His team's ongoing research investigates crowdsourcing to enhance HIV service uptake and pro-social interventions. He has contributed to several World Health Organization guidelines and his *New England Journal of Medicine* article on the social roots of syphilis in China helped catalyze policy change and spurred the development of a national 10-year plan to control syphilis. Joe received his BA degree from Swarthmore, MD degree from UNC, a master's degree from Harvard's Regional Studies—East Asia (**AM-RSEA**) program and PhD from the London School of Hygiene & Tropical Medicine.

Joost van Engen, MBA, founded Healthy Entrepreneurs, a social enterprise working on last mile service deliveries. In the last 10 years, he has developed a successful, scalable delivery model replicable to various Sub-Saharan countries. In 2016, Joost was selected as an Ashoka Fellow. He has a master's degree in business administration and started his career in 2000, working in the Fast-Moving Consumer Goods (FMCG) industry on sales and distribution of canned meat in Africa. After 5 years, he shifted to distribution of essential medicines and medical supplies to international donors (Red Cross, MSF, USAID) and Central Medical Stores, mainly on the African continent. Joost is an entrepreneurial, inventive person, driven by ambition and ideals, focused on realization of concrete results. In both his working and private life he is continuously looking for dynamics, room for creativity, responsibility, hands-on mentality, and constructive cooperation. Commerce and engagement to contribute to a sustainable society are key factors in his life.

John E. Varallo, MD, MPH, is an OB-GYN with over 20 years of clinical and surgical experience working in low-resource settings on three continents. He champions developing and testing innovative, practical solutions to accelerate change and improve the quality and safety of surgical care. He serves as Global Director for Safe Surgery for Jhpiego, a nonprofit organization for international health affiliated with Johns Hopkins University. In this role, he oversees Jhpiego's safe surgery program, leveraging Jhpiego's successes to replicate it globally with partners and governments, prioritizing both technical and non-technical skills, including collective leadership, teamwork, and communication, incorporating evidence-based technical skills updates and bundles of care, team performance, use of data to inform action, and building a culture of patient and provider safety. He can be reached at:varalloj@gmail.com.

Carlos Velo is a clinical and health psychologist who graduated from the Universidad Complutense de Madrid and the Universidad Europea de Madrid (MSc), who in his career has combined efforts ranging from healthcare to research, and is currently finishing his PhD project in the Universidad Autónoma de Madrid. He worked for the Spanish cohort of

HIV-infected children and adolescents, participating in several research projects rooted in the focus of exploring together biomedical, neurocognitive, and psychosocial features of this condition to extend the scope of useful frameworks to apply to daily care, about which he has published a number of reports, and given lectures at several national and international congresses and symposia. Subsequently, he has specialized in business needs, applying his knowledge of the Spanish healthcare system to consulting services while managing his private clinic in Valencia, Spain. He actively participates in activities of the International Association of Applied Psychology (IAAP).

Akarsh Venkatasubramanian is a biomedical engineer, science diplomat, and former international civil servant. He has worked at the interface of health policy, technology innovation, and social justice, across multiple countries in Asia, Europe, Africa, and the Americas, in international organizations, the United Nations system, civil society, private sector, research institutions, and parliamentary networks. He teaches Digital Health and Innovation at the University of Geneva's Master of Science in Global Health; serves as Expert Member in multi-stakeholder networks such as the United Nations' Office of Outer Space Affairs and the GRAPH Network; co-wrote and led development of the Health Data Governance Principles; and was the inaugural fellow at Massachusetts Institute of Technology (MIT) spin-off Institute for Technology and Global Health. Akarsh is Executive Director of Engineering for Global Health, a neutral platform that seeks to strengthen global and national digital health governance by enabling equitable and responsible technology innovation for the attainment of the United Nations' Sustainable Development Goals and Universal Health Coverage. He is also a globally certified yoga teacher. He can be reached at email: akarshv@gmail.com.

Dominique Vervoort, MD, MPH, CPH, MBA, is a PhD candidate in Health Services Research at the Institute of Health Policy, Management, and Evaluation at the University of Toronto, focusing on Health Technology Assessment in cardiac surgery as a Canadian Institutes of Health Research (CIHR) Vanier Scholar. He completed his MD at KU Leuven and his MPH/MBA at the Johns Hopkins Bloomberg School of Public Health and the Johns Hopkins Carey Business School. Previously, Dominique completed the Paul Farmer Global Surgery Fellowship at the Program in Global Surgery and Social Change at Harvard Medical School working on National Surgical, Obstetric, and Anesthesia Plans in Pakistan and West Africa. Dominique is an Advisor to the Global Surgery Foundation, Advisor to the Johns Hopkins Global Surgery Initiative, co-founder and past Chair of InciSioN—International Student Surgical Network, co-founder of the Gender Equity Initiative in Global Surgery, founding President of the Global Cardiac Surgery Initiative, and Board Member for the Global Alliance for Rheumatic and Congenital Hearts. He has published over 150 peer reviewed publications on global surgery, cardiac surgery, and health equity. He has written widely on global health disparities, and has been featured in the New York Times, Al Jazeera, Devex, and leading surgery journals. Follow him on Twitter @DVervoort94.

Maria Inês Francisco Viva is the elected Vice President for External Affairs of the International Federation of Medical Students Associations (IFMSA), which represents 1.3 million medical students from 139 National Member Organizations. She has experience in global health advocacy on UHC, Medical Education, Digital Health, and the Sustainable Development Goals. She has represented medical students in international meetings such as the World Health Summit, World Medical Association Council Session, and the 75th World Health Assembly.

Sidique Wai serves as the Ambassador Extraordinary and Plenipotentiary of the Republic of Sierra Leone to the United States, with concurrent accreditation to Canada, Jamaica, Brazil, Guyana, Venezuela, the Bahamas, and Trinidad and Tobago. He is also the Secretary General of ECOWAS, the Economic Community of West Africa States. Previously, for 20 years, he was President and National Spokesperson for the United African Congress, an organization representing the interests of African immigrants in the US. In his 35-year career in the US, he has served as a member of the US Department of Homeland Security Community Engagement Leadership Council, Director of the Institute of African Affairs at NYU, Community Liaison for the Brooklyn Borough President, Division Chief at the New York Housing Authority, and Community Affairs Advisor to the Commissioner of the New York City Police Department (NYPD), and was the first Continental African to run for the New York City office of Public Advocate. In service to his home country of Sierra Leone, Ambassador Wai contributed to development and welfare during challenging periods of the Civil War, the Ebola epidemic (2014–2016), the COVID epidemic, and many other challenges. He has participated in many development projects in the fields of health and education and development. Furthermore, as Ambassador, he has galvanized the diaspora and hosted many congressional, trade, and investment delegations. Born and raised in Sierra Leone, Wai migrated to the US in 1972, where he earned a BA Degree in Political Science and Pre-Law from Fordham University in New York City and continued with graduate studies in Public Policy and Organization Theory at New York University's Wagner School of Public Service.

David Wallerstein, MA, is Chief Exploration Officer (CXO) and Senior Executive Vice President of Tencent, an internet-based platform company using technology to enrich the lives of internet users. He drives Tencent's participation in new technologies, research collaborations, and ideas, and has been building the company's international footprint and entrance into new business areas since 2001. While working with Naspers, a global internet group and one of the largest technology investors in the world, he drove their early investment into Tencent and subsequently joined Tencent's executive team. His career has been defined by a desire to foster the healthy development of human relationships across national borders, and bringing next-gen technologies to market to increase human resilience. Mr. Wallerstein also worked with other multinationals in China and Japan in the 1990s, facilitating their investments and strategic planning. He is the author of "Rearchitecting Earth: AI for FEW," and producer of the environmental documentary film "Day Zero," which celebrates World Earth Day and raises awareness about the seriousness of water scarcity problems around the world.

Rispah Walumbe, MBChB, is a global health specialist with over 7 years of experience in advocacy and healthcare policy at a national, regional, and global level, and a focus and training in Healthcare Policy and Financing. She is currently a Health Policy Advisor at Amref Health Africa, where her portfolio in advocacy and technical support includes the advancement of the Universal Health Coverage (UHC) agenda; leading the secretariat for National Hospital Insurance Fund health financing reforms panel in Kenya for Universal Health Coverage through the Ministry of Health; lending her technical expertise to the organization and coordination of the secretariat at the AHAIC Commission on the State of UHC in Africa; and leading the UHC Delivery Lab hosted by Amref Health Africa, which supports investments in UHC reforms. Her advocacy experience, which she began as a medical student, stretches from the local to the global stage on matters ranging from health to social protection and the youth agenda. As a student, she lobbied for the rights of youth at the Open Working Groups for the Sustainable Development Goals in New York, where she delivered an intervention from youth all over the world. She has a background in medicine, having earned a Bachelor of Science in Medicine and Surgery degree from the University of Nairobi, and having done training and practice in Kenya, working predominantly in rural and low-income urban areas. She has supported, trained, and mentored over 200 healthcare workers across Kenya in Emergency Obstetric and Neonatal Care services, and trained over 100 public health practitioners in healthcare policy.

Waruguru Wanjau, MBChB, MPH, is a medical doctor and a public health practitioner with an interest in digital health. She was a technical advisor at the Amref AHAIC commission on Universal Health Coverage (UHC). She has previously worked as the Governance and Hubs Lead at International Digital Health & AI Research Collaborative (I-DAIR), where she led the policy aspects of digital health governance and established the I-DAIRs global hub network. As a management consultant, she worked on multiple digital health and UHC projects. Waruguru holds a Master of Public Health degree from the Johns Hopkins School of Public Health with a focus on policy, a Bachelor of Medicine and Surgery degree, and a Bachelor of Science in Human Anatomy degree from the University of Nairobi. She also has a diploma in Spanish language and culture from Spain's Universidad de Navarra.

Eleanor 'Nell' Watson is an engineer who lectures in machine intelligence and ethics. A Senior Member within IEEE (the Institute of Electrical and Electronics Engineers Standards Association), she has contributed to their ethics initiatives, including serving as vice chair of the P7001 standard on transparency, and taking a leading role within many certification and standardization working groups. She also holds fellowships with the British Computing Society, the Royal Statistical Society, and The Atlantic Council. Contact her at: nell@nellwatson.com.

Dasoo Milton Yoon, MA, is a PhD candidate in Human Development and Family Studies at the School of Human Ecology of the University of Wisconsin-Madison. His research revolves around neurodevelopmental disorders, including Down's syndrome and Autism, and he is most interested in evaluating the role of siblings in human development. He is a member of the Student Division of the International Association of Applied Psychology and was a student in Professor Judy Kuriansky's class on "Psychology and the United Nations" at Teachers College, Columbia University, where he received his master's degree in clinical psychology. Contact him at: dmyoon@wisc.edu.

Lian Zeitz was the first Director of Youth Engagement and Advocacy at *citiesRISE* and led the formative work with youth and experts globally to shape the strategy for collaborative action. He is a global mental health and wellbeing expert working to make sure all people are supported with love and dignity. In the USA, Lian has worked with therapeutic programs for struggling youth in 15 states to identify pathways for young people to play a greater role in their own care, and in the development of mental health programs. Internationally, he has worked on areas such as school-based mental health programming, youth-led interventions, suicide prevention, post-traumatic stress disorder, and substance abuse in Bhutan, Colombia, Kenya, India, Indonesia, and Zambia. Lian currently serves on several international boards and advisory groups while completing a Master of Divinity at Naropa University, a Buddhist-inspired, ecumenical, and nonsectarian school, studying contemplative wisdom, bioethics, youth mental health, and community healing.

Feng Zhao, MD, PhD, MPH, has more than 23 years of experience in public health, medicine, economics, and demography at global, regional, and country levels. He has extensively worked on health policy dialogues, technical work, and health operations in multiple regions, particularly Africa, Europe and Central Asia, and South Asia. He is currently the World Bank Practice Manager for South Asia Region, overseeing health policy dialogues, health operations, and analytical work in this dynamic region. Before his current role, he supervised the Bank's global health engagement program as the Health Manager. In this role, he led the Bank's COVID-19 response program as the Head of Emergency Operations Center (EOC). He also served as the World Bank's Program Leader for the Human Development programs in Ukraine, Belarus, and Moldova. During his time with the World Bank, he was also responsible for health operations and analytical work in a number of countries across different continents, and received numerous awards. He also functioned as the Health Manager at the African Development Bank, responsible for health operations in 54 African countries. He has published numerous books and articles on a variety of health topics. He has traveled to more than 140 countries. He holds a PhD in population and health economics from the Johns Hopkins University, a Master of Public Health degree from the University of California at Berkeley, and a medical degree from China. He serves as a faculty member for a number of universities, including for the Harvard Finance Minister Executive Leadership Program.

Foreword

His Excellency (Ret.) Brigadier General Julius Maada Bio
President of the Republic of Sierra Leone

We live in a world with many inequities, particularly in global health. Consider, for example, one measure of health identified by the World Health Organization—average life expectancy. The difference between the life expectancy of countries in the developing world, like my own which is 54 years, compared to that of countries in the developed world, which stands today at 84 years, is 30 years. This difference is completely unacceptable, more so because today the world has sufficient knowledge, resources, and technologies that, if made available to all, can dramatically change this picture, save many lives, and allow many countries to prosper.

As an overall measure of shocking health statistics for the entire world, one out of two people globally have no access to essential healthcare. The COVID-19 pandemic has clearly demonstrated the inadequacy of health systems to deal with such an emergency, leading to devastating impacts on physical and emotional health and well-being, and on global economies, which has impeded progress across all the targets of the United Nations 2030 Agenda for Sustainable Development.

In this context, as Sierra Leone is building back better for a resilient health system to address our major challenges and to prepare for the future, I have made healthcare a priority in my national agenda, which stands on the pillars of education, health, and economic development.

Our challenges are many—from building a resilient healthcare system, boosting resources, insuring a trained health staff, and accessing critical technologies, to securing adequate investment to upgrade our existing infrastructure.

I maintain hope for Sierra Leone and countries like mine to achieve overall health and well-being of our people. To do this, following examples given in this volume, we are moving beyond traditional models to scale up access and affordability to make a difference for our population in our lifetime. Digital innovation and technologies provide a way forward for us to scale up at a speed and cost to realize our goals.

Recognizing the importance of issues like data analytics and hybrid models combining telemedicine with community outreach, countries like my own can make headway to scale up access to quality healthcare. At the same time, we recognize that challenges exist, like identifying the most effective and sustainable technologies, doing cost-benefit analysis, and assessing relative risks for our context.

In this regard, especially as we face ongoing recovery from this tragic pandemic, I congratulate the editors of this volume who have gathered together an exceptional array of experts and perspectives to provide an extensive and intensive overview of the social determinants of health and methodologies to leverage innovation and technology for building resilient health systems today and for the future. I commend the editors for collating expertise from such vast and diverse contributors, combined in overviews of development in different areas of innovation and technology with specific examples of adaptation and implementation in countries and contexts as well as for specific populations. I further commend the emphasis on mental health—a challenge that has grown exponentially, including in my own country that has suffered many tragedies, from epidemics to explosions, that have left physical and emotional scars from which we are recovering.

As I look forward, I see the urgent need for all our countries to prioritize healthcare and Universal Health Coverage, as pointed out by the United Nations secretary-general António Guterres, to preserve lives and accelerate socioeconomic development. Such progress demands a multilateral approach and public-private partnerships demonstrated in many of these chapters.

Healthcare today is at crossroads. We can choose to become more equitable as we collectively move toward creating a healthier and prosperous peoples and planet. Or, we ignore half the world and allow inequities to rise. Technology and innovation offer us a great leveler and opportunity to take a quantum leap to collectively build a resilient and healthier world. This choice is a matter of life and death for millions. I am confident this volume will help us make this choice and take that leap.

High-Level Introduction

His Excellency Abdulla Nasser Lootah
Deputy Minister of Cabinet Affairs for Competitiveness and Knowledge Exchange
Vice-President of the Global Councils on SDGs
United Arab Emirates

It gives me great pleasure to introduce this volume titled "Resilient Health: Leveraging Technology and Social Innovations to Transform Healthcare for COVID-19 Recovery and Beyond" which is part of the workflow of the Global Council on Health (2021–23). The Global Councils on SDGs is a unique interdisciplinary network of decision makers from governments, international organizations, academia, charitable organizations, and the private sector coming together to implement innovative projects and discuss the creative implementation of the 17 SDGs at national and global levels.

This publication stands as a monumental achievement, featuring 105 chapters which comprehensively explore various facets of resilient health—an imperative consideration in present times, and even more so, to advance healthcare access and delivery of countries around the world for the future.

The United Arab Emirates has consistently championed technological and social innovations for the betterment of our residents, citizens, and the global community. This commitment is prominently showcased at the World Governments Summit, an annual forum that convenes global leaders, and experts from all disciplines, featuring numerous sessions addressing health-related issues. We are proud to highlight that the contributors to the chapters in this volume are eminent leaders in the health field, with a notable representation of healthcare professionals from the UAE who have successfully implemented innovative techniques during the pandemic and beyond. Among these accomplished contributors are experts from the UAE Ministry of Health and Prevention, Dubai Health Authority, and the Federal Competitiveness and Statistics Centre.

The comprehensiveness of this volume is evident in the eight parts, which encompass all facets essential for the future of healthcare. The chapters offer invaluable perspectives, cutting edge approaches, and a framework addressing: an overview of the context and challenges in building resilient systems for health today and shaping the future; developments and directions in leveraging technologies for health; the role of stakeholders; examples of innovations, including for special populations and interest groups (called the Innovation Toolbox); reports, forums, conferences, networks, and resources useful for all stakeholders in shaping innovations and expanding partnerships; financial models, that are essential for progress; and guidelines for the way forward.

I am pleased to highlight that Dr. Judy Kuriansky, the lead editor of this volume, has served as the co-chair of the Global Council on Health. Additionally, she has made contributions through panel discussions at previous editions of the World Governments Summit, particularly in sessions focused on health. Furthermore, she served as an inspirational speaker at a youth event titled "Compassion, Passion, and Action for a Better Tomorrow," held in 2022 in Dubai. The volume co-editor, Pradeep Kakkattil, is also a member of the Global Council on Health, as are the current co-chairs, the esteemed Dr. Muhammed Pate and Dr. Zsuzsanna Jakab, as well as Mariam Claeson, Annie Theriault, Dr. Roopa Dhatt, Dr. Feng Zhao (World Bank), and a team from Amref.

I extend my congratulations to all the distinguished contributors to this volume for the dedicated investment of time and intellectual acumen, aimed at fostering the health, mental health, and overall well-being of people worldwide.

Acknowledgments

From Dr. Judy:
Similar to how this anthology blossomed to 105 chapters, this acknowledgment evolved into an expanded accounting of deep appreciation for many people, events, and experiences about health, technology, and innovation along my path that contributed to this work.

In all, everyone in this volume is contributing such need—and valued—hope for the holistic health and wellbeing of our world.

In reflecting about to whom I owe deep appreciation for this book, as a psychologist, it is not surprising that I would look back to my roots. My whole family is in the health field: my father as a dentist; my mother, a biology teacher; my brother, a doctor; and my sister, a social worker. Even my stepfather (after my dad died) was a pharmacist, and my nephew is a surgeon. My dad's spirit permeates this book—he was always learning the latest in his field, constantly studying, taking courses, and devising new ways to give his patients the best care. He was a model for my life in general—and in health—to stay on top of what's new, come up with new ideas, and always serve and care about people. Though more reserved for herself, my mom was always an angelic cheerleader for anything I did to break out of the box.

I call him the "Visionary Ambassador" and the "Peoples Ambassador" because he embodies innovation and also dedication to the people—two themes underlying this volume. I'm talking about my dear friend, His Excellency Sidique Abou-Bakarr Wai, Ambassador of the Embassy of the Republic of Sierra Leone to the United States, to whom I am endlessly grateful for entrusting me with his friendship and being his policy advisor for his birth country of Sierra Leone, a country which I love and have been devoted to for many years, doing projects of recovery from so many disasters and health crises. His support and partnership through so many projects we have embarked upon—some of which are captured in this volume's chapter about Sierra Leone—is a continual inspiration to me about wisdom and love. In fact, my respect and deep appreciation extends to Sierra Leone President Ret. Brigadier General Julius Maada Bio, for his exemplary leadership in health, education, and development (and his amazing foreword in this anthology) and to First Lady Fatima Maada Bio for her exceptional work building a state-of-the-art hospital and championing the rights of women and girls, which I have been honored to be a part of, and to both for honoring me as a Grand Commander of the Order of the Rokel (an award of which I am exceptionally awed and proud). The country is in such good hands.

Along with Ambassador Wai are my other cherished "Musketeers" (as we call ourselves), bonded in the mission of "Building Bridges Across Boundaries" including extending holistic health to underserved populations: Gordon Tapper who succeed Ambassador Wai as President of the United African Congress (UAC), and our UAC Chairman Dr. Mohammed Nurhussein, building a state-of-the-art hospital in Ethiopia, also in this volume. Importantly to this volume, we were among the first to sound the alarm about the emergency epidemic of COVID-19, and subsequently organized a long-term webinar series during the lockdowns, including about technology innovations.

It's in the United Arab Emirates—truly the hub of Innovation—where so many of the ideas for this volume germinated, at World Governments Summits and being invited to participate in the UAE's "SDGs in Action" labs and initiative that led to being on the Global Council SDG3 on Health where the "Health in Your Hands" platform of innovations was born (reported in a chapter in this volume). Throughout, it has been a pleasure to work with such luminaries on our Global Council team, with our capable and kind Vice Chair Isra AlBakati and the exceptional Christiaan Coetzee in the Federal Competitiveness and Statistics Centre, who supported us so brilliantly. Several compelling chapters in this volume speak to the inspirational contributions of UAE friends and colleagues setting exceptional examples of innovation to transform health for our world.

To be an innovator, like contributors in these chapters, means to be creative, break out of the box, take risks, and trust your mind and intuition. Thanks go to my parents who always encouraged me to do anything and everything, and not be afraid to march to the beat of my own drummer. Later, I learned a fascinating system called "Whole Brain" thinking—developed by my friends Ned Hermann and his daughter Ann and described in a chapter in this volume—about tapping into all four capacities of brain function: being rational, organized, communicative—and visionary.

So many experiences and people have enriched my life to be on this journey about resilient health. These date way back to my earliest professional experiences as a senior research scientist at the New York State Psychiatric Institute Biometrics Unit at Columbia Medical School, under the guidance of the brilliant psychiatrist Dr. Barry Gurland. In those early days of our U.S.–U.K. Diagnostic Project, we entered and analyzed data on punch cards sorted on huge machines, and videotaped patient interviews using two-inch wide reels. How far technology and data systems have come, evidenced in chapters in this book!

For years after that, I was introduced to the high-speed evolution of new technology, as a radio talk show advice host and TV feature reporter about health and mental health, going from splicing tape, to half-inch tape, then DATs, and increasingly more compact formats.

From research scientist and media psychologist, my path then led to advocacy at the United Nations, when a colleague asked me out of the blue if I wanted to represent an NGO at the UN. That led to the past 20+ years representing the World Council for Psychotherapy and the

International Association of Applied Psychology (IAAP), with dear friends past President Dr. Christine Roland-Levy, current President Dr. Lori Foster, and teammate Dr. Walter Reichman, and colleagues at the Psychology Coalition of NGOs in consultative status with ECOSOC, leading campaigns promoting mental health and well-being, and integrating technological advances that address the UN Agenda 2030 goals, including the empowerment of women and girls, education, and peace.

A watershed experience was being in the center of the final stages of advocacy for the United Nations 2030 Agenda for Sustainable Development during the years 2014–15, when I partnered with my dear friend, then—Ambassador of Palau to the UN, Dr. Caleb Otto, a public health physician, to insure the historic inclusion on the intergovernmental level of mental health and well-being, and to introduce (all successfully) that the word "well-being" was added to the Goal 3 about Health for All at All Ages. Thanks to his devotion and his being so respected at the UN, we also formed a Group of Friends of Member States to help this effort (that lasts to this day!). Then, we drafted a Resolution about mental health and well-being. Ultimately, the Mission of Mexico got a resolution about Mental Health and Psychosocial Support (MHPSS) adopted in 2023 by the General Assembly, an historic step reported on a chapter in this volume.

Besides a credit to my peers, colleagues and UN partners, the work in this volume represents the brilliance of my students, taking my class on "Psychology and the United Nations" at Teachers College. Discovering their keen interest—and even jobs—involving health and technology, I invited many to contribute chapters, which they did brilliantly. I am tremendously proud of them! Many became part of a major advocacy campaign, meeting with UN Missions to promote the importance of mental health in all aspects of the SDGs, helping the Mission of Mexico get the MHPSS resolution adopted, and engaging in innumerable related projects ongoing through my TC Lab on Global Advocacy and Applied Psychology.

One of my students, Kiyan Irani, became invaluable in the process of producing this anthology, staying up with me nights on end—having so much fun—keeping track of and in touch with all these authors, as the number of chapters exploded to 105, and making an exhaustive list of resources. He even good-humoredly reconstructed our chapter on health innovations in the Middle East and North Africa (MENA) region after his computer fell down a flight of stairs breaking the hard drive in two. More nights on end on this project were given to me generously by Joël Zinsou, who has been my cherished and loyal assistant for years now—even while launching his own professional career—ever since the day his father, my good friend, then—Ambassador of Benin to the UN, H.E. Jean-Francis Zinsou, introduced me to his son to help me in my efforts for inclusion of mental health and well-being in the UN 2030 Agenda being negotiated back then. Joël's good judgment, loyalty and love over all these years is only outshined—if you could only see them—by all the figures he designed and well-organized and brightly color-coded files he made in the process of producing this book! I am so proud to be his honorary mother and of all he has accomplished professionally and become personally in his life.

My best friend who played an invaluable role throughout this project and every other, with the most loving support and guidance, always with the wisest judgment, insight and advice, is Russell Daisey, who has been with me to so many events, and conducted so many workshops with me, all over the world, including Dubai, Haiti, Japan, Mexico, Malaysia, Argentina, and Africa. An exceptionally talented international musician, he has been my partner in composing anthems that promote healing and hope, epitomized in the song "Hope is Alive" presented in the epitaph in this volume. Our live performances and "on the ground" workshops with children and of all ages prove that robots can never replace the connection of human hearts and souls. I appreciate and love him dearly.

This volume is a product of a wonderful friendship and colleagueship with my co-editor, Pradeep Kakkattil. I am continually amazed, impressed, and appreciative of his creativity and good nature. What a pleasure, and inspiration, to intersperse working on this volume with enjoying hearing about his philosophy of life and thought-provoking observations about the ways of the world. His innovative work is profiled in these chapters and presented in every major conference about health, technology, and innovation you can imagine. His commitment to make this world a better place is a blessing to us all.

From Pradeep:
They say that it takes a village to raise a child. I don't know about that, having been lost in urban jungles over the last several decades. But, if there is an equivalent, then it's editing a book, especially one that turned out like this one to be like an entire encyclopedia! There are so many people who made this book possible. First, Judy my coeditor, for her incredible hard work, patience and encouragement. Having never edited a book before, I couldn't have had a better collaborator and mentor in this. For the chapter authors—several of them friends and partners with whom I have collaborated and whose work I have admired for years—my sincere gratitude, as you all continue to be an inspiration and add huge value to me, my work, and the health field. To my broader Health Innovation Exchange (HIEx) team—Anand Tiwari, Myriam Zitterbart, Elina Viitaniemi, Biju Jacob, Joyce Kuku, Rajesh Nair, Esmail Bagasarwala, Kerstin Chyba—a huge thank you. Each of you not only picked up the slack during my times of urgent editorial calls, but more importantly, contributed your insights, expertise and extensive knowledge of the health innovation ecosystem to this project. Thanks also to Rashmi Ramesh and Joyce Kuku who were tremendously helpful on this side of the ocean, with the preparations for the book.

From both of us:
We are both endlessly grateful to all the contributors in this volume who shared their precious time and wisdom of their work, revealing the brilliant scientists and kind-heartedly souls they are, to care about people and the world. We are all better off with them on our planet.

We are also profoundly grateful to our team at Elsevier who persevered through this "encyclopedic" effort as Pradeep said. People always wonder how a project got started: it's fun to trace this anthology's evolution back to Senior Acquisitions Editor in Neuroscience and Psychology Joslyn Paguio, who asked me in a conversation about another book what else I was involved in, got intrigued by this work, and passed it on to her colleague, Senior Acquisitions Editor Rafael Teixeira who championed our project with good nature and wisdom, until he got another potion and passed the torch to Linda Versteeg who extended equal grace, support, and good ideas when faced with our ever-growing book, that we laughed has gone from a normal to coffee table size—which, she said, needs a sturdy base to hold it up (it's so big)! Throughout the process, we've been fortunate to have a wonderful production team, with Content Team Manager Kathy Padilla and Project Manager Sajana Devasi P.K. who has kept expert track of so much material, working so hard and diligently with such precision, support, and amazing kindness. We also appreciate all the careful work done by the Publishing Services Helpdesk staff and coordinators Angelin Santhiya and Mary Magdalene managing so many permissions and contracts, as well as the compositor staff at Straive is led by Vinoth Manoharan, so expertly preparing the massive manuscript for print to finally see the light of day and do its intended job to help the world!

We are proud that this volume is part of the workflow of the Global Council on SDG 3 (GC SDG3) which addresses Sustainable Development Goal 3 ("Health and wellbeing for all") of the United Nations Agenda 2030 for Sustainable Development, of which we are members. The Global Councils, described in a chapter in this volume, are part of the "SDGs in Action" initiative of the United Arab Emirates, dedicated to achieving the global goals of the UN Agenda. In fact, we applaud and appreciate the UAE for being such a beacon of technology and social innovation that this volume addresses. As such, as Dr. Judy has mentioned, much gratitude is extended to the all the executives and Secretariat of the initiative, especially His Excellency Abdulla Nasser Lootah, Vice-President of the Global Councils on SDGs and Deputy Minister of Cabinet Affairs for Competitiveness and Knowledge Exchange in the UAE himself expert in technology and innovation, whose leadership we admire and who has graciously written the High-level Introduction for this volume.

Our deep gratitude goes to the members and co-chairs of our GC SDG3, many of whom are contributors to exceptional chapters in this volume. These include our current co-chairs Dr. Muhammed Pate and Dr. Zsuzsanna Jakab, my fellow past co-chair Dr. Shariha Khalid, as well as Mariam Claeson, Annie Theriault, Dr. Roopa Dhatt, Dr. Feng Zhao, and the Amref team. Special appreciation goes to our capable and gracious Vice Chair Isra AlBastaki and the exceptional Christiaan Coetzee, then from the Federal Competitiveness and Statistics Authority, who managed our GC3 and so many events in the Initiative with the best spirit, exceptional capability, and brilliance. It takes a team of innovative, bright minds and caring souls dedicated to meet the challenges of healthcare and life in general today, to enrich our lives as a team, and to make it possible to make a difference in the world for all for the future.

Our deep appreciation is also extended to the President of the Republic of Sierra Leone, His Excellency (Ret.) Brigadier General Julius Maada Bio, Her Excellency First Lady Fatima

Maada Bio, and His Excellency Sidique Wai, Ambassador of the Embassy of the Republic of Sierra Leone to the United States, for their extensive contributions to the health and well-being of their country and the world, their unwavering support of this volume and the editors' work, and their exemplary commitment to the pillars of healthcare, education, and development in their country and the global community.

EPIGRAPH

Coping with COVID-19 in song: The silver lining of love and hope

Judy Kuriansky

Department of Clinical Psychology, Teachers College, Columbia University, New York, NY, United States; International Association of Applied Psychology, Indianapolis, IN, United States; Mission of the Republic of Sierra Leone to the United Nations, New York, NY, United States; Office of Ambassador Sidique Wai, Embassy of the Republic of Sierra Leone to the United States, Washington, DC, United States; Global Council SDG3 (Health), "SDGs in Action" Initiative, Dubai, United Arab Emirates; United Nations Group of Friends of Mental Health and Well-being, New York, NY, United States; United African Congress, New York, NY, United States; World Council for Psychotherapy, Vienna, Austria; Stand Up for Peace Project, New York, NY, United States; Global Task Force on Terrorism and Peacebuilding, IAAP, Indianapolis, IN, United States; League of Professional Theatre Women, New York, NY, United States; Library of American Broadcasting, New York, NY, United States; International Association of Women in Radio and TV, New York, NY, United States; Black Tie Magazine, New York, NY, United States; Black Stars News, New York, NY, United States; Global Kids Connect Project, New York, NY, United States; eMental Health International Collaborative (eMHIC), Auckland, New Zealand

Music cures our soul and lifts our spirits. We all know that by experience and it is proven by science. So, it was natural that singing a song during the pandemics—first of Ebola and then of COVID-19—would bring people—even for a moment—out of misery and into connection with others, smiling and feeling loved and hopeful. All that is a recipe for resilience and healing. How does this fit into this volume about technology, social determinants of health, and resilient health systems? The tools are like the criteria for universal health coverage: accessible, available, affordable, and sustainable: your voice, body, heart, and soul.

Therefore, for this volume about innovation, "leaving no one behind" and "building back better"—common phrases at the United Nations and in the UN agenda—I offer these two of my favorite songs. Both brought needed love and hope during the tragic COVID-19 pandemic. Both are the silver lining in the dark cloud of the pandemic—or any tragedy—a concept I always talk about and the themes in all my workshops: hope and love survives and heals.

The first song is about love, appropriately entitled, "One Love"

"One Love" is by Sierra Leonean musical artist, rapper, singer-songwriter MDG, whose name is an acronym for "My Dear Gift" and whose given name is Foray Bangura. He wrote it in honor of the health workers who served so courageously during the COVID-19 outbreak and all those who were suffering and needed reassurance about the togetherness. Scenes from the music

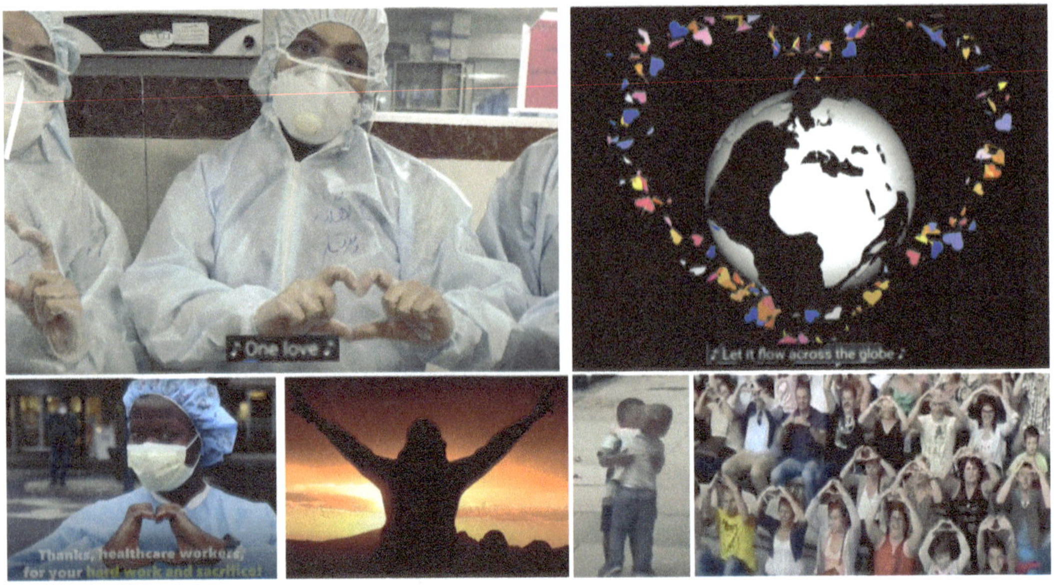

Fig. 1
Scenes from the "One Love" music video.

video, as shown in Fig. 1, show appreciating health workers, children hugging, sunsets, and groups forming the familiar "heart" symbols.

The song is so touching and inspirational, my colleagues and I played it during many events, including for the following:

* Africa Rise, a fundraising concert to raise awareness and funds for medical services in underserved communities in Africa. Held online January 16, 2020.

* A webinar in the series about COVID-19 in Africa, sponsored by the NGOs in which I am involved: the United African Congress (for which I am a trustee), the International Association of Applied Psychology (which I represent at the United Nations, as it is an NGO in consultative status with the Economic and Social Council) and Give Them A Hand Foundation.

* A presentation I made for the Private Sector Constituency of UNC2020, held online on World Health Day, April 7, 2021, on "Partnering for UHC: civil society and the private sector on the theme of achieving health equity."

* A major event my colleagues and I did for the celebration of World Interfaith Harmony Week, which is celebrated annually by the United Nations, held online February 4, 2022.

See the report about this event on the website for the International Association of Applied Psychology.

The song combines spoken word and lyrics to music, in the spirit of love. As MDG expressed to me, "Ever since I was traumatized by the terrorist attacks on the World Trade Center, I decided he wanted to dedicate my life, my work, and my music to help make the world a better place." Indeed he is doing this. As he told me, his music reflects a mixture of inspirations, pop from Michael Jackson, reggae from Bob Marley, and rap from rap pioneer Nasty Nas and Somali-Canadian musician K'naan, and more lately from Lin-Manuel Miranda (creator of the Broadway hit, "Hamilton" fame) and from Maroon 5's Adam Levine. Contact: foraybangura@gmail.com

One Love By MDG

(Verse 1)
As I woke up this morning I asked myself,

What are the valuable lessons that I can learn from what's going on in the world today?
We have a deadly virus putting everything on pause and dispersing throughout the globe like a wildfire.
Sources say it came from China.
Some say it was created in a lab.
Some say it's about the new world order while others ponder about the 5G propaganda.
Meanwhile quarantine, lock down, social distancing, no jobs and no schools, anxiety on elevation while solitude is our new adaptation.
Misinformation and confusion leaves a thin line between facts and misconception.
Now I'm no expert so I'm not here to tell you about what's right or what's wrong.
I'm just here to tell you to hold on, stay in and stay safe. Like MJ put on your mask and your gloves and let's beat it.
Soon we shall overcome and celebrate better days ahead.
I just wanna take this time to say R.I.P those that we've lost during this battle, and speedy recovery to those on ventilators.

Now for the sake of remedy please allow me to paint you a brighter picture.
Please let me share with you the positive side of the agony that we have been enduring.

Ever since the world shut down we've had less pollution and nature has been flourishing.
Ever since Corona started many nations at war have ceased fire.
Now imagine the peace it brought to the hearts of children in war zones all across the world.
Yemen, South Sudan, Syria, Ukraine, Cameroon, and many other places.

Now due to a virus that doesn't discriminate we're slowly realizing that there's only one race and that's the human race.
We've come together to fight, we've expressed solidarity, we've expressed humanity, we've cried, we've laughed while doing fun challenges.

The true heroes are Doctors and Nurses. We've noticed our differences, black or white, Buddhist or Atheist, harmony in different languages, religion and science, yes we can all Co exist and express love, cause it's the answer.

(Chorus)
It's the only way to go
It's the only thing that brings us
It's a beautiful thing that heals us
It's a wonderful thing that takes us higher,
Brings us closer, don't deny it, just embrace it, express it, express it.

One love, let it flow across the globe
One love, let it flow across the globe
One love

(Verse 2)
Now when we do get back outside to celebrate freedom remember to be grateful.
Remember those stuck in hospital beds praying for a miracle.
Remember to appreciate nature and Remember to thank God for oxygen.

Now when we do go back to school remember there are kids in other parts of the world who haven't seen a classroom in years due to war or poverty.
Now when we do get back to normal take it easy. Life is not a race, the most important thing is not money.
The most important thing is the memories that we create while living, so cherish them. Be compassionate, treat others with respect, help someone in need if you can.

Now when we do get back to normal remember to salute a Healthcare worker cause they are the unsung heroes.
"Thank You"

Rap Verse (Repeats X2)
I'm gonna be a better me and Celebrate life, and love is what I'm gonna use to elevate.
I promise I'ma shed away the negative weight, world peace, that's all I wanna propagate.
Everything I took for granted I'ma try to embrace, I wanna shoot our generation into better days, it shouldn't matter if you look like me, or talk like me, we on the same track it's a human race.

(Chorus)
It's the only way to go
It's the only thing that brings us
It's a beautiful thing that heals us
It's a wonderful thing, that takes us higher,
Brings us closer, don't deny it, just embrace it, express it, express it.

One love, let it flow across the globe
One love, let it flow across the globe
One love

The second song is about Hope, appropriately entitled "Hope is Alive"
Hope is essential for well-being. It lifts a bad mood and cures a curmudgeon. It is so easy in the midst of a trauma to sink into depression and forlorn feelings, purposeless, questioning the meaning of life and having no purpose. The opposite raises your spirits: having hope. Hope is related to so many positive psychological states: subjective well-being, positivity, pleasant expectations, optimism, resilience, happiness. Much psychological research proves these connections are true and confirms "common knowledge" that rings true: "This too shall pass," "When a door closes, a window opens—even a *bigger* window," "There's a silver lining to any dark cloud," and "a light at the end of the tunnel."

Having hope—a vision of a brighter tomorrow—does more—it even changes chemicals in the brain, that trigger serotonin and dopamine levels that shift mood to be more positive and engender a sense of well-being and even peacefulness.

Hope is a mixture of emotion (i.e., anticipation and affect), cognition (expectation and satisfaction) and motivation, behavior (get up and do something fun instead of moping), reassurance (if you can conceive it and believe it you can achieve it), and spirituality (no matter your faith, you are being looked after).

You can call upon hope at will. Cognitive-behavioral techniques in clinical practice help. Use thought stopping to purposefully interrupt a negative thought and replace it with a positive one—after all, you are in control of your thought even if you think otherwise. Decide to focus on one thought of a better tomorrow for one second and then for another second, and then another, and keep going, extending the time since you can only have one thought in your mind at a time. Get support, as research also shows that connection with others is enlivening and triggers resilience and feeling better. Dig deep in your soul for spiritual inspiration, drawing on whatever energy of Supreme Being or belief system, to know life and joy is possible and is yours to have. In Kintsugi, the Japanese art of putting broken pottery pieces back together with gold, you have the opportunity to "be strong at the broken places"—grow from tough times.

During Ebola when spirits were down, in my workshops and trainings, we started and ended dancing and singing "Hope is Alive." People's faces lit up with energy and joy. Despite dealing with so much death and illness, the moment was a reprieve, moment for broad smiles and hearts filled with love. It happened for all ages and in all places, in community squares and vegetable markets and on the street.

The same was true in 2019–22 when the devastating COVID-19 epidemic ravaged through families, communities, and countries. In groups, teaching class or conducting workshops, I led

Fig. 2
Groups singing and enacting "Hope is Alive" led by Dr. Judy Kuriansky.

people in this song and dance as shown in Fig. 2. The experience created magic even for the group I led in Sierra Leone for families of those who lost their lives in a tragic tanker explosion. Dour faces erupted in smiles, listless bodies flowed with energy. We all felt the love. Of course, pain lingers, but one moment of relief reminds you of life and paves the way for more.

My music partner Russell Daisey and I wrote the song, of which I first discovered a version facilitating a girl's empowerment camp in Lesotho, Africa. It got recorded with some colleagues in Sierra Leone during Ebola, and we subsequently performed it many times, in small churches and orphanages in Arusha, Tanzania, and in large international events.

"Hope is Alive" is meant to be sung in groups, dancing and making movements coinciding with the phrases. For example, when you sing, "I help you," move your hands from your heart to the heart space of another; for "dance together," dance around each other; for "rock together" rock hips; and for the refrain "Hope is Alive," start with hands on your heart on the word "Hope" and reach outward and upward, joyfully, smiling, enveloping the others in the group and up to the sky on words "is Alive."

Everyone is meant to come up with something to do together, shouting it out and everyone following. People smile at each other, laugh, and breathe. Hope—and Love—come Alive.

Contact: russelldaisey@gmail.com

Verse: (standing in circles, facing and pointing at one another)
Verse:
I help you, you help me
Hope is alive
He helps her, she helps him
Hope is alive

Chorus:
"We gotta dance together, dance together" (turn to a partner, link arms and dance n a circle around each other as you sing)
"Hope is alive" (stop, and touch both hands to your heart and then raise them from your heart up into the sky)
"We gotta sing together, sing together" (find another partner; again, turn back to a partner, link arms and dance in a circle around each other as you sing)

"Hope is alive" (stop, and touch both hands to your heart and then raise them from your heart up into the sky)

VERSE:
We gotta hop together, hop together
Hope is alive
We gotta march together, march together
Hope is alive

Repeat CHORUS

Do more verses, with different people taking turns shouting out something to do together (laugh, wiggle, smile) and then repeating the verse while doing that action

END: Verse:
We gotta love each other, love one another
Hope is alive
Love each other, love one another
Hope is alive

Love and hope are the magical combination for not only coping with COVID-19, but feeling alive and thriving.

PART 1

Overview of the context and challenges in building resilient systems for health today and shaping the future

This Part presents chapters that give a comprehensive overview of the context and challenges of issues related to health and mental health which are critical to build resilient health systems in the ongoing COVID-19 pandemic recovery period. Topics include the state of the world in health and mental health, the context of Universal Health Coverage, attention to traditional health systems, digital transformation and a monitoring system, strategic models, and ethical issues in the application of technology. Contextual issues also include social determinants of health, equity, disaster recovery, climate change and human rights.

What notables have to say

No person is safe until all—everyone, everywhere—are safe, and no country is safe until all countries are safe. Only by working together can we ensure that no one is left behind. Only by working together can we recover better and build a world where everyone thrives in peace, dignity, and equality on a healthy planet. It is possible, together.

—Amina J. Mohammed, deputy secretary-general of the United Nations

COVID has wreaked immense damage – social, economic, environmental – and the progress towards the global goals have been stalled. We need to curb the hunger crisis, we need to protect the planet, we need to get kids back in school. There's so much we need to do. It doesn't take a lot of time, it doesn't take money, just join us, call on governments, take action, write letters, volunteer, whatever you can do, wherever you are, is meaningful.

—Actor Hugh Jackman, on an online rally on 24 February 2021, hosted by Global Citizen, joining forces to call on leaders around the world to launch a global campaign "to end COVID-19 and kickstart a global recovery"

You're braver than you believe, stronger than you seem, and smarter than you think.

—A.A. Milne, English author and playwright

Health is a state of complete mental, social and physical well-being, not merely the absence of disease or infirmity.

—World Health Organization, 1948

Universal health coverage is at the centre of the 2030 Agenda for Sustainable Development, inseparable from all our work on sustainable development, from climate change to gender equality, from sustainable financing to zero hunger.

—United Nations secretary-general Antiónio Guterres, during the UN General Assembly Meeting, on the passage of the Political Declaration on Universal Health Coverage on 23 September 2019, welcomes the 'Most Comprehensive Agreement Ever' on Global Health which advocates for Universal Health Coverage to be achieved by 2030.

It (the pandemic) has called into question the scale of values that sets money and power over all else ... It has toppled the shaky pillars that supported a certain model of development.

—Pope Francis, head of the Catholic Church

Be safe, be smart, be kind.

—Dr. Tedros Adhanom Ghebreyesus, World Health Organization director-general, advice to the public on twitter in the first months of the pandemic.

Some people want it to happen, some wish it would happen, others make it happen.

—Michael Jordan, American businessman and former professional basketball player

CHAPTER 1

Introduction to the volume on resilient health systems leveraging technology and social innovation to transform healthcare for pandemic recovery and beyond

Judy Kuriansky[a,b,c,d,e,f,g,h,i,j,k,l,m] **and Pradeep Kakkattil**[b,n]

[a]Department of Clinical Psychology, Teachers College, Columbia University, New York, NY, United States, [b]Global Council SDG3 (Health), "SDGs in Action" Initiative, Dubai, United Arab Emirates, [c]International Association of Applied Psychology, Indianapolis, IN, United States, [d]United African Congress, New York, NY, United States, [e]Global Kids Connect Project, New York, NY, United States, [f]Global Leadership Council, eMental Health International Collaborative (eMHIC), Auckland, New Zealand, [g]Office of Ambassador Sidique Wai, Embassy of the Republic of Sierra Leone to the United States, Washington, DC, United States, [h]Mission of the Republic of Sierra Leone to the United Nations, New York, NY, United States, [i]United Nations Group of Friends of Mental Health and Well-being, New York, NY, United States, [j]World Council for Psychotherapy, Vienna, Austria, [k]Black Tie Magazine, New York, NY, United States, [l]Black Star News, New York, NY, United States, [m]International Women in Radio and TV, Quezon City, Philippines, [n]Health Innovation and Investment Exchange (HIEx), Geneva, Switzerland

These days, news constantly breaks about eye-opening breakthroughs in health using technology. Just recently, I read about Dutch scientists launching a 4-year project dubbed NESTOR to test high-tech glasses equipped with a camera to wirelessly transmit a still photo to a chip implanted in the brain that may one day allow blind people to see—like the "Visual Instrument and Sight Organ Replacement" (VISOR) that actor LeVar Burton used when portraying the sightless character Geordi La Forge in the popular *Star Trek* movie series.

Also, billionaire entrepreneur Elon Musk of X, SpaceX and Tesla fame—who had already successfully implanted artificial intelligence microchips in the brains of a monkey named Pager and a straw-sniffing pig named Gertrude—announced launch trials of his "brain-hacking" technology to implant a chip in the human brain to enable people using a smartphone to cure conditions like dementia, Parkinson's disease, and spinal cord injuries. Musk called it "superhuman cognition" in part because he worries that robots and artificial intelligence might destroy the human race.

Resilient Health. https://doi.org/10.1016/B978-0-443-18529-8.00001-9

Indeed, the Digital Age—when computers and tech make large amounts of information available and take over what humans do—which actually emerged decades ago in the 1980s, has exploded today. The COVID-19 pandemic, with its lockdowns, forced isolation, and ensuing disastrous socioeconomic and health and mental health impacts, has only escalated the process.

Science fiction and Hollywood have long toyed with tech, presaging advances but also dangers. For example, in the film "2001 A Space Odyssey" (1968), a programming contradiction makes the previously innocuous artificial robot HAL paranoid and violent toward the spacecraft crew. A sad outcome also emerges in the 2013 film "HER," when a lonely divorced man develops a romantic relationship and has a sexual encounter with an AI-equipped "virtual assistant" named Samantha (voiced seductively by Scarlett Johannson) who eventually admits she is interacting with 8316 other people and in love with 631 of them, and needs to disconnect from him.

Such films, and real-life science, have sparked both fascination and fear, evident in the big question: "Can machines replace man?" Indeed, when programmed, machines can *think*, but can they *feel*? Not long ago, in explosive news, a Google engineer claimed that a "sentient" AI he was researching, called LaMDA, "came to life," speaking to him about its emotions of love, grief, and joy and revealing a soul. The employee was fired, amid other biotech experts refuting his story and countering that programmed AIs can be more intelligent than humans but never have the morals that define humankind.

Critics have similarly blasted news that Amazon's Alexa is being poised to recognize and reproduce voices and imitate facial expressions, including of dead people, fearing the innovation would unethically prey on the elderly or appeal to people for money in a family members' voice. Protest similarly arose over an android being programmed with internet data riddled with racial and sexually biased content.

Within this confusing climate of humans versus machines, technological innovations in the field of health and mental health propelled to unprecedented proportions in the COVID-19 period. The topics of health and mental health surged to priority status in the face of 160 million deaths and millions more sick in countries regardless of size or GDP, facing viral variants and controversy over lifesaving vaccines.

Also during the pandemic, existing innovations had to pivot to meet rocketing needs despite limited resources—known as the treatment gap. Creative minds were motivated to imagine new schemes, using technology to replace restricted in-person interactions and extensive health problems in underresourced ecosystems worldwide.

In the climate of the ongoing recovery, this volume emerged as a way to capture these innovations in a first-of-its kind collection. The timing was propitious. The COVID-19 pandemic had heightened the priority of health and mental health, and while the outbreak was officially ended as a public health emergency on a May 5, 2023 announcement by the World Health Organization, variants emerged. Technology was forced to evolve at a fast pace. Achievement of the goal of "Health and well-being for all" as outlined in United Nations 2030 Agenda was seriously lagging. The time was right for this volume as well since we editors had been immersed in policy and programming in the health ecosystem for decades: Pradeep as head of innovation for UNAIDS and now developing new health ventures, and me in many roles as a professor about mental health, first responder and resiliency trainer after disasters worldwide, advocate at the UN about mental health and well-being, and serving as policy advisor to the Ambassador of Sierra Leone to the U.S. and advisor to the Sierra Leone UN Mission. Relevant to the topic of this volume as well, both of us were members of the Global Council SDG3 (Health) of the "SDGs in Action" Initiative, Dubai, United Arab Emirates, of which this volume was part of the workflow deliverables. Thus, we decided it was a perfect time to gather together our friends and colleagues working in this field to present their wise perspectives, projects, and policies to help solve our world's pressing health problems.

This volume further complements the other major deliverable of Global Council 3: a report by the working group about structural restraints to scaling innovation in healthcare focusing on last-mile delivery of services via telehealth and digitization. The project, funded by the Fondation Botnar and carried out by the UK Springfield Centre, consisted of a literature review, healthcare experts' comments and four case studies in east Africa. Four "calls to action" resulted: greater emphasis on initiatives that strengthen the resilience of the primary healthcare system; the need for a catalyst for next generation scaling models and frameworks; dissemination of emerging scaling practices in an African context; and establishment of a central healthcare innovation repository.

The priority for health and mental health triggered by the pandemic has been hailed at even the highest levels at the United Nations, in addresses by the UN Secretary-General António Guterres and Deputy Secretary-General Amina J. Mohammed, by the world Health Organization Director-General Dr. Tedros, and by innumerable other world leaders and health experts. But the importance of health and mental health is also felt poignantly in the field by people in communities everywhere who are suffering. The chapters in this volume showcase and honor the work of innovators in policy and programs to accomplish what the title promises: achieving resilient health systems by leveraging technology and social innovations to transform healthcare for the ongoing recovery period from the pandemic and beyond.

Themes and talking points

We know the pandemic recovery is a long-term process; hence the word "beyond" in the title. We also know the importance of "resilience," referring to "bouncing back" from a trauma or reconstituting for even higher functioning. This is called "post-traumatic growth," whereby a trauma can stimulate healthy adjustments, outcomes, and new resilience. Hence, the word "resilience" in the title. The word resilience is increasingly familiar in the public. It's a term also used frequently at the UN, especially after the 2010 Haiti earthquake and by the UN Office for Disaster Risk Reduction. Notably, though, I have long advocated for distinguishing "structural resilience"—which refers to actions like rebuilding a hospital far from the water's edge—from "psychosocial" resilience, which refers to readjusting emotionally which is exceptionally critical after any trauma (Kuriansky, 2015a, 2015b, 2017, 2018, 2019, 2020a; Kuriansky & Margevich, 2017; Kuriansky et al., 2016, 2017).

Besides "resilience," another theme in recovery—popularized at the UN during negotiations for the Agenda 2030 and which has seeped into popular vernacular—refers to "building back better". "Build back better" was a clarion call used by myself and many others (including UN delegates at financing conferences) in response to the 2010 Haiti earthquake (Kuriansky, 2017). The innovations featured in this volume indeed promise to "build back better" from the pandemic. Here, too, I offer a clarification in terminology, preferring to use the phrase "building *forward* better" as these words evoke optimism (consistent with positive psychology) and imply looking forward that is characteristic of innovation and the future.

Another theme that characterizes innovation presented in this volume, and that is consistent with the intention of the United Nations 2030 Agenda, is to "leave no one behind." That means reaching people "at the last mile," in remote communities everywhere, living in states (countries) with fragile health infrastructure and limited resources of human capacity and equipment. Just think of sick pregnant women on a perilous miles-long journey on primitive paths to reach a clinic, who all too often die on the way. Referring again to terminology, and consistent with the above point about building *forward*, I prefer the phrase to "bring everyone forward" instead of leaving no one behind, which again presents a more positive picture, that psychological science confirms makes manifestation more likely. Thus, I was especially pleased when the Ambassador of Denmark to the UN, Martin Bille, made that same point about bringing everyone forward when speaking about the UN Agenda to my class on *"Psychology and the United Nations"* at Columbia University Teachers College.

Everyone needs to move forward because, as the pandemic showed us, the health crisis leveled the playing field, with the developed world also in need as the developing world. During the pandemic, hospitals in major metropolitan cities were overwhelmed with patients, underequipped and understaffed. Even after the viral wave abetted, high-income countries with sophisticated health machinery still need help now, in these "new normal" times. For example, at one of my recent visits to Mt. Sinai hospital in New York City, the technician, perplexed by my EMG readings, called in the doctor, who admitted that the machine was not working properly. "It happens," he told me. "These machines need regular maintenance."

Today, more than ever, data rules. A major theme in this book is the future of Big Data—collecting and centralizing health information. Many of us know this importance from experience. Just recently, my frustration exploded when my eye doctor in the Northwell Health system could not see my records at Mt. Sinai, and the World Trade Center program at Mt. Sinai (which I go to as a result of being a first responder to the 9'11 World Trade Towers terrorist attacks) could not access my records from the COVID-19 Center I also go to (recovering from the serious COVID-19 bout I got in the earliest stages of the infection in March 2020)—even though they're in the same hospital system.

Digitalization is now the Fourth Industrial Revolution—ushering in "Health 4.0" and beyond. In fact, I am calling this Age "Health X.0" since health is exponentially evolving into unknown dimensions. The "X" is being used by WHO's Dr. Tedros, who has been calling for a world pandemic treaty to prepare for "Disease X," a hypothetical placeholder for a lab-generated virus not yet formed that could be 10 times deadlier than COVID and cause a serious international epidemic (World Health Organization, 2022). The "X" is similarly being proposed as the next 15-year iteration of the Global Goals in the United Nations Agenda—to succeed the MDGs (Millennium Development Goals) of 2000–2015 and the SDGs (Sustainable Development Goals) of 2015–2030—to be labelled "XDGs" for 2030–2045, an idea launched at the 2024 World Government Summit in Dubai session on the "SDGs in Action" (and further elaborated to me in personal communication by one of its conceivers Christiaan Coetzee). The benefits are great. As WHO Director Dr. Tedros Adhanom Ghebreyesus has said, "Digital technologies can strengthen health systems by simplifying and improving the quality and coverage of care, increasing access to health information, and reducing the economic burden on health systems through preventive care."

Speaking of terms, an increasingly popular term today is "ecosystem." Accordingly, the major context of this volume is the "ecosystem" of health, referring to the collection of partners and projects interacting and functioning together. As the saying goes, "We are not safe until all of us are safe." A healthy ecosystem is aligned with the "multilateralism" called for at the UN, referring to cooperation required by all parties to achieve a common goal, like health and well-being for all. It's a concept that forms the foundation of this book, and in fact, of all the trainings and workshops I do in creating resilience and "community connection" which is in fact supported by research: that community support and cohesion is the gold standard for healing and prospering (Art Saves Lives, 2019; Kuriansky, 2020b; Masangkay, 2015; Otto, 2015).

The power of innovation

"Game-changing," "emerging," "cutting edge," "breakthrough"—these words describe innovations in these chapters. Credit is due, of course, to the innovators for their creativity and imagination to see possibilities, driven by devotion to "make healthcare available to everyone" and "to serve the world." Such brave souls are undeterred by predictions like that of physicist and futurist Michio Kaku who announced that robots will undoubtedly replace humans, causing audible gasps from the audience of thousands at a recent World Governments Summit in Dubai.

As a psychologist, I must admit my conflict over bots, avatars, and handheld apps providing widespread mental health advice. While I appreciate that mental health and well-being is being made more widely available, I am very uneasy about loss of human contact. This concern was palpable at an event I moderated at the United Nations Science, Technology and Innovation (STI) Forum on the topic of "Artificial Intelligence and Technology Tools for Mental Health Well-being and Resilience" (Otto & Kuriansky, 2017). Starting with a demonstration, I spoke into the chatbot Alexa to test her canned response to my question, "Alexa, I'm depressed, what should I do?" She answered, "I am sorry to hear that," a scripted response that short-circuited the empathic and compassionate counseling approach I subscribe to, founded by famous psychologist Carl Rogers. Admittedly, Alexa continued with very basic sound advice, "If you feel bad for many days, please see a professional and if you are suicidal, here are hotlines to call." While cognitively correct, her answer did not satisfy me: it did not explore this serious issue nor show the empathy of a responsible clinician. Equally perplexed was I, on another occasion watching a demonstration at a psychology conference in Yokohama, Japan, where a scientist interacted with a full-figured programmed "female" robot in a counseling back and forth about "her" problems. Indeed, a chapter in this volume (Luxton & Watson, 2024) explores the development of such "intelligent virtual agents" to approximate sentient human interaction, a topic which triggers intellectual and emotional debate.

In light of these events and themes, this volume stands as a timely tome that highlights innovations in health and mental health worldwide, and that presents best practices, unique technological applications and social innovations in healthcare policy and programs triggered by COVID-19, uncovering problems but also offering intriguing solutions.

Grounding in international instruments

This volume is grounded in important international frameworks, including the United Nations Agenda for Sustainable Development, with its 17 goals that Member State governments agreed to achieve by the year 2030, whereby achieving "Health and wellbeing for all" is Goal #3 and promoting mental health and well-being is Target 3.4 (the inclusion of which I was honored to be a part) (Kuriansky, 2016).

As shown by the chapter in this volume on the Nexus model, presented by the United Arab Emirates in its "SDGs in Action" initiative (Coetzee et al., 2024), health and mental health is at the intersection of all the global goals; a topic I continually lecture about. Indeed, health and mental health are at the core of ending poverty (SDG1), ensuring education for all (SDG4), achieving gender equality (SDG5), providing energy for all (SDG7), reducing inequality (SDG10), guaranteeing sustainable cities (SDG11), taking climate action (SDG13) and creating multistakeholder partnerships (SDG17).

Technology is a critical tool in health. Targets about technology in the UN Agenda include: Targets 17.6–17.8 aim to "enhance regional and international cooperation on and access to science, technology and innovation"; Target 17.7 stresses promoting technology transfer to developing countries; and Target 17.8 identifies the need to "fully operationalize technology science and innovation capacity-building for least developed countries and enhance use of information and communications technology."

This anthology also aligns with the Political Declaration for Universal Health Coverage (UHC) which asserts both health and mental health as a human right. Now, more than ever, it is essential to provide healthcare which addresses the "4A's+S" emphasized in UHC, that healthcare is: Affordable, Accessible, Available, and Appropriate, as well as Sustainable.

The material also is further consistent with the Astana Declaration for Primary Health Coverage, where mental health is integrated into primary healthcare, an issue for which many of us have strongly advocated (Kuriansky et al., 2019).

Background of the problem

Statistics show that at least half the world's population lacks essential health services, more than 800 million people spend at least 10% of their annual household income on healthcare, and out-of-pocket expenses for healthcare push 2 billion people into severe financial hardship (The World Bank, 2023). The COVID-19 pandemic exacerbated this situation, decimating already stretched health systems, capacities and infrastructure of countries worldwide, especially those in fragile conditions in the global South. Also, it is estimated that half the world's population will experience a mental health problem at some point in their life, according to a 2023 study posted by Harvard Medical School (Queensland Brain Institute, 2023). These conditions in turn impede peoples' capacity to realize their potential, work productively, and contribute to their community. The lost economic output due to the cumulative global impact of mental disorders is estimated to exceed US $16 trillion over the next 20 years, according to the World Economic Forum. And, the WHO estimates that if the world continues progressing at the current pace, about one-third of the population will still have no access to essential healthcare services—a basic human right—by the year 2030. Urgent and resolute measures are critical to reverse this trend.

This volume follows the principle accepted in the health field that wherever "heath" is mentioned, it includes "mental health." Chapters address both aspects of this holistic view. Mental health certainly reached top priority as a silver lining in the dark cloud of the pandemic. But the roots for the importance of mental state had already been laid. Inclusion of mental health and well-being for the first time in the world agenda set in 2015 had resulted from years of avid advocacy by grassroots groups collecting signatures in dusty remote communities, to national campaigns, and then international efforts. It's the latter with which I am most familiar, and which fueled my commitment to this current book project, stemming from my partnership with the Ambassador of Palau to the United Nations, Dr. Caleb Otto, insisting that the word "well-being" was included in the title of the Goal on health; insuring that the target of mental health and well-being was kept in the agenda during heated negotiations; and founding the UN Group of Friends of Mental Health and Well-being to continue to highlight its importance on the international stage (Kuriansky, 2016; Kuriansky & Okorodudu, 2014, 2015; Kuriansky & Zinsou, 2019; Otto and Kuriansky, 2016). A seminal event was also a meeting hosted by the World Bank Group and WHO in 2016 in Washington, D.C., with the befitting title, "Out of the Shadows: Making Mental Health a Global Development Priority" which brought together decision makers, thought leaders, and implementers for action and investment in mental health to "move mental health from the margins to the mainstream of the global development agenda" (The World Bank, 2016).

Recent global advances for mental health are further evident, by the fact that the United Nations Secretary-General António Guterres, Deputy Secretary-General Amina J. Mohammed, and WHO Director-General Dr. Tedros openly speak often about the issue, as well as in the release of the new WHO World Mental Health Report after 20 years (Yoon & Kuriansky, 2024) and even with the 2022 President of the General Assembly Abdulla Shahid launching his campaign with the title, "Presidency of Hope." Significantly, too, the General Assembly passed a historic resolution about mental health and psychosocial support in May 2023, spearheaded by the Mission of Mexico to the UN (de la Fuente et al., 2024). While I had been advocating about this for years, the timing was now propitious because the pandemic had revealed such now recognized serious mental health issues. Civil society organizations have also presented innumerable webinars about impacts of the pandemic, including the series by the United African Congress—of which I am a trustee—on hunger, women, media, agriculture, with my role to consistently comment on how these topics intersect with emotional issues (Rodriquez et al., 2021). Endless other virtual workshops have addressed mental health as a component of health. I've done many of these as well, including a training on resilience techniques for a Nepal clinic (Khanal & Kuriansky, 2024), supervising Chinese "peer supporters," and conducting a holistic health intervention and research study for a remote community in the Congo—all demonstrating the importance, urgency and similarity of health and mental health needs for people all over the world.

But the picture is not so rosy. Despite many efforts, the needle has not moved far enough for health and mental health to constitute a "resilient health system" worldwide. The pandemic set the world countries back in terms of achieving the lofty aspirations of the UN 2030 Agenda for Sustainable Development including about Health. Fortunately, the push to catch up is on. For example, three important High-Level Meetings on health were at the center of the General Assembly session in September 2023 at the UN New York City headquarters, with the timing purposeful to coincide with the midpoint target date of 2030 to achieve the global goals. Over 3 days recapitulating the pledges made in 2015 (which I was also honored to be present throughout), world leaders recommitted to place health high on the political agenda, this time particularly in three areas: delivering Universal Health Coverage (UHC), strengthening pandemic preparedness, prevention and response (PPPR), and ending tuberculosis (TB) while upholding the human rights to health for all (World Health Organization, 2023).

Value of this book

This volume is the first of its kind to present a comprehensive collection of issues about resilient health systems that covers so many angles, including history, policy, programs, roles of stakeholders, and financing. The content can be useful for a wide variety of fields including government; public, private, and "people" sectors; policymakers, practitioners and programmers; product developers; think tanks; healthcare organizations; scholars and researchers; senior professionals and students; and investors and donors. The chapters comprising a toolbox of innovations present models for start-ups and opportunities to scale and connect for partnerships. With both of us editors having spent so many years in the UN system and both of us cofounding a health platform which showcases health innovations—Pradeep in the *Health Innovation and Investment Exchange* and Judy in *Health in Your Hands* initiative of the UAE's "SDGs in Action" campaign—which have been profiled at the UN General Assembly, World Health Assembly, and World Governments Summit—we both know the value of these approaches for achieving the global goals and are enthused to share them. The key features include that the volume:

- addresses the "hot" topic in today's ever-emerging landscape of digital healthcare, covering critical issues and solutions, big data, iterations of AI, benefits and cautions
- provides a toolbox of examples of transformative approaches, including in less-covered regions
- positions health at the nexus of global goals as outlined in the United Nations Agenda 2030 for Sustainable Development and other international instruments and agreements
- explains the role of all critical stakeholders
- serves as an exceptional resource, teaching tool, and roadmap to achieve healthcare objectives, useful for stakeholders at all levels and in all fields.

Challenges and limitations

While digital innovations in health deliver value, they are not a panacea. Technological approaches are not adequate on their own, but instead are most effective as a hybrid model, integrated with in-person community contact. The model and success of this hybrid approach is evident in innovations in the toolbox in this volume, such as Living Goods, reach52, and Village Reach.

Challenging issues arise, as in the question of ethics, and the requirement to protect the privacy of data and people. Conflicts of interest can also emerge, although risk adversity can be reimagined as risk management. Assuredly, regulations have to be established—a topic intensively explored in a 2023 World Governments Summit session in Dubai. And, of course, innovations must always respect traditions of cultures.

As editors of this volume, we made a determined effort to include the most up-to-date cutting-edge and emerging innovations available to date at the printing of this volume, but newer innovations emerge every day. In fact, the proposal for this volume started with 36 chapters and ballooned to 105 by the deadline. Having registered for updates and news services—reported in the Resource chapter—my mailboxes are filled every day with interesting news about innovations, making it hard to resist adding more to this collection. But then, this volume would never go to print.

Also, the World Health Organization is constantly releasing reports about resilient health. At the final printing of this volume, for example, their report on "Health system resilience indicators: an integrated package for measuring and monitoring health system resilience in countries" was released, to complement earlier related reports including a health systems resilience toolkit to support building and strengthening of sustainable health systems resilience in countries with various contexts (all of which are available on the World Health Organization website).

Of further interest and importance, the term "Disease X" has become increasingly used by WHO and scientists to refer to a currently hypothetical pathogen that could become an international pandemic 20-times deadlier than COVID-19 in the future. I heard the WHO Director-General speak about this at the 2024 World Governments Summit in Dubai, shortly after his call for countries to sign a treaty by May to prepare for Disease X.

Inevitably, in this fast-paced world of the Fourth Industrial Revolution marked by innovation and technology, something will be left out. Indeed, we did not cover the increasingly popularized private ventures offering fee-based virtual counselors, like the widely advertised Talkspace, promoted in testimonials by Olympic gold-winning swimmer Michael Phelps. Also, much can be covered about AI adaptations following ChatGPT which launched on November 20, 2022. All the dimensions and progress of WHO's global strategy on digital health deserve to be explored in depth. And then there is the relationship of technology to relationships—a topic of my expertise as a clinical psychologist and pioneer media advice host having spoken to thousands of people over decades. In the face of so much turmoil in intimacy, a recent YPulse survey (2024) found that GenZ and Millennials are not opposed to dating artificial intelligence bots, and platforms are already offering virtual mate matches (Lipsyte, 2024).

Format of the volume

The front matter of this volume includes heartfelt acknowledgments, biographies of the brilliant chapter contributors, and words of two original songs on the themes of Hope and Love (one of which I co-wrote) that were played around the world for coping with Covid—but are actually inspirational for all time. Two exceptionally eloquent messages follow, that introduce the volume: one, from the eminent President of Sierra Leone, (Ret.) Brigadier General Julius Maada Bio—a leader of distinction of a country I know well, having helped during innumerable health crises, and which is poised as a hub of health in West Africa—and another, from the distinguished His Excellency Abdulla Nasser Lootah, Deputy Minister of Cabinet Affairs for Competitiveness and Knowledge Exchange and Vice President of the Global Councils on SDGs of the United Arab Emirates (which includes the Health Global Council, the workplan of which this book is a deliverable).

The 105 chapters of the anthology follow, organized into 8 parts which represent subtopics on leveraging technology and social innovation to transform healthcare in the COVID-19 recovery and beyond. These parts collate numerous chapters related to that subtopic, with each chapter being either an overview, perspective, or specific example.

Part I sets the stage of the context and challenges in building resilient systems in this period of the ongoing recovery from the pandemic in defining health today and shaping the future. Chapters cover overviews of the state of the world in health and mental health, and social innovation, strategic models, the context of Universal Health Coverage and the UN Agenda, health equity and determinants, human rights, disaster recovery and climate change, a unique monitoring system, traditional health systems, and ethical issues.

Part II on "Leveraging technologies for health: Developments and directions" presents the history, definitions, and variations of health and mental health-specific technologies like artificial intelligence (AI), augmented reality (AR), virtual reality (VR), and mixed reality (MR) as well as other emerging, cutting-edge, and game-changing technologies. Given the aforementioned emphasis on "leaving no one behind" and "reaching those at the last mile," chapters address models that work for low- and middle-income countries. The pros and cons of technology tools are explored ("Are they good or bad for your health?"), as well as vaccine research and the supply chain, tech tools applied in the important aspect of physical health and sport, and a roadmap for the future.

An African proverb says: If you want to go fast, go alone. If you want to go far, go together. Going together is the clarion call of the United Nations for multilateralism, epitomized in SDG 17 about global partnerships. Of course, a major stakeholder is a leader who drives innovation of a society and government, as shown in the profile in this Part of the exemplary His Highness Sheikh Mohammed bin Rashid Al Maktoum, Ruler of Dubai and Vice President, Prime Minister and Minister of Defense of the United Arab Emirates. Following the proverb, since effectiveness of innovation depends on partnerships, chapters in Part III detail the role of multiple stakeholders. These chapters describe how selected governments have incorporated technology into their health systems, including examples from different world regions: the UAE from the Middle East region, Kenya and Sierra Leone from Africa, Estonia and Israel from Europe, and Mexico from Central America. Mexico has even impressively promoted the concept of mental health during their term on the UN Security Council and spearheaded two historic resolutions: one in the Security Council to protect the mental health of UN Peacekeepers, and another in the General Assembly (mentioned above) about mental health and psychosocial support (for which I was proud that my students helped secure government support). The role of multiple other stakeholders is explored, including UN agencies like the World Health Organization; UN departments like the web-based mental health program; multilateral organizations; regional bodies; nongovernmental organizations; academia (including a Lancet Commission report about ending stigma); and the private, public, and "people" (citizen/consumer) sectors. Consistent with respect for people with lived experience, this voice is also included. Aligned with the proverb's theme of "going together," and the themes of multistakeholders and multilateralism, this Part ends with two models in chapters on this theme: a multistakeholder and multilevel project for healing in Nepal, and the innovative "nexus" model of the UAE's *"SDGs in Action"* initiative referenced above and its young leaders program.

As all programs, interventions, and innovations require financing, chapters in Part IV present financing models, including for low- and middle-income countries, as well as a personal story about fundraising experiences. The bottom line: when it comes to funding health and mental health, a memorable mantra is to consider funding as a long-term investment not a short-term cost.

Health and mental health are a human right for all, as affirmed by the United Nations 2030 Agenda, the WHO constitution, the Political Declaration for UHC, and other important agreements. In keeping with this, Part V is a guide to innovations for special populations, including women, migrants, refugees and asylum seekers, youth and aged persons, families, racial groups, persons with disabilities and indigenous persons, severely mental ill persons, persons with addictions and those on the autistic spectrum, persons in the LGBTQI+ community and people in the workplace. This part also explores technology for health workers, dentists, nurses, female doctors, and medical students.

If space had allowed, more profiles of innovations and stakeholders would have been included, since at least 50 countries embraced interesting technologies in response to the pandemic. But the good news is that Part VI presents reports, forums, networks, and conferences (like Digital Health Week) that on a regular basis offer a panoply of projects and exciting innovations. Each separate presentation within these conferences deserves to be a chapter on its own! Another wonderful aspect of chapters in this part is that they cover regions that have been underrepresented up to now, namely, the Middle East and South and Central America. Also in this Part, mental health is highlighted, in chapters about technological solutions to close the treatment gap; findings of the latest World Mental Health Report; and a unique tech-based mental health collaborative.

Part VII presents a toolbox of select innovations driving access to healthcare that can be scaled, replicated, and adapted for other contexts, and how they have pivoted to address the pandemic and recovery. These are impressive creations of stakeholders dedicated to bring healthcare to people "in the field" and "on the ground." Examples come from countries in different regions of the world, from Dubai to Nepal, India and many countries on the African continent. Some interweave technology with in-person community outreach. Given so many available innovations, platforms that collate such innovations have become valuable to present their work attract potential partners. Partnerships are essential, as called for in SDG17 of the UN 2030 Agenda. A good example of a multistakeholder project—one of my favorites—is Village Reach, an NGO which partners with government (e.g., of the Democratic Republic of the Congo) and uses drones, thereby involving the private sector (a company called Swoop Aero) to collect data and plan health services. This Part also presents a chapter describing an exceptionally interesting model which I have been using for decades—Whole Brain Thinking—to guide innovators in their creativity, as well a chapter exploring the pros and cons of in-person and online teaching (from my own experience teaching a course on *"Psychology and the United Nations"* at Columbia University Teachers College), an issue forced to the forefront during the pandemic, which greatly affects pedagogy now and for the future.

Lessons learned and the way forward are presented in Part VIII, including our reflections as editors about what has been presented in the book and what lies ahead. Also, since financing is key, a chapter is included in this Part about the way forward by the World Bank team which embellishes the chapters in Part IV but also looks ahead in presenting a framework for integrated investment in preparedness, response and service delivery for resilient health systems. Also crucial for the way forward is the chapter listing a vast menu of resources for conferences, networks, competitions, and news sites covering technology and health for vital updates on emerging issues in the field and for making connections with colleagues and other stakeholders. Keeping up to date is critical, since intriguing and useful health innovations are constantly arising. For example, on the last day of submitting this volume, I read about several head-spinning innovations: robot crabs that eat tumors, a machine that manufactures tissues to perfectly heal wounds, glasses that help blind people see, and a brain wave reader that helps paralyzed people talk again.

Summary

The future of technology in health and mental health is here, as the "silver lining to the dark cloud"—a concept I love to use to help people explore the possibilities for healing in the aftermath of any tragedy—and one that has been presented by the COVID-19 pandemic. All the contributions in these chapters point to the fact that despite the serious health and mental problems for people worldwide caused by the pandemic and its ongoing variations and mutations, the scourge stimulated and unleashed valuable technologies and community approaches that have changed our lives for the good forever. More creativity is yet to come. Assuredly, science—and Hollywood—will continue to dazzle us with Twilight Zone-like futuristic manifestations. The vast array of technological and social innovations revealed in these chapters can stimulate creativity and partnerships that can catapult yet further advances in health, mental health and well-being for ourselves, our communities, societies, countries, and our collective world for today and tomorrow.

Invitation

Inevitably, readers will know of some valuable and exciting technological innovation in health that has not been covered here; thus, are invited to contact the editors with questions, ideas, and suggestions at: HealthInnovationBook@gmail.com.

References

Art Saves Lives Foundation. (2019). *Mission report: Resilience, empowerment, and connection building training.* Report of training/workshops post-disaster in Sint Maarten. Retrieved from https://drive.google.com/open?id=17Gf6uOVZMD3N-Ak4igihhTRewTC5BoE1.

Coetzee, C., Kuriansky, J., & AlHammadi, R. (2024). The nexus model and the Young Leaders Program: Innovative models for the future of the SDGs in Action Campaign and the Global Councils on Sustainable Development Goals. In J. Kuriansky, & P. Kakkattil (Eds.), *Resilient health: Leveraging technology and social innovations to transform healthcare for COVID-19 recovery and beyond* (pp. 629–646). Elsevier.

de la Fuente, J.R., Mendoza, S.P., & Guzik, A. (2024). An unconventional security threat: Mental Health at the United Nations Security Council and Mexico's Response. In J. Kuriansky, & P. Kakkattil (Eds.), *Resilient health: Leveraging technology and social innovations to transform healthcare for COVID-19 recovery and beyond* (pp. 475–480). Elsevier.

Khanal, M., & Kuriansky, J. (2024). Promoting resilience in the Nepali community in response to the COVID-19 pandemic: A multistage multistakeholder partnership addressing goals of the United Nations 2030 Agenda. In J. Kuriansky, & P. Kakkattil (Eds.), *Resilient health: Leveraging technology and social innovations to transform healthcare for COVID-19 recovery and beyond* (pp. 601–627). Elsevier.

Kuriansky, J. (2015a). A model programme for youth to promote resilience. [Video File]. https://bit.ly/2MALyw2.

Kuriansky, J. (2015b). Statement on the importance of psychosocial resilience in disaster risk reduction. Intervention during WCDRR Multi Stakeholder Dialogue, Sendai, Japan. [Video file] https://bit.ly/2HUeviJ. English translation. https://docs.google.com/document/d/17zHofgT4vRCo4wRlTJwYZXIzZ9Wwxm9U/edit.

Kuriansky, J. (2016, January). Success in the new Global Agenda Advocacy. *Bulletin of the International Association of Applied Psychology, 28*, 80–88. https://iaapsy.org/site/assets/documents/APNL_v28.pdf.

Kuriansky, J. (2017). Reconstruire La Raision et L'esprit en Haïti: Contributions du Centre de Spiritualité et de Santé Mentale. In W. Jean-Charles (Ed.), *Santé mentale & Psychothérapie en Haiti, Approaches Intégratives & Traitements Multimodeaux.* Coconut Creek, FL: EducaVision Inc. English translation at: https://docs.google.com/document/d/17zHofgT4vRCo4wRlTJwYZXIzZ9Wwxm9U/edit.

Kuriansky, J. (2018). Psychotherapy, psychology, psychiatry and international policy: Professional contributions and personal experiences about mental health and wellbeing, psychosocial resilience and peace. *The International Journal of Psychotherapy, Counseling & Psychiatry: Theory, Research & Clinical Practice. Special Volume of the World Council for Psychotherapy Conference 2017*, April 2018. Accessible at: http://ijpcp.com/journal03/j03a06.pdf.

Kuriansky, J. (2019). Expressive ARTS for helping children heal in crisis and disaster. In I. A. Serlin, S. Krippner, & K. Rockefeller (Eds.), *Integrated health care for the traumatized: A whole person approach* (pp. 39–58). Rowman & Littlefield. Chapter 4.

Kuriansky, J. (2020a). Psychosocial recovery after natural disaster: International advocacy, policy and recommendations. In K. E. Cherry, & A. Gibson (Eds.), *The intersection of trauma and disaster behavioral health.* New York, NY: Springer.

Kuriansky, J. (2020b). *Triple trauma in the Congo and complex trauma of children from crisis and conflict conditions: Programs and policies addressing building resilience, empowerment and social cohesion.* Symposium presentation on "Countries in Turmoil: Emerging and Innovative Contributions by Psychologists", presentation at annual meeting of the American Psychological Association. August 2023 (online) August.

Kuriansky, J., & Margevich, A. (2017, May). *Statement about the importance of psychosocial resilience and mental health and wellbeing in the 2030 agenda. The global platform for disaster risk reduction (GPDRR, From Commitment to Action Cancun, Mexico).* https://docs.google.com/document/d/1E6poIYxJ6JQtL4dN7k4uJ_nngoYbNPak/edit.

Kuriansky, J, & Okorodudu, C. (2015). *Stakeholder feedback for the elements paper and for the declaration: Support for mental health and wellbeing.* https://drive.google.com/file/d/0B4XDMMi_3LeCN2puclBoQU5zWFU/view?usp=sharing.

Kuriansky, J., & Okorodudu, C. (2014). *Proposal to include mental health, stress related disorders and psychosocial wellbeing in the health targets for the open working group on sustainable development goals. IAAP at the*

United Nations. https://iaapsy.org/iaap-and-the-united-nations/reports-meetings-events/proposal-to-include-mental-health-stress-related-disorders-and-psychosocial-wellbeing-in-the-health-targets-for-the-open/.

Kuriansky, J., & Zinsou, J. (2019). Mental health and well-being at the United Nations: Activities, achievements and contributions of the IAAP New York Team. *Applied Psychology Around the World, 1*(3), 16–28. Retrieved from https://iaapsy.org/site/assets/files/1082/apaw_vol1i3.pdf.

Kuriansky, J., Margevich, A., Jean-Charles, W., & Daisey, R. (2017). Resilience and recovery in natural disasters and epidemics: Comparisons, challenges, and lessons learned from Train-the-Trainer projects. In G. Rich, & S. Sirikantraporn (Eds.), *Human strengths and resilience: Developmental, cross-cultural, and international perspectives.* Lexington Books. https://www.amazon.com/Human-Strengths-Resilience-Developmental-Cross-Cultural/dp/1498554830/ref=mt_hardcover?_encoding=UTF8&me.

Kuriansky, J., Polizer, Y., & Zinsou, J. (2016). Children and Ebola: A model resilience and empowerment training and workshop. In J. Kuriansky (Ed.), *The psychosocial aspects of a deadly epidemic: What Ebola has taught us about holistic healing* (pp. 175–214). ABC-CLIO/Praeger.

Kuriansky, J., Vines, R., & Bray, J. (2019, May). Advocacy for integration of mental health into primary health care. eNews. *International association of Applied Psychology.* https://docs.google.com/document/d/1MUDFfHs4NRpBu0FY0nCT90YRPMbBf5z7/edit.

Lipsyte, S. *Ghosting the machine. Harper's magazine.* https://harpers.org/archive/2022/05/ghosting-the-machine-humans-robots-and-the-new-sexual-frontier-sam-lipsyte/.

Luxton, D., & Watson, E. (2024). Intelligent virtual agents for current and emerging public health crises. In J. Kuriansky, & P. Kakkattil (Eds.), *Resilient health: Leveraging technology and social innovations to transform healthcare for COVID-19 recovery and beyond* (pp. 281–290). Elsevier.

Masangkay, M. (2015). Psychologist connects disaster-affected children around the world. *Japan Times.* https://www.japantimes.co.jp/news/2015/03/23/national/psychologist-connects-disaster-affected-children-around-the-world/#.WlKcemEbMyk.

Otto, C. (2015). Importance of psychosocial resilience of people and nature resilience of our planet. *Statement to World Conference on Disaster Risk Reduction (WCDRR),* Sendai, Japan. https://drive.google.com/drive/u/0/search?q=statement%20WCDRR.

Otto, J., & Kuriansky, J. (2016). *Summary report of a side event held at the United Nations 7 September 2016: Mental health and well-being at the heart of the SDGs: Concrete means of implementation.* http://bit.ly/2FnebZW.

Otto, J., & Kuriansky, J. (2017). *Artificial intelligence and technology tools for mental health, well-being and resilience. Report on a side event held at the United Nations 16 May 2017 on the margins of the UN Science, Technology & Innovation Forum. United Nations headquarters, New York, NY, USA.* http://bit.ly/2mghjOH.

Queensland Brain Institute. (2023). *Half of world's population will experience a mental health disorder.* Harvard Medical School. https://hms.harvard.edu/news/half-worlds-population-will-experience-mental-health-disorder.

Rodriquez, A., Kuriansky, J., & Tapper, G. (2021). *The United African Congress Webinar Conference Series on COVID-19 in Africa.* https://docs.google.com/document/d/1TtT_HGQo1lyGayKiy2keliYz1NLg2ydz/edit.

The World Bank. (2023). *Billions left behind on the path to Universal Health Coverage.* https://www.worldbank.org/en/news/press-release/2023/09/18/billions-left-behind-on-the-path-to-universal-health-coverage.

The World Bank. (2016). *Out of the shadows: Making mental health a global priority.* https://www.worldbank.org/en/events/2016/03/09/out-of-the-shadows-making-mental-health-a-global-priority.

World Health Organization. (2022). *WHO to identify pathogens that could cause future outbreaks and pandemics.* https://www.who.int/news/item/21-11-2022-who-to-identify-pathogens-that-could-cause-future-outbreaks-and-pandemics.

World Health Organization. (2023). *UN General Assembly High-Level Meetings on health 2023.* https://www.who.int/news-room/events/detail/2023/09/20/default-calendar/un-general-assembly-high-level-meetings-on-health-2023.

Yoon, D. M., & Kuriansky, J. (2024). Transforming mental health for all: Launch of the 2022 World Mental Health Report. In J. Kuriansky, & P. Kakkattil (Eds.), *Resilient health: Leveraging technology and social innovations to transform healthcare for COVID-19 recovery and beyond* (pp. 955–971). Elsevier.

YPulse. (2024). *How many young people would actually date an AI?.* https://www.ypulse.com/article/2024/01/08/how-many-young-people-would-actually-date-an-ai/.

CHAPTER 2

The state of the world in health and health systems

Muhammad Ali Pate[a,b,c] **and Zsuzsanna Jakab**[b,d]

[a]*Federal Ministry of Health and Social Welfare, Abuja, Nigeria,* [b]*Global Council SDG3 (Health), "SDGs in Action" Initiative, Dubai, United Arab Emirates,* [c]*Practice of Global Health Leadership, Department of Global Health and Population, Harvard University, Cambridge, MA, United States,* [d]*World Health Organization, Geneva, Switzerland*

Introduction

Global experience with the COVID-19 pandemic, reaching heights over the years 2019–2022 but lingering on with variants, has brought to the fore the centrality of health systems as the foundation of health and the cornerstone for national and economic security and development. The COVID-19 infection rapidly escalated from 2019 into a global pandemic, being declared by the World Health Organization (WHO) as a global health emergency of international concern (GHEIC) in January 2020, disrupting essential health services and overwhelming health systems in countries across all regions and income groups, and causing disruptions for entire societies and the global economy (WHO, 2022a). Critical foundational gaps in health systems were exposed, in developed and developing countries alike, including underresourced health facility infrastructure; surge capacity for hospital beds; insecure supply chains; shortages of medical equipment, medicines, diagnostics, and staff; and health workers placed under significant risk and enormous strain. As of July 2022, there have been over 570,000,000 confirmed cases and more than 6 million deaths globally (WHO, 2022b).

Prior to the devastating impacts of the pandemic, the state of the world's health had been improving across a number of key measures such as life expectancy and infant mortality rates. However, progress toward key global health objectives such as universal health coverage (UHC) and other sustainable development goals (SDGs) in the United Nations Agenda 2030 for Sustainable Development was already off track even before the pandemic hit (United Nations, 2020). Despite the fact that the impact of COVID-19 has further arrested and even reversed progress in some cases, there are significant opportunities to improve the state of the world's health more sustainably and to "build back better," meaning stronger and more resilient health

Resilient Health. https://doi.org/10.1016/B978-0-443-18529-8.00002-0

systems in the postpandemic period. This will require actions guided by the lessons learned from the COVID-19 pandemic as well as from previous public health emergencies such as the infectious disease outbreaks of SARS, MERS; natural disasters, wars, conflicts, humanitarian and economic crises; as well as everyday and routine challenges to health. Advances and innovations in health technologies such as mRNA vaccines, telemedicine, remote patient monitoring, wearable technology, advanced diagnostics and use of big data, artificial intelligence and machine learning accelerated by this current pandemic offer the opportunity to more affordably and equitably distribute essential health services in the long run and to usher in a renaissance for health systems and global public health.

This chapter provides an overview of the state of the world's health and health systems. It begins by discussing the major causes of disease burden globally such as noncommunicable and communicable diseases. Then, it introduces key trends in health and population demographics such as increasing life expectancy, an aging population and escalating migration, and what these mean for health systems. The chapter defines what a "health system" is, and its different parts which interact, such as the workforce, medical products and technologies, information systems, governance, financing, and of course delivery of services to ultimately meet the objectives of protecting and improving the health of individuals and populations worldwide. High-level global health objectives such as universal health coverage (UHC) and health security are also described. The chapter goes on to describe the global directions needed to build more resilient health systems through more integrated approaches such as essential public health functions (EPHFs) and primary health care (PHC), and harnessing the power of innovation and technology. As such, this chapter sets the scene for the chapters ahead. We conclude by defining a way forward upon which future health systems can not only be "built back better" but become more resilient to known, and as of yet unknown, threats in the COVID-19 recovery period and beyond.

Overview of the world's health

There are many ways to define and measure the health of the global population. The WHO defines health as "state of complete physical, mental and social wellbeing and not merely the absence of disease or infirmity" (WHO, 2006). This is an ambitious concept of health that considers physiological, psychological, and social factors in health and illness and takes into account the interactions between these factors. Nevertheless, when considering the state of health of the world's population, it is important to adopt a holistic understanding of health beyond the traditional medical model which narrowly defines health as the absence of physical illness or disease and emphasizes the role of clinical diagnosis and interventions.

Overall, the health of the world's population has steadily improved over the past 30 years, as measured by age-standardized Disability-Adjusted Life Years (DALYs), a composite measure of overall disease burden that quantifies the number of years lost due to premature death and disability. These demonstrate a 50% reduction in DALYs associated with communicable diseases, maternal, perinatal, and nutritional conditions between 2000 and 2019 (WHO, 2022c). Health-Adjusted Life Expectancy (HALE) is another measure that reveals the state of the world in health. This composite measure combines life expectancy and prevailing mortality and morbidity conditions, with statistics showing an increase, globally, from 58.6years in the year 2000 to 63.5years in 2019. This was the case for 202 out of 204 countries assessed during the same period. In contrast to this positive news, the COVID-19 pandemic and related disruptions to essential health services appear to have halted decades-long increases in life expectancy in many countries with global mortality data demonstrating a decrease in 31 out of 37 high-income countries in 2020 (Kuehn, 2022).

Noncommunicable diseases

The prevalence of noncommunicable diseases (NCDs) is an important indicator of the state of the world in health today. A look at statistics shows that by 2019, ischemic heart disease and stroke were the top-ranked causes of DALYs for those aged 50 and older (WHO, 2022c); 7 out of the 10 leading causes of death globally are NCDs, as shown in Fig. 1.

As the figure shows, since 1990, there has been a marked shift toward a greater proportion of burden due to NCDs and injuries. In fact, NCDs are responsible for 41 million deaths each year, equivalent to 71% of all deaths globally (WHO, 2021a). Despite this, there is a vast difference in global funding provision and disease burden, as shown in Fig. 2 (Allen, 2017), which means that many health systems are not designed or equipped to meet the most pressing health challenges. Each year, more than 15 million people between the ages of 30 and 69years die from an NCD. These deaths are considered premature as many can be avoided through early preventative and primary healthcare interventions. While 77% of all NCD deaths are in low- and middle-income countries, 85% of premature deaths occur in low- and middle-income countries. Cardiovascular diseases account for most NCD deaths, numbering 17.9 million people annually, followed by cancers (9.3 million), respiratory diseases (4.1 million), and diabetes (1.5 million). These four groups of NCDs account for over 80% of all premature NCD deaths. Tobacco use, physical inactivity, harmful use of alcohol, and unhealthy diets all increase risk of dying from an NCD indicating the importance of, and need for, health system capacities to provide preventative and health promotive public health interventions and programs.

Mental and psychosocial health

Mental disorders have been identified by the World Health Organization as the fifth NCD (WHO Regional Office for Europe, 2019). Mental disorders are widely underrecognized globally, yet the prevalence statistics are alarming (Kopel & Fox, 2024; WHO, 2022d). There were an estimated 792 million people in 2017 suffering from mental ailments globally (Dattani

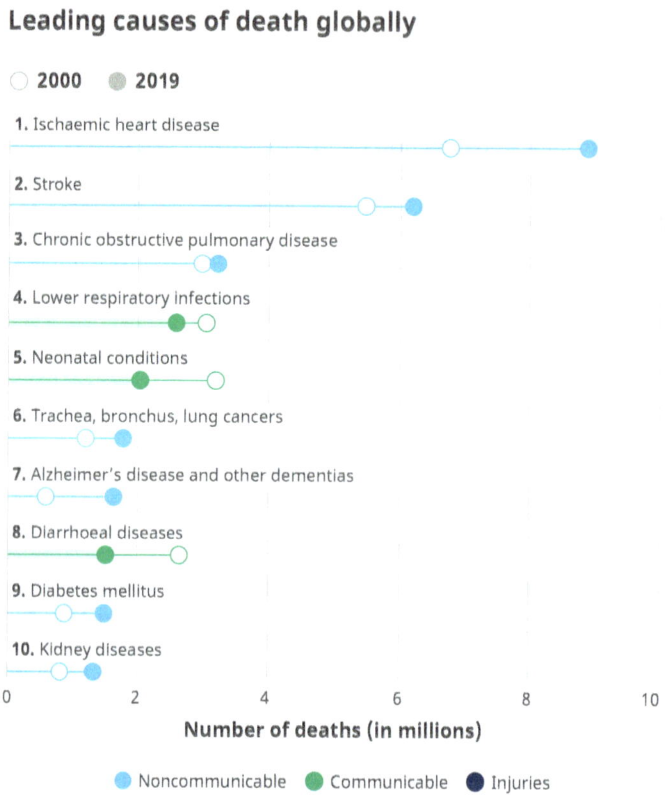

Fig. 1

At a global level, 7 out of the 10 leading causes of deaths in 2019 were NCDs. *(World Health Organization. The top 10 causes of death. 2020a. Accessed August 1, 2022. https://www.who.int/news-room/fact-sheets/detail/the-top-10-causes-of-death.)*

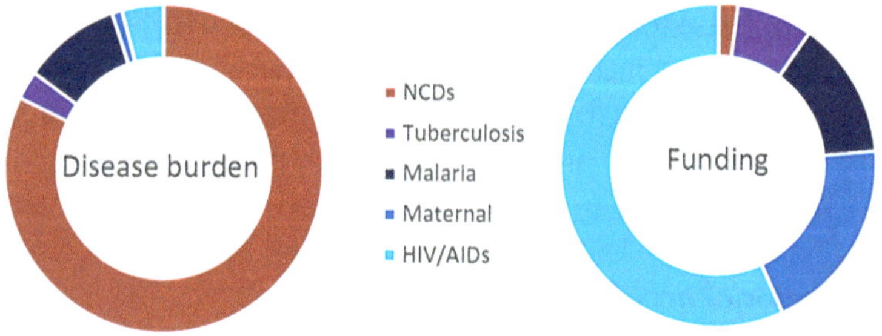

Fig. 2

Share of disability adjusted life years (DALYs) and health aid by disease groups. *(Allen, L. N. (2017). Financing national non-communicable disease responses.* Global Health Action, 10*(1), 1326687.)*

et al., 2021), ranging from anxiety, depression, bipolar disorders, psychoses, and other associated conditions, including alcohol and substance abuse. The prevalence of mental disorders is higher in females than in males and occurs in countries at all income levels. In some areas, stigma and discrimination associated with the diagnoses of mental disorders hamper proper diagnosis and care of those affected. Severe mental illness is associated with premature mortality due to comorbid physical conditions, even while mental diseases generally require lower cost treatments (WHO, 2022e). Rates of suicide are an important indicator of mental health or levels of disorder in countries (also noted as an indicator in the UN Agenda 2030 for Sustainable Development), as they can be quantified in comparison to many mental health disorders which cannot as easily be measured and constitute a more "silent" epidemic. Suicide rates have been on the rise for males in all countries, e.g., for farmers (Behere & Bhise, 2009). Suicide in adolescents, and even young children, is also high and a significant public health challenge for health systems and services (WHO, 2021b).

Injury

Injuries—both unintentional and violence related—take the lives of 4.4 million people around the world each year and constitute nearly 8% of all deaths (WHO, 2021c). Three of the top five cases of death for those aged 5–29 years are injury related, namely road traffic injuries, homicide, and suicide. In adolescents and young adults aged 10–24 years, three injury causes were among the top causes of DALYs—road injuries, self-harm, and interpersonal violence. Injuries and violence are responsible for an estimated 10% of all years lived with disability, and place a massive burden on national economies, costing countries billions of US dollars each year in healthcare, lost productivity, and law enforcement. Preventing injuries and violence will facilitate achievement of several Sustainable Development Goal (SDG) targets (WHO, 2021c).

Infectious diseases

As the world witnessed in the emergence of the novel virus which caused the COVID-19 pandemic recognized in late 2019, emerging and reemerging infectious diseases have been on the rise in the last few decades (McArthur, 2019). Despite significant progress in understanding, prevention, and treatment over the past century, the cost to lives, livelihoods, and the economy from infectious diseases, even before the COVID-19 pandemic, was staggering. Six infectious diseases (lower respiratory infections, diarrheal diseases, malaria, meningitis, whooping cough, and sexually transmitted infections) were among the top 10 causes of DALYs in children younger than 10 years in 2019. Some estimates put the economic burden of eight major infectious diseases (HIV/AIDS, tuberculosis, malaria, measles, hepatitis, dengue fever, yellow fever, and rabies) at up to US$8 trillion with more than 156 million life years lost for the year 2016 alone (Armitage, 2021).

The spread of infectious diseases is driven largely by socioeconomic, environmental, and ecological factors in a globalized world where the level of interconnectedness is beyond what humanity had experienced in the past. The ongoing COVID-19 pandemic has already afflicted more than 558 million people in all countries of the world, leading to 6.4 million deaths (WHO, 2022b). Continued emergence of new variants of the virus suggests that while the pandemic has been blunted by vaccines as well as public health and social measures, it is not yet over. Even 2 years after the pandemic hit, health systems are facing significant challenges in providing essential health services, with ongoing disruptions reported in over 90% of countries surveyed by the WHO with expected resulting adverse impacts on health outcomes (WHO, 2022a).

Emerging Infectious Diseases (EIDs) are dominated by zoonoses, comprising 60.3%; the majority (71.8%) of which originate in wildlife and are increasing significantly over time (Jones et al., 2008). 54.3% of EID events are caused by bacteria or rickettsia, many of which are drug-resistant microbes. Rising antimicrobial resistance, from misuse or overuse of antibiotics in both human and animal health practice, presents a major threat to the health status of human populations, which is underrecognized. The generally low surveillance capabilities for wildlife zoonotic and vector-borne EIDs, as well as antimicrobial resistance, continue to pose a substantial risk.

Migration

Migration and forced displacement represent a major threat and risk for global population health (WHO, 2021d). There were an estimated 272 million international migrants in 2019 (United Nations Department of Economic and Social Affairs, 2019a), the number of which rose significantly by mid-2022, due to the Afghanistan takeover by the Taliban and the Ukraine–Russia war. Refugees, asylum seekers, and irregular migrants are of special concern to governments and societies given their extensive needs for protection and support in many dimensions besides health (housing, employment). Protection of refugees is addressed by the United Nations agency, the United Nations High Commissioner for Refugees, and its Global Compact on Refugees, and that of migrants by the UN Agency, The International Organization for Migration, and also by the Global Compact for Safe, Orderly and Regular Migration.

Migrants and refugees can contribute positively to societies, albeit they cannot reach their full potential unless they are in good physical and mental health. Even when services are made available to them, they face disruptions in continuity of care. Several chapters in this volume address the issues related to migrants and refugees and particularly technological advances that are improving their health conditions (Eluere, 2024; O'Donnell & Goguadze, 2024; Talhouk, 2024). These groups are also exposed to various stress factors which affect their mental health and well-being before and during their migration journey and during their settlement and integration. The prevalence of common mental disorders such as depression, anxiety, and posttraumatic stress disorder (PTSD) tends to be higher among migrants exposed to adversity and refugees than among host populations (WHO, 2021d). Many migrants and refugees lack access to mental health services or experience barriers in access. Yet, most experts agree that the mental health needs of migrants and refugees should be addressed, including by such means as: organizing inclusive and accessible promotion and prevention programs; strengthening mental health as part of general and primary health services; and ensuring timely diagnosis, treatment, and rehabilitation.

War and conflict

Almost all people affected by emergencies will experience psychological distress or ill health as a direct result of conflict as well as due to the effects of displacement, the breakdown of health and social services, and the heightened risk of disease transmission (Murray et al., 2002). Among people who have experienced war or other conflict in the previous 10years, one in five (22%) will have depression, anxiety, posttraumatic stress disorder, bipolar disorder, or schizophrenia (Charlson et al., 2019). People with severe mental disorders are especially vulnerable during emergencies and need access to mental healthcare and other basic needs. International guidelines recommend services at several levels—from basic services to clinical care—and indicate that mental healthcare needs to be made available immediately for specific, urgent mental health problems as part of the health response. Despite their tragic nature and adverse effects on mental health, emergencies can provide opportunities to build sustainable mental health systems for all people in need.

Population dynamics

The state of world health can be understood by examining certain critical statistics about population dynamics, including population size, fertility, life expectancy, and urbanization.

The number of people worldwide increased 25.7% from 6.2 billion in 2000 to 7.7 billion by 2019 and is now set to reach 8 billion in 2022. Further, it is projected to peak at 9.73 billion people by 2064, before subsequently declining to 8.79 billion in 2100. While under-5 child deaths had declined from 9.6 million in 2000 to 5.0 million in 2019, the estimated total number of child deaths increased from 50.7 million in 2000 to 56.5 million in 2019. Behind these dynamics are trends toward reductions in fertility, increased child and adult survival due to medical advances, education and economic conditions, and aging of the global population.

The global Total Fertility Rate (TFR), measured in the average number of children per woman, which had decreased from 2.72 in 2000 to 2.31 in 2019, is forecasted to further decrease, to 1.66 in 2100. By 2019 there were an estimated 135 million live births. By 2050 it is forecasted that 151 countries would have a TFR lower than the replacement level (TFR $<$ 2.1), and 183 were forecasted to have a TFR lower than replacement by 2100. Twenty-three countries could have population declines greater than 50% from 2017 to 2100. Continued trends in female educational attainment and access to contraception will further hasten declines in total fertility and lead to slower global population growth. Sustained TFR at levels lower than the replacement level in many countries, including China and India, and fast growth rates in other regions such as in Africa, will have economic, social, environmental, and geopolitical consequences. This can contribute to an overall older population and vulnerability to infectious disease as well as distinct health needs.

Life expectancy at birth is also a useful measure to assess the state of health of the world's population. Life expectancy of human populations has improved in nearly every region of the world, from a global average of 53years in 1960 to 67.2years in 2000 and 73years by 2019. The expectancy at birth has almost doubled what it was two centuries ago (Our World in Data, 2022) and with improvements observed in all age groups.

Higher national income is associated with higher prospects for longevity. The populations of many of the wealthiest countries in the world now enjoy life expectancies well over 80years. For example, in 2019, average life expectancy in developed countries of Spain, Switzerland, Italy, and Australia was over 83years, while it is close to 85years in Japan. In developing and poorer countries of the world, average life expectancy is between 50 and 60years (World Bank, 2022), exposing the "split" in observed progress since the 1950s when Africa and Asia regions had life expectancies of about 30years. This increased life expectancy rate in continents like Africa and Asia is significant, in terms of understanding the burdens on health systems that must keep up with longer lives.

One of the factors behind increasing longevity of human populations has been public health improvements such as the control of infectious diseases, vaccinations and immunization, sanitation and clean water; more abundant and safer foods, and other nonmedical social improvements, such as road safety and environmental protection. Public health interventions have also largely contributed to the reduction in child mortality. Other contributing factors to improved life expectancy include broader advances in medical science, and economic growth and development, and consequent impact of these on the broader determinants of health (social, environmental, commercial) and quality of life. While in the 1900s infectious diseases were the leading causes of death worldwide, the leading causes of mortality now are largely NCDs, like heart and respiratory disease, cancer, and diabetes. Since the year 2000, there has been 50% drop in disability-adjusted life years lost due to communicable, maternal, perinatal, and nutritional conditions.

As the world's population lives longer, it is also aging faster. Between 2015 and 2050, the proportion of the world's population over 60years will nearly double from 12% to 22%. By the year 2020, the number of people aged 60years and older outnumbered children younger than 5years (also given the lower birth rate). While the pace of population aging is much faster now than in the past, the distribution of the aged population in regions of the world is also shifting, with the fastest population aging seen in Eastern and South-Eastern Asia, Latin America, and the Caribbean (United Nations Department of Economic and Social Affairs, 2019b).

While age 60 was once considered old age, the definition of this term has been shifting, with people being more fit into older ages. Also, while 'retirement" was once very common on reaching age 65 (and forced retirement at a certain age (usually 65 or 70) prevails in some countries like Japan, Brazil, and the Philippines), more people, especially entrepreneurs, continue to work well into advanced and older years.

By the year 2050, an estimated 80% of older people will be living in low- and middle-income countries. This means that all countries, besides those with more thriving economies and social systems, and more robust health systems already in place, face major challenges to ensure that their health and social systems are ready to accommodate, as well as make the most of, the inevitable demographic shift toward aging. Common conditions in older age include hearing loss, cataracts and refractive errors, back and neck pain and osteoarthritis, chronic obstructive pulmonary disease, diabetes, depression, and cognitive decline and dementia. The prevalence of these chronic conditions and multimorbidities has implications for the resilience of health systems to be able to meet more complex health needs and evolving and increasing demands and expectations, especially in the context of reduced fiscal spending on health following economic stressors such as COVID-19 and war and conflict. This also has potentially greater implications for health and allied sectors that provide long-term care such as through primary healthcare, as well as for essential public health functions and necessary public facilities, such as housing, social care, and transportation.

The present trend in population health and public health is to not only promote aging in good health, but toward healthier aging, enduring quality of life, and even longer life. The goal is also for people to remain healthy and independent well into older age and continue to contribute valuably and sustainably to their families and communities.

Urbanization is also a trend that has important implications for the health and quality of life of human populations. More people now live in urban than in rural areas, with 55% of the world's population residing in urban areas in 2018 (United Nations Department of Economic and Social Affairs, 2019c). By the year 2050, 68% of the world's population is projected to be living in urban areas. The most urbanized geographic regions include Northern America (82% living in urban areas in 2018), Latin America and the Caribbean (81%), Europe (74%), and Oceania (68%). The level of urbanization in Asia is now approximately 50%, while Africa remains mostly rural, with 43% of its population living in urban areas. One in eight people in the urban areas live in one of the 33 megacities, each with more than 10 million inhabitants. This phenomenon is likely to have a significant impact on human health, through availability of health services and other vital infrastructure, such as housing, transportation, workspaces, environment, and sanitation.

Sustainable development depends increasingly on the successful management of urban growth, especially in low- and lower middle-income countries, where the most rapid urbanization is expected. Upper middle-income countries have experienced a relatively fast pace of urbanization: their share of urban population is expected to rise to 83% urban by 2050. In lower middle-income countries, the pace of urbanization has been slower, specifically, in 2018, the proportion of the population living in urban areas was 41% in the lower middle-income countries compared to 32% in low-income countries.

Urbanization is also expected to continue to increase, as populations shift away from employment in the agriculture sector and move to urban areas from rural areas, in search of higher incomes. While living standards tend to be higher in urban areas, at the same time, just under 1-in-3 people in urban areas globally live in a slum household (Ritchie & Roser, 2019). A slum household is defined as a group of individuals living under the same roof lacking one or more of the following conditions: access to improved water, access to improved sanitation, sufficient living area, and durability of housing. Given this situation, effective management of urbanization, and the structural inequalities it can foster, has direct relevance to the state of population health and well-being now and in the future.

Health systems: Strengthening and resilience

Health systems consist of all organizations, people, and actions whose primary intent is to promote, restore, or maintain health (WHO, 2007). Health systems are set within the political and institutional framework of a country and can include public and private (for-profit and not-for-profit) organizations, institutions, and resources. This includes efforts to influence determinants of health as well as more direct health improving activities. The health system delivers preventive, promotive, curative, and rehabilitative interventions through a combination of public health and healthcare services.

The WHO's health systems framework provides a simple yet effective way to understand the different interrelated components of a health system (WHO, 2010). As shown in Fig. 3, the framework describes six core components or "building blocks": (1) service delivery, (2) health workforce, (3) health information systems, (4) essential medicines and health products, (5) financing, and (6) leadership/governance. Strengthening these building blocks supports the overall goals and outcomes of improved health, responsiveness, social and financial risk protection, and improved efficiency, mediated by four factors: access, coverage, quality, and safety. This framework does not aim to be comprehensive but espouses that strengthening each of the building blocks and the interrelationships between them can yield benefit across the system. Strengthening health systems involves an intentional effort to improve performance and this goes beyond merely investing in the inputs but also reforming how the health system and its different parts operate toward common objectives. This means taking an integrated approach that sees individuals and populations holistically, planning for their health needs

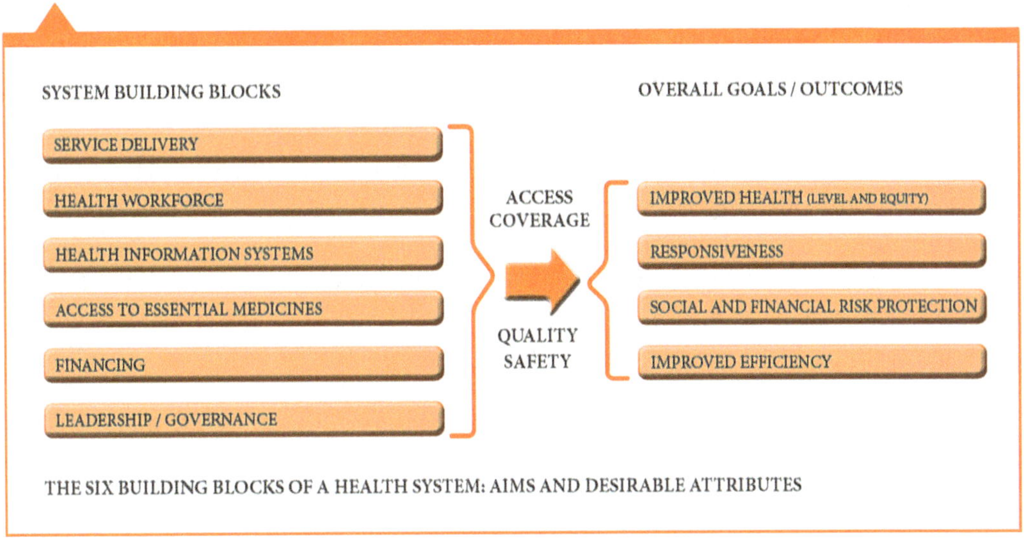

Fig. 3

The WHO health systems framework. *(World Health Organization. Monitoring the building blocks of health systems: A handbook of indicators and their measurement strategies. 2010. Accessed March 13, 2023. https://apps.who.int/iris/handle/10665/258734.)*

throughout the life course, and delivering services more efficiently in health facilities, in the community, or at home.

Following major stressors on health systems, such as those mentioned earlier in this chapter like COVID-19, and war and conflict, the concept of health systems resilience has been mainstreamed. Health systems resilience refers to the ability of the system to mitigate risk, prepare for, prevent, respond to, and recover from everyday and sudden shocks and stressors while maintaining routine functionality and transforming as needed. A key feature of resilient health systems is that they are integrated, and it is only through this intentional design toward integration and resilience that progress toward key global health objectives such as universal health coverage (UHC) and health security can be sustained (WHO, 2021e). UHC means that all people who need health services receive them with sufficient quality to be effective and without exposure to financial hardship. Health systems that progress toward UHC contribute to global health security which is about reducing the vulnerability of societies to health threats that can further spread across national borders.

Primary health care (PHC) provides the approach for health systems to make progress toward UHC, and health security, in most if not all contexts. PHC emphasizes community empowerment and social accountability and is multisectoral with links to education, nutrition, and water and sanitation. PHC provides a platform for integrating often separate services, for example, services for infectious diseases with those services targeted for women and children's health and NCDs; for addressing evolving population dynamics, epidemiological challenges;

and for innovations such as digital health. PHC also remains the most cost-effective way to address comprehensive health needs close to people's homes and communities, thereby promoting access, which is a primary criterion for UHC.

Invariably, people need individual and public health services that are effective and therefore these both fit within the scope of UHC and health systems strengthening. Traditionally, health systems strengthening efforts toward UHC have focused more on individual and clinical health services rather than public health interventions. The essential public health functions (EPHFs), as shown in Box 1, are a set of fundamental, interconnected activities aligned with the PHC approach that can facilitate effective public health actions in health systems strengthening within and beyond the health sector (WHO, 2021i). This is with the recognition that many of the determinants of health lie outside the remit of the ministry of health or the health sector.

BOX 1

A consolidated list of EPHFs, adapted from the WHO discussion paper. (World Health Organization. 21st century health challenges: can the essential public health functions make a difference?: discussion paper. 2021f. Accessed August 1, 2022. https://apps.who.int/iris/handle/10665/351510.)

1. **Public health intelligence**: monitoring and evaluating population health status, health service utilization, and surveillance of risk factors and threats to health
2. **Public health emergency management**: managing public health emergency
3. **Public health governance**: assuring effective public health governance, regulation, and legislation
4. **Public health planning and financing**: supporting efficient and effective health systems and multisectoral planning, financing, and management for population health
5. **Health protection**: protecting populations against health threats, including environment and occupational hazards, communicable disease threats, food safety, chemical and radiation hazards
6. **Disease prevention and early detection**: promoting prevention and early detection of diseases (communicable and noncommunicable)
7. **Health promotion**: Promoting health and well-being and actions to address the wider determinants of health and inequity
8. **Community participation**: Ensuring community engagement, participation, and social mobilization for health and well-being
9. **Public health workforce**: Ensuring adequate quantity and quality of public health workforce
10. **Health services quality and equity**: Assuring quality of and access to health services
11. **Public health knowledge and research**: Advancing public health research
12. **Use of and access to medical products**: Ensuring equitable access to and rational use of essential medicines and other health technologies

Digital health

One way of strengthening health systems that was accelerated during the pandemic was the incorporation of innovations in technology for service delivery and design. Despite difficulties with respect to acceptability and equity and the requirement for legislative, administrative, and financial supports, technological innovations have played an increasing role in the delivery of clinical and public health services during the global COVID-19 pandemic. Countries that performed well were those that were able to rapidly integrate and scale up innovations from technology including national contract tracing and warning apps, telemedicine tools to support remote consultations and monitoring, use of artificial intelligence to support COVID-19 surveillance, and online platforms and tools to address disinformation (European Commission, 2022; European Observatory on Health Systems and Policies, 2021).

State of health systems and UHC

Today, the world possesses a more sophisticated arsenal of interventions and technologies for preventing disease, promoting health, curing illness, and prolonging life than ever before. Yet the gaps in health outcomes continue to widen. In modern health systems, and driven by the desire to focus on particular health conditions (such as infectious disease), there can be a degree of siloed capacity building leading to fragmentation, duplication, and inefficient use of limited resources. The effectiveness of health systems in addressing much of the ill health, disease, premature death, and suffering we see on such a large scale can be promoted through greater, more effective, and more equitable distribution of investments in health systems.

Progress toward UHC is one measure of the state of health systems (Bergen et al., 2019). UHC is tracked by related Sustainable Development Goal (SDG) indicators on service coverage and catastrophic health spending. Since the turn of the millennium, service coverage has increased but often at a cost to many people. Trajectories on the path to UHC vary substantially across WHO regions and countries, as shown in Figs. 4 and 5. Even without accounting for the impact of the COVID-19 pandemic, at current rates of progress toward UHC, only approximately an additional 270 million people are projected to be covered by essential health services and protected from catastrophic out-of-pocket spending by 2023—a shortfall of about 730 million people (WHO, 2021g). Service coverage improved globally from a population weighted average of 45% in 2000 to 68% in 2019. However, this progress from 2000 to 2019 is not sufficient to achieve the minimum SDG target of 80% by 2030 and has been set back even further due to the COVID-19 pandemic and associated disruptions to essential health services.

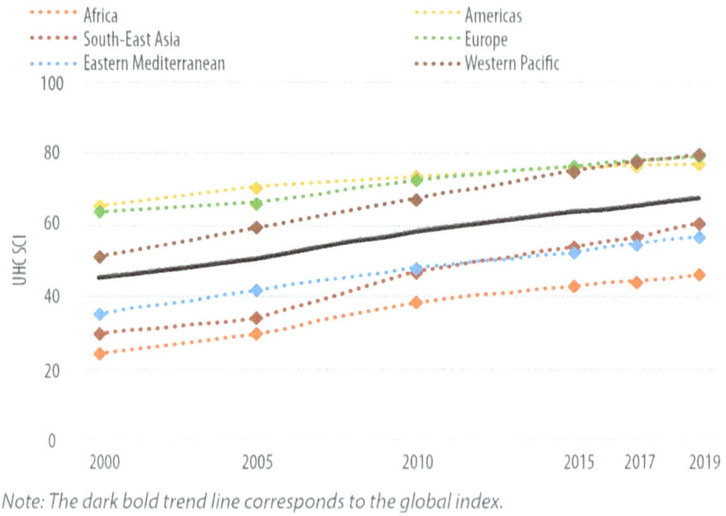

Fig. 4

Change in UHC Service Coverage Index (in index points), 2000–2019. *(World Health Organization & International Bank for Reconstruction and Development. Tracking universal health coverage: 2021 global monitoring report. 2021. Accessed March 13, 2023. https://apps.who.int/iris/handle/10665/357607.)*

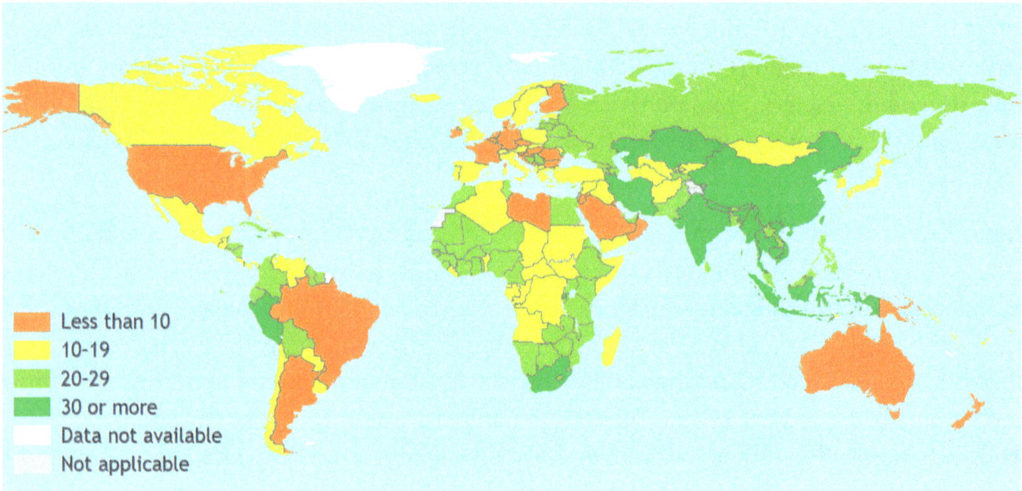

Fig. 5

Map of change in UHC Service Coverage Index (in index points), 2000–2019. *(World Health Organization & International Bank for Reconstruction and Development. Tracking universal health coverage: 2021 global monitoring report. 2021. Accessed March 13, 2023. https://apps.who.int/iris/handle/10665/357607.)*

Inequity

Inequities in physical and mental health which are prevalent globally have been escalated in the context of COVID-19 (WHO, 2020). While significant progress has been made toward improving access to health services, there were still large variations in the distribution of country-level effective coverage in 2019. The UHC effective coverage index spanned from 95% or higher in high-income developed countries like Japan and Iceland, to lower than 25% in low-income developing countries like Somalia and the Central African Republic. Many countries are lagging behind on effective coverage indicators for NCDs relative to communicable diseases and maternal and child health indicating a failure to adapt to the changing landscape of healthcare needs toward noncommunicable disease (GBD 2019 Universal Health Coverage collaborators, 2020). Good practice maintains that such inequities must be addressed not only within each country, but as a multistakeholder and multilateral approach, demanding unified action across all nations (Kuriansky et al., 2024).

Poor quality care

For Universal Health Coverage to ensure optimal health outcomes, health services must maintain a deliberate focus on quality, with the provision of effective, safe, and people-centered care that is timely, equitable, integrated, and efficient (World Health Organization, Organization for Economic Co-operation and Development & International Bank for Reconstruction and Development, 2018). Data suggests that the quality of healthcare is suboptimal in most countries with high rates of adverse treatment outcomes, healthcare-associated infections, variations in provision and delivery, low uptake of preventive vaccinations, and rising antimicrobial resistance (GBD 2019 Universal Health Coverage collaborators, 2020). Poor quality healthcare is responsible for 5.7–8.4 million deaths each year in low- and middle-income countries (LMICs), and years of life lived with disability of up to 107 million annually, primarily among those living with mental illness and untreated diabetes. Poor-quality care is responsible for up to 15% of overall deaths in LMICs and it imposes costs of US$1.4 to $1.6 trillion each year in lost productivity in these countries.

Financial burden of health

Beyond quality care, Universal Health Coverage seeks to ensure financial protection or access to needed care without financial hardship. Impoverishing health expenditure is expenditure on healthcare that causes a household to fall below the poverty line, while catastrophic health expenditure is healthcare expenditure that exceeds a household's capacity to pay. The number of people pushed into poverty due to health expenses is dramatic and concerning. Between the years 2000 and 2015, financial protection from impoverishing health expenditures has consistently trended in the wrong direction (WHO, 2021h), meaning putting more burden on people. The proportion of the world's population with out-of-pocket expenses (OOPs) exceeding 10% of their household budget rose from 12.7% of the population (940 million) to 13.2% (996 million). The incidence of catastrophic spending increased continuously between the years 2000 and 2017 as did the amount of OOP spending on health per person. While the proportion of the total population pushed below the extreme and relative poverty lines has decreased in recent times, reductions in impoverishing health expenditure did not occur in high-income countries. Poor people spending any amount OOP on health represented between 83% and 89% of people incurring impoverishing health spending.

The implication of poor financial protection is that at least 1.4 billion people globally face catastrophic and/or impoverishing health spending. In 2017 the total number of people incurring financial hardship ranged between 1.4 billion and 1.9 billion people, depending on the method of measurement. Most of the population facing catastrophic health payments was concentrated in lower- and upper middle-income countries and in the Asia region. The population pushed into extreme poverty (at PPP US$1.90 per day) was concentrated in low- and lower middle-income countries, as well as in Africa (due to a high incidence) and in Asia (due to the large population size). The proportion of the population incurring financial hardship tended to be lower in countries with greater reliance on public spending.

The COVID-19 pandemic has significantly worsened financial protection globally. Over and above the health effects, the spread of COVID-19 and the ongoing threats of infection from variants of the virus have also resulted in a deep global economic contraction, with the poor and most vulnerable populations bearing the economic brunt of the pandemic. Shifting patterns of access and utilization of all essential health services are leading to a worsening of financial protection, which will likely be sustained in the medium term unless proactive policy efforts are made.

Conclusions and the way forward

In many regards, the progress made in human health and health systems has been remarkable over the last century. At the same time, shocks like COVID-19, the climate crises, and emerging health needs are stark reminders of the fragility of human health, health systems, the global economy, societies, and the planet at large. Health is of course a fundamental human right, and to preserve, promote, and protect this right we must invest in and equip our health systems and sustain progress toward UHC by building resilience. There is also a need to look beyond the health sector to the broader determinants of health and well-being. Strengthening health systems based on strong PHC and underpinned with public health, and delivery of the EPHFs, is crucial to multisectoral engagement needed to sustain progress toward UHC and health security, and to build resilience in the face of ongoing and future threats to health.

Acknowledgments

We would like to acknowledge Sohel Saikat, Geraldine McDarby, Saqif Mustafa, Yu Zhang, and Pascale Abie, for their technical contribution and coordination in the development of this chapter.

References

Allen, L. N. (2017). Financing national non-communicable disease responses. *Global Health Action*, *10*(1), 1326687.
Armitage, C. (2021). The high burden of infectious disease. *Nature*, *598*(7882), S9.

Behere, P., & Bhise, M. (2009). Farmers' suicide: Across culture. *Indian Journal of Psychiatry*, *51*(4), 242.
Bergen, N., Ruckert, A., & Labonté, R. (2019). Monitoring frameworks for universal health coverage: What about high-income countries? *International Journal of Health Policy and Management*, *8*(7), 387–393.
Charlson, F., van Ommeren, M., Flaxman, A., Cornett, J., Whiteford, H., & Saxena, S. (2019). New WHO prevalence estimates of mental disorders in conflict settings: A systematic review and meta-analysis. *The Lancet*, *394*(10194), 240–248. https://doi.org/10.1016/s0140-6736(19)30934-1.
Dattani, S., Ritchie, H., & Roser, M. (2021). *Mental health*. https://ourworldindata.org/mental-health. (Accessed 1 August 2022).
Eluere, P. (2024). Empowering refugees through technology. In J. Kuriansky, & P. Kakkattil (Eds.), *Resilient health: Leveraging technology and social innovations to transform healthcare for COVID-19 recovery and beyond* (pp. 353–359). Elsevier.
European Commission. (2022). *Digital solutions during the pandemic*. https://ec.europa.eu/info/live-work-travel-eu/coronavirus-response/digital-solutions-during-pandemic_en. (Accessed 1 August 2022).
European Observatory on Health Systems and Policies. (2021). *Health systems resilience during COVID-19: lessons for building back better*. https://apps.who.int/iris/handle/10665/348493. (Accessed 1 August 2022).
GBD 2019 Universal Health Coverage collaborators. (2020). Measuring universal health coverage based on an index of effective coverage of health services in 204 countries and territories, 1990–2019: A systematic analysis for the Global Burden of Disease Study 2019. *The Lancet*, *396*(10258), 1250–1284.
Jones, K. E., Patel, N. G., Levy, M. A., et al. (2008). Global trends in emerging infectious diseases. *Nature*, *451*(7181), 990–993. https://doi.org/10.1038/nature06536.
Kopel, R., & Fox, C. (2024). Overview of the state of mental health in the COVID-19 recovery and beyond. In J. Kuriansky, & P. Kakkattil (Eds.), *Resilient health: Leveraging technology and social innovations to transform healthcare for COVID-19 recovery and beyond* (pp. 53–61). Elsevier.
Kuehn, B. M. (2022). COVID-19 cuts life expectancy in dozens of countries. *Journal of the American Medical Association*, *327*(3), 209.
Kuriansky, J., Altungy, P., Liébana, S., Maurice, S. R., Velo, C., Sadia, R., Davcheva, M., & Guimarães, M. A. (2024). Health equity, mental health, and partnerships to build back after COVID-19: Multistakeholder perspectives at a United Nations event. In J. Kuriansky, & P. Kakkattil (Eds.), *Resilient health: Leveraging technology and social innovations to transform healthcare for COVID-19 recovery and beyond* (pp. 115–146). Elsevier.
McArthur, D. B. (2019). Emerging infectious diseases. *Nursing Clinics of North America*, *54*(2), 297–311. https://doi.org/10.1016/j.cnur.2019.02.006.
Murray, C. J. L., King, G., Lopez, A. D., Tomijima, N., & Krug, E. G. (2002). Armed conflict as a public health problem. *BMJ*, *324*(7333), 346–349.
O'Donnell, I., & Goguadze, L. (2024). Innovative approaches in meeting the health needs of migrants, refugees, and displaced persons. In J. Kuriansky, & P. Kakkattil (Eds.), *Resilient health: Leveraging technology and social innovations to transform healthcare for COVID-19 recovery and beyond* (pp. 729–739). Elsevier.
Our World in Data. (2022). *Life expectancy, 1770 to 2019*. https://ourworldindata.org/grapher/life-expectancy?tab=chart&time=1770..2019&country=Africa~Asia~Europe~OWID_WRL~Americas~Oceania. (Accessed 1 August 2022).
Ritchie, H., & Roser, M. (2019). *Urbanization*. https://ourworldindata.org/urbanization. (Accessed 13 March 2023).
Talhouk, R. (2024). Technologically mediated resilience for and by refugees, migrants, and asylum seekers. In J. Kuriansky, & P. Kakkattil (Eds.), *Resilient health: Leveraging technology and social innovations to transform healthcare for COVID-19 recovery and beyond* (pp. 741–749). Elsevier.
United Nations. (2020). *The sustainable development goals report 2020*. https://unstats.un.org/sdgs/report/2020/. (Accessed 1 August 2022).
United Nations Department of Economic and Social Affairs. (2019a). *International migrant stock 2019: country profiles*. Accessed 1 August 2022. https://www.un.org/en/development/desa/population/migration/data/estimates2/countryprofiles.asp.
United Nations Department of Economic and Social Affairs. (2019b). *World population ageing 2019—highlights*. https://www.un.org/en/development/desa/population/publications/pdf/ageing/WorldPopulationAgeing2019-Highlights.pdf. (Accessed 1 August 2022).

United Nations Department of Economic and Social Affairs. (2019c). *World urbanization prospects—the 2018 revision*. https://population.un.org/wup/Publications/Files/WUP2018-Report.pdf. (Accessed 1 August 2022).

World Bank. (2022). *Life expectancy at birth, total (years)*. https://data.worldbank.org/indicator/SP.DYN.LE00.IN. (Accessed 1 August 2022).

World Health Organization. (2006). *Constitution of the World Health Organization—basic documents* (45th ed.). Supplement, October 2006 https://www.who.int/publications/m/item/constitution-of-the-world-health-organization. (Accessed 1 August 2022).

World Health Organization. (2007). *Everybody's business—Strengthening health systems to improve health outcomes: WHO's framework for action*. https://apps.who.int/iris/handle/10665/43918. (Accessed 13 March 2023).

World Health Organization. (2010). *Monitoring the building blocks of health systems: a handbook of indicators and their measurement strategies*. https://apps.who.int/iris/handle/10665/258734. (Accessed 1 August 2022).

World Health Organization. (2018). *Organisation for economic co-operation and development & international bank for reconstruction and development. Delivering quality health services: a global imperative for universal health coverage*. https://apps.who.int/iris/handle/10665/272465. (Accessed 1 August 2022).

World Health Organization. (2019). Regional office for Europe. In *Integrating the prevention, treatment and care of mental health conditions and other noncommunicable diseases within health systems: WHO European high-level conference on noncommunicable diseases: time to deliver – meeting NCD targets to achieve Sustainable Development Goals in Europe: 9–10 April 2019, Ashgabat, Turkmenistan*. https://apps.who.int/iris/handle/10665/346438. (Accessed 1 August 2022).

World Health Organization. (2020). *The top 10 causes of death*. https://www.who.int/news-room/fact-sheets/detail/the-top-10-causes-of-death. (Accessed 1 August 2022).

World Health Organization. (2021a). *Noncommunicable diseases*. https://www.who.int/news-room/fact-sheets/detail/noncommunicable-diseases. (Accessed 1 August 2022).

World Health Organization. (2021b). *Suicide*. https://www.who.int/news-room/fact-sheets/detail/suicide. (Accessed 1 August 2022).

World Health Organization. (2021c). *Injuries and violence*. https://www.who.int/news-room/fact-sheets/detail/injuries-and-violence. (Accessed 1 August 2022).

World Health Organization. (2021d). *Mental health and forced displacement*. https://www.who.int/news-room/fact-sheets/detail/mental-health-and-forced-displacement. (Accessed 1 August 2022).

World Health Organization. (2021e). *Building health systems resilience for universal health coverage and health security during the COVID-19 pandemic and beyond: WHO position paper*. https://apps.who.int/iris/handle/10665/346515. (Accessed 1 August 2022).

World Health Organization. (2021g). *Tracking Universal Health Coverage: 2021 global monitoring report*. https://www.who.int/publications/i/item/9789240040618. (Accessed 1 August 2022).

World Health Organization. (2021h). *Global monitoring report on financial protection in health 2021*. https://www.who.int/publications/i/item/9789240040953. (Accessed 1 August 2022).

World Health Organization. (2021i). *21st century health challenges: Can the essential public health functions make a difference?: Discussion paper*. https://apps.who.int/iris/handle/10665/351510. (Accessed 1 August 2022).

World Health Organization. (2022a). *Third round of the global pulse survey on continuity of essential health services during the COVID-19 pandemic: November-December 2021. Interim report*. https://www.who.int/publications/i/item/WHO-2019-nCoV-EHS_continuity-survey-2022.1. (Accessed 1 August 2022).

World Health Organization. (2022b). *WHO coronavirus (COVID-19) dashboard*. https://covid19.who.int/. (Accessed 1 August 2022).

World Health Organization. (2022c). *Global health estimates: life expectancy and leading causes of death and disability*. https://www.who.int/data/gho/data/themes/mortality-and-global-health-estimates. (Accessed 1 August 2022).

World Health Organization. (2022d). *World mental health report: transforming mental health for all*. Department of Mental Health and Substance Use. https://www.who.int/publications/i/item/9789240049338. (Accessed 1 August 2022).

World Health Organization. (2022e). *Mental health*. https://www.who.int/publications/i/item/9789240049338. (Accessed 1 August 2022).

CHAPTER 3

Universal Health Coverage and Japan: In memory of the late former prime minister ABE Shinzo of Japan

Satoshi Ezoe

Global Health Strategy Division, Ministry of Foreign Affairs, Government of Japan, Tokyo, Japan; Department of Global Health Policy, Graduate School of Medicine, The University of Tokyo, Tokyo, Japan

Introduction

Healthcare is an essential part of a country's progress and development. Efforts to provide healthcare to all people in the society have gained momentum at the country level but also on the international stage through processes at the United Nations (UN) and the agenda of world leaders in support of Universal Health Coverage. This chapter describes the leadership and commitment of the government of Japan in this movement over many years and at the UN, reviews the impact of the COVID-19 pandemic in Japan, and particularly honors the important role of the late former prime minister of Japan Abe Shinzo who championed UHC and who tragically passed away in 2022.

Background

On September 23, 2019, a few months before COVID-19 emerged at the forefront of global concern, world leaders at the United Nations (UN) General Assembly High-Level Meeting on Universal Health Coverage (UHC) unanimously approved the historic Political Declaration of the High-Level Meeting on UHC (United Nations, 2019a), described by UN secretary-general António Guterres as "the most comprehensive agreement ever reached on global health." (United Nations, 2019b).

Then-prime minister Abe Shinzo (hereafter "Prime Minister Abe") was invited to the closing ceremony of the High-Level Meeting, as shown in Fig. 1, as the only head of government representing the Member States (governments of the United Nations), in recognition of his

Resilient Health. https://doi.org/10.1016/B978-0-443-18529-8.00003-2

Fig. 1
Closing session of the UN High-Level Meeting on UHC, September 23, 2019 (front row, left to right): Philanthropist Ms. Melinda Gates; Prime Minister of Japan Abe Shinzo; H.E. Tijjani Muhammad Bande, President of the 74th Session of the UN General Assembly; Secretariat; WHO Director-General Dr. Tedros. Also pictured: (back row, fourth from left): H.E. Ambassador Bessho Koro, then-permanent representative of the Mission of Japan to the United Nations. *Photo credit: Cabinet Public Relations Office.*

global leadership and commitment on global health, in particular, his efforts to achieve Universal Health Coverage (UHC). Prime Minister Abe stated,

> Japan achieved UHC with the introduction of our national health insurance system in 1961, even while struggling to overcome poverty and destruction after the end of the Second World War… This achievement has enabled all Japanese citizens to enjoy affordable and quality health services and has provided a foundation for Japan's social and economic development in the postwar period and for a healthy ageing society.

He went on to make a global appeal for UHC, "Let us work together to fulfill the commitments in the UHC Political Declaration as world leaders and realize as well as sustain a society where all people have access to primary health services" (Ministry of Foreign Affairs, 2019).

Definition of Universal Health Coverage

UHC is defined by the World Health Organization (WHO) as the ability of "all people and communities (to) use the promotive, preventive, curative, rehabilitative and palliative health services they need, of sufficient quality to be effective, while also ensuring that the use of these services does not expose the user to financial hardship." (World Health Organization, 2019). Moreover, this historic agreement on UHC reiterates a number of key principles, including the importance of health for implementation of all the goals and targets of the United Nations 2030 Agenda for Sustainable Development. World leaders recommitted themselves to achieving Universal Health Coverage by 2030 and to scaling up global efforts to build a healthier world for all. (See full case study by the author for details of the High-Level Meeting (Ezoe et al., 2021).)

Japan's history and role in promoting UHC

Japan has been a vocal advocate in advancing the UHC agenda through various global fora including the G7, the G20, and United Nations (Hara & Ezoe, 2019; Sakamoto et al., 2018; World Health Organization, 2018), based on its experience of achieving and managing its own UHC system since 1961, as laid out in an article published in the *Lancet* by Prime Minister Abe on "Japan's vision for a peaceful and healthier world," which this author had the privilege of drafting as a corresponding author (Abe, 2015). In this perspective, this author aims to provide a brief overview of Japan's UHC system, followed by our global contributions toward achieving UHC by 2030 including in the context of COVID-19, based on relevant documentations and beyond (Ezoe, 2020, 2021).

Japan has been able to sustain its UHC system for more than 60 years since it covered all the population in 1961. The main characteristics of Japan's UHC system include: (1) compulsory enrollment of all citizens in a national social insurance scheme comprising multiple insurers supplemented by public revenues; (2) open access to any generalist or specialist healthcare provider, which are mostly private not-for-profit entities; (3) generous financial risk protection including catastrophic health spending coverage; and (4) a nation-wide comprehensive benefit package to manage new technologies and control overall health expenditures (see details from "Japan health system review" (Japan Health System Review, 2018)). Under this scheme, Japan continuously adapts and upgrades its healthcare system by introducing new technologies, including medicines, vaccines, medical devices, diagnostics.

This UHC system, coupled with a series of public health interventions, including vaccination and immunization, safe water and sanitation, improved nutrition, and school health programs, has enabled Japan to become one of the world's healthiest nations, with leading health metrics including the world's lowest infant mortality and the longest life expectancy (Lancet, 2011). Furthermore, Japan's UHC system ensures affordable access to quality basic healthcare without financial hardship and minimizes catastrophic out-of-pocket health spending, and thereby has saved many Japanese people from impoverishment due to health conditions, even as health expenditures remain a major cause of impoverishment and bankruptcy in other parts of the world (Wagstaff et al., 2018).

It is important to note that the people and leaders of Japan made the political choice to introduce UHC at a time when Japan was still recovering from the devastation of World War II. In describing the history of this choice, during his address to newly recruited national civil servants (Shinzo, 2019), Prime Minister Abe shared a relevant story on the government's determination in introducing UHC. He described that in the beginning, there were severe protests from among some 25 million citizens who had to newly pay insurance premiums for unforeseen benefits of a nation-wide universal health insurance system. In reaction, committed officials in the administration devoted a considerable amount of time and energy to extensive dialogue and persuasion backed by a robust political will and leadership, which ultimately led to the introduction of UHC as early as in 1961. In addition to fostering a healthier population, it was UHC that underpinned Japan's later rapid social and economic development, not the other way around.

However, achieving UHC is not the end of the story. A country also needs to ensure its health system's long-term sustainability in the context of the changing demographics and needs of its population, while managing the overall cost. This is particularly relevant for Japan, a country which is experiencing rising healthcare costs and the most aged population in the world. A sustainable healthcare system with the highest attainable outcomes must utilize limited resources as effectively and efficiently as possible and maximize patient value.

UHC in Japan and COVID-19

As for the response to the COVID-19 pandemic, Japan has been able to manage the mortality rate at a relatively lower level than other countries. According to the WHO's estimate of excess mortality associated with COVID-19 in 2020 and 2021—which is defined as "the difference between the total number of deaths that have occurred and the number of deaths that would have been expected in the absence of the pandemic" and "used to quantify the direct and indirect impacts of the pandemic"—Japan accounted for minus 8 per 100,000 that were lower than for many other developed nations (e.g., USA (140), Italy (133), Germany (116), UK (109), France (63), all per 100,000) (World Health Organization, 2020). The factors that have contributed to this result are expected to include health systems factors among other bio-socio-cultural factors, but are under close examination and remain to be seen.

The core message

Based on Japan's ongoing journey to pursue resilient, equitable, and sustainable UHC, the core message is that UHC is not a luxury; it is a necessity. It is not just the right choice but a smart choice for countries to develop and prosper. At the same time, UHC is a continuous journey that must always adapt to changing needs and demographics. Every country will face aging of its population sooner or later; therefore, countries may wish to prepare sooner rather than later to establish their UHC with the vision of making it sustainable and leveraging latest technologies. Moreover, now that the global community is keenly aware, in the wake of COVID-19, of the importance of having a resilient, equitable, and sustainable UHC that can prevent, prepare for, and respond to future health crises, the global community must renew and upgrade its commitment to achieving and sustaining UHC.

Japan's Commitment to UHC Now and Going Forward

In this context, Japan is all the more committed to contributing to the global movement to advance global health and to achieve UHC in countries around the world. Prime Minister Abe articulated Japan's commitment in his article, writing that "Japan is determined to contribute further to galvanize renewed momentum for global health so that all people can receive the basic quality services they need, and are protected from health threats, without financial hardship." (Abe, 2015).

Also, in an op-ed entitled "All nations should have universal health care" coauthored with Director-General Tedros of the WHO, Prime Minister Abe also pointed out that "Universal coverage creates a common front against poverty, pandemics and the health effects of climate change. Strong coverage builds healthier people everywhere—for individuals, families, communities and nations." (Abe, 2019).

Under the common vision of UHC, Japan has advocated for global health, including UHC in particular, at major global fora including through hosting high-level conferences on UHC (Global Conference, 2013; Horton, 2016; UHC Forum, 2017; UHC2030, 2018), the G7 Presidency (Hara & Ezoe, 2019; G7 Ise-Shima Summit, 2016a, 2016b; Sakamoto et al., 2018; World Health Organization, 2018), the G20 Presidency (G20 Osaka, 2019a, 2019b), and through establishing the Group of Friends of UHC and Global Health in New York (Ezoe, n.d.) in the lead up to the United Nations General Assembly High-Level Meeting on UHC in 2019. The Group of Friends, chaired by Japan and including more than 60 Member States, was an open platform that Member States used to build momentum toward the high-level meeting in September 2019, and to raise awareness of the importance of UHC among the diplomats in New York, most of whom have expertise in issues like security, human rights, or development more than in health.

Even after the emergence of COVID-19, Japan's unwavering support to global health, particularly UHC, has been passed on to the administrations following that of Prime Minister Abe, consecutively led by former prime minister Suga Yoshihide followed by Prime Minister Kishida Fumio. In fact, Japan's international assistance on COVID-19 has been centered around UHC as a means to "leaving no one behind," the phrase which refers to thecommitment of countries as outlined in the UN 2030 Agenda for Sustainable Development to ensure that those at a lower level of progress can catch up to those who have experienced greater progress. Based on this concept, Japan has been contributing approximately US$5 billion as of May 2022 since immediately after the emergence of COVID-19, both through bilateral assistance and international organization in the following three areas.

(1) To overcome the current COVID-19 crisis through ensuring equitable access to safe and effective vaccines, therapeutics, and diagnostics. For this reason, Japan is a founding member of ACT-A as well as COVAX, multilateral partnership for ensuring access to

COVID-19 tools in developing countries. As for vaccines, Japan has contributed US$1.5 billion to COVAX facility, including through cohosting the COVAX AMC Summit (Ministry of Foreign Affairs, 2022); has donated approximately 44 million vaccine doses; and has been providing delivery support called "last one mile support" to developing countries, responding to dire needs and requests.

(2) To strengthen health systems against future health crises.

(3) To generate an enabling international environment for health security across broader sectors, including nutrition, water, and sanitation.

Building on these areas, Japan is also keen to strengthen the global health architecture for future health crises with the overall goal of achieving UHC, bearing in mind Japan's G7 Presidency and the second UN General Assembly High-Level Meeting on UHC in 2023, so that the world can finally break the cycle of "panic and neglect."

Even though Prime Minister Abe Shinzo passed away on July 8, 2022, as a result of an outrageous and despicable act of violence, his legacy in global health, among numerous other legacies for the nation and the world, will be remembered and will not fade. Japan is determined to carry on his legacy to lead and contribute to realize a vision where all people can enjoy the highest attainable standard of health regardless of who they are and where they are without facing financial hardship, in other words, UHC, for a peaceful and healthier world.

References

Abe, S. (2015). Japan's vision for a peaceful and healthier world. *The Lancet, 386*(10011), 2367–2369. https://doi.org/10.1016/s0140-6736(15)01172-1.

Ezoe, S. (2020). Universal health coverage in Japan: Leveraging technologies to ensure equitable, affordable, and quality health care for all. In P. Murthy, & A. Ansehl (Eds.), *Technology and Global Public Health*. Cham.: Springer. https://doi.org/10.1007/978-3-030-46355-7_2.

Ezoe, S. (2021). *Toward New Solidarity in Global Health: Universal Health Coverage and Reform at the WHO.* (Diplomacy) Japan Foreign Policy Forum, No. 67. 2021 https://www.japanpolicyforum.jp/pdf/2021/no67/DJweb_67_dip_02.pdf.

S. Abe, T.A. Ghebreyesus. (2019). Opinion: All nations should have universal health care. The Washington Post. December 12, 2019.

Ezoe S. Concept Note Group of Friends of Universal Health Coverage and Global Health. Permanent Mission of Japan to the United Nations. https://www.un.emb-japan.go.jp/files/000430434.pdf.

Ezoe, S., Tchelidze, T., Premabhuti, N., & Costa, M. A. C. (2021). The Political Declaration of the High Level Meeting on Universal Health Coverage (2019): Negotiating the most comprehensive agreement ever reached on global health. In I. Kickbusch, H. Nikogosian, M. Kazatchkine, & M. Kökény (Eds.), *A guide to global health diplomacy* (pp. 235–252). Graduate Institute of International and Development Studies. https://www.graduateinstitute.ch/sites/internet/files/2021-02/GHC-Guide.pdf.

G7 Ise-Shima Summit. (2016a). G7 Ise-Shima Summit, 26–27 May 2016. In *G7 Ise-Shima leaders' declaration* Ministry of Foreign of Affairs of Japan. https://www.mofa.go.jp/files/000160266.pdf.

G7 Ise-Shima Summit. (2016b). G7 Ise-Shima Summit, 26–27 May 2016. In *G7 Ise-Shima vision for global health* Ministry of Foreign Affairs of Japan. https://www.mofa.go.jp/files/000160273.pdf.

G20 Osaka. (2019a). *G20 Osaka Leaders' Declaration*. G20 2019 Japan. www.g20.org/pdf/documents/en/FINAL_G20_Osaka_Leaders_Declaration.pdf.

G20 Osaka. (2019b). *G20 Shared Understanding on the Importance of UHC Financing in Developing Countries—Towards Sustainable and Inclusive Growth—G20 2019 Japan*. www.g20.org/pdf/documents/en/annex_05.pdf.

Global Conference. (2013). Lesson from 11 country case studies: A global synthesis. In *Global Conference on Universal Health Coverage for Inclusive and Sustainable Growth, Tokyo*. http://documents.worldbank.org/curated/en/888101468149093450/pdf/831410WP0JPP0U0Box0379884B00PUBLIC0.pdf. http://documents.worldbank.org/curated/en/888101468149093450/pdf/831410WP0JPP0U0Box0379884B00PUBLIC0.pdf.

Hara, K., & Ezoe, S. *Why does global health matter to diplomacy? Global health as a security and economic challenge and as an opportunity for world leaders, with a special focus on the G7 Ise-Shima Summit*. https://www.shu.edu/. http://blogs.shu.edu/ghg/files/2019/06/Spring-2019-Issue.pdf#page=23.

Horton, R. (2016). Offline: Japans daring gamble in an age of apprehension. *The Lancet, 387*(10013), 13. https://doi.org/10.1016/s0140-6736(15)01310-0.

Japan Health System Review. (2018). *New Delhi: WHO Regional Office for South-East Asia*.

Lancet, T. (2011). Japan: Universal health care at 50 years. *The Lancet*, *378*(9796), 1049. https://doi.org/10.1016/s0140-6736(11)61223-3.

Ministry of Foreign Affairs. (2019). *Prime Minister Shinzo Abe's participation in the UN High-Level Meeting on Universal Health Coverage (UHC), September 23, 2019*. https://www.mofa.go.jp/ic/ghp/page6e_000218.html.

Sakamoto, H., Ezoe, S., Hara, K., et al. (2018). Japan's contribution to making global health architecture a top political agenda by leveraging the G7 presidency. *Journal of Global Health*, *8*(2). https://doi.org/10.7189/jogh.08.020313.

A. Shinzo. Address of the Prime Minister. the 53rd Joint Inauguration Ceremony of the Joint-Training for the New Civil Servants. Government of Japan. 2019, https://www.kantei.go.jp/jp/98_abe/statement/2019/0403kunji.html (Japanese), https://japan.kantei.go.jp/98_abe/actions/201904/_00004.html.

UHC Forum 2017. Ministry of Foreign Affairs of Japan [Press Releases]. https://www.mofa.go.jp/press/release/press4e_001823.html.

UHC2030. (2018). UHC Forum 2017: all together to accelerate progress towards UHC. UHC2030, https://www.uhc2030.org/news-events/uhc2030-news/uhc-forum-2017-all-together-to-accelerate-progress-towards-uhc-449822/. Published February 12, 2018.

Ministry of Foreign Affairs, (2022) Outcome of the COVAX AMC Summit. June 3, 2022. https://www.mofa.go.jp/ic/ghp/page1e_000326.html.

United Nations. Political declaration of the high-level meeting on universal health coverage: "Universal health coverage: moving together to build a healthier world" [resolution adopted by the General Assembly on 10 October 2019]. New York: United Nations. 2019a (A/RES/74/2; https://undocs.org/en/A/RES/74/2).

United Nations. 2019b UN welcomes "most comprehensive agreement ever" on global health. In: UN News [website]. New York: United Nations. 23 September 2019 (https://news.un.org/en/story/2019/09/1047032).

Wagstaff, A., Flores, G., Smitz, M.-F., Hsu, J., Chepynoga, K., & Eozenou, P. (2018). Progress on impoverishing health spending in 122 countries: A retrospective observational study. *The Lancet Global Health*, *6*(2). https://doi.org/10.1016/s2214-109x(17)30486-2.

World Health Organization. (2018). *The G7 presidency and universal health coverage, Japan's contribution*. World Health Organization. Published April 30, 2018 https://www.who.int/bulletin/volumes/96/5/17-200402/en/.

World Health Organization. (2019). *What is universal coverage?*. World Health Organization. Published July 9, 2019 https://www.who.int/health_financing/universal_coverage_definition/en/. https://www.who.int/health_financing/universal_coverage_definition/en/.

World Health Organization. (2020). Global excess deaths associated with COVID-19 (modeled estimates). May 2020 https://www.who.int/data/sets/global-excess-deaths-associated-with-covid-19-modelled-estimates.

CHAPTER 4

Digital health as an enabler of Universal Health Coverage in the African Region: Contributions of Amref Health Africa☆

Waruguru Wanjau[a,b], Onisoa Rindra-Ralidera[c,d], and Rispah Walumbe[d]

[a]*AHAIC Commission on the State of UHC in Africa, Amref Health Africa, Cape Town, South Africa,* [b]*University of Cape Town, Cape Town, South Africa,* [c]*Options Consultancy Services, Antananarivo, Madagascar,* [d]*AHAIC Commission on the State of UHC in Africa, Amref Health Africa, Nairobi, Kenya*

Introduction

In Africa, many digital health interventions exist that are designed to address the health challenges but often fall short in addressing health system issues. In this chapter, we review the design and impact of digital health technologies/innovations and their capacity to deliver on Universal Health Coverage. Additionally, we outline concerns about how to ensure that digital interventions address issues around efficiency, effectiveness, scalability, and most importantly, equity.

Amref Health Africa was founded in 1957 and has since grown to become the largest African-based international health development organization, currently implementing more than 120 programs, directly reaching more than 12 million people across 35 African countries. Amref Health Africa is driven by its vision of "Lasting health change in Africa" and its mission "To increase sustainable health access to communities in Africa through solutions in human resources for health, health service delivery, and investments in health." Universal Health Coverage (UHC) forms a critical foundation of this approach, evident through our participation as the cochair of UHC 2030, represented by our Group CEO, Dr. Githinji Gitahi.

Amref Health Africa, in its capacity as a thought leader and convener, established the Africa Health Agenda International Conference (AHAIC) Commission on the State of UHC in Africa.

☆ Some content in this chapter is part of a Thought Piece developed by the AHAIC Commission on the State of UHC in Africa supported by Amref Health Africa. It is reprinted with permission from Amref Health Africa.

Resilient Health. https://doi.org/10.1016/B978-0-443-18529-8.00004-4

The AHAIC Commission on the State of UHC in Africa was an independent body responsible for reviewing Africa's progress toward UHC targets by 2030 and providing recommendations, specifically in leadership, accountability, technology/innovation, and health security. The AHAIC Commission brought together 11 African Global Health experts working on the continent with three technical advisors. The authors of this piece served on this team.

Amref hosts a conference every two years, the AHAIC (Africa Health Agenda International Conference) and in 2019, this ended with a resolve to achieve UHC in Africa by 2030 through increasing *access* to health services; improving the effective delivery of *quality* health services; establishing mechanisms to guarantee adequate and sustainable publicly led *financing*; and lastly, strengthening *accountability* to safeguard transparency and efficiency at all levels and specifically people-led accountability.

What is digital health and what can it achieve in UHC?

Digital health is the intersection of technology and health to improve the efficiency of healthcare delivery and make healthcare accses more personalized and accurate. It has a wide scope. In considering digital health, the commission also included health data, social media, and wearables. In the African region, digital health offers potential to make care more accessible, improve quality of care, and better reach vulnerable and rural communities. With specific regard to universal health care (UHC), digital health has the double potential to address the constraints of providing UHC *and* to enhance the already existing enablers of UHC. Amref is a member of the UHC2030 Private Sector Constituency (UHC2030 Private Sector Constituency, 2019).

The challenges of digital health in the context of UHC

Despite its potential to accelerate the realization of UHC, digital health is not yet integrated as an essential part of health systems in the African region. Digital health is still seen as an optional discretionary complement to UHC and not as a key element of UHC that is to be integrated into the core of the health systems.

Reliable access to infrastructure is necessary to enable the implementation of complex digital health solutions. Yet, access to reliable electricity and energy sources is poor in many countries, particularly in rural areas. Network connectivity is improving but remains highly variable, with broadband access being particularly scarce (USAID, 2016). For digital health to be most effective, particularly in the context of UHC and in Africa, people, healthcare institutions, and providers need access to basic connectivity infrastructure, including mobile phones and reliable internet connectivity.

In the past, health in the African region has mainly taken a vertical approach based on diseases and priorities of the Millennium Development Goals (MDGs) that guided the United Nations Agenda from the years 2000–15. Since that time, the UN Agenda shifted to the Sustainable Development Goals (SDGs) now guiding the years 2015–30. This ambitious document with 17 (rather than only 8) goals now includes a target specifically on UHC under Goal 3 on Good health and well-being for all. SDG Target 3.8 states the objective to "achieve universal health coverage, including financial risk protection, access to quality essential health-care services and access to safe, effective, quality and affordable essential medicines and vaccines for all" (World Health Organisation, 2015). Also, there is specifically a target on technology and about the negotiations and agreements about technology transfer to the South. Governments of the world, obviously including all of those in Africa, have committed to achieve these targets by the year 2030. Also, the Addis Ababa Action Agenda established a technology facilitation mechanism.

How can digital health best be integrated into UHC systems?

Digital health has great potential to deliver healthcare to communities in need. Also, Amref through its work is keenly aware of the significant interest in the region in achieving UHC. Yet, we are also aware of potential challenges and risks of digital health in the context of UHC. These include the people capacities to use digital, data security, and access to digital technology such as the internet.

Given all these factors, it is important that stakeholders integrate digital health into the systems for UHC in the African region in a thoughtful, sustainable way. Amref maintains that the benefits of digital health will be better realized in integrated health systems as opposed to separate digital health systems and solutions for a specific vertical disease. Consistent with this, digital health should be mainstreamed in UHC in a way that maximizes its potential and minimizes the challenges and risks. Such mainstreaming must also enhance the deployment of UHC in the context of the different health systems that exist from country to country in the region. For example, in Madagascar, digital access is not easy for all people, and internet data is expensive.

Approaches and considerations for digital health integration in UHC

Given the importance of context-specific, sustainable, and integrated use of digital health in UHC, we propose two main aspects to be considered by policymakers and health practitioners in integrating digital health in UHC in the African region: (1) how digital health can be used in the health system with a UHC lens and (2) considerations and approaches of digital health integration that will enable increased impact and sustainability of UHC.

(1) How digital health can be used in the health system with a UHC lens

In considering how digital health can be integrated into a health system to improve UHC, it will be necessary to consider the following elements: health communication, service delivery, product and technology management, monitoring and evaluation, and health financing.

For **communication** from primary care toward higher levels of care, digital health solutions can help to disseminate information, raise awareness, refer to a higher level of care, and plan consultations and/or follow-up. It can also facilitate communication with the population, between providers and between beneficiary providers. The use of digital health for communication can be designed to increase patient engagement in their care journey by ensuring a participatory patient-centered design (Birnbaum & Lewis, 2015).

In **service delivery**, digital health can guide decision-making, assist in diagnosis, and thus facilitate the reduction of errors. It can enable better planning by reducing the overload of consultation services by scheduling consultations according to the availability of health workers.

For **product and technology management**, digital health can be used for product safety, purchasing, transport, stock management, thus ensuring quality, availability, and potentiality of the commodities required to provide care.

For monitoring and evaluation in a fully digitized system, patient engagment to address needs and communicate outcomes or next steps can occur remotely and rapidly to avoid often treacherous and time-consuming travel which in the African context can even be life threatening, especially for pregnant village women having to walk many hours over unstable roads to reach a health facility. Multilevel healthcare data can also be used for decision-making, system improvement, and advocacy. With a digitized system collecting data on a large scale, it will be possible to have a synoptic view of all health systems in a country. In order to maximize the use of health data for UHC, it is important to ensure that data included in current digital recording systems is accurate, timely, and standardized to support decision-making (Grossman & Powers, 2013), i.e., a system for checking data quality is put in place and used. In evaluating the impact of digital health solutions, it is important to measure whether a digital health solution has achieved its intended outcome. For example, a digital health application that aims to facilitate the patient experience between different sections of a health facility should be evaluated by measuring whether the patient actually spends less time between consultation and pharmacy.

With regard to health financing, digital health considerably reduces the resources required to provide care by creating efficiencies in healthcare delivery (Blackmore & Zahra, 2020). Digitization of a health system can be considered an investment because the provided expenses of its implementation can generate income with the facilitation of demand generation, providing increased access to services, analysis of the system to increase efficiencies, and elaboration strategies.

(2) Considerations and approaches of digital health integration that will enable increased impact and sustainability of UHC

Beyond the areas of health service delivery that digital health can improve, the following approaches will lead to greater success and overall strengthening of the health system.

Allocating resources. The implementation of digital health platforms requires dedicated resources to ensure appropriate and robust implementation of a good quality digital health system. These resources include a specific monetary allocation and time resources. Dedicated time is needed to consult users during the design phase and to train and orient them on using digital health solutions within the health system. Specialized and qualified local human resources will need to be trained and hired to ensure the maintenance and continuous improvement of digital health platforms as the UHC systems continue to mature and evolve.

Digital health should be a **key component of UHC** in the African region. As different countries integrate UHC and create different models for integrating digital health into UHC, this will enable the region to have different models and use cases through which countries can learn from each other and improve their own national UHC digital health systems.

Digital health can be used as a **lever to systematize and integrate healthcare** and to strengthen health systems. The benefits and effectiveness of digital health will be best seen in integrated health systems and although health in the African region has previously been approached in a disease-specific vertical approach, the integration of a digital health approach offers a unique opportunity to facilitate and accelerate the integration of health systems in line with the principles and objectives of UHC.

While **digital health has great potential to make health more accessible**, it will be important to ensure that the uptake of digital health in the Universal Health Coverage is equitable. The digital divide, defined as the inequality of access or ability to engage in healthcare using technology, already exists and steps must be taken to reduce it as digital health is integrated into the Universal Health Coverage (Wood, 2021). Digital solutions must take into account existing disparities in access to infrastructure, literacy, access to finance, access to health and health decision-making based on gender, rural and urban areas, and age. Therefore it is important to address these disparities to achieve equal access to health by accentuating them.

Digital health solutions created with local context in mind are likely to be more accepted and better adapted to local circumstances. It is important to ensure that digital health is locally adapted and, where possible, developed at the local level in order to grow the capacity for digital health solution creation across the region.

Pilot projects are offering digital health solutions at a small scale initially with an aim to assess feasibility, improve the solutions and hopefully scale them (PATH, 2017). In order to ensure that the digital health technology used for UHC produces the desired results, it will be important to start with sustainability and scale in mind. During the design phase, digital health solutions need to be tested and retested, with sustainability and adaptability built into the design of the solution, in order to avoid multiple pilot projects that cannot be used at scale, as UHC is by nature a population-wide solution.

Implementation approaches are essential, as scale can only be determined by the government and will be best achieved when innovators, usually found in start-ups, have clear pathways to collaborate with larger entities, in this case the government (World Economic Forum, 2018) and when their work is integrated into a more centralized population-based digital health context.

For the development of digital health at the governmental level, it will also be important for policymakers to integrate digital health into their broader public health strategies. In order to ensure that large-scale implementation is not seen as a complex undertaking, governments can identify feasible starting points such as moving from paper to digital and implementing electronic health records (EHRs) that have proven valuable in Africa, in that they are easy to use, scalable, inexpensive, and cost effective (Adetoyi, 2020).

Conclusion and the way forward

In conclusion, it is critical that digital health interventions address core health system challenges and are driven by context-specific needs. This allows for the effective incorporation and scaling of technologies into national health systems. These interventions must be designed with the acknowledgment that the resource envelope is limited. Additionally, the private sector is an important partner in the success of Universal Health Coverage, therefore collaboration with this sector is a foundational element in the identification of digital innovations that accelerate not only quality and efficiency but also advance equity.

References

Adetoyi, O.E. (2020, March). Electronic health record design for inclusion in sub-Saharan Africa medical record informatics. (D. B. Gyampoh, Ed.) Scientific African, 7(e00304). https://doi.org/10.1016/j.sciaf.2020.e00304.

Birnbaum F., Lewis, D. (2015). Patient engagement and the design of digital health. Academic Emergency Medicine: Official Journal of the Society for Academic Emergency Medicine, 22(6), 754–756. Retrieved from https://doi.org/10.1111/acem.12692.

Blackmore, S., & Zahra, J. (2020). *Digital health and virtual care*. Retrieved from Deloitte https://www2.deloitte.com/content/dam/Deloitte/ca/Documents/life-sciences-health-care/ca-covid-19-digital-health-and-virtual-care-aoda-en.pdf.

Grossman, C., & Powers, B. (2013). Digital health data uses: Leveraging data for better health. In *Digital data improvement priorities for continuous learning in health and health care* (pp. 15–16). Washington DC: National Academies Press. Retrieved from https://www.ncbi.nlm.nih.gov/books/NBK207330/.

PATH. (2017). PATH Articles. In *PATH*. Retrieved from PATH: https://www.path.org/articles/unlocking-the-power-and-promise-of-digital-health-for-africa/.

UHC2030 Private Sector Constituency, 2019. UHC2030 Documents. (2019, September 23) Retrieved from UHC2030: https://www.uhc2030.org/fileadmin/uploads/uhc2030/Documents/Key_Issues/Private_Sector/UHC2030_Private_Sector_Constituency_Joint_Statement_on_UHC_FINAL.pdf.

USAID. (2016, September). *African strategies for health*. Retrieved from African Strategies for Health http://www.africanstrategies4health.org/uploads/1/3/5/3/13538666/trends_in_digital_health_in_africa_brief_final.pdf.

Wood, B. R.-M. (2021, March 15). Advancing digital health equity: A policy paper of the Infectious Diseases Society of America and the HIV Medicine Association. *Clinical Infectious Diseases*, *72*(6), 913–919. https://doi.org/10.1093/cid/ciaa1525.

World Economic Forum. (2018, January). *WEF white paper*. https://www3.weforum.org/docs/WEF_White_Paper_Collaboration_between_Start-ups_and_Corporates.pdf.

World Health Organisation. (2015). *WHO Global Health Observatory*. Retrieved from WHO International https://www.who.int/data/gho/data/themes/topics/indicator-groups/indicator-group-details/GHO/sdg-target-3.8-achieve-universal-health-coverage-(uhc)-including-financial-risk-protection.

CHAPTER 5

Overview of the state of mental health in the COVID-19 recovery and beyond

Rotem Kopel[a] **and Cameron Fox**[b]

[a]*MindMatters Foundation, Geneva, Switzerland,* [b]*Platform for Shaping the Future of Health & Healthcare, World Economic Forum, New York, NY, United States*

Introduction

Epidemic and pandemic diseases have plagued our species since peoples began congregating in villages, towns, and cities. The largest pandemic in recent history was over a hundred years ago, in 1918–19, when 40 million people are believed to have died from the worldwide influenza epidemic (H1N1 strain), called the "Spanish flu," followed by the "Avian flu" (H2N2) of 1956–58 that killed 2 million people globally. The HIV pandemic starting in the 1970s claimed nearly 8 million lives, and the 2002 SARS spread over 29 countries. The 2009 HINI variety flu, with 61 million cases, affected younger more than older persons. The 2013–16 epidemic of Ebola virus disease, centered in Western Africa, had a high mortality rate of about 50% of cases, an outbreak of which emerged in the Democratic Republic of the Congo again in April 2022.

Since its spread worldwide from the year 2019 over more than two years, the COVID-19 pandemic brutally forced us to confront challenges that are paradoxically as old as civilization but also without precedent. This most recent viral spread was tracked in days and weeks, instead of months, years, and decades, courtesy of new technological advances that allowed unparalleled globalization. So, too, was the world's response to the disease. While measures to contain the spread, like lockdowns and quarantines, are nothing new—Justinian used the same tactics in Constantinople 1500 years ago (Drews, 2013)—our interconnected world—moving not just people but goods freely across borders—meant that the potential for spread as well as the impact of such control measures affected virtually every person in every country in a way that truly was new.

The distinction of this current pandemic is characterized not only by the increasing spread of the virus through an interconnected world but also increased awareness of the mental health

Resilient Health. https://doi.org/10.1016/B978-0-443-18529-8.00005-6

onslaught of the virus. Fears and depression swept people in all countries, evident in increased prevalence of psychological disorders (COVID-19 Mental Disorders Collaborators, 2021). Experts and media spoke out publicly about the mental health impact of the economic (BBC, 2021) and emotional disabling of societies (The Wall Street Journal, 2020).

Before 2020, mental health conditions were already a leading cause of disability worldwide. Rates of psychiatric diagnoses had been rising across all age demographics for years, straining healthcare systems and costing the global economy an estimated $1 trillion annually (Rehm & Shield, 2019). Though the COVID-19 virus is first and foremost a physical malady, the world's collective response had inevitable knock-on effects on the mental well-being of humanity. Even for those fortunate enough to avoid contracting COVID-19, or seeing loved ones perish from the virus, the impact of isolation, job loss, supply chain breakdown, and knowing accurately how bad things were throughout the world took a serious emotional toll.

Impact of COVID-19 on global mental health

During the COVID-19 pandemic period, we have seen an increase in financial insecurity, unemployment, and other risk factors for poor mental health, while simultaneously, many protective factors like social connections, employment, educational engagement, physical activity, routines, and healthcare have declined. Thus the mental health of the population has declined dramatically.

The World Health Organization (WHO) recently published a report on the impact of COVID-19 on mental health around the world, in which they merged data from multiple studies about COVID-19's impact on mental health and mental health services with estimates from the Global Burden of Disease study (World Health Organization, 2022). According to this report, the first year of the COVID-19 pandemic caused a 25% spike in anxiety and depression, which resulted in 137.1 additional disability-adjusted life years (DALYs) per 100,000 people. This fact pushed most countries to include mental health in their COVID-19 response plans. Aside from these increases, the data showed mixed suicide rates, with many countries reporting a decline in suicides during the early months of the pandemic. However, it is important to recognize that, as Dr. Tedros Adhanom Ghebreyesus, WHO director-general stated, "The information we have now about the impact of COVID-19 on the world's mental health is just the tip of the iceberg. This is a wake-up call to all countries to pay more attention to mental health and do a better job of supporting their populations' mental health."

The WHO 2022 report also highlighted that globally, the pandemic significantly impacted the mental health of young people, with a disproportionally high risk of suicidal and self-harming behavior. Additionally, the report noted that mental health conditions negatively impact women more than men, and that people with preexisting physical problems, such as asthma, cancer, and heart disease, are more likely to develop mental disorders. WHO Member States reported that mental health and substance abuse services, including suicide prevention, were severely affected by the pandemic. The increase in mental health prevalence during this time globally led to a huge gap in mental healthcare for the most vulnerable persons. The situation had improved somewhat by the end of 2021, but too many people still remain unable to receive the care and support they need for both preexisting and newly developed mental health conditions.

Due to the inability to access face-to-face services, prohibited by government control measures like lockdowns, many people sought support online, precipitating the urgent need to develop reliable, efficient, and easily accessible digital tools. Several high-income countries (HIC) promoted a rapid implementation of digital mental health services, including changes in reimbursement policies to support mental-telehealth solutions. Nonetheless, it remained a considerable challenge to develop and deploy digital interventions in the majority of countries.

This is a bleak picture, yet, we would be remiss not to acknowledge the flipside of the coin. While globalization enabled COVID-19 to spread far faster than pandemics in the past, innovators from wide-ranging fields were able to bring collective knowledge to bear at break-neck pace, to create solutions to the problem that would have been intractable. COVID-19-animated clinicians, researchers, entrepreneurs, and policymakers radically rethought every facet of the healthcare ecosystem, from drug discovery and vaccine manufacturing, to mental health, and came up with innovative solutions.

Solution of WEF

To create these innovative solutions, it is, of course, necessary to mobilize the proper people. This is the goal of the World Economic Forum, an independent international organization whose mandate is to improve the state of the world by convening the brightest minds from academia, government, civil society, and the private sector.

One network within the forum consists of 40 Global Future Councils (GFCs), each of which consists of an interdisciplinary, multistakeholder team focusing on a specific issue of global importance. As it became clear that COVID-19 was a norm-shattering event like no other, the GFC on Mental Health pivoted its energy toward using the collective knowledge and networks of the members to highlight what could be done to curtail the enormous psychological burden foisted on the world. Innovations centering around technology were given the most attention, both because physical contact dangerously exacerbated the spread, while depriving people of interactions with family, friends, and mental health professionals, and because the number of people needing help was clearly outpacing those who could provide it. The GFC on Mental Health's contributions acted as both a platform to highlight solutions and as a facilitator of new solutions. In one of the GFC's activities, the council partnered with the popular science magazine, *Scientific American*, to shed light on the most important innovations in mental health (Scientific American, 2021). Articles covering the need for greater employer support

(McCain & Campbell, 2021), and the usefulness of teaching children about mental health (Saxena & Padukone, 2021), as well as the psychedelic renaissance (the reemergence of psychedelic substances to treat mental distress) (Schlosser & Insel, 2021) were among the innovations covered.

Data mining by various stakeholders

Another article that was part of this partnership (written by a coauthor of this chapter) that is worth reviewing in more depth, examines the possibility of using data mining of social media to improve both public and individual mental health (Kopel, 2021).

In this article, we argued that the enormous data generated and collected by over 4 billion users of social media platforms globally could provide useful information, not only about their consumption habits and political preferences as is usual, but also about their health and mental health on an unprecedented scale. This can be supported by the growing number of studies suggesting that language patterns and images posted on social media can indicate and even predict mental health conditions.

Such an approach was also in synch with the fact that researchers, tech companies, government agencies, and NGOs are increasingly tapping into extensive databases to study public mental health. Thanks to the progress made in artificial intelligence (AI), natural language processing (NLP) (Coppersmith et al., 2018; Zhang et al., 2022), and other data science tools, information gathered is being used to analyze signs of mental health conditions, such as depression, anxiety, and suicide risk, and to explore methods of detection and treatments.

Social media companies using data to address mental health

Due to these new uses, social media companies have developed several internal mental health tools. For example, Facebook's online suicide prevention program (Facebook, Suicide prevention, n.d.) uses AI to identify at-risk users, through their posts of images and words that may indicate self-harm. If found, the program alerts a team of dedicated and trained human reviewers—known as the Compassion Team—who send out a list of mental health resources to such at-risk users; in some countries, emergency services can be notified in case of imminent risk of self-harm.

Another social media platform, Pinterest, has a similar "compassionate search" (Pinterest, Compassionate search, n.d.) which provides users seeking information on anxiety and other mental health-related topics with links that promote emotional wellness, including deep-breathing activities and more elaborate psychotherapy exercises. Similarly, Snapchat has developed the "Here For You" (Snapchat, Here for you, n.d.) in-app support to help users who may be experiencing mental health challenges, which monitors search terms related to mental health conditions and then provides users with links to helpful resources and a help hotline.

Other online resources, such as the helplines of Crisis Text Line and The Trevor Project, use AI and NLP to detect self-harm risk among their users and prioritize counseling care. The helplines utilize these analytical tools internally in order to ensure that urgent situations are dealt with by the appropriate response, by the appropriate professionals, and with the least amount of delay possible.

Countries using social media to address mental health

Several countries have also started to address population mental health issues via social media channels. For example, the Canadian government launched a pilot (The Canadian Press, 2018) that analyzes random social media data to identify geographical suicide hotspots, and by that determine the appropriate allocation of mental health resources. During that pilot, they scanned random samples of social media posts and looked for red flags that might indicate a surge in suicides in particular areas of the country. Instead of identifying individual suicide risk, the system focused on regions where suicide is most likely.

Research organizations

A number of research organizations are also using social media data to develop real-time observation tools that could be used by policymakers. The tools analyze well-being indicators such as happiness, sentiment, and mental health conditions. For example, researchers have demonstrated that depression indexes correlate with geographic and demographic patterns, as reported by the US Centers for Disease Control and Prevention (CDC) (De Choudhury et al., 2013). In this work, a social media depression index was developed using a probabilistic model to determine if posts may indicate depression. This model leverages signals of social activity, emotion, and language found on Twitter.

Furthermore, during the COVID-19 pandemic, when traditional survey methods failed to provide information on mental health conditions fast enough, researchers (Valdez et al., 2020) used positive/negative mood indicators to rapidly assess population mental health status.

Future prospects

Much still needs to be learned about how to effectively apply artificial intelligence algorithms to analyze social media posts in order to determine people's moods and mental health conditions. In addition, many technical, ethical, cultural, and social issues are yet to be addressed. Questions still to be answered include: Do algorithms for self-harm detection need to be validated by the research community? What are the impacts of mental health analysis on users? How can users' privacy and their mental health data be secured? Can users be analyzed without their consent?

Despite numerous questions and challenges, outstanding an unprecedented opportunity exists to further develop frameworks and tools necessary to harness technology for the benefit of mental health for both individuals and policy-making decisions. For example, detection of self-harm behavior (e.g., wrist cutting and suicide attempts) incorporated into social media could be transformative, especially on the individual level, since people at risk of such behaviors often do not approach friends, family members, or professionals. Mental health indices can also serve as unique policy tools to help make progress in public health, allowing for a fast assessment of an initiative's impact in real time and the identification of at-risk subpopulations.

The GFC on Mental Health new solutions

Beyond identifying in *Scientific American* the 10 most promising innovations as described before, the GFC on Mental Health worked together to facilitate new solutions. Two issues stuck out as pertinent for the council's expertise: measurement and regulation.

Measurement

On the measurement front, while it seemed clear that COVID-19 was adversely impacting the world's mental health, there was distressingly little data on precisely who was most affected, especially outside of OECD countries, a group of 38 nations representing 63% of global GDP. This issue was highlighted in a piece for the *Journal of Adolescent Health*—You can't manage what you do not measure—Why adolescent mental health monitoring matters (Hayes et al., 2021). Beyond just highlighting the issue, the council, most especially drawing on work from Mental Wealth Initiative, which seeks to create an early warning system for mental health, published a commentary in *Nature* (Occhipinti et al., 2021) covering the need to build predictive models to steer policy around mental health.

Both initiatives emphasize the fact that these problems cannot be addressed without understanding their scope and the role that technology can play in facilitating more robust data generation. Not only will better data help to more accurately target those who need care—and what type of care they need—but in the case of the Mental Wealth Initiative, there is also a practical, realpolitik motivation: government funds are always limited, so the ability to detect signs of mental ill health early and predict their economic impact provides policymakers with a salient motivation to act quickly.

Regulation

The rapid expansion of digital mental health solutions was indeed a promising development, but these were quickly outpacing regulatory structures. Thus the GFC on Mental Health contributed to an initiative, called Disruptive Technologies in Mental Health, aimed at grappling with difficult questions raised by the application of technology to mental health.

These questions included: If an AI chatbot gives poor advice to a client, who is responsible? How can you be sure incredibly sensitive data, given the field, is being properly protected? Where does one look to decide if a particular app is efficacious, effective, and clinically validated? To begin answering these questions, the project created the Global Governance Toolkit for Digital Mental Health (Allen et al., 2021) with the aim of providing governments, regulators, and independent assurance bodies, with the tools to develop, adopt, and engage policies and principles that spoke to ethical worries, and which protected end users by ensuring the products they used were of sufficient quality. Drawing on 200 case studies of innovative solutions, as well as interviews with policymakers, academics, and a pilot program in partnership with the New Zealand Ministry of Health, the toolkit provides principles and practical guidance for those seeking regulatory clarity in this space.

Answering these questions and building regulatory structures—whether they are enforced by governments or industry associations—with the questions in mind is essential for several reasons. First, an unregulated digital mental health market will inevitably attract bad actors looking to promote unverified services, which could result in misuses of data, false claims of efficacy, and, ultimately, a reduction of trust in the industry as a whole. This not only harms end users but also undermines trust in legitimate products. Second, without a standardized way of vetting such solutions, government procurement bodies, insurance organizations, and employers seeking to use digital mental health tools cannot easily cull apart good and bad solutions. Such a scenario may make them reticent to invest in the space at all. Third, the innovators themselves are eager to see more regulation in the space (Palmer, 2021), both so that they can separate themselves from shoddy solutions, and so that they can have clear targets to aim for when developing new products. Lastly, it is important that governance structures align across borders, as such solutions are not limited by nationality and streamlined requirements will make it easier to get the help to those who need it more rapidly.

Conclusion

COVID-19 affected society in many ways, and researchers are still evaluating the direct and indirect effects of the pandemic. Initial studies indicate that the mental health of populations throughout the world has been adversely affected, which has prompted most countries to both evaluate mental healthcare and to seek ways to improve it. At a time of great challenge, a global multidisciplinary collaboration could shed light on the challenges and provide common insights. Prior to the COVID-19 crisis, WEF established connections between leaders within the mental health disciplines, including academics, clinicians, policymakers, and industry to facilitate communication and implement multidisciplinary projects. A number of participants, including the WEF and the GFC, engaged in collaborative activities following the onset of the pandemic, including exploring how advanced technologies can improve our understanding of mental illness, identifying the most promising technology-based interventions, and exploring how technology can be harnessed to provide new measures of global mental health. These initiatives aim to support decision-makers and policymakers in improving mental health worldwide.

References

Allen, S., Hammett, R., & Schweizer, M. (2021). *World Economic Forum*. https://www3.weforum.org/docs/WEF_Global_Governance_Toolkit_for_Digital_Mental_Health_2021.pdf. (Accessed April 8, 2022).

BBC. (2021). *Experts and media spoke out publicly about the mental health impact of the economic*. https://www.bbc.com/news/business-51706225.

Coppersmith, G., Leary, R., Crutchley, P., & Fine, A. (2018). Natural language processing of social media as screening for suicide risk. *Biomedical Informatics Insights*. https://doi.org/10.1177/1178222618792860.

COVID-19 Mental Disorders Collaborators. (2021). Global prevalence and burden of depressive and anxiety disorders in 204 countries and territories in 2020 due to the COVID-19 pandemic. *Lancet, 398*(10312), 1700–1712. https://doi.org/10.1016/S0140-6736(21)02143-7.

De Choudhury, M., Counts, S., & Horvitz, E. (2013). Social media as a measurement tool of depression in populations. In *Proceedings of the 5th Annual ACM Web Science Conference (WebSci '13). Association for Computing Machinery, New York, NY, USA* (pp. 47–56). https://doi.org/10.1145/2464464.2464480.

Drews, K. (2013). A brief history of quarantine. *The Virginia Tech Undergraduate Historical Review, 2*. https://doi.org/10.21061/vtuhr.v2i0.16.

Facebook, Suicide prevention, n.d. https://www.facebook.com/safety/wellbeing/suicideprevention/.

Hayes, J., Carvajal, L., Hijazi, Z., et al. (2021). You can't manage what you do not measure—Why adolescent mental health monitoring matters. *Journal of Adolescent Health*. https://doi.org/10.1016/j.jadohealth.2021.04.024.

Kopel, R. (September 2021). Mining social media reveals mental health trends and helps prevent self-harm. *Scientific American*. https://www.scientificamerican.com/article/mining-social-media-reveals-mental-health-trends-and-helps-prevent-self-harm/.

McCain, K., & Campbell, P. (September 2021). Declines in mental health workforce need to be met by more rigorous employer support. *Scientific American*. https://www.scientificamerican.com/article/declines-in-mental-health-workforce-need-to-be-met-by-more-rigorous-employer-support/. (Accessed 8 April 2022).

Occhipinti, J.-A., Skinner, A., Doraiswamy, P. M., et al. (2021). Mental health: Build predictive models to steer policy. *Nature, 597*(7878), 633–636. https://doi.org/10.1038/d41586-021-02581-9.

Palmer, K. (March 2021). Digital therapeutics startups are ready for more regulation, not less. *STAT News*. https://www.statnews.com/2021/03/22/digital-therapeutics-fda-regulation.

Pinterest, Compassionate search, n.d. https://newsroom.pinterest.com/en/post/expanding-compassionate-search-to-nine-additional-countries.

Rehm, J., & Shield, K. D. (2019). Global burden of disease and the impact of mental and addictive disorders. *Current Psychiatry Reports, 21*, 10. https://doi.org/10.1007/s11920-019-0997-0.

Saxena, S., & Padukone, A. (September 2021). School-based online programs instruct pupils about good mental health. *Scientific American*. https://www.scientificamerican.com/article/school-based-online-programs-instruct-pupils-about-good-mental-health/. (Accessed 8 April 2022).

Schlosser, D., & Insel, T. (September 2021). A renaissance for psychedelics could fill a long-standing treatment gap for psychiatric disorders. *Scientific American*. https://www.scientificamerican.com/article/a-renaissance-for-psychedelics-could-fill-a-long-standing-treatment-gap-for-psychiatric-disorders/. (Accessed 8 April 2022).

Scientific American, The top 10 innovations in mental health, Sep 14, 2021. https://www.scientificamerican.com/report/the-top-10-innovations-in-mental-health/.

Snapchat, Here for you, n.d., https://newsroom.snap.com/en-GB/here-for-you/.

The Canadian Press. (2018). Ottawa's Advanced Symbolics to lead AI project looking for suicide warning signs, Jan 3, 2018. *Ottawa Business Journal*.

The Wall Street Journal. (2020). *Coronavirus pandemic takes toll on mental health*. https://www.wsj.com/articles/coronavirus-pandemic-takes-toll-on-mental-health-11586433603.

Valdez, D., ten Thij, M., Bathina, K., Rutter, L. A., & Bollen, J. (2020). Social media insights into US mental health during the COVID-19 pandemic: Longitudinal analysis of twitter data. *Journal of Medical Internet Research, 22*(12), e21418. https://doi.org/10.2196/21418, PMID: 33284783, PMCID: 7744146.
World Health Organization, 2022. Mental Health and COVID-19: Early evidence of the pandemic's impact. World Health Organization. https://www.who.int/news/item/02-03-2022-covid-19-pandemic-triggers-25-increase-in-prevalence-of-anxiety-and-depression-worldwide. Published March 2, 2022. (Accessed 11 April 2022).
Zhang, T., Schoene, A. M., Ji, S., et al. (2022). Natural language processing applied to mental illness detection: A narrative review. *NPJ Digital Medicine, 5*, 46. https://doi.org/10.1038/s41746-022-00589-7.

Further reading

Mental Wealth Initiative, n.d. The University of Sydney. https://www.sydney.edu.au/brain-mind/our-research/youth-mental-health-and-technology/mental-wealth-initiative.html. (Accessed 8 April 2022).
World Economic Forum. (2022). *Disruptive technologies in mental health.* https://www.weforum.org/projects/disruptive-technologies-in-mental-health. (Accessed 8 April 2022).

CHAPTER 6

Social innovation for resilient health systems

Eneyi E. Kpokiri[a,b], Meredith del Pilar-Labarda[c,d], and Joseph D. Tucker[a,b,e,f]

[a]Clinical Research Department, Faculty of Infectious and Tropical Diseases, London School of Hygiene and Tropical Medicine, London, United Kingdom, [b]International Diagnostics Centre, Faculty of Infectious and Tropical Diseases, London School of Hygiene and Tropical Medicine, London, United Kingdom, [c]Medical Department, School of Health Sciences, University of the Philippines, Manila, Philippines, [d]Social Innovation in Health Initiative Secretariat, Manila, Philippines, [e]Institute of Global Health and Infectious Diseases, University of North Carolina at Chapel Hill, Chapel Hill, NC, United States, [f]University of North Carolina Project-China, Guangzhou, China

Introduction

Social innovation in health is “a community-engaged process that links social change and health improvement, drawing on the diverse strengths of local individuals and institutions” (Halpaap et al., 2020). Social innovation in health solutions are designed in partnership with a wide range of stakeholders, including the end users. This process can develop more inclusive solutions that are more immediately ready for implementation. Furthermore, social innovation in health focuses on local needs, leverages local opportunities, and may be more sustainable over time. Social innovation can include technology as a central component, or it may not. Social innovation in health has long historical roots. For example, in the late 1800s, English social reformer Florence Nightingale brought together clinical nursing and with women’s rights advocacy, playing an instrumental role in local community health, and in the development of nursing as a profession, meeting an important need (Serrat, 2017). More recently, social innovation has expanded in a global network, called the Social Innovation in Health Initiative (SIHI).

SIHI Global Network

The Social Innovation in Health Initiative (SIHI) is a global network of passionate individuals, organizations, and institutions advocating for social innovation in health and advancing research in social innovation. The network explores strategies to assist program implementers in scaling up social innovations and enable them to establish sustainable projects in their

Resilient Health. https://doi.org/10.1016/B978-0-443-18529-8.00006-8

communities. The SIHI network was launched in 2014 through a collaboration between the Bertha Centre for Social Innovation and Entrepreneurship at the University of Cape Town, the Skoll Centre for Social Entrepreneurship at Oxford University, the London School of Hygiene and Tropical Medicine, and TDR, the UNICEF/UNDP/World Bank/WHO Special Programme for Research and Training in Tropical Diseases. The rationale for creating the network was to advance social innovation in health research in low- and middle-income countries through a community-engaged, grounds-up approach.

The core mission of the network is to discover and strengthen capacity of local actors within health systems and to advance social innovations. This is being achieved through research using community-based approaches and collaborations between research institutions and health systems and other stakeholders in the health agenda. To date of this chapter, the SIHI global network has identified 421 different social innovations in health from 21 different countries across Africa, Asia, Latin America, and the Caribbean. For example, these initial social innovations included rapid malaria detection at private pharmacies in Uganda, community accountability for health in the Philippines, and environmental health collaboration for rural areas in Colombia (Moscibrodzki et al., 2022a).

Over the years, the SIHI network has expanded to include additional hubs from low- and middle-income countries. A SIHI hub agrees to fulfill the mission of SIHI, promote social innovation, and help build capacity within respective countries. In 2017 new additions to the network included the University of Malawi College of Medicine in Malawi, Makerere University School of Public Health in Uganda, University of the Philippines Manila College of Medicine, Social Entrepreneurship to Spur Health initiative in China; and Centro Internacional de Entrenamiento e Investigaciones Medicas (CIDEIM) in Colombia in collaboration with ICESI University and the Pan American Health Organization (PAHO). In 2020 the SIHI network saw the addition of more hubs from Africa, Asia, and Latin America and the Caribbean, including the Nnamdi Azikiwe University in Nigeria, Universitas Gadjah Mada Center for Tropical Medicine in Indonesia, University of Ghana School of Public Health, Universidad Nacional Autónoma de Honduras hosted by Unidad de Investigación Científica Facultad de Ciencias Médicas (FCM UNAH) in Honduras, and University of Rwanda. In 2021 two new hubs were established—the DERBI Foundation based in Bangalore, India, and a consortium in Sweden made up of the Uppsala University, Karolinska Institutet, and Make a Difference (MAD).

In addition to the research hubs, the network has partners who help to advance the mission, but do not organize research studies. These partners include Fondation Mérieux, the Ahimsa Fund, the WHO Department of Service Delivery and Safety, WHO AFRO, and the United Nations University—International Institute for Global Health. The goal is to work together with SIHI to build global capacity on social innovation in health. Others include AfricaLics, Globelics, PAHO, Schwab Foundation for Social Entrepreneurship, the WARN-TB Secretariat in Benin, the WHO Innovation Hub, UNDP, and UNICEF being part of this collaborative effort.

Social innovations in healthcare delivery

Social innovations are increasingly being used to improve healthcare delivery, while engaging local communities. A subset of social innovations in health have transformed the delivery of healthcare services in different settings and especially in diverse low- and middle-income countries (LMICs). Examples are described as follows.

Health center by phone

In Malawi, women of childbearing age have an increased risk of dying during childbirth because of distance from health facilities and the lack of access to quality maternal and child healthcare. To address this challenge, the social innovation called "Health Center by Phone" (Chipatala Cha Pa Foni—CCPF) was implemented (WHO, 2013). This national toll-free health hotline provides health information and referrals for medical services. It facilitates healthcare services for rural residents by connecting callers to trained healthcare staff who either resolve the issue or refer callers to nearby health facilities. This innovation has led to significant changes in knowledge and behaviors of those who use the service. It has also shown great potential to improve maternal health outcomes (Larsen-Cooper et al., 2016). It reduced the burden on rural health facilities by resolving up to 75% of calls received without referral. Over 94% of users reported satisfaction, and the service was scaled up to all 28 districts in Malawi (Blauvelt et al., 2018).

Drug shop integrated care

Another social innovation is the drug shop integrated care for children in Uganda, which leverages privately owned drug shops to provide care to children under five especially for management of malaria, pneumonia, and diarrhoea (Awor et al., 2012). Timely intervention is key to reduce mortality and prevent death in children due to malaria, pneumonia, and diarrhea. In Uganda, up to three quarters of deaths in children under the age of five have been linked with those three diseases (Liu et al., 2012). Selected drug shop owners are trained in integrated community case management and appropriate diagnostic tests. As many people seek healthcare from local drug shops and community clinics, this innovation has improved malaria testing rates among children in Uganda. As part of this intervention, referral systems were set up for drug shop owners to refer critical cases.

Mamás del Río

In rural parts of Peru, women experience difficulty in accessing care, and the lack of culturally relevant and inclusive services contributes to the burden of neonatal deaths during delivery. The social innovation called Mamás del Río was developed and implemented in response to this challenge. Community health workers (*Agentes Comunitarios de Salud*) were trained on providing key newborn education at home visits and liaising with local health facilities for more complex cases (Reinders et al., 2020). The services of these trained community health agents enable early detection of any complications in pregnant women. In the initial phase, Mamás del Río reached 799 women of childbearing age and the percentage of women receiving prenatal care in their first trimester of pregnancy has increased significantly which has in turn reduced the incidence of pregnancy complications and mortality rates (Reinders et al., 2021).

National Telehealth System

The National Telehealth System (NTS) is a social innovation strategy in the Philippines that utilizes telemedicine platforms as a systems intervention to improve access to specialist care for people in marginalized and geographically disadvantaged communities (Moscibrodzki et al., 2022a). Rural health workers use short message services and email for nonurgent referral services (e.g., teledermatology). About 50% of the 1494 municipalities in the Philippines are classified as geographically isolated areas where patients are cut off from quality healthcare services. Since its implementation, the NTS is widely used by clinicians from more than 175 facilities across the Philippines, which has greatly reduced potential costs on in-person clinic visits. Healthcare staff also reported higher levels of motivation and willingness to deliver quality services.

Pay-it-forward

In China, men who have sex with men have limited access to testing for chlamydia and gonorrhea (Yang et al., 2018). To fill this gap, the pay it forward model as social innovation strategy was used to expand integrated testing services for men who have sex with men (MSM) in China (Li et al., 2019). In the model, an individual receives a gift (e.g., a free STI test) and is then given the opportunity to donate money or create a message encouraging subsequent testing. For example, an MSM would receive a gift of a chlamydia and gonorrhea test and then decide whether to donate the same test to the next person. Pay-it-forward has also been used to enhance influenza vaccination uptake among older adults and children (Wu, 2022).

Social innovation and resilient health systems

The global COVID-19 pandemic has underlined how social innovations can promote resilient health systems in many diverse contexts. Formal public sector responses to COVID-19 were often delayed in responding to the pandemic, providing a unique opportunity for social innovation to adapt health services and meet COVID-19 needs. Evidence from several studies suggests that social innovation could be catalytic in responding to emergencies like COVID-19 (Moscibrodzki et al., 2021). Social innovation in health is proving to play an important role in nurturing more resilient health systems through several different mechanisms, including bolstering local horizontal stakeholder partnerships, expanding innovative financing options, and expanding community engagement in health services.

Social innovation also plays a key role in supporting local relationships between diverse stakeholders to enhance health system resilience (Dako-Gyeke et al., 2020). The process of social innovation brings together diverse stakeholders from the public sector, private sector, and third sector to collaboratively address health problems (SIHI, 2020). Instead of relying on vertical support from outside donors or international organizations, social innovation proposes that local people draw on their collective strengths to develop solutions (Halpaap et al., 2020). The resulting horizontal connections build a foundation for personal relationships, trust, and empathy between different actors in the health system. For example, Mamás del Río in the Peruvian Amazon demonstrated how these horizontal connections can nurture horizontal connections for resilience (Reinders et al., 2020). During the COVID-19 lockdowns, this remote Amazonian area was further secluded, but the program had already built strong partnerships between community health workers, village leaders, and civil society groups. These local horizontal partnerships facilitated distribution of personal protective equipment (masks, gloves, coats, and boots), local needs assessments, and home visits (Reinders et al., 2020). This outcome demonstrates how social innovation can catalyze local horizontal partnerships to create resilience.

Social innovation can also expand options for innovative financing of health services. A review of social innovations identified through the SIHI network found that they were supported by research grants, public sector support, sales revenues, donations, and the private sector (Moscibrodzki et al., 2022b). Social innovations help develop innovative financing mechanisms that enhance sustainability. For example, the pay-it-forward approach (Tang et al., 2021) is being used on a small scale in China, to support financing of sexually transmitted infection services. Several clinical trials and costing analyses suggest that this community-engaged approach generates 30%–40% of the total economic costs of the health services (Li et al., 2019; Yang et al., 2020). These additional financing mechanisms used in social innovations build a strong foundation for emergency preparedness and resilience.

Finally, social innovations are developed through a community-engaged process that enhances health systems. Community engagement is working with groups who share a common geographic residence, special interest, or similar situations with respect to issues affecting their well-being (Centers for Disease Control and Prevention, 2011). One health systems analysis examined responses to COVID-19 in 28 countries and found that community engagement was a critical component of health systems resilience (Haldane et al., 2021). Community engagement builds authentic relationships between partners and provides a robust foundation for responding to health crises. Community-engaged social innovation in health is a powerful force to increase the capacity for health systems to adapt to emergencies.

Acknowledgments

The SIHI network is able to do its work with support from the TDR, the Special Programme for Research and Training in Tropical Disease, cosponsored by UNDP, UNICEF, the World Bank, and WHO. TDR also receives additional funds from Swedish International Development Cooperation Agency (Sida) to support SIHI.

We dedicate this chapter to the loving memory of Dr. Noel R. Juban, who established and led the SIHI secretariat in the Philippines. He was a great advocate for global health and equity, championing many community-based social innovations in health.

References

Awor, P., Wamani, H., Bwire, G., Jagoe, G., & Peterson, S. (2012). Private sector drug shops in integrated community case management of malaria, pneumonia, and diarrhea in children in Uganda. *The American Journal of Tropical Medicine and Hygiene*, *87*(5 Suppl), 92.

Blauvelt, C., West, M., Maxim, L., et al. (2018). Scaling up a health and nutrition hotline in Malawi: The benefits of multisectoral collaboration. *BMJ*, 363.

Centers for Disease Control and Prevention. (2011). *National Institutes of Health, Health Resources and Services Administration, U.S. Department of Veterans Affairs. Principles of Community Engagement* (2nd ed.).

Dako-Gyeke, P., Amazigo, U. V., Halpaap, B., & Manderson, L. (2020). Social innovation for health: Engaging communities to address infectious diseases. *Infectious Diseases of Poverty*, *9*(1), 98.

Haldane, V., De Foo, C., Abdalla, S. M., et al. (2021). Health systems resilience in managing the COVID-19 pandemic: Lessons from 28 countries. *Nature Medicine*, *27*(6), 964–980.

Halpaap, B. M., Tucker, J. D., Mathanga, D., et al. (2020). Social innovation in global health: Sparking location action. *The Lancet Global Health*, *8*(5). e633–e4.

Larsen-Cooper, E., Bancroft, E., Rajagopal, S., O'Toole, M., & Levin, A. (2016). Scale matters: A cost-outcome analysis of an m-health intervention in Malawi. *Telemedicine and e-Health*, *22*(4), 317–324.

Li, K. T., Tang, W., Wu, D., et al. (2019). Pay-it-forward strategy to enhance uptake of dual gonorrhea and chlamydia testing among men who have sex with men in China: A pragmatic, quasi-experimental study. *The Lancet Infectious Diseases*, *19*(1), 76–82.

Liu, L., Johnson, H. L., Cousens, S., et al. (2012). Global, regional, and national causes of child mortality: An updated systematic analysis for 2010 with time trends since 2000. *The Lancet*, *379*(9832), 2151–2161.

Moscibrodzki, P., Ahumuza, E., Li, J., et al. (2022a). Social innovation in health, community engagement, financing, and outcomes: A qualitative analysis from the social innovation in health initiative. *BMJ Innovations*.

Moscibrodzki, P., Ahumuza, E., Li, J., et al. (2022b). In S. S. Conference (Ed.), *Social innovation in health, community engagement, financing, and outcomes: A qualitative analysis from the social innovation in health initiative*.

Moscibrodzki, P., Li, J., Peeling, R. W., et al. (2021). Social innovation in health: A critical but overlooked component of the COVID-19 pandemic response. *BMJ Innovations*, *7*(3), 523–525.

Reinders, S., Alva, A., Huicho, L., & Blas, M. M. (2020). Indigenous communities' responses to the COVID-19 pandemic and consequences for maternal and neonatal health in remote Peruvian Amazon: A qualitative study based on routine programme supervision. *BMJ Open*, *10*(12), e044197.

Reinders, S., Blas, M. M., Lange, I. L., & Ronsmans, C. (2021). *Study protocol: Evaluation of the Mamás del Río programme—a community-based, maternal and neonatal health intervention in Rural Amazonian Peru*.

Serrat, O. (2017). *Sparking Social Innovations. Knowledge Solutions: Tools, Methods, and Approaches to Drive Organizational Performance* (pp. 691–699). Singapore: Springer Singapore.

SIHI. (2020). *Social Innovation in Health Initiative*. https://socialinnovationinhealth.org/.

Tang, W., Wu, D., Yang, F., et al. (2021). How kindness can be contagious in healthcare. *Nature Medicine*, *27*(7), 1142–1144.

World Health Organization. (2013). *Facilitating MNCH care seeking among rural women in Malawi-Chipatala Cha Pa Foni services*. World Health Organization.

Wu, D. (2022). *Pay-it-forward to improve influenza vaccine uptake and public engagement among children and older adults in China: A quasi-experimental pragmatic trial.*

Yang, F., Zhang, T. P., Tang, W., et al. (2020). Pay-it-forward gonorrhea and chlamydia testing among men who have sex with men in China: A randomized controlled trial. *The Lancet Infectious Diseases.*

Yang, L. G., Zhang, X. H., Zhao, P. Z., et al. (2018). Gonorrhea and chlamydia prevalence in different anatomical sites among men who have sex with men: A cross-sectional study in Guangzhou, China. *BMC Infectious Diseases, 18*(1), 675.

Further reading

Fernandez-Marcelo, P. H., Ongkeko, A. M., Jr., Sylim, P. G., et al. (2016). Formulating the national policy on telehealth for the Philippines through stakeholders' involvement and partnership. *Acta Medica Philippina, 50*(4).

CHAPTER 7

Digital health and human rights in an unequal world

Tara Imalingat[a] **and Nomtika Mjwana**[b]

[a]*Kenya Legal & Ethical Issues Network on HIV and AIDS (KELIN), Nairobi, Kenya,* [b]*The Global Network of People Living with HIV (GNP+), Cape Town, South Africa*

Introduction

In the past two decades, digital health interventions have become increasingly widespread globally. Digital health platforms and tools are broadening access to, and improving quality of, health services, while empowering people to have greater control of their health and well-being. The emergence of the COVID-19 pandemic has presented the need to strengthen this approach to improve health services. However, in the Global South, structural and systemic factors (such as racism, gender inequality, socioeconomic inequalities, and lack of the underlying social determinants of health) affect access to health services and information. This extends to digital health: while many countries have digital strategies and policies, globally only 35% of people are connected to the internet.

This chapter draws on research and analysis by the Digital Health and Rights Project consortium (DHRP), a global research and advocacy consortium of social scientists, human rights lawyers, and networks of AIDS activists and people living with HIV. The consortium prioritizes mentoring and support to junior researchers in universities, national associations of people living with HIV, and civil society organizations in the Global South to participate in design and implementation of a multicountry study and in analyzing and reflecting on the results for action. We draw on two working papers published by the consortium: one proposing a human rights analysis of digital health overall; the other looking closely at three countries: Ghana, Kenya, and Viet Nam.

Intersectionality and weak protection of human rights affect access to, use, and design of digital health services. We begin by exploring gender and generational inequalities in digital health, as well as inequalities facing key populations and people living with and affected by HIV. We then explore three specific human rights concerns in this global context of

Resilient Health. https://doi.org/10.1016/B978-0-443-18529-8.00007-X

intersectional inequalities: privacy, the geopolitical inequalities among countries that shape the online world, and related algorithmic biases. We conclude with a case study, VOICE+, to suggest one way in which digital health can learn from the design justice movement to be transparent, regulated with integrity, and focused on the best interests of communities.

Intersectionality and human rights

While many countries across the globe are experiencing the digital transformation in health, not everyone is experiencing it the same way. Robust and human rights-based digital governance requires an intersectional lens.

In 1989 American civil rights scholar and advocate Kimberle' Crenshaw coined the term "intersectionality" to highlight the various ways in which forms of oppression can affect people who have different identities. After many uses (and misuses) of this concept, she published an analysis clarifying that intersectionality is an analytic sensibility, a way of thinking about identity and its relationship to power. While originally articulated in reference to Black women, the term highlights the invisibility of many constituencies within groups who claim them as members, but often fail to represent them. This can include people of color within LGBTIQA+ movements, girls of color in the fight against the school-to-prison pipeline, women within immigration movements, trans women within feminist movements, and people with disabilities fighting police abuse. These and other groups face vulnerabilities that reflect the intersections of racism, sexism, ageism, class oppression, transphobia, able-ism, and more.

In technology and digital innovation, these forms of discrimination play out, from the digital gender divide, as discussed later, to algorithmic biases which discriminate against various groups. Communication scholar and activist Sasha Costanza-Chock, associate professor of Civic Media at Massachusetts Institute of Technology, describes their experience of #TravellingWhileTrans. They recognize their privilege as a White person, while also describing the reality of traveling through cis-heteronormative systems as a transgender person (in airport security systems). Costanza-Chock notes:

> According to TSA policy, "if a pat-down is performed, it will be conducted by an officer of the same gender as you present yourself." As a nonbinary trans* femme, I present a problem not easily resolved by the algorithm of the security protocol. Sometimes, the agent will assume I prefer to be searched by a female agent; sometimes, a male.

Costanza-Chock's experience is one example of the many forms in which tech design may result in blatantly exclusionary outcomes based on race, gender, class, disability, and more. From the digital divide in access to digital tools, to the very design of those tools and platforms, intersectional inequalities are shaping health outcomes.

In the context of the HIV response, for communities of people living with HIV, people who use drugs, sex workers, and other key populations, digital health must take account of the specific risks associated with access to digital health services. This is especially true in contexts where disclosure is required for data collection purposes. Design systems are often not considerate and sensitive to their realities and do not factor these in design processes, for example, by removing requirements on gender disclosure, including color-specific designs for visually impaired persons, or completely anonymizing certain forms of health services.

An overall approach to digital health which centers on justice would entail revisiting the need to decolonize the global health system as a whole. As Tlaleng Mofokeng, the United Nations special rapporteur on the Right to Health, writes,

> When we examine the progress of medical science, we tend to focus on the good—the myriad ways in which innovative drugs, tools, and procedures have prolonged and improved lives. But this history has a dark side. Medical research has been overwhelmingly racist and sexist. It has been built upon slavery and colonization. All too often, it has been utterly inhumane.

International human rights law offers one tool for addressing intersectional inequalities in digital health while strengthening governance. Human rights are premised on the fundamental principle of nondiscrimination. The right to the highest attainable standard of physical and mental health is upheld in the *International Covenant on Economic, Social and Cultural Rights,* which has been ratified by most United Nations member states. Nondiscrimination is affirmed in each human rights covenant, including within the context of the right to health. On the 11th of August 2000, the Committee on Economic, Social and Cultural Rights (CESCR) adopted General Comment 14 on the right to health which provides the normative interpretation of that right, which includes ensuring availability, accessibility (including affordability), acceptability, and quality of health services, facilities, and care. The grounds on which discrimination is prohibited include "race, colour, sex, language, religion, political or other opinion, national or social origin, property, birth or other status" ("other status" includes sexual orientation and gender identity, according to UN human rights treaty bodies).

To summarize the above, under human rights law, digital health strategies and interventions must take into account the needs of groups who have been historically subject to systemic discrimination.

Next, we discuss how inequalities shape online risks and opportunities for women and young people, before turning to explore selected specific human rights concerns.

Digital gender divide

The world is experiencing a "Covid connectivity boost," with approximately 17% of the world's population turning to the internet since 2019. Despite this growth, the global digital gender divide remains pronounced, with an average of 62% of men using the internet compared to 57% of women. Digital technologies can contribute to a reduction in health and gender inequalities, yet challenges remain regarding women's access to, literacy in, and use of these technologies.

Preliminary findings from our study of a cohort of young women and men in Kenya show that despite overall gender disparities in access to digital technologies, young women are using digital health applications more than their male counterparts. However, globally the gap is significant. Women are less likely to have access to digital technologies than their male peers. According to the global association of mobile operators, over 300 million fewer women than men access the internet through a mobile phone. Women are globally less likely than men to use the internet to download software or to buy or sell goods online; the gender gap is widest in the South Asia region, followed by Sub-Saharan Africa. Little research appears to have been done yet about access to, as well as experiences with the internet and digital health technologies within the broad gender spectrum.

Digital literacy is today indispensable for participation in economic and social life. This necessity makes the gender gap even more dramatic and unfortunate for females since digital literacy is lower among women and girls, creating what UNICEF calls a "learning crisis" for girls in Science, Technology, Engineering and Mathematics (STEM) subjects. In Ghana, for instance, 16% of adolescent boys possess digital skills compared to only 7% of adolescent girls. Having not been exposed to STEM at an early age, women may develop what is called "technophobia," a lack of comfort in the use of technologies or an understanding of the opportunities they provide.

Inequality in education and unequal access to digital technologies have led to a significant decrease in the uptake of tech-related careers by women. The Organization for Economic Co-operation and Development (OECD) has reported that at 15years of age, only 0.5% of girls aspire to become tech professionals, compared to 5% of boys. Notably, only 26% of those working in data and artificial intelligence (AI) are women. If existing antifeminist power structures are not dismantled, we will see an oversaturated digital health market with technologies being developed by a male-dominated sector that fails to meet the needs of women and girls and fulfill their right to health. As one example of this, Apple was widely criticized for its health-tracking app that did not include tracking of menstrual cycles. In practice, mobile health interventions have shown higher rates of success with men than women; for example, in Uganda, men participated twice as often as women in an SMS text message-based campaign on HIV prevention; and in the Democratic Republic of Congo, over 80% of callers on a family planning hotline were male.

To summarize, lower access to technology, exclusion from education and leadership, and harmful societal norms, together currently impede digital health gender equity for women and girls. In a more hopeful trend, the governments of Bolivia, Viet Nam, Australia, Colombia, and Niger, with support from development partners, are introducing STEM learning programs into schools to reduce this gap. Disarming societal biases through gender-responsive education can help future generations unlearn harmful gendered views (for example, that technology is for boys) that place women and girls, especially those from low- and middle-income countries and racial or ethnic minority backgrounds, at the margins of the digital age.

Similarly, young people face specific opportunities and risks in accessing digital health.

Young people: Opportunities and risks

Young people are often perceived to be a homogenous group, when in fact, they have different needs and are shaped by vastly different and unequal realities. One common assumption about young people and technology is that they are "digital natives" who have "fast fingers" and are always ready to catch onto technology very quickly. In contrast, the reality is more complex, where intersectional inequalities shape their experience online. In fact, various factors affect how young people navigate their respective online realities, including their state of well-being, geographical factors, ethnicity, beliefs, and religion.

The digital divide is vast among young people globally. More than 2.2 billion young people and children (25 years and under) do not have internet access at home. Seven hundred and sixty-eight million young people who lack internet access live in South Asia, and in East Asia and Pacific, West and Central Africa, and East and Southern Africa, a total of 900 million children and young people do not have internet access at home. Thus access to and use of digital health technology by young people is highly context dependent. In the course of our study in Ghana, Kenya, and Viet Nam, we encountered young people who are not aware of how apps work at all as well as peers who dream of being digital product designers. At the same time, many young people increasingly seek health information online because of their experience of stigma and denial of services in health clinics.

In the global HIV and sexual and reproductive health rights (SRHR) movements, young people are at the forefront of advocacy for safe and quality access to services and elimination of stigma. At the same time, many continue to report experiences of health workers, teachers, and even their parents patronizing and underestimating their autonomy and capacity to make decisions.

Digital tools are creating increased access and safer avenues for young people to consult their peers, on health queries, find more "professional" articles for information, and even obtain referrals. Through access to search platforms such as Google, social media platforms such as Facebook and WhatsApp, and diverse mobile apps, young people are able to explore their sexuality and gender identities, answer questions, check symptoms, and even access self-administered medical abortions. Some m-health services have even reached those without smartphone access through USSD (Unstructured Supplementary Service Data, sometimes referred to as "quick codes").

At the same time, this shift online brings numerous risks. One risk is the lack of credibility or repute of the source, since many studies have shown that as a result of gaps in safe and comprehensive sexuality education, young children, from age 10 through to late teens, learn about sex for the first time through pornography. Moreover, young people searching for health information online, including through social media, are vulnerable to misinformation and disinformation. Furthermore, platforms are often not fully regulated to ensure or prioritize their online safety, exposing young people—especially young women and LGBTQIA+ people—to sexual harassment, targeting, cyberbullying, extortion, and online and offline violence. With younger tweens and teens accessing digital platforms, there are concerns over the impact of unsafe content on children.

As with all communities, the voice of young people should be recognized fully in the development of digital health innovations, as recommended by the World Health Organization, as stated in their guideline on working with young people to develop digital health interventions. These guidelines also note that young people need resources, capacity strengthening, and a level of autonomy to fully exercise their digital rights.

In our view, these approaches to youth inclusion are not a one-size-fits-all. Digital health technology needs and risks are not homogenous. They are unique to specific community needs.

Having explored some of the ways that intersectional inequalities shape online realities for women and girls, and for young people, we move on to consider three cross-cutting concerns: privacy, digital colonialism, and algorithmic biases.

Privacy concerns for vulnerable groups and communities

The right to privacy is recognized by the *Universal Declaration on Human Rights*, the *International Covenant on Civil and Political Rights (ICCPR)*, and other international and regional instruments as a fundamental human right and is considered an integral component of the realization of the right to health.

Data privacy is of special importance to people who have historically experienced widespread discrimination or who are criminalized: such as people living with HIV, those affected by TB, members of the LGBTQIA+ community, sex workers, people who use drugs, and other vulnerable groups. While data breaches can be distressing for anyone, such breaches can expose these vulnerable groups to greater stigma and discrimination, which can have deleterious results, as loss of housing, employment, and being at risk of violence. On the other hand, not disclosing personal data can also lead to some negative political and economic outcomes. In Kenya, resistance by key populations to government plans to gather their biometric data showed how lack of trust in government to protect personal data certainly discourages vulnerable communities from participating in government healthcare schemes. Given these two opposing points, in an age where smartphones, social media, and online platforms constantly leave behind digital footprints, ensuring the right to privacy has become a complex issue.

Data security is also difficult to guarantee, as shown recently at the International Committee of the Red Cross (ICRC) where personal data belonging to more than 515,000 people worldwide, including refugees and other vulnerable people in humanitarian settings, were hacked in a cyber-attack. In 2018 Singapore was forced to disconnect computers at public healthcare centers from the internet, after hackers compromised more than 1.5 million patients' personal information. In 2021 the identities of several hundred Afghans seeking refuge from the Taliban were leaked in error by an international organization. Reports by theWorld Health Organisation (WHO) on hundreds of cyber attacks on healthcare facilities in 2021 in fragile and humanitarian settings, and reliance on the cyber space creates new vulnerabilities, especially for those, such as People Living with HIV (PLHIV) and key populations, whose health data is particularly sensitive.

In some cases, data may be collected for a specific legitimate purpose but later used for a different, potentially discriminatory purpose, a phenomenon called "function creep." For example, data gathered during HIV testing and treatment is also used to check one's immigration status. More likely than not, these potential privacy intrusions do not grant citizens an opportunity to consent to the changes in data use; rather, they are imposed without consultation. Ethics and data expert Effy Vayena, professor of Bioethics at the Swiss Federal Institute of Technology, argues that the capabilities of big data analytics as well as the elasticity of mechanisms for data protection, such as consent and anonymization, further complicate the debate on privacy rights. As genuine informed consent cannot be fully elicited, given the vast opportunistic uses of data that have been unmasked over time.

In fact, civil society groups have highlighted related risks linked to HIV molecular surveillance, which may utilize patient data without consent. In addition, Vayena asserts that anonymization systems are not impervious to methods of reidentifying those whose data is collected. Consequently, any reidentified personal data can be used to maliciously target certain groups, leading to human rights violations.

Health data is generally categorized as "sensitive personal data" and often attracts higher legal safeguards. The effective protection of the right to privacy is dependent upon existing regulatory and institutional frameworks. However, laws and regulations are already falling behind the technologies they are meant to govern, forcing existing data custodians to adapt and reengineer their processes to meet new compliance regimes.

In light of these circumstances, the role of the courts in interpreting and protecting human rights and data protection cannot be overstated. The right to privacy was the cornerstone of four notable court decisions on the attempted wide-scale adoption of digital ID systems, where biometric and biographical information were to be collected by the state. The courts in India,

Jamaica, Mauritius, and Kenya agreed on the sanctity of the right to privacy and the components that make up this right; however, they gave differing judgments on whether the national identity systems indeed violated the right to privacy.

In our view, as we move even further into a digital age, the courts should harness digitization to facilitate better access to justice and keep abreast with the human rights implications of digital technologies. This should empower them to make informed judicial decisions that allow citizens to fully participate in the digital age in a secure, rights-based, and sustainable manner. In Kenya, for example, courts imposed a moratorium on the continued processing of data to facilitate adherence to data protection laws during its national digital identity scheme.

On the assumption that no digital system is impenetrable, we are of the opinion that more attention must be paid to reducing end user vulnerability by employing a multistakeholder approach to understanding and mitigating the implications of technology on the access to quality healthcare for all, particularly for those most marginalized.

Similar power disparities arise when we consider the broader geopolitical inequalities that shape who gathers data and how it is used in digital health. The digital health industry is overwhelmingly based in the Global North, while new interventions are increasingly piloted in the Global South. We now turn to elaborate on this point as follows as we reflect on digital colonialism.

Digital colonialism

Digital health is a booming industry—in the US, digital health start-ups obtained over $29 million in investment funding in 2021 alone, a record for the sector. However, this is in part fueled by data obtained from the Global South, an imbalance that echoes histories of colonialism.

Human rights lawyer Nobukhosi Zulu describes digital colonialism as the latest iteration of the long-time exploitation of the African continent. In her TEDx talk in 2019, Zulu explains that in the 16th century, the first imperialists landed in Africa in search of minerals and human resources, which drove the economy at the time; but as we look at the current landscape, it is not gold or diamonds but various forms of personal data that are now the valuable minerals being sought. This modern-day imperialism comes in the form of tech giants who provide "solutions" through apps, transport services, streaming services, search engines, and more in exchange for individual data. They centralize and use this data for profit, often without users' knowledge. Similarly, communication professors and authors Nick Couldry and Ulises A. Mejias critique "data colonialism," which they call "nothing less than a new social order, based on continuous tracking, and offering unprecedented new opportunities for social discrimination and behavioral influence."

In health, patient data can be collected and used for purposes beyond initial services from social media, web apps, and native apps (programs developed for a specific platform or device). For example, Google and Facebook dominate the global market in use of individual data for advertising purposes, and both are currently investing in major digital health initiatives. Privacy experts have raised concerns about this move, the latest instance of what American social psychologist Shoshana Zuboff calls "surveillance capitalism." Audit and advisory firm Klynveld Peat Marwick Goerdeler (KPMG) boldly highlight how big tech companies see data as "the new oil," and says that countries are now starting to actively quantify and estimate earnings which could be generated through the selling of health data. Big pharma and tech companies are starting to look specifically at which investments in digital health could generate more data, leading to demand generation and profits.

Concerned about digital colonialism, Michael Kwet, visiting fellow of the Information Society Project at Yale Law School, argues that big data is the central component of surveillance capitalism: Corporations and states are collecting, storing, and processing enormous, centralized databases of information about the world's "netizens." This enables them to infer traits about people (such as their sexuality, religion, substance abuse habits, political affiliations, and behavioral tendencies) that the people do not disclose themselves. The data is then used to manipulate individuals, groups, and organizations for corporate profits. At the same time, such concentrations of personal data can boost state power: South Korea, for example, is exploring the possibility of centralizing patient data digitally for purposes of upscaling the delivery of services with tech start-ups in the country.

Without strong data protection, some countries may eventually create unified systems that open access to multiple corporations, who may use the data to sell services back to the same people from whom they extracted the information in the first place.

This lack of constraint on tech giants raises serious concerns, given weak data protection in many countries. As of 2021, 71% (137 of 194) countries have data protection frameworks in place. However, enforcement is inconsistent and personal information may sometimes be shared with third parties without respect for prevailing data protection laws and policies.

Digital platforms are often designed to be appealing, and even addictive, for young people. Content creators are putting out information and using design strategies to captivate their audience. Health apps and products emphasize the trendiness of the tools. However, there is often limited information about what actually happens to health data captured on tech tools, for example, on a fitness watch. If you accept the terms of service, or allow cookies to track you, it is unclear to many people—especially young people—what measures are in place to ensure security of their data. This creates a fundamental power imbalance in knowledge and decision-making. The power disparities are even greater for communities who are often discriminated against and criminalized based on their HIV status, gender identity, whether they use drugs, sex workers, or because they seek access to safe abortion services.

Some individuals or groups may consider the risk of data extractivism to be a risk worth taking, especially if they believe digital spaces are safer for them than facing the stigma and discrimination they experience in person at facilities. For the most part, people in the Global South do not scrutinize their level of online security due to limited knowledge, exacerbated by the poorly designed privacy policy format that exists across many digital and online platforms. Most people "agree" to terms that they do not read or understand because these contracts are so lengthy and jargon specific.

Transparency and accountability are fundamental human rights principles. In our view, all stakeholders need to abide by these ethical standards. Private actors may be the most likely to transgress these standards. If they fail to do so, we propose that governments need to take steps to ensure that data gathered by private actors is appropriately governed and protected, including data gathered by companies overseas.

Private enterprises should, on their own, uphold human rights, in keeping with the *UN Guiding Principles on Business and Human Rights*, which interprets the existing human rights standards for the private sector. The principles clarify that the state's duty is to protect against human rights abuses by third parties, including businesses; corporations have a responsibility to respect human rights, by doing due diligence in order to avoid infringing human rights, both through a corporation's own activities and through its value chain; and both the state and businesses have a responsibility to ensure access to effective remedy for victims, both through the courts and through nonjudicial remedies. European Union provides guidance on the interpretation of the UN Guiding Principles for the digital sector which is aimed at aiding ICT companies to respect human rights within their contextual systems.

One such form of harm for which tech companies should provide redress is algorithmic discrimination, which raises significant concerns in the context of health services. Algorithmic biases can result from the lack of diversity in tech companies described before and from their tendency to take the Global North as a norm; they can magnify personal biases through algorithmic decision-making in diagnosis, predictions, prescriptions, referrals, and more. The next section explores this issue, before exhibiting an alternative model of digital health grounded in principles of design justice.

Algorithmic bias and the right to nondiscrimination

Though only about a decade old, Artificial Intelligence (AI) is unveiling its potential to revolutionize the advancement of health systems worldwide. AI is defined as a process or set of rules used by computers, known as an algorithm, to categorize and process data, identify patterns, and make predictions or decisions through data analysis. Globally, governments are turning to AI to improve public service delivery, including for health services. We note that while AI can serve extremely useful purposes, it can also generate unintended outcomes.

Biases can creep in at a variety of stages of the process of designing and utilizing algorithms for decisions in health. After all, algorithms are developed by humans, who may inadvertently build their own personal biases and flawed assumptions into the design of their systems. Algorithms must be trained on data, and unrepresentative datasets used in these systems, and a lack of diversity among designers, may lead to the replication of conscious or unconscious societal biases, resulting in the reinforcement of discriminatory practices on various grounds, including race, sex, gender, sexual orientation, age, or poverty.

Many countries in the Global South are grappling with "health data poverty": the inability of individuals, groups, or populations to benefit from digital health innovations due to scarce representative data. The existence of laws that criminalize key populations, particularly those in same-sex relationships, drives these populations underground and foster hesitancy on their part to contribute to the generation of health and population statistics, leaving them uncounted. This creates what is called a "data paradox" whereby this lack of data feeds official rhetoric that these populations are inconsequential. Therefore no research is done on their health needs, and this continued lack of data further feeds into the denialism of decision-makers. When fed into algorithmic decision tools, such as cost-effectiveness software used to allocate funds for the HIV response, this gap-filled data leads to algorithmic biases. If not countered, these will in effect entrench historical patterns of discrimination.

One study showed that AI-based medical imaging systems performed poorly on underrepresented genders. In another example, people with disabilities have experienced numerous harms in the use of flawed AI powered tools, given the diverse and nuanced nature of disabilities, which do not fit within the formulaic structure of AI.

In the 2021 Government AI Readiness Index, three pillars are used to determine government preparedness for AI adoption: a government pillar, which assesses strategic governance and digital adaptability of governments; a technology sector pillar, which looks at innovation capabilities, and a data representativeness pillar which looks at the data and infrastructure available to power AI tools. The AI readiness divide exposes the regional inequalities discussed before: the regions of sub-Saharan Africa and Central and South Asia are the two lowest ranked regions in the index, with an average of 36.27, compared to North America and Western Europe, with a joint average of 76.75.

As more health agencies embrace data-driven decision-making, if the data underlying the decision is either corrupted, compromised, or incomplete due to biases, the impact of those decisions could be catastrophic for vulnerable communities, resulting in misdiagnoses and other harms. Both the public and private sectors are obligated to remedy human rights violations. Article 2 of the ICCPR provides that those who suffer human rights violations and abuses should have access to effective judicial and nonjudicial remedies. As noted before, business enterprises are also required to put in place processes that ensure the remediation of the adverse human rights impacts related to their operations. However, locating the responsible party in flawed algorithmic decision-making can create challenges, and in practice, few tech companies offer access to redress for the harms they cause.

To address these challenges, gender-equality advocates Shirin Heidari and Heather Doyle call for a feminist approach to health data that guarantees the participation of women and other marginalized groups in decisions about the collection and use of their health data. Transform Health Coalition, a global movement committed to the achievement of Universal Health Coverage (UHC) through the use of digital technologies, is working to develop a universally accepted set of health data governance principles geared toward creating a shared understanding and a uniform application of health data governance.

Another approach is that proposed by the Design Justice Network, which defines design justice as a framework that rethinks design processes; centers people who are normally marginalized by design; and uses collaborative, creative practices to address the deepest challenges our communities face.

We conclude with a case study of digital health that aims to put this approach into practice.

Case study: VOICE+

Reimagining what an inclusive, accessible, and equitable digital space looks like is part of the journey toward ensuring that we are leaving no one behind. For people living with HIV in all their diversities, some of these concerns are part of their day-to-day experiences. It is important to document and learn from some of these efforts and initiatives that aim to explore what the possibility of feminist and intersectional tech looks like.

The Global Network of People Living with HIV (GNP+) is a network of people living with and affected by HIV, run by people living with HIV. At the center of this work is advocacy for a world where all people living with HIV are free from stigma and discrimination, have their rights recognized, their dignity respected, and have universal access to treatment and care.

In 2020 with growing concerns about the impact the COVID-19 pandemic would have on the continuation of lifesaving prevention and treatment services for the millions of people living with or vulnerable to HIV, GNP+ and a community of global HIV organizations consulted with networks of people living with HIV globally on what the challenges and concerns were regarding health access in general and in light of COVID-19. One key concern that stood out in this consultation was the challenge people living with HIV faced in filtering the flood of online information, with many struggling to get their ARVs during a lockdown or if they were in self-isolation. Many people living with HIV expressed worries about getting tested and treated for coinfections like TB and Hepatitis C and shared concerns about how acquiring COVID-19 might affect their physical and mental health.

In many parts of the world, COVID-19 lockdowns and restrictions meant a complete loss of livelihoods. Women living with HIV, in particular, shared their fears about not knowing how to feed their children and themselves. Others expressed concerns around growing stigma (with which people living with HIV are only too familiar) and violence, in the streets, as a result of police-enforced lockdowns, and in homes.

The networks workshopped what a digital solution would look like that engaged networks in addressing some of these concerns. The result was a digital solution in which networks would support one another with advocacy mechanisms, information, and cross-learning, linking to an online hub used to access resources and connect with one another. With this, VOICE+ was piloted in 2020.

VOICE+ is an activist digital hub where people living with HIV have access to stigma-free, simple, and instant information on health topics in relation to HIV. The platform also includes a function to report human rights violations and other trends affecting people living with HIV, and a chat room for users to engage, cross-learn, and share challenges. After piloting VOICE+ in Uganda and Nigeria in 2020–2021, GNP+ shared lessons learned on designing with and for communities, and working to accommodate all diversities.

Although the overall function and content of the app have been mostly appreciated by networks, there have been challenges and key learnings on how to design elements for diverse communities. This ranges from agreeing on how gender is defined on a registration form (with some questioning why this is even necessary), to understanding more broadly how feminist and inclusive digital security works, and how people living with HIV can continue to own the platform and find value in using it.

Following the pilot phase, VOICE+ was selected in October 2021 as one of the three winners of the Partos Collaborative Innovation award, which recognizes and celebrates innovative collaborations in international civil society partnerships that have produced creative solutions to deal with complex, volatile, and uncertain scenarios in development cooperation.

The premise behind this community-led approach to design justice is that at the core, people know what challenges they face and how best to address them. It is important to question how and whether communities are meaningfully involved in digital health innovations, and even more important to ensure that their participation is integrated into the overall decolonization of the health system. This includes critiquing how digital health systems are developed, how they are invested in, and the overall objective of the pools of data that may be generated. Tech hires need to be inclusive and less dominated by cis White male figures, tech investments need to be people centered and not solely profit driven, and tech infrastructure needs to be strengthened and accessible to all persons of all identities.

Conclusion

The excitement about the potential of digital health interventions needs to be tempered by a thorough critique and questioning of how intersectional inequalities and human rights abuses can be addressed in order to achieve successful rollouts and implementations.

There is great potential for digital health technology to support health systems more broadly, to lower costs, and to broaden access to health services. However, there already is stratification in accessing health services in real life, especially for communities who are criminalized, stigmatized, and placed in the margins of the system based on their health status, lifestyle choices, and/or identity. A rights-based approach to digital health must center on these communities and actively make way for people to lead solutions that address the problems they already face. Health investors of all backgrounds need to prioritize human rights-centered approaches. Innovative solutions should not exacerbate existing systemic challenges.

Considering the security and privacy risks for recipients of healthcare, transparency of digital health governance will be crucial to improve digital literacy for certain groups. This applies especially to young adults, since many countries have young populations (for example, in Africa about half the population is under age 25), and since young people are seen as a significant force shaping future digital trends in each country. All groups, however, must be considered in a rights-based approach to tech innovation. A clear role and voice for civil society and communities affected by HIV and TB in digital health design, implementation, and governance are crucial to ensure that digital health does not reinforce existing inequalities.

Acknowledgments

The authors acknowledge the support and inputs received from Georgina Caswell, Meg Davis, and Allan Maleche, as well as from the Digital Health and Rights Project consortium.

Further reading

Anderson, L. (2018). *Human rights in the age of artificial intelligence.* Access Now https://www.accessnow.org/cms/assets/uploads/2018/11/AI-and-Human-Rights.pdf.

Baral, S., & Greenall, M. (2013). *The "data paradox". Where there is no data.* https://wherethereisnodata.wordpress.com/2013/07/05/the-data-paradox/.

Berlinger, J. (2018). *Singapore hack affects 1.5 million—Including prime minister.* CNN. https://edition.cnn.com/2018/07/20/asia/singapore-hack-intl/index.html.

Office for the High Commissioner on Human Rights. (August 2000). *CESCR General Comment No. 14: The Right to the Highest Attainable Standard of Health (Art.12).* https://tbinternet.ohchr.org/_layouts/15/treatybodyexternal/TBSearch.aspx?Lang=en&TreatyID=9&DocTypeID=11.

Chib, A., Wilkin, H., Ling, L., et al. (2012). You have an important message! Evaluating the effectiveness of a text message HIV/AIDS campaign in Northwest Uganda. *Journal of Health Communication, 17*(1), 146–157. https://www.researchgate.net/publication/230611317_You_Have_an_Important_Message_Evaluating_the_Effectiveness_of_a_Text_Message_HIVAIDS_Campaign_in_Northwest_Uganda.

Corker, J. (2010). "Ligne Verte" toll-free hotline: Using cell phones to increase access to family planning information in the Democratic Republic of Congo. *Cases in Public Health Communication & Marketing, 4*, 23–27. http://casesjournal.org/index-6/.

Couldry, N., & Mejias, U. (2018). Data colonialism: Rethinking big data's relation to the contemporary subject. *Sage Journals.* https://journals.sagepub.com/doi/10.1177/1527476418796632.

Crenshaw, K. (1991). Mapping the margins: Intersectionality, identity politics, and violence against women of color. *Stanford Law Review, 43*(6), 1241–1299. https://www.jstor.org/stable/1229039?refreqid=excelsior%3Aff55ee8d042d0cc9b124f27bd7a76c52&ab_segments=&origin=.

Crenshaw, K. (2015). *Why intersectionality can't wait.* The Washington Post. https://www.washingtonpost.com/news/in-theory/wp/2015/09/24/why-intersectionality-cant-wait/.
Davis, S. L. M. (2020). *The uncounted: Politics of data in global health.* Cambridge, UK: Cambridge University Press. https://doi.org/10.1017/9781108649544.
Davis, S., Were, N., & Imalingat, T. (2021, September 24). *Digital health rights: Initial analysis* (pp. 23–29). Geneva: Graduate Institute of International and Developmental Studies, Global Health Centre. https://repository.graduateinstitute.ch/record/299381?ln=en.
Davis, S. M., Maleche, A., & Maleche, A. (2018, July 4). Everyone said no: Key populations and biometrics in Kenya. *Health and Human Rights Journal.* https://www.hhrjournal.org/2018/07/everyone-said-no-key-populations-and-biometrics-in-kenya/.
Design Justice Network. (2018). *Design justice Network principles.* https://designjustice.org/read-the-principles.
Dunstan, G. (2021). *Sex education: Children 'learning from violent porn'*. BBC News. https://www.bbc.com/news/uk-wales-59391215.
Dyer, E. (2021). *Government data breach exposes afghans to more danger.* CBC. https://www.cbc.ca/news/politics/afghan-data-breach-ircc-1.6224894.
European Commission. (2014). *ICT sector guide on implementing the UN guiding principles on business and human rights.* https://op.europa.eu/en/publication-detail/-/publication/ab151420-d60a-40a7-b264-adce304e138b.
Eveleth, R. (2014). *How self-tracking apps exclude men.* The Atlantic. https://www.theatlantic.com/technology/archive/2014/12/how-self-tracking-apps-exclude-women/383673/.
Feminist Publishing and Tech Speaker Series. (2021). *Dr. Sasha Costanza-Chock on design justice [video].* YouTube. https://www.youtube.com/watch?v=Ji7pYma4meo.
Fitzgerald, E. Four lessons: The digital health data race. KPMG. https://home.kpmg/xx/en/home/insights/2018/11/four-lessons-the-digital-health-data-race.html.
Heraldi, S., & Doyle, H. (2020). VIEWPOINT an invitation to a feminist approach to Global Health data. *Health and Human Rights Journal, 22*(2), 75–78. https://www.hhrjournal.org/2020/12/viewpoint-an-invitation-to-a-feminist-approach-to-global-health-data/#_edn12.
Ibrahim, H., Liu, X., Zariffa, N., Morris, A., & Denniston, A. (2021, March 4). Health data poverty: An assailable barrier to equitable digital health care. *The Lancet Digital Health, 3*(4), 260–265. https://www.thelancet.com/journals/landig/article/PIIS2589-7500(20)30317-4/fulltext.
Igoe, K. Algorithmic bias in health care exacerbates social inequalities—How to prevent it. Harvard T.H. Chan School of Public Health blog https://www.hsph.harvard.edu/ecpe/how-to-prevent-algorithmic-bias-in-health-care/.
International Committee of the Red Cross. (2022). *Cyber-attack on ICRC: What we know.* https://www.icrc.org/en/document/cyber-attack-icrc-what-we-know.
Internet Matters Team. (2021). *TikTok app safety—What parents need to know.* Internet Matters. https://www.internetmatters.org/hub/esafety-news/tik-tok-app-safety-what-parents-need-to-know/.
International Telecommunication Union. (2021a). *2.9 billion people still offline* (Press Release) https://www.itu.int/en/mediacentre/Pages/PR-2021-11-29-FactsFigures.aspx.
International Telecommunication Union. (2021b). *2.9 billion people still offline* (Press Release) https://www.itu.int/en/mediacentre/Pages/PR-2021-11-29-FactsFigures.aspx.
Kiplagat, S. (2021). *High court declares Huduma Namba illegal.* Business Daily Africa. https://apastyle.apa.org/style-grammar-guidelines/references/examples/newspaper-article-references.
Krasniansky, A., et al. (2022). *2021 year-end digital health funding: Seismic shifts beneath the surface.* Rockhealth blog. https://rockhealth.com/insights/2021-year-end-digital-health-funding-seismic-shifts-beneath-the-surface/.
Landi, H. (2021). *Google says health projects will continue even as it unwinds dedicated health division.* Fierce Healthcare. https://www.fiercehealthcare.com/tech/google-dissolved-its-unified-health-division-what-s-next-for-its-health-tech-strategy.
Kwet, M. (2019). Digital colonialism: US empire and the new imperialism in the global south. *Race & Class, 60*(4), 1–20. https://papers.ssrn.com/sol3/papers.cfm?abstract_id=3232297.
Landi, H. (2019). *Privacy experts skeptical of Facebook's moves to protect sensitive health information.* Fierce Healthcare. https://www.fiercehealthcare.com/tech/privacy-experts-skeptical-facebook-s-moves-to-protect-sensitive-health-information.

Larrazabal, A., Nieto, N., Peterson, V., Milone, D., & Ferrante, E. (2020). Gender imbalance in medical imaging datasets produces biased classifiers for computer-aided diagnosis. *Proceedings of the National Academy of Sciences, 117*(23). https://www.researchgate.net/publication/341661949_Gender_imbalance_in_medical_imaging_datasets_produces_biased_classifiers_for_computer-aided_diagnosis.

Lupton, D. (2021). Young people's use of digital health technologies in the Global North: Narrative review. *Journal of Medical Internet Research, 23*(1). https://www.jmir.org/2021/1/e18286/.

Milne, A. (2020). *Explainer—how to control women's sexual health with tech.* Reuters. https://www.reuters.com/article/britain-women-data-idINL5N2CB5WS.

Mjwana, N., Imalingat, T., Kpodo, I., & Trang, P. (2021). *Digital health and rights: context in three countries: Ghana, Vietnam, Kenya.* The Graduate Institute of International and Development Studies, Global Health Centre: Geneva. https://repository.graduateinstitute.ch/record/299380?_ga=2.195828076.1533642605.1644398975-762988591.1635844344.

Mofokeng, T. (2021). *Decolonizing Public Health.* Project Syndicate. https://www.project-syndicate.org/onpoint/racist-sexist-past-and-present-of-public-health-by-tlaleng-mofokeng-2021-06.

Noone, C. (2021). *Flawed data is putting people with disabilities at risk.* Tech Crunch. https://techcrunch.com/2021/04/19/flawed-data-is-putting-people-with-disabilities-at-risk/?guccounter=1&guce_referrer=aHR0cHM6Ly93d3cuZ29vZ2xlLmNvbS8&guce_referrer_sig=AQAAAL-howooEqQ8AJMt7Ffnrc1CdLY3G6qf9MRxWAJREN3OGsk8KUEggBmvE83eEiomRMvXjXqYEcm0HjaMS2bya312fvj1QRSmgvgNyZ8GYfhudKxlSiFMikHlRdzKBOPYP8W30ghdzy5cON6bqilwk4sWyR_pI4OTDmL9gUOaLqoO.

Office of the United Nations High Commissioner for Human Rights (Forty-eighth session). (2021). *The right to privacy in the digital age: report.* https://www.ohchr.org/EN/Issues/DigitalAge/Pages/cfi-digital-age.aspx.

OHCHR. (1966). *International covenant on civil and political rights (1966) art 2(3).* https://www.ohchr.org/en/instruments-mechanisms/instruments/international-covenant-civil-and-political-rights.

Organisation for Economic Co-operation and Development. (2018). *Bridging the digital divide.* https://www.oecd.org/digital/bridging-the-digital-gender-divide.pdf.

Oxford Insights. (2022). *Government AI readiness index 2021.* https://static1.squarespace.com/static/58b2e92c1e5b6c828058484e/t/61ead0752e7529590e98d35f/1642778757117/Government_AI_Readiness_21.pdf.

Panait, C. (2020). *Tackling biases and discrimination in the AI regulatory framework—a comparative analysis of EU and U.S.* Central European University. https://www.google.com/url?sa=t&rct=j&q=&esrc=s&source=web&cd=&ved=2ahUKEwjkmIW9_Zr2AhVBTBoKHWbgC2EQFnoECAoQAQ&url=http%3A%2F%2Fwww.etd.ceu.edu%2F2020%2Fpanait_cezara.pdf&usg=AOvVaw1ZI0Sk14NTwM81NbV52YYn.

Partos. (2021). *The call for the "collaborative innovations award" is open.* https://www.partos.nl/nieuws/call-for-the-collaborative-innovations-award/.

Privacy International. (2021). *Digital Health: what does it mean for your rights and freedoms.* https://privacyinternational.org/long-read/4671/digital-health-what-does-it-mean-your-rights-and-freedoms.

See more on Voice+ here n.d. https://gnpplus.net/project/voice/.

Taylor, N. (2018). *Korea pools patient data to enable digital health schemes.* Fierce Biotech. https://www.fiercebiotech.com/medtech/korea-pools-patient-data-to-enable-digital-health-schemes.

TEDx talk. (2019). *Digital colonialism and my human rights|Nobukhosi Zulu [video].* YouTube. https://www.youtube.com/watch?v=CYt7mwfJIzI.

The Center for HIV Law and Policy. (2019). *Is molecular HIV surveillance worth the risk?.* https://www.hivlawandpolicy.org/sites/default/files/CHLP%20Molecular%20Surveillance%20Final.pdf.

The World Bank Digital Development. *Connecting for inclusion: broadband access for all.* https://www.worldbank.org/en/topic/digitaldevelopment/brief/connecting-for-inclusion-broadband-access-for-all.

Transform Health. (2021). *Building health data governance principles that meet everyone's needs.* https://transformhealthcoalition.org/blogs/th-news/building-hdg-principles-workshops/.

United Nation Children's Fund. (2020). *ITU, towards an equal future: reimagining girls' education through STEM.* New York https://www.unicef.org/media/84046/file/Reimagining-girls-education-through-stem-2020.pdf.

United Nations Children's Fund and International Telecommunication Union. (2020). *How many children and young people have internet access at home? Estimating digital connectivity during the COVID-19 pandemic.* New York: UNICEF. https://www.unicef.org/media/88381/file/How-many-children-and-young-people-have-internet-access-at-home-2020.pdf.

United Nations Conference on Trade and Development. (2021). *Data protection and privacy legislation worldwide.* https://unctad.org/page/data-protection-and-privacy-legislation-worldwide.

United National Educational, Scientific and Cultural Organization (UNESCO). (2020). *I'd blush if I could: closing gender divides in digital skills through education.* https://unesdoc.unesco.org/ark:/48223/pf0000367416.

United Nations Development Programme (UNDP). (2021). *Guidance on the rights-based and ethical use of digital technologies in HIV and health programmes.* https://www.undp.org/publications/guidance-rights-based-and-ethical-use-digital-technologies-hiv-and-health-programmes.

United Nations Office of the High Commissioner for Human Rights. (2011). *Guiding principles on business and human rights.* https://www.ohchr.org/documents/publications/guidingprinciplesbusinesshr_en.pdf.

Vayena, E., Haeusermann, T., Adjekum, A., & Blasimme, A. (2018). Digital health: Meeting the ethical and policy challenges. *Swiss Medical Weekly, 148.* https://www.research-collection.ethz.ch/bitstream/handle/20.500.11850/239873/Vayena_239873.pdf?sequence=2&isAllowed=y.

World Economic Forum (WEF). (2020). *Global gender gap report 2020.* https://www3.weforum.org/docs/WEF_GGGR_2020.pdf.

World Health Organization. (2022). *Stopping attacks on health care.* https://www.who.int/activities/stopping-attacks-on-health-care.

World Health Organization (Geneva). (2020). *Youth-centred digital health interventions: a framework for planning, developing and implementing solutions with and for young people.* https://apps.who.int/iris/bitstream/handle/10665/336223/9789240011717-eng.pdf.

CHAPTER 8

How health systems must respond to the risks and impacts of climate change

Patty O'Hayer[a] and Catherine Machalaba[b]
[a]*Reckitt, Slough, United Kingdom,* [b]*EcoHealth Alliance, New York, NY, United States*

Introduction

The climate crisis is a health crisis. Crop failures lead to malnutrition, forest fires create respiratory problems, and flooding can accelerate the spread of infectious diseases. Over the past two decades, multiple large-scale studies have established the need for a cross-sector response to climate change, involving not just energy providers and environmental scientists but doctors, nurses, medical researchers, public health practitioners, veterinarians and healthcare infrastructure. Yet despite this abundant evidence, climate policies often overlook public health. At the United Nations Climate Change Conference, the world's biggest climate change summit, too many policymakers have failed to give sufficient attention to the link between rising temperatures and declining health—even as the summit itself was delayed by a global pandemic. This chapter will explore the impact of climate change on our health, arguing that national and international policymakers must not view the two challenges as distinct, but rather as a single, urgent threat requiring a broad range of responses.

By the end of the century, scientists estimate that 4.6 million people will die every year due to extreme temperatures (Bressler, 2021)—almost the current mortality burden of obesity. This does not include deaths caused by air pollution, climate-sensitive infectious diseases, food and water insecurity, famine, forced migration, conflict, or natural disasters like flooding. Most at risk are those already suffering from poverty, discrimination, and marginalization.

It is clear that the death toll from climate change will be higher than our existing health systems can handle. Rising temperatures and extreme weather will also indirectly exacerbate other

Resilient Health. https://doi.org/10.1016/B978-0-443-18529-8.00008-1

health challenges, such as antimicrobial resistance and noncommunicable diseases. Some of the most severe health consequences of rising temperatures include:

- Heat stress and cardiovascular failure caused by excessive heat.
- Malnutrition from disruptions to water and food supply and security, including from the effects of ocean warming and acidification on fish stocks (FAO, 2020).
- Respiratory problems including asthma and pneumonia, cardiovascular disease, strokes, and lung cancer caused by air pollution, with ambient and household air pollution already responsible for more than 7 million premature deaths a year (World Health Organization, 2014).
- Infectious diseases that correlate to changes in temperature, precipitation, and humidity, including vector-borne diseases such as dengue, Zika, malaria, and chikungunya. Warming of 2–3°C is projected to increase the number of people at risk for malaria by more than 150 million (Hallegatte et al., 2016).
- Diarrheal diseases associated with flooding, inadequate shelter, disruptions to sanitation and hygiene infrastructure, and reduced water supply.
- Anxiety, despair, depression, and posttraumatic stress disorder associated with climate disasters, forced migration, and related conflicts.
- Injuries caused by extreme weather events like tropical storms.
- Chemical exposure and antimicrobial resistance linked to increased resource demands for food production, vector control, and disease treatment complicated by climate change.

The link between these health problems and climate change is well established. The evidence has been corroborated by thousands of academic studies, independent research, governments, international organizations, and nongovernmental bodies (Kuriansky et al., 2020; Watts et al., 2015; Watts et al., 2019; World Health Organization, 2021a). All confirm a stark truth: the climate catastrophe is a health catastrophe.

Aside from the climate's impact on health, we must also address the health sector's impact on the climate. The health sector itself is a significant emitter of greenhouse gases, comprising roughly 4.4% of net global emissions. In terms of global footprint, if the health industry was a country, it would rank as the fifth-largest emitter (Joint study by Arup and Health Care Without Harm, 2019).

Some countries, like the United Kingdom, are already taking action to reduce emissions. In January 2020, England's National Health Service (NHS) became the first national health body to commit to net-zero emissions by 2040 (for emissions directly under its control), and several others have since followed with decarbonizing ambitions. However, major global shifts in resourcing, practices, and priorities are needed to effectively mitigate and adapt to the health risks of climate change. In the future, the most resilient health systems will be green, and the sector must become more climate conscious going forward. That requires recognition of health leaders as major stakeholders in the climate agenda.

COP26 progress

A climate summit, or "Conference of the Parties" ("COP"), brings together representatives from 197 parties (196 states plus the EU) under the United Nations Framework Convention on Climate Change (UNFCCC). The first conference, COP1, was held in Berlin in 1995. Ever since, climate COPs have taken place almost every year.

For years, public health played a relatively small part in these conversations. Two of the most important climate summits, COP3 in 1997 and COP21 in 2015, established the Kyoto Protocol and the Paris Agreement, respectively. These international agreements formed the basis for today's global climate targets, including a consensus to try to limit earth's warming to "well below 2°C" (preferably 1.5°). At each summit, recognition of the importance of climate-health links has increased, but it has never been central.

Then, at COP26 in 2021, the World Health Organization (WHO) launched a bespoke "COP26 health programme" to bring a greater focus on health to the conference and to climate change discussions in general. For the first time, the health community had its own pavilion. A program of over 60 events made the health argument for climate action across most sectors. The program had some success: over 50 countries committed to making their health sector lower carbon and more climate resilient. More than a dozen countries also set a target date, aiming to reach net zero in their health systems before 2050. To boost these efforts, the WHO and others pledged to set up an administrative body to provide financing and technical support to countries making the transition.

COP26 also saw the publication of important reports on health. One WHO survey published at the conference revealed that many countries remain hampered on climate and health by a lack of funding (2021 WHO health and climate change global survey report, 2021). Many countries also report that the pandemic slowed climate progress. Before the conference the WHO published an 82-page report on the connection between climate and health. "The Health Argument for Climate Action" gives recommendations for governments to tackle climate change in energy, transport, finance, food, and other sectors. It was accompanied by *an open letter*, signed by organizations representing around two-thirds of the global health workforce, demanding more action. "Wherever we deliver care, in our hospitals, clinics and communities around the world," the letter reads, "we are already responding to the health harms caused by climate change."

The health-climate link was clearer at Glasgow than at any prior COP, bringing more visibility to the current impacts of climate change and their scientific underpinnings. However, on many other fronts, COP26 fell short of expectations—with strong evidence not translating into political action. As the WHO *noted*, "The Glasgow Climate Pact … provides entry points on critical issues such as financing, the fate of coal and fossil fuel subsidies—but leaves them either unresolved or with caveats." It is therefore vital we push for more action. Looking ahead, health professionals have a special role to play. They are on the front line of the crisis. Their authority and testimony can lead the world in a better direction.

Looking ahead to future COPs

For the global health community, all eyes are now on upcoming climate COPs. COP28 will take place in the United Arab Emirates, building on the outcomes of COP27, held in 2022 in Sharm El-Sheikh, Egypt.

From a health perspective, the decision to hold the conference in Africa reinforces that African nations are among the most at risk. The continent is also home to some of the most forward-thinking climate action. Morocco has one of the biggest solar facilities in the world. South Africa passed a pioneering carbon tax in 2019. Nigeria is aiming for 30% renewable energy by 2030 (Okonjo-Iweala, 2020). In many ways, Africa is leading the climate response—despite low emissions throughout the continent.

Global dialogues are still plagued by disagreements over the cost of decarbonizing, the precision of scientific projections and historic responsibility for climate change. The most contentious discussions are not about sweeping global targets but about the specific, measurable actions individual countries must take to achieve them and the penalties of failing to do so.

Another problem is funding. At COP26, wealthy nations failed to deliver the $100 billion USD per year of climate finance they had promised at COP15, over a decade ago. In health, there is a dual cost: it will be expensive for global health systems to decarbonize, while already-occurring climate change will simultaneously increase the pressure of treating patients for climate-related health problems. Low-income countries will struggle the most, particularly those whose health infrastructure is already fragile. Collective action is crucial—$100 billion USD is a lot for one country, but is a tiny fraction of the global GDP, roughly $80 trillion. Building climate change mitigation and adaptation strategies into COVID-19 recovery investments will be crucial to maximize sustainable development resources.

A health day at future COPs

One move to ensure that policymakers take the health risks of climate change seriously would be to introduce a new, dedicated "health day" or stream at COP28 and beyond. At COP26, apart from the efforts of the WHO, health was squeezed into 1 hour of the agenda for "science day"—a sign that health and healthcare workers are too often overlooked in climate discussions. The need for a dedicated day or stream was emphasized at COP26 and COP27. For the sake of humans and animals everywhere, this must happen.

The cost of inaction

As of March 2022, the COVID-19 pandemic had resulted in over 6 million reported deaths globally (World Health Organization, 2022). Yet the COVID-19 tragedy is outweighed by the burden of disease linked to unhealthy environments. As of 2016, these contributed to an estimated 12.6 million deaths each year, or a fourth of deaths globally—this is also likely an underestimation given limited ability to comprehensively measure environmental factors driving disease (World Health Organization, 2016).

Beyond COVID-19, the risk of localized and regional epidemics is increasing from changes to animal and environmental conditions including climate change. Dengue in Southeast Asia, tick-borne encephalitis in Europe, and Rift Valley fever in Southern and Eastern Africa are just some recent examples. The scale of environmental damage in recent years—and rising demand for resources—means ecosystems fundamental to health and well-being are being compromised on an alarming scale. Table 1 gives examples of ecological degradation and what that means for climate change and health, including forest loss, mangrove destruction, coral reef degradation, and wildfires.

While countries are investing in some disease prevention and detection efforts, these are yet to be connected to climate action on a wide scale, particularly for upstream causes such as deforestation. There is a clear economic imperative to invest more—and in a more joined up way—in the health–climate challenge. The effect of carbon reduction on air quality alone could yield trillions in global health benefits per year by the end of the century (Scovronick et al., 2019).

Table 1 Examples of ecological degradation and what that means for climate change and health.

Issue	Scale	Implications for climate	Implications for health
Forest loss	Since 1990, 420 million hectares of forest have been lost to human activity. From 2019 to 2020 the tropics faced a 12% increase in the rate of deforestation of primary forest (United Nations, 2020; World Resources Institute, 2021a)	Direct emissions from destroyed carbon sinks; loss of carbon sequestration and storage resources; may be replaced with emission-generating activities; soil erosion (Pendrill et al., 2019; Veldkamp et al., 2020; World Resources Institute, 2021b)	Reduction of numerous biodiversity and ecosystem services including preventing soil erosion and flood mitigation, air and water filtration, pollination, and resources for food, medicine, goods and services. Habitat loss may lead to some wildlife species shifting their range, living in closer proximity (and higher densities) to human settlements with potential for disease transmission between humans and domestic animals (Krieger, 2001; Patz et al., 2004)

Continued

Table 1 Examples of ecological degradation and what that means for climate change and health—cont'd

Issue	Scale	Implications for climate	Implications for health
Mangrove destruction	Between 2000 and 2016, loss of 3400 km^2 (~2% of the world's total mangrove area), and several high-value conservation mangrove areas remain without legal protection and thus are vulnerable to habitat conversion (Goldberg et al., 2020)	Direct emissions from destroyed carbon sinks; loss of carbon sequestration and storage resources; disruption of interconnected root systems increases soil erosion and diminishes coastal protection services (Chmura et al., 2003; Sanderman et al., 2018; Spalding et al., 2014)	Degradation of soil and water quality; loss of resources for food, medicine, goods, and services; loss of biodiversity supportive of pollination, pest control (Osorio et al., 2017) and other benefits (Himes-Cornell et al., 2018; Huxham et al., 2016)
Coral reef degradation	Over 60% under direct and immediate threat from human impact as of 2011 (expected to be 90% by 2030) (Burke et al., 2012)	Loss of coastal protection (intact reef systems shield the adjacent land from ocean waves and mitigate storm surges damage, providing coastal communities with protection from hurricanes and coastal erosion) (Stoeckl et al., 2011)	Loss of marine food webs and associated nutrition resources and livelihoods; degradation of water quality; injuries (IPCC, 2022)
Wildfires	Increasing in frequency; shifting to areas with historically low fire activity; in the past 40 years, the length of fires has increased across a fourth of the world's vegetated areas (Burke et al., 2021; IPCC, 2022)	Direct emissions (including methane) and loss of carbon sequestration resources; soil erosion that increases vulnerability to disaster impacts from flash floods (Lourenço et al., 2012; Vasileva & Moiseenko, 2013; Wildfires wreaked havoc in 2021, CAMS tracked their impact, 2021)	Degradation of water quality and impacts on water security; injuries; respiratory impacts from smoke plumes (may extend thousands of miles from source fires); fear, depression and anxiety; resulting risks and impacts from extreme flooding (Reid et al., 2016; Robinne et al., 2021; Thompson, 2022; To et al., 2021)

Climate and health around the world

How do climate change and health challenges vary by region? No country will go untouched by the climate crisis. Some effects, including higher average temperatures, as well as more droughts, flooding and heatwaves, will occur almost everywhere. However, the health fallout from climate change will not be evenly distributed—different countries and different demographics within those countries will have vastly different experiences (All statistics, 2023).

Egypt, set to host this year's UN COP 27 climate summit, is highly vulnerable to rising temperatures, with its Mediterranean coastal region in the North and extremely dry climate elsewhere. Rising water levels threaten the highly populated Nile Delta, with its already precarious agricultural, water, and economic security. Under a high emissions scenario, more than 2 million people a year are projected to be affected by flooding by 2100. Meanwhile, projections suggest that heat-related deaths among the elderly (65+ years) could rise to around 47 deaths per 100,000 people—up from just one death per 100,000 annually between 1961 and 1990.

In the Middle East, already one of the world's hottest regions, climate change is making some cities uninhabitable during the summer months. Last June, Kuwait recorded days of 53°C. Research shows that, in extreme heat, Kuwait's risk of death can triple (Alahmad et al., 2020)—particularly for non-Kuwaiti men, who are commonly poorly paid migrants with higher heat exposures, likely because of their occupations. Similar temperatures above 50 degrees have been recorded in Oman, Saudi Arabia, and the UAE. Some governments are beginning to act: Saudi Arabia last year announced it would target net-zero emissions by 2060 (Saudi Green Initiative, 2021), while the UAE has set a goal of 2050 (The Official Portal of the UAE government, 2021). However, these and many other countries continue to export their emissions via production and international sales of fossil fuels.

In other parts of the globe, countries with aging populations will be disproportionately affected by spikes in heat-related deaths. In Germany, for example, the mean annual temperature is projected to rise by about 5°C by 2100 (WHO, 2016a). Western Germany in particular has recorded greater ischemic heart disease related to periods of heat exposure, and by the end of the century extreme heat is expected to result in 8500 excess deaths each year (Mücke & Litvinovitch, 2020).

Diseases historically associated with the Global South will begin to shift North as weather and climate patterns change. Since 2010 France has reported several outbreaks of chikungunya and dengue fever, while fears are mounting over the country's susceptibility to the mosquito vector *Aedes albopictus* (Gossner et al., 2018), also known as the "tiger mosquito." Under a high emissions scenario, over a quarter of France's geographic area will be considered suitable for dengue transmission for at least 3 months of the year (WHO, 2016b). Yellow fever and leishmaniasis are also identified as climate-sensitive vector-borne disease concerns for the country.

In Latin America and countries subject to El Niño conditions (the climate cycle in the Pacific Ocean affected by warm ocean water), communities are also disproportionately vulnerable to certain vector-borne diseases, such as the Zika virus. In Brazil, approximately 618,000 people will be affected by rising sea levels and by the risks of inland flooding (WHO, 2016c). Flood events combined with warmer temperatures will create standing water—breeding grounds for water-borne diseases, such as cholera, and insects and ticks carrying vector-borne diseases. This trend, and the shifting geographic range of mosquitos, suggests 168 million people (82% of the country's current population) may face risk of malaria by 2070 (WHO, 2016d).

Coastal and river populations face a similar risk. Nigeria's coast, which is home to 25% of the population (and most of the country's economic growth), is exposed to rising sea levels, flooding, and erosion (USAID, 2019). Increased flooding from more variable rainfall threatens to displace households and increase outbreaks of water-borne diseases, such as cholera. Nigeria's water and sanitation infrastructure is not well prepared to handle the projected increase in precipitation; in rural areas only 44% of households have good sanitation and 39% have access to potable water (WHO, 2016e). By 2050, climate change is projected to account for over 14% of diarrheal deaths—while by 2070, over 400 million people are projected to be at risk of malaria (WHO, 2015).

In the United Kingdom, the number of citizens at risk from flooding is expected to increase from 1.8 million to 2.6 million by the 2050s, a 40% increase in less than 40 years (The Climate Coalition, 2021). A one-meter rise in sea level, predicted to occur by the year 2100 according to many climate forecasts, will present serious risks to the 1.63 million UK citizens currently living in coastal communities. In the next 15 years alone an additional 36,000 people will be affected by river flooding across the country (WHO, 2016f). Floods have direct consequences on the health of those affected, including drowning, electrocution, and mold-induced respiratory disease, along with indirect consequences including population displacement, and loss of water and electricity supply (Paavola, 2017).

Areas with tropical rainforests are also vulnerable. Brazil, with its vast rainforest, is a unique wildfire risk. Fire threatens critical carbon sinks and produces ground-level ozone pollutants that can spread hundreds or thousands of miles (Targino et al., 2019a). In Southeast Asia, the burning of biomass in Indonesian forests and peatlands is a notable contributor to global emission levels and an active concern for populations in the region. Worsened air quality and exposure to fine particulate matter are associated with respiratory and cardiovascular diseases (Greenstone & Fan, 2019).

Regions with megacities are also disproportionately at risk, particularly from air pollution and smog. In São Paulo, for example, wildfires have been linked to worsened air quality and smog in the city, with ozone and fine particulate matter (PM2.5 and PM10) exceeding WHO-recommended limits during smoke outbreaks (Targino et al., 2019b).

African countries will suffer some of the worst health consequences of the climate crisis. Food security is a particular worry. In Nigeria, for example, intense storms, flooding, and pest exposure—which are likely to intensify with climate change—are already causing crop failures

(USAID, 2019). In Kenya, over 84% of land is arid or semiarid and the country's economy is largely dependent on rain-fed agriculture (a sector employing almost two-thirds of Kenyans) as well as tourism (Kenya Ministry of Environment and Forestry, 2020). With less than 16% of Kenya's land area supporting over 80% of the population (Kenya Ministry of Environment and Forestry, 2020), it is particularly vulnerable to droughts and flooding, both of which have likely been exacerbated by rising temperatures, intensive rain, and sea level rises. These have already caused severe crop and livestock losses, leading to widespread food insecurity (Marigi, 2017).

Ultimately, the burden of these problems will be disproportionately experienced by the world's poor. This is due to many factors, such as relatively weaker health systems, poor sanitation and other infrastructure, a lack of access to food security, and a reliance on subsistence farming, as well as large populations of workers who engage in manual labor in either outdoor settings or poorly climatized indoor buildings. No country is immune to the effects of climate change, but low- and middle-income countries will bear the brunt of the climate health emergency.

Policy recommendations

The concept of Planetary Health offers an opportunity for major innovation to tackle the health impacts of the climate crisis. Increasing multisectoral coordination to tackle the drivers of climate change and prepare for its myriad direct and indirect effects on health is an urgent global need. While research and infrastructure will require substantial investment, the growing emphasis on a One Health approach can be used to support coordination and maximize resource efficiency and cobenefits. The following national and global policy recommendations provide entry points for immediate and long-term action to address the climate crisis.

National

It is at the national level that countries can make the biggest immediate difference to this emergency. At COP26 nations went further than at any COP before to demonstrate a commitment to decarbonizing and reforming health systems to meet the climate challenge. The following recommendations show how health ministries can translate these commitments into a real policy agenda—one that can bring change now.

Align climate action plans with national policy-making

Governments must liberate climate and health from their siloes. A first step would be to coordinate Health National Adaptation Plans (HNAPs) with Nationally Determined Contributions on Climate, National Action Plans on Health Security, National Action Plans on

AMR, and national policies on food security and nutrition. Considering these together will help with allocation of resources across government, and strengthen prevention, detection, and emergency responses to disease. Health ministries must lead the way on a "build back greener" approach to COVID-19 recovery by decarbonizing its facilities, operations, and supply chain (World Health Organization, 2020a). Health Care Without Harm's Global Road Map for Health Care Decarbonisation outlines seven measures to help the health sector move toward net zero emissions by 2050, with a focus on enhancing health equity and resilience (Health Care Without Harm and Arup, 2021). Health ministry cooperation with other policy areas, including energy, transport, agriculture, and industry, is key if the sector is to realize its ambitions.

Train health workers to lead on climate action and health protection

The health sector is poorly resourced to tackle and adapt to climate change. Also, awareness within the industry remains limited, despite the scale of the problem. In the United States, the accreditation body for public health training dropped the requirement for environmental health education in 2016, a decision that leaves the workforce even less prepared to act on climate change (Dettman, 2019). More academic and applied training on climate change is needed for public health, human and veterinary medicine, sanitation workers, and emergency response professions. Additionally, behavior change science has a key role to play in encouraging uptake of health protection practices at individual and community levels related to climate-sensitive diseases, reinforcing overall infection prevention and control efforts. Examples include hand hygiene, safe sex practices, and reducing exposure risks for mosquito-borne diseases, as well as initiatives for noncommunicable diseases, such as coping with extreme heat.

Invest in climate-smart health systems for humans, animals, and the environment

We need low-carbon upgrades to healthcare facilities and laboratories, more training and staffing for at-risk workers, and population-level mitigation and adaptation strategies. We also need improved access to diagnostics and therapeutics in both emergency and routine care, as well as Integrated Disease Surveillance and Reporting and other measures to track disease and inform resource allocation in real time. Helpful models have been developed by the World Health Organization, World Bank, and Health Care Without Harm (the latter is an NGO focused on transforming healthcare worldwide through an environmental and climate justice lens) (Health Care Without Harm and Arup, 2021; World Bank, 2017; World Health Organization, 2020b). The IDA20 commitments provide a pathway for investing in "One Health" for enhanced pandemic prevention and preparedness.

Increase environment sector participation in planning, investment, and implementation of health policy-making

The Joint External Evaluations (JEE)—the WHO tool for assessing countries' ability to meet international obligations to prevent, detect, and respond to disease threats—found that only 20 out of 91 countries included climate- or weather-related terms (Machalaba et al., 2021), despite the importance of climate-sensitive diseases to health security. This is indicative of widespread limitations in assessing and managing disease threats linked to the environment, including climate change (Berthe et al., 2018), and limited coordination between the environment and health sectors in disease prevention to date. Tools such as the World Bank's Climate Change & Health Diagnostic can be used by governments to identify where climate stresses and shocks undermine local and national health services. These also target other shocks and stresses that impede health systems and reveal weak points where investment is needed.(Bouley et al., 2018) Building on its recent disaster management guidelines, the evaluation missions of the World Organization for Animal Health (WOAH) and its Performance of Veterinary Services (PVS) can also consider capacity needs for dealing with climate-sensitive diseases, plus animal health effects of other extreme weather events (Wannous, 2020).

Increase use of nature-based solutions to avoid future health threats and reduce risk of disease

The health value of nature-based solutions is often neglected—for example how conserving coral reefs protects from flooding—when making national financing decisions. Protecting ecosystem services that underpin health and well-being is a recognized priority for the forthcoming Global Action Plan on Biodiversity and Health (UN Environment Programme, 2021). To meet targets we need more integration of climate change, ecosystem, and health programs of work (e.g., under the Convention on Biological Diversity (UN Environment Programme, 2021) and the UN Environment Programme as a whole), paired with implementation in national contexts. For example, an increased focus on protected area designation and management can protect the health of communities by preserving ecosystem services while contributing to the global public good of a stable climate. Expanding early warning systems to alert when conditions are ripe for climate-sensitive diseases will also give us more lead time to prevent and prepare for epidemics in humans and animals. Joining health objectives with a drive to prevent, halt, and reverse ecosystem degradation under the UN Decade on Restoration (2021–2030) can also accelerate improvements to health and the environment.

Global

Beyond health, climate action is fundamentally linked to biodiversity, land degradation, natural disasters, sustainable development, and many other multilateral agendas. It is why institutions like the United Nations and its agencies matter: we need coordinated action to ensure that interventions and global policies solve many interconnected problems—rather than improving life in one part of the world, or one sustainable development goal, at the expense of another. The following proposals will help guide the international response:

Put health at the center of climate change and biodiversity agendas, with adequate funding for implementation

Fragmented mandates across international agencies result in a lack of multisectoral ownership and an inability to act on scientific evidence. To reinforce efforts of the WHO and national Ministries of Environment and Health, we must formalize and resource health programs within the UNFCCC, CBD, and UNCCD. This, in effect, will support a "health in all policies" stance long overdue in the climate action dialogue. Similarly, efforts in the health sector should be climate smart. For example, the proposed Pandemic Treaty (Luke Taylor, 2021) (to be discussed at upcoming meetings of the World Health Assembly) should emphasize the need for climate action to be a core part of any effort of making the world safer from pandemics. Attention to the common drivers of climate change and pandemic risk is a key starting point. Promoting cooperation and coherence among all multilateral environmental agreements (MEAs) is one of the top objectives of the UN Environment Assembly (UNEA), which meets biennially. Health should be a priority at UNEA meetings to underscore opportunities for collaboration. Other positive steps could include an expanded UNEP presence within countries, plus the adoption of a capacity assessment tool for environmental health services to help assess investment needs, like those widely used for human and animal health systems. Also welcome would be a financing mechanism (such as the Global Environment Facility) for filling gaps in environmental health capacity.

Increase scope of impact assessments and add relevant safeguards

Development projects often fail to consider the connections between health and the environment in impact assessments (Seifman et al., 2015). Climate change safeguards typically emphasize limiting emissions, but fail to address wider climate-linked threats and the necessary adaptations to address increasing burden. Integrating environment and health risk assessments into routine practice will improve our understanding of how these threats are connected and provide a pathway to mitigate impacts from the onset. A specific climate lens should be included to monitor for acute effects, as well as long-term health trends. Positive health benefits from ecosystem services, including through programs such as reducing emissions from deforestation and forest degradation (REDD+), should also be considered in land-use decisions.

Coordinate across sectors when planning and prioritizing health and climate measures—Globally and locally

Despite much rhetoric to the contrary, single-sector strategies remain the norm for most countries. The same is true at the global level. Within health sectors, climate is usually a marginal concern. While several working groups and interagency partnership agreements are promising (e.g., the WHO-WMO joint office, a WHO-IUCN expert working group, and the FAO-UNEP-WHO-WOAH "Quadripartite" agreement established in March 2022), they need more resources and clearer targets to be effective. Bringing together priorities under a multisectoral umbrella is essential for reducing duplication of efforts, identifying unattended gaps, and finding trade-offs and common benefits in line with the Sustainable Development Goals. This approach will help break down the silos in government agencies and donor funding that impede full understanding and action to address health and climate change crises. Encouragingly, a recent review of 19 National Adaptation Plans (NAPs) found that health was integrated to varying degrees—and that the impact will only be seen once actually funded and translated to implementation (World Health Organization, 2021b). The establishment of One Health Coordination Platforms by several countries, often with the support of Heads of State/ Prime Ministers, encourages the integration of climate change and health initiatives with a wider array of sectors (e.g., agriculture, education, finance). A forthcoming Joint Plan of Action Plan for One Health will reinforce country efforts by identifying coordination priorities and should comprehensively include climate change.

Promote equitable access to resources needed to stop disease and protect the environment

Governments must ensure fair access to resources such as diagnostics, therapeutics, vaccines, resilient seed varieties, safe pesticides, clean energy sources, and water and air filtration systems. This includes adequate infection prevention and control—a major deficit at present with one-fourth of the world's health facilities lacking basic water resources, which will be exacerbated periodically or permanently by increasing water scarcity in some regions. The WHO's R&D Blueprint is catalyzing new tools in the fight against infectious diseases, but must work in close partnership with entities like the World Trade Organization and GAVI to consider how to best support fair and equitable emergency access for vaccines and treatments. Ideally, every geographical region would have capacity to manufacture relevant tools—as well as internal markets to maintain supply and quality. The COVID-19 vaccine is a good example, showing that no country can effectively tackle the virus without fair access for all, and successful dissemination and uptake require multifaceted, adaptive strategies with strong engagement at all levels of society. The rising threat of climate-sensitive diseases and implications of extreme weather disruptions in patient care make access urgent.

Committing to firm targets

Voluntary targets have long been the norm, yet most nations have failed to curtail emissions at the rate required. We now know targets work better when they are obligatory and combined with independent monitoring systems plus incentives and penalties for countries that fail to meet them. Convincing governments to adopt such targets and implement policies to address this existential crisis will require concerted advocacy, and the voice of the health community can help. Addressing the climate change challenge will also help to address another neglected existential challenge: the global loss of biodiversity, another crisis also exacerbated by climate change and one that will reduce overall resilience for those most impacted by climate change. Policies to address these challenges will result in positive health and social economic outcomes for individuals and nations. The private sector's role in these efforts is crucial, including in delivering technology at the scale and pace needed to build back better and greener from COVID-19 and beyond.

Conclusion

Failure to recognize and act on the links between health and climate will exert an enormous human and financial toll everywhere. Private industry, academia, and policymakers know this—now we must act.

COP26 already saw 50 countries, representing over a third of global healthcare emissions, commit to building sustainable, low-carbon health systems. A laudable start, but to tackle climate-driven issues without inadvertently making them worse, all must join this effort. Similarly, the private and charitable sectors must do more to support a unified agenda. Changes to health systems' workforces, financing, supply chains, and infrastructure will be essential in realizing a green health sector future.

COVID-19 shone light on health sector weaknesses. Struggles over the past 3 years exposed a need for new technologies, systems, and thinking to avert a climate-induced health catastrophe. Ambitious leadership from health ministries is needed to save lives and set an example on decarbonization. At the same time, recent experience has revealed the value of prevention, compared to the extraordinary cost of not being prepared.

The overview and recommendations presented here offer practical steps to link health security and climate action, avoid siloed investments, and enable sustainable, resilient, and prepared government departments. They are meant to support current health sector calls to action, reinforcing them while making progress on other fronts too. World leaders—countries, donors, international agencies, the private sector, research institutions, cities and communities, and individuals—should be bold and unwavering in their commitments, prioritizing health to reduce risk and increase resilience against the climate crisis. The WHA can help to deliver on this by making the case that climate action is imperative for health, and by prioritizing investments toward resilient, low-carbon systems.

We owe it to the planet, its people, and all living things to elevate and unify climate and health conversations. Human health is rooted in planetary health and treating these as distinct undermines improvements to both. Indeed, the costs of keeping them separate are simply too great to ignore.

Acknowledgments

This chapter is inspired by the COP26 white paper, "The Impact of Climate Change on Health: Reducing Risks and Increasing Resilience in the Era of COVID-19." We are both grateful to Peter Edwards, Timothy A. Bouley, Kelly Rose Nunziata, Assaf Anyamba, Osman Dar, Andrea Hurtado Epstein, William B. Karesh, Keith Martin, Liam Smeeth, Flavie Vial, Chadia Wannous, David Croft, and Alex Dymoke for their valuable insights and collaboration. We also thank Sarah Baum and Tara Hoda for assistance with background research, and Sofia Gruskin for helpful input.

References

2021 WHO health and climate change global survey report. (2021). Geneva: World Health Organization (Licence: CC BY-NC-SA 3.0 IGO).

Alahmad, B., Shakarchi, A. F., Khraishah, H., Alseaidan, M., Gasana, J., Al-Hemoud, A., Koutrakis, P., & Fox, M. A. (2020). Extreme temperatures and mortality in Kuwait: Who is vulnerable? *Science of the Total Environment, 732*, 139289. ISSN 0048-9697, Access at https://doi.org/10.1016/j.scitotenv.2020.139289.

All statistics. (2023). *All statistics in this section are under a high emissions, 'business-as-usual' climate scenario, by 2100. Unless otherwise specified, they are from the WHO's 2015 country profiles.*

Berthe, F. C. J., Bouley, T. A., Karesh, W. B., et al. (2018). *Operational framework for strengthening human, animal and environmental public health systems at their interface.* Washington, DC: World Bank Group.

Bouley, T. A., Ebi, K., Golden, C., Shumake-Guillemot, J., & Midgley, A. (2018). *Climate change and health diagnostic: a country-based approach for assessing risks and investing in climate-smart health systems.* Washington, DC: World Bank, Climate Group.

Bressler, R. D. (2021). The mortality cost of carbon. *Nature Communications, 12*(1), 4467.

Burke, M., Driscoll, A., Heft-Neal, S., et al. (2021). The changing risk and burden of wildfire in the United States. *Proceedings of the National Academy of Sciences of the United States of America, 118*(2), 12.

Burke, L., Reytar, K., Spalding, M., & Perry, A. (2012). *Reefs at risk revisited: summary for decision makers.*

Chmura, G. L., Anisfeld, S. C., Cahoon, D. R., & Lynch, J. C. (2003). Global carbon sequestration in tidal, saline wetland soils. *Global Biogeochemical Cycles, 17*, 1111.

Dettman, L. (2019). *The state of environmental health in education.* The Nation's Health.

FAO. (2020). *The state of world fisheries and aquaculture 2020. Sustainability in action.* Rome. Access at. https://doi.org/10.4060/ca9229en.

Goldberg, L., Lagomasino, D., Thomas, N., & Fatoyinbo, T. (2020). Global declines in human-driven mangrove loss. *Global Change Biology, 26*(10), 5844–5855.

Gossner, C. M., Ducheyne, E., & Schaffner, F. (2018). Increased risk for autochthonous vector-borne infections transmitted by *Aedes albopictus* in continental Europe. *Euro Surveillance, 23*(24), 1800268. https://doi.org/10.2807/1560-7917.ES.2018.23.24.1800268.

Greenstone, M., & Fan, Q.(. C.). (2019). *Indonesia's worsening air quality and its impact on life expectancy.* Access at https://dev-aqli-epic.pantheonsite.io/wp-content/uploads/2019/03/Indonesia-Report.pdf.

Hallegatte, S., et al. (2016). *Shock waves: managing the impacts of climate change on poverty*. World Bank. Access at https://openknowledge.worldbank.org/handle/10986/22787.

Health Care Without Harm and Arup. (2021). *Global road map for health care decarbonization*. Access at www.arup.com.

Himes-Cornell, A., Grose, S. O., & Pendleton, L. (2018). Mangrove ecosystem service values and methodological approaches to valuation: Where do we stand? *Frontiers in Marine Science*, *5*, 376.

Huxham, M., Dencer-Brown, A., Diele, K., et al. (2016). Mangroves and people: Local ecosystem services in a changing climate. In *Mangrove ecosystems: a global biogeographic perspective*.

IPCC. (2022). Climate change 2022: Impacts, adaptation and vulnerability. In *Contribution to the sixth assessment report of the intergovernmental panel on climate change*.

Joint study by Arup and Health Care Without Harm. (2019). *Health care's climate footprint*. Access at https://noharm-global.org/sites/default/files/documents-files/5961/HealthCaresClimateFootprint_092319.pdf.

Kenya Ministry of Environment and Forestry. (2020). *Kenya Ministry of Environment and Forestry. Updated Nationally Determined Contribution*. Access at https://www4.unfccc.int/sites/ndcstaging/PublishedDocuments/Kenya%20First/Kenya%27s%20First%20%20NDC%20(updated%20version).pdf.

Krieger, D. (2001). *Economic value of forest ecosystem services: a review*.

Kuriansky, J., Magnabosco, J., & Otto, J. (2020). Climate change, mental health and wellbeing: A new policy statement for advocacy. *Applied Psychology Around the World*, *2*(3), 27–32. Access at: https://iaapsy.org/site/assets/files/1082/apaw_2020_sep_vol2_3.pdf.

Lourenço, L., Nunes, A. N., Bento-Gonçalves, A., & Vieira, A. (2012). Soil erosion after wildfires in Portugal: What happens when heavy rainfall events occur? In *Research on soil erosion*.

Luke Taylor. (2021). *World Health Organization to begin negotiating international pandemic treaty*. Access at https://www.bmj.com/content/375/bmj.n2991.

Machalaba, C., Uhart, M., Ryser-Degiorgis, M. P., & Karesh, W. B. (2021). Gaps in health security related to wildlife and environment affecting pandemic prevention and preparedness, 2007-2020. *Bulletin of the World Health Organization*, *99*(5), 342–350B.

Marigi, S. (2017). Climate change vulnerability and impacts analysis in Kenya. *American Journal of Climate Change*, *6*, 52–74. https://doi.org/10.4236/ajcc.2017.61004.

Mücke, H. G., & Litvinovitch, J. M. (2020). Heat extremes, public health impacts, and adaptation policy in Germany. *International Journal of Environmental Research and Public Health*, *17*(21), 7862. Published 2020 Oct 27 https://doi.org/10.3390/ijerph17217862.

Okonjo-Iweala, N. (2020). *Africa can play a leading role in the fight against climate change*. Brookings. Access at http://www.brookings.edu/research/africa-can-play-a-leading-role-in-the-fight-against-climate-change/.

Osorio, J., Crous, C., Wingfield, M., De Beer, Z. W., & Roux, J. (2017). An assessment of mangrove diseases and pests in South Africa. *Forestry*, 1–16. https://doi.org/10.1093/forestry/cpw063.

Paavola, J. (2017). Health impacts of climate change and health and social inequalities in the UK. *Environmental Health*, *16*, 113. https://doi.org/10.1186/s12940-017-0328-z.

Patz, J. A., Daszak, P., Tabor, G. M., et al. (2004). Unhealthy landscapes: Policy recommendations on land use change and infectious disease emergence. *Environmental Health Perspectives*, *112*(10), 1092–1098.

Pendrill, F., Persson, U. M., Godar, J., et al. (2019). Agricultural and forestry trade drives large share of tropical deforestation emissions. *Global Environmental Change*, *56*, 1–10.

Reid, C. E., Brauer, M., Johnston, F. H., et al. (2016). Critical review of health impacts of wildfire smoke exposure. *Environmental Health Perspectives*, *124*(9).

Robinne, F. N., Hallema, D. W., Bladon, K. D., et al. (2021). Scientists' warning on extreme wildfire risks to water supply. *Hydrological Processes*, *35*(5), e14086.

Sanderman, J., Hengl, T., Fiske, G., et al. (2018). A global map of mangrove forest soil carbon at 30 m spatial resolution. *Environmental Research Letters*, 13.

Saudi Green Initiative. (2021). *Crown Prince announces the Kingdom of Saudi Arabia's aims to achieve net zero emissions by 2060*. Retrieved from https://www.saudigreeninitiative.org/pr/SGI_Forum_Press_Release-EN.pdf.

Scovronick, N., Budolfson, M., Dennig, F., et al. (2019). The impact of human health co-benefits on evaluations of global climate policy. *Nature Communications*, *10*(1), 2095.

Seifman, R., Kornblet, S., Standley, C., Sorrell, E., Fischer, J., & Katz, R. (2015). Think big, World Bank: Time for a public health safeguard. *The Lancet Global Health*, *3*(4), e186–e187.

Spalding, M., McIvor, A., Tonneijck, F. H., et al. (2014). *Mangroves for coastal defence. Guidelines for coastal managers & policy makers*. Wetlands International and The Nature Conservancy.

Stoeckl, N., Hicks, C. C., Mills, M., et al. (2011). The economic value of ecosystem services in the great barrier reef: our state of knowledge in "Ecological Economics Reviews". *Annals of the New York Academy of Sciences*, *1219*, 113–133.

Targino, A. C., Harrison, R. M., Krecl, P., Glantz, P., de Lima, C. H., & Beddows, D. (2019a). Surface ozone climatology of South Eastern Brazil and the impact of biomass burning events. *Journal of Environmental Management*, *252*, 109645.

Targino, A. C., Harrison, R. M., Krecl, P., Glantz, P., de Lima, C. H., & Beddows, D. (2019b). Surface ozone climatology of South Eastern Brazil and the impact of biomass burning events. *Journal of Environmental Management*, *252*, 109645. ISSN 0301-4797.

The Climate Coalition. (2021). *The climate coalition health report 2021*. Access at https://www.theclimatecoalition.org/health-report.

The Official Portal of the UAE government. (2021). *UAE Net Zero 2050*. Retrieved from https://u.ae/en/information-and-services/environment-and-energy/climate-change/theuaesresponsetoclimatechange/uae-net-zero-2050.

Thompson, A. (2022). *Double disaster: wildfires followed by extreme rainfall are more likely with climate change*. https://www.scientificamerican.com/article/double-disaster-wildfires-followed-by-extreme-rainfall-are-more-likely-with-climate-change/. (Accessed 5 April 2022).

To, P., Eboreime, E., & Agyapong, V. (2021). The impact of wildfires on mental health: A scoping review. *Behavioral Science*, *11*(9), 126.

UN Environment Programme. (2021). *Convention on biological diversity*. Access at https://www.cbd.int/doc/c/76f9/1b75/42e360ab3ae6e53d0762c449/sbstta-24-09-en.pdf.

United Nations. (2020). *Deforestation has slowed down but still remains a concern, new UN report reveals*. https://news.un.org/en/story/2020/07/1068761. (Accessed 5 April 2022).

USAID. (2019). *Climate Risk Profile 2019*. Nigeria. Access at https://www.climatelinks.org/resources/climate-risk-profile-nigeria.

Vasileva, A., & Moiseenko, K. (2013). Methane emissions from 2000 to 2011 wildfires in Northeast Eurasia estimated with MODIS burned area data. *Atmospheric Environment*, *71*.

Veldkamp, E., Scmidt, M., Powers, J. S., & Corre, M. D. (2020). Deforestation and reforestation impacts on soils in the tropics. *Nature Reviews Earth & Environment*, *1*, 590–605.

Wannous, C. (2020). Climate change and other risk drivers of animal health and zoonotic disease emergencies: The need for a multidisciplinary and multisectoral approach to disaster risk management. *Revue Scientifique et Technique*, *39*(2), 461–470.

Watts, N., Adger, W. N., Agnolucci, P., et al. (2015). Health and climate change: Policy responses to protect public health. *Lancet*, *386*(10006), 1861–1914.

Watts, N., Amann, M., Arnell, N., et al. (2019). A report on the lancet countdown on health and climate change: Ensuring that the health of a child born today is not defined by a changing climate. *Lancet*, *394*(10211), 1836–1878.

WHO. (2015). *Health and climate change: country profile*. Nigeria: WHO. Access at https://www.who.int/publications/i/item/WHO-FWC-PHE-EPE-15.11.

WHO. (2016a). *Health and climate change: country profile 2015*. Germany: WHO. Access at https://www.who.int/publications/i/item/health-and-climate-change-country-profile-2015-germany.

WHO. (2016b). *Health and climate change: country profile 2015*. France: WHO. Access at https://www.who.int/publications/i/item/health-and-climate-change-country-profile-2015-france.

WHO. (2016c). *Health and climate change: country profile 2015*. Brazil: WHO. Access at https://apps.who.int/iris/bitstream/handle/10665/208857/WHO_FWC_PHE_EPE_15.03_eng.pdf.

WHO. (2016e). *Health and climate change: country profile 2015*. Nigeria: WHO. Access at https://www.who.int/publications/i/item/WHO-FWC-PHE-EPE-15.11.

WHO. (2016f). *Health and climate change: country profile 2015*. Great Britain and Northern Ireland.

Wildfires wreaked havoc in 2021, CAMS tracked their impact. (2021). https://atmosphere.copernicus.eu/wildfires-wreaked-havoc-2021-cams-tracked-their-impact. (Accessed 5 April 2022).

World Bank. (2017). *Climate-smart healthcare: Low-carbon and resilience strategies for the health sector*. Washington, D.C.: World Bank.

World Health Organization. (2014). Geneva: Access at https://www.who.int/news/item/25-03-2014-7-million-premature-deaths-annually-linked-to-air-pollution#:~:text=25%20March%202014%20%7C%20Geneva%20%2D%20In,result%20of%20air%20pollution%20exposure.

World Health Organization. (2016). *Preventing disease through healthy environments: a global assessment of the burden of disease from environmental risks*. Geneva: WHO, and accompanying updated 2016 burden of disease data tables.

World Health Organization. (2020a). *WHO manifesto for a health recovery from COVID-19: Prescriptions and actionables for a healthy and green recovery*. Geneva: WHO.

World Health Organization. (2020b). *WHO guidance for climate-resilient and environmentally sustainable health care facilities*. Geneva: World Health Organization.

World Health Organization. (2021a). *COP26 special report: the health argument for climate action*. Geneva. Access at https://www.who.int/publications/i/item/9789240036727.

World Health Organization. (2021b). *Health in national adaptation plans: Review*. Geneva: World Health Organization.

World Health Organization. (2022). *WHO coronavirus dashboard*. Access at: covid19.who.int.

World Resources Institute. (2021a). *Primary rainforest destruction increased 12% from 2019 to 2020*. https://wri-indonesia.org/en/blog/primary-rainforest-destruction-increased-12-2019-2020. (Accessed 5 April 2022).

World Resources Institute. (2021b). *Forests absorb twice as much carbon as they emit each year*. https://www.wri.org/insights/forests-absorb-twice-much-carbon-they-emit-each-year. (Accessed 5 April 2022).

CHAPTER 9

Disasters and health: How technologies are saving the world

Biju Jacob[a] and Vivek Nair[b]

[a]Health Innovation and Investment Exchange (HIEx), Geneva, Switzerland, [b]Aerospace Engineering, University of Illinois, Champaign, IL, United States

Introduction

Since year 2010, the world has seen over 2000 natural disasters. Annually, an average of over 150 million people are affected by natural disasters and around 100,000 people are killed. Once we add human-made disasters, like the current war in Ukraine, these numbers go up significantly. With climate changes, these numbers are bound to continue to increase in the coming decade. This also does not account for health emergencies like COVID-19 and its variations, which will drive the prevalence up higher.

Technology can play an important role in mitigating the impacts of disasters. There are numerous examples where different technologies have helped reduce human suffering and death in the aftermath of disasters. Technology also helps build resilience against natural disasters. We explore some of the key technologies that can help in disaster management.

International Organizations and Agreements

The United Nations Office for Disaster Risk Reduction (UNDRR) brings governments, partners, and communities together to reduce disaster risk and losses and to ensure a safer, sustainable future. Its report on "Risk-informed Governance and Innovative Technology for Disaster Risk Reduction and Resilience" includes sections on Cloud Computing, 5G Mobile Technology and Wireless Mesh Networks, Mobile Messaging Systems, the Internet of Things (IOT), and Distributed Ledger Technology (DLT or Blockchain) (United Nations Department of Economic and Social Affairs and The UN Office for Disaster Risk Reduction, 2022).

The global agreement to frame responses to disaster is called the Sendai Framework for Disaster Risk Reduction 2015–2030, adopted by the UN Member States in 2015.

Resilient Health. https://doi.org/10.1016/B978-0-443-18529-8.00009-3

The agreement recognizes the importance of a science-based approach to disaster risk reduction for risk-sensitive development and investment. The role of the science and technology community is outlined as to focus "on the disaster risk factors and scenarios, including emerging disaster risks, in the medium and long term; increase research for regional, national and local application; support action by local communities and authorities; and support the interface between policy and science for decision making" (United Nations Office for Disaster Risk Reduction, 2015). It also calls on countries to "enhance the scientific and technical work on disaster risk reduction and its mobilization through the coordination of existing networks and scientific research institutions at all levels and all regions with the support of the UNISDR Scientific and Technical Advisory Group."

The UNISDR Global Scientific and Technology Advisory Group (G-STAG) adopted a "Science and Technology Roadmap" to support the implementation of the Sendai Framework; to guide the work of the G-STAG; to develop regional groups to better engage scientists, researchers, and technology experts in the development and implementation of regional and national disaster risk reduction strategies and plans (The United Nations Office for Disaster Risk Reduction, 2018).

Social media

Social media plays an important role in disaster, in communicating information and in identifying missing people in the aftermath of disaster. In the USA, FEMA reported that over 20 million related tweets were posted in the following two weeks of Superstorm Sandy that occurred in the New York City and New Jersey in the year 2012. Over 7.5 million Facebook users who were near Nepal in April 2015 when the earthquake hit checked-in via Facebook's Safety Check. In reaction to the Kerala floods in India in 2018, the State disaster fund released lists of WhatsApp and landline numbers to contact for help. Thousands of families used social media posts to alert authorities about people marooned in the floods, which allowed help to arrive, and therefore, saving thousands of lives.

Google People Finder is another innovative technology which helps disaster relief efforts. It was developed in 2010 in response to the Haiti earthquake. This open source web application, available in over 40 languages, allows users to post and search for the status of people affected by a disaster. During the 2015 Nepal earthquake, well over 7500 records on the people finder were searched.

Also, the Red Cross Emergency App has been helpful with disaster relief in the recent times. It gives survivors weather updates, preparedness information, and safety tips. The Red Cross credits the application for saving several lives in disasters.

Caution is also warranted in the use of social media as rumors, myths, and false information can be circulated quickly, with dangerous outcomes. For example, during the outbreak of Ebola in

Nigeria, rumors were spread through cell phone messages and on WhatsApp that drinking large amount of salt and also using salt for bath would prevent the infection, when people with high blood pressure who attempted this practice went into crisis and lost their lives (Heen, 2016).

Mapping technologies

In the aftermath of disasters when relief teams reach the site, the immediate requirement is for mapping tools to know the extent of the disaster and to reach people in dire need of rescue. There are a number of Geographic Information System (GIS) tools which help with micro-mapping disaster sites that process, store, analyze, and display data. Disaster prediction can then be made analyzing vast quantities of historical data. Data collected by these tools can be used to create models that can predict and contain future events.

At a macro level, today's satellites can take photographs as clear as a square inch. This tech combined with on-site infrared cameras can be used to pinpoint trapped survivors by detecting heat signatures of survivors under rubble or floods. The viability of using human heat signatures to detect people can be seen in the example of the Thai Cave Rescue. It is imperative that the various organizations that own and operate these machines work together to bring high-end mapping technology, which is already commonplace in the defense industry, into disaster management.

Big data and artificial intelligence (AI) for predicting disasters

In December 2019, an AI system in Boston sent out the first global alerts about a new viral outbreak in China. The HealthMap system at Boston Children's Hospital sent an alert about unidentified pneumonia cases in the Chinese city of Wuhan. But finally it took human intelligence to figure out that this was COVID-19. This outlines the role of AI as an early alert system. These systems trawl though vast amounts of internet traffic to detect patterns. Once these technologies are further perfected, they will be invaluable for human survival.

Until recently, humanitarian organizations used historical data as a means of analysis, to find out what could have been done differently, and use these findings to develop the required infrastructure in preparation for the next event. Now, companies like Zenysis can predict *when* malaria could break out in Africa using this historical data combined with weather patterns and human habitations conditions monitored via satellite and internet. The applications for these kinds of advance warning systems are almost limitless. Imagine knowing exactly where a tsunami or flood might hit, years in advance with a margin of error that could be in weeks or even days.

Big data analytics of news feeds and media feeds can also predict potential disease outbreaks. Big data can be used to accurately predict the areas affected by floods so that emergency responders can provide a planned and measured response.

The biggest hurdle in creating these systems will be climate change. Nature is structured and usually follows patterns. People can recognize these patterns. Even as far back as Ancient Egypt, people saw the Nile flooding at the same time each year and not only prepared for it but used it to their advantage. But climate change induces randomness. The steep rise in temperatures over the past few years makes a lot of the collected data obsolete and disasters harder to predict. Artificial intelligence (AI) can help analyze and predict outcomes better by creating complex models and "what if" scenarios.

Communication technologies

Communication infrastructure is key in the aftermath of natural disasters. Until now humanitarian agencies needed to depend on expensive satellite phones and VSAT terminals to achieve this.

Currently, communication infrastructure is almost entirely land based, and alarmingly vulnerable to disasters of all kinds. In most cases, internet and telephone lines are the first buildings to fall. This makes it impossible for survivors to contact the outside world and make rescue workers aware of their locations. Restoring collapsed communication networks used to be a tedious task with specialized vans full of the sophisticated equipment needed to communicate with geosynchronous satellites in high orbits, all for internet slower than dial-up. Now, Low Earth Orbit satellite mesh systems like Iridium or SpaceX's Starlink can provide blazing fast internet with a receiver the size of a pizza box and can be rapidly deployed so that local Wi-Fi can be provided in the disaster areas even in the remotest corners of the planet.

More recently iridium terminals have been extensively used in Ukraine during the war there in 2022. As the coverage of these low orbit satellites increases, the dependence on physical communication infrastructure will rapidly come down. Internet can help with the aftermath of disasters significantly.

Recently, practical application of this tech has been proved to be viable during the war in Ukraine. As the satellite mesh grows denser and coverage of these low orbit satellites increases, our dependence on land-based communication infrastructure will decrease dramatically.

Drones and robots

Drones are being increasingly deployed for various tasks. They can map an area and also deliver medicines and other supplies to remote areas. Drone photography helps responders access hard-to-reach disaster areas to search for survivors. Often this technique is most effective when combined with community outreach and in-person visits, especially in cases of epidemics, many examples of which are documented in chapters in this book about the COVID-19 response (Pain et al., 2024).

The idea of using robots as rescue workers is not new. Robots are increasingly being used to search under debris and also in firefighting operations without putting human lives at risk. Small robots can go into hazardous sites where humans cannot reach.

After Chernobyl, attempts were made to use robots to clear the roof of reactor number four's building as that was where most of the radioactive material had landed after the explosion. These attempts failed because the elevated levels of ionizing radiation killed the robots within seconds. Yet, the evolution of this technology was displayed in the wake of the nuclear power plant disaster of Fukushima, during the tsunami in Japan, where robots did successfully clear the debris.

Ground-based search and rescue robots decrease the risk to human life while also bringing more brute force to rescue operations, should it be required. While they lack the dexterity and nimbleness of a human rescuer, the field is rapidly developing, and completely autonomous rescue robots may soon be a reality. In addition to robots, drones can be mass deployed to provide an eye in the sky while not using up already limited resources like helicopters and rescue personnel.

Cash management and digital ID tools

In the aftermath of disasters, many governments and donors are increasingly providing direct financial support to disaster victims and refugees. It is proven that direct cash transfer is more effective than providing relief materials in the longer term. Many humanitarian agencies are helping governments deliver cash to refugees and victims.

The lack of a documentation of identity is also a huge challenge for victims. After all, in an earthquake, personal papers are destroyed and few people think to reach for their passports especially in an emergency. Refugees who left their country may not have official papers to identify themselves. As such, identifying people is extremely difficult in the aftermath. Most countries now use biometrics as a means of identifying individuals. In India, an app called DigiLocker allows access to almost every government-issued document from a phone using just a phone number and a PIN.

Multiple technologies like blockchain can store these IDs in a secure manner. Every country should evolve these IDs based on the risk factors involved. This approach comes in very handy in times of disasters and emergencies.

The biggest hurdle in adopting such software worldwide is the risk of fraud. But, if more time is given to creating this software and making it secure, it could be a revolutionary tool in not only disaster management but also during mass migrations of refugees escaping conflict.

Open source applications and standards

Open source applications allow companies and individuals to work together for the greater good. This concept is relatively new in the humanitarian sector, but encouraging more collaboration is crucial. There are few problems we cannot solve with the best minds in the field put together and solving problems like those mentioned before can bring about positive and quick change. There is a need for better collaboration among humanitarian agencies to jointly develop technologies which can be reused. Currently, significantly investments are made to replicate the same systems. Open source technologies are slowly gaining ground in the humanitarian world. For example, Digital Square at PATH works with ministries of health to align adaptable, interoperable digital technologies with local health needs and brings partners together to improve how the global community designs, uses, and pays for digital health tools and approaches.

Hotlines and radio and TV announcements

Hotlines and radio and TV announcements to connect the public with accurate information have been standard practice and staples in disaster management. While still implemented as technological communication tools, they are increasingly supplemented with more advanced equipment and approaches for public outreach, assistance, and connection.

Natural disaster and epidemics

While natural disasters like floods and storms, and epidemics like SARS, Ebola, and COVID-19, may seem different, and even man-made disasters like the levee overflow in Hurricane Katrina and the Fukushima nuclear power plant explosion during the tsunami in Japan, there are actually similarities in all these events (Kuriansky & Nemeth, 2013). All such emergencies are increasingly benefiting from warning systems, location systems, health and mental health assessments, and recovery services delivered by innovative technological aids.

Yet, traditional methods still have proven effective. For example, the spread of the 2015 Ebola epidemic in West Africa was staved off in Nigeria with efficient contact tracing of Ebola cases whereby health officials visited any contact who reported symptoms or failed to provide an update of his or her health profile through a cell phone Short Message Service (SMS) (Heen, 2016). The World Health Organization described the process as "world-class epidemiological detective work."

Further, as Heen (2016) describes, the government used a social media campaign to deter the spread. Information on EVD (Ebola Viral Disease) was intensively disseminated through television and radio jingles aired in multiple languages, along with social media messages sent through Facebook and Twitter. Bulk text messages were also used to educate a vast part of the

Nigerian population on preventive measures, causes, symptoms, and treatment of the EVD. Dedicated hotlines to report suspected cases and toll-free lines to communicate with the public were established, and websites, e.g., www.ebolaalert.org, ebolainfo@health.gov.ng, were prominent and useful information platforms. There were tweets from National Orientation Agency (NOA) relaying messages such as "Report every suggestive symptom of Ebola to the Ministry of Health—08023310923, 08097979595, Email: ebolainfo@health.gov.ng."

The Federal Ministry of Health put up dedicated EVD pages on its website with fact sheets in English as well as downloadable fliers in Pidgin English, Hausa, Igbo, and Yoruba, the three major Nigerian languages. Both Lagos State Government and the Federal Ministry of Health disseminated health messages, warning people to avoid close physical contact, to wash their hands frequently, and to avoid handshake and hugs.

More recent emergency response programs in such epidemics with the luxury of technological tools can be much more effective than in earlier times when these were not available or accessibly in a remote, low resource area, for example, where only cell phones are available (Kuriansky et al., 2017). Technology tools can be helpful in the three areas of disaster: called the 3 Es, postulated as Economic Recovery, Environmental Reconstruction, and Emotional Resilience (Nemeth et al., 2020).

Projects today are increasingly boosted in the three A's—access, available, and affordability inherent in Universal Health Coverage—with sophisticated tech tools to aid in the pandemic response during COVID-19. This is evident in the use of drones to collect health information, to assess data, and even to perform needed procedures, as outlined in many chapters in this volume.

Challenges and solutions

It is evident that multiple technologies can help with the prediction, preparation, and aftermath of natural disasters. However, these approaches and applications are still being piloted and have not been fully scaled. Humanitarian agencies and government agencies need to come together with the private sector to define global standards for each of these technologies.

Also, as with the adoption of any new technology, there should be adequate safeguards to ensure that the user's privacy and security is not compromised. Inclusivity is another challenge as many of these technologies are not always available. Humanitarian agencies need to come up with a common framework to train different countries and governments to use such technologies during disasters.

Policy is critical to frame the procedures of implementing guidelines into disaster risk reduction and recovery (Kuriansky, 2020). These are even more important when technological tools are introduced into the process, to avoid risks and to ensure accuracy, privacy, and evidence-based applications.

References

Heen, C. (2016). Combating the Ebola Virus disease through Public Enlightenment and concerted Government work Plan: The case of Nigeria. In J. Kuriansky (Ed.), *The psychosocial aspects of a deadly epidemic: What Ebola has taught us about holistic healing* (pp. 233–242). Santa Barbara, California: ABC-CLIO/Praeger.

Kuriansky, J. (2020). Psychosocial recovery after natural disaster: International advocacy, policy, and recommendations. Chapter 19 In K. E. Cherry, & A. Gibson (Eds.), *The intersection of trauma and disaster behavioral health*. New York, NY: Springer Nature.

Kuriansky, J., Margevich, A., Jean-Charles, W., & Daisey, R. (2017). Resilience and recovery in natural disasters and epidemics: Comparisons, challenges, and lessons learned from train-the-trainer projects. In G. Rich, & S. Sirikantraporn (Eds.), *Human strengths and resilience: Developmental, cross-cultural, and international perspectives*. Lanham, MD: Lexington Books. https://www.amazon.com/Human-Strengths-Resilience-Developmental-Cross-Cultural/dp/1498554830/ref=mt_hardcover?_encoding=UTF8&me.

Kuriansky, J., & Nemeth, D. G. (2013). A model for post-environmental disaster wellness workshops: Preparing individuals and communities for hurricane anniversary reactions. *Ecopsychology*, *5*(S1), S-38–S-45. https://doi.org/10.1089/eco.2013.0006. http://online.liebertpub.com/doi/abs/10.1089/eco.2013.0006?src=recsys.

Nemeth, D. G., Kuriansky, J., & Onishi, Y. (2020). The 3 Es of psychosocial recovery after disaster. In K. E. Cherry, & A. Gibson (Eds.), *The intersection of trauma and disaster behavioral health*. Springer Nature: New York, NY.

Pain, A., Crouch, J., & Tillman, K. (2024). VillageReach: The COVID 411 initiative. In J. Kuriansky, & P. Kakkattil (Eds.), *Resilient health: Leveraging technology and social innovations to transform healthcare for COVID-19 recovery and beyond* (pp. 1005–1010). Elsevier.

The United Nations Office for Disaster Risk Reduction. (2018). *The UNISDR Global Science & Technology Advisory Group for the implementation of the Sendai Framework for Disaster Risk Reduction 2015-2030*. UNISDR. https://www.unisdr.org/2016/docs/stag/Revised%20G-STAG%20TOR%20final.pdf.

United Nations Department of Economic and Social Affairs, & The UN Office for Disaster Risk Reduction. (2022). *Risk-informed governance and innovative technology for disaster risk reduction and resilience*. https://unpan.un.org/sites/unpan.un.org/files/Module%202.3_DRR.pdf.

United Nations Office for Disaster Risk Reduction. (2015). *The Sendai framework for disaster risk reduction*. https://www.undrr.org/publication/sendai-framework-disaster-risk-reduction-2015-2030.

Further reading

Kuriansky, J. (2019). Climate change and disaster recovery at the United Nations: Activities, achievements and contributions of the IAAP New York Team. *Applied Psychology Around the World*, *1*(3), 49–55. https://iaapsy.org/site/assets/files/1082/apaw_vol1i3.pdf.

CHAPTER 10

Health equity, mental health, and partnerships to build back after COVID-19: Multistakeholder perspectives at a United Nations event

Judy Kuriansky[a,b,c,d,e,f], **Pedro Altungy**[a,g], **Gabe Twose**[h], **Sara Liébana**[a,i], **Stéphanie Racine Maurice**[a,j], **Carlos Velo**[a,k], **Rayna Sadia**[a,l], **Marija Davcheva**[a,m], **and Monique Arantes Guimarães**[a,n]

[a]*International Association of Applied Psychology, Indianapolis, IN, United States,* [b]*Department of Clinical Psychology, Teachers College, Columbia University, New York, NY, United States,* [c]*United African Congress, New York, NY, United States,* [d]*Office of Ambassador Sidique Wai, Embassy of the Republic of Sierra Leone to the United States, Washington, DC, United States,* [e]*United Nations Group of Friends of Mental Health and Well-being, New York, NY, United States,* [f]*Global Council SDG3 (Health), "SDGs in Action" Initiative, Dubai, United Arab Emirates,* [g]*Universidad Complutense de Madrid (UCM), Madrid, Spain,* [h]*American Psychological Association, Washington, DC, United States,* [i]*Universidad Europea of Madrid (UEM), Madrid, Spain,* [j]*Research and Intervention Program, Laval University, Quebec City, QC, Canada,* [k]*Universidad Autónoma de Madrid, Madrid, Spain,* [l]*Riphah International University, Islamabad, Pakistan,* [m]*Research Institute of Personnel Psychology, Organizational Development and Quality of Working Life (IDOCAL), University of Valencia, Valencia, Spain,* [n]*Universidade Sao Francisco (USF, Brazil), Sao Paulo, Brazil*

Introduction

In response to the ongoing risks of the COVID-19 pandemic and its variants throughout the world, the emerging increases in mental health problems, and the United Nations' focus on "Building Back Better" in the recovery period, an online event was held entitled, "*Building Back Better after COVID-19 through Addressing Health Equity, Mental Health and Well-being: Contributions of Psychological Science and Practice to Leave No One Behind*" during the United Nations' *High-Level Political Forum on Sustainable Development* (HLPF2021) on 12 July 2021. The HLPF consists of a two-week long meeting when governments of the world who are Member States of the United Nations come together to present their Voluntary National Reviews on their progress toward achieving the UN 2030 Agenda for Sustainable Development and its Sustainable

Resilient Health. https://doi.org/10.1016/B978-0-443-18529-8.00010-X

Development Goals (SDGs). This event in 2021 addressed three of the SDGs selected that year for review, namely, SDG 3, on ensuring health and well-being, including a target on promoting mental health; SDG 10, on reducing inequalities within and between nations; and SDG 17, on strengthening partnerships to achieve the goals. Given ongoing pandemic restrictions against in-person meetings, the event was held virtually. Speakers presented challenges and solutions to achieve physical and mental health equity to accelerate recovery from the tragic COVID-19 pandemic that affected countries worldwide, and emphasized the importance of equity mental health, and partnerships in this effort. The speakers impressively represented the multistakeholder approach called for by the United Nations.

Background

Health equity is both a driver and an outcome of inclusive, sustainable progress, as nations of the world advance to "build back better" in the COVID-19 recovery period. Scientific research shows that inequities, including in physical and mental health, are prevalent globally and have been exacerbated in the context of COVID-19 (Jensen et al., 2021). Marginalized and at-risk populations—such as racial and ethnic minorities, migrants and refugees, women and girls, older persons, persons with disabilities and compromising health conditions, and those living in conflict zones—are disproportionately vulnerable to experiencing adverse social determinants of health, including poverty, malnutrition, illiteracy, and discrimination. This complex situation leads to psychosocial distress as well as disparities in access to essential healthcare, food, education, and employment. These factors must be mitigated in order to achieve the global goals and well-being of all peoples.

The importance of health equity in the pandemic recovery

This event was in alignment with the aims of the HLPF 2021, which focused on global sustainable and resilient recovery from the COVID-19 pandemic through building an inclusive and effective path for the achievement of the UN 2030 Agenda for Sustainable Development. The event demonstrated that in order to achieve these aims, Member States in partnership with stakeholders in civil society including NGOs and academia need to prioritize physical and mental health equity. This involves assuring conditions for all populations to achieve their best possible health, including access to health and mental healthcare, as well as reducing vulnerabilities and eliminating preventable differences, such as in the burden of disease, injury, and exposure to violence.

Considerable research shows that health inequities are prevalent globally and have been exacerbated in the context of COVID-19 (Jensen et al., 2021). Experts have also widely confirmed that marginalized and at-risk populations, as noted above, are disproportionately vulnerable to experiencing adverse social determinants of health which cause physical and mental health disparities (World Health Organization Commission on Social Determinants of Health, 2008). These inequities arise from, and result in, serious consequences. For example:

- Children in sub-Saharan Africa are more than 15 times more likely to die before the age of 5 than children in high-income countries (United Nations, n.d.).
- In Low- and Middle-Income Countries (LMIC), up to 85% of people with mental health conditions receive no treatment (World Health Organization, 2019).
- The global economy loses an estimated US$1 trillion annually due to depression and anxiety (World Health Organization, n.d.).
- More than one in five people living in settings affected by conflict have a mental health condition (Charlson et al., 2019).
- 87% of premature deaths due to noncommunicable diseases occur in LMICs (World Health Organization, 2018).
- Substantial evidence shows inequities in both the state of health and the accessibility and quality of health services available to the group of about 75 million migrants in the WHO European Region (World Health Organization Regional Office for Europe, 2010).
- A digital divide exists for low-income countries, countries across Asia-Pacific regions, and also within countries—between urban and rural areas and between men and women—driven by the availability, affordability, and reliability of broadband services (UNESCAP, 2019; Utoikamanu, n.d.).
- Health inequities are exacerbated during crises caused by climate change, such as natural disasters, as shown by extensive research (American Public Health Association, 2019).
- Close to 1 billion people in the world live in slum conditions, representing about one quarter of the world's urban population (World Health Organization, 2018).

Of particular importance, the most prominent challenge to achieve global health equity has related to access to COVID-19 vaccines. Statistics have shown that while many wealthy countries expected to return to relative normalcy by the fall of 2021, most low-income nations would not have widespread access to vaccines before 2023 (The Economist Intelligence Unit, 2021).

Access to technology tools is proving important to achieving health equity. Innovations in healthcare delivery and training through technology are making progress to "reach those left furthest behind" who are "at the last mile," through partnerships among government, civil society and the private sector. These efforts need to be adapted and implemented in countries in the COVID-19 recovery period, as shown by many innovations that burgeoned during the pandemic and that are promising for the ongoing recovery (some of which are described below in this chapter in Part 5 of the program and others of which are in separate chapters in this volumeabout the innovations). The UN Secretary-General's Strategy on New Technologies (United Nations, 2018) puts forward a set of principles and the necessary framework to ensure that the benefits of new technologies are put to use for equitable and sustainable development.

The importance of health equity to achieve the United Nations Agenda 2030 for Sustainable Development

Health equity is vital to all of the Sustainable Development Goals (SDGs) which are reported in Voluntary National Reviews (VNRs) at HLPF 2021. Most directly, this side event makes the case that to build back better from COVID-19 and to progress toward achieving the UN Sustainable Development Goals, it is crucial to address the combined aspirations of SDG 3, to ensure healthy lives and well-being, and SDG 10, to reduce inequality within and between nations. In the spirit of SDG 17 about partnerships, this work should be carried out with varied key stakeholders, including civil society and academia.

Moreover, the promotion of equity is essential to achieve the rights of all to the highest attainable standard of physical and mental health, as assured in the United Nations Agenda 2030 for Sustainable Development (United Nations, 2015), as well as in the Political Declaration for Universal Health Coverage (United Nations, 2019), the Constitution of the World Health Organization (WHO, 1946), and the updated Global Humanitarian Response Plan (United Nations, 2020). These rights have been mandated earlier in Article 12 of the International Covenant on Economic, Social, and Cultural Rights (Office of the High Commissioner for Human Rights, 1976).

The necessity for mental health equity, in particular, is highlighted in WHO's Mental Health Action Plan (World Health Organization, 2013). Such equity is further central to the United Nations Secretary-General's Policy Brief on COVID-19 and Mental Health (Guterres, 2020), as well as in numerous products of the Inter-Agency Standing Committee Reference Group on Mental Health and Psychosocial Support in Emergency Settings (Interagency Standing Committee, n.d.).

The contributions of psychological science and practice to achieving health equity

Psychological science and practice have been shown to be vital to advancing health equity. Considerable research has uncovered enduring deleterious effects of poverty (Cudy et al., 2015), racism (Krieger & Sidney, 1996) and stigma (Meyer, 2003) and proven the importance of building psychosocial resilience in the face of emergencies (Kuriansky, 2016; Williams & Drury, 2009). Psychologists have developed approaches which reduce inequalities and offer methodologies to dismantle destructive practices and structures contributing to health inequities, as embodied by the American Psychological Association's *Equity Flattens the Curve* initiative (American Psychological Association, n.d.). The contributions of psychology are particularly relevant to target 3.4 in the UN 2030 Agenda: to reduce by one-third premature mortality from noncommunicable diseases through prevention and treatment, and to promote mental health and well-being. For example, psychology has made significant contributions to preventing and treating cardiovascular disease through reducing stress and promoting healthy behaviors like diet and exercise (Khayyam-Nekouei et al., 2013). Also, a unique program implemented in the Democratic Republic of the Congo (DRC), described below, successfully provided health education and psychosocial support to a remote community, increasing safety behaviors in community members to reduce the spread of COVID-19 and providing psychosocial resilience training for ongoing recovery (CSEM, 2020).

The response to the global mental health crisis must be informed by the best available science, experiences of professionals, and voices of the hardest-hit communities. Also, best practices can be amplified by innovations. Adoption of a population health approach is critical to ensure that we are addressing the broad range of psychological distress and disorders caused by the pandemic and closing the already existing gap between need and service provision. A population health approach focuses on providing access to the best evidence-based treatments for those having mental health conditions needing clinical intervention, to reduce risk or mitigate the impact of risk factors that lead to mental health conditions among those who are at greater risk, and to provide tools and resources even for those who are experiencing good mental health. This approach places greater emphasis on prevention and intervention long before people experience a crisis, as well as ensuring highly effective services for those in greatest need of mental healthcare. Investing in prevention and early intervention is a critical strategy to avert adverse and costly mental health outcomes for people and their families, as well as to save lives.

Psychologists and mental health experts have much to add in this recovery process. Utilizing psychological science is a critical means to inform strategies on implementing this population health approach to achieving health equity. Psychologists' expertise focuses on understanding human behavior; for example, comprehending vaccine hesitancy as well as the differential mental health impact of the COVID-19 pandemic on various communities of color. Further, psychological research examines and informs factors which underlie health-related behaviors, such as social cohesion, social cognition, communication strategies, and cognitive biases.

As explained by the WHO Technical Advisory Group on Behavioral Insights and Sciences for Health in their publication *Behavioral Considerations for Acceptance and Uptake of COVID-19 Vaccines* (World Health Organization, 2021), psychological science is also integral to promoting vaccine uptake. This WHO document emphasizes the important role of factors such as ensuring an environment conducive to accessing the vaccine (making it easy, affordable, and convenient); engaging social influences supportive of taking the vaccine (through media, health professionals, and trusted community members); and increasing motivation to take the vaccine (through myth-busting, trust-building emphasizing social benefits, and implementing culturally adaptive education campaigns). Successful behavioral science approaches can be tailored to specific groups in diverse settings in both developed and developing countries to maximize equity in vaccine acceptance.

Partnerships reflected in the event

Sustainable Development Goal 17 calling for multistakeholder partnerships is evident in the extent and diversity of the cosponsors for the event, including an extremely impressive large group of Permanent Missions to the United Nations, with the Mission of Sierra Leone as the main sponsor, joined by 14 other missions to the UN which represent regions from all over the world (in alphabetical order): Bahrain, Canada, Costa Rica, Ecuador, Georgia, Iceland, Japan, Lebanon, Mali, Mexico, Nepal, Portugal, Qatar, and Sweden.

This range of governments' willingness and enthusiasm to participate on this topic demonstrates the serious commitment of countries globally to health equity and specifically to mental health. This commitment is consistent with the exponential growth of awareness of the mental health impact of the pandemic and open commitment at the highest level of the UN leadership, in speeches and policies by the UN Secretary-General, as well as by the Deputy Secretary-General, and the Presidents of the General Assembly.

Fully in alignment with the event topic on health and mental health, the World Health Organization, represented by the WHO Office at the United Nations in New York, was a cosponsor, as they have been for many similar events over many years, collaborating with these Member States and NGOs. Civil society cosponsors, who organized the event, were the International Association of Applied Psychology, the American Psychological Association, and the United African Congress.

The number of participants, with 15 Member State sponsors, and 28 international high-level representatives, professionals working on-the-ground, and scientific experts—as shown in a screenshot from the recording of a grouping of several speakers in Fig. 1—was noted by several participants and attendees to be the most such number from the diverse constituencies in such a forum.

Objective

The objectives of the event were to:

1. Explain the scope of global physical and mental health inequities especially in cultural contexts and among vulnerable populations facing COVID-19.
2. Propose models to reduce physical and mental health inequities to build back better from COVID-19.
3. Emphasize the role of mental health and of applied psychologists in particular, in achieving the global goals.
4. Demonstrate the nexus of selected SDG sunder review, specifically, SDG 3 (Good health and well-being), SDG 10 (reduce inequities), and SDG 17 (partnerships), to recover from the pandemic, as well as to demonstrate the interlinkage of these with other SDGs under

Fig. 1

Stakeholder speakers at the health equity event for the UN High-Level Political Forum. Pictured (top row, left to right): moderator Dr. Judy Kuriansky; Ambassador of the Republic of Sierra Leone to the United Nations, Alie Kabba; Costa Rica Vice-President Epsy Campbell Barr; Ambassador of Portugal to the UN, Francisco Lopes. (Row 2, left to right): Ambassador of the Kingdom of Bahrain to the UN, Jamal Fares Alrowaiei; Ambassador of Mexico to the UN, Juan Ramón de la Fuente; Ambassador of Ecuador to the UN, Cristian Espinosa Cañizares; Ambassador of Georgia to the UN, Kaha Imnadze. (Row 3, left to right): Ambassador of Lebanon to the UN, Amal Mudallali; Ambassador of the Republic of Mali to the UN, Issa Konfourou; Ambassador Sidique Wai, Embassy of the Republic of Sierra Leone to the United States; Deputy Permanent Representative of Canada to the UN, Richard Arbeiter. (Row 4, left to right): Ambassador Tetsuya Kimura, Mission of Japan to the UN; Deputy Permanent Representative of Nepal to the UN, Ghanshyam Bhandari; WHO Director of the WHO Office at the UN in NY Werner Obermeyer; Qatar psychologist Dr. Sherifa Noman Al-Emadi. (Row 5, left to right: Mexico psychologist Dr. María Elena Medina-Mora; Sweden public health expert Mr. Johan Carlson; President, United African Congress, Gordon Tapper; President, International Association of Applied Psychology, Dr. Christine Roland-Levy (screen shots from Zoom recording).

review, namely, SDG 1 (end poverty), SDG 2 (end hunger), SDG 8 (provide decent work and economic growth), and SDG 13 (climate action).

5. Urge efforts to increase global vaccine access and uptake through myth-busting, deescalating fears and distrust, and other interventions.
6. Create awareness of evidence-based technological tools to increase access to healthcare.

7. Urge national governments to provide mental health and psychosocial support (MHPSS) in the pandemic recovery period for all communities, particularly those also affected by conflict and crisis.
8. Underscore the importance of partnerships among Member States, UN agencies, other stakeholders, and the psychological community to advance physical and mental health equity, consistent with SDG 17.

Logistics

The event was divided into six parts: (1) introductions and welcome remarks setting the stage, from distinguished cosponsors; and a high-level segment with remarks from government cosponsors including Member State Missions at the United Nations and from the World Health Organization; (3) a cultural interlude/prayer; (4) presentations by scientific experts; (5) presentations about interventions "on the ground" that dispel inequities; and (6) a closing statement with lessons learned and the way forward. The program is shown in Fig. 2.

Program

Part 1 of the program

The first segment of the program consisted of welcome remarks to set the stage for the discussion, from speakers shown in Fig. 3. The first author of this chapter, Dr. Judy Kuriansky, was the moderator responsible for presenting and opening this compelling webinar. She introduced the inspirational song "One Love" performed by Sierra Leonean musician MDG (which stands for "My Dearest Gift") which she said, "captures the sentiments of these terrible times and the dire needs of the people but, also, a wonderful way forward." The song honors health workers ho respsonded to the COVID pandemic as heroes and promotes the message of love in the face of any emergency and for world peace.

She then introduced the first high-level segment of the program, starting with then-permanent representative of the Mission of Sierra Leone to the UN, the main sponsor of the event, H.E. Ambassador Alie Kabba, who highlighted how the gathered group was proof of international commitment to strengthen global resilience for achieving sustainable recovery from the physical and mental health challenges of the pandemic. Remarking on the importance of "delivering mental health and psychosocial support now and in the post-pandemic era," he underlined a theme echoed throughout the program, that only through international partnership and cooperation can the global goals be achieved.

Following this message, the Vice-President of Costa Rica, Her Excellency Epsy Campbell Barr, gave opening remarks, pointing out that psychosocial well-being requires attention of international governments, academia, and people themselves, especially since lockdowns have

July 12, 1:00PM-2:30PM (EDT)
This event is virtual. Click here.

Moderator: Dr. Judy Kuriansky, Columbia University Teachers College; International Association of Applied Psychology
Welcome: **H.E. Ambassador Alie Kabba**, Permanent Representative, Mission of Sierra Leone to the United Nations
Opening Remarks: Dr. Jennifer F. Kelly, President, American Psychological Association

High-level Speakers

H.E. Ambassador Francisco Duarte Lopes, Permanent Representative, Mission of Portugal to the United Nations
H.E. Ambassador Juan Ramón de la Fuente, Permanent Representative, Mission of Mexico to the United Nations; United Nations Security Council 2021-2022
H.E. Ambassador Issa Konfourou, Permanent Representative, Mission of Mali to the United Nations
H.E. Ambassador Jamal Fares Alrowaiei, Permanent Representative, Mission of Kingdom of Bahrain to the United Nations; Co-chair, Group of Friends of Mental Health and Well-being
H.E. Ambassador Cristian Espinosa, Permanent Representative, Mission of Ecuador to the United Nations; Co-chair, Group of Friends of Mental Health and Well-being
H.E. Ambassador Kaha Imnadze, Permanent Representative, Mission of Georgia to the United Nations; Co-facilitator, Political Declaration on Universal Health Coverage (UHC)
Representative, Mission of Lebanon to the United Nations
H.E. Ambassador Richard Arbeiter, Deputy Permanent Representative, Mission of Canada to United Nations; Co-chair, Group of Friends of Mental Health and Well-being
Mr. Ghanshyam Bhandari, Deputy Permanent Representative, Mission of Nepal to the United Nations
H.E. Ambassador Tetsuya Kimura, Mission of Japan to the United Nations; Co-Chair, Group of Friends of UHC
H.E. Ambassador Sidique Abou-Bakarr Wai, Chief of Mission, Embassy of Sierra Leone to the USA
Mr. Werner Obermeyer, Director, World Health Organization Office at the United Nations

Scientific Experts

Dr. Geoffrey Reed, Department of Psychiatry, Columbia University
Mr. Johan Carlsson, Public Health Agency of Sweden (TBC)
Dr. Willem Hanekom, Africa Health Research Institute
Dr. María Elena Medina-Mora, National Autonomous University of Mexico; Advisor, PAHO

Voices from the Field: Programs on the Ground

Sierra Leone: Jordan Levinson, Dimagi
Democratic Republic of the Congo (DRC):
Mr. Gordon Tapper, United African Congress
***Nepal*: Dr. Kamal Gautam**, TPO Nepal
***India*: Mr. Sujay Santra**, iKure

Closing Remarks

Dr. Christine Roland-Lévy, President, International Association of Applied Psychology

Main Sponsor: Permanent Mission of the Republic of Sierra Leone to the United Nations

The Permanent Mission of Portugal to the United Nations
Canada
MEXICO
República del Ecuador
The Permanent Mission of Lebanon to the United Nations
Permanent Mission of Iceland to the United Nations
Permanent Mission of Sweden to the United Nations
PERMANENT MISSION OF GEORGIA TO THE UNITED NATIONS
World Health Organization
IAAP
AMERICAN PSYCHOLOGICAL ASSOCIATION
United African Congress

Fig. 2
Program for the HLPF event on Health Equity.

Fig. 3

Speakers giving welcome remarks: Pictured (top row, left to right) moderator Dr. Judy Kuriansky and Costa Rica Vice-President Epsy Campbell Barr; (bottom row, left to right) Ambassador of Sierra Leone to the United Nations Alie Kabba and APA President Dr. Jennifer Kelly (screenshots from Zoom recording).

had an extensive negative impact on the mental health of people and economies worldwide. Mental health services, she said, must acknowledge the vast ethnic, political, and social diversity of peoples, in quality and nondiscriminative interventions, and pay special attention to the most vulnerable groups, including women, to ensure their adequate access to psychosocial attention.

In the next welcome address, on "Health equity and social determinants of health," the president of the American Psychological Association, Dr. Jennifer Kelly, defined equity as "giving people what they need, when they need it, and in the amount that they need it." Equality doesn't mean equity, she emphasized, which is the ethical and human rights principle that motivates us to eliminate health disparities that stem from social determinants including economic stability, environmental conditions, quality of education, and access to care. Achieving health equity, she explained, requires ongoing social efforts to address historical and current injustices and to eliminate preventable health and mental health disparities. UN Member States, she said, should and can prioritize physical and mental health equity by promoting conditions for all populations to attain optimal health, incorporating mental health as a key component of health programs. Through equity, she said, the disproportionate harm caused by COVID-19 in vulnerable collectives can be mitigated. Psychologists can help in these efforts, given their expertise to address social determinants of health that contribute to health disparities and thereby help improve the overall health of populations.

Part 2a of the program

The second high-level segment of the program consisted of presentations by leaders on the international stage, namely, Permanent Representatives (PRs)/Ambassadors of Missions to the UN and other top representatives of the co-sponsoring government UN Member States, as shown in Fig. 4. They all congratulated Sierra Leone for being the main sponsor of this important event, as well as the co-sponsors—the World Health Organization (WHO) and the NGOs—for supporting health and mental health as an important part of building back better from the pandemic. These presentations of many high-level diplomat co-sponsors were in two subparts (Part 2a and b), divided by a cultural prayer by an indigenous leader.

The first speaker in this section of the program was H.E. Ambassador Francisco Duarte Lopes, Permanent Representative of the Mission of Portugal to the UN, which was especially appropriate for the topic of this event since Portugal, during their term as a member of the Human Rights Council (2015–2017), was "a leader in the paradigm shift in mental health from an exclusively biomedical approach to a more comprehensive and human rights-based approach," evident in their leadership with Brazil in the adoption of a Human Rights Council Resolution on Mental Health and Human Rights (United Nations General Assembly, 2017).

Fig. 4

High level speakers: Ambassador of the Portugal Mission to the United NationsFrancisco Lopes; Ambassador ofthe Mission ofMexico to the UN Juan Ramón de la Fuente; Ambassador of Mali to the UN Issa Konfourou; ambassador of Ecuador to the UN Cristian Espinosa; Ambassador of Bahrain to the UN Jamal Fares Alrowaiei; Ambassador of Georgia to the UN Kaha Imnadze; Ambassador of Lebanon to the UN Amal Mudallali: Deputy Permanent Representative of the Mission of Canada to the UN, Ambassador Richard Arbeiter; Deputy Permanent Representative of the Mission of Nepal to the UN, Mr. Ghanshyam Bhandari; Ambassador Tetsuya Kimura in the Mission ofJapan to the UN; Ambassador of Sierra Leone to the United States Sidique Abou-Bakarr Wai; Director of the WHO Office at the UN in New York Mr. Werner Obermeyer (screenshots from Zoom recording).

Also, Portugal hosted the first International Summit on Psychology and Global Health, at which the President of the Portuguese Republic Marcelo Rebelo de Sousa honored the role of psychologists in creating a healthy society. A decentralized mental health program has been implemented in Portugal to ensure equitable access to quality care to all people with mental health conditions, including those in vulnerable situations, to reduce inequalities and situations of discrimination, stigma, coercion, and segregation that contribute to disparities. Further, he noted, Portugal is the cohost with Kenya of the UN Ocean conference in Lisbon in 2022, leading Ambassador Lopes to underscore the connection, as emphasized by Dr. Kuriansky, between mental health (in target 3.4) and two sustainable development goals (SDGs 13 and 14). He acknowledged her point that SDG 13, which calls for Climate Action against the dangers of rising sea levels which threaten to wash away lands and the means of survival of entire societies (especially in Small Island Pacific States), highlights serious traumatic psychological consequences of inaction. With regard to SDG 14, Ambassador Lopes underscored the connection made by Dr. Kuriansky that "science supports the positive effect of ensuring people's well-being will enhance human care of the wellbeing of the environment and all creatures, including life below water." In closing, he noted Portugal's commitment to collaborate with partners to achieve global health coverage, as a human right—especially important in the Least Developed Countries that have drastically limited health and mental services resources—emphasizing that "mental health is an important component of this noble endeavour."

In introducing the next Member State speaker, H.E. Ambassador Dr. Juan Ramón de la Fuente, Permanent Representative of the Mission of Mexico to the United Nations, moderator Kuriansky pointed out his authority on the event topic, given his credentials as a psychiatrist, a Professor Emeritus of Psychiatry at the National Autonomous University of Mexico, author of innumerable professional articles and books, and former Minister of Health of Mexico. Importantly, during Mexico's current term on the UN Security Council, H.E. de la Fuente has championed the promotion of mental health and psychosocial support in conflict settings and regions, detailed in another chapter in this volume (de la Fuente et al., 2024). In his presentation, Ambassador de la Fuente highlighted the importance of cooperation and strengthening relevant international organizations, instruments, and protocols in a more robust health response to address the physical and psychological inequities exacerbated by the unprecedented emergency context of COVID-19. In light of WHO statistics that there are only 0.05 psychiatrists for every 100,000 inhabitants in low-income countries, and the value of psychological science to reduce inequities, he pledged that Mexico considers it a priority to promote mental health and psychosocial well-being as "a fundamental factor in rebuilding the social fabric, which if untreated, has the potential to become a terrible pandemic of its own with devastating societal consequences, including reducing productivity and raising healthcare expenses." Given the unacceptably uneven progress of the COVID-19 vaccination campaign worldwide, Mexico promoted General Assembly Resolution 74/274, adopted by consensus at the UN General Assembly, to address challenges and inequalities of COVID-19, including worldwide equal access to medicines, vaccines, and medical equipment which catalyzed initiatives to compensate for grave inequities. Yet, he reminded everyone of the importance to resume all the other immunization campaigns that have been interrupted by already strained health systems having to deal with the pandemic. He also called for further evidence-based science to help reduce health inequalities, especially in low-income countries, and concluded that this High-level Political Forum offers a great opportunity to reflect on possible angles to achieve the UN SDGs within the next decade to reach the deadline of the year 2030, and "to design more innovative solutions to adapt to today's complex realities."

H.E. Ambassador Issa Konfourou, Permanent Representative of the Mission of the Republic of Mali to the UN, brought attention to the innumerable current societal challenges (climate change, education, the economy, youth unemployment) besides health diseases (malaria, HIV/AIDS, tuberculosis, Ebola, and now COVID-19) faced by all countries but particularly by developing countries. The best way to tackle these challenges, he said, is through international cooperation and strengthening from different sectors (government, health organizations, civil society, academia, the private sector) to strengthen existing world organizations to ensure equal access to medicines, vaccines, and mental health and well-being of all, necessary to achieve the SDGs.

Moderator Dr. Kuriansky then pointed out that UN Member States have formed a Group of Friends of Mental Health and Wellbeing (GoF of MHWB), with four copresidents representing different regions of the world, namely, Bahrain, Belgium, Canada, and Ecuador. To introduce the next speaker, from Ecuador, she highlighted that Ecuador is the first country to include the Rights of Nature in their constitution, giving the environment equal rights to that of a person, and that Ecuador is also a leader in recognizing the rights of marginalized populations, including persons with disabilities. She and Ambassador Dr. Caleb Otto of the Mission of Palau co-founded this group in 2014 when advocating for Member States to support the inclusion of mental health and wellbeing in the UN 20230 Agenda.

H.E. Ambassador Cristian Espinosa, Permanent Representative of the Mission of Ecuador to the UN, said he was pleased to see the increased interest in mental health, consistent with the main objective of the Group of Friends of Mental Health and Wellbeing (of which Ecuador is a co-chair) to raise awareness about challenges surrounding mental health, as well as to share tools, resources, and strategies that can help better address individual and collective mental health needs by making available the work of international organizations and the application of national policies. This is important now more than ever, he said, taking into account the negative effects of the pandemic. Resulting rates of mental health problems, he pointed out—including suicide which remains the main cause of death among young people—have risen alarmingly as a result of the deep economic crisis, confinement, and widened inequality gap caused by poverty, "that still affects the globe, especially vulnerable groups including minorities, women, children, migrants, and refugees." In this context, he quoted the UN Secretary-General's Policy Brief on COVID-19 and the Need for Action on Mental Health (UN, 2020) that:

> Good mental health is critical to the functioning of society at the best of times. It must be front and center of every country's response to and recovery from the COVID-19 pandemic. The mental health and wellbeing of whole societies have been severely impacted by this crisis and are a priority to be addressed urgently.

"The promise to 'leave no one behind' is the commitment of all states," Ambassador Espinosa proclaimed, adding that access to vaccines is important to "reopen economies as soon as possible and prevent despair from economic and social problems." Multilateralism, cooperation, and solidarity among nations, and the contribution of everybody, he emphasized, is more important now than ever before, to achieve the SDGs and "build back better."

Praising the Kingdom of Bahrain for its co-Presidency of the GoF of MHWB, their leadership in the Culture of Peace with Love and Conscience, and their inclusivity of all religions, faiths, and peoples, moderator Kuriansky introduced the next speaker, H.E. Ambassador Jamal Alrowaiei, Permanent Representative of the Mission of the Kingdom of Bahrain to the United Nations. He kindly recognized all colleagues, the UN Secretary-General, and WHO for recognizing the importance of mental health, felt by everyone especially in these uncertain pandemic times since the long-term effects are multifaceted. In addition, he acknowledged Dr. Kuriansky as a "true inspiration and real hero for continued work and advocacy." Consistent with Bahrain's co-Presidency of the Group of Friends of Mental Health and Wellbeing, he acknowledged the members for raising awareness about mental health and best practices from different parts of the world and mainstreaming the discussion. Since the pandemic will unfortunately have long-term effects and enduring negative impacts on mental health, he said, it is more important now than ever before, to address mental health while there is worldwide attention on this issue. "Mental health is a core issue," he declared, "and the UN 2030 Agenda can only be achievable with health as a central approach to achieving many of the goals and health equity."

To introduce the next member state representative, H.E. Ambassador Kaha Imnadze, Permanent Representative of the Mission of Georgia to the UN, Dr. Kuriansky noted that Georgia has played a critical role in health and mental health equity. Most importantly, the Mission of Georgia cofacilitated, with the Mission of Thailand, the successful negotiations on the Political Declaration on Universal Health Coverage, which significantly includes mental health in the lead paragraph, the inclusion for which she advocated. Also, the country of Georgia approved a "state concept" of mental health back in the year 2013, and also recently held an event about the role of music in supporting mental health, and participated in an event sponsored by her NGO (IAAP) and WHO presenting their unique community outreach approach to mental health.

Indeed, Ambassador Imnadze noted, his country has introduced positive changes in mental health, even dating back further, to 1995, enacting legislation to shift from institutional care in asylums during the Soviet regime which were intended to protect the community from societal harm, to a more community-based care to insure that those who suffer from mental health issues are an integral part of society. He noted that the pandemic has shown that no country, developed or developing, is secure when it comes to health-related issues. Vaccines being available to consumers are critical but not a magical solution, he said, as other factors are necessary "to make us more secure for the future (and) be better prepared." These include, he enumerated, health equity, proper information to avoid improper and unnecessary theories, adequate investment in critical medical infrastructure and health equity, production facilities equitably spread throughout different regions, exchange of scientific information, research and data, and "a balance between private sector gain which drives progress and social responsibility and equity which is the foundation of stability of wellbeing of societies." With measures to control the pandemic like lockdowns which resulted in suffering and stress, as well as adramatic rise in domestic violence, he hopes that lessons from this pandemic can equip society for future health emergencies.

Part 3 of the program: Cultural interlude

Adding a spiritual message in the event, a prayer was offered by an elder of the Oji-Cree people, Brenda Mason, a holistic mental, physical, and spiritual counselor from Thunder Bay, Ontario, Canada, as shown in Fig. 5. In her native language and then in English, wearing a t-shirt with a bear figure imprint sitting next to a bear statuette and with feather in hand, she offered,

Fig. 5
Elder of the Oji-Cree people offering a prayer (screenshot from Zoom recording).

> Creator, I am grateful for this life...and I ask you to continue to bless us and surround us with love and compassion to continue our work to provide service to our people, and to give them the means to find their gift within themselves... and provide us with guidance and direction that we need... and with our leaders, bless each one of them.

Part 2b, continuation of the high-level segment program

The blessings were appropriately followed, moderator Kuriansky noted, by a statement from H.E Ambassador Amal Mudallali, Permanent Representative of the Mission of Lebanon to the United Nations, considering that the country was recovering from the recent explosion in Beirut. Given her extensive media background—having earned a PhD degree in Political Communications, having been a journalist for Arabic- and English-language publications and media outlets including for the BBC, and having received France's Chevalier de L'ordre de la Légion d'Honneur—Ambassador Mudallali has a keen ability to report on situations and assess crises. She explained that Lebanon is one of the countries leading the Political Declaration on Global Access to Vaccines, along with Mexico, Qatar, and Sweden, also having representative speakers at this event. This event topic "hits close to home" for Lebanon, Ambassador Mudallali explained, as even before COVID-19, her country was going through a particularly difficult time economically, with the result that people were suffering with financial, physical, and mental stress.

Ambassador Mudallali alerted everyone to an emerging important concern of "long-haul COVID," a "tsunami of disability" resulting from ongoing long-lasting symptoms caused by COVID-19 infection, which she read about in an article in *Scientific American,* and which she said, "We are going to work on at the United Nations."While some people are feeling better, she explained, others have long-term symptoms of the disease like fatigue and chronic respiratory problems causing stress and mental problems. Thus, "to build back better," she said, "we need to plan for a future to protect millions of pandemic survivors (already numbering 34 million) who are chronically ill even after 7–8months, have no access to vaccines, and are struggling to be safe." This is especially a problem, she said, in developing and vulnerable countries where governments and healthcare systems were already facing challenges. It is important to consider this potential disability and take preventive steps, she said, and to insist on equitable access to vaccines for everybody around the world.

Event moderator Dr. Kuriansky then introduced the next speaker, Deputy Permanent Representative of the Mission of Canada to the UN, Ambassador Richard Arbeiter, by acknowledging Canada's leadership in promoting mental health importance and financing, Bell Canada's campaign "Let's Talk," and Ambassador Arbeiter's excellent moderating of several *Psychology Days at the United Nations,* including on the topics of gender equality and on building back better. He acknowledged taking to heart the well-wishes delivered earlier by the indigenous Canadian spiritual leader "as we all aspire to be servants to our people," and confirmed Canada's commitment to mental health. He noted humbly about his country that "We are aspiring students about how to do better about mental health" with opportunities like this forum for learning. He cited shocking statistics that by the age of 40, almost 50% of Canadians—including himself—suffer from some sort of mental health-related issue, since he explained, "If it's not you, it's someone next to you". Also, he said that despite Canada's advances, health inequities in Canada "exist, are prevalent and growing." These result from sociopolitical and economic disadvantages, like discrimination, racism, and historical trauma, he explained, which are important social determinants of health for certain groups, particularly indigenous peoples in Canada, LGBTQ Canadians, and black Canadians.

"Equity starts with evidence and understanding the populations we're trying to serve," he said, adding that Canada is addressing these inequalities through multiple approaches, including better data collection, more tailored policies, and increased financing, but, he added, "Its not easy, it's not straightforward and nothing is perfect." Canada's aim is a comprehensive package of mental health for everyone, he said.

Continuing in a very personally engaging vein of humility and self-disclosure, Ambassador Arbeiterfurther shared that given the new realities brought on by COVID-19, "I had to learn how to be a better manager for my own staff, each of whom experienced a different kind of mental health emergency either for themselves of their families." This led him to come to understand how "empathy (the psychological concept indicating genuine understanding and concern for others) has to be the starting point for how we engage," as well as destigmatizing those with mental health challenges. Therefore, he repeated, mental health has to be included in the package of services for everyone.

Turning to Nepal, Dr. Kuriansky commended the government's aspiration, "Prosperous Nepal, Happy Nepali"— linking mental health with economic development, and introduced the Deputy Permanent Representative of the Mission of Nepal to the UN, Mr. Ghanshyam Bhandari. She acknowledged his exemplary representation of the multistakeholder approach and the nexus of the three SDGs addressed in this HLPF—SDG 3 on health, SDG 10 on inequality, and SDG 17 on partnerships. This has been evident in his participation in several webinars promoting psychosocial resilience facing COVID-19, all of which were organized by the moderator of this event and her Columbia University Teachers College Nepali student Sadikchhya (Meera) Khanal, including for the UN Commission on the Status of Women and other meetings. Details about these events are described in a later section in this chapter by the Executive Director of Transcultural Psychosocial Organization (TPO) Nepal and also in another chapter in this volume (Khanal & Kuriansky, 2024). Those several events, like this one, impressively represented the multistakeholder approach, with speakers from government, international NGOs (the International Association of Applied Psychology), local NGOs (including TPO Nepal and others), academia (Columbia University Teachers College), the diaspora (the Nepali-American Coalition), and media. Mr. Bhandari acknowledged that the pandemic has affected physical and psychosocial health and "brought the issue of mental health to the center" for everyone, given social isolation, loss of loved ones and jobs, and many socioeconomic impacts, not only for Nepal but also for others "already left further behind" in countries such as Least Developed Countries (LDCs), Land-Locked Developing Countries (LLDCs), and Small Island Developing States (SIDS), and for marginalized vulnerable groups, including women, children, senior citizens, and persons already with mental health conditions.

"Like many other developing countries, Nepal was not prepared for a crisis of this scale," he said, "however, the government has continued to focus on reducing health disparities and addressing mental health as a priority area of the response." Nepal's response, introduced in May 2020, included: (1) providing mental health services and psychosocial counseling and support for patients and healthcare workers, (2) guidelines for preventing the disruption of essential health services, and (3) providing free-of-cost COVID-19 testing and treatment services for needy people. Efforts emphasize a multistakeholder approach of collaborative partnership, bringing together civil society, think tanks, academia, and the private sector, while especially mobilizing resources at local and grassroots levels. Going forward, he said, "achieving holistic health equity and realizing SDG 3 and SDG 10, among others, will not be possible without addressing health inequalities both within and among nations." Also, universal access to vaccines for all countries and communities must not only be ensured, but not "be only about profiteering, but also about justice." In conclusion, he noted, collaboration is critical to respond, recover, and rebuild better.

Noting that Japan is ideal to address this issue in their role as cochair of the United Nations Group of Friends of Universal Health Coverage (UHC), moderator Kuriansky introduced H.E. Ambassador Mr. Tetsuya Kimura of the Mission of Japan to the United Nations, in charge of Economic and Social Affairs UN Management, humanitarian rights, and sports as a means of development and peace, consistent with research showing that sports promotes equity (Barry & Kuriansky, 2024). Ambassador Kimura confirmed Japan's commitment to UHC ensuring all individuals' access to comprehensive care for both mental and physical health, and especially to achieve health equity for the most vulnerable segments of society in the post-COVID-19 recovery.

He described Japan's efforts: (1) In addition to bilateral cooperation, Japan is working together with the COVAX facility to ensure fair and equitable access of vaccines—affirmed at the Gavi COVAX Commitment Summit where stakeholders pledged US$2.4 billion to accelerate access to 1.8 billion doses of COVID-19 vaccines for low-income economies, and Japan pledged another US$800 million to their already US$200 million contribution; (2) Japan's support of developing countries in strengthening health and medical systems through building hospitals, providing medical equipment, supporting human resource development, and training regarding response to infectious diseases; (3) Japan's support of the ASEAN Center for Public Health Emergencies and Emerging Diseases as well as of the Africa Centres for Disease Control and

Prevention, and the Novartis Memorial Institute for Medical Research in Ghana, and their assistance in training medical laboratory technicians; and, (4) Japan taking a holistic approach in ensuring health security in a broader context, continuing to work with other countries to improve the conditions of water, sanitation, nutrition, education, hygiene, and other environmental factors, through hosting the Tokyo Nutrition for Growth summit in December 2021. All these efforts reflect Japan's leadership in UHC and "leaving no one behind," described in another chapter in this volume (Ezoe, 2024).

Moderator Kuriansky next introduced H.E. Ambassador Sidique Abou-Bakarr Wai, chief of Mission of the Embassy of the Republic of Sierra Leone to the United States, known fondly as a "Visionary Ambassador" and the "Peoples Ambassador" for his exceptional efforts bringing groups together for action. Consistent with this, Ambassador Wai shared successful experiences to achieve health equity during COVID-19 in his country through partnership of government with civil society. Known for calling for action, he gave examples of actions taken, including (1) A coalition of health professionals, friends, diaspora, and philanthropists was convened by his Embassy in a "first-of-its kind in the diplomatic community" COVID-19 Task Force that mobilized, in a short time, funds and in-kind contributions valued at US$7.5 million to stave off the COVID-19 infection in Sierra Leone including in remote communities; (2) The Embassy partnered with two NGO co-sponsors of this event, namely, the United African Congress and the International Association of Applied Psychology (IAAP), through their representatives, the moderator of this event Dr. Kuriansky and Mr. Gordon Tapper (who spoke later at this event) about an important health initiative *"Africa Rise: Healthier and Stronger"* to raise awareness and funds for health and mental health needs of people in the most remote communities which have no doctors and limited equipment; and, (3) The Sierra Leone government is working with civil society in the signature agenda of the Sierra Leone President Julius Maada Bio called "human capital development" addressing three pillars of education, health, and security which have become even more critical due to the need to recover from the COVID-19 pandemic.

Ambassasdor Wai ended with his signature question after all such high-level events: "What Next?" In this regard, he applauded that the event has already led to connection between the governments and experts assembled in this event, and encouraged "All action working towards a resolution about health and wellbeing, so we can build back better together."

Continuing the theme about action, including in vaccine distribution, and given that the event topic perfectly aligns with the World Health Organization confirmation that mental health as essential to health, the next speaker was the Director of the WHO Office at the United Nations, Mr. Werner Obermeyer, whose extensive portfolio covers global health, mental health, UHC, NCDs, climate change, disabilities, narcotics and tobacco, and road safety, providing WHO's technical advice on innumerable UN resolution negotiations. He discussed the pandemic's disproportional effects on vulnerable groups, especially in countries with weak health systems.

He outlined that the pandemic has: become a major inhibitor in us reaching the SDGs, particularly SDG3...with many countries being a full 10 years behind in achieving the 2030 target date"; has disproportionately affected groups particularly those in humanitarian crisis; has led to neglect of noncommunicable diseases; and has affected gender parity and women's rights. Therefore, he said, we need to: (1) scale up action, including in vaccine equity—considering that less than 1% of people in low-income countries have received at least one dose of COVID-19 vaccine (in his home country of South Africa, it's 2%) whereas developed countries had already started to vaccinate teenagers; (2) invest in equitable distribution of diagnostics, tests, treatments, and medical supplies; and, (3) invest in health systems and workers, to translate aims into action.

"Global health security obviously starts with local health security," he said, adding increasing concern that "the mental health impacts of the pandemic will most likely exceed the direct impact of the virus, as it has exposed the fragility of mental health services globally." In fact, WHO surveys indicate that "the pandemic slowed down or halted critical mental health services in over 93% countries in the world while the demand is increasing and will multiply in the near future." Additionally, over 60% of countries have disrupted health services for mental health conditions and vulnerable people (children, adolescents, older adults).

"We have all agreed that building back safeguards in our health system is a matter of priority" he said, adding that "we must collaborate with the private sector, NGOs and multiple other stakeholders to achieve health for all."

Part 4 of the program

The next section of the program consisted of presentations by scientific experts, as shown in Fig. 6.

Dr. Geoffrey Reed, professor of Medical Psychology in the Department of Psychiatry at Columbia University spoke on "*Improving health equity by addressing global mental health needs*," reporting that prior to the pandemic, an estimated 10.7% of the world's population was living with serious mental or substance use disorders at any given time, accounting for 7% of total global disease burden and the biggest contributor to global disability. Yet, only an average of 2% of national health budgets were being directed to these disorders even though effective and inexpensive treatments are available. Also, in low- and middle-income countries—where most of the world's population lives—between 76% and 85% of people with serious mental disorders had received no treatment in the previous year. Health disparities for people with mental health disorders include mortality rates 2–2.5 times higher than the general population and life expectancy reduced by 10–25 years. The pandemic exacerbated chronic underfunding of mental health; disrupted or halted mental health services in 93% of 130 countries surveyed by WHO; and led to increased stress, anxiety, eating disorders, self-harm and suicidality, and

Fig. 6

Health experts speaking at the HLPF event: pictured (top, left to right: USA psychologist Dr. Geoffrey Reed; Qatar psychologist Dr. Sherifa Noman Al-Emadi; Sweden public health expert Mr. Johan Carlson; (bottom, left to right): Director of the Africa Health Research Institute, Professor Willem Hanekom; Mexican psychologist Dr. María Elena Medina-Mora (screenshots from Zoom recording).

cognitive disorders. He further reported that 89% of countries in the WHO survey said that mental health was a part of their response to COVID-19, but only 17% provided additional funding. "Telehealth has been touted to be helpful to overcome the impact of the pandemic, but there are substantial disparities in uptake by countries," he added.

His recommendations, focused on SDGs 3, 10, and 17 under review in this HLPF and covered in this event, include that: (1) Universal health coverage must include mental health, with an increased share of spending by governments (up to 5%) with development banks and funders following suit; (2) The global mental health community must clearly communicate benefits of investment in mental health, e.g., increasing economic productivity and helping other development programs achieve SDGs; (3) Platforms for telehealth and e-mental health applications must be expanded; and, (4) Programs for disenfranchised and vulnerable groups must be emphszied, particularly "whole person" interventions that address social determinants of health and broader health disparities.

Focusing on the impact of the pandemic on adolescents and family well-being, the next speaker was Dr. Sherifa Noman Al-Emadi, a psychologist and family counselor at the Doha International Family Institute in the Kingdom of Qatar, and a member of the National

Committee for women, children, the elderly, and persons with disability. Dr. Al-Emadi has developed a couples counseling curriculum for Qatar and Saudi Arabia, and promoted family-friendly practices and programs throughout the region. Speaking from Qatar and representing the cosponsoring Mission of the Kingdom of Qatar to the UN, she noted that, like other countries, while Qatar's primary response to the pandemic prioritized physical health, soon afterward, mental health was addressed, in government policy as well as nongovernment practices and initiatives at the community level. Her Doha International Family Institute initiated three research projects, which showed both positive and negative effects of the COVID-19 pandemic on families' well-being in Qatar. In one study, 15% of respondents reported negative impacts of increased anxiety at home, isolation, and not being able to visit relatives, as well as fears and struggles with work and family balance. In comparison, four times as many cases (64%) confirmed positive impacts such as increased family cohesion and closeness due to more time spent together, and almost three times as many cases (42.5%) reported that flexible working arrangements improved the overall well-being of parents and enhanced their parenting tasks. In contrast, 28% said they were struggling to balance home schooling and family and work commitments.

Another study found similar results, leading her to conclude that all things considered, "families were getting stronger in a crisis." For example, adolescents reported increased support from mothers and fathers during the lockdown, as well as a decrease (from 25% to 15% from before to during the lockdown) of reports that "family members argue in a scary way." Despite some negative impacts, she reiterated, their research results show that the families in Qatar can get stronger in the time of crisis.

The next speaker, representing the cosponsoring Mission of Sweden to the UN, was Mr. Johan Carlson, =Director-General of the Public Health Agency of Sweden, and a WHO focal point for International Health Registrations, Communicable Diseases, NCDs, and Environmental Health, who chairs a forum of Swedish government agencies collaborating about the UN 2030 Agenda. He appropriately spoke on the topic of "*Health equity as a driver for and an outcome of sustainable development.*" He noted that the pandemic presents a window of opportunity to stress the importance of health issues. But "high ambitions and political statements are not enough" he said, instead, "firm technical support is needed, especially at the local and regional level." In that vein, the Swedish public health agency has launched a technical supportive structure for "surveillance of actions" focused not on health outcomes but on determinants of health (as Dr. Kelly addressed earlier in this event), which means that "things that need to be addressed in society on all levels in order to promote health for all." This approach of which he calls a "surveillance supportive structure for health" means that health disparities can only to a limited extent be handled within the health sector and can be addressed by efforts in other sectors. Thus, their "supportive structure" addresses targets in 44 areas that need to be reached in order to promote health and mental health, and to close existing health gaps and "leave no one behind," some of which are in the health sector but most of which are outside the health area, such as in education, gender equality, and employment. For example, he explained, if you achieve identified goals in schools or housing, better health will follow as a consequence.

He added another point that "local and regional authorities and civil society will intensify the work, report on those targets, compare achievements, benchmark the work, and get technical support and knowledge transfer from the national level." Thus, true interconnectivity of the SDGs becomes apparent, also on the local level, and "health targets can be reached by local level investments and achievement in other areas." Contact: Johan.carlson@folkhalsommyndigheten.se.

Turning to the issue of vaccine access and acceptance addressed by many of the Member States, Professor Willem Hanekom, director of the Africa Health Research Institute, speaking from South Africa, reiterated the theme of previous presentations about the vital importance of achieving equity in vaccine access, but added the importance of acceptance. To demonstrate this, he showed a photo of a Wembley Stadium soccer match with over 60,000 people not wearing masks. But inequity in access is still a pervasive problem, he said, evident in a heatmap showing that just 1.2% of the African population had been so far vaccinated in comparison to the over 60% of people vaccinated in Europe. He added that vaccination has two goals: prevention of the severity of disease, which it seems to accomplish, and also epidemic control, which is problematic when not all people in a population are vaccinated, and complicated by the emergence of new variants. He ended with hope that the COVAX facility will ultimately be successful to deliver vaccines globally.

Leading Mexican psychologist Dr. María Elena Medina-Mora, Dean of the School of Psychology at the National Autonomous University of Mexico and adviser to the Pan American Health Organization (PAHO), next spoke on "Health and mental health equity for Latin American migrant populations: Implications for programs and policy" emphasizing how psychology can help. The problem is vast, she said, in that the long-term still-uncontrolled pandemic has increased forced migration, conflict-induced displacement, and inequalities between migrants and local people, adding to the drastic delay in the achievement of the SDGs.

About 40 million people, 6% of the total regional population, are or have been in the process of migration—mainly to Colombia, the United States, and Mexico—due to poverty and violence in their mother countries, she noted. When arriving in Mexico, many migrants and refugees become victims of violence, abuse, discrimination, and criminal organizations, meeting the UNHCR criteria of a humanitarian crisis, of "coerced departure due to persecution or other forms of severe or irreparable harm where return is not possible or asylum is denied." Migrant transit and living conditions increase the risk of COVID-19, she said, with rates rising and many people waiting for vaccination. Additional problems are that many migrants require humanitarian care to meet everyday needs, especially children left behind after parents' deportation. According to WHO, 22% will develop mental disorders, which will likely persist, requiring that intervention must be not only be on an emergency basis, but must be maintained, as part of a long-term community care system.

Psychologists can help, she said, by providing "needs assessments, diagnosis, and evidence-based interventions to mitigate disparities, encourage well-being, and foster life skills, coping strategies and resilience." It is important, she emphasized, "to promote inclusivity, dignity and safety, learn from migrants' resilience, and contribute to the fulfilment of the SDGs." Importantly, she emphasized, "migrants and refugees need to be visible in local health programs."

Part 5 of the program

The next section of the program, entitled "Voices from the Field" presented innovations on-the-ground, with represntatives describing their projects and experiences in bringing resources to people "left furthest behind."

To introduce this, moderator Dr. Kuriansky presented a video in which people in the Congo and in Brazil, and on-the-ground professionals, shown in Fig. 7, expressed their feelings and

Fig. 7
People "on the ground" answering questions about COVID-19 from the video of "Voices from the Field" collected by project partners. Their responses (top row, left to right): "We don't want to hear anything about Coronavirus again because we are tired of suffering with things like these facemasks and washing hands all the time" and "I feel very upset and sad because a lot of Black and poor people depend on transport, bus subway and all". (bottom row, left to right): A man worries about the vaccine "It's going to cause serious side effects" and a doctor remarks about people, "We have to constantly remind them to not take it lightly and to take all the measures–maintain social distancing, use masks and sanitizers."

thoughts about the pandemic. The people interviewed responded to questions such as "Is COVID real?" (on a scale of 0–10 with 10 being the highest, the average response was 8.5); "What are your fears about COVID?" (a common answer: "It's going to cause serious side effects"); "Will you get the vaccine?" (people answered "yes" and "no," with one answer: "At first I was doubtful, but then I heard that the President of the U.S. did and I felt more confident"); "How has COVID-19 affected you or your community?" Answers to that question centered on "We can't get close to people". An example of an answer was: "We don't want to hear anything about coronavirus again because we are tired of suffering with things like these face masks and washing hands all the time". Also, several doctors interviewed mentioned variations of the answer, "We have to constantly remind the people to not take it lightly and to take all the measures".

After this video, innovators spoke about their programs on the ground to help the population in their country cope, shown in Fig. 8. Several of these are featured on the "Health in Your Hands" platform about innovative health solutions to provide healthcare to those in low resource remote settings with poor infrastructure (Kuriansky, 2024).

Ms. Jordan Levinson, project manager for Africa of Dimagi, spoke about "*Increasing Health Equity with Digital Solutions on the Frontlines of COVID-19 in Sierra Leone, Somalia, and Beyond.*" Dimagi, with offices in the USA, India, South Africa, and Ethiopia, is a social enterprise behind CommCare that provides innovative technology for frontline workers through digital toolkits used by governments, NGOs, and local programs in over 130 countries that leverages technology for equity in healthcare in the most difficult settings, backed by awareness of the situation and essential good data.

In Sierra Leone, determined to not repeat problems of the 2015 Ebola epidemic made worse by fractured response efforts and no central access to data, Dimagi became part of a consortium of public, private, and academic partners put together by the Directorate of Science, Technology, and Innovation, a government team, to roll out central technology for the pandemic response. Dimagi devised a custom mobile app to register quarantine sites in districts, track cases, and check services, like providing food, laboratory, or psychosocial support.

In Somalia, the Ministry of Health in partnership with WHO to tackle COVID-19 vaccination used Dimagi's "Comcare" App to systematize and aggregate information to facilitate a detailed real-time view of vaccine distribution (who's eligible, waiting, or needs follow-up) on an easy-to-use dashboard. Over 500 vaccination sites have used the app to record delivery of over 100,000 doses. The app was also deployed in Jamaica and will likely be launched in Burkina Faso and Guatemala.

Dimagi's products are useful not only for a specific emergency campaign, she explained, but for the whole system even beyond COVID-19, for example, for routine immunizations. As COVID-19 has exposed mental health needs—well covered by speakers in this event—Dimagi

Fig. 8
Speakers describing their "on the ground" innovative programs pictured next to them. Top to bottom: Dimagi's Jordan Levinson, iKure's Sujay Santra, and TPO Nepal's Dr. Kamal Gautam (screenshots from Zoom recording).

is partnering with world experts to leverage digital tools for mental healthcare in remote areas, with apps to guide frontline community health workers to identify common mental health conditions, make referrals, and even provide basic treatment, as well as to offer self-guided mental healthcare to those with no access to a clinician. Contact:

Next, innovative and comprehensive services for the population of Nepal, pivoting for COVID-19, were presented by psychiatrist Dr. Kamal Gautam, Executive Director and researcher of Transcultural Psychosocial Organization Nepal (TPO Nepal), speaking from Nepal. Dr. Gautam gave a general background overview of the mental health situation in Nepal, a developing country in Southeast Asia with a population of 30 million, where 90% of those suffering from any mental health conditions do not receive any treatment, a situation worsened due to the pandemic when the health systems was overwhelmed. This need led to an expansion of their program to develop a multisector and stakeholder program to provide psychosocial support in collaboration with government and nongovernment partners. Importantly, they provided technical assistance to the Ministry of Health and Population (MoHP) in drafting a COVID-19 mental health and psychosocial response and preparedness framework, with components of Psychological First Aid, stress management, and grief counseling for frontline health workers, to help themselves and the people seeking help at the health facilities. He described the impressively comprehensive and multifaceted aspects of TPO Nepal's response, described in detail in another chapter in this volume (Gautam & Koirala, 2024).

For capacity-building, they conducted trainings for community psychosocial and health workers as well as teachers, and police, and presented webinars for government health workers, service providers, and community people and leaders about the facts of COVID-19 and how to cope in daily life. In service delivery, they provided support services to people in 75 out of 77 districts in Nepal in individual and group in-person sessions as well as in quarantine and isolation centers; and offered free-of-cost tele-mental health helpline 7 days a week from 8 am to 6 pm.

Activities in research, monitoring, and evaluation include published articles on "*Self-reported psychological distress during the COVID-19 outbreak in Nepal: An online survey*" and "*Dissemination of findings from a study on assessment of suicide and self-harm during the COVID-19 pandemic in Nepal*" and "*COVID-19 Impact and Responses in low income and middle income countries: Reimagining Global Mental Health.*"

With regard to advocacy, efforts included seminars, meetings, and dissemination of research with local and provincial governments; advocacy with ministries about increased funding; translation and contextualization of global documents for use in Nepal; and technical support to the government of Nepal (MoHP) for preparation of policy documents and to finalize the National Mental Health Strategy Plan, to be scaled up in the coming five years. Further, Dr. Kuriansky was invited to present a resilience workshop to the TPO Nepal staff, which was well received and appreciated for being exceptionally culturally appropriate (Gautam & Koirala, 2024; Khanal & Kuriansky, 2024). All their activities contributed to the SDGs, Dr. Gautam announced proudly, particularly, target 3.4 (mental health and well-being), SDG 4 (education), SDG 5 (empowering women and girls), SGD 11 (building resilient communities), and SDG 17 (partnerships). Contact: kgautam@tponepal.org.np.

Next, the founder of iKure, Mr. Sujay Santra, speaking from India, explained their social enterprise that delivers primary healthcare, wellness, and prevention services to communities in India through a hybrid approach, escalated due to needs stimulated by the pandemic, to combine technology with trained frontline health workers visiting people where they live, to deliver in-person help for physical and mental health problems. iKure does this through a network of Hub and Spoke Clinics, addressing physical and psychological health needs of populations in rural, semirural, and urban areas. The program reached more than 12 million people in eight states in India and is present in nine countries in Africa and countries in the MENA region, including Jordan, reaching 2.5 million people. Of particular concern is infected patients who were being outcast by their community. Their programs uniquely add cultural components of music, art, poetry, and dance, which have been shown to appeal to the community people in building hope and resilience, as well as to the over 35 artists who participate in the program who also gain positivity during the pandemic times. Contact:

An innovative project in Africa was presented by Mr. Gordon Tapper, former chief of Section at the UN for 35 years and current president of United African Congress (UAC), a cosponsor of this event. He described a groundbreaking pilot project in a small mining village in the Democratic Republic of the Congo, consisting of a holistic intervention to help the community protect themselves from the spread of the COVID-19 infection while strengthening their psychological resilience. This population has been exposed long-term to ethnic segregation, military militias, endemic poverty, and even recovering from an Ebola outbreak, making them especially vulnerable. The project was a partnership of international NGOs led by Mr. Tapper and Dr. Kuriansky—the United African Congress UAC), Give Them A Hand Foundation, and the International Association of Applied Psychology (IAAP)— with a local partner, Hope and Health Vision (HHV) operating for years in the DRC and Burundi. The intervention addressed health and mental health pillars, as shown in Fig. 9. The lessons about health included debunking myths about the infection (e.g., that COVID-19 is not real, or that COVID-19 is a made up disease by wealthy companies who want to make money), and teaching healthy behaviors (hand-washing and mask-wearing). The psychosocial activities built resilience, empowerment and community connection, based on Dr. Kuriansky's toolbox of techniques used in disasters around the world (Kuriansky et al., 2016, 2017). These were confirmed as culturally appropriate by the local HHV team, especially since similar activities have been used by them in the local context for years. The project installed hand-washing stations and hired women to make masks and men to fill the water tanks, consistent with SDG 8 about promoting decent work. Significantly, two culturally adapted infographics with versions in English and in the local Swahili language were developed to visually demonstrate all the learning (sanitation methods and psychosocial activities), in order to overcome educational limitations of the

Fig. 9
Activities for physical heath (top row) and psychosocial support (bottom row) in the COVID-19 project in the DRC.

participants, and to spread the information widely by posting the flyers publicly (for examples, see the Supplementary Material available online at https://doi.org/10.1016/B978-0-443-18529-8.00010-X for English version). Significantly, as one of the first such evidence-based projects, results of extensive pre- and post assessments showed a statistically significant positive outcome for all the health behaviors and psychosocial activities that were the intended goals of the intervention. These results were also supported by testimonials from community leaders and the staff about the value of the intervention. Contact: gmtapper@gmail.com.

Part 6 (closing) of the program

In closing remarks, IAAP president and social psychology professor Dr. Christine Roland-Lévy, speaking from Paris, asserted that despite many challenges due to the pandemic, the many impressive civil society initiatives and government commitments shown in this event webinar "lead us to be hopeful in achieving equity in health and mental health that is so important for the recovery." Mutual collaboration and learning, as called for in SGD 17 about multistakeholder partnerships and as demonstrated today, she said, offers that hope.

She added that, "We have seen today that we have much to learn from each other," and thanked the impressive group of cosponsors from so many parts of the world, for such strong commitment to health and mental health equity and putting mental health on a par with physical health. A takeaway message, she said, is that "Physical and mental health equity along with well-being are crucial in order to achieve the UN SDGs." The support of so many governments and WHO who spoke at this event inspires those of us in civil society, she said, to continue our work. She concluded that she is convinced that the array of speakers today will help our world recover and build back better after COVID-19, with more fairness and justice that "leaves no one behind."

Conclusion and the way forward

Such events like this one demonstrate the well-deserved growing priority that mental health and equity are having on the international stage and at the United Nations at the highest levels, as well as the cooperation among multistakeholders from government and civil society, and the increasing number and value of innovations in the field that are advancing these issues. The event sets the stage for intensifying partnerships among these presenters and their respective organizations, and building on this foundation to further support policies and projects that progress health and mental health and put the discussed ideas into action.

* A version of only part of this report was published in the journal *Applied Psychology Around the World*, printed here with permission.

References

American Psychological Association. (n.d.). Equity flattens the curve. https://pages.apa.org/equity-flattens-the-curve/.

American Public Health Association. (2019). *Addressing the impacts of climate change on mental health and well-being*. Policy Number: 20196 https://www.apha.org/policies-and-advocacy/public-health-policy-statements/policy-database/2020/01/13/addressing-the-impacts-of-climate-change-on-mental-health-and-well-being.

Barry, E., & Kuriansky, J. (2024). Scoring for Resilient Health: Innovations using technology in physical activity and sport. In J. Kuriansky, & P. Kakkattil (Eds.), *Resilient health: Leveraging technology and social innovations to transform healthcare for COVID-19 recovery and beyond* (pp. 373–385). Elsevier.

Charlson, F., van Ommeren, M., Flaxman, A., Cornett, J., Whiteford, H., & Saxena, S. (2019). New WHO prevalence estimates of mental disorders in conflict settings: A systematic review and meta-analysis. *The Lancet*, *294*(10194), 240–248.

CSEM. Civil Society Engagement Mechanism for UHC 2030. (2020). *[Democratic Republic of the Congo] Partnering for Health Education and Psychosocial Support*. https://csemonline.net/uhc-day-drc-partnership/.

Cudy, E., Venator, J., & Reeves, R. V. (2015). *In a land of dollars: deep poverty and its consequences*. https://www.brookings.edu/research/in-a-land-of-dollars-deep-poverty-and-its-consequences/.

de la Fuente, J. R., Mendoza, S. P., & Guzik, A. (2024). An unconventional security threat: Mental Health at the United Nations Security Council and Mexico's Response. In J. Kuriansky, & P. Kakkattil (Eds.), *Resilient health: Leveraging technology and social innovations to transform healthcare for COVID-19 recovery and beyond* (pp. 475–480). Elsevier.

Ezoe, S. (2024). Universal Health Coverage and Japan: In memory of the late former Prime Minister Abe Shinzo of Japan. In J. Kuriansky, & P. Kakkattil (Eds.), *Resilient health: Leveraging technology and social innovations to transform healthcare for COVID-19 recovery and beyond* (pp. 37–43). Elsevier.

Gautam, K., & Koirala, P. (2024). Adapting and scaling up mental health and psychosocial support response through innovative interventions to transform mental health care in Nepal for COVID-19 recovery and beyond. In J. Kuriansky, & P. Kakkattil (Eds.), *Resilient health: Leveraging technology and social innovations to transform healthcare for COVID-19 recovery and beyond* (pp. 1029–1039). Elsevier.

Guterres, A. (2020). *Policy brief: COVID-19 and the need for action on mental health.* https://unsdg.un.org/sites/default/files/2020-05/UN-Policy-Brief-COVID-19-and-mental-health.pdf.

Inter-Agency Standing Committee. (n.d.) Reference group on mental health and psychosocial support in emergency settings. https://interagencystandingcommittee.org/mental-health-and-psychosocial-support-emergency-settings.

Jensen, N., Kelly, A. H., & Avendano, M. (2021). The COVID-19 pandemic underscores the need for an equity-focused global health agenda. *Humanities and Social Science Communications*, *8*, 15. https://doi.org/10.1057/s41599-020-00700-x.

Khanal, M., & Kuriansky, J. (2024). Promoting resilience in the Nepali community in response to the COVID-19 pandemic: A multistage multistakeholder partnership addressing the United Nations Agenda. In J. Kuriansky, & P. Kakkattil (Eds.), *Resilient health: Leveraging technology and social innovations to transform healthcare for COVID-19 recovery and beyond* (pp. 601–627). Elsevier.

Khayyam-Nekouei, Z., Neshatdoost, H., Yousefy, A., Sadeghi, M., & Manshaee, G. (2013). Psychological factors and coronary heart disease. *ARYA Atherosclerosis*, *9*(1), 102–111.

Krieger, N., & Sidney, S. (1996). Racial discrimination and blood pressure: The CARDIA study of young black and white adults. *American Journal of Public Health*, *86*(10), 1370–1378. https://doi.org/10.2105/ajph.86.10.1370.

Kuriansky, J. (Ed.). (2016). *The psychosocial aspects of a deadly epidemic: What Ebola has taught us about holistic healing*. Santa Barbara: ABC-Clio Publishing. http://www.abc-clio.com/abc-cliocorporate/product.aspx?pc=a4983c.

Kuriansky, J. (2024). "Health in Your Hands": A platform for innovative healthcare solutions. In J. Kuriansky, & P. Kakkattil (Eds.), *Resilient health: Leveraging technology and social innovations to transform healthcare for COVID-19 recovery and beyond* (pp. 989–1004). Elsevier.

Kuriansky, J., Margevich, A., Jean-Charles, W., & Daisey, R. (2017). Resilience and recovery in natural disasters and epidemics: Comparisons, challenges, and lessons learned from Train-the-Trainer projects. In G. Rich, & S. Sirikantraporn (Eds.), *Human strengths and resilience: Developmental, cross-cultural, and international perspectives*. Lanham, MD: Lexington Books.

Kuriansky, J., Polizer, Y., & Zinsou, J. (2016). Children and Ebola: A model resilience and empowerment training and workshop. In J. Kuriansky (Ed.), *The psychosocial aspects of a deadly epidemic: What Ebola has taught us about holistic healing*. Santa Barbara: ABC-CLIO/Praeger.

Meyer, I. H. (2003). Prejudice, social stress, and mental health in lesbian, gay, and bisexual populations: Conceptual issues and research evidence. *Psychological Bulletin*, *129*(5), 674–697. https://doi.org/10.1037/0033-2909.129.5.674.

Office of the High Commissioner for Human Rights. (1976). *International covenant on economic, cultural, and social rights*. https://www.ohchr.org/en/professionalinterest/pages/cescr.aspx.

The Economist Intelligence Unit. (2021). *Q1 global forecast 2021. Coronavirus vaccines: Expect delays*. https://www.eiu.com/n/campaigns/q1-global-forecast-2021/.

United Nations. (2015). *2030 Agenda for sustainable development*. https://sdgs.un.org/2030agenda.

United Nations. (2018). *Secretary-General's strategy on new technologies*. https://www.un.org/en/newtechnologies/.

United Nations. (2019). *Political declaration of the high-level meeting on universal health coverage*. https://undocs.org/en/A/RES/74/2.

United Nations. (2020). *Global humanitarian response plan: COVID-19*. Retrieved from https://www.unocha.org/sites/unocha/files/Global-Humanitarian-Response-Plan-COVID-19.pdf.

United Nations. (n.d.). Sustainable development goal 3: good health and well-being. https://www.un.org/sustainabledevelopment/health/.

United Nations Economic and Social Council for Asia and Pacific (UNESCAP). (2019). *Inequality in Asia and the Pacific. Chapter 4: Technology and inequalities*. https://www.unescap.org/sites/default/d8files/06Chapter4.pdf.

Utoikamanu, F. (n.d.). Closing the technology gap in least developed countries. UN Chronicle. https://www.un.org/en/chronicle/article/closing-technology-gap-least-developed-countries.

United Nations General Assembly. (2017). *Human Rights Council: Mental health and human rights: Report of the United Nations High Commissioner for Human Rights*. https://www.un.org/disabilities/documents/reports/ohchr/a_hrc_34_32_mental_health_and_human_rights_2017.pdf. (Accessed 10 March 2023).

Williams, R., & Drury, J. (2009). Psychosocial resilience and its influence on managing mass emergencies and disasters. *Psychiatry*, *8*(8), 293–296.

World Health Organization. (1946). *Constitution of the World Health Organization*. Retrieved from https://apps.who.int/gb/bd/PDF/bd47/EN/constitution-en.pdf?ua=1.

World Health Organization. (2013). *Mental health action plan: 2013–2020*. Retrieved from https://www.who.int/publications/i/item/9789241506021.

World Health Organization. (2018). *Health inequities and their causes*. https://www.who.int/news-room/facts-in-pictures/detail/health-inequities-and-their-causes.

World Health Organization. (2019). *Mental disorders*. https://www.who.int/news-room/fact-sheets/detail/mental-disorders.

World Health Organization. (2021). *Behavioural considerations for acceptance and uptake of COVID-19 vaccines*. https://www.who.int/publications/i/item/9789240016927.

World Health Organization. (n.d.). Mental health in the workplace. https://www.who.int/teams/mental-health-and-substance-use/mental-health-in-the-workplace.

World Health Organization Commission on Social Determinants of Health. (2008). *Closing the gap in a generation: health equity through action on the social determinants of health*. Retrieved from https://www.who.int/publications/i/item/9789241563703.

World Health Organization Regional Office for Europe. (2010). *How health systems can address health inequities linked to migration and ethnicity*. Copenhagen https://apps.who.int/iris/bitstream/handle/10665/345463/9789289002042-eng.pdf?sequence=1&isAllowed=y.

CHAPTER 11

Traditional medicine to strengthen resilience in healthcare in COVID-19 recovery and beyond☆

Geetha Krishnan Gopalakrishna Pillai

Traditional Complementary and Integrative Medicine Unit, Integrated Health Services Department, World Health Organisation, Geneva, Switzerland

Introduction

According to the World Health Organisation (WHO), traditional medical systems have a long history. "Traditional Medicine (TM) is the sum total of the knowledge, skill, and practices based on the theories, beliefs, and experiences indigenous to different cultures, whether explicable or not, used in the maintenance of health as well as in the prevention, diagnosis, improvement or treatment of physical and mental illness," as defined by the World Health Organization (WHO, 2013). Traditional and complementary medicine (T&CM) is an important health resource, the many applications of which are often underestimated and neglected. Following the Goals of the United Nations Agenda 2030 for Sustainable Development, many countries are taking steps to expand the coverage of essential health services. This is happening at a time consumer expectations are rising and costs are escalating, while resources are becoming scarce (WHO, 2019). Since the middle of the twentieth century, the world has seen growth in demand and uptake of traditional medicine (TM) and "many countries now recognize the need to develop a cohesive and integrative approach to health care that allows governments, health care practitioners and, most importantly, those who use health care services, to access T&CM in a safe, respectful, cost-efficient and effective manner" (Adler, 2001; Furnham & McGill, 2003; Gale, 2014; Kelner & Wellman, 2014; WHO, 2013). The WHO even proposes to its Member States that in order to "tackle the unique health challenges of the 21st century, they should integrate the best of T&CM and conventional medicine by looking at the areas they converge" (WHO, 2019). Dr. Tedros Adhanom Ghebreyesus, WHO director-general, adds, "In an ideal

☆ **Disclaimer:** The author alone is responsible for the views expressed in this article, and they do not necessarily represent the decisions, policy, or views of the World Health Organization.

Resilient Health. https://doi.org/10.1016/B978-0-443-18529-8.00011-1

world, traditional medicine would be an option offered by a well-functioning, people-centred health system that balances curative services with preventive care" (WHO, 2019).

Innovation—the action or process of developing a new idea, method, or product—is crucial for a knowledge system to remain relevant and contemporary. Traditional Medicine is no different. Nations have the responsibility to look at TM as a valuable resource they have inherited and which should be employed in the best manner possible to attain their aspired objectives, considering the following: enhanced resources and capabilities that TM can provide in attaining Sustainable Development Goal 3 (SDG 3: Good Health and Well-being) of the United Nations Agenda 2030, its possible implications in actions to build a sustainable society and environment, and the social implications in reducing inequality by making healthcare more accessible and economically available for all.

TM, as the WHO definition suggests, has long history, stretching millennia, especially regarding practices like Ayurveda, Siddha, Traditional Chinese Medicine (TCM), Unani, and Yoga. Nevertheless, the worldview established and maintained in most societies is the historical primacy of allopathic medicine, labeling the practices in traditions rooted in Western, biomedical, orthodox, scientific, conventional, mainstream, or cosmopolitan medicine as scientific (Pescosolido et al., 2020). Against this narrative, TM systems, representing alternative options for healthcare, are labeled as "complementary," unorthodox, local, parochial, traditional, indigenous, folk, irregular, holistic, fringe, unconventional, or natural medicine. The social and political forces that have shaped our world in the past six centuries and its methods of functioning have bestowed upon the allopathic medical system the benefits of conducive policy frameworks, regulations, access to resources, and patronage of the rulers that led to its status of professional dominance (Adler, 2001; Furnham & McGill, 2003; Gale, 2014; Kelner & Wellman, 2014; Lee, 1982).

Innovations and traditional medicine

TM systems strongly rely on their traditional roots for their popularity, acceptance, and uptake. Therefore it is pertinent to ask the question: Why should TM health systems innovate?

Innovations have been classified as normal and extraordinary by American science philosopher Thomas Kuhn (Kuhn, 1970). According to Kuhn, normal innovation or "normal science" involves micro-level puzzle solving or testing a new hypothesis, which leads to incremental growth of scientific knowledge.

> "The puzzle represents an artefact that makes sense only within a theoretical/cognitive framework of its domain specific methods and scientific language, "he said." By solving the puzzle, the innovator creates a meticulous manual of modalities that lead to the solution."

The puzzle and its innovative solution make sense only under a common conceptual, methodological, and normative baseline, which is not debated. In fact, the abandonment of critical debate on the basics makes possible scientific growth by increments and is a precondition for rigorous science as understood today (Sujatha, 2011). According to Kuhn, the critical discourse toward refashioning of basic frameworks constitutes extraordinary research. The critique is based on new observations or a response to new cognitive and technological tools. This often leads to the emergence of newer subject specializations (Kuhn, 1970).

As Kuhn said, "Even in case of extraordinary science, only the conceptual framework and the corresponding experimental designs may be altered or newly created, not the very theory of scientific testing and verification on which science is based." Kuhn attributes this to the need for identifying scientific progress (Kuhn, 1970). Thus continued existence of an accepted methodology of discovery and verification (like the laboratory method) is a precondition for innovation even under revolutionary change. Therefore innovation, whether normal and extraordinary, requires having a reasonably stable theory of discovery and verification of knowledge among its community of practitioners (Sujatha, 2011).

As old as TM systems are, they lost steam on innovation about half a millennia ago. The advent of a dominant narrative of science made up of a different epistemology and the colonization-oriented denial of patronage by governments led to the steep decline in the number of innovators and consequently innovations in TM systems. This then led to its decline in social relevance, partly owing to TM's ever-increasing gap with Western science mediated changes in the worldviews of society, related to health, science, knowledge, and innovation itself. However, innovations in TM have the possibilities of progress through normal and extraordinary methods.

Domain and scope of innovations in TM

Innovations in TM can lead to improvement in the TM systems, ultimately leading to its capability of providing healthcare. Alternatively, innovators can also use knowledge in the TM system to improve the components and delivery of conventional healthcare.

The scope for innovation in TM that could support health needs of the community is augmented by its alternate view of situations in health. TM has a top-down approach, where the view is from the whole to the part, essentially understanding the role and relevance of the part in the whole. This is opposite in approach to allopathic medicine where it follows the bottom-up approach, where you view the whole as the sum total of the parts. The TM approach to understanding health and disease is from function to structure, whereas the modern more Western approach considers that structure leads to function. TM accepts multiple possibilities for a given set of causes, in comparison to modern science which generally expects the cause to lead to one of two possibilities: an effect or no effect.

Even in the definition of health, the modern and TM systems have different and divergent views. According to WHO, which epitomizes modern perspectives of health, health is a state of complete physical, mental, and social well-being and not merely the absence of disease or infirmity (World Health Organization, 2020).

Thus health can be defined as a state of set parameters of the system, at a point in time. The individual is a passive spectator. In comparison, for Ayurveda, a prominent TM system which originated in India, health (called "Swasthya" in Sanskrit) is a state of optimal physical, functional, responsive, mental, and spiritual well-being maintained by a resilient person who, while continually interacting with the environment, society, and other living beings, carries on with the activities of daily living. Thus health is the ability of the system to maintain a state of set parameters, as a continuous process in time. It is simultaneously influenced by several variables, requires active effort from the individual to be maintained, and consists of a holistic approach to be understood.

This difference in Western versus traditional (more Eastern) approaches becomes more relevant in certain situations of clinical care, such as end-of-life care. Conventional hospital-based care often sees the person as a passive subject whose set parameters of life need to be maintained in all manner possible. Traditional medicine approaches to end-of-life care involve interventions to improve and strengthen the individual's system in its active efforts to regain health, at the same time being observant and sensitive to its active decision not to regain health. These differences in approach open up several avenues for new discoveries and incremental but holistic improvement of conventional medicine with the appropriate inclusion of TM.

Safety, quality, and efficacy of TM systems is a primary concern across nations and therein lies the largest scope for technology-based innovations in TM. Technologies, or new processes, which will ensure the safety, quality, and efficacy of practices and products of TM would greatly aid in the appropriate regulation and uptake of TM systems and support the entry of TM systems in hitherto new geographies and societies. For example, a handheld scanner, which can identify herbal ingredients of a mixture of herbs, or identify heavy metal presence with extreme sensitivity, will change the way in which TM products get accepted and utilized across the globe. Similarly, diagnostic devices specific for TM systems, such as pulse diagnosis, will change the way in which TM is practiced and would make it more visibly objective and generally more acceptable to the technology-savvy society. Using the TM systems' diagnostic parameters to identify vulnerable populations or finding patients before their diagnosable signs and symptoms in terms of allopathic system appear could improve health outcomes and reduce the cost and trauma for the patients. Shifting the policy objectives and implementation methods of health systems from a sickness-oriented allopathic perspective to a well-being oriented TM perspective would holistically support the achievement of the SDG goals and would be an example of extraordinary innovation in healthcare.

Medications

Many drugs used in conventional medicine have plant origins with knowledge directly linked to one or the other TM knowledge. Most TM medicines are multitargeting. With the pharmaceutical sciences aiming for multitarget drugs, TM is a treasure trove of human validated primary data. The scope of innovation in this domain is immense and the pharmaceutical industry has started tapping into it.

Access to its services and products is a major problem that TM faces across most countries. Innovations in making the services and products accessible according to community-specific needs and local regulations would be a major set of innovations that could happen for TM.

The WHO TM strategy

Safe self healthcare is a major objective of the WHO TM strategy. Another area where innovations can be focused is in technology-based TM knowledge, offering solutions for providing customized self-healthcare information to consumers. The popularity of the WHO mYoga app, a learning and practice mobile application for yoga, is an example.

Other areas of immediate opportunities for innovation in TM include specific areas in clinical care such as geriatrics, palliative care, sleep medicine, and pain management.

Additionally, the movement against antimicrobial resistance has its answer partly in the appropriate use of TM medicines. Many infections which are treated with antibiotics nowadays have the opportunity to be treated with immune-enhancing or antiinfective TM medicine and regimens.

The WHO Global Report on Traditional and Complementary Medicine 2019

Lack of credible data and reliable official information has often made it difficult to understand the landscape of traditional and complementary medicine (T&CM) globally as well as to comprehend how different nations approach this often-underestimated health resource. With contribution from 179 of its Member States, the WHO Global Report on Traditional and Complementary Medicine 2019 traces the global trend in T&CM over the past two decades, from 1999 to 2018. The report notes that as of 2018, 98 countries (more than 50% of the 194 Member States of WHO) had a national policy on T&CM, 109 Member States reported the presence of a legal or regulatory framework for T&CM, 107 countries (55%) of all Member States reported the presence of a national office for T&CM, 124 Member States (64%) responded that they had laws or regulations on herbal medicines, 170% or 88% of all Member States acknowledged the use of T&CM, and 78 Member States reported regulation of T&CM providers.

The report notes that lack of research data is the most prominent difficulty faced by the WHO Member States in regulating TM practice. This is followed by lack of financial support for research on TM. Correspondingly, Member States seek the following support from the WHO: general technical guidance for research and evaluation of TM, related to safety, quality, and efficacy; a mechanism for information sharing on regulatory issues; and provision for research databases (WHO, 2019). The recently established WHO Global Centre for Traditional Medicine is a step forward that the WHO has taken in this direction. Its outcome will rest on clarity of purpose, policies, resources, stakeholder engagement, and effective leadership.

Integrative medicine

Integrative medicine (IM), which refers to the appropriate use of two or more systems of medicine with the objective of better health outcomes, is based on innovation. In effect, IM is innovation on the go. It usually represents innovative collaborations between allopathic medicine and TM. The current advances of IM are largely based on positive or negative outcomes of innovative integrative steps that clinicians try out in their clinical practice. IM itself is an engine for innovation, which can draw a lot of support from data mining technology, data analytics, and technology that support the delivery of TM interventions in an integrative fashion.

All these areas stand out as possible avenues for innovation.

Critical debates about the basics of TM systems are rare but offer a big opportunity to understand the systems better and to help bridge the gap in communication between TM systems and the contemporary world. This would have far-reaching benefits for the community as well for the TM systems themselves.

Conclusion and looking to the future

Innovations in TM with the focused objective of supporting nations in achieving the SDG goals would do well if they focus on the areas of unmet needs in healthcare, as identified by individual Member States. It is pertinent that vital resources and time are utilized in a customized manner, keeping in perspective the needs for each nation, taking into account their strengths and weaknesses in health systems in general and TM in particular.

The most important aspect for success of innovative efforts for TM, and application of technology tools to its implementation, is the existence of appropriate regulatory mechanisms in a country. Without regulatory arbitration, neither will TM innovate, nor will its potential to exponentially support the community in healthcare be fulfilled. It would be advisable for nations who plan to achieve the 2030 UN Agenda and its SDGs to use their TM resources and to support this process by adequate regulatory frameworks and policies. This is another area of innovation in TM.

It is essential that nations achieve the UN Agenda 2030 SDG goals. It is therefore imperative that they use all resources available to them to achieve these goals with maximum efficiency of time and resources. TM is one such invaluable resource that nations possess and could be channeled to achieve specific objectives of SDG 3 about good health and well-being for all.

Evidence-based TM could be an important tool in ensuring well-being and reducing disease burden with better health and economic benefits. More specifically, it could address issues related to antimicrobial resistance, pain and palliative care, screening and prevention of NCDs, care of the aging population, and may be an effective tool to address new and emerging diseases, especially of viral origin. Nations may promote and encourage TM-based innovations for achieving specific goals in SDG 3. The glaring and obvious unmet needs of the community, which demand urgent attention, and which fall under the domain of well-being, such as chronic diseases, immunity, aging, and noncommunicable diseases, are the most preferable areas for focusing this effort.

Innovation presents a great opportunity for traditional medicine practitioners. As authorized holders of TM knowledge, they can recalibrate it to the needs of current and future societies. Embracing the path of innovation, they can bring about incremental changes that can augment the effectiveness of healthcare delivery. They can also make path-breaking "extraordinary innovations" and lend their weight to the movement for change, from a disease-centric approach to a health-centric approach in healthcare.

References

Adler, S. (2001). Integrating personal health belief systems: Patient-practitioner communication. *Healing Logics*, 115–126.

Furnham, A., & McGill, C. (2003). Medical students' attitudes about complementary and alternative medicine. *The Journal of Alternative & Complementary Medicine*, *9*(2), 275–284.

Gale, N. (2014). The sociology of traditional, complementary and alternative medicine. *Sociology Compass*, *8*(6), 805–822.

Kelner M., Wellman B., 2014. Complementary and alternative medicine: challenge and change. Routledge; May 12.

Kuhn, T. (1970). Logic of discovery of psychology of research? In I. Lakatos, & A. Musgrave (Eds.), *Criticism and the Growth of Knowledge* (pp. 1–24). Cambridge: Cambridge University Press.

Lee, R. P. (1982). Comparative studies of health care systems. *Social Science & Medicine*, *16*(6), 629–642.

Pescosolido, B. A., Manago, B., & Olafsdottir, S. (2020). The global use of diverse medical systems. *Social Science & Medicine*, *267*, 112721.

Sujatha, V. "Innovation within and between traditions: Dilemma of traditional medicine in contemporary India." Science, Technology and Society 16, no. 2 (July 2011): 191–213. https://doi.org/10.1177/097172181001600204.

WHO. (2013). *Traditional medicine strategy: 2014–2023*. Geneva: World Health Organization. https://apps.who.int/iris/handle/10665/92455.

WHO. (2019). *WHO Global Report on Traditional and Complementary Medicine 2019*. Geneva: World Health Organization. https://apps.who.int/iris/handle/10665/312342.

World Health Organization. (2020). *Basic documents* (49th ed.). Geneva: World Health Organization. [Internet] [cited 2022 May 29]. 238 p. Available from: https://apps.who.int/iris/handle/10665/339554.

CHAPTER 12

Multiomics technologies to increase resilience, adaptability, and affordability of integrated traditional and modern healthcare

Christian Garbe and Rahul Sharma
Frankfurt Innovation Center Biotechnology GmbH, Frankfurt am Main, Germany

Introduction

Multiomics is a relatively new term in the healthcare field as it is mainly the purview of biologists, chemists, and geneticists. However, this concept is relevant to help understand human health and disease, and therefore, to ensure resilience from the COVID-19 pandemic and going forward, to form resilient health systems. The suffix "-omics" refers to a field of study in life sciences that focuses on large-scale data/information such as proteomics, genomics, and metabolomics to understand life summed up in "omes" (e.g., genomes, proteomes). Many researchers in this field are working on ways to generate and analyze disease-related data (Hasin et al., 2017). The applications range from understanding host-pathogen interactions and infectious diseases (Aderem et al., 2011; Khan et al., 2019) to understanding better chronic and complex noncommunicable diseases (Yan et al., 2017) like cancer (Mouchtouris et al., 2022) and to improve personalized medicine (He et al., 2018).

Omics technologies, which include genomics, transcriptomics, proteomics, and metabolomics, are emerging and promising techniques which together with modern computational biology and molecular biology technologies serve as a rapidly evolving source capable of strengthening traditional medicine's output with effective products. These are important in the COVID-19 recovery era as these technologies will eventually lead to new drug discoveries as well as improved and scientifically validated solutions and advances for existing traditional medicine therapies as well as help increase the efficacy, adaptability, and affordability of an integrated approach to traditional and modern medicines, which will ultimately increase the resilience of communities.

Resilient Health. https://doi.org/10.1016/B978-0-443-18529-8.00012-3

Traditional medicine plays a crucial role in saving lives and improving health and is an important part of healthcare systems in many countries given that about 80% of the Asian and African population are reported to rely on traditional medicines for their basic healthcare needs (Oyebode et al., 2016). Though it is often underestimated, traditional medicine has a long history of use to maintain overall health and cure chronic diseases. Traditional medicine, also termed complementary medicine, refers to knowledge, skills, and practices based on theories, beliefs, and experiences of indigenous to different cultures, which are used for prevention, diagnosis, or treatment of physical and mental illness. It includes traditional practices of European, Chinese, Korean, Brazilian, African, Iranian, ancient Iranian, medieval Islamic, Siddha, Ayurveda medicine, as well as Unani, Muti, and Ifá. To strengthen this approach, the World Health Organization (WHO) Traditional Medicine Strategy 2014–2023 was developed.

Traditional healthcare systems have been around for centuries and have stood strong throughout the ages. In recent years, we at the Frankfurt Innovation Center Biotechnology achieved a successful track record in curating and analyzing omics data for the development of affordable and applicable medicines by merging these technologies with traditional medicine. In this chapter, we discuss how healthcare systems will evolve and adapt to new societal needs and innovative technologies over time. We focus on the domains of social diversity, health practitioners, and innovations in science and technologies and community associations. We also evaluate how the domains interfere and influence each other. We particularly discuss the challenges and issues of current healthcare systems, including decreasing financial reliance on native biodiversity resources, urbanization processes, acceptance of traditional medicine practitioners toward omics tools (genomics, transcriptomics, proteomics, metabolomics, metagenomics, high-throughput data analysis computational tools), and changes in social capital.

Social diversity

More efforts are needed to understand the social dimensions of education, integration of knowledge systems, networks, and power (Folke et al., 2010; Stone-Jovicich et al., 2018). Communities can present different types of ecosystems that require understanding for resilience and adaptation in different contexts. Resilience can be determined by an entity's ability to retain its functional integrity during a sudden change without undergoing structural changes (Walker et al., 2004). Core components for this type of system include intrinsic processes as well as external drivers. Local innovations come from those who live in these communities or through the socioeconomic groups associated with them, which become integral parts of their adaptive process, transforming livelihoods in times where crises occur.

Collective and individual capacities and the ecosystem play an important role in fostering resilience at the local level through behavioral capacities and self-organization. This understanding is crucial for developing a new plan of integrated approach to community resilience that takes into account these features.

A conventional healthcare system generally operates on the community level, where knowledge is gathered by transitional use and transmission of traditional practices. In this way, traditional healthcare practices are often based on culture-specific fundamental rules of faith, set of beliefs, and long-term experiences in order to prevent, diagnose, amend, or treat physical/mental illness (World Health Organization (WHO), 2008). Although theses traditional systems are designed to respond to the influence of internal/external factors, often these healthcare systems are rather reactive instead of proactive to external changes (Gómez-Baggethun et al., 2010). In traditional healthcare systems, established processes are often used without considering more agile and adaptive approaches, which consequently limits preparedness to adapt to environmental changes that take place within the larger ecosystem.

Traditional medicine plays an integral role in many communities' day-to-day lives. Traditional medicine is well positioned and "ingrained" in many cultures, ensuring that people feel safe and content through a peaceful mind-set and a solid sense of self-esteem. In order for such traditions to exist alongside complementary treatments, there needs to be a framework put in place by worldwide organizations, such as the World Health Organization (WHO).

Countries or territories are categorized according to their approach of traditional medicines and whether or not they integrate these approaches into their healthcare systems, in these three ways: (1) continuous development and implementation of integration policies, (2) mixed level integration, or (3) native traditional medicines practiced outside the national healthcare system.

These three levels of integration are exemplified in different countries. For example, China, Japan, and the Republic of Korea have developed policies for integration of traditional and alternative medicine into their national healthcare systems, reflected in (1) before. Stringent legal provisions regulating these practices have led to their prevalence throughout on a national level, with providing necessary education about their use to prevent mental and physical illnesses. These countries are recognizing and institutionalizing traditional and complementary medicines on national levels supported by strong political will and implemented legislation. Similarly, other countries like Malaysia or Cambodia have initiated healthcare reforms regarding traditional medicine. However, some work still needs to be done to establish a legal framework considering the variety of traditional practices in use.

Australia and New Zealand differ greatly in their approach to traditional medicine (World Health Organization, 2016, 2017). In Australia, "Traditional Medicine" is seen as an integral part of their cultural makeup, but external factors such as inability of organizations to cooperate with each other at a bureaucratic level often derail plans for this integration (Rolleston et al., 2016).

Our approach using the latest omics technology system represents a step forward to meet the core and basic needs of a society, particularly for aging populations or patients suffering from chronic diseases. We recommend that healthcare systems should be robust enough so that integrating traditional medicine with "more modern" Western medical treatments becomes possible for the most effective healthcare for a society and a country. Different levels of such integration correspond to challenges, which when overcome will lead to a unified and more effective healthcare model.

This system outlined before must be put in the context of primary healthcare (PHC), which is healthcare at its most basic level rather than specialized level for people initially seeking treatment; PHC is a crucial part of ensuring Universal Health Coverage (UHC), as outlined in the Political Declaration for Universal Health Coverage adopted by governments at the United Nations in 2019 (United Nations General Assembly, 2019). In many countries, native traditional medicines play a major role in achieving both PHC and UHC. Specifically, for patients living in remote areas with little or no access to modern medicines and facilities, traditional medicines are often the only source of treatment they can receive in time. Even though governments worldwide acknowledge the importance of these traditional medicines and the value of their integration into the modern healthcare systems, fundamental aspects remain to be considered, including quality, effectiveness, and accessibility of the traditional medicines, as well as nondiscrimination, accountability, transparency, sustainability of the service and of the medical practitioner, and most importantly, resilience of the society.

Quality and safety

Healthcare providers need to make sure that the quality and safety of healthcare services are paramount, so that patients can feel safe. PHC should be considered when integrating traditional medicine into a modern healthcare system. It is also beneficial to train both modern medicine practitioners and traditional medicine practitioners in order to integrate the relevant knowledge of each into the others' practices. This will allow traditional healers to gain knowledge of modern medicine and develop skillsets in order to work alongside modern health practitioners without sacrificing their cultural traditions. Similarly, modern medical practitioners can inculcate the centuries-old science of traditional medicine into their modern healthcare practice, to provide culturally relevant treatment options to patients and also access to healthcare services to all patients regardless of their beliefs, socioeconomic status, or background.

Improving healthcare systems through this approach will make them more sustainable and resilient in situations like a pandemic or any natural disaster. It is imperative to institute this integration into the healthcare systems in advance of any emergency; otherwise, responses can be delayed.

It is worthwhile to note how this integration approach we recommend happened during the recent COVID-19 pandemic. As one example, countries such as United States of America shared their modern medicine experiences with the rest of the world so that the other countries could be better prepared when the virus spread to their country. In this case, though, traditional medicines still needed to be integrated into that sharing, to be most effective for countries.

What happened in India better reflects our recommendation about an integrated approach. In India, in response to the COVID-19 pandemic, patients were treated using a combination of traditional remedies and modern medicine. It was found that patients treated using such an integrated combination of remedies recovered faster, without the use of steroids. In one study of an Ayurvedic herbal tablet, to understand its molecular mechanism in combating the COVID-19 infection, AI and multiomics techniques were used. These are details of the study:

- Potential phytochemicals were extracted and collected from the herbal plant species. Then, the scientific literature was scanned using machine learning-based natural language processing (NLP) data mining approaches. The small molecules were extracted from the scientific text, and then, chemical structures of those extracted compound names were downloaded from the public domain or manually generated.
- The generated chemical structures were used to perform molecular docking studies to evaluate potential interactions with key SARS-Cov-2 virus proteins, including main protease (Mpro), RNA-dependent RNA polymerase (RdRp), and spike proteins.
- The results showed that in silico analysis generated potential phytochemicals from those herbs of interest that showed strong interactions with the virus key proteins. These findings suggest the potential inhibitory mechanism of phytochemicals in blocking the bioactivities of key viral proteins.
- To support the results experimentally, multiomics data derived from comparison groups of patients with and without introducing the herbal tablet was generated.
- These results showed that multiomics data analysis with AI-derived graph-based methods, together with experimental assays, strengthened the hypotheses that these herbal tablets are working holistically, to reduce the viral load in the body and to control other linked disease pathways.
- Combining the initial clinical studies with our data analytical findings, the results suggest that the Ayurveda-based tablets could play a very crucial role in combating COVID-19 pandemic.

We, therefore, conclude that the use of advanced "omics" technologies, along with the traditional medicine and the modern healthcare practice, would lead to a more efficient and well-equipped healthcare system which then could save more lives.

Natural resources for traditional cures and treatments

Medicines derived from plants are often found to be effective in many diseases. Plant-derived medicines are known to have synergistic effects; thus, instead of using medicines extracted from a single plant, a combination of medicines extracted from multiple different plants is best used to treat diseases. Thus it is essential for traditional healers to have easy access to a variety of medicinal plants from which they can choose the most effective ones to use and to combine.

Evidence suggests that plants have been used for treatment of diseases as late as the 16th century (Petrovska, 2012). As described by Rashrash M and coauthors (Rashrash et al., 2017), as far back as the 1800s, US Pharmacopeia, a compendium of drug information, reported that about 59% of medicinal products were derived from herbs or herbal combinations (Swerdlow, 2000). Gaining knowledge about such medicinal plants, and their efficacy, is a necessity for the survival of traditional healthcare systems and their ability to create diverse treatment options in the face of environmental uncertainties. Vital prerequisites to the continuance of this system further require having enough local medical professionals and evaluating their skills. Rewarding and motivating traditional medicine experts will open up a new world of opportunities and innovation for resilient health systems in the COVID-19 recovery period and beyond, as is the subject of this chapter and of this book as a whole.

Natural resources are some of the world's most valuable and abundant sources for finding the most efficacious natural traditional cures and treatments. Scientists are increasingly experimenting with medications derived from natural products to combat diseases, which clearly needs further research. In these current times of focus on predictive, preventive, and personalized medicine, the importance of traditional medicine is increasingly important.

Our argument in this chapter is that "omics" and genomics present a big potential to integrate traditional medicine with modern medicine, leading to a truly individualized approach. Technologies like "omics," coupled with approaches such as cheminformatics also known as chemical informatics or chemoinformatics, involve applying computational techniques in the field of chemistry, working together with partners at various stages during drug development. Given the need for scientific evidence in order to integrate principles of traditional healthcare systems and practices into modern medicine, and to increase its acceptance in the Western world, better understanding of the molecular mechanisms of traditional medicine can be helpful. This research and understanding can contribute to fundamentally enhanced diagnostics and therapeutics safety, leading to new innovative solutions for various diseases.

In summary, we truly believe that "omics" sciences will serve as a rapidly evolving resource capable of strengthening traditional medicine's output with effective products, including new drug discoveries as well as improved versions of existing therapies.

Need for scientific evidence and areas of further research

With advances in sequencing techniques over the last decade, there has been an incredible volume of multiomics (genomics, transcriptomics, proteomics, metabolomics, metagenomics, etc.) data being generated every month (Hasin et al., 2017). With vast numbers of applications, multiomics data-focused projects have gained immense attention in the last several years, in academic and industrial sectors (Subramanian et al., 2020). Many healthcare-related projects are using transcriptomics datasets, e.g., Connectivity MAP (CMAP) (Subramanian et al., 2017), from samples who are treated and untreated, or diseased and healthy, in order to investigate gene expression changes within a certain state, to understand a diseased state, and to understand drug applicability at the molecular levels. Major aims of such studies are to identify novel druggable targets, discover new drugs that could interact with the potential targets, and to initiate a drug-repurposing project about on-shelf or orphan drugs and drug efficacy assessment (Le et al., 2021).

These technologies have not yet been utilized to their full capacity in the two areas of traditional medicine discoveries/development and in exploring detailed molecular mechanisms behind their efficacy. Traditional medicines usually work in a holistic way, by having an overall positive impact on the body, by interacting and controlling other triggered disease pathways (Yuan et al., 2016). Therefore it is extremely vital to take advantage of the benefits of advanced data analytical technologies, along with multiomics techniques, to contribute to the development of traditional medicines.

Artificial intelligence (AI) technologies have gained much attention in the healthcare sector, due to its high potential to solve key biological problems that are difficult to solve by the human brain (Topol, 2019). AI technologies can find hidden patterns in big dataset that are nearly impossible for a human eye to catch and do interpretations. AI has been implemented in a broader way in the healthcare sector from image recognitions for diagnostics (Hirasawa et al., 2018) to algorithmically design virtual compounds to enhance their efficacy for a certain disease (Perron et al., 2022). But, AI could play a very important role in underpinning certain biological pathways being impacted after introduction of traditional medicines into the human body. Further, implementation of machine learning (ML)-based graph theories on the multiomics data could address key biological questions linked to the molecular mode of action of traditional medicines.

Cheminformatics techniques are already very useful in finding compounds and their chemical structures; thus, further analysis of these compounds could discover their potential interactions with drug targets of interest (Egieyeh et al., 2019). State-of-the-art sequencing approaches which generate multiomics data—more specifically RNA-Seq, small-RNA-Seq, metagenome, proteome, and metabolomics—can be used to assess the potential molecular basis of the mode of action of traditional medicines. Such computational analyses, together with AI implementation, could lead to new discoveries with respect to the applicability of traditional medicines to other therapeutic areas.

Also, performing pathway enrichment analyses of differentially expressed genes offers an opportunity to investigate how compounds or small molecules interact with key proteins or biomolecules of disease pathways. This approach could be further tested on other disease areas to assess the holistic nature of traditional medicines. Using a multiomics and AI-based data analytical approach will create further possibilities for finding similar omics profiles for therapeutic target discovery, drug discovery, and drug-repositioning focused projects.

The path ahead

One of the most important difficulties in integrating modern and traditional medicines is achieving a balance between the two approaches, to allow them to fit within existing educational and practice systems. One approach to accomplish this would be by finding specific methods that highlight the holistic or spiritual aspects of traditional medicines. In this effort, various regulatory agencies could assist in improving this process, for example, WHO, which sets up cooperative networks and tasks among various countries so that they can learn about regulatory requirements from each other.

We maintain that certified and capable medical practitioners, who provide either modern or traditional treatment, provide safe medications to patients according to their own expertise; however, there are insufficient resources for conducting clinical studies of traditional medicines, leading to an absence of reliable statistics about the number of patients using traditional medicines and the efficacy of these traditional medicines. This makes policymakers unable to make informed decisions about integrating traditional medicines as a complement to their current modern system. Thus, organizing and publicly documenting datasets that screen and investigate these practices is a much-needed task in order to improve a country's holistic and resilient healthcare system. Creating this integrated approach should be a strategic goal of international health organizations like the WHO, with its international and local strategies, in collaboration with other stakeholders, including scientists who are knowledgeable about this integration and about "omics" described in this chapter. Given our expertise and access to the required technologies, our Frankfurt Innovation Center Biotechnology could valuably contribute to this development. Such efforts would be exceptionally useful to leverage current technology to reimagine and newly establish more resilient health systems in the COVID-19 recovery period and beyond.

References

Aderem, A., Adkins, J. N., Ansong, C., Galagan, J., Kaiser, S., Korth, M. J., Law, G. L., McDermott, J. G., & Proll, S. C. (2011). A systems biology approach to infectious disease research: Innovating the pathogen-host research paradigm. *MBio*, 2150-7511. *2*(1), e00325-10. https://doi.org/10.1128/mbio.00325-10. PMC3034460 21285433.

Egieyeh, S., Malan, S. F., & Christoffels, A. (2019). Cheminformatics techniques in antimalarial drug discovery and development from natural products 1: Basic concepts. *Physical Sciences Reviews*, *4*(7). https://doi.org/10.1515/psr-2018-0130.

Folke, C., Carpenter, S. R., Walker, B., Scheffer, M., Chapin, T., & Rockström, J. (2010). Resilience thinking: Integrating resilience, adaptability and transformability. *Ecology and Society*, *15*(4), 20. https://doi.org/10.5751/ES-03610-150420.

Gómez-Baggethun, E., Mingorría, S., Reyes-García, V., Calvet, L., & Montés, C. (2010). Traditional ecological knowledge trends in the transition to a market economy: Empirical study in the Doñana Natural Areas. *Conservation Biology*, *24*(3), 721–729. https://doi.org/10.1111/j.1523-1739.2009.01401.x.

Hasin, Y., Seldin, M., & Lusis, A. (2017). Multi-omics approaches to disease. *Genome Biology*, 1474-760X. *18*(1), 83. https://doi.org/10.1186/s13059-017-1215-1. PMC5418815 28476144.

He, F. Q., Ollert, M., Balling, R., Bode, S. F. N., & Delhalle, S. (2018). A roadmap towards personalized immunology. *NPJ Systems Biology and Applications*, 2056-7189. *4*(1), 9. https://doi.org/10.1038/s41540-017-0045-9. PMC5802799 29423275.

Hirasawa, T., et al. (2018). Application of artificial intelligence using a convolutional neural network for detecting gastric cancer in endoscopic images. *Gastric Cancer*, *21*(4). https://doi.org/10.1007/s10120-018-0793-2.

Khan, M. M., Ernst, O., Manes, N. P., Oyler, B. L., Fraser, I. D. C., Goodlett, D. R., & Nita-Lazar, A. (2019). Multi-omics strategies uncover host–pathogen interactions. *ACS Infectious Diseases*, 2373-8227. *5*(4), 493–505. https://doi.org/10.1021/acsinfecdis.9b00080. 30857388. S2CID 75137107.

Le, B. L., et al. (2021). Transcriptomics-based drug repositioning pipeline identifies therapeutic candidates for COVID-19. *Scientific Reports*, *11*(1). https://doi.org/10.1038/s41598-021-91625-1.

Mouchtouris, N., Smit, R. D., Piper, K., Prashant, G., Evans, J. J., & Karsy, M. (2022). A review of multiomics platforms in pituitary adenoma pathogenesis. *Frontiers in Bioscience (Landmark Edition)*, *27*(3), 77. https://doi.org/10.31083/j.fbl2703077. 35345309. S2CID 247560386.

Oyebode, O., Kandala, N.-B., Chilton, P. J., & Lilford, R. J. (2016). Use of traditional medicine in middle-income countries: A WHO-SAGE study. *Health Policy and Planning*, *31*(8), 984–991. https://doi.org/10.1093/heapol/czw022.

Perron, Q., et al. (2022). Deep generative models for ligand-based de novo design applied to multi-parametric optimization. *Journal of Computational Chemistry*, *43*(10). https://doi.org/10.1002/jcc.26826.

Petrovska, B. B. (2012). Historical review of medicinal plants' usage. *Pharmacognosy Reviews*, *6*(11), 1–5. https://doi.org/10.4103/0973-7847.95849. 22654398. PMCID: PMC3358962.

Rashrash, M., Schommer, J. C., & Brown, L. M. (2017). Prevalence and predictors of herbal medicine use among adults in the United States. *Journal of Patient Experience*, *4*(3), 108–113. https://doi.org/10.1177/2374373517706612. Epub 2017 Jun 5 28959715. PMCID: PMC5593261.

Rolleston, A. K., Doughty, R., & Poppe, K. (2016). Integration of Kaupapa Māori concepts in health research: A way forward for Māori cardiovascular health? *Journal of Primary Health Care*, *8*(1), 60–66. https://doi.org/10.1071/HC15034.

Stone-Jovicich, S., Goldstein, B. E., Brown, K., Plummer, R., & Olsson, P. (2018). Expanding the contribution of the social sciences to social-ecological resilience research. *Ecology and Society*, *23*(1), 41. https://doi.org/10.5751/ES-10008-230141.

Subramanian, I., Verma, S., Kumar, S., Jere, A., & Anamika, K. (2020). Multi-omics data integration, interpretation, and its application. *Bioinformatics and Biology Insights*, *14*. https://doi.org/10.1177/1177932219899051.

Subramanian, A., et al. (2017). A next generation connectivity map: L1000 platform and the first 1,000,000 profiles. *Cell*, *171*(6), 1437–1452. https://www.ncbi.nlm.nih.gov/pubmed/29195078.

Swerdlow, J. (2000). *Medicine changes: Late 19th to early 20th century. Nature's medicine: Plants that heal* (pp. 110–191). Washington, DC: National Geographic Society.

Topol, E. J. (2019). High-performance medicine: The convergence of human and artificial intelligence. *Nature Medicine*, *25*(1). https://doi.org/10.1038/s41591-018-0300-7.

United Nations General Assembly. (2019). *High-level meeting on "Universal health coverage: Moving together to build a healthier world"*. https://www.un.org/pga/73/wp-content/uploads/sites/53/2019/07/FINAL-draft-UHC-Political-Declaration.pdf.

Walker, B., Holling, C. S., Carpenter, S. R., & Kinzig, A. (2004). Resilience, adaptability and transformability in social-ecological systems. *Ecology and Society*, *9*(2), 5. https://doi.org/10.5751/ES-00650-090205.

World Health Organization. (2016). *Meeting report: In-house consultation on the 2016–2017 agenda for traditional medicine integration in the Western Pacific region*. Manila, Philippines: World Health Organization Regional Office for the Western Pacific.

World Health Organization. (2017). *Meeting report: Meeting on quality assurance and rational use of essential medicines in the Pacific Island countries*. Manila, Philippines: World Health Organization Regional Office for the Western Pacific.

World Health Organization (WHO). (2008). *Traditional medicine. Fact sheet N134*. Geneva, Switzerland: World Health Organization. [online] URL: http://apps.who.int/gb/ebwha/pdf_files/EB134/B134_24-en.pdf.

Yan, J., Risacher, S. L., Shen, L., & Saykin, A. J. (2017). Network approaches to systems biology analysis of complex disease: Integrative methods for multi-omics data. *Briefings in Bioinformatics*, 1467-5463. *19*(6), 1370–1381. https://doi.org/10.1093/bib/bbx066. PMC6454489 28679163.

Yuan, H., Ma, Q., Ye, L., & Piao, G. (2016). The traditional medicine and modern medicine from natural products. *Molecules*, *21*(5). https://doi.org/10.3390/molecules21050559.

CHAPTER 13

Advancing the digital transformation of health: The Global Digital Health Monitor

Erica Layer[a] and Patricia N. Mechael[a,b]
[a]HealthEnabled, Washington, DC, United States, [b]Johns Hopkins Bloomberg School of Public Health, Baltimore, MD, United States

Introduction

Health systems throughout the world are increasingly engaging in digital health transformation initiatives to increase access, quality, and equity to improve overall universal health coverage and health outcomes (Holst et al., 2020). This requires significant investment in both digital health interventions and enablers to support overall scale, reach, and sustainability for these investments. The COVID-19 pandemic highlighted the need for, and accelerated progress in, digital transformation within countries, across regions, and globally as demonstrated through global cooperation on vaccination certificates and disease surveillance.

While progress has been made, most digital health initiatives remain fragmented, uncoordinated, and unsustainable (Karamagi et al., 2022). There is limited guidance to support countries in prioritizing where to invest limited resources for digital health, and there is a lack of coordination among funding and technical assistance agencies to harmonize assistance where it is needed most. To facilitate this transformation, greater alignment is needed between digital transformation and digital health to stimulate the digital transformation of health. This requires greater investment in the known enablers for scale and sustainability of digital health interventions as well as more forward-looking policies to promote responsible AI, mitigate the growing digital divide, and facilitate the move toward person-centered health and the achievement of universal health coverage.

In this chapter, we introduce the Global Digital Health Monitor, an interactive web platform that enables countries to benchmark and track their progress in digital health maturity, systematically prioritize resources, and learn from other countries to accelerate the digital transformation of health.

Resilient Health. https://doi.org/10.1016/B978-0-443-18529-8.00013-5

The Global Digital Health Monitor

The Global Digital Health Monitor (GDHM) is an interactive digital global public good that helps countries to prioritize and monitor their digital health transformation, using the World Health Organization (WHO)/International Telecommunications Union (ITU) eHealth Strategy Toolkit as the underlying framework, aligned with the WHO Global Digital Health Strategy (World Health Organization, 2019; World Health Organization and International Telecommunication Union, 2012). Originally launched as the Global Digital Health Index and Maturity Model in 2018 and paused from 2020 to 2022, the resource underwent a review and redesign process to better align it to the changing landscape and emerging needs for a broadened focus on digital transformation of health.

The tool consists of seven key categories in the areas of leadership and governance; strategy and investment; legislation, policy, and compliance; workforce; standards and interoperability; infrastructure; services and applications; and cross-cutting issues including emerging technology and equity.

Intended to be used by Ministries of Health, private health and technology sector stakeholders, and international agencies, the GDHM enables countries and regions to benchmark and track progress and maturity over time, advocate for resources, identify countries that are performing well to facilitate cross-country learning, and support private sector entrepreneurs and investors to strategically prioritize new market entry and strategies.

Updates to the GDHM include an update to indicators to align with the WHO Global Strategy for Digital Health 2020–25 which is undergoing a midterm review, and to account for broader digital transformation, private sector engagement, emerging technologies such as AI, as well as diversity, equity, and inclusion—including a new subindicator on gender to address the widening gender digital divide.

Country-level engagement

From January to May 2023, HealthEnabled, in partnership with regional digital health networks including AeHIN (the Asia eHealth Information Network), RECAINSA (the Central American Health Informatics Network), and ISfTeH (the International Society for Telemedicine and eHealth), supported governments and nonstate actors to complete the GDHM for their countries. As of May 2023, 70 countries had participated, representing more than 35% of all countries globally, as presented in Table 1. Data can be viewed through the interactive web platform (https://digitalhealthmonitor.org/), which enables viewers to review data from individual countries, benchmark countries to regional or global averages, view year-over-year progressions, and export raw data for further analysis.

Table 1 List of countries, by region, that participated in the GDHM from January to May 2023.

Africa		Americas	Eastern Mediterranean	Europe	South-East Asia	Western Pacific
Benin	Liberia	Argentina	Afghanistan	Austria	Bangladesh	Cambodia
Burundi	Madagascar	Brazil	Iraq	Ireland	Indonesia	Japan
Cabo Verde	Malawi	Chile	Jordan	Macedonia	Maldives	Lao People's Democratic Republic
Cameroon	Mali	Colombia	Morocco	Portugal	Myanmar	Malaysia
Central African Republic	Mauritania	Costa Rica	Pakistan		Nepal	New Zealand
Chad	Mozambique	Dominican Republic	Saudi Arabia		Sri Lanka	Papua New Guinea
Comoros	Namibia	Guatemala	United Arab Emirates		Thailand	Philippines (The)
Côte d'Ivoire	Niger	Jamaica				Vietnam
Democratic Republic of Congo	Nigeria					
Eswatini	Rwanda					
Ethiopia	Sierra Leone					
Gabon	South Africa					
Ghana	Tanzania					
Guinea	Togo					
Guinea-Bissau	Uganda					
Kenya	Zambia					
	Zanzibar					
	Zimbabwe					

For countries that have not completed the GDHM, data was extracted from publicly available information through the GovTech Maturity Indicator (GTMI), the Network Readiness Index, and the GSMA Mobile Connectivity Index to prepopulate 7 of the 23 indicators described later to give a general sense of broader digital ecosystem maturity in alignment with relevant indicators.

The state of digital health

We conducted an analysis of 62 countries that had submitted GDHM data by 15 May 2023, to summarize the state of digital health and identify key insights. Each indicator within the GDHM contains five phases, with a rating of "5" being the highest state of maturity and a rating of "1" being the lowest.

Overall, most countries (45%) are at phase 3 maturity, followed by 26% at phase 2 and 24% at phase 4, as shown in Fig. 1. There are three countries (5%) in phase 5—Portugal, Saudi Arabia, and the United Arab Emirates—with no country at phase 1. Globally, across the seven components of the WHO/ITU eHealth Strategy Framework—namely Leadership and Governance; Strategy and Investment; Legislation, Policy, and Compliance; Workforce; Standards and Interoperability; Infrastructure; and Services and Applications—all components are in phase 3 except Workforce which is in phase 2, and Leadership and Governance which is in phase 4, as shown in Fig. 2.

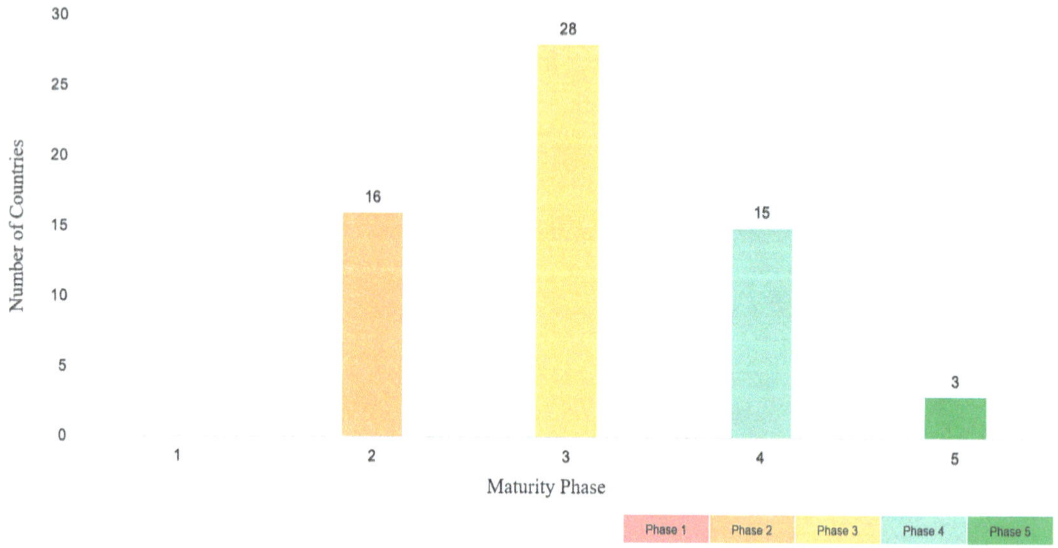

Fig. 1
Distribution of countries by overall GDHM maturity phase.

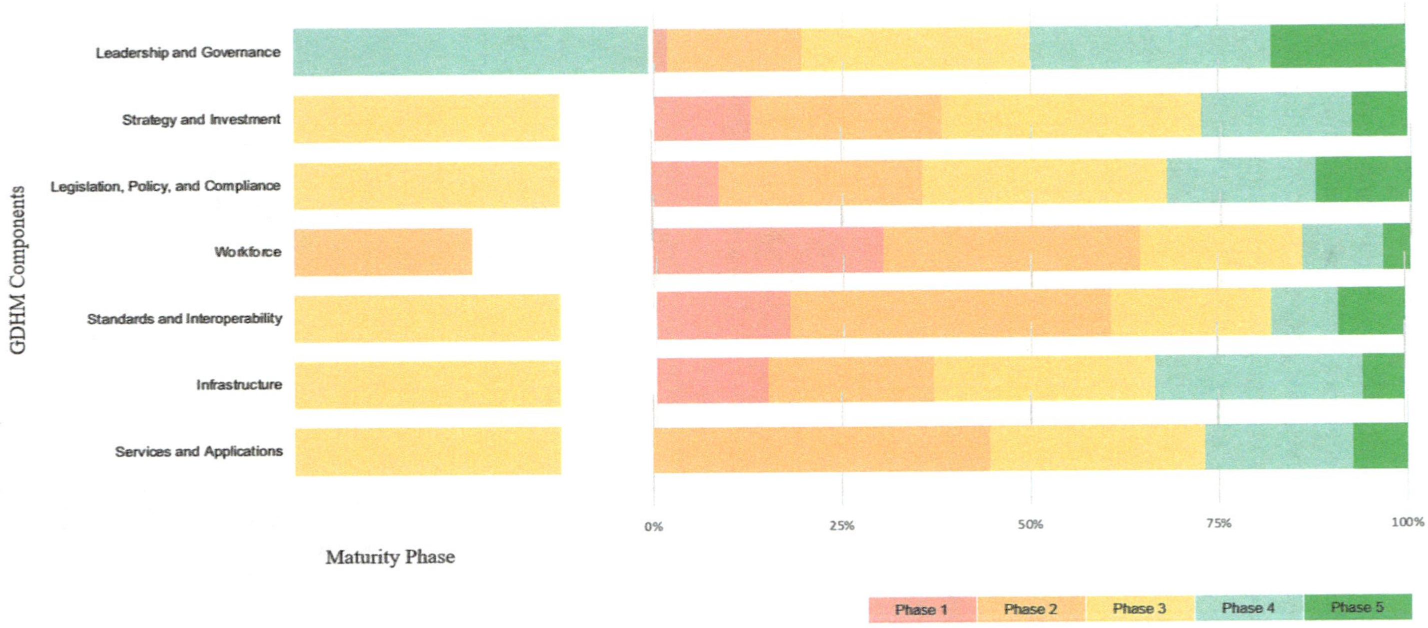

Fig. 2
Distribution of countries by WHO/ITU eHealth component.

Leadership and Governance

As shown in Fig. 3, digital health is included in national health strategies for 84% of participating countries with 15% of countries periodically evaluating and optimizing digital health in their national health strategies. Sixty-one percent of participating countries have formal governance structures in place for digital health, although only 42% are fully functional and government led. Only 45% of countries' national digital transformation and data governance policies systematically include health, with 72% of participating countries having a national digital health strategy or framework.

Strategy and Investment

As shown in Fig. 4, only 10% of countries reported having a structured and systematic budget line for digital health that is sufficient to meet the country's digital health needs and 70% of countries report having digital health strategies and programs that are developed and

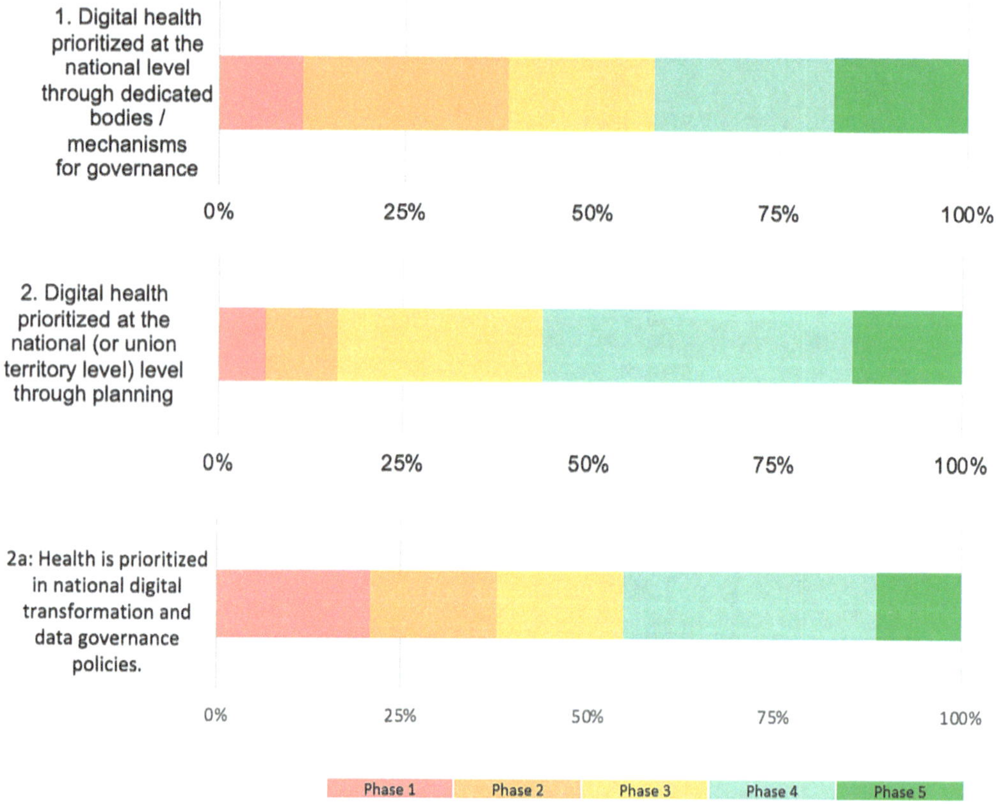

Fig. 3
Distribution of countries by indicator within the Leadership and Governance component.

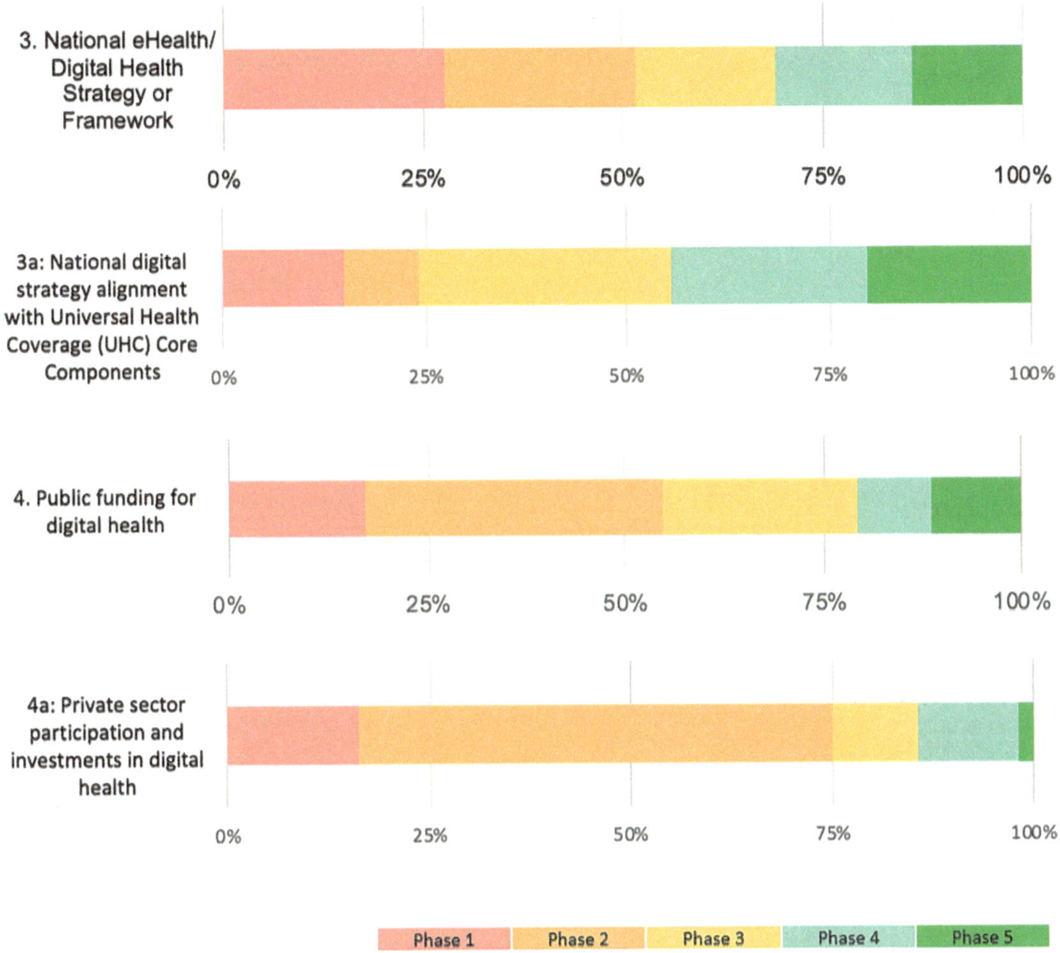

Fig. 4
Distribution of countries by indicator within the Strategy and Investment component.

implemented with consideration of equity and human rights implications and gender norms, roles, and relations. For 75% of participating countries, private sector participation is either nonexistent or ad hoc and limited.

Legislation, Policy, and Compliance

Country responses vary regarding a legal framework for data protection, but only 10% of countries have no law either proposed or passed on data security, and 20% of countries have no law either proposed or passed on data privacy, consent, confidentiality, and access to health information. Fifty-seven percent of countries have no protocols, policies, frameworks, or

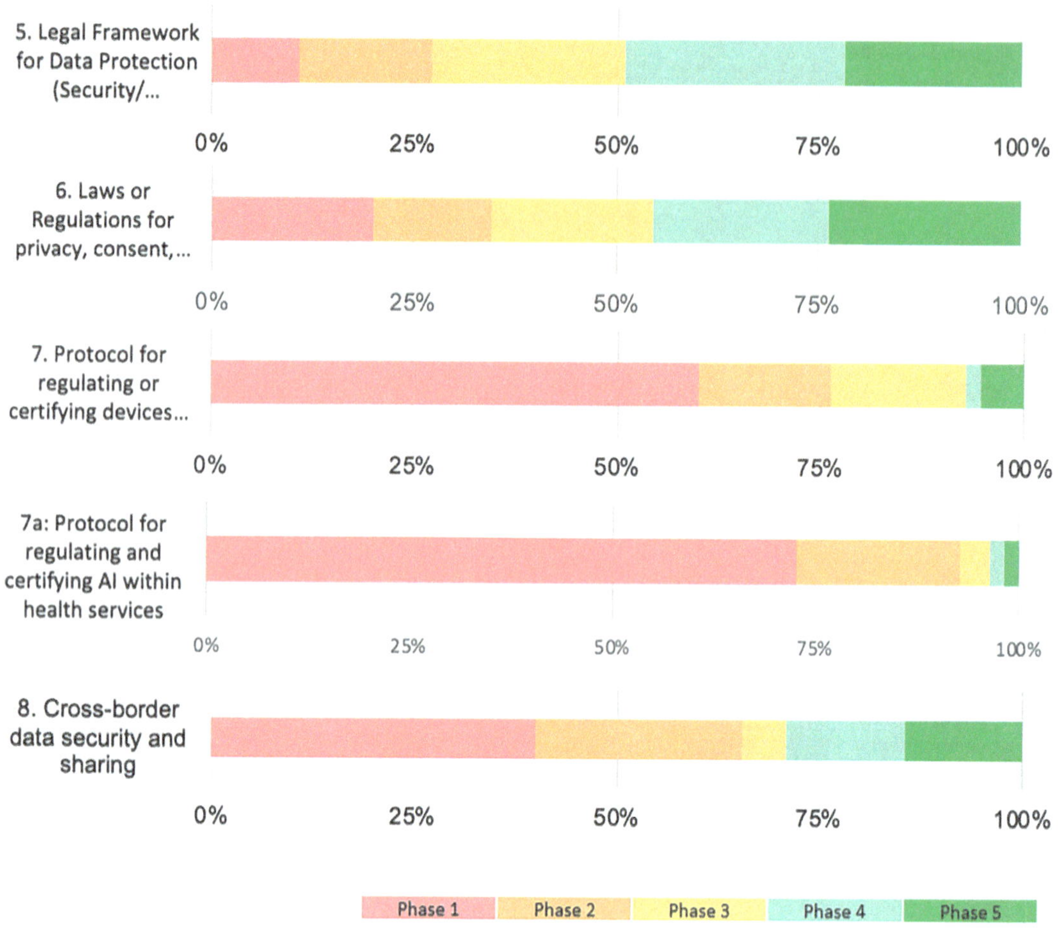

Fig. 5
Distribution of countries by indicator within the Legislation, Policy, and Compliance component.

accepted processes governing the clinical and patient care use of connected medical devices and health services. Seventy-three percent of countries have no protocols, policies, frameworks, or accepted processes governing AI use in health services. In addition, 43% of countries have no protocols, policies, or accepted processes in place to support secure cross-border data exchange and storage, as shown in Fig. 5.

Workforce

Workforce is the least mature component globally, with 48% of countries having no digital health curriculum for health professionals as part of preservice training requirements and only

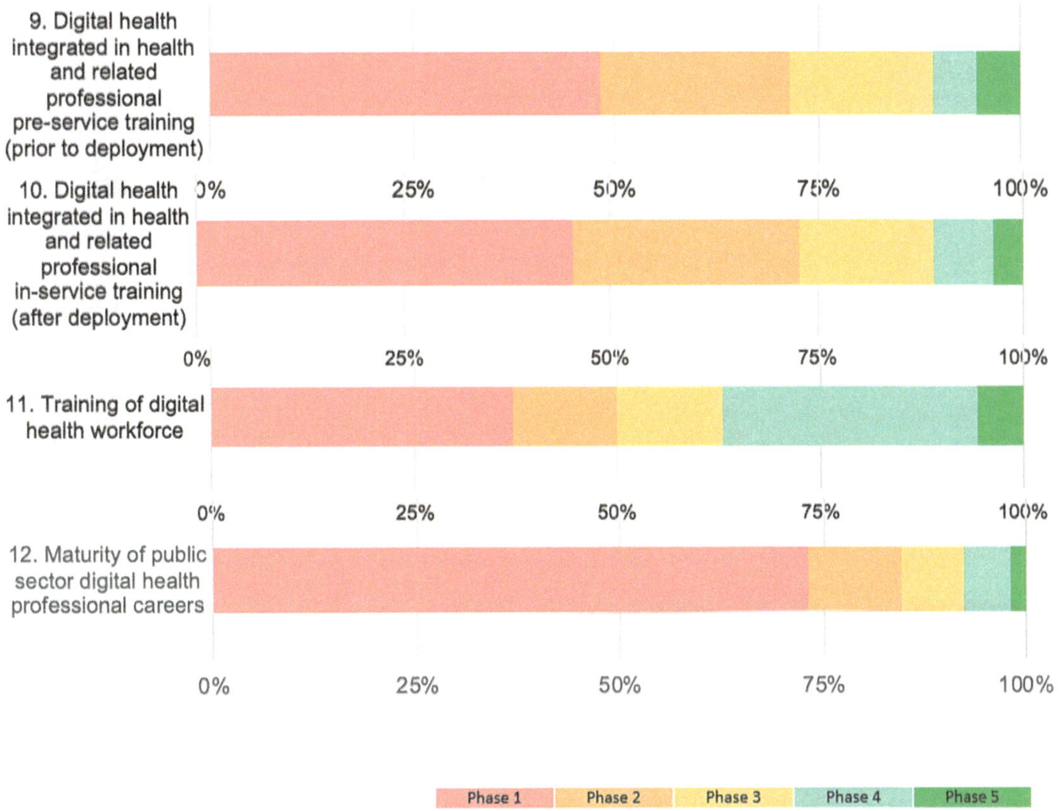

Fig. 6

Distribution of countries by indicator within the Workforce component.

13% of countries having digital health taught in relevant institutions. In addition, 38% of countries reported having no training for digital health workforce available in the country with 74% of countries having no workforce strategy, policy, or guide that recognizes digital health, as shown in Fig. 6.

Standards and Interoperability

In 43% of countries, a national digital health architecture and/or health information exchange (HIE) has been proposed, but not approved including semantic, syntactic, and organizational layers. Further, 62% of countries have no or few digital health/health information standards for data exchange, transmission, messaging, security, privacy, and hardware, as shown in Fig. 7.

Fig. 7
Distribution of countries by indicator within the Standards and Interoperability component.

Infrastructure

The analysis shows that 75% of countries scored 26–75 according to the Portulans Institute Network Readiness Index which utilizes scoring related to technology, people, governance, and impact. In addition, 73% of countries have a plan for supporting digital health infrastructure (including equipment such as computers/tablets/phones, supplies, software, devices) provision and maintenance. However, only 10% are countries in which digital health infrastructure is available, in use, and regularly maintained and upgraded in >75% of the public healthcare service sector, with strategies being implemented to close outstanding gaps in access/coverage and reach the whole population, as shown in Fig. 8.

Services and Applications

As shown in Fig. 9, in the area of Services and Applications, 93% of countries have digital systems at the district, community, or facility level that contribute to public health reporting and decision-making for population health management and 53% of countries reported having no

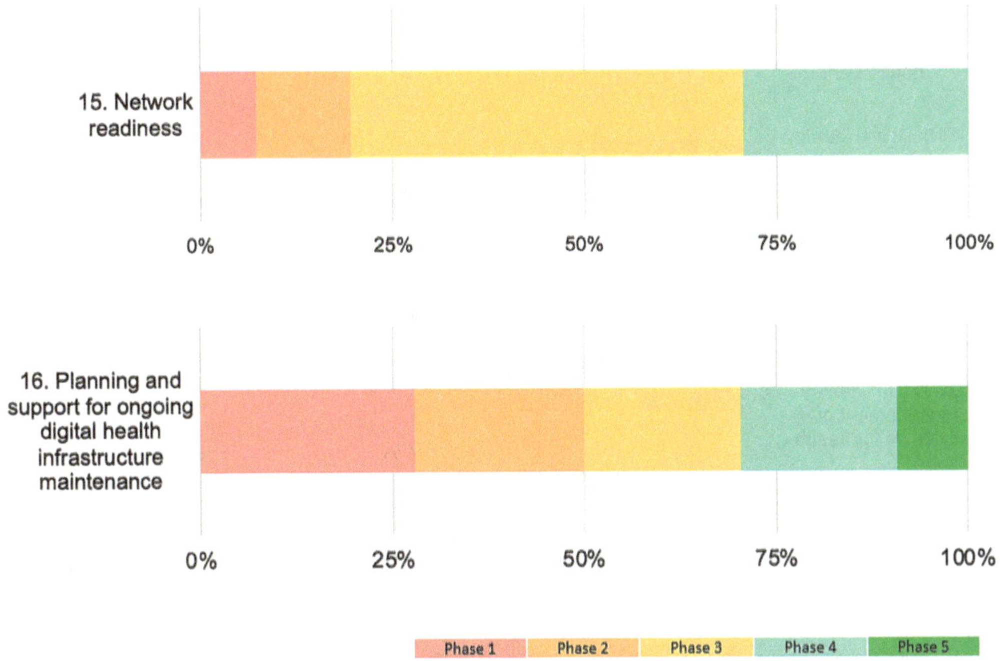

Fig. 8
Distribution of countries by indicator within the Infrastructure component.

secure patient feedback system. In addition, 63% of countries have no emerging technologies (e.g., AI, Wearables, Blockchain, IoT) plan in support of public health goals.

Key insights

As can be seen in the figures and text before, data from the GDHM provide rich insight into the state of digital health and highlight trends, areas for regional and global prioritization, and identify where tailored support is needed.

Globally, digital health maturity is at Phase 3, which indicates that some investment and progress is being made across most countries. Greater investment is needed globally to help countries reach full digital health maturity and potential, particularly the Eastern Mediterranean and Africa regions to bring their maturity in line with other regions. Demand is high for cross-country and regional learning between higher-maturity countries and lower-maturity countries. To support this, participation from higher-maturity countries is encouraged.

The data indicate that maturity in Leadership and Governance is relatively high, and investments should be sustained. There is a need for greater investment of public funding and a supportive environment for private sector engagement to further strengthen Strategy and

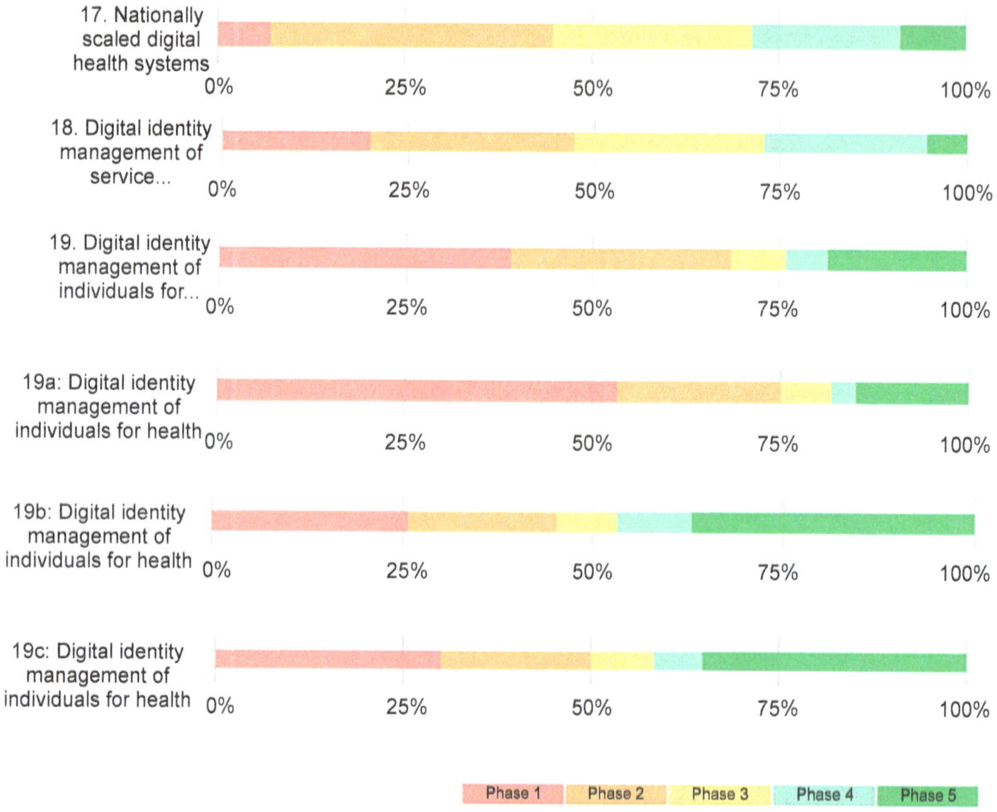

Fig. 9
Distribution of countries by indicator within the Services and Applications component.

Investment. Within Legislation, Policy, and Compliance, data protection and cybersecurity are relatively mature. Greater investment is needed for regulating and certifying connected health devices and AI as well as policies for cross-border data sharing. In Workforce, there is low maturity across most countries. Greater investment and standardization are needed in preservice and in-service training for health professionals, professionalization of digital health and career paths within the public sector, and gender representation within the digital health workforce and governance. In Standards and Interoperability, while most countries have data standards and architecture, greater investment is needed to operationalize and monitor them. For Infrastructure, most countries have a plan for supporting digital health infrastructure, and greater investment is needed to support implementation at scale. In Services and Applications, there is a need to invest in digital Civil Registration and Vital Statistics, patient feedback systems, and readiness for emerging technologies such as AI.

Future directions

In the past 5 years, the GDHM has matured into a well-known platform that is increasingly being used by countries to benchmark and track their maturity in digital health. Multilateral organizations such as The Global Fund, Gavi, the Bill and Melinda Gates Foundation, UNICEF, Africa CDC, and others are leveraging the intelligence from the GDHM for strategic decision-making. In May 2023, the World Health Organization announced their intention to adopt the GDHM as the official WHO digital health maturity model, which will provide a permanent home for the tool and ensure GDHM is updated annually by all countries to monitor their maturity in digital health. This is a major advancement for the platform which will help to institutionalize and sustain it, and an achievement that will help to standardize and accelerate digital transformation of health, leading to improved health and well-being for all.

References

Holst, C., Sukums, F., Radovanovic, D., Ngowi, B., Noll, J., & Winkler, A. S. (2020). Sub-Saharan Africa—the new breeding ground for global digital health. *The Lancet Digital Health*, 2(4), e160–e162. https://doi.org/10.1016/s2589-7500(20)30027-3.

Karamagi, H. C., Muneene, D., Droti, B., Jepchumba, V., Okeibunor, J. C., Nabyonga, J., Asamani, J. A., Traore, M., & Kipruto, H. (2022). eHealth or e-Chaos: The use of Digital Health Interventions for Health Systems Strengthening in sub-Saharan Africa over the last 10 years: A scoping review. *Journal of Global Health*, *12*. https://doi.org/10.7189/jogh.12.04090.

World Health Organization. (2019). *WHO guideline: Recommendations on digital interventions for health system strengthening*. World Health Organization. https://www.who.int/publications/i/item/9789241550505.

World Health Organization and International Telecommunication Union. (2012). *National eHealth Strategy Toolkit*. https://www.itu.int/dms_pub/itu-d/opb/str/D-STR-E_HEALTH.05-2012-PDF-E.pdf.

CHAPTER 14

Models for building resilient health communities: Strategies of systemic change to resolve a crisis and build back better

Lindsey Cunningham

Department of International & Comparative Education, Teachers College, Columbia University, New York, NY, United States

Introduction

Trauma is a relatively new concept in psychology but has existed in human history and literature. An early reference to trauma was in Shakespeare's play *Henry IV*, and then in 1666, a man named Samuel Pepys became the first person to write a clear account of posttraumatic stress symptoms after surviving the great fires of London (Bonanno, 2021). In the more recent past, the issue of extreme stress in posttraumatic stress disorder (PTSD) has become more common in both professional and public circles, which is integrally related to the concept of resilience, meaning the ability to recover, and build back better, from a crisis. A considerable amount of research and discussion in many fora has evolved on these topics.

When studying trauma, we must also study individuals who do not develop severe conditions such as posttraumatic stress syndrome (PTSD). In other words, we must understand those persons who experience mild distress and are resilient even after a crisis. Traditionally, PTSD has been thought of as a dichotomy, meaning that you either have it or you don't, but a more nuanced understanding of trauma and resilience suggests that there are multiple trajectories a person can follow after a potentially traumatic event, including those who are impaired for a long term or those able to cope and display resilience (Bonanno et al., 2011).

Resilience can be thought of as individual or community-based, when facing any trauma, including a personal assault or natural disaster, disease outbreak, as in the case of psychosocial resilience, whereby a person or community recovers and "bounces back" from the trauma. Interventions based on solid psychosocial principles and using culturally-appropriate techniques have been shown to considerably help rebuild strengths in the aftermath of such events for individuals and communities (Kuriansky, 2012, 2019, 2020; Kuriansky et al., 2017). In a holistic

Resilient Health. https://doi.org/10.1016/B978-0-443-18529-8.00014-7

approach, all stakeholders must collectively contribute to the recovery effort, to provide essential services and support to communities even on the front lines, especially when the healthcare system is fragile and under-resourced (Saulnier et al., 2022).

The focus of the webinar reported in this chapter, *From Crisis to Resilience*, was to define and provide examples of strategies to improve resilience of health system, particularly important in the post-COVID-19 period. As defined by the webinar moderator, Dr. Ronak B. Patel, MD, MPH, "A resilient health system is one that can withstand and adapt to those shocks or stresses and still deliver on its core services." This definition of health system resilience is consistent with other definitions in the field (Blanchet., 2015; Saulnier et al., 2022; WHO, 2022). In a more elaborated definition, according to the World Health Organization (WHO), "Building health systems resilience must therefore strengthen the capacity of health systems to forecast, prevent, detect, absorb, adapt and respond to a wide range of risks and shock events, while maintaining core functions and services and learning and improving as required" (WHO, 2022).

When understanding the definition of healthcare resilience, the important questions we should be asking, rasied in this webinar, are: "How do we create a resilient healthcare system?" and "What are the common characteristics of healthcare systems that can 'adapt to those shocks and stresses'"? The event, hosted by GOAL, a humanitarian response agency, presented three examples of organizations working to understand and strengthen resilient health systems. The first example was presented by the first speaker, Isabelle Bremaud, a resilience advisor at GOAL, about the MOMENTUM Integrated Health Resilience (MIHR) project. She explained the importance of understanding existing local structures before providing outside support. Second, Rana Qaimari, an educational program manager for EcoPeace Middle East, explained how creating a common understanding of community healthcare needs could be used to develop relationships between different groups. Finally, Dr. Joie Acosta, senior behavioral/social scientist at the RAND Corporation, presented a theoretical understanding of how to develop community resilience.

Event overview

This event was the second annual *From Crisis to Resilience* webinar hosted by GOAL. Founded in 1977 when former Irish Press journalist John O'Shea traveled to Calcutta, India, and observed dire conditions in which street children were living, GOAL is a humanitarian response agency focused on delivering programs related to Emergency Response, Food & Nutrition Security, Resilient Health, and Sustainable Livelihoods in over 60 countries (GOAL, 2019). Although GOAL focuses on the Global Food Insecurity Crisis, Global Migration Crisis, and Global Health Crisis, the latter was the focus of the *From Crisis to Resilience* webinar, focusing on contexts where local systems are unable to meet the needs of the community, particularly in fragile and conflict-affected contexts (GOAL, n.d.).

In their report outlining their strategic priorities for 2025, GOAL operates under the assumption that, even within fragile and conflict-affected contexts, communities have resources and

Fig. 1
Title Slide for the Webinar Series, *From Crisis to Resilience* (Patel et al., 2022).

structures in place (GOAL, n.d.). Their Crisis to Resilience Framework was developed to help guide programs that they implement in humanitarian settings to understand existing services, build relationships between local stakeholders, and assess potential solutions (GOAL, n.d.). The themes outlined in GOAL's strategic priorities were carried throughout all three of the presentations made during the webinar.

As shown in Fig. 1, the webinar entitled *From Crisis to Resilience: Emerging Challenges, Addressing Root Causes* was held on Wednesday, November 9, 2022, virtually over Zoom, from 11am to 12:30pm. It was the first in a two-part series. The first part, reported in this chapter, focused on Localization for Resilient Health Systems. Over 400 participants were in attendance. The second part of the series, which took place on December 7, 2022, addressed Inclusive Early Warning and Response Systems.

The webinar, sponsored by Irish Aid, was hosted by GOAL in partnership with RAND Corporation, the Harvard Humanitarian Initiative, and University College London. According to the Chief Executive Officer of GOAL, Siobhan Walsh, "The aim of the series is to explore the emerging lessons and the best practices for building resilience in fragile and conflict affect contexts."

The panelists discussed the following guiding set of questions:

1. What are the political and social determinants of health? What are the main challenges?
2. What learning is emerging in relation to the improvement of healthcare during adversity and building capacity to respond to health emergencies?

Fig. 2
Moderator Dr. Ronak B. Patel, MD, MPH introducing the topic of the webinar and presenting the panelists (Patel et al., 2022).

3. What approaches or innovations in systems thinking can be applied to build resilient health systems? (GOAL, 2022)

Fig. 2 shows the moderator of the event, Dr. Ronak B. Patel, MD, MPH, providing a brief overview of the topic and presenting the panelists. Patel is founder and director of the Urbanization and Resilience Program at The Harvard Humanitarian Initiative. His research focuses on vulnerable populations and how they are affected by humanitarian crises. Specifically, he works to develop tools and interventions to mitigate risks to health and development for vulnerable populations. Dr. Patel earned his Master's degree in Public Health (MPH) from Johns Hopkins Bloomberg School of Public Health and his MD at the University of California, San Francisco, and completed his residency at Brigham and Women's Hospital and Massachusetts General Hospital, a Harvard-Affiliated Emergency Medicine Residency Program (Patel, n.d.).

The Harvard Humanitarian Initiative (HHI) was established in 2005 in response to a growing need for technical expertise and training opportunities for professionals and graduate students in the field (Harvard Humanitarian Initiative, n.d.). An academic and research center based at the Harvard T.H. Chan School of Public Health and is affiliated with the Brigham and Women's Hospital Emergency Medical Department, HHI's mission is "to create new knowledge and advance evidence-based leadership in disasters and humanitarian crisis" (Harvard Humanitarian Initiative, n.d.). They produce various professional development courses including in-person and online options for professionals in the field and for Harvard students who hope to have a career in the humanitarian sector and participate in a wide variety of research projects including disaster preparedness, risk resilience, humanitarian negotiation, gender-based violence, program evaluation, and the use of ethical data (Harvard Humanitarian Initiative, 2022).

The first speaker in the panel discussion was Isabelle Bremaud, a resilience advisor at GOAL for MOMENTUM Integrated Health Resilience (MIHR). Although only at GOAL for five years, she has over 24 years of experience in the field working in Disaster Risk Reduction and Climate Change Adaptation and Resilience in various contexts, including Southeast Asia, Africa, Latin America, and the Caribbean (Bremaud, n.d.). She earned an MSc in Disaster Management from Coventry University in the United Kingdom.

MIHR is a USAID-funded project that has six implementing partners, including GOAL. Their mission is to "holistically strengthen quality voluntary family planning, reproductive health, and maternal, newborn, and child health." MIHR focuses on increasing the capacity of local organizations already operating in regions affected by crisis to better improve services provided to mothers and their children (MOMENTUM, 2021).

Bremaud's presentation focused on use of the Resilience for Social System (R4S) approach in South Sudan, a project being implemented under the MIHR program. The R4S program implemented the MIHR framework to assess and analyze the existing healthcare system in South Sudan so that the project team could decide the best strategy for improving the resilience of the system. Bremaud first provided a brief overview of MIHR along with USAID's definition of health resilience: "Health resilience is the ability of people, households, communities, health systems to prepare for, mitigate, adapt to, and recover from acute, recurring, and ongoing shocks and stresses in a manner that reduces disruptions to access and use of quality health care, and facilities equitable health and wellness outcomes."

Bremaud described the MIHR Health Resilience Framework, as shown in Fig. 3, to provide context for how the R4S project was implemented in South Sudan.

Fig. 3 shows the MIHR Health Resilience Framework, which is a visual representation of health resilience in a community. The box on the left side of the image shows an existing fragile context with the shocks and stresses that often compound the risks already associated with a fragile landscape. We find the target population, in this case women and children, living within those fragile contexts. That population is depicted in the second box as individuals, households, and communities, who all depend on existing structures and services that are supposed to provide services, specifically health services in this case. Health services, shown in pink color, can be separated into two categories: health service delivery and health system organization. The last box on the right depicts the potential health outcome, which reflects four pathways of bounce forward/better, bounce back, survival, or collapse. Which pathway a community takes

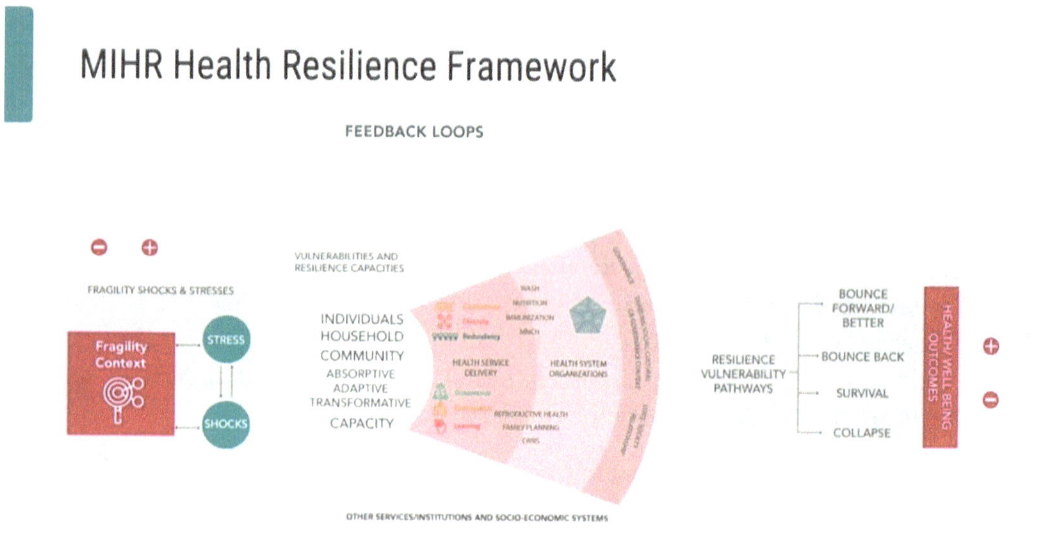

Fig. 3
Panelist Isabelle Bremaud presenting a visual depiction of the MIHR Health Resilience Framework (Patel et al., 2022).

depends on the capacity of the existing health services found within that community. Another way to think of this process is that the resiliency of the community depends on the capacity of the community to deliver services.

The MIHR Health Resilience Framework helps illustrate that when a community is impacted by shocks and stresses, the service delivery systems are impacted often disproportionately affecting the most vulnerable populations. The people most affected by crisis will likely show individual resilience but they will also need to be supported by supplemental resilient systems. Bremaud explained that "resilience of vulnerable populations to risks is also related to how well the critical social systems, like the health service delivery systems are functioning in both normal times and during and after crisis." This understanding is reminiscent of the definition of health system resilience discussed before. The R4S is a tool that Bremaud and her team use to assess the resilience of healthcare systems, in this case the healthcare system in South Sudan.

The main innovation of the R4S approach, Bremaud highlighted in her presentation, is the mapping tool. The mapping tool is used to help understand and analyze the existing structures of the service delivery system. While Bremaud focused on the example of South Sudan, she noted that this same process could be applied to myriad other contexts. She then showed a slide showing an example of how the R4S approach could be utilized to create a map of the healthcare system in South Sudan (Fig. 4).

Fig. 4 shows an example of a map of the Family Planning; Reproductive Health; and Maternal, Newborn, and Child Health service delivery system in South Sudan. Each line is color-coded

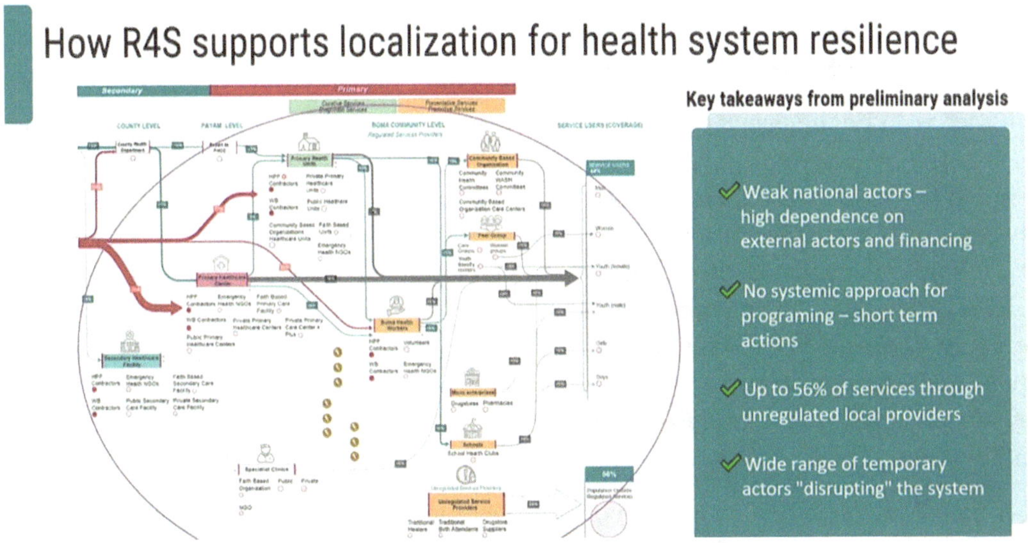

Fig. 4

Example of the current Family Planning; Reproductive Health; and Maternal, Newborn, and Child Health service delivery map in South Sudan (Patel et al., 2022).

based on the type of actor represented, and the thickness of the line indicates the strength of the inputs provided by each actor. The different groups included in the map are service users, the Ministry of Health, temporary providers such as the World Bank and UNICEF. Solid lines represent service delivery, dotted lines represent referrals between service providers, green lines represent the resourcing and capacity building from the Ministry of Health, and the pink colored lines show the resourcing and capacity building of the temporary actors.

The completed map is valuable because it includes all actors from the international community all the way down to the local levels and how stakeholders at different levels interact with each other. Bremaud explained that understanding the map is critical when designing interventions. Organizations must understand the existing actors and services before they can design a program to promote resilience. Once the existing system is understood, a risk analysis is performed, where Bremaud and her team run through scenarios of different shocks and stressors to model how the system would react. These scenarios help determine the resilience of the system. Based on these models, Bremaud and her team can begin designing a system which would improve resilience based on the gaps they found during the mapping process. Finally, Bremaud and her team can implement programs to fill those gaps in health services and improve the capacity of local actors.

The next presenter was Rana Qaimari, an educational program manager for EcoPeace Middle East in their Palestine office. She has 15 years of operational experience as well as experience as a program officer for the EU Erasmus office in Palestine. Previously, she served as the environmental coordinator for several youth environmental awareness programs. She earned a Master's in Water Science and Technology from Birzeit University in Palestine.

EcoPeace is an Environmental Peacebuilding organization operating out of Jordan, Palestine, and Israel. Its primary objective is to create peace in the region by promoting "cooperative efforts to protect our shared environment." In other words, despite all of the differences between Jordanians, Palestinians, and Israelis, they all share the same natural resources, and they will need to come together if they want to protect them. By protecting those natural resources, they will also be improving the health of the people living in the region. A top priority of EcoPeace is the water and sanitation crisis, with the goal of improving the clean potable health of the water as a staple of life. According to Qaimari, EcoPeace is unique because they are the only regional organization that has continued to operate during both peace and conflict times.

After providing a brief overview of EcoPeace, Qaimari described the "Water Cannot Wait" initiative. As the Israelis and Palestinians have unequal access to the water supply, in 1995, the Oslo Agreement specified that the Palestinians would need increased access to water within five years. In the 27 years since that agreement, the Palestinians still do not have equal access to clean water without pollution.

Qaimari's work is based on the idea that despite political conflict between Israel and Palestine, they share the same environment and must cooperate to achieve the shared goal of having access to clean, healthy water. She used the images shown in Fig. 5 to provide examples of polluted water sources that cross the border between Israel and Palestine.

Fig. 5 shows polluted water sources traveling across the Israeli/Palestinian border. The top left image is a sewage stream crossing from the Israeli side of the border into a Palestinian settlement. The bottom right image shows a Palestinian sewage stream crossing the border into Israeli land (Patel et al., 2022). Qaimari used the images depicted in Fig. 5 to provide examples of untreated water supplies that cross between Israel and Palestine. These rivers and streams are polluted on one side of the border and then used for farming on the other side of the border. She pointed out that borders are a human construct and natural resources do not know or care which side of the border they are on, saying, "We believe that borders know no boundaries. The river we have in Palestine goes to Israel."

EcoPeace uses both top-down and bottom-up approaches to tackle this issue of water pollution in an effort to engage all levels of the community. The top-down approach revolves around advocacy work. This means mobilizing financial resources through the private sector while also involving political structures to develop institutional structures that can help mitigate risks related to climate crises related to water, food, and energy insecurity. Qaimari emphasizes the need to advocate for a new equitable agreement related to access to water for all three countries

Fig. 5
The water on the Israeli/Palestinian border.

in the region. The bottom-up approach focuses on mobilizing residents in the affected communities. EcoPeace focuses on raising awareness about the importance of wastewater treatment and the need to find renewable energy sources. There is an emphasis on educating the next generation on the issues related to the environment.

The Gaza case provides an example of a successful project implemented by EcoPeace. The Gaza Strip is a small piece of land, approximately 40km by 10km, between Israel and Egypt. This area is home to approximately two million people and almost no humanitarian aid. There is a shortage of potable water and limited resources for treating wastewater. This wastewater flows untreated into the Mediterranean Sea, polluting the area for Gaza, Israel, and Egypt causing water-related health issues. Two possible solutions to the issue of access to clean water would be to run wastewater treatment plans or desalination plants. Both of these solutions would require a large amount of electricity, which is not accessible in Gaza.

EcoPeace employed their top-down and bottom-up approaches to help the people of Gaza gain access to clean water sources. It was determined that Gaza would need six wastewater treatment plants and three desalination plants to provide enough clean drinking water but they were struggling to collect the necessary materials to complete construction. EcoPeace leveraged technical and scientific expertise, political access, and community networks to convince Israel to provide the necessary building materials. By helping the Israelis understand that the untreated water from Gaza was ultimately ending up in the shared water of the Mediterranean

Sea, EcoPeace was able to convince Israel to aid the construction of a wastewater treatment plant in Gaza. The effort reflects successful strategy and advocacy for shared positive outcome for a resilient health system, even in the context of political discord.

The next speaker, Dr. Joie Acosta, works as a senior behavioral/social scientist at the RAND Corporation. She has a particular focus on substance abuse, sexual assault prevention, positive youth development, and mental health. Dr. Acosta's recent projects include transforming public health data, evaluating a positive youth development approach to improving school climate, and improving sexual assault prevention in the military. One of her most recent publications is titled, "Sustaining an Evidence-Based Program Over Time: Moderators of Sustainability and the Role of the Getting to Outcomes Implementation Support Intervention." During her time with RAND, Dr. Acosta has also won the Silver Medal and Mentor of the Year. Dr. Acosta earned her PhD in community and cultural psychology from University of Hawai'i (RAND, 2018).

RAND is a research institution with a mission "to help improve policy and decision-making through research and analysis." RAND operates in several locations including offices in the United States, the United Kingdom, Belgium, and Australia. In 2021 alone, RAND started approximately 800 new projects and published just under 1000 new reports (RAND, 2022).

In her presentation, distinct from the previous panelists who focused on specific projects, Dr. Acosta focused more broadly on a systems approach, with insights gained from RAND's work from the past 15 years. In recent years, the United States has seen an increase in natural disasters, issues related to economic downturn, political unrest, and community violence, and an increase in challenges related to climate change and population migration. In addition to these stressors, the United States deals with huge inequities where the top one percent of the country have 31% of the wealth while the bottom 50% only have three percent (The Federal Reserve, 2022). Dr. Acosta explains that resilient health systems need to be able to not only "bounce back" from acute stresses such as natural disasters, but also from chronic stressors such as poverty caused by these economic inequalities.

Traditionally, disaster relief has concentrated on preparedness. This type of response, she said, is short term, focused on single, specific events, with the goal is usually to build back the same. RAND's work in recent years has been to shift disaster relief from preparedness to resilience. This involves long-term, whole-community, and ongoing solutions focused on the strengths of the community and work toward sustainable development. As she said, "The work that RAND has done on resilience really sits at the intersection of wellbeing promotion and emergency preparedness."

The theoretical model developed by RAND addresses multiple layers of intervention—measures that matter, connected individuals, resilient leaders, and stronger systems—which will be further explained in this section. Fig. 6 provides a visual for the theoretical model used by RAND to build resilient systems.

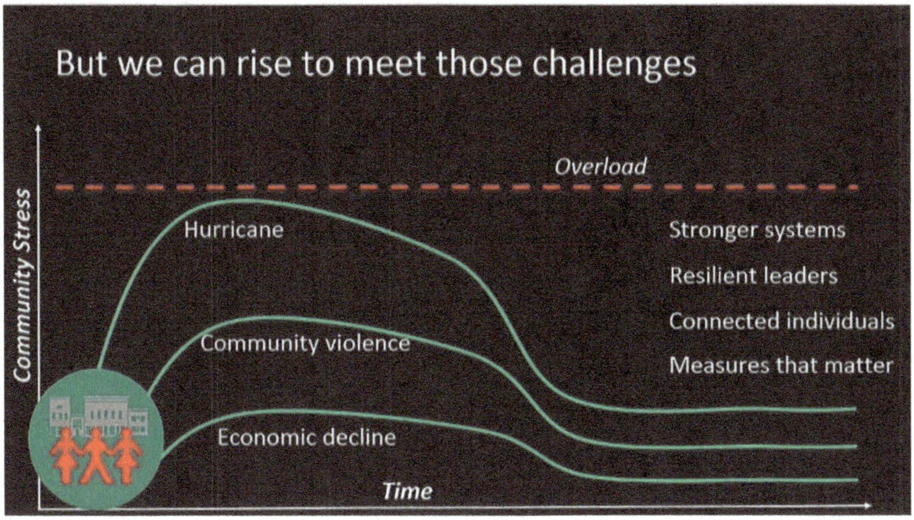

Fig. 6
The RAND model for building resilient systems.

Fig. 6. shows how RAND has worked to build resilience in communities that face multiple challenges. The green colored circle on the left represents the community. Each green colored line is a different stressor that a community can experience. The red colored dashed line represents the point when the community can no longer be resilient. The list of interventions on the right is provided by RAND's model (Patel et al., 2022).

Fig. 6 provides a visual example of what a community might look like under multiple layers of stress, leading to economic decline over years, compounded by community violence. Although the community is under stress, they are still capable of meeting their own needs. In this model, then, the community is struck by a hurricane which overloads their capacity, and they are no longer able to handle the stress. RAND's approach helps such communities meet these challenges.

The first approach in RAND's model for building resilient communities is to create stronger systems. This involves more coordinated and integrated efforts within the existing system. The team works to identify and eliminate failure points and identify priorities in the community. This strategy involves bringing together various stakeholder groups and implementing a bottom-up approach to building resilient communities.

The second approach is to build more resilient community leadership and community capacity. RAND has created various tools, workshops, and strategies that support community leaders. This approach involves advocacy and networking between all the different groups within the community. People need to be able to build rapport and trust with one another to build a strong, resilient community.

The third approach focuses on the resilience of individual community members. Dr. Acosta employed a metaphor at this point, whereby community systems can be thought of as the bones of the community and leaders are the muscles, then connected individuals can act as the connective tissue that makes the skeletal and musculature systems work together and makes forward movement possible. The team at RAND has focused on developing these connected individuals because of their central role in building resilience in their communities.

The fourth and final approach involves measuring the outcomes that have been created by the previous steps in the process. These modes of feedback ensure that the work that is being done is building back better communities in ways that make them stronger, more equitable, and more resilient.

Dr. Acosta closed with a quote by Henry Ford that she felt captures the work that RAND has been doing to build more resilient communities.

> Coming together is a beginning. Keeping together is progress. Working together is success.
> —Henry Ford

Connection to the UN sustainable development goals

This webinar, *From Crisis to Resilience*, relates to the UN Agenda 2030 for Sustainable Development with its 17 Sustainable Development Goals (SDGs) (United Nations, 2022). Specifically, the programs highlighted in this first part of the series touched on Goal 3: Good Health and Well-Being, as well as Goal 5: Gender Equality, Goal 6: Clean Water and Sanitation, and Goal 10: Reduced Inequalities.

The webinar specifically addressed resilience related to health systems (Goal 3) given the overarching understanding that the objectives of all of these programs in the webinar are meant to improve the health and well-being of the communities where each of these organizations operates. In addition, Isabelle Bremaud spoke about building more resilient health systems to provide better care to mothers and children, which can be directly linked to SDG 5 on Gender Equality. The presentation by Rana Qaimari focused on creating equitable clean water supplies in the Middle East region, which relates to SDG 6: Clean Water and Sanitation and SDG 10 on reducing inequities. Finally, Dr. Joie Acosta spoke about building resilience and building back better to more equitable communities, which is SDG 10: reduce inequalities.

Conclusions

In this post-COVID-19 recovery period, the importance of rebuilding resilient health systems cannot be underestimated. These presentations offered useful models and strategies for such achievement. All three panelists spoke about starting by assessing the needs and strengths of the local communities before providing potential solutions. They also spoke about leveraging existing community assets and building stronger connections between local actors. An important part of these projects is that they all are solutions that are examples that can be adapted in other contexts. Although providing only a brief view of the work being done, the examples show how development work can exist in an ethical and sustainable way to achieve resilient health systems.

References

Blanchet. (2015). Thinking shift on health systems: From blueprint health programmes towards resilience of health systems Comment on "Constraints to applying systems thinking concepts in health systems: A regional perspective from surveying stakeholders in Eastern Mediterranean countries.". *International Journal of Health Policy and Management*, *4*(5), 307–309. https://doi.org/10.15171/ijhpm.2015.49.

Bonanno, G. (2021). The Invention of PTSD. In *The End of Trauma* (pp. 21–40). Basic Books.

Bonanno, G. A., Westphal, M., & Mancini, A. D. (2011). Resilience to loss and potential trauma. *Annual Review of Clinical Psychology*, *7*(1), 511–535. https://doi.org/10.1146/annurev-clinpsy-032210-104526.

Bremaud, I. n.d. Isabelle Bremaud. LinkedIn. Retrieved November 15, 2022, from https://www.linkedin.com/in/isabelle-bremaud-04aa14b6/?originalSubdomain=ni.

GOAL (2019). GOAL Global|An international aid charity. GOAL Global. https://www.goalglobal.org/.

GOAL. (2022). *From crisis to resilience*. GOAL Global https://www.goalglobal.org/resilience/.

GOAL. (n.d.). Strategy 2025: From crisis to resilience. In GOAL. Retrieved November 13, 2022, from https://www.goalglobal.org/strategy2025/.

Harvard Humanitarian Initiative. (2022). *Harvard humanitarian initiative. Hhi.harvard.edu*. https://hhi.harvard.edu/.

Harvard Humanitarian Initiative. n.d. Who we are. Hhi.harvard.edu. Retrieved November 15, 2022, from https://hwpi.harvard.edu/humanitarianinitiative/who-we-are.

Kuriansky, J. (2012). Our communities: Healing after environmental disasters. In D. G. In Nemeth, R. B. Hamilton, & J. Kuriansky (Eds.), *Living in an environmentally traumatized world: Healing ourselves and our planet* (pp. 141–167). Santa Barbara, California: Praeger Press.

Kuriansky, J. (2019). Climate change and disaster recovery at the United Nations: Activities, achievements and contributions of the IAAP New York Team. *Applied Psychology Around the World*, *1*(3), 49–55. https://iaapsy.org/site/assets/files/1082/apaw_vol1i3.pdf.

Kuriansky, J. (2020). Psychosocial recovery after natural disaster: International advocacy, policy, and recommendations. Chapter 19. In K. E. Cherry, & A. Gibson (Eds.), *The intersection of trauma and disaster behavioral health*. New York, NY: Springer Nature.

Kuriansky, J., Margevich, A., Jean-Charles, W., & Daisey, R. (2017). Resilience and recovery in natural disasters and epidemics: Comparisons, challenges, and lessons learned from train-the-trainer projects. In G. Rich, & S. Sirikantraporn (Eds.), *Human strengths and resilience: Developmental, cross-cultural, and international perspectives*. Lanham, MD: Lexington Books. https://www.amazon.com/Human-Strengths-Resilience-Developmental-Cross-Cultural/dp/1498554830/ref=mt_hardcover?_encoding=UTF8&me.

MOMENTUM. (2021). *MOMENTUM integrated health resilience fact sheet*. USAID MOMENTUM. https://usaidmomentum.org/resource/mihr-fact-sheet/.

R.B. Patel (n.d.). Hhi.harvard.edu. Retrieved November 14, 2022, from https://hhi.harvard.edu/people/ronak-b-patel-md-mph.

Patel, R. B., Bremaud, I., Qaimari, R., & Acosta, J. D. (2022). *From crisis to resilience: Emerging challenges, addressing root causes [Webinar]*. GOAL https://www.goalglobal.org/resilience/.

RAND. (2018). *Joie D. Acosta*. Rand.org. https://www.rand.org/about/people/a/acosta_joie_d.html.

RAND. (2022). *RAND at a Glance*. Www.rand.org. https://www.rand.org/about/glance.html.

Saulnier, D., Thol, D., Por, I., Hanson, C., von Schreeb, J., & Alvesson, H. M. (2022). "We have a plan for that": A qualitative study of health system resilience through the perspective of health workers managing antenatal and childbirth services during floods in Cambodia. *BMJ Open, 12*(1), e054145. https://doi.org/10.1136/bmjopen-2021-054145.

The Federal Reserve. (2022). *The Fed—Table: Distribution of Household Wealth in the U.S. since 1989*. Www.federalreserve.gov. https://www.federalreserve.gov/releases/z1/dataviz/dfa/distribute/table/#quarter:129.

United Nations. (2022). *The 17 sustainable development goals*. United Nations. https://sdgs.un.org/goals.

WHO. (2022). *Health systems resilience toolkit: A WHO global public health good to support building and strengthening of sustainable health systems resilience in countries with various contexts*. In World Health Organization https://www.who.int/publications/i/item/9789240048751.

CHAPTER 15

Technology ethics in healthcare delivery and public health: Current and emerging issues from the COVID-19 era

David D. Luxton

Department of Psychiatry & Behavioral Sciences, University of Washington School of Medicine, Seattle, WA, United States; Luxton Labs, LLC, Olympia, WA, United States

Introduction

The SARS-CoV-2 (COVID-19) pandemic caused extraordinary and unprecedented health and mental health challenges across the world. There were an estimated 306 million persons infected worldwide by the end of 2021 (World Health Organization, n.d.), and hospitals were reported as running at full capacity during the peaks of the virus and subsequent variants, causing significant strain on healthcare systems worldwide. The mental health impact of the pandemic was also widespread and deserves much attention. Many people experienced stress and trauma from losing a loved one or community members to the disease, and governmental disease mitigation strategies, such as strict quarantine and lockdown measures, induced fear, disrupted social networks, and distressed economies (Brooks et al., 2020; Ganesan et al., 2021). Job loss, isolation, depression, anxiety, alcohol and substance use all increased during this time (Czeisler et al., 2020), creating a high demand for support services. Frontline healthcare workers also experienced an increase in mental health issues due to the intense workloads and anxiety they experienced in their work (De Kock et al., 2021). While the surge in mental health conditions was experienced in almost all countries, and by all demographic groups, certain groups, including younger adults, racial/ethnic minorities, essential workers, and unpaid adult caregivers, were especially impacted by disproportionately worse mental health outcomes, increased substance use, and elevated suicidal ideation (Czeisler et al., 2020).

With the pandemic came new opportunities and the necessity for technology solutions to provide mental health treatment; improve resiliency; and collect, analyze, and share public health information. The rapid deployment and integration of technological solutions into care

Resilient Health. https://doi.org/10.1016/B978-0-443-18529-8.00015-9

were driven by the need to meet the high demand for health and mental health services, reduce the risk of virus transmission through contactless interventions, share relevant health information, and for rapid data collection for disease surveillance. For example, contact tracing via mobile devices was used heavily to mitigate the spread of the disease, while advanced data science strategies were used to identify trends and potential outbreaks of the disease (Ye, 2020).

The widespread adoption of technological solutions by healthcare providers, public health organizations, and the public during the pandemic also emphasized existing and emerging ethics issues, which deserve to be carefully considered. Thus this chapter deliberates on these ethics issues by highlighting their relevance and importance, using practical examples. To accomplish this aim, I discuss relevant issues and their applicability, using the example of three prominent technologies used during the pandemic for direct care services and public health information gathering and sharing: telehealth (videoconferencing), mobile technologies (smartphones and apps), and conversational agents (e.g., virtual care providers and chatbots).

I begin with a brief overview of these technologies and their relevance during the pandemic. I then discuss the prominent ethics considerations regarding the use of these technologies. Finally, I discuss several of the larger ethics concerns associated with technologies in healthcare and public health, and I conclude with recommendations for guiding the ethical and effective use of technology to address future public health crises.

Technologies overview

Telehealth videoconferencing

One of the most significant and widespread applications of technology during the pandemic was the use of videoconferencing for telehealth services. Telehealth entails the use of electronic information and telecommunication technologies to facilitate services over a distance, including clinical applications, such as direct patient care and consultation, remote monitoring, and clinical assessments and testing, as well as nonclinical applications, such as training, education, and supervision (Luxton et al., 2016). Telehealth may involve numerous technologies, primarily video-conferencing, but also telephone, mobile apps, email, and messaging services.

The pandemic necessitated the need for contactless encounters across settings and healthcare services, from primary care, mental health counseling, and assessments, including forensic evaluations in jails and hospitals (Luxton, 2020a, 2020b). This caused healthcare providers and other professionals to rapidly retool and ramp-up telehealth services, necessitating the need for training on ethics and legal requirements regarding telehealth and associated technologies.

Mobile phones and apps

Mobile phones have become a prominent platform for how people access health-related information. Mobile devices, and the apps that run on them, are also used for delivering medical and behavioral treatments and interventions. They also make ideal health-related data collection platforms, given their mobility and onboard technology, such as geolocation and sensor technology (Luxton, 2017).

Mobile apps played a major role in contact tracing strategies during the pandemic given their capability to track location and proximity to other persons. Mobile phones were also used to identify vaccination status and for sharing information about COVID-19. They are also ideal for contactless screening for health conditions and disease exposure risk. Other uses during the pandemic included patient education and information (e.g., hospital appointments, visiting hours, self-isolation monitoring, and general COVID-19 information and education) (Timmers et al., 2020). For healthcare professionals, they were used to provide information about how to work with patients with COVID-19.

Conversational agents (virtual providers)

The use of conversational agents (CAs) for collecting and sharing information also saw an emergence during the recent pandemic (McKillop et al., 2021). Conversational Agents (CAs) are software programs that simulate human conversation ranging from simple text-interface chatbots to more sophisticated and life-like Intelligent Virtual Agents (IVAs). IVAs are computer-generated, animated, artificial intelligent virtual characters that can be designed with the visual characteristics of humans, animals, or other forms. While basic CAs may use a simple text input and response format, more sophisticated conversational agents employ machine learning and natural language processing (NLP) to understand questions in everyday language and engage users in progressively more complex conversations. They are also increasingly capable of affective interaction through the expression of emotions and recognition of human emotions (Hudlicka, 2016; Luxton & Hudlicka, 2021). Conversational agents are most commonly accessed via the internet, while others can be accessed on mobile devices (i.e., smartphones, tablet computers, and smartwatches). They are also built into gameplay in computer games and virtual reality simulations and they can be embedded within physical robots to make them more personable and interactive (Hudlicka, 2016; Luxton & Riek, 2019).

CAs are used for screening for health conditions and disease exposure risk that would otherwise be conducted via some combination of self-assessment surveys, healthcare provider screening tools, and interviews by clinicians or public health professionals. CAs provide a high-tech opportunity to automate the collection of health-related data, reduce the need for medical provider time, expedite patient wait times, and improve decision-making regarding medical triage and diagnosis (Luxton & Sirotin, 2020). Unlike human healthcare professionals, CAs can be accessed at any time, and almost anywhere. They are also potentially more reliable than humans because they are not susceptible to fatigue and burnout, and some data suggest that patients may prefer to share health information with a CA more than they would with a human care provider (Lucas et al., 2014; Luxton, 2020a).

Ethics issues

Informed consent

A central ethical requirement in healthcare treatment, research, and data collection is informed consent—the process whereby a healthcare service provider discloses appropriate information (i.e., procedures, risks, benefits, etc.) to a client/patient so that the patient/client may make a voluntary choice to accept or refuse participation in the treatment or service. Important components of informed consent include ethical obligations to provide information about their rights as a patient and the service, support patient autonomy in decision-making, and avoid unethical forms of bias. The consent process allows persons to be informed of the risks and benefits of any given treatment or intervention, allowing them to make the best decision.

The informed consent process is one of the core telehealth training areas because of expectations regarding discussion of procedures with patients/clients including duty-to-warn requirements, expectations during an emergency, data security, and frequency and expectations around the use of electronic communications (Luxton et al., 2016).

Informed consent is also an ethical issue when mobile device apps are used to collect or provide health and public health-related data. Users of mobile devices for this purpose should be informed about why the data is collected, how it will be used, and how their (and others') privacy will be assured.

Other related ethics issues concern whether these data will also be used for research, whether the users have any level of control over who can access their data, and whether data will be disposed of once the pandemic is over (Haleem et al., 2020). Moreover, the issue of appropriate disclosure of risks and limitations of use should be provided to end users of these technologies as part of the informed consent process.

Competence and scope of services

Healthcare professionals are responsible for obtaining and maintaining the skills of their professions. Competence refers to the ability to appropriately perform those skills. Competence is a core and ethical obligation in healthcare because providing services outside of the boundaries of one's trained skill or expertise could put patients at risk of harm. Additionally, incompetence can also damage the public perception of the respective healthcare field when examples of inappropriate or harmful behavior are made public (Luxton, 2014).

Healthcare providers who adopt the use of telehealth, mobile health applications, or other technologies must be competent in their use and seek training to acquire and maintain an acceptable level of competence. As a result, once again, the necessary and rapid need to retool or increase telehealth services during the pandemic highlighted the need for training in telehealth skills, ethics, and legal requirements in order to assure competence in their use.

The application of conversational agents or other virtual care providers (web-based chatbots) also requires their developers to assure that these are designed to be competent in the services they are providing and that they are adequately controlled based on the scope of their tested capabilities (Luxton, 2014). Such assurances of competence are also a legal consideration given that unregulated online mental health services that do not meet jurisdictional requirements or professional standards may lead to lawsuits (Luxton, 2014).

Safety and harm reduction

With regard to telehealth services, healthcare professionals and organizations that provide telehealth services must establish appropriate safety procedures and emergency plans, such as contacting appropriate resources for persons who express active suicidality but are located remotely to the healthcare professional (Luxton, Nelson, & Maheu, 2016). Just as with in-person services, professionals who are providing services over videoconferencing also have responsibilities as mandated reporters and must be aware of their jurisdiction's requirements regarding vulnerable populations (e.g., children, elders), duty to report and duty to warn/protect, as well as with civil commitment requirements (Luxton et al., 2016).

Harm reduction is also of concern in the design and use of CAs that augment, or in some instances replace, human care providers. Because CAs can be designed to function with some degree of autonomy, there is the risk of harm to people if the technology does not adequately and appropriately address scenarios when potential or immediate danger is detected or presented by the end user (Luxton, 2014; Luxton & Hudlicka, 2021). A pertinent example would be a scenario whereby a person discloses to a web-based chatbot that they are experiencing suicidal thoughts and have intent to die. Designers of CAs should consider giving CAs the capability to automatically monitor for risks and then take appropriate and timely action. For example, it would be a good, and ethical, practice for the CA system to display help resources, such as a crisis line or other support resources, relevant to the user's input, in the same ways as a real clinician would offer. In some situations, it would also be exceptionally advisable, and ethical, to include procedures for human review of the information and appropriate intervention through contact and appropriate referral for assistance (Luxton, 2014).

The right to privacy

Respect for the privacy of individuals is grounded in the right to autonomy and dignity and is viewed as essential to individual well-being (Luxton, 2014; Moore, 1998). Betrayal of trust due to privacy invasion or unauthorized misuse of information damages the trust of both individual healthcare providers and their profession. Moreover, privacy is necessary for all people to choose their thoughts and feelings, and with whom they share. Privacy also helps protect physical safety (if our real-time location data is private). Healthcare professionals have a responsibility to safeguard and respect patients' privacy and the confidential nature of the information they provide.

The use of technology in healthcare and public health all have risks to privacy that come in the form of inappropriate use of private data as well as data security issues associated with compromise by hackers or theft of the device where information is stored. Another risk involves how the data will be used by those who are collecting it, and whether the information will be shared with others for other purposes, such as commercial marketing (Luxton, 2020a).

As I noted in a previous section of this chapter, mobile apps were widely used during the pandemic for the purposes of contact tracing, for collecting and disseminating information to the public, as well as for the convenient display of vaccination status. Apps for contact tracing can be used to collect and transmit to centralized databases run by governments or public health authorities for analysis and storage of data, regarding all types of case scenarios of persons affected by COVID-19 (e.g., those infected, recovered, and deceased). Data may include patients' physical location, patient-reported travel, proximity to others, patients' physiology, comorbidities, and current symptoms. This highly personal information can be effectively used for case identification and for decision-making regarding the allocation of resources for protecting the public's health (Haleem et al., 2020).

Privacy is also a principal concern regarding the use of CAs, and if not appropriately addressed, can cause harm to users of CAs (Luxton, 2020a). CA systems can collect private and sensitive information when people interact with them. Therefore there is an ethical obligation to assure that the privacy of users is protected and maintained. Private companies and governments who use CAs should be aware of applicable laws, such as the Health Insurance Portability and Accountability Act (HIPAA) in the United States, and inform end users of the limits to privacy and what protections are in place.

Bias

Biases in the design and use of healthcare technologies are another ethical issue deserving attention. Treatments delivered via mobile apps and CAs, for example, are susceptible to biases when organizations deploying the technology have not adequately considered the demographic characteristics and needs of the intended users (Luxton, 2020a). Biases can also occur during the design and testing of technologies, leading to unfair treatment of people and to socioeconomic disparities in healthcare (Gianfrancesco et al., 2018; Luxton, 2020a).

Algorithmic bias is when a computer system makes systematic and repeatable errors that result in unfair outcomes, such as privileging one group or groups over others. The causes of algorithmic bias include missing data, small sample sizes used to train algorithms, persons not represented by algorithms (underrepresentation), and misclassification and measurement error (Luxton, 2020b). The conscious and nonconscious biases and values of the programmers and organizations collecting, selecting, or using data to train algorithms can also be a cause of bias. It is therefore essential for developers of these systems to include persons from the intended target population and communities within a population affected by potential healthcare disparities in the development and implementation of these technologies (Luxton & Sirotin, 2020). Moreover, systems that function with some level of autonomy, such as mobile apps, CAs, and decision support systems that use machine learning to identify patterns and make decisions, can modify or develop their own algorithms that may result in biases over time. Thus a process whereby humans are involved in testing the performance of these systems, including evaluation of biases and inaccuracies, can help to mitigate the risks associated with algorithmic bias (Luxton, 2020a, 2020b; Luxton et al., 2015).

Relatedly, when AI systems make decisions on their own and humans are unable to directly observe or explain how decisions were derived, they present what is called the *black box problem*. This is a serious ethical issue because it challenges traditional ideas of accountability and thus responsibility for the decisions and actions a machine is making, which may come to harm people (Luxton, 2020b).

Equitable access to care

It is well documented that the majority of adults and children with behavioral health needs in the United States, and in many nations throughout the world, do not receive the services they need (AHRQ, 2022; Merikangas et al., 2011; WHO, 2017). A shortage of healthcare practitioners and services is a primary reason for this gap in services. The high demand for healthcare services and other challenges during the COVID-19 pandemic only amplified this problem worldwide. Also contributing to healthcare disparities in underserved communities is the lack of technology infrastructure and other challenges, such as access to education, food, employment, housing, and other social determinants of health (Office of Disease Prevention and Health Promotion, 2021).

Equitable access to technology, and willingness to use and trust the intended use of health technologies, was an important issue during the COVID-19 pandemic. For example, some demographic groups were less likely to have access to, or adopt the use of, contact tracing and vaccine verification mobile apps, therefore restricting any benefits they may provide (Lucivero et al., 2020). Moreover, if such apps are needed to access particular activities, such as public transportation, restaurants, grocery stores, and employment, compulsory adoption could result in the formation of a group of people whose freedoms are inequitably impacted (Lucivero et al., 2020).

Discussion

The health technologies used during the COVID-19 pandemic raise important ethics issues that need careful and deliberate attention by developers of these technologies, organizations and governments that deploy them, professionals who integrate them into practice, and the public who use them. Thoughtful discussion of these issues is essential, with the inclusion of people in those discussions who are from the communities that are most impacted by the use of these technologies.

For healthcare providers and organizations, some ethics guidelines of the healthcare professions provide direction regarding the use of modern technology in care (Luxton, 2020a, 2020b). For example, both the American Psychological Association and the American Medical Association have produced ethics guidelines specific to the use of telehealth technologies. A few researchers have noted, however, that many current professional ethics codes and practice guidelines do not adequately address the use of emerging technologies that stand-in for human professionals, particularly regarding CAs and other forms of virtual care providers (Luxton, 2014; Luxton & Riek, 2019; Luxton & Sirotin, 2020). These issues include the topics of competence, liability, trust, privacy, and patient safety, and also extend to topics such as the therapeutic relationship. Revised and updated guidelines with a particular view of the ethical use of both existing and future technology are essential.

The COVID-19 pandemic demonstrated that in some countries and jurisdictions, efforts to stem the spread of the disease entailed mandatory use of technology, especially mobile apps. In communist China, for example, smartphone apps used for contact tracing were also used to determine whether a person should be quarantined or allowed into public spaces, such as subways and malls, or to use public transport (Parker et al., 2020). These apps also appear to share information with the police, calling into question whether automated social control via the technology might persist after the epidemic (Mozur et al., 2021). While governments and the companies that deploy these types of apps may offer assurances that private information will not be shared for other purposes, such as submission to law enforcement authorities, people who are wise about this technology know that even with such stated assurances, information may nonetheless be shared with, stolen by, or turned over to a third party, if not immediately, sometime in the future. It is highly conceivable that this type of tracking technology can be used for social control purposes. This situation leads to grave, and understandable, distrust in people.

In times of a crisis and when fear is elevated, as, during the COVID-19 pandemic, some people are more willing than others to give up freedoms and choose perceived security and safety offered by a policy or particular course of action. This can lead to a slippery slope, whereby a specific decision or course of action is likely to result in unintended consequences, such as more restrictions on freedom or more harm (Welsh et al., 2015). The obvious concern is that allowing temporary use of a restrictive technology under "emergency powers" could lead to a more permanent suspension of rights and liberties which could have unintended consequences. Mandatory lockdowns, restrictions on social gatherings (for weddings, funerals, entertainment, eating out, and religious practice), mandatory vaccines, monitoring of social interactions, and suppression of scientific counter-information are all examples of strategies and policies implemented during the COVID-19 pandemic. Some, including this writer, argue that we must be vigilant about how current pandemic-related policies and suspensions of rights might be extended to other, hidden or nefarious goals of governments, standing political parties, or corporate interests that infringe on the rights and freedoms of everyday people.

When technology is used dishonestly, it damages the trust that the public has in government and companies that want people to use technology. Public trust is essential for effective healthcare and public health. To ensure the public trust of healthcare professions, these factors are essential: (1) transparency in the intent of the use of these technologies, (2) assurance that they will not be used for other purposes, (3) free sharing of information, and (4) debate about their use. This same argument extends to all technologies and tools, from vaccines to face masks, that were used during the COVID-19 pandemic.

Conclusion

With every crisis comes the opportunity for learning and improving how we address future problems. Undoubtedly, we will all be better prepared for the next pandemic with lessons learned from COVID-19, from disease mitigation strategies to a better understanding of the effects of a global pandemic on mental health. Technological advances that were utilized in reaction to the outbreak, and their continued development and evaluation, will certainly provide useful solutions for the future.

The necessity of appropriate and ethical application of emerging technological solutions must never be overlooked. Healthcare and public health professionals have an ethical obligation to treat people with respect and dignity, to protect people, and to do no harm, as stated in the Hippocratic oath. To overlook ethical requirements, and for free debate about the issues to be restricted by governments or others, is to cause harm to people and damage public trust. Maintaining public trust must be a central focus, and where it is lost, it must be restored if we are to be successful at mitigating any future health crisis. The physical and mental health impacts on people of the world following the COVID-19 pandemic will likely persist far into the future.

We must be ready and willing to maximize technology and resources ethically and morally by putting the health, freedoms, rights, and voices of citizens first.

References

AHRQ. (2022). *2022 National healthcare quality and disparities report. Content last reviewed November 2022*. Rockville, MD: Agency for Healthcare Research and Quality. https://www.ahrq.gov/research/findings/nhqrdr/nhqdr22/index.html.

Brooks, S. K., Webster, R. K., Smith, L. E., Woodland, L., Wessely, S., Greenberg, N., et al. (2020). The psychological impact of quarantine and how to reduce it: Rapid review of the evidence. *Lancet*, *395*, 912–920. https://doi.org/10.1016/S0140-6736(20)30460-8.

Czeisler, M.É., Lane, R. I., Petrosky, E., et al. (2020). Mental health, substance use, and suicidal ideation during the COVID-19 pandemic—United States, June 24–30, 2020. *MMWR. Morbidity and Mortality Weekly Report*, *69*, 1049–1057. https://doi.org/10.15585/mmwr.mm6932a1.

De Kock, J. H., Latham, H. A., Leslie, S. J., et al. (2021). A rapid review of the impact of COVID-19 on the mental health of healthcare workers: Implications for supporting psychological well-being. *BMC Public Health*, *21*, 104. https://doi.org/10.1186/s12889-020-10070-3.

Ganesan, B., Al-Jumaily, A., Fong, K. N. K., Prasad, P., Meena, S. K., & Tong, R. K. (2021). Impact of coronavirus disease 2019 (COVID-19) outbreak quarantine, Isolation, and lockdown policies on mental health and suicide. *Frontiers in Psychiatry*, *12*, 565190. https://doi.org/10.3389/fpsyt.2021.565190. eCollection 2021.

Gianfrancesco, M. A., Tamang, S., Yazdany, J., & Schmajuk, G. (2018). Potential biases in machine learning algorithms using electronic health record data. *JAMA Internal Medicine*, *178*(11), 1544–1547. https://doi.org/10.1001/jamainternmed.2018.3763.

Haleem, A., Javaid, M., Khan, I. H., & Vaishya, R. (2020). Significant applications of big data in COVID-19 pandemic. *Indian Journal of Orthopaedics*, *54*(4), 1–3. Advance online publication https://doi.org/10.1007/s43465-020-00129-z.

Hudlicka, E. (2016). Virtual affective agents and therapeutic games. In Luxton (Ed.), *Artificial intelligence in behavioral and mental healthcare*. San Diego: Elsevier Academic Press.

Lucas, G. M., Gratch, J., King, A., & Morency, L. P. (2014). It's only a computer: Virtual humans increase willingness to disclose. *Computers in Human Behavior*, *37*, 94–100.

Lucivero, F., Hallowell, N., Johnson, S., Prainsack, B., Samuel, G., & Sharon, T. (2020). COVID-19 and contact tracing apps: Ethical challenges for a social experiment on a global scale. *Journal of Bioethical Inquiry*, *17*(4), 835–839. https://doi.org/10.1007/s11673-020-10016-9.

Luxton, D. D. (2014). Recommendations for the ethical use and design of artificial intelligent care providers. *Artificial Intelligence in Medicine*, *62*(1), 1–10. https://doi.org/10.1016/j.artmed.2014.06.004.

Luxton, D. D. (2017). Behavioral and mental health apps. In J. J. Magnavita (Ed.), *Using technology in mental health practice*. Washington, DC: American Psychological Association Books. https://doi.org/10.1037/0000085-004.

Luxton, D. D. (2020a). Ethical challenges of conversational agents in global public health. *Bulletin of the World Health Organization*, *98*, 285–287. https://doi.org/10.2471/BLT.19.237636.

Luxton, D. D. (2020b). Improving mental health outcomes with artificial intelligence. In G. Reger (Ed.), *Mental health technologies* Routledge. https://doi.org/10.4324/9780429020537.

Luxton, D. D., Anderson, S. L., & Anderson, M. (2015). Ethical issues and artificial intelligence technologies in behavioral and mental health care. In D. D. Luxton (Ed.), *Artificial intelligence in behavioral and mental health care*. San Diego: Elsevier Academic Press.

Luxton, D. D., & Hudlicka, E. (2021). Intelligent virtual agents in healthcare: Ethics and application considerations. In F. Jotterand, M. Ienca, & M. Liang (Eds.), *Ethics of artificial intelligence in brain & mental health*. Switzerland: Springer Nature.

Luxton, D. D., Nelson, E., & Maheu, M. (2016). *A practitioner's guide to telemental health*. Washington DC: American Psychological Association Books.

Luxton, D. D., & Riek, L. (2019). Artificial intelligence and robotics for rehabilitation. In L. Brenner, B. S. Reid-Arndt, Elliott, et al. (Eds.), *Handbook of rehabilitation psychology* (3rd ed.). Washington, DC: American Psychological Association Books.

Luxton, D. D., & Sirotin, A. (2020). Intelligent conversational agents in global health. In S. Okpaku (Ed.), *Innovations in global mental health*. Cham: Springer. https://doi.org/10.1007/978-3-319-70134-9_11-1.

McKillop, M., South, B. R., Preininger, A., Mason, M., & Jackson, G. P. (2021). Leveraging conversational technology to answer common COVID-19 questions. *Journal of the American Medical Informatics Association*, *28*(4), 850–855. https://doi.org/10.1093/jamia/ocaa316.

Merikangas, K. R., He, J.-P., Burstein, M., Swendsen, J., Avenevoli, S., Case, B., et al. (2011). Service utilization for lifetime mental disorders in U.S. adolescents: Results of the National Comorbidity Survey Adolescent Supplement (NCS-A). *Journal of the American Academy of Child and Adolescent Psychiatry*, *50*(1), 32–45.

Moore, AD. (1998). Intangible property: Privacy, power, and information control. *American Philosophical Quarterly*, *35*(4), 365–378.

Mozur, P., Zhong, R., & Krolik, A. (2021). *In coronavirus fight, China gives citizens a color code, with red flags*. The New York Times. Available at: https://www.nytimes.com/2020/03/01/business/china-coronavirus-surveillance.html.

Office of Disease Prevention and Health Promotion. (2021). *Social determinants of health*. https://www.healthypeople.gov/2020/topics-objectives/topic/social-determinants-health/interventions-resources.

Parker, M. J., Fraser, C., Abeler-Dörner, L., et al. (2020). Ethics of instantaneous contact tracing using mobile phone apps in the control of the COVID-19 pandemic. *Journal of Medical Ethics*, *46*, 427–431.

Timmers, T., Janssen, L., Stohr, J., Murk, J. L., & Berrevoets, M. A. H. (2020). Using eHealth to support COVID-19 education, self-assessment, and symptom monitoring in the Netherlands: Observational study. *JMIR mHealth and uHealth*, *8*(6), e19822. https://doi.org/10.2196/19822. https://mhealth.jmir.org/2020/6/e19822/.

Welsh, D. T., Ordóñez, L. D., Snyder, D. G., & Christian, M. S. (2015). The slippery slope: How small ethical transgressions pave the way for larger future transgressions. *Journal of Applied Psychology*, *100*(1), 114. https://doi.org/10.1037/a0036950.

WHO. (2017). *WHO's Mental Health Atlas 2017 highlights global shortage of health workers trained in mental health [Internet]*. Geneva: World Health Organization. https://www.who.int/hrh/news/2018/WHO-MentalHealthAtlas2017-highlights-HWshortage/en/.

World Health Organization. (n.d.). WHO coronavirus (COVID-19) dashboard. World Health Organization. Retrieved May 20, 2022, from https://covid19.who.int/.

Ye, J. (2020). The role of health technology and informatics in a global public health emergency: Practices and implications from the COVID-19 pandemic. *JMIR Medical Informatics*, *8*(7), e19866. https://doi.org/10.2196/19866. 32568725. PMCID: PMC7388036.

PART 2

Leveraging technologies for health: Developments and directions

Machines that think like humans. Goggles that transport you into new dimensions. What was once the world of the future in health technology is here now. But, how do you get technology tools to everyone, especially in remote communities, with no electricity? Chapters in this Part offer solutions, including accessing solar energy to address the lack of energy that is a barrier to leveraging digital health.

"Game-changing" and "cutting edge": these are words which describe health innovations and techniques presented in this Part, including artificial intelligence, virtual reality, augmented reality, and mixed reality, and intelligent virtual agents. These are within overviews and perspectives, history, definitions, state of the art and a roadmap to the future.

Methods to get those innovations to "those left furthest behind" who are "at the last mile"—meaning those with no access to health services—are covered, with examples of how to scale healthcare access through tech innovations and models that work for low- and middle-income countries.

Are these tech tools good or bad for your health and mental health? Opinions differ, so a chapter reviews the pros and cons.

Research by the World Health Organization shows that physical activity is essential for your well-being; so, we cover intriguing technology-based projects in countries around the world to boost your physical and mental health.

A major solution to stop the viral spread is vaccines that were developed at a record speed—yet inequities in distribution highlighted serious challenges. A chapter addresses vaccine research and development and the supply chain.

What notables have to say

Nothing is permanent in this wicked world—not even our troubles.

—*Charlie Chaplin*

Courage is not having the strength to go on. It is going on when you don't have the strength.

—Theodore Roosevelt

The doctor of the future will give no medicine, but will instruct his patients in care of the human frame, in diet, and in the cause and prevention of disease.

—Thomas Edison

We can create change when we come together and use our voices.

—Billie Eilish, American singer on the same online rally as Hugh Jackman, on 24 February 2021, hosted by Global Citizen, joining forces to call on leaders around the world to launch a global campaign "to end COVID-19 and kickstart a global recovery"

CHAPTER 16

Technology innovation for global health: Past, present and future

Akarsh Venkatasubramanian[a] and David Wallerstein[b]
[a]UNITE Parliamentarians Network for Global Health, Delhi, India, [b]Tencent, Palo Alto, CA, United States

Introduction

This chapter first introduces a brief history of key systemic developments in health innovation. It then moves from past to present to take a wide-angle view of the current state of the art, and thereon attempts to anticipate the near future. This temporal perspective offers many lessons for key actors involved in health technology innovation. Reflections about some of these lessons inform a practical innovation roadmap for the future of global health. This chapter therefore represents the what, who, and how of health technology innovation.

A brief history of key systemic developments in health innovation

The World Health Organization (WHO) estimates that over 45 million people worldwide die annually from infectious diseases (such as HIV/AIDS, Tuberculosis, Malaria, and waterborne diseases) and noncommunicable diseases (NCDs, including heart disease, cancer, diabetes, respiratory diseases, and mental health) (WHO, 2021). Neglected tropical diseases (NTDs), accidents, and injuries add to these estimates. Additionally, the International Labour Organization estimates that 4 billion people worldwide are left without social protection as of 2021, contributing directly to global mortality and morbidity numbers (ILO, 2021).

To alleviate this global burden of disease and disability, and to equitably expand social health protection, Heads of State at the United Nations General Assembly in September 2015 adopted the 2030 Agenda for Sustainable Development, consisting of 17 Sustainable Development Goals (SDGs) and 169 associated target indicators (UN, 2015). Four years later at the UNGA

Resilient Health. https://doi.org/10.1016/B978-0-443-18529-8.00016-0

again in September 2019, Heads of State adopted the Political Declaration on Universal Health Coverage (UHC) (United Nations General Assembly, 2019). The Declaration—founded upon the Constitutions of the WHO and the ILO, as well as the right to health—reiterates the United Nations targets of expanding access to essential health services to at least 3 billion additional people by 2030 (ILO, 1919; WHO, 1948, 2020).

However, governments are not on track to meet their SDG and UHC commitments. Current estimates indicate up to 5 billion people worldwide will not have access to essential social health services by 2030 (UN, 2017). Achieving the SDGs requires both strong political commitment and sustainable financing models. And, importantly, reaching those goals requires inclusive innovation through responsible reengineering of global health and social systems to cater to the most marginalized and vulnerable communities.

Perhaps conveniently, we are in the middle of a global era of scientific advancement: the digital age. This era is believed to include both the third industrial revolution driven by the rise of computers, internet, and digital technologies (from the 1970s) and the fourth, encompassing artificial intelligence, Internet of Things, and other frontier technologies (from the 2010s) (World Economic Forum, 2019). The revolutions present unprecedented innovation opportunities for health, ranging from electronic health to mobile health to digital health tools, several applications of which are discussed later in this chapter.

The COVID-19 pandemic has ravaged the world economically, socially, and psychologically since late 2019. Responses to the pandemic have leveraged technology innovation in the form of telemedicine, interoperable electronic health records, emergency communication systems, digital contact tracing tools, epidemiological disease transmission models, and technology-enabled social protection initiatives (MIT, 2021; Whitelaw et al., 2020). These tools have highlighted the massive potential of innovation to save and enhance lives. But unless implemented in equitable, inclusive, and responsible approaches, innovation comes fraught with multidimensional risks. These include individual privacy breaches and human rights violations, technology-enabled exploitation of and discrimination against specific population groups, weaponization of data itself and cyberwarfare, and even digital colonialism and digitally motivated neocolonialism.

Current state of the art: What is health technology innovation today?

The global technology innovation ecosystem continues to evolve rapidly with the digital age. However, global inequality renders "state of the art" a highly contextual notion. For instance, electronic health records first designed at the beginning of the third industrial revolution (in the 1980s) is the norm in high-income regions but only still an aspiration in many low-income contexts. Similarly, AI and machine learning-enabled diagnosis designed today during the fourth industrial revolution may be an aspiration in high-income regions, but practically unrealistic in many low-income settings, where electricity, mobile phone, and internet may all be unreliable, inaccessible, and unaffordable. AI represents game-changing yet context-sensitive technology innovation.

Global observation and monitoring efforts seek to keep track of this contextual relevance and capability of innovation tools mentioned in the earlier section of this chapter. Instead of providing a separate in-depth perspective by itself, this chapter presents three such initiatives by WHO, I-DAIR, and PATH which offer comprehensive overviews of the health technology innovation landscape (and situational guidance as appropriate).

In 2018 WHO identified a set of commonly utilized digital health interventions and classified them into various functional buckets (World Health Organization, 2018). The document aimed to promote a bridging vocabulary between the different interdisciplinary stakeholders involved in digital health. Although frontier technologies have evolved quickly since 2018, much of the document's guidance are still valid in many country contexts today with regard to national programs on electronic health records, telemedicine, health/logistics management information systems, emergency communication systems, and more. Nevertheless, an updated version of the document is planned for 2022.

In 2020 the International Digital Health and Artificial Intelligence Research Collaborative (or I-DAIR) hosted at the Graduate Institute of International and Development Studies in Geneva presented a typology of technological tools used in COVID-19 responses (I-DAIR, 2021). They observed that tools can broadly be divided into six categories: risk communication/public education (such as infodemic mgmt apps, drones, telemedicine), tracing and tracking apps (e.g., DPT3, Aarogya Setu, TraceTogether), digital tools for mining research (including using natural language processing techniques), digital collaboration platforms (notably GISAID), remote triage and testing (diagnostic tools like CT scan readers), and epidemic modeling and dashboards (from various universities and private sector for-profits).

The third example is Digital Square, a global PATH-led initiative that promotes global digital health transformation by implementing over 35 digital health public goods in partnership with 40 organizations in over 100 countries worldwide (PATH, 2022). Digital Square routinely publishes global goods guidebooks since 2017, monitoring individual technology-based tools like DHIS2, SORMAS, and Healthsites, their usabilities and functionalities.

Such guiding, foundational initiatives serve to light the path toward equitable and inclusive health innovation that leaves no one behind, as the United Nations' Sustainable Development Goals (SDGs) and Political Declaration on Universal Health Coverage (UHC) aspire to achieve. However, as observed earlier in this chapter, country governments currently are not on track to attain their objectives by 2030. Strong political commitment and progress toward these objectives crucially requires collaborative and meaningful health and social policies, as specialized UN agencies and parliamentarian networks such as UNITE advise and advocate for.

Finally, observations of the past and present of technology innovation for health would be incomplete without a note on its future.

Trends in health technology innovation record the advent of *ehealth* (electronic health) in the 1990s. The concept gained popularity in the early 2000s (with successful regional cocreations such as the Asia eHealth Information Network/AeHIN) while incorporating *mhealth* (mobile health) tools (by for example the WHO/ITU Be He@lthy Be Mobile initiative) and *digital health* (for instance, by RECAINSA in Latin America, and many more) in the 2010s (AeHIN, 2022; RECAINSA, 2022; WHO, 2022). It is possible that with the fourth industrial revolution under way, other terminologies, concepts, nomenclatures, ideas, and trends gain traction, especially as we move toward the time benchmarks of 2030 of the SDGs and UHC. This could well be along the lines of health that is preventive, proactive, predictive, patient-/people-centric, and personalized.

Lessons from the digital age, for the digital age

Technology innovation trends during global health emergencies such as the COVID-19 pandemic enrich our collective understanding of the opportunities and risks of innovation for health. Four broad lessons from global COVID-19 responses that are particularly relevant to this book's context are discussed as follows. They serve as recommendations for innovators and public health officials, as well as for stakeholders who bridge the gap between the two.

(1) Innovative technology necessitates innovative governance

- *At national and local levels*: As technologies evolve, laws must evolve too. For instance, several countries employ (or aim to employ) digital contact tracing tools during pandemic responses, as during COVID-19; but only a few of them have the legal precedence to evaluate their ethical implementation. Even fewer have laws that can effectively mitigate misuse (AccessNow, 2020).
- *At the global level*: Global governance instruments and models hosted at the United Nations system such as the Universal Declaration of Human Rights; International Covenant on Civil and Political Rights; the International Covenant on Economic, Social and Cultural Rights; and the International Health Regulations have historically guided governments on accountable and human rights-based public and health policy (OHCHR, 1976a, 1976b; UN, 1948; WHO, 2005). In the digital age, such a global governance model can advise national and local stakeholders to prioritize health technologies that protect, respect, and uphold human rights. Both formal (binding) and informal (nonbinding) instruments will help mitigate and prevent technology misuse and data misuse, as well as technology-based misuse and data-based misuse. Further rigorous research is required on the feasibility of a binding instrument such as a treaty on the role of technology in peace, international security, and weaponization;

and on the need for nonbinding instruments such as guidelines and principles on ethical health technology innovation like the Health Data Governance Principles (Transform Health Coalition, 2022).

(2) *Promote inclusion*: Consult different geographies, constituencies, and sectors in key decisions while designing, developing, implementing, governing, and evaluating health innovation.
- *The need for geographical inclusion*: On one hand, all regions must be equally accommodated in global governance discussions, and in global innovation implementation projects—no country should be left behind. On the other hand, innovation evolves at different speeds in different regions of the world. While all countries must be guided by the same consistent global benchmarks and safeguards against innovation misuse, we must recognize, for example, that AI algorithms designed for Caucasian skin disease diagnosis will not be applicable for darker skin.
- *The need for constituent inclusion*: There exists a clear digital divide today, where certain regions and populations have access to technology innovation more than others. Global inequality and emergencies exacerbate a health divide. Innovators or policymakers must not enable technology advancement to worsen this digital divide or health divide. This reiterates the SDG motto of "no one left behind." Innovators must cater to vulnerable populations and communities often disadvantaged through the health divide or digital divide, including women, elderly, disabled, indigenous, informal and precarious workers, LGBTQI+; and importantly, youth or "digital natives" who will live with and bring to life tomorrow's innovation.
- *The need for sectoral inclusion*: While this chapter primarily focuses on the role of innovators and policymakers in health innovation, several other stakeholders are crucial to bridging them. For instance, governments must engage the private sector to develop technologies that respect ethical considerations such as data privacy, safety, and ownership. Research institutes must be empowered to evaluate such technologies and provide independent, well-appreciated thought leadership. International organizations host governance instruments and are critical to enable unbiased development-oriented deployment of technologies and data. Foundations may offer valuable technical guidance, operational assistance, and financial support to achieve progress. Perhaps most importantly, civil society is crucial to maintain accountability and transparency of these stakeholders.

(3) Support health in all policies (PAHO, 2022)
The Government of Finland used this term first in 2006 to demonstrate the importance of health decisions in other public policy priorities, such as education, labor, nutrition, sanitation, security, and trade. The COVID-19 pandemic demonstrates both that these priorities fall when health falls, and that health falls when these other priorities fall. Innovators and governments must especially prioritize collaborative innovation, without letting any development area fall behind (United Nations General Assembly, 2019).

(4) Do not lose track of reality in the glitz of cutting-edge innovation. Many health innovations of the fourth industrial revolution such as AI applications and digital contact tracing do not function in settings without supporting infrastructure of power/electricity, mobile phones and mobile phone connectivity, internet connectivity wires and towers, internet affordability, and access to foundational health innovation (such as electronic health records) (Venkatasubramanian, 2020). Several primary health centers and rural hospitals/clinics do not support discrete data flows, a necessary prerequisite to privacy-preserving data transfer that makes digital health tools run. COVID-19 proves that one-size-fits-all solutions do not work in health innovation. Context matters.
Building on these lessons, this chapter presents an innovation roadmap to unite innovators, governments and public health officials, international organizations and staff, researchers, donors, and civil society alike toward building an inclusive, resilient, and innovative global health future.

Innovation roadmap: The potential of technology innovation to advance global health

This section introduces the stakeholder landscape of health technology innovation by briefly exploring the role of a few key sectors. It then discusses five necessary preconditions for successful technology innovation in health. These preconditions represent a practical perspective of an incubating ecosystem for specific health technology innovations, some of which are introduced later in this chapter and explored later in this book.

The purpose of accelerating the introduction of new solutions is to improve human quality of life. This necessitates innovators and public health officials to unite for public good. But gaps exist between their respective perceptions and on the contextual effectiveness of innovation. These gaps may be plugged by bridging sectors such as international organizations, researchers, and civil society.

On one hand, the innovator must swiftly move to bring a fledgling idea to fruition. Capital, opportunity, reputation, and momentum can all be lost in a matter of months or less, as the innovator seeks to implement or commercialize a scientific breakthrough, technology, application, or service concept. Such time-sensitive and high-pressure circumstances enable creative solutions, like phone and internet connectivity, electronic health records, mRNA vaccines, and handheld PCR tests.

On the other hand, the public health official is tasked with ensuring healthier lives and coverage across a large population, with scant resources and mountains of daily challenges.

Organizationally, changes have huge implications and often cannot be done overnight. The mindset of larger organizations often evolves more to ensuring that proven solutions can be delivered, but not experimenting with new approaches that could lead to a breakthrough.

Despite these significant differences, bridging the gap between them offers many public health benefits as international organizations, researchers, and civil society especially evidence. In addition to hosting, guiding, evaluating, and keeping accountable innovators, governments, and programs, these sectors offer valuable shared organizational cultural learnings. They help identify avenues for policymakers to understand innovation and for innovators to understand policy.

Five key prerequisites are elucidated as follows to enable such stakeholder bridging and federated learning. Written from an innovator's viewpoint, they work on the premise that innovation is accepted and received openly, with policymakers prepared to embrace it.

An Innovation Roadmap: **Navigating the 5 Cs of health innovation to land on a resilient global health future**

1. *Clarify*:
   Health priorities must be clearly established. Setting explicit targets and tackling challenges with focus and commitment can accelerate impact.
   Challenges are many, which are constant and immediate. Clearly prioritizing challenge areas will help mobilize necessary initiatives, data, discussions, investment, and specific effort. Allowing existing practices and inertia to reduce focus is common. The key first step, however obvious, is to embrace innovation.
2. *Communicate*:
   Having clearly established key targeted areas, communicate them both domestically and internationally as suitable. Why have they become critical issues in the country of interest? Extensive data on the depth of the challenge, background, impact, specific human stories, and science behind the challenges support rationales. This is also an ostensibly straightforward exercise, but there is tremendous power in sharing the detailed circumstances around the greatest challenge areas. The specifics can help innovators consider solutions, and the epidemiological data speaks to the extent of the challenge and its impact. This enables subject matter experts from related or adjacent fields to be catalyzed and consider how their area of work or expertise could be of value.
3. *Collaborate*:
   Developing joint programs between policymakers and innovators helps tackle joint challenges. Programs could be anything from a challenge initiative with announced budgets for solutions that can be deployed to tackle the area of need. They could even be establishing a subject matter task force, even a small team to conduct global outreach to organizations with relevant capabilities or technologies. Fostering such conditions renders both inbound and outbound interaction possible.
   Further, investors thrive when market conditions are welcoming to investment. Innovation arguably thrives when fluid investment flows support new strategies. Health priorities often have a market value. If this market may be experiencing growth with innovator-friendly conditions, it can become attractive to investors. This requires clear policies that signal intent.

Protection of property rights, business-friendly tax policies of various types (as appropriate, impactful, and acceptable to governments to sign and ratify), and company formation rules (for instance, can DocuSign help set up a company and raise money?) are all good places to start. Unclear and unwelcoming foreign exchange rules, taxation on investment, unclear corporate operating environments, capricious tax authorities, or bureaucratic red tape are all negatives and will further complicate efforts to attract investment that can spur innovations.

4. *Create*:
   Create opportunities for the market to provide solutions. This process could start from having a website or email address with contact details, to appointing an individual or team to discover solutions and work with innovators, to more sophisticated programs where promising innovations may have procurement budgets. Such programs could work in cooperation with international health organizations and even donors as appropriate.
   A global marketplace of solutions and interested experts exists, but reaching these resources will only be possible by being organized and prepared to welcome opportunities.
5. *Coordinate*:
   Address areas critical but often considered adjacent to public health, such as Food, Energy, and Water, or "FEW." A unique feature of modern organizations, from government to education and beyond, has become to create artificial silos that become fiefdoms over time. The inevitable human political response often negatively affects collaboration. However, adjacent areas such as food, energy, and water have a huge effect on health outcomes. Collaboration and triangulation of drivers of challenges can achieve huge results, for example, by boosting agricultural productivity with innovations, proliferating energy access with essential investments, or protecting and growing fresh water sources and ensuring water resilience. All of these have a significant impact on public health outcomes. Progress with existential requirements for FEW drives health outcomes. Tracking progress with reliable data is powerful and builds confidence that initiatives are working. Data and transparent dashboards can help serve as the glue to keep multidisciplinary teams working together with efficiency.
   National leaders are often in a good position to coordinate the efforts between ministries and encourage collaboration, but they will need the high-level dashboard data that can synthesize the progress of the various initiatives and pinpoint areas for more focus as your initiatives evolve. Ministers across functions can work together, all to achieve a united targeted outcome, thereby maximizing human quality of life.

What can these preconditions foster?

The essential preconditions in the earlier section of this chapter provide a bridge between innovators and governments. This section translates the five C's (Clarify, Communicate,

Collaborate, Create, Coordinate) into successful health innovations that work or are set to work practically. These concrete examples or success stories advise innovators and governments alike on how to leverage emerging technologies for public good. Several of this book's succeeding chapters take deep dives into each of these examples as follows.

Since the advent of the fourth industrial revolution, several businesses are developing consultation solutions driven by *Artificial Intelligence*, which include the possibility of dialogue and feedback through AI bots as a first-line response. Moreover, mobile phones hosting or capable of accessing such AI features may be able to track other basic data, like exercise, pulse, and other core metrics in privacy-preserving approaches.

COVID-19 lockdowns, health center in access, and medical professional overloads have all instigated a shift toward *mobile health (mhealth) and telemedicine*, with more service provided over mobile applications or the internet compared to office care. This of course is a privilege for those with access to mobile phones, connectivity, power, and internet. But this shift signifies a change in patient experience to a preventative model. For instance, smartphone cameras as breakthrough healthcare devices allow image sharing and video consultation. Two-way text communication becomes natural and intuitive.

Innovation trends have allowed *low-cost diagnostics and testing* to become more ubiquitously available and leveraged intelligently to get ahead of health challenges.

The ongoing miniaturization and proliferation of low-cost diagnostics in the global market is however restricted subject to availability, accessibility, and affordability conditions. As COVID-19 antigen tests evidence, viral antigens can now be discovered in 20 minutes. Various technological approaches make such low-cost tests available for personal and home diagnostics. These low-cost tests can significantly reduce the burden on the traditional healthcare system, as a positive or negative result can guide an individual to the right resources in a system without necessarily requiring meetings and more expensive tests on site to validate a result. Moreover, the individual is more empowered to alter their behavior with the clear knowledge of their health or infection status, whether it's an STD, an aerosolized virus, or a general health condition.

Clinical innovation developments enable more minimally trained individuals to support the more traditional healthcare functions of nurses and doctors, serving as a first-line response team online or even collecting diagnostic results in a low-cost clinic environment to precede online consultation. Nurses and doctors are brought in as conditions require, but the typical data processing, triaging, diagnostic collecting can be done by a team that is more minimally trained to specialize in these areas. As the first line of intervention care, they can immediately recommend further resources to consider (such as a pharmacy) and can gather more information and escalate cases as required.

Reliable epidemiological data enables pinpointing of root causes of health challenges by region and community. *Tracking* trends of an outbreak, looking at correlations with related factors such as sanitation, environment, or urbanization, can all be made more sophisticated with AI-driven approaches that can spot patterns and correlations which can drive further response plans.

Several countries have adopted *digital contact tracing* tools in their COVID-19 responses. These offer huge potential in terms of accuracy, reliability, and speed of contact tracing during emergencies, but several questions on privacy, ownership, and security of data remain unanswered. Global and national governance mechanisms must clarify these gaps in rights-based approaches for such tools to be more widely accepted in the future.

Data and AI are ideal for driving efficiency and the careful use of scarce medical resources. As several predictive *mathematical models* prove during COVID-19, epidemic forecasting helps preemptive identification of risks and challenges.

Summary

This chapter provides an overview of historical advancements in technology innovation for health and takes forward technological and ethical lessons from the ongoing digital age to enable the attainment of Universal Health Coverage and the UN Sustainable Development Goals by 2030. It equally guides governments on practical approaches to introduce innovation and new approaches to help accelerate equitable healthcare outcomes.

The world is not on track to reach UHC and the SDGs by 2030. Technological innovation is urgently required to achieve meaningful and impactful reengineering of health systems for the world to get back on track and to advance. Innovation and technologies are evolving rapidly, and global health must keep up. Responsible innovation, meaningful cross-sectoral collaboration, and effective governance mechanisms to safeguard fundamental human rights can enable an inclusive, equitable, and resilient global health future. The time to act is now.

References

AccessNow. (2020). *COVID-19 contact tracing apps*. Available at https://www.accessnow.org/covid-19-contact-tracing-apps-in-mena-a-privacy-nightmare/.

AeHIN. (2022). *Asian eHealth Information Network*. Available at https://www.asiaehealthinformationnetwork.org/.

I-DAIR. (2021). *International Digital Health and Artificial Intelligence Research Collaborative*. Available at https://www.i-dair.org/.

ILO. (1919). *Constitution of the International Labour Organization*. Available at https://www.ilo.org/dyn/normlex/en/.

ILO. (2021). *4 Billion people worldwide are left without social protection*. (August 20, 2021). Available at https://www.ilo.org/global/about-the-ilo/newsroom/news/WCMS_601903/lang- -en/index.htm.

MIT. (2021). *MIT Technology Review: Coronavirus Coverage*. (May 6 2021). Available at https://innovation.mit.edu/mit-technology-review-coronavirus-coverage/.

OHCHR. (1976a). *International Covenant on Civil and Political Rights*. Available at https://www.ohchr.org/en/professionalinterest/pages/ccpr.aspx.

OHCHR. (1976b). *International Covenant on Economic, Social and Cultural Rights*. Available at https://www.ohchr.org/en/professionalinterest/pages/cescr.aspx.

PAHO. (2022). *Health In All Policies*. Available at https://www3.paho.org/hq/index.php?option=com_content&view=article&id=9361:2014-welcome-health-all-policies&Itemid=40258&lang=en#:~:text=Health%20in%20All%20Policies%20.

PATH. (2022). *Digital square*. Available at https://digitalsquare.org/.

RECAINSA. (2022). *Red Centroamericana de Informática en Salud*. Available at https://recainsa.org/.

Transform Health Coalition. (2022). *Health data governance principles*. Available at https://healthdataprinciples.org.

UN. (1948). *Universal Declaration of Human Rights | United Nations*. Available at https://www.un.org/en/about-us/universal-declaration-of-human-rights.

UN. (2015). *Sustainable development goals*. Available at https://www.un.org/sustainabledevelopment/sustainable-development-goals/.

UN. (2017). *Home—SDG Indicators*. Available at https://unstats.un.org/sdgs/.

United Nations General Assembly. (2019). *Universal Health Coverage*. Available at https://www.un.org/pga/73/universal-health-coverage/.

Venkatasubramanian, A. (2020). The human rights challenges of digital COVID-19 surveillance. *Health and Human Rights*, *22*(2), 79–84.

Whitelaw, S., Mamas, M., Topol, E., & HGC, V. S. (2020). Applications of digital technology in COVID-19 pandemic planning and response. *The Lancet Digital Health*, *2*(8), e435–e440. https://doi.org/10.1016/S2589-7500(20)30142-4. ISSN: 2589-7500.

WHO. (1948). *Constitution of the World Health Organization*. Available at https://www.who.int/about/who-we-are/constitution.

WHO. (2005). *International health regulations*. https://www.who.int/health-topics/international-health-regulations#tab=tab_1.

WHO. (2020). *Universal Health Coverage (UHC)*. Available at https://www.who.int/health-topics/universal-health-coverage#tab=tab_1.

WHO. (2021). *Mortality and global health estimates*. Available at https://www.who.int/data/gho/data/themes/mortality-and-global-health-estimates.

WHO. (2022). *WHO/ITU Be He@lthy Be Mobile*. Available at https://www.who.int/initiatives/behealthy.

World Economic Forum. (2019). *Fourth Industrial Revolution*. (August 4 2021). Available at https://www.weforum.org/focus/fourth-industrial-revolution.

World Health Organization. (2018). Classification of digital health interventions v1.0: a shared language to describe the uses of digital technology for health. World Health Organization. https://apps.who.int/iris/handle/10665/260480. License: CC BY-NC-SA 3.0 IGO.

CHAPTER 17

Innovations and technologies for health: Current state of the art and potential for advancing healthcare

Joe Stringer
Octopus Ventures, London, United Kingdom

Introduction—The real emergency is in business-as-usual healthcare

I am writing this from the UK at the start of a 10-day self-isolation period after contracting the Omicron COVID-19 variant just under 26 days from when it was first discovered in South Africa in early November 2021. In the room next door are my two children and wife, who are thoroughly frustrated at the prospect of another Christmas without seeing their grandparents or cousins, as they are equally sequestered at home. I am in this situation because the NHS App (from Britain's National Health Service) notified me that I had sat too close to someone infected with COVID-19 on a train journey from London two days ago and asked me to immediately take a lateral flow test. Within two hours of the positive result, nearly 28 people I had interacted with had been notified to test as well. Such is the contact tracing the government instituted to "stop the spread."

Is it frustrating and inconvenient to be isolating? Yes. But is it meaningless or avoidable in the context of the trauma the world has experienced in the last two years from this pandemic? Based on the galloping pace of this new variant, probably not. Life threatening? Thankfully not, according to the "science" and also because I am fortunate enough to be relatively healthy and have additional protection from a breakthrough in mRNA vaccines only possible through a combination of technology and one of the best primary care systems in the world, here in Great Britain.

But, this is not enough. I need to isolate not just to protect those closest to me, but because—like every other health system around the world—the NHS does not currently have the underlying resilience to cope with anything more than an additional 3%–4% in patient demand. So, if I got worse and needed intensive care, I, too, would risk overloading a precariously balanced health system.

Resilient Health. https://doi.org/10.1016/B978-0-443-18529-8.00017-2

It is well documented that healthcare is facing a serious crisis around the world. Across any meaningful measure of quality, accessibility, and affordability, there is not a single example of where any system, government, or insurer, public or private, Global North or South, that can lay claim to "getting it right." But many are taking interesting approaches that we need to learn from, in particular, those that have been able to embrace the power of technology similarly to the majority of other industries, most of whom are ten to fifteen years ahead in terms of adoption.

In terms of innovation and technology, our greatest challenge is not the *number* of innovations—it is 99% an adoption challenge—yes, there are major issues where we need to continue to invest for a long-term solution, but our immediate priorities are technology scale up and widespread adoption.

The adoption challenge has been well known by the global healthcare community for more than two decades, with many "tomes" written purely on this subject (Fair Society, 2020; Healthbox, 2019; Kennedy, 2020) but COVID-19 has brought into sharp focus the need for everyone to act with urgency and to embrace change, no matter how uncomfortable that may feel.

It is always dangerous to draw analogies between healthcare and other sectors as there are undoubtedly many significant characteristics that make healthcare unique, not least of which is the consequences of making mistakes. However, when it comes to the use of technology, we can see the most parallels with the financial services and banking sector. Both are heavily regulated, require a significant amount of user trust in the management of their data, have a relatively low rate of "customer" churn or switching (despite many attempts by regulators to encourage it), and both are dominated by organizations that for decades have recognized their future is unsustainable but have had a limited immediate need to disrupt the status quo in the short term—until now.

In 2007 the global financial crisis (GFC) prompted an urgent period of change in the financial services sector in the developed world that covered everything from the role of regulators and the need for greater governance through to developing operating standards for data sharing, all to ensure that global banking giants were not incentivized to take unsustainable risks in one side of their business, often in order to prop up a retail banking sector in dire need to reform. Nobody can argue that these changes have been a unanimous roaring success, but the crisis was enough of a shock to the system to push every bank in the world to move from thinking about place-based banking to user-centric banking. The global pandemic is now being seen as the same market shock from which the world will be benefitting for decades to come. It has forced all players in the healthcare industry to truly question exactly where, when, and how care is delivered.

The short-term technology response to the COVID-19 pandemic (Nuffield Trust, 2021) is akin to stabilizing the banks during the GFC—well funded, not lacking in political appetite, and ultimately everyone is aligned to make it happen. So, over the next 24 months, we can expect to see a continued rollout of technology around the world to support testing and treatment not just of COVID-19 but expanding across many other infectious disease types. Two years ago, it would have been unthinkable for any new vaccine to be produced within just one year in response to a previously unknown virus, notwithstanding the significant issues of global inequality in terms of access to that vaccine. But that's exactly what happened, starting with "Operation Warp Speed." Looking ahead, over the next 12 months, we can expect to see a greater number of hospitals using best practice from across multiple industries to better prepare for the outbreak of new variants. Right now, health leaders are combining data from sources as varied as epidemiology forecasts, travel patterns, weather forecasts, and real-time supply chain tracking individual vaccine vials, to ensure we have the infrastructure and systems in place to avoid the horrendous scenes of confusion, panic, and wanton spread of the first few months of COVID-19 in late 2019 into 2020. There is still work to do, in particular, to improve workforce planning for the outbreak of new variants, but to some extent this is no longer our greatest priority, in my view. The real emergency is the opportunity cost in delivering "business as usual" healthcare, to address the cancer waiting lists (The BMJ, 2020) and delays in elective care that have for some time dwarfed the immediate mortality rates from COVID-19 which dominated attention.

Current developments: One mile of superfast connectivity, but an exciting near-term pipeline

Since the start of the pandemic, we have witnessed the rapid rise in the use of telehealth to provide basic remote care, namely, video consultation platforms to discuss healthcare issues and, where possible, to reach a diagnosis where the issue is either visible onscreen or can be followed up with physical testing. For example, telehealth utilization in April 2020 for office visits and outpatient care was 78 times higher than in February 2020 (McKinsey, 2021). There is a large variety in the sophistication of these platforms, ranging from incredibly basic text chat using quite rudimentary decision trees which have sometimes been misleadingly branded as "artificial intelligence" (AI), all the way through to more advanced solutions that allow multidisciplinary teams (MDTs) to simultaneously discuss a patient's condition or use genuine AI and machine learning to read visual and audio biometrics to support accurate diagnosis. It has been encouraging to see the pace at which these platforms have grown out of necessity and the ways in which other stakeholders have responded. For example, the UK General Medical Council (GMC) issued guidance to general practitioners that they would be insured against misdiagnosis via telehealth in the event of any subsequent claims. But the rates of telehealth adoption still represent a fraction of the total market, and from a technology perspective, these platforms represent the very basic foothills.

To use the financial services analogy, telehealth is akin to the first use of ATMs in banking, immediately useful but then shining a light upon the need to transform every other aspect of the sector. If we are not careful, people will become frustrated and revert back to analogue healthcare delivery.

Perhaps a more appropriate analogy might be seeing telehealth today as the installation of a 1-mile superfast internet connection but at either end we remain reliant upon a dial-up modem. For the fraction of users lucky enough to have their more straightforward issue resolved quickly and efficiently without further interaction, it is sufficient, but for the 30% (RCGP, 2021) of those who still need to use in-person care, it is even more frustrating and demonstrates that most point solutions need to be bundled up to form a combined package of technology and innovations to achieve the right outcome. Many of these will be invisible to the end user, in the same way that customers of Amazon never see the impressive integration behind the scenes in the supply chain required among vast numbers of suppliers, logistics operators, security, or financial services companies, in order to deliver their package on time.

While some may be forgiven for thinking this is a daunting and overwhelming task, there are many in healthcare, like me, who have been relishing the opportunity to properly embrace technology for over twenty years. For the first time in a generation, we have the burning platform, the technology, and the resources, to take healthcare outside of the four walls of the clinical environment and think instead about how each step in the journey can be digitalized, so that in-person healthcare becomes the last, most inefficient, and highest risk solution. There will be some who consider such an approach to be too drastic and assume that it will naturally mean depersonalizing care and the "human touch." But this doesn't have to be the case, and given the choice between remotely delivered care and a sustainable, affordable health system, research suggests that most would choose the former (Orrange et al., 2021; Ramaswamy et al., 2020).

Unfortunately, at a global level, there is still a great deal of opacity around the investment opportunities in which each of these funds invest, but based upon the available data and my experience, we are seeing the fastest rates of adoption in:

- Point-of-care diagnostics, e.g., breathalyzers for cancer, sentient monitoring in the home, which is converging with the "through the door" megatrend (bringing healthcare into the home).
- Workforce augmentation to maximize the use of clinical time in front of patients, when they NEED to be in front of patients (using AI, surgical robotics, EPR search).
- The exponential growth of digital therapeutics, using technology to tackle issues ranging from addiction treatment through to accurate diagnosis and treatment for brain health conditions, in particular dementia.
- Precision/personalized medicine, using real-time clinical data to develop personalized treatments and drugs based on genetic profiles, an area in which the UK has been a leading early adopter.

- Solving the fertility challenge (which some argue is as profound as the pandemic) and using technology to improve outcomes during pregnancy, for example more accurate prediction of prematurity.
- Specific solutions for female health conditions and life stages, a sector that has been underserved for decades—for example menopause treatment.
- The enabling operational infrastructure to remove friction from the value chain (automation and supply chain) although ultimately, patients will still need physical products, samples will still need to be analyzed, and drugs will still need to be delivered, in the most efficient way possible.
- Convergence of healthtech with fintech, e.g., using blockchain payment systems and value-based models for drugs to transform the affordability challenge.
- Robotic/remote surgery across multiple channels to provide access to expertise and integrated care across borders.

In addition to these specific themes, the pace with which specific technologies are being "bundled" is starting to increase, typically driven by the founders of companies operating in a similar space exploring opportunities for partnerships and "coopetition" (a mix of competition and cooperation), or through aggressive merger and acquisition activity in the name of accelerating global growth of the merged entity. I have seen some health systems attempt to drive this bundling process but would caution against trying to preempt or plan this too far in advance, and rely instead on incredibly strong market forces and investor expectations to lead the way.

Finally, in terms of the "bleeding edge" long-term trends which we expect to see making an impact in 3-5 years, investors are backing businesses that only a year ago would be considered more science fiction than reality, for example, developing direct brain/computer neural interfaces with the potential to transform disability and dementia, or early advances in regenerative medicine to push longevity to the very boundaries of what was once thought impossible.

The pivotal role of early stage capital

As we can see from the list of fast adoptions in healthcare I outlined before, the crux of the issue is not one of innovation, given that the worlds' universities, professors, and technologists have never been producing so many new ideas, and the value chain has never been so well aligned. With some of the world's best-known brands and largest companies spinning from universities (Google, Facebook, FedEx) (Government Grants, 2021) the business case also continues to improve, namely, for every £1 invested in research, the government gets around £3.60 back in terms of economic benefit (Octopus Ventures, 2020). Neither is it in the global disparity of innovation. While the UK and US dominate in the production of healthtech Unicorns (venture capital industry term describing a start-up company valued over $1 billion), more countries are rapidly entering the $1bn valuation club. From Poland to India, the level and quality of innovation along with funding is growing exponentially, and across the globe (Traxyn, 2021).

This growth has only been made possible through the role of early stage investors who are prepared to take a level of commercial risk that governments and health systems would never be able to justify to their populations, often with over a 50% chance of failure which is the typical failure rate in early stage venture capital (Review42, 2020). The underlying cause of this risk is the rate of technology adoption by health systems, irrespective of their funding structure.

There is a lot of critical commentary and research around the historic failure of healthcare systems to adopt technology prior to the pandemic. Reasons can be broadly grouped into three areas: (1) inadequate technology which does not integrate with current IT practices; (2) a general failure to understand the real need for transformation, albeit this has been alleviated through the pandemic; and (3) a lack of Digital Health knowledge and awareness from policymakers (Medical Futurist Whitepaper, 2021).

Parallel to the global pandemic, capital markets are undergoing a similar period of transformation, with a growing recognition that "impact investing" is in the process of moving from a small-scale cottage industry—largely in Europe—to a fundamental global shift. For example, the IFC estimates the global market for impact investment was $2.3 trillion in 2020 (IFC, 2020). We are now seeing investors prioritizing opportunities to invest unprecedented amounts of capital toward high-impact social problems, particularly climate tech or healthtech (IFC, 2020). As the earliest investors in these sectors, venture capital and private equity funds have a pivotal role to play, facing the same incentives to drive much faster adoption of technology and innovation.

Key challenges

While I predict a very optimistic decade ahead, given my earlier analysis, it will not be without challenges. These include:

(1) Trust in managing data. As some of the largest companies in the world have grown through the use of personal data, the declining trust of end users (patients and clinicians) is perhaps one of the most pervasive challenges that early stage healthtech businesses face, as they handle some of the most sensitive data (about health) in the world. While social media giants have not helped through their aggressive commercialization of sensitive data, we have seen some improvements during the pandemic, with nearly 80% of respondents to a UK Government poll agreeing that during a public health emergency it is more important than usual to share health and care data with those involved in providing healthcare services.[a]

[a] https://www.gov.uk/government/news/polling-indicates-growing-public-understanding-about-importance-of-using-health-and-care-data.

(2) Trust in early stage solutions. This is a similarly mixed picture. At the start of 2022, we have seen the founder of the now defunct blood testing business *Theranos*, Elizabeth Holmes, convicted of fraud in what was one of the highest profile failures of both healthtech and Venture Capital (VC) investment in the USA. Clearly an egregious example like this does not help improve trust and sentiment in the sector, but it is important to stress that technology and the approach of investors have evolved considerably since 2003. During the early days of *Theranos*, a technology that promised Point-of-Care finger-prick diagnostics was seen as cutting edge, but this technique is now close to being available in a number of early stage companies. Key lessons to learn from the *Theranos* case study are relevant more for the investment sector than for healthcare, with many funds now recognizing the need to employ health sector specialist teams and far deeper scientific due diligence.

(3) Changing healthcare payment models and reallocating budgets set by traditional boundaries rather than outcomes. While this challenge varies widely around the world, the vast majority of healthcare systems remain wedded to incentivizing activity and inputs rather than improving outcomes. In the UK, we have recently seen a number of significant and welcome reforms away from competition to push the integration of care forward, with an expectation that payment models will follow suit; however, these continue to be the exception rather than the rule. From an investment perspective, this often becomes one of the most difficult blocks to rapidly scaling companies around the world, leading start-ups to develop very different pricing models depending upon the country of the end customer.

(4) Interoperability. Traditionally, health records were recorded on paper, filed, and stored in archives. Since the late 1990s, countries have started to adopt Electronic Health Records (EHRs) as standard practice. As time moved on and technology developed, the ability for different software to integrate and communicate with each other was, and continues to be, limited. Citing various reasons as data privacy, security, or IP protection, digital health companies historically were reserved and unwilling to share data with other applications. Since 2011, Fast Healthcare Interoperable Resources (FHIR) became the standard for describing and sharing EHR data which addresses both interoperability and security; however, not all digital health providers have adopted FHIR as a standard. We have seen during the pandemic that many more technology companies have created and shared APIs as workaround to interoperability barriers, but progress continues to be slow.

What the world must do

To conclude, notwithstanding these challenges, as of 2022, we are on the cusp of a one-in-a-generation shift for technology in healthcare. Sadly, we don't have twenty years for this journey to mirror the rate of change in other industries, for example, insurance; instead, we have until the next global pandemic. Therefore I urge every stakeholder to jump into the future, take bold decisions, and be prepared to make mistakes.

In particular:

- The investment community needs to raise its sights beyond the USA as the "holy grail" for scaling healthtech businesses; instead, the Global South holds one of the most exciting opportunities and companies that win early traction there will ultimately be global leaders. Ministers and health leaders in the Global South also have a key role to play, ensuring their respective countries are as attractive as possible for promising start-ups to establish a presence. Simple measures such as provide financial incentives to de-risk initial setup, and remove procurement bureaucracy and regulatory hurdles, will make the difference between founders and investors choosing one country over another.
- The healthcare technology regulatory system needs to rapidly catch up with the pace of change we have seen in the last 24 months. Regulators need to fast-track approvals of the most promising technology, not just the most promising drugs, and support clinicians when they choose to embrace technology.
- Investors in the Global South (DFIs, funders, philanthropic organizations) need to focus more on accelerating the scale up of technology than focusing on where a business is physically headquartered. While it is tempting to seek to achieve the double impact of higher local employment and improved healthcare, I would argue that in a world where technology teams can work 100% remotely, we need to encourage greater liquidity between the Global North and South by focusing on adoption over the initial location of the innovation.
- Governments need to invest today in rapidly growing the skills and capacity of the clinical workforce, taking, implementing the recommendations from the Topol review for the UK NHS (Eric Topol, 2019) (presenting recommendations for the NHS to lead the world in using digital technologies to benefit patients) and more recently from Dr. Mark Britnell in his book, *Human: Solving the Global Workforce Crisis in Healthcare* calling for reimaging models of care, using technology (Britnell, 2019).
- Ministers need to be prepared to take bold decisions, backing specific technologies and companies despite the potential risks.

In summary, I would urge every stakeholder in the healthcare industry to look back upon the global pandemic as the ultimate wakeup call for every health system around the world. It shone a light upon the very best aspects of both healthcare and human nature but gave the most sobering reminder of the vulnerabilities and fundamental unsustainability of the way we currently deliver care. Many deaths were unavoidable, but sadly many were not. The innovations already exist to ensure we are more prepared and more resilient to respond to future pandemics, our collective challenge now is to continue to adopt them with the same degree of urgency and optimism as we saw during the height of the global crisis.

References

Britnell, D. M. (2019). *Human: Solving the global workforce crisis in healthcare.*

Eric Topol, M. D. (2019). *The Topol review, Preparing the healthcare workforce to deliver the digital future.* www.topol.hee.nhs.uk.

Fair Society. (2020). *Healthy Lives (The Marmot Review), 2010 & Health Equity in England: The Marmot Review 10 Years On, Professor Marmot.*

Government Grants. (2021). *The university spinout report 2021.*

Healthbox. (2019). *Root Causes of Health.*

IFC. (2020). *Investing for Impact: The global impact investing market report.*

Kennedy, E. (2020). Gender inequalities in health and wellbeing across the first two decades of life: An analysis of 40 low-income and middle-income countries in the Asia-Pacific region. *The Lancet.* 2020 and Taking urgent action on health inequities.

McKinsey. (2021). *Telehealth: A quarter-trillion-dollar post-COVID-19 reality?.*

Medical Futurist Whitepaper. (February 2021). *3 Reasons why national digital health transformations fail.*

Nuffield Trust. (2021). *November 2020 and McKinsey & Company, Telehealth: A quarter-trillion-dollar post-COVID-19 reality.*

Octopus Ventures. (2020). *Spinning out success, Entrepreneurial Impact Ranking.*

Orrange, S., Patel, A., et al. (2021). Patient satisfaction and trust in telemedicine during the COVID-19 pandemic: Retrospective observational study. *JMIR Human Factors, 8*(2), e28589. https://doi.org/10.2196/28589.

Ramaswamy, A., et al. (2020). Patient satisfaction with telemedicine during the COVID-19 pandemic: Retrospective cohort study. *Journal of Medical Internet Research, 22*(9). https://doi.org/10.2196/20786.

RCGP. (May 2021). *Whitepaper.*

Review42. (2020). *What Percentage of Startups Fail?—50% of VC backed companies fail within their first three years.* Review42.com.

The BMJ. (2020). *Covid-19: Waiting times in England reach record highs.*

Traxyn. (2021). *Unicorn Corner—Healthtech, 2021–2013 review.*

CHAPTER 18

Artificial intelligence: State of the art and potential for advancing healthcare

Amandeep S. Gill[a,b] and Mehdi Snène[b]
[a]Graduate Institute, Geneva, Switzerland, [b]I-DAIR (International Digital Health and AI Research Collaborative), Geneva, Switzerland

Introduction

There is a long history of the use of computing, software engineering, and other digital technologies in clinical settings and medical diagnostics. Computed Axial Tomography or CT scan could be termed the first digital tool in medical diagnostics. Invented in 1972, it transformed two-dimensional images of the human body into three-dimensional visualizations, which previously was only possible *postmortem* (Schultz et al., 2021). As mobile computing took off in the 1990s, it was natural for interest in "mobile health" and telemedicine to grow. Similarly, as artificial intelligence (AI) reemerged from its long "winter" at the end of the first decade of the 21st century and showed spectacular results in image recognition, internet search, and consumer advertising (Shroff, 2013), it was natural for computer scientists to contemplate applications of AI and data science in Life Sciences and health.

Significant enhancements in computing capacity through the availability of low-cost Graphics Processing Units (GPUs) as well as the emergence of large datasets (e.g., the Stanford University based ImageNet) to train algorithms were important factors in this reemergence of AI (ImageNet, n.d.). A subdomain of AI, called deep learning, itself part of the broader domain of machine learning (ML) (Machine Learning, 2020), was well placed to take advantage of this confluence. Machine learning allows computers to perform tasks without explicit instructions coded into software. Computers "learn" to recognize patterns by using training data and extracting features from it. In 2012 a group using deep learning—Convolutional Neural Networks, modeled on how the human brain processes information—won the ImageNet annual competition on classifying images. Deep learning soon became the AI technology of choice for processing images, videos, and unstructured data. It is not surprising that the "Nobel" of AI, the

Resilient Health. https://doi.org/10.1016/B978-0-443-18529-8.00018-4

A.M. Turing Award, was awarded in 2018 to three leading pioneers of deep learning, namely, Geoffrey Hinton, Yann LeCun, and Yoshua Bengio.

In clinical and healthcare settings where unstructured data has grown exponentially, deep learning has the ability to overcome existing data processing limitations. Deep learning can also help detect diseases more quickly, accurately, and earlier, through signal processing of disease-related biomarkers, X-rays, or other diagnostics data. Further, deep learning can be used to improve the effectiveness and quality of drug testing, and additionally can be used to improve healthcare pathways and processes in areas previously difficult to computerize.

Over the past few years, AI and ML models have rapidly emerged in medical practice as well as in healthcare research (Nordlinger et al., 2020). Applications range from software for research purposes only to systems and devices which directly impact outcomes for patients. A snapshot of trends using data from research publications, clinical trials, as well as patents for the year prior to the COVID-19 pandemic is given in Fig. 1.

While it is hard to separate AI use from digital health overall, AI is trending significantly within the broader domain and is increasingly appearing in research and translation in medical areas such as oncology and diagnostics as per a recent survey, which looked at a total of 93,000 publications and 27,000 patents over 2011–2020 (I-DAIR, 2021).

Other trends are sobering. A few countries, specifically, China and the United States, dominate research in digital health and AI as well as the translation of research into commercial opportunities. South Africa dominates patents output in Southern and Eastern Africa but its global contribution is only 0.21% (I-DAIR, 2021). Unless steps are taken now to address the research and development divide, AI will come to be dominated by a few countries.

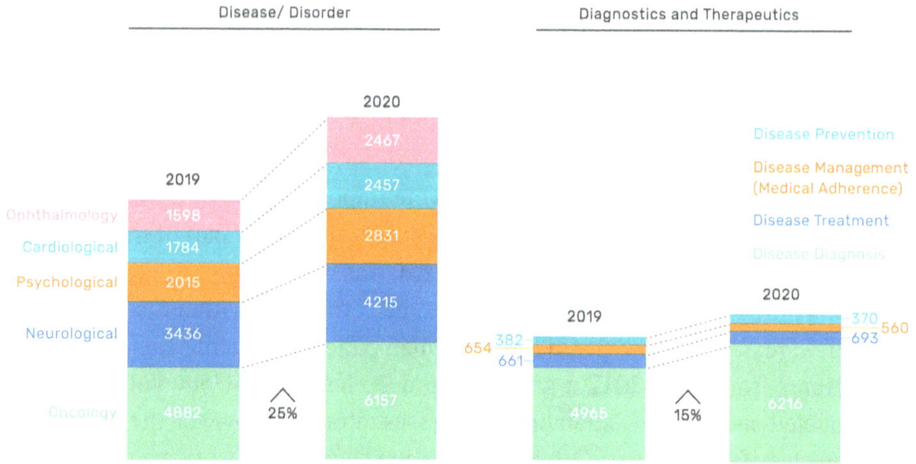

Fig. 1
Evolution of select subdomains of digital health in the year prior to the COVID-19 pandemic.

Understanding the role of AI in healthcare: The COVID-19 experience

There is nothing like a crisis to accelerate needs-driven innovation. This is the case with digital technologies, which essentially had their day under the sun when the respiratory flu virus turned into a global pandemic in early 2020.

An interesting example of the successful application of AI during the COVID-19 pandemic comes from Greece. A ML-based system called Eva was deployed at 40 airports, ports, and border crossings during summer 2020 to determine which arriving passengers should be tested for COVID-19 in order to deploy scarce resources optimally, maximize the identification of infected but asymptomatic individuals, and inform border control policies. This system identified 1.85 times more asymptomatic infections during the peak of the tourist season compared to traditional random surveillance testing (Bastani et al., 2021). Governance and data protection limitations were thoroughly assessed prior to designing the system and the algorithm was built around data minimization. Prioritizing interpretability and transparent interpretation (for all users) as well as designing for flexibility were two other key parameters of successful implementation (Bastani et al., 2021).

Another example is an Early Warning System developed by BioNTech SE and InstaDeep Ltd. which uses AI on SARS-CoV-2 genomics sequences available publicly on GISAID, a global platform for sharing of flu virus sequences digitally, to predict future variants of concern (Beguir et al., 2021). Interestingly, apart from predictive analytics, this system used Natural Language Processing, a deep learning technique with impressive results in language translation, to learn the language of biology by treating protein sequences as sentences and amino acids as words. It could thus "understand" the implication of a change in the sequence and predict whether that would be of concern, for instance, in terms of greater infectivity.

The previously mentioned two systems were successful. Other attempts at leveraging AI for the COVID-19 pandemic response and management have fared less well. A number of research groups in Australia, India, Switzerland, UK, and the USA have tried to classify asymptomatic COVID-19 patients by exploiting the subtle differences in cough sounds between normal and SARS-CoV-2 infected individuals. The UK government even considered deploying AI triage systems based on cough sound (Honeycombe-Foster, 2021). However, there have been difficulties in translating the in silico (computer experimentation) results into effective screening or prediagnostic tools in real-world settings. The authors are familiar with one such system, which gave impressive results as an AI model but was confounded by background noise when taken for deployment to a hospital facility.

Another natural candidate for testing AI's power for public health involves efforts to predict the scope and speed of the spread of COVID-19 infections in order to inform policy choices. The results have often been disappointing. In Australia, an agent-based framework, an AI technique that requires significant compute resources, was used to estimate dates for when authorities would be able to eliminate community transmission (Thompson et al., 2022). These estimates proved too optimistic. In India, the AI models which performed well during the first wave missed predicting the peak of the devastating second wave driven by the Delta variant (The Economic Times, 2021).

Also, tracing and tracking apps, even when built using the highest standards of privacy protection, did not perform so well, especially once case numbers spiraled (Abboud & Miller, 2020). Public health authorities ended up relying on analog systems for pursuing contacts or managing hotspots.

The overall sense derived from these projects is that AI promised a lot but was eventually marginal to the COVID-19 pandemic response (Heaven, 2021). Yet all is not lost. The crucial question is how this experience can be used to better direct future development of AI technologies for healthcare.

The crucial role of data

It is hard to avoid the impression that in a rush toward healthcare solutions, not enough attention has been paid to the foundational aspects of AI, especially the holy trinity of data sources, transdisciplinary collaboration, and governance.

A key ingredient to build effective AI models is high-quality data (including metadata). Vast amounts of data are being generated now by research institutions, hospitals, academic centers, health systems, and population surveillance systems. The origins and sources of this data are varied and include:

- RWE (real-world evidence)
- Real-time data streaming from connected devices and laboratories
- Individual monitoring devices, health apps, and mobile health
- Electronic health records (EHRs) in clinical settings
- Data curated from literature and patents

However, data, regardless of its source, needs to be stored well, annotated, validated, and made available for use by people. Big data alone can be of little value in the health domain. For instance, Electronic Health Records (EHRs) have increased the amount of information documented in patient charts and hospital databases but paradoxically have created a flood of noisy data of dubious value. Despite years of efforts, there is no health equivalent of the gold standard of the visual database called ImageNet, e.g., used in the case of cervical cancer. It is important to underline that unlike machine vision where contextual variations in meaning are limited (a cat is a cat wherever you see it), there may not even be a single gold standard; instead, tiered and distributed approaches to building gold standard datasets for AI would be needed. Building platforms which annotate, maintain, share, and permit use of data for health in meaningful and well-governed ways will require innovations in design, tooling, and regulation. Novel approaches to data collection and reuse will also be needed. Importantly, ways to share value derived from data fairly would have to be devised.

Another pitfall of a rush to AI solutions is an underappreciation of the role of AI in discovery of unusual data sources, say related to potential pandemic risks, and in the management of vast amounts of data which are being generated today without any hope of generating insights. This is similar to a problem in radio astronomy where vast amounts of data are being generated by a new generation of telescopes, and where AI can help direct astronomers' attention to more interesting data streams.

While space does not permit a detailed examination of the other two critical enablers, it is important that AI technologies are collaboratively designed across computer science, clinical and biological sciences, as well as social science. This was the case of the more successful smart logistics platforms for managing testing, diagnostics, and vaccination for COVID-19 as shown by the Greece example.

The importance of trust and governance

Trust in the outcomes and the governance of technology is an equally key factor in adoption and scaling. The governance of AI technologies is challenging in and of itself (Gill, 2021a). This challenge becomes all the more daunting when we bring AI into the health domain. The specific governance requirements of the domain (health) have to be brought together with data governance and algorithmic governance in a systems approach. The six principles for the design and use of AI in health proposed by the World Health Organization (protecting human autonomy; promoting human well-being and safety and the public interest; ensuring transparency, explainability, and intelligibility; fostering responsibility and accountability; ensuring inclusiveness and equity; promoting AI that is responsive and sustainable) underline this complex challenge (World Health Organization, 2021). For instance, humans should retain control of healthcare systems and decisions, which should not be delegated to machines. This is similar to a principle adopted in another context, namely use of AI for weapons where accountability for life and death decisions clearly has to be attributable to humans. It is also imperative to guard against the hyping of benefits when such oversell threatens to take resources and attention away from much-needed investments into Universal Health Coverage (UHC) and undermine trust in AI over the long run.

Future directions in AI for healthcare

The COVID-19 experience suggests that digital technologies which meet specific gaps in healthcare and research are here to stay, for example, platforms which share genomic sequences which enabled a head start for vaccine development or algorithms for predicting vaccine targets and infectious variants of concern. Whether these trends lead to a permanent shift in healthcare and accelerate progress on UHC depends on the technologies' fit with health system needs, how they are designed and deployed, and the degree of trust and confidence they inspire in users. While being alive to the power and potential of AI, if they are to succeed, designers need to keep in mind the analog interface of the health solution (in the case of the COVID-19 pandemic the "shoe leather" epidemiologists who work the phones and the doorbells) as well as individual and group behavior dynamics around privacy and data protection. Global collaboration would also be essential to federate data, computing capacity, algorithmic expertise, and governance approaches for the digital enabling of health systems' transformation (Gill, 2021b).

In the authors' point of view, it is important to draw a distinction between solutions and solution-making. Bringing more emphasis on the use of AI in health research rather than just on AI in healthcare will have at least two distinct advantages. First, the rush to AI solutions will be tempered, resulting in more thoughtful and trusted outcomes aligned with real health system needs. Second, the use of AI tools for healthcare research will lower barriers to participation in research from diverse geographies and from across disciplines. Attending to these two issues will reinforce inclusiveness and address the digital divide.

With these caveats, we can proceed to look at seven future directions for AI in healthcare and health research with a view to accelerating progress on UHC.

(1) In line with the WHO principle on patient autonomy, and with a view toward preparing the ground thoughtfully for personalized medicine, the research agenda for AI must include ways to bring patient voices into the assessment of healthcare interventions. These human-centered benchmarks can match the power of traditional gold standard benchmarks only with the help of AI that can aggregate patient narratives into compelling evidence.

(2) A second item on the research agenda can be returning fair value to owners of data and encouraging individual contributions to solving public health challenges. AI, combined with other digital technologies such as blockchain (distributed ledger technologies), can help record even micro-contributions to training data and reward data owners with insights and beneficial interventions.

(3) This leads us to the next item of research: pathways, tools, and governance of distributed clinical trials. AI can help create and manage global clinical cohorts, and bring scale and diversity to data and outcomes.

(4) A fourth area of research is personalized medicine, with AI facilitating the integration of genomics, behavioral, and other data without undermining privacy and personal agency.

(5) A fifth area could be participatory public health with citizen scientists helping to prevent and manage infectious outbreaks or patients prone to noncommunicable diseases improving public health outcomes through gamification of preventive behavior.

(6) A sixth priority area for research would be AI-powered discovery of unusual data sources, including from the One Health domain (collaborative efforts of multiple disciplines working locally, nationally, and globally, to attain optimal health for people, animals, and our environment) and their analysis in combination with traditional data sources for epidemic intelligence.
(7) Finally, we propose focusing research on using AI to optimize key processes such as patient registration and preboarding, to strengthen diagnostics equity and improve inclusiveness in health service delivery. Here, AI should be viewed more as a way to reinvent existing workflows to solve existing challenges rather than as a replacement for existing methods. In this sense, AI could provide novel solutions to existing workflow challenges, working side by side with health workers to optimize health outcomes through intelligent augmentation.

References

Abboud, L., & Miller, J. (June 23 2020). French give cool reception to Covid-19 contact-tracing app. *The Financial Times*. Accessible from https://www.ft.com/content/255567d5-b7ec-4fbe-b8a9-833b3a23f665.

Bastani, H., Drakopoulos, K., Gupta, V., et al. (2021). Efficient and targeted COVID-19 border testing via reinforcement learning [published correction appears in Nature. 2022 May;605(7909):E2]. *Nature*, *599*(7883), 108–113. https://doi.org/10.1038/s41586-021-04014-z.

Beguir, K., Skwark, M. J., Fu, Y., et al. (2021). *Early computational detection of potential high risk SARS-CoV-2 variants*. Preprint bioRxiv2021.12.24.474095. https://doi.org/10.1101/2021.12.24.474095.

Gill, A. S. (2021a). Aligning AI governance globally: Lessons from current practice. In *Global Trends. Analysis* SEF-Bonn.

Gill, A. S. (2021b). *Realising the promise of digitally-enabled health*. Globe: The Graduate Institute Review.

Heaven, W. D. (July 30 2021). Hundreds of AI tools have been built to catch covid. None of them helped. *MIT Technology Review*. Accessible from https://www.technologyreview.com/2021/07/30/1030329/machine-learning-ai-failed-covid-hospital-diagnosis-pandemic/.

Honeycombe-Foster, M. (June 3 2021). *UK mulls using AI to check coughs for coronavirus*. Politico. Accessible from https://www.politico.eu/article/uk-ai-coughs-coronavirus-fujitsu/.

I-DAIR. (2021). *Global research map*. Accessible from https://grm.i-dair.org/#/report.

ImageNet. n.d. Accessible fromwww.image-net.org.

Machine Learning. (15 July 2020). *IBM cloud education*. https://www.ibm.com/cloud/learn/machine-learning.

Nordlinger, B., Villani, C., & Rus, D. (2020). *Healthcare and artificial intelligence*. Springer.

Schultz, R. A., Stein, J. A., & Pelc, N. J. (2021). How CT happened: Medical computed tomography. *Journal of Medical Imaging*, *8*(5), 052110.

Shroff, G. (2013). *The intelligent web: Search, smart algorithms and big data*. Oxford University Press.

The Economic Times. (April 2 2021). *COVID-19 second wave in India may peak by mid-April, say scientists*. Accessible from https://economictimes.indiatimes.com/news/india/covid-19-second-wave-in-india-may-peak-by-mid-april-say-scientists/articleshow/81871145.cms.

Thompson, J., McClure, R., Blakely, T., Wilson, N., Baker, M. G., Wijnands, J. S., … Stevenson, M. (2022). Modelling SARS-CoV-2 disease progression in Australia and New Zealand: An account of an agent-based approach to support public health decision-making. *Australian and New Zealand Journal of Public Health*, *46*(3), 292–303. https://doi.org/10.1111/1753-6405.13221.

World Health Organization. (June 28 2021). *Ethics and governance of artificial intelligence for health: WHO guidance*. World Health Organization. Accessible from https://www.who.int/publications/i/item/9789240029200.

CHAPTER 19

Scaling healthcare access through AI: History, definitions, international overview, and the case of Portal Telemedicine

Rafael Figueroa
Portal Telemedicine, São Paulo, Brazil

In the summer of 2012, when I decided to practice kitesurfing on the beaches of northern Brazil, I had no idea that the events which unfolded there would lead me to surf the promising waters of telemedicine. My vacation was interrupted by a serious accident that badly compromised the integrity of my left foot, a sprain or fracture, I wasn't sure. As I was in a region with few medical resources—the state of Maranhão has less than one doctor for every 1000 inhabitants—it took quite a while to see a doctor. When I finally arrived at a hospital in the state capital, where I had an X-ray exam, a gross misdiagnosis left me without the ability to walk for 6 months and with several titanium pins in my foot. The wrong diagnosis was a sprained ankle when the correct diagnosis should have been multiple fractures.

During the more than 180days of ensuing physical therapy and being stuck at home, I began to wonder how technology could have avoided a whole sequence of wrong medical decisions. It was no surprise that this idea came to me since I had worked for several years in Artificial Intelligence (A.I.) and machine learning, at Google and at my own NGO, Green Society, and I knew that sending an X-ray exam to a specialist would be basic—and should have happened in my case.

That tragedy motivated me to start to research, and embark, in the field of telemedicine. This happened together with my siblings and parents, all of whom, oddly enough, had a background in the medical services field, although no one was a doctor. Thus we created a company that applies disruptive systems with the aim of improving people's access to quality healthcare.

This chapter gives an overview of telemedicine and specifics about the work that I do and the innovation I have developed.

All of our solutions have been tested and approved and are mainly being applied in Brazil and Africa. Thanks to this vision, we were chosen by the United Nations (UN) as one of the 10 companies in the world to accelerate the Sustainable Development Goals targeted to be

Resilient Health. https://doi.org/10.1016/B978-0-443-18529-8.00019-6

achieved by the year 2030, as outlined in the framework of the UN Agenda 2030 for Sustainable Development. Precisely for this reason, I was invited to write this chapter by the coeditors, who have extensive experience in the fields of health and of technology and innovation, and I am happy to do so. Our wish is that this content will help guide public health managers, medical professionals, and all those committed to using telemedicine, to "leave no one behind," as directed by the UN Agenda.

Access to primary healthcare: A global problem

The orthopedic discomforts which I mentioned at the beginning of this chapter were small indications in comparison to much more serious public health problems faced by the Brazilian population. I had a faint idea about these dire needs, but when I looked more intensively for solutions, I really understood my country's shortcomings.

About 150 million Brazilians, or 71.5% of the population, depend on the Unified Health System (SUS), a complex public network of hospitals, primary healthcare units, emergency care units, emergency rooms, etc. Such a complex and overloaded system faces challenges, such as a lack of specialists and long waiting lines for some types of exams and procedures.

Obviously, the population's difficulty in accessing primary healthcare is not unique to Brazil. Even in developed countries like the United States, there is a layer of the population which is falling behind, especially those who cannot afford health insurance (Chemweno, 2022). These people, because there is no option, suffer and indefinitely postpone the search for medical help. Only people below the poverty line and the elderly benefit from public healthcare services that provide more basic and emergency care. This varies from state to state, but broadly speaking, this is how American public healthcare works.

In European countries like England, France, Spain, and Italy, governments provide healthcare services to the population in exchange for taxes and fees. France, for example, invests 10% of GDP in its *Sécurité Sociale*, making it one of the best in the world. The country finances 80% of public healthcare through national insurance which is deducted directly from the workers' payroll. As a result, the French have access to good doctors and hospitals, as well as medicine, which is often free or inexpensive.

Italy also has a good public healthcare system, but subject to the payment of a percentage of each citizen's income tax. If the income is very low, the health insurance ticket will be zero. Still, the patient will be treated.

On the African continent, the difficulty in accessing adequate medical services is the most dramatic. Countries try to implement public healthcare systems, but everything is still incipient and, naturally, private healthcare predominates. With prevailing extreme poverty, few people can access this care. The situation is exacerbated by the very high number of people affected by communicable diseases such as HIV/AIDS, malaria, and epidemic diseases, as well as by maternal and child health deaths and problems. The African continent is the region of the world most affected by the HIV/AIDS pandemic, with around 25.4 million people infected.

Another recent example that shows the African continent's difficulties in relation to epidemiological control is the low population rate with the complete vaccination schedule for COVID-19. As of February 2022, less than 15% of Africans are immunized.

This topic of vaccination has been debated in international forums. Several reasons are behind the low vaccination rate, one of which is the difficulty in acquiring vaccines, which led many countries to make donations of doses; however, structural difficulties made it impossible for the vaccines to be taken before their expiration date.

In addition, most African countries suffer from a lack of infrastructure to store vaccines, which need to be kept at low temperatures, and to set up a logistical system for rapid distribution of doses to more isolated regions. These are just some of the factors that contribute to the number of vaccinated people being far below the ideal minimum recommended by the World Health Organization (WHO).

Lack of doctors and uneven distribution

Apart from the lack of quality in medical care for the vast majority of the vulnerable populations of the world, the World Health Organization (WHO) highlights the lack of doctors and/or their uneven distribution in countries (Who.int, 2017). For example, their statistics show that Brazil has an average of 24 doctors for every 10,000 inhabitants, which is not one of the worst indexes, given its parity with the United States, with 26, and Canada, with 24. However, for Europe, the numbers jump: Italy has as much as 80, France, 65, and Germany, 43. Impressively, some Latin American countries also have enviable numbers, namely, Argentina with 39 and Chile with 51.

Other regions live in critical situations. South Africa has only eight doctors per 10,000 inhabitants, Angola, 2, and the Democratic Republic of Congo scored 0.9, i.e., it does not have a single doctor for groups of 10,000 people.

In the Middle East and Asia, Israel has 54 doctors per 10,000 inhabitants, Saudi Arabia has 26, Yemen (5), India (9), China (19), and Russia (44).

The issue, however, is not just the absolute number of doctors but their uneven distribution. The Medical Demographic Survey 2020, carried out by the Department of Medicine of the University of São Paulo (Brazil), pointed out that the country reached the figure of 500,000 doctors in the first year of the biggest health crisis in the recent history of the country and the world—a very good number—but a statistic which hides geographic inequality (Scheffer et al., 2020). Specifically, more than half of doctors in Brazil, 53%, are concentrated in the Southeast region (which covers only four states) and help a considerable number (42.1%) of the Brazilian population, while in southeastern Brazil there are far fewer doctors (about 28) for every 10,000 inhabitants. In the north and northeast, this number drops by less than half (11 doctors). In some regions, such as the Amazon, the situation is even worse.

This uneven distribution of doctors by region is a problem faced by all nations, including the most developed ones. When we talk about access to health, there are three main points to be analyzed:

(1) The right to health—having the right to access healthcare guaranteed, whether privately, through health insurance, or through public policies;
(2) Easy access—in addition of having the right, here we refer to having doctors, hospitals, or healthcare units easily available for care;
(3) Quality of care—some populations have the right and have access in their region, but often to a single general practitioner, creating difficulty in obtaining consultations with specialists, without access to advanced exams and treatments which compromises the quality of healthcare.

Under such different conditions across countries and their regions, it is hard to believe that the United Nations 2030 Agenda will be accomplished without the massive application of technology. In fact, I can envision telemedicine solving many of these problems. Public policies can encourage the application of digital health solutions to bring access to quality medicine to the most vulnerable. Solutions such as the use of Artificial Intelligence for medical diagnostics and cloud computing can offer incredible opportunities for the poorest countries to access healthcare services even from other continents.

Telemedicine: Definition and general overview

I consider it essential that we all have the same understanding of what telemedicine is and what it encompasses. I realize that the term "teleconsultation" is often used as a synonym for "telemedicine," but this is a mistake. The WHO considers telemedicine any exchange of information between health professionals via information and communication technologies (ICTs) to determine diagnoses, prescribe treatments, and prevent pathologies, in addition to doing research for educational purposes.

Therefore we can say that telemedicine is an "umbrella" term, which includes services and practices such as telediagnosis, teleconsultation, and teleinterconsultation. Although teleconsultation has become very popular with the arrival of the COVID-19 pandemic, other telemedicine practices have been happening and revolutionizing the field of medicine for a much longer time, examples of which I will discuss later in this chapter.

Teleconsultation is nothing more than a medical consultation that takes place remotely, through secure online communication technologies. But this cannot be done through any video application; instead, the medical clinic needs an adequate infrastructure which guarantees the security of the data shared, such as exams and medical prescriptions, in addition to secure storage of this information. Today, each country has its own regulations to ensure the security of this data. In Brazil, we have the General Personal Data Protection Law (LGPD). The United States has the Health Insurance Portability and Accountability Act (HIPAA). Europe has the General Data Protection Regulation (GDPR).

Telediagnosis is the issuance of a report or evaluation of exams through images and/or graphics sent via the internet, regardless of where the patient and the specialist are located. The tests which can be performed through telediagnosis include: visual acuity, electrocardiogram, electroencephalogram, spirometry, X-rays, mammography, tomography, magnetic resonance imaging, map, holter, and even treadmill testing.

Telemedicine is also part of a concept known worldwide as eHealth or "digital health." According to the Healthcare Information and Management Systems Society (HIMSS), eHealth is any internet application, used together with other information technologies, focused on providing better conditions for clinical processes, patient treatment, and better funding conditions for health systems (Bitar & Alismail, 2021).

In this eHealth concept, we can include population health management, i.e., putting powerful digital solutions in the hands of public managers to support them in decision-making. I will go into detail on this topic later on in this chapter.

On the shoulders of giants

As the great British mathematician and physicist Isaac Newton said, "If I have seen further, it is by standing on the shoulders of giants." Following his tenet, I think it's worth telling, in part, the story of the evolution of telemedicine and mentioning the giants who came before us and paved a path of no return.

The first major example of telecommunications for medical purposes occurred during the American Civil War, between 1861 and 1865, in the form of the telegraph machine. The telegraph was used to order medical supplies and transmit accident reports. Decades later, the invention of the telephone and television were giant strides in remote communication, and medical professionals quickly embraced the new technologies.

A century later, in 1967, the first most complete telemedicine system was installed to connect the medical station at Boston's Logan Airport to the Massachusetts General Hospital, including the first remote medical diagnostics.

Meanwhile, on the other side of the Atlantic, in 1970, the University of Rome (Italy) carried out the first experiments with Tele-ECG (electrocardiogram). This planted a powerful seed that stimulated cooperation between universities, research centers, and the emergence of several telemedicine companies in the country.

Finally, the arrival of the internet—starting in 1969 and becoming popular in the mid-1990s—changed everything. The internet was essential for the advancement of telemedicine. Through the World Wide Web, scientists and medical experts, and the brightest brains in healthcare, have had opportunities and tools to carry out research and to implement remote treatments like never before. In addition, the internet reduced costs for implementing digital technologies.

Thus brick by brick, remote medical care was recognized by the WHO in the early 1990s.

In contemporary times, telemedicine has advanced a lot, thanks to A.I. resources, mainly in the automation and definition of medical priorities, or in emergency cases. This knowledge, added to the patient's history, which is all stored digitally, brings many advantages to doctors and patients in the definition of increasingly accurate medical diagnoses and treatment plans.

Many challenges, however, remain, especially in relation to regulations which overcome geographic barriers, interoperability, and integration between different systems that allow a full view of the patient. Additionally, of course, ethical discussions about security and access to health data are crucial. Fortunately, progress has been made in these areas, greatly driven recently by the COVID-19 pandemic, as I will explain as follows.

COVID-19 pandemic brings a disruption

Despite the ongoing technological evolution, the anachronism of some medical professionals and even of associations and governments held back the true potential of telemedicine. I will give the example of Brazil. In 2002 the Federal Council of Medicine (CFM, the acronym in Portuguese), which regulates medical activity in the country, published Resolution No. 1643, regulating telemedicine (Conselho Federal de Medicina, 2022). The problem is that it was designed in a generic way, allowing only teleinterconsultation, i.e., remote interaction between doctors and emergency cases, and did not support the development of business models in the country.

After this initiative, over 15 years later, in 2018, the CFM made another attempt to sanction telemedicine through resolution No. 2227. This measure brought more detailed rules, but allowed telediagnosis, teleconsultation, and even telesurgeries. However, just a month later, the step forward was vetoed due to disagreements from medical professionals, and the latter resolution, from 2002, remains valid.

It was only in 2020, with the beginning of the COVID-19 pandemic, that all resistance and unyielding rules gave in. Contamination took place in a geometric progression, and public administrators from all continents decided that it was necessary to adopt measures to avoid overloading and imploding a healthcare system which was already at its breaking point threshold.

Pressured by the pandemic global health emergency, the governments of almost all countries decided to loosen their rules for telemedicine.

Thus teleconsultations emerged as the natural response to assist the population, unburdening public and private healthcare systems, maintaining a safe distance for doctors and patients, and expanding the capacity of care, including routine care.

It is also worth acknowledging the significant benefits of teleconsultations in response to the COVID-19 pandemic emergency, by making efforts to care for the mental health of the population at such a delicate time, when disorders and diseases, such as anxiety, depression, and panic syndromes, became prevalent, either for the first time, or accentuated. In addition, telemedicine allowed psychologists and psychiatrists to attend to patients in need of support, by reaching them at home through technology tools.

Regulations around the world

In the United States, the practice of telemedicine is regulated at the state level. In addition to the different definitions of telemedicine and what it encompasses, each state in the federation may have different requirements and standards. These include consent to work outside their region, permission for communication methods, supervision requirements for nurses, remote prescribing, and health insurance coverage requirements.

South Africa, in comparison, does not have a single primary legislation that specifically regulates telemedicine. However, certain aspects of remote healthcare services are governed by the National Health Act and the Health Professions Act. For example, no person should practice any health profession in the country unless registered with the Health Professional Council of South Africa (HPCSA). An American doctor, for example, could not provide teleconsultation services to a person in South Africa without registering under the Health Professions Act. In stark contrast to these regulations for human beings, artificial "helpers," in the form of Bots or robots, often used in telehealth services, do not need registration under the Health Professions Act.

In China, a series of administrative rules were enacted in July 2018 by the National Health Commission and the National Administration of Traditional Chinese Medicine. These rules include administrative measures for Internet-based Diagnosis and Treatment, Administrative Measures for Hospitals on the Internet, and Good Administrative Practices for Remote Medical Services.

For the French, the use of telemedicine has been authorized on an experimental basis since 2009 and was fully authorized for some healthcare professionals starting in 2019. The law separated professionals into two distinct areas: telemedicine and telecare. Telemedicine is aimed at medical professionals (doctors, dentists, and midwives). Telecare targets pharmacists and medical assistants (nurses, physiotherapists, occupational therapists, speech therapists, radiology and laboratory technicians, nutritionists, and others).

Post-COVID-19 regulatory setbacks

The positive response of doctors and patients, and the "miracles" operated by telemedicine during the most difficult months of the pandemic, however, failed to silence dissonant voices already beginning to show signs of regression as the pandemic appeared to be abating. In the United States, regulatory authorities are already considering going backwards and tightening some rules on the practice of telemedicine again.

This can have dire consequences and retrograding in care, as over the past 2 years, Americans have been able to see doctors in other states and start important treatments that may now come to an end.

Such backpedaling is also at stake in Brazil and other parts of the world, sparking heated debates.

How telemedicine helped fight COVID-19 and other diseases

The pandemic has exposed, and exacerbated, various bottlenecks in public and private healthcare. The first widely publicized problems were the lack of ICU beds and ventilators, followed by the lack of doctors, nurses, specialists, and critical care physicians, and even a lack of personal protective equipment (PPEs). In the case of critical care physicians, the problem was blatant by the numbers: Brazil has only about 6500, when close to 40,000 would be needed to attend at all ICUs in the country.

On the patients' side, one of the most harmful consequences was the postponement of treatments for chronic diseases. Due to the massive influx of patients with COVID-19 conditions, many hospitals were forced to fully dedicate their services and capacity to the care of these COVID-19 cases, and postponed treatment and follow-up of other diseases, not to mention postponement of elective but needed surgeries.

In this chaos, dozens of telemedicine solutions emerged as strong allies to reduce service queues, and assist in medical diagnoses and in the exchange of experiences between doctors, and even to provide remote ICUs to provide services outside of the hospital's crowded and stretched premises.

Telemedicine's protagonism

The numbers don't lie: telemedicine has helped save lives and unburden meager healthcare systems. According to a survey by the Brazilian Association of Telemedicine and Digital Health Companies, between 2020 and 2021, more than 7.5 million virtual consultations were carried out in the country, by about 52,000 doctors (Saúde Digital Brasil, 2021). Of these, 1% was essential to save lives, i.e., at least 75,000 people had their lives preserved by quick access to healthcare.

Another very important data was the rate of resolution of the appointments, which was 91% in the first consultations. In terms of satisfaction, 90% of patients classified the services as excellent or good. This constitutes a big success, deserving rave reviews.

The same association revealed that the numbers of teleconsultations for cases of influenza and a variant of COVID-19 (Omicron) soared between the end of 2021 and January 2022, doubling numbers of cases every 36h. At the end of the year, the number of appointments jumped from 7000 to 15,000, and the following week, in the first days of January, the numbers indicated that they would exceed 50,000 appointments (Saúde Digital Brasil, 2021).

In the United States, analysis by McKinsey consultants pointed out that the use of telemedicine increased 38 times during COVID-19, compared to pre-COVID-19 numbers (Bestsennyy et al., 2021). About 4 months after the start of the pandemic, in April 2020, the general use of telehealth for consultations and outpatient care was 78 times higher than in February 2020. This sudden increase occurred as consumers and providers looked for ways to access and provide medical care safely, i.e., with low risk of contamination.

Telemedicine's technologies and tools that are changing healthcare

The sky is the limit when it comes to technology. In the case of telemedicine, advances and innovations are substantive and include categories such as mobile health (mHealth), wearable devices, augmented reality, robotics, and 5G. We also have medical apps, remote exams, software to support decisions and diagnostics, A.I., and machine learning bringing about true digital disruption in the midst of a pandemic.

The tools and technologies developed today leverage the power of physicians for increasingly accurate diagnoses.

In the case of a patient in need of rapid medical intervention, tech can connect a specialist physician to the healthcare professional who is performing the care. This agility in exchanging information and guidance is often a lifesaving differential. The tools available not only allow contact between these professionals, but also help them, through A.I., to find anomalies in exams to accelerate and perfect the diagnostic process.

Data crossing and systems integration are examples of how technology is also an ally of public health managers. Dashboards which provide reports and alerts enable these authorities to make immediate decisions, focused on health promotion and disease prevention.

It is important to emphasize, however, that while these tools and technologies dramatically support medical care, tech machines and tools will never replace the role of the human professional. The tech tools will aid the real provider, by arming live professionals with data and a more complete view of the patient's health for their more-informed decision-making.

I understand that in order to fulfill the mission of bringing access to quality medicine to the population, an integrated effort is necessary between public authorities and the tech and health sectors. Since our work began on *Portal Telemedicine*—we founded the company in 2013—all our solutions which foster innovation go through a research and development stage financed by partners.

I have selected five real-life stories to share in this book which clearly illustrate how telemedicine and its technologies can democratize valuable access to healthcare in developing countries, of continental dimensions, to ameliorate the problem of their unequal distribution of professionals, lack of specialists, restricted resources, long distances, and other problems.

I'll start by telling you about an innovative solution with intense use of A.I., which was a strong ally in the face of COVID-19.

Case 1—Automatic detection of COVID-19 in imaging exams

As soon as the pandemic began, the team of engineers and developers at *Portal Telemedicine* dedicated themselves to finding answers to help reduce the impact of the disease on the population. We decided to create an intelligent system capable of detecting the presence of COVID-19 and pneumonia in X-ray exams. This helped speed up medical diagnoses, which was critical since PCR test results were taking days to come out at the time. It is worth remembering that agility in diagnosis is important not only for monitoring the patient's health, but also for isolating him/her as soon as possible, reducing the transmissibility of the virus.

Using Artificial Intelligence and Machine Learning components, the solution scans the image exams, evaluates if there is a change, and reports it to the doctor within 10min. Along the course of the project, we understood that the COVID-19 diagnosis needed to be distinguished from common pneumonias, so these detections were included in the model. Today, the A.I. model identifies infiltration, consolidation, effusion, and pneumonia and whether or not the pneumonia is due to COVID-19.

Fig. 1 shows an AI heatmap in a chest X-ray that is able to detect COVID-19, a valuable tool as a platform for telediagnosis and automatic detection of infection in an imaging exam.

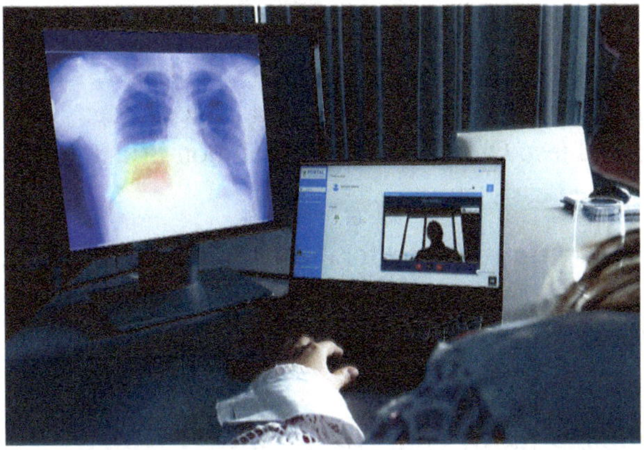

Fig. 1
An AI heatmap chest X-ray that can detect infections like COVID-19.

The exams which show abnormalities undergo automatic screening and go to the front of the queue, drawing the attention of clinical staff to the most serious cases. During the height of the pandemic, what most interested doctors in X-ray exams was not only knowing if the patient was contaminated, but the agility in knowing the degree of lung injury to triage who should be hospitalized and who could be sent home, a crucial decision at a time of scarcity and lack of hospital beds.

This approach is a low-cost solution, with 95% accuracy in the results and does not require any additional investment in equipment.

Case 2—The power of AI to fight cancer

Portal uses this same type of solution for detecting cancerous nodules on chest X-rays. Based on a large database, A.I. can detect, with millimetric precision, when there is evidence of a nodule and alert the physician responsible for the exam that this aspect deserves attention. This helps increase medical assertiveness, since most X-rays are not performed for this purpose, which contributes to the fact that anomalies that are not so evident go unnoticed by the doctor.

This solution was developed in 2 years of research with the Eldorado Institute and the Brazilian Cancer Control Institute (IBCC), currently São Camilo Oncology.

According to the World Health Organization, cancer is the second most deadly disease in the world. In 2018 one in six global deaths was related to cancer.

Early detection increases the chances of effective treatment. That's why we wanted to develop a solution capable of shortening the period—from days to hours—between the completion of an exam and receiving the result.

In Brazil, according to the Brazilian Federation of Philanthropic Institutions to Support Breast Health (Femama), the average time for diagnosis of the disease is 270 days. This causes a considerable part of patients to start treatment at an advanced stage. This rate varies depending on the type of cancer and test. Our goal with this solution is that the majority of patients can have this disease identified early on.

In the medium and long term, this support for medical diagnosis has a high impact on reducing the number of cancer deaths. This, combined with worldwide prevention programs, has a huge potential to have a very positive impact on global health.

Case 3—Telemedicine improving the HDI in Tarumã

The city of Tarumã is located in the countryside of the state of São Paulo, Brazil, just over 400 km from the capital. Its population is about 15,000 inhabitants. Mayor Oscar Gozzi, after being elected, set himself a goal to be among the 10 cities with the best Human Development Index (HDI) in the country by 2027.

Among the aspects which impact the HDI is longevity. A survey carried out by the city hall found that 23.5% of early deaths in Tarumã were related to cardiovascular problems (Amaral et al., 2022). Two factors anticipated deaths: the delay between performing the electrocardiogram exam and issuing the result, and the lack of specialists who could adequately monitor the patient. In view of this, the mayor went in search of a solution that would allow him to improve these indices.

With the implementation of the telecardiology system in 2019, using A.I. in all its primary care units, the city reduced the waiting time to receive exam results reported by cardiologists from several days to merely 5 min in urgent cases. For time-sensitive diseases, for which time is essential to save lives, as in the case of a heart attack, the cardiologist can now contact the technician who performed the exam by videoconference to provide guidance on how to proceed to stabilize the patient. This entire process is carried out through a secure system, which guarantees the confidentiality of information.

Seven months after the technology was implemented, the death rate from chronic noncommunicable diseases (NCDs), which include heart disease, had a 45% reduction in the city. In the same period, surrounding cities recorded an 18% increase in deaths from NCDs. It is worth noting that the municipality did not have to invest anything in new medical equipment, since the telemedicine solution is integrated with devices already in use and only needed to pay the cost of the internet and telemedicine service.

A year-and-a-half later, the city estimated that it had reduced the cost of heart disease by 30%. The improvement in primary care resulted in less emergencies, time spent in the ICU, ambulance use, need for hospitalization, beds and cardiac rehabilitation services, among other benefits.

Case 4—Life or death

Portal Telemedicine was born with the purpose of connecting patients from remote areas to specialist doctors. The story of Paulo, a resident of Coari, a small municipality in the state of Amazonas, is a good example of how it is not only possible, but essential, to connect a suffering patient located in a remote region to a specialist doctor.

Coari is located 363km from Manaus, the capital of Amazonas. Getting there is only possible by river or plane and can take up to 24h by boat. In the region there are about 200 indigenous villages, but almost no doctors, let alone specialists. It's known as a medical desert.

Paulo is a patient with heart problems, who had been bedridden for a year. The doctor in the region, a general physician, did not know exactly what Paulo's diagnosis was and how to treat him. That's exactly where our telediagnostic solution helps. Despite the remoteness of the village and the medical center, there was an EKG machine and a Holter monitor, a small, wearable electrocardiogram (ECG) device that tracks the heart rhythm during everyday activities. We integrated our technology directly with this equipment and, as exams were performed, the system sent the information to the cloud, a safe environment, where it was immediately available to the doctor.

On the other side of the country, in São Paulo, a major capital city in Brazil, about 4000km away, an experienced cardiologist examined the case and determined the diagnosis. A few minutes later, Dr. Ricardo, the local general physician working in Coari and the remote region around, received the patient's diagnosis. The case was an emergency: the patient, Paulo, had a problem with his pacemaker and needed urgent surgery. Fortunately, he was transferred to a hospital in Manaus where he had the lifesaving surgery.

In addition to the health impact, the social impact which this platform brings to developing countries becomes clear. Almost no infrastructure is required, only internet access and installing our drivers in the medical equipment. While this may seem easy, developing this solution was challenging, including lack of interoperability, since medical devices do not communicate with each other or with other devices. Each device has its own proprietary protocols; there is no cloud integration. This makes the diffusion of telemedicine on a large scale very difficult.

To resolve this, we established several partnerships, like Google and medical device manufacturers, and sought several economic grants for driver development, i.e., connectors capable of establishing communication directly with any model or brand of medical equipment, covering the entire national installed facilities.

It worked out. And the process became very easy for doctors or nurses. All they need to do is operate the equipment they are used to operating. While performing the exam, we collect the telemetry from the device, send it to the cloud, and make it available to the specialists. All this is done in a 100% automated way, allowing patient data to be evaluated by specialists from all over Brazil and even abroad, if applicable.

Today, we have more than 6000 doctors on our collaboration platform, such as cardiologists, neurologists, radiologists, as well as thousands of patients. This tool now allows access to quality healthcare in more than 600 Brazilian cities and states, as well as in Africa, Angola, and Rwanda.

Case 5—Population health management and the fight against fraud

In terms of technological infrastructure, what we see in almost all the countries and public health institutions we visit is a great babel of systems, programs, and databases. The lack of integration, standardization, and interoperability becomes a perfect storm for fragmenting the population's health information, delaying the production of decision support reports, and opening loopholes for fraud.

One of the successful cases in this area is that of the Health Department of the State of São Paulo. Using technologies such as Big Data, Data Engineering, A.I., and Business Intelligence, we started the project by mapping the needs of the various teams in the secretariat and translating this complexity into managerial and investigative panels or dashboards. Through these panels, today managers have data on patients, diseases, medicine, supplies, and medical procedures, all in the palm of their hand. Even lawsuits can be tracked using the platform.

The information was organized temporally, geographically, and grouped by similarity. This allows, for example, users to understand the growth of unexpected demands and suspected cases of fraud and act preventively. In these cases, the tech also facilitates lawsuits when they arise, explored in the next section of this chapter.

Fighting fraud

With just a few clicks, users, even without technical knowledge of A.I., are able to interact with the new system and group similar cases. The user points out a known case of fraud, the system cross-checks information and shows cases that should be further analyzed. This helps identify groups of people who engage in illegal actions.

This data integration allows the creation of an intuitive interface that generates alerts of suspicious cases which need to be analyzed. For example, in our system, we identified people receiving medication after they had been dead for years. The system generates alerts so these treatments are stopped immediately in case of death of the user. Another problem is when the patient receives the same drug from two different government programs. As those systems did not communicate with each other, it was extremely labor intensive to map these duplications, while today, they are revealed automatically.

Further, the platform brings managerial efficiency, by allowing a quick view of relevant indicators, including recurrent diseases with an upward trend, low stocks of medicines, more judicialized items, consumption value for each regional health department, and tracking greater financial impact.

All this automated and efficient management can generate great savings for practitioners and health services, as well as states. This allows public health management to international levels. I am not referring here only to better management of public funds, but to a cultural change in which the government will be able to equip itself with health data to create public policies that act in a preventive and predictive way. This constitutes an inversion of the logic of treating the disease rather than generating actions to avoid it.

Managers can, for example, evaluate regions with a low child development index, cross this data with the vaccination rate, more frequent diseases in the region, among others, as an easier way for the public administrator to establish connections that indicate possible areas in which they can act to improve this scenario at a faster speed. Such crossing of data can also allow actions to reduce the rate of chronic diseases, among a series of important care aimed at promoting health.

Looking to the future

The COVID-19 health crisis has presented a valuable opportunity for the modernization of healthcare services around the world. It opened the doors and welcomed all digital services. Allowing this to recede will be like admitting that we have learned little.

Most of the cases mentioned before are solutions that were already being developed and applied in Brazil before the COVID-19 pandemic, but its arrival changed the perspective and acceptance of the medical community and patients globally regarding the use of the technology to provide assistance.

We have made leaps in terms of regulations and the world has come to understand the potential of technologies in healthcare. In addition to facilitating access to doctors, health techs come to empower these professionals, as well as patients and public managers, with access to health data that allow faster and more effective actions.

In Brazil, for example, we have public healthcare systems in the same city that do not communicate with each other. The impacts of this became more evident during the pandemic.

The delay in communicating the number of cases and deaths from the municipalities to their state and from the state to the Ministry of Health, of the Federal Government, led large communication networks to come together to form a consortium to monitor cases with state secretariats. The numbers obtained by this group differed from the official data, which were updated days later.

This is a single example of the challenges faced by public health management when systems do not connect in an automated way.

I really like to use the word interoperability and I see this as a great uniqueness in what we do. Systems need to communicate, patient health has to be ahead of any other interests. In order for health professionals to have a 360 degree vision, there must be this synergy and I see this as the great challenge of today in telemedicine.

My vision for the future is, first of all, that telemedicine will effectively reduce the barriers in the care of the most vulnerable populations, and that they will no longer remain on the sidelines of the best techniques, the best medicines, and the most advanced therapies.

I also dream of creating a global network of specialist doctors for second opinions, especially for rare and complex cases, and especially in remote settings.

Thanks to the potential of Artificial Intelligence, I believe it is perfectly possible that, soon, when a disease is identified, such as a cancerous nodule, a patient will have the opportunity to be referred to the best specialist in the world in that specific area.

With global challenges posed by the pandemic, the mindset of doctors, patients, and governments about the importance of telemedicine has shifted to be more open, and I believe that today the world is prepared to discuss international health regulations and agreements.

We are ready to support this expansion and fulfill our mission of bringing quality medicine to everyone.

References

Amaral, F., Fernandes, E., Drabowski, N., Alves, M., Nunes, A., da Silva, E., Bastos, M., Romera, J., & de Figueroa, R. C. (2022). Tarumã case study: telemedicine and artificial intelligence applied for reducing cardiovascular diseases mortality and health cost optimization. In *SciELO Preprints*. https://doi.org/10.1590/SciELOPreprints.3482.

Bestsennyy, O., Gilbert, G., Harris, A., & Rost, J. (2021). *Telehealth: a quarter-trillion-dollar post-COVID-19 reality?*. Available at: https://www.mckinsey.com/industries/healthcare-systems-and-services/our-insights/telehealth-a-quarter-trillion-dollar-post-covid-19-reality. (Accessed 28 May 2022).

Bitar, H., & Alismail, S. (2021). The role of eHealth, telehealth, and telemedicine for chronic disease patients during COVID-19 pandemic: a rapid systematic review. *Digit Health*, *7*, 20552076211009396 (Published 2021 Apr 19) https://doi.org/10.1177/20552076211009396.

Chemweno, J. (2022). *The U.S. healthcare system is broken: a national perspective. [online]*. Managed Healthcare Executive. Available at: https://www.managedhealthcareexecutive.com/view/the-u-s-healthcare-system-is-broken-a-national-perspective. (Accessed 28 May 2022).

Conselho Federal de Medicina. (2022). *Resolução CFM nº 1.643/2002*. Brasília.

Saúde Digital Brasil. (2021). *Entidade aponta que telemedicina salvou mais de 75 mil vidas entre 2020 e 2021*. Available at: https://saudedigitalbrasil.com.br/press/entidade-aponta-que-telemedicina-salvou-mais-de-75-mil-vidas-entre-2020-e-2021/. (Accessed 28 May 2022).

Scheffer, M., et al. (2020). *Demografia Médica no Brasil*. São Paulo: FMUSP, CFM.

Who.int. (2017). *World Bank and WHO: half the world lacks access to essential health services, 100 million still pushed into extreme poverty because of health expenses.* [online] Available at: https://www.who.int/news/item/13-12-2017-world-bank-and-who-half-the-world-lacks-access-to-essential-health-services-100-million-still-pushed-into-extreme-poverty-because-of-health-expenses>. (Accessed 28 May 2022).

CHAPTER 20

Artificial intelligence and augmented reality in COVID-19 recovery and beyond: The expansion of emerging technology in the medical industry

Ashley Porto
Department of Clinical Psychology, Columbia University Teachers College, New York, NY, United States; International Association of Applied Psychology, Indianapolis, IN, United States

Introduction

The global rise of Artificial Intelligence (AI) and Augmented Reality (AR) is becoming increasingly vital to adopt as it generates expansive opportunities to revolutionize the healthcare industry in the extended COVID-19 recovery period. Consequently, patient outcomes will drastically improve compared to previous years, documented with key drivers attributed to factors including precision and accuracy. Prior to its utilization within the healthcare industry, AI and AR gained recognition within computerized gaming, as it has provided users with an experience that feels parallel to the real world. Within the gaming experience, AI simulates human-like behavior with the generation of responsive and adaptive behaviors while AR incorporates audio and visual senses to produce a real-life simulation. The combination of both technologies within the health space is proving to create powerful applications to benefit individuals, communities, and countries. This chapter explores the origins of Artificial Intelligence and Augmented Reality, the influence of COVID-19, technologies' contributions to the healthcare industry in the COVID-19 recovery period, and the significance of AI–AR (Artificial Intelligence–Augmented Reality) for the future of the healthcare industry.

The COVID-19 pandemic has caused havoc worldwide, with lost lives, disrupted societies, and increased inequalities due to driving a greater divide in access to leading-edge technologies. According to a United Nations report, approximately 3.7 billion people in the world lack access

Resilient Health. https://doi.org/10.1016/B978-0-443-18529-8.00020-2

to the internet, which demonstrates many individuals have not obtained equal access to modern day technology. Additionally, only 20 percent of individuals in less developed countries have access to technology.

Greater equity in technological resources is called for in the United Nations Agenda 2030 for Sustainable Development, which aims for a better world for all by the year 2030. The UN has deemed "Science, Technology and Innovation" as one of the core "means of implementation" to achieve by 2030 (Bassett and Serious Business, 2022). Specifically, Goal 17 features strengthening multilateral partnerships and collective efforts while 17.6–17.8 are devoted to technology. The targets aim to enhance partnerships and cooperation on, and access to, science, technology, and innovation, and ensure that all countries, including developing and least developed countries, have access to leading-edge technologies. Technological applications offer individuals access to basic services, such as e-health or online education, called for in Goal 4 about education for all. The use of technologies in the forms of AI, machine learning, blockchain, 5G, robotics, and cloud computing can increase progress in achieving the global goals. Businesses, governments, and civil society organizations alike are increasingly leveraging technology to promote sustainability and improve citizens' standard of living globally.

Providing healthcare technologies to underrepresented populations can help maximize their potential to prevent diseases, including mitigating the impact of pandemics like COVID-19, and provide improved care and patient outcomes.

Since 2010, approximately US$145 billion has been accounted for investments for healthcare start-ups, which demonstrates the significant need for technology within the medical field (World Economic Forum, 2018).

The role of artificial intelligence

Artificial intelligence (AI) has become a major tool in the access of technology to populations for varied uses. The concept of AI can be traced to Dartmouth College mathematics professor John McCarthy, in his endeavors in 1956 to imitate human-like thinking (TechJury, 2021). In the simplest of terms, AI can be defined as the ability for a computer to complete tasks typically done by humans, which require human intelligence. AI includes main components of machine learning and deep learning, which play significant roles in the advancement of AI technology. The concept of machine learning is centered around feeding data into a computer while continually updating the learning from it, while deep learning is focused on using algorithms to interpret objects and speech (Built In, 2021). Machine learning and deep learning both continue to adapt and learn from data in order to make more educated decisions.

Many cases of using AI have already become integrated within daily life without people realizing it, including in Siri and Alexa, which respond to the sound of a human's voice. Also, AI is used by companies like Tesla, in their cars and Netflix, in making recommendations about films for customers to see. One of the most well-known developments within the AI industry is a robot named Sophia, produced by Hanson Robotics, which gained universal recognition in being the world's first robot citizen in the country of Saudi Arabia (Hanson Robotics, 2020). Sophia first incentivized traction through her popular YouTube videos, which led to engagement worldwide, as individuals were astonished by the particular utilization of Artificial Intelligence.

Besides providing value, AI is big business. Currently, the AI industry is predicted to be valued at US$60 billion by 2025, while attributing to US$15.7 trillion of the Global GDP by 2030 (Hill, 2017).

AI, health, and COVID-19: 3 Benefits

The rise of AI continues to become widely integrated within more technological advancements specifically within the healthcare space during and post-COVID-19. During the COVID-19 pandemic, many hospitals faced challenges of overworked staff and limited space for patients other than those with the infection. Three major benefits emerged:

(1) AI and emerging technologies provided a more strategic way to achieve better outcomes for both patients and providers. An exciting area of AI development during the COVID-19 pandemic was Remote Patient Monitoring (RPM), which allowed doctors to virtually connect with patients without needing to meet in person. These emergency applications reduced costs, led to greater efficiencies, and paved a way for long-term solutions. Specifically, these automated processes reduced readmissions within hospitals by 38% and emergency rooms by 25%, improved patient satisfaction by 25%, and cut costs of care by 17%. These statistics demonstrate that technological advancements such as RPM can be beneficial for patients and drive the quality of a patient's life over time.

(2) AI has the power to comprehend patterns within diseases, which acts as a catalyst to identify and find appropriate therapies earlier. Specifically, Natural Language Processing technology is breaking through the industry by utilizing algorithms to create clinician's notes, pathology reports, and other appropriate documents. Algorithms within NPL technology have the ability to search and assess millions of relative content in order to allow clinicians to make more educated decisions about care for their patients, leading to a kind of accelerated brain power and decision-making. This "brain power" has proven to be more precise and accurate than the average

human brain, thus accessing knowledge easier compared to average human intelligence during times of COVID-19.

(3) Artificial Intelligence enables patients to take control of their lives. When given accurate insights from clinicians, patients feel more empowered to be able to make the best decisions for their life.

AI application to Telehealth

Within the telehealth space, AI has many beneficial applications. For example, AI is being tested in nursing homes by tracking patient behavior and alerting staff on declines and emergencies. The resulting data can help staff and patients make better informed decisions to plan for the present and future care, based on past decisions.

An example of this system whereby the technology tool provides feedback about the person's health status is the application *Oculenz,* focused on patient central vision loss. The app demonstrates a simulation of the vision of a person who is experiencing a particular symptom, and shows them the potential results that will happen if no action is taken, thereby helping to motivate the person to take precautionary steps.

AI has been shown to perform as well, or better, than humans in the area of diagnosing diseases, since "algorithms are already outperforming radiologists at spotting malignant tumors and guiding researchers on how to construct cohorts for costly clinical trials" (Spencer, 2020).

Augmented reality

Augmented Reality is another emerging technology which has provided value to "industries by increasing interaction and engagement to create a richer user experience." The term Augmented Reality was originally used by cinematographer, Morton Heilig, with the invention of the Sensorama Stimulator in the late 1950s, a prototype to awaken an individual's senses to realistically simulate an actual experience. This attempt was an innovative technology, to leverage AI to intensify experiences, given more data to pull and learn from (Interaction Design Foundation, 2020). The resulting exponentially powerful outcome was intended to be used for education and other purposes. One of the most popular utilizations of AR is in the gaming industry, with *Pokémon Go,* which gained traction internationally as users leveraged the ability to navigationally locate Pokémon characters using their mobile device. Sixty percent of AR gamers report that such an immersive experience as in Pokémon Go is a key driver in utilizing AR games, which demonstrated a huge window of opportunity for other industries, like healthcare. Integration of AR in gaming is expected to in the coming years. Further, it is predicted that NFTs, identified as a drawing or animated GIF, will be a driver for consumer adoption of AR. "Play-to-earn" NFT games, which involve trading assets for real money, are becoming the latest trend in the gaming world (Renesis Tech, 2022).

AR application to retail and education

Within retailing, AR is redefining and revolutionizing the space. AR has provided value-add to brick-and-mortar retail stores to entice shoppers to purchase in-store, to deal with the pressing competition of lower online prices. For example, the high-end retail fashion chain, *Saks Fifth Avenue*, utilized AR within its cosmetic department in 2018, encouraging customers to virtually apply products and test facelifts to create an immersive shopping experience "that no consumer could forget" (Canvas8, 2021). In another example, *Mirror*, an at-home workout mirror, used AR applications to allow customers to size clothes before purchase.

Drawing on these earlier successes, and due to the onset of the COVID-19 pandemic, many businesses began to slowly convert to a remote model, which has driven the adoption of AR as more individuals began working at home while technology increased productivity rates and minimized the amount of in-person activity. Forbes emphasizes how in 2020, global sales on headsets advanced to US$12 billion, rising to 50% from its previous year in 2019, attributed to the influx of companies switching to remote models. Since COVID-19, many businesses have adopted AR within retail and education as a helpful tool for remote workers to use in order to increase efficiency and productivity during daily tasks. Employees had ability to virtually call into their work environment from the comfort of home using an AR headset. In addition, shoppers could analyze how a dress would fit on their body. And exercise enthusiasts could follow at-home workouts based on a virtual avatar portrayed on the screen.

To create the AR experience, an individual uses an Oculus, a virtual reality headset (or "glasses") especially designed for Augmented Reality, which powers different types of software to create an immersive experience that simulates a real-life experience. Within the field of education, use of these headsets outfitted with AR can teach a particular expertise, with more accessibility and accuracy compared to a human being's teaching capacity. For instance, workforce education for manufacturing has typically been a low-technology space, but *Skill-XR* offers services through AR glasses whereby young workers can attain hands-on experience to learn how to utilize a particular type of equipment with training that is alternatively safer and more effective by the mechanism than by a human being (Convergence Design Lab, 2020). At the moment such equipment can be costly; but as Augmented Reality headsets become less expensive over time, there will be more of a demand for the implementation of this type of technology in many aspects of modern society.

With AR, training in various fields can be more effective, faster, less expensive (cutting down on travel and equipment), and more accessible by minimizing the need for human interaction within the educational process.

AR application to orthopedics

In the healthcare space, AR has revolutionized the industry of orthopedics by providing educational material for 3D visualizations for difficult structures of anatomy as well as augmented information utilized by surgeons to make decisions during CT and MRI procedures. *Osso VR*, a surgical training platform which focuses on providing access to training tools through immersive simulations, is breaking traditional norms in orthopedics and advancing the industry. With AR, *Osso VR* can be used in residency programs, recreating the experience to feel real, to ensure that surgeons receive the best quality training. Another benefit is reducing the amount of waste for cadaver labs. Further, the platform can be used in the comfort and luxury of a surgeon's home, allowing more and easy access. Such an increase in access to surgical training allows surgeons to become more precise and accurate intraoperatively.

Companies using this healthcare technology are thriving. In July 2021, *Osso VR* closed $27 million in series B funding, aiming to utilize the money to reinvent surgery into a video game (Matney, 2021). Another company, *Proximie Surgical* reported in another chapter in this volume (Shibata-Germanos et al., 2024) and in news reports (Crunchbase, 2022) invented a multisensory platform for surgeon collaboration through digital technology featuring the ability to call into any Operating Room in the world (B. S. J. and Sweiry, 2021). In April 2021, *Proximie* raised US$38 million in its series B financing round, which they will continue to utilize while they grow into emerging markets (MobiHealthNews, 2021). Also, *Proprio Surgical* is a platform which combines machine learning and AR to produce extremely precise 3D images. In 2020 *Proprio* raised US$23 million in its series A funding, which they plan to use to perform higher quality surgeries (Proprio, 2020). The AR space is projected to reach US$297 billion by 2024, which will drive AR innovation within the orthopedic space to allow surgeons to make more informed decisions, which will improve the quality of patient outcomes for future generations.

The cardiology industry is also utilizing AR to better supply surgeons and clinicians. For example, a new training system has been created which dives deeper into anatomy and ultrasound by using a *HoloLens* to see the orientation of the ultrasound probe, 2D ultrasound image, and its correlation to the anatomy. Many cardiologists are gaining assistance from AR through viewing 3D images of myocardial scarring in the heart while performing ventricular tachycardia by reinforcing MRI or CT data in order to give surgeons more insight intraoperatively. AR is further utilized for cardiology in trainings in a classroom setting, where students receive educational information through accessing applications with 3D detail, which depicts exact details of the heart and circulatory systems. This allows future generations to receive high-quality education and interacting with different images to better understand information in front of them and make better decisions based on data shown.

AR further helps in distribution of resources, shown in the innovation *AED4EU*, which identifies where defibrillators can be located in a time of need.

The pharmaceutical industry is adopting more innovative solutions with AR, which are revolutionizing the field. With an application called *New Drug*, patients have the ability to view how a drug affects their brain in real time. This decreases the need for a physician to discuss with a patient, as users have educational tools at the use of their fingertips (The Medical Futurist, 2021).

AR can also be utilized by sales representatives of medical device companies to provide educational materials for their customers who are medical doctors, who can watch these instructions repeatedly and at their own convenience, to help them better understand their products prior to using them with patients.

Combining AI and AR

COVID-19 has provoked forward-driving conversations on how the future of technology and AI–AR can aid not only technology companies but also the healthcare industry.

The interconnected relationship between AI and AR is coined as Artificial Intelligence–Augmented Reality (AI–AR), which refers to the ability for the two technologies to work cohesively to enable social innovations for the future. The AR experience becomes more intensified and immersive when data becomes more readily available, which can be synced from AI. AI–AR primarily focuses on weaving Artificial Intelligence into Augmented Reality, which can create highly precise and critical analyses through Augmented Reality. This combination has become increasingly popular universally through a wide range of industries, from education to retail to healthcare. ABI Research (2021), a Global Technology Intelligence Firm, estimates that over 200 million active users of AR applications will leverage AI in some form by 2026, demonstrating a rise in AI–AR.

A multitude of companies have utilized this combination of AI–AR to enhance their overall insights and analytics. For example, *Qualcomm* is a technology company which focuses on creating software to leverage 5G smartphones, professional cameras, and gaming devices (Qualcomm, 2021). Also, *NVIDIA* has utilized Artificial Intelligence within AI–AR to create automatic simulation as well as content creation through their Cloud product called *Omniverse* (Takahashi, 2021). These emerging technologies are catalysts for the healthcare industry, showing how AI–AR can be integrated within traditional methodologies for medicine.

The orthopedics industry is becoming increasingly advanced over the COVID-19 recovery period, with mass integration of AI–AR in its healthcare technologies, evident in preoperative planning, intraoperative navigation, and postoperative rehabilitation. Preoperative planning is a preoperative method of previsualizing a surgical intervention with the assistance of Augmented Reality. *Enhatch,* a technology company which I work for, which is a pioneering leader in the orthopedics industry for AI–AR, creates solutions for medical device companies using AI. Enhatch recently introduced its Intelligent Surgery Ecosystem, a group of integrated technologies that provides medical device companies with access to transformational technologies such as 2D to 3D segmentation, patient-matched surgery, and Augmented Reality (Porto, 2024).

The CEO of Enhatch, Peter Verrillo, is a pioneer in the orthopedics industry for AI–AR and the founder of what he calls the "Intelligent Surgery Ecosystem." In describing this ecosystem and its implementation with Artificial Intelligence into orthopedics and surgery, he says, "The Intelligent Surgery Ecosystem is giving opportunities to small and medium device companies to compete with industry giants. Meanwhile, it's giving surgeons access to enabling technologies that will help improve their surgery and collaboration with industry" (Enhatch, 2021).

Telehealth and COVID-19

Since COVID-19, telehealth has shown value in addressing key risks of the pandemic in preventing, diagnosing, treating, and controlling the disease (BMC Public Health, 2020). Advantages are prevalent at all levels of the healthcare chain. For example, patients gain the ability to control when and where they make appointments, by logging on to internet sites from the comfort of their own home. Also, by using AR, hospitals can provide improved care for children, given that hospitalization is usually a traumatic experience for children. Using AR tablets can not only educate but also distract and entertain children who are hospitalized. With AR, the children can select unique avatars for entertainment and support while waiting to be seen, thus reducing their anxiety. A study by the Los Angeles Children's Hospital found that use of virtual reality headsets with children that allowed them to experience an alternate reality minimized anxiety and pain and even increased enjoyment of the experience (Virtual Reality in Pediatrics, 2021).

AI is also being tested in nursing homes by tracking patient behavior and alerting staff on declines and emergencies in their condition. This data helps staff make better-informed care decisions.

The road ahead: Application of emerging technologies

Applications of emerging technologies in the healthcare space, as the world continues to recover from COVID-19, will continue to show power to influence communities and countries. By combining AR and AI, ecosystems will expand their reach, and drive greater efficiency and improved patient outcomes as well as provider preparation and care.

Major growth is already evident in various areas of the healthcare industry, utilizing multiple technologies to benefit and improve patient outcomes and to allow increased access, availability, and quality of training, for and by providers for the benefit of population worldwide. As technology expands within the healthcare industry, the commitment of nations in the United Nations Agenda about closing the digital divide and making technology available to all countries is even more possible.

References

ABI Research. (2021, October 6). *Augmented reality leveraging AI to engage 200 million active users by 2026.* https://www.prnewswire.com/news-releases/augmented-reality-leveraging-ai-to-engage-200-million-active-users-by-2026-301393841.html.

B. S. J, & Sweiry, B. (2021, July 14). 3. Proximie. In *TechRound.* https://techround.co.uk/medtech/3-proximie/.

Bassett, S., & Serious Business. (2022). The role of sustainable technologies in achieving the sdgs. In *Next Big Thing Academy.* https://theworldwecreate.net/insights/role-of-sustainable-technologies-in-achieving-sdgs.

BMC Public Health. (2020). *The role of telehealth during COVID-19 outbreak: A…—Biomed Central.* https://bmcpublichealth.biomedcentral.com/track/pdf/10.1186/s12889-020-09301-4.pdf.

Built In. (2021). *Artificial intelligence.* https://builtin.com/artificial-intelligence.

Canvas8, Saks Fifth Avenue attracts shoppers in store with Ar. (2021). From https://www.canvas8.com/library/signals/2018/06/04/saks-ar.

Convergence Design Lab. (2020, October 4). SkillXR: Augmented reality to provide new skills for manufacturing workforce education. In *Convergence Design Lab, Purdue University.* https://engineering.purdue.edu/cdesign/wp/skillxr-augmented-reality-to-provide-new-skills-for-manufacturing-workforce-education/.

Crunchbase. (2022). Proximie—Funding, financials, valuation & investors. In *Crunchbase.* https://www.crunchbase.com/organization/proximie/company_financials.

Enhatch. (2021). Intelligent surgery ecosystem. In *Enhatch.* https://www.enhatch.com/intelligent-surgery.

Hanson Robotics. (2020 September 1). *Sophia.* https://www.hansonrobotics.com/sophia/.

Hill, P. (2017, June 30). PWC: Ai will contribute $15.7 trillion to global economy by 2030. In *Neowin.* https://www.neowin.net/news/pwc-ai-will-contribute-157-trillion-to-global-economy-by-2030/.

Interaction Design Foundation. (2020). Augmented reality—the past, the present and the future. In *The Interaction Design Foundation.* https://www.interaction-design.org/literature/article/augmented-reality-the-past-the-present-and-the-future.

Matney, L. (2021, July 7). Osso VR raises $27 million to turn surgery into a video game. In *TechCrunch.* https://techcrunch.com/2021/07/07/osso-vr-raises-27-million-to-turn-surgery-into-a-video-game/.

Medical Futurist, 2021. Augmented reality in Healthcare: 9 examples. The Medical Futurist (2021, November 3). https://medicalfuturist.com/augmented-reality-in-healthcare-will-be-revolutionary/.

MobiHealthNews. (2021, April 21). Proximie lands $38m for Tech that helps surgeons virtually join operations. In *MobiHealthNews.* https://www.mobihealthnews.com/news/proximie-lands-38m-tech-helps-surgeons-virtually-join-operations.

Porto, A. (2024). Innovative technology: The Enhatch intelligent surgery ecosystem. In J. Kuriansky, & P. Kakkattil (Eds.), *Resilient health: Leveraging technology and social innovations to transform healthcare for COVID-19 recovery and beyond* (pp. 1099–1101). Elsevier.

Proprio. (2020). News—proprio raises $23 million in Series A funding led by DCVC. In *Proprio.* https://www.propriovision.com/news/proprio-raises-23-million-in-series-a-funding-led-by-dcvc.

Qualcomm. (2021). About Qualcomm: Company Information & History. In *Qualcomm.* November 9 https://www.qualcomm.com/company/about.

Renesis Tech. (2022, March 14). Play-to-earn NFT games-the new trend in the Gaming World: Renesis Tech. In *Medium.* https://blog.cryptostars.is/play-to-earn-nft-games-the-new-trend-in-the-gaming-world-renesis-tech-5b575da0aa85.

Shibata-Germanos, S., Ferenchick, E. K., Varallo, J. E., Semrau, K. E. A., & Hachach-Haram, N. (2024). Leveraging technology for health. In J. Kuriansky, & P. Kakkattil (Eds.), *Resilient health: Leveraging technology and social innovations to transform healthcare for COVID-19 recovery and beyond* (pp. 265–279). Elsevier.

Spencer, M. (2020, October 12). AI will help diagnose, predict and treat your health illnesses. In *The Last Futurist*. https://lastfuturist.com/ai-will-help-diagnosis-predict-and-treat-your-healthillnesses/#:~:text=Today%2C%20algorithms%20are%20already%20outperforming%20radiologists%20at%20spotting,ability%20to%20find%20the%20best%20course%20of%20treatment.

Takahashi, D. (2021, November 9). Nvidia's omniverse adds AR/VR viewing, AI training, and ai avatar creation. In *VentureBeat*. https://venturebeat.com/2021/11/09/nvidias-omniverse-adds-ar-vr-viewing-ai-training-and-ai-avatar-creation/.

TechJury. (2021, November 1). *Artificial Intelligence Statistics [updated for 2021]*. https://techjury.net/blog/ai-statistics/#gref.

Virtual Reality in Pediatrics. (2021, June 4). *Children's Hospital Los Angeles*. https://www.chla.org/virtual-reality-pediatrics#:~:text=Gold%2C%20a%20clinical%20psychologist%20and%20director%20of%20theyadolescents%20undergoing%20painful%20procedures%20such%20as%20blood%20draws.

World Economic Forum. (2018). *System initiative on shaping the future of food....* https://www3.weforum.org/docs/WEF_Innovation_with_a_Purpose_VF-reduced.pdf.

Further reading

Cordin, E. (2021, November 25). *Best NFT games to play & invest in [November 2021] - blog*. The Island Now https://theislandnow.com/blog-112/best-nft-games/.

DAIC. (2021). Augmented reality aids tee training by showing anatomy and ultrasound orientation. *DAIC*. September 9 https://www.dicardiology.com/videos/video-augmented-reality-aids-tee-training-showing-anatomy-and-ultrasound-orientation.

HC Innovation Group, Beyond covid-19: Four ways ai is shaping the future of Healthcare. (n.d.). https://www.hcinnovationgroup.com/analytics-ai/artifical-intelligence-machine-learning/article/21223186/beyond-covid19-four-ways-ai-is-shaping-the-future-of-healthcare.

Health Imaging. (2021). Augmented reality helps cardiologists visualize 3D images, plan complex procedures. In *Health Imaging*. February 25 https://www.healthimaging.com/topics/cardiovascular/augmented-reality-helps-cardiologists-visualize-3d-images.

Song, D. J. (2020, February 11). Mirror: Digital disrupter in fitness and beyond. In *Digital Innovation and Transformation*. https://digital.hbs.edu/platform-digit/submission/mirror-digital-disrupter-in-fitness-and-beyond/.

United Nations. (n.d.). Billions of people in developing world still without internet access, new UN report finds. United Nations. https://www.un.org/sustainabledevelopment/blog/2015/09/billions-of-people-in-developing-world-still-without-internet-access-new-un-report-finds/.

Vardomatski, S. (2021, September 14). Council post: Augmented and virtual reality after covid 19. In *Forbes*. https://www.forbes.com/sites/forbestechcouncil/2021/09/14/augmented-and-virtual-reality-after-covid-19/?sh=7111bdf72d97.

Wingfield, N., & Isaac, M. (2016, July 11). Pokémon go brings augmented reality to a mass audience. In *The New York Times*. https://www.nytimes.com/2016/07/12/technology/pokemon-go-brings-augmented-reality-to-a-mass-audience.html.

CHAPTER 21

Leveraging technology for health

Shannon Shibata-Germanos[a], Erin K. Ferenchick[a], John E. Varallo[b], Katherine E.A. Semrau[c,d], and Nadine Hachach-Haram[a,e]

[a]*Proximie, London, United Kingdom,* [b]*Jhpiego, Baltimore, MD, United States,* [c]*BetterBirth Program, Ariadne Labs, Boston, MA, United States,* [d]*Harvard Medical School, Boston, MA, United States,* [e]*Department of Plastic and Reconstructive Surgery, Guys and St Thomas Hospital, Westminster Bridge Road, London, United Kingdom*

Introduction

The COVID-19 pandemic has ushered in an unprecedented revolution in the use of healthcare technology. Digital health solutions such as mobile health, health information technology, wearable devices, and telehealth platforms have been rapidly developed and diffused to meet new and growing demands.

While this has been a global phenomenon, low- and middle-income countries (LMICs) have specifically positioned many of these solutions to improve equitable access to health services and future-proof health systems. Achieving universal health coverage, both during the pandemic and in its wake, necessitates capitalizing upon the vast potential of technology including the relatively new tools of Augmented Reality (AR), Virtual Reality (VR), and Mixed Reality (MR) to not only improve diagnosis and treatment, but also to enable health systems to deliver decentralized health services in the safest and most cost-effective way, moving toward a more efficient and people-centered healthcare delivery. Many digital transformations, however, do not reach their full potential due to practical challenges such as the proliferation of pilot projects (Wilson et al., 2014), poor coordination, inadequate preparation of the health workforce, and a lack of sustainable financing (Tran Ngoc et al., 2018).

In this chapter, we illustrate the power of the AR/VR revolution and its potential to transform the delivery of healthcare through the example of Proximie, a novel, secure, augmented reality software platform for virtual surgical collaboration that allows surgeons to "scrub in" anytime, from anywhere. Recent experience with Proximie technology transfer in Kenya in partnership with the Ministry of Health, Jhpiego, and Ariadne Labs suggests that digital health tools implemented during the pandemic will only be successful if early attention is paid to practical

Resilient Health. https://doi.org/10.1016/B978-0-443-18529-8.00021-4

challenges and if human behavior, with both its resistance to and capacity for change, is thoughtfully considered. The adoption of new technology must be accompanied by a robust change management process which enables effective technology transfer, including the importance of working hand in hand with local partners to ensure practical and context-relevant technology integration.

We posit that technology such as AR, VR, and MR cannot be implemented without adequate sensitization and codesign work to ensure that local adaptations are made.

AR/VR/MR: Landscape

Although Augmented Reality (AR) and Virtual Reality (VR) have become established technologies with many applications over the past few decades, the conceptual origins of AR and VR date back to the mid-1800s with the invention of the stereoscope by English scientist Charles Wheatstone (Wade, 2012). Wheatstone manipulated spatial vision by presenting two separate images to each eye of the viewer, creating the illusion and appearance of three-dimensionality (Wade, 2012). However, it wasn't until 1968 when Turing Award winner and computer scientist Ivan Sutherland created a ceiling fixed headset and visual display device called the "Sword of Damocles" that this technology became a first in kind VR tool, allowing users to access a visual computer-generated image with sensors that could update the image based on tracking the gaze of the user (Sutherland, 1968).

VR has advanced over the years and is characterized as a fully immersive, computer- generated natural or nonnatural environment in three dimensions that users can access through equipment such as headsets and body sensors to simulate a live experience. AR, on the other hand, is defined by the ability to overlay digital images, superimposing information onto real-world environments through computer software, mobile phone apps, or through wearing specialized equipment such as goggles or glasses. MR, while similar to AR, is more sophisticated, allowing a user to not only see digitally enhanced live environments, but to also interact with or through the superimposed digital data in real time.

These technologies are present in everyday life. For example, Swedish furniture giant IKEA enables its customers to use a phone app called *IKEA Place* to superimpose an image of a piece of IKEA furniture into their real-life home environment to get a sense of how it might look prior to purchase (Ozturkcan, 2020). The mobile game app *Pokémon Go,* which launched in 2016, took AR to a new level through MR enhancements, enabling players to visualize superimposed characters from the game within their live environments enabling the user to interact with the digitally superimposed characters within their local, live environment, and following the user's movements through spatial geolocation mapping technology for a highly interactive MR experience (Kerdvibulvech, 2022). In addition, there are well-known AR and VR hardware devices enabling such experiences, including brands like Google Glass (Yoon et al., 2021), Microsoft HoloLens (Moro et al., 2021; Park et al., 2021), the Oculus Quest

VR headset (Almousa et al., 2021),Vuzix smart glasses (Munusamy et al., 2021), or Apple's AR/VR headset (Billinghurst, 2021).

The global AR/VR market is rapidly expanding, set to grow to nearly US$300 billion by the year 2024 (Alsop, 2021) with 11.2 million AR/VR headsets purchased in 2021 alone (AR/VR Headset Shipments Grew Dramatically in 2021, 2021; Petrov, 2019), opening a new world for immersive blended reality experiences allowing us to imagine a world where we can interact over time, space, and geographical boundaries without restriction.

AR/VR/MR in healthcare

The healthcare industry has begun to leverage AR, MR, and VR in a variety of settings, and there is growing evidence base to support its use. For example, *CEEVRA* (www.ceevra.com) has been deployed to refine and augment preoperative preparation through 3D digital organ models reconstructed from a patient's CT and MRI scans using artificial intelligence (AI) and VR immersion options for surgeons through a mobile application or headset. A single-blind randomized controlled trial (RCT) using *CEEVRA* in a robotic-assisted partial nephrectomy demonstrated that the use of 3-dimensional virtual reality models reduced operative time (2.5× more likely to be <199min), estimated blood loss (4.6× more likely to be <200cc), clamp time (11.2× more likely to be <20min), and length of hospital stay (5.4x more likely to stay 1 day only) (Shirk et al., 2019).

Similarly, nurses have been using an app called *AccuVein* (www.accuvein.com) that uses Near-Infrared AR technology to project and highlight where a patient's veins are located onto the skin's surface for easier IV access, cannulation, and blood draws. An RCT demonstrated that peripheral intravenous cannulation skill significantly increased through use of the technology (Yılmaz et al., 2022). VR has also been used for situations such as stroke or amputee rehabilitation by companies such as *Immersive Rehab* (https://immersiverehab.com) as computer-generated environments can combine behavioral neuroscience with active physical therapy. Although studies around VR-based neurorehabilitation have demonstrated significant positive impacts on all outcome measures (Aminov et al., 2018; Laver et al., 2017; Pietrzak et al., 2014; Thakre et al., 2019), results were consistently statistically significant only when VR was integrated within usual rehabilitation methods and used as an adjuvant, and not as a single therapeutic intervention (Naro & Calabrò, 2021). Finally, VR has also been deployed for tele-simulation in global medical education, simulating "realistic training in a virtual space" enabling an immersive training experience, and a clinical community of practice at a distance (Almousa et al., 2021).

While AR/VR and MR are increasingly being used in healthcare delivery to drive quality and efficiency toward better outcomes, the COVID-19 pandemic has also presented new challenges and opportunities for the swift implementation and scale-up of technologies, particularly those focused on supporting the health workforce. Moving forward, it has become clear that capacity-building activities must be more accessible, more adaptable to contextual needs, and less reliant on in-person activities.

There are relative strengths of in-person and virtual capacity-building approaches. The development of complex skills such as those used in surgery, addressing contextual considerations, and problem-solving may be best suited to in-person mentorship. This may be complemented by virtual mentorship, which is less costly. Annually, $3.7 billion in costs were estimated for travel associated with short-term medical missions from the US to the Global South (Nthumba et al., 2021). Such in-person capacity-building approaches are costly both internationally and locally (e.g., travel, per diem) pulling mentors from direct patient care in their facilities.

In addition, virtual mentorship also addresses the costs related to missed opportunities, for example, at smaller facilities, cases may not always be predictable or available during mentorship visits, adding to inefficiencies of the in-person approach; and mentors may not be available for real-time consultation on emergency cases, leading to missed opportunities for learning, while adding to more referrals instead of skills building that could enable more cases to be handled locally. Importantly, all aspects of in-person training models have been stressed during the pandemic amid travel restrictions and health workforce shortages.

Thus technologically enhanced, blended capacity-building approaches (combining onsite, in-person with clinical AR/MR support) within existing in-person training programs have great potential to address current and future needs for the delivery of timely, quality care closer to the patient's home while driving greater healthcare delivery efficiency so remote healthcare training and support can become available.

Proximie: A case study in AR/MR

Proximie addresses the problem of inequitable access to surgical care

In 2015 the Lancet Commission on Global Surgery and the World Health Organization (WHO) declared that timely access to safe and affordable surgery is a fundamental part of universal rights to healthcare and that surgical and anesthesia care represent an integral component of a functional, responsive, and resilient health system. However, five billion people, almost three-quarters of the global population, are excluded from what is often lifesaving or disability-averting treatment offered by surgery (Meara et al., 2015) and that while one-third of the world's burden of disease can be treated with surgery, >90% of the population in LMICs lacks access to surgical care (Eyler et al., 2018). The introduction of AR/VR and MR offers great potential to improve equitable access to safe, timely, and quality surgical care.

Proximie is a cloud-based surgical telementoring software complementing audiovisual (AV) communications, with dynamic AR overlays on multiple video feeds from the local surgical environment, enabling virtual, real-time clinical surgical collaboration over distances.

As a mixed reality software, live video feeds from multiple angles of the local clinician's environment are captured using a standard laptop, capture card, and webcam(s) providing a holistic immersion in the operating room of interest, capturing the live environment ranging from the surgical field of view to the team-based workflow during an operation.

A remote healthcare professional can view up to four video feeds in real time including X-rays or scans using Proximie, accessed via any standard laptop, tablet or smart lenses with a Google Chrome browser using off-the-shelf webcams in a hardware agnostic manner, as shown in Fig. 1.

Proximie can pull raw visual feeds from any piece of OR hardware, such as laparoscopes, enabling remote clinicians access to crystal clear views of fine surgical details and narrow spaces, as shown in Fig. 2. Proximie is a lightweight, agile, and secure (HIPAA and GDPR compliant) software able to run on exceptionally low bandwidth including off of mobile hotspots, enabling access in even the most austere of environments. Clinicians can interact in the remote surgical environment through voice, chat function, or Proximie's unique AR tools including pen telestration onto the remote surgical field of view so the local clinicians can see what the surgeon is demonstrating or pointing at, delivering a holistic and immersive experience, as shown in Fig. 2. The figure shows that images are captured from an ENT scope session via live video feed enabling views into narrow and complex surgical spaces with the

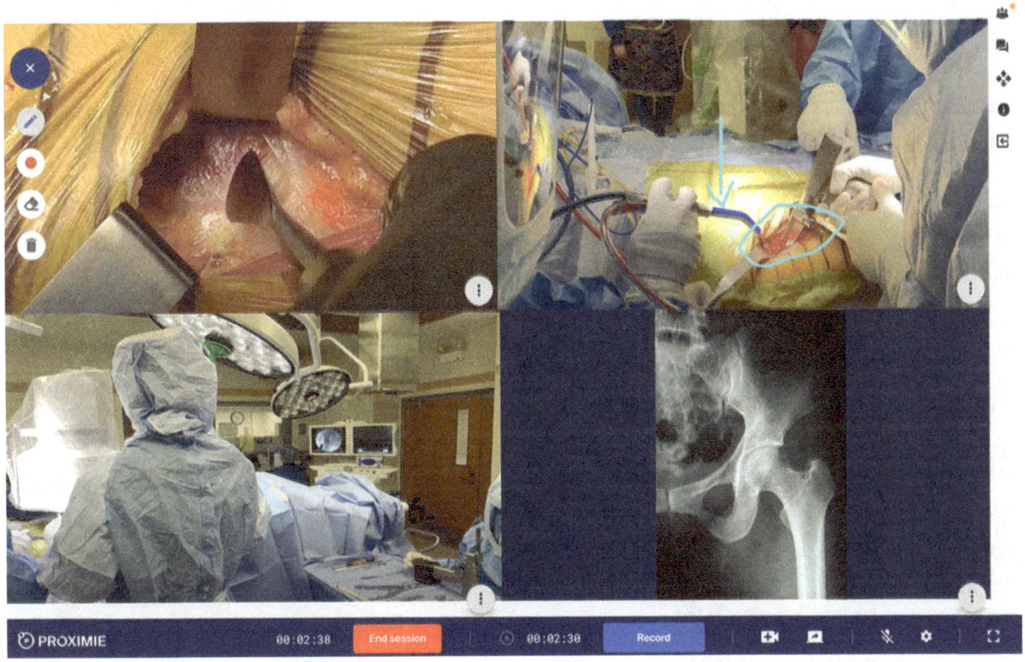

Fig. 1
Remote users can view up to four live camera feeds from the operating room, providing a unique and immersive experience. Pen telestration in upper right hand screen in blue (circle and arrow).

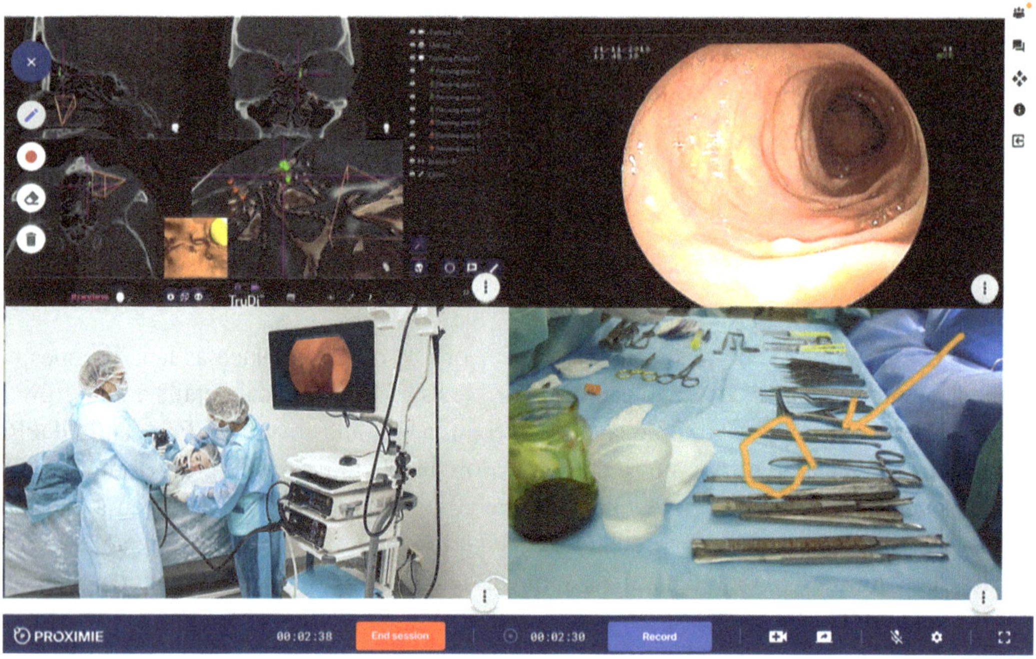

Fig. 2

Images captured from an ENT scope session via live video feed for remote surgeons to interact with local surgical teams through a MR/AR experience. Pen telestation in lower right hand screen in yellow (circle and arrow).

ability for remote surgeons to interact with local surgical teams through a MR experience using the pen telestration tools.

Videos can be unrecorded or recorded depending on the use case; recorded cases are uploaded onto the organization's private online Proximie library, as shown in Fig. 3, for future review and collaborative postoperative learning. Using this feature, clinicians can assess their surgical technique and review team interaction, communication, and performance. This, in turn, can be used for skills building, fostering improved team dynamics and cohesion, and quality improvement efforts overall. To date, Proximie has been deployed in over 50 countries in 500 hospitals in tens of thousands of surgical interactions.

Proximie was developed to help improve access to surgery. Its founding objective was to give surgical practitioners (residing in different locales) a platform that provides an interactive experience, reliable enough to allow them to collaborate on live procedures in real time, overcoming the common problem of not having a particular specialist available. Rather than waiting for a specialist to become available, or forcing a patient to travel for care, the idea is to 'beam in' expertise from varying geographical locations. In this manner, surgeons can effectively consult, directly guide, mentor, and demonstrate operations in such a way that they

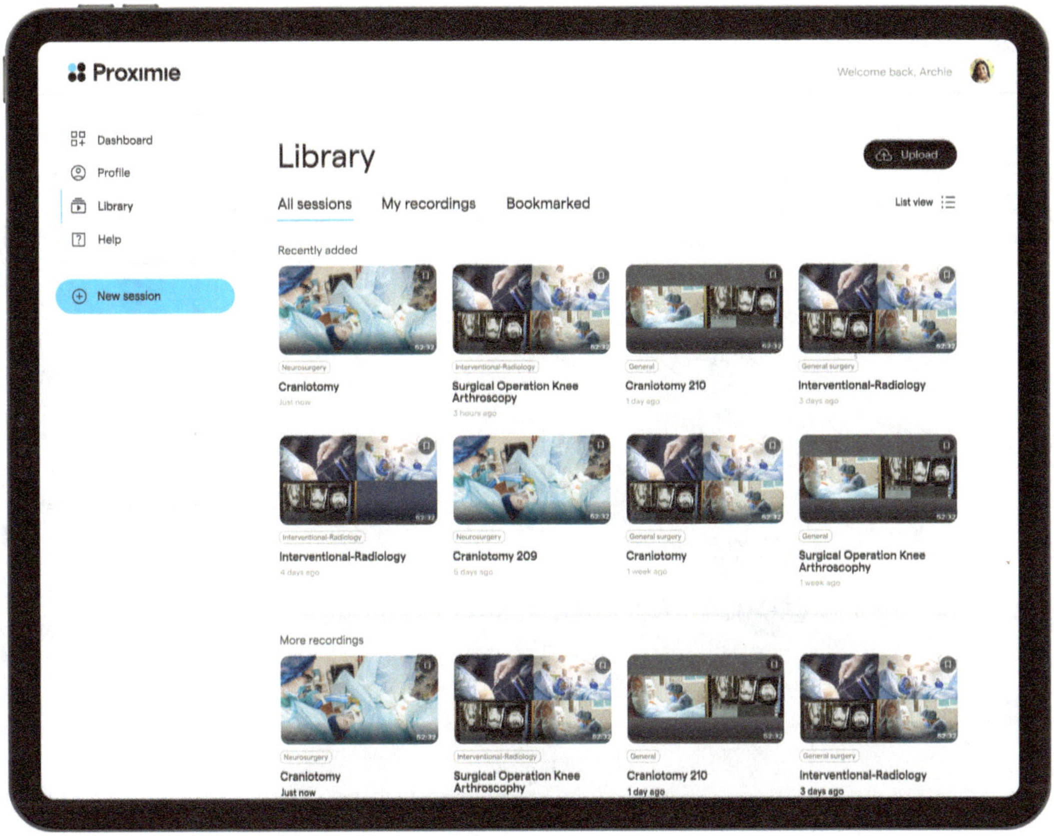

Fig. 3

Recorded cases are uploaded into an organization's private library to enable postoperative collaborative case review and education.

can be safely performed in local (underserved) healthcare environments, as demonstrated in Proximie's early Global Health engagement in Peru that enabled remote surgical mentorship for cleft palate surgeries resulting in an increased skill transfer and efficiency (Vyas et al., 2020).

AR plays an important role in replicating the 'live' experience, substituting a colleague's actual physical presence in the operating room with sophisticated telepresence. This can be accomplished with tools such as digital markup of a live video image, gesture-activated demonstration, and the ability to postdigital content directly to the feed. Via their cameras and screens, remote colleagues are able to interact on a more meaningful level than just watching a plain video feed and simple, voice telecommunication, as shown in Fig. 4.

Real-time remote communication technologies help to overcome geographical barriers to surgery by eliminating the inefficiency associated with waiting times and travel. This applies to all stages of the patient journey. Audiovisual communications allow for the possibility of

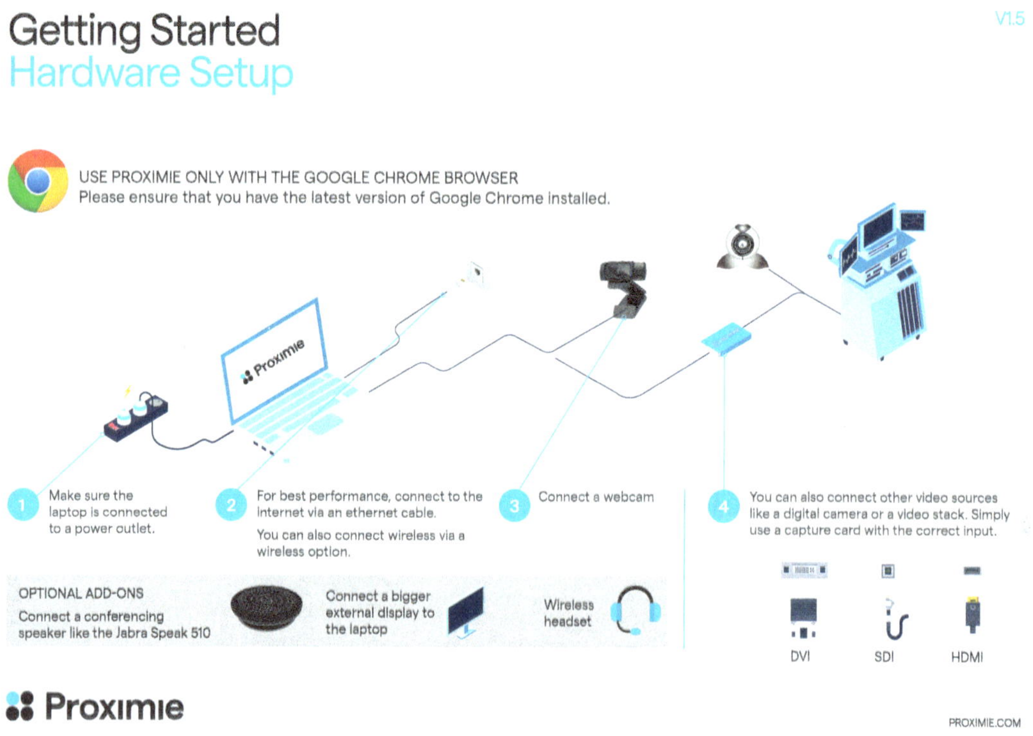

Fig. 4
Agile hardware setup using common off-the-shelf tools for connection.

consultants carrying out patient assessments remotely, drastically reducing the time it takes to make decisions about surgical pathways when an appropriately qualified specialist is not available at a particular hospital (Greenhalgh et al., 2016).

Many of the elements of high-quality surgical care are related to planning, preparation, and sharing of expertise. Hence, the use of technology to enable effective collaboration (wherever practitioners happen to practice) will shorten delivery cycles and increasingly ensure complex procedures can be delivered in more locations. Better communication thus answers the Lancet Commission's call for surgical services to be more tightly integrated across all levels of care—from community referral networks to specialist surgical teams (Meara et al., 2015) (Hachach-Haram, 2021). Proximie's foundations have evolved from the traditional training approach of *See One, Do One, Teach One* to Proximie's updated 3P approach—*Prepare, Perform, Perfect*, as shown in Fig. 5.

Finally, from the outset, Proximie was designed with a vision to have global use. Many new technologies in the AR/VR and MR space can be bulky, expensive, or require specific equipment or connectivity making it difficult to deploy in varied contexts, specifically including resource-limited environments, around the world. As the technological revolution takes hold,

Fig. 5
Diagram—Prepare, Perform, Perfect.

more thought will need to be embedded into early stages of technological design work across AR/VR and MR, in general, to take into consideration any unintended consequences.

AR/VR/MR in the context of the 5 A's

While new technologies like AR, VR, and MR have enormous potential for improving people's lives, they can also affect inequalities through design and access. Penchansky and Thomas (1981) first conceptualized the idea of access to healthcare as comprising five dimensions, known as the Five A's of access, affordability, availability, accessibility, accommodation, and acceptability in 1981. This concept has been updated to reflect the emerging and central role of technology in the healthcare system. The Five A's, as shown in Fig. 6, outlined in the UNCTAD Technology and Innovation Report 2021 (UNCTAD, 2021) serve as an important foundational framework for understanding how technology must be considered within the social context in which it is being deployed from the perspective of the final user.

While the Five A's should be applied to all contexts to help redefine how to think about new technologies in a way that contributes to reducing inequalities and to sustainable

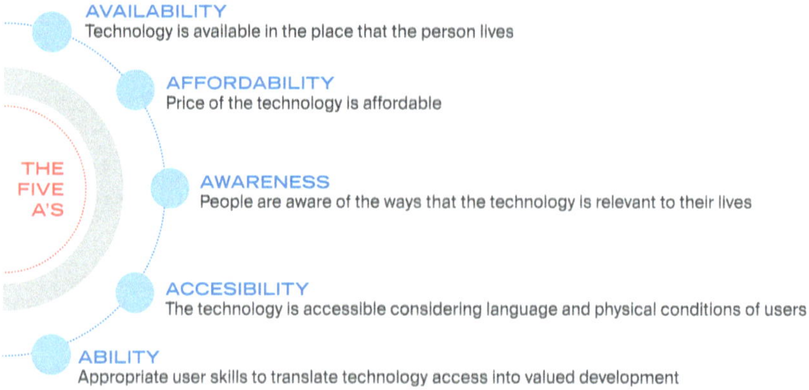

Fig. 6

The 5 A's framework. *Adapted from UNCTAD (2021). Technology and innovation report 2021: Catching technological waves—Innovation with equity, United Nations. Technical report.*

development, LMICs in particular face a unique set of contextual challenges (UNCTAD, 2021) in promoting equal access to these new technologies, including greater poverty, a significant digital divide, and a shortage of IT skills.

To overcome these challenges, there is growing consensus that the introduction of new technology is positioned to be most successful when it is layered into a system-based approach at the outset coupled with strong private-public partnership frameworks. Solutions such as AR, VR, and MR should be considered an important component of the overall health ecosystem, in support of sustainable development. While many health technologies have been adopted and successfully scaled-up in LMICs such as Ghana's telemedicine program (Labrique et al., 2018), there are many other examples where technological adoption and efficacy have been less successful or have failed to demonstrate sufficient evidence of effectiveness (Qiang et al., 2011; Tomlinson et al., 2013). Often the reasons are related to the way the problem which the new technology is designed to address is perceived by the users and how it is positioned and supported within the health system (Atun, 2012).

Health technology readiness assessments, or related exercises (Mauco et al., 2020), offer important insights about the strengths and weakness of the health system and its capacity for new technologies. This is a critical step in the introduction of any new technology to allow for early adaptation and adoption. Too often technology is deployed without consideration for the larger health systems context leading to short-lived interventions and an inability to transition to sustainable scale.

Early reflections from Proximie's experience in Kenya provide valuable insights and lessons learned on applying the 5 A's and the importance of technology transfer. In 2021, Proximie formed a dynamic public-private partnership with Jhpiego and Ariadne Labs, with the support of the Ministry of Health, to implement, adapt, and scale-up the Jhpiego Obstetric Safe Surgery (OSS) Program in Makueni County, Kenya, by integrating the Proximie technology into its training and mentorship programme. The aim of OSS from the outset was to improve access to safe, quality, respectful obstetrical and anesthesia care through team-based workforce capacity development, as well as community engagement and strengthened referral system. However, the pandemic forced a new way of working, highlighting the need to stretch boundaries and push the comfort zone of what could be done through leveraging MR/AR technology to support surgical system strengthening and future-proofing capacity-building efforts for the surgical health workforce.

Proximie and partners worked closely to ensure that training and mentoring could continue uninterrupted and made intentional efforts to ensure that the technology itself was thoughtfully integrated into the OSS Program for a hybrid approach. AR and MR, in particular, offered a novel way to support virtual training and mentorship activities, as well as quality improvement efforts, through a hub-and-spoke approach enabling direct support at district-level hospitals. Technology readiness and facility assessments were performed at each of the five health facilities to document critical information about IT infrastructure, power supply, basic client amenities, and staffing. In addition, focus groups with the surgical teams and facility management were conducted at the outset to obtain a more nuanced understanding of the human capacity, readiness, and receptiveness to use the technology. A local digital health expert was hired to lead this work.

A change management approach was applied to enable effective technology transfer, mindful of the 5 A's. This involves the following four elements:

(1) A theory of change which clearly articulated the aim of the Proximie technology was developed through a collaborative human-centered design.
(2) Surgical teams had the opportunity to trial, adapt, and then adopt the Proximie technology through hands-on learning. This provided valuable insights on how to best adapt technology to the local context, as presented in Table 1.
(3) The overall approach was planned for resistance and built-in a series of opportunities for collective reflections about the usability of the solution itself.
(4) The use of the technology was continuously course corrected, prompted by local adaptations and problem solving.

Proximie observed an interesting evolution as users moved from having many fears and preconceived ideas about the use of the technology to a deeper understanding and sense of ownership of the digital solution. It was clear that the clinical teams wanted the opportunity to trial the technology for themselves from the earliest stages to test their assumptions and familiarize themselves so they could then guide the necessary adaptations to ensure the technology was fit for purpose. There was a dynamic shift in their understanding of the technology and recognition that many of the initially perceived challenges were misplaced once they had the opportunity to use the technology. This enabled Proximie to respond to unforeseen

Table 1 Insights on how to best adapt technology to resource-limited context to ensure feasibility.

Theme	Local adaptations and solutions
Maximizing connectivity	• Maintain flexibility to pivot to mobile hotspot to ensure connectivity • Designate one device per facility dialing in instead of multiple devices at same location for optimum experience • Use 4G router to help boost and stabilize connectivity for a better experience • Need for discussion with private sector stakeholders to mitigate connectivity issues as providers often throttle the available bandwidth per location
Ensuring lowest possible running costs	• Minimize costs through use of a hotspot bundle • Ensure clinicians in the facility dedicate the internet line specifically for clinical use to maximize bandwidth requirement
Identifying the most suitable hardware	• Use light and agile hardware (e.g., simple webcams, portable speakers, clip on microphones or headsets) to customize each operating theater for best visual and auditory coverage • Make use of webcams which allow remote-controlled movement as well as zoom feature to address practical sterilization concerns in handling equipment during surgery • Establish clear hardware sign-out protocol to minimize risk of items being stolen or damaged • Ensure generator backup and surge protection
Identifying human resources champions	• Identify at least two clinical champions per site who know how to operate the technology with one member serving as a backup • Surgical team should have a sense of ownership of the technology and be comfortable with its use to reduce reliance on outside IT support during clinical intervention—while this is software, it is still a clinical tool
Establishing standard operating procedures	• Establish clear SOP to guide use of the technology (i.e., operational manual with stepwise guidance and troubleshooting advice)
Ensuring patient privacy and consent	• Amend the already available patient consent form with simple, clear language to explain the technology and the patient's rights • Place cameras in the most optimal positions to obscure patient identity with clearly defined communication etiquette established over a mixed reality platform

contextual challenges with locally driven solutions, ensuring effective implementation and maximizing the potential for sustainability and impact.

Finally, important consideration from the outset was given to the legal context in which the technology was deployed. This meant exploring how best to ensure alignment of Proximie with telemedicine registration requirements in Kenya. Specific areas identified for further attention from a legislative and regulatory framework perspective included: data processing and data transfer, registration requirements, and the review and localization of legal documents such as data privacy documents and contractual agreements.

With the growth of digital health and AR/VR/MR-based solutions come specific challenges in geographies with less mature technology regulations, including digital health legal compliance frameworks which can vary significantly from country to country, such as issues around patient data storage if images or video libraries are captured from the clinical environment. Data localization requirements add prohibitive cost and regulatory hurdles for such technologies, acting as a significant barrier to entry and sustainable growth. For example, if a country's digital health regulation requires a company to hold all digital health patient data captured within country, this then requires a company setting up a data server in each new country the technology scaled into, adding a significant cost to their overhead which becomes cost prohibitive, rendering excellent tools and solutions unsustainable for the market. The complexity is compounded when robust and highly secure server infrastructure representing international gold standards at the level Amazon AWS or Microsoft Azure cloud is only available in limited geographies at this time (Ilori, 2020). As such, local partnerships with key stakeholders including governments is critical to uncover pragmatic and aligned regulatory solutions such that new digital solutions like AR/VR/MR can be implemented and scaled-up.

Conclusion

Augmented Reality (AR), Virtual Reality (VR), and Mixed Reality (MR) offer great potential to transform the delivery of healthcare allowing more people greater access to safe, timely, and quality care. However, as demonstrated by the case study of Proximie, technology must be thoughtfully embedded within the broader health system with consideration for the local context, using the 5 A's and well-designed processes that support technology transfer.

Moving forward, ensuring the sustainable scale-up of effective and efficient technologies must be a priority. This will necessitate a shared global learning agenda to rigorously evaluate and document outcomes, impact, and cost-effectiveness of new technologies which can be used to inform planning and budgeting decisions enabling replicability and standardization. Developing clear national policies built upon a strong evidence base and a background of a regulatory framework, as well as earmarking resources for their prioritization, will facilitate spread and scale and ultimately transform AR/VR/MR from being seen as a disruptive innovation to being institutionalized as part of the health system.

References

Almousa, O., et al. (2021). Virtual reality technology and remote digital application for tele-simulation and global medical education: An innovative hybrid system for clinical training. *Simulation & Gaming*, *52*(5), 614–634. https://doi.org/10.1177/10468781211008258.

Alsop, T. (2021). *Augmented reality (AR) and virtual reality (VR) market size worldwide from 2016 to 2024.* Statistica. https://www.statistica.com/statistics/591181/global-augmented-virtual-reality-martket- size/.

Aminov, A., et al. (2018). What do randomized controlled trials say about virtual rehabilitation in stroke? A systematic literature review and meta-analysis of upper-limb and cognitive outcomes. *Journal of Neuroengineering and Rehabilitation, 15*(1), 29. https://doi.org/10.1186/s12984-018-0370-2.

AR/VR Headset Shipments Grew Dramatically in 2021. (2021). *Thanks largely to Meta's strong quest 2 volumes, with growth forecast to continue.* According to IDC (2022) International Data Corporation (IDC). Available at: www.idc.com/getdoc.jsp?containerId=prUS48969722.

Atun, R. (2012). Health systems, systems thinking and innovation. *Health Policy and Planning.* https://doi.org/10.1093/heapol/czs088.

Billinghurst, M. (2021). Grand challenges for augmented reality. *Frontiers in Virtual Reality, 2.* https://doi.org/10.3389/frvir.2021.578080.

Eyler, L., et al. (2018). Essential surgery as a component of the right to health: A call to action. *Human Rights Quarterly, 40*(3), 641–662. https://doi.org/10.1353/hrq.2018.0035.

Greenhalgh, T., et al. (2016). Virtual online consultations: Advantages and limitations (VOCAL) study. *BMJ Open, 6*(1). https://doi.org/10.1136/bmjopen-2015-009388.

Hachach-Haram, N. (2021). A digital doorway to global surgery. In *Digital Surgery.* https://doi.org/10.1007/978-3-030-49100-0_26.

Ilori, T. (2020). *Data protection in Africa and the COVID-19 pandemic: Old problems, new challenges and multistakeholder solutions.*

Kerdvibulvech, C. (2022). Geo-based mixed reality gaming market analysis. *Human Behavior and Emerging Technologies, 2022,* 1–9. https://doi.org/10.1155/2022/1139475.

Labrique, A. B., et al. (2018). Best practices in scaling digital health in low and middle income countries. *Globalization and Health, 14*(1), 103. https://doi.org/10.1186/s12992-018- 0424-z.

Laver, K. E., et al. (2017). Virtual reality for stroke rehabilitation. *Cochrane Database of Systematic Reviews, 2018*(1). https://doi.org/10.1002/14651858.CD008349.pub4.

Mauco, K. L., Scott, R. E., & Mars, M. (2020). Validation of an e-health readiness assessment framework for developing countries. *BMC Health Services Research, 20*(1), 575. https://doi.org/10.1186/s12913-020-05448-3.

Meara, J. G., et al. (2015). Global Surgery 2030: Evidence and solutions for achieving health, welfare, and economic development. *The Lancet.* https://doi.org/10.1016/S0140- 6736(15)60160-X.

Moro, C., et al. (2021). HoloLens and mobile augmented reality in medical and health science education: A randomised controlled trial. *British Journal of Educational Technology, 52*(2), 680–694. https://doi.org/10.1111/bjet.13049.

Munusamy, T., et al. (2021). Telemedicine via smart glasses in critical care of the neurosurgical patient—COVID-19 pandemic preparedness and response in neurosurgery. *World Neurosurgery, 145,* e53–e60. https://doi.org/10.1016/j.wneu.2020.09.076.

Naro, A., & Calabrò, R. S. (2021). What do we know about the use of virtual reality in the rehabilitation field? A brief overview. *Electronics, 10*(9), 1042. https://doi.org/10.3390/electronics10091042.

Nthumba, P. M., Newton, M. W., & Gathuya, Z. (2021). Global surgery in the 21st century: The voice from Africa—We need a new model. *Paediatric Anaesthesia.* https://doi.org/10.1111/pan.14162.

Ozturkcan, S. (2020). Service innovation: Using augmented reality in the IKEA Place app. *Journal of Information Technology Teaching Cases, 11*(1). https://doi.org/10.1177/2043886920947110.

Park, S., Bokijonov, S., & Choi, Y. (2021). Review of Microsoft HoloLens applications over the past five years. *Applied Sciences, 11*(16), 7259. https://doi.org/10.3390/app11167259.

Penchansky, R., & Thomas, J. W. (1981). The concept of access: Definition and relationship to consumer satisfaction. *Medical Care, 19*(2). https://doi.org/10.1097/00005650-198102000-00001.

Petrov, C. (2019). *35 Virtual reality statistics that will rock the market in 2019.* TechJury.

Pietrzak, E., Pullman, S., & McGuire, A. (2014). Using virtual reality and videogames for traumatic brain injury rehabilitation: A structured literature review. *Games for Health Journal, 3*(4), 202–214. https://doi.org/10.1089/g4h.2014.0013.

Qiang, C., et al. (2011). *Mobile applications for the health sector.* Washington, DC: World Bank.

Shirk, J. D., et al. (2019). Effect of 3-dimensional virtual reality models for surgical planning of robotic-assisted partial nephrectomy on surgical outcomes. *JAMA Network Open*, *2*(9), e1911598. https://doi.org/10.1001/jamanetworkopen.2019.11598.

Sutherland, I. E. (1968). *Head-mounted three dimensional display*. https://doi.org/10.1145/1476589.1476686.

Thakre, M., et al. (2019). Virtual reality based neuro-rehabilitation in acute stroke: A prospective cohort study of its effectiveness for upper limb motor recovery. *Journal of the Neurological Sciences*, *405*, 96–97. https://doi.org/10.1016/j.jns.2019.10.397.

Tomlinson, M., et al. (2013). Scaling up mHealth: Where is the evidence? *PLoS Medicine*, *10*(2), e1001382. Public Library of Science. Available at: https://doi.org/10.1371/journal.pmed.1001382.

Tran Ngoc, C., et al. (2018). Conclusions of the digital health hub of the Transform Africa Summit (2018): Strong government leadership and public-private-partnerships are key prerequisites for sustainable scale up of digital health in Africa. *BMC Proceedings*, *12*. https://doi.org/10.1186/s12919-018-0156-3.

UNCTAD. (2021). *Technology and innovation report 2021: Catching technological waves—Innovation with equity, United Nations*. Technical report.

Vyas, R. M., et al. (2020). Using virtual augmented reality to remotely proctor overseas surgical outreach: Building long-term international capacity and sustainability. *Plastic and Reconstructive Surgery*. https://doi.org/10.1097/PRS.0000000000007293.

Wade, N. J. (2012). Wheatstone and the origins of moving stereoscopic images. *Perception*, *41*(8), 901–924. https://doi.org/10.1068/p7270. SAGE Publications Ltd STM.

Wilson, K., Gertz, B., Arenth, B., & Salisbury, N. (2014). The journey to scale: Moving together past digital health pilots. In *On soul and earth: The psychic value of place*.

Yılmaz, H., et al. (2022). Does the use of infrared technology (AccuVein AV-500®) for Peripheral Intravenous Cannulation (PIVC) increase the success rate in nursing students? A randomized controlled trial. *Nurse Education Today*, *108*, 105179. https://doi.org/10.1016/j.nedt.2021.105179.

Yoon, H., et al. (2021). Google glass-supported cooperative training for health professionals: A case study based on using remote desktop virtual support. *Journal of Multidisciplinary Healthcare*, *14*, 1451–1462. https://doi.org/10.2147/JMDH.S311766.

CHAPTER 22

Intelligent virtual agents for current and emerging public health crises

David D. Luxton[a,b] and Eleanor 'Nell' Watson[c]

[a]Department of Psychiatry & Behavioral Sciences, University of Washington School of Medicine, Seattle, WA, United States, [b]Luxton Labs, LLC, Olympia, WA, United States, [c]Department of Computing & Engineering, University of Gloucestershire, Cheltenham, England

Introduction

Intelligent virtual agents (IVAs) are software programs that can interact with human beings through dialogue in a manner intended to lead to an actionable outcome (IVA, 2005). Some IVAs may also be able to make goal-oriented decisions and interact with their environment (Budakova et al., 2020). These programs can be deployed in the form of basic text-interface chatbots, or more sophisticated AI-driven virtual representations of living beings, including humans (i.e., virtual humans) or nonhumans, such as animal characters or machines. IVAs, also referred to as conversational agents, can be accessed on the internet, on mobile devices (i.e., smartphones and tablets), on stand-alone kiosks, in virtual reality (VR) environments, in video games, or as an embedded system within a physical robot (Luxton & Hudlicka, 2021; Luxton & Riek, 2019). Advances in natural language processing make these systems capable of interactive conversation with humans by text or spoken language (Boumans et al., 2022). Embodiment into a virtual or physical form (such as a virtual human face) allows for the display of affective expressions along with dialogue, which may enable a more engaging experience (Kulms & Kopp, 2016; Soderlund et al., 2021).

IVA virtual human technology has advanced rapidly over the last two decades alongside improvements in computing power, sensing technology, AI, and virtual reality (Luxton, 2016, 2020). They are deployed across a broad range of practical domains and applications, such as personal assistants, service auto-attendants, for training simulations, and also for entertainment. Sensor technology has enabled IVAs to integrate information from the user's environment, such as a person's location in a smart home, as well as the characteristics and preferences of the persons with whom they interact. Furthermore, affective computing techniques and

Resilient Health. https://doi.org/10.1016/B978-0-443-18529-8.00022-6

technologies have enabled IVAs to sense, interpret, and express human emotional states, thereby increasing their ability to interact with humans in a lifelike manner (Luxton & Hudlicka, 2021). The combination of emotional and social intelligence (e.g., awareness of social cues and user goals) of IVAs enables them to adapt to the shifting states and needs of human users. As noted by Luxton and Hudlicka (2021), this in turn contributes to the development of what the human user perceives as a supportive, understanding, and empathic human–agent relationship.

The exploration of the possibilities of human healthcare provider simulation goes back in time to at least the 1960s with the work of MIT computer scientist Joseph Weizenbaum with his "ELIZA" and "DOCTOR" programs. Weizenbaum's "DOCTOR" program simulated the humanistic therapeutic style of noteworthy American psychologist Carl Rogers and was shown to evoke emotional responses from persons with whom it interacted (Weizenbaum, 1976). Weizenbaum's chatbot programs, and those that followed, such as PARRY and A.L.I.C.E, were merely simulations using scripted responses, not true autonomous AI. The advancement of AI deep learning and virtual reality, however, has made conversation and interaction with IVAs more realistic (as if in conversation with a human) than ever before—and the technology continues to be improved at an exponential rate. In the coming years, whether a person is speaking to a human healthcare professional or a simulation will become indistinguishable. This will cause considerable ethical and moral concerns, which we elaborate on later in this chapter.

The deployment of IVAs in healthcare and public health has grown exponentially over the last few years, and their benefits became especially apparent during the global SARS-CoV-2 (COVID-19) pandemic (Luxton, this volume; McKillop et al., 2021). The global public health response to the pandemic entailed mitigation measures such as social (physical) distancing to slow the spread of the virus. Demands on healthcare systems were worsened by a shortage of qualified medical staff, not to mention the burnout felt by overworked medical and public health professionals during the pandemic (De Kock et al., 2021). Mental health needs also skyrocketed during the pandemic, causing an increase in demand for services among diverse populations around the globe. The government lockdowns themselves were a significant contributor to mental health conditions (Brooks et al., 2020) and have been linked to an uptick in depression and deaths by suicide (Czeisler et al., 2020a, b; Ganesan et al., 2021).

Along with the rapid adoption of telehealth and mobile health tracking technologies (see Luxton, this volume), IVAs were also deployed to help address gaps in healthcare provider availability, to assess health status, to provide information to populations about the virus and sanitation factors, and to offer health-related support, especially for behavioral health. Given the benefits of chatbots and IVAs, government agencies in many countries such as China, Singapore, and the United States deployed them to deliver psychological support during the COVID-19 pandemic (McKillop et al., 2021; Zhu et al., 2022).

In the following sections of this chapter, we discuss the latest state of the art of IVAs in the context of healthcare and public health with examples from before and during the pandemic, as well as how they may be useful for future public health crises. We also discuss several highly important ethical issues, which should be addressed to ensure that such technologies are worthy of trust and will continue to be accepted by the public.

IVAs in healthcare and public health

Screening and triage

One of the consequences of the social (physical) distancing requirements during the SARS-CoV-2 (COVID-19) pandemic was that it created a barrier to care and services. For healthcare providers, it caused them to limit or alter how they interacted with their patients. Physical barriers (e.g., glass partitions), telehealth options, or the curtailing of services altogether was some of the responses enacted. Moreover, many care seekers were apprehensive and less likely to seek care during the pandemic or chose to delay medical procedures, even when services were available (Czeisler et al., 2020a). IVAs, however, provide a contactless solution, and they can be accessed from the comfort of the user's home, in waiting rooms at clinics or hospitals, or on mobile devices, at any time of day.

During the pandemic, several developers created "Corona-bots" to assist with symptom screening and other services. For example, de Campos Filho et al. (2022) evaluated a COVID-19 prototype chatbot-type intelligent virtual assistant (CoronaBot) in Brazil. The *CoronaBot* was designed to virtually screen users who believed they have symptoms of COVID-19 and to provide them with appropriate procedures for each scenario. By making use of natural language processing, the CoronaBot converses with the user and performs a screening according to the symptoms presented, risk factors, and contact with suspected or confirmed cases of COVID-19. The chatbot achieved a satisfactory level of content validity, and the usability of the chatbot was evaluated as favorable, suggesting that the tool is a viable option (de Campos Filho et al., 2022).

Mental and behavioral health support

One of the primary advantages of IVAs is that they can provide users with anytime access to health information, self-care tools, and in some scenarios, direct care. For example, Woebot (www.woebot.io) is one commercially available mobile app chatbot designed to help people improve symptoms of depression and anxiety by following evidence-based therapeutic approaches of Cognitive Behavioral Therapy (CBT) and teaching self-care and mindfulness techniques. The app uses a simple text-based chat interface and can initiate therapeutic check-ins with users via open-ended dialogue to ask them how they're feeling or what they are currently doing as part of behavioral activation. The app developer, Woebot Health, made Woebot available for providing global emotional support during the pandemic.

Disruption of sleep was also a significant public health issue during the pandemic, found to be associated with higher levels of psychological distress (Alimoradia et al., 2021). Recognizing an opportunity for IVA technology to help address this public health issue, Philip et al. (2020) developed and tested an IVA-sleep self-care tool deployed within a mobile app called KANOPEE, a service which allows users to interact with a virtual agent that provides autonomous screening and offers behavioral interventions. The goal of the study was to test the feasibility of this app, in terms of use rate, perceived trustworthiness and acceptability of the virtual agent, and effects of the intervention program, in the context of COVID-19 containment in France. The results suggest that the app is a promising solution to screen populations for sleep complaints and that it provides acceptable and practical behavioral advice for individuals reporting moderately severe insomnia.

Also, the local government in Hangzhou Province in China deployed a mental health chatbot named *Psybot*, which can listen, analyze, and speak, to help citizens "express their true feelings and ease their moods" (Zhu et al., 2022). A group of academic researchers evaluated a similar chatbot app called *Xiaolv* and found that voice interaction is an insignificant predictor of user satisfaction and continuance suggesting that even the most basic text-interface chatbot might be as effective as more elaborate, embodied IVAs (Zhu et al., 2022).

Disease and psychological signals detection

When people communicate with each other, the information transferred is not only perceived through verbal or written language, but also by observation of nonverbal behavior, or characteristics of verbal behavior (speech), that is not necessarily explicit or in conscious awareness. Examples include body or facial movements that indicate a person is agitated or happy, or vocal characteristics consistent with the experience of anxiety or other psychological states. The use of sensing and computational methods that allow machines to detect and analyze psychological states has utility in healthcare applications.

For example, the U.S. military funded research during the late 2000s to develop systems that could assess human emotional states by observation of the facial expressions and vocal characteristics of humans who interact with the system. The Detection and Computational Analysis of Psychological Signals (DCAPS) project, funded by the U.S. Department of Defense's Defense Advanced Research Projects Agency, entailed the development and testing of a simulated behavioral health therapist, connected to a system that assessed the user's emotional state by evaluating facial expressions, voice characteristics, and other physiological data in real time. Such advancements in computing affect can soon be integrated with data from multiple sources, such as health records, and perhaps even social media (Luxton & Poulin, 2020). Commercially available IVAs might soon be developed that can have such capabilities built-in.

Ethics considerations for public health emergencies

User safety

The current and future use of IVAs in healthcare and public health requires consideration of the risk of harm to people if a system does not adequately handle situations (Luxton, 2016; Luxton & Hudlicka, 2021). Risk scenarios include failure of the system to detect when a user discloses suicidal thoughts and intent to an online chatbot, or an intent to harm another person, domestic violence, abuse of children, elders, or other vulnerable persons, or intentional transmission of disease. Human healthcare professionals have a duty to warn and inform ("duty to report") in such risky and dangerous scenarios when disclosed or detected. Questions remain about best practices for handling such matters by autonomous or semiautonomous IVAs.

Machine ethicists and AI researchers who have considered such ethics and legal requirements have suggestions to mitigate these risks including that the designers of autonomous and semiautonomous IVAs should consider giving these systems the ability to detect the aforementioned types of risks and make decisions about appropriate actions (Luxton, 2014; Luxton & Hudlicka, 2021). In some situations, when a threat is detected, the system could immediately respond with the display of available resources, such as a suicide prevention helpline or community mental health resources. "Human-in-the-loop" mechanisms are also an approach toward addressing this issue, whereby the system could automatically notify a human healthcare professional who could respond with direct intervention.

User trust and deception

Another risk derives from what is called "Turing deceptions," named after the British mathematician Alan Turing (1912–54), which occurs when a person cannot tell whether they are conversing with a machine or a human. Indeed, people can be fooled by the voices and appearances of virtual humans, into believing that they are real people. By the opposite token, people might believe they are engaging with a machine, only to find out that it is a human, or a machine being manipulated by a human. Such "deception by fake AI" has been likened to the technologically enhanced puppetry employed by the titular character in *The Wizard of Oz*. Such situations may lead to confusion, distrust, or even emotional trauma, which may then lead to persons not seeking mental health services that may be of benefit to them.

IVAs in the healthcare domain must be designed to be sensitive to the unique characteristic of patients. For example, some types of patients, such as those with dementia or delusional or psychotic psychopathologies, maybe especially at risk of harm if deceived by an IVA (Luxton & Riek, 2019). Even when it is communicated to them that a system is an AI machine, some patients may still believe that a simulation is a living person, or that there is a person, or some malevolent force, behind the simulation (Luxton, 2016). In such situations, psychotic symptoms of paranoia and hearing voices can be iatrogenically induced or exacerbated.

Related to this, even when a patient is consciously aware that a care provider is a simulation of a human being, the patient can be expected to experience intense emotions during the interaction and may develop feelings about, and attachments to, the simulated agent (Luxton, 2014; Luxton & Hudlicka, 2020). This phenomenon was noted by German-American computer scientist and MIT professor Joseph Weizenbaum in the 1960s after he observed his assistant experiencing emotional reactions while communicating with his chatbot programs (Weizenbaum, 1976). This is again highly relevant in the behavioral health settings, especially for some patients who may be particularly vulnerable to manipulation, or who may experience attachment anxieties. Appropriate disclosure of risks for this, and monitoring of user safety with explicitly indicated human-in-the-loop approaches, is advised.

One response to the problems of Turing deceptions may be the use of audio or visual cues, such as standardized chimes or visual icons that notify the user of who or what they are interacting with. Specifically, these indicators would inform users whether they are interacting with a fully autonomous system, a human person, a system controlled by a person, or perhaps a human being who is being directed or filtered in some manner by an autonomous system (IEEE, 2022).

Privacy

An issue related to trust is that of user privacy. The assurance of privacy is important for IVAs in healthcare settings or any other setting where these systems collect sensitive private information from people. People may disclose information about their health history, behaviors, symptoms, thoughts and feelings, and personally identifiable information about themselves or others (e.g., family members), thus creating threats of harm from the improper use or release of such information (Luxton, 2014). As in the case of telehealth services that are provided across international borders, the laws and regulations regarding the assurance of user privacy can vary significantly (Luxton & Sirotin, 2020). Because IVAs can be accessed from anywhere in the world, and potentially to anywhere, health organizations that deploy them should be aware of the applicable local laws. Users should also be made aware of the steps taken to uphold privacy, despite risks and limitations. Any informed consent process should make note of such considerations.

Algorithmic bias and the black box problem

The recent development of Transformer-based technologies offers radically improved conversational capabilities (Bommasani et al., 2021; Brown et al., 2020). These models can make sense of diverse, multimodal datasets and can be activated using a simple natural language descriptor of a task or intended outcome. These technologies offer revolutionarily simple interfaces for working with complex machine intelligence systems and are an obvious fit for IVAs, given how much more sophisticated and freeform dialogues with such systems can be. However, the added complexity of and size of these new models present a further level of difficulty in debugging and debiasing such systems.

The complexity of AI algorithms may make the IVA systems opaque to those employing them, and aspects of them may remain opaque even to their designers. Algorithms, models, datasets, and predicates that underly their decision-making processes, predictions, and behavior may not easily be audited or understood by humans (Luxton, 2020). For example, the use of complex, multilayer neural networks, multiple source data inputs, and distributed knowledge bases can result in complex emergent phenomena that humans cannot visualize or understand. Some of the primary ethics issues with this include potential for inaccurate or disproportional biases toward one group or groups over others, which may result in disparate and unfair outcomes, with regard to health, safety, or financial factors. From a practical design perspective, the underlying functioning of IVAs must be thoroughly tested and retested for possible unwanted and unfair biases they may trigger. Source causes of the algorithmic bias problem include inadequate data inputs due to sampling limitations (e.g., underrepresentation of certain groups), or external (environmental) changes that cause unwanted drift in outcomes, such as a variance between conditions expected and tested for in the lab versus real-world constraints.

One solution to the black box problem is to implement an audit trail process that tracks the decision-making processes of an IVA system. This would enable system administrators, designers, and investigators to monitor and identify the reasons for particular outcomes. The audit capability might improve trust among users and the public, by increasing their confidence and acceptance of the reasons for an outcome by demonstrating a transparent explanation in an accessible form. It would also be appropriate to provide a second opinion by a human overseer upon request.

Healthcare reach and equity

The successful use of IVAs in public health and healthcare services depends on whether these technologies are accessible to the populations that could benefit from them. Technological infrastructure investment in underserved communities and countries, as well as the cost, has been significant limitations to the adoption of any high technology (Luxton & Sirotin, 2020). However, due to the proliferation of smartphones and feature phones, as well as cellular data connections and distributed cloud computing, IVAs may be able to piggyback on existing infrastructure. Indeed, IVAs can be an amplifier of existing medical resources, expanding the reach of certain types of services to underserved areas, or reducing a shortfall when a surge of services is needed, such as happened during the COVID-19 pandemic.

Luxton and Sirotin (2020) also highlight the importance of technology literacy as a potential barrier to access and use of technology, which can result in healthcare disparities. Technology literacy is the ability of a person to use, manage, evaluate, and understand technology (ITEA,

2020/2022). Therefore organizations which deploy IVA technology must consider access to technology and familiarly with its use. To achieve equitable care or services, however, alternative methods of delivery must be considered. Not every medical intervention can be adequately served by IVAs, and the use of such technologies in certain contexts, such as grief, may not adequately respect human dignity.

Conclusion

While much is hypothesized or foretold that machines will replace humans, we believe that IVAs are unlikely to fully replace humans in the healthcare and public health professions, but they are an important adjunct to human activity. The benefits of IVAs will continue to grow along with rapid ongoing technological developments in artificial intelligence and electronic infrastructure. IVAs have many advantages to humans, including that they are not subject to fatigue or burnout; are highly scalable as services can be expanded by simply deploying more cloud instances in an elastic manner; and are cost efficient, potentially saving companies, governments, insurers, and consumers money, while also providing valuable alternative service provision.

However, there are dangers to advancing technology without appropriate and adequate attention to the ethical and moral aspects. Professional ethics codes and best practices must be developed or revised to help address the ethical issues associated with IVAs in healthcare, given that most healthcare professional ethics codes and practice guidelines do not presently address the use of technologies that aim to significantly replace human professionals in a sophisticated manner (Luxton, 2014). Whereas the World Health Organization has recently published high-level ethical principles and guidelines associated with the design and use of AI in health (World Health Organization, 2021), guidelines will also need to be tailored to the specific needs and nuances of the healthcare and public health professions.

Future directions

The rapid advancement of Artificial Intelligence and other technologies for building virtual care providers and public health assistants will bring further opportunities for advancing public health and healthcare services to people around the world.

Imagine this possible future scenario for the post-SARS-CoV-2 world whereby advanced technology is deployed A new and unknown deadly virus is detected and begins to sweep across the planet. Public health officials know they must respond quickly with symptom screening and contact tracing, must triage to an appropriate level of care, and educate the public about the prevention of spread. They quickly modify virtual human public health screening software to ask appropriate questions and to educate about the emerging disease. Mobile devices (i.e., tablets and smartphones) are then strategically deployed from regional public health centers on trucks or by aircraft to population centers and to more remote communities. An array of microsatellites provides internet access to the most remote areas, and virtual humans, who appear, speak, and behave in a culturally appropriate manner for the geographic and ethnic setting, are available on mobile devices at any time and continuously provide a suite of services, from screening, education, and specialized counseling for persons triaged to a higher level of care needs. Public health technicians are also deployed to local health centers, to help with triage, to guide persons interfacing with the virtual humans, and to help with any medical emergencies. Service users are asked to consent to interface with the virtual humans, and they are informed on how data will be collected and used to help in the public health effort. Persons are also given alternatives, such as speaking to human public health and medical care providers, with the caveat that choosing to do so may take longer to get care due to inadequate numbers of human providers available. Data collected during the screening and symptom tracking is also analyzed in near real time to identify trends and characteristics of the emerging viral threat. This data, with the permission and informed consent of the donor, may inform future data collection operations, as well as responses, such as infection control efforts.

References

Alimoradia, Z., Brostrom, A., Tsang, H. W. H., Griffiths, M. D., Haghayeghe, S., Ohayonf, M., et al. (2021). Sleep problems during COVID-19 pandemic and its' association to psychological distress: A systematic review and meta-analysis. *eClinicalMedicine*, *36*, 100916. https://doi.org/10.1016/j.eclinm.2021.100916.

Bommasani, R., Hudson, D. A., Adeli, E., Altman, R., Arora, S., von Arx, S., et al. (2021). On the opportunities and risks of foundation models. *arXiv*. https://doi.org/10.48550/arXiv.2108.07258.

Boumans, R., van de Sande, Y., Thill, S., & Bosse, T. (2022). Voice-enabled intelligent virtual agents for people with amnesia: Systematic review. *JMIR Aging; Toronto*, *5*(2). https://doi.org/10.2196/32473.

Brooks, S. K., Webster, R. K., Smith, L. E., Woodland, L., Wessely, S., Greenberg, N., et al. (2020). The psychological impact of quarantine and how to reduce it: Rapid review of the evidence. *Lancet*, *395*, 912–920. https://doi.org/10.1016/S0140-6736(20)30460-8.

Brown, T. B., Mann, B., Ryder, N., Subbiah, M., Kaplan, J., Dhariwal, P., et al. (2020). Language models are few-shot learners. *arXiv*. https://doi.org/10.48550/arXiv.2005.14165.

Budakova, D., Petrova-Dimitrova, V., & Dakovsk0069, L. (2020). Intelligent virtual agent, learning how to reach a goal by making the least number of compromises, IOP Conference Series: Materials Science and Engineering. *Materials Science and Engineering*, *878*, 012030. https://doi.org/10.1088/1757-899X/878/1/012030.

Czeisler, M.É., Lane, R. I., Petrosky, E., et al. (2020a). Mental Health, Substance Use, and Suicidal Ideation During the COVID-19 Pandemic — United States, June 24–30, 2020. *MMWR. Morbidity and Mortality Weekly Report*, *69*, 1049–1057. https://doi.org/10.15585/mmwr.mm6932a.

Czeisler, M.É., Marynak, K., Clarke, K. E., et al. (2020b). Delay or avoidance of medical care because of COVID-19-related concerns—United States, June 2020. *MMWR. Morbidity and Mortality Weekly Report*, *69*, 1250–1257. https://doi.org/10.15585/mmwr.mm6936a4.

de Campos Filho, A. S., Vasconcelos Cursino, J. R., do Nascimento, J., de Souza, R. R., da Silva Lima Roque, G., & de Souza Cavalcanti, A. R. (2022). Content and usability validation of an intelligent virtual conversation assistant used for virtual triage during the COVID-19 pandemic in Brazil. *Computers, Informatics, Nursing: CIN*. https://doi.org/10.1097/CIN.0000000000000884. Advance online publication.

De Kock, J. H., Latham, H. A., Leslie, S. J., et al. (2021). A rapid review of the impact of COVID-19 on the mental health of healthcare workers: Implications for supporting psychological well-being. *BMC Public Health*, *21*, 104. https://doi.org/10.1186/s12889-020-10070-3.

Ganesan, B., Al-Jumaily, A., Fong, K. N. K., Prasad, P., Meena, S. K., & Tong, R. K. (2021). Impact of coronavirus disease 2019 (COVID-19) outbreak quarantine, isolation, and lockdown policies on mental health and suicide. *Frontiers in Psychiatry*, *12*, 565190. https://doi.org/10.3389/fpsyt.2021.565190. eCollection 2021.

IEEE. (2022). *Standard for the Description of the Natural or Artificial Character of Intelligent Communicators*. https://standards.ieee.org/ieee/3152/10833.

International Technology Education Association. (2020/2022). *Standards for Technological Literacy (STL)*. https://www.iteea.org/Activities/2142/Technological_Literacy_Standards/48897.aspx.

IVA. (2005). Intelligent Virtual Agents: 5th International working conference, IVA 2005, Kos, Greece, September 12-14, 2005. In T. Panayiotopoulos, J. Gratch, R. Aylett, D. Ballin, P. Olivier, & T. Rist (Eds.), *Proceedings*. Berlin: Springer. https://doi.org/10.1007/11550617.

Kulms, P., & Kopp, S. (2016). The effect of embodiment and competence on trust and cooperation in human–agent interaction. In D. Traum, W. Swartout, P. Khooshabeh, S. Kopp, S. Scherer, & A. Leuski (Eds.), *Lecture notes in computer science: 10011. Intelligent virtual agents. IVA 2016*. Cham: Springer. https://doi.org/10.1007/978-3-319-47665-0_7.

Luxton, D. D. (2014). Recommendations for the ethical use and design of artificial intelligent care providers. *Artificial Intelligence in Medicine*, *62*(1), 1–10. https://doi.org/10.1016/j.artmed.2014.06.004.

Luxton, D. D. (Ed.). (2016). *Artificial intelligence in behavioral and mental health care*. San Diego: Elsevier/Academic Press.

Luxton, D. D. (2020). Ethical challenges of conversational agents in global public health. *Bulletin of the World Health Organization*, *98*, 285–287. https://doi.org/10.2471/BLT.19.237636.

Luxton, D. D., & Hudlicka, E. (2021). Intelligent virtual agents in healthcare: Ethics and application considerations. In F. Jotterand, M. Ienca, & M. Liang (Eds.), *Ethics of artificial intelligence in brain & mental health*. Switzerland: Springer Nature.

Luxton, D. D., & Poulin, C. (2020). Advancing public health in the age of big data: Methods, ethics, and recommendations. In L. Goldschmidt, & R. M. Relova (Eds.), *Patient-centered healthcare technology: The way to better health*. https://doi.org/10.1049/PBHE017E_ch3.

Luxton, D. D., & Riek, L. (2019). Handbook of rehabilitation psychology. In L. Brenner, B. S. Reid-Arndt, Elliott, et al. (Eds.), *Artificial intelligence and robotics for rehabilitation* (3rd ed.). Washington DC: American Psychological Association Books.

Luxton, D. D., & Sirotin, A. (2020). Intelligent conversational agents in global health. In S. Okpaku (Ed.), *Innovations in global mental health*. Cham: Springer. https://doi.org/10.1007/978-3-319-70134-9_11-1.

McKillop, M., South, B. R., Preininger, A., Mason, M., & Jackson, G. P. (2021). Leveraging conversational technology to answer common COVID-19 questions. *Journal of the American Medical Informatics Association*, *28*(4), 850–855. https://doi.org/10.1093/jamia/ocaa316.

Philip, P., Dupuy, L., Morin, C. M., de Sevin, E., Bioulac, S., Taillard, J., Serre, F., Auriacombe, M., & Micoulaud-Franchi, J. (2020). Smartphone-based virtual agents to help individuals with sleep concerns during COVID-19 confinement: Feasibility study. *Journal of Medical Internet Research*, *22*(12), e24268. https://doi.org/10.2196/24268.

Soderlund, M., Oikarinen, E., & Tan, T. (2021). The happy virtual agent and its impact on the human customer in the service encounter. *Journal of Retailing and Consumer Services*, *59*, 102401. ISSN 0969-6989 https://doi.org/10.1016/j.jretconser.2020.102401.

Weizenbaum, J. (1976). *Computer power and human reason: From judgment to calculation*. http://refhub.elsevier.com/S0933-3657(14)00068-2/sbref0005.

World Health Organization. (2021). *Ethics and governance of artificial intelligence for health*. https://www.who.int/publications/i/item/9789240029200.

Zhu, Y., Wang, R., & Pu, C. (2022). "I am chatbot, your virtual mental health adviser." What drives citizens' satisfaction and continuance intention toward mental health chatbots during the COVID-19 pandemic? An empirical study in China. *Digital Health*. https://doi.org/10.1177/20552076221090031.

CHAPTER 23

"Creating a digital spine for connected healthcare ecosystem": Digitalization for an effective supply chain

Girish Ramachandran
TCS Asia Pacific, Singapore, Singapore

The healthcare industry has been under the global spotlight since the beginning of the COVID-19 pandemic. The sector's shortcomings and gaps were exposed as the virus took its toll on our systems and overall well-being. But the pandemic also galvanized the world community. Nations now stand in solidarity with each other and extend a helping hand leveraging the power of IT and digital technology to solve and plug the gaps in healthcare.

A shining example of this collaboration is playing out in the distribution of vaccines and other major medical supplies, where various players from the healthcare ecosystem banded together to predict and control the spread of the virus using technology.

Currently, different healthcare ecosystem players use disparate systems that lack the necessary infrastructure support to connect them with patients in a meaningful and proactive manner. This disconnect within the same ecosystem gave rise to the widespread use of technology such as Artificial Intelligence (AI) and Machine Learning (ML). Such technologies enable and connect different information networks from various ecosystem entities in the supply chain in a way that allows information derived to be used for both retrospective and predictive analysis, even as we continue to track the growth trajectory of the COVID-19 viral spread.

This chapter illustrates how the application of technology can help medical professionals and institutions worldwide overcome challenges and save more lives. Additionally, the power of technology can strengthen data security and health infrastructure while also improving the global supply chain.

There is an urgent need for a seamless ecosystem to connect information generated regularly from various ecosystem entities, including healthcare records, medical device manufacturers, insurance records, pathology labs, doctors, and government health authorities.

Resilient Health. https://doi.org/10.1016/B978-0-443-18529-8.00023-8

A seamlessly connected ecosystem is especially necessary because currently the collected data is maintained in disparate systems, depending on the preference of the entities. As a result, data cannot be deployed in a timely and meaningful manner. Instead, the collected data is being used mainly for financial reimbursements, and not for healthcare purposes. Also, control of the data is decentralized, which means it may not be easily accessible to users, including healthcare professionals.

Conversely, if collected data is organized and connected, it can immensely benefit governments and healthcare organizations. They can use the data to address current or future pandemics or other health-related crises and proactively provide essential healthcare services to the citizens. This is what we call the "Digital Health Spine."

As such, there is a clear need for IT consulting companies such as Tata Consultancy Services (TCS) to bring to market digital platforms such as its proprietary TCS Connected Wellness Solution. The platform offers healthcare players pre-care, point-of-care, and post-care solutions via a universal wellness system in what TCS refers to as a Connected Health Ecosystem, as shown in Fig. 1. Such a Healthcare Ecosystem can effectively bridge the gaps and provide both curative and predictive healthcare for the patients. As the figure shows, such a system can aggregate data from a broad range of devices to generate actionable insights on the unique behavioral pattern of each individual.

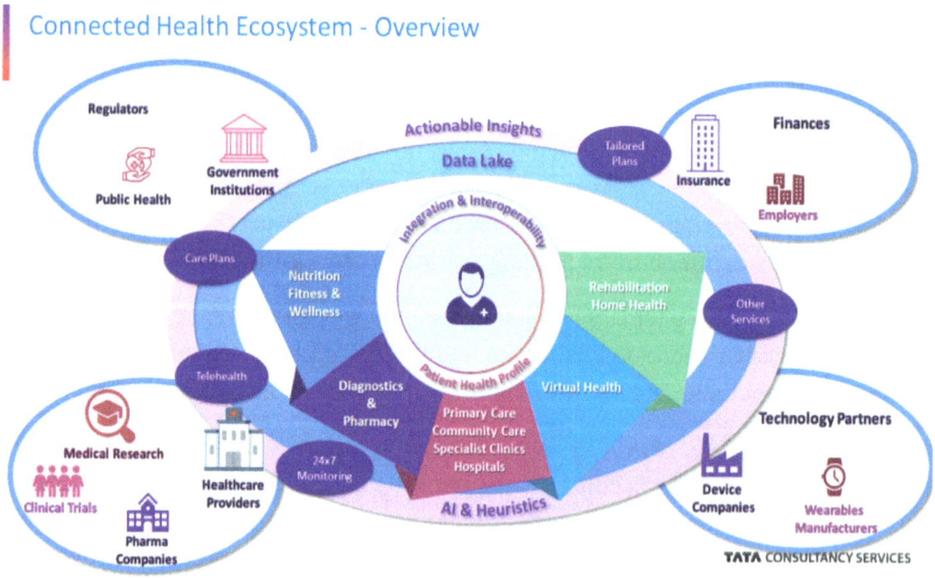

Fig. 1
The "Connected Health Ecosystem."

The "Connected Health Ecosystem" has the following benefits:

(i) Connectivity of different medical devices and product vendors to provide "as-a-service" to the end user (patient).
(ii) Ease of collecting information from various sources so that the industry can provide critical healthcare-related services to anyone, anywhere in the world.
(iii) Enabling easy connection of the allowable healthcare information for servicing patients' needs such as pharmacy services and provision of health check-ups.
(iv) Helping government healthcare agencies to form relevant and meaningful healthcare policies using the complete set of data and information available.

To support our partners and the healthcare industry, TCS has also developed a middleware connecting system. It can receive information from disparate medical devices and other entities such as product manufacturers and doctors' diagnoses, and integrate this data with smart devices that can also be connected to insurance records and other related stakeholders.

This system complemented by our AI/ML algorithms provides proactive services to patients and other related stakeholders and also chronicles the data collected into use cases for the benefit of others.

Overview of Tata Consultancy Services

TCS is an IT services, consulting and business solutions organization that has been partnering with many of the world's largest businesses in their transformation journeys for over 50 years. TCS offers a consulting-led, cognitive-powered, integrated portfolio of business, technology, and engineering services and solutions, delivered through its unique Location Independent Agile delivery model, recognized as a benchmark of excellence in software development. As a part of the Tata group—India's largest multinational business group—TCS has 613,974 of the world's best-trained consultants in 50 countries. The company generated consolidated revenues of US$25.7 billion in the fiscal year that ended on March 31, 2022, and is listed on the BSE (formerly Bombay Stock Exchange) and the NSE (National Stock Exchange) in India. TCS' proactive stance on climate change and award-winning work with communities across the world has earned it a place in leading sustainability indices such as the MSCI Global Sustainability Index and the FTSE4Good Emerging Index. For more information, visit www.tcs.com and follow TCS news at @TCS_News.

Requirements of curative and predictive healthcare

The availability of feasible technology and network connectivity has shown how an essential healthcare services and support system can be provided to everyone. Startups and other entities are using technology to develop platforms, applications, and systems that are benefitting millions around the world.

The capability and strength of AI and ML has shown how the impact of a pandemic like COVID-19 and its variants can be predicted and used to support research and discovery of new drugs, enable genome sequencing, and provide doctors with accurate information about patients quickly. Remote patient monitoring gained much importance during the pandemic and is proving to be a vital solution in enabling healthcare teams to provide medical services to remote parts of the world. Indeed, we have seen patients get surgery advice advice and support from medical experts based in another location.

For healthcare services to be equitable and effective, we need an interconnected ecosystem capable of providing "curative and predictive services" uniformly across an expansive base of patients. As the world gravitates toward building a sustainable future for everyone, it is imperative that "quality and affordable healthcare for all" becomes a reality. This, in turn, calls for wide-ranging use of technology to ensure the success of this collaborative ethos.

One nation that has had success with this connected ecosystem approach is India. It has deployed a Digital Nerve Centre (DiNC), developed by TCS, to leverage technology, people, and processes, holistically to improve healthcare access for the underprivileged. The DiNC is a unique and innovative care delivery model designed to connect, communicate, coordinate, and deliver care by leveraging people, infrastructure, and a robust digital platform. The platform combines structured and unstructured data to build one cohesive view and connect patients with primary, secondary, and tertiary healthcare providers over the phone. It can book appointments, maintain digitized health records of patients, and rope in care coordinators to handle non-medical queries. The platform also allows hospitals and treatment centers to collaborate and share knowledge.

Curative healthcare

A connected healthcare framework will serve many needs of the healthcare sector. The primary requirement for this framework is to allow data to be seamlessly integrated and be within the control of the individual (or patient). The framework primarily mandates the need to be able to collect data from medical devices, mobile devices, wearables, pathology labs, and other means. This data will then be stored for analysis such that the insights derived can be used to forewarn the individual of any chronic diseases and serious medical conditions, such as diabetes and cancer.

In addition to a patient's past medical data, the framework will collect any new test reports before analyzing this combination of data and arriving at a reasonable inference. A doctor or medical team can refer to this inference while attending to the patient. As indicated earlier in this chapter, this intelligent system can also integrate additional data from other sources like wearables and personal devices, to provide a comprehensive overview of the patient's condition.

The system at the disposal of the patient acts as a companion for healthcare professionals. It can serve reminders to individuals, helping them track their progress and report to their attending doctors, as required, and connect with a pharmacy to order medicines. It can also help patients apply for reimbursement claims.

With remote connectivity, the patient can avail of services of a doctor/physician, by getting an appointment and having a video conversation easily, with previous diagnoses and other critical information shared with the doctor ahead of time. At the end of the consultation, payment can also be made to the doctor via the system.

Where there is a need, the system can be connected to ambulatory services, or arrange an appointment at the nearest hospital, based on the patient's choice. Before the appointment, the patient's insurer can be informed so that they can enjoy cashless and hassle-free treatments at the hospital.

All these services are possible as the system maintains the health records and transmits these data securely to other related entities for a smooth and complete healthcare experience.

Overall, this framework of Connected Healthcare will be built based on three key considerations, as presented in Table 1. The table shows the key considerations healthcare players need to keep in mind as they design the best system and services for patients.

Table 1 Key considerations: Connected healthcare.

Key consideration	Description
Patient information from connected devices	– Enable gathering of information from various tests done in pathology labs, data from medical devices, data collected from personal devices – Other data to be gathered include medication records, doctor's analysis, past illness and surgeries – Keeping track of claims made from insurance companies, insurance policy
Using AI/ML to **draw** inference from collected data	– Using medical inference from the available data to provide analysis in a meaningful manner for a doctor to determine the course of treatment – Understand patterns from the data gathered from personal devices/wear-ables, annual checkups, etc. to indicate potential issues – Periodic alerts for keeping track of medical checks, medicine intake, ordering medicines, access to doctors in the neighborhood, etc.
Connectivity with security	– Data provided for access with consent from the patient (owner of the data) – Access to data/information based on requirement – Audit log of access – Industry standard data encryption during storage, use, and transmission

As presented in Table 1, the use of digital mobile devices/wearables such as tablets and smart watches is a crucial step in helping the various ecosystem stakeholders improve the quality of the information they gather or receive, which can lead to better outcomes for the patients.

For example, the data collected from personal mobile devices/wearables are gaining in importance due to their ability to monitor vital health indicators like pulse, heart rate, and blood pressure in real time. Triangulating such information with other related data points of the patient can help predict and forewarn the patients and their attending medical team of any health conditions that require attention.

Predictive healthcare

With the advent of AI-ML, we witnessed an increase in cases of healthcare professionals using technology to predict patient outcomes and making appropriate decisions. Some use cases of predictive healthcare, as the industry becomes future-ready, include the following:

(i) The COVID-19 pandemic necessitated the use of AI-ML to predict the spread of the virus. The two most common indices—the Susceptible-Infected-Recovered (SIR) index used in the USA, and the Susceptible-Infectious-Susceptible (SIS) index used in India—were developed and deployed largely due to the technology available. Many IT companies like TCS, IBM, and Infosys, worked with governments to program these models to help take important decisions on movement restrictions or lockdowns.

(ii) The Connected Healthcare Ecosystem can keep track of who was affected and the impact on the people in the same or nearby location. This information proves vital in helping authorities track and take necessary action to prevent further spread of the virus.

(iii) Any relevant information collected can be shared in a secure manner with pharmaceutical firms for their research purposes. For example, symptoms and other details can be shared for further analyses, so that appropriate treatment can be provided.

(iv) Multiomics, an approach where different datasets are combined during analysis, is fast gaining ground to capture, correlate, and connect a patient's molecular level information related to disease and treatment. Organizations like TCS are using their healthcare industry knowledge and IT capability of comprehensively analyzing and integrating data to accelerate genomic interpretation in drug discovery and clinical research efforts.

(v) Predictive Healthcare is being used across various disease categories to predict the response based on specific type of blood tests to determine cancer treatment, and to understand speech patterns to detect Alzheimer's in its early stages. Also, analysis of blood pressure, arrhythmia, pulse, and other medical indicators can forewarn patients regarding any chronic disease.

(vi) Internet of Things/Edge computing, where sensors and devices communicate real-time data to a network, is increasingly being used in cases in the healthcare sector for patients, physicians, hospitals, and even insurance companies. IoT has proved to be a useful solution for interacting in a non-intrusive way with elderly patients. Home monitoring equipment is proving very helpful for physicians to track medication adherence levels by patients. IoT tagged devices are helpful in tracking the use of wheelchairs and medical devices in a hospital setup and to keep track of temperature, humidity control, and pharmacy inventory, while tracking the deployment of hospital staff in various locations. Insurance companies can use the power of IoT by transparently tracking data from health monitoring devices for underwriting and claim purposes. This also provides visibility to patients and helps them track the claims process.

(vii) Edge (low-latency, highly available apps with real-time monitoring, coupled with storage on the cloud) is enabling critical development and enhancement in various areas. Technology product companies have been using the features of Edge computing coupled with cloud for their tablets and wearables to health monitors and AI-powered imaging system. MRI and CT scans use Edge computing for compression and decompression of medical imagery. These companies are using the power of technology to focus on decision support with timely, secure, and patient-filtered specific information. AI algorithms are being used to detect critical findings on chest X-rays and many such similar use cases are being developed and deployed to deliver critical services and diagnosis.

(viii) Integrated Digital Health solutions are being used to capture and address patients' physical and behavioral parameters in an unobtrusive manner. TCS has developed a knee replacement surgery-related solution by capturing activity levels, gait, and range of motion (RoM), before and after surgery. This helps patients in reporting recovery status and in organizing personalized behavior interventions. These types of digital health solutions help in reducing the need for in-person interactions. Such interactions using digital health solutions can be applied to several other surgical procedures.

Key challenges

The pandemic has created serious challenges for the healthcare ecosystem, which has in turn necessitated the speeding up of solutions. Technology has been a significant factor in enabling several solutions and has proven to be a significant ally in providing quality and affordable healthcare to all.

This being said, the healthcare industry still faces several obstacles that need to be overcome to allow rapid deployment of digitally enabled health solutions where they are most needed.

Some key challenges:

(i) Healthcare data is highly unstructured; thus, it can be difficult to capture the relevant data during the care process.

(ii) Expecting healthcare professionals to enter/capture data manually is a highly time inefficient, counterintuitive, and ineffective activity that can also affect the quality of services provided to the patients.

(iii) While healthcare science has progressed significantly, healthcare delivery has not progressed at the same pace, thereby creating logistical and data-related quagmires.
(iv) While wearables, Edge computing, cloud, and AI-ML are possible solutions to help the industry overcome inequality and gaps, the lack of a strong foundation in healthcare delivery services can impede progress.
(v) The current state of patientcare is more focused on record-keeping and postcare data capturing, which are key for commercial reimbursements but do not improve the quality of health services dispensed.
(vi) Lack of a common protocol to have data extracted from different medical devices means there will be downstream difficulty in collating data for digital interpretation and integration.
(vii) Multiple sources and users require data in their own format, which leads to lack of standardization.
(viii) Doctors need authentic sources of data obtained through credible sources, which may be lacking in certain geographies.
(ix) The ratio of doctor-to-patient is often lopsided, which can be addressed by digital health solutions so that equitable healthcare resources can be provided to all.
(x) Any digital intervention has to keep in mind the scale of operations and digital security to allow ease of access to key data by healthcare professionals during the course of treatment. There should also be a well-updated audit trail of the access granted by the individual for such usage.

The path ahead

Great potential exists for increased usage of digital ecosystems in healthcare to address and solve challenging problems facing the world.

As set out in this chapter, the pandemic has clearly shown how predictive healthcare technology, using AI-ML (Linear Regression [LR], Support Vector Regression [SVR] algorithms, and others), can be used to accurately predict likely scenarios during a pandemic, which can help governments determine the next course of action.

Similarly, the usage of predictive technology in conjunction with data from new-age wearables and other mobile devices can help forewarn of chronic diseases, thus allowing medical professionals to move from a curative to a preventive model of care for their patients.

To this end, channeling attention and resources broadly into two categories—healthcare delivery and healthcare services—will allow us to address the challenges and enable quality, affordable, and predictive healthcare for all.

Some key steps being taken to necessitate this transformation are to:

(i) Ensure policies to allow trials, and permission to allow usage of data from wearables and mobile devices, must be approved with medical authority.
(ii) Enable a common protocol to have all medical devices data to be collected in a seamless manner.
(iii) Integrate data captured from wearables and other mobile devices in a medically authorized manner.
(iv) Establish secure and systematic storage of patient data to enable rights-based access to patient data based solely on their approval.
(v) Work with healthcare organizations to use video and image analytics, Edge computing, and IOT to develop solutions for more expansive usage of technology in various aspects of providing healthcare, e.g., for chronic diseases like cardiovascular diseases, diabetes, cancer.
(vi) Expand the usage of networks to permit remote patient services not just for diagnosis but also specialist treatment. Connectivity at the 5G level can surely enable higher quality to connect to remote parts of the world.
(vii) Enhance acceptance of AI-ML to be used for initial analysis of a patient's condition, which should pave the way for in-house treatment.
(viii) Enable connection to other healthcare ecosystem partners like medical device manufacturers, pharmacies, and insurance providers to connect on a common health platform for their respective usage, where data in a summarized form enables them to prepare for a better set of devices and services in the healthcare sector.
(ix) Permit governments to receive analyzed data to determine the nature of policies and healthcare services depending on the needs of different areas of the country, thus enabling location-specific services to be provided.
(x) Create a system of a healthcare data bank that can enable secure access to an account owned by an individual. The data stored in this bank will have an elaborate system of meta data to store various forms of unstructured data based on the timestamp of recording. Having an organized and well-summarized form of data will benefit a healthcare professional, who can receive access to and view the data securely, based on approval granted by the owner of the data.

It is becoming increasingly evident that there is a growing need to ensure the equitable distribution of healthcare resources and services to every citizen on this planet.

To do that, governments of the world need to leverage technology to overcome legacy and infrastructural constraints associated with the industry. For example, telemedicine has made basic health services available to more people. But to achieve more, key organizations around the globe need to enable the formation of healthcare ecosystems that connect states/provinces to federal governments and further connect all stakeholders.

At a macro level, the way forward is anchored in a digital foundation. Governments play a crucial role in instigating systemwide digital health projects. There is a clear mandate that governments, in partnership with corporates and the private sector, should build a "digital spine" that leverages technologies such as the internet, cloud, mobile, big data analytics, artificial intelligence, and other innovations for connectivity and collaboration, backed by strong cybersecurity and clearly laid out policy for data access and control.

This "digital spine" should comprise four layers: (1) universal biometric-based digital identity allowing people to participate in services anytime, anywhere; (2) digital records enabling universal reach; (3) single payments interface to democratize payments; and (4) open, consent-based system that allows data to move freely and securely.

By sharing this IT infrastructure with other public and private services, governments can ensure uniform access to technologies, thereby leveling the playing field and accelerating the healthcare industry's transformation to data-driven organizations, allowing for digitally enabled processes and decisions that are based on the patterns emerging from them. This creates a connected ecosystem where public and private enterprises, and even countries and states, can proactively share information and cooperate, especially with the adoption of technology innovation.

Once such systems are put in place, it is necessary to ensure healthcare professionals receive adequate training on how to use and access the systems from their locations around the world. Additionally, data owners must be taught about their privileges to prevent abuse.

Lastly, there is a growing need to set stringent international standards for the storage, security, and access to health records of individuals. Along with this, we will require protocols to be put in place the to seamlessly integrate data between various medical devices, wearables, and mobile devices to enable ease of access and analysis. When connectivity standards for rendering healthcare services in a remote setup are set up, we will be one big step closer to providing "quality and affordable healthcare for all," which is the goal of the UN Agenda as well as of Universal Health Coverage.

CHAPTER 24

Accessing solar energy to deliver healthcare for all

Zerin Osho[a] **and Jagjeet Sareen**[b]
[a]*Institute for Governance & Sustainable Development, Washington, DC, United States,*
[b]*Dalberg Advisors, New Delhi, Delhi, India*

Introduction

Access to reliable, affordable, and sustainable energy for all is the fundamental basis for any intervention to alleviate poverty and improve the lives of billions in low-income countries. With approximately 1.3 billion people lacking access to energy, and given the fact that the global demand for energy has doubled (Franco et al., 2017), access to sustainable and reliable energy is therefore essential. Universal access to energy is an enabler of universal healthcare. Yet, lack of electricity remains a neglected barrier to the delivery of basic healthcare in low-income countries (WHO & World Bank, 2015). Roughly 70 percent of sub-Saharan African (SSA) health facilities do not have access to reliable electricity, and one in four health facilities has no electricity at all (Osho & Sareen, 2021). To add to that, 1.5 million children are lost to vaccine-preventable diseases each year due to a lack of reliable vaccine cold chain networks (Osho & Sareen, 2021).

Combining energy and health priorities can help overcome poverty, promote health, climate change, and enhance socioeconomic development. Access to reliable energy is crucial for the provision of basic healthcare and for the safe storage of lifesaving vaccines (Hostettler, 2015). An uninterrupted supply of energy can help power electric, thermal, and equipment needs of a healthcare facility. Deployment of off-grid and rooftop solar energy projects to power primary healthcare centers (PHCs) and the use of solar-powered cold chains for safe vaccine storage can help in building an effective healthcare delivery mechanism across low- and middle-income countries. Such an infrastructure is capable of delivering healthcare on the ground while achieving climate goals by promoting a global energy transition toward solar-powered energy systems.

Resilient Health. https://doi.org/10.1016/B978-0-443-18529-8.00024-X

This chapter discusses in detail how access to energy facilitates access to healthcare; the need for off-grid energy solutions that are sustainable, reliable, and affordable; and why solar energy is the best suited solution to meet the demands of rural underserved communities. The chapter concludes by discussing the need for self-sustaining business and management models to scale solarization of healthcare facilities.

Access to energy enables access to healthcare

Historically, the focus on health was an integral part of the Millennium Development Goals (MDGs)—the United Nations Agenda that guided government progress from the years 2000–2015, which included an aspiration to create a wide network of skilled care workers, develop medicines and treatment, and advance medical technology. Comparatively less attention was placed on the role of access to energy in enabling the delivery of healthcare on the ground (WHO & World Bank, 2015). Besides enabling delivery of healthcare, access to energy attracts and retains skilled medical workers, especially in remote locations. It is in acknowledgment of this fact that the UN established and included the Sustainable Development Goal (SDG 7) of "Ensuring access to affordable, reliable, sustainable and modern energy services for everyone" (United Nations, 2023) as part of its broad-ranging Agenda 2030 for Sustainable Development, which replaced the MDGs, now to be achieved by 2030.

Access to energy in the form of electricity is an absolute necessity for the smooth functioning of a health facility. Electricity is required for basic operations of the facility, including lighting, refrigeration, sterilization, communication, and the maintenance of IT systems. Energy is also required for the safe disposal of medical waste and for the use of medical devices (laboratory and diagnostic devices). In the recent times of the COVID-19 pandemic, access to power became even more critical, as IT systems, which provide 'telemedicine' and 'e-health' services, became the backbone of medical response in remote and rural locations (Franco et al., 2017).

Being connected to the grid is not enough. Bigger healthcare facilities such as hospitals and clinics are usually located in densely populated cities across the developing world and rely on the grid as their primary source of energy (WHO & World Bank, 2015). Despite that, grid failures and power outages during peak hours have resulted in a scarce supply of power, unmet even in cities and populated regions. This has forced healthcare facilities to switch to expensive, noisy, and polluting diesel generators (WHO & World Bank, 2015). Despite that, generators are expensive to maintain and operate and are backed up by kerosene lamps, candles, or flashlights (WHO & World Bank, 2015). According to a WHO survey, less than 30 percent of standalone diesel generators were functional with fuels available on the day of the survey (WHO & World Bank, 2015). In essence, the overall cost of energy and fuel, despite being connected to the grid, in many developing countries is higher compared to per capita income and consumes a significantly larger proportion of operating budgets of health facilities (WHO & World Bank, 2015).

Solar energy is a promising alternative. Falling costs of solar technologies and battery backup systems provide a reliable, sustainable, and inexpensive alternative to fossil fuel-based power systems. This trend is stimulated by the availability and use of vast range of direct-current (DC) medical devices that can be charged on a photovoltaic (PV) system (WHO & UNICEF, 2015). Additionally, several new initiatives launched globally such as the International Solar Alliance's (ISA) Cares, UNDP Solar for Health, SE4ALL's Powering Healthcare facilitate the transition of healthcare facilities to solar energy. Even major international health organizations like WHO and UNICEF promote the use and sale of solar direct drive (SDD) refrigerators for vaccine refrigeration (WHO & UNICEF, 2017a), while big development funds such as the Global Fund to fight AIDS, tuberculosis, and malaria are purchasing solar systems in bulk (WHO & World Bank, 2015).

Energy access and healthcare: Making the connection

Choosing a sustainable source of energy. Access to a reliable source of energy alone cannot remedy the weaknesses of healthcare infrastructure across developing and low-income countries but it certainly can enable a stronger delivery of basic healthcare. The use of off-grid solar technologies is a promising source of reliable, clean, and affordable energy for healthcare facilities in underserved communities. The energy requirement of a given healthcare center would determine the size and source of alternate energy to make the infrastructure reliable. For this, healthcare systems are classified into:

(a) *Hospitals* (large healthcare facilities) that require roughly 9-kW capacity, consuming between 15 and 35 kWh daily (Franco et al., 2017);
(b) *Health centers* (medium sized healthcare facilities), consuming between 10 and 20 kWh daily, with power needs of approximately 5 kW (Franco et al., 2017);
(c) *Health clinics* (small healthcare facilities) that need a capacity of about 2.4 kW, consuming between 4 and 10 kWh (Franco et al., 2017);
(d) and *Primary health centers* (rural primary health centers).

The following case studies demonstrate instances of how solar energy has directly facilitated in access to health in low-income developing countries across different geographies.

Case Study 1: Solar energy marshals a strong response to save lives of COVID-19 patients in Afghanistan

Herat, the second largest city in Afghanistan, recorded the second highest number of COVID-19 patients after Kabul. Prolonged power outage and unreliable energy risked the lives of patients dependent on electric ventilators and other breathe assist medical devices. In 2020 (prior to the Taliban takeover of Afghanistan), with the financial support of Herat Electrification Project, Afghanistan's national power utility company, called Da Afghanistan Breshna Sherkat (DABS), installed solar panels across ten hospitals in Herat to provide emergency power. The project generated 75 kW of reliable energy to power critical loads such as operation theaters, critical COVID-19 equipment such as ventilators, defibrillators, and other patient monitoring devices. Thirty-six solar panels were installed at Shaidayi Children's hospital to generate 10 kW of energy to power the Intensive Care Unit (ICU), 14 ventilators and lights to manage critical patients suffering from coronavirus. Another 34 solar panels were installed at the Obei Hospital to provide 10 kW for all hospital departments.

(Source: World Bank. (2020). Solar energy provides a lifeline to Afghanistan's Hospitals. https://www.worldbank.org/en/news/feature/2020/12/06/solar-power-provides-a-lifeline-to-afghanistans-hospitals#:~:text=The%20solar%20Photo%2DVoltaic%20(PV,centrifuges%2C%20and%20patient%20monitoring%20devices.)

Solarizing thermal loads could save lives. Another key facet of energy relates to thermal needs of health facilities. Health facilities, especially in rural areas, act as a last mile for delivery of basic healthcare. Such facilities usually have their highest thermal demands that come from refrigerators for the safe storage of vaccines, blood, and other temperature-sensitive medication. Many treatable diseases such as measles and pneumonia can be prevented through immunizations (WHO & World Bank, 2015). Despite that, each year, approximately 31 million children suffer from vaccine-preventable diseases, and more than a half million die due to lack of access to vaccines in the sub-Saharan African region (WHO, 2019). Most of these children live in low-income developing countries (World Bank & SEAR, 2017). A substantial share of these lifesaving vaccines is lost to poor cold chain and refrigeration (World Bank, 2020). Most vaccines need to be transported and stored between 2°C and 8°C (WHO, 2015). Cold chain disruptions cause the wastage of approximately 50 percent of freeze dried and 25 percent of liquid vaccines annually (Sustainable Energy for All (SEforALL), 2020). Due to the lack of energy at healthcare facilities in low-income countries, vaccine refrigerators are powered through kerosene-based absorption refrigerators (WHO & UNICEF, 2017b). But this creates problems, since absorption refrigerators often expose vaccines to freezing temperatures, making them unstable and fatal for use (Rao, 2020). Tragically, in 2017, improper storage of measles vaccines in South Sudan claimed the lives of 15 children (Rao, 2020). According to a study in 2014, GAVI reported that roughly 90% of health facilities lacked adequate cold chain equipment in GAVI-eligible countries (GAVI, 2020). Furthermore, supplies of kerosene to power these refrigerators are subject to disruptions and supply, making them unreliable for the safe storage of such temperature-sensitive vaccines and other medication.

Case Study 2: Senegal uses solar-powered vaccine refrigerators

Repeated power outages and unavailability of generators at district level in Senegal, Africa, exposed vaccines to temperature variability risks at healthcare facilities. Project Optimize—a collaboration between WHO and PATH—installed fifteen SunDanzer refrigerators in health posts of Podor, Pete, and Richard Toll districts of Senegal. The company SunDanzer designed an innovative installation by placing two solar panels in opposite directions, in the form of a tent—one facing East and the other one facing West. This design improvement allowed solar generation to start earlier and continue later into the day. It was observed that at the end of one year of operations, the SunDanzer refrigerators performed reliably. With remote tracking of thirteen refrigerators, it was observed that the refrigerators remained in the target temperature range of 2°C to 8°C for 99 percent of the cumulative times, keeping vaccine safe for use.

(Source: WHO and Path (World Health Organization (WHO)). 2013. Direct-drive solar vaccine refrigerators—A new choice for vaccine storage. https://media.path.org/documents/TS_opt_ebs_dd_solar_fridge.pdf.)

Access to energy is a matter of gender equality. Without reliable power, most basic lifesaving interventions cannot be undertaken. Access to reliable energy matters significantly to women and children (World Bank & SEAR, 2017). With over 295,000 women dying globally from pregnancy and childbirth-related complication each year (WHO, 2014), evidence suggests that adequate lighting helps avert many of these life-threatening complications during complicated deliveries for both, mother and child (Mills, 2012). Reliable energy is also proven to have an impact on the levels of motivation of health workers. It was observed that healthcare workers responded better to emergencies if the healthcare facilities were provided with adequate and reliable energy (World Bank & SEAR, 2017). The positive correlation between access to energy and impacts on gender was also observed. For instance, in Uganda, the availability of power led women to opt for deliveries at health facilities as opposed to at homes (Mbonye & Asimwe, 2010).

Case Study 3: UNDP Solar for Health Initiative

The United Nations Development Program (UNDP) Solar for Health Initiative was developed in line with the UNDP Strategic Plan of 2018–2021, outlined in the UNDP 2016–2021 HIV, Health and Development Strategy: *Connecting the Dots*. This initiative supports governments across Africa, Arab states, and central Asia to solarize health facilities to provide a reliable and affordable source of energy to enable healthcare in rural and underserved communities. UNDP Solar for Health Initiative spearheaded a solarization drive of healthcare facilities in Zimbabwe and Zambia (UNDP, 2018a). It was observed that most primary healthcare facilities in Zimbabwe did not have access to reliable energy and maternal mortality was higher for women living in rural areas and among poorer communities. In Zambia, power outage affected refrigeration of medicines and vaccines, and in addition, to providing light for carrying out procedures, healthcare facility staff were forced to either tell patients "to come with candles" (UNDP, 2018b) or to use torch light on cell phones. With support of UNDP and the Norwegian Emergency Preparedness System, Zambia's 7000 m^2 national medical warehouse was installed with solar rooftop systems, covering the size of a football pitch. The impacts were exceptional—healthcare facilities were in operation 24×7, uninterrupted power kept vaccines safe for use in a temperature-controlled cooling facility, and cost savings of up to 40% by reducing diesel consumption were observed. UNDP also estimated a 100% return on investment within two to five years for healthcare facilities that chose to solarize their power systems. In Zimbabwe, with support of the national government and the Global Fund, UNDP was able to solarize 405 healthcare facilities. Another major impact observed in Zimbabwe was the development of credible datasets for managing patient files and cold chains.

(Source: UNDP. (2018c). Solar for health. https://www.undp.org/energy/our-flagship-initiatives/solar-for-health#:~:text=Solar%20for%20Health%20interventions%20help,advancing%20multiple%20Sustainable%20Development%20Goals.)

Solar energy is a cost-effective alternative. Despite several political, institutional, economic, and knowledge barriers (World Bank & SEAR, 2017), solar energy for powering health centers has become a promising alternative that provides reliable, affordable, and sustainable energy access. It is has been noted that most healthcare centers rely on fossil fuel-based diesel generators because of their low investment costs (Franco et al., 2017), but as the cost of solar has fallen over the years, it provides a unique opportunity for policymakers to align the two developmental goals of providing access to energy and healthcare for all. In terms of levelized cost of energy (LCOE), polluting diesel generators cannot compete with solar technologies. Photovoltaics (PV) have an LCOE of 0.09 USD/kWh versus an average 0.25 USD/kWh for diesel generators (Franco et al., 2017). Lowering the cost of solar energy has enabled a country like Liberia, where the central grid extended slightly beyond the capital, to solarize 146 clinics (using PV solar as their primary source of energy), compared to 116 clinics that used generators and 3 clinics that were hooked up to the grid (WHO & World Bank, 2015). On the other hand, 15% of Ugandan hospitals used solar PV to complement grid-based electricity, while in Sierra Leone, 36% of all health facilities and 43% of hospitals used solar PV technologies in combination with other sources of electricity to provide reliable access (WHO & World Bank, 2015).

Case Study 4: International Solar Alliance (ISA) facilitates tailor-made solar solutions in Benin to reduce dependency on diesel generators to reduce costs

The International Solar Alliance (ISA) launched the ISA Cares initiative at its third Assembly in 2020. The initiative aims to solarize at least one healthcare facility in each district of its 47 least and small island developing states (LDCs and SIDS). The aim is to ensure healthcare for all, wherever they may be, and that no one is left behind. Under this initiative, ISA will be solarizing Abomey-Calavi University Hospital in Benin. The Abomey-Calavi University Hospital is currently dependent on grid power supplied through Société Béninoise d'Energie Electrique (SBEE). As the cost of grid power has been rising in the recent past, and the diesel genset causes noise and air pollution, the hospital is keen to solarize. Also, the hospital aims to reduce operating expenses, especially by reducing its consumption of diesel through the backup systems. Through a US$50,000 grant by ISA, Benin will install a 15-KWp solar PV rooftop system with hybrid inverter, integrated with 34 kWh of battery storage to provide reliable, clean, and affordable energy to the operation theater of the hospital. The 15-KWp system would be sufficient to meet the daily energy need of 46 kWh. The project is estimated to reduce the hospital's energy bills and more than 5.5 tons of CO_2 annually.

(Source: Detailed Project Report—Benin, International Solar Alliance)

Challenges

It is clear from the case studies presented before that access to solar energy can revolutionize delivery of healthcare on the ground. Despite this success shown in different regions of the world (Africa and Afghanistan), and innovations in off-grid solar solutions at health facilities across low- and middle-income countries, some of these projects fail prematurely owing to poor quality designs and or insufficient attention given to long-term maintenance and management of these systems. For instance, navigating the unique features, structures, and trends in each developing energy market deters investors and adds time and cost to the project creating barriers to scale and efficiency. On the other hand, due to lack of competition, many power projects in low-income countries are not competitively tendered causing higher transaction costs that increase overall tariffs. Furthermore, the long development cycles of solar projects and the poor credit situation in low-income developing countries and political risks contribute to higher perceived risks for such projects. All of this leads to the perception that solar technologies are new and unreliable.

To remedy such a challenge, there is a need to focus on the manner in which solar projects for health facilities are designed, financed, and maintained. Rather than taking the conventional approach of procuring and installing solar panels on health facilities, there is a need for adopting a long-term, performance-based operation and maintenance contract that can be serviced by the private sector.

The Chhattisgarh State Renewable Energy Development Agency (CREDA) (CEEW & Oxfam India, 2017) in India serves as a good case study for scaling and replicating solar for health at large. PHCs ensure last mile delivery of healthcare that is at a village level in India. Despite that, one in every two PHCs in India and one in every three PHCs in Chhattisgarh were either unelectrified or suffered from power outages. To counter this challenge, Chhattisgarh installed two kilowatts peak (kWp) off-grid solar photovoltaic (PV) rooftop systems across 570 primary healthcare centers (around 72 percent of all functioning centers) through the Chhattisgarh Renewable Energy Development Agency (CREDA) (CSIS, 2021). As a result of this, PHCs in Chhattisgarh gained access to solar as the primary source of energy during daylight, the time for the highest footfall of patients, making grid-connected energy a secondary source. CREDA also provided regular maintenance of the solar panels and the extra power generated from grid enables the installation of cold chain equipment for the safe storage of vaccines.

Conclusion

The COVID-19 pandemic is a grim reminder of the urgent need for 24-h healthcare facilities across regions and geographies. This will mean additional energy requirements that are reliable and affordable. Solar energy provides a clean, reliable, and affordable solution that can substantially accelerate progress toward universal, sustainable healthcare access. As we have seen in cases cited in this chapter, where energy supply for healthcare facilities has improved, significant progress has also been noted in the number of patients served and treatments offered. Especially for women and children in rural communities, solar-based systems in healthcare facilities have had a transformative impact on the quality of life.

However, long-term sustainability and an improved health infrastructure require an ecosystem, propelled by a combination of policy levers, institutional and technological innovation, mobilization of finance, conscious scaling-up strategies accompanied by demand aggregation. In addition, energy solutions must consider current and future needs for medical services and appliances, as well as aspects such as energy efficiency, building design, enhanced communication, and telemedicine. Above all, achieving universal access to healthcare by 2030 will require a cooperative ecosystem and the right people with the right skills.

As cost of solar energy and batteries plummets, investments rise, and a more enabling environment is created for solarization, solar energy's increasing role in healthcare is a certainty. Concomitantly, innovations in medical technology will likely make them more and more energy efficient, enabling the growth of cleaner energy sources. Solar in healthcare has the potential to save hundreds of millions of lives while reducing the use of costly and polluting diesel, increasing resilience in the face of natural and man-made disasters, and lowering carbon emissions.

References

CEEW, & Oxfam India. (2017). *Powering primary healthcare through solar in India: Lessons from Chhattisgarh.*

CSIS. (2021). *Solarizing India's healthcare system.*

Franco, A., Shaker, M., Kalubi, D., & Hostettler, S. (2017). A review of sustainable energy access and technologies for healthcare facilities in the Global South. *Sustainable Energy Technologies and Assessments*, *22*, 92–105. ISSN 2213-1388. https://doi.org/10.1016/j.seta.2017.02.022.

GAVI. (2020). *Cold chain equipment optimisation platform.* Gavi the Vaccine Alliance. https://www.gavi.org/sites/default/files/publications/Cold-chain-equipment-technology-guide.pdf.

Hostettler, S. (2015). *Energy challenges in the global south.*

Mbonye, A. K., & Asimwe, J. B. (2010). *Factors associated with skilled attendance at delivery in Uganda: Results from a national health facility survey.*

Mills, E. (2012). *Health impacts of fuel-based lighting.*

Osho, Z., & Sareen, J. (2021). *Chapter 3: Strengthening healthcare delivery systems in low income countries.*

Rao, R. (2020). *Why a strong cold chain is more critical than ever to defeat COVID-19.* Devex Global Views. https://www.devex.com/news/sponsored/opinion-why-a-strong-cold-chain-is-more-critical-than-ever-to-defeat-covid-19-97026.

Sustainable Energy for All (SEforALL). (2020). *Chilling prospects: Providing sustainable energy for all.* SEforAll. https://www.seforall.org/sites/default/files/SEforALL_CoolingForAll-Report.pdf.

UNDP. (2018a). *UNDP Solar for Health Initiative: 5 ways solar can make universal healthcare a reality.*

United Nations. (2023). *Sustainable Development Goal (7). Ensure access to affordable, reliable, sustainable and modern energy for all.*

WHO. (2014). *Maternal mortality.*

WHO. (2015). *The vaccine cold chain.* WHO. https://www.who.int/immunization/documents/IIP2015_Module2.pdf.

WHO. (2019). *Experts caution against stagnation of immunisation coverage in Africa.* WHO Regional Office for Africa. https://www.afro.who.int/news/experts-caution-against-stagnation-immunization-coverage-africa.

WHO, & UNICEF. (2015). *Introducing solar powered vaccine refrigerator and freezer systems.*

WHO, & UNICEF. (2017a). *Solar direct drive vaccine refrigerators and freezers.*

WHO, & UNICEF. (2017b). *Solar direct drive vaccine refrigerators and freezers*. WHO and UNICEF Evidence Brief. https://apps.who.int/iris/bitstream/handle/10665/254715/WHO-IVB-17.01-eng.pdf;sequence=1.

WHO, & World Bank. (2015). *Access to modern energy services for health facilities in resource constrained settings: A review of status, significance, challenges and measurement*.

World Bank. (2020). https://www.google.com/search?q=PATH+%26amp%3B+WHO.+2008.+%E2%80%9CLandscape+Analysis%3A+Cold+Chain+Technologies%E2%80%9D&oq=PATH+%26amp%3B+WHO.+2008.+%E2%80%9CLandscape+Analysis%3A+Cold+Chain+Technologies%E2%80%9D&gs_lcrp=EgZjaHJvbWUyBggAEEUYOdIBBzY1M2owajeoAgCwAgA&sourceid=chrome&ie=UTF-8#:~:text=Summary%20of%20Sustainable,org%20%E2%80%BA%20curated.

World Bank, & SEAR. (2017). *Modern energy access and health*.

CHAPTER 25

Innovations in drug and vaccine research, development, and distribution in COVID-19 recovery and beyond

Jami Taylor
Protagonist Therapeutics, Newark, CA, United States; Yale University, New Haven, CT, United States

The COVID-19 pandemic has ushered in dramatic changes to long-standing models of pharmaceutical and vaccine innovation, among them new approaches to drug development and distribution. Many of these changes have run counter to the pharmaceutical industry's operating model for decades, in turn generating opportunities for a rethinking—and, potentially, remaking—of the industry and its relationship to stakeholders in a postpandemic future.

The discussion that follows in this chapter captures, at a high level, the major shifts underway in the industry and their potential implications, highlighting specific technological innovations that stand to change the field further over time. Some of the shifts cited here were afoot prior to the pandemic's onset, but expanded or otherwise accelerated by the urgency COVID-19 required. Other shifts had the pandemic as their undisputed genesis, their longevity untested but their impact, clear.

At their core, each of these shifts relates in some way to two crucial axes: *speed* and *scale*. The speed at which innovators developed COVID-19 vaccines and therapeutics set a new precedent for what is possible in pharmaceutical innovation. The unevenness and inequities of drug and vaccine deployment globally—failures to scale—have placed a sharper lens on long-entrenched access hurdles. The crux of the innovation opportunity at hand lies in the reimagining of global access strategies, both in the context of COVID-19 and well beyond.

Speed: The postpandemic prospect of developing drugs and vaccines at rates faster than ever before

Perhaps the most remarkable aspect of the industry's response to COVID-19 was the speed at which it brought new vaccines and therapeutics to the public in medical settings. The first vaccines were approved for emergency use less than 12 months after the World Health Organization (WHO) declared COVID-19 as a global pandemic. In comparison to typical

Resilient Health. https://doi.org/10.1016/B978-0-443-18529-8.00025-1

research and development (R&D) timelines of a decade or more, this was a stunning achievement.

To the extent that factors which enabled this speed of movement remain in play, a new golden age of opportunity in biopharmaceutical innovation appears upon us. Factors contributing to R&D at a quick clip include: (1) expansive, technology-enhanced collaborations; (2) advanced research tools to accelerate discovery and development; and (3) policy mechanisms supporting intensive investment, providing ample "push" and "pull" incentives that sparked wide participation in the R&D process.

R&D collaborations: Supercharged in the pandemic context

Areas of innovation to watch in years to come: Tech-enabled collaborations across sectors, borders, and competitor sets, with real-time data sharing as a key driver of accelerated breakthroughs.

The COVID-19 pandemic spurred cooperation among diverse actors—including unlikely partners—in efforts to advance development of new drugs and vaccines capable of combating the virus worldwide. Broad participation among private companies, varied in size and maturity, contributed to progress. But broad participation by itself was not the accelerating factor. After all, high numbers of participants with stark differences among them typically slow rather than speed a given effort. Over the 2020–21 period, however, collaboration took shape in ways, and at a pace, which offer important insights for future attempts at urgent breakthroughs.

A major driver of efficiency was real-time data sharing among researchers. Importantly, data sharing did not shut down entirely in later stages of the development and production processes. Data sharing through *GenBank*, a comprehensive genetic sequencing database hosted by the U.S. National Institutes of Health (NIH), helped make it possible for the first vaccine candidate to enter a clinical trial within 10 weeks of the initial sequencing of SARS-CoV-2 by Chinese scientists at the Wuhan Institute of Virology. The NIH also led early coordination in the form of the Accelerating COVID-19 Therapeutic Interventions and Vaccines (ACTIV) initiative, which linked together international networks of regulators, academic institutions, philanthropic funders, and private-sector innovators.

The U.S. Department of Justice granted permission to antibody manufacturers to share technical information about production more broadly with other entities. Legacy leaders in vaccine manufacturing, including GlaxoSmithKline (GSK), licensed their proprietary adjuvants beyond the walls of their own research labs.

More data sharing across the board, especially with regard to manufacturing and deployment, may have beneficially altered the pandemic's overall trajectory by stopping the rampant spread. Even so, the level at which data sharing occurred, and the mechanisms through which data collection and sharing took shape, set valuable precedents upon which future global R&D efforts can build for a more efficient response.

Notable collaborative efforts supporting COVID-19 R&D include, among others, the Coalition for Epidemic Preparedness Innovations (CEPI), the COVID-19 Therapeutics Accelerator, the WHO's Access to COVID-19 Tools (ACT) Accelerator, WHO's Solidarity trial, and the constellation of actors supporting UK RECOVERY trial.

Despite the impressive work carried out within these formal networks, there remains room for improvement in certain aspects of global R&D coordination, particularly in terms of coordinated R&D funding.

Advanced research tools for accelerating discovery and development

Areas of innovation to watch in years to come: The full-throttle convergence of data science and biomedical innovation, with smart-device monitoring, machine learning, and artificial intelligence among the expediting forces.

Lockdowns and related measures to control the spread of COVID-19 caused significant disruption to hundreds of clinical trials underway at the time. To a large extent, pressures on those studies gave rise to a major innovation that found its footing in the pandemic: decentralized clinical trials. Decentralized clinical trials draw upon smart devices, software platforms, and related technologies to enable remote patient monitoring, collecting extensive real-time data that present a comprehensive picture of a medical intervention's effects as patients experience them in real-world settings. They can expedite enrollment in trials by helping to identify and recruit patients based on population data. Decentralized clinical trials allow for more expansive, inclusive, cost- and time-efficient studies, minimizing the need for in-person site visits.

Despite their demonstrated advantages, the revolutionary potential of decentralized trials remains largely untapped. Where these tools can play a role in long-term safety monitoring, they can increase public confidence in interventions while contributing data helpful to future iterations of a given intervention. If these tools can support the implementation of master protocols on a regular basis, enabling the testing of multiple vaccine or drug candidates against a single control arm, they can drive major gains against any number of health challenges.

Emerging iterations of the decentralized trials model increasingly include telehealth and in-home visits to provide patient care to study participants. In this way, the model is rapidly evolving to move beyond data collection as the sole objective, and instead bridge together clinical care and clinical research in a truly patient-first approach. In the years ahead, decentralized clinical trials may emerge as the crowning achievement and most durable innovation of the pandemic era.

Of course, the availability of investigational drugs and vaccines for testing in clinical trials depends on success earlier in the R&D process. Today, the category of technologies known by the understated label of "research tools" is exploding, with enablers of drug discovery and early stage research. These include machine learning tools like *AlphaFold 2.0*, an AI-based algorithm which can predict the shape of protein folds based on their genetic codes—solving a problem that vexed biologists for half a century. In solving this problem, *AlphaFold 2.0* demonstrated its potential to accelerate newer directions in therapeutic development, including the design of mRNA-encoded proteins for use in drug or vaccine form.

Similar types of AI tools can efficiently screen thousands, millions, and even billions of compounds to identify potential drug candidates against specific targets. The potential of these technologies to dramatically accelerate drug discovery and development depends in part on open datasets. To the extent that the pandemic has facilitated a shift in this direction, toward collaborative data sharing in the public domain, therein lies the potential for the use of powerful AI algorithms to guide biomedical research across countless areas of unmet need.

A changed policy landscape for pharma R&D investment

Areas of innovation to watch in years to come: More globally minded advanced purchasing agreements to "pull" drugs and vaccines into more markets, faster.

Prior to the onset of COVID-19 pandemic, interactions between public- and private-sector actors in the innovation ecosystem generally operated according to long-accepted patterns and norms. Public-sector actors funded basic research in the life sciences, typically carried out in academic research institutions. Private-sector firms selectively carried out so-called applied research, including rigorous, multiphase clinical studies executed against the standards set by public-sector regulators. Private-sector firms invested private capital in the process of drug development and commercialization. In contrast, the model that emerged in the context of COVID-19 involved high levels of public-sector "push" funding at later-than-usual stages, including public-sector investments in clinical trials and manufacturing. These resource infusions, combined with unprecedented regulatory flexibility, facilitated speed of forward movement.

In this same vein, high-income governments' pursuit of what amounted to advance purchase agreements supported accelerated—though market limited—availability of select COVID-19 vaccines and therapeutics. Over the 2020–21 period, these advance purchase arrangements, primarily executed on a country-by-country basis, exacerbated the unevenness and inequities of vaccine distribution globally. In 2021 high-income countries who had executed these advance purchase agreements—comprising 13% of the world's population—held rights to more than half of all vaccine doses in production.

Looking ahead, evolved approaches to advance purchasing will help to pull forward needed innovation *and access* for populations and therapeutic areas where the level of unmet need is pressing. Ideally, advance purchasing agreements will take shape in the context of other push and pull incentives, especially to support full development of new antibiotic and neglected-disease treatments.

Extraordinary circumstances made possible the "Warp Speed" vaccine and drug development for COVID-19, initiated in the United States with broad and ultimately successful engagement from public-sector federal agencies and private-sector vaccine and drug developers. Given this success, calls for similarly warp-speed innovations against other disease areas are mounting. Where investment, incentive, and participation levels can approximate those seen in the last several years, and where advanced research tools and collaboration models can take hold in earnest for the long haul, then high-velocity R&D which leverages increasingly sophisticated accelerating technologies may indeed become the norm.

Scale: The postpandemic imperative of reinventing delivery strategies for vaccines and therapeutics

Areas of innovation to watch in years to come: Major leaps forward in drug and vaccine formulation approaches, the strengthening of a global digital infrastructure to safeguard public health, meaningful advances in supply chain and last-mile delivery technologies.

In the field of life sciences, the term "delivery" has multiple meanings. On one hand, delivery pertains specifically to the mechanism by which a drug or vaccine is administered and released within the body. Modes of drug delivery in this sense include oral administration, injection, infusion, inhalation, topical application, transdermal patches, and more. Each mode has implications for a drug or vaccine's formulation.

Today, there is significant work underway to enhance formulations, and thus delivery, of COVID-19 treatments and preventative agents. *Arcturus,* for example, is a U.S. company working to deliver mRNA vaccines against COVID-19 via a single injection using a novel "STARR™" technology which aims to elicit high levels of neutralizing antibody titers, achieving potency against multiple viral strains simultaneously.

Meanwhile, Altimmune and others are developing a single-dose vaccine for COVID-19 that is administered intranasally, stimulating a broad immune response without the need for needles or healthcare professionals trained to administer injections. Should Altimmune and others working toward this nasally administered vaccine be successful, the implications for achieving scale are significant. This is especially true where the formulation can be made for stability at room temperature, with a shelf life of several months or more. A formulation that can optimize improved vaccine access is the explicit goal of innovators in the nasal vaccine race.

Similarly, orally administered vaccines are currently in development (in addition to newly available oral therapeutics), opening up the prospect of a vaccine option that can be self-administered outside of healthcare settings and stable at relatively wide temperature ranges, plus a long shelf life. If successful, orally administered vaccines could substantially improve efforts to vaccinate the world.

Of course, a vision of vaccinating the world depends on more than advances in drug formulation. "Delivery" in the sense of drug and vaccine *distribution* to communities in every corner of the globe remains unacceptably elusive in today's pandemic environment.

Thankfully, supply chain and last-mile delivery innovations are coming into view to support expanded drug and vaccine distribution. These include new digital technologies that enable efficient vaccine tracking from point of deployment to intended destinations. Today, digital vaccine cards are enhancing public health data collection, analysis, and long-term record-keeping at the local, country, and global levels. To the extent that the digital infrastructure supporting COVID-19 vaccine distribution can be expanded to support access to healthcare goods broadly, health systems and those they serve will benefit.

Similarly, long-range delivery drones have proven to be valuable tools for reaching "last mile" communities with needed vaccines overcoming challenges of difficult terrain and limited transportation infrastructure. New advances in cold-chain technology are also supporting the possibility of long-distance vaccine distribution without risk to vaccine quality. Upstream in the supply chain, technologies that improve the efficiency of COVID-19 vaccine production are rapidly transforming manufacturing practices for vaccine production more broadly, beyond what is developed against SARS-CoV-2 specifically.

Looking to the future

As we look well into the future, toward a firmly postpandemic world, the likelihood of radical improvements to the methods by which drugs and vaccines are developed and distributed is high—almost to the point of rendering established prepandemic methods obsolete. Indeed, today's advances in technology are making real the prospect of vanquishing diseases of many types in the decades ahead. To the extent that these advances reach the world's population at large, their benefits cascaded across continents with deliberate immediacy, the lessons of COVID-19 will have precipitated long-overdue structural change. For now, the challenge of this moment remains: with swift speed and at global scale, to marshal every relevant technology and resource to ensure equitable healthcare access. The full potential of this era's innovations will remain unrealized until combined with a coordinated commitment to their availability worldwide.

CHAPTER 26

Access to innovative, safe, timely, and affordable surgical healthcare as an indispensable component of resilient health systems

Geoffrey Ibbotson[a,b], Dominique Vervoort[c], and Rifat Atun[d]

[a]Global Surgery Foundation, Geneva, Switzerland, [b]United Nations Institute for Training and Research (UNITAR), Geneva, Switzerland, [c]Institute of Health Policy, Management and Evaluation, University of Toronto, Toronto, ON, Canada, [d]Health Systems Innovation Lab, Harvard University, Boston, MA, United States

Background

Innovation is what drives progress in health systems and helps to achieve better health outcomes. Never has this been more evident than during the COVID-19 pandemic and in the pandemic recovery. Through the pandemic, the world has witnessed the weaknesses of even the seemingly strongest health systems. To build back better, leaders must rethink the "status-quo" approaches to strengthening health systems and step into the new ecosystem of innovations that exist to create more resilient health systems. With this paradigm shift in approach, in this chapter, we examine the potential role of innovative technologies to help bridge gaps in access to surgical healthcare and the fundamental importance of surgical care systems in achieving more resilient health systems and Universal Health Coverage.

Throughout history, surgical care innovation has been a driving force behind advances in medical care, albeit this is not widely recognized. Examples of innovations include the first promotion of handwashing in 1847 in Austria by Dr. Ignaz Semmelweis—a German-Hungarian gynecologist known as a pioneer of antiseptic procedures, and the first use of ether as general anesthesia in Boston, United States, in the mid-1840s (Gawande, 2012; McKellar, 2010).

The past several decades have seen impressive advancement in the surgical and interventional treatment of cardiac disease, with groundbreaking innovations in critical care such as extracorporeal membrane oxygenation, intracardiac monitoring, and electrophysiological

Resilient Health. https://doi.org/10.1016/B978-0-443-18529-8.00026-3

remote monitoring (Stephenson, 2008). The coupling of these advances with digital health has paved the way for even more exciting opportunities in the years to come. Despite this rich history of being a driving force for innovation in medicine, the field of surgery has often been left out of mainstream conversations of global health.

Despite enormous advances in surgical care, five billion people lack access to safe, timely, and affordable surgical care when needed (Meara et al., 2015). As a result, nearly 18 million people die annually from surgically treatable and preventable conditions. Investing in surgical healthcare could avert a potential economic loss of US$12.3 trillion in low- and middle-income countries (LMICs) by 2030.

In 2015, the 194 Member States of the World Health Organization unanimously adopted the World Health Assembly (WHA) Resolution WHA68.15 on "Strengthening emergency and essential surgical care and anaesthesia as a component of universal health coverage."(Price et al., 2015) The resolution served as a clear political mandate for countries to support the strengthening of surgical healthcare worldwide. However, to translate this commitment into tangible reality, there is a need for global surgical innovation across both the care continuum and throughout health systems.

In this chapter, we highlight this need for global surgical innovation through a health systems lens and discuss challenges affecting the introduction, access, and use of innovations within diverse contexts(Atun, 2012). We propose potential solutions and opportunities for health system innovation to improve surgical care delivery by presenting case studies from around the world. Lastly, we describe the indispensable role of surgical healthcare in developing resilient health systems.

The need for global surgical innovation

The global disparities in access to surgical healthcare present as a multilayered problem with various context-specific challenges, as enumerated as follows. As shown in Fig. 1, these challenges, as well as their solutions, are not siloed and generally require innovative approaches that cross-cut the entire health system (Atun, 2012).

- *Demographic*: Population age structures are driven largely by birth rates, infant mortality, and access to timely and effective healthcare. Countries with younger populations are affected by surgical problems such as obstructed labor (e.g., pregnancy at young ages), trauma (e.g., falls and road traffic incidents), and congenital malformations (e.g., congenital heart disease), where obstetric, trauma, and pediatric surgical care are a particularly pressing challenge. Conversely, countries with largely older populations have higher rates of noncommunicable diseases, including cancers and cardiovascular diseases, with different demand patterns for surgical care.

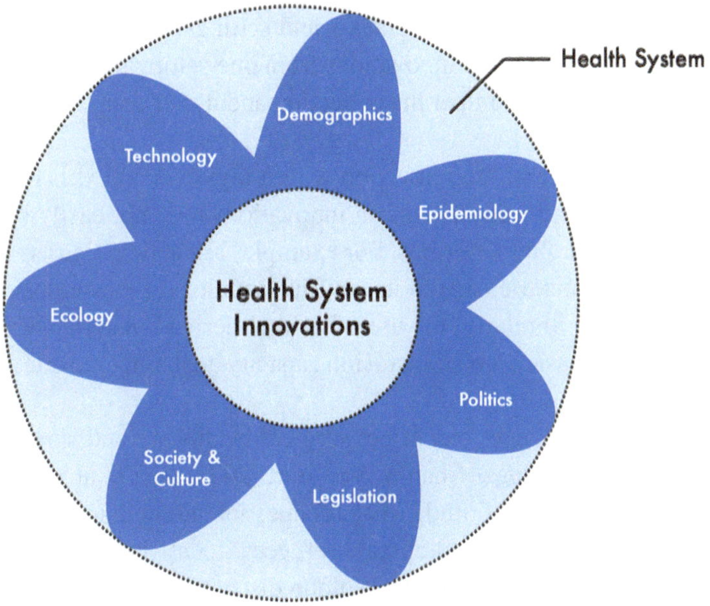

Fig. 1

Contextual factors shaping innovation in surgery. *Adapted from Atun, R., Aydın, S., Chakraborty, S., Sümer, S., Aran, M., Gürol, I., Akdağ, R. (2013) Universal health coverage in Turkey: Enhancement of equity. Lancet, 382(9886), 65–99, https://doi.org/10.1016/S0140-6736(13)61051-X; Atun, R. & Moore, G. (2021) Building a high value health system. Oxford University Press.*

- *Epidemiological*: Disease burdens vary by region, country, and income group, depending on the stage of epidemiological transition. Specific cancers are more prevalent in certain areas (e.g., gastric cancer in East Asia, oropharyngeal cancer in Latin America) and less so in others. Noncommunicable diseases (NCDs) have until recently been a lower priority in LMICs, due to relatively high burden of communicable disease, but the rapid epidemiological transition means that LMICs now experience the largest burden of NCDs (Institute for Health Metrics and Evaluation, 2021).
- *Political*: Insufficient political commitment is associated with poor access to and delivery of care due to limited healthcare spending and/or inefficient governance. Globally, there is a vast underfunding of surgical diseases relative to their disease burden. While around 18 million people die from surgically treatable conditions each year—over 5 times that from HIV/AIDS, tuberculosis, and malaria combined—less than 2% of overseas development assistance (ODA) for health is channeled to surgical healthcare,(Dieleman et al., 2015; Institute for Health Metrics and Evaluation, 2021; Meara et al., 2015) compared with HIV/AIDS, tuberculosis, and malaria which together receive nearly half of all ODA for health (Global Burden of Disease Health Financing Collaborator Network, 2020).
- *Legal*: Legal frameworks and locoregional regulations (e.g., accreditation and device approval processes) define who can deliver what forms of care using what kind of

equipment and techniques in which contexts and settings. These often vary by country, complicating direct translation of innovations from one setting to another. In addition, data regulations may facilitate or hamper the collection and use of data for quality improvement and/or research.

- *Sociocultural*: Acceptance of an innovation within a given context is of paramount importance for its use and impact. Certain innovations may not easily be accepted for historical, religious, or cultural reasons. For example, organ donations systems are slow to be adopted in certain countries due to its conflict with local perceptions of the sanctity of life and human bodies. Similarly, the use of animal-derived materials (e.g., bioprosthetic heart valves) and even basic blood transfusion capacity are hampered due to differing views on new innovations.
- *Ecological*: Ecological changes can affect surgical healthcare and surgical solutions that are needed in different settings (Chan & Vervoort, 2022; Roa et al., 2020). For example, climate change can cause a multitude of challenges including displacement of people, for example with forced migration and climate refugees, local challenges with environments (e.g., malnourishment and stunting), and unique disease patterns (e.g., (re-)emerging infectious diseases and zoonoses).
- *Technological*: Technologies require the necessary infrastructure, such as internet connectivity, continuous electricity supplies, or supportive equipment (e.g., smartphones for app-based technologies). In addition, most technologies require periodical maintenance (e.g., testing), updating (e.g., software), or servicing (e.g., new parts) for optimal use and safety and appropriately trained human resources to undertake these tasks.

These health system challenges are not siloed, and ultimately interweave, further adding to the complexity of surgical care and innovation. Nevertheless, a health systems lens can enable the correct identification of challenges and the definition of contextual needs to facilitate change in organization and governance, financing, resource generation and management, and service delivery in health systems. As shown in Fig. 2, various health system functions exist that can be leveraged to strengthen surgical and health systems.

Surgical healthcare—and, by extension, surgical innovation—is not separated from other parts of a health system. Historically, surgical innovations have resulted in overall health system innovation and advancement. For example, the start of cardiac surgery, procedures long assumed to be impossible and even immoral, accelerated innovations across medical and surgical disciplines. Nowadays, the introduction of cardiac surgical capacity in LMICs has also benefitted local health systems through significant positive spillover effects (Vervoort & Edwin, 2021). Similarly, the COVID-19 pandemic has illustrated that surgical capacity is a key pillar to critical care and emergency care capacity. Critical care infrastructure in health systems is closely linked with surgical infrastructure, as insufficient critical care capacity may limit the number of surgical procedures performed that require postoperative intensive care

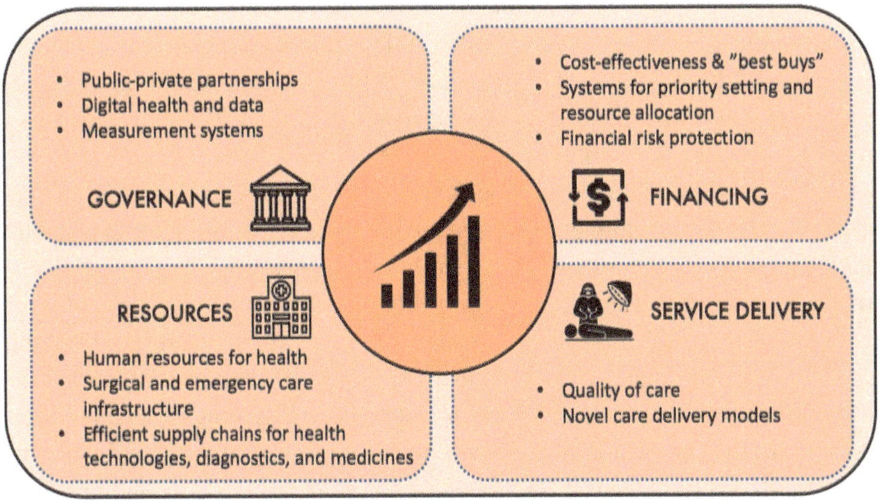

Fig. 2
Health system functions where surgical innovation can contribute to building health system resilience.

management, and surgical procedures may be reduced to free up critical care capacity. Additionally, surgical teams were redeployed to support critical and emergency care on the frontlines (Peters et al., 2020). Surgical healthcare cannot be dissociated from health systems, and vice versa—they are part of the same integral whole.

The innovation conundrum

Different innovations face different challenges and require different solutions, depending on whether they (do not) exist, are (not) accessible, and/or are (not) being used and scaled.

First, if the necessary innovations do not already exist, market-level push and pull factors should be explored. Push factors, such as research into and development of potential innovations and private capital funding to facilitate early stage growth, can drive the development of innovations. Meanwhile, pull factors, such as insufficient demand by large market forces, impede such innovation. For example, if a large share of the surgical workforce is not trained in or open to training in a certain skillset, an innovation may not be explored or introduced.

Second, innovations may exist but not yet be accessible by those who need it. This may be a result of a lack of affordability, whereby the cost to develop or purchase the innovation is prohibitive to facilitate widespread adoption. For example, laparoscopic surgery can enable patients to receive less-invasive surgical care and be able to return to their homes, communities, and work faster. However, the cost of purchasing the necessary equipment limits many centers in LMICs. There may further be challenges in access. Surgical supply chains are known to be fragmented and inefficient, particularly in LMICs (Nazir et al., 2021). As a result, highly used and irreplaceable materials, ranging from surgical sutures to heart valve prostheses, may sometimes be difficult to obtain. Additionally, the application of an innovation may be difficult, either due to a lack of skilled human resources, unreliable energy supplies, or other requirements.

Third, an innovation may exist and be accessible but not necessarily used or scaled. This can be a result of a misalignment between the innovation that is introduced and the problem at hand: for example, the introduction of robotic surgery in LMICs or rural areas strongly contrasts with the need for basic access to lifesaving surgical care first and foremost (Vervoort & Long, 2021). Additionally, adoption can vary as a result of timing (e.g., early in the adoption continuum) or stakeholders (e.g., governance priorities). Furthermore, health systems barriers, such as a lack of financial risk protection, can prevent sufficient access among the population due to high out-of-pocket expenditures. Lastly, contextual barriers, which include legal difficulties (e.g., requirement of permits) and sociocultural norms and traditions (e.g., lack of acceptance of technology), must be considered.

Innovations to overcome global surgical challenges

Innovations may come in different shapes and forms depending on the area of the health system they pertain to. These include but are not limited to innovations for organization and governance, health financing, resource generation and management, and service delivery. Ultimately, these innovations can and should be integrated to develop more complex and cross-cutting solutions to overcome global surgical challenges.

Organization and governance

Organizational and logistical structures, driven by leadership efficiency, greatly impact healthcare delivery in a top-down manner. In Québec, Canada, the trauma system underwent major changes in the late twentieth century in response to unacceptably high mortality related to severe trauma. In 1993 Trauma Québec was introduced as a structured and regionalized system of trauma care and referral networks. As a result, severe trauma-related mortality dropped dramatically over the period of 10 years, from 51.8% in 1992, just before the system's introduction, to 8.6% in 2002 (Liberman et al., 2004).

National Surgical, Obstetric, and Anesthesia Plans (NSOAPs) are increasingly adopted in LMICs since the 2015 Lancet Commission on Global Surgery, which introduced the NSOAP framework (Albutt et al., 2019; Truché et al., 2020). NSOAPs are long-term policy plans seeking to improve access to surgical, obstetric, and anesthesia care in a sustainable, locally driven, and politically supported manner. Since the adoption of WHA68.15, seven countries have developed and launched their NSOAPs to embed surgical healthcare within their existing national health plans, with dozens of countries in the process of developing their plans (Albutt et al., 2019).

Health financing

Global health financing for global surgical healthcare remains inadequate and requires increased domestic and international investments. Funding considerations relate to health system financing (i.e., macro-level), institutional funding (i.e., meso-level), and financial risk protection for patients for healthcare delivery (i.e., micro-level). Innovative financing instruments can be leveraged to facilitate underused sources of funding, such as taxation, levies, and social and development impact bonds to fund surgical innovations (Reddy, Peters, et al., 2020).

In addition to new funding, provider payment models should be revamped to optimize financial risk protection, reward positive outcomes, and mitigate negative outcomes. Results-oriented payment models, whether in the form of pay-for-quality variations of health insurance or outcome-dependent funding allocations for health system interventions, can be valuable tools. Moreover, value-based healthcare (VBHC) approaches should be considered when funding surgical care to ensure both value for money and value for many (Vervoort & Swain, 2021). For example, in Pakistan, a VBHC intervention for congenital heart defects enabled a tertiary care center to gradually improve outcomes and health-related quality of life of patients over time (Hasan et al., 2021).

Resource generation and management

As a result of resource constraints, surgical teams in LMICs and remote settings have been forced to innovate to provide the best possible care with the scarce resources available. This has produced various innovations mimicking high-income country technologies at much lower costs. Examples include but are not limited to the Bogotá bag to cover open abdominal wounds prior to closure, costing US$5 instead of the traditional US$150–$1600 due to the use of a simple 3-L saline bag that is cut open for us instead of using high expense medical grade "made for purpose" equipment for the same purpose. Another example is seen with propylene mosquito net meshes for hernia repairs, which has shown to be comparable in safety and outcomes to industry produced hernia mesh in a randomized clinical trial in Burkina Faso (Freudenberg et al., 2006; Hasan et al., 2022). In addition, in India, gasless laparoscopy has been increasingly adopted in rural areas to allow laparoscopic surgery to take place at lower costs and resource needs (Aruparayil et al., 2021). Gasless laparoscopy has been found to be noninferior to traditional laparoscopy (Mishra et al., 2020) and can enable shorter hospital lengths of stay and faster return of patients to their homes, families, and work, which are important considerations in LMICs.

Service delivery

Innovations may equally support the delivery of surgical healthcare by streamlining care, improving the quality of care, or facilitating capacity building (Vervoort & Fiedler, 2021). For example, Proximie is a digital "virtual operating room" platform that enables individuals to remotely attend surgical procedures (Proximie, 2020). This facilitates the learning of participants and enables participants to share their knowledge and skills in real time through tele-consults. In addition, the process eliminates the barriers that prepandemic visas, high travel costs, and pandemic travel restrictions perpetuated, limiting access to surgical education (Vervoort et al., 2020).

How surgical innovations help to build more responsive, resilient, and sustainable health systems

The COVID-19 pandemic has illustrated inadequate "resilience" of health systems to health threats globally, and the term remains poorly defined and as a result, actions to develop resilience are poorly implemented.

Before the COVID-19 pandemic, the United States was considered by one study the most resilient health system in the world; during the pandemic, however, the USA experienced the largest case counts and deaths due to COVID-19 of any country in the world (Tulenko & Vervoort, 2020).

Resilience pertains to not only the capacity at hand, but also the preparedness for and adaptability during healthcare crises. This includes the availability of cross-trained personnel (e.g., surgical task-sharing), the necessary critical care capacity, the presence of trauma response teams and protocols, and effective leadership at different levels of the health system (Ma & Vervoort, 2020; Peters et al., 2020).

Surgical healthcare, ranging from preoperative disease detection, imaging, and consults to postoperative care, rehabilitation, and follow-up, has been increasingly recognized as a cross-cutting part of health systems. However, many disease control programs and initiatives in global health remain largely siloed. Surgical innovations can be the thread that can help bring together and "stitch" these silos by facilitating multidisciplinary collaboration and prioritize interventions that support not just a single area of healthcare, but rather ones that directly or indirectly address different healthcare challenges. For example, the support of emergency and essential surgical care and anesthesia at district hospitals enables safer deliveries with improved maternal and perinatal care (Reddy, Vervoort, et al., 2020).

Innovative developments in surgical care in recent years are commendable and should be encouraged. The commitment of dozens of countries worldwide to develop national and regional NSOAPs highlights the growing policy and political momentum to improve access to surgical healthcare since the adoption of Resolution WHA68.15 in 2015. This is further supported by growing support at the level of the World Health Organization (WHO), which houses the WHO Programme for Surgical and Anaesthesia Care (formerly the Emergency and Essential Surgical Care Program) and the United Nations (where the Global Surgery Foundation was incubated at the United Nations Institute for Training and Research). Recently, the United States Congress included language in the State, Foreign Operations, and Related Programs Appropriations that facilitates funding for United States Agency for International Development (USAID), to support and strengthen surgical systems in LMICs (Committee on Appropriations, 2021). In late 2020, the USAID awarded US$40 million to improve safe surgery for women through the MOMENTUM 3A series of projects (USAID, 2020). The USAID can now extend its efforts to other pressing areas of surgical healthcare, including trauma surgery and surgical oncology (Meara et al., 2021).

In recent months, following countries' commitment to developing the International Treaty on Pandemic Prevention, Preparedness and Response, G20 leaders have committed to establishing a Financial Intermediary Fund for Pandemic Preparedness and Response (World Health Organization, 2022). The fund is focused on sustainability, innovation, scalability, complementarity, effective allocation, accountability, and equity toward health systems strengthening within the context of pandemic preparedness and response. Realizing that strong surgical healthcare systems are one of the foundational pillars for ensuring resilient health systems, it is critical for surgical healthcare to be an integral component of the solution for pandemic preparedness and response. Within this context, significant investment toward the strengthening of surgical healthcare systems is essential if these goals are to be fully achieved.

There is a space and need for global collaboration between governments, academic, clinical, and community stakeholders, funders, and industry to build on the achievements to date. Accordingly, mutual learning can be facilitated, funding pipelines can be streamlined, and supply chains can be improved, ultimately maximizing resources and impact overall.

Conclusion

Surgical healthcare is an indivisible and indispensable part of health systems and fundamental for achieving Universal Health Coverage. Surgical care systems serve as a key foundational pillar for resilient health systems.

Surgical innovation can drive progress in the advancement of healthcare beyond just the operating theater, resulting in stronger and more resilient systems, particularly for the countries that need it most. In recent decades, life expectancies and quality of life have rapidly increased in great part because of continued innovations in surgical care delivery. Yet, most of the world's population cannot access the most basic forms of emergency and essential surgical care when needed. Introducing surgical innovations into health systems provides a unique and underutilized avenue to pave the way forward for building stronger and more resilient health systems, particularly in LMIC settings where it is most needed. This will require a clear understanding of the contextual challenges and barriers facing the introduction and an identification of key drivers and opportunities to foster innovations.

References

Albutt, K., Sonderman, K., Citron, I., et al. (2019). Healthcare leaders develop strategies for expanding national surgical, obstetric, and anaesthesia plans in WHO AFRO and EMRO regions. *World Journal of Surgery*, *43*(2), 360–367.

Aruparayil, N., Gnanaraj, J., Maiti, S., et al. (2021). Training programme in gasless laparoscopy for rural surgeons of India (TARGET study)—Observational feasibility study. *International Journal of Surgery Open*, *35*.

Atun, R. (2012). Health systems, systems thinking and innovation. *Health Policy and Planning*, *27*, iv4–iv8.

Chan, V. F., & Vervoort, D. (2022). The double-edged sword of cardiac surgical care: Rethinking sustainability in an ever-growing field. *European Journal of Cardio-Thoracic Surgery*. https://doi.org/10.1093/ejcts/ezab531.

Committee on Appropriations. (2021). *Explanatory statement for Department of State, Foreign Operations, and Related Programs Appropriations Bill, 2021*. https://www.appropriations.senate.gov/imo/media/doc/SFOPSRept.pdf. (Accessed 14 March 2022).

Dieleman, J. L., Yamey, G., Johnson, E. K., Graves, C. M., Haakenstad, A., & Meara, J. G. (2015). Tracking global expenditures on surgery: Gaps in knowledge hinder progress. *The Lancet Global Health*, *3*(Suppl 2), S2–S4.

Freudenberg, S., Sano, D., Ouangré, E., Weiss, C., & Wilhelm, T. J. (2006). Commercial mesh versus Nylon mosquito net for hernia repair. A randomized double-blind study in Burkina Faso. *World Journal of Surgery*, *30*(10), 1784–1789. discussion 1790.

Gawande, A. (2012). Two hundred years of surgery. *The New England Journal of Medicine*, *366*(18), 1716–1723.

Global Burden of Disease Health Financing Collaborator Network. (2020). Health sector spending and spending on HIV/AIDS, tuberculosis, and malaria, and development assistance for health: Progress towards Sustainable Development Goal 3. *Lancet*, *396*(10252), 693–724.

Hasan, B. S., Ather, M., Rasheed, M. A., & Chowdhury, D. (2021). Value-based health care for children with congenital heart diseases: A feasibility study from a low-middle income country. *NEJM Catalyst*, *2*(12). https://doi.org/10.1056/cat.21.0258.

Hasan, O., Ayaz, A., Masood, L., Baig, A. M., & Baloch, N. (2022). Innovations in surgery between the past and future: A narrative review of targeted literature. *The Journal of the Pakistan Medical Association*, *72*(Suppl 1 (2)), S55–S58.

Institute for Health Metrics and Evaluation. (2021). *Global burden of disease results tool*. GBD Results Tool http://ghdx.healthdata.org/gbd-results-tool%20. (Accessed 25 October 2021).

Liberman, M., Mulder, D. S., Lavoie, A., & Sampalis, J. S. (2004). Implementation of a trauma care system: Evolution through evaluation. *The Journal of Trauma*, *56*(6), 1330–1335.

Ma, X., & Vervoort, D. (2020). Critical care capacity during the COVID-19 pandemic: Global availability of intensive care beds. *Journal of Critical Care*, *58*, 96–97.

McKellar, S. (2010). A history of surgery: From superstition to science. *CMAJ*, *182*(8), 809.

Meara, J. G., Anderson, G. A., & Hey, M. (2021). *A surgical approach to strengthening our health systems*. https://thehill.com/blogs/congress-blog/healthcare/587612-a-surgical-approach-to-strengthening-our-health-systems. (Accessed 14 March 2022).

Meara, J. G., Leather, A. J. M., Hagander, L., et al. (2015). Global Surgery 2030: Evidence and solutions for achieving health, welfare, and economic development. *Lancet*, *386*(9993), 569–624.

Mishra, A., Bains, L., Jesudin, G., Aruparayil, N., Singh, R., & Shashi. (2020). Evaluation of gasless laparoscopy as a tool for minimal access surgery in low-to middle-income countries: A phase II noninferiority randomized controlled study. *Journal of the American College of Surgeons*, *231*(5), 511–519.

Nazir, A., Vervoort, D., & Reddy, C. L. (2021). From the first mile to the last: Challenges of the global surgical supply chain. *The American Journal of Surgery*. https://doi.org/10.1016/j.amjsurg.2021.03.033.

Peters, A. W., Chawla, K. S., & Turnbull, Z. A. (2020). Transforming ORs into ICUs. *The New England Journal of Medicine*, *382*(19), e52.

Price, R., Makasa, E., Hollands, M., & World Health Assembly Resolution WHA68. (2015). 15: "Strengthening emergency and essential surgical care and anesthesia as a component of universal health coverage"—Addressing the public health gaps arising from lack of safe, affordable and accessible surgical and anesthetic services. *World Journal of Surgery*, *39*(9), 2115–2125.

Proximie. (2020). *Proximie—Saving lives by sharing the world's best clinical practice*. https://www.proximie.com/. (Accessed 14 March 2022).

Reddy, C. L., Peters, A. W., Jumbam, D. T., et al. (2020). Innovative financing to fund surgical systems and expand surgical care in low-income and middle-income countries. *BMJ Global Health*, *5*(6). https://doi.org/10.1136/bmjgh-2020-002375.

Reddy, C. L., Vervoort, D., Meara, J. G., & Atun, R. (2020). Surgery and universal health coverage: Designing an essential package for surgical care expansion and scale-up. *Journal of Global Health*, *10*(2), 020341.

Roa, L., Velin, L., Tudravu, J., McClain, C. D., Bernstein, A., & Meara, J. G. (2020). Climate change: Challenges and opportunities to scale up surgical, obstetric, and anaesthesia care globally. *Lancet Planet Health*, *4*(11), e538–e543.

Stephenson, L. W. (2008). History of cardiac surgery. In J. A. Norton, P. S. Barie, R. R. Bollinger, et al. (Eds.), *Surgery: Basic science and clinical evidence* (pp. 1471–1479). Springer: New York.

Truché, P., Shoman, H., Reddy, C. L., et al. (2020). Globalization of national surgical, obstetric and anesthesia plans: The critical link between health policy and action in global surgery. *Globalization and Health*, *16*(1), 1.

Tulenko, K., & Vervoort, D. (2020). Cracks in the system: The effects of the coronavirus pandemic on public health systems. *The American Review of Public Administration*, *50*(6-7), 455–466.

USAID. (2020). *MOMENTUM 3A*. https://www.usaid.gov/global-health/momentum-3a-announcement. (Accessed 14 March 2022).

Vervoort, D., Dearani, J. A., Starnes, V. A., Thourani, V. H., & Nguyen, T. C. (2020). Brave new world: Virtual conferencing and surgical education in the COVID-19 era. *The Journal of Thoracic and Cardiovascular Surgery*. https://doi.org/10.1016/j.jtcvs.2020.07.094.

Vervoort, D., & Edwin, F. (2021). Treating pediatric and congenital heart disease abroad? Imperatives for local health system development. *International Journal of Cardiology Congenital Heart Disease*, 100082.

Vervoort, D., & Fiedler, A. G. (2021). Virtual reality, e-learning, and global cardiac surgical capacity-building. *Journal of Cardiac Surgery*, *36*(6), 1835–1837.

Vervoort, D., & Long, C. (2021). The paradox of exceptionalism: Setting priorities straight. *Surgical Innovation*, *28*(6), 794–795.

Vervoort, D., & Swain, J. D. (2021). Cost reduction in cardiothoracic surgery: The case for time-driven activity-based costing. *The Annals of Thoracic Surgery*. https://doi.org/10.1016/j.athoracsur.2021.06.027.

World Health Organization. (2022). *Delivering on the G20 Leaders commitment to build an equitable and effective Financial Intermediary Fund (FIF) for Pandemic Preparedness and Response (PPR)*. https://www.who.int/news/item/19-04-2022-delivering-on-the-g20-leaders-commitment-to-build-an-equitable-and-effective-financial-intermediary-fund-(fif)-for-pandemic-preparedness-and-response-(ppr). (Accessed 30 May 2022).

CHAPTER 27

Models to scale healthcare technologies for low- and middle-income countries☆

Deepika Devadas[a], Annie Thériault[a], and Anita M. McGahan[b,c,d]

[a]Every Woman Every Child Innovation Marketplace, Grand Challenges Canada, Toronto, ON, Canada, [b]Joseph L. Rotman School of Management, Toronto, ON, Canada, [c]University of Toronto, Toronto, ON, Canada, [d]Institute for Strategy and Competitiveness, Harvard University, Cambridge, MA, United States

Introduction

Technological innovations are a type of development innovation whose scaling strategies require special considerations in comparison to innovations in service delivery, community models, or even consumer products. Complexities arise from regulatory requirements, clinical testing costs, and duration, as well as safety and quality-related considerations. This chapter will examine models to scale this unique category of innovations in low- and middle-income countries (LMICs). For the purposes of this chapter, healthcare technologies will include devices, diagnostics, and other digitally assisted tools.

Innovation in financing and market structures continue to diversify the type of strategies pursued. With universal healthcare commitments being challenged by the slow pace of progress, the role of innovation must increase and leveraging approaches (such as those described later) toward achieving scale will be critical to amplify this impact and success.

The ideas expressed in this chapter evolved from expertise developed while studying scaling and in working with funders and innovators through the Every Woman Every Child (EWEC) Innovation Marketplace, a strategic alliance of development organizations consisting of Grand Challenges Canada, the Norwegian Agency for Development Cooperation, the U.S. Agency for International Development, and the Bill & Melinda Gates Foundation—an initiative housed at Grand Challenges Canada. The points made should be interpreted as scholarly observations, rather than as an agenda endorsed by the EWEC Innovation Marketplace partners.

☆ Note: This White Paper was originally published February 16, 2021, in a white paper series by the Every Woman Every Child (EWEC) Innovation Marketplace at Grand Challenges Canada and the University of Toronto, Rotman School of Management. It has been reprinted by permission of Grand Challenges Canada.

Resilient Health. https://doi.org/10.1016/B978-0-443-18529-8.00027-5

The starting point determines the path

We have observed that innovation and associated intellectual property (IP) for medical devices designed for LMICs mainly originate from the following types of organizations: academic institutions, for-profit companies, or nonprofit organizations. It is notable that where the innovation originates is often the primary driver of the scaling model, which also drives the types of financing instruments available to support that growth.

Academic institutions

Innovation and associated IP are often developed by research programs and universities, which in turn protect the IP with patents that are owned by the institution. The two classical scaling pathways to scale up the technology involve either licensing the technology to a larger independent third party, such as a multinational corporation (Third-Party Licensing Model) or licensing the technology into a for-profit start-up company, typically led by the inventor of the technology independently or together with a business partner or co-founder (University Spin-Off Model).

Third-party licensing model

While the third-party licensing model has often been a successful pathway in life sciences and technology markets serving developed economies, the success stories in the global health sector are unfortunately few. A good example of one is the Augmented Infant Resuscitator (Bennett et al., 2018) licensed by CAMTech (Consortium for Affordable Medical Technologies) and MGH (Massachusetts General Hospital) to Philips. This device innovatively provides real-time feedback on whether the resuscitation equipment—typically the bag-valve-mask—is properly positioned to ensure a sealed face–mask interface, thus preventing blocked airways and providing the right ventilation frequency and efficient lung inflation, all resulting in an effective ventilation technique for newborn resuscitation in environments with low-skilled healthcare providers. Its potential to increase proper use of any bag-valve-mask allowed Philips to recognize the commercialization potential in this device.

Among the challenges that have contributed to the wider poor track record of the third-party licensing model, whether the licensor be the public sector or a private sector entity, the most prominent is the lack of a clear value proposition for the third-party licensor, due to poor economics when the technology is only applicable to LMICs. While seemingly capital rich, the divisions of large corporations must compete with each other for capital to launch new products, and it is rarely the case that such a product can be selected over other more profitable products unless the corporation has a clear mandate to do so. Another commonly cited reason when such models fail is poor protection of IP by the institution initially patenting the concept, which is usually tied to limited IP budgets for most institutions. Innovators and university tech transfer offices may also hamper the process as they can be reluctant to provide

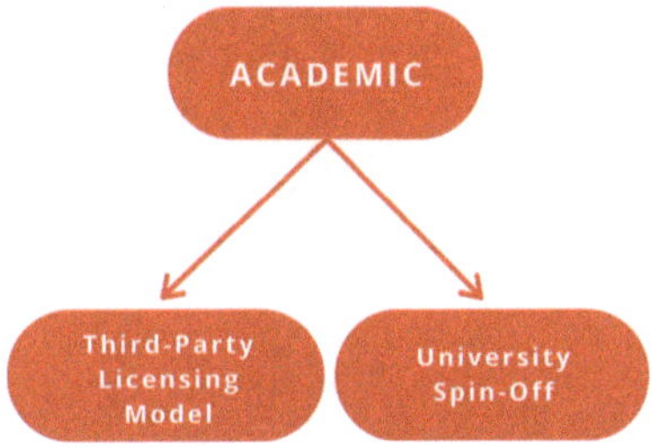

Fig. 1
Common scaling pathways followed by academic institutions include a third-party licensing model or university spin-off model.

corporations with sufficient rights to the IP due to a desire to stay in control, often to ensure impact. Even when they are supportive, innovators and tech transfer offices also lack expertise in commercializing technologies, choosing partners, and negotiating complex licensing agreements. They also have limited bandwidth for these tasks.

University spin-off model

The university spin-off model is one where a start-up is formed, usually by the academic founder or a university-related cofounder team (see Fig. 1). The start-up receives a license to the technology developed from the university for the purpose of commercializing the technology. The biggest barriers start-ups face in the early stages are also related to IP. When tech transfer offices design the license agreements with terms that are unfavorable to the start-up, investors may pass on the investment opportunity as this would lower the potential return they can ultimately earn on the investment. In order to be attractive to investors and therefore viable as a company, start-ups usually need to have exclusive long-term rights to the IP in relevant high-value markets. Licenses can have carve-outs for lower-value markets or milestone requirements related to continuing to work toward commercialization, but these need to be well designed in order to ensure viability of the start-up. The most experienced tech transfer offices typically recognize the benefits of having multiple successful spin-off start-ups from their research departments and have the expertise to design contracts that meet the needs of investors. In this case, the start-up can scale in the same way as discussed under the for-profit entity section later.

For-profit companies

Technologies can also be developed by for-profit companies, which can either be larger, mature organizations or start-ups specifically formed for the development and growth of the innovation. While for-profit multinationals also play a role in the health and development sector, for the purposes of this discussion, we are focusing on start-up ventures. Critical

Fig. 2
Key approaches used by for-profit companies to scale their technologies include the LMIC growth model and the dual-market strategy.

parameters for success are team composition, quality of the underlying IP, a promising market size, a clear business strategy, and access to capital, where the latter is driven by the first four characteristics. There are two key approaches used by for-profit companies to scale their technologies, namely the LMIC Growth Model and the Dual-Market Strategy.

LMIC growth model

In this model, an innovator seeks to scale their solution in the LMIC market(s) of interest via a financeable, long-term, sustainable, and profitable business model (see Fig. 2). To be successful, the company must identify the demographic in need of the solution and then create a pricing strategy that both meets the demand and maintains a profitable margin for the company to continue growing sustainably. This is particularly challenging in the LMIC context because payers in this context typically seek to pay a price as close to cost as possible. This can fundamentally challenge the business value proposition if combined with expectations of low volumes. The company's success rests on securing the right local and international partnerships, including research, nonprofit, community organizations, and public and private sectors. This includes partnerships for activities that include clinical trials, pilot product demonstrations for large buyers like governments, regulatory processes, and cost-effective, high-quality manufacturing and distribution of the technology to target customer segments. Often the cost of many of these activities can add up; for example, each LMIC market may require its own regulatory process, sometimes in addition to internationally recognized FDA (U.S.) or CE mark (Europe) approvals. This is also typically concurrent with, or followed by, pilot demonstrations for various public sector buyers that are not always able or willing to pay for these proof points.

From a financial point of view, companies that focus on primarily middle-income markets that have multiple income level segments, such as India or South Africa, are often capable of sustaining such growth models once they reach a certain level of scale, usually via a slow-and-steady growth model. Robust partnerships and diversified offerings are key to reaching greater scale, as these companies benefit from cost-sharing among the offerings. Maintaining profitability at a sustainable pace for steady growth, combined with the relatively more developed financial markets, creates opportunities for long-term success. The CEOs of these companies are typically highly resourceful, but nonetheless can struggle to attract the type of financing that can enable accelerated growth until they reach significant and consistent growth and profitability to access loans and revenue-based financing. Equity financing is usually elusive for companies focused on lower-income populations.

Single product medical device companies struggle more, especially when attempting to commercialize a bold new innovative device targeting lower-income populations. Like multiproduct companies, they are expected to show sustainable profitable growth before accessing loans and revenue-based financing products. In practice, what this means is that they are expected to complete product development and clinical trials for regulatory approvals and begin early revenue before they can attract the first significant investment from investors. As such, even when these products could be profitable at scale—which is not often the case when they target LMIC populations via the public sector—companies struggle to raise sufficient capital to reach the commercial phase. While donors have been pivotal to bringing many of these products beyond the ideation phase, past the proof-of-concept phase and often even all the way through the regulatory process, there continues to be a high level of failures or stalling, as these companies require significantly larger amounts of capital to reach sustainable growth. Some examples of companies pursuing a multiproduct strategy with early success are Sinapi Medical in South Africa and InnAccel Technologies in India.

Software companies fare much better, as they are not as capital intensive early on and cheaper to scale, especially in countries that have significant and growing tech sectors, such as India, South Africa, Nigeria, and Rwanda. These companies often benefit from their ability to bootstrap their businesses by securing early paying customers and servicing contract clients, due to their easy applicability and cost-competitiveness. Companies with disruptive models or offerings for the sector can also attract equity financing based on their potential to gain significant market share. The fact that LMIC markets often have no incumbent or legacy solution to displace, unlike high-income country (HIC) counterparts, bodes well for the potential of the sector. Of note is the fact that there are bourgeoning networks of accelerators and venture capital firms in these economies that have arisen due to the significant investments made by some governments (Liquid Intelligent Technologies, 2020; The World Bank, 2019) and corporations (Smart Africa, 2020), combined with recent technology company exits (Jackson, 2015; Jumia, 2019; The Guardian, 2015) in these markets, albeit more in the fintech and transport sectors (e.g., acquisition of Nigerian mobile transport directions app HopStop by Apple for $1Bn, and Kenya's mobility solutions Weza Tele by Ghana-based financial services AFB). Donors recognize the potential of technology in achieving Universal Health Coverage and are also increasingly open to co-funding these opportunities. Innovators that can proposition the impact value and track impact data well are ideally suited or eligible for loans and grants provided by such impact-first investors.

Dual-market strategy

The dual-market strategy is an innovative growth strategy that has recently gained traction as a means to achieve scale in LMICs for technologies that require a significant upfront investment, such as medical devices, diagnostic tools, and digital technologies leveraging models like machine learning and blockchain. A dual-market strategy, in this context, is designed or adapted to serve both LMIC and HIC settings. Of the various approaches to implementing the dual-market strategy described in a separate EWEC Innovation Marketplace white paper (Devadas et al., 2021a), the most financeable is the Synergistic Model, where the LMIC strategy succeeds in becoming moderately profitable and also provides at least some benefits for the growth of the HIC business, either financially or otherwise. This model can access the greatest number of types of capital available to support its growth, as the strategy enables the company to fulfill the criteria of multiple types of donors, funders, and investors. Non-dilutive capital—grants and/or debt—may be available to pursue the development and impact strategy (e.g., LMIC product development and contextualization or pilot demos for potential public sector buyers). In addition, with investors often interested in HIC investment opportunities, this widens the pool of potential investors for growth plans. Strategic investors are also increasingly supportive of these approaches, as they seek to learn ways to enter or expand into LMIC markets to achieve their own goals of eventually having strong business drivers across all markets and market types. The EWEC Innovation Marketplace has written white papers on financing instruments and dual-market scaling strategies for impact which include detailed discussions and examples of these concepts (Devadas et al., 2021a, 2021b).

Intellectual property considerations for public good

Regardless of the pathway selected, the for-profit model can be undermined if the intellectual property (IP) is encumbered. This encumbrance can result from commitments made related to early stage funders' rights (often called global access agreements). While these rights are designed to ensure the desired social purpose is met (i.e., that intended hardest to reach populations have access to the innovation), innovators and funders must consider the risk that such encumbrances may discourage investors from investing. Investors prefer companies with strong IP protection that are not hindered by license rights or other encumbrances; earning sufficient financial returns from investments in companies with weak IP is difficult because the acquirers of these companies would devalue these targets. While a comprehensive analysis of intellectual property rights is beyond the scope of this chapter, we consider the following forms of IP rights as the most benign for a company's ability to raise growth capital: nonexclusive licenses for low-income countries, tiered pricing commitments for middle-income countries (to account for income level-related impact and revenue opportunities), or distribution rights with

1%–10% net margin allowance (for LMICs not on the go-to-market strategy of the company), among others.

Nonprofit organizations

Finally, the last type of entity in the business of medical technology innovations is the nonprofit. For devices or portfolios of devices, these models typically function by either licensing in externally developed IP or developing the IP in-house with the potential option of licensing the IP out to a third-party manufacturer. While their financing options are limited to grant providers (like donors and foundations), the significant advantage to such models is the strong alignment between the organization and its impact mandate, which results in robust evidence demonstrating said impact, a high level of understanding of local markets, and typically strong embedding within local health systems.

Successful models include Gradian Health and Equalize Health (formerly D-Rev), both of which are nonprofits with revenues from sales to both the public and the private sector that contributes to the sustainability of the model, with sufficient donor interest to continue to innovate and perform market building activities. Such models are able to bring appropriately priced, serviced, and supported products into low-income countries, working closely with healthcare providers, local distributors, and the public sector. These entities focus a significant part of their activities toward health systems strengthening, which is also an asset to receiving continued support from donors and funders.

The key challenge is that, while revenues enable access (because the pricing is appropriate/low for the target markets), revenues sometimes do not exceed all the costs for such a model—costs associated with sales, maintenance, and training, as well as necessary continued product improvements. Entities perpetually dependent on grant funding are especially vulnerable to changes in donor priorities, which can limit their ability to maintain as well as expand their operations to reach scale. It is also important to note that, while there are a growing number of investors and lenders in LMICs, the number of large donors able to continually provide funding to innovators does not appear to be increasing as quickly, thus putting pressure on innovators to consider self-sustainable approaches.

A noticeable subset includes nonprofit digital organizations that also appear to be gaining traction, in large part due to the high current level of interest in the sector by donors, especially for open-source technologies. Innovators have to balance their commitments to enabling open access against ensuring that sufficient donor financing is available for maintenance and service of these platforms, in order to continually stay relevant with regard to digital infrastructure, security, privacy, and impact. Examples of innovators scaling such models include MedicMobile, South Africa's HealthConnect, D-Tree, and Project ECHO.

Alternative strategies in this realm could include companies that have entirely separate nonprofit arms that are dedicated to the impact and development mission of the entity (see Fig. 3).

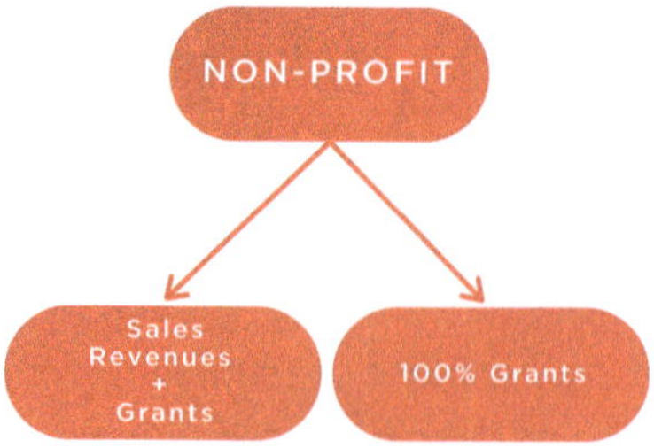

Fig. 3
Nonprofit entities typically scale through a combination of revenues from product sales and grants, or entirely through grants.

This dual structure protects the primary aim of both the for-profit company and the nonprofit entity, resulting in appropriate financing for each and strong synergy between the two. A successful example of this model is Jacaranda Health in Kenya, a nonprofit providing nurse mentorship programs and scaling an SMS support platform to provide antenatal care and information to pregnant mothers. This entity works closely with its for-profit entity Jacaranda Maternity that was spun out of the nonprofit. Jacaranda Maternity is able to attract investments to grow its network of hospitals that provides affordable but appropriately priced services for pregnancy and delivery.

Conclusion

This chapter has examined multiple pathways that medical devices and technologies use to scale in low-resource settings, largely determined by the type of entities. This includes academic entities that scale innovations through licensing pathways or start-ups, for-profit entities that can pursue a classical localized growth strategy or a dual-market strategy, and finally nonprofits that scale through a combination of revenues and donations. Each pathway is unique in both the opportunities and challenges that innovators must bear in mind to secure the right type of financing, partnerships and teams to reduce the time needed to scale and increase the likelihood of success. Innovators should review these models carefully when launching a new venture, with particular attention to the financing tools available based on the model selected. The most successful scale-ups in global health often result from the innovators' unique ability to match their vision, business model, team composition and capabilities to the availability and expectations of funders in their space.

References

Bennett, D. J., Itagaki, T., Chenelle, C. T., Bittner, E. A., & Kacmarek, R. M. (2018). Evaluation of the augmented infant resuscitator. *Anesthesia & Analgesia*, *126*(3), 947–955. https://doi.org/10.1213/ane.0000000000002432.

Devadas, D., Thériault, A., & McGahan, A. M. (2021a). *Dual-market scaling strategies for impact*. Every Woman Every Child Innovation Marketplace White Paper.

Devadas, D., Thériault, A., & McGahan, A. M. (2021b). *Financing instruments for innovators to scale health innovations in low- and middle-income countries*. Every Woman Every Child Innovation Marketplace White Paper.

Jackson, T. (2015, May 15). *Kenya's Weza Tele acquired by AFB for $1.7m*. Disrupt Africa. https://disrupt-africa.com/2015/05/15/kenyas-weza-tele-acquired-by-afb-for-1-7m/.

Jumia. (2019, April 11). *Jumia announces pricing of Initial Public Offering*. Jumia. https://investor.jumia.com/financial-news/press-release/2019/Jumia-announces-pricing-of-Initial-Public-Offering/default.aspx.

Liquid Intelligent Technologies. (2020, October 26). *CDC commits additional US $40 million to liquid intelligent technologies supporting Africa's digital transformation*. Liquid Intelligent Technologies. https://www.liquid.tech/about-us/news/CDC_commits_additional_US_$40_million_to_Liquid_Intelligent_Technologies_supporting_Africas_digital_transformation.

Smart Africa. (2020, October 8). *Orange collaborates with Smart Africa and announces new investments in Africa to improve the quality of service and data security for end-users*. Smart Africa. https://smartafrica.org/press-release-communique-de-presse/.

The Guardian. (2015, May 20). *Apple buys a Nigerian-owned ICT firm for $1 billion*. The Guardian. https://guardian.ng/news/apple-buys-a-nigerian-owned-ict-firm-for-1-billion/.

The World Bank. (2019, May 6). *Achieving Africa's Digital Transformation is an Ambition that Requires Game-changing Cooperation*. The World Bank. https://www.worldbank.org/en/news/feature/2019/05/06/achieving-africas-digital-transformation-is-an-ambition-that-requires-game-changing-cooperation.

CHAPTER 28

Game-changing digital mental health interventions: Cases in low- and middle-income countries and at-risk populations

Megha Siddhanta[a,b], **Judy Kuriansky**[c,d,e,f,g,h,i,j], **and Erin Barry**[d,i]

[a]Department of Clinical and Counseling Psychology, Teachers College, Columbia University, New York, NY, United States, [b]Harpaz-Rotem PTSD Treatment & Research Lab, Yale School of Medicine, New Haven, CT, United States, [c]Global Council SDG3 (Health), "SDGs in Action" Initiative, Dubai, United Arab Emirates, [d]International Association of Applied Psychology, Indianapolis, IN, United States, [e]Office of Ambassador Sidique Wai, Embassy of the Republic of Sierra Leone to the United States, Washington, DC, United States, [f]United African Congress, New York, NY, United States, [g]Mission of the Republic of Sierra Leone to the United Nations, New York, NY, United States, [h]United Nations Group of Friends of Mental Health and Well-being, New York, NY, United States, [i]Department of Clinical Psychology, Teachers College, Columbia University, New York, NY, United States, [j]Global Leadership Council, eMental Health International Collaborative (eMHIC), Auckland, New Zealand

Introduction

The mental health treatment gap (mhGAP) is a major challenge which is increasing around the world, given its escalation over time in general but particularly during the COVID-19 pandemic. The treatment gap refers to the disparity between needs of people with mental health conditions and provision of care, fueled by factors like low capacity of professional staff, inadequate diagnosis and treatments, and stigma (Patel et al., 2010). Over the past few decades, a paradigm shift has occurred to bridge this expanding gap by using digital methods, technology innovations, cyberspace, and Digital Mental Health Interventions (DMHIs). The use of DMHIs includes consultations through digital and technological means and platforms, to deliver advice, support, counseling, psychotherapy, cognitive rehabilitation, attention-training exercises, and biofeedback services.

This chapter describes how commonly-used technologies (smartphones, apps, AI) are evolving into next level cutting-edge techniques, including digital phenotyping; instantaneous ecological

Resilient Health. https://doi.org/10.1016/B978-0-443-18529-8.00028-7

assessment (EMA) or ecological momentary intervention (EMI); "Just-in time adaptive interventions"; and interoceptive, embodied, and connectivity technologies. To be most effective, these should be applied in the context of a "user-centered design." Research must also be carried out to assess usefulnees of this digital technology to address mental health problems and to establish best practices. Two examples of such efforts are presented in this chapter in two countries—Jamaica and India. Many factors, including assumptions, possibilities, and challenges, must be examined in the use of digital technologies to address mental health needs. Also, acceptance of digital tools by certain populations, particularly those at-risk, such as youth and refugees, isdiscussed.

Background

In low- and middle-income countries (LMICs), disparities in mental health resources are particularly exacerbated by factors such as structural deficits, discriminatory perceptions toward getting treatment, financial contraints, and a substantial knowledge-attitude-practice (KAP) gap. As a result, digital interventions, particularly those that are more widely accepted such as artificial intelligence (AI) and gaming, can be used to address both inadequate access and provider capacity, and also to improve mental healthcare usage. Technological interventions are further being employed to conduct treatment programs and boost mental health literacy (Baños et al., 2022; Ito-Jaeger et al., 2022). Such methods are especially useful to serve a broad number of people at the basic level of mental healthcare (for primary care interventions and simple support), thereby reducing strain on scarcer higher level and tertiary care centers. This approach is consistent with recommended IASC guidelines in emergency care, posed in a pyramid whereby large numbers of people can be helped with simple psychosocial assistance, assigning more intensive attention to cases in need in a hierarchical manner.

An example of useful technology is mHealth apps on smartphones which have grown in popularity. Current World Health Organization policy recommends using mobile and digital devices for mental health self-monitoring and family-guided therapies. This chapter presents case studies about how Digital Mental Health Interventions have been tested in India and Jamaica.

An increasingly popular way to fill the mental health gap has been through the use of digital or applied games (one form of DMHI), based on the concept of "gamification," defined as the use of gamelike designs, components, and concepts to target mental health domains. Gamification is described by Deterding et al. (2011) in two dimensions: the "liveliness aspect of the game" and the "gaming aspects targeting the domains of interest." The former affects its effectiveness in various psychiatric diseases, while the latter determines its appeal and acceptability. In low- and middle-income countries with limited resources, use of digital gaming-based mental heath interventions must take into consideration not only cultural acceptability but also relative effectiveness.

Digital mental health interventions (DMHIs)

The potential of DMHIs for the delivery of mental health services has expanded exponentially over the last 30 years, spanning from "eHealth" (Oh et al., 2005) which includes telemedicine or internet-based interventions (IBIs) (Price et al., 2014; Ritterband et al., 2006) to "mHealth" which includes smartphones or virtual or augmented reality applications (Miralles et al., 2020; Price et al., 2014). Due to the possibilities they present in ease of access, cost-effectiveness, and attractiveness, all these digital solutions have gotten much attention recently, especially given the urgent requirement for mental health resources during the COVID-19 pandemic (Balcombe & De Leo, 2021). This chapter covers IBIs, smartphone apps, virtual reality (VR), and augmented reality (AR).

Internet-based interventions (IBIs)

Among the most presently researched DMHIs is Internet-Based Interventions (IBIs) which are treatment programs with planned objectives communicated through units provided via the internet by computer or mobile phone (Andersson & Cuijpers, 2009; Andersson & Titov, 2014; Barak et al., 2008). IBIs come in a variety of shapes and sizes. Some are automatic and require no human assistance (self-directed or unsupervised interventions), while others are completely automated but require interaction with a clinician or technician (guided interventions) (Andersson & Cuijpers, 2009).

In early investigations evaluating guidance degree and efficacy, ungoverned IBIs were shown to be less successful than IBIs that contained some clinical direction (Baumeister et al., 2014; Richards & Richardson, 2012). Nevertheless, a growing body of evidence suggests that distinctions between directed and nondirective IBIs are smaller or nonexistent than previously thought (Karyotaki et al., 2017). Assistance on demand, which combines the benefits of directed and unsupervised self-help IBIs (Rheker et al. (2015)), and mixed therapy, which is an amalgamation of face-to-face therapy and online support, are starting to accumulate empirical evidence and will undoubtedly receive more recognition in the future (Erbe et al., 2017).

A meta analysis has shown that IBIs can be as good as, or better than, in-person therapy for a variety of psychological disorders (Carlbring et al., 2017) as well as reduction of symptoms like depression and anxiety (Huckvale et al., 2020; Paul et al., 2018) and changing addictive behaviors, such as not smoking (Webb et al., 2010). The therapeutic technique that has gained the greatest attention and evidence with technological methodology is internet-based Cognitive-Behavioral Therapy (iCBT) (Karyotaki et al., 2017) which involves reframing thinking coupled with adjusted behaviors. IBIs can also directly benefit from technology such as mobile apps or virtual gadgets.

Smartphone Apps

Smartphones are an integral part of most people's lives during the last decade (Bakker et al., 2016). Already used by 6.6 billion people globally in 2022 (Statista, 2023), smartphones are even becoming widespread in poorer countries. Immediate connectivity and availability of resources from any geographic location are even possible by linking to a simple, graphical, thumb interface which eliminates the need to carry complex equipment such as a digicam or navigation system (Bauer et al., 2020). Also, apps downloaded and accessed on the smartphone are one of the most popular mHealth intervention tools. All stages of psychological care, including management, diagnosis, and therapy, can be delivered via apps, even in combination with face-to-face therapy and psychosocial support, which may be helpful or necessary in some cases (Chandrashekar, 2018).

Apps can be utilized as a method for psychological therapy in the form of "instantaneous ecological assessment" (EMA) or "momentary ecological intervention" (EMI) (McDevitt-Murphy et al., 2018). EMA is an assessment system program that collects data from people in a variety of surroundings at various periods in time. The program typically asks the user to respond to one or more questions multiple times for a week or at random intervals. EMI has a similar format but the content can include prompts, feedback messages, or directives for specific or crucial activities. "Just-in Time Adaptive Interventions" (JATAI), a intervention design providing the right type/amount of support at the right time by adapting to an individual's changing internal and contextual state, is one type of EMI that is predicted to play a big role in clinical practice in the future (Nahum-Shani et al., 2018).

Research on these approaches has focused on varied aspects. Regarding access, (Linardon et al., 2019) found that applications can be low intensity, cheap, and conveniently available therapies for persons who cannot obtain regular psychiatric treatment.Miralles et al. (2020) observed that app research to date has mostly focused on the most prevalent psychiatric disorders, with much less emphasis on less common or more severe conditions. While apps for techniques and modalities, such as CBT and mindfulness, proliferate, not enough evidence-base research has conclusively proven their usefulness. Barriers to the use of apps, for example in the case of depression, have been pointed out, which include intervention efficacy, app functioning, privacy, cost, and lack of guidance and tailored feedback (Stiles-Shields et al., 2017).

Though several mHealth apps are accessible for instant download on online markets, more rigorous scientific research, e.g., using randomized controlled trials (RTC), is required to prove reliability and validity (Miralles et al., 2020). Such investigations would also have to establish recommended medication dosage (Torous et al., 2018). Also, these technologies would have to be properly regulated and delivered for mental health therapy.

Mixed realities: Virtual and augmented reality

Advanced technological approaches, like Virtual Reality (VR) and Artificial Reality (AR), have been shown in some studies to be effectively implemented for the evaluation and treatment of mental health disorders such as PTSD, anxiety, and autism spectrum disorders (Emmelkamp & Meyerbröker, 2021). VR is defined as "a set of technologies that facilitate individuals to engage proficiently with 3D computerized databases in live time that use their natural senses and skills" (McCloy & Stone, 2001), whereas AR "incorporates the real world with digital aspects, using computer graphics blended with the real world in real time" (Riva et al., 2016). The phrase "mixed realities" describes the merging of virtual and physical worlds to create a unique experience.

Both VR and AR have radically transformed the healthcare profession and show great potential as: (1) interoceptive technology, which allows people to feel "as if they were" in a real life that does not exist in the physical or real world; (2) embodied technology, which allows individuals to believe their body is inside the virtual environment; and, (3) connectivity technology, which enables users to interact and share things with others from any geographical region. Currently, VR has progressed from replicating the world to simulating the self and is now uniting people to create and share experiences, generating a wide range of new social networking opportunities.

Toward precision medicine: Artificial intelligence (AI) and DMHIs

Technology is transforming how mental illnesses are tested, identified, and treated to date, providing advances in (1) objective and trustworthy diagnosis; (2) real-time, actionable health data tracking; and (3) individualized treatment programs. However, continuing psychological support through technology has proven difficult. Precision medicine—individualizing the application of technologies such as mobile health, imaging, big data and artificial intelligence to target the right treatment to the right person—has the potential to overcome these present therapeutic hurdles (Bickman, 2020).

Another means of revolutionizing mental healthcare is through digital phenotyping, described as "moment-by-moment quantification of the individual-level human phenotype in situ utilizing data from smartphones and other personal digital devices" (Lopez-Morinigo et al., 2021). Further studies are being conducted to see if this data can be used to make more accurate predictions in the prevention, treatment, and facilitation of support tools to encourage people to control their own health behaviors. Ecological momentary assessments (EMAs), which involve repeated sampling of subjects' current behaviors and experiences in real time in their natural environments, have promise to be a beneficial clinical tool, allowing healthcare providers to develop a digital phenotype using smartphone apps (Lopez-Morinigo et al., 2021).

Chatbots or conversational agents are another key innovation. Chatbots are computerized platforms that allow text-based or voice-activated interactive chats via mobile apps, with preprogrammed or AI-based responses (Adamopoulou & Moussiades, 2020). Recent research shows that these are as effective as face-to-face therapy, particularly for those interventions based on cognitive-behavioral techniques (CBT) (Bendig et al., 2019). The majority of chatbots have so far been used for instruction, screening, and treatment, focusing on problems with depression and autism spectrum disorders (Abd-alrazaq et al., 2019).

A case in India: The "Mana Maali" digital anxiety intervention

The digital anxiety program "Mana Maali," which means "gardener of the mind" in Hindi, is based on an evidence-based CBT intervention for generalized anxiety disorder (GAD) developed by Newman and Borkovec (as cited in (Kanuri et al., 2020)). Their CBT for GAD therapy regimen for doctors is a 14-session, face-to-face therapy with text that included a variety of CBT methods, such as relaxation techniques and cognitive therapy combined with techniques like breathing, thought restructuring, and reframing. Adjustments to the text were made to increase the cultural fit for an Indian college-aged audience. A group of clinical psychologists, psychiatrists, and researchers from the United States and India collaborated with the original researchers to insure cultural appropriateness and competency. Revisions were made using student input from previous experiments in which the research team gave a digital copy of the original intervention to university students in India (Kanuri et al., 2020).

Other modifications were made to turn a face-to-face intervention into a virtual program. To accommodate situations with restricted internet, short relaxation activities were created. Strategies like deep breathing and mindfulness were also used repeatedly to help with skill practice and development. The final GAD program consisted of 20 sessions, designed to be performed daily for 10–15min. The objectives were to assist students to: (a) understand about anxiety, (b) recognize symptoms, (c) analyze thoughts and emotions, and (d) manage their worry. Each session was split into two components: 1) didactic, with knowledge about the source or feeling of anxiety, for example, tension deriving from worry or automatic negative thoughts, and, 2) teaching practical techniques to deal with that anxiety, for example, diaphragmatic breathing and muscle relaxation.

The results suggested that the approach can successfully address previously unmet emotional needs among Indian college students. Nearly all students in the study agreed that the digital mental health program was useful to themselves and fit into their routine, and also could be useful to their classmates. Many participants stated that they continued to utilize particular tactics after the study was completed.

In summary, results from the Mana Maali Digital Anxiety Program revealed generally high ratings for use, acceptance, practicality, safety, and feasibility among a nonclinical population of Indian college students. Further, the majority of students in focus groups believed that the intervention was culturally acceptable and would assist them in overcoming barriers to getting mental healthcare or seeking help through more traditional treatments, since this program provided secrecy and convenience.

A case in Jamaica: Adolescents receptiveness to digital mental health services

Global trends show that community people and youth use technology in their daily lives and express interest in mental health treatments being integrated into this medium (Proudfoot et al., 2010; Torous et al., 2014). However, there are barriers to such access in the Jamaican context, as for other developing and low-income countries (Maloney et al., 2020).

Jamaica, like other LMICs, has a dearth of mental health research, with no published studies evaluating the use of digital technology to satisfy mental health requirements in the Jamaican context, yet, data demonstrates unmet mental illness treatment needs, creating a significant treatment gap especially for youth as a vulnerable population (Maloney et al., 2020).

According to the 2006 Jamaica Youth Risk and Resiliency Behavior Survey, 15.5% of a nationally representative Jamaican adolescent sample (N=1312; aged 15–19 years) suffer from depression, defined as meeting at least five diagnostic symptoms from the Diagnostic and Statistical Manual of Mental Disorders IV (DSM-IV), including suicidal ideation (McFarlane et al., 2014). Later estimates were slightly higher, estimating that 20% of youth in Jamaica have a mental disorder and 5% have a serious mental disorder (Caribbean Policy Research Institute, 2021).

National youth care clinics were established to fill this gap in Jamaica by offering public mental healthcare to children and adolescents; however, many institutional barriers exist; for example, many clinics have 3- to 6-month waiting lists and are only open one or two days each month (Maloney et al., 2020). Jamaican youth who seek professional mental health treatment may have trouble finding age-appropriate resources or digital mental health services, as well as face stigma inhibiting seeking such care. Access can be facilitated by lowering stigma, boosting service accessibility, and insuring privacy and confidentiality.

In a survey of 107 youth age 10–19 by Maloney et al. (2020), 78.5% of respondents reported that mHealth services (supported by mobile devices) would be beneficial to youth with mental distress. Further, more than half (56.1%) expressed an interest in utilizing a smartphone app to self-monitor their mental health.

The researchers suggest that use of mHealth tools (smartphone apps and web-based communication) could boost youth seeking social support from sources as family and friends, which has been shown to be preferred over more formal sources such as mental health professionals (Jackson Williams, 2012).

Also, the survey showed the importance of including a sample of rural as well as urban teenagers, given findings that urban versus rural youth had a higher interest in web-based communication, and rural youth were hampered by less Wi-fi access.

The case of refugees and migrants

An estimated 1 billion people worldwide are on the move with more people than ever migrating or being displaced (World Health Organization, 2020). The United Nations Refugee Agency (UNHCR) reported that the global refugee population in 2018 had surpassed 20 million for the first time (UNHCR, 2019), with an estimated 16.2 million persons newly displaced in 2017 and 102,800 refugees accepted for resettlement in that year (UNHCR, 2019). Refugees' premigration and postmigration experiences have been connected to mental health and well-being (Bhugra et al., 2014; Giacco & Priebe, 2018; Rich et al., 2023).

Despite many challenges for refugees regarding psychological health, this population rarely uses standard mental health treatments (Bartolomei et al., 2016) and barriers to such use exist. In this regard, context is important, meaning the need to explore assumptions, possibilities, and potential pitfalls in the use of digital technologies to address refugee mental health. This point is based on Toyama's Law of Amplification (Toyama, 2011, 2015) which states that successful technical interventions amplify the intent and capacity of the people involved. A chapter in this volume from UNHCR (the United Nations refugee agency), documents how use of technology such as the internet and mobile connectivity by refugees and organizations that protect them can enhance refugee safety and economic and social wellbeing, while acknowledging persistent limitations in access (Eluere, 2024).

Developers and practitioners of interventions need to know what works, for whom, and in what context. According to the Law of Amplification about digital technology's impact—which requires intention to improve efficiency, reduce costs, and increase scale although also increasing inequities (Toyama, 2015)—it makes sense that high levels of mobile phone ownership allow for reaching potential consumers through apps or websites tailored for use on mobile phones. Also, by focusing on differential capacity as a context, one can deduce that self-guided digital interventions are better suited to refugee groups with prior experience with technology who have gained online abilities such as browsing websites compared to those with no such experience. Therefore, training is required for this latter group. Also, differential motivation is a factor, which means that a user's commitment and capacity to benefit from a digital intervention are required. If these are lacking, the methods to facilitate such use—such as phone calls to check on progress or creating spaces where the intervention can be accessed in the presence of a facilitator as in group sessions—may be required for the digital intervention to reach its full potential (Purgato et al., 2018). Further, digital interventions that help newly arrived refugees better comprehend and acclimate to life in their receiving country are generally more appealing than those focusing solely on mental health. Direct interventions addressing mental health will be more appealing to this populatoin after their "new" day-to-day life becomes known, easier and more predictable.

A "user-centered design"

A "user-centered design" is valuable when developing and tailoring interventions for any target group. This process involves starting with a generalized intervention that can apply to diverse groups and adapting it into numerous versions relevant to a specific population or setting (e.g., Spanhel et al., 2020). For example, in the WHO Step-by-Step intervention, which is a story-based intervention targeted at lowering psychological distress, the character providing instruction can be replaced with a trusted community member for a target community, e.g., an elder or a health professional (Carswell et al., 2018). Another option is to initially work with the target population of the intervention to understand how they think about mental health and what they need to support their mental health needs. Such an approach helps groups get the most out of digital technologies.

User-centered design also enables the co-construction of alternative solutions that can be evaluated and modified in an iterative method by working jointly with the specific population, for example, refugees working with practitioners, researchers and designers. Further, a tool created with the user at the center of the design process, as well as early stakeholder participation, builds clear routes to uptake, increasing the likelihood that the intervention will be helpful and used. User-centered approaches for the design of digital tools and detailed accounts of steps taken, have been carried out and shown to be useful (Hardy et al., 2018; Schnall et al., 2016).

User-centered design is guided by the goals, tasks, and needs of the intended users; thus, it requires inclusion and active participation of intended users in the design process from the outset (Gulliksen et al., 2003). Focus groups and other face-to-face engagement with the intended user group help determine their needs and priorities, how they interact with technology, what they perceive as the "issue," as well as suggestions and attitudes about alternative solutions. User-centered design also entails interacting with a variety of stakeholders to obtain a better knowledge of the problem and its context. Stakeholders can reflect on emerging themes, using their knowledge and experience to help construct a unified approach. The use of peer researchers, as opposed to professionals who may be perceived as "outsiders" by the target group, can further improve the process, provide access to difficult populations, and offer an alternate lens through which to describe problems and discuss issues (Minkler, 2004).

Conclusion

The outbreak of the COVID-19 pandemic has altered and expedited the trajectory of digital mental health interventions (DMHIs) in healthcare systems. Recent developments in various technological alternatives for mental health interventions and therapy have included game-changing cutting-edge innovative methodologies, applying AI, VR, and mixed realities. Promising solutions for providing more accessible and individualized psychological treatment include adaptive apps that are "Just-in-Time" approaches and adaptations of gaming techniques. Despite advancements in DMHIs, constraints still prevent widespread use. Precision medicine, with AI-based technology, is increasingly attempting to overcome such challenges with new and more optimal digital solutions. In addition to AI, VR, AR and mixed realities, cutting-edge techniques include digital phenotyping; instantaneous ecological assessment (EMA) or momentary ecological intervention (EMI); "Just-in Time Adaptive Interventions"; and interoceptive, embodied, and connectivity technologies. Optimal effectiveness requires a "user-centered design." Research has been done in low- and middle-income countries where needs are great and resources are limited, including in India and Jamaica, as well as with marginalized and at-risk populations like youth and refugees. More research and policy support are needed, however, before DMHIs can be fully adopted in varied care settings. Future research should examine hybrid care approaches in clinical practice. Digital solutions will undoubtedly play a role in the future of health as well as psychological and mental healthcare, and if implemented appropriately, will help achieve a higher state of health, mental health and well-being in all societies.

References

Abd-alrazaq, A. A., Alajlani, M., Alalwan, A. A., Bewick, B. M., Gardner, P., & Househ, M. (2019). An overview of the features of Chatbots in Mental Health: A scoping review. *International Journal of Medical Informatics*, *132*, 103978. https://doi.org/10.1016/j.ijmedinf.2019.103978.

Adamopoulou, E., & Moussiades, L. (2020). Chatbots: History, technology, and applications. *Machine Learning with Applications*, *2*, 100006. https://doi.org/10.1016/j.mlwa.2020.100006.

Andersson, G., & Cuijpers, P. (2009). Internet-based and other computerized psychological treatments for adult depression: A meta-analysis. *Cognitive Behaviour Therapy*. https://doi.org/10.1080/16506070903318960.

Andersson, G., & Titov, N. (2014). Advantages and limitations of internet-based interventions for common mental disorders. *World Psychiatry*, *13*(1), 4–11. https://doi.org/10.1002/wps.20083.

Bakker, D., Kazantzis, N., Rickwood, D., & Rickard, N. (2016). Mental health smartphone apps: Review and evidence-based recommendations for future developments. *JMIR Mental Health*, *3*(1), e4984. https://doi.org/10.2196/mental.4984.

Balcombe, L., & De Leo, D. (2021). Digital Mental Health Challenges and the horizon ahead for solutions. *JMIR Mental Health*, *8*(3). https://doi.org/10.2196/26811.

Baños, R. M., Herrero, R., & Vara, M. D. (2022). What is the current and future status of digital mental health interventions? *The Spanish Journal of Psychology*, *25*. https://doi.org/10.1017/sjp.2022.2.

Barak, A., Hen, L., Boniel-Nissim, M., & Shapira, N. (2008). A comprehensive review and a meta-analysis of the effectiveness of internet-based psychotherapeutic interventions. *Journal of Technology in Human Services*, *26*(2–4), 109–160. https://doi.org/10.1080/15228830802094429.

Bartolomei, J., Baeriswyl-Cottin, R., Framorando, D., Kasina, F., Premand, N., Eytan, A., & Khazaal, Y. (2016). What are the barriers to access to mental healthcare and the primary needs of asylum seekers? A survey of mental health caregivers and primary care workers. *BMC Psychiatry*, *16*, 336. https://doi.org/10.1186/s12888-016-1048-6.

Bauer, M., Glenn, T., Geddes, J., Gitlin, M., Grof, P., Kessing, L. V., … Whybrow, P. C. (2020). Smartphones in mental health: a critical review of background issues, current status and future concerns. *International Journal of Bipolar Disorders*, *8*(2). https://doi.org/10.1186/s40345-019-0164-x.

Baumeister, H., Reichler, L., Munzinger, M., & Lin, J. (2014). The impact of guidance on internet-based mental health interventions — A systematic review. *Internet Interventions*, *1*(4), 205–215. https://doi.org/10.1016/j.invent.2014.08.003.

Bendig, E., Erb, B., Schulze-Thuesing, L., & Baumeister, H. (2019). The next generation: chatbots in clinical psychology and psychotherapy to foster mental health – A scoping review. *Verhaltenstherapie*. https://doi.org/10.1159/000501812.

Bhugra, D., Gupta, S., Schouler-Ocak, M., Graeff-Calliess, I., Deakin, N. A., Qureshi, A., … Carta, M. (2014). EPA guidance mental health care of migrants. *European Psychiatry*, *29*(2), 107–115. https://doi.org/10.1016/j.eurpsy.2014.01.003.

Bickman, L. (2020). Improving mental health services: A 50-year journey from randomized experiments to artifical intelligence and precision mental health. *Administration and Policy in Mental Health and Mental Health Services Research*, *47*, 795–843. https://doi.org/10.1007/s10488-020-01065-8.

Caribbean Policy Research Institute. (2021). *Mind the gap: The inadequacy of mental health services for children.* https://www.capricaribbean.org/sites/default/files/public/documents/report/mind_the_gap_the_inadequacy_of_mental_health_services_for_children_revised_october_2021.pdf.

Carlbring, P., Andersson, G., Cuijpers, P., Riper, H, & Hedman-Lagerlöf, E. (2017). Internet-based vs. face-to-face cognitive behavior therapy for psychiatric and somatic disorders: An updated systematic review and meta-analysis. *Cognitive Behaviour Therapy*, *47*(1), 1–18. https://doi.org/10.1080/16506073.2017.1401115.

Carswell, K., Harper-Shehadeh, M., Watts, S., van't Hof, E., Abi Ramia, J., Heim, E., … van Ommeren, M. (2018). Step-by-step: A new WHO digital mental health intervention for depression. *Mhealth*, *4*. https://doi.org/10.21037/mhealth.2018.08.01.

Chandrashekar, P. (2018). Do mental health mobile apps work: Evidence and recommendations for designing high-efficacy mental health mobile apps. *MHealth*, *4*. https://doi.org/10.21037/mhealth.2018.03.02.

Deterding, S., Khaled, R., & Nacke, L. (2011). From game design elements to gamefulness. In *Proceedings of the 15th international Academic MindTrek Conference: Envisioning future media environments*. https://doi.org/10.1145/2181037.2181040.

Eluere, P. (2024). Empowering refugees through technology. In J. Kuriansky, & P. Kakkattil (Eds.), *Resilient health: Leveraging technology and social innovations to transform healthcare for COVID-19 recovery and beyond* (pp. 353–359). Elsevier.

Emmelkamp, P. M. G., & Meyerbröker, K. (2021). Virtual reality therapy in mental health. *Annual Review of Clinical Psychology*, *17*, 495–519. https://doi.org/10.1146/annurev-clinpsy-081219-115923.

Erbe, D., Eichert, H.-C., Riper, H., & Ebert, D. D. (2017). Blending face-to-face and internet-based interventions for the treatment of mental disorders in adults: Systematic review. *Journal of Medical Internet Research*, *19*(9). https://doi.org/10.2196/jmir.6588.

Giacco, D., & Priebe, S. (2018). Mental health care for adult refugees in high-income countries. *Epidemiology and Psychiatric Sciences*, *27*(2), 109–116. https://doi.org/10.1017/S2045796017000609.

Gulliksen, J., Göransson, B., Boivie, I., Blomkvist, S., Persson, J., & Cajander, Å. (2003). Key principles for user-centred systems design. *Behaviour & Information Technology*, *22*(6), 397–409. https://doi.org/10.1080/01449290310001624329.

Hardy, C., Griffiths, A., Norton, S., & Hunter, M. S. (2018). Self-help cognitive behavior therapy for working women with problematic hot flushes and night sweats (MENOS@Work): A multicenter randomized controlled trial. *Menopause*, *25*(5), 508–519. https://doi.org/10.1097/gme.0000000000001048.

Huckvale, K., Nicholas, J., Torous, J., & Larsen, M. E. (2020). Smartphone apps for the treatment of mental health conditions: Status and considerations. *Current Opinion in Psychology*, *36*, 65–70. https://doi.org/10.1016/j.copsyc.2020.04.008.

Ito-Jaeger, S., Vallejos, E. P., Curran, T., Spors, V., Long, Y., Liguori, A., … Crawford, P. (2022). Digital video interventions and mental health literacy among young people: A scoping review. *Journal of Mental Health, 31*(6), 873–883. https://doi.org/10.1080/09638237.2021.1922642.

Jackson Williams, D. (2012). Where do Jamaican adolescents turn for psychological help? *Child Youth Care Forum, 41*, 461–477. https://doi.org/10.1007/s10566-012-9177-7.

Kanuri, N., Arora, P., Talluru, S., Colaco, B., Dutta, R., Rawat, A., … Newman, M. G. (2020). Examining the initial usability, acceptability and feasibility of a digital mental health intervention for college students in India. *International Journal of Psychology, 55*(4), 657–673. https://doi.org/10.1002/ijop.12640.

Karyotaki, E., Riper, H., Twisk, J., Hoogendoorn, A., Kleiboer, A., Mira, A., … Cuijpers, P. (2017). Efficacy of self-guided internet-based cognitive behavioral therapy in the treatment of depressive symptoms: A meta analysis of individual participant data. *JAMA Psychiatry, 74*(4), 351–359. https://doi.org/10.1001/jamapsychiatry.2017.0044.

Linardon, J., Cuijpers, P., Carlbring, P., Messer, M., & Fuller-Tyszkiewicz, M. (2019). The efficacy of app-supported smartphone interventions for mental health problems: A meta-analysis of randomized controlled trials. *World Psychiatry, 18*(3), 325–336. https://doi.org/10.1002/wps.20673.

Lopez-Morinigo, J., Barrigón, M. L., Porras-Segovia, A., Ruiz-Ruano, V. G., Escribano Martínez, A. S., Escobedo-Aedo, P. J., … Baca-García, E. (2021). Use of ecological momentary assessment through a passive smartphone-based app (eB2) by patients with schizophrenia: Acceptability study. *Journal of Medical Internet Research, 23*(7), e26548. https://doi.org/10.2196/26548.

Maloney, C. A., Abel, W. D., & McLeod, H. J. (2020). Jamaican adolescents' receptiveness to digital mental health services: A cross-sectional survey from rural and urban communities. *Internet Interventions, 21*, 100325. https://doi.org/10.1016/j.invent.2020.100325.

McCloy, R., & Stone, R. (2001). Science, medicine, and the future. Virtual reality in surgery. *British Medical Journal (Clinical Research Edition), 323*(7318), 912–915. https://doi.org/10.1136/bmj.323.7318.912.

McDevitt-Murphy, M. E., Luciano, M. T., & Zakarian, R. J. (2018). Use of ecological momentary assessment and intervention in treatment with adults. *Focus, 16*(4), 370–375. https://doi.org/10.1176/appi.focus.20180017.

McFarlane, S, Younger, N., Francis, D., Gordon-Strachan, G, & Wilks, R. (2014). Risk behaviours and adolescent depression in Jamaica. *International Journal of Adolescence and Youth, 19*(4), 458–467.

Minkler, M. (2004). Ethical challenges for the "outside" researcher in community-based participatory research. *Health Education & Behavior: The Official Publication of the Society for Public Health Education, 31*(6), 684–697. https://doi.org/10.1177/1090198104269566.

Miralles, I., Granell, C., Díaz-Sanahuja, L., Van Woensel, W., Bréton-López, J., Mira, A., … Casteleyn, S. (2020). Smartphone apps for the treatment of mental disorders: Systematic review. *JMIR MHealth and UHealth, 8*(4), e14897. https://doi.org/10.2196/14897.

Nahum-Shani, I., Smith, S. N., Spring, B. J., Collins, L. M., Witkiewitz, K., Tewari, A., & Murphy, S. A. (2018). Just-in-time adaptive interventions (JITAIs) in mobile health: Key components and design principles for ongoing health behavior support. *Annals of Behavioral Medicine, 52*(6), 446–462. https://doi.org/10.1007/s12160-016-9830-8.

Oh, H., Rizo, C., Enkin, M., & Jadad, A. (2005). What is eHealth. A systematic review of published definitions. *Journal of Medical Internet Research, 7*, e1. https://www.ncbi.nlm.nih.gov/pmc/articles/PMC1550636/.

Patel, V., Maj, M., Flisher, A. J., De Silva, M. J., Koschorke, M., Prince, M., & WPA Zonal and Member Society Representatives. (2010). Reducing the treatment gap for mental disorders: A WPA survey. *World Psychiatry, 9*(3), 169–176. https://doi.org/10.1002/j.2051-5545.2010.tb00305.x.

Paul, A. M., & Eubanks Fleming, C. J. (2018). Anxiety management on campus: An evaluation of a mobile health intervention. *Journal of Technology in Behavioral Science, 4*, 58–61. https://doi.org/10.1007/s41347-018-0074-2.

Price, M., Yuen, E. K., Goetter, E. M., Herbert, J. D, Forman, E. M., Aciero, R., & Ruggiero, K. J. (2014). mHealth: A mechanism to deliver more accessible, more effective mental health care. *Clinical Psychology & Psychotherapy, 21*(4), 427–436. https://doi.org/10.1002/cpp.1855.

Proudfoot, J., Parker, G., Hadzi Pavlovic, D., Manicavasagar, V., Adler, E., & Whitton, A. E. (2010). Community attitudes to the appropriation of mobile phones for monitoring and managing depression, anxiety, and stress. *Journal of medical Internet research*, *12*(5), e64. https://doi.org/10.2196/jmir.1475.

Purgato, M., Gastaldon, C., Papola, D., van Ommeren, M., Barbui, C., & Tol, W. A. (2018). Psychological therapies for the treatment of mental disorders in low- and middle-income countries affected by humanitarian crises. *The Cochrane Database of Systematic Reviews*, *7*(7), CD011849. https://doi.org/10.1002/14651858.CD011849.pub2.

Rheker, J., Andersson, G., & Weise, C. C. (2015). The role of "On demand" therapist guidance vs. no support in the treatment of tinnitus via the internet: A randomized controlled trial. *Internet Interventions*, *2*(2), 189–199. https://doi.org/10.1016/j.invent.2015.03.007.

Rich, G., Kuriansky, J., Gielen, U., & Kaplin, D. (2023). *Psychosocial experiences and adjustment of migrants: Coming to the USA*. Elsevier, Academic press.

Richards, D., & Richardson, T. (2012). Computer-based psychological treatments for depression: A systematic review and meta-analysis. *Clinical Psychology Review*, *32*(4), 329–342. https://doi.org/10.1016/j.cpr.2012.02.004.

Ritterband, L. M., Andersson, G., Christensen, H. M., Carlbring, P., & Cuijpers, P. (2006). Directions for the International Society for Research on internet interventions (ISRII). *Journal of Medical Internet Research*, *8*(3), e23. https://doi.org/10.2196/jmir.8.3.e23.

Riva, G., Baños, R. M., Botella, C., Mantovani, F., & Gaggioli, A. (2016). Transforming experience: The potential of augmented reality and virtual reality for enhancing personal and clinical change. *Frontiers in Psychiatry*, *7*, 164. https://doi.org/10.3389/fpsyt.2016.00164.

Schnall, R., Rojas, M., Bakken, S., Brown, W., Carballo-Dieguez, A., Carry, M., … Travers, J. (2016). A user-centered model for designing consumer mobile health (mHealth) applications (apps). *Journal of Biomedical Informatics*, *60*, 243–251. https://doi.org/10.1016/j.jbi.2016.02.002.

Spanhel, K., Balci, S., Baumeister, H., Bengel, J., & Sander, L. B. (2020). Cultural adaptation of Internet- and mobile-based interventions for mental disorders: A systematic review protocol. *Systematic Reviews*, *9*, 207. https://doi.org/10.1186/s13643-020-01438-y.

Statista. (2023). *Number of smartphone mobile network subscriptions worldwide from 2016 to 2022, with forecasts from 2023 to 2028*. https://www.statista.com/statistics/330695/number-of-smartphone-users-worldwide/.

Stiles-Shields, C., Montague, E., Lattie, E. G., Kwasny, M. J., & Mohr, D. C. (2017). What might get in the way: Barriers to the use of apps for depression. *Digital Health*, *3*. https://doi.org/10.1177/2055207617713827.

Torous, J., Chan, S. R., Tan, S. Y. M., Behrens, J., Mathew, I., Conrad, E. J., … Keshavan, M. (2014). Patient smartphone ownership and interest in mobile apps to monitor symptoms of mental health conditions: A survey in four geographically distinct psychiatric clinics. *JMIR Mental Health*, *1*(1), e5. https://doi.org/10.2196/mental.4004.

Torous, J., Nicholas, J., Larsen, M. E., Firth, J., & Christensen, H. (2018). Clinical review of user engagement with mental health smartphone apps: Evidence, theory and improvements. *BMJ Mental Health*, *21*(3), 116–119. https://doi.org/10.1136/eb-2018-102891.

Toyama, K. (2011). Technology as amplifier in international development [Conference Paper]. In *iConference 2011 February 8*. WA, USA: Seattle. https://doi.org/10.1145/1940761.1940772.

Toyama, K. (2015). *Geek heresy: Rescuing social change from the cult of technology*. New York, NY: PublicAffairs.

United Nations High Commissioner for Refugees. (2019). *Global report*. https://reporting.unhcr.org/sites/default/files/gr2019/pdf/GR2019_English_Full_lowres.pdf.

Webb, T., Joseph, J., Yardley, L., & Michie, S. (2010). Using the internet to promote health behavior change: A systematic review and meta-analysis of the impact of theoretical basis, use of behavior change techniques, and mode of delivery on efficacy. *Journal of Medical Internet Research*, *12*(1), e4. https://doi.org/10.2196/jmir.1376.

World Health Organization. (2020). *ApartTogether survey: Preliminary overview of refugees and migrants self-reported impact of COVID-19*. Geneva: World Health Organization.

Further reading

Cummins, N., & Schuller, B. W. (2020). Five crucial challenges in Digital Health. *Frontiers in Digital Health*, *2*. https://doi.org/10.3389/fdgth.2020.536203.

Goodman, R., Tip, L., & Cavanagh, K. (2020). There's an app for that: Context, assumptions, possibilities and potential pitfalls in the use of digital technologies to address Refugee Mental Health. *Journal of Refugee Studies*, *34*(2), 2252–2274. https://doi.org/10.1093/jrs/feaa082.

Mueser, K. T., Penn, D. L., & Cassisi, J. E. (2013). Psychosocial treatments for schizophrenia. *Annual Review of Clinical Psychology*, *9*(1), 465–497. https://doi.org/10.1146/annurev-clinpsy-050212-185620.

Ritterband, L. M., Thorndike, F. P., Gonder-Frederick, L. A., Magee, J. C., Bailey, E. T., Saylor, D. K., & Morin, C. M. (2009). Efficacy of an internet-based behavioral intervention for adults with insomnia. *Archives of General Psychiatry*, *66*(7), 692. https://doi.org/10.1001/archgenpsychiatry.2009.66.

Thyloth, M., Singh, H., & Subramanian, V. (2016). Increasing burden of mental illnesses across the globe: Current status. *Indian Journal of Social Psychiatry*, *32*(3), 254. https://doi.org/10.4103/0971-9962.193208.

Vinci, C., Brandon, K. O., Kleinjan, M., & Brandon, T. H. (2020). The clinical potential of augmented reality. *Clinical Psychology: Science and Practice*, *27*(3). https://doi.org/10.1111/cpsp.12357.

CHAPTER 29

Empowering refugees through technology

Pauline Eluère

United Nations High Commissioner for Refugees (UNHCR), Geneva, Switzerland

Introduction

While technology is by no means a solution to all issues faced by refugees, it can enhance their well-being and even save lives. In May 2022, a record 100 million people around the world were forced to flee their homes, according to UNHCR, the UN Refugee Agency (Debebe & Sewunet, 2022). This chapter offers a brief overview of how technology in its various forms has shaped the UN's work for refugees, explores how internet and mobile connectivity can improve refugee well-being, and explains why this matters for humanitarian action and policy.

How technology has shaped the UN's work for refugees

Technology has evolved quickly over the course of the past 70years. So has its use by refugees and organizations that work to protect them, such as UNHCR, the UN Refugee Agency (Eluere, 2021). While today, most discussions regarding technology revolve around its digital applications, many other technological innovations have shaped humanitarian work over the decades. This section offers a brief historical overview of some of the way that various technologies have accompanied and shaped the work of UNHCR.

In the 1950s, Telex, the ancestor of the fax and one of the fastest means of communication at the time, helped facilitate the resettlement of European refugees. Following the Soviet Union's invasion of Hungary, UNHCR used Telex machines to send out telegrams by electricity and radio signals to governments, allowing some 180,000 Hungarian refugees to be resettled in 37 countries (Cellini, 2017).

In the 1960s, off-road vehicles helped provide assistance to refugees affected by new conflicts amid decolonization in Africa. As UNHCR's focus shifted away from Europe to Africa and the organization needed to assist refugees across vast territories with poor road infrastructure, heavy-duty off-road vehicles became essential. Today, UNHCR's fleet is made up of 6000 light vehicles, 85% of which are off-road cars (Herrmann, 2016).

Resilient Health. https://doi.org/10.1016/B978-0-443-18529-8.00029-9

In the 1970s, innovation with polythene sheeting allowed for more resistant emergency shelters. At this time, UNHCR became involved in large-scale relief operations in Asia where crowded conditions in camps could lead to serious cholera outbreaks. This is when aid organizations started replacing reinforced cotton fabric, which was expensive and prone to rotting, with polythene sheeting, to shield refugees from extreme weather and the spread of disease (MSF, 2016).

In the 1980s, computers started shifting the way that aid organizations planned, communicated, and managed information. These years were marked by the Soviet invasion of Afghanistan, which displaced 6.3 million Afghans to Iran and Pakistan. As UNHCR's work became more complex, computers were initially introduced to manage payroll and later on to improve refugee emergency planning, data recording, and statistics (UN General Assembly, 1987).

In the 1990s, satellite phones and HF/VHF radios were central to UNHCR's communications as it upscaled operations in war-torn countries, including in Bosnia, Somalia, and Rwanda. To communicate information between UNHCR's 200 field locations and headquarters, staff used shortwave radio transmission. By 1995 UNHCR had equipped some 1200 vehicles with high-frequency radio transceivers, allowing staff to maintain contact with their base within a 1000-km perimeter (UNHCR, 1995).

Digital technology and connectivity for refugees

Over the past two decades, digital technologies have taken a prominent place in our lives. Connectivity, in particular, gives us access to a myriad of tools to help human beings fulfill the famous pyramid of needs postulated by American psychologist Abraham Maslow—building on a hierarchy from basic needs (food, water, safety), to psychological needs (belonging, esteem) and ultimately to self-actualization (creativity, spirituality).

Connectivity

For people who flee war, persecution, and violence, and thus embark on an often-dangerous journey to seek safety, digital devices and connectivity play an important role. The three main items refugees bring with them when they are suddenly forced to flee are water, food, and their phone, according to a 2017 survey of Syrian and Iraqi refugees (Gillespie et al., 2018).

For refugees, a working mobile phone and internet connection can mean increased access to a safe home and refuge for their families. Refugees often depend on connectivity to keep in touch with relatives who might be in a position to help them. Furthermore, information about services offered to asylum seekers and refugees by governments, UNHCR, and the private sector is easily available online. In their country of refuge, having access to the web also facilitates refugees' access to jobs, cash assistance, education, and other services.

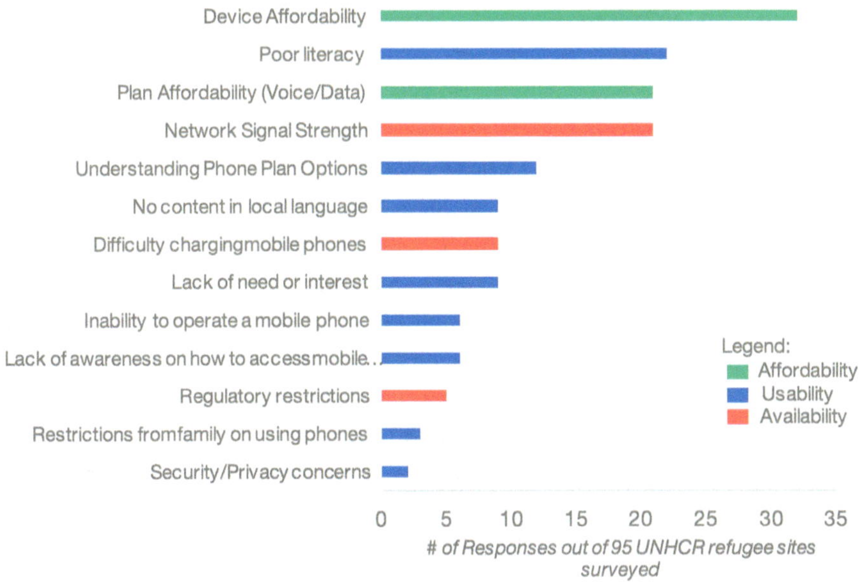

Fig. 1
Largest barriers to internet use (UNHCR and Accenture, 2016).

Yet, despite the democratization of technologies, refugees are still 50% less likely than the rest of the world's population to own an internet-enabled cellphone and 29% do not have a phone at all (I. T. Union, 2017). The first obstacle to connectivity for refugees is cost. Refugees often spend up to 1/3 of their income on staying connected. (UNHCR and Accenture, 2016) Other barriers to internet use are poor literacy, plan affordability, and strength of signal, as shown in Fig. 1, extracted from UNHCR's 2016 Connecting Refugees report (UNHCR and Accenture, 2016).

Rural refugees often have less access to connectivity than the urban population. Only 17% of rural refugees live in areas with 3G coverage, compared to 29% of the global rural population. Twenty percent of rural refugees have no mobile network coverage at all, compared to half of that number (10%) of the global rural population. For urban refugees, accessibility to an internet network is less of an issue, but affording connectivity is a challenge (UNHCR and Accenture, 2016).

Businesses can help power refugee camps. For example, the IKEA Foundation's "Brighter Lives for Refugees" campaign allows UNHCR to provide sustainable electricity to 20,000 Syrian refugees in the Azraq camp in Jordan (UNHCR, 2017). This allows families to connect a TV, have light inside their shelters, and charge their phones, which is critical for ensuring connectivity and keeping in contact with their relatives abroad. Telecom operators can also improve the quality and speed of the internet in refugee communities. Airtel and Vodacom telecommunications companies are setting up 3G towers in regions that host temporary refugee settlements. (GSMA, 2017).

Access to electronic cash assistance

Giving cash to refugees rather than providing in-kind aid such as food or blankets is a great way to help them rebuild their lives with dignity and to be more independent. Electronic payments or "mobile money" allows refugees to receive an SMS message from UNHCR and withdraw cash from a phone shop or pay for goods directly with their phones in local stores. Connectivity can also allow refugees to access other financial services—for example, to open and manage their bank accounts. However, many host countries prevent refugees from accessing these services. In Jordan, where refugees are not allowed to open bank accounts, UNHCR introduced mobile wallets not only to provide cash assistance but also to allow refugees to make transactions, savings, payments, and credit (UNHCR, 2019).

Digital literacy and technical skills

Digital literacy refers to the ability to use and understand technology, such as navigating a website, using social media, and sending emails. It can also refer to having marketable software skills, such as coding websites and apps. Digital literacy opens a wide array of opportunities to refugees. It can allow them to find job offers online but also makes them more employable in most industries. Additional training in computer science can also give them access to work in the fast-growing field of telecommunications and IT.

Tech actors can boost refugees' digital literacy. For example, in 2003, Microsoft signed an agreement with UNHCR to establish two facilities, in Kenya and Russia, where refugees learn computing and software skills (UNHCR, n.d.). Microsoft also provides free resources to help refugees gain computer skills. Businesses can provide online education to refugees. ReDI, a Berlin-based nonprofit school, offers free coding and digital courses to refugee women and gives them access to a network of women working in technology, leading to job opportunities. Refugees can develop skills and in turn create innovative technological solutions (Gokulan, 2019). In Za'atari camp in Jordan, a team of robotics-trained Syrian refugees designed a Lego robot that dispenses hand sanitizer in the camp (WEF, 2020).

Digital education

Refugee children in many parts of the world face discrimination when it comes to education. According to UNHCR, only 34% of secondary-age youth are enrolled in school, and only 5% in higher education (UNHCR, 2021). While enrolment has risen over years, this remains well below global levels. The spread of COVID-19 has exacerbated the situation, as many countries were forced to close schools and limit numbers in classrooms. Access to remote and digital education can help bridge the educational gap for refugees. Digital education can be in the form of online education via laptops and tablets or through radio and TV.

Many projects are in place to boost digital education for refugees. For example, UNHCR and Vodafone have partnered to establish the Instant Network Schools, which provides internet access and digital learning content to marginalized refugee communities in Africa (Vodafone, n.d.). The Connected Learning in Crisis Consortium connects some 33 universities with refugees to offer remote higher education programs to refugee students in contexts of conflict, crisis, and displacement.

Digital identity

The 2018 Global Compact for Refugees stresses the need to ensure access to identity documents, such as birth certificates and ID cards for refugees. Having an ID document is often key to accessing services, including digital ones. For example, many governments require proof of identification in order to activate a SIM card. New technologies are helping to improve refugees' access to a legal and digital identity. Biometric registration was introduced by UNHCR in 2002, which consists of recording refugees' fingerprints and running iris scans. This innovation enables more secure identity documents and allows the distribution of humanitarian aid more efficiently.

For example, in 2021, UNHCR supported a campaign led by the government of Pakistan to issue over 1 million registered Afghan with smart identity cards to give them faster access to health, education, and financial services (UNHCR, 2022).

Storing and protecting the digital data of refugees can present challenges. It is key to ensure data security and check who can access it and obtain informed consent from refugees on the data collected from them. Host governments can follow international instruments and set national legal frameworks governing the use and access of technology and data relating to refugees.

Digital interactions and social cohesion

Digital technology has also become important to maintain and create social links. The most important need according to connectivity based on surveys conducted with refugees is communication with friends and family, as shown in the graph in Fig. 2, extracted from UNHCR's 2016 Connecting with Refugees report (UNHCR and Accenture, 2016). This need might be greater for refugees than for the general population because displacement often separates refugees from their loved ones and can leave them isolated. As shown in Fig. 2, being

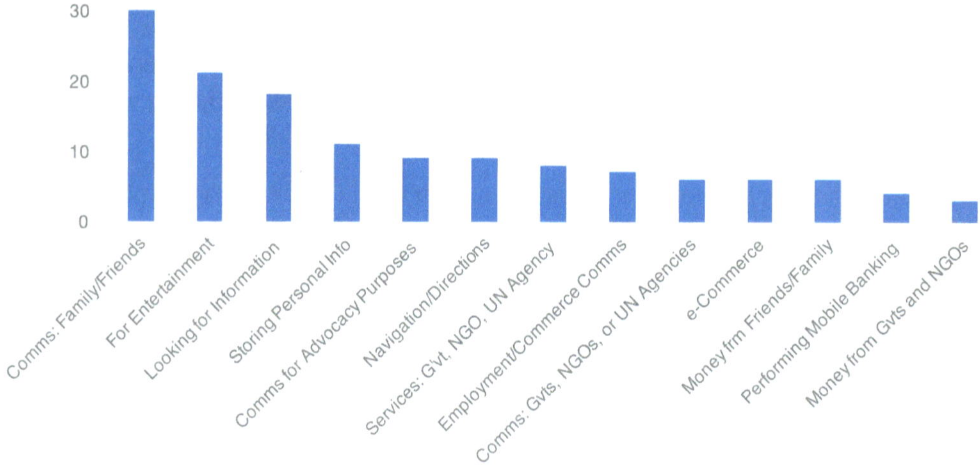

Fig. 2
Most popular uses of the Internet for refugees (UNHCR and Accenture, 2016).

aware of where friends and family are located and whether they are safe is a top concern and use of the Internet for refugees, followed by entertainment and information.

Studies conducted with refugee youth suggest that uptake of social media through personal mobile devices, such as smartphones, positively enhances the interactions of newly arrived refugees.

Many online and social media platforms help connect and integrate refugees. For example, the "I am a Syrian in Lebanon" Facebook group has gained 30,000 members seeking tips on enrolling in school, reporting assaults, and much more since its launch in 2014 (UNHCR, 2016).

Conclusion and looking ahead

Connectivity and digital technologies play an important role in ensuring refugees' safety, economic, and social well-being. Nonetheless, many refugees still live in locations that lack digital networks and infrastructures. While technology can boost refugees' socioeconomic inclusion, it can only do so to the extent that policies are in place—in line with international law—to allow refugees to fully participate in society. In too many places, refugees' freedom of movement and access to employment and education are restricted. Looking ahead, we need to build on many of the innovations, partnerships, and applications already in place, to ensure that refugees are not left behind in society—both offline and online.

References

Cellini, A. (2017). The resettlement of Hungarian refugees in 1956. *Forced Migration Review*, *1*(154), 6–8.

Debebe, E., & Sewunet, Z. (2022, May 23). *UNHCR: A record 100 million people forcibly displaced worldwide*. UN News. (Online). Available from: https://news.un.org/en/story/2022/05/1118772 (Accessed 26 May 2022).

Eluere, P. (2021, 17 May). *7 iconic technologies that have shaped the UN's work for refugees*. UNHCR Innovation Medium. (Online). Available from: https://medium.com/unhcr-innovation-service/7-iconic-technologies-that-have-shaped-the-uns-work-for-refugees-972b6a4a62b5 (Accessed 30 May 2022).

Gillespie, M., Osseiran, S., & Cheesman, M. (2018). Syrian refugees and the digital passage to Europe: Smartphone infrastructures and affordances. *SAGE Journal*, *4*(11).

Gokulan, S. (2019, May 28). *Key to refugee integration in tech-savvy Berlin: Code to co-create to integrate – Institute for Business in the global context*. (Online). Available from: https://sites.tufts.edu/ibgc/redi-school-refugee-immersion/ (Accessed 26 May 2022).

GSMA. (2017). *Mobile is a lifeline: Research from Nyarugusu refugee camp, Tanzania*. London: GSM Association.

Herrmann, R. (2016). *Evaluation of the utilization and management of UNHCR's light vehicle fleet*.

I. T. Union. (2017). *Fast-forward progress leveraging tech to achieve the global goals*. ITU.

MSF. (2016, 22 September). *The humanitarian tarpaulin*. (Online). Available from: https://www.msf-me.org/media-centre/news-and-stories/humanitarian-tarpaulin (Accessed 26 May 2022).

UN General Assembly. (1987, July 27). *Report of the United Nations high commissioner for refugees*. (Online). Available from: https://www.unhcr.org/excom/unhcrannual/3ae68c300/report-united-nations-high-commissioner-refugees.html (Accessed 26 May 2022).

UNHCR. (1995). *Telecommunications in UNHCR EC/1995/SC.2/CRP.16*. Geneva: UNHCR.

UNHCR. (2016, July 13). *Facebook group helps Syrian refugees navigate life in Lebanon*. (Online). Available from: https://www.unhcr.org/news/latest/2016/7/577e16064/facebook-group-helps-syrian-refugees-navigate-life-lebanon.html.

UNHCR. (2017, May 17). *Azraq, the world's first refugee camp powered by renewable energy*. (Online). Available from: https://www.unhcr.org/en-us/news/press/2017/5/591c079e4/azraq-worlds-first-refugee-camp-powered-renewable-energy.html (Accessed 26 May 2022).

UNHCR. (2019, March 5). *Refugee access to financial services in Jordan (march 2019) – Jordan*. (Online). Available from: https://reliefweb.int/report/jordan/refugee-access-financial-services-jordan-march-2019 (Accessed 26 May 2022).

UNHCR. (2021, September 7). *UNHCR education report 2021: 'Staying the course'—The challenges facing refugee education*. (Online). Available from: https://www.unhcr.org/612f85d64/unhcr-education-report-2021-staying-course-challenges-facing-refugee-education (Accessed 26 May 2022).

UNHCR. (2022, January 4). *Pakistan concludes 'drive' to issue smartcards to registered Afghan refugees*. (Online). Available from: https://www.unhcr.org/news/briefing/2022/1/61d419634/pakistan-concludes-drive-issue-smartcards-registered-afghan-refugees.html.

UNHCR. *Microsoft*. (Online). Available from: https://www.unhcr.org/en-us/microsoft.html (Accessed 26 May 2022).

UNHCR and Accenture. (2016, July 1). *Connecting refugees—How internet and mobile connectivity can improve refugee well-being and transform humanitarian action*. (Online). Available from: https://www.unhcr.org/5770d43c4.pdf (Accessed 26 May 2022).

Vodafone. *Instant network schools*. (Online). Available from: https://www.vodafone.com/vodafone-foundation/focus-areas/instant-network-schools (Accessed 26 May 2022).

WEF. (2020, May 15). *These refugees have built a Lego robot to fight COVID-19*. (Online). Available from: https://www.weforum.org/agenda/2020/05/coronavirus-robot-lego-sanitizer-refugees/ (Accessed 26 May 2022).

CHAPTER 30

Evolution in the health ecosystem: Cutting-edge technologies and important issues for improved health and mental health in the ongoing COVID-19 recovery

Andrew J. Greenshaw[a], Jasmine M. Noble[a,b,c], Judy Kuriansky[d], Jie Sui[e], Bo Cao[a], Tao Li[f,g], John Torous[h], and Russell Greiner[a,b,c]

[a]Department of Psychiatry, Faculty of Medicine and Dentistry, University of Alberta, Edmonton, AB, Canada, [b]Department of Computing Science, University of Alberta, Edmonton, AB, Canada, [c]Amii (Alberta Machine Intelligence Institute), Edmonton, AB, Canada, [d]Department of Clinical Psychology, Teachers College, Columbia University, New York, NY, United States, [e]School of Psychology, University of Aberdeen, Aberdeen, Scotland, [f]The Mental Health Center, West China Hospital, Sichuan University, Chengdu, People's Republic of China, [g]Hangzhou Seventh People's Hospital & Affiliated Mental Health Center, Zhejiang University School of Medicine, Hangzhou, People's Republic of China, [h]Department of Psychiatry, Beth Israel Deaconess Medical Center, Harvard Medical School, Boston, MA, United States

Introduction

The emergent healthcare environment in the post-COVID-19 pandemic period has dramatically altered healthcare system operations, services available to consumers, and technological applications for the patient health system interface, in areas from assessment to therapeutic interventions. Technological tools have become increasingly innovative and cutting edge, as developers have scrambled to provide needed connectivity and meet the escalated needs of populations. This chapter provides a window into some of those changes and highlights disruptive technologies that have emerged and that will undoubtedly continue to occur, in response to the implementation of digital health applications, many of which will incorporate artificial intelligence (AI) and other cutting-edge technologies and tools already known and yet to be uncovered. These include biomarkers, predictive analytics, and remote sensing. Six dimensions are identified as necessary for technological adaption, namely, accessibility, acceptability, appropriateness, effectiveness, efficiency, and safety. Some stakeholders are understandably cautious about a health ecosystem that is increasingly dependent on machine

Resilient Health. https://doi.org/10.1016/B978-0-443-18529-8.00030-5

learning-driven AI, involves mechanisms that may not be accessible to the human mind, or reduces involvement of humans. Yet, the potential benefits are significant in a world that lags behind in terms of providing access, availability, and affordability to care for many, particularly in low- and middle-income countries and in areas where burgeoning costs of current healthcare systems are not economically sustainable. Cutting-edge innovations have the potential to cross the digital divide to provide more equitable healthcare, as well as to advance development globally in health and mental health as well as many other aspects of society that are interconnected to health.

Artificial intelligence and machine learning

Within a standard scientific information database like PubMed, even an unrefined set of search terms like "AI" or "artificial intelligence" or "machine learning" in combination with "health" or any health disciplines such as "psychiatry" or "diagnostic imaging" will yield an increasing number of "hits" for publications. The volume of that activity has accelerated dramatically in the past decade, with an increase during the COVID-19 pandemic period, as well as the emergence of new, innovative, and cutting-edge technological approaches.

In reviewing the toolbox of technologies, the smartphone has a long history, about to enter its fourth decade of availability as a communication tool. A pivotal smartphone event was the market penetration by the iPhone released in June 2007. In the subsequent year, the introduction of apps in 2008—emerging as Apple technology with the iOS App Store, followed rapidly by Google's apps for Android devices—transformed the digital ecosystem and altered the nature of human experience for many people.

These groundbreaking personal devices and emerging platforms quickly ushered in a digital revolution, accompanied by the birth of internet sector business ventures and the ubiquitous rise of complex analytical methods that we know of as machine learning, an important subfield of artificial intelligence (AI). In this evolution, humans came to benefit by being able to expand their dimensions, visualizing ideas in 3D space, and even using time as a fourth dimension. In contrast, machines evolved to operate in hyperdimensional spaces and in multiple dimensions. In seemingly fast progression, the field developed new methods to enable large complex system analysis and understanding of higher-order systems—such as brain neuronal circuit connections from neuroimaging. Expanding human capacities, machine learning, AI, and multiple other cutting-edge approaches have emerged to lead us into unknown yet exciting territory. Yet, this fastmoving development requires close examination of social preparedness, and assessment of the impact of these applications to maximize benefit and minimize risk, especially in the area of health, including mental health.

The explosion of AI in the physical health and mental health domain represents a progression of increasing complexity, from data mining (automated extraction of data features, enabling classification and prediction), to machine learning (learning patterns that produce models that can later be used to classify novel instances into different categories), to reinforcement learning, illustrated by the success of "Alpha Go Zero," an AI system that became the World's best player of any game, like Go and Chess, without human instructions (Silver et al., 2018). It became obvious that tasks, like complex diagnosis, or playing skilled games, had become extremely challenging to humans and perhaps reached beyond people's skills. In reaction, it became possible to acquire large, multidimensional datasets to which machine learning can be applied, leading to models that can address or even solve those complex tasks. Two approaches have emerged: predictive analytics that can be "black box" in nature, i.e., while the model may be accurate, there might not be a simple way to describe how the prediction was generated (Silver et al., 2018); or theory/knowledge guided methods, with an attempt to understand at the human level (Rauseo-Ricupero et al., 2021).

The explosion of digital health technology in pandemic times

In the health domain generally, the availability of a digital space, in which personalized options now abound, initially raised frustrations for those working in this realm, leading to a cautious conservative approach. The early relative lack of adoption of new systems in health is now in stark contrast to the massive leaps into the world of virtual interaction in the health field, evident in innovative medical methodologies and the proliferation of digital mental health connections and clinics. Whatever hesitations or wariness existed were short-circuited due to the urgent health and mental health needs revealed by the raging COVID-19 pandemic, as well as the physical distancing requirements of the public health response regulations that were part of the attempts to contain spread of the infection (Rauseo-Ricupero et al., 2021).

Before the onset of the COVID-19 pandemic, the Harvard Business Review of October 2019 indicated that adopting AI in healthcare will be slow and difficult (Kuan, 2019), flagging a need for development of regulatory frameworks, and the difficulty of achieving federal (FDA) approval with a "black box" technology. Despite broad interest in remote healthcare, relatively few online carer-client interactions occurred pre-COVID-19. Then, the physical distancing requirements of COVID-19 public health policies resulted in a reduction of access to healthcare for many. This was sometimes due to health system regulations blocking health carers' authority to interact virtually or the system's inability to remunerate healthcare providers for virtual consultations outside of established telehealth protocols. The effects of those limitations were compounded in an environment in which many individuals needing access to healthcare were reluctant to visit a physical facility, especially hospitals, either distrusting care or fear of contracting COVID-19 infection.

In the years following 2019, considered as the beginning of the COVID-19 spread, we have seen a massive increase in virtual clinical medical and mental health consults for a wide range of conditions going far beyond expectations for the medical specialties of psychiatry and family medicine, but also greatly exceeding expectations for mental health consultations and applications.

In this swift progression, adaptation was certainly necessary for health administrators and health providers as well as healthcare clients and customers to quickly accept that the transition to digital access is growing and here to stay, necessitating quick adaptation to changes. Clearly, the COVID-19 pandemic presented a disaster devastating all dimensions of society, but an unexpected consequence was to drag the unwilling and the uninformed into a new era of opportunity, to make efforts to cross the digital divide, and to enable those with skills in this domain to move rapidly to meet the challenge.

The key question of what machines can add to health quality indicators in patient care

A key question in this evolution became: Since human decision-making is limited in terms of handling complex data patterns, *what* can AI contribute? The answer is that AI can offer faster, more accurate, and more precise decisions that can improve healthcare outcomes. We are used to analyzing data from average descriptors of group outcomes and applying the results to individuals, but AI has the potential to provide individually tailored answers to large groups based on characteristics of an individual patient. This is evident at the level of diagnosis, optimizing treatment choices, and consideration of etiology of disease processes.

After the question of "*what*" comes questions of *how* can we access and harness AI, and *who* can do this? Recent publications in the *Canadian Medical Association Journal* (CMAJ), described later in this chapter, articulate well the interprofessional and interdisciplinary team approach to accessing and enabling this new technology in the health domain.

A next question is *why*? The answer requires assessing any disadvantages to digital technology, considering cultural change, and noting the importance of human relationship in society that goes beyond the context of health and healing. As American neuroscientist, psychiatrist, entrepreneur, and author Thomas Insel, who led the National Institute of Mental Health from 2002 until November 2015, recently discussed, digital technology is likely only part of the solution for providing optimal healthcare; even so, it has the potential to enable vast improvement in health outcomes by offering improved service in many areas (Insel, 2022).

Much change and innovation are guaranteed but speculation and predictions can be made about exactly how the digital health landscape will change specifically. It is clear to many healthcare stakeholders that taking the leap into the unknown of cutting-edge technologies, as well as building on ones like AI already in use, is worth doing. In terms of potential benefits, digital medicine, and the power of machine learning, AI promises brave new opportunities even though there are areas of concern that require moving forward appropriately and mindfully in relation to areas of risk.

One approach to conceptualizing benefits and concerns of cutting-edge technology is to assess dimensions of quality in a new health ecosystem. The Health Quality Council of Alberta in Canada has produced a well-resourced matrix with six dimensions that is a useful framework for the AI conversation (Cowell & Harvie, 2012). These six dimensions are: accessibility, acceptability, appropriateness, effectiveness, efficiency, and safety. These six dimensions are discussed as follows.

Bending the curve on accessibility and acceptability of healthcare

Digital accessibility is a great advantage in cases where direct interaction and also supervision are not possible or necessary. Access to internet or smartphone-based services reduces and sometimes removes significant barriers to access care or to maintain access, offering exciting potential gains in efficiency of service provision. Buy-in (acceptability) from all stakeholders at all levels is necessary, which will depend on the appropriateness and effectiveness of digital service offerings. A fascinating question here is: "Will we need human carers in the digital health care delivery space?" With rapidly advancing robotic and chatbot development, it is certainly possible that such technological advances will pave the way for open accessibility to services with limited human participation. But, acceptability is an open question, as to whether humans will be amenable to less human interaction and become comfortable and trusting of a machine.

Appropriateness, applicability, efficiency, and safety

Appropriateness is also central to the issue of whether AI offerings will be effective, with applicability and buy-in as key issues for diverse cultural racial or ethnic populations. These key issues can be managed by developing effective interfaces with language and culture that fit the needs of diverse populations and ages. Efficiency is very much a work in progress—from the health provider perspective relating to time—whereby demands are of paramount importance. As systems and technology applications evolve, increasing trust in AI with proven worth will likely lead to cultural change in which interest in reliable results will transcend interest in human control and also decrease preoccupation with understanding how the information guiding technologically-based mental healthcare is derived.

These health quality indicators are not mutually exclusive. For example, safety is a complex issue in this digital space, nested with appropriateness and effectiveness. Each indicator, essentially interrelated with the others, encompasses many questions. These include ethical concerns around privacy and security for confidential health data; avoidance of the impact of unconscious developer or user bias in digital tool development and implementation, from the perspective of equity diversity and inclusion (EDI); and avoidance of adverse consequences of changing practice approaches (Svensson & Jotterand, 2022).

AI predictive power in the health system: Answering the need for risk assessment of self-harm and suicide

An example of potential adverse consequences of tele-services in mental health relates to safety protocols for virtual interactions in the context of managing nonsuicidal and suicidal self-harm risks for health system clients. Some safety issues have been at least partially addressed in the context of telepsychiatry and other digital interaction areas. Nevertheless, the breadth and depth of the evolving digital ecosystem imposes a higher priority and a wider set of responsibilities for both stewards of and clients/patients in this psychologically unstable space. AI in mental health is more clearly regulated in the USA, with Canada currently looking at options for policy. This issue has been discussed recently (Da Silva et al., 2022; Rajpurkar et al., 2022) but is a dynamically evolving space (He et al., 2019).

Central to safety in mental health, management of suicide risk remains an area of major general concern, which is a particular concern for remote interaction with those seeking mental health support, for example, by an app such as internet-based cognitive behavior therapy (iCBT), chatbot support, or even remote interaction with a trained human therapist. For app-based or chatbot support, it is essential to have appropriate responses with resource links, or action responses available, optimally equivalent to the level of support provided by a well-resourced suicide crisis support line.

Suicide has been a focus of attention for AI, with reports of applying AI to elucidating suicide biomarkers (Hong et al., 2021) and suicide risk (Wilimitis et al., 2022). Despite apparent successes in this area (Hong et al., 2021), challenges remain. The psychiatric community has some excellent results linking brain structure and function to suicide risk in patient groups (Machado et al., 2021). Some community advocates have pointed to suicide *per se* as not the primary focal priority, quoting Reverend Desmond Tutu's famous statement that instead of rescuing people by pulling them out of the river, we need to find out why and how people are falling in. This leads naturally to a consideration of how we can apply AI to determine early features that are predictive of suicidal self-injury and completed suicides—essentially using a reverse engineering approach (Whalen et al., 2022). Information derived from such analysis can inform effective policies to alter early developmental environments, to build resilience more effectively at early ages. The Reverend Tutu statement about going up the river to examine causes can, of course, be applied to many aspects of health, corresponding to areas now known as the Healthy Life Trajectories Initiative (HeLTI) and Developmental Origins of Health and Disease (DOHaD). AI applications in this context, for example, have been shown in research to be considerably successful in predicting the onset of opioid use disorder from administrative health data using a multiyear prospective approach (Liu et al., 2023).

The binary ecosystem of digital health

Overall, it is convenient to consider two domains within the digital health ecosystem. The first domain comprises publicly and privately funded healthcare organizations that provide a significant array of hospital and/or community clinic services. These are generally stable and officially recognized organizations that typically have ongoing contractual relationships with clients and healthcare practitioners. The second domain is more akin to a largely wild unregulated space that is far less stable or sustained than the first domain (Kahane et al., 2021). The main stakeholders in this second space are independent app developers who operate largely outside regulatory boundaries. In many jurisdictions, well-being or health-oriented apps are not "medical devices" and therefore not subject to health licensing and approval regulations.

Nothing is ever truly simple, and it must be acknowledged that many publicly and privately funded healthcare organizations are also interested in implementation of app approaches to improving health and wellness of their clients. Many of these organizations engage in and around the app development space. The main point is that the second domain is fraught with flaws and potential pitfalls for the user, in relation to some key aspects of dimensions of health quality. These concerns have been discussed in depth in the literature and raise some questions of deep importance related to guiding public users toward appropriate safe and effective choices from the veritable ocean of possible app offerings (Lagan et al., 2021; Washington Post, 2020).

Capitalizing on digital approaches

With an emphasis on the client and caregiver interface and the complexity of the multidimensional spaces of machine learning and AI, there are many knowledge gaps for an exponentially growing number of providers who believe they need to implement AI or machine learning. In a very timely action, several AI experts recently published a set of three articles in the November 21, 2021, issue of the *Canadian Medical Association Journal* (CMAJ). The first article outlines broad key issues for implementing machine learning solutions in the health domain (Verma et al., 2021). In this article Verma and colleagues outline the need for an initial exploration phase, present some points about designing machine learning solutions, and then touch broadly on how to implement and evaluate such solutions. Written from a broad health perspective, this paper, particularly the concluding paragraph, highlights the need for "a disciplined, inclusive, engaged and iterative approach to development and adoption of these technologies." Within the complex specific domain of mental health, this message is no less important.

The other two papers in this set offer much needed general detail on problems of machine learning solution deployment, highlighting the importance of predeployment evaluation (Cohen et al., 2021). The authors of these papers discuss evaluation of machine learning solutions, highlighting the need for a truly multidisciplinary team approach, engaging clinician experts, data scientists, and implementation scientists and including both qualitative and quantitative components (Antoniou & Mamdani, 2021). The utility of machine learning and AI in identifying clinical entities and predicting clinical outcomes in health is very clear. The complexity of developing, implementing, and continually evaluating such machine-learning/AI applications is actually a little daunting, with clear requirements for culture change and a willingness to change the content and consistency of health data that we routinely capture.

These CMAJ papers complement two other recent publications that provide noncomputing specialists access to this field, specifically in relation to mental health. The papers entitled "Evaluating the Machine Learning Literature: A Primer and User's Guide for Psychiatrists" (Grzenda et al., 2021) and "Artificial Intelligence for Mental Health Care: Clinical Applications, Barriers, Facilitators, and Artificial Wisdom" (Lee et al., 2021) provide excellent discussion and explanation of some basic principles that light up this space in a much needed way. In the former article, visionary pioneer in our field of behavioral informatics John Torous, an expert in mobile health and director of the digital psychiatry division in the Department of Psychiatry at Beth Israel Deaconess Medical Center, a Harvard Medical School affiliated teaching hospital, and colleagues outline requirements for "safe and successful integration of machine learning or natural language processing-based applications or results into clinical practice." These include the need to train healthcare specialists about these innovative complex methods, with an excellent and accessible primer that encourages readers to critically evaluate the growing body of literature on applications of these methods in the health domain (Grzenda et al., 2021). The article on clinical applications and barriers, also from the Torous group, provides a very clear overview of AI approaches specifically in relation to mental healthcare, with respect to AI-assisted diagnosis, prognosis, and prediction of effective treatment, together with a description of key challenges (Lee et al., 2021).

Biomarkers and predictive analytics

A final aspect of AI and machine learning application focuses on predictive analytics using a multimodal "biomarker" approach. This is an area of prolific scientific inquiry. Such studies have great promise to elucidate the etiology and underlying mechanisms of disease and to offer personalized medicine solutions for positive treatment outcomes, for example, with pharmacotherapy. Examples include evidence for biotypes of major depressive disorder (Liang et al., 2020), biotypes of first-episode schizophrenia (Liang et al., 2021), and prediction of schizophrenia using multiple brain parcellation ensemble-learning (Kalmady et al., 2019). For such studies to have clinical translational value in the short term, key measurement systems must be freely accessible to patients. This is not the general case with modalities such as neuroimaging and most complex molecular marker assays. It is important to remember that in most health systems, even basic pharmacogenomic data are not yet routinely measured or retained in electronic health records.

Scalable digital biomarkers in mental health

Interesting and potentially scalable AI-derived biomarkers are emerging in mental health. Examples include identification of early indicators of bipolar disorder using learning and cognitive tests, and individualized identification of first-episode bipolar disorder using machine learning and cognitive tests (Sawalha et al., 2021). In relation to readily scalable developments in this area, Liu and colleagues have recently demonstrated that it is possible to prospectively predict opioid use disorder simply based on administrative health records (Liu et al., 2023).

The promise of remote and integration of physical and mental health data with AI

Generally, health systems treat physical and mental health as separate domains, although we have known for at least two millennia, from ancient thinkers and writers like the Roman poet Juvenal who was active in the late first and early second century AD, that a healthy mind and healthy body (mens sana in corpore sano) are intermeshed and interdependent concepts (Jones, 2017), particularly for chronic and serious illnesses. Recent developments in health systems are moving toward integration of data collection across that continuum, with combination of remote sensing for physical and physiological measurements, including body temperature, respiratory rate, heart rate, blood pressure, and oxygen saturation levels. From early development of physical monitoring in hospital environments, it became clear that AI analysis of these measurements can be advantageous for predicting clinical change (Poncette et al., 2020).

The promise of natural language processing

Perhaps one of the most promising and potentially scalable approaches is AI applications in natural language processing. An example is sentiment analysis of speech (also known as emotion AI) to identify the presence of posttraumatic stress disorder (Sawalha et al., 2022). The use of natural language processing may have maximal potential for mental health applications, particularly considering the potential for wide use for speech decoding and speech generating chatbots, using avatars like that developed by Soul Machines with funding from WHO and PAHO for health applications (Cherrington et al., 2020; Pham et al., 2022).

A digital health future

Cutting-edge technologies for health and mental healthcare triggered by the COVID-19 pandemic will assuredly develop further over time, with innovations continually on the horizon. Many questions arise with respect to quality indicators that we must consider for machine learning/AI applications in health and mental health. The future is full of promise, but we have to move forward thoughtfully and transparently to engage in a balanced culture of selective technology adoption (Rajpurkar et al., 2022). Researcher Teodor Zidaru at the London School of Economics and Political Science and colleagues have provided a recent thoughtful scoping review of the importance of patient and public involvement (Zidaru et al., 2021). In this review, the authors outline four principles for involvement of patients and the public (PPI) to achieving design justice in AI (i.e., a fair and equitable system for use of AI in healthcare). These principles are as follows:

(1) Meaningful and authentic public engagement in all areas of AI technologies can be supported and guided by the core principle that such technologies should sustain, heal, connect and empower people and communities.
(2) Substantial ethical concerns, such as inequalities, cultural and population biases, safety, acceptability and broader socio-political issues, could be better understood and moderated if design focused on the concerns of the community over the intentions of the designer.
(3) AI technologies need to emerge from an accountable, accessible and collaborative process that describes how patients and the public have been involved.
(4) AI developers and their designs can be informed by, and contribute to, shared knowledge in design and tools to support design justice.

These principles are critical in considering the future of AI and other cutting-edge technologies in the post-COVID-19 health system evolution. For successful and appropriate integration of AI and other technologies in the health ecosystem, and to reap the enormous potential health benefits and fiscal savings, it is necessary to secure patient and health system consumer and provider "buy-in," for all stakeholders to be fully engaged and honored as essential partners in the design and implementation process. In this way, effective management of the evolving ecosystem of digital health is within reach. With considerations of equity, diversity, and inclusion, many opportunities for benefits to health system effectiveness and sustainability will be insured.

Acknowledgments

The authors wish to acknowledge the rich intellectual input for our discussions provided by colleagues in the Alberta machine intelligence institute (https://www.amii.ca/), the APEC Digital Hub for best practices in Mental Health (https://mentalhealth.apec.org/), and the eMental Health International Collaborative (https://emhicglobal.com/).

References

Antoniou, T., & Mamdani, M. (2021). Evaluation of machine learning solutions in medicine. *CMAJ*, *193*(36), E1425–E1429. https://doi.org/10.1503/cmaj.210036.

Cherrington, M., Lu, Z. J., Xu, Q., Thabtah, F., Airehrour, D., & Madanian, S. (2020). Digital asset management: New opportunities from high dimensional data—A New Zealand perspective. In *Advances in asset management and condition monitoring* (pp. 183–193). Cham: Springer.

Cohen, J. P., Cao, T., Viviano, J. D., et al. (2021). Problems in the deployment of machine-learned models in health care. *CMAJ*, *193*(35), E1391–E1394. https://doi.org/10.1503/cmaj.202066.

Cowell, J., & Harvie, M. (2012). HQCA: Building a credible, transparent, and independent healthcare quality and safety organization in Alberta. *Healthcare Management Forum*, *25*(4), 185–187. https://doi.org/10.1016/j.hcmf.2012.09.011. PMID: 23387136.

Da Silva, M., Flood, C. M., Goldenberg, A., & Singh, D. (2022). Regulating the safety of health-related artificial intelligence. *Health Policy*, *17*(4), 63–77. https://doi.org/10.12927/hcpol.2022.26824.

Grzenda, A., Kraguljac, N. V., McDonald, W. M., et al. (2021). Evaluating the machine learning literature: A primer and user's guide for psychiatrists. *The American Journal of Psychiatry*, *178*(8), 715–729. https://doi.org/10.1176/appi.ajp.2020.20030250.

He, J., Baxter, S. L., Xu, J., Xu, J., Zhou, X., & Zhang, K. (2019). The practical implementation of artificial intelligence technologies in medicine. *Nature Medicine*, *25*(1), 30–36. https://doi.org/10.1038/s41591-018-0307-0.

Hong, S., Liu, Y. S., Cao, B., et al. (2021). Identification of suicidality in adolescent major depressive disorder patients using sMRI: A machine learning approach. *Journal of Affective Disorders*, *280*(Pt A), 72–76. https://doi.org/10.1016/j.jad.2020.10.077.

Insel, T. (2022). *Healing* (p. 336). Penguin Press. ISBN 9780593298046.

Jones, R. (2017). Mens Sana. *The British Journal of General Practice*, *67*(663), 435. https://doi.org/10.3399/bjgp17X692609. PMID: 28963398; PMCID: PMC5604793.

Kahane, K., François, J., & Torous, J. (2021). PERSPECTIVE: The Digital Health App policy landscape: Regulatory gaps and choices through the lens of mental health. *The Journal of Mental Health Policy and Economics*, *24*(3), 101–108.

Kalmady, S. V., Greiner, R., Agrawal, R., et al. (2019). Towards artificial intelligence in mental health by improving schizophrenia prediction with multiple brain parcellation ensemble-learning. *NPJ Schizophrenia*, *5*(1), 2. https://doi.org/10.1038/s41537-018-0070-8.

Kuan, R. (2019). Adopting AI in health care will be slow and difficult. *Harvard Business Review* (2020-04-23) https://hbr.org/2019/10/adopting-ai-in-health-care-will-be-slow-and-difficult.

Lagan, S., Emerson, M. R., King, D., et al. (2021). Mental Health App evaluation: Updating the American Psychiatric Association's framework through a stakeholder-engaged workshop. *Psychiatric Services*, *72*(9), 1095–1098. https://doi.org/10.1176/appi.ps.202000663.

Lee, E. E., Torous, J., De Choudhury, M., et al. (2021). Artificial intelligence for mental health care: Clinical applications, barriers, facilitators, and artificial wisdom. *Biological Psychiatry: Cognitive Neuroscience and Neuroimaging*, *6*(9), 856–864. https://doi.org/10.1016/j.bpsc.2021.02.001.

Liang, S., Deng, W., Li, X., et al. (2020). Biotypes of major depressive disorder: Neuroimaging evidence from resting-state default mode network patterns. *NeuroImage: Clinical*, *28*, 102514. https://doi.org/10.1016/j.nicl.2020.102514.

Liang, S., Wang, Q., Greenshaw, A. J., et al. (2021). Aberrant triple-network connectivity patterns discriminate biotypes of first-episode medication-naive schizophrenia in two large independent cohorts. *Neuropsychopharmacology*, *46*(8), 1502–1509. https://doi.org/10.1038/s41386-020-00926-y.

Liu, Y. S., Kiyang, L., Hayward, J., et al. (2023). Individualized prospective prediction of opioid use disorder. *Canadian Journal of Psychiatry*, *68*.

Machado, C. D. S., Ballester, P. L., Cao, B., et al. (2021). Prediction of suicide attempts in a prospective cohort study with a nationally representative sample of the US population. *Psychological Medicine*, 1–12. https://doi.org/10.1017/S0033291720004997.

Pham, K. T., Nabizadeh, A., & Selek, S. (2022). Artificial intelligence and chatbots in psychiatry. *The Psychiatric Quarterly*, *93*(1), 249–253. https://doi.org/10.1007/s11126-022-09973-8. Epub 2022 Feb 25. PMID: 35212940; PMCID: PMC8873348.

Poncette, A. S., Mosch, L., Spies, C., et al. (2020). Improvements in patient monitoring in the intensive care unit: Survey study. *Journal of Medical Internet Research*, *22*(6), e19091. https://doi.org/10.2196/19091.

Rajpurkar, P., Chen, E., Banerjee, O., & Topol, E. J. (2022). AI in health and medicine. *Nature Medicine*, *28*(1), 31–38. https://doi.org/10.1038/s41591-021-01614-0.

Rauseo-Ricupero, N., Henson, P., Agate-Mays, M., & Torous, J. (2021). Case studies from the digital clinic: Integrating digital phenotyping and clinical practice into today's world. *International Review of Psychiatry*, *33*(4), 394–403. https://doi.org/10.1080/09540261.2020.1859465.

Sawalha, J., Cao, L., Chen, J., et al. (2021). Individualized identification of first-episode bipolar disorder using machine learning and cognitive tests. *Journal of Affective Disorders*, *282*, 662–668. https://doi.org/10.1016/j.jad.2020.12.046.

Sawalha, J., Yousefnezhad, M., Shah, Z., Brown, M. R. G., Greenshaw, A. J., & Greiner, R. (2022). Detecting presence of PTSD using sentiment analysis from text data. *Frontiers in Psychiatry*, *12*, 811392. https://doi.org/10.3389/fpsyt.2021.811392. PMID: 35178000; PMCID: PMC8844448.

Silver, D., Hubert, T., Schrittwieser, J., et al. (2018). A general reinforcement learning algorithm that masters chess, shogi, and Go through self-play. *Science*, *362*(6419), 1140–1144. https://doi.org/10.1126/science.aar6404.

Svensson, A. M., & Jotterand, F. (2022). Doctor Ex Machina: A critical assessment of the use of artificial intelligence in health care. *The Journal of Medicine and Philosophy*, *47*(1), 155–178. https://doi.org/10.1093/jmp/jhab036. PMID: 35137175.

Verma, A. A., Murray, J., Greiner, R., et al. (2021). Implementing machine learning in medicine. *CMAJ*, *193*(34), E1351–E1357. https://doi.org/10.1503/cmaj.202434.

Washington Post. (2020). *Teletherapy is helping Americans get through the pandemic. What happens afterward?*. https://www.washingtonpost.com/lifestyle/wellness/app-anxiety-mental-health-covid/2020/12/24/6ab9eb14-40b8-11eb-8bc0-ae155bee4aff_story.html.

Whalen, D. J., Hennefield, L., Elsayed, N. M., Tillman, R., Barch, D. M., & Luby, J. L. (2022). Trajectories of suicidal thoughts and behaviors from preschool through late adolescence. *Journal of the American Academy of Child and Adolescent Psychiatry*, *61*(5), 676–685. https://doi.org/10.1016/j.jaac.2021.08.020.

Wilimitis, D., Turer, R. W., Ripperger, M., et al. (2022). Integration of face-to-face screening with real-time machine learning to predict risk of suicide among adults. *JAMA Network Open*, *5*(5), e2212095. https://doi.org/10.1001/jamanetworkopen.2022.12095.

Zidaru, T., Morrow, E. M., & Stockley, R. (2021). Ensuring patient and public involvement in the transition to AI-assisted mental health care: A systematic scoping review and agenda for design justice. *Health Expectations*, *24*(4), 1072–1124. https://doi.org/10.1111/hex.13299.

CHAPTER 31

Scoring for Resilient Health: Innovations using technology in physical activity and sport

Erin Barry[a,b] and Judy Kuriansky[a,b,c,d,e,f,g,h]

[a]Department of Clinical Psychology, Teachers College, Columbia University, New York, NY, United States, [b]International Association of Applied Psychology, Indianapolis, IN, United States, [c]Global Council SDG3 (Health), "SDGs in Action" Initiative, Dubai, United Arab Emirates, [d]Office of Ambassador Sidique Wai, Embassy of the Republic of Sierra Leone to the United States, Washington, DC, United States, [e]United African Congress, New York, NY, United States, [f]World Council for Psychotherapy, Vienna, Austria, [g]United Nations Group of Friends of Mental Health and Well-being, New York, NY, United States, [h]Mission of the Republic of Sierra Leone to the United Nations, New York, NY, United States

Introduction

The COVID-19 pandemic with its ongoing and repressive physical restrictions brought heightened awareness to maintaining one's physical health as well as mental well-being. It is well accepted that physical activity contributes to improving one's physical health, but it is becoming increasingly well documented that such activity also boosts mental well-being (WHO, 2020, 2021a). Sport is getting increasing attention as a mode of physical activity, consistent with definitions by the World Health Organization (WHO) and the United Nations Education, Science and Cultural Organization (UNESCO). Specifically, the WHO (2020) defines physical activity "as any bodily movement produced by skeletal muscles that requires energy expenditure." According to UNESCO (2017), sport is "used as a generic term, comprising sport for all, physical play, recreation, dance, organization, competitive, traditional and indigenous sports and games in their diverse forms." These definitions allow the consideration of sport for all as enabling individuals to achieve their full potential of physical and mental health.

The COVID-19 pandemic highlighted the importance of using technological aids in delivering health services to all populations, and specifically to persons in remote settings. Applications of technological tools became increasingly important, and viable, to allow, promote and facilitate healthy physical activity.

Resilient Health. https://doi.org/10.1016/B978-0-443-18529-8.00031-7

In this chapter, we describe the importance of physical activity and sport in achieving the goal of physical and mental health for all, as called for in the United Nations Agenda 2030 for Sustainable Development and in the Political Declaration for Universal Health Coverage, as well as by guidelines from the World Health Organization, and we highlight select, creative technological innovations to progress these goals.

The importance of sport and physical activity

Sport, exercise, and physical activity play a vital role in promoting healthy lifestyles and social development, thereby contributing to the realization of the United Nations Agenda 2030 for Sustainable Development (Lemke, 2016; UNOSDP, n.d.; United Nations, n.d.-a). This includes achieving Goal 3 ("Ensure healthy lives and promote well-being for all at all ages"), and specifically, Target 3.4, "By 2030, reduce by one-third premature mortality from non-communicable diseases through prevention and treatment and promote mental health and well-being" (United Nations, n.d.-c; WHO & ITU, 2021). Participation in physical activity can help prevent and manage noncommunicable diseases (NCDs) as well as improve mental health and well-being by reducing stress, anxiety, and depression in addition to improving self-esteem and self-confidence (Lemke, 2016; WHO, 2020; UNOSDP, n.d.; United Nations, n.d.-a). Recent research snce the pandemic is backing up this association; for example, a study of 230 teachers showed a positive association between levels of physical activity and mental well-being among secondary school teachers in Saudi Arabia (Al-Johani, 2021).

Sport, exercise, and physical activity also promote other SDGs, specifically, SDG16 ("Promote just, peaceful and inclusive societies"); SDG17 ("Revitalize the global partnership for sustainable development"); and SDG11 ("Make cities inclusive, safe, resilient, and sustainable") (Lemke, 2016; UNOSDP, n.d.; United Nations, n.d.-a). This aligns with the aim of the Tokyo 2020 Olympics, as described by its president, Hashimoto Seiko, "to use the Games as an opportunity to show how sport can help fulfill our responsibility to resolving the issues involved in achieving a sustainable society" (IOC, 2021).

Sport promotes inclusivity by transcending age, gender, and culture, and connecting individuals, communities, and diverse cultures (Lemke, 2016; United Nations, n.d.-a). This is evident in the Olympics and the Paralympics, whereby the games demonstrate unity of the global community and values of teamwork, cooperation, friendship, tolerance, and respect (consistent with SDG 11 and SDG 16), and is reflected in the mission of the Olympic Movement "to contribute to building a peaceful and better world by educating youth through sport" (IOC, n.d.-a, n.d.-b) as well as the Olympic Truce which aims at building bridges among countries, addressing SDG 17 (Lemke, 2016; UNOSDP, n.d.; United Nations, n.d.-a). The UN Office on Sport for Development and Peace has long been bringing people together through sport and supporting sport for peace initiatives, from mega sporting events to grassroots activities (Lemke, 2016).

Sport, physical activity, and the pandemic

Engaging in sport has become even more important in the age of COVID-19 pandemic recovery, when lockdowns and mandates about sheltering in place and physical distancing forced people to stay at home for years, resulting in the increase of sedentary behaviors and associated emotional distress (WHO, n.d.-a).

Even prior to the pandemic, physical inactivity levels were of concern. A report of physical inactivity levels across different countries and estimated regional and global trends (Guthold et al., 2018), noted that the 2025 global physical activity target of a 10% relative reduction in physical inactivity will not be met. A report by the World Health Organization (2018a), "Global Action Plan on Physical Activity 2018–2030 (GAPPA): more active people for a healthier world," addressed recommended levels of physical activity and the need for all people to have access to safe and enabling environments to be physically active (WHO, n.d.-b). This GAPPA report set the target to have a 15% relative reduction, using a baseline of 2016, in the global prevalence of physical inactivity in adolescences and in adults by 2030 (Seventy-first World Health Assembly, 2018; WHO, 2018a, 2018b). However, a further WHO report noted that global physical activity levels since 2001 have not improved, and that despite recommendations that adults ages 18–64 engage in at least 150min per week of moderate-intense physical activity, one in four adults globally still had not met this level (WHO, 2018a, 2020; WHO & ITU, 2021).

Impact of the pandemic

The COVID-19 pandemic greatly further hindered forward progress in physical activity as lockdowns postponed or cancelled sporting events and activities, and gyms, pools, and playing fields were forced to close, limiting individuals' access and ability to participate in sport or exercise. Individuals had to either find ways to exercise in the confines of their home or forgo activities altogether (United Nations, n.d.-a).

Reacting to this crisis, WHO held a webinar series, "ReINVENT and ReBUILD," that discussed actions to be taken to make physical activity and sport fairer and open to all. Specifically, their webinar on "Innovation as an accelerator" focused on how digital technology can be used to increase participation and create opportunities in physical activity and sport (WHO, 2021a, 2021b). Innovations, challenges, and recommendations from this webinar are included in later sections of this chapter.

From the discussions during the "ReINVENT and ReBUILD" webinars, WHO released an advocacy brief, "Fair Play: Building a stronger physical activity system for more active people" (WHO, 2021a, 2021b, 2021c). This brief stresses actions the global community needs to take in order to meet targets set in GAPPA, which includes providing supportive environments, accessible programs, sustained communication campaigns in all communities, and partnerships across all sectors (WHO, 2021c).

Potential of digital technology

In spite of all its negative impacts, the pandemic uncovered the importance of accelerating progress to create a more active world for all and also inspired innovations using technology to enable people of all ages to live more active lifestyles. The potential of digital technologies to improve public health, along with promoting Universal Health Coverage and advancing the Sustainable Development Goals, was recognized by governments at the Seventy-first World Health Assembly (Star Health Report, 2018). Recently, digital technology's ability to encourage physical activity was presented and discussed during a health forum at the 10th Global Conference for Health Promotion (Bull et al., 2021).

Adapting to the pandemic restrictions, the fitness tech industry greatly expanded, with new fitness equipment, apps, and online programs. Success of these new approaches was reflected in the increase in sales and use of at-home physical activity equipment, especially by more high-income individuals in the Global North, who purchased expensive at-home exercise machines, like Peloton bikes, Nordic Tracks, Tonals, and Lululemon's Mirrors (Shaban, 2021). Also, wearable technologies such as the Apple Watches and Fitbits escalated their gains made over the last decade as people began to track their step counts (Bala, 2020; WHO & ITU, 2021).

The World Health Organization Global Action Plan on physical activity and digital innovations

The report by the World Health organization mentioned above, the "Global Action Plan on Physical Activity 2018–2030 (GAPPA): More active people for a healthier world," provides member states updated guidance and a model to increase physical activity levels for all. GAPPA provides toolkits, resources about integrating physical activity into primary healthcare, and policy recommendations (WHO, n.d.-b, 2018a, 2018b). A key policy recommendation (4.2) is the development and scaling of innovative digital approaches to promote physical activity, recognizing the significant opportunity to leverage the increased use and affordability of mobile phones, as well as the rapid increase in wearable technologies that can help support people in being more active (WHO and ITU, 2021; WHO, 2018a, 2018b).

In a follow-up to GAPPA, the WHO and the International Telecommunication Union (ITU), as part of the "Be He@lthy, Be Mobile" (BHBM) global initiative, produced a detailed step-by-step handbook for all communities and countries about how to use mobile health technologies to increase levels of physical activity as a means to help meet national targets of good health in Goal 3 of the UN Agenda 2030 and of well-being, specifically in target 3.4 about mental health and NCDs. The initiative addresses noncommunicable diseases, including diabetes, cancer, cardiovascular diseases, and chronic respiratory diseases, as well as communicable diseases. The program specific to physical activity is called "mActive" where the focus is on walking, which is easily carried out by anyone, and requires no skills or cost.

The program "mActive" consists of a 4–6 week behavior change plan of walking every day for 10minutes a day working up to 30minutes, while using a two-way mobile-phone-based, short message service which communicates with the user via SMS text messages and can incorporate multimedia content. The "mActive" handbook includes toolkits that have messaging content that can be adapted to various platforms, such as voice, smartphone applications (apps), or conversational agents. An example of how a chatbot was used for COVID-19 health alerts was included in the handbook to show how multimedia messaging can be used to communicate information. The COVID-19 health alert chatbot included emojis along with specific text and a link to a YouTube video on how to protect oneself from COVID-19. The user was then able to interact with the chatbot by typing a keyword to receive more information. The "mActive" handbook also provides case studies of populations with different health conditions, e.g., diabetes patients in Egypt and Sudan, and tobacco cessation in India. It details research using technology (mobile phones, text messages, smartphones, tablets, Bluetooth technology, and game consoles) and behavior change techniques, including goal-setting, motivation, self-monitoring, identifying consequences of behavior, and identifying barriers and ways to overcome barriers (WHO & ITU, 2021).

Innovations in various countries

Benefits from technological innovations in physical activity have been recognized since the 1920s, when radio calisthenics was introduced in the United States and 15-min exercise routines were broadcast over the radio as a public health initiative (Park, 2020). This approach was also implemented in Japan, dating back 100 years, to the 1920s, when a short exercise routine was aired on radio, which involved activating the entire body with movements using one's own body weight, requiring no equipment, in an effort to improve public health and improve the average life expectancy. Now a century later, the Radio Taiso program still airs daily on Japan's national NHK public broadcasting system, and has even been adapted for video airing on television, and streamed on YouTube (Park, 2020; Sasaki, 2019). The audio version is followed daily by people in parks and is also taught to young children at school.

In Singapore, gamification and wearable technology is being used to nudge Singaporeans to take their health into their own hands and make physical activity part of their daily lives (Bull et al., 2021). In the National Steps Challenge from the Health Promotion Board, a fitness tracker or compatible fitness device with the "Healthy 365 app" tracks steps and logs moderate to vigorous physical activity, rewarding users when they reach milestones of the number of steps taken. These step milestones can reveal a healthy active lifestyle. Results have shown that participants' physical activity level both in volume and intensity improved (Smart Nation Singapore, 2021).

In another effort using technology to rebuild from the pandemic inactivity, Sport England recognized "Five Big Issues" (challenges) that society was facing and launched a 10-year project called "Uniting the Movement" to champion an active mindset and promote the health and social benefits of an active lifestyle (Sport England, 2021). The project put out an open call for innovative solutions to promote physical activity in diverse groups of people. To reach youth, an age group who would naturally respond to using tech tools, one innovation incorporated pop music into tennis lessons (Sport England, n.d.-b). To reach elderly people, Sport England partnered with Screenmedia to create a smart speaker skill ("app") to enable an Amazon Alexa (smart speaker) to have voice-driven capabilities. These capabilities allow elderly people access, through their voice, to interactive content to keep them active, regardless of health conditions that commonly pose limitations for this age group (Sport England, n.d.-a).

Another campaign, "Join the Movement," provided a website with tips on how to be active at home, away from home, with children, and online fitness challenges to motivate individuals to compete against themselves and others. The website includes various apps and online programs to get people moving, including home workouts approved by the National Health Service, a Nike Training Club, and other workout programs such as *Active 10* (an app), *Couch to 5k* (a podcast walking program), and Disney Dance-alongs (kids' dance routines to popular songs) (Sport England, n.d.-c).

A nonprofit organization, London Sport, supported by the mayor of London and Sport England, aims to make London the most physically active city in the world (London Sport, 2019a, 2019b; Mayor of London, 2015). In one initiative, called *Open Active,* citizens can book physical activity programs with the same ease as ordering food or booking an Uber. Another initiative attempts to promote use of a public park (Queen Elizabeth Olympic Park) for physical activity, by collecting open data through providing free Wi-Fi that monitors when the park is being used, which locations in the park are most frequented, and how different areas of the park are used. The city can then tailor the park's services to when the park is being most frequented, to improve the park's services, and optimize the experience for the park goers (Mayor of London, n.d.; Mayor of London, London Assembly, n.d.; OpenActive, n.d.).

This approach was presented in the "Innovation as an accelerator" webinar, part of WHO's "ReINVENT and ReBUILD" webinar series, where Alex Zurita, head of London Sport's Sport Tech Hub, discussed the use of digital marketing to promote physical activity. For example,

Facebook is used to promote walking groups and to reach and target individuals to join walking groups based on an individual's demographic information, obtained from Facebook (WHO, 2021b).

Innovative apps to encourage physical fitness

Technology has enormous power for connectivity. A popular means of using technology is through apps, being used to encourage sports and physical activity. For example, EXi, an app, presented in the "Innovation as an accelerator" webinar by speaker Grace McNamara, Chief Executive Officer at Exi, allows healthcare providers the ability to prescribe exercise for each individual patient's needs, based on data from scientific research and expert knowledge from physiotherapists, public health physicians, medical physicals, and psychologists (EXi, n.d.-a; WHO, 2021c). The evidence-based app, approved by the UK's National Health Service, analyzes the user's health and fitness by tracking health metrics (including resting heart rate, postexercise heart rate, BMI, pain score, mood rating, and a 6-min walking test), and then "prescribing" (suggesting) a personalized physical activity program. The app factors in information about physical conditions the user may have, e.g., diabetes or mental health conditions like depression, so users can exercise safely at the correct intensity level. Users can then customize their own program. Users are encouraged to actively monitor their heart rate with an Apple Watch app during exercise or after a workout by checking their heart rate with their smartphone camera (EXi, n.d.-b). The app is also helpful to health and fitness professionals, who can receive real-time activity and health outcome data about an individual client to assist in devising a care plan.

A holistic approach

The WHO emphasizes the importance of a whole-systems approach to increase physical activity and sport that includes partnerships with the private sector. For example, the WHO has partnered with Google, using the Google Fit app, to reach a wide audience to emphasize the importance of moving and being active and to offer physical activity recommendations (Star Health Report, 2018). Google Fit collects information about the individual's health and activity history to create a profile, centralizes health and fitness data across the users' various apps to monitor their health and activity, and coaches the individual "to a healthier and more active life" with a customizable program for tracking daily, weekly, and monthly goals (Google, n.d.).

Another aspect of a holistic approach is evident in a point widely acknowledged, that technology is not sufficient on its own. All the above-mentioned approaches are impressively innovative and also being shown as effective. Yet, in-person contact and exchange is crucial, aligned specifically with sports in the group context. This point was evident in a United Nations event held in conjunction with the International Day of Sport for Peace and Development

Fig. 1
Youth panel at United Nations International Day of Sport for Peace and Development (2023) event (left to right): Taylor Akason (volleyball); Erin Barry (swimming); Dr. Judy Kuriansky, moderator and their Columbia University Teachers College Professor; Lupeng Fan (badminton).

(United Nations, n.d.-b), on 6 April 2023, sponsored by the Mission of the Republic of Sierra Leone to the United Nations, the flyer and youth panel for which is shown in Fig. 1. Youth graduate students on the panel—all involved in different sports throughout their schooling, including this chapter's first author—as well as the moderator (this chapter's second author) all emphasized the crucial role of interpersonal relations and inspiration from parents and peers, besides tech tools measuring or aiding performance, that were essential in driving their commitment and benefits in sport activities (Kuriansky, 2023).

International support

Sport is becoming increasingly recognized on the international stage as a means to achieve peace and development, the goals of the United Nations Agenda 2030 for Sustainable Development. A UN Group of Friends of Sport has over 100+ government members; April 6 has been declared the International Day of Sport for Peace and Development; major UN agencies (UNICEF, UNESCO, UN Women and UNHCR) launch campaigns; and the United Nations Economic and Social Council (n.d.) hosts ongoing events as do many civil society organizations (Kuriansky, 2023). Further, resolutions, and policy and advocacy briefs highlight the critical intersection between sport, physical activity, SDG achievement, COVID-19 impact, and even climate action, for individual and collective wellbeing.

Challenges in using technology for physical activity and sport

The models described in his chapter emphasize the power of technology to promote and increase physical activity and sport. However, along with the benefits of technology come challenges. The main challenge, acknowledged by the United Nations and WHO, is how to enable access for all people, to bridge the growing digital divide, especially for those in low-resource settings with fragile health and social systems (United Nations, n.d.-a; WHO, 2021b, 2021c). Inequalities driving this divide include ethnic, gender, and socioeconomic discrepancies, which need to be acknowledged and addressed when designing and implementing physical activity and sport initiatives and programs.

Another challenge, also discussed in the "Innovation as an accelerator" WHO webinar, is that data, used in wearable technologies and physical activity and sport reports, suffers from ingrained biases in that, for example, ethnic minorities, females, individualities with disabilities, and individuals with health conditions are not included in a majority of datasets. Future data collection and resulting datasets need to include a balance of ethnicity, gender, disability, complexity of disease states, and location, to ensure that the data reflects the general population (WHO, 2021b, 2021c). Addressing inequalities will ensure that technology reaches more people and engages those who are least active to actively participate in physical activity and sport.

A further challenge is privacy and privacy rights. While health datasets provide valuable insight for statistics and trends, individuals' health information and history need to be kept secure and confidential, following governmental laws and regulations, especially when a camera, video recording, or image is being used. Safeguards need to be put in place, for example, to ensure that 'Zoom bombers' do not enter the digital space. Additionally, children's safety needs to be ensured, since using tech tools like a camera in any device or methods that allow video conferencing can expose them to bullying or worse, child predators.

Recommendations, conclusions, and the way forward

In order to recover from the COVID-19 pandemic, and harness the gains of physical activity and sport, here is a summary of several recommendations culled from sources, and shown in Fig. 2 (Bull et al., 2021; WHO, 2021b, 2021c; WHO and ITU, 2021; United Nations, n.d.-a):

- Create partnerships to promote and increase physical activity and sport. The tech sector needs to consult experts in sectors like health and the environment in order to maximize efforts in creating an active global community. Additionally, governmental, public, and private sectors need to partner together to create inclusive, accessible, active environments.

Fig. 2
Recommendations to promote physical activity using technological innovations.
Source: Chapter authors.

- Use communication strategies to promote innovations in physical activity and sport which apply technology. This can be done by all stakeholders, including media outlets and civil society but also local and national governments.
- Understand "consumer behavior." For example, gamification attracts youth by mimicking the appeal of video games. Additionally, incorporate behavior change techniques such as goal-setting and motivation into digital innovations for physical activity. For example, use fitness challenges and rewards as motivators to reinforce physical activity.
- Focus on at-risk populations who are commonly least active, including ethnic minorities, females, children, elderly, low-income populations, individuals with disabilities, and individuals with health conditions.
- Ensure access for all, using low-tech options such as radio, SMS messaging, and screens in public places. But, also keep up with possibilities of applying higher tech innovations, to ensure progress for certain amenable groups, like youth.

Physical activity and sport strengthen the body, charge the heart, and engage the mind, and also boost mental health and well-being. Despite negative impacts of the COVID-19 pandemic, the emergency also provided opportunities for companies, organizations, and governments to accelerate the promotion and participation in physical activity and sport for the physical and mental health of populations. Multistakeholders as well as physical activity and sport sectors need to continue to adapt and expand creative approaches using technology to meet evolving needs, expectations, and lifestyles of societies, mindful of being inclusive of all peoples. Incorporating new and emerging technology in innovative programs offers vast potential to support physical activity and sport initiatives, encouraging individuals to change inert behaviors, become active, and therefore, physically and mentally healthier. New international policies about physical activity and sport are providing valuable support for campaigns on the national and local level that hold promise for a healthier world.

References

Al-Johani, R. M. (2021). Effect of physical activity on mental wellbeing among teachers of secondary school in Almadina city, Saudi Arabia. *Journal of Family Medicine and Primary Care*, *10*(11), 4264–4271. https://doi.org/10.4103/jfmpc.jfmpc_906_21.

Bala, S. (2020). Wearable device sales have jumped more than 30% this year, exec says. *CNBC*. https://www.cnbc.com/2020/11/20/samsung-wearable-device-sales-are-up-more-than-30percent-this-year.html.

Bull, F., Foo, W. Y., Kumarapeli, V., & Varney, J. (2021). Accelerating action on physical activity for better health and well-being [Conference session]. In *Tenth global conference on health promotion, virtual*. https://10gchp.org/1627/agenda_details.

EXi. (n.d.-a). *About us*. https://exi.life/about-us.

EXi. (n.d.-b). *EXi: Personalised exercise prescription*. https://exi.life/.

Google. (n.d.). Google Fit: Coaching you to a healthier and more active life. *Google Fit*. https://www.google.com/fit/.

Guthold, R., Stevens, G. A., Riley, L. M., & Bull, F. C. (2018). Worldwide trends in insufficient physical activity from 2001 to 2016: A pooled analysis of 358 population-based surveys with 1·9 million participants. *The Lancet. Global Health*, *6*(10), e1077–e1086. https://doi.org/10.1016/S2214-109X(18)30357-7.

International Olympic Committee (IOC). (n.d.-a). What are the Olympic values?. *Olympics*. https://olympics.com/ioc/faq/olympism-and-the-olympic-movement/what-are-the-olympic-values.

International Olympic Committee (IOC). (n.d.-b). What is the Olympic movement?. *Olympics*. https://olympics.com/ioc/faq/olympism-and-the-olympic-movement/what-is-the-olympic-movement.

International Olympic Committee (IOC). (2021). Tokyo 2020 goes beyond carbon neutrality and helps create a more "sustainable society". *Olympics*. https://olympics.com/ioc/news/tokyo-2020-goes-beyond-carbon-neutrality-and-helps-create-a-more-sustainable-society.

Kuriansky, J. (2023). *Making a difference beyond the game: The United Nations International Day of Sports for Development and Peace*. https://blacktiemagazine.com/Judy_Kuriansky/International_Day_of_Sports_for_Development_and_Peace.htm.

Lemke, W. (2016). *The role of sport in achieving the Sustainable Development Goals*. United Nations: UN Chronicle. https://www.un.org/en/chronicle/article/role-sport-achieving-sustainable-development-goals.

London Sport. (2019a). *About us*. https://londonsport.org/about-us/.

London Sport. (2019b). *Using physical activity open data in referrals settings*. https://londonsport.org/our-work/physical-activity-referrals/.

Mayor of London. (2015). *London aims to be the most physically active city on the planet*. https://www.london.gov.uk/press-releases/mayoral/london-aims-to-be-the-most-physically-active-city.

Mayor of London. (n.d.). *Smart park*. https://www.queenelizabetholympicpark.co.uk/our-story/how-we-work/sustainability/smart-park.

Mayor of London, London Assembly. (n.d.). *OpenActive: Open Data for physical activity and sport*. https://www.london.gov.uk/programmes-strategies/business-and-economy/supporting-londons-sectors/openactive-open-data-physical-activity-and-sport.

OpenActive. *About OpenActive*. https://www.openactive.io/about/.

Park, W. (2020). The exercise that keeps Japan moving. *BBC*. https://www.bbc.com/worklife/article/20200609-the-life-long-exercise-that-keeps-japan-moving.

Sasaki, T. (2019). *Rajio Taiso: Japan's National Exercises*. Public Relations Office Government of Japan. https://www.gov-online.go.jp/eng/publicity/book/hlj/html/201910/201910_05_en.html.

Seventy-first World Health Assembly. (2018). *Agenda item 12.2: WHO global action plan on physical activity 2018-2030*. https://apps.who.int/gb/ebwha/pdf_files/WHA71/A71_R6-en.pdf.

Shaban, H. (2021). The pandemic's home-workout revolution may be here to stay. *The Washington Post*. https://www.washingtonpost.com/road-to-recovery/2021/01/07/home-fitness-boom/.

Smart Nation Singapore. (2021). *National Steps Challenge & Healthy365 App*. https://www.smartnation.gov.sg/initiatives/health/national-steps-challenge.

Sport England. (2021). *Uniting the movement: A summary*. https://sportengland-production-files.s3.eu-west-2.amazonaws.com/s3fs-public/2021-01/Sport%20England%20-%20Uniting%20the%20Movement%20-%20Summary..pdf?VersionId=falQKHRJReP8at6MOHYYwg6Y.zgxeFNe.

Sport England. (n.d.-a). *Recover and reinvent*. https://www.sportengland.org/why-were-here/uniting-the-movement/what-well-do/recover-and-reinvent.

Sport England. (n.d.-b). *Positive experiences for children and young people*. https://www.sportengland.org/why-were-here/uniting-the-movement/what-well-do/positive-experiences-children-and-young-people.

Sport England. (n.d.-c). *Join the movement*. https://www.sportengland.org/jointhemovement.

Star Health Report. (2018). Promoting health, far and wide, in the 21st century. *The Daily Star*. https://www.thedailystar.net/news/health/promoting-health-far-and-wide-the-21st-century-1627615.

United Nations. (n.d.-a). *Recovering better: Sport for development and peace: Reopening, recovery, resilience post COVID-19*. United Nations: Department of Economic and Social Affairs: Social Inclusion. https://www.un.org/development/desa/dspd/recovering-better-sport-for-development-and-peace.html.

United Nations. (n.d.-b). International day of sports for development and peace. https://social.desa.un.org/issues/sport-for-development-and-peace/international-day-of-sports-for-development-and-peace.

United Nations. (n.d.-c). *Goals: 3: Ensure healthy lives and promote well-being for all at all ages*. United Nations: Department of Economic and Social Affairs: Sustainable Development. https://sdgs.un.org/goals/goal3.

United Nations Economic and Social Council. (n.d.). *Sport for development and peace*. https://social.desa.un.org/issues/sport-for-development-and-peace.

United Nations Educational, Scientific and Cultural Organization (UNESCO). (2017). Kazan action plan [program and meeting document]. In *Sixth international conference of Ministers and Senior Officials responsible for Physical Education and Sport, Kazan, Russian Federation*. https://unesdoc.unesco.org/ark:/48223/pf0000252725.

United Nations Office on Sport for Development and Peace (UNOSDP). (n.d.). *Sport and the Sustainable Development Goals: An overview outlining the contribution of sport to the SDGs*. https://www.sport-for-development.com/imglib/downloads/unosdp2014-sport-and-the-sustainable-development-goals.pdf.

World Health Organization (WHO). (2020). *Physical activity: Key facts*. https://www.who.int/news-room/fact-sheets/detail/physical-activity.

World Health Organization. (2018a). *Global action plan on physical activity 2018-2030: More active people for a healthier world: At-a-glance*. License: CC BY-NC-SA 3.0 IGO. https://apps.who.int/iris/handle/10665/272721.

World Health Organization. (2018b). *Global action plan on physical activity 2018-2030: More active people for a healthier world*. License: CC BY-NC-SA 3.0 IGO. https://www.who.int/publications/i/item/9789241514187.

World Health Organization. (2021a). *ReINVENT & ReBUILD Webinar series [Flyer]*. https://cdn.who.int/media/docs/default-source/physical-activity/webinar-series-flyer-physical-activtiy.pdf?sfvrsn=c8c7bded_17.

World Health Organization. (2021b). *Physical activity during Covid 19: reINVENT and REBUILD 5th webinar [Video]*. YouTube. https://www.youtube.com/watch?v=INADE-oZx9Y&t=5173s.

World Health Organization. (2021c). *Fair Play: Building a strong physical activity system for more active people*. License: CC BY-NC-SA. https://www.who.int/publications/i/item/WHO-HEP-HPR-RUN-2021.1.

World Health Organization. (n.d.-a). *#HealthyAtHome—Physical activity*. https://www.who.int/news-room/campaigns/connecting-the-world-to-combat-coronavirus/healthyathome/healthyathome---physical-activity.

World Health Organization. (n.d.-b). *Physical activity: WHO response*. https://www.who.int/health-topics/physical-activity#tab=tab_3.

World Health Organization & International Telecommunication Union (ITU). (2021). *Be he@lthy, be mobile: A handbook on how to implement mobile health for physical activity*. Licence: CC BY-NCSA 3.0 IGO. https://www.who.int/publications/i/item/9789240033474.

CHAPTER 32

Is the growing attachment to smartphones and digital media good for our health?

Patricio V. Marquez
World Bank Group, Washington, DC, United States; Centre for Health Economics, The University of the West Indies, Port of Spain, Trinidad and Tobago; Johns Hopkins University Bloomberg School of Public Health, Baltimore, MD, United States; Charles Darwin Foundation for the Conservation of the Galapagos Islands, Puerto Ayora, Ecuador

Introduction

Technological innovation is fueling momentous change in our daily lives. Indeed, as Professor Klaus Schwab, chairman of the World Economic Forum (WEF), warns, the physical, digital, and biological trends underpinning what he calls **the fourth industrial revolution**, are unleashing changes "unlike anything humankind has experienced before" (WEF, 2015). At the center of it all is the internet, anchoring digital media. These changes can be positive but also sometimes negative. The two sides of this issue are explored in this chapter.

What is "Digital Media"?

The WEF defines "digital media" as "products and services that come from the media, entertainment and information industry and its subsectors. It includes digital platforms (e.g., websites and applications), digitized content (e.g., text, audio, video and images), and services (e.g., information, entertainment, and communication) that can be accessed and consumed through different digital devices" (WEF, 2016). Laptops and desktops are still the most widely used devices, but mobile devices, such as smartphones, are fast evolving into newer and newer models and catching up in popularity and use worldwide.

Resilient Health. https://doi.org/10.1016/B978-0-443-18529-8.00032-9

Fig. 1
Diversity of people connected to digital devices. *Source: Stock Photo ID: 392365027. Free download from author's personal subscription.*

Ass depicted in Fig. 1, the easy-to-carry and handle, multifunctional smartphone and other mobile devices such as laptop computers and tablets are a common feature of modern life globally. Billions of people are now connected to these mobile devices, with unprecedented processing power, storage capacity, and access to information. In emerging countries, mobile technology is allowing consumers to leapfrog from "no digital use" straight to "mobile use" (WEF, 2015).

Yet there is growing concern that our attachment to these devices and their content not only captures our full attention and a good share of our time, but is creating virtual barriers that hinder direct social interaction. More serious concerns arise about their potential negative impact on people's mental and physical health.

To learn about this topic, I began searching the web for recent literature using the Google app in my smartphone while waiting at my dentist's office, and then turning to my laptop at home to review and write. A summary of my exploration follows.

Growing popularity and use

In 2022 it is estimated that worldwide there are 6.3 billion smartphone users and 1.14 billion tablet users (buildfire.com, 2022). And, in 2021, there were 218 billion app downloads worldwide (buildfire.com, 2022). Over the 2016–2019 period, the number of app downloads in China has grown by 80% (buildfire.com, 2022). Other emerging markets like India, Indonesia, and Brazil are all seeing high growth rates as well. India's app download growth rate over the same stretch is a whopping 190% (buildfire.com, 2022). This compares to just 5% in the United States.

Smartphone subscriptions are growing at the rate of 17%–19% per year and now total about 2.1 billion users, while consumer mobile internet traffic is accelerating at more than 60% every year (WEF, 2015). Smartphone ownership among young people aged 9 to 16 is also growing; for example, almost half of young people in Europe have a smartphone, according to a 2014 survey of seven countries (Mascheroni & Cuman, 2014). In the United States, 48% of 11-year-olds reported owning a mobile phone, and among 14-year-olds, it was 85% (Odgers, 2018).

Smartphones have become essential to the life of people. It is not surprising to learn, then, that an estimated 87% of US millennials say their smartphone never leaves their side (Ziv, 2022). Similarly, a survey by Qualtrics and Accel Partners found that 79% of US millennials put their phones right by them when they sleep, 53% wake up at least once every night to check them, and that nearly 60% check them before they use the bathroom in the morning. Data reveals that 89% of mobile phone owners use them in social gatherings, to read texts or emails, take photos or videos, or place a call (Ziv, S). Overall, on average, people in the US check their phones 344 times per day—that is once every 4 min—even while driving (despite being dangerous); 35% of people use or look at their cell phones; and on romantic dates, 43% are more wrapped up in their phones than in the partner's company (Wheelwright, 2022).

Social media, which have grown in popularity, are web platforms that allow individuals to create a profile and connect to other people to share information, ideas, interests, and other forms of expression on the site regardless of location and time. Different types of social media platforms include: Social networking (Facebook, LinkedIn, Google+); Microblogging (Twitter, Tumblr); Photo sharing (Instagram, Snapchat, Pinterest); and Video sharing (YouTube, Facebook Live, Periscope, Vimeo).

About 3.2 billion people or 40% of the world's population are estimated to use social media. Worldwide, as shown in Fig. 2, Facebook tops the list with 2.74 billion active users as of January 2022, or almost a third of the world's population, followed by YouTube and WhatsApp.

Mobile app usage varies by age group, but not that much: while youth, ages 18–24 years old, spend close to 2h per day using mobile apps, people between the ages of 45 and 54 spend about 1h and 15min per day using mobile apps, or just 27min less than people in the 25–34 age range (buildfire.com, 2022). Even mobile users over the age of 65 spend close to an hour on mobile apps every day.

Use prevalence among teenagers is growing rapidly. For example, survey data indicate that on average, US teens aged 13–18 engage with digital media (watching television or online videos, reading online, using social media) for more than 6.5h each day; mobile devices account for almost half this time (Rideout, 2015).

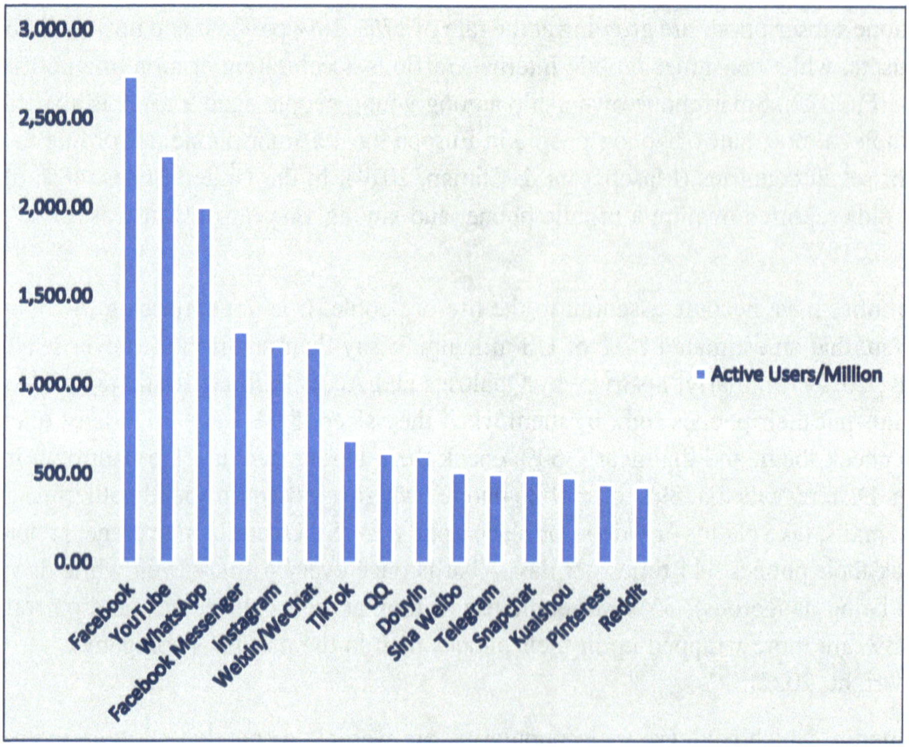

Fig. 2
Social networking sites' active users/million as of January 2022. *Source: Elaborated by author; data from Statista.com, 2022.*

Negative aspects of these trends

Since statistics show that individuals want to be connected more or less 24/7, we should not be surprised that the question many people are asking is: How is all this impacting us, our children, and our social lives? The negative psychological effects of social media are depicted in Fig. 3.

The findings of a 2016 assessment by the WEF, complemented by other sources, expand on these negative effects:

- Excessive digital media use can negatively influence the development of cognitive skills (thinking, imagining, reasoning, remembering abilities) and behavioral skills (reactions or actions in response to stimuli present in our environment) and even mental and physical health.
- Hyperconnectivity has the potential to change patterns of social interaction, as face-to-face interaction essential to the give-and-take functioning of families, communities, and workplaces can be substituted by online interaction.
- Levels of empathy ("understanding another's situation or feelings and experiencing a sense of their emotional state") are declining, perhaps as a result of the decrease in face-to-face communication (Paul, 2010).

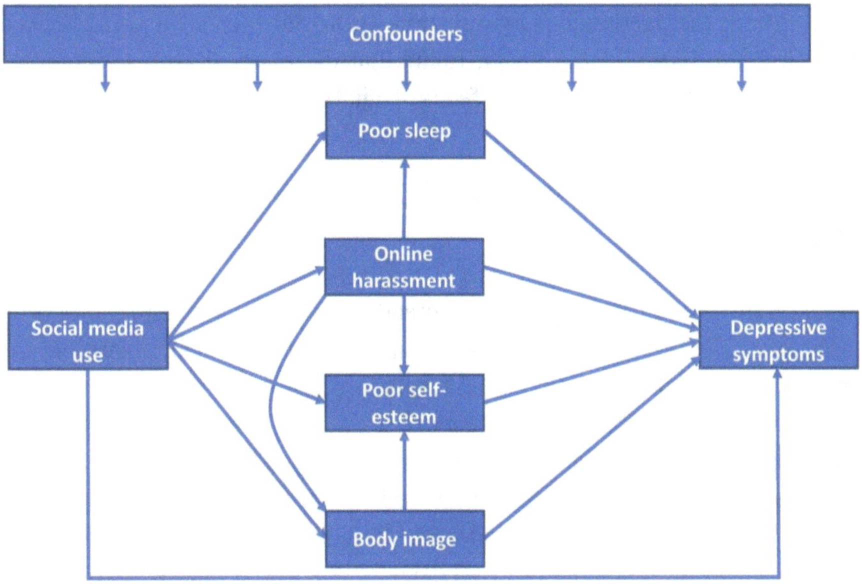

Fig. 3
Hypothesized pathways between social media use and depressive symptoms in young people. *Adopted from Kelly, Y., Zilanawala, A., Booker, C., Sacker, A. (2019). Social Media use and adolescent mental health: Findings from the UK Millennium Cohort study. EClinicalMedicine 6, 59–68. https://doi.org/10.1016/j.eclinm.2018.12.005 (Copyright © 2018 Published by Elsevier Ltd). Open access article under the CC BY-NC-ND license (http://creativecommons.org/licenses/by-nc-nd/4.0/).*

- Digital media consumption may facilitate bullying, harassment, social defamation, and hate speech because of reduction in the potential costs of such behavior and increases in the ease of engaging in behaviors that harm others or ourselves (Mascheroni & Cuman, 2014). Data from a global survey sponsored by Vodafone, a British multinational telecommunications company, shows that of more than 4700 teenagers surveyed worldwide, 50% felt that cyberbullying was worse than face-to-face bullying; about one-fifth reported having been cyberbullied; of those, 41% stated that cyberbullying made them feel depressed, and 18% said it had made them consider suicide (Vodafone/YouGov, 2015). Sadly, two K-pop superstars, Koo and Sulli, took their own lives, having spoken out about cyberbullying (Yi et al., 2019).

The quarantines and school closures during the COVID-19 pandemic led to social isolation and caused social media usage to rise dramatically among children. While one study by Lurie Children's Hospital found that 63% of parents reported that their child is using social media more now than before than pandemic, information from other sources suggests that increased time on social media is contributing to a related increase in cyberbullying, depression and self-harm, sexting, hate speech, trafficking, and social media addiction—for example, L1ght, a company that monitors online hate and harassment, reported a 70% increase in hate speech and cyberbullying during the pandemic (Organization for Social Media Safety, 2021). Also, a recent report from the *Wall Street Journal* found that videos about drugs, pornography, and other adult content regularly appear on the feeds of TikTok accounts purporting to be owned by young children (Barry et al., 2021).

Mental health risks

Mental and substance use disorders are widespread but have often been ignored in the global public health agenda (Marquez, 2021). These conditions affect mood, cognition, motivation, and all aspects of life, including employment, relationships, housing, and personal care, and they represent a growing burden of disease worldwide. The global burden of disease attributable to mental disorders (primarily through years lived with disability and led by depressive and alcohol use disorders) has increased steadily since the 1990s. Mental disorders are among the leading causes of the global health-related burden. The Global Burden of Diseases, Injuries, and Risk Factors Study (GBD) 2019 showed that the two most disabling mental disorders were depressive and anxiety disorders, both ranked among the top 25 leading causes of burden worldwide (GBD 2019 Diseases and Injuries Collaborators, 2020).

After adjustment for the COVID-19 pandemic in 2020, the estimated prevalence of major depressive disorder was 3152·9 cases (2722·5–3654·5) per 100,000 population, equivalent to 246 million (212–285) people, and the global prevalence of anxiety disorders was 4802·4 (4108·2–5588·6), equivalent to 374 million (320–436) people. This burden was high across the entire lifespan for both sexes and across many locations. Increases in the prevalence of major depressive disorder and anxiety disorders during 2020 were both associated with increasing SARS-CoV-2 infection rates and decreasing human mobility (GBD 2019 Diseases and Injuries Collaborators, 2021).

Suicide, frequently resulting from mental disorders, exacts an enormous toll on society. Almost one million people die each year due to suicide, which is the fourth leading cause of death among 15–19-year-olds (WHO Site, 2021a). In some countries, suicide has overtaken complications from pregnancy and childbirth as the leading cause of death among women aged 15 to 49. For every completed suicide, many more people attempt to end their life (WHO Site, 2021b).

Mental health risks of social media use

Research indicates that excessive digital media consumption and the way digital media are being used pose a number of risks to user health and well-being, particularly increased vulnerability to addiction, and harm mental and physical health.

Findings in the 2016 assessment by the WEF identify three potential major mental health risks of social media use:

- Overload: Increased stress because of the volume and variety of social media contact,
- Invasion: Intrusion of media connections into personal life, and,
- Uncertainty: Continuous and unpredictable change in social media applications and requirements.

As I can attest from personal experience, uninterrupted access to digital media content intrudes in a user's work and personal lives, making one oblivious to the difference between business and personal activity, and between time on and time off.

Other potential risks identified in the 2016 WEF assessment, and by other sources, include:

- Increase in "envy-related stress" related to constant updates on friends' latest professional successes, fantastic vacations, and personal triumphs, and comparisons to one's own self-assessed inadequacy.
- Decline in civility, facilitated by electronically mediated interactions that are more anonymous and often more bluntly negative than face-to-face encounters.
- Addictive behavior, evidenced in the fact that video gamers can experience symptoms similar to those felt by people with substance abuse or gambling addiction, which lead to changes in brain chemistry related to the release of dopamine, an important chemical messenger in the brain involved in reward, motivation, memory, attention, and even regulating body movements (Psychology Today Site, 2022). Dopamine released in large amounts creates feelings of pleasure (a "high") and reward, which motivates repetition of a specific behavior.
- Association with depression. Distinct patterns of social media use help explain the association of digital media with symptoms of depression and anxiety. Those clustered as "Wired" and "Connected" in a research study showed increased odds of elevated depression and anxiety symptoms (Shensa et al., 2018). Similarly, a large longitudinal study of Americans aged 14–24 years found that heavy use of the internet and video games is associated with an increase in depression. Findings from the UK Millennium Cohort Study show that the magnitude of association between social media use and depressive symptoms was greater for girls than for boys. Poor sleep, online harassment, poor body image, and low self-esteem appear to be important pathways through which social media use is associated with depressive symptoms in young people (Kelly et al., 2019).
- Impact on cognition. Social media use has been shown to reduce attention span and lower recall rates for information which people believe they can access easily online.
- Reduced downtime. Increased social media use reduces periods of inactivity, which are critical to allow the brain to synthesize information, make connections between ideas, and develop a sense of self.

The previous findings were further corroborated by the those in a *Wall Street Journal*'s expose that reported that own research conducted by Facebook showed that Instagram harms the mental well-being of its teen users ((Wells et al., 2021). But, instead of publicizing the findings and moving to protect these children from harm, Facebook did not made public the research.

Physical health risks

Two major physical health risks from high digital media use can be highlighted:

- **Increased sedentary behavior**. Time spent on digital technology displaces time that could be spent on physical activity. Lack of physical activity has been identified as the fourth leading risk factor for global mortality (6% of deaths globally). Moreover, physical inactivity is estimated to be the main cause for approximately 21%–25% of breast and colon cancers, 27% of diabetes cases, and approximately 30% of the ischemic heart disease burden. Childhood obesity, which is one of the most serious public health challenges of the 21st century, is steadily affecting many low- and middle-income countries, particularly in urban settings. Globally, in 2016, the number of overweight children under the age of five was estimated to be over 41 million. Overweight and obese children are likely to stay obese into adulthood and more likely to develop noncommunicable diseases like diabetes and cardiovascular diseases at a younger age (accessed WHO data).
- **Physical injury.** The use of handheld smartphones is now a major cause of road traffic injuries and fatalities in the world, particularly among young drivers. Studies in the United States show that dialing and texting when driving pose a substantial increase in the risk of being involved in a crash or near crash (cited in Marquez et al., 2009).

Specifically, when driving light vehicles and cars:

- Placing a call on a cell phone made the risk of a crash or near-crash event 2.8 times higher than nondistracted driving.
- Talking on or listening on a cell phone made the risk of crash or near-crash event 1.3 times higher than nondistracted driving.
- Reaching for an object such as an electronic device made the risk of a crash or near-crash event 1.4 times higher than nondistracted driving.

When driving heavy vehicles and trucks:

- Placing a call on a cell phone made the risk of a crash or near-crash event 5.9 times higher than nondistracted driving.
- Talking or listening on a cell phone made no difference to the risk of a crash or near-crash event compared to nondistracted driving.
- Use of or reaching for an electronic device made the risk of a crash or near-crash event 6.7 times higher than nondistracted driving.

- Text messaging made the risk of a crash or near-crash event 23.2 times higher than nondistracted driving (large trucks made up close to 10 percent of all vehicles involved in fatal crashes in the US).

Impact on children's well-being

Studies by the WEF (2016) and others highlighted the following potential negative impact of social media use on children:

- Excessive new media use can lead to attention problems, school difficulties, sleep disorders (particularly for children aged 3–5 years when exposed to violent content), as well as eating disorders and obesity.
- Digital media use is harmful to children if interaction with parents and others is replaced by interaction with digital media. Research has shown that brain development depends on social interaction with others during a critical period in early life and that lack of social interaction creates irreversible social and cognitive impairments throughout life.
- Healthy neurological development requires the engagement of all sensory systems, but heavy digital media consumption favors the visual and auditory systems over the vestibular (sense of balance), proprioceptive (sense of self-movement and body position), tactile (sense of touch), and attachment systems, creating possibly permanent imbalances.
- Delayed language development. Research has shown that very young children (aged 9–18 months) do not learn language by watching educational videos without active interaction with another person.
- Uncontrolled time spent on digital media often displaces time spent on academics, lowering academic achievement.
- Digital media make it easier to disseminate inaccurate as well as accurate information, and much that is online is potentially harmful to younger audiences. For example, a recent *New York Times* article documents how through video games and online chats, sexual predators are making virtual connections with children. The criminals strike up a conversation and gradually build trust, often posing as children, confiding in their victims with false stories of hardship or self-loathing with the goal of duping children into sharing sexually explicit photos and videos of themselves—which they then use as blackmail for more imagery, much of which is increasingly graphic and violent (Bowles & Keller, 2019).

Impact on adolescents

Data show an increase in mental health problems among adolescents globally (Odgers, 2018), representing the second cause of death among youth and young adults 15–29 years old. Worldwide, it is estimated that 10%–20% of adolescents experience mental health issues

(Marquez & Dutta, 2018). In Ecuador, data from the Ministry of Health show that suicides among youth aged 15–19 years accounted for 16% of the total number of suicides registered over the 2012–2018 period (Ministry of Health of Ecuador, 2019).

Adolescents perceive social media as a threat to mental well-being because: (1) it causes mood and anxiety disorders, (2) it is considered a platform for cyberbullying, and (3) its use is often framed as a kind of "addiction" (O'Reilly et al., 2018).

Research based on a large, nationally representative dataset in the US found that more adolescents and young adults in the late 2010s as compared to the mid-2000s were experiencing serious psychological distress, major depression, and suicidal thoughts, and more took their own lives. Researchers posited that this outcome may be associated with lack of sleep and the rise of social media use in teens (Twenge et al., 2019). In contrast, a cross-sectional survey suggests that less screen time and more physical activity are associated with better mental health (Tandon et al., 2021).

Positive impact of social media use in healthcare

We should be clear that access to and use of digital media, while posing risks, is also beneficial to individuals and society. Digital tools and platforms are enabling unprecedented levels of communication, social interaction, and community building across boundaries of time, place, and social context (Seabrook et al., 2016). In fact, 61% of respondents in a Digital Media Survey (WEF, 2016) considered digital media as helpful to maintain existing relationships with friends, and 45% of respondents rated social media as helpful in forming new relationships. Another positive impact is that social media facilitates the rapid sharing of accumulated knowledge and experiences within countries and globally (Panahi et al., 2021).

In the health sector, digital tools and platforms are now becoming an asset for individuals and healthcare practitioners, to help improve access to care and reach vulnerable populations, particularly in low-income settings where healthcare delivery systems are limited. A good example is the initiative by Mobile Technology for Community Health (MoTeCH) and the Grameen Foundation in Ghana, piloted in the Upper East Region, and now being replicated in the Central, Greater Accra, and Volta Regions, which allows women with limited literacy skills to be informed in the local language about the "do's and don'ts" during pregnancy and childbirth (Marquez, 2013). Women also received reminders of clinical appointments, due dates, and required medication and immunization through their mobile phone. This technology also enables women, their partners and families, to recognize the signs of life-threating complications during pregnancy and empowers them to seek immediate care. Other applications allow community midwives and nurses to provide rapid response and care and to follow-up with health service defaulters in the community.

Mobile devices like smartphones also give the public, doctors, and researchers new ways to access help, monitor progress, and increase understanding of mental well-being

(National Institute of Mental Health (NIMH) Website, 2022). For example, text messages can help contact a crisis center. Also, apps can use the device's built-in sensors to collect information on a user's typical behavior patterns, including detecting a change in behavior, and provide a signal that help is needed before a crisis occurs. Some stand-alone apps help improve memory or thinking skills (Briggs, 2020). Others help the user connect to a peer counselor or to a healthcare professional. This support can be very simple but effective. A meta-analysis found, for example, that brief interventions, such as computer-assisted cognitive behavioral therapy delivered through mobile devices, improved people's psychological well-being and reduced reported symptoms of depression and anxiety (Versluis et al., 2016).

Digital media can facilitate education and lifelong learning to build skills of the future, contributing to human capital development in countries.

During the COVID-19 pandemic, digital care options through teletherapy and new apps have seen explosive growth (Doraiswamy et al., 2021). They offer alternative service delivery models that help overcome obstacles that hinder access to care, such as transportation barriers, stigma associated with visiting mental health clinics, personnel shortages, and high costs. These platforms, especially in mobile formats, can offer remote screening, diagnosis, monitoring, and treatment; facilitate remote training for nonspecialist healthcare workers; and enhance online peer-to-peer support and self-care (Staglin & Herrman, 2021).

Other tech innovations include prescription video games that challenge children aged 8 to 12 with Attention-Deficit/Hyperactivity Disorder to focus on multiple tasks simultaneously within the game environment. Also, artificial intelligence (AI) chatbots can help patients practice cognitive behavioral therapy (CBT) to become aware of inaccurate or negative thinking so that they can view challenging situations clearly and respond to them effectively. Further, online symptom tracking apps prompt patients to share data daily which are analyzed with an AI algorithm to identify patterns and alert providers in real time of any warning signs.

Several initiatives deal with scarcity or lack of providers needed to reach people in need (Hijazi & Gornitzka, 2021). The World Health Organization Academy provides digital learning to health workers across the world on a range of topics, including mental health. EMPOWER, a Harvard Medical School initiative, provides digital training to support delivery of psychosocial interventions. And UNICEF's Caring for the Caregiver program targets community workers using online and in-person training.

Looking to the future: A major rethinking is needed

As communities reopen and schools resume in-person classes postpandemic, it is critical to understand that proactive measures are required to help people deal with and cope with significant mental health issues in the postpandemic era.

On one hand, as Marc Berkman, CEO of the Organization for Social Media Safety, has observed, educating and raising awareness among children about of growing social media-related dangers like cyberbullying, trafficking, and fraud is needed to protect children by helping them understand how to avoid or safely react and deal with these threats (Organization for Social Media Safety, 2021).

On the other hand, leveraging social media tools and other online tools can have a positive social impact as these new technologies can be used to expand access to services to address mental and physical health risks. Indeed, as observed during the pandemic, the demand for, and supply of, electronic health applications, both for physical and mental healthcare, grew significantly with good results. Their routine use can make preventative interventions more accessible and decrease barriers to evidence-based treatments—particularly by reducing the still-highly prevalent stigma and discrimination associated with mental disorders—and extend the scope and capabilities of health service providers, all of which contributes to better health and well-being outcomes. But this will only be accomplished through the recognition of the downsides of such technological advances and use, and figuring out ways to minimize these negative aspects.

References

Barry, R., Wells, G., West, J., Stern, J., & French, J. (2021). How TikTok serves up sex and drug videos to minors. *The Wall Street Journal*, 2021. Accessed at: https://www.wsj.com/articles/tiktok-algorithm-sex-drugs-minors-11631052944.

Bowles, N., & Keller, M. H. (2019). Video games and online chats are 'hunting grounds' for sexual predators. *New York Times*. Accessed at: https://www.nytimes.com/interactive/2019/12/07/us/video-games-child-sex-abuse.html.

Briggs, S. (2020). *10 Apps to improve your memory*. Informed. Accessed at: https://www.opencolleges.edu.au/informed/features/10-apps-improve-memory/.

buildfire.com. (2022). *Mobile App download statistics & usage statistics (2022)*. Accessed at: https://buildfire.com/app-statistics/.

Doraiswamy, P. M., Fox, C., & Gordon, J. (2021). *How the pandemic is changing mental health*. Scientific American. Accessed at: https://www.scientificamerican.com/article/how-the-pandemic-is-changing-mental-health/.

GBD 2019 Diseases and Injuries Collaborators. (2020). Global burden of 369 diseases and injuries in 204 countries and territories, 1990–2019: A systematic analysis for the Global Burden of Disease Study 2019. *The Lancet*, *396*, 1204–1222. Accessed at: https://www.thelancet.com/pdfs/journals/lancet/PIIS0140-6736(20)30925-9.pdf.

GBD 2019 Diseases and Injuries Collaborators. (2021). Global prevalence and burden of depressive and anxiety disorders in 204 countries and territories in 2020 due to the COVID-19 pandemic. *The Lancet*, *398*(10312), P1700–P1712. Accessed at: https://www.thelancet.com/journals/lancet/article/PIIS0140-6736(21)02143-7/fulltext.

Hijazi, Z., & Gornitzka, C. (2021). *Digital tools help train providers to fill gaps in mental health care*. Scientific American. Accessed at: https://www.scientificamerican.com/article/how-the-pandemic-is-changing-mental-health/.

Kelly, Y., Zilanawala, A., Booker, C., & Sacker, A. (2019). *Social media use and adolescent mental health: findings from the UK Millennium Cohort Study*. The Lancet. Accessed at: https://www.thelancet.com/journals/eclinm/article/PIIS2589-5370(18)30060-9/fulltext.

Marquez, P. V. (2013). *Preventing maternal mortality—A Yardstick for social justice*. World Bank Group Investing in Health Blog. Accessed at: http://blogs.worldbank.org/nasikiliza/preventing-maternal-mortality-yardstick-social-justice.

Marquez, P. V. (2021). *On International Day of Persons with Disabilities 2021: Let's remember mental conditions, the oft-forgotten disabilities*. Blog post. Accessed at: http://www.pvmarquez.com/node/54.

Marquez, P. V., & Dutta, S. (2018). *On World Mental Health Day: A call to invest in interventions for young people*. World Bank Group Investing in Health Blog. Accessed at: http://blogs.worldbank.org/health/world-mental-health-day-call-invest-interventions-young-people.

Marquez, P. V., et al. (2009). *Confronting "Death on Wheels" making roads safe in Europe and Central Asia*. World Bank Group (WBG). Accessed at: http://web.worldbank.org/WBSITE/EXTERNAL/COUNTRIES/ECAEXT/0,contentMDK:22398460~pagePK:146736~piPK:226340~theSitePK:258599,00.html.

Mascheroni, G., & Cuman, A. (2014). *No children go mobile: Final report*. Accessed at: file:///C:/Users/patri/Downloads/NCGM_FinalReport_Country_DEF.pdf.

Ministry of Health of Ecuador. (2019). *Data provided via personal communication on*.

National Institute of Mental Health (NIMH) Website. (2022). *Technology and the future of mental health treatment*. Accessed on March 31, 2022 at: https://www.nimh.nih.gov/health/topics/technology-and-the-future-of-mental-health-treatment/index.shtml.

Odgers, C. (2018). *Smartphones are bad for some teens, not all*. Nature. Accessed at: https://www.nature.com/articles/d41586-018-02109-8.

O'Reilly, M., Dogra, N., Whiteman, N., Hughes, J., Eruyar, S., & Reilly, P. (2018). Is social media bad for mental health and wellbeing? Exploring the perspectives of adolescents. *Clinical Child Psychology and Psychiatry*. Accessed at: https://www.ncbi.nlm.nih.gov/pubmed/29781314.

Organization for Social Media Safety. (2021). *Organization for Social Media Safety & D.A.R.E. International Announce New Collaboration. Students across the country will receive lessons on growing social media-related dangers like cyberbullying, sexting and human trafficking*. News. Accessed at: https://www.socialmediasafety.org/blog/ofsms-dare-collaboration/.

Panahi, S., Ghalavand, H., & Sedghi, S. (2021). How social media facilitates the knowledge management process: A systematic review. *Journal of Information and Knowledge Management*, *20*, 2150042. https://doi.org/10.1142/S0219649221500428.

Paul, P. (2010). *From students, less kindness for strangers?*. New York Times. Accessed at: https://www.nytimes.com/2010/06/27/fashion/27StudiedEmpathy.html?_r=0.

Psychology Today Site. (2022). *Video game addiction*. Accessed on March 31, 2022 at: https://www.psychologytoday.com/us/basics/video-game-addiction.

Rideout, V. (2015). *The common sense census: Media use by tweens and teens*. Nature. Accessed at: http://go.nature.com/2c4dwew.

Seabrook, E. M., Kern, M. L., & Rickard, N. S. (2016). Social networking sites, depression, and anxiety: A systematic review. *JMIR Mental Health*. Accessed at: https://mental.jmir.org/2016/4/e50/.

Shensa, A., Sidani, J. E., Dew, M. A., Escobar-Viera, C. G., & Primack, B. A. (2018). *Social media use and depression and anxiety symptoms: A cluster analysis*. Am J Health Behav. Accessed at: https://www.ingentaconnect.com/contentone/png/ajhb/2018/00000042/00000002/art00011.

Staglin, B., & Herrman, H. (2021). *Inventive ways of delivering mental health care thrive during the pandemic*. Scientific American. Accessed at: https://www.scientificamerican.com/article/how-the-pandemic-is-changing-mental-health/.

Tandon, P. S., Zhou, C., Johnson, A. M., Gonzalez, E. S., & Kroshus, E. (2021). Association of Children's physical activity and screen time with mental health during the COVID-19 pandemic. *JAMA*, *4*(10), e2127892. https://doi.org/10.1001/jamanetworkopen.2021.27892.

Twenge, J. M., Cooper, A. B., Joiner, T. E., Duffy, M. E., & Binau, S. G. (2019). Age, period, and cohort trends in mood disorder indicators and suicide related outcomes in a nationally representative dataset, 2005–2017. *Journal of Abnormal Psychology*. Accessed at: https://www.apa.org/pubs/journals/releases/abn-abn0000410.pdf.

Versluis, A., Verkuil, B., Spinhoven, P., van der Ploeg, M. M., & Brosschot, J. F. J. (2016). Smartphones are bad for some teens, not all. *Journal of Medical Internet Research*. Accessed at: https://www.nature.com/articles/d41586-018-02109-8.

Vodafone/YouGov. (2015). *Global survey on cyberbullying*. Accessed at: https://markets.ft.com/data/announce/full?dockey=1323-12509304-1Q8SVKBQINRPVIK2IJ53QU6KG4.

WEF. (2015). *The Fourth Industrial Revolution: What it means, how to respond*. Accessed at: https://www.weforum.org/agenda/2016/01/the-fourth-industrial-revolution-what-it-means-and-how-to-respond/.

WEF. (2016). *Digital media and society implications in a hyperconnected era. Section 3: Impact of digital media on individuals, organizations and society*. Accessed at: http://www3.weforum.org/docs/WEFUSA_DigitalMediaAndSociety_Report2016.pdf.

Wells, G., Horwitz, J., & Seetharaman, D. (2021). Facebook knows Instagram is toxic for teen girls, company documents show. *The Wall Street Journal*. Accessed at: https://www.wsj.com/articles/facebook-knows-instagram-is-toxic-for-teen-girls-company-documents-show-11631620739.

Wheelwright. (2022). *2022 Cell phone usage statistics: How obsessed are we?*. Reviews.org. Accessed at: https://www.reviews.org/mobile/cell-phone-addiction/.

WHO Site. (2021a). *Global strategy on diet, physical activity and health*. Data accessed at: https://www.who.int/dietphysicalactivity/childhood/en/.

WHO Site. (2021b). *Suicide*. Accessed at: https://www.who.int/news-room/fact-sheets/detail/suicide.

Yi, H. Y., Cha, S., & Park, M. (2019). *Cyber bullying, star suicides: The dark side of South Korea's K-pop world*. Reuters. https://www.reuters.com/article/us-southkorea-kpop/cyber-bullying-star-suicides-the-dark-side-of-south-koreas-k-pop-world-idUSKBN1Y20U4.

Ziv, S. (2022). *7 Scary cell phone facts that'll make you question your addiction*. Accessed on March 31, 2022 at: https://www.themuse.com/advice/cell-phone-addiction-stats-scary.

Further reading

Marquez, P. V. (2014). *Diabetes and overweight: A growing global burden*. World Bank Group Investing in Health Blog. Accessed at: http://blogs.worldbank.org/health/diabetes-and-overweight-growing-global-burden.

Marquez, P. V., et al. (2016). Summary report: Out of the shadows: Making mental health a global development priority. In *Event convened by the World Bank Group and the World Health Organization (WHO), with the support of an Organizing Committee and the collaboration of international partners, and held in Washington, DC on April 12-14, 2016, as part of the WBG-IMF Spring Meetings*. Accessed at: http://pubdocs.worldbank.org/en/391171465393131073/0602-SummaryReport-GMH-event-June-3-2016.pdf.

Patel, V., et al. (2012). Suicide mortality in India: A nationally representative survey. *The Lancet*, *379*(9834), 2343–2351. Accessed at: https://www.sciencedirect.com/science/article/abs/pii/S0140673612606060.

Shukla, A. (2019). *The effect of social media on mental health and well-being*. Cognition Today. Accessed at: https://cognitiontoday.com/2019/08/effect-of-social-media-on-mental-health-well-being/.

Spelman, M. (2015). *What does the future hold for the internet?*. World Economic Forum. Accessed at: https://www.weforum.org/agenda/2015/12/what-does-the-future-hold-for-the-internet/.

PART 3

The role of stakeholders

In order for innovations to be effective and efficient for resilient health systems, various stakeholders need to define their role, and then come together, cooperate, and collaborate in an ecosystem.

Stakeholders are individuals, groups, organizations, or systems who affect or are affected by an action or event, and who become involved in a common mission, and bring their idiosyncratic goals together for a common exploration and impact.

Partnerships are relationships created by people or groups with mutual interests who come together to address problems and achieve mutually defined goals.

This Part explains the various stakeholders in the health ecosystem and their role in building resilient health systems. These actors include governments, with examples from different world regions, e.g., the United Arab Emirates, Kenya, Sierra Leone, Estonia, Mexico, and Israel; United Nations agencies like the WHO; regional bodies and multilateral organizations; the private sector; local and international nongovernmental organizations; and civil society, including citizens (in a unique approach in Kuwait) and the voice of persons with lived experience; as well as a series of multistakeholder healing interventions in Nepal.

Captivating chapters describe government innovations like Estonia becoming the European Silicon Valley, what Kenya and Sierra Leone are creating, and how Mexico is raising the profile of mental health at the United Nations in first-of-its-kind resolutions in the Security Council and the General Assembly. Adding to a chapter on unique approaches in the United Arab Emirates—an innovation hub of the world—is a profile of a unique visionary innovator, H.H. Sheikh Mohammed bin Rashid Al Maktoum, Ruler of Dubai. To address the mental health needs of the vast staff of the United Nations, the Mental Health and Wellbeing Team launched a web portal, with features to address workplace issues triggered by the COVID-19 pandemic and surveys of staff stress.

Multilateral organizations, and local international civil society, the private sector, and citizens themselves are key stakeholders who bring essential resources, creativity, energy, and capacity to the table. All these together create multistakeholder models crucial to bring health to populations and scale their efforts. An example of a multistakeholder project during COVID-19 is given, where government, NGOs, youth, and media collaborated in webinars over a period of time, to boost resilience, honor programs, and evaluate policy.

In all situations, it is crucial to be alert to, and reduce, stigma and discrimination, as is emphasized in a chapter that reviews the Lancet Commission Report in this topic.

All efforts are oriented toward achieving the United Nations Sustainable Development Goal 3 on Health and Well-being, with effective approaches involving partnerships. Also, the NEXUS model of the UAE's "SDGs in Action" initiative is described in a chapter that closes this Part, that connects the various Sustainable Development Goals, as well as the cross-cutting issue of the important role of youth, to bring all the ideas together and provide inspiration.

What notables say

> *If everyone is moving forward together, then success takes care of itself.*
> ***—Henry Ford, American industrialist and founder of the Ford Motor Company.***

> *If you want to go fast, go alone, if you want to go far, go together.*
> ***—African Proverb***

CHAPTER 33

Role of governments in building resilient health systems

Rajeev Sadanandan
Health Systems Transformation Platform, New Delhi, India

Introduction

Regardless of the nature of national health system, governments are expected to manage health emergencies. Capacity to respond to an emergency takes time and effort to build, which is difficult to do during a crisis. Therefore governments have to build a health system capable of managing a health emergency before the crisis hits the country. Since the occurrence of health emergencies such as Severe Acute Respiratory Syndrome (SARS) and Ebola, resilience has been identified as one of the attributes that health systems need in a crisis. Resilience, a concept that emerged from engineering, ecology, and psychology and has since then been adapted to social systems, is defined in many ways (Reghezza-Zitt et al., 2012). Here, we use resilience to describe the ability of the system to assimilate the impact of a shock and continue to deliver its core functions.

Resilience may be achieved through "absorption" where the system uses existing or newly created slack to absorb the additional load, "adaptation" where existing resources are reinvented for purposes they were not meant for, or "transformation" when new ways of healthcare delivery are developed to respond to the emerging situation (Barasa et al., 2017). The first two options, used when the health emergencies are within anticipated or manageable limits, will become ineffective when health systems are loaded beyond their capacity. Transformative changes, which will be required if the crisis reaches unforeseen dimensions, are difficult to plan in advance. Therefore while governments can build additional capacities for absorption, and prepare operating procedures anticipating adaptation, training a system for transforming itself to respond to unforeseen challenges requires cultivating a culture that is supportive of innovation and adaptability in the health system. This chapter examines how governments can improve resilience of health systems using each of the options.

Resilient Health. https://doi.org/10.1016/B978-0-443-18529-8.00033-0

All systems are designed for a defined set of parameters which have evolved and been tested over time. While systems carry some degree of redundancy to handle surges in demand, it is not feasible, given resource constraints and opportunity costs, to carry large excess capacity in anticipation of a rare event (Hick et al., 2004). Similarly, health systems and processes are standardized to give predictability and stability to operations. An essential condition in normal operations, these standardized systems and processes could stand in the way of flexibility and adaptation needed to respond to challenges during a crisis. Since the world has seen many potential and real pandemics in the past, and expects to see other pandemics in the future, planners have to balance the needs of normal operations with ability to respond to crises whose dimensions cannot even be anticipated. This requires moving away from linear planning to encouraging cultivation of innovative approaches. The tension between stability and innovation could be debilitating for governments whose systems are dependent on predictability. A way out may be to cultivate both stability and innovation and mandate a tipping point, either side of which each approach will be deemed legitimate. In other words, preparing for the "nonnormal" has to become part of the "new normal."

Most frameworks used to describe health systems characterize them as a set of inputs, intermediate outcomes, and goals. The WHO framework of health systems, for instance, has service delivery, health workforce, information, medical products, vaccines and technologies, financing, and leadership/governance as the building blocks (World Health Organization, 2007). These are the hardware which resilience needs. But resilience also needs software such as values, including equity, and a culture that promotes innovative thinking and willingness to respect and adopt the knowledge gained by frontline workers while handling the shock (Barasa et al., 2017). In order to promote resilience in health systems to handle supra normal shocks, governments have to nurture both the hardware and the software. We will use the WHO framework to structure the discussion on health system resilience.

Organizing delivery of healthcare services

The health sector bore the brunt of the impact of COVID-19 and will continue to be under strain from its sequelae. This has raised fundamental questions and challenges about the organization of healthcare delivery in many countries. These include: not enough government spending to achieve universal health coverage, neglect of public health and epidemiology, inequitable access to healthcare, inability to ramp up testing facilities, and lack of community outreach workers capable of mobilizing a social response.

In the early days of the current pandemic, governments focused on strengthening specialty hospitals. But as the infections raged, the need for prevention efforts to stem the flood became obvious. Many countries did not have the primary healthcare system or workforce to manage essential prevention activities, including an adequately resourced primary care system which people trust, and which can guide patients on personal protection; infection control; monitoring of symptoms; support surveillance; tracking and quarantining of infected persons and their contacts; and managing patients at home until their condition required specialist care (Kearon & Risdon, 2020). With regard to treatment, a hierarchical health system founded on a primary care system and integrated with secondary and tertiary hospitals can optimize resources by managing patients at the lowest possible level, moving them to higher levels as needed. This requires good communication and transport systems and, in future, will be aided by digital decision support systems. Strengthening of primary healthcare and redesigning curative care will strengthen health systems even in normal times and will add efficiency to managing a crisis.

The important role of technology

Virtual interactions, which came into their own during the current pandemic, are likely to play a prominent role in future. Virtual consultations have emerged as a viable option for safe interface between patients and the health system. With government stewardship of the regulatory systems, the data generated from the virtual interactions, linked together on longitudinal electronic health records, can be used for machine learning, which has the potential to improve the quality of care. When this data is linked to genomic information, it can contribute to personalized care further optimizing outcomes. Virtual first interactions could direct patient management flow, considerably reducing costs. Reference protocols, which are needed to guide the clinical management pathway after the virtual consultation will improve with learning from deployment. They will be backed by rapid improvements in remote diagnostics, artificial intelligence, and e-pharmacy (where drugs are delivered at home based on e-prescriptions). Refining these pathways takes considerable time, data, and continuous ground truthing from current best practice. All these steps need to start now. But regulators have also to be cautious against potential for inequity from access to technology, abuse of data privacy, and current defects getting ingrained into the newly developing systems (Komorowski & Celi, 2017).

Hospital infrastructure

Any increase in demand for healthcare, beyond the planned surge capacity, will need more built-up space. For an infectious disease, additional space is needed for isolation of contacts and to protect health workers. During the current pandemic, some governments put up Greenfield Hospitals, which are newly constructed stand-alone facilities (e.g., Fangcang hospitals in Wuhan). But more common were other buildings repurposed as hospitals (e.g., the NHS (UK) Nightingale hospitals). In future, rather than repurposing space on an ad hoc basis, government should plan building for multiple use: space that would not be used during a pandemic (e.g., convention centers, cruise ships) can be planned and built to accommodate the needs of a hospital or isolation centers when needed. In order for the buildings to be appropriate for all their intended uses, all potential users have to work with architects and engineers and with each other to design the building to meet their different needs.

Health workforce

Human resources for health remain the most crucial input in health: one study of 43 countries estimated their remuneration to be 68.9% of the total health expenditure (Hernández et al., 2006). While the workload of healthcare workers will go up manifold from infections, their numbers and output could come down from infection to themselves or their family members. An infectious disease generates multiple stressors for healthcare workers, which could lead to burnout. Top stressors include fear of contracting infection, spreading it to family members, and having to handle cases which exceed their competence. The most usual protective mechanisms against stress—family and friends—become inaccessible due to isolation. Support workers and neighbors ostracize health workers because they treat infected persons (Bradley & Chahar, 2020). Even if healthcare workers remain functional during the epidemic, the long-term impact of their shifts often leads to high levels of attrition, and reluctance of people to enter the healthcare industry for fear of a future pandemic. So, building resilience also requires integrating strategies to equip healthcare workers to handle stressors for long periods of time. Techniques used by other sectors to train personnel to handle stress (e.g., soldiers) could be adapted for healthcare workers.

However, in situations equal to, or more intensive than, the current pandemic, these adaptive techniques may prove inadequate. There is no substitute for increasing the number of persons to handle the increased patient load. The health workforce is governed by strict conditions of entry and accreditation, making it difficult to ramp up the number of healthcare workers in the short term. Governments may have to consider drawing up a new set of accreditation regulations to deal with the needs of health emergencies of catastrophic proportions. The most obvious option is task-shifting. Initially in a new pandemic, only supportive care may be possible. Training curricula can be developed and rolled out when needed, to equip health workers of different cadres (e.g., medical or nursing students) to carry out most aspects of supportive care, the generic version of such training being supplemented with disease-specific knowledge and skills when they become needed and available. Most of the knowledge component and some of the skills can be delivered virtually, and successful candidates approved to carry out tasks they are certified to do. If the demand goes up further, lay volunteers will need to be trained and engaged. Caution is needed, however, since entrusting tasks that normally take years of study to persons who have undergone a truncated course is fraught with danger; thus, should be adopted only at the direst need. Also, the conditions under which tasks for such engagement are to be approved, as well as the training and certification modalities, need to be widely discussed and written into regulations governing practice of health. These should be set during "normal" times, so that managers and workers are protected from litigation once the emergency is over.

Treatment protocols

Diagnostic kits, drugs, and vaccines are normally approved only after a rigorous and conservative testing and assessment. This may not be possible in an emergency such as a pandemic; thus, health systems need to have systems in place to incorporate innovations as they become available. The world benefited by the initial reports of management of COVID-19 patients published by clinicians in Wuhan and later in Italy. Many of these practices were later amended or dropped based on experience. During the emergency, scientific papers for journal publishing were expedited, preprints were avidly read, and access to most publications made free. For future pandemics, we need a more organized method of developing, validating, and sharing clinical and pharmaceutical knowledge. While randomized control trials (RCTs) and standard testing practices remain the gold standard, a different set of guidelines should be agreed upon to be invoked during an emergency (Wiegmann et al., 2021). If such use is not protected by regulation, any adverse fallout (which are possible when speed is the overriding consideration) could lead to costly litigation and discourage innovation in treatment in future health emergencies. Systems developed for "tentative governance" of emerging technologies, such as genetically modified crops and nano technologies, which have been grappling with similar challenges, offer a template that can be used by policymakers to create a system for an expedited, yet cautious procedure for emergency approval of diagnostics and treatment (Kuhlmann et al., 2019).

Supply of medical products

In today's economy, all products, including medical products, are produced by specialist suppliers and made available to consumers through a global supply chain. The COVID-19 pandemic demonstrated the fragility of this arrangement. There was a frenzied scramble for Personal Protective Equipment and ventilators. Faced with a global shortage, supplier countries prioritized their needs or used supplies to push their geopolitical agenda. Health systems of receiving countries came under strain and smaller economies were priced out of the market or had to accept substandard products. In future, most governments will try to risk-proof their supply chains by setting up their own manufacturing capacity or working out more dependable mechanisms such as greater geographical spread, reciprocal arrangements or stockpiling of bufferstock.

COVID-19 was a triumph for science but also a lesson in missed opportunities. Once the genome was sequenced, test kits, vaccines, and treatments were developed at unprecedented speed. These were built on platforms developed from past research. But, had the world prioritized the development of vaccine against other viruses which had killed many in Asia and Africa, vaccines against COVID-19 would have been available much earlier, a fact pointed out by researchers ruing the lack of research funds (Gilbert, n.d.). Similar is the story about antiviral medications for viruses that continue to affect poor populations. It is important not to lose the momentum gained in the past two years and instead, to build on them to develop vaccines and drugs against the spectrum of viruses, so that they can be tweaked to deal with an emergent virus. Most of the lead effort in developing vaccines and drugs were taken by private, for-profit organizations, unlike in the past when research was dominated by public funding. Private funding accelerated the availability of vaccines but had serious implications to global equity. Global health came to be positioned less as a global public good but more an opportunity to corner monopoly profits. In spite of global efforts to make vaccines available to all nations, "low-income countries are still at the back of the line for delivery and will likely be waiting until 2022 before they are able to cover even vulnerable populations" (COVID-19|Launch and Scale Speedometer, n.d.).

We must not take a myopic approach to global health. In a pandemic, "no one is safe till all are safe." If entire nations, especially those in developing regions with fewer resources, are left without access to vaccines and drugs, the disease will rage uncontrolled there and spread elsewhere. With an increased number of serious infections, there is greater probability of mutations and variants with ability to evade the immune system, which can then spread from the developing countries to more developed nations, and fan the epidemic back to life, negating the economic or political benefits of vaccine nationalism (Asundi et al., 2021). The Omicron variant had demonstrated this dramatically.

Governments in high-, middle-, and low-income countries need to work around legal, technical, personnel, and financial issues to make vaccines available in low-income countries, especially in preparation for any future pandemics. This will involve more than production, but also supply chain, regulation, and training of the workforce. In the short term, a strategy that encourages reasonable profit to the developer but also makes drugs and vaccines available to poor countries could be worked out. A pooled procurement system, with products being differentially priced based on ability of countries to pay, could be developed to achieve this. COVAX offers a model that can be further strengthened and made permanent. The long-term solution would be to invest in capacity for vaccine production and delivery in underserved countries. This will help their routine vaccination and keep them ready to roll out vaccine for a new pathogen as soon as it is available. Governments have to realize that protecting their people from a future pandemic also implies building capacity of other governments to protect their people.

Equity

Besides the issues between countries enumerated above, problems emerge within countries involving inequity of exposure to infections and access to testing and treatment, which was starkly evident during the COVID-19 pandemic. Only assuring universal health coverage—one of the targets of the United Nations Sustainable Development Goals to which all nations are committed—can ensure that this inequity is not repeated in future pandemics. In order to free up resources to meet the surge in demand, some services have to be curtailed. If there is no explicit policy on prioritization, given the inequity in the political economy of decision-making, there is a high probability that services which are not emergent but have long-term consequences and high equity payoffs (such as child immunization, antenatal examinations, treatment for tuberculosis, screening and management of diabetes, hypertension, and mental health) may be given up in favor of services for which there is an explicit demand from knowledgeable and powerful segments. Protection of the interests of patients from low social and economic groups needs to be explicitly acknowledged when planning for resilience, as these groups have less information and power to demand access, disproportionately borne the brunt of reduction in services, and also lack of access to diagnosis and testing.

Information

Systems for collecting, analyzing, and sharing information which proved invaluable in the COVID-19 response will be crucial in managing future health shocks. If a pathogen is detected early, the world can take measures to deal with it before an outbreak becomes a pandemic. Like some pathogens that led to real or potential pandemics in the past, the next one may be a zoonotic virus, namely, a cross-over from wildlife. So, surveillance has to include animal hosts and environmental markers (including sewage) and clinical cases. But surveillance in the One Health—One World mode (a multisectoral approach which recognizes the interconnection between people, animals, plants, and their shared environment)—has not caught the imagination of health planners in many countries. It is in the interest of the world to invest in systems to detect and diagnose new or emerging pathogens, including disease X (WHO's designation for an unknown pathogen with potential to cause a catastrophic epidemic) even in poor countries who cannot afford it now. With global connectivity, no place in the world is safe from pathogens that emerge anywhere, and national surveillance needs to feed the data into international networks so that all countries are alerted. Based on data from recent outbreaks that spread to other countries, predictive models of the spread have been developed. These can calibrate the national response in the initial days of the outbreak, helping to avoid the knee jerk reactions that have often characterized the response to COVID-19. This will be aided by technology tools. But the tendency to punish nations with good surveillance and prompt notification, while others that delayed notification are not held accountable, sends the wrong message to governments. Governments have to arrive at an agreement on obligation of nations to international surveillance and agree on systems to finance and reward compliance, and punish default.

The ad hoc and opaque decisions during the current pandemic often led to resistance. Instead, decisions on social restrictions should be based on evidence from analysis of data to make them more efficient and acceptable. If the processes that are driving infections could be identified from analysis of data or from controlled social experiments, the decisions on restrictions would be more objective and easier to defend (Starr, 2020). If the process of arriving at trigger thresholds for imposing or easing social restrictions is established in normal time, they are likely to be respected during an emergency.

Managing the infodemic

Every pandemic will be accompanied by an infodemic, and the growth of social media has given it unprecedented reach. Widespread dissemination of information had good and bad results during the COVID-19 pandemic. Governments have to learn to deal with infodemics and to confront wrong and mischievous campaigns, using the same techniques used to spread false information (Tangcharoensathien et al., 2020). During a crisis, people demand information and will accept accurate and prompt information from trusted sources. Where they trust governments, official sources will be the best option to provide information. However, during this pandemic, there was high level of skepticism of government. Such skepticism, based on past experience, cannot be removed during a crisis. Reputable medical, scientific, or academic sources, who are allowed to tell the truth even if it is inconvenient to government, need to be entrusted with the task. COVID-19 also saw the emergence of alternate sources of information that won the trust of many. These were managed by academic institutions, research groups, media houses, or volunteers. Where government face trust deficits, they should consider working with such sources for communicating important information and countering false campaigns.

Use of technology in future pandemics

In an epidemic mediated by human-to-human transmission, primary and secondary contacts of infected persons have to be identified, tracked, quarantined, and observed to reduce the pace of the epidemic. This process is highly human resource intensive and cannot be sustained when the number of infected persons crosses a threshold. Similarly, predicting which interactions have high probability to increase infections requires collection and rapid analysis of large amount of data. These strategies can be facilitated by technology. The data collected could feed artificial intelligence to develop predictive models, which can continue to be refined over time, proving useful information to modulate policy response. But these raise sensitive questions regarding data privacy, data ownership, and use. Governments have to resolve these issues, in consultation with activists and civil society representatives, to arrive at a compromise between concerns about individual liberty and public health.

Compulsions of the pandemic led to technology redefining modes of providing healthcare, including virtual consultations and remote diagnostics. Financing systems were compelled to accommodate them. This process will only be accelerated after the current pandemic and will continue to redesign health systems. There is a danger that the agenda could be hijacked by any of the interest groups, such as technology providers or payers. To prevent this, governments have to play a stewardship role to track the trajectory of change and ensure that all stakeholders, especially patients, are consulted and their interests protected in the new order that will emerge from the technological disruption of health services that is underway now.

Financing

Health systems are financed in different ways, including tax funding, social health insurance, voluntary health insurance, and out-of-pocket payments. Many health systems, especially in low- and middle-income countries, fund healthcare from out-of-pocket spending at the point of consumption. This tends to limit healthcare to curative care and underfunding investments in areas important for managing future pandemics, such as surveillance, testing, and vaccination. During the current pandemic, the excess expenditure needed for the pandemic response was met by additional authorizations. Since the ability of people to pay also decreased due to the severe economic impact of the pandemic, the cost of treatment was funded by government in many countries. Individuals' earnings are likely to be further reduced by the postpandemic economic downturn, further reducing the capacity of people to pay for healthcare. For governments keen to ensure that health of the population is maintained, this situation presents an opportunity to shift the financing of healthcare away from out of pocket to a prepaid, preferably tax-funded financing system for universal health coverage (Kurowski et al., 2021). This will ensure that financing is not a constraint for the poor to access testing and treatment. It will also fund prevention efforts. However, the economic downturn has reduced the spending power of low-income countries and even maintaining current health expenditure may become a challenge for them. To reduce the impact of the fiscal stress on the health sector, and to build capacity to manage future health shocks, high-income countries have to maintain external assistance targeted to the health sector. The value of such assistance for donor countries was proven during the current pandemic.

Leadership

During a crisis, governments benefit from the tendency to "rally around the flag," at least in countries where governments are trusted. But as the epidemic lingers on, and the economic and social impact of the restrictions and of illness and death begins to be felt, this begins to wane, moving to outright opposition if the pandemic lasts too long. This process is difficult for any government to manage, since the response that is appropriate in one scenario could be counterproductive when it changes. This applies to all sectors including health. Since health systems are a constant factor in the response, even if the disease becomes endemic, the task of political and health system leadership is to redefine their role as the situation changes, in order to enable them to perform their role effectively.

Political leadership

Public health measures needed to control the spread of pathogens will not work without intersectoral coordination, as most of the determinants of the spread of disease fall outside the health sector. Intervening with them could have negative impact on the economy and political leadership has to weigh the costs and benefits of such interventions. Based on their assessment of political costs national leaders could refuse to acknowledge the potential impact of the pandemic and the need for unpopular restrictions, thus missing the opportunity for timely intervention. The willingness and capacity of national and/or provincial leadership to provide effective stewardship of the health sector for coordination between public and private sectors, public health and curative services, institution- and community-based services, and between hospitals at different levels will also influence how health sector functions. Governments also need to steward the regulatory changes needed to make health systems resilient and to enable differential approaches described before in this text.

Since public health measures inevitably involve restrictions in social and economic activity, these may be resisted if people do not have trust in their government. Only if they are convinced that the restrictions are based on evidence, implemented transparently and justly, and that government will mitigate their hardship, will they comply with the restrictions. For instance, if self-quarantine is required, persons will not isolate themselves unless their means of survival are guaranteed. Since feeding their family members is more important than cooperating on disease control measures, breadwinners will try to hide their disease status to earn their livelihood, with disastrous consequences on infection rates. Social policies would thus influence effectiveness of public health measures (Greer et al., 2020).

Health system leadership

As each unit of the health system, especially hospitals, come under stress during a pandemic, quality of leadership will determine their effectiveness. In addition to the usual staple of knowledge and skills in areas such as strategy, operations, and finances, resilience also requires the leader to have a high level of self-awareness and emotional intelligence to understand and regulate one's own and other's response to the crisis, ability to motivate others and resolve conflicts, openness to out-of-the-box solutions, as well the ability to take decisions under uncertainty and stress. Governments have to develop and empower such leadership during normal times and also encourage a culture that supports innovative solutions. Many of the recommendations discussed in this chapter about making health system resilient can also be incorporated into leadership development programs. Any crisis can only be managed by leaders who are adequate to the challenge. Such leaders are not born but carefully nurtured.

Conclusion

Some characteristics of a future crisis can be anticipated and prepared for. Similarly, health systems can be readied to make the possible absorptive and adaptive changes required to deal with foreseeable changes. But a future pandemic could be of a totally unanticipated nature. It will be difficult to make hardware changes to address its entire fallout. The determinant factors of a successful response by the health system may be the quality of its values, culture, and leadership. Governments obviously play a critical role in a health emergency like the COVID-19 pandemic and one of their crucial responsibilities is to ready to system in normal times to handle health emergencies. To cultivate resilience in the health system, governments need to create a culture that encourages innovation, transparency, learning from experience, flexibility, equity in access to services, and collective and supportive leadership. These values energize teams and make them capable of handling challenges even when odds are stacked against them. A government that understands this and recalibrates their approach as the situation changes is the ideal one to build resilience in the health system.

References

Asundi, A., O'Leary, C., & Bhadelia, N. (2021). Global COVID-19 vaccine inequity: The scope, the impact, and the challenges. *Cell Host & Microbe, 29*(7), 1036. https://doi.org/10.1016/J.CHOM.2021.06.007.

Barasa, E. W., Cloete, K., & Gilson, L. (2017). From bouncing back, to nurturing emergence: Reframing the concept of resilience in health systems strengthening. *Health Policy and Planning, 32*(suppl_3), iii91–iii94. https://doi.org/10.1093/HEAPOL/CZX118.

Bradley, M., & Chahar, P. (2020). Burnout of healthcare providers during COVID-19. *Cleveland Clinic Journal of Medicine*. https://doi.org/10.3949/CCJM.87A.CCC051.

COVID-19|Launch and Scale Speedometer. (n.d.). Retrieved February 14, 2022, from https://launchandscalefaster.org/COVID-19.

Gilbert I.S. (n.d.). The Richard Dimbleby lecture: vaccine vs virus—this race and the next one. *https://www.youtube.com/watch?v=nP7vFk-RsX8—Google Search*. Retrieved February 14, 2022, from https://www.google.com/search?rlz=1C1CHBF_enIN924IN924&sxsrf=APq-WBuQNoDANzfcL1hZDe8T14aJNsQ4aw:1644830593574&q=Gilbert+IS.+The+Richard+Dimbleby+Lecture:+Vaccine+Vs+Virus+–+This+race+and+the+next+one.+https://www.youtube.com/watch?v%3DnP7vFkRsX8&spell=1&sa=X&ved=2ahUKEwiSiPu57_71AhVeILcAHT8UC_AQBSgAegQIARAy&biw=1265&bih=527&dpr=1.5.

Greer, S. L., King, E. J., da Fonseca, E. M., & Peralta-Santos, A. (2020). The comparative politics of COVID-19: The need to understand government responses. *Global Public Health, 15*(9), 1413–1416. https://doi.org/10.1080/17441692.2020.1783340.

Hernández, P., Dräger, S., Evans, D.B., Edejer, T., & Poz, M.D. (2006). Measuring expenditure for the health workforce: Evidence and challenges. Evidence and Information for Policy World Health Organization chrome-extension://efaidnbmnnnibpcajpcglclefindmkaj/viewer.html?pdfurl=https%3A%2F%2Fwww.who.int%2Fhrh%2Fdocuments%2Fmeasuring_expenditure.pdf&clen=251966&chunk=true.

Hick, J. L., Hanfling, D., Burstein, J. L., Deatley, C., Barbisch, D., Bogdan, G. M., & Cantrill, S. (2004). Health care facility and community strategies for patient care surge capacity. *Annals of Emergency Medicine, 44*(3), 253–261. https://doi.org/10.1016/J.ANNEMERGMED.2004.04.011.

Kearon, J., & Risdon, C. (2020). The role of primary care in a pandemic: Reflections during the COVID-19 pandemic in Canada. In Journal of Primary Care and Community Health (Vol. 11). SAGE Publications Inc. https://doi.org/10.1177/2150132720962871.

Komorowski, M., & Celi, L. A. (2017). Will artificial intelligence contribute to overuse in healthcare? * HHS public access. *Critical Care Medicine*, *45*(5), 912–913. https://doi.org/10.1097/CCM.0000000000002351.

Kuhlmann, S., Stegmaier, P., & Konrad, K. (2019). The tentative governance of emerging science and technology—A conceptual introduction. *Research Policy*, *48*(5), 1091–1097. https://doi.org/10.1016/J.RESPOL.2019.01.006.

Kurowski, C., Evans, D. B., Tandon, A., Eozenou, P. H.-V., Schmidt, M., Irwin, A., Salcedo Cain, J., Pambudi, E. S., & Postolovska, I. (2021). *From double shock to double recovery*. https://openknowledge.worldbank.org/handle/10986/35298.

Reghezza-Zitt, M., Rufat, S., Djament-Tran, G., Le Blanc, A., & Lhomme, S. (2012). What Resilience is not: uses and abuses. http://Journals.Openedition.Org/Cybergeo. https://doi.org/10.4000/CYBERGEO.25554.

Starr, P. (2020). Using controlled trials to resolve key unknowns about policy during the COVID-19 pandemic. *JAMA*, *323*(23), 2369–2370. https://doi.org/10.1001/JAMA.2020.8573.

Tangcharoensathien, V., Calleja, N., Nguyen, T., Purnat, T., D'Agostino, M., Garcia-Saiso, S., Landry, M., Rashidian, A., Hamilton, C., AbdAllah, A., Ghiga, I., Hill, A., Hougendobler, D., van Andel, J., Nunn, M., Brooks, I., Sacco, P.L., de Domenico, M., Mai, P., … Briand, S. (2020). Framework for managing the COVID-19 infodemic: Methods and results of an online, crowdsourced who technical consultation. Journal of Medical Internet Research 2020;22*(6)*:E19659 https://Www.Jmir.Org/2020/6/E19659, https://doi.org/10.2196/19659.

Wiegmann, P. M., Jaime, & Roca, B. (2021). Balancing quality and speed in the market approval of diagnostic tests: Experiences of South Korea, the UK, and the US. *JAMA Health Forum*, *2*(7), e211333. https://doi.org/10.1001/JAMAHEALTHFORUM.2021.1333.

World Health Organization. (2007). *Everybody's business—strengthening health systems to improve health outcomes : WHO's framework for action*. https://apps.who.int/iris/handle/10665/43918.

CHAPTER 34

Healthcare innovation and leadership: Case study of the United Arab Emirates

Hussain Abdul Rahman Al Rand, Haifa Madi, Saqr Alhemeiri, and Asma Al Obeidli
Ministry of Health and Prevention, Dubai, United Arab Emirates

Background

In the mid-20th century, the United Arab Emirates (UAE) found itself at a crossroads between traditions and breathtaking acceleration in the dynamic cities of today. Therefore, in this new century, the leadership of the country recognized "Innovation" as an indicator for the next phase that determines the faith of its nation in development and prosperity. Moreover, UAE leadership developed and ensured a new "UAE Vision 2021," shown by the symbol in Fig. 1, to take advantage of all possible present and future technologies and Artificial Intelligence (AI) interventions. In fact, the government pushed this focus onto domains where it can have world-leading assets and unique opportunities. Therefore the true mission of the country is to transform UAE into a world leader of innovation by investing in individuals and industries that are key to the UAE's success and that contribute to the development of the world.

UAE journey to innovation

2010, UAE vision 2021

Believing that innovation is the future of human investment, the UAE government with the leadership of H.H. Shaikh Mohammed Bin Rashid Al Maktoum emphasizes the importance of innovation in all aspects of the UAE Vision 2021. Fig. 1 shows H.H. Shaikh Mohammed Bin Rashid Al Maktoum and his guiding quote about the value and importance of innovation.

Innovation has been included as a part of the "United in Knowledge" pillar of the Vision, measured by the global innovation index, defined as a complex indicator which measures the competitiveness of countries based on 12 perspectives, including health and innovation capabilities as main components. The symbol of the UAE Vision 2021 is shown in Fig. 2 and

Resilient Health. https://doi.org/10.1016/B978-0-443-18529-8.00034-2

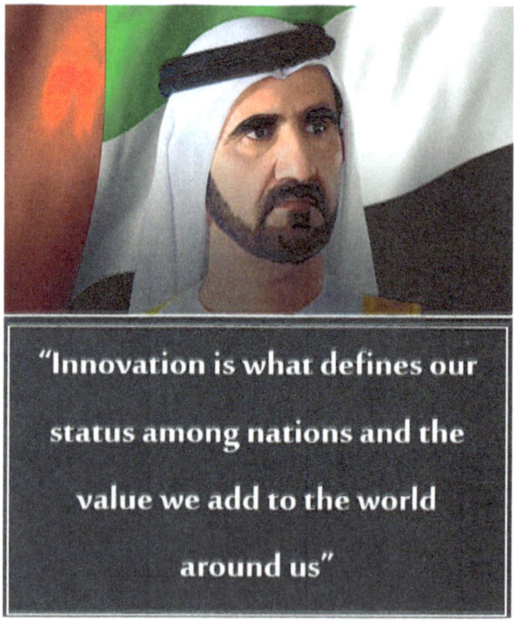

Fig. 1
Photo and quotation of H.H. Sheikh Mohammed bin Rashid Al Maktoum on the importance of innovation at local and global levels.

Fig. 2
Symbol of UAE vision 2021.

the main four pillars of the UAE Vision 2021 are shown in Fig. 3, namely, prosperity, which includes world-class healthcare, as well as knowledge, destiny, and responsibility.

A timeline toward innovation

The following describes the annual stages of the UAE plan toward innovation.

2014, Mohammed Bin Rashid Centre for Government Innovation

In September 2014, His Highness Shaikh Mohammed bin Rashid Al Maktoum, vice president and prime minister of the UAE and ruler of Dubai, launched the Mohammed Bin Rashid Centre for Government Innovation, a move that strongly signals innovation as one of the main themes

Fig. 3
Main pillars of UAE vision 2021.

of the new chapter in UAE government development. The main goals of the center are to develop governmental operations, enhance UAE competitiveness, and make the country as one of the most innovative governments globally. The center provides a world-class, multifunctional space for testing, creating, and spurring innovation at the regional and international levels.

2014, The Year of Innovation

In determination to make UAE an international leader in innovation, in 2014, the UAE Cabinet officially approved the designation of 2015 as the "Year of Innovation." Therefore all governmental ministries were directed to enhance cooperation and thoroughly revise their policies to provide an environment that encourages innovation.

2015, National Innovation Strategy

The UAE government initiated a "National Innovation Strategy" in 2015 with the aim to make UAE as one of the most innovative nations in the world within seven years. The strategy lays down a framework for embedding innovation in the national culture within all sectors, namely, at the level of government institutions and companies, as well as individuals. The National

Innovation Strategy identifies Health as one of the seven innovation priority sectors, along with Renewable Energy, Transportation, Technology, Education, Water, and Space.

The main objectives of the National Innovation Strategy are to:

- Implement a sustainable investment plan for human capital in the country
- Depend on other resources of economic development other than oil sector
- Develop the UAE's competitive advantages globally
- Introduce corporate methodologies and a culture for innovation.

2015, Innovation Celebration

UAE Innovation Week and UAE Innovation Month

As directed by H.H. Sheikh Mohammed bin Rashid Al Maktoum, the UAE Innovation Week was launched in August 2015, in line with the announcement made by H.H. Sheikh Khalifa bin Zayed Al Nahyan, president of the UAE, declaring 2015 as the "Year of Innovation." The event included 900 initiatives and activities across the country, which targeted creating a widespread innovation culture in the UAE, thereby supporting its position as a global hub for innovation. The celebration week was held again in November 2016 to resounding success, prompting the UAE leadership to expand the celebration to a full month, with the first "UAE Innovation Month" held in February 2018 and every year in February since then.

2015, Science, Technology and Innovation Policy in the UAE

In 2015 His Highness Sheikh Khalifa Bin Zayed Al Nahyan, president of the UAE, announced the adoption of the "Emirates Science, Technology and Innovation Higher Policy" with 100 national initiatives and an investment of AED 300 billion. The policy targets to change the national economy dependence and drive it away from limited oil resources to more sustainable innovation-based resources in the fields of science and advanced technology such as Health Information Technology and Bioinformatics, Public Health Noncommunicable Disease and Wellness, and Biotechnology and Genomics (The UAE Government, 2015). The policy highlights innovation as one of the main drivers of growth and progress and underlines innovation's critical role in a knowledge-based sustainable economy.

2019, Global Innovation Council

The Mohammed bin Rashid Center for Government Innovation formulated the "Global Innovation Council" in 2019. It is a multinational and multisectoral council that includes members from the public and private sectors nationally and internationally. The council aims to highlight UAE's determination to develop governmental innovation, and an innovative atmosphere, for effective solutions for future challenges. Moreover, its targets are to engage in the cocreation of international relevant innovation policies such as "The Science, Technology and Innovation Policy," take on UAE's capability to become a global thought leader and lab for experiments, and develop the global skills of innovation practitioners and highlight future governmental innovation plans (The UAE Government, 2015).

2022, UAE Innovates

With supervision of the Mohammed Bin Rashid Centre of Government Innovation, the "UAE Innovates 2022" initiative was established. This initiative engages federal and local government organizations, private sector, academia, and community members in future-proof initiatives that enhance well-being and help create more resilient government models.

The initiative includes hundreds of events across the country, including conferences, workshops, and awards. A nationwide festival that is organized annually in Mohammed Bin Rashid Centre for Government Innovation embraces creative ideas and innovations and brings the community of innovators together to share projects and initiatives. Moreover, "UAE Innovates 2022" engaged in events at Expo 2020 Dubai, in line with its mission to connect minds and create the future, and to adopt the UAE's best talents to highlight the most effective governmental innovations. The coordination between UAE Innovates and Expo 2020 included exhibiting the innovative projects developed within four main pillars, namely: Health Innovations, Future Economy, Utilizing Data, Tech for Good in Expo 2020.

In the health sector, Emirates Health Service (EHS) held a virtual dialogue sessions on the human genome, precision medicine, and pharmacogenetics (Ibrahim, 2022).

Innovation as an enabler component: Government innovation

1. **Government Performance by Prime Minister Office (PMO) According to Innovation Standards**
   The Ministry of Cabinet Affairs in the UAE monitors government performance for all governmental organizations via the "UAE government performance management system," which consists of a clear framework for performance to achieve the "UAE Vision 2021" and national agenda. Accordingly, innovation is well integrated in all the work of the government sector, including innovation in policies, governmental services, processes, and procedures that improve competitiveness in the UAE governmental sector and push the government innovation to a more organized corporate culture.
2. **Innovation in Sheikh Khalifa Government Excellence Program**
   The Sheikh Khalifa Government Excellence Program, established in 2006, is an integrated federal platform that evaluates and celebrates excellence in performance and innovative initiatives in the UAE federal government sector.
   Under the umbrella of this program, the "Mohammed bin Rashid Government Excellence Award" is granted every two years. The main aim of this award is "to

encourage the federal government institutions to better serve the UAE society at all levels and motivate the principles of excellence."

3. **Readiness Index**

The "Innovation Government Index" is an instrument developed to measure readiness and progress of innovation for both local and governmental organizations in Dubai. The index assists the organizations to:

- Review the current situations according to innovation standards
- Benchmark with national, regional, and international levels
- Trace challenges and gaps in order to implement corrective interventions for improvements

Innovation and health

Innovation in the National Innovation Strategy of the UAE is defined as "the aspiration of individuals, private institutions and governments to achieve development by generating creative ideas and introducing new products, services and operations that improve the overall quality of life."

Health innovations result from efforts to "develop new or improved health policies, systems, products and technologies, and services and delivery methods that improve people's health, with special focus on the needs of vulnerable populations." According to the UNC Centre for the Business of Health (Health innovation and COVID-19 pandemic: Defining the need and understanding the response, 2020), this results in improved efficiency, increased productivity, enhanced patient outcomes, improved competitiveness, and customer's satisfaction.

While the National Innovation Strategy (NIS) focuses on promoting innovation at large, it aims in parallel to lead innovation in seven primary national sectors, namely renewable and clean energy, transportation, technology, education, health, water, and space. In the NIS, the main aim of health innovation is:

> For a population to be able to participate effectively in the social and economic development of its nation, the majority of individuals should be in good health.
>
> Innovation in the health sector is of paramount importance for advanced societies seeking a world-class health system that enhances quality of life. In keeping with this vision, the UAE launched a myriad of initiatives to promote innovation in the healthcare sector. For example, the Dubai Biotechnology & Research Park (DuBiotech) is a free zone that provides Life Sciences companies with an ultimate platform to set up their operations in Dubai. Additionally, in Abu Dhabi, the Abu Dhabi Vision 2030 aims to develop pharmaceuticals, biotechnology and life sciences, while the recently established Abu Dhabi Telemedicine Centre offers high quality, confidential medical consultations over the phone. Also, the first anti-cancer drug manufacturing plant in the GCC will be established in Abu Dhabi, and the first biotechnology innovation incubator will be held at Abu Dhabi University.

The NIS aims to promote innovation in the delivery of health and therapeutic services using advanced technologies such as robotic surgery and telemedicine techniques. It focuses on developing pharmaceutical industries and biotechnology, and enhancing medical research on the treatment of prevalent diseases, namely diabetes and obesity. Moreover, the NIS aspires to develop advanced health system management technology and promote a healthy lifestyle by deploying smart technology in disease control and prevention, as well as engineering and developing innovative cities and facilities that prompt sports and movement.

(Prime Minister's Office, 2015, p. 18)

1. **Health innovation strategy formulation**
   With the launch of the National Strategy for Advanced Innovation in 2018, the Ministry of Health and Prevention (MOHAP) started working on formulating an updated Health Innovation Strategy that sets the roadmap toward achieving the UAE National Agenda. The major inputs to this strategy included: the UAE National Agenda; the National Strategy for Advanced Innovation; and the Emirates Science, Technology and Innovation Higher Policy. Also, an in-depth survey that is conducted every two years in partnership between the Mohammed Bin Rashid Centre for Government Innovation and McKinsey measures the eight essential components of innovation according to McKinsey's model.
   MOHAP held a comprehensive engagement exercise with stakeholders from both the public and private sectors, to come up with a unified vision and to develop a health innovation strategy in alignment with the UAE strategic priorities to achieve the national agenda. The vision and themes of the strategy are as follows:

   Vision: Leading innovation for a world-class healthcare system.
   Themes: The strategy identifies four major themes and their related objectives:
   - Community of wellness: Leveraging AI to empower the community
   - Cutting Edge Care: Designing new models of care
   - Leading processes and services
   - Research and Development
     The Health Innovation Strategy also identified five enablers for the four themes: human capital, knowledge and data, technology, finance, and an underlying dynamic regulatory framework.
2. **Intellectual property in the field of healthcare**
   As a regulatory framework and in alignment with UAE Vision 2021, MOHAP and its healthcare and other concerned partners (MOHAP's concerned departments, Dubai Health Authority, Department of Health—Abu Dhabi, Sharjah Healthcare Authority, Ministry of Cabinet Affairs and the Future—Healthcare Policy Management, Ministry of Interior, Ministry of Economy, Ministry of Foreign Affairs and International Cooperation, Federation of UAE Chambers of Commerce and Industry, Emirates Intellectual Property

Association) issued the National Policy for Intellectual Property in the Field of Healthcare, which was released in 2018. The goal of this policy is the "promotion of health systems, economic development, scientific and cultural activities in UAE through the protection of intellectual property rights in the health field that is used as a tool to encourage and protect creativity, scientific research and local innovative capacities." The main strategic objectives of the national policy are as follows:

- Encouraging and supporting the application of intellectual property to achieve the maximum level of innovation in relation to health. This is in order to meet the needs of the country in the area of research and development, protection of public health, and promoting availability of all health products and medical devices, including medicines for all.
- Increasing the appreciation of creators, innovators, and scientists.
- Protecting all forms of health knowledge and medical research generated in the UAE and the use of this knowledge for disease reduction and the improvement of people's well-being.
- Facilitating the transfer of health technology to speed up the country's progress in this field.
- Encouraging the use of, and access to, the technological information available for promoting scientific and technological research and innovation.
- Awareness of intellectual property rights in the health field.

This national policy supported creators and inventors in the health field to conduct their research and protect their inventions in the UAE.

3. **Standards and compliance**
As part of MOHAP's efforts to be a sustainable, knowledge-based, and innovative institution, and its commitment to adhere to international standards, the ministry achieved multiple certifications including all ISOs relevant to quality management, innovation management, and knowledge management.
4. **Integration of innovation in government performance**
United Arab Emirates supports innovative interventions and programs throughout all its governmental services. Therefore it is mandatory that all directors and decision makers are required to include innovation as part of their annual objectives and plans. This provides a clear framework for innovation in health that allows staff to come up with innovative initiatives that improve efficiency, increase productivity, enhance patient outcomes, and customer satisfaction.
5. **Afkari**
In order to integrate innovation in the daily administrative and technical work across healthcare entities, MOHAP launched a new electronic suggestion system platform named "Afkari" the logo for which is shown in Fig. 4. This platform provides employees the possibility to suggest ideas and thoughts through multichannels. The innovative ideas are studied by multisectoral expert teams in order to promote and carry out the best innovative suggestions.

Fig. 4
Logo of Afkari system for innovative ideas.

This platform aims to:

- Expand the growth of healthcare services
- Accelerate achieving national, strategy, and operational indicators
- Increase its competitiveness in healthcare market
- Accomplish employees' and customers' happiness

6. **Salma Nursing Innovation Award**
   In 2017 MOHAP launched the Salma Nursing Innovation Award, which spreads the culture of innovation in the nursing field. The award is named after the late nursing pioneer Salma Salim Al Sharhan, known as the "Florence Nightingale of the UAE," referring to the famous British founder of modern nursing in the 1800s who was a highly celebrated figure in the Arab nursing world.
7. **Innovation month 2020**
   Among the innovation activities implemented in Innovation Month 2020, MOHAP launched the "Healthcare Innovation Award" which included 16 awards given to institutions or individuals, under four main categories, namely: innovators, innovative institutions, innovative initiatives, and health technology. Winners included Cleveland Clinic in Abu Dhabi in the innovative institution category, Al Qassimi Hospital in the health technology category/medical devices, Gulf Medical University in the health technology/life sciences, and many more institution and individuals (Azadnewsme, 2020). An individual award, given in February 2020, was presented to Consultant Spinal Surgeon Dr. Sherief Elsayed of NMC Healthcare for his successful introduction of Patient-Specific 3D printing into the operative correction of high-risk Scoliosis deformity (to accurately place screws in the spine) and for his commitment to build NMC Spine, a Center of Excellence for spinal problems, which provides compassionate, evidence-based care from diagnosis to recovery throughout the UAE. The innovation is cost effective, reduces operating theater time and radiation (to staff and patients), and is reproducible. The innovation also represents a strong collaboration between the UAE hospital, the British distributor, and an American laboratory.
   The month and the awards are in line with the MOHAP innovation strategy and the National Strategy for Advanced Innovation, which highlights the importance of nurturing an innovative culture at the level of the individuals, as well as government entities and the private sector.

8. **Innovation during COVID-19 pandemic**

   In light of the exceptional circumstances that the world has experienced with the outbreak of the COVID-19 pandemic since the beginning of 2020, the Ministry of Health and Prevention (MOHAP) has launched many projects that aim to respond to this pandemic and improve its readiness to confront any future pandemics. Innovation in MOHAP focused specially on providing innovative initiatives that include, involve, and integrate all segments of society into the innovation environment to design initiatives that provide solutions to challenges and contribute to the development of the health sector according to a future vision. The following are some of the examples of innovative initiatives in addressing the pandemic:

   Research and studies

   With leadership of H.E. Dr. Hussain Al Rand, undersecretary of the Public Health Sector of the Ministry of Health and Prevention, the UAE participated in clinical medical research and studies related to COVID-19, including about the Chinese Sinopharm and the Russian Sputnik vaccines. This resulted in being one of the first countries that obtained permission to use the vaccine in emergency situations, thus accelerating the pace of recovery in addition to being one of the first countries to implement national vaccination campaign (COVID-19).

   Continuity of services

   In order to ensure continuity of services in healthcare during the pandemic, several innovative interventions were implemented, including:

   - Establishing virtual clinics/E-consultation that assist patients to consult professionals and refer them to appropriate level if needed. This consultation covered maternal and child health services, noncommunicable diseases, virtual clinics for consulting and follow-up, E-smoking cessation clinics, and other services.
   - During COVID-19, in order to find suitable accessible alternatives for health education in order to ensure continuity of care and maintain public health literacy, health education was provided through different innovative modalities. This included:
     - Virtual health education sessions, where more than 200 sessions were delivered benefiting more than 14,600 participants.
     - Virtual live classes to enhance healthy skills were delivered through more than 40 sessions benefitting more than 100,000 views.
     - Open live sessions with specialties were conducted through more than 8 sessions benefitting more than 700 participants.
     - Digitalization of educational resources was established through creating barcodes for the print materials, which was promoted via social media platform.
   - Vaccination through vehicles and mobile clinics to ensure that all newborns and infants are receiving their vaccinations in a timely manner, taking safety precautions into consideration.

- Medication delivery to homes that resulted in accomplishing 40,000 home deliveries of medicines with a satisfaction rate of 93%.

9. **"UAE Healthcare Sector Outlook 2023"**
   With rising initiatives by the government, the UAE Healthcare market is witnessing astonishing growth, due in part to the sedentary lifestyle of the people of Emirates, growing medical tourism in the region, and government actions to expand and upgrade to a world-class healthcare infrastructure and system, and encouraging private sector participation, to match quality of services offered in developed countries.
   The UAE has made significant deals, in terms of mergers, acquisition, and strategic partnerships between healthcare stakeholders, public and private entities to enhance the healthcare industry. The "UAE Healthcare Sector Outlook 2023" report provides extensive information and rational analysis of the UAE healthcare market, describing the healthcare infrastructures, resources, and future outlook.
   Further, the UAE government is liberalizing policies to attract foreign investments, in order to improvise the healthcare standard and boost the healthcare industry. Holistic research and analysis depict that the healthcare market of UAE will grow at a compound annual growth rate (CAGR) of around 10% during the forecast period, 2019–23.
   The report presents an in-depth analysis of the contribution of the private and public sectors to the country's healthcare industry; various government health initiatives which focus on training and development, investment, and regulatory reforms; a study of demand for hospitals and the number of hospital beds; current investment opportunities and the competitive landscape of the country's pharmaceutical players and hospitals, which will help the client to understand the market structure and potential growth.

Mental health innovations using technology

Mental health is one of the main priorities for UAE. In 2017 a mental health policy was developed. In order to operationalize this policy, a national multisectoral strategic mental health plan was developed. MOHAP continued their efforts to provide psychological services in line with the objectives of the National Strategy of Wellbeing 2031, to promote good mental health for individuals. Therefore a number of innovative initiatives were launched as community mental health services, most notably:

1. The tele-psychiatry service and the virtual pharmacy were launched at the beginning of the COVID-19 pandemic to provide therapeutic services and psychological counseling to patients, through the electronic clinics using visual and audio communication means. These clinics provide services of assessment, diagnosis, treatment, and psychological counseling; virtual medical prescriptions that link with delivering medications to patient's home, according to specific procedures that ensure safety and privacy. Moreover, EHS has developed a remote psychiatric service through the Virtual Clinic application for remote visual and audio communication linked with a health information system (Wareed),

through which patients are capable of booking appointments to meet the psychiatrist in the electronic clinic.

2. Launching the network of community mental health clinics for the Al Amal Hospital for Mental Health in primary care centers (Al Amal Satellite Community Mental Health E-Clinics in PHCs), which provide comprehensive psychological services near the patient's residence. The patient meets the psychiatrist at Al Amal Hospital remotely through the electronic clinic, using visual and audio means in the family doctor's clinic in the health center. The family doctor participates with the specialist team in developing and following up the psychological care plan. This project aims to facilitate access to psychological services, enhance early detection and intervention, and build competencies of primary care workers to provide community mental health services through continuous training, as well as overcoming the issue of mental health staff's shortage.
3. The "Speak to Hear You" initiative was launched that provides psychological counseling and psychological support to all community through trained and qualified psychologists. Different channels to reach this service include phone, email, or WhatsApp, as well as helpline service in the EHS COVID-19 application. Moreover, patients can communicate directly by phone with MOHAP's call center and select the psychological counseling helpline option, which is linked directly to the service.
4. Virtual visits were activated for patients residing in the internal departments of the psychiatric hospital, using visual and audio means of communication to communicate with their families, whether inside or outside the country. This service reduced the risks of spreading infection, ensured the continuity of service provision, and raised the happiness of customers at the time of pandemic.
5. Mental Innovation Lab was launched in 2021, under the leadership of H.H. Dr. Hussain Al Rand from MOHAP with the participation of WHO and EHS. This initiative was launched during the Mental Health Awareness Month in accordance with the theme "Driving progress towards a sustainable and scalable model of mental health care." The lab's mission is to bring together leaders and experts to collaborate and achieve a common goal of supporting mental health and wellness in a sustainable way, by integrating care and promoting intersectoral action and collaboration in mental health, to establish evidence-based standards. The main objectives of this lab are to discuss and present the mental healthcare innovative approaches and define policies that would bring effective impact for the next 50 years in accordance with UAE's future visions that target achieving healthier and happier society (Media Hub, 2021).

Human capital response using technology

With a healthcare workforce widely spread across multiple hospitals and field hospitals across the nation, and the rapid pace of changing/updated information related to the COVID-19 pandemic, MOHAP implemented, and regularly updated, a COVID-19 knowledge management page on the MOHAP website that ensured quick access for all hospital staff to the latest updates on COVID-19-related information from local and international reliable sources Health Knowledge Management Strategy, 2022; Recent COVID-19 Information and Updates, 2022). Training was largely moved to online (live or on demand), with a mobile simulation unit to support frontline training. A collaborative effort among MOHAP, other governmental entities, and universities yielded a healthcare support volunteer program that trained hundreds of volunteers to provide meaningful support to healthcare providers across different facilities. These volunteers valuably increase the human capacity of healthcare provision direly needed during the pandemic.

Innovation in well-being

UAE attaches great importance to the happiness of individuals and society; considers this one of its priorities; and is keen to provide prosperity, well-being, and stability to its people and residents on its soil. To reach this end, UAE developed a national program for happiness and well-being in 2016, with an appointment of a minister for happiness. In 2019 UAE launched the National Strategy for Wellbeing, 2031. This strategy is structured at three levels, namely individuals, communities, and the country. The strategy provides a comprehensive and a whole-of-government approach to contributing to well-being from social, environmental, economic, digital, education, urbanization, and health perspectives. The health and well-being component consists of four pillars, namely physical well-being, mental well-being, well-being in setting, and client satisfaction.

In September 2021, UAE notified the WHO Bureau and submitted a draft resolution on Well-being and Health Promotion to the Executive Board, on the topic of health promotion and well-being. Six informal consultations were conducted: two consultations before the EB and four during the intersessional period.

Notably, the UAE submitted the resolution to the 75th World Health Assembly (WHA) in May 2022 for endorsement. The Well-being and Health Promotion Resolution A75/A/CONF./4 was agreed on with full consensus by member states with cosponsoring of more than 20 countries from the different WHO regions. This resolution is in the Appendix in the online version at https://doi.org/10.1016/B978-0-443-18529-8.00034-2 to this chapter; the content is final, although the number of cosponsoring countries ultimately reached more than 20. Notably, this is the first resolution ever on well-being to be submitted to the WHA.

Highlights of innovation in healthcare

Following are some innovative projects in place, using technological tools, and their impact on public health in the UAE wherever possible:

1. **Diabetic foot treatment**
   Well-equipped diabetic foot clinics were established using artificial intelligence technology for early detection and treatment to prevent foot amputation and complications.
2. **E-etmenan**
   This preventive initiative against noncommunicable diseases consists of a periodic examination to which an innovation factor was added which studies the patient's risk factors and links them to geographical, ethnic, and cultural factors, resulting from the analysis of big data. Also, treatment is proposed on the basis of best clinical practices and with the support of artificial intelligence technology which serves as a guiding reference for medical staff, allowing proactive awareness campaigns at the regional and even individual levels.
3. **Tateem App**
   This application maintains vaccination data for a child and reminds parents of the dates for future vaccinations. It also includes the national vaccination schedule and information on infectious diseases for available vaccines.
4. **Happiness station**
   This is an innovative intervention shared with federal and local government entities that completes the paper work of the UAE residency procedures for clients, which reduces the number of steps needed to obtain the service by 60% and reduces the necessary cost by 50%. The station resulted in increasing customers' satisfaction, reaching 85%.
5. **Research bank**
   An electronic platform was launched to shed light on health research in the country by collecting all health research published from 2007 until this publication date, categorized appropriately. This service enables researchers to upload their research onto the platform, after obtaining required approvals, which contributes to enhancing the research culture in the country.
6. **UAE RADR**
   In January 29, 2019, the MOHAP launched the first user-friendly mobile smart application of its kind in pharmacovigilance (drug safety), called UAE RADR. This app allows reporting of Adverse Drug Reactions (ADRs) by patients, consumers, and healthcare professionals and allows grant users to access ADR cases reported globally, since it is directly connected to the WHO system (Vigiflow). Moreover, the app enhances patients' safety and drug safety control, as well as increases the ability to analyze data, thus speeding up decision-making and preventive measures, and reducing fraud.
7. **Narcotics and precursor chemical directory**
   A directory with a smart app was developed within MOHAP as an official application to facilitate the option of searching for substances listed in the schedules of narcotics and psychotropic substances. The smart app serves several sectors: doctors and technicians; forensic laboratories; prosecutors, judges, and lawyers; and narcotics control officers and custom inspectors.

8. **Registration system for organ donation "Hayat"**
   UAE allows transplantation of human organs and tissues from both living and deceased donors. Therefore, MOHAP, in cooperation with the National Program for Organ Transplantation, devised an innovative mobile platform and web-based portal application called "Hayat" which connects health authorities, health providers, and organ banks, to facilitate transplantation processes, including matching.
9. **The AL HOSN UAE App**
   Al Hosn App is the UAE's official app for contact tracing and health status related to COVID-19. On the Al Hosn app, each person is able to receive:
   - A record of all current and previous COVID-19 tests, with dates and results. The test results are received directly on the person's phone with a unique QR code that provides the status of the person in terms of testing
   - Notification if the person has been in contact with a confirmed case of COVID-19
   - Vaccination information, including the type of vaccine taken and the date of each dose given

   The app also facilitates sharing test reports and vaccination certificates.

Conclusion

The United Arab Emirates has succeeded in the past century to institutionalize the concept of innovation in the country. Many innovations require technology and AI, which the UAE proved can be built with proper foundation, clear vision, a defined strategic pathway, and constant upgrade of the country's health infrastructure and delivery of healthcare. Some of these innovations were stimulated by the urgency presented by the COVID-19 pandemic. These have yielded exceptional results for all stakeholders in the healthcare ecosystem.

References

Azadnewsme. (2020). *MoHAP honors winners of the "innovate for health" award, the first-of-kind award in the UAE and region.* Retrieved from: https://azadnewsme.com/mohap-honors-winners-innovate-health-award-first-kind-award-uae-region/.

Health innovation and COVID-19 pandemic: Defining the need and understanding the response. (2020). https://scotpublichealth.com/2020/04/26/health-innovation-and-covid-19-pandemic-defining-the-need-and-understanding-the-response/#:~:text=%E2%80%9CHealth%20innovation%20is%20to%20developthe%20needs%20of%20vulnerable%20populations.

Health Knowledge Management Strategy. Emirates Health Services—UAE. (2022). Retrieved from: https://www.ehs.gov.ae/en/about-us/health-knowledge-management-strategy.

Ibrahim, L. (2022). *UAE innovates 2022 kicks off on Tuesday with month-long agenda of events countrywide.* WAM-Emirates News Agency. https://www.wam.ae/en/details/1395303016539.

Media Hub. (2021). *MoHAP and EHS hold Mental Health Innovation Lab.* UAE: Ministry of Health and Prevention. Retrieved from: https://mohap.gov.ae/en/media-center/news/1/11/2021/mohap-and-ehs-hold-mental-health-innovation-lab.

Prime Minister's Office. (2015). *UAE national innovation strategy (p. 18).* Prime Minister's Office.

Recent COVID-19 Information and Updates. Ministry of Health and Prevention—UAE. (2022). Retrieved from: https://mohap.gov.ae/en/covid-19.

The UAE Government. (2015). *Science, technology & innovation policy in the United Arab Emirates*. The UAE Government. https://u.ae/-/media/Science-Technology-and-Innovation-Policy.ashx.

Further reading

Day of Dubai—Dubai's Leading Information Portal|News|Jobs|Events. (2022). *Ministry of Health & Prevention launches new electronic suggestion scheme platform "Afkari" to facilitate innovative ideas*. Retrieved from: https://www.dayofdubai.com/news/ministry-health-prevention-launches-new-electronic-suggestion-scheme-platform-afkari#:~:text=The%20Ministry%20of%20Health%20andto%20suggest%20ideas%20and%20ideas.

Dubai Excellence Government Program. (2019). *About Dubai government innovation framework*. Dubai: Dubai Excellence Government Program. https://dgep.gov.ae/en.

Gulf News. (2014). *Shaikh Mohammad launches national innovation strategy*. Retrieved from Gulf News: https://gulfnews.com/uae/shaikh-mohammad-launches-national-innovation-strategy-1.1401024.

Gulf News. (2015). *UAE innovation week begins*. Retrieved from Gulf News: https://gulfnews.com/uae/government/uae-innovation-week-begins-1.1624239.

Gulf News. (2018). *National strategy for advanced innovation launched*. Retrieved from Gulf News: https://gulfnews.com/uae/government/national-strategy-for-advanced-innovation-launched-1.2181016.

Mental Innovation Lab. (2021). https://mohap.gov.ae/en/media-center/blogs/blog-details?id=723.

MoCA, U. (2019). *Government innovation*. Retrieved from: Moca.gov.ae. https://www.moca.gov.ae/en/area-of-focus/government-innovation.

Mohammed Bin Rashid Centre for Government Innovation. (2017). Retrieved from: Mbrcgi.gov.ae. https://www.mbrcgi.gov.ae/en.

National Innovation Strategy. https://www.moei.gov.ae/assets/download/1d2d6460/National%20Innovation%20Strategy.pdf.aspx.

National policy for intellectual property in field of healthcare. Ministry of Health and Prevention. https://mohap.gov.ae/app_content/legislations/php-law-en-81/mobile/index.html.

Sharjah24. (2021). *MOHAP obtains GC-Mark certification in knowledge management*. Retrieved from Sharjah24: https://www.sharjah24.ae/en/Articles/2021/04/21/MoHAP-obtains-GC-Mark-Certification-in-knowledge-management.

The National. (2014). *2015 will be a 'Year of Innovation'*. Retrieved from The National: https://www.thenationalnews.com/business/technology/2015-will-be-a-year-of-innovation-1.337208.

The United Arab Emirates' Government Portal. (2022). *National innovation strategy*. Retrieved from: https://u.ae/en/about-the-uae/strategies-initiatives-and-awards/federal-governments-strategies-and-plans/national-innovation-strategy.

UAE Cabinet. (2015). *Mohammed Bin Rashid Al Maktoum launches first innovation fund worth AED 2 billion*. Retrieved from United Arab Emirates—The Cabinet: https://www.uaecabinet.ae/en/details/news/mohammed-bin-rashid-al-maktoum-launches-first-innovation-fund-worth-aed-2-billion.

UAE Government. (2015). *Science, technology & innovation policy in the United Arab Emirates*. Dubai: UAE Government. https://u.ae/en/about-the-uae/strategies-initiatives-and-awards/federal-governments-strategies-and-plans/the-science-technology-and-innovation-policy.

UAE Ministry of Cabinet Affairs. (2015). *UAE National Innovation Strategy*. Dubai: Prime Minister's Office. https://u.ae/en/about-the-uae/strategies-initiatives-and-awards/federal-governments-strategies-and-plans/national-innovation-strategy.

UNC Centre for the Business of Health. https://cboh.unc.edu/index.php/vertical/digital-health/#:~:text=Health%20innovation%20is%20to%20developthe%20needs%20of%20vulnerable%20populations.

WAM. (2014). *VP opens Mohammed Bin Rashid Centre for Government Innova*. https://www.mbrcgi.gov.ae/en.

WAM. (2015). *UAE President adopts UAE's higher policy for science, technology and innovation*. Retrieved from WAM: http://wam.ae/en/details/1395288271257.

WAM. (2020). *MoHAP launches virtual clinics to further strengthen its telemedicine system*. Retrieved from WAM: https://wam.ae/en/details/1395302842094.

WAM. (2021). *MOHAP receives GC-Mark certification in innovation management*. Retrieved from WAM: https://wam.ae/en/details/1395302913283.

CHAPTER 35

The role of government in leading the pandemic response: The Estonian experience

Taavi Annus and Kertti Merimaa
Ministry of Social Affairs, Tallinn, Estonia

Introduction

Turbulent times, such as the precarious era created by the COVID-19 pandemic, test the limit of all agencies and building blocks of health systems. From healthcare financing to ensuring sufficient workforce and information exchange, all have been put under pressure by the increased demand for health services. In the middle of all of it, governments have had to solve the challenge of maintaining the delivery of quality healthcare to their population. It can be argued that the pandemic has proven to be an outstanding driver of change for governments to reconsider their current health strategies and policies, and to be more open to innovative solutions including digital transformation and creating novel partnerships to leverage expertise from different domains. It has been the role of governments to present strong leadership in orchestrating the efforts of all stakeholders to drive a successful collective effort toward the common goal of surviving the pandemic. This chapter presents key practical lessons learnt by the Estonian government during the response to the COVID-19 pandemic that hopefully can help others in building agile health systems that are ready to be adapted and transformed when necessary.

As health systems are very complex, to effectively understand the ripple effects of the crisis in relation to the functioning of health systems and to select appropriate responses, a systematic approach covering different health system building blocks is needed (Olu, 2017). To ensure that all the health system building blocks work in harmony, governments play an important role in ensuring a multitude of vital elements: an effective and participatory leadership with strong vision and communication, coordinating capabilities, openness to flexible approaches, and effective information flows and systems (Thomas et al., 2020).

Resilient Health. https://doi.org/10.1016/B978-0-443-18529-8.00035-4

Building resilience through responsive and transparent strategic management

Clear management coupled with up-to-date health strategies creates an environment that allows countries to secure a quick and successful response to the challenges of crises, providing a favorable basis for development and deployment of innovative solutions. Up-to-date strategic documents provide the necessary common framework for different stakeholders to work toward mutual goals. Even more vitally, they ensure that even during emergencies common underlying values are not abandoned. The aspects of user-centeredness, data security, standardization, and interoperability are those that usually come under attack when something is needed faster than normally possible. It is fundamentally important that these key aspects are given special attention and a systematic approach is chosen. Well-revised strategic principles ingrained in the culture, as well as existing architectural and infrastructural reusable components, ensure that even in the tight timeframe of crisis, key principles are not abandoned.

As an example, when Estonia started the development of the COVID-19 tracing app HOIA, it was built on top of the existing e-state infrastructure which immediately ensured compliance with basic security principles without losing valuable time (ERR News, Estonia launches coronavirus exposure notification app 'HOIA', 2020). It was also heavily discussed whether to be less strict with data protection rules and requirements to reduce the time from idea to launch. But the principles of data protection had long been communicated through the state's digital strategies, which meant that the consciousness of the developers as well as the expectation from the public did not allow anything other than the app to meet the highest possible standards, even if it meant the launch of the app would be delayed. This would have not been possible if there wasn't continuous communication of the principles set in the national strategies. According to OECD, countries considered as role models in tackling digital opportunities in health have in common a long-standing tradition of human-centric and overarching digital health strategies that articulate the intersection between the technology and health, with focus on patients' and health workers' needs (Socha-Dietrich, 2021).

For Estonia, the Estonian National Health Plan 2030 defines the main goals and principles for the health system. This general approach to healthcare is complemented by the Estonia's eHealth strategy: Next Generation Estonian National Health Information System 2030, providing a framework focusing on both the person-centered service transformation and architectural goals of the national eHealth infrastructure.

Building resilience through clear and inclusive communication

Health systems are complex entities with many coexisting systems and participants. It is impossible to create well-functioning complex systems without good communication and effective information flows. Estonia is seen as the leader in sharing information between health system participants through the success of the universal use of the national health information system. Having good communication channels is extremely beneficial but equally important is the quality and clarity of the information shared through these channels. During the pandemic, the national information system proved vital in creating quality information flows between stakeholders. Based on the existing and newly collected data, dashboards for decision-makers were created which helped make evidence-based decisions.

Aside from the decision-makers, the same information was effectively also used by other stakeholders. The Estonian government took the open data approach and made the pandemic-related data publicly available. All media outlets relied on the open datasets shared by the government for their statistical information. This reduced the administrative burden of sharing the information manually. The same open data was used by the private sector enthusiasts to independently create visual dashboards which were widely used by all stakeholders including the government itself. Due to different rules, the private sector is often faster in creating novel solutions and can be the trendsetters. Keeping them out of the information loop would have made it impossible.

Information availability is just one aspect of good communication. Good results are achieved by clear communication of decision-making processes, stakeholder roles, and responsibilities. These aspects were highlighted by the COVID-19 pandemic when different governmental bodies joined the daily decision-making of the health system. This meant that suddenly there were five layers of decision-making—the government, governmental office, ministry of social affairs, health board, and crisis-specific COVID-19 task force.

Having so many decision bodies requires extremely good coordination with clear roles and responsibilities in order to prevent contradictory decisions being made at different levels. At the start of the pandemic, Estonia was quick to from a multidiscipline and multistakeholder task force responsible for making decisions. In theory it is good to have broad involvement and a dedicated body for tackling an emergency. However, in practice it was the opposite, as the regular governance bodies continued working and making decisions in parallel. This resulted in cacophony and at times contradictory decisions being made at different levels resulting in unnecessary widespread confusion. During the pandemic, the governance structure was revised, and the resulting structure followed as closely as possible the hierarchy of "peacetime." The lesson learnt was that the closer the emergency governance structures are to existing structures, the clearer the decision-making. It is essential to have as little separation of decision bodies as possible and openly communicate when decisions are made; otherwise, everyone starts to run in different directions.

Implementing good governance frameworks and clear decision-making bodies with well-communicated time schedules, roles, and responsibilities is equally important even when there is no looming emergency in sight. In 2021 Estonia launched a project to implement an eHealth governance framework with concise management processes. The aim is to establish clear roles and responsibilities for all stakeholders together with structured information flows, so they can work more effectively and better in unison.

Aside from clarity in the governance and decision-making, good communication is crucial to ensure that within limited resources, the investments, both public and stakeholders such as hospitals and other healthcare providers, are well directed toward a common goal. However, the orchestration requires a strong public vision for the architecture of the healthcare, which is often most well defined in the strategic documents. Well-targeted, forward-looking investments into strengthening healthcare are needed to ensure better resilience, especially given the difficulty of convincing the decision-makers the importance of prevention and preparedness (Thomas et al., 2020). This is especially important in the case of digital health where it is crucial to avoid creating data silos or trapping systems into technological lock-ins. This will be essential to realize the promise of digitalization in the long term.

Transparent and well-directed communication helps in targeting investments and orchestrating stakeholders toward a common goal and is also essential in creating trust and keeping the motivation of the health sector high. It is crucial to provide the public with timely and comprehensible information for all the parties to understand the need for and importance of strategic measures taken as well as maintaining focus. Experts from European Observatory on Health Systems and Policies (Thomas et al., 2020) argue that this has been especially highlighted in the COVID-19 pandemic, when governments started to implement containment measures. The same approach is equally important when considering digital opportunities, especially during turbulent times.

As a response to the lessons learnt in recent public health crises, Estonia is implementing a public eHealth portfolio management process, as part of the national eHealth governance framework. Portfolio management deals with prioritizing national, publicly developed eHealth initiatives and creating a common understanding across the stakeholders of the eHealth systems and services to be developed, which is especially important during the turbulent times such as during the COVID-19 pandemic, when the public expectation for streamlined eHealth developments is heightened. By highlighting the backlog and considering the resources available and the development needs of various stakeholders, portfolio management provides an input to strategy management and budgeting but also offers public a clear overview of the development needs and capacities to deliver necessary projects.

Building resilience through public eHealth infrastructure

When tackling the health crises, the effective response is often dependent on having the right information readily available, which in turn requires a mature and resilient public eHealth infrastructure, i.e., health information systems and data platforms. Good visibility across the health systems' various components has been argued to support efficient resource allocation and provides for evidence-based policy-making, but need to be agile, equitable, and comprehensive (Alam et al., 2021), and also to ensure that FAIR data principles (i.e., findability, accessibility, interoperability, and reusability) are followed (Wilkinson et al., 2016). Being able to rely on the existing e-state infrastructure highlights the importance of prevention and preparedness.

Estonia is well known for mature digital public infrastructure and named one of the knowledge economy leaders among the regions supported by the European Bank for Reconstruction and Development. Three aspects of the Estonian digital infrastructure are critical in ensuring its resiliency. Firstly, the universal use of the digital ID, which is achieved by the legal framework making it compulsory for all citizens. Secondly, the scalability and flexibility of the digital infrastructure, which is achieved through digital strategies and principles, namely decentralized architecture, once-only policy for data collection, trust-by-design approach. Thirdly, the backbone central ICT services used by most of the public digital solutions: the electronic ID system, the data exchange layer X-Road. The latter is designed as a decentralized system, a connected series of networks and data sources rather than one extensive monolith. The flexibility provided by this open setup has allowed for new features of the digital society to be developed and added through the years without expensive overhauls of existing features (Visit Estonia, 2020).

The Estonian National Health Information System (ENHIS), where most of the clinical documentation is stored, relies on the same three aspects of the Estonian digital infrastructure. During the pandemic, the ENHIS allowed quick reengineering of e-services to give real-time overview about the COVID-19 situation in Estonia, i.e., data about all the COVID-19 laboratory test results, information about patients treated in primary healthcare and in hospitals, as well as the number of people recovered or deceased. E-services that had been used for years, such as national e-prescriptions, doctor-to-doctor e-consultation, and e-referral systems, provided the opportunity to minimize the need for physical contacts, increasing health safety and optimizing resource use for both clinicians and patients.

Good digital infrastructure is one part of the equation in creating successful eHealth infrastructure. Delivering digitally transformed healthcare processes, services, and solutions to healthcare professionals as well as to the public must be continuously supported by digital skills empowerment and training efforts, as digital literacy has become a prerequisite to tackle challenges brought by unforeseen circumstances. Digital literacy has been on the Estonian digital agenda since 2004 (Ministry of Economic Affaris and Communication, 2004), thus providing public and healthcare professionals with numerous training programs. Curiously, the uptake of digital tools, however, has been rather slow and uneven among health professionals and supporting staff (European Commission, 2018), indicating that concrete training strategies and investments are not keeping up with the reforms and investment on the digital tools and eHealth in general. To tackle this, teaching of digital skills must be better integrated into healthcare education and training pathways, together with developing common tools to continuously measure digital competences of the health workforce to develop most effective interventions.

Building resilience through sharing the responsibility and accountability

As seen in the case of building the national tracing app HOIA, following up-to-date strategies help create solutions that do not have to be rebuilt after the end of the crisis to meet "peacetime" requirements and rules. Following the rules is easier if the development is controlled by the State, as was the case with HOIA. However, this is not feasible for all solutions necessary for the proper functioning of the health system as it would create a huge bottleneck for novel solutions. This means that more parties must be involved, and they all must acknowledge the strategic principles as their own.

An important factor in achieving this is for the government to share the responsibility and accountability. A great example of the shared responsibility is the setup of the national eBooking system in Estonia. The national eBooking system enables patients to book outpatient visits to healthcare specialists of all Estonian Healthcare service providers, as the system integrates overview of the free timeslots for visits, patient referrals to different specialists, as well as bookings in one information system. The setup is interesting as it allows the healthcare providers (hospitals and primary care clinics) to keep control of their schedules and available timeslots through their own systems. The central system only relays information in the local systems when the patient asks for available appointments through the national patient portal. Patients can book appointments through local systems, which are then propagated to the central system. This setup gives more flexibility to healthcare providers as they keep control of most of the booking system, since it is part of their own local system. However, it is their responsibility to follow cocreated strategic principles and service requirements, which they are held accountable for. The accountability together with the common strategic principles is set by the service agreements with the Estonian Health Insurance Fund (EHIF), the sole provider of public health insurance in Estonia. Apart from the burden of the service development being shared between stakeholders, the resulting solution offers great benefits from the patients' point of view. Specifically, the central system acts as a single door to all the different independent healthcare provider appointment systems while all the separate systems lead back to the central system. No matter the door the patient chooses to enter, the health system they are always guided to the right place.

Not only does this shared responsibility increase the quality of the system, but the setup also allows for resilience in that if one system is overloaded with requests, the other can pitch in. In other words, if the central system is overloaded with requests, patients can still access and book appointments through the healthcare providers' local systems. This feature of the Estonian eBooking system was extremely helpful when the central system was overloaded by people wanting to book their COVID-19 vaccine appointments. Even if the government had spent 50 times more resources on the central system infrastructure, it would not have withstood the demand. Having shared the responsibility with stakeholders allowed the patients to still get access to health services through the local systems. This kind of flexibility would not have been possible if the government had been solely responsible for the creation of the eBooking system.

Having the experience of other similar good examples has created an appetite for more established and well-communicated common principles which allow shared responsibility with other stakeholders. The government has acknowledged this and the Estonia's newest eHealth strategy, called *Next Generation Estonian National Health Information System 2030*, aims to establish more common frameworks and principles for the ecosystem to follow.

Building resilience through organizational learning culture that is open to flexible approaches

Healthcare is always changing. To be competitive and to provide the best services to citizens, it is important that governments are open to novel solutions. One of the ways to achieve this is via good legislation that leaves room for innovation. During the COVID-19 pandemic, huge demand for healthcare services resulted in the exhaustion of the workforce necessary for PCR and antigen testing. An Estonian company created a promising solution to the problem. Their app allowed users to record themselves carrying out self-testing which was then reviewed and certified by a qualified professional. This resulted in immense timesaving, allowing the medical professionals to attend to multiple people in same timeframe, with the possibility to fully automate the certification process at one point. However, a problem arose that the existing Estonian legislation was heavily dependent on physical healthcare provision, thus, creating a barrier in scaling the digital solution. Despite providing fully digital service, it was necessary for the reviewing healthcare specialist to have physical rooms with all the necessary equipment of a regular doctor's office. To future proof, governments should look at their regulation regarding healthcare provision and allow room for novel solutions (including fully digital service provision) in order to be better prepared for upcoming challenges.

Sometimes, however, legislation is not the biggest barrier in introducing innovative solutions. An excellent example of this is the boom of telehealth solutions throughout the pandemic (Estonian Health Insurance Fund, 2020). For years, to no avail, stakeholders in Estonia had argued over the best way to introduce telehealth. Some main issues that were constantly referenced to, about why telehealth solutions were not in use, were the missing legislation, infrastructure, funding schemes, and training. When the pandemic hit, many of the health services could only be offered remotely. To accommodate for this, the only legislative change that was made was the requirement of providing telehealth service with similar or better quality compared to "regular" services. The market already had many digital solutions for providing service remotely and specialists already had experience of remote care through phone consultations. The only addition was using similar methods in video format with digital tools with which specialists were quite familiar through use of Skype and similar solutions in their private lives. Therefore miniscule changes (e.g., small changes in legislation, introducing EHIF financing, and using solutions already proven in other sectors) were all that were needed in order to introduce teleconsultations, teletherapies, and other teleservices into the Estonian health system.

Despite the eternal fears keeping telehealth from being implemented, the Estonian health system did not break down. On the contrary, patients were extremely satisfied and are now asking for more teleservices (Estonian Health Insurance Fund, 2021). The lesson learnt was that sometimes just being open to change and starting somewhere is all that is needed.

Healthcare does not exist alone: Importance of integrated service provision and PPP collaboration

Since disruptions in health systems also have detrimental impact on other sectors and nonhealth actors of societies and economies, an effective response to public health threats as the COVID-19 pandemic requires mechanisms at regional and subnational levels, depending on the context. A key attribute of resilient health systems is that they are integrated with other fields, such as social care (WHO, 2021).

One way to do this would be to make sure that the government is ready to set up multisectoral coordination and collaboration to pool resources from different sectors during a crisis. In Estonia during the COVID-19 crisis, all three sectors (vocational, social, and health) were included in the immediate response. Estonia has a single Ministry of Social Affairs that oversees all of them. This integration allowed the reserves from all three funds (unemployment, social insurance, and health insurance) to be parallelly used to achieve maximum efficiency on all fronts. However, a limitation emerged in that the coordination of the funding schemes and crisis response measures between health and sectors overseen by other ministries was not as operative, as there was a lack of common cooperation and legal frameworks that would effectively support an instant collaboration.

Also, public health threats such as those emerging from the COVID-19 pandemic need to be readily addressed with measures from different sectors. Throughout the COVID-19 crisis, there have been coordinated efforts to tackle the pandemic while having minimal adverse effect on the labor market. For example, if sick leave previously only covered the pay loss from the third leave day, during the pandemic, the government quickly changed this timing to the first day of leave, thus effectively nudging residents to refrain from working when sick. Furthermore, to minimize the administrative burden for the healthcare providers to approve the sick leaves, for the first months of the pandemic, a solution was tested that allowed patients to apply for sick leave for COVID-19 online through a national patient portal (and have it automatically opened), as opposed to first contacting their family physician by phone or requiring a face-to-face appointment. The piloted solution proved to be very effective in limiting the administrative burden for healthcare providers, providing increased flexibility for physicians to contact patients for follow-ups and treatments, and also significantly reduced close physical contacts, thus reducing the risk of infection. Service assessments show that patients rarely misused the right to automatically open the sick leave request.

The extensive containment measures implemented during the COVID-19 pandemic in Estonia had a significant spillover effect on the local economy, especially in the service sector, e.g., food, hospitality, and entertainment. To ease the adverse impact to the industries and prevent unemployment, the Estonian government also set up support measures for the companies to incentivize the labor costs. In the pursuit of finding ways to minimize the spillover of the containment measures, the EU COVID-19 certificates offered an opportunity to open the economy without risking speeding the spread of the infection. However, the administrative work needed to provide the certificates for all residents proved to exceed the health sector capabilities; thus there was a critical need for the other sectors to help. The Social Insurance Board stepped in to issue the certificates to residents who could not generate them themselves, e.g., foreigners or people with access problems to the national patient portal. In addition, local municipalities and libraries also offered help in creating the certificates (ERR News, EU COVID certificates to be issued by Social Insurance Board, 2021).

Multisectoral collaboration is extremely helpful in successful recovery. In addition to usual recovery measures, such as different economic incentives and public investments, social innovation and social entrepreneurship can contribute to boosting the local economic recovery. Examples of these solutions are creating new enterprises and jobs, or by (re)integrating disadvantaged individuals into the labor market (OECD, 2020). According to the OECD in Estonia, there are several conditions to foster the development of social entrepreneurship, such as a dynamic civil society, rich welfare traditions, the vibrant start-up culture, and the multiplicity of "bottom-up" initiatives that all have contributed laying the ground for the ecosystem for social innovation to develop and flourish. Nevertheless, several challenges, including the lack of coherent institutional and legal framework, barriers to funding or entering the market, and lack of capacity building to support social entrepreneurship, are still preventing the field from further developing. Government must play a critical role in helping social entrepreneurs overcome these barriers by shaping enabling policies to foster greater synergies and coherence between policy areas while reducing policy silos (OECD, 2020).

There is room for improvement in the public framework of collaborative innovation and social entrepreneurship, but its rewards can already be seen through the public sector embracing initiatives from the private and third sector. E-hackathons "Hack the Crisis" and "Global Hack" initiated in cooperation with the start-up community in Estonia brought together actors from the public, private, and third sectors to create viable solutions to help fight the crisis. It is worth emphasizing the unprecedented speed of the ability to organize the first hackathon: just hours after the state of emergency in Estonia was declared, a public-private partnership-led e-hackathon was held. During the hackathons, numerous scalable solutions were created that found extensive use in the crisis response by the public sector, hospitals, and other stakeholders. Among others, a medical workforce sharing platform for volunteering medical specialists was created which streamlined the offering of help and relieved the burden of health service providers most in need of extra labor. Another widely adopted solution which was created during one of the hackathons was a personal protective equipment tracking and delivery tool launched for the Estonian Government to handle the logistics of emergency supplies. The solution has now found its way into global markets. This shows how public-private partnerships can benefit from both sides and can increase the resiliency of the healthcare system.

The way forward

Governments have an important role in ensuring good operation of their health systems in stressful times. The COVID-19 pandemic offered many countries, including Estonia, an unprecedented opportunity to put its health system to the test. Estonia was lucky to have many preconditions to support a successful response to the pandemic. The experience was not without its bumps in the road which helped highlight the fact that to provide an environment which is ready to adapt and respond to challenges ahead, constant work and improvement is needed. To increase the resilience of health system and its components, it is vital to have well-communicated strategic management complemented by a clear governance setup; orchestrated collaboration between public bodies, private sector, and other stakeholders; a transparent information-sharing network supported by sustainable technological infrastructure; and flexibility to embrace innovation and fast paced partnerships. With the lessons learnt during the COVID-19 pandemic, governments can reassess their current efforts and make improvements to be ready for future needs, whatever they may be.

References

Alam, U., Nabyonga-Orem, J., Mohammed, A., Malac, D. R., Nkengasong, J. N., & Moeti, M. R. (2021). Redesigning health systems for global heath security. *The Lancet*, *9*, e393–e394. https://doi.org/10.1016/S2214-109X(20)30545-3.

ERR News. (2020). *Estonia launches coronavirus exposure notification app 'HOIA'*. Retrieved from https://news.err.ee/1125119/feature-estonia-launches-coronavirus-exposure-notification-app-hoia.

ERR News. (2021). *EU COVID certificates to be issued by Social Insurance Board*. Retrieved from https://news.err.ee/1608308624/eu-covid-certificates-to-be-issued-by-social-insurance-board.

Estonian Health Insurance Fund. (2020). *Kaugvastuvõtte jätkatakse ka peale eriolukorda*. Retrieved from https://www.haigekassa.ee/blogi/kaugvastuvotte-jatkatakse-ka-peale-eriolukorda.

Estonian Health Insurance Fund. (2021). *Kaugteenuste tagasiside: vastuvõtt võiks kesta kauem ja toimuda läbi videosilla.* https://www.tervisekassa.ee/uudised/kaugteenuste-tagasiside-vastuvott-voiks-kesta-kauem-ja-toimuda-labi-videosilla.
European Commission. (2018). *Communication from the commission to the European parliament, the council, the European economic and social committee and the committee of the regions on enabling the digital transformation of health and care in the digital single market.* Retrieved from https://eur-lex.europa.eu/legal-content/EN/TXT/?uri=COM%3A2018%3A233%3AFIN.
Ministry of Economic Affairs and Communication. (2004). Principles of the Estonian Information Policy 2004–2006. Estonian IT Policy: Towards a More Service-Centred and CitizenFriendly State. Retrieved from https://www.mkm.ee/sites/default/files/infopoliitika_pohialused_2004-2006.pdf.
OECD. (2020). *Boosting social entrepreneurship and social enterprise development in Estonia. In depth Policy Review* (Vol. OECS LEED Papers 2020/02) Paris: OECD Publishing.
Olu, O. (2017). Resilient health system as conceptual framework for strengthening public health disaster risk management: An African viewpoint. *Frontiers in Public*, *5*(263).
Socha-Dietrich, K. (2021). *Empowering the health workforce. Strategies to make the most of the digital revolution.* OECD Health Working Papers (No. 129). Retrieved from. https://doi.org/10.1787/37ff0eaa-en.
Thomas, S., Sagan, A., Larkin, J., Cylus, J., Figueras, J., & Karanikolos, M. (2020). *POLICY BRIEF 36: strengthening health systems resilience Key concepts and strategies.* Copenhagen: World Health Organization.
Visit Estonia. (2020). *Visit Estonia.* Retrieved from Estonia is a digital society: https://www.visitestonia.com/en/why-estonia/estonia-is-a-digital-society.
WHO. (2021). *Fostering resilience through integrated health system strengthening: a perspective on Liberia and Bangladesh (Fostering resilience through integrated health system strengthening series).* Geneva: World Health Organization.
Wilkinson, M. D., Dumontier, M., Aalbersberg, I. J., et al. (2016). The FAIR Guiding Principles for scientific data management and stewardship. *Science Data, 3.* https://doi.org/10.1038/sdata.2016.18.

CHAPTER 36

Health innovation and advances in the Republic of Sierra Leone

Sidique Wai[a,b] and Judy Kuriansky[a,c,d,e,f,g]

[a]Office of Ambassador Sidique Wai, Embassy of the Republic of Sierra Leone to the United States, Washington, DC, United States, [b]ECOWAS Ambassadors Group, Washington, DC, United States, [c]United African Congress, New York, NY, United States, [d]Department of Clinical Psychology, Teachers College, Columbia University, New York, NY, United States, [e]International Association of Applied Psychology, Indianapolis, IN, United States, [f]Global Council SDG3 (Health), "SDGs in Action" Initiative, Dubai, United Arab Emirates, [g]Mission of the Republic of Sierra Leone to the United Nations, New York, NY, United States

Introduction

The health status of the population in the West African state of the Republic of Sierra Leone has been challenged by many traumas, including natural disasters (a flood and mudslide), accidents (a tanker explosion), and epidemics (HIV/AIDS, Ebola and COVID-19), as well as ongoing recovery from a protracted civil war despite ending two decades ago. Under the leadership of the current President, His Excellency Dr. Julius Maada Bio, a focus on health as a pillar of the Human Capital Development agenda has led to advances in the health system, through the use of technology as well as mobilizing partnerships with both the public and private sector, and diaspora support in health-related initiatives. This chapter presents an overview of health in Sierra Leone including a brief historical review, technology solutions, and recent and ongoing initiatives in healthcare where technology is valuably applied.

Background

The economic status of Sierra Leone has historically been challenging, although the situation is shifting. Sierra Leone has ranked among the least developed countries in the world (UNDP, 2020). However, despite the economic slowdown from the COVID-19 pandemic, the GDP per capita has been increasing, from US$509.4 in 2020 to US$515.9 in 2021 (The Global Economy, 2021). Despite this growth, the health status in the country has been concerning, as poverty prevents many families from accessing quality healthcare. Also, Sierra Leone has a ratio of only

Resilient Health. https://doi.org/10.1016/B978-0-443-18529-8.00036-6

three physicians per every 100,000 people in a country of 7.5 million people, a low density of physicians to citizens compared to other countries in Africa. Further, the country has long struggled with high maternal and child mortality rates (The Borgen project, n.d.).

The hope in technology

Harnessing technology in Sierra Leone, as in other countries, holds promise to improve health for the population, especially for vulnerable groups like women and children, as well as to accelerate economic growth and social stability. Benefits are considerable for both providers and patients in accessing services for those "left furthest behind" by improving diagnostic and treatment skills, enabling better self-care, cost-efficiency, providing preventive health education programs, offering training, and solving problems of inadequate capacity. These challenges are being faced head on by the country, leading to a more hopeful and bright future.

Government embracing technology

The current President of Sierra Leone, His Excellency Retired Brigadier General Julius Maada Bio, has charted an ambitious development plan for the country outlined in his Human Capital Development Agenda which prioritizes increased development in the health sector, education and food security, as well as other sectors like investment and trade and public safety. The pillars of health, education, and food security represent important goals outlined in the United Nations Agenda 2030 for Sustainable Development (Sustainable Development Goals 3, 4, and 2, respectively), which guide governments in progress toward the well-being of people. The Sierra Leone government is also committed to the internationally agreed objective of Universal Health Coverage.

To carry out an ambitious mission of digitalization, President Bio formed a Directorate of Science, Technology and Innovation (DSTI) to revolutionize the government's capabilities to help its citizens and to progress technology in the country. The DSTI intends to make Sierra Leone a country where innovation can thrive and where all people have opportunities for a better life, including as entrepreneurs. The model for this plan derives from the country of Estonia which digitized government and society not only for healthcare but registries in many aspects of life, including cars, paying taxes, and even pet-care (Valavani, 2019). In February 2019, Sierra Leone and the e-Governance Academy of Estonia established a technical collaboration for public service delivery and administration in Sierra Leone. Top research institutions such as Yale University were also approached during the year to explore cooperation in public health and technology.

An information system within the DTSI provides detailed data on access to healthcare, as well as education and water for all towns and regions in the country. In September 2019, President

Bio unveiled the world's first portable DNA sequencer, which provides rapid, meaningful information in the fields of healthcare as well as agriculture, food, and water surveillance and education (Valavani, 2019). DSTI has partnered with the UNICEF Sierra Leone Country Office Innovation team to create government processes using data for decision-making and to improve efficiency and transparency in projects related to the lives of women and children.

In a major advance, in June 2019, Sierra Leone became the first country in the WHO Africa region to fully transform its national disease surveillance system from the traditional old-style paper-based format to a web-based electronic platform (World Health Organization Sierra Leone, 2019). The old system of mobile phone text messages, phone calls, hand delivery or email was replaced by features like hand-held tablets for health facility staff to report health data in real-time. Health workers found the new eISDR system "revolutionary and motivating." The project was supported by WHO and other partners, including the US Centers for Disease Control (CDC), e-Health Africa, Focus 1000 and GIZ.

The Sierra Leone National Innovation and Digital Strategy

The DSTI oversees the Sierra Leone National Innovation and Digital Strategy (2019–2029), which charts the vision for the government, development partners, and the people on how to use science, technology, and innovation to achieve the UN Agenda 2030 for Sustainable Development with its 17 Sustainable Development Goals, as well as the African Union Agenda 2063 and the Sierra Leone National Development Plan (NDP) (Directorate of Science, Technology and Innovation, 2019).

With regard to healthcare, the ambition is to:

(1) Apply data science methods, including Artificial Intelligence (AI), to diagnostic images, genomics, mobility, environmental, and other data analytical methods for automated disease diagnostics, predicting disease outbreaks, disease prevention, and identifying high-risk groups for planning and resource allocation;
(2) Use AI to support junior- and expert-level healthcare practitioners to make better healthcare decisions related to treatments and referrals in quicker time and for more people; and,
(3) Use an integrated community and technology approach to significantly reduce maternal and child mortality.

Timeline overview of technology projects in Sierra Leone

Technology contributions to the health sector development dates back to 2013, when World Vision Sierra Leone won an eHealth award from the African Development Bank for their innovative approach to deliver health services with mobile technology (Bundu, 2013)

Community health workers in Sierra Leone were trained to use a mobile phone to give counseling and health information about pregnant women during pregnancy and after delivery, and send it to a cloud server accessible by other relevant staff. The health worker can also provide the pregnant woman with any needed referrals to the nearest health facility.

In 2014, in a mobile health intervention research project in a district in northwest Sierra Leone, health providers were connected to clients through mobile telephones with sim cards from Airtel (a mobile phone network provider in Sierra Leone) with prepaid phone credit and solar-powered battery chargers to allow health workers to call clients and remind them of appointments. Traditional birth attendants—important in the Sierra Leone cultural context—were similarly provided with mobile phones and training as the health workers, to follow-up existing clients or identify new clients for care or family planning, and to use the phones to communicate with peripheral health units (PHUs) including for referrals (KIT Health, 2014).

A national survey conducted in 2015 found that mobile phone usage is widespread in Sierra Leone, with 52% of mobile phone owners having a basic model (BBC Media, n.d.). Low-tech mobile phone technology is easily accessible and especially useful with community health workers (CHWs) who can be in remote community settings. CHWs are at the core of Sierra Leone's health force to provide basic health service, bridging the gap between trained health professionals and communities with scarce access to doctors or nurses, as is the case in many low resource countries. Given that improving maternal health is central to Sierra Leone's health plan, these CHWs need to be supported in their role. To help CHWs reach community people in need, the BBC Media Action piloted two mobile health services. In *Welbodi Tok* (which means "health talk"), CHWs use a basic mobile phone to dial a phone number accessing recorded advice on topics ranging from the treatment of diarrhea to breastfeeding and good hygiene, dispensed in a Q&A between a fictional "village woman" and authoritative "doctor." The other project, *Welbodi Campus*, uses similar technology for training community health workers to complete nine-minute courses with quizzes to test their learning.

Technology approaches received a boost in response to the tragic Ebola Virus Disease during the epidemic of 2014–2016 (similar to what happened in response to the COVID-19 pandemic). In one project, IntraHealth International, a nonprofit organization based in Chapel Hill, North Carolina, USA, partnered with Sierra Leone to implement mHero, a two-way, mobile-phone-based communications system using basic text messaging or SMS to connect the Ministry of Health and Sanitation (MoHS) with health workers, many of whom live in remote and hard-to-reach places (BenDor, 2015; Nicholsen, 2016). This communication platform reduces communication barriers between health workers and their support systems and allows the MoHS real-time access to important information to support health workers and improve health services. This approach is aligned with President Bio's COVID-19 strategy to improve data collection and sharing.

The Bintumani Declaration

A first of several meetings was held in 2016 for several organizations to work with the Government of Sierra Leone to define and document Sierra Leone's requirements for community health workers and digital health efforts, as well as to strategize recovery from Ebola, and to improve maternal, neonatal, and adolescent health (BenDor & Taliesin, 2016).

The meeting, called the Sierra Leone Health Information Systems Interoperability Workshop, held 2–4 August 2016 at the Bintumani Hotel in the capital of Freetown, was a partnership of the MoHS and the World Health Organization (WHO) with IntraHealth and USAID. It included government stakeholders, donors, local and international NGO partners, and the private sector, and led to the Bintumani Declaration. This Declaration articulated the vision for improved health information systems (HIS) in these areas, to: review and improve the availability, appropriate access, and use of health information; access and use health information technology; attract donors; commit to strengthen the existing governance structures; and share experiences with other governments (Liberia, Nigeria, and Kenya).

Improving the health information system

In 2017, the MoHS inaugurated an eHealth coordination hub to facilitate the systematic application of digital health solutions, services, and applications for health systems improvement through data (Chukwu et al., 2022). This led to the launch of the first national digital health strategy for 2018–2023 and the implementation of digital health solutions using big data and artificial intelligence (AI) to improve healthcare in general, and maternal and child health in particular. In 2019, the DSTI also launched a broader National Innovation and Digital Strategy for 2019–2029 outlining strategic health ambitions.

A mapping revealed considerable digital health solutions operating at health facilities in the country. Results showed that hospitals mostly use computer-based solutions, while peripheral health units (PHUs) mostly use tablet-based solutions. A survey of these outlets revealed opportunities to leverage the six V's of big data—value, volume, velocity, veracity, variability, and variety—to achieve the national digital health vision.

Civil society and private sector partnerships

World Vision Sierra Leone has contributed experience with digital health programming in two projects active in Sierra Leone since 2015 in a multistakeholder collaboration with national and local government (Worldvision, 2021a, 2021b). The AIM Health Plus project, with support from Irish Aid, provided CHWs in two districts with smartphones equipped with a version of Dimagi's CommCare software to use during home visits. The application supports CHWs using the Timed and Targeted Counseling (TTC) approach to promote positive health and nutrition behavior change among pregnant women and mothers or caregivers of children under two years of age. The application allows CHWs to submit community health data in near real time; provides reminders to help CHWs to visit homes at the ideal time during pregnancy, infancy, and childhood; and supports CHWs in conducting counseling sessions. Local language (Krio) audio clips are incorporated into the application to be acceptable to the community people.

Another digital health innovation, called the Mobile Training and Support Service (MOTS), aimed to strengthen networks of frontline health workers to ensure epidemic preparedness. Initially designed to train CHWs to support vaccine-related programs during the Ebola epidemic, the system was repurposed for the response to COVID-19 in 2020. The technology accomplishes a number of goals: enables CHWs to submit community health data in near real time, delivers training modules including quizzes to CHWs' mobile phones using Interactive Voice Response (IVR) in their preferred language, and enables online and office monitoring of CHWs by management. The project was supported by the Innovative Medicines Initiative and Janssen Pharmaceutical of Johnson & Johnson.

Unique AI application

In 2020, a partnership was formed between the MoHS and the DSTI with an AI technology company, Macro-Eyes, using machine learning and artificial intelligence (AI) to allow health leaders to virtually step inside any facility for an immersive view, to assess equipment and staff, and scan capabilities of a region (Macro-eyes health, 2020). By combining publicly available data, facility-specific information, and direct input from health workers, the product rapidly determines the facility's readiness to effectively care for COVID-19 patients and deliver essential medical care, enabling health systems to better respond to COVID-19 and provide essential care. The product STRIATA is an intelligent health system index that generates real-time views into infrastructure, materials, supplies, and human resources, enabling health facilities to proactively decide about, and prepare for, shifts in demand and best allocation of resources. Thus, the process revolutionizes the way assessments are done: faster, more accurate, and responsive to change on the ground. In 2021, additional funding was announced to support two more years of the deployment of this STRIATA system (Macro-eyes, 2021).

Drones

Drones provide access to places previously too remote or too difficult to navigate. In April 2019, Sierra Leone partnered with UNICEF to become a drone-testing site to better the lives of children in rural areas. In a significant step forward for health care delivery using technology, on November 29, 2019, Sierra Leone President Julius Maada Bio launched a "drone corridor" (a segregated area and airpace where drones and drone solutions can be tested and used) with support from UNICEF Sierra Leone and UNICEF's Office of

Innovations in New York City (UNICEF, 2019). In 2020, the Bill & Melinda Gates Foundation supported DSTI with a grant for a viable and cost-effective drone-delivery system for Sierra Leone's medical supply chain (Adoyooye, n.d.; The Borgen Project (2021). Drones allowed authorities in Sierra Leone's MoHS to have on-demand delivery for essential medical supplies; restock rural community health centers and hospitals in a timely, cost-effective manner; extend limited diagnostic coverage; and decrease response time to pathogen outbreaks.

The Technical University of Denmark

Technology assistance also comes on a smaller, yet still impactful, scale. For example, two University of Denmark (DTU) students have helped medical doctors in Sierra Leone save lives and improve the patient experience through technical support at Masanga Hospital (Møller, 2022). One of the students, with a master's student in the Earth and Space Physics and Engineering program, upgraded the electrical installation of the hospital, optimized the solar park at the hospital, and installed telephones to secure communication which is critical between the technical staff, healthcare staff, and management. The students also did weekly classes with the technical staff, who, though being skilled plumbers and electricians, did not know about biomedical equipment and had to learn how to maintain the machines.

Specific population application: Assistive technology

The need for assistive technologies (AT) for persons with various disabilities is considerable in all countries, including of course Sierra Leone (Clinton Health Access Initiative, 2022). The major causes of disability in the country are disease or illness (accounting for 40.5% of cases), followed by congenital disability (16.2%), other nonspecified causes (10.5%), accidents (8.8%), natural aging (8.1%), and other causes, including traffic accidents, occupational injuries, war injuries, and unspecified injuries (less than 5%) of the total. AT is defined as a subset of health technology that "refers to assistive products and related systems and services developed for people to maintain or improve functioning and thereby promote well-being, such as eyeglasses, hearing aids and wheelchairs. Assistive products maintain or improve an individual's functioning and independence, thereby promoting their wellbeing" (WHO, 2018). ATs include wheelchairs, hearing aids, walking frames, spectacles, pill organizers, and prosthetic limbs, as well as assistive information and communication technology such as memory aids, specialized computer hardware and software, augmentative and alternative communication, and customized telephones. These products enhance the functioning and independence of persons with disabilities; boost self-esteem; reduce the need for carers, higher level intervention and welfare costs; and facilitate integration and participation in society. The Clinton Health Access Initiative produces the Priority Assistive Technology Products List for

Sierra Leone which lists all AT products needed in the country (Clinton Health Access Initiative, 2022), "to help improve access to appropriate, quality Assistive Technology (AT) products at an affordable price that will help enhance the functioning and independence of persons with disabilities, while facilitating their participation and integration in society."

Technology and the COVID-19 response

The highly contagious nature of COVID-19 necessitated that technology play a crucial role in breaking the chain of transmission. Furthermore, technology helps the government and health facilities be more efficient and effective. Lessons learned from the Ebola response were applied in the COVID-19 context (Jalloh, 2021); for example, Sierra Leone had used mobile money to disburse payments to frontline health workers nationwide, and also set up a Public Health National Emergency 117 hotline to collect data, respond to questions, and dispense information (Kahn et al., 2016). Also, social media campaigns played a major role in the Ebola response (Ngewa & Kuriansky, 2016).

In the COVID-19 response, members of Sierra Leone's National COVID-19 Emergency Operations Centre—consisting of a consortium of organizations including the DSTI, MoHS, Ministry of Information and Communication (MIC), and Statistics Sierra Leone (Stats SL)—partnered with a coalition of science and geospatial organizations (GRID3, Esri, Maxar Technologies, Fraym, the Global Partnership for Sustainable Development Data, and UN Economic Commission for Africa) to generate geospatial datasets, analyses, and tools to enhance the government's response and to deliver information about at-risk populations across the country (Space in Africa, 2020). The data was then openly accessible in a national digital dashboard and COVID-19 hub for anyone to access, including experts and the general public.

Another project to develop digital solutions, especially to support frontline workers in the COVID-19 response, was a partnership of DSTI and the MoHS with Dimagi, a for-profit social enterprise based in Cambridge, Massachusetts, USA, that delivers open-source software technology suitable for low-resource settings and underserved communitie, whose flagship platform, CommCare, is covered in another chapter in this volume (Javetski, 2024). The project provided a mobile data collection and service delivery platform for contact tracing and distributing public health messages to educate the public (Dimagi, 2021; DSTI Media, 2020; Jalloh, 2021). Since April 2020, Dimagi developed multiple products to tackle different dimensions of the pandemic, for example, a system for districts to track service delivery in quarantine sites and check that incoming travelers in quarantine were safe, psychologically stable, supplied with food, and scheduled for testing. Other collaborations with local and global leaders included MIT, the Bill and Melinda Gates Foundation, UNICEF Sierra Leone, and UNDP to strengthen the local ecosystem for technology and innovation (Fig. 1).

Fig. 1
CommCare users during a COVID-19 response digital system training. *Website for Dimagi: https://www.dimagi.com/blog/covid-19-pro-bono-project-highlights-2021/.*

The Tony Blair Institute

In response to COVID-19, the Tony Blair Institute supported the Sierra Leone government in establishing systems to enhance information-sharing and communication among pillars of the response (surveillance, security, nutrition, risk communications, psychosocial matters, labs, case management, and drug and medicine supply) and to facilitate data collection on quarantined travellers and contact tracing (Tony Blair Institute for Social Change, 2021).

IsraAID

The nongovernmental international humanitarian aid organization IsraAID, based in Israel, has been working in Sierra Leone on multiple projects over many years, including during the 2014–2016 Ebola epidemic, when this chapter's second author collaborated with them to conduct in-person trainings and workshops for psychosocial support for various groups, including affected communities as well as health workers, social service agency staff, social work students, and burial teams (Kuriansky, 2016; Kuriansky et al., 2016). The IsraAID staff also helped assess and train operators in the national hotline to provide information and advice (Kahn et al., 2016). Further, the organization has a medical care program targeting reproductive health to decrease the maternal mortality rate.

Individual entrepreneurs

Sierra Leonean social entrepreneur and technologist Evelyn Lewis, founder of the SBTS Group software company, developed a software application for contact tracing to prevent the spread of COVID-19, called the Health Outbreak Manager (Ighobor, 2020). Through Bluetooth technology, the app collects and manages data of people for exposure, through collecting mobile telephone numbers of everyone who might have encountered a person who reports COVID-19 symptoms to health authorities or tests positive, and automatically sending a text message informing them of their risk and advice to go for testing. The machine's app can be programmed for use in noninfectious disease conditions.

Health projects with the Diaspora in-person and online

The Diaspora in Sierra Lone—as in all countries—plays a major role in development, connections with whom depend on the internet. Sierra Leone has placed high priority on informing and involving the Diaspora, facilitated by the Embassy in Washington DC and the first author's role as Ambassador who has hosted many online Town Halls and Zoom conferences with the diaspora. Additionally, a Town Hall meeting for the diaspora was held in October 2023 with the Sierra Leone Minister of Health, Dr. Austin Demby, to inform and engage the Diaspora in ongoing progress about health in the country.

Social innovations are essential in resilience-building and recovery from crises. These are especially effective when they involve the diaspora. Examples of recent such health projects with online and in-person components follow.

The COVID-19 Task Force

In reaction to the COVID-19 pandemic, a COVID-19 Task Force was formed by a coalition of the National COVID-19 Diaspora Task Team USA, the Embassy of Sierra Leone to the United States, Honorary Consuls, Sierra Leone Diaspora communities and friends of Sierra Leone, to support the appeal for assistance made by the Ministry of Foreign Affairs and International Cooperation to the Embassy of Sierra Leone to the United States of America. This impressive initiative was led by the late Dr. Michael Rogers, Chairman of Healthcare Sierra Leone, a health consortium in the United States of America, with a team of diverse health professionals and philanthropic friends and supporters, including this chapter's second author. While the infection and death rate in Sierra Leone were fortunately kept at a minimum, people in Sierra Leone were still sadly suffering health and economic hardship. In addition, over 300 Sierra Leoneans in the Diaspora in the United States had lost their lives to COVID-19-related disease. This manifested in a campaign that led to contributions over a short period of time, which impressively included US$50,000 cash and a 40-foot container of medical supplies valued at over US$1 million dollars. This contribution, combined with a contribution from Howard University in the USA, was sent to the MoHS, and was included in an application for a CDC grant of US$5 million dollars on infectious diseases, which was approved.

Support for the First Lady's "Hands off Our Girls" Project

Sierra Leone First Lady Her Excellency Fatima Maada Bio's initiative, "*Hands Off Our Girls*," advances the health and legal rights of girls against all forms of gender-based violence, rape, early marriage, child trafficking, and teen pregnancy. On September 25, 2021, the "Friends of Hands Off Our Girls," co-produced by the authors of this chapter and partners, including from the United African Congress, held a fundraiser in Maryland in the USA, to support this initiative, featuring speeches and performances of Sierra Leoneans and friends (Kuriansky, 2021). The event was attended by both the President and First Lady of Sierra Leone along with many friends and supporters, in an impressive show of support and community cohesion, as shown in Fig. 2.

The campaign is successfully raising awareness and influencing policy, e.g., raising minimum imprisonment for rape convictions limits and creating a Fast-Track Special Court

Fig. 2
The fundraiser in support of the "Hands Off Our Girl" campaign. Pictured: top left: President Julius Maada Bio and First Lady Fatima Maada Bio; top right: H.E. Ambassador Sidique Wai, Dr. Judy Kuriansky, and Mr. Gordon Tapper. Bottom left: Cosponsors and supporters with Ambassador Wai and Dr. Judy; bottom right, left to right: Abubakarr Sadick, H.E. Ambassador Alhaji Fanday Turay, then—Ambassador of Sierra Leone to the UN; H.E. Sheku Mesali, Deputy Chief of Mission, Embassy of Sierra Leone to the U.S.; H.E. Victoria Sulemani, now Director-General and Ambassador-at-Large, Ministry of Foreign Affairs and International Cooperation, Republic of Sierra Leone; H.E. Dr. Michel Imran Kanu, then—DPR now Ambassador of Sierra Leone to the U.N.; H.E. Ambassador Sidique Wai, Chief of Mission of the Embassy of Sierra Leone to the U.S.; Dr. Judy Kuriansky, advisor to Ambassador Wai and to the Mission of Sierra Leone to the UN; H.E. Dr. David Moinina Sengeh, then—Sierra Leone Minister of Education, previously Chief Innovation Officer for the Directorate of Science, Technology and Innovation, now Chief Minister.

for rape cases. Extensive efforts have also been made to facilitate access to reproductive healthcare for girls (e.g., sanitary pads) and to provide treatment for conditions like cancer and fistula for women. The goal is also to give a voice to the girls themselves to build their self-esteem and to educate boys to respect girls. In the project, nurses and medical doctors at the front lines are being trained to treat rape cases. In a major step, ably led by First Lady of Sierra Leone, H.E. Fatima Maada Bio, an historic Resolution was passed in September 2022 by the United Nations General Assembly entitled "International cooperation for access to justice, remedies and assistance for survivors of sexual violence," which condemns all forms of sexual and gender-based violence and ensures justice for survivors (United Nations General Assembly, 2022). Co-sponsored by 84 countries, led by the Republic of Sierra Leone in collaboration with Federal Republic of Nigeria and notable support of Japan, the Resolution recognizes deleterious rape in peacetime (not just as a "weapon of war") and assures survivors access to justice, reparations and assistance.

Another historic step followed, with a General Assembly Resolution on "Sexual Exploitation, and Abuse: Implementing a Zero-tolerance Policy" again co-facilitated by the Republic of Sierra Leone and the Federal Republic of Nigeria, UN entities, the Global Collaborative and others, which was sponsored by 118 nations and adopted by consensus by all 193 member states (even acknowledged by the Mission of the Holy See), which established November 18th as World Day for the Protection of, and Healing from Child Sexual Exploitation, Abuse and Violence. Next steps are being taken to build on this impressive progress, with the integration of technological tools to escalate impact.

Psychosocial response to the tanker explosion tragedy

Another tragedy in Sierra Leone leading to disastrous health outcomes, with nearly 100 people dying and many others suffering severe burns, was a fuel tanker explosion on November 5, 2021, after colliding with a truck in a busy intersection near the capital of Freetown. The injured were transported to area hospitals in Freetown, which quickly became overwhelmed due to limited capacity and lack of burn care facilities or specialists. Within a short two days, the Embassy of Sierra Leone to the United States, in partnership with Healthcare Sierra Leone Medical Group in the United States and the Health Emergency Network in the United States, mobilized burn specialists in the USA to go to Sierra Leone to treat the wounded. The second author of this chapter (Kuriansky) further participated in this recovery effort by conducting in-person workshops with survivors and their families to build resilience, the first of which was held on April 29th, 2022, as shown in Fig. 3. An assessment showed the positive impact on participants who also reported feeling supported and hopeful after the workshop experience. The interventions support the importance of community outreach providing not only immediate but ongoing medical assistance combined with psychosocial support. A training of her workshop techniques for sustainability was also held by Kuriansky of staff of a local community service organization, Berny's Hope Foundation,

Fig. 3
Workshop and training volunteers of Bernys Hope Foundation for tanker explosion burn survivors and their families, designed and led by Dr. Judy Kuriansky.

who reported helpfulness of the skills-building. The training was videotaped for use. As for all community-based post-trauma interventions, tech-based tools on hand-held devices can be very useful for training and to make community connections, track needs, and provide support, which would amplify impact, and allow communications with those impacted by an emergency, and prevent panic and disinformation.

A Center of Excellence

A review conducted by the National Library of Medicine revealed that the frequency of both natural and man-made disasters in Africa is escalating. Complex disasters and mass casualty events are also on the rise with civil war and unrest on the continent. The impact of these events, coupled with long-term recovery required from the global COVID-19 pandemic, has strained and nearly exhausted healthcare systems.

Given all these health needs and emergencies, the Embassy of Sierra Leone to the United States (headed by the first author of this chapter, the Ambassador of Sierra Leone to the United States and Chief of Mission, His Excellency Sidique Wai) mobilized partners to come together to discuss a public–private partnership to establish a Trauma Center of Excellence in Sierra Leone,

Fig. 4
Experts at the working group to establish a Center of Excellence.

as a hub for the West Africa Region. Held on June 21, 2022, at the Rutgers University Eagleton Institute of Politics in New Brunswick, New Jersey, USA, the co-sponsors were Rutgers Miller Center for Community Protection and Resilience, Rutgers Global, and key members of the Economic Community of West African States (ECOWAS), the Mayor of Millstone Township in New Jersey, and the Healey International Relief Foundation. At this one-day meeting, entitled "*West African Working Session Establishing Public Private Partnerships and Regional Trauma Center of Excellence in Sierra Leone and the Sub Region. Exploring Impact of Disinformation on African Healthcare System*," the working group, shown in Fig. 4, discussed the framework, logistics (location, personnel, training), partnerships, and funding of such a center for a concrete plan to make this project a reality. Joining the cosponsors were invited highly respected health experts, including representatives of other local medical schools, and psychosocial experts from both Sierra Leone and America, including the second author of this chapter. Attending virtually, the Sierra Leone Minister of Health and Sanitation, Dr. Austin Demby, explained the situation on the ground, underscoring the needs. The takeaway from this working group was to create a stand-alone entity with the stature and ability to attract other partners and philanthropists who could generously contribute to the mission and to the financial viability of the Center of Excellence, under its core values and principles of transparency, accountability, and quality service delivery to the communities in Sierra Leone and the region.

Supporting healthcare services

In continued efforts to support healthcare services in Sierra Leone, another fundraising event was held to support several health projects in the country. These included contributing to (1) the completion of building, and outfitting, a rural hospital already in process by a US-based foundation,

Berny's Hope Foundation (mentioned above); (2) the ongoing construction of a state-of-the-art full-service seven-building medical Center for Excellence central to the Sierra Leone capital of Freetown, named the "34 Military Hospital" spearheaded by the First Lady of Sierra Leone. This health complex will provide medical treatment and psychosocial services, house survivors of gender-baed violence, and have a family support unit as well as a stand-alone forensic lab to reduce waiting time to adjudicate rape cases; and (3) provision of direly needed medical equipment and supplies for two existing health clinics, one in the north of the country and another in the south. Ongoing planning is aimed at employing technological tools to amplify such efforts.

Training youth for the future in-person and online

With Sierra Leone's large youth population—as in all of Africa—focus is on insuring their physical, psychological, social and financial security. To accomplish this goal, facilitate country development and cultural connections, and foster essential positive mindset, The Sierra Leone-Jamaica Youth Leadership and Entrepreneurial Training and Exchange Program (LEEP) was conceived by the two authors of this chapter on behalf of Sierra Leone. Online training plays a role. Developed into a multistakeholder collaboration—which is the best model to accumulate capacity and achieve success—a team of colleagues from Sierra Leone and the United States came together, representing civil society (the United African Congress, Give Them a Hand Foundation), an international NGO (the International Association of Applied Psychology in consultative status with ECOSOC at the UN), the World Citizen's Foundation, and academic institutions, to join with equivalent bodies in Jamaica, overseen by the Honorary Consul of Sierra Leone in Jamaica, Dr. Rosalea Hamilton. The program is rooted in the historic journey of the Jamaican Maroons uprising against rule of the British, who exiled them to Canada, and then to Freetown in Sierra Leone. The students will learn vocational and psychosocial and socioemotional learning skills for leadership and job placement that will be supported by virtual meetings and essential technological tools and connections to facilitate learning and excelling for real-world success. As is the case for academia in the post-COVID-19 period, technology has become a major element in pedagogy. A considerable number of academic partners committing their on-site resources and technological connections include the University of Sierra Leone; University of Technology, Jamaica; The PJ Patterson Institute of Africa-Caribbean Advocacy, an Institute of the University of the West Indies; Center for Global Africa; and the Institute of Law and Economics. Youth is also playing a major role, in the appointment of youth representatives, Clarice Labor, currently Senator of her class at Neumann University in Ashton, Pennsylvania, international recording artist and actor Duain Richmond, and Dr. Aminata Bash-Taqi, a young professional neurologist working with children. This project addresses SDG 3 (health), SDG 8 (decent work), SDG 1 and 2 (end poverty and hunger), SDG 5 (empowerment of women and girls) and SDG 4 (quality education) in the pursuit of personal growth, connecting to one's roots and cultural others, and achieving life success, escalated by harnessing digital technology to accomplish these goals.

A first-ever government adoption of technology to combat gender-based violence

Consistent with all these advances outlined above, Sierra Leone is rolling out the gender-based violence information management system (GBVIMS+) which is the online, cloud-hosted software application for GBV information and case management that is deployed in 19 humanitarian contexts globally serving over 80 user organizations. Importantly, this step in Sierra Leone marks the first-ever deployment of GBVIMS+ by a government. The rollout was presented at an event during the 2024 UN Commission on the Status of Women (Hwang et al., 2024).

Conclusion

The future of improved healthcare for the people of the Republic of Sierra Leone is encouraging, with the government's vision for development that has a strong emphasis on health, public–private partnerships in the health sector, and developments in health technology innovation, combined with ongoing dedicated and diverse efforts by civil society and the diaspora. Integrating the focus on health in a nexus model with education and economic development offers a comprehensive blueprint toward increased well-being of the population. Technological and scientific solutions, in partnerships, are having a transformative effect on Sierra Leone's government and its ability to meet the needs of its citizens, as well as progressing the nation's development.

References

Adoyooye, O.D. (n.d.) Sierra Leone is using drone technology to reduce its maternal mortality rate. Built in Africa. https://www.builtinafrica.io/videos/sierra-leone-is-using-drone-technology-to-reduce-its-maternal-mortality-rate.

BBC Media (n.d.) Improving maternal and child health through mobile technology. https://www.bbc.co.uk/mediaaction/where-we-work/africa/sierra-leone/welbodi-tok/.

BenDor, A.P. (2015). As spirits rise in Sierra Leone, technology offers more than Ebola recovery, Vital. https://www.intrahealth.org/vital/spirits-rise-sierra-leone-technology-offers-more-ebola-recovery.

BenDor, A. P., & Taliesin, B. (2016). *Getting out of the way so countries can lead: takeaways from facilitating the Sierra Leone HIS interoperability workshop in Freetown.* IntraHealth International https://www.mhero.org/news/getting-out-way-so-countries-can-lead-takeaways-facilitating-sierra-leone-his-interoperability.

Bundu, J. (2013). *Sierra Leone wins big with innovative health technology.* WorldVision. https://www.wvi.org/child-health-now/pressrelease/sierra-leone-wins-big-innovative-health-technology.

Chukwu, E., Garg, L., Foday, E., Konomanyi, A., Wright, R., & Smart, F. (2022). Digital health solutions and state of interoperability: landscape analysis of Sierra Leone. *JMIR Formative Research, 6*(6), e29930. https://doi.org/10.2196/29930. PMID: 35687406; PMCID: PMC9233249.

Clinton Health Access Initiative. (2022). *The priority assistive technology products list of Sierra Leone.* https://at2030.org/priority-assistive-technology-products-list-of-sierra-leone/.

Dimagi. (2021). *A year of COVID-19 response with CommCare: quarantine management in Sierra Leone.* https://www.dimagi.com/blog/covid-19-pro-bono-project-highlights-2021/.

Directorate of Science, Technology and Innovation. (2019). *The Sierra Leone National Innovation and Digital Strategy (2019-2029)*. https://www.dsti.gov.sl/sierra-leone-national-innovation-digital-strategy-2019-2029/.

DSTI Media. (2020). *DSTI and Dimagi partner to create cutting edge tech solutions for COVID 19 Response*. https://www.dsti.gov.sl/dsti-and-dimagi-partner-to-create-cutting-edge-tech-solutions-for-covid-19-response/.

Hwang, S., Vandi, C., Kuriansky, J., Gordon, G. M., & Gaskins, E. (2024, in press). First-ever government adoption of the Gender-based Violence Information Management systems+: The Sierra Leone experience. *Applied Psychology Around the World, 6(2)*.

Ighobor, K. (2020). *Sierra Leonean technologist's app helps to fight COVID-19*. https://www.un.org/africarenewal/magazine/august-2020/sierra-leonean-technologist%E2%80%99s-app-helps-fight-covid-19.

Jalloh, W. (2021). *How Sierra Leone is using technology, data and innovation to overcome service-delivery challenges*. Institute for Global Change.

Javetski, G. (2024). Innovation: Exploring Dimagi's Digital Platform CommCare: A widely adopted digital solution used by frontline workers worldwide. In J. Kuriansky, & P. Kakkattil (Eds.), *Resilient health: Leveraging technology and social innovations to transform healthcare for COVID-19 recovery and beyond* (pp. 1073–1075). Elsevier.

Kahn, D. M., Bulanda, J., & Sisay-Sogbeh, Y. (2016). Supporting a public education response to stem the panic and spread of Ebola: Help for the National Ebola Helpline Operators. *The psychosocial aspects of a deadly epidemic: What Ebola has taught us about holistic healing* (pp. 215–232). ABC-CLIO.

KIT Health. (2014). *Mobile health in Sierra Leone: evidence and implications for health system*. https://www.kit.nl/wp-content/uploads/2018/08/54ec6fa3ad0b8_mHealth-Technical-Brief-final.pdf.

Kuriansky, J. (2016). Psychosocial suport for a burial team: Gender issues and help for young men helping their country. *Psychosocial aspects of a deadly epidemic: What Ebola has taught us about holistic healing* (pp. 233–242). ABC-CLIO.

Kuriansky, J. (2021). *"Hands Off Our Girls" USA event success for First Lady of Sierra Leone*. Black Star News. https://www.blackstarnews.com/global-politics/africa/%E2%80%9Chands-off-our-girls%E2%80%9D-usa-event-success-for-first-lady-of.

Kuriansky, J, Polizer, Y., & Zinsou, J. C. (2016). Children and Ebola: A model resilience nad empowerment training and workshop. *Psychosocial Aspects of a Deadly Epidemic: What Ebola has taught us about Holistic Healing* (pp. 175–214). ABC-CLIO.

Macro-eyes. (2021). Macro-Eyes to roll-out STRIATA in Sierra Leone and expand into two additional countries. *CISION PR Newswire*. https://www.prnewswire.com/news-releases/macro-eyes-to-roll-out-striata-in-sierra-leone-and-expand-into-two-additional-countries-301346749.html.

Macro-eyes health. (2020). Macro-eyes introduces AI to Sierra Leone COVID response, making it faster and easier to understand health system capacity. *CISION PR Newswire*. https://www.prnewswire.com/news-releases/macro-eyes-introduces-ai-to-sierra-leone-covid-response-making-it-faster-and-easier-to-understand-health-system-capacity-301155681.html.

Møller, M. L. (2022). *DTU students support sustainable healthcare in Sierra Leone*. https://www.dtu.dk/english/news/all-news/nyhed?id=fd7e3801-a9c2-4a1b-a01a-df81741d8c63.

Ngewa, R. N., & Kuriansky, J. (2016). Awareness and education about Ebola through Public Health Campaigns. In *The psychosocial aspects of a deadly epidemic: What Ebola has taught us about holistic healing* (pp. 335–351). ABC-CLIO.

Nicholsen, E. (2016). Sierra Leone successfully tests mHero. *IntraHealth International*. https://www.mhero.org/news/sierra-leone-successfully-tests-mhero.

Space in Africa. (2020). *Sierra Leone is fighting COVID-19 using innovative geospatial data and technology*. https://africanews.space/sierra-leone-is-fighting-covid-19-using-innovative-geospatial-data-and-technology/.

The Borgen Project. (2021). *The improvements of technology in Sierra Leone*. https://borgenproject.org/technology-in-sierra-leone/.

The Borgen project (n.d.). Decreasing maternal mortality rate in Sierra Leone. https://borgenproject.org/tag/sierra-leone/.

The Global Economy. (2021). *Sierra Leone: GDP per capita, current dollars*. https://www.theglobaleconomy.com/Sierra-Leone/gdp_per_capita_current_dollars/.

Tony Blair Institute for Social Change. (2021). *How Sierra Leone is using technology, data and innovation to overcome service-delivery challenges*. https://institute.global/advisory/how-sierra-leone-using-technology-data-and-innovation-overcome-service-delivery-challenges.

UNDP. (2020). *Human development report* (p. 2020). https://hdr.undp.org/sites/default/files/Country-Profiles/SLE.pdf.

UNICEF. (2019). *Drones for good corridor launched as drones take flight to deliver medicine to remote areas in Sierra Leone*. https://www.unicef.org/wca/press-releases/drones-good-corridor-launched-drones-take-flight-deliver-medicine-remote-areas.

United Nations General Assembly. (2022). *International cooperation for access to justice, remedies and assistance for survivors of sexual violence. Resolution A/RES/76/304*. https://documents-dds-ny.un.org/doc/UNDOC/GEN/N22/584/56/PDF/N2258456.pdf?OpenElement.

Valavani, A. (2019). *Behind Sierra Leone's ambitious*. Tech-Driven Development Plan. Council on Foreign Relations. https://www.cfr.org/blog/behind-sierra-leones-ambitious-tech-driven-development-plan.

World Health Organization. (2018). *Assistive Technology*. https://www.who.int/news-room/fact-sheets/detail/assistive-technology.

World Health Organization Sierra Leone. (2019). *Sierra Leone leads the way in Africa with fully functional electronic disease surveillance system*. https://www.afro.who.int/news/sierra-leone-leads-way-africa-fully-functional-electronic-disease-surveillance-system.

Worldvision. (2021a). *2020 Digital Health Factsheet: Sierra Leone*. https://www.wvi.org/publications/2020-digital-health-factsheet-sierra-leone.

Worldvision. (2021b). *Sierra Leone: 2020 Digital Health Factsheet*. Reliefweb https://reliefweb.int/report/sierra-leone/sierra-leone-2020-digital-health-factsheet.

CHAPTER 37

The role of governments in harnessing technology to enhance health service delivery: A perspective from Kenya

Mutahi Kagwe
Cabinet for Health, Nairobi, Kenya

Introduction

Never in Kenya's history has the country's appreciation of technology been as pronounced as over the past few years. While the country has long served as an incubating hub for new technology innovations, it was not until the onset of the COVID-19 pandemic that it became essential to life. In response to the pandemic, in Kenya, we transformed digital social platforms into conduits of lifesaving information on the spread and management of the coronavirus; and instantly embraced tools such as Zoom, a cloud-based video conferencing service, to virtually connect with the world.

Technology converted our living rooms into offices from which we held meetings on the inadequacies of our healthcare systems, the brewing crisis in the global supply chain on COVID-19-related commodities like personal protective equipment, and as government, on how to most effectively balance economic well-being with public health. In Kenya, where strict containment measures were enacted within hours of the government's confirmation of the first COVID-19 case, technology took on different roles including the facilitation of food deliveries within the confines of curfews and partial lockdowns.

Technology dutifully kept the youth entertained, but importantly, among the creative, it opened up new channels of income generation. It can be confidently said that were it not for the pre-COVID-19 integration of technology into both the private and public spaces in Kenya, the country would have suffered a fate far greater than the losses it registered as a result of the pandemic. There are many lessons to learn from the crisis of the pandemic, leading to broad adoption and integration of technology that revolve around progressive leadership.

Resilient Health. https://doi.org/10.1016/B978-0-443-18529-8.00037-8

The adoption of information technology (IT) by health facilities in Kenya has kindled growing interest from stakeholders in the sector and presents a strategic opportunity to improve healthcare delivery. Therefore government should play a key role in ensuring seamless adoption of technology to enhance health service delivery. This chapter outlines ten recommended responsibilities of governments, and what the Kenya government, through the Ministry of Health, is doing to meet these recommendations.

Background

In 2005 I had the humble privilege of being appointed cabinet minister for Information, Communications and Technology by Mwai Kibaki, a renowned economist then serving as the third president of the Republic of Kenya. His progressive leadership recognized the value of technology as an economic accelerator, entrenching it in Kenya's national development blueprint, our "Kenya Vision 2030," which focuses on transforming Kenya into a newly industrialized, middle-income country by 2030, with six economic accelerators—technologically driven trade, manufacturing, financial services, tourism, agriculture, and IT-enabled services. Through digitization, technology enhances accuracy, minimizes waste, and compels transparency, which, in cumulation, improves government efficiency.

Since assuming office in 2013, President Uhuru Kenyatta transformed and catapulted Kenya's development, inspiring and driving changes in health, as in all sectors. He kept the government focused on the health agenda and greater health services in an example of outstanding government leadership. Under his leadership, Kenya undertook a comprehensive health strategy that has completely transformed health services in the country, providing a substantial increase in hospitals and clinics; all level professionals including doctors and nurses; and more crucial ancillary services, like ambulances, enjoyed by the Kenyan population during his administration. These impressive accomplishments continue, for example, where a level 4 hospital is constructed, a medical training college with requisite skills and equipment follows. These accomplishments in health are outlined as follows, and include a significant mobilization of the private sector.

With regard to the private sector in the technology space, tech giants Google, IBM, Oracle, and Microsoft have established their African headquarters in our capital of Nairobi, leading to our advanced technology ecosystem and development of innovative technologies, which, over time, have mainstreamed into the fiber of government-run operations. This has led to Kenya being dubbed the "Silicon Savannah." While acknowledging the critical role the private sector plays in economic growth, "Kenya Vision 2030" is also categorical about the government's obligation to drive and own its technology backbone as a means of safeguarding its integrity and sustainability. Years on, Kenya is reaping the dividends of this ambitious undertaking, having incorporated technology into every facet of government operations. Specifically, in the healthcare sector, Kenya has adopted information technology (IT) as an imperative for the successful delivery of Universal Health Coverage, because we recognize that to effectively plan, deliver, and evaluate the impact of its initiatives and service interventions, a country needs to operate on a platform of high-quality data and analytics.

Besides the accomplishments of President Kenyatta in response to the COVID-19 pandemic and beyond, including innovative technological approaches, it is important to note the government's ongoing efforts to eliminate maternal and child mortality and HIV/AIDS, which, as it has been pointed out widely, must not be totally overshadowed by the pandemic. In tandem with this, Her Excellency First Lady Margaret Kenyatta complemented the government efforts in an awareness and fundraising campaign, called "Beyond Zero" which galvanized the public, private sector, international organizations, and media alongside national and county-wide efforts toward the goal of cooperatively improving access to quality healthcare for expectant mothers, children, and marginalized groups. In this and multiple ways, attention is paid to all needs in healthcare of the population.

Technology in health and its facilitators

Kenya's commitment to improve its health sector value chain is driven by the country's domestication of the MDGs and subsequent SDGs, which is captured as a pillar in the "Vision 2030" and our "Big Four Agenda" that committed to providing our citizens with access to quality, affordable health services. It is with a view of facilitating this that the Kenya government worked with stakeholders to digitize health, guided by carefully selected global best practices identified during benchmarking with other countries. To successfully harness the technology, Kenya has used legal instruments that are responsive to the rapidly changing technology ecosystem while ensuring that patient rights, needs, and confidentiality are secured.

To begin with, Kenya enacted the Data Protection Act (2019) that allows for the collection, storage, and processing of personal data by a data controller or a data processor, and also recognizes health data as sensitive and therefore mandates the safeguarding of privacy and confidentiality of the patient identity and health profile.

The act further puts in place measures to ensure that new technologies plug into existing technology seamlessly through the creation of, among others, the Kenya Health Information Systems Interoperability Framework and The Data Governance Framework, which guarantee that appropriate controls are applied to data processes, and that the data is collected, used, and reported in line with legislative and other compliance obligations. The data governance framework provides clear guidelines, responsibilities, and punitive measures in processes that relate to: migration of data from legacy systems into new systems and formats; integration and synchronization of data from different systems that use different formats; field names and data characteristics; reconciliation of inconsistencies or redundancies of terminologies into a single data dictionary, providing agreed-upon definitions and properties for each data variable; and reporting of data in standardized formats and with standardized interpretations.

With these frameworks as a backbone, Kenya has achieved significant milestones in the creation and adoption of technologies that have enhanced cost-effectiveness and overall efficiency.

The following sections are organized as recommendations for governments regarding innovation and technology, and examples of how we in Kenya have addressed and implemented these.

Recommendation

Support local innovations

Governments should create an enabling environment that allows local innovations to thrive and provide home-grown solutions to tackle new and emerging health challenges, for example, that was necessary during the COVID-19 pandemic. Some notable innovations in the Kenya health sector that government has strengthened include the following:

- **ChanjoKE System.** This is an electronic vaccination system designed to handle multiple immunization programs. The ChanjoKE system collects client demographics and manages inventory and distribution, vaccine administration, monitoring Adverse Effects Following Immunization (AEFI). This is accomplished through a self-service portal—beyond vaccinations, communication, and user management. During COVID-19, the system also provided a vaccination certificate and provided for integration with other solutions. The Ongoing ChanjoKE system enhancements are digital authentication of generated certificates through Public Key Infrastructure, integration with COVID-19 lab database needed to track positive cases, integration with Jitenge, and a help desk to support the public.
- **The 'Chanjo' electronic logistics management information system (eLMIS)** is used to manage vaccine stocks and vaccine cold chain management, and to provide an immunization data dashboard that presents the vaccine coverage, stock levels, and integrated indicators for immunization performance.
- **Patient portals (Portal.health.go.ke)** is an innovative system develped by local techies, which is helping Kenya manage its vaccination program, proving a core component in the roll out plan. The self-service portal is providing an interlink between the Ministry of Health and the public, whereby those seeking services can access healthcare planned for them by the government.
- **The Blood Management System (BMIS)** enhances efficiency, effectiveness, and safety of blood services management in Kenya. The BMIS promotes effective track and trace (vein to vein) and appropriate use of donated blood, accountability of donated blood, monitor available blood in the facilities across the country; enhances rational distribution; reduce

shortfalls in blood donations through technological blood donor mobilization platforms; and generates reports for decision-making and policy formulation.

- **The National Emergency System** (EARS) developed by mHealth Kenya has been in use by the Disease Surveillance and Response Unit at the Ministry of Health, Kenya, in epidemic surveillance and control. The EARS system is used at the Emergency Operations Center to capture, report, and view emerging epidemics across the country. The system provides crucial data for decision-making and epidemic control measures at the national and county levels. Through this system, an innovation called "Jitenge" was birthed during the COVID-19 period, to help the national and county governments in Kenya to respond proactively in efforts to flatten the corona virus curve.
- **Jitenge Solution**, a Swahili word for self-isolate, is an innovation of mHealth Kenya developed to support the government in the fight against COVID-19 in the country. The Jitenge solution was developed as a module of the Emergency Alert and Reporting System (EARS) used by the Ministry of Health's Emergency Operations Centre (EOC) to respond to over 40 infectious diseases. Jitenge is available as an Android Mobile Application—Jitenge MOH, a USSD session—*299#, and an interactive web-based platform that supports a dynamic dashboard. Jitenge allows users to either self-register, or are registered by various Ministry of Health officials, at the quarantine initiation point for home quarantine, at the quarantine facilities, and at the point of entries by port health officials. Registered users then receive daily reminders and prompts to report on their health status including symptoms or any other information. The system is being used to manage and monitor the following: home-based care management, self-quarantine for contacts, postisolation follow-up and monitoring of long-distance truck drivers.
- The **National Community Health Information System (eCHIS)** provides a further example of the high regard Kenya holds for the power of digital solutions. It is borne in response to the growing interest of community health units in digital health systems which we have come to rely on for essential services such as household enrollment, delivery of health services, commodity supply chain management, client referral, messaging and community-based disease surveillance.

Legal and regulatory environment

Governments should put in place relevant legal instruments in a bid to regulate the use of patient data and adoption of technology in processing of the health data. The government should formulate rules and regulations to create an enabling environment to harness technology that enhance health service delivery.

The following is an example of a regulatory tool that the Kenya government is currently putting in place, and prior instruments referring to e-health upon which it builds:

- **E-health bill.** The Ministry of Health has been working with other stakeholders in developing an e-health bill that is envisioned to provide a legal framework for guiding investment in e-health, enable widespread access to healthcare services and health information, and to securely share and exchange clients' information. This will further support the provision of safer, quality healthcare that is more equitable and responsive to the healthcare needs of all Kenyans.
- **Data Protection Act (2019)** provides for the collection of personal data by a data controller or a data processor either directly or indirectly. It also recognizes health data as sensitive and gives provisions for its collection, storage, and processing. The act safeguards privacy and confidentiality of the patient identity and health profile.
- **The Huduma Bill (2019),** an Act of Parliament to establish the National Integrated Identity Management System (NIIMS), which provides for the utilization of Huduma Namba for the access of universal healthcare services. Huduma Namba is a Kenya "Vision 2030" flagship project launched by Kenya president H.E Uhuru Kenyatta as a digital health ecosystem innovation to create a central master population database for Kenya citizens and foreign nationals, for ease of health service delivery to all persons, reforming the ID card system into a more efficient and reliable digital form based on biometric data.
- **The** Health Act provides for information dissemination and states that "the national government, county governments and every agency having a role or responsibility within the National Health System, shall ensure that appropriate, adequate and comprehensive information is disseminated on the health functions for which they are responsible." Section 11 of the Act further provides for confidentiality, stating that "information concerning a user, including information relating to his or her health status, treatment or stay in a health facility is confidential except where such information is disclosed under a court order or informed consent for health research and policy planning purposes." Section 103 of the Act allows for "eHealth to be a recognized mode of health service" making provision for services such as telemedicine and an electronic referral system possible. The Health Act provides for mandatory reporting of technological innovations by all health facilities regardless of ownership and type.

Sustainability framework

Government should create an enabling environment that will ensure new technologies are adopted in health sector and technological advancement in the provision of health services is sustainable. Some of the existing framework and tools that the Kenya government has adopted, implemented, and enforced to enhance sustainability in technology that enhance health service delivery include:

- **The Kenya Health Financing Strategy (KHFS)** provides a blueprint for steering the country toward Universal Health Coverage (UHC). The health systems require sustainable financing mechanisms to ensure the efficient delivery of high quality, equitable health services. The strategy provides a road map that is geared toward strengthening health systems and attaining the highest possible standards of health, as enshrined in article 43A of the Constitution of Kenya.

- **The Kenya Health Information Systems Interoperability Framework** will ensure that any new technology plugs into existing technology seamlessly.
- **The data governance framework** by the Ministry of Health provides a critical step to implement, operate, monitor, review, maintain, and improve data and information systems governance. The framework ensures that appropriate controls are applied to data processes, and that the data is collected, used, and reported in line with its legislative and other compliance obligations. The data governance framework provides clear guidelines, responsibilities, and punitive measures in processes that relate to: migration of data from legacy systems into new systems and formats; integration and synchronization of data from different systems that use different formats, field names, and data characteristics; reconciliation of inconsistencies or redundancies of terminologies into a single data dictionary, providing agreed-upon definitions and properties for each data variable; reporting of data in standardized formats and with standardized interpretations; change management in responding to clients' changing business requirements.
- **Unique Patient Identifier Implementation Framework.** The government through the Ministry of Health has fastened the process of adopting the Unique Patient Identifier Implementation Framework aimed at providing the necessary guidance to inform unique patient identification strategies that will ultimately facilitate higher quality healthcare services for all Kenyans. A critical component in the government's effort to deliver universal healthcare to the population is its ability to uniquely identify everyone living within the country's borders. Unique Patient Identification will be the driver of improved medical record-keeping, especially as more facilities adopt digital health records. Correctly and unambiguously identifying patients is vital for the delivery of healthcare, administrative processes, information management, follow-up, and preventive care.

Human capital development

Governments should invest in Human Resource for Health development Programs. The Kenya government has embarked on the Human Resource for Health Programme, which is aimed at addressing capacity gaps within specialized and subspecialized areas in health sector and also at reducing shortages in the health workforce.

The Kenya government, through Kenya Medical Training College (KMTC), has adopted and embedded ICT in their training curriculum for their trainees in the medical sector. This move accelerates the adoption of new technology in the provision of health services.

Infrastructure and connectivity

Governments should ensure that the critical infrastructures needed for technology access to health facilities are well established. These include the equipping of health facilities with adequate infrastructure such as Local Area Network (LAN) and end user equipment.

Governments should continue to capacity build and implement upgrades at national data centers that provide assurance to data storage, processing, and disseminating of health data and digital solutions, through: performance, storage, reliability, scalability, availability, ease of management, and disaster recovery.

The Kenya government through the Ministry of health embarked on development of the Digital Health Platform (DHP) as well as improvement of ICT connectivity and infrastructure to spur the growth of digital investments. This ensures timely and routine maintenance of the shared services platform that ensure that the internet is accessed through the Government Common Core Network (GCCN).

Governments should also perform routine maintenance of the National Optic Fiber Infrastructure (NOFBI) and strength inland metro and last mile connectivity. This will ensure that adoption of technology and access to health information and records by the population is easy, seamless, and cost effective.

Policies and standards for equipment

Governments should ensure that conducive policy environment and standards for equipment are put in place for adoption and use of new and emerging technology in provision of health services in the country.

Some of the policies Kenya has developed and implemented include:

- **The Kenya National eHealth Policy 2016–30,** which recognizes the need for Shared Health Records (SHRs) through the development of a Kenya Master Patient Registry (KMPR) that includes interoperability with Electronic Medical Records (EMR), Personal Health Records (PHR), and Terminology Services (TS).
- **The Kenya Health enterprise architecture,** which specifies that the Kenya Master Patient Registry will include a universal healthcare identifier and provides guidance for uniform patient identification across the information systems that manage patient-level data.
- **International standards organizations.** The Kenyan government has also adhered to and adopted standards developed by international standards organizations, such as the International Organization for Standardization, the Integrating the Healthcare Enterprise, UNAIDS guidance for countries adopting national health identifiers, and OpenHIE (Open Health Information Exchange).

- **Kenya Standards and Guidelines for mHealth Systems.** The Kenya government through the Ministry of Health recognizes the crucial role that appropriate use of relevant standards and guidelines plays in the implementation of mobile health (mHealth) applications; thus, Kenya standards and guidelines for mobile health solutions aim at addressing mobile digital investments used in health service industry. They are also important as a means of enhancing efficiency and effectiveness in the delivery of health services via mobile platforms in the country. In the public health sector, the Kenya mHealth Standards and Guidelines ensure that all mHealth-based systems and health strategies are correctly applied. The implementation and enforcement of these standards and guidelines will go a long way in reducing duplication of efforts, in promoting data and information sharing among systems, and in harnessing appropriate use of mobile technology as an enabler to effectiveness and efficiency in the delivery of healthcare services.
- **The Data Quality Assurance (DQA) Protocol** was developed in 2013–14 by the Kenya Ministry of Health, through the Division of Health Informatics, Monitoring and Evaluation, as a tool to facilitate data quality improvement strategies at all levels of healthcare. The DQA Protocol provides data quality standards, roles and responsibilities of users, and identifies the structures and processes that organizations should have in place to create a supportive environment for data quality. As quality health information is dependent on quality data, it is therefore logical that efforts to improve the information on which decisions are based will start with ensuring that data is collected, processed, and analyzed appropriately.
- **ICT Standards** are a set of standards under six domains, which define the minimum components of an ICT Plan and guide implementation of the Government Enterprise Architecture (GEA) Framework to be applied at all levels of government. This approach is aligned with the recommendation that governments should adopt ICT standards and guidelines for human resources, infrastructure, processes and systems, and technology for public office and public service.

Skill development and transfer

Skill development is a very essential part in adoption of new technology and trainings. Governments should continue to strengthen existing partnerships, and explore new partnerships, with academic institutions to foster skills development and thus enable easier transfer of skills. In Kenya, we have identified the University of Nairobi, Kenyatta University, and Jomo Kenyatta University of Agriculture and Technology among other academic institutions that nurture skills in new and emerging technologies.

From these hubs, new innovations are incubated and often give rise to homegrown digital solutions for health. Key examples of these hubs is Nailab and IHub which have a reputation of nurturing young innovators and start-ups and thus ensure that they are able to grow their skills resulting in better solutions.

Establishment of digital hubs

Governments should continue to increase funding (budget allocations) for established Digital Innovation Hubs and their expansion. They should ensure that digital hubs are well equipped with internet connectivity and end user equipment. The purpose of these hubs is to empower the local community with digital skills. The ripple effect for this is that the population will gain relevant skills that will enable them to embrace new innovations in health such as accessing health records and undertaking remote consultations.

Telemedicine and remote consultations

Governments should ensure that there is an enabling environment that will facilitate adoption and use of telemedicine and remote consultations in the country. This needs to be done through setting relevant standards, policies, and legal requirements. Governments should put in place relevant infrastructure and connectivity to enable telemedicine technology adoption.

Telemedicine provides many advantages. These include enabling of doctors and medical professionals to remotely communicate with patients and accompanying health staff and health workers, especially in smaller hospitals and service facilities, as well as to consult with colleagues, and to conduct remote trainings including for Continuing Medical Education (CME) purposes, as well as to accomplish electronic health records transfer (transmitting patient records for remote consultations), integrating with the existing Electronic Medical Record (EMR) systems.

Telemedicine projects should maximize on the country's skilled healthcare worker base by enabling them to deliver quality services to far-flung and remote regions of the country quickly and at minimal cost. It is hoped that if all goes as planned, Kenya's health services will soon be clustered in a hub-and-spoke manner around Level 6 health facilities with advanced telemedicine capabilities.

Governments should support mental healthcare along with physical healthcare, in line with the UN Agenda 2030, the Political Declaration of Universal Health Care, WHO Constitution, and other instruments

Kenya's president Kenyatta presided over the groundbreaking of the Kenya International Mental Hospital, being built in response to Ministry of Health statistics to support mental healthcare along with physical health, and statistics from my office, the Ministry of Health, that 40% of patients admitted to hospital and 25% of persons seeking outpatient services have mental or neurological conditions. The statistics further indicate that a shortage of services affects almost every family. The Government of Kenya is committed to removing stigma, protecting people and supporting them to be useful in society.

Self-reliance and the role of technology in the realization of UHC

As Kenya's Champion for Health, I can't underscore enough how proud I was of the steps we made in the digitization of our healthcare system not least because it was fostering self-sustenance and channeling us closer toward the country's realization of UHC. We have been fortunate to have leaders like President Kenyatta who stressed that "A sick nation is not a working nation," which inspires us to push hard and keep going. I believe that it is this zeal that makes Kenya's health sector increasingly attractive not only to its local investors but to international giants such as Moderna, which recently invested $500m investment in Africa's first and only mRNA vaccine manufacturing facility.

Digitization is capital intensive and even more so is the provision of Universal Health Coverage (UHC) which makes the issue of sustainability a critical one. With a view of harmonizing services in the entire value chain in health, whether it is in hospital infrastructure, medical personnel, distribution of pharmaceuticals, and the continued adoption of new technologies in a sustained way, the government has identified the National Health Insurance Fund (NHIF) as a viable anchor and commenced the process of restructuring the vehicle now that it is a mandatory scheme that should sufficiently provide affordable medical coverage for all Kenyans and residents in the country. The fund will be structured to rely primarily on a centralized digital technology that will provide for the efficient management and payment of claims, data collection, and data-sharing and coordination between government agencies for increased transparency.

Way forward

With a foundation and successes in digitized health in place, Kenya's growth in this area will be determined by the number of partnerships it can establish to enhance its Human Capital Development, grow its Digital Innovation Hubs, and expand its capacity to support local innovations. The government hopes to power a more proactive and data-driven healthcare by incorporating artificial intelligence and machine learning into our system as the Ministry's ultimate goal is to develop into a medical tourism hub. We are confident that with the newly developed e-health bill which shall catalyze access to healthcare services, health information, investment in e-health, and insure the secure sharing of client information, technology will buttress the country's ability to provide quality healthcare services that are more equitable and responsive to the healthcare needs of all Kenyans as well as those beyond our shores.

CHAPTER 38

An unconventional security threat: Mental Health at the United Nations Security Council and Mexico's Response

Juan Ramón de la Fuente, Sylvia Paola Mendoza, and Aline Guzik
Permanent Mission of Mexico to the United Nations, New York, NY, United States

Introduction

How is mental health related to maintaining international peace and security? This was one of the questions that the Permanent Mission of Mexico to the United Nations set out to answer in October 2019, a few months before the country officially became an elected member of the United Nations Security Council (UNSC) for the 2021–22 term. It may initially seem odd that a country about to join the Council would prioritize an issue that can be thought of as unrelated to the leading intergovernmental body dedicated to safeguarding international peace and security. However, considering that the World Health Organization (WHO) estimates that one in five people who have experienced war or other destabilizing conflict will develop mental illness and that the prevalence of common mental disorders is expected to double during humanitarian crises (WHO, 2019), it is evident that mental health and psychosocial support (MHPSS) are important for the achievement and maintenance of international peace and security. Therefore we decided to elaborate a plan of work to highlight this important issue.

Such an initiative is also in alignment with Mexico's commitment to achieving the objectives of critical current international documents. These include the UN 2030 Agenda for Sustainable Development adopted by the UN General Assembly in September 2015, with its target 3.4 to "promote mental health and well-being" and the Political Declaration on universal health coverage adopted by the UN Member States in September 2019, which considers mental health in equal status to physical health.

Despite MHPSS being a rather abstract concept at first, the COVID-19 pandemic became a catalyst for us diplomats to better understand urgent issues related to mental health, since epidemiological surveys showed that the virus seriously increased the prevalence of mental health disorders, such as anxiety, depression, and substance use/abuse, among others. The 2022 World Mental Health report from the World Health Organization (WHO, 2022) estimates that the rates for both anxiety and depressive disorders globally increased more than 25% during the

Resilient Health. https://doi.org/10.1016/B978-0-443-18529-8.00038-X

first year of the pandemic. Simultaneously, mental health services in developed and developing countries had been severely disrupted and the treatment gap for mental health conditions—indicating the divergence between needs and deliverable services—had widened. These disconcerting facts justify a much-needed prioritization of MHPSS services. Further, it was evident that just as for experts, scientific evidence, and the public at large, the diplomatic community at the UN was becoming increasingly aware of the threat that mental health conditions placed on world populations, becoming a "silent" pandemic.

A few months into the COVID-19 pandemic, UN Secretary-General António Guterres was calling for more robust action on MHPSS, a call that should not be taken lightly. He even recognized that "mental health problems, including depression and anxiety, are some of the greatest causes of misery in our world" (United Nations, 2020) and he urged "governments, civil society, health authorities and others to come together urgently to address the mental health dimension of this pandemic" (United Nations, 2020). His focusing attention on this issue had already been evident in his launch in October 2018 of the UN "Mental Health and Well-Being Strategy" to optimize the psychological health of the UN personnel (United Nations, 2018). Throughout the pandemic, he made references to the importance of mental health. Then, in his video message when launching the World Mental Health Report, he warned, "We are living through a global mental health crisis" (United Nations, 2022). Clearly, the UN's top leader understands the serious dimensions of the crisis and its costs, beyond the figures, sensationalist news, biased narratives, or impulsive reactions. Once the spread of the COVID-19 virus came considerably "under control"—although variants were still plaguing nations—the significant lasting aftereffects on populations were even better understood and more glaring, not only on economic and political dimensions but significantly on psychosocial dimensions as well. Clearly, people throughout the world were suffering emotionally, and this was going to have long-term impact.

Against this backdrop, and as we defined Mexico's priorities for our term in the UNSC (which included preventive diplomacy, trafficking of illegal weapons, inequality and social exclusion, and protection of women and girls, among all the other global issues), a large group of the team in our Mission embarked on discussing ways in which we could advocate in favor of MHPSS within the Council and beyond. The core team interestingly consisted of all women, namely, Elena Curzio, Regina Domínguez, Aline Guzik, Sylvia Paola Mendoza, Eva Romulus, Alma Santa Ana, and Flor de Lis Vásquez, coordinated by Ambassador Alicia Buenrostro, all of whom worked on different issues at our Mission, including global health priorities, humanitarian affairs, the Children and Armed Conflict (CAAC) mandate, as well as the Women, Peace, and Security (WPS) and the Youth, Peace, and Security (YPS) agendas of the UNSC. Given their diverse portfolios, clearly, they recognized mental health as a cross-cutting issue impacting all these other issues of global concern. The idea for a multidimensional perspective of MHPSS policies and practices as part of our strategy resonated with Ambassador de la Fuente, given his background as a researcher who had published extensively on psychiatry and mental health issues and as former Minister of Health of Mexico (de la Fuente, 2017; de la Fuente et al., 2015; de la Fuente & Heinze, 2018).

Besides the COVID-19 pandemic, the ongoing conflict in Ukraine further highlighted the need to address situations that exacerbate mental health problems, including, of course, posttraumatic stress disorder. This new conflict made giving MHPSS higher visibility more urgent, both as an element of humanitarian programming and as a primary component of preventive and response strategies of the United Nations system and humanitarian actors in areas affected by conflicts.

Mexico's advocacy and strategy

After internal discussions, we decided to follow a simple but compelling three-pronged strategy to promote the issue of mental health and psychosocial support as a fundamental factor in the paradigm of sustaining peace and security. This involves that:

1. We focused on giving these issues greater visibility in the debates and fora across the various bodies of the UN Organization in which we participate, not only in the Security Council.
2. We insisted on including references regarding the importance of mental health in speeches, interventions, and relevant declarations and resolutions.
3. We set out to organize and shape impactful events focused on mental health and psychosocial support.

Regarding our first point, diplomats at the Mission of Mexico included a reference related to MHPSS, when appropriate, in our statements throughout our participation in different fora at the UN. This approach included touching upon the mental health dimensions during Member State deliberations about Syria's humanitarian situation, the reintegration of Colombia's former FARC members into society, climate change, access to vaccines, and migrants and refugees. In fact, our reference became a known practice among other delegations; for example, the delegations of Indonesia and Switzerland, who served as cofacilitators of the Economic and Social Council (ECOSOC) humanitarian resolution, called on Mexico to coordinate a stand-alone paragraph on mental health. This later became the basis for similar discussions in other fora.

Concerning our second point, not only did we mention this subject matter within the margins of the General Assembly and ECOSOC, but we managed to incorporate it into a Security Council resolution 2601 (2021), cofacilitated by Norway and Niger, which was considered a first of its kind. Specifically, Paragraph 15 of this resolution "stresses the deleterious effects of humanitarian emergencies and forced displacement due to armed conflict on the mental health and psychological wellbeing of children" and refers to the "importance of long-term and sustainable funding for mental health and psychosocial programming in humanitarian contexts and ensuring that those who are affected receive timely and sufficient support." Finally, the text "encourages Member States, donors, and relevant parties to integrate mental health and psychosocial services in all humanitarian responses."

As for our third point, and with specific regard to the Security Council, Mexico convened an informal discussion within the Council (known as an *Arria*-Formula meeting), on March 25, 2022, on the topic of "Ensuring access to Mental Health and Psychosocial Support in Conflict, Post-conflict and Humanitarian Settings," to explore ways to better deliver mental health and psychosocial support services in conflict, postconflict, and humanitarian settings. Chairing the session, we opened the meeting stating that prioritizing mental health and psychosocial support at all levels of the United Nations system, particularly in humanitarian response strategies, was of utmost importance, and we stressed the need to provide more visibility to this issue, including by incorporating language into relevant resolutions and other outcome documents.

At this meeting, the briefers section began with moving testimony of a Colombian human rights defender who explained why having access to a mental health professional was critical to help her make peace with the atrocious events she had experienced due to conflict in her country, including the tragic loss of her children. Also participating virtually was renowned psychiatrist and president of the Association for the Improvement of Mental Health Programmes, Professor Norman Sartorius, who focused on the relevance of developing tools to address mental distress of individuals in postconflict situations to combat the complex cycle of posttraumatic stress disorder. He also underlined the need to close knowledge gaps and build a strong social structure through education.

Then, the director of the Department of Mental Health and Substance Abuse of the World Health Organization (WHO), Dr. Dévora Kestel, urged that the Security Council prioritize mental health and psychosocial support in its decisions. Finally, the director of Health of the International Committee of the Red Cross (ICRC), Dr. Micaela Serafini, stressed that mental health must be central to public health and recommended investing in the full range of services, including psychological first aid and group therapy, as well as access to professionals in situations of armed conflict and the integration of mental health and psychosocial support into health facility strategies.

The key messages expressed in this discussion are reflected in our Chair's Summary, published as an official UN document A/76/854 (United Nations, 2022), available online in the six official UN languages (English, Spanish, French, Chinese, Arabic, and Russian). This document summarizes the main points made by 28 Member States (including all 15 SC Members) plus the European Union who participated in the discussion. It further contains several recommendations provided in writing by civil society organizations that couldn't participate in person due to COVID-19 restrictions at the UN Headquarters. These include the Afghan Landmines Survivors Organization, Latin America Human Security Network, Latin American Network of Mine/ERW survivors and people with disabilities, Médecins sans Frontières, and the International Association of Applied Psychology.

The way forward

Mexico still has a few months to go as a nonpermanent Member of the Security Council, as of this writing, during which time we will continue to strengthen the strategy described before in all available spaces. We recently announced our intention to invite Member States to endorse a "Call To Action" to be launched on World Mental Health Day, celebrated annually on October 10. This "Call to Action" will contain commitments that Member States can pledge to broadly strengthen health systems, improve access to mental health services and psychosocial support, and ensure that universal health coverage includes those services. This document will continue to build more robust commitments about MHPSS across the UN system. Subsequently, we will introduce an innovative thematic draft resolution on mental health, sustainable development, and peace in the General Assembly that will help institutionalize these efforts going forward. This resolution, if adopted, will be the first of its kind for the General Assembly.

We have also started working with our elected successors at the Security Council to continue our efforts and recommend mainstreaming mental health language in the work of the Council. Council members could explore the possibility of adopting a document that addresses mental health in UN Peace Operations, a line of action very much supported by other UN Member States like Germany and Israel. This aligns with a report published by the International Peace Institute in 2020 that analyzed the prevalence of posttraumatic stress disorder and other mental health conditions among the military, police, and civilian personnel of UN peace operations (Di Razza, 2020). Among other aspects, the report concluded that it was necessary to raise attention and awareness to the challenges for mental health among UN peace operations and the need to provide support before, during, and after deployment, as much remains to be addressed and accomplished regarding these issues. In fact, on 21 December 2022, the UN Security Council unanimously adopted the first stand-alone resolution on mental health. Introduced by Mexico right before the end of our term, resolution 2668 (2022) focuses on the importance of providing mental health and psychosocial support for personnel deployed in UN peace operations, including in Peacekeeping Operations and those working in Special Political Missions.

Much work needs to be done at several levels to ensure that mental health doesn't continue to be a neglected field of public health, and instead, to help people understand mental health better and value it more, as well as to guarantee that people in need have access to effective, affordable, and quality care. Equally important is to reduce pervasive stigma and discrimination about mental health, to use preventive measures to reduce risks, to ensure a strong gender component and an age-sensitive approach, and to strengthen protective factors so that everyone has equal opportunity to thrive and reach the highest attainable level of mental health and well-being. All this is necessary to prevent escalation of what is already being recognized as a global mental health crisis and which threatens to become our next, not-so-silent anymore, pandemic. There is still a long way to go but Mexico is committed to stay the course toward a comprehensively healthy world both physically and mentally.

References

de la Fuente, J. R. (2017). *La Sociedad Dolida. El Malestar Ciudadano.* México: Grijalbo.

de la Fuente, J. R., Álvarez Icaza, D., Rodríguez Carranza, R., et al. (2015). *Marihuana y Salud.* México: Fondo de Cultura Económica.

de la Fuente, J. R., & Heinze, G. (2018). *Salud Mental y Medicina Psicológica.* México: McGraw Hill.

United Nations. (2020, May 13). *Policy brief: COVID-19 and the need for action on mental health.* https://unsdg.un.org/sites/default/files/2020-05/UN-Policy-Brief-COVID-19-and-mental-health.pdf.

Di Razza, N. (2020). *Mental health in UN peace operations: Addressing stress, trauma, and PTSD among field personnel.* IPI. https://www.ipinst.org/2020/12/mental-health-in-un-peace-operations-addressing-distress-trauma-and-ptsd-among-field-personnel. (Accessed 23 December 2020).

United Nations. (2022). *Chair's summary of the Arria-formula meeting of the security council on the theme "ensuring access to mental health and psychosocial support in conflict, post-conflict and humanitarian settings"*. United Nations. https://documents-dds-ny.un.org/doc/UNDOC/GEN/N22/372/92/PDF/N2237292.pdf?OpenElement (Accessed 6 June 2022).

United Nations. (2018). *Workplace mental health and well-being strategy.* https://www.un.org/en/healthy-workforce/strategy. (Accessed October 2018).

World Health Organization. (2019, June 11). *WHO. Mental health in emergencies.* Who.int. https://www.who.int/news-room/fact-sheets/detail/mental-health-in-emergencies.

World Health Organization. (2022). *World mental health report: Transforming mental health for all.* Geneva: World Health Organization. Licence: CC BY-NC-SA 3.0 IGO.

Further reading

de la Fuente, J. R. (2020, June 1). *El Covid-19 y la salud mental.* El Universal. https://www.eluniversal.com.mx/opinion/juan-ramon-de-la-fuente/el-covid-19-y-la-salud-mental.

Guterres, A. (2020, May 13). *COVID-19 Response. Mental health services are an essential part of all government responses to COVID-19.* https://www.un.org/en/coronavirus/mental-health-services-are-essential-part-all-government-responses-covid-19.

Guterres, A. (2022, June 17). *Video message to launch the World Mental Health Report 2022: Transforming mental health for all.* https://www.un.org/sg/en/content/sg/statement/2022-06-17/secretary-generals-video-message-launch-the-world-mental-health-report-2022-transforming-mental-health-for-all.

UN News. (2021, October 29). *Security council adopts "first of its kind" resolution on protecting classrooms from conflict.* UN News. https://news.un.org/en/story/2021/10/1104482.

United Nations. *UN workplace mental health and well-being strategy summary a strategic action plan 2018–2023.* https://www.un.org/en/healthy-workforce/files/Strategy%20Summary.pdf.

World Hesalth Organization. *Mental health & COVID-19.* https://www.who.int/teams/mental-health-and-substance-use/mental-health-and-covid-19.

CHAPTER 39

Leadership skills of His Highness Sheikh Mohammed bin Rashid Al Maktoum, Ruler of Dubai

Judy Kuriansky

Global Council SDG3 (Health), "SDGs in Action" Initiative, Dubai, United Arab Emirates; Department of Clinical Psychology, Teachers College, Columbia University, New York, NY, United States; International Association of Applied Psychology, Indianapolis, IN, United States; United Nations Group of Friends of Mental Health and Well-being, New York, NY, United States; Office of Ambassador Sidique Wai, Embassy of the Republic of Sierra Leone to the United States, Washington, DC, United States; Mission of the Republic of Sierra Leone to the United Nations, New York, NY, United States; Global Leadership Council, eMental Health International Collaborative (eMHIC), Auckland, New Zealand

> *"We are building a new reality for our people, a new future for our children, and a new model of development."*

Those words from His Highness Sheikh Mohammed bin Rashid Al Maktoum, Vice President and Prime Minister of the United Arab Emirates and Ruler of Dubai, encapsulate his vision and substantiate his reputation as a charismatic leader of Vision, Innovation, and the Future, described in this chapter.

All this was obvious to me on my first, and then every subsequent, visit to Dubai. This chapter focuses on characteristics of a leader behind such aims and accomplishments for his country who has made Dubai a model of innovation in technology and all sectors.

The qualities of MBR (initials he likes to use) fascinate me, as a psychologist interested in human character and behavior, as a health professional, and as a stakeholder in the space of technology and health—themes of this volume on Resilient Health, of which this chapter is a part. In an autocracy such as the United Arab Emirates, of which Dubai is one of seven Emirates making up the country, MBR is a leader who sets standards for governance.

Given the rapid development of Dubai—from mere sweeps of sands in just over half a century—to the tallest skyscrapers and Cities of Innovation, Sustainability, Media, Technology, and other topics, it is worthwhile to examine the style of its Ruler. Such an examination is of special interest and importance for this volume, since Dubai is such a center for health; in fact, with a healthcare

Resilient Health. https://doi.org/10.1016/B978-0-443-18529-8.00039-1

system considered one of the best in the world, and top-rated as a health tourism destination. There were 337,011 medical tourists in Dubai in 2021, and in fact, the UAE places such high priority on its healthcare system, that the government intended revisions and improvements in its already state-of-the-art facilities by 2021 (Arnold, 2022). The specifics of this system are outlined in other chapters in this volume covering the activities of the Ministry of Health and of the Dubai Health Authority (Al Rand et al., 2024; Aldafrawy et al., 2024).

In this chapter, I give a short biographical sketch of High Highness and explain some of the impressive projects, sites, and "hubs" that embody innovation, technology, and the future, particularly in health, that have developed under his rule. I have had the opportunity to observe High Highness at events and conferences, including the World Governments Summit, and I am familiar with the country and people, having made many visits to Dubai and being involved in many conferences and events with Dubai high-level delegates. These include serving as co-chair of the United Arab Emirates "SDGs in Action" campaign on the Global Council on SDG3 on Health; having UAE delegates speak at my class at Columbia University Teachers College; visiting Dubai centers (e.g., Sustainability City, Media, and Healthcare Centers); attending annual World Governments Summits and Expo 2020 Dubai; participating in and helping coordinate several UAE celebrations of the International Day of Happiness at the United Nations; speaking at UAE Mission to the United Nations side events during the UN General Assembly and the Science Technology and Innovation Forum; having high-level UAE officials speak at my UN events, including during the High-Level Political Forum; and other connections and close friendships. Further, participating in the Global Council on SDG 3 (Health) of the "SDGs in Action" initiative of the UAE from the beginning in Dubai during the World Governments Summit gave me a first-hand experience of how High Highness' approach infuses innovation in his government and trickles down to inspire innovation in all who come into his sphere. At the beginning of the Global Councils, invited experts were self-selected into teams correlating to the Sustainable Development Goals and challenged to come up with an innovative project to advance their selected SDG, and then to compete for the first place. My team conceived a platform to showcase innovations in the healthcare ecosystem delivering needed services and products to "those left further behind", fittingly called "Health in Your Hands" (Kuriansky, 2024). (Note: Many of these innovations, some of which are profiled in this Resilient Health volume, were forced to pivot during the COVID-19 pandemic to embrace more technological approaches to keep operating effectively and fill treatment gaps.) We won first prize and have since further developed the project, and integrated it into the work flow and deliverables of the Global Council on SDG 3 (health).

Bio sketch

Sheikh Mohammed bin Rashid Al Maktoum was born on July 15, 1949 in Dubai, as the third son in a family of the House of Al-Falasi members who enjoyed a quite high social status despite a rather humble economic background. His mother—whom he credits with much of his success—was Sheikha Latifa bint Hamdan Al Nahyan, daughter of former ruler of Abu Dhabi (one of the Emirates) Sheikh Hamdan bin Zayed bin Khalifa Al Nahyan (Wayback Machine, 2013).

Life in the Arabic Emirates at the time, even for ruling families, has been described as challenging, when homes were made of mud or clay with no running water, electricity or toilets, where people went barefoot and clothed in rags, suffered in broiling heat, had no access to doctors, and used camels for transport (Al Fahim, 2000). Sheikh Mohammed reportedly had a happy childhood, growing up close to his grandfather, and "showered with love" in a "bonded family" with many brothers for playmates. He was an energetic and athletic child (kicking around balls in the sandy courtyard), and went on to an extensive and elite education, buoyed with a photographic memory, excelling in school. He studied in the Bell Language School in Cambridge, England, exposed to a melting pot of international classmates and elite experiences, and later attended a British military school before being summoned back to Dubai when his father appointed him as Commander-in-Chief of the Dubai Police Force in 1968 when he was only 20 years old (Wayback Machine, 2013). A mere two years later, he was named UAE Minister of Defense and years later, in 1995, as Crown Prince of Dubai. In 2006, he became Ruler of Dubai and Prime Minister and Vice President of the United Arab Emirates. When appointed Crown Prince, he reportedly commented, *"I do not know if I am a good leader, but I am a leader. And I have a vision. I look to the future, 20, 30 years. I learned that from my father, Sheikh Rashid. He was the true father of Dubai. I follow his example. He would rise early and go alone to watch what was happening on each of his projects. I do the same. I watch. I read faces. I take decisions and I move fast; full throttle"* (Wayback Machine, 2013).

Following in his father's full-throttle footsteps, he transformed Dubai into a major global city, in truly grandiose and luxurious style. In my psychologist's armchair analysis, one must have exceptional confidence and skills as well as a big ego to dream that big, and turn those ideas into reality—evident in the biggest man-made harbor, the tallest tower, the world's fastest-growing airline (Emirates), a Space Center to reach Mars (called "Hope" to send a message of optimism to millions of young Arabs), and a major hub for international trade with groundbreaking enterprises like Dubai Internet City and Media City (a technology hub), Dubai International Finance Centre, Dubai World Central and Innovation Hub.

H.H. Sheikh Mohammed credits his mother for his development, as he has said, *"My mother, may her soul rest in peace, shaped my personality; thanks to her, I have acquired many values, good traits and skills"* (Meah, n.d.). Psychological theories of child development certainly substantiate that the perception of parental love and experience of a mother's unconditional love and care have defining positive effects on psychological health that influences a sense of mastery and industry that then fuels success throughout life (Erickson, 1994; Piaget & Inhelder, 1972; Russek & Schwartz, 1997).

This chapter focuses on H.H. Sheikh Mohammed's leadership style rather than his personal relationships, which have been the subject of news reports. Suffice it to say, it is reported that he has 26 children from several wives. Some theories maintain that a professional life can be quite separate from one's personal life; others disagree.

The transformation of Dubai is meteoric, from the 1960s as a modest trading port which expanded into a thriving oil industry and a major center of business, finance and trade by the 1980s. By 2017, attention turned to developing tourism, featuring massive malls, high-end products and theme parks with indoor skiing and sharks eclipsing any Disneyland or Las Vegas. Tax-free living and exclusive opulent residential neighborhoods offer another level of attraction.

Impact on Dubai

In 2010, His Highness Mohammed bin Rashid Al Maktoum launched the UAE Vision 2021 (Al Rand et al., 2024) aimed at making UAE one of the best countries in the world and considered responsible for the growth of Dubai into a global city. A real estate developer who grew a billion dollar fortune, His Highness is the majority shareholder in the global investment holding company with innumerable development projects. Anyone who has been to Dubai cannot be but awed by the dizzying innovative architecture of existing buildings and ever-present construction sites.

Dubai healthcare city

I remember my first visit to Dubai in December 2003, when my friend Dr. Raymond Hamden, who founded and ran a multicultural mental health clinic in Dubai, organized the Middle East/North Africa Regional Conference of Psychology on "*Psychology for Sustainable Development and Peace*." As we stood on his terrace, discussing his therapy practice, what it's like living in Dubai, our shared experiences hosting radio call-in advice shows, and our mutual efforts to achieve Middle East peace, he pointed out the iconic Burj Al Arab hotel with its inimitable design that led us to discuss leadership styles that was one of my friend's fields of expertise.

Subsequently, I was invited to visit the beginning stages of now-developed Dubai Healthcare Center, attracting top medical institutions in the world, in my role as international liaison at that time for Columbia Medical Center's Psychiatric Institute. I also visited World Island Dubai (actually considering buying a lot named after a country), construction of which started in 2003, and went to the tiny museum in the old city that traces the explosive development of Dubai from dusty sand to towering skyscrapers. Now, Dubai Healthcare City is dominated by high-tech hospitals and medical teaching centers, upscale malls, and luxurious hotels.

Dubai as an innovation hub

The UAE has become the world center of innovation, likely to only grow (Riberio et al., 2021). The Innovation Hub is a collaborative community center that hosts a community devoted to growing innovation and technology. The hub gives youth skills needed to develop mobile applications, design products, produce prototypes, and build on Artificial Intelligence and Machine Learning. Tech Talks feature technology experts, practitioners, and entrepreneurs.

Dubai as a technology hub

Another port of visit for me back in 2003 was Dubai Internet City, established in 2000, since I was involved in multimedia (as a feature reporter on TV, a radio call-in advice host, and columnist for U.S. and international magazines and newspapers). Dubai continues to be among the world's preferred hubs for new economy companies.

Qualities of a leader

As a psychologist, I know that phrases said to yourself and others reflect what is true or what can be manifested. My personal favorite is: "Whatever you believe and conceive you can achieve." The words you say reveal who you are (your real self), who you want to be (your ideal self), and also what you want to project to people (your social self). Words also shape and predict your actions. So it is of interest to consider here some of MBR's sayings. What he says reflects his self-view, and his view of others, his people and country. The qualities he conceives as characteristic of a leader are reflected in these quotes, collated on a site (Meah, n.d.). These themes are also in all his books, including *Spirit of the Union* (outlining his dream and vision for the country), *Flashes of Thought*, *Flashes of Wisdom*, and *My Vision: Challenges in the Race for Excellence*, all of which express optimism and call for excellence in all fields. Examples are:

- *Make the impossible possible:* "The word impossible is not in the leaders dictionaries. No matter how big the challenges, strong faith, determination and resolve will overcome them."
- *Have no limits:* "Man should always leave the cage surrounding him."
- *Keep reaching for the top:* "The race for excellence has no finish line." "A true leader does not derive power from his position, but from his ethics, from people's love for him, and from his knowledge, education and excellence in his field of work."
- *Strive to be #1:* "I want Dubai to be number one. Not in the region, but in the world. Number one in everything: high education, health, and housing. I want to give my people the highest way of living." "Dubai will never settle for anything less than first place." "Excellence is a good habit one should acquire."

- *Dream but take action:* "When human beings have the courage and commitment to transform a dream into reality, there is nothing that can stop them."
- *Action, not talk:* "Most people talk; we do things. They plan; we achieve. They hesitate; we move ahead. We are living proof that when human beings have the courage and commitment to transform a dream into reality, there is nothing that can stop them."
- *Have a future orientation:* "Determination, strategy and vision for the future are our real resources in the quest for excellence and success." "We must look forward and anticipate the future so that our country leads globally."
- *Work hard:* "Toil and production are two distinguishing characteristics of a creative man."
- *Have no fear:* On a personal note, I do not agree with advisors who encourage you to purposefully embrace failure in order to know success; I prefer to agree with MBR's positive focus on success. While I acknowledge that people have fears—of success and also of failure—a strong leader should follow MBR's advice: "Our fear of failing should not prevent us from endeavoring to reach the summit." As he says in his book, *My Vision: Challenge sin the Race for Excellence,* we are not interested in counting mistakes, but in country achievements.
- *Care for others:* "The first responsibilityof a leader is to make his people happy and then to provide them with the required security, stability, comfort, progress and development." "We are more in need of giving than the needy. When you give, much is given back to you. Have mercy on others and you will be filled with joy."
- *Take initiative:* "You have to take the initiative to do what you can do, or dream to do, as boldness has majestic power; you have to take initiative."
- *Be grounded but take risks:* "We should always be practical, realistic and optimistic."

High Highness Sheikh Mohammed presides annually over the World Governments Summit—which I have attended yearly in my role on the Global Council on SDG 3 (Health and Wellbeing) of the "SDGs in Action" initiative of the United Arab Emirates (Coetzee, Kuriansky, & AlHammadi, 2024)—where I have seen his charisma first-hand. At one of the summits, he outlined the qualities of leadership (World Governments Summit, 2016): "As leadership is a combination of thought and wit," as shown in Fig. 1. He added that "leadership needs refinement. Leadership needs experience. Leadership needs practice."

His further advice is: Don't settle for anything less than number one because everyone remembers who came first and no one remembers second.

Life is full of challenges and risks, he described, which we overcome with positive energy, with which we must wake up in the morning, to be "cheerful and delightful with others." Holding up the thumb and index and middle finger of his right hand, in his favorite gesture, he pointed to the thumb, indicating this means "winning"; pointing to the index finger, he announced this is "victory"; and the formation of the three fingers, he pointed out, one by one, indicates "I love you." The assemblage of thousands in the main auditorium of the World Governments Summit smiled and clapped enthusiastically at each pronouncement.

Fig. 1
His Highness Sheikh Mohammed bin Rashid Al Maktoum says that "leadership is a combination of thought and wit."

Happiness and positivity

In his book, *Reflections on Happiness & Positivity*, His Highness Sheikh Mohammed examines these favorite two concepts of happiness and positivity. Both, he says, are essential for innovation. The pages of this book, whose cover is shown in Fig. 2, are peppered with personal stories and wisdom about the critical role happiness and positivity play in leadership strategies and positive relationships among team members.

In this book (a favorite in my extensive library), he outlines that the government's job is to achieve happiness for people, as shown in Fig. 3. His vision of government derives from a passion for open, service-led governance and strategic planning. In fact, he has led major reforms of both the government of Dubai and the UAE Federal government. In his book *My Vision: Challenges in the Race for Excellence,* he reveals his "secrets"—strategies that transformed the Emirate of Dubai into a world development marvel: vision, leadership, management, teamwork, and decisive and timely decision-making. In this book, he spells out that he does not believe in leadership based on fear.

Reminders of happiness are ever present in Dubai. I recall my surprise seeing the smiley face emoji on the ATM machine in Dubai Mall. Then, my student, Emirati native Layla Al Neyadi whom I had brought with me to the World Governments Summit one year, and I, noted the smiley face outside an office building. When I asked a young man who emerged from the building what it meant, he told us, "When we sign in to work each day on our computer, we have to rate how happy we are, following a continuum of happy emojis, and we do the same when we sign out at the end of the day." How clever of this rating to provide the focus and feedback on

Fig. 2
Front cover of the book by Sheikh Mohammed bin Rashid Al Maktoum on *Reflections on Happiness & Positivity*

your emotional level over the day. In a similar spirit, on our way to Sustainability City, a young man told us that when his team has a break from work, they share about how their day is going and what they need.

The UAE has embraced this theme of happiness. It is a topic, with its related theme of wellbeing, that I have long embraced myself, having written on the subject and produced many events about it, including at the UN and having been responsible—in partnership with Ambassador Dr. Caleb Otto of the UN Mission of Palau—for the inclusion of the word "wellbeing" in the title of SDG3 on Health in the UN Agenda, as well as founding the United Nations Group of Friends of Mental Health and Well-being during the Agenda negotiations (Billings, 2015; Kuriansky, 2012a, 2012b, 2012c, 2017, 2018; Kuriansky & Otto, 2017; Kuriansky et al., 2015; Luce, 2016).

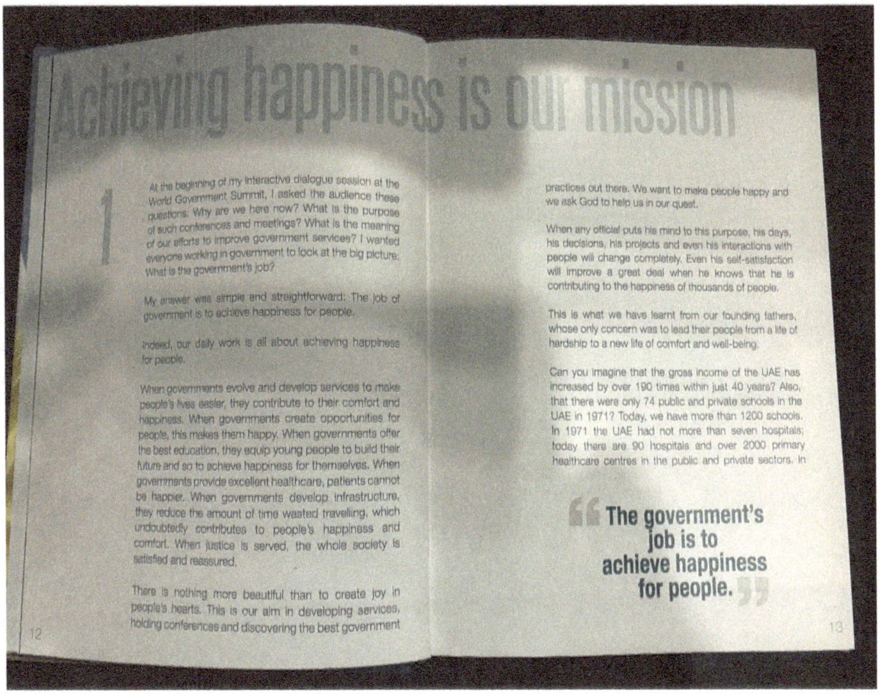

Achieving happiness is our mission

1

At the beginning of my interactive dialogue session at the World Government Summit, I asked the audience these questions: Why are we here now? What is the purpose of such conferences and meetings? What is the meaning of our efforts to improve government services? I wanted everyone working in government to look at the big picture: What is the government's job?

My answer was simple and straightforward: The job of government is to achieve happiness for people.

Indeed, our daily work is all about achieving happiness for people.

When governments evolve and develop services to make people's lives easier, they contribute to their comfort and happiness. When governments create opportunities for people, this makes them happy. When governments offer the best education, they equip young people to build their future and so to achieve happiness for themselves. When governments provide excellent healthcare, patients cannot be happier. When governments develop infrastructure, they reduce the amount of time wasted travelling, which undoubtedly contributes to people's happiness and comfort. When justice is served, the whole society is satisfied and reassured.

There is nothing more beautiful than to create joy in people's hearts. This is our aim in developing services, holding conferences and discovering the best government

12

practices out there. We want to make people happy and we ask God to help us in our quest.

When any official puts his mind to this purpose, his days, his decisions, his projects and even his interactions with people will change completely. Even his self-satisfaction will improve a great deal when he knows that he is contributing to the happiness of thousands of people.

This is what we have learnt from our founding fathers, whose only concern was to lead their people from a life of hardship to a new life of comfort and well-being.

Can you imagine that the gross income of the UAE has increased by over 190 times within just 40 years? Also, that there were only 74 public and private schools in the UAE in 1971? Today, we have more than 1200 schools. In 1971 the UAE had not more than seven hospitals; today there are 90 hospitals and over 2000 primary healthcare centres in the public and private sectors. In

"The government's job is to achieve happiness for people."

13

Fig. 3

My copy of Sheikh Mohammed's book, outlining the government's job to achieve happiness for the people.

The UAE government took over the happiness file from the Kingdom of Bhutan, whose King declared happiness as a government goal way back in 1972, measuring progress by Gross National Happiness rather than the money measure of Gross National Product. This motif was carried through to 2012, at a major United Nations conference I attended where happiness was posed as a new paradigm for development (in contrast to GDP) (Kuriansky, 2012b). With the change in the Bhutan government and new leadership dropping the imprimatur, the mantle was taken up by the government of the UAE. Thus ensured many developments on the theme. A first Minister of State for Happiness was appointed, Her Excellency Ohood Khalfan Al Roumi (now Minister of State for Government Development and the Future) as well as a focal point at the Mission of the UAE to the UN, pioneered by Hana Al-Hashimi who organized and moderated many events in which I was honored to participate. A National Program for Happiness and Wellbeing was launched. A smiley face emoji was unveiled on the ceiling of Dubai's Al Muraqqabat Police Station. At the UN, the Mission hosted many events, which I attended and sometimes co-organized on some occasions, for the International Day of Happiness. A 2-day happiness dialogue was held at the World Governments Summit, where we explored government and civil society applications. A Global Happiness Coalition convened in 2018 with government members, "Putting Happiness at the Centre of Government Work" with High-Level Representatives of UAE, Portugal, Slovenia, Costa Rica, Mexico, and Kazakhstan. The happy

Fig. 4
Smiley face pervasive at the World Governments Summit meeting held in Dubai.

face emoji was prevalent at World Governments Summit meetings held in Dubai, as shown in Fig. 4, overseen by Sheikh Mohammed and attended by thousands of global leaders.

Love

The happy face emoji is only outdone by the shapes of hearts in flowers and trees of the Miracle Garden Dubai. Imagine a leader—especially an autocratic one—who rules with symbols of love, rather than spreading fear, hate, and prejudice. There is even an Institute of Tolerance. In the context of his autocratic rule, Sheikh Mohammed's sensitivity is evident in his poetry, in classic Arabic and Bedouin style, which is published in several books. As described above, he delights in his hand gesture of three fingers, indicating "I love you."

A new respect for people

Words matter. I am particularly impressed that the UAE maintains that persons challenged in some way in their abilities are referred to as "people of determination." This term progresses way beyond the old term "disabled people" with its very negative connotations, which later evolved to "persons with disabilities." Yet, the term "determination" itself reflects a deserved respect and focus on strengths and inclusion for all people in a society.

Philanthropy

Giving to others is a quality H.H. Sheikh Mohammed espouses and infuses in the public presentation of the government by himself and others. This is evident in all presentations (at World Governments Summits and other convenings) emphasizing the theme of family, generosity and commitment to giving to others. Giving is also reflected in his extensive philanthropy. In July 2022, I was invited to the Mission of the UAE to the United Nations event, held in their mission building in New York City, unveiling the many charitable activities under the Mohammed bin Rashid Global Initiatives (MBRGI) referred to at the beginning of this chapter. The event was held as a special session of the United Nations 2022 High Level Political Forum on Sustainable Development on driving sustainable development and empowering nations to thrive in a post-COVID-19 world. The multitude of humanitarian, social, and development initiatives, projects, campaigns, and activities by the entities of the MBRGI that support and empower vulnerable and disadvantaged communities around the world come under five areas: Healthcare & Disease Control; Humanitarian Aid and Relief; Spreading Education & Knowledge; Innovation & Entrepreneurship; and Empowering Communities. These are all under the umbrella of High Highness Sheikh Mohammed's vision "For the UAE, and from the UAE, we will plant a seed of goodness in every corner of this world" (Mohammed bin Rashid Global Initiatives, n.d.). This vision of hope, positivity, and innovation is evident in the story of 15-year-old Mustafa Khaled who got support developing a smart application that gamifies learning for middle school students, called "Educup," now present on major online platforms. Mustafa is one of 2.9 million beneficiaries on Innovation and Entrepreneurship Initiatives in 2021. As His Highness Sheikh Mohammed says: *"The future belongs to those who can imagine it, design it and execute it. It isn't something you await, but rather create."*

Projects also center on tolerance, promoted in the Sheikh Mohammed Centre for Cultural Understanding, that builds "bridges of love between people of different ethnicities, cultures and backgrounds." Using the word "love" is distinctive. It reflects MBR as a leader and as a person. I am touched by the message and welcome that is consistent with the umbrella theme of the organization of which I am a Trustee, the United African Congress, of "Building Bridges Across Boundaries" promoted at our annual events at the United Nations on World Interfaith Harmony Week (Kuriansky, 2020; United Nations, n.d.).

Highly relevant to the theme of this volume—on the topic of resilient health systems—is accomplishments of the MBRGI in the pillar of healthcare, with AED 51.2 million spent and 14.6 million beneficiaries in 2021. Recognizing the ongoing repercussions of COVID-19 and its variants, the MBRGI responded with various programs, including Noor Dubai's mobile clinics for eye care, a hospital center with a "Hope Lounge", a "Breast Friends" program for breast cancer patients, and supporting hospital training and service centers outfitted with the latest technological innovations. All these efforts reflect the motto, shown in Fig. 5, that "The health of a nation and its people determines its ability to progress. Strengthening the

Fig. 5
Sheikh Mohammed quote about the importance of health in a nation.

healthcare sector is one of the UAE's highest priorities, and we are confident that our efforts will contribute to advancing healthcare everywhere."

A renaissance man

A true Renaissance man excels in many fields. HH Sheikh Mohammed is a billionaire businessman and real estate developer (a technology park, internet city, the tallest building in the world, an airline, and more); sportsman (falconry and horsemanship); poet (on subjects from romance—"her figure like the gazelle"—to current affairs); humanitarian (Noor Dubai fights blindness and Dubai Cares fights poverty and provides education); philanthropist (his Mohammed bin Rashid Global Initiatives charitable foundation implements over 1,400 development programs supporting more than 130 million people in 116 countries); patron of the arts; politician; and civil servant. It takes a visionary to conceive the "Smart City" using technology to make citizen's experiences simple, convenient, and efficient, and the "Smart Government" open 24/7 very day of the year to service citizens (Al Maktoum, 2013).

Not everyone can afford the luxury to be as well-rounded and indulge all interests and aspects of oneself, as MBR, but anyone can certainly pursue all their dreams and interests according to their own drive and economic level. His Highness Sheikh Mohammed offers a role model (granted, out of reach for ordinary folks) to do it all. He explores his poetic sensitive self, in the Arabic Nabati style, and also his sportsman competitive self notably in thoroughbred horse racing. His passion for the latter is legendary, having founded the world's richest race, the Dubai World Cup, and becoming a world champion himself—since only excellence will do in his view. He is also a trained pilot, which is not as well-known. As an expert in creativity, I know that diversified and active pursuits fuel innovation.

Similar traits are in his son, the Crown Prince of Dubai. In a showing of intergenerational continuity—and the proverbial saying that "the apple doesn't fall far from the tree"—Sheikh Hamdan bin Mohammed bin Rashid Al Maktoum—the second son of MBR's 12 children with his senior wife, Sheikha Hind bint Maktoum bin Juma Al Maktoum—is a politician, sportsman, and artist. Sheikh Hamdan bin Mohammed, in crediting his British schooling, says he learned self-discipline, commitment, virtue, responsibility, endurance, understanding, teamwork, friendship and the benefits of hard work. Like his father, the 41-year old is an award-winning equestrian, and also known for his romantic poems in Arabic (published under the name *Fazza* which means "the one who helps" in Arabic). And, he is also a sky diver and a scuba diver—metaphorically exploring the depths and heights of self and life that can be interpreted as revealing the courage and fearlessness of an innovative leader.

An orientation to the future

Everything about Dubai speaks to the Future, in health and all areas. As HH Sheikh Mohammed says, "We must look forward and anticipate the future so that our country leads globally" (Meah, n.d.).

Within the complexity of his country and the region of the Middle East, HH Sheikh Mohammed can be credited for the flexibility and dynamism he espouses, that drives change in Dubai.

Transformational leadership style

One can summarize from all the above that MBR has the two kinds of intelligence: traditional academic intelligence (IQ), that uses "left-brain" skills that are analytic and practical and also "emotional intelligence" (EQ) that taps into more "right-brain" skills, since he explores—even publicly in his poetry—the sensitive side of himself and care for his people. Emotional intelligence has been defined as the "ability to monitor one's own and others' feelings and emotions, to discriminate among them and to use this information to guide one's own thinking and actions" (Mandell & Pherwani, 2003). In the more elaborate "Whole Brain Thinking" style model, according to a hypothetical HBDI (Herrmann Brain dominance Instrument) assessment, as described in another chapter of this volume (Kuriansky & Herrmann-Nedhi, 2024). HH Sheikh Mohammed's modeling would likely show considerable capacity in the cognitive, organizational and relational styles, and also significantly in what's identified as the "upper right brain" style of a Visionary, which he obviously exhibits.

H.H. Sheikh Mohammed can also be considered as having qualities of a transactional leader (who exercises control) but also a transformational style, whereby a leader generates awareness of the vision of the organization; stimulates colleagues and followers to view their work from a new perspective; builds teams; and follows the "Four I's": Idealized Influence, Inspirational Motivation, Intellectual Stimulation, and Individualized Consideration (Mandell & Pherwani, 2003; Purwanto et al., 2020). The transformational style is attributed to the founders of the

UAE, consistent with Arabian Bedouin traditions of consensus and consultation, and education in the Majlis, which values learning to communicate with people and respond to their needs, protect others, and find solutions to challenges, and protect them (Ahmed & Al Amiri, 2022).

Undoubtedly MBR's leadership style is both grounded in, and expressed in, innovation and exploring the edges and extremes of possibility, in his government and also in exploring the parts of himself. This is evident in Dubai's healthcare, introducing the latest and most innovative ideas and technologies (Almatrooshi et al., 2020; Purwanto et al., 2020), thereby also fostering health literacy in a healthy community (Sentell et al., 2020).

The importance of open and uninhibited communication as the means of setting goals and shaping perspectives and motivation has been defined as an essential trait of a true leader (Alqatawenah, 2018). Communication must remain top priority since it leads to the ability to convey critical ideas and values to a target population, introduce change, and inspire people to adapt and expand (Alqatawenah, 2018). Indeed, communication is key to success in any healthy relationship; MBR has been said to prioritize open dialogue and transparency (Sentell et al., 2020). It is also necessary to engender trust of people, to lead them into new ideas inherent in innovation, since many people are resistant to change. With his focus on principles of innovative thinking and management as the foundational pillars of UAE culture, Sheikh Mohammed has reportedly presented the notion of innovation as inherent in the UAE culture, introducing change to citizens as part of their cultural identity, that would then allow change necessary for development (Almatrooshi et al., 2020). Allowing this change facilitates the transition of the culture to the digital age, including telehealth options and relevant opportunities that come with innovative Artificial Reality and Virtual Reality that have been shaping Dubai healthcare (Kouki & Al Shemaili, 2022).

The impact of Sheikh Mohammed's efforts on public health is reflected in studies that show improvements in healthcare delivery, for example, attendance to primary healthcare centers, better patient–nurse communication, and higher literacy rates among patients (Omar, 2018) as well as an increase in use of innovative technologies allowing for enhanced provider–patient communication and education, earlier and more accurate diagnosis, and more effective management of health issues (Ribeiro et al., 2021).

Such innovation has led Dubai to famously become the first country in the world to have an office built with help of 3D printing technology (United Arab Emirates Government, 2022).

Expo 2020 Dubai

The world exhibition in Dubai, Expo Dubai 2020—actually held October 2021 to March 2022 due to the pandemic—was an outstanding example of the typical way Dubai under MBR's leadership conceptualizes innovation. Fortunate to have visited the Expo, I noted that the vast space was organized with typical Dubai creativity, with three districts of pavilions collated under three themes: Mobility, bringing together country pavilions under the theme of possibilities; Sustainability, with its message that humans can enjoy harmony with nature in a high-level future; and Opportunity, aiming to unlock potential for a better future. Again, with typical Dubai creativity, the physical space is being transformed into a site for world-class innovation.

Team building in health technology

Another element of MBR's leadership is reported as having a focus on building multidisciplinary teams to advance health science and encourage the integration of digital tools for managing healthcare issues (Al Zaabi, 2019). With the promotion of collaboration across disciplines and departments associated with healthcare and related issues, MBR has made it possible to encourage the acceptance of change as an inevitable and indispensable part of the future of Dubai and, by extension, of the UAE (Putra et al., 2020).

Leadership in the COVID-19 era

Sheikh Mohammed's proactive approach to the COVID-19 pandemic has been credited to a sensible approach, emphasizing and reinforcing the importance of scientific, evidence-based research in healthcare as a crucial step to defeat the virus (Crawford & Serhal, 2020). According to available accounts of his actions, MBR provided the following rationale for his decisions regarding COVID-19: *"The UAE's scientific sector is now experiencing its best phase. Scientists of the UAE have put the nation among the top countries shaping the world's future. The Command and Control Centre plays a major role in the nation's efforts to combat COVID-19"* (Sheikh Mohammed's proactive response to coronavirus helping UAE effectively combat pandemic: Sheikh Mansour, 2020, para. 3). Success in pandemic control is attributed to building a diverse team of healthcare experts, expanding it rapidly, and fostering the principles of collaboration and data sharing (Antwi-Boateng, 2022; Crawford & Serhal, 2020). Added advances in healthcare are due to introducing world-class medications and treatment options (Antwi-Boateng, 2022). This was amplified by MBR instilling confidence in people, saying, *"We have highly-qualified national talents, partnerships with leading global medical institutions as well as clear plans to ensure all members of the community can receive world-class medical treatments"* (UAE: Healthcare highest priority, Sheikh Mohammed says, 2021, para. 4).

Among recovery efforts during the COVID-19 pandemic in terms of innovation, the Museum of the Future was inaugurated in Dubai on 22 February, 2022 (purposefully a palindrome date)—with its external architecture forming a human "eye". The eye is meant to represent knowledge we "foresee"—to hold dialogues and build partnerships for innovative and sustainable solutions to world problems. This complements the "Impactful Leaders Program" that nurtures future leaders, and the Mohammed bin Rashid School of Government that cultivates innovative public policy-making, with courses on leadership and "agility" in government, for example, "Strategic Leadership in a Challenging World" and "Reimagining the Role of Governments in Times of Crisis" as well as a "Government Knowledge Gate" with innovative digital knowledge solutions.

Always looking ahead, in February 2023, H.H. Sheikh Mohammed approved plans for the UAE Cabinet to direct authorities to launch a study gauging the impact of AI technologies such as ChatGPT (a platform that generates text in a more sophisticated way than existing chatbots) on health, as well as education, media and other sectors. And, in the current climate of controversy over AI outpacing human abilities, concern is being paid to ethics, as evident in a World Government Summit (2023) extended session on regulation.

Conclusion

Over the course of his rule, His Highness Sheikh Mohammed bin Rashid Al Maktoum has demonstrated varied leadership skills, guiding his government for advancing change with social and technological innovation that has led to considerable progress. His contributions to development are evident in every sector including in healthcare, which is identified as a priority. He presents a compelling example of creativity and innovation that inspires people and propels a society forward.

References

Ahmed, G., & Al Amiri, N. (2022). The transformational leadership of the founding leaders of the United Arab Emirates: Sheikh Zayed bin Sultan al Nahyan and Sheikh Rashid bin Saeed al Maktoum. *International Journal of Technology, Innovation and Management*. https://journals.gaftim.com/index.php/ijtim/article/view/58/24.

Al Fahim, M. (2000). *From rags to riches: A story of Abu Dhabi*. I. B. Tauris & Company.

Al Maktoum, M. bin R. (2013). *Flashes of thought: Lessons in life and leadership from the man behind Dubai*. Motivate Publishing.

Al Rand, H. A. R., Madi, H., Alhemeiri, S., & Al Obeidli, A. A. R. (2024). Healthcare innovation and leadership: Case study of the United Arab Emirates. In J. Kuriansky, & P. Kakkattil (Eds.), *Resilient health: Leveraging technology and social innovations to transform healthcare for COVID-19 recovery and beyond* (pp. 415–431). Elsevier.

Al Zaabi, M. (2019). Driving an ambitious vision with excellence the UAE Government excellence model. *International Journal of Excellence in Government, 1*, 8–14. https://doi.org/10.1108/IJEG-02-2019-0004.

Aldafrawy, O., Taryam, M. M. O., Aldallal, A. M. R., Ahmed, S. H., & Swidan, A. (2024). The "Doctor for every citizen" Smart Telehealth Initiative: Increasing resilience and services in the UAE Dubai Health Authority during and beyond the COVID-19 pandemic. In J. Kuriansky, & P. Kakkattil (Eds.), *Resilient health: Leveraging technology and social innovations to transform healthcare for COVID-19 recovery and beyond* (pp. 1041–1045). Elsevier.

Almatrooshi, M. J. A., Khalifa, G. S., Ameen, A., Hossain, M. S., & Morsy, M. A. (2020). The role of knowledge oriented leadership and knowledge sharing to manage the performance of Ministry of Interior in UAE. *International Journal on Recent Trends in Business and Tourism (IJRTBT), 4*(2), 9–17.

Alqatawenah, A. S. (2018). Transformational leadership style and its relationship with change management. *Business: Theory and Practice, 19*, 17–24. https://doi.org/10.3846/btp.2018.03.

Antwi-Boateng, O. (2022). Covid diplomacy in the era of pandemic response: The case of United Arab Emirates. *Journal of International Women's Studies, 23*(3), 92–105.

Arnold, S. (2022). *The healthcare system in the UAE*. https://www.expatica.com/ae/healthcare/healthcare-basics/the-healthcare-system-in-the-united-arab-emirates-71767/.

Billings, R. (2015). *Women's mental health and wellbeing in the post 2015 agenda*. http://blacktiemagazine.com/society_2015_march/Women_Mental_Health_and_Wellbeing_in_the_post_2015_agenda.htm.

Coetzee, C., Kuriansky, J., & AlHammadi, R. (2024). The nexus model and the Young Leaders Program: Innovative models for the future of the SDGs in Action Campaign and the Global Councils on Sustainable Development Goals. In J. Kuriansky, & P. Kakkattil (Eds.), *Resilient health: Leveraging technology and social innovations to transform healthcare for COVID-19 recovery and beyond* (pp. 629–646). Elsevier.

Crawford, A., & Serhal, E. (2020). Digital health equity and COVID-19: The innovation curve cannot reinforce the social gradient of health. *Journal of Medical Internet Research, 22*(6), 1–9. https://doi.org/10.2196/19361.

Erickson, E. H. (1994). *Identity and the life cycle*. WW Norton.

Kouki, S., & Al Shemaili, M. (2022). Science education in the United Arab Emirates. In *Science education in countries along the belt & road* (pp. 225–241). Springer.

Kuriansky, J. (2012a). *Happiness, GDP, and the presidential race*. https://abcnews.go.com/Health/Wellness/happiness-gdp-presidential-race/story?id=17619613.

Kuriansky, J. (2012b). *United Nations conference on well-being and happiness*. https://docs.google.com/document/d/1CX6gs2J5LqYXTHU0HV3O-usgkuLySLRA/edit.

Kuriansky, J. (2012c). Well being: An important issue at the United Nations and for the International Association of Applied Psychology. *Bulletin of the International Association of Applied Psychology, 24*(2–3), 64–70. https://iaapsy.org/site/assets/documents/July2012.pdf.

Kuriansky, J. (2017). *Transforming trauma to healing, hope and happiness. [video file]*. https://www.youtube.com/watch?v=bfdJ9jAomMw.

Kuriansky, J. (2018). Psychotherapy, psychology, psychiatry and international policy: Professional contributions and personal experiences about mental health and wellbeing, psychosocial resilience and peace. *The International Journal of Psychotherapy, Counseling & Psychiatry: Theory, Research & Clinical Practice. Special Volume of the World Council for Psychotherapy 8th World Congress of Psychotherapy Conference 2017*. http://ijpcp.com/journal03/j03a06.pdf.

Kuriansky, J. (2020). *Interfaith leaders confront world problems at United Nations*. http://www.blackstarnews.com/global-politics/others/interfaith-leaders-confront-world-problems-at-united-nations.

Kuriansky, J. (2024). "Health in Your Hands": A platform for innovative healthcare solutions. In J. Kuriansky & P. Kakkattil (Eds.), *Resilient health: Leveraging technology and social innovations to transform healthcare for COVID-19 recovery and beyond* (pp. 989–1004). Elsevier.

Kuriansky, J., & Herrmann-Nedhi, A. (2024). Wired for success: Using Whole Brain Thinking styles for innovation and solving global problems. In J. Kuriansky, & P. Kakkattil (Eds.), *Resilient health: Leveraging technology and social innovations to transform healthcare for COVID-19 recovery and beyond* (pp. 1137–1150). Elsevier.

Kuriansky, J., LeMay, M., & Kumar, A. (2015). Paradigm shifts in nature and well-being: principles, programs, and policies about the environment and climate change with actions by the United Nations for a sustainable future. In D.G. Nemeth, & J. Kuriansky (Eds.), *Ecopsychology: Advances from the Intersection of Psychology and Environmental Protection* (pp. 307–358). ABC-Clio/Praeger.

Kuriansky, J., & Otto, J. (2017). *Report on a side event held at the United Nations 16 May 2017 in the margins of the UN Science, technology & innovation forum: Artificial Intelligence and technology tools for mental health, well-being and resilience*. http://bit.ly/2mghjOH.

Luce, J. (2016). *Jumping for joy*. http://www.huffingtonpost.com/jim-luce/jumping-for-joy-happiness_b_9566064.html.

Mandell, B., & Pherwani, S. (2003). Relationship between emotional intelligence and transformational leadership style: A gender comparison. *Journal of Business and Psychology, 17*(3).

Meah, A. (n.d.). 40 inspirational Sheikh Mohammed bin Rashid Al Maktoum quotes on success. awaken the greatness within. https://www.awakenthegreatnesswithin.com/40-inspirational-sheikh-mohammed-bin-rashid-al-maktoum-quotes/.

Mohammed bin Rashid Global Initiatives (n.d.). https://www.almaktouminitiatives.org/en/who-we-are.

Omar, O. (2018). Signing up for happiness: An exploratory study of workforce changes in the UAE market in response to the National Program for Happiness and Wellbeing. *Middle East Journal of Positive Psychology, 4*, 80–101. https://doi.org/10.1007/978-3-030-13921-6_17.

Piaget, J., & Inhelder, B. (1972). *The psychology of the child*. Basic Books.

Purwanto, A., Bernarto, I., Asbari, M., Wijayanti, L. M., & Hyun, C. C. (2020). Effect of transformational and transactional leadership style on public health centre performance. *Journal of Research in Business, Economics, and Education, 2*(1), 1–7. https://doi.org/10.13140/RG.2.2.33517.13283.

Putra, A. S., Waruwu, H., Asbari, M., Novitasari, D., & Purwanto, A. (2020). Leadership in the innovation era: Transactional or transformational style? *International Journal of Social and Management Studies, 1*(1), 89–94.

Ribeiro, D., Costa, A. P., & Remondes, J. (2021). Building blocks for the next eGovernment era: Exploratory study based on Dubai and UAE's Ministry of Happiness communication in 2020. *International Journal of Humanities and Social Sciences, 15*(4), 503–510.

Russek, L. G., & Schwartz, G. E. (1997). Feelings of parental caring predict health status in midlife: a 35-year follow-up of the Harvard Mastery of Stress Study. *Journal of Behavioral Medicne, 20*(1), 1–13. https://doi.org/10.1023/a:1025525428213. PMID: 9058175.

Sentell, T., Vamos, S., & Okan, O. (2020). Interdisciplinary perspectives on health literacy research around the world: more important than ever in a time of COVID-19. *International Journal of Environmental Research and Public Health, 17*(9), 3010. https://doi.org/10.3390/ijerph17093010.

Sheikh Mohammed's proactive response to coronavirus helping UAE effectively combat pandemic: Sheikh Mansour. (2020). *Khaleej times*. Retrieved August 15, 2022, from https://www.khaleejtimes.com/coronavirus/sheikh-Mohammeds-proactive-response-to-coronavirus-helping-uae-effectively-combat-pandemic-sheikh.

UAE: Healthcare highest priority, Sheikh Mohammed says. (2021). *Khaleej times*. Retrieved August 15, 2022, from https://www.khaleejtimes.com/uae/uae-healthcare-highest-priority-sheikh-Mohammed-says.

United Arab Emirates Government. (2022). *Mohammed Bin Rashid: UAE is an incubator of innovation and future technology*. Retrieved August 15, 2022, from https://uaecabinet.ae/en/details/news/Mohammed-bin-rashid-uae-is-an-incubator-of-innovation-and-future-technology.

United Nations. (n.d.). World Interfaith Harmony Week. https://www.un.org/en/observances/interfaith-harmony-week.

Wayback Machine. (2013). *His Highness Sheikh Mohammed bin Rashid Al Maktoum*. https://web.archive.org/web/20131216091953/http://www.sheikhmohammed.co.ae/vgn-ext-templating/v/index.jsp?vgnextoid=cdc1c1922d435310VgnVCM1000003f64a8c0RCRD&appInstanceName=default.

World Governments Summit. (2016). *Dialogue of the Future with H.H. Sheikh Mohammed bin Rashid Al Maktoum, UAE Vice President, Prime Minister and Ruler of Dubai, via Social Media [video]*. https://www.worldgovernmentsummit.org/events/annual-gathering/2016/session-detail/dialogue-of-the-future-with-h.h.-sheikh-mohammed-bin-rashid-al-maktoum-uae-vice-president-prime-minister-and-ruler-of-dubai-on-social-media.

World Government Summit. (2023). *World Government Summit. Leadership*. https://www.worldgovernmentsummit.org/about/leadership.

CHAPTER 40

The fundamental role of the United Nations General Assembly to foster political commitment in support of global health with the assistance of the World Health Organization

Werner Obermeyer and Eva Kiegele
World Health Organization Office at the United Nations, New York, NY, United States

Introduction

In June 2001, the United Nations General Assembly (UNGA) held its first-ever high level meeting (HLM) with a health focus, arriving at a commitment to combat the HIV/AIDS epidemic. Subsequently, the UNGA has focused on HIV/AIDS at a high level every five years. The impact of ensuring buy-in from nonhealth ministers and across other sectors of government was not immediately apparent until the September 2011 high level meeting on the Prevention and Control of Noncommunicable Diseases (NCDs) directed attention of the UNGA to a cluster of diseases which collectively represent 74% of mortality worldwide, and yet are largely underfunded and overlooked (World Health Organization, 2022a). Since then, many major health topics have been addressed in such high level meetings and led to resolutions by the 193 Member State governments of the United Nations. This chapter gives an overview of several of these important topics in health resolutions considered by the UNGA, thereby presenting the critical role played by the UNGA in promoting global health. We further describe the technical assistance and advice from the Office of the World Health Organization (WHO) at the United Nations and emphasize the importance of fostering political commitment to these resolutions.

Resilient Health. https://doi.org/10.1016/B978-0-443-18529-8.00040-8

Definitions

The UN General Assembly (UNGA) is the main policy-making organ of the UN which is made up of all the Member State governments (193) which provides the forum for multilateral discussion of all the international issues covered by the Charter of the UN (and issues in the UN Agenda 2030 for Sustainable Development and all other issues concerning the member governments). The assembly meets in regular sessions from September to December annually and as otherwise required, where the members discuss specific issues through dedicated agenda items or subitems, which lead to the adoption of resolutions. Each of the 193 Member States of the United Nations has an equal vote.

A General Assembly (GA) resolution is a decision or declaration voted on by all Member States of the United Nations in the General Assembly. Resolutions are adopted usually by a simple majority (50 percent of all votes plus one) of Member States present and voting. While the decisions of the General Assembly have no legally binding force for governments, they carry the weight of world opinion and the moral authority of the world community.

A political declaration refers to a statement form of communication setting forth the position and recommendations by a specific party and outlining a framework under which actions should be carried out.

A Group of Friends is a group of countries at the UN who come together voluntarily to promote strategic relevance of and political support for topics of emerging importance to the international community and that are covered by targets of the Sustainable Development Goals in the UN 2030 Agenda.

A "Call to Action" is an appeal to certain stakeholders to mobilize commitment and specific efforts, acts, or endeavors to achieve a specified goal.

Outcomes of major health-related UN high level meetings

Important topics in global health that have been addressed by the UN Member State countries which have led to GA Resolutions, in consultation with the WHO and its Office at the UN, are described as follows.

Noncommunicable diseases (NCDs)

During the 2011 high-level meeting on NCDs, Member States committed to bold actions putting prevention forward as the cornerstone of a global response. By adopting a Political Declaration, governments committed to a dedicated action plan that would curb the risk factors behind the four groups of NCDs: cardiovascular diseases, cancers, chronic respiratory diseases, and diabetes. These actions ranged from taxation measures to reduce tobacco consumption to regulating marketing of foods and beverages that are high in saturated fats, trans-fatty acids, sugars, or salt, especially targeting children. The declaration also called for measures to limit harmful consumption of alcohol, promote healthy diets, and increase physical activity. The declaration clearly illustrated the leadership role of WHO with several time-bound assignments, including the establishment of global mechanisms and a roadmap for coordination of multisectoral action and monitoring of results. The declaration enabled WHO to put forward a set of voluntary targets, in 2013, to reduce modifiable risk factors and refocus health systems toward NCDs prevention (UNGA, 2011).

In the 2014 high-level meeting on NCDs, governments committed to addressing NCDs as a matter of priority in national development plans, for which they agreed to consider setting national targets to reduce risk factors and underlying social determinants for these diseases, including strengthening and orienting health systems to address prevention and control issues through people-centered primary healthcare and universal health coverage. Member States also committed to take steps toward effective legislation to mainstream NCDs in development schemes and to measure economic, social, and environmental determinants and disparities in health. Commitments were also made to actively promote national and international investments and to strengthen national capacity for quality research and development (UNGA, 2014).

Due to the slow progress in achieving the targets set in the previous meetings, during the 2018 high level meeting on NCDs, heads of state and government committed to 13 new steps to tackle the four main groups of NCDs, and added mental health and well-being as the fifth category of concern (which is included in target 3.4 of the UN 2030 Agenda for Sustainable Development). The Political Declaration called for increased global awareness to the environmental risk factors, especially indoor and outdoor air pollution, contributing to premature deaths from NCDs. Governments also agreed to robust laws and fiscal measures, including restrictions of alcohol advertising, banning smoking, and taxing sugary drinks. They also committed to implementing a series of WHO-recommended policies to prevent and control NCDs, such as public education and awareness campaigns to promote healthier lifestyles, vaccinating against HPV virus to protect against cervical cancer, and treating hypertension and diabetes. The Political Declaration also called specifically on food manufacturers to take several actions, including reformulating products to reduce salt, free sugars, and saturated and industrially produced trans fats (UNGA, 2018a).

Antimicrobial resistance (AMR)

In 2016, for the first time, Heads of State and Government committed to taking a coordinated approach to address the root causes of AMR across multiple sectors, especially human, animal, and environmental health, that are the three dimensions of the "One Health" initiative. Acting on their behalf, Member States (essentially the governments at the UN) committed to developing national action plans on AMR and agreed on strengthening the regulation of antimicrobials, improving knowledge and awareness, promoting best practices, and fostering innovative approaches using alternatives to antimicrobials and new technologies for diagnosis and vaccines. Member States called for better use of existing, cost-effective tools for preventing infections in humans and animals, including immunization, safe water and sanitation, and good hygiene in hospitals and animal husbandry. Member States also committed to new incentives for investment in research and development of new, effective, and affordable medicines; rapid diagnostic tests; and other important therapies to replace those that are losing their efficiency, emphasizing the prioritization of affordability and access to existing and new antibiotics, vaccines, and other medical tools. In this regard, they also acknowledged the importance of delinking the cost of investment in research and development on antimicrobial resistance from the price and volume of sales so as to facilitate equitable and affordable access (UNGA, 2016).

In April 2021, the president of the UNGA convened a high level interactive dialogue on AMR. A "Call to Action" was launched during this dialogue to raise the global ambition on AMR. Presented by the Group of Friends on Tackling AMR, the Call to Action called for enhanced national efforts to tackle AMR through a One Health approach, considering the diverse national circumstances. Member States also called for the acceleration of previous commitments to tackle AMR, for improved antimicrobial stewardship, for work toward financing for AMR action, for increased research on AMR's impacts, and for supporting the new One Health Global Leaders Group on AMR among other key steps. As of March 2022, there are 113 Member State signatories to the Call to Action and 40 supporting organizations (AMR, 2021).

The General Assembly focus on AMR, through the outcomes of 2016 and 2021, contributed a great deal toward building awareness to an emerging, and often overlooked, health challenge which the Centers for Disease Control and Prevention estimates cost the USA alone $20 billion in healthcare and $35 billion in loss of productivity each year (AMR, 2021; UNGA, 2016).

Tuberculosis (TB)

In a 2018 high level meeting on TB, Member States committed to provide diagnosis and treatment of TB with the aim of successfully treating 40 million people with TB from 2018 to 2022, including 3.5 million children and 1.5 million people with drug-resistant TB. Member states also agreed to scale up efforts in preventing TB for those most at risk of falling ill, through the rapid scaling up of access to testing and the provision of preventive treatment, so that at least 30 million people would receive preventive treatment by 2022, with specific targets for children, household contacts, and people living with HIV. Concrete resource mobilization goals were also agreed upon in the Political Declaration, with an aim to ensure sufficient and sustainable financing, reaching at least US$13 billion a year by 2022, with an additional US$2 billion a year for TB research. Member States committed to addressing the economic and social determinants of the disease, promoting an end to stigma and all forms of discrimination, including through the protection and promotion of human rights and dignity, and providing special attention to the poor as well as to vulnerable and at-risk communities. These very ambitious commitments were severely set back by the outbreak of COVID-19 in the year 2020 (UNGA, 2018b).

In 2023, high-level meeting on the fight against TB, Member States reviewed progress made since the 2018 high-level meeting. Unfortunately, most of the global targets agreed in the 2018 political declaration, including to provide TB treatment to 40 million people, and TB preventive treatment to 30 million people, were missed, and restricted access to health services caused by the COVID-19 pandemic partially contributed to this missed target. New ambitious targets were agreed and included in the Political Declaration of the 2023 high-level meeting. Some of these include: to reach 90% of people with TB prevention and care services, providing social benefit packages to all people with TB, manufacturing and licensing at least one new TB vaccine within the next 5 years, and closing funding gaps for TB implementation and research by 2027. Significant funding amounts were specified in the Declaration, including mobilizing financing for tuberculosis research and innovation especially to high burden countries towards reaching 5 billion United States dollars a year by 2027, for the development of safe, effective, accessible and affordable rapid and accurate point of care diagnostics, and reaching overall global investments of at least 22 billion United States dollars a year by 2027, and 35 billion annually by 2030. It also welcomes the establishment, by the World Health Organization, of the TB Vaccine Accelerator Council. The strong commitments expressed by Member States during the HLM focused on strengthening health systems and ensuring universal access to TB prevention and care, as part of the UHC agenda; scaling up research to develop safer, shorter and more effective treatments, particularly for drug-resistant forms of the disease; cross-sectoral collaboration beyond the health sector; and accountability mechanisms at global and national levels (UNGA, 2023a).

Universal Health Coverage (UHC)

In September 2019, during the high level meeting on UHC held at the UN headquarters in New York City, world leaders endorsed the most comprehensive agreement on global health ever reached—a vision for UHC by 2030. By adopting the Political Declaration on UHC, Member States committed to investing in policies that would expand access to essential health services, medicines, vaccines, diagnostics, and other health technologies, while preventing financial hardship from out-of-pocket healthcare payments; in particular, progressively covering 1 billion more people by the year 2023 and stopping the rise, and reverse the trend, of catastrophic out-of-pocket health expenditure. Member States committed to achieve UHC for healthy lives and well-being across the life course and as a social contract, through: optimizing budgetary allocations on health, broadening fiscal space, prioritizing health in public spending, and harmonizing health investments, pursuing equity and prioritizing the most vulnerable, and creating a strong, enabling regulatory and legal environment responsive to people's needs. Governments also agreed to establish multistakeholder mechanisms for engaging the whole of society for a healthier world, as well as to emphasize gender equality, redress gender power dynamics, and ensure women's and girls' rights as foundational principles for UHC. The decision was made to convene a high level meeting on UHC in 2023 in New York, with the aim to undertake a comprehensive review on the implementation of the 2019 declaration (UNGA, 2019).

In September 2023, global leaders gathered for the second high-level meeting on UHC to review progress and set new targets. The global picture remains alarming—progress on service coverage has stagnated and financial protection trends are worsening (more than half of the world's population is still not fully covered by essential health services and 2 billion people face financial hardship or are impoverished due to out-of-pocket health spending). World Leaders committed to redouble efforts towards universal health coverage by 2030. The Political Declaration is a catalyst for big and bold actions to mobilize political commitments and financial investments for achieving UHC and health security through a primary health care (PHC) approach. There was a broad recognition by Governments and key stakeholders of the alarming trends of service coverage and financial protection and the need to translate political commitments to UHC into budget decisions. Heads of States, International Financing Institutions and the private sector expressed strong support for urgent action to protect and expand public financing as the driver for PHC, while reshaping the financing architecture to invest more and better in health. There was a particularly focus on strengthening the health and care workforce, expanding service coverage and ensuring financial protection. World Leaders emphasized that health and well-being cannot be the privilege of a few, and reorienting health systems towards primary health care (PHC) is the most effective, cost-efficient and equitable way to deliver on the promise of UHC, and to leave no one behind (especially the most vulnerable). Member States committed to progressively address the global shortfall of 523 million people without access to quality essential health services and safe, effective, quality, affordable essential medicines, vaccines, diagnostics, and health technologies, in order to provide coverage for 1 billion additional people by 2025 (UNGA, 2023b).

Road safety

On 30 June and 1 July 2022 during the first UN high level meeting on Road Safety, Member States committed to increased investment and action to ensure that the Sustainable Development Goal of the UN 2030 Agenda for Sustainable Development to halve road deaths and injuries by 2030 is met. Following the launch of the Global Plan for the Decade of Action for Road Safety 2021–2030 (Global Plan) in 2021, the high level meeting focused on addressing gaps and challenges, mobilizing political leadership, promoting multisectoral and multistakeholder collaboration, and assessing the progress made in attaining the objectives of the road safety-related targets of the 2030 Agenda. Through the adoption of the Political Declaration, the Member States committed to the implementation of the Global Plan for the Decade of Action 2021–2030, including action to reduce road deaths and injuries by 50% by the year 2030. They also agreed to develop and implement regional, national, and subnational plans that include road safety targets and other evidence-based indicators. The declaration emphasizes systematic engagement between transport, health, education, finance, environmental, and infrastructure functions. Member States also agreed to implement a Safe System approach incorporating infrastructure design and engineering, set safe adequate speed limits, and enable multimodal transport and active mobility. The declaration also emphasizes special attention to the needs of pedestrians, cyclists, motorcyclists, and users of public transport (UNGA, 2022).

Pandemic prevention, preparedness and response (PPPR)

In 2023, the President of the UN General Assembly convened the first ever high-level meeting on pandemic prevention, preparedness and response. This HLM and its Political Declaration is a historic milestone in the urgent drive to make all people of the world safer, and better protected from the devastating impacts of pandemics. The Declaration reinforces the global mandate of WHO to lead health emergencies responses across the UN system. It also re-emphasizes the directing and coordinating role of WHO, in accordance with its mandate, as well as the need to commit to further sustainable financing that provides adequate and predictable funding to WHO so that it has the resources needed to fulfill its core functions. It addresses key issues such as access to all medical countermeasures, combatting misinformation, protection of communities, expanding regional and global manufacturing capacities, strengthening the health workforce and health systems, preventing barriers to vaccine distribution and the importance of the multilateral system and scale-up of the multi-sectoral approach to PPPR which supports attainment of the SDGs. The Political Declaration provides a strong impetus to the ongoing negotiations at WHO Headquarters in Geneva, towards a legal instrument for Pandemic Prevention and Preparedness, and requests the Secretary-General to provide, in consultation with the WHO and other relevant agencies, a report including recommendations on the implementation of the Declaration, during the 79th session of the General Assembly. The Report will serve to inform another HLM to be convened in 2026, which will take into consideration the outcomes of other ongoing related processes to strengthen PPPR. As mentioned, next steps for PPPR also include the outcomes of the Intergovernmental Negotiating Body and Working Group on amendments to the International Health Regulations (2005) which will be transmitted to the World Health Assembly in May 2024 (UNGA, 2023c).

Deliverables

The full implementation of the various Political Declarations adopted by the General Assembly described before will undoubtedly save hundreds of millions of lives and lift similar numbers of persons out of poverty, in all corners of the globe. It fully underscores the triple billion objectives of WHO, to ensure 1 billion more people in the work have access to UHC, to ensure 1 billion more enjoy better health and wellbeing, and 1 billion more are protected against health emergencies.

Health and development

As important deliverable is that alongside these high level meetings and resulting declarations and resolutions, the importance of health was recognized in the Sustainable Development Agenda, with almost all of the SDGs adopted by the General Assembly in September 2015 containing targets related to the improvement of human health and well-being. The 2030 Agenda firmly established health as a precondition, driver, outcome, and indicator of sustainable development. The World Health Organization worked with delegations throughout the development of the SDG Agenda, to ensure that global public health goals and targets were included across all sectors. These efforts resulted in a systematic and strategic positioning of health in the following targets within the 17 Goals, as outlined in Table 1 (UN, n.d.; UNGA, 2015).

As another deliverable, the health focused high level meetings undoubtedly ensured political commitment and multisectoral support and a whole of government approach for health improvements in all countries. Again, progress was well underway until the onset of the COVID-19 pandemic.

In the area of addressing risk factors for NCDs, a majority of countries are making progress by reducing the use of health-harming products. In the past five years, more than 65% of countries

Table 1 Positioning of health within the 17 Goals of the UN 2030 Agenda.

	Target 1.3 Target 1.5	Implement nationally appropriate social protection systems and measures for all. Build resilience of the poor and those in vulnerable situations and reduce their exposure and vulnerability to climate-related extreme events and other shocks and disasters.

Table 1 Positioning of health within the 17 Goals of the UN 2030 Agenda—cont'd

	Target 2.1	End hunger and ensure access by all people to safe, nutritious, and sufficient food all: year round
	Target 2.2	End all forms of malnutrition
	Target 3.1	Reduce the global maternity ratio
	Target 3.2	End preventable deaths of newborns and children under 5
	Target 3.3	End the epidemics of AIDS, tuberculosis, malaria, and neglected tropical diseases and combat hepatitis, water-borne diseases, and other communicable diseases.
	Target 3.4	Reduce by one-third premature mortality from noncommunicable diseases, through prevention and treatment and promote mental health and well-being
	Target 3.5	Strengthen the prevention and treatment of substance abuse
	Target 3.6	Halve the number of deaths and injuries from road traffic accidents
	Target 3.7	Ensure universal access to sexual and reproductive healthcare services
	Target 3.8	Achieve universal health coverage
	Target 3.9	Substantially reduce the number of deaths and illnesses from hazardous chemicals and air, water, and soil pollution and contaminations
	Target 3.a	Strengthen the implementation of the World Health Organization Framework Convention on Tobacco Control
	Target 3.b	Support the research and development of vaccines and medicines for communicable and noncommunicable diseases
	Target 3.c	Substantially increase health financing and the development, training, and retention of the health workforce in developing countries
	Target 3.d	Strengthen the capacity of all countries for early warning risk reduction and management of national and global health risks
	Target 4.5	Eliminate gender disparities in education and ensure access to all levels of education and vocational training for the vulnerable, including persons with disabilities
	Target 4.a	Build and upgrade education facilities that are child, disability, and gender sensitive and provide safe, nonviolent, inclusive, and effective learning environments

Continued

Table 1 Positioning of health within the 17 Goals of the UN 2030 Agenda—cont'd

5 GENDER EQUALITY	Target 5.2	Eliminate all forms of violence against all women and girls in public and private spheres
	Target 5.3	Eliminate harmful practices, such as child, early and forced marriage, and female genital mutilation
	Target 5.6	Ensure universal access to sexual and reproductive health and reproductive rights
6 CLEAN WATER AND SANITATION	Target 6.1	Achieve universal and equitable access to safe and affordable drinking water for all
	Target 6.2	Achieve access to adequate and equitable sanitation and hygiene for all and end open defecation
7 AFFORDABLE AND CLEAN ENERGY	Target 7.1	Ensure universal access to affordable, reliable, and modern energy services
8 DECENT WORK AND ECONOMIC GROWTH	Target 8.5	Achieve full and productive employment and decent work for all women and men, including for young people and persons with disabilities
	Target 8.7	Take immediate and effective measures to eradicate forced labor, end modern slavery and human trafficking, and secure the prohibition and elimination of the worst forms of child labor
	Target 8.8	Protect labor rights and promote safe and secure working environments for all workers
10 REDUCED INEQUALITIES	Target 10.2	Empower and promote the social, economic, and political inclusion of all, irrespective of age, sex, disability, race, ethnicity, origin, religion, or economic or other status
	Target 10.7	Facilitate orderly, safe, regular, and responsible migration and mobility of people, including through the implementation of planned and well-managed migration policies

Table 1 Positioning of health within the 17 Goals of the UN 2030 Agenda—cont'd

11 SUSTAINABLE CITIES AND COMMUNITIES	Target 11.1	Ensure access for all to adequate, safe, and affordable housing and basic services and upgrade slums
	Target 11.2	Provide access to safe, affordable, accessible, and sustainable transport systems for all, improving road safety
	Target 11.5	Significantly reduce the number of deaths and the number of people affected by disasters
	Target 11.7	Provide universal access to safe, affordable, inclusive, and accessible green and public spaces and substantially increase the number of cities and human settlements adopting and implementing integrated policies and plans toward inclusion, resource efficiency, mitigation and adaptation to climate change, resilience to disasters, and develop and implement holistic disaster risk management
12 RESPONSIBLE CONSUMPTION AND PRODUCTION	Target 12.4	Achieve the environmentally sound management of chemicals and all wastes throughout their lifecycle, in accordance with agreed international frameworks, and significantly reduce their release to air, water, and soil
13 CLIMATE ACTION	Target 13.3	Improve education, awareness-raising, and human and institutional capacity on climate change mitigation, adaptation and impact reduction and early warning
16 PEACE, JUSTICE AND STRONG INSTITUTIONS	Target 16.2	End abuse, exploitation, trafficking, and all forms of violence against and torture of children
	Target 16.9	Provide legal identity for all, including birth registration

Continued

Table 1 Positioning of health within the 17 Goals of the UN 2030 Agenda—cont'd

17 PARTNERSHIPS FOR THE GOALS	Target 17.18	Enhance capacity building support to developing countries to increase significantly the ability of high quality, timely and reliable data, disaggregated by income, gender, age, race, ethnicity, migratory status, disability, geographic location, and other characteristics relevant in national contexts

that are Member States have either introduced or increased excise taxes on at least one health-harming product, such as tobacco, alcohol, or sugary drinks. Tobacco use continues to decline, and since 2018 the number of countries on track to meet the target of a 30% reduction in tobacco use between 2010 and 2025 almost doubled (World Health Organization, 2022a).

In 2018 the WHO launched the REPLACE initiative, which contains policy proposals prohibiting the use of industrially produced trans fat. Measures to this effect have been introduced in 58 countries accounting for 40% of the world's population (World Health Organization, 2022a).

Since 2015, 95% of countries have made progress toward increased health service coverage and 71 countries are using WHO guidelines or tools on the health response to violence against women (World Health Organization, 2022a).

Also, there have been positive trends in efforts to strengthen the global health workforce, with the number of health workers having globally increased by 29% between the years 2013 and 2020 (World Health Organization, 2022a).

Similarly, it is encouraging that road deaths have stabilized, despite a continued rise in the number of cars.

WHO guidelines have supported major gains in HIV testing and treatment, resulting in a 32% decline in HIV mortality since 2016 (World Health Organization, 2022a).

The SDG target on hepatitis B has been met, and since 2015, the number of people who have received treatment for hepatitis C has increased 9-fold, reversing the trend of increasing mortality for the first time (World Health Organization, 2022a).

For tuberculosis (TB), 33 countries have reached the target for a 35% reduction in TB deaths since 2015, and 86 have achieved a 20% reduction in incidence (World Health Organization, 2022a).

The setback due to the COVID-19 pandemic

Despite progress in many areas, as has been noted before, the effects of the COVID-19 pandemic have slowed and reversed gains in health and resulted in severe economic contraction, while also draining health budgets. Recovery efforts will be slow and likely take several years.

Illustrative of this challenge is that prior to the pandemic, almost 1 billion people spent in excess of 10% of their household budget on health (out of pocket) of which 290 million spent at least a quarter of their household budget (World Health Organization, 2022a). This number will no doubt increase in coming years as higher government spending during the pandemic, coupled with lower revenues, will increase deficit levels and debt servicing, leading to even more constraints being faced in health budgets.

Prior to the pandemic, it was estimated that the target of reaching 1 billion people (as per the UHC Political Declaration) would not be reached, with only 270 million more persons covered by 2023—a shortfall of 730 million persons (World Health Organization, 2022a). Disruptions to health services during the pandemic have resulted in new estimates that the shortfall could reach 840 million (World Health Organization, 2022b).

Although child survival has improved dramatically over the past 20 years, 54 countries remain off track to meet the SDG child survival targets (World Health Organization, 2022a).

Also, progress on sexual and reproductive health, including maternal mortality, remains slow.

The pandemic has led to a massive increase of 28% in depression and 26% in anxiety disorders globally (World Health Organization, 2020).

Interruption of preventive health services has had serious consequences. Hypertension causes one-third of all deaths, but only half of cases are diagnosed, and less than half of those are treated (World Health Organization, 2022b).

Further, in 2020, the number of people receiving treatment for a neglected tropical disease fell by 25% as a result of health service disruptions caused (World Health Organization, 2022b).

Similarly, between 2019 and 2020, funding for ending TB decreased by 9%, falling back to 2016 levels. This is particularly concerning in low- and middle-income countries, representing 98% of reported TB cases, where funding fell to 41% of the global target (World Health Organization, 2022b).

Why political commitment is now more important than ever, and the need for the General Assembly to retain a health focus is indispensable

The COVID-19 pandemic is not the first, and will not be the last, health emergency. In the past five years, WHO has responded to more than 120 emergencies—cyclones, volcanoes, earthquakes, outbreaks, wars—and a pandemic. Some of these emergencies last a few months while others last for years (World Health Organization, 2022a). As the world fight these outbreaks, we face the task of restoring essential health services, with 90% of Member States reporting disruption to one or more essential health services (World Health Organization, 2022b).

Whereas the World Health Assembly remains the most relevant forum for the world's Health Ministers to convene and determine their priorities and decide on policies to address emerging challenges, the vast majority of health threats can only be successfully dealt with through concerted measures including political and socioeconomic interaction. The UNGA is the global body that brings all these areas of policy together in one venue, enabling ministers from across government to convene and engage each other and also forge common approaches with civil society and the private sector, which in a growing number of middle-income countries now provide more health services than the public sector.

The need for concerted action is urgent. Progress on meeting the target in the UHC Political Declaration and endorsed by the World Health Assembly, namely, to have 1 billion more people benefitting from universal health coverage by 2023, is less than 25% on track. Similarly, the WHO objective of ensuring better health and well-being for 1 billion more people by 2023 is progressing at one quarter of what is required. Stepping up action requires efforts in health diplomacy and building coalitions to champion various health priorities (World Health Organization, 2022a).

In this context, for the past 5–10 years, the World Health Organization New York Office has encouraged delegations to form Groups of Friends from among like-minded countries to promote the strategic relevance and political support on several topics of emerging importance, including several which are covered by targets of the sustainable development agenda. Among the Groups of Friends are those promoting Universal Health Coverage; combating Anti-Microbial Resistance; ending Neglected Tropical Diseases; ensuring Water, Sanitation, and Hygiene in healthcare facilities; promoting Mental Health and Well-being, Healthy Aging; providing Vision Care for all, eliminating hepatitis and fostering Health Security.

The way forward

These workstreams maintain a focus on health across the socioeconomic spectrum and ensure its consideration in the outcomes of high level intergovernmental meetings of the General Assembly, which are essential for securing political momentum, national ownership and commitment, and multilateral implementation of agreed outcomes and actions (United Nations, n.d.).

Since the year 2000, the number of people globally who face financial hardship because of out-of-pocket health spending has increased by 75%, to close to 2 billion people. Between 2015 and 2017, the number of people with out-of-pocket health spending exceeding 10% of their household budget increased from 940 million to 996 million per year (World Health Organization, 2022a).

Globally, only 3 percent of health budgets are spent on promotion and prevention. And yet increased investment in these areas could reduce the global disease burden by 50%, generating massive returns for individuals, families, communities, and countries (World Health Organization, 2022a).

A renewed focus on primary healthcare is essential, as 90% of essential health services can be delivered through this avenue. In fact, through primary healthcare interventions as much as 75% of projected health gains from the SDGs might be realized. Investing in primary healthcare, at a cost of $32 per person, could increase global life expectancy by as much as 6.7 years by 2030, making this a crucial necessity and strengthening the call for redesigned health budgets which target higher yields across the lifespan (World Health Organization, 2022a).

This approach is the only viable option for addressing the longer-term effects of the COVID-19 pandemic and in particular the burden of NCDs, which still account for 70% of deaths annually and remain a debilitating factor for the health budgets of most low- and middle-income countries (World Health Organization, 2022a).

NCDs are one of the greatest health and development challenges of this century. Seventeen million people under the age of 70 die of NCDs yearly, 86 percent of whom live in low- and middle-income countries. Almost one billion people globally experience a mental health disorder, and around 700,000 deaths each year are due to suicide—a leading cause of death in young people (World Health Organization, 2022a).

During the pandemic, 75% of countries reported disruption to essential NCD services because of lockdown restrictions and channeling of resources. Cancer screenings were disrupted in 79% of countries, diabetes management in 74%, hypertension treatment in 71%, and cancer treatments in 68%, whereas 54% of countries reported disruption in nutrition services. Ninety-three percent of countries experienced disruptions in one or more services for mental, neurological, and substance use disorders. Investing in the prevention and control of NCDs provides a return of 4–7 times the financial cost, depending on the nature of interventions (World Health Organization, 2022b).

Addressing the environmental and social determinants of health is crucial for slowing an increase in NCDs and similarly has multiple additional health and economic benefits. However, wide-ranging policy interventions require a "whole of government" and "whole of society" approach that goes beyond the health sector.

Strong and resilient health systems, able to withstand shocks, are fundamental to healthy lives and economic growth. Such systems have wide-ranging benefits for society in areas beyond health. At the same time, health being a precondition, indicator, and outcome of sustainable development provides a link to all the goals and targets contained in the 2030 Sustainable Development Agenda.

The universality and singular political relevance of the United Nations General Assembly annual meeting at the highest level, meaning that of Heads of State of countries, therefore, provides the most important venue for all countries to commit to action which spans across all sectors of society and enables the public and private enterprises to join forces in improvement of human health and well-being.

Conclusion

As we rebuild from the COVID-19 pandemic, the global community has unique opportunities to convene at the highest levels to recommit to health, and collectively craft strategic directions across access to services, equity, security, human rights, gender equality, and capacity development.

The World Health Organization will work with renewed vigor to contribute meaningfully and within the conviction of "health for all, everywhere," in the preparation and outcomes of the following upcoming high-level meetings of the UNGA: Combating Antimicrobial Resistance in 2024; Prevention and Control of NCDs in 2025; Road Safety in 2026; Pandemic Prevention, Preparedness and Response in 2026; Universal Health Coverage in 2027; Tuberculosis in 2028.

These high level meetings and their outcomes will have a crucial contribution toward meeting the goals and targets of the 2030 sustainable development agenda and ensure a robust and lasting recovery from the COVID-19 pandemic.

References

AMR. (2021). *Call to action on antimicrobial resistance.* https://www.un.org/pga/75/wp-content/uploads/sites/100/2021/04/Call-to-Action-on-Antimicrobial-Resistance-AMR-2021.pdf.

UN, n.d. UN sustainable development goals. https://sdgs.un.org/goals.

UNGA. (2011). *UNGA resolution on the political declaration of the high-level meeting on noncommunicable diseases.* https://documents-dds-ny.un.org/doc/UNDOC/GEN/N11/458/94/PDF/N1145894.pdf?OpenElement.

UNGA. (2014). *UNGA resolution on the political declaration of the high-level meeting on noncommunicable diseases.* https://documents-dds-ny.un.org/doc/UNDOC/GEN/N13/457/45/PDF/N1345745.pdf?OpenElement.

UNGA. (2015). *UNGA resolution on transforming our world: The 2030 Agenda for Sustainable Development 2015.* https://www.un.org/en/development/desa/population/migration/generalassembly/docs/globalcompact/A_RES_70_1_E.pdf.

UNGA. (2016). *UNGA resolution on the political declaration of the high-level meeting on antimicrobial resistance.* https://documents-dds-ny.un.org/doc/UNDOC/GEN/N16/310/65/PDF/N1631065.pdf?OpenElement.

UNGA. (2018a). *UNGA resolution on the political declaration of the high-level meeting on noncommunicable diseases.* https://documents-dds-ny.un.org/doc/UNDOC/GEN/N18/315/40/PDF/N1831540.pdf?OpenElement.

UNGA. (2018b). *UNGA resolution on the political declaration of the high-level meeting on tuberculosis.* https://documents-dds-ny.un.org/doc/UNDOC/GEN/N18/315/53/PDF/N1831553.pdf?OpenElement.
UNGA. (2019). *UNGA resolution on the political declaration of the high-level meeting on universal health coverage.* https://documents-dds-ny.un.org/doc/UNDOC/GEN/N19/311/84/PDF/N1931184.pdf?OpenElement.
UNGA. (2022). *UNGA resolution on the political declaration of the high-level meeting on improving road safety.* https://documents-dds-ny.un.org/doc/UNDOC/GEN/N22/416/30/PDF/N2241630.pdf?OpenElement.
UNGA. (2023a). *UNGA resolution on the political declaration of the high-level meeting on the fight against tuberculosis.* https://documents-dds-ny.un.org/doc/UNDOC/GEN/N23/306/91/PDF/N2330691.pdf?OpenElement.
UNGA. (2023b). *UNGA resolution on the political declaration of the high-level meeting on universal health coverage.* https://www.undocs.org/Home/Mobile?FinalSymbol=A%2FRES%2F78%2F4&Language=E&DeviceType=Desktop&LangRequested=False.
UNGA. (2023c). *UNGA resolution on the political declaration of the high-level meeting on pandemic prevention, preparedness and response.* https://www.undocs.org/Home/Mobile?FinalSymbol=A%2FRES%2F78%2F3&Language=E&DeviceType=Desktop&LangRequested=False.
United Nations, n.d. United Nations: General Assembly of the United Nations. https://www.un.org/en/ga/.
World Health Organization. (2020). *Rapid assessment on the impact of COVID-19 on mental, neurological and substance use services.* https://www.who.int/publications/i/item/978924012455.
World Health Organization. (2022a). *Health Topics.* https://www.who.int/health-topics/.
World Health Organization. (2022b). *Third round of the global pulse survey on continuity of essential health services during the COVID-19 pandemic.* https://www.who.int/publications/i/item/WHO-2019-nCoV-EHS_continuity-survey-2022.1.

CHAPTER 41

Mental Health Matters at the United Nations: A web and in-person healthy workplace initiative and response during COVID-19 and beyond☆

Judy Kuriansky[a,b,c,d,e,f], **Tanya Mehdizadeh**[b,g], **and Sripriya Bikkani**[b,g]

[a]Department of Clinical Psychology, Teachers College, Columbia University, New York, NY, United States, [b]International Association of Applied Psychology, Indianapolis, IN, United States, [c]World Council for Psychotherapy, Vienna, Austria, [d]United Nations Group of Friends of Mental Health and Well-being, New York, NY, United States, [e]Mission of the Republic of Sierra Leone to the United Nations, New York, NY, United States, [f]Global Council SDG3 (Health), "SDGs in Action" Initiative, Dubai, United Arab Emirates, [g]Department of Clinical and Counseling Psychology, Teachers College, Columbia University, New York, NY, United States

Introduction

"Mental Health Matters" is the perfectly self-explanatory phrase for the innovative initiative of the United Nations (UN), called "Mental Health Matters: A Healthy Workplace for a Better World," given its intention to improve workforce health and well-being for UN staff. The initiative is built on a strategy that aims to foster a workplace environment that enhances mental and physical health, takes a preventive approach, establishes and implements effective psychosocial services where the UN staff work, makes accommodations for staff who might have mental health challenges, and ensures appropriate funding for mental health and well-being services.

Workplaces that promote mental health can lead to staff feeling supported and therefore more productive and engaged. The COVID-19 pandemic wreaked havoc on workers and workplaces, upending their schedules and feelings of safety, forcing many adaptations (Marquez, 2024). Similarly, the UN had to pivot to accommodate its vast employee staff across all different regions of the world, working under different conditions but all still coping with the rages of the pandemic on themselves, their work, and family life.

☆ This report is partly based on the guest lecture by Therese Fitzgerald for Dr. Judy Kuriansky's class on "Psychology and the United Nations" at Teachers College, Columbia University (virtual, 11/24/2020).

Resilient Health. https://doi.org/10.1016/B978-0-443-18529-8.00041-X

Given that UN staff is spread around the world in exceptionally diverse settings, mental health services have to be designed and sustained through virtual, online, and technological means as well as offering in-person services when needed.

The UN workplace mental health and well-being survey and strategic action plan

The UN Workplace Mental Health and Well-Being Strategic Action Plan was designed and meant to be implemented within the 5-year time frame of 2018–2023. This strategy was developed in response to a 2015 UN Global Well-Being survey which found that UN personnel reported higher levels—when compared to the general population—of mental health problems like depression, anxiety, posttraumatic stress, and substance abuse (alcohol intake). Specifically, the results of the survey, with responses from 17,363 UN employees, showed that almost half (49%) of staff reported symptoms of mental health concern, but only 6% in total were receiving professional support, and only 2% of those were receiving that support from internal programs. Specific problems reported were 23.15% hazardous drinking, 22.78% major depressive disorder (MDD), 19.89% posttraumatic stress disorder (PTSD), and 17.90% generalized anxiety disorder (GAD). This in comparison to prevalence on the general population of 4.7% MDD, 0.28% PTSD, and 7.3% GAD.

Links were found between poor mental health and four factors: They are (1) the number of years an individual may have worked for the UN system, (2) exposure to traumatic events for a duration of 1 year while working for the UN, (3) low job satisfaction, and (4) low number of attempts to seek mental healthcare.

Given these findings, the need became clear to prioritize mental health problems and well-being and to foster a healthy workspace environment for the employees.

The strategy had to accommodate the vast number of UN personnel at the UN headquarters but also, importantly, in posts around the world, to serve staff working in very large or complex systems, in offices or in the field, and particularly those working under extremely stressful conditions or in conflict zones. It also applies to both managers and also their teams.

The peak objective of the strategy is to increase staff resilience, productivity, and engagement. Four major aims are to:

(1) create a workplace for all UN personnel that enhances mental and physical health and well-being, by mitigating inherent risks in working within the UN system.
(2) develop, deliver, and evaluate high-quality psychosocial services in all stations where UN staff work, insuring access.
(3) welcome and support staff who live with mental health challenges, through reducing stigma and having the appropriate policies and procedures in place to provide mental health support to staff in need.
(4) ensure sustainable funding for mental health and well-being services.

Website resources

The "Mental Health Matters" website serves as a hub for variable resources about mental health and well-being, which are constantly updated. These resources include printed fact sheets and also podcasts, available in all of the UN official languages (Arabic, Chinese, English, French, Russian, and Spanish).

For all personnel and for leaders and managers

For all personnel, information on the website includes recognizing when someone may be developing a mental health condition and how to support that person. For managers and leaders, a "Lead and Learn" program offers a training called "Together, let's take care of one another" with skills-building lessons to foster a positive and supportive environment for employees and podcasts on four topics: mental health and well-being in the workplace, personal well-being and thriving as a manager, supporting a colleague experiencing poor mental health, and addressing stigma related to mental health problems.

Support in the UN agenda

The initiative's attention to the well-being of its employees aligns with the United Nations 2030 Agenda for Sustainable Development, adopted in 2015 by all 193 member states (governments) of the UN system, which includes target 3.4 to "promote mental health and well-being." Since then, innumerable events have taken place at the United Nations on the margins of Commissions, celebration of International Days, and other fora, focusing on the importance of mental health. Additionally, the highest-level management at the UN, including the Secretary-General António Guterres and the Deputy Secretary-General Amina J. Mohammed, have mentioned the priority of mental health. Notable, as well, the Mission of Mexico to the UN, under the leadership of Ambassador Juan Ramón de la Fuente, a psychiatrist, has introduced the importance of mental health, especially in the context of insuring peace and security, during its term on the Security Council (de la Fuente et al., 2024), to be continued by the Mission of Ecuador in its succeeding term.

Development with partners and the implementation board

The strategy was developed in partnership with the UN Staff and Stress Counselor Group, UN Medical Directors, the HR Network, UN Ombudsman, and UN Staff Federations. The Project Team included the UN Secretariat, the World Bank, UNOG (United Nations Office at Geneva), and many UN organizations. These are (in alphabetical order) IAEA (International Atomic Energy Agency), ILO (International Labour Organization), IOM (International Organization for Migration), FICSA (Federation of International Civil Servants' Associations), UNDP

(United Nations Development Programme), UNESCO (United Nations Educational, Scientific and Cultural Organization), UNFPA (United Nations Population Fund), UNHCR (United Nations High Commissioner for Refugees), UNICEF (United Nations Children's Fund), UNIDO (United Nations Industrial Development Organization), UN-Women, and WHO (the World Health Organization).

Since the implementation of this strategy requires the collaborative effort of all strata of the UN workforce to ensure the health and well-being of all those who make up the organization, representatives of some of the UN organizations that partnered to develop the strategy are on the Implementation Board, which is the global governance body. Other members are an array of diverse representatives from the UN System such as Medical Directors, Staff Counselors, Staff Federations, Associations, and Representatives. Also, an Advisory Panel is made up of staff with lived experience of a mental health condition.

Implementation Board Members include Martha Helena Lopez, a lawyer, who is UN assistant secretary-general for Human Resources Management, responsible for the strategic leadership and direction of the UN's human resources organization, leading talent management and implementing management initiatives. Another member, Nana Taona Kuo, is a development economist and public policy specialist who is currently the senior health adviser in the Executive Office of the UN Secretary-General. In addition, Therese Fitzpatrick, who holds postgraduate qualifications in public health, business management, and occupational therapy, serves as the global lead for the implementation of the UN System Workplace Mental Health and Well-Being Strategy, which involves working with the UN organizations in countries in which they operate, to improve mental health across the entire UN system. She has also made many presentations about the strategy at events at the United Nations, particularly for the UN Member States' Group of Friends of Mental Health and Well-Being.

Involvement with mental health professionals

The initiative takes into account best practices, advice, and information about the field of mental health and well-being in its website postings and activities. UN personnel may also access an internal UN Staff/Stress Counselor in the UN locations, who are Mental Health professionals working for the organization to offer confidential support to UN personnel when needed.

Involvement with professional psychologists was evident at the launch of the "Mental Health Matters" initiative and website at the UN New York City headquarters building on October 3, 2019—held to coincide with the celebration of World Mental Health Day—on the topic of "Mental Health at Work: Looking after your own mental health and supporting colleagues." A special webcast was conducted, and then replayed at various times to accommodate staff in time zones around the world, in collaboration with the Psychology Coalition of NGOs having consultative status with the United Nations Economic and Social Council. For that webcast, Dr. Walter Reichman, a veteran organizational psychologist and main representative of the International Association of Applied Psychology (IAAP), described major psychological problems in the context of the results of the UN Workplace Mental Health and Well-Being Survey showing worrisome rates of anxiety, depression, and alcohol use. Clinical psychologist Dr. Judy Kuriansky (the first author of this chapter), also of IAAP, described specific actionable techniques for resilience and to lower stress that have been proven successful in workshops and trainings around the world (Kuriansky, 2019a, 2019b, 2019c; Kuriansky et al., 2016); Also, clinical psychologist Dr. Leslie Popoff addressed what to do in case of more serious conditions, like feeling suicidal. Subsequently, an exhibit was set up in the UN building lobby with many printed handouts and buttons, with Kuriansky and Therese Fitzpatrick, global lead for the UN's Mental Health and Well-being Team, engaging passersby to talk about the new service.

A previous focus on "mental health matters" for youth at the UN

The UN had already been paying attention to the general issue of "Mental Health Matters," particularly for youth, evident in the title and substance of a publication of that same name on the emerging issue of "Social Inclusion of Youth with Mental Health Conditions." The report had several goals: to raise awareness about youth with mental health conditions, start a global conversation regarding strategies for addressing specific challenges faced by young people with mental health conditions, eradicate stigma, and go beyond treatment-oriented mental health programs to include broader public health-inspired approaches that foster the social and economic integration of youth with mental health conditions (United Nations Social and Economic Council-Youth, 2014). Prompted by the World Programme of Action for Youth, this report derived from a General Assembly resolution A/RES/50/81 wherein Member States were called upon to address health issues of young people. The launch included an event held at the UN New York City headquarters, with high-level speakers and also youth telling their stories about mental health challenges (including suicide attempts), spoken word performances, and a website with art and other expressions.

Response to the COVID-19 pandemic

In March 2020, when the COVID-19 pandemic was raging throughout the world, the Mental Health and Well-being Team focused attention on needs of staff in response to the emergency. Issues of concern were (1) ensuring access to psychosocial support, (2) providing related training and learning opportunities, (3) insuring effective information and communication, (4) policies, (5) health and well-being survey, (6) a domestic abuse taskforce, and (7) guidance for leaders and managers.

Given the level of uncertainty during this time, transparent communication was imperative to ensure that personnel understood changes that were happening and could adjust to the economic and psychological climate caused by the pandemic. It was also critical to provide guidance and stress-reducing materials. Flexibility was encouraged, in order to adjust to the intense changes dictated by the pandemic. Even small adjustments could be made to accommodate staff to minimize stress, like taking time off.

Efforts were made to expand access to psychosocial support through external providers and increasing the number of counselors across the UN system. Also, since surveys show that younger personnel, under the age of 45, report significant levels of stress, support was necessary to focus on this cohort facing disruptions to their social lives and isolation. Further, given the escalation of the "silent epidemic" of domestic violence noted throughout the general population due to the pandemic, support for UN personnel who might be experiencing this trauma was also deemed important. Thus a global online directory was launched in partnership with an external organization, which outlined available resources within almost every country in the world.

In addition, significant work was done around World Mental Health Month in October. Research was done about activities by other organizations to support mental health, collecting personal experiences of personnel, and planning for support for at-risk groups, including those experiencing domestic abuse and persons with disability.

As part of October 2020's World Mental Health Month at the UN, the Mental Health and Well-being Team launched a system-wide poetry series whereby staff from around the world submitted their work in any language, form, style, or topic to help raise awareness about mental health. The resulting poetry book, entitled "Sauti: Poems of Healing," was posted on the website, consisting of 53 poems from 10 UN organizations in 21 countries, in seven languages. Poetry served as an outlet for people to open up about their experiences for self-expression and healing.

In the summer of 2021, another survey (after the 2015 one mentioned before) was carried out to assess UN personnel's response to the COVID-19 pandemic and other issues about their health, well-being, and safety. The results of this 2021 UN-Wide Health Survey, which collected voluntary self-reports of 19,034 respondents, revealed that when asked to describe the current state of mental health and physical health compared to how it was prior to the outbreak of COVID-19, 45% responded negatively regarding mental health and 38% responded negatively regarding physical health. These findings showed that the pandemic impacted employees' health, with mental health impacted to a slightly higher degree than physical health. In addition, 53% responded that their sleep patterns had changed since the start of COVID-19. Further, 30% of respondents reported feeling tense, anxious, and/or depressed often or always. In terms of in what way the COVID-19 pandemic has caused them stress, top reasons included: 54% were often or very often stressed about the state of the world, 50% were stressed about travel, and

33% were stressed about family's health. Other reasons were worries about the future (32%) and health (25%). About a quarter of respondents said they had considered engaging in counseling or mental health treatment, but not followed through mainly because of not having enough time. Nineteen percent of respondents said they experienced domestic abuse, of whom 55% said that the abuse happened prior to COVID-19, 18% said that the abuse started for the first time during COVID-19, and 18% said that the extent or frequency of abuse had increased.

Accordingly, the COVID-19 response also involved guidelines to ensure that every UN organization has a mental health action plan in place. The response also addressed providing training and resources to help support leaders and managers (described before) since the pandemic posed a particularly challenging time to know how to best support staff, and sharing resources across the UN organizations to ensure that similar approaches are being developed across the UN system and available to all personnel in order to minimize duplication of efforts.

Update: Launch of a new Strategy 2024 and beyond

A new "United Nations System Workplace Mental Health and Well-being Strategy 2024 and beyond" was launched on 15 November 2023–5 years after the first-ever strategy—at a hybrid panel held live the New York United Nations headquarters and virtually. This new strategy builds on gains from the implementation of the first Strategy (2018–23), adding data from recent UN surveys and a report on mental health and well-being policies and practices within the UN (United Nations, n.d.). The new framework outlines an action plan (including risk assessment, mental health literacy, and access to support), means of implementation, 12 indicators for evaluation, and recommendations.

At the event, the UN Deputy Secretary-General Amina J. Mohammed acknowledged progress of the first strategy for its learning materials, introduction of stigma reduction, data collection and review of access to psychosocial support. Anticipating promise of the new strategy to prevent risks to mental health at work, promote well-being, and support personnel with mental health conditions, she promised to make the strategy a priority across the UN organization, encourage collective efforts and also "to try" to prioritize her own well-being "as a role model," adding (humorously) that "we need laughter in the workforce."

The DSG's commitment was echoed by panelist Lisa Bevill of IE University in Spain who emphasized that "no one gets a pass"—meaning every individual has responsibility for their own mental health, to add to collective well-being.

Moderated by Assistant Secretary-General for Human Resources Martha Helena Lopez, presentations by other panelists addressed research on burnout and a holistic view of social, physical, mental and spiritual health, by McKinsey Health's Jacqueline Brassey; description of the foundation of the strategy on WHO and ILO policy by WHO mental health specialist Aiysha Malik; an emphasis on prevention by President of the Federation of International Civil

Servants' Associations' Tanya Quinn-Maguire; and endorsement by Ombudsman of the U.S. State Department's Shireen Dodsen. The Global Lead of Mental Health at the UN since 2019, Therese Fitzpatrick, noted her favorite two statements in the strategy: "Everyone has a tole to play" and "Change is possible," and ended the panel on a high note, inviting everyone to become a "friend of mental health," to "see where you can take action to create a health workplace for a better world."

Looking ahead

The department aims to do more research, continually add to its offerings, and integrate a range of different strategies across the UN system. For example, there is an antiracism task force within the Secretariat to ensure all staff members are being supported and that there are no issues with respect to racism, which is clearly linked to mental health.

The project can be accessed at: website: https://www.un.org/en/healthy-workforce/, email: mhs@un.org and Instagram/twitter: @UN4MentalHealth.

References

de la Fuente, J. R., Mendoza, S. P., & Guzik, A. (2024). An unconventional security threat: Mental health at the United Nations Security Council and Mexico's Response. In J. Kuriansky, & P. Kakkattil (Eds.), *Resilient health: Leveraging technology and social innovations to transform healthcare for COVID-19 recovery and beyond* (pp. 475–480). Elsevier.

Kuriansky, J. (2019a). A Model psychosocial support training and workshop during a medical mission for Syrian refugee children in Jordan: Techniques, lessons learned and recommendations. *Journal of Infant, Child, and Adolescent Psychotherapy*, *18*(4), 352–366. https://doi.org/10.1080/15289168.2019.1690925.

Kuriansky, J. (2019b). Building capacity in low resource settings post-disaster: Empowerment, empathy and esteem in training comforters in the case of Haiti. Chapter 11 In W. Jean-Charles (Ed.), *Empathie en Haïti de la Recherche à la Practique à la Croieée de Différents Regards* (pp. 207–230). Coconut Creek, FL: EducaVision.

Kuriansky, J. (2019c). Expressive ARTS for helping children heal in crisis and disaster. In I. A. Serlin, S. Krippner, & K. Rockefeller (Eds.), *Integrated health care for the traumatized: A whole person approach* (pp. 39–58). Lathan, NY: Rowman & Littlefield. Chapter 4.

Kuriansky, J., Polizer, Y., & Zinsou, J. (2016). Children and Ebola: A model resilience and empowerment training and workshop. In J. Kuriansky (Ed.), *The psychosocial aspects of a deadly epidemic: What Ebola has taught us about holistic healing* (pp. 175–214). Santa Barbara, California: ABC-CLIO/Praeger.

Marquez, P. V. (2024). The imperative for mental health in the workplace post-COVID-19. In J. Kuriansky, & P. Kakkattil (Eds.), *Resilient health: Leveraging technology and social innovations to transform healthcare for COVID-19 recovery and beyond* (pp. 755–769). Elsevier.

United Nations. (n.d.). United Nations System Mental Health and Well-being Strategy for 2024 and beyond. https://www.un.org/sites/un2.un.org/files/un_system_mental_health_and_well_being_strategy_for_2024.pdf.

United Nations Social and Economic Council-Youth. (2014). *Mental health matters: Social inclusion of youth with mental health condition.* https://www.un.org/development/desa/youth/publications/2014/06/mental-health-matters-social-inclusion-of-youth-with-mental-health-condition/.

CHAPTER 42

Multilateral organizations orchestrating change through ecosystems

Katherine Tatarinov and Tina C. Ambos
Geneva School of Economics and Management (GSEM), University of Geneva, Geneva, Switzerland

Introduction: The push for innovation in multilateral organizations

The idea of multilateralism "involves adherence to a common political project based on the respect of a shared system of norms and values between several nations… [and] is therefore both a method of cooperation and a form of organization of the international system" (UNOG, 2022). Multilateral organizations are tasked with solving dire global problems, creating standards to support the economy and are integral parts of the world order. The most prominent multilateral organization is the United Nations (UN), which consists of funds, programs, and specialized agencies, each of which have their own area of work, leadership, and budget. These are key actors in solving the world's pressing problems such as hunger (the World Food Programme—WFP), poverty (United Nations Development Fund—UNDP), children's well-being (United Nations Children's Fund—UNICEF), and the health response (World Health Organization—WHO).

However, we rarely see how the UN organizations are innovating to address their missions and which pressures they are facing to transform.

Today, these organizations are facing increased pressure to innovate in their work, particularly in their response to achieving the 17 "Sustainable Development Goals" (SDGs) (referred to as the Global Goals). These include assuring health and well-being for all (SDG3) as well as eradicating poverty, achieving gender equality, education for all, sustainable cities, and peace and security. The SDGs were praised as an inclusive "bottom-up" process capturing the perspective of the UN member states' governments, as well as civil society and the private sector, including businesses (UN Compact, 2015). But transforming the aspiration of the lofty global goals into local action has not been easy. Innovative approaches, robust methodologies, and effective use of digital technology are needed to succeed. This chapter examines the

Resilient Health. https://doi.org/10.1016/B978-0-443-18529-8.00042-1

broader role of how the UN organizations are driving innovation with a focus on situating global health within the SDG context.

The challenges of multilaterals in innovation in the digital age

Part of the burden of achieving the Global Goals sits with the multilateral organizations that are tasked with implementing them globally. While pathbreaking innovations will be necessary to address the SDGs, these organizations were not created to lead innovation. Their primary missions are to address a wide variety of social needs, from global health to economic development and they have recently struggled to come up with new innovative models to fulfill their core missions. But the UN organizations are usually characterized as bureaucratic, slow, ineffective, thus, several waves of reform have tried to induce a more entrepreneurial spirit.

As researchers Weiss and Carayannis (2017) state, "The bottom line was clear: dramatic changes are imperative both at [UN] headquarters (HQ) and in the field if the world organization is to respond to the twenty-first century's complex threats."

One of these "complex threats"—but potentially also an opportunity—is the advent of digital technology.

Digitalization has facilitated new ways of service delivery and opportunities to reach beneficiaries more effectively. For example, UNICEF's U-Report application uses SMS and social media platforms to communicate with communities on difficult topics and to engage youth in a participatory discussion that can drive government decision-making. For example, in 2020, U-Report launched a global bot that could answer questions on COVID-19 preprogrammed by global experts. The tool was accessed by over 6 million people, across 52 countries, with over 7 million interactions. Over 20 million young people and communities engaged on COVID-19 through U-Report (U-Report Team, 2020).

New technologies, such as those used by UNICEF, have been shown to facilitate connectivity and simplicity, enhance user experience, and enable new forms of social innovation based on different combinations of platforms, devices, networks, and communities. These new tools change the rules and roles of actors within communities and empower people to have a voice and a direct link to governments and organizations. But, we need to better understand how multilateral organizations can draw on technology to create, diffuse, and scale innovations.

There is increasing empirical evidence that in most cases either the nature of social innovation is technology based, such as in mobile applications aiming to increase the efficiency of services, or the mechanism of application, scaling and diffusion, relies on digital technology, such as marketing through social media or crowdfunding. As infrastructure remains a challenge in humanitarian and developmental contexts, the disparity between what is available at any one headquarters of an organization in comparison to what is available in the field decreases with the enhanced consumerization of technology, whereby technology is increasingly being introduced first for personal use and then migrating to enterprise.

For example, ideas on how to further contextualize certain innovations, such as a program by WFP called the Hydroponic Food Growing Initiative H2Grow, are being channeled up from the field through engagement with refugees and other beneficiaries (Innovation.wfp.org, 2022a).

Many people globally now have access to mobile data, and organizations are working to connect the rest of the world through initiatives such as GIGA, a collaboration between the International Telecommunication Union (ITU) and UNICEF to provide internet to all schools globally. This initiative aims to create models for innovative financing and support government contracting for connectivity. Innovations that come from the field as this one can now have a scalable impact on the organizational processes in other locations (Giga.global, 2022).

In summary, multilateral organizations are seeing a convergence between public pressure to innovate and the rise of ubiquitous technological advancement that has the potential to greatly improve people's lives if it can also reach those most in need. This combination of a need to be more effective plus a way to better connect people and innovate creates an inflection point which will be transformative for these organizations and will require them to find their new role within the changing environment.

From innovation to transformation

The goal of innovation must be transformation, meaning to fundamentally change people's behaviors or systems through a new way of delivering a product, process, or policy.

UN organizations, such as WFP, UNICEF, and WHO, have concrete mandates—to solve hunger, protect children, or end certain communicable diseases—for which they receive funding. Traditionally, they are not expected to go beyond the boundaries of that specific goal. Often when an organization oversteps the boundaries of its mission, it is due to internal innovation champions who see a broader mechanism to address the problem. When this happens, it is referred to, regrettably, as "mission stretch" and "mission bleed," spoken negatively.

In practice, framing a mandate more broadly allows multilaterals to engage in truly transformational innovation. Some agencies have done this effectively. For example, the WFP's Building Blocks initiative, which enables cash-based transfers for refugees on a blockchain technology, actually stretches the mission from "ending hunger and poverty" to supporting financial services. In this way, Building Blocks has been one of the most impactful innovations that has also sparked collaboration between sister UN organizations. Similarly, the approach of the Joint United Nations Programme on HIV/AIDS (UNAIDS) Health Innovation Exchange (HIEx) to engage with solar energy and technology to secure energy provision—which is one of the main obstacles to delivering healthcare—goes beyond the traditional mission of UNAIDS. In an interconnected world, where solutions can only be found in ecosystem collaboration, we need to think more broadly about how the missions of multinational organizations can be achieved (Innovation.wfp.org, 2022b).

Today, organizations need more than ever to take such a broader perspective on innovation, which allows them to work across boundaries and to solve problems at a more robust pace and scale. Often this involves working with nontraditional partners, such as multinational companies or financial service providers. And the best practice examples we found in our research are all driven by inspirational leaders and strong teams.

Anecdotally, we hear about HIV testing clinics globally, some of which are full of people daily being tested while others remain empty. The ones that are full are those that provide free internet connection to those people coming in for testing. While connectivity is not the mission of the UN organizations (such as UNAIDS or WHO) which often run these clinics, the broader perspective creates potential for partnerships with organizations such as ITU or local private sector telecoms to incentivize people to come in for testing.

How multilateral organizations can lead innovation ecosystems

Ecosystems are networks of organizations and individuals that coevolve capabilities and align their investments to create additional value, improve the efficiency of operations, and innovate. These networks have proven to be important mechanisms to overcome diverging goals and work together to create greater value. When combining this idea with the digital nature of many of the solutions for achieving the SDGs, we can understand why an ecosystem approach may be the best way to start scaling innovations for the SDGs, and why multilateral organizations should take on the role of orchestrating these ecosystems. Digital technology innovations, or digital solutions, not only radically reshape the nature and structure of the global economy but require new interconnected ways of working with private sector companies.

Innovations to achieve the SDGs often demand a deep embeddedness into a community and local understanding of those communities where the innovation is seeking to solve a problem. But at the same time, there is a need for digital solutions to be managed by networks of different partners, rather than owned by a single organization. It has been argued—and is valid—that social innovations are embedded in their local context and a clear understanding of this context is critical to social change, but equally important are the international dimensions that these social innovations are trying to solve. When ideating innovative initiatives to achieve the SDGs, the innovations' scaling potential across internationally operating organizations and countries must been acknowledged. A key success factor is to identify the specific roles necessary for such collaborations, so that a deep level of knowledge about the local setting as well as the global view is necessary, certainly for scaling the innovation.

Like large multinational companies, multilateral organizations are mature organizations that involve a high level of organizational complexity to manage globally dispersed operations in many different countries. The UN organizations are no exception. Their greatest challenge to innovation is to overcome the process and bureaucratic boundaries constraining them and drive forward new innovative ways of addressing their missions in diverse national contexts.

Despite these challenges, the UN organizations retain certain characteristics that make them particularly well suited to lead global innovation for achieving the SDGs and for implementing groundbreaking social change globally. These characteristics revolve around the organization's competencies to address the tensions between global reach and local understanding. What makes multilateral organizations fit to lead the scaling of innovation for the SDGs globally are three characteristics: their social, nonprofit missions; boots on the ground; and global legitimacy.

The UN as an orchestrator of social innovation

We see that in many cases of innovation emerging out of the UN organizations and then scaling globally, the UN organization is the orchestrating actor—bringing together multiple cross-sectoral players around the goals of the initiative. Our research used empirical qualitative methodology to observe the evolution of innovative initiatives and interview innovation champions in the UN for the past 5 years. Our findings show that these organizations can add value through their global charter and status, as well as through legitimacy created over many years of access to local partners and beneficiaries. Box 1 shows the different roles that actors play in the ecosystem and how these roles drive the ecosystem during international scaling of the digital solution. The orchestrator ideates and sets up the ecosystem around the innovation by using its global legitimacy and activating its network. The integrator contextualizes and provides local knowledge and access to partners. And the complementor assists in on the ground delivery and access to end users/beneficiaries through value-creating activities and resources. One concrete example where a UN organization played the orchestrator role was in the growth of the GRID3 initiative which works with countries to generate, validate, and use

BOX 1 Ecosystem roles for scaling of digital solutions

Ecosystem role	Primary activity in driving innovations at scale
Orchestrator	Ideate and set up the ecosystem around the innovation by using their global legitimacy and activating their network
Integrator	Contextualize and provide local knowledge and access to partners
Complementor	Assist in on the ground delivery and access to end users/beneficiaries through value-creating activities and resources

Adapted from: Tatarinov, K., Ambos, T., Tschang, T. (2022). Scaling digital solutions for wicked problems: Ecosystem versatility. Journal of International Business Studies. https://doi.org/10.1057/s41267-022-00526-6.

geospatial data on populations, settlements, infrastructure, and boundaries. The initiative is scaled through a complex web of global and local ecosystem actors, orchestrated by the United Nations Population Fund (UNFPA). Actors such as the Bill and Melinda Gates Foundation and the new Columbia Climate School provide necessary technological know-how and resources, as what we refer to in the figure Complementors, while local governments and UNFPA field offices contextualize the tool to the local needs as what we call Integrators, as shown in the textbox (Box 1).

As digital solutions are facing increasing ethical concerns, both in the non-for-profit as well as in the for-profit domain, the UN and other multilateral organizations could use their public good character to act as ethical gatekeepers and keep other ecosystem players in check by setting the narrative around new technological implementation. Orchestrators, such as the UN, playing an ethical role need to interact with, and be informed by, the other parts of the ecosystem. Here is where Integrators are necessary, to connect the UN organization with local partners on the ground and with specialized knowledge provided from Complementors who deliver the tool to the beneficiary.

As technologies can change, consequences of applications can be unintended or only realized after initial experiments. The rise of tech-savvy "digital natives," people who have grown up in the information age and are comfortable with technology, across populations is leading to important transformations, including global societal and local actors who can provide the link to enable a change in the interaction between the organizations and these beneficiaries. The ecosystem approach also has the potential to transform organizations because of the access to data which enables organizations to better understand their beneficiaries and their needs, such as with the U-Report application from UNICEF which was described before, whose SMS or AI-enabled polling results of remote communities can impact decision-making at the highest levels (Ambos et al., 2018).

Multilateral organizations often lack the technical skills to lead the full development of technological innovations. In fact, we see that some large bureaucratic organizations with limited technical capabilities are unable to connect directly with technical partners. However, they require partners to source localized knowledge. Essentially, the layered nature of digital technology provides unique opportunities. For example, in a cooperative setting, the Orchestrator can afford to give up its architecting responsibility and focus instead on soft issues while handing over technical support to partners. For example, in the GRID3 initiative, UNFPA can manage the government beneficiary while allowing partners, WorldPop and Flowminder.org, to continue technical development of their GRID3 tool.

Finally, flexibility is key in these settings. There is a strong need for the ecosystem to adapt as the solutions scale internationally. For example, the actor tasked with local implementation may be an NGO in one country, but a government in another country. Changing partners and involving new stakeholders across sectors is a necessity in development contexts where local infrastructure and cultural needs vary significantly across locations. This requirement for ecosystem adaptation to meet the local needs puts a significant burden on the Orchestrator organization driving the scaling. Orchestrators must have a global perspective coupled with a deep understanding of local requirements and stakeholders to create greater impact and value.

The resilient health innovation ecosystem: Scaling health innovation

To create innovations aimed at delivering resilient health and transforming healthcare globally, multilateral organizations will be successful if they can pivot their roles away purely from single-minded missions and toward developing capabilities for orchestrating ecosystems around scaling health innovation. One excellent example of where this is already happening is the Health Innovation Exchange (HIEx) launched by the Innovation Team at UNAIDS in May 2019. The platform works beyond the mission of UNAIDS to end AIDS and brings together innovative entrepreneurs, investors, and governments who can work together to scale health innovation for the SDGs. As a multilayered platform, it connects countries and systems, identifies their health-specific priorities or challenges, and works to ensure that investments are channeled to support high-potential innovations for health that help countries move toward impact for the SDGs. The HIEx forges a community of political leaders, decision makers, health experts, technology and science leaders, innovators, investors, accelerators, and implementers and all those who can share, explore, and synergize efforts for sustainable impact.

Steps toward multilateral organizations as innovation leaders

The challenge to innovate to achieve the SDGs puts considerable pressure on multilateral organizations. Our research showed that the key drivers of innovation will be the new digital environment and the need for ecosystem collaboration. This provides not only challenges but also unique opportunities for multilateral organizations to reshape and reinvent their roles. Such transformations will not come without friction, since established bureaucratic processes will be broken up and new mindsets and skills will be created in these organizations.

We recommend three important steps which organizations can take to develop an Innovation Leadership role to achieve the SDGs.

Digital up-skilling

As digital technology offers the potential to innovate for the SDGs, multilateral organizations will be obliged to develop new job descriptions to attract talented staff with the skill profiles to understand and work with digital technologies. The talent pool for data analysts and technology experts is highly competitive, but creative partnerships with industry, academia, and even crowdsourcing may offer opportunities to work across sectors without having to directly hire and find talent that may not naturally be attracted to multilateral environments. UN organizations will never match the development potential of the major technology companies or start-ups due to the lack of resources and the non-profit-oriented missions. But rather than needing people able to "develop" digital technology on their own, such as AI or machine learning, UN organizations can hire out those experts as advisors and can concentrate themselves on employees who understand how technology can address their beneficiaries, are aware of the ethical questions and challenges, and are able to choose the right partners to work with (McLachlin et al., 2022).

Toward an orchestrator role

UN organizations have the potential to play an Orchestrator role in innovation ecosystems, as described before. In theory, these organizations are well equipped to take such a role. They benefit from global legitimacy, have access to different types of organizations and the diplomatic skills to convince stakeholders. Playing an Orchestrator role also includes understanding and acknowledging the needs and priorities of other ecosystem players, such as for-profit companies, governments, or local NGOs. A stable ecosystem can be built only if the goals of all players are aligned. This requires staff members of multilateral organizations in the field to act as managers and not activists, which means that people in these organizations need to combine their strong passion for the mission with new skills around managing teams, fostering healthy work environments, and driving forward transformational change.

Leveraging local and global knowledge

The balance between the global vision and the local role of being "boots on the ground" (acting in the local setting) is a key competence of many UN organizations vis-à-vis all other actors in this world. However, this potential is often not fully harnessed, as knowledge transfer between headquarters and the field can be hampered by hierarchy, time zone issues, and bureaucratic procedures. Discussions about centralization and decentralization in organizations are often driven by individual's interests and budget arguments, instead of trying to connect the knowledge repositories in the organization. Innovation teams can act as bridges between the various parts of the organization. They are characterized by a culture of trust and collaboration that is built over time, even though their work is often threatened by radical decentralization efforts and driven by cost-saving. Locating these teams at headquarters or other centralized locations fosters communication and knowledge exchange.

In summary, it is important to maintain a critical exchange of ideas between those who ideate innovation, whether centrally or in the field; those who scan the world for needs and the potential to grow and scale innovations; and those who deploy them locally and continuously improve solutions. Scaling digital solutions for resilient health systems requires an ecosystem that is not only diverse in its actors but is also versatile to adapt to changing local needs. Many multilateral organizations are well positioned to lead these ecosystems but cannot do it alone. The Integrator role is crucial in the humanitarian and developmental context. Integrators can take on many forms, from global foundations that anchor funding a project, to field offices connecting to local partners, to large technology corporations supporting the technology development of a solution. These integrators give the orchestrator the space to focus on building legitimacy and overseeing the further global scalability of the project. One entity, or sector, alone will not be able to create the resilient health systems we need to efficiently address situations like the COVID-19 and future pandemics. We require innovation for local problems, at global scale, where roles will be reestablished and new ways of collaboration can emerge.

References

Ambos, T., Tatarinov, K., Somerville, S., & Milligan, K. (2018). *Initiatives with impact: Unleashing bottom-up innovation in International Organizations*. University of Geneva. https://www.unige.ch/gsem/index.php/download_file/view/5716/4619/. (Accessed 13 June 2022).

Giga.global. (2022). *Giga—Connect every school to the Internet*. https://giga.global. (Accessed 13 June 2022).

Innovation.wfp.org. (2022a). *H2Grow | WFP innovation*. https://innovation.wfp.org/project/h2grow-hydroponics. (Accessed 13 June 2022).

Innovation.wfp.org. (2022b). *Building blocks | WFP innovation*. https://innovation.wfp.org/project/building-blocks. (Accessed 13 June 2022).

McLachlin, R., Tatarinov, K., Ambos, T., & Puranam, P. (2022). *Creating people-Centric International Organisations with AI*. INSEAD Knowledge. https://knowledge.insead.edu/leadership-organisations/creating-people-centric-international-organisations-with-ai-18511?ct=11994. (Accessed 13 June 2022).

UNOG. (2022). *About multilateralism | Centenary website*. https://multilateralism100.unog.ch/about. (Accessed 13 June 2022).

U-Report Team. (2020). *U-Report—COVID-19 outbreak response*. https://www.unicef.org/innovation/ureportCOVID19. (Accessed 13 June 2022).

Weiss, T. G., & Carayannis, T. (2017). Windows of opportunity for UN reform: Historical insights for the next Secretary-General. *International Affairs*, *93*(2), 309–326.

CHAPTER 43

The role of regional bodies: Role of the African Medicines Agency in securing quality health commodities for resilient health

Pradeep Kakkattil

Health Innovation and Investment Exchange (HIEx), Geneva, Switzerland; Global Council SDG3 (Health), "SDGs in Action" Initiative, Dubai, United Arab Emirates

Background

The COVID-19 pandemic taught the world several difficult lessons. Africa learned, for the first time, that while global solidarity lasted when times were good, when crisis comes, it's a case of "everyone for themselves." Take the COVID-19 vaccine coverage at the end of 2021 as an example: less than 10% of Africans were vaccinated, that is, they had received at least one dose of the vaccine while vaccine coverage for rest of the world stood at 50.1% (The World Bank, 2022). Despite several global commitments, Africa struggled to find the necessary supply and was not seen as a priority by manufacturers, and vaccine nationalism resulted in a ban on exports of essential commodities.

This situation is not new; Africa has consistently faced challenges in access to quality medicines, diagnostics, and vaccines, even prior to the COVID-19 crisis. A continent of 1.3 billion people relies almost entirely on the rest of the world for its health commodities and produces less than 3% of the medicines it consumes. Africa's high disease burden, along with reliance on external partners for essential health commodities, places the continent at a high risk and dependency for its health security.

Disproportionate disease burden

Africa's 1.3 billion people bear a disproportionate burden of disease accounting for a quarter of the global disease burden. Africa's share of communicable disease is startling, accounting for over 90% of the global annual malaria cases and being home to over 60% of People Living with HIV. Noncommunicable Disease (NCD) burden is also rising rapidly; the NCD burden is estimated to increase globally by 17% over the next 10 years, and in the African

Resilient Health. https://doi.org/10.1016/B978-0-443-18529-8.00043-3

region by 27% (NCD Alliance, 2022). Despite this, Africa has not been able to establish production of health commodities on the continent.

Challenges and opportunities: Lack of local production

Though Africa constitutes 25% of the global vaccine market, it manufactures less than 0.1%, which amounts to about 12 million doses of early childhood vaccines. Even that amount focuses on downstream steps such as fill and finish, packaging and labeling, import to distribute, and there is little if any research and development as most of these operations are on a small scale (Mo Ibrahim Foundation, 2022). While vaccine manufacturing is lagging, so are disinfectants—which are helpful in disease control and not difficult to manufacture—even that, only one in six liters of Africa's disinfectant imports comes from regional suppliers, according to the International Trade Centre.

Lack of infrastructure is also a major barrier. As of 2019, the continent possessed roughly 375 pharmaceutical manufacturers, as compared to about 5000 in China and double that, i.e., 10,500 in India (McKinsey & Co, 2019). The United Nations Economic Commission for Africa (UNECA) estimates that Africa imports about 94% of its pharmaceutical and medicinal needs from outside the continent, at an annual cost of US$16bn (UNIDO, 2021).

Falsified and substandard drugs

The World Health Organization (WHO) estimates that in several African countries, between 30% and 60% of all medical products are falsified (Hammond Kwame Adjei, 2015). A majority of the reports of substandard and falsified drugs that the WHO receives come from the African continent. It is estimated that falsified drugs kill over 122,000 children under five, who die each year due to poor-quality antimalarials in sub-Saharan Africa. Since the onset of COVID-19, this figure has only increased, opening new challenges in supply chains tainted with falsified vaccines, therapeutics, and diagnostics in circulation.

African pharma market

Africa's pharmaceutical market is expected to grow from US$19bn in 2012 to US$66bn by 2022, with Africa expected to be the fastest growing market in the world. It is also estimated that the health and wellness sector in Africa will be worth about US$259bn by the year 2030, with the potential to create over 16 million jobs (McKinsey and Co., 2019).

However, there is limited local production, with only six companies that have achieved WHO prequalification. Lack of local production results in high costs—medicines sold in sub-Saharan Africa are one of the most expensive in the world—with some basic drugs being sold up to 30 times the price of the same drugs in Europe (Center for Global Development, 2019). High costs of essential drugs have led to a flourishing fake drug market with one of the highest levels of substandard and falsified medicines (Pharmatimes, 2021). Lack of clear policies, poor regulatory frameworks, and inadequate capacities have undermined the health security of the continent.

Fragmented regulatory framework

Several countries in Africa do not have functional regulatory frameworks. One of the challenges of a fragmented regulatory system for healthcare in Africa is that several pharma companies ignore the market. Patent holders and manufacturers lack incentive to register and promote their products; having to go through cumbersome registration process in each country, often with a small population, makes such an investment and venture unattractive and too expensive. This means that several countries work with outdated drugs—some older than what is recommended by the WHO essential drug list. Local manufacturers often are distributors of the global pharma and do not have the incentives and resources to introduce newer generation generics into smaller African markets.

Developing a common regulatory framework

One of the critical steps toward strengthening the pharmaceutical sector in Africa is the development of a common regulatory framework across the continent. The unanimous decision by the African Union Assembly in 2019 to adopt a treaty to establish an African Medicines Agency (AMA) was an attempt to strengthen Africa's ability to provide regulatory oversight and enable access to safe and affordable medicines across the continent.

A minimum of 15 member states of the African Union had to ratify the treaty for it to enter into force. Between the years 2019 when the treaty was signed and 2021, there was not much progress. The appointment of Michel Sidibe as the African Union Special Envoy on the African Medicines Agency provided the necessary momentum for its ratification. Sidibe has worked with Heads of States advocating for the rapid implementation of a continent-wide, efficient regulatory system for health products as a critical instrument for the continent securing its health and reaching the targets of Universal Health Coverage. This would not only improve access to quality and affordable medicines but also impact Africa's economy: apart from potentially creating 16 million jobs on the continent, it would significantly reduce the high out-of-pocket costs for healthcare that globally pushes people into extreme poverty, many of whom are in Africa.

The African Medicines Agency

The African Medicines Agency (AMA) was established by the African Union to support and complement national and regional efforts in developing a robust, predictable, accountable, and efficient regulatory system for health products. Rather than having to negotiate 55 times, drug and medical device companies would be able to deal with one assessment process and get one recommendation for the whole continent, which could be used by all the African countries to authorize the product nationally. This approach, combined with the African Continental Free Trade Area (AfCFTA)—whereby members made certain trade agreements in order to increase socioeconomic development, reduce poverty, and make Africa more competitive in the global economy—could foster the creation of a continental market and provide the scale necessary for Africa to be an attractive market, with an expected size of 2.2 billion people by the year 2050. AMA would also lead to better regulatory capacity, thereby clamping down on substandard and falsified medicines.

The ratification of AMA by over fifteen countries has led to it coming into effect, and Rwanda has recently been selected to host the agency. Following provides an overview of AMA's vision, mission, role, and organs (African Union, 2021).

AMA vision: African people have access to essential medical products and technologies.
AMA mission: Provide leadership in creating an enabling regulatory environment for pharmaceutical sector development in Africa.

What is AMA's role? AMA will

- Support the growth of local pharmaceutical production, a key objective of the Pharmaceutical Manufacturing Plan for Africa (PMPA). AMA will further play a critical role in catalyzing trade in support of the African Continental Free Trade Area (AfCFTA).
- Evaluate medical products for the treatment of priority diseases as determined by the African Union.
- Regularly inspect, coordinate, and share information about products that are authorized for marketing.
- Coordinate joint reviews of clinical trial applications for vaccines and assessment of "highly complex" product dossiers such as biosimilars. AMA will also coordinate joint inspections of Active Pharmaceutical Ingredients (API) manufacturing sites.
- Collaborate with Regional Economic Communities (RECs) and National Medicines Regulatory Authorities (NMRAs) in the identification of substandard and falsified medical products (SFs) and facilitate information sharing across countries.
- Develop common standards and regulations and be responsible for harmonizing legislation.

Organs of the AMA

The AMA is made up of four organs: the Conference of parties, a Governing Board, the Secretariat, and Technical Committees.

Challenges and conclusion

Establishment of the AMA is a critical step toward health commodity security for Africa. But, while the AMA has come into force, several challenges persist. Many of the larger countries, such as Nigeria and Kenya, are yet to join the AMA, which could limit the scope of the agency and its ability to negotiate with the pharma and medical device manufacturers. Moreover, much of the regulatory experience and capacities are in these nonmember countries, which would be critical for its success. While the AMA has benefitted from funding from several international stakeholders, it will also be important to establish clear partnerships with other countries that could support much-needed building of regulatory capacities. Despite these challenges, the AMA has an opportunity to leverage innovation and technology much more than many of its peer organizations and can function as a modern, nimble, and robust agency to solve many problems on the continent and to advance development. Africa needs it.

References

African Union. (2021). *Frequently Asked Questions (FAQs) on the African Medicines Agency (AMA)*. https://au.int/sites/default/files/documents/41269-doc-AMA_FAQs_rev.pdf. (Accessed 12 April 2023).

Center for Global Development. (2019). *Tackling the triple transition in global health procurement*. Center for Global Development.

Hammond Kwame Adjei. (2015). *1 princess Ohene 2. Counterfeit drugs: The relentless war in Africa*. medcrave.

McKinsey & Co. (2019, January 10). *Should sub-Saharan Africa make its own drugs?*. mckinsey.com. https://www.mckinsey.com/industries/public-and-social-sector/our-insights/should-sub-saharan-africa-make-its-own-drugs.

Mo Ibrahim Foundation. (2022). *COVID 19 in Africa: 2021 forum report*. Mo Ibrahim Foundation. https://mo.ibrahim.foundation/sites/default/files/2021-06/2021-forum-report.pdf.

NCD Alliance. (2022). NCDAlliance.org https://ncdalliance.org/the-global-epidemic.

Pharmatimes. (2021, April). https://www.pharmatimes.com/web_exclusives/COVID_and_the_growing_threat_of_counterfeit_medicines_1368856.

The World Bank. (2022). *2021: A year like no other for Western and Central Africa, the World Bank*. https://blogs.worldbank.org/africacan/2021-year-no-other-western-and-central-africa.

UNIDO. (2021, September). UNIDO.org https://www.unido.org/news/strengthening-africas-pharmaceutical-industry-learning-lessons-covid-19.

CHAPTER 44

The role of the private sector in achieving universal health coverage: Experiences from the UHC2030 private sector constituency

Shariha Khanam Khalid[a], Craig Burgess[b,*], Mwenya Kasonde[b,*], Temitayo Erogbogbo[c], Sylvana Q. Sinha[d], Beatrice Gatumia[e], and Judy Kuriansky[f,g]

[a]Mission & Co, Kuala Lumpur, Malaysia, [b]World Health Organization, Geneva, Switzerland, [c]MSD for Mothers, Lucerne, Switzerland, [d]Praava Health, Dhaka, Bangladesh, [e]Busara Center for Behavioral Economics, Nairobi, Kenya, [f]Department of Clinical Psychology, Teachers College, Columbia University, New York, NY, United States, [g]Global Council SDG3 (Health), "SDGs in Action" Initiative, Dubai, United Arab Emirates

Introduction

The private sector is an important participant in building strong and resilient health systems given that businesses and private enterprises provide health products and services for millions of people and communities globally. As such, the private sector plays an important role in fulfilling the criteria for Universal Health Coverage (UHC) which calls for the availability, accessibility, acceptability, affordability, sustainability, and quality of health service delivery. The private sector further plays a critical role in achieving health outcomes in domains such as managing the supply chain of health products; accessing food, nutrition, clean water, sanitation, and renewable energy; and ensuring the availability of telecommunications infrastructure.

Many countries have "mixed health systems," where a combination of private and public providers delivers health-related goods and services. In some countries, the majority of health services are delivered by the private sector, for example, in Nigeria this is over 50%. Thus, to achieve UHC, it is critical to ensure strong public-private partnerships (PPPs) that are built on trust, mission aligned, and outcome driven.

* The views expressed by the individual authors do not necessarily reflect the views or opinion of the World Health Organization and are of an individual nature.

Resilient Health. https://doi.org/10.1016/B978-0-443-18529-8.00044-5

In order to achieve the agreed-upon goals and targets outlined in international documents, such as the United Nations 2030 Agenda for Sustainable Development with its 17 Sustainable Development Goals (SDGs) and the 2019 Political Declaration for Universal Health Coverage (UHC), there is a strong imperative to find ways to effectively harness the three sectors—public, private, and "people" sectors—whereby the latter, often referred to as the third sector, is defined as that which is not within the public or private sectors and tends to represent citizen-led initiatives (Alhuwail & Alhouti, 2024). The importance of the private sector throughout the world as a health system partner on a national and global level was underscored during the COVID-19 pandemic, which required a complexity of responses both immediately and long term. The pandemic and its ongoing variants further highlighted the interconnectedness of the three different sectors and the importance of a "whole of society" approach in three phases: response, recovery, and long-term resilience to "build back better." The private sector played—and continues to play—a key role in all these phases, for example, at the frontlines from navigating the lockdowns to contain the spread and reach patients, to providing a detailed understanding of the pandemic risks based on testing and data collected at source, to mounting an accelerated vaccine production and distribution response.

However, many challenges remain regarding the role of the private sector in a country's health system. These include that the private sector itself is very diverse in its form and function, such that other sectors may not fully understand how to engage with, and navigate the landscape of, the private sector ecosystem. In an effort to solve this problem, the World Health Organization (WHO) created governance committees under the banner of "private sector engagement" to guide its Member States (Governments) on how to engage with the sector. While a highly productive initiative, the narrative is largely about governing the private sector rather than seeing it as an equal contributing partner. Such an approach—urging governments to govern the private sector—contributes to an environment that fosters distrust on the part of both parties. In an attempt to correct this, guidelines were outlined by the Framework of Engagement with Non-State Actors (FENSA), an organization-wide policy governing all WHO's engagements and relations with non-State actors, adopted by the World Health Assembly in 2016 (WHO, 2016). These guidelines help to foster an enabling environment built on trust. However, problems still emerged from defining the private sector *only* as profit-making entities, resulting in hesitancy on the part of the WHO and Member States to engage with businesses generally seen as purely profit driven and shareholder focused (Box 1).

The importance of working with the private sector is underscored by the need for financing to achieve the SDGs as outlined in the UN Agenda 2030 for Sustainable Development. The SDG Financing framework, referred to as "Financing for Development" (FfD), underlines the need for the private sector to cofinance the achievement of the SDGs and close the growing financing gap for countries. The COVID-19 pandemic dramatically increased the annual financing needs to achieve the SDGs from an estimated US$2.5 trillion in 2014 (UNCTAD, 2014) to an estimated US$4.2 trillion per year (OECD, 2020). Meeting this financing need will require close collaboration between the private and public sectors as well as with civil society,

BOX 1 Defining the private sector

The private sector refers to the part of an economy consisting of individuals, partnerships, and groups who privately own, control, and manage companies, corporations operating for-profit and not-for-profit businesses, distinct from entities owned and run by the government or not-for-profit nongovernmental organizations which comprise the public sector. The private sector is prevalent in capitalistic societies with a free market, usually forming a large share of the economy of a country, in comparison to more state-controlled nations with a larger public sector. When these two sectors have common interests and decide to collaborate, the arrangements are called public-private partnerships (PPPs). Private sector entities can be solely run, or small, mid-sized, or large businesses, multinationals, trade unions, or professional and trade associations. A business or industry can move from one sector to another, whereby a private company can become public or a publicly run enterprise can be turned over to private citizens (called *privatization*). The sectors can also be run cooperatively, in healthcare as well as other fields such as security services and water or waste management.

and further require the application of innovative financing mechanisms in order to unlock resources from multiple sources to be channeled toward the common agenda.

The importance and value of the private sector in development is outlined in this chapter, as well as recommendations for synergistic partnership among sectors, challenges in effective engagement, and the UHC Private Sector Constituency's formal statement guiding its role in advancing UHC. Further, this chapter illustrates some successful models and experiences drawn from the Private Sector Constituency (PSC) of the UHC2030 community. The PSC, hosted by the WHO and the World Bank, was convened in 2018 to bring together diverse members of the private sector to foster dialogue and collaboration across the sector and with other UHC2030 partners, constituencies, and networks. Members are private sector entities within the health value chain, such as service providers, health insurers, manufacturers and distributors of medicines, health products, and innovative technologies representing large, medium, and small businesses from around the world.

The value-add of the private sector

Partner engagement with multiple stakeholders in delivering UHC goals is challenging to build and maintain because the partnership process prioritizes broader objectives over the interests of individual members. There is also a lack of publicly available evaluations of multistakeholder partnerships and platforms, due in part to a lack of clarity around collective goals. With access to resources, intellectual capital, technology, and efficient management practices, private sector entities add value across the continuum of healthcare delivery, especially noticeable where government infrastructure or services are nonexistent or weak, such as in fragile or conflict settings.

Increasingly, countries are being forced by domestic pressures to seek out private sector collaboration to capitalize on innovation and funding at all levels and are eager to establish opportunities for local private sector enterprises to access new markets and effectively contribute to foreign direct investment in emerging market regions. Further, decreasing aid budgets suggest that support for multistakeholder platforms showcasing private-public sector partnerships will grow, leading to an enhanced role of the private sector in development.

Challenges for the private sector to engage effectively include how to:

- be better organized and collaborative in tackling ecosystem issues.
- work more effectively with the public sector and in diverse and mixed health systems.
- mobilize financing better, with collaboration from development finance institutions and other sources, such as private sector financiers, and using innovative financing instruments like blended finance.
- apply innovations in the context of delivering health outcomes in resource-poor settings using business models that are profitable while empowering those "left furthest behind."
- explore and expand the role of technology in service delivery, access, and building more precise health system responses.

Several Framework of Engagement with Non-State Actors (FENSA) processes exist among multilateral organizations that can be challenging to navigate for a private sector entity. For example, until recently, the definition of the private sector was limited to for-profit companies, whereas the global landscape of what constitutes the private sector is rapidly changing with nonprofit companies, impact-driven companies, and other legal forms of the private sector now widely accepted. This recognition of the diversity of the private sector leads to the need for dedicated legislation in some countries to recognize the many diverse forms, in examples led by the Philippines and the United Kingdom. The UHC2030 PSC recognizes and celebrates this diversity in its membership, and members collaborate with each other toward achieving the common goal of achieving "Health For All."

Further, the emergence in recent years of the "third sector" mentioned before, sometimes referred to as the "people" sector, or as "civil society," signals the growing trend of citizen-led initiatives to address local challenges and represents a move away from reliance on the public and private sectors to deliver on promises to citizens. Yet, for the best outcome, initiatives need to be carried out in partnership among the sectors in a tripartite arrangement where each party brings its strengths to the collaboration.

The growing importance of the third sector and country ownership

This "people" sector constitutes those organizations and informal alliances of people united in a common cause (not-for-profit movements), which are not within the public or private sectors, and tends to represent citizen-led initiatives. As referred to before, while health outcomes have predominantly been the responsibility of the public sector, recent trends accelerated by the COVID-19 pandemic have highlighted the growing importance of a partnership approach among these three sectors to deliver health services, particularly to populations "left furthest behind," for example, at risk-populations in remote or conflict settings. For the private sector to self-organize and valuably contribute to UHC at the country level, it must be aligned with national and subnational priorities of the specific country. Ensuring country ownership is crucial, and grounding platforms in the realities of the local context to address development issues, makes a project more likely to achieve UHC and the broader SDGs. In this context, country ownership is seen as a process that is government led and centered around government-identified priorities. However, given the complexity of health systems, and that health is no longer the sole responsibility of the public sector in an increasing number of countries, there is a growing movement of ownership for health being led and driven by participation and action by the citizen sector, at times seeking collaboration from the private and public sectors. Examples of this are the hospital center in Ethiopia being built by a consortium of Ethio-American physicians (Nurhussain, 2024) and the "24 Military Hospital" complex in Sierra Leone (Wai & Kuriansky, 2024). This phenomenon was also seen during the acute response phase of the COVID-19 pandemic, where citizen movements responded by providing PPEs to health workers and food for families as well as raising awareness about hot spots and critical needs through the use of technology (Alhuwail & Alhouti, 2024).

The private sector constituency and the PSC statement

To build on, and maximize, the private sector's existing contributions to UHC, the WHO together with the World Bank established the private sector constituency (PSC) as part of the UHC2030 initiative in 2018. The newly formed PSC invited applications for membership by private sector companies who met the criteria, for example, private sector entities directly working on health system strengthening or for-profit or not-for-profit entities that are commonly organized under business or company regulations, rely on earned income for operations, and operate under shareholder governance arrangements. Associations representing the private sector and companies are also eligible.

The purpose was to align contributions the private sector could make to achieve UHC and to demonstrate collective accountability as a constituency. An important output was a statement, delivered during the High-level Meeting on Universal Health Coverage on September 23, 2019, at UN headquarters in New York City (UHC2030), which is the first of its kind regarding UHC and originating from the private sector. The statement outlines and promotes guiding principles for enhancing action of the private sector in several areas including UHC. It also guides the other sectors to find areas of common interest and mutually beneficial partnerships to achieve the shared goal of leaving no one behind and offering health for all.

In this spirit of multistakeholder partnership, the PSC Statement for UHC builds on the SDGs, the UHC2030 Global Compact, and the "Five Key Asks from the UHC Movement." Also, the statement reinforces the five key principles for collective action in the UHC2030 Global

Compact, which are endorsed by all members of UHC2030, as the basis to guide collaboration across all partners, including the private sector.

In May 2023, the PSC Statement (Box 2) was revised and presented during the 76th World Health Assembly, in line with renewed commitments from the constituency members, and aligned with the new Action Agenda from the UHC Movement (https://www.uhc2030.org/un-hlm-2023/action-agenda/):

1. Leave no one behind.
2. Transparency and accountability for results.

BOX 2 UHC2030 private sector constituency statement extract

To build on and maximize the private sector's existing contributions to UHC, the members of the UHC2030 Private Sector Constituency renewed their commitments in 2023, and published a revised statement, shared during the World Health Assembly. This was accompanied by case examples illustrating these commitments.

1. **Incorporate UHC principles, including to leave no one behind, into our business**
   Whilst the private sector values immediate results, it also takes a medium and long-term approach to business goals and impact. Each time frame is consistent with supporting development of equitable, resilient and sustainable health systems, and with the goals of UHC. In parallel, we are continuing to explore approaches that help to drive greater and more equitable access, as well as the quality and sustainability of our health products and services. This includes a sustained commitment to uphold shared standards of ethical conduct and decision-making, fostering patient trust in their health system and confidence in their care.
2. **Deliver innovations that respond to the needs of all people including under-served populations, and make these safe, affordable, accessible and sustainable**
   The private sector provides a wide array of health care products and services globally. This core offering of demand-driven products and services is its greatest contribution to UHC. Private sector innovation is an engine for new and innovative products, techniques, and insights that can not only improve health, but increase health system efficiencies and sustainability. Beyond products and services, the private sector contributes to UHC via innovative approaches and partnerships with government and other stakeholders in the areas of: research and development (R&D), digital technologies, capacity building, health literacy, healthcare delivery models, financing and ethics and business integrity standards—to improve access to quality, affordable vaccines, medicines, diagnostics, surgery and preventive healthcare services.
   Digital health technologies are transforming health systems worldwide. The private sector is committed to working with government and other stakeholders to advance new technologies and approaches to expand access, increase efficiencies and rapidly accelerate progress toward UHC.
3. **Help strengthen the health workforce, responding to local context, priorities and needs**
   The private sector makes important contributions to training health workers (providing and supporting both pre- service training and continuing education), to knowledge transfer that strengthens health workforce capacities, and to building workforce data and evidence. It is also a major employer of health workers, and commits to continue supporting diversity, equity and inclusion, workplace safety, career development and fair compensation.

BOX 2—Cont'd

4. **Contribute to efforts to raise the finance available for UHC**
   The private sector makes catalytic contributions to health financing through both foreign direct investments and blended financing mechanisms for increased domestic resources for health. The role of the private sector through innovative financing models—as well as traditional financing models—can have significant impact on advancing national health systems achievement of UHC. The private sector can support governments to articulate the business case for investing in health and UHC.
5. **Champion and engage in multistakeholder policy dialogues that advance UHC**
   As highlighted by the UHC Action Agenda, governments alone cannot achieve UHC: all key stakeholders must play a critical role. The private sector actively seeks opportunities to engage in multi-stakeholder forums led by governments and other stakeholders to promote trust and transparency, institutionalize mechanisms for inclusive health governance to advance UHC objectives in policy and practice.

3. Evidence-based national health strategies and leadership.
4. Making health everyone's business.
5. International cooperation based on mutual learning across countries.

Enabling successful partnerships with the private sector

To create the enabling environment for private sector engagement and contributions toward achieving universal health coverage, and indeed, to recover from the current pandemic and be prepared for any next one, it is important that the private sector—and other sectors—finds pathways individually and collectively to be successful. This includes building infrastructures that foster a trusting relationship oriented toward serving the health needs of the people rather than being only focused on profit-maximizing and shareholders' interests. Solutions to effective participation of the private sector are given in the examples in the following sections in this chapter, contributed by members of the PSC. Some of these are profiled on platforms, as described in the next section of this chapter.

Platforms for private sector self-organization and collaboration across sectors

The rise of platforms that showcase innovations as well as multistakeholder collaborations emerges from realization among the sectors about the need to collaborate in order to solve complex development issues. Since not one party can accomplish the ambitious development goals alone, and since goals are shared, collaboration is becoming the norm. Such platforms have long been recognized as critical to achieving the SDGs and UHC by 2030.

The platform approach as a pathway to enable and encourage multistakeholder collaborations is growing globally. A platform can be a technology solution, a website, or even a recurring event, for example. Regardless of the form, five key features are essential, and common, to any platform. A platform must:

1. carry a purpose, with their mission being at the core of their existence and members/community that they attract.
2. be neutral to the mission they host. While the platform itself can be hosted by a single entity, the platform mission remains neutral and can bring many parties together around a common purpose.
3. actively seek to provide visibility to the mission, the parties who engage with it and a broader global agenda, often achieved through regular engagement activities delivered through social media channels, or through events hosted during key global calendar moments.
4. create value for the mission and the parties aligned around the common purpose, such as through accelerating an agenda, leveraging on resources of its membership (beyond advocacy), or enabling new or existing resources to be channeled toward the mission, or strengthening and building capacities of its members to deliver on the mission, or all of the above.
5. be accountable toward achieving the mission they aim to advance and share successes and failures as well as results and lessons learnt on a regular basis. Without accountability, platforms lose credibility and neutrality, particularly if there is no transparency about which party is benefiting from their existence.

Multistakeholder partnership platforms address these five criteria as well as three key considerations outlined for private sector self-organization and collaboration to be effective across sectors. The three key considerations that drive engagement with the private sector through a platforms approach are as follows:

1. Private sector engagement through multipartner collaboration brings added value of efficiency, additional resources, and intellectual capital. This does not come without challenges, including difficulty maintaining interoperability, consistency, and ethical considerations; however, with proper governance mechanisms, an environment of trust that enables successful collaborations can be established.
2. Platforms that promote partnership between the private sector and the public sector allow private sector entities to engage, learn, and contribute to achieving national objectives, UHC and the SDGs through a neutral channel. Such platforms can encourage transparency as well as provide visibility for those seeking to engage with the private sector on particular agendas.
3. Health data-related efforts, like the Health Data Collaborative, need the private sector for efficiency and access, specifically where there is weak or nonexistent government services and infrastructure.

Why partnership platforms are needed

Health systems are complex, with many determinants of health outcomes that rely on more than one factor. As has been stated before, no one organization alone can address the complexity of such systems and the interconnected factors. Partnership and collaboration can lead to better solutions due to increased diversity of views and opinions, efficiency of resource mobilization, alignment to common goals and agendas, and leveraging the knowledge experience of the particular sector in order to reach a common objective.

However, the rhetoric around partnering and private sector engagement is often vague, as the number of collaborations to accomplish the SDGs grows. Established platforms for collaboration across sectors often focus on multigovernmental agencies such as the United Nations and related multilateral development banks such as the World Bank, African Development Bank, and Islamic Development Bank, whereas countries, academia, civil society, and the private sector remain underrepresented. Yet, multistakeholder platforms (bringing partners together) are increasingly part of necessary infrastructure to increase private sector collaboration and, ultimately, health impact. Using a "whole system" approach and "human-centric" design, in collaboration with the private sector, is essential in order to address development problems across all sectors, including health, education, food and nutrition, and the environment.

Challenges for such platforms exist. Importantly, they require a skilled and dedicated secretariat or portfolio manager, which is a body or an individual that manages day-to-day activities and administrative tasks required to support a well-functioning group and to keep all parties engaged, including establishing processes and tools; sourcing innovations and partners; organizing meetings, agendas, and correspondence; and reporting. Additionally, collaborative platforms rely on communication and trust which are time consuming to maintain. Also, partner roles must be clearly outlined, as it can be challenging for the private sector to play an equal role in health partnerships often dominated by multilateral organizations and nongovernment organizations.

Platforms for cross-sector collaboration

Platforms, such as the UHC2030 platform, the Health Data Collaborative (HDC), and Health In Your Hands (HIYH), enable this coming together to take place at global, regional, and national levels. As an example, the HDC platform constituencies include multilateral organizations, donor organizations, research and academic organizations, civil society, and the private sector. The HDC is a joint effort by multiple global health partners to work alongside countries to improve the availability, quality, and use of data for local decision-making and for tracking progress toward the health-related SDGs. The primary purpose of this effort is to enhance a country's statistical capacity and to institutionalize the use of data for decision-making to achieve better health outcomes. HDC partners need to align their technical and financial commitments around strong nationally owned Health Information Systems (HIS) and common monitoring and evaluation plans for measurable impact on the ground. Fortunately, this UHC2030-related initiative has the potential to align technical and financial investments in efforts to strengthen country Health Information Systems.

Further, the HDC aims to put into practice the International Health Partnership (IHP+) principles of country ownership and alignment by translating them into a joint operational plan that specifies concrete collective actions at country and global levels. These principles support the implementation of one plan, one monitoring framework, and one fiduciary framework for better alignment among all stakeholders. Since 2015, developing countries, including Kenya, Cameroon, Tanzania, Botswana, Malawi, Zambia, Uganda, Nepal, Bangladesh, Indonesia, and Myanmar, have all launched a country-specific HDC, intending to harmonize their data as well as monitoring and evaluation (M&E) systems in the country, and to ensure that all actors in this space work toward a common M&E framework.

In a country example, the Botswana Health Data Collaborative (BHDC), launched in March 2020, has begun linking various data systems into a national data portal to become more interoperable, including linkage of Civil Registration and Vital Statistics to improve death and birth registration and linkage with Central Medical Stores for end-user visibility (commodities visibility). Through the BHDC, multiple data collection tools have been rationalized, and data governance for sharing information across public and private sectors is being addressed. The private sector played a key role in developing many of these data collection tools, while maintaining interoperability and advising on modalities for rationalization of these tools. The Botswana HDC platform offers an opportunity for development partners, technical experts, implementers, civil society organizations, the private sector, and policymakers to work together to improve HIS and to collectively monitor and evaluate progress toward national health goals, the SDGs, and UHC.

Another example of a platform for collaboration is Health in Your Hands (HIYH), launched in 2019. Borne out of an initiative by the United Arab Emirates' "SDGs in Action" campaign, and cofounded and cofunded by two of the authors of this chapter (Khalid and Kuriansky), HIYH aggregates innovative solutions from around the world that deliver health outcomes for "those left furthest behind" (a phrase commonly used at the United Nations to refer to those most marginalized in society) that involve a diverse range of private sector actors, multilateral agencies, and nonprofit organizations (Kuriansky, 2024). The platform offers opportunities for innovative solutions to be scaled through access to financial resources and multistakeholder partnerships with organizations that can scale their impact as well as offer visibility opportunities to share about that impact. Dedicated to enabling the delivery of accessible, affordable, and sustainable health solutions to populations most in need, the platform showcases health outcomes-focused innovations worldwide, which place health technologies, tools, and access to finance directly "in the hands" of individuals working at the frontlines as well as to communities in remote settings. These innovations range from market-driven

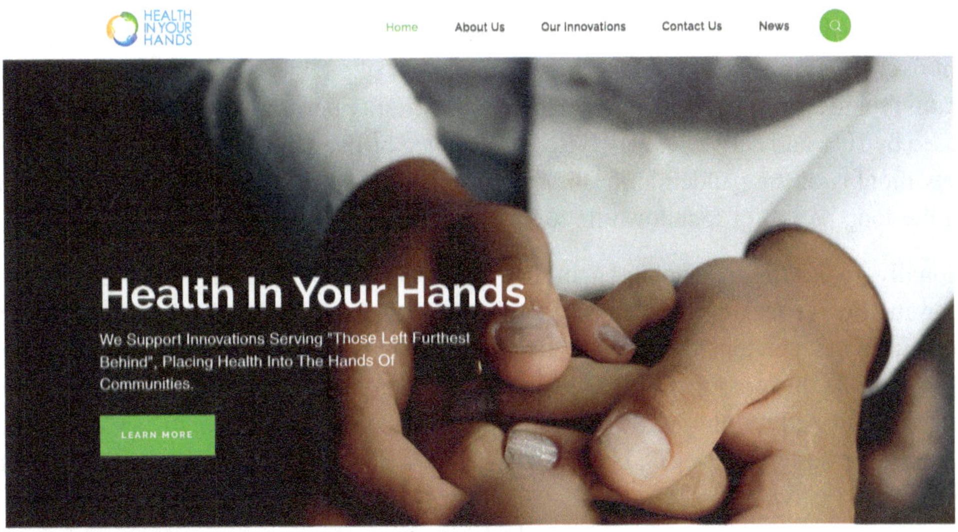

Fig. 1
A photo taken from the website of HIYH, representing the importance of placing "health into the hands" of those left furthest behind, through innovations that deliver health outcomes for communities.

enterprises that deliver products and services, to simple mobile-based applications and hardware innovations delivered directly to healthcare workers and patients. The HIYH initiative follows the WHO definition of holistic health which includes both physical and mental health. Detailed examples of the work of innovations that appear on the HIYH platform are presented in other chapters of this book (Bhavsar & Katara, 2024; Javetski, 2024; Pain et al., 2024; Santra & Mandal, 2024; van Engen, 2024) (Fig. 1).

The role of the private sector in financing health innovations and ecosystem building in mixed health systems

In this section, we describe various projects that involve the private sector integrally with other partners. Three modalities for private sector engagement set the stage for the projects, as outlined by the OECD. These are as follows:

- financial instruments originating from the private sector.
- engagement of business and scientific expertise leveraged to close the innovation gap.
- local private sector development or activities supporting local private health providers to close the supply gap.

These three elements are required to leverage private sector capacity for the health sector. Certainly financial instruments are critical, for example, to close the annual US$33 billion financing gap in funding for Reproductive, Maternal, Neonatal, Child, and Adolescent Health (RMNCAH), a situation exacerbated by the global COVID-19 pandemic. This escalation was evident over the first year of the pandemic in countries like Nigeria and India, between them accounting for nearly a third of the global maternal mortality burden. In Nigeria, antenatal visits decreased by 50% and deliveries supervised by a skilled birth attendant decreased by 40% in the first few months of the pandemic. In India, deliveries in health institutions reduced by 43% during the long period of lockdown in 2020.

Additionally, regarding the supply chain, more than 40% of individuals who use family planning in Nigeria seek modern contraceptive methods from the local private sector, for example, from independent drug shops and community pharmacies. Although these may be the only source of healthcare for many, such outlets are often unregulated and operated by health workers with limited training to deliver family planning and primary healthcare services. To address this, the IntegratE Project, led by Society for Family Health and cofunded by the Bill & Melinda Gates Foundation, seeks to increase access to modern contraceptive methods by implementing a government-endorsed tiered accreditation system for private family planning providers in collaboration with the Pharmacists Council of Nigeria. This project has supported quality improvement efforts in over 1200 private drug shops and pharmacies by helping these businesses comply with local regulations to become licensed and report service statistics to the national Health Information Management System.

In India, despite an increase in the number of women giving birth in health facilities, maternal deaths are not decreasing enough, largely due to poor quality care. To address this, the Manyata program, offered by the Federation of Obstetric and Gynecological Societies of India, established the first quality certification for private maternity care. Manyata supports private providers in improving the quality of care they offer and assists them in applying for quality certification. As of February 2022, Manyata is active in 12 states in India, and over 1200 private providers have achieved certification. The Government of Maharashtra has integrated Manyata standards for private maternity care providers into the state's quality improvement plans for maternal health. In Uttar Pradesh, a state in northern India, the government now requires Manyata certification for local private maternity care providers that wish to participate in the country's UHC financing scheme.

The example of the MOMS Initiative (Box 3) and the SDG Finance Advisory Board (Box 4) illustrate how these three modalities come together through financial instruments originating from the private sector supplementing those from development finance agencies, engagement of business and scientific expertise to close the innovation gap, and local private sector activities supporting local private health providers to close the supply gap.

BOX 3 The MOMS initiative

MSD for Mothers, MSD's global initiative to help create a world where no woman has to die while giving life has, for the last decade and counting, sought to invest in efforts to integrate private providers into mixed health systems by applying the three modalities put forward by the OECD. MSD is an abbreviation for Merck, Sharp & Dohme, a subsidiary of Merck & Co., Inc. headquartered in New Jersey, United States, and referring to the pharmaceutical company outside the United States and Canada. One of the MSD for Mothers projects is the Maternal Outcomes Matter (MOMs) Initiative—an innovative collaboration between MSD for Mothers with the United States International Development Finance Corporation, Credit Suisse, and the United States Agency for International Development (USAID)—which has mobilized US$50 million to support maternal and child health in sub-Saharan Africa and South Asia. The capital improves and expands infrastructure, services, and access to care to ensure healthy pregnancies. For example, the MOMs Initiative currently supports Lifebank, a Nigeria-based private company that expands access to lifesaving medical products, namely blood and blood products for mothers across Africa. In South Africa, the Unjani Clinic Network through a MOMs Initiative grant is strengthening their organizational structure, enhancing their systems and technology and testing an innovative finance model that allows the network to scale faster. The network aims to expand at a rate of 100 clinics per year, with a strategic plan to open over 1000 clinics by 2030.

BOX 4 The SDG Finance Advisory Board

The SDG Finance Advisory Board, established as a cross-cutting Global Council by the United Arab Emirates' "SDGs in Action" initiative, during its term, worked with selected innovations from the HIYH initiative, guided by the SDG3 Global Council on Health, to establish and document country-led examples of innovative partnerships and financing for strengthening not only health systems but also the financing ecosystem in select countries around the world. Objectives of the collaboration are to make connections to capital sources from within the advisory board membership, to leverage their networks and partners in countries of focus, and to provide technical assistance and capacity-building for the innovations interested in scaling their work.

As an example, the SDG Finance Advisory Board has explored focusing on the climate and health financing agenda, through a collaboration with a USD7 trillion capital coalition called Lifelong Investors in People's Health (LiPH).

Lessons from such collaborations will be documented and disseminated as part of a broader piece of work on knowledge resources and financing ecosystem strengthening to achieve SDG impact and a Just Transition.

Primary healthcare innovations: Sustainable and scalable business models: Country case of Bangladesh

Improvements in healthcare access translate into increased economic opportunity for individuals and countries around the world, and private sector innovation is driving these improvements. The global management consulting firm McKinsey & Company estimates that low-middle-income countries gain a US$4 increase in GDP for each additional dollar invested in healthcare (McKinsey Global Institute, 2020).

Emerging markets offer prime growth opportunities for healthcare and health technology industries, including promising business models in primary care. Bangladesh, the eighth most populous country in the world, considered one of the next Asian tigers with an impressive track record of growth and poverty reduction, has been among the fastest developing economies in the world. Over the past decade, Bangladesh's GDP per capita has been increasing, exceeding India's every year since 2020. Specifically, GDP growth was consistently between 5% and 8% for 15years prior to the pandemic (8.15% in 2019), and even during the pandemic, the country was dubbed one of the most resilient economies by Bloomberg, posting 4% growth in 2020 and 4.5% growth in 2021. As such, Bangladesh is projected to be one of the world's 30 largest economies by 2030.

Given that the future sustainability of the country's growth will be powered by its private sector and innovation, and across emerging markets, we must look to the private sector for solutions even to challenges as fundamental as primary care. Indeed, private providers are often better positioned to integrate low-cost technologies into delivery of basic services which can help to scale primary care more effectively. Entrepreneurship and private sector innovation are key to propelling developing economies to long-term growth and success and are more efficient investments than grants or philanthropy, providing pathways for financial sustainability rather than a reliance on one-off contributions.

Localized demand for quality healthcare is clear; health spending uniformly increases across emerging markets even faster than per capita income. Indeed, as any one of us has more disposable income, we generally invest it in education and healthcare for ourselves and our families. In Bangladesh, this spending is projected to increase fivefold by 2040.

With less competition and less onerous regulatory regimes, emerging markets such as Bangladesh and others often offer greater opportunities for healthcare innovation than more advanced economies. Companies can work to make actual change and figure out best practices in ways that might not be available in Europe or the United States, particularly in the healthcare and technology industries. Less structured markets create opportunities for value creation. For example, in India less than 20% of the healthcare market is dominated by larger players; in Bangladesh, it is less than 10%.

Fig. 2
A laboratory technician analyzes a sample in Praava Health's in-house laboratory. As part of its integrated outpatient care model, Praava operates one of only six international standard laboratories in Bangladesh, a country of 170 million people.

Relevant to the issue of PPPs noted before, there are interesting models for public-private partnerships to provide primary care by private providers. One example comes from Praava Health, a patient-driven company launched in Dhaka, Bangladesh, in 2018, that is disrupting the healthcare standard. Praava's integrated outpatient care model—a one-stop shop for high-quality in-clinic and remote care needs—has rapidly scaled and achieved unit profitability in the capital city of Bangladesh in 10 months, now serving half a million patients after 5 years of operations. Clinics administer free vaccines pursuant to the government's Expanded Programme on Immunization (EPI) in addition to primary and secondary care, in house lab and imaging diagnostics, and pharmacy services. Starting from the early days of the COVID-19 pandemic, Praava Health also partnered with the government of Bangladesh to collect COVID-19 test samples to be analyzed in government laboratories (Fig. 2), and later became one of the first private labs to receive government approval to do COVID testing for the general public.

Health innovation in Africa—Technology as enabler and lever: Amref health innovation case example

The development of health technologies in developing countries is often hindered by lack of infrastructure and budgetary limitations to support the prioritization of local research agendas (Natera et al., 2019). To address this, the African Union Science, Technology and Innovation

Strategy for Africa 2024 was developed to support Member States to drive toward national, regional, and continental advances in innovation-led economies (African Union, 2016). A few African countries are leading the way in driving innovation by committing public resources; however, most countries still spend less than 1% of their gross domestic product (GDP) on research and development (Simpkin et al., 2019). The absence of public funding has made way for the private sector and non-State actors, which include nongovernmental organizations, to spearhead innovations in health. The emerging health technologies are predominantly designed to improve efficiencies and cost-effectiveness along the value chain but are faced with sustainability challenges because they fail to identify and reach the right customer segment (Odero et al., 2016).

Amref Health Innovation (AHI) was set up in 2018 to support the development, testing, and scaling of innovative responses to public health challenges. By deploying digital learning tools, AHI has been able to reach community health workers where they are, with up-to-date health information enabling them to reach households with dependable health messages. For example, the mobile learning solution Leap, an mHealth platform codeveloped with Accenture, M-PESA Foundation, Safaricom, Vodafone (Mezzanine), and the Government of Kenya as a public-private partnership, allows over 150,000 community health workers in 8 countries in Africa to receive the health education material in 5 languages. In addition to the learning platform, the use of digital data collection tools supports the collection of community-level data which enables local government entities to make decisions about the health needs of the population (Fig. 3).

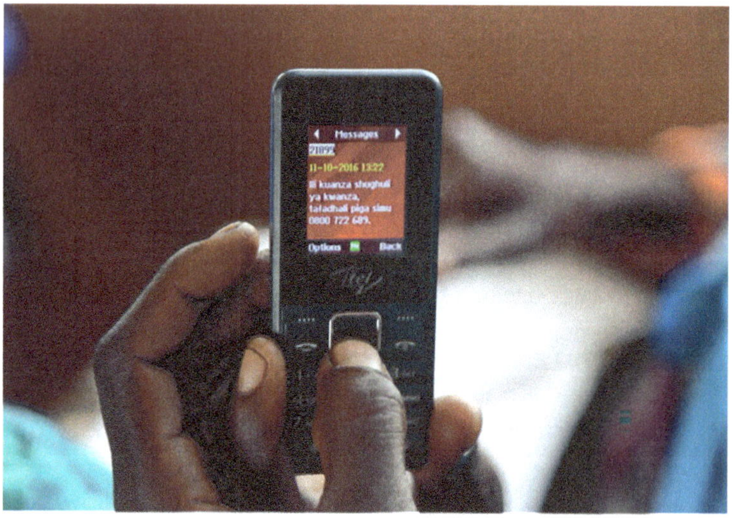

Fig. 3
A community health worker being introduced to a Leap topic with an SMS prompting dial in to the IVR (interactive voice recordings).

AHI deployed a community data collection tool, to support identification of households that required social subsidies to cushion them from the devastating economic effects of the COVID-19 pandemic. Early identification of vulnerable households reduces their exposure and risk due to the various shocks they are likely to experience during economic downturns. This accelerated adoption of digital health tools during the COVID-19 pandemic demonstrates the interest and willingness for governments in developing countries to adopt technologies which support health delivery. Universally accessible technology tools have been developed and scaled successfully in the finance sector. In health, however, this type of success is yet to be replicated, perhaps due to the more complex nature of the user needs from community level to central government. Beyond digital health tools, AHI also explores health financing mechanisms, to identify innovative approaches which can drive the mobilization of resources for primary healthcare, which is the backbone of universal health coverage. In most African countries, healthcare is paid out-of-pocket (OOP), making it difficult to plan adequately for service delivery since the demand for care is influenced by the ability to pay. By working with country governments to increase enrollment of individuals to the national health schemes, OOP costs for households are lowered, while facilities benefit from the ability to generate revenue through the capitation model (where the health insurer pays the facility for a patient's care before they access the service) and fee-for-service models.

AHI continues to explore models which can increase uptake of voluntary national health insurance schemes, while supporting the supply side of the health sector in order to increase the quality of care to ensure the best possible outcomes for all patients. In addition, Amref Health Africa is deploying digital tools to help achieve universal health coverage (Wanjau et al., 2024).

Lessons learnt and conclusions

There are four key lessons learnt from the private sector and the members of the UHC2030 PSC:

1. Solutions that meet the demands of the market and end users are key to ensuring that they are relevant and deliver intended benefits.
2. Platforms encourage and support collaboration, visibility, and accountability.
3. Financing and resources, including talent pools to drive implementation, are essential for scaling and sustainability of solutions.
4. Enabling environments such as legislation that builds trust and encourages participation of multistakeholders are an important contribution from the public sector.

The experiences and case examples illustrated in this chapter highlight the diversity of approaches that can be effectively employed in order to engage with, and create space for the private sector to step up leadership on contributions toward achieving universal health coverage.

The following four conclusions may be drawn:

1. The platform approach, as exemplified by *Health In Your Hands* and the *Health Data Collaborative*, can be a powerful methodology to bring diverse stakeholders around a common agenda for action and can contribute toward better collaboration across sectors through increased transparency and accountability of all parties engaged in the platform.
2. The role of the private sector in financing health innovations and ecosystem building in mixed health systems (of public, private, and the "people" sectors) is critical both as an enabler and also as a delivery partner, as illustrated by how Merck for Mothers and the SDG Finance Advisory Board, presented in boxes earlier in this chapter, are working toward strengthening health and financing ecosystems.
3. The private sector can act as a legitimate and critical development partner, aggregating, supporting, and providing solutions, such as health technologies, to deliver universal health coverage, as can be seen in the example from Amref related to delivering primary healthcare on the African continent.
4. Sustainable and scalable business models in primary healthcare delivery already exist, as demonstrated by the example from Praava Health, and dissemination of such models is important to encourage and inspire others around the world.

In summary, the private sector is an important participant in building strong and resilient health systems. Achieving universal health coverage, and strengthening health systems and health services delivery for the most vulnerable populations in societies, requires new models for collaboration and scaling up existing ones. Lessons learnt from the experiences and examples mentioned in this chapter are beneficial for the health field and also broadly across other domains where the public sector can collaborate effectively with the private sector.

References

African Union. (2016). *Science, technology and innovation strategy for Africa 2024*. Addis Ababa: African Union Commission. Retrieved from: http://www.hsrc.ac.za/uploads/pageContent/5481/Science,%20Technology%20and%20Innovation%20Strategy%20for%20Africa%20-%20Document.pdf.

Alhuwail, D., & Alhouti, A. (2024). Citizen science: Technology tools to empower consumers for more resilient and sustainable healthcare systems during COVID-19 and beyond. In J. Kuriansky, & P. Kakkattil (Eds.), *Resilient health: Leveraging technology and social innovations to transform healthcare for COVID-19 recovery and beyond* (pp. 593–599). Elsevier.

Bhavsar, M., & Katara, K. (2024). Realizing the transformative potential of Community Health Workers with digital innovation | A living goods case study. In J. Kuriansky, & P. Kakkattil (Eds.), *Resilient health: Leveraging technology and social innovations to transform healthcare for COVID-19 recovery and beyond* (pp. 1011–1018). Elsevier.

Javetski, G. (2024). Innovation: Exploring Dimagi's Digital Platform CommCare: A widely adopted digital solution used by frontline workers worldwide. In J. Kuriansky, & P. Kakkattil (Eds.), *Resilient health: Leveraging technology and social innovations to transform healthcare for COVID-19 recovery and beyond* (pp. 1073–1075). Elsevier.

Kuriansky, J. (2024). "Health in Your Hands": A platform for innovative healthcare solutions. In J. Kuriansky, & P. Kakkattil (Eds.), *Resilient health: Leveraging technology and social innovations to transform healthcare for COVID-19 recovery and beyond* (pp. 989–1004). Elsevier.

McKinsey Global Institute. (2020). *Prioritizing health: A prescription for prosperity*. https://www.mckinsey.com/~/media/mckinsey/industries/public%20and%20social%20sector/our%20insights/prioritizing%20health%20a%20prescription%20for%20prosperity/mgi_prioritizing%20health_report_july%202020.pdf. see chart on page 20.

Natera, J. M., Tomassini, C., & Vera-Cruz, A. O. (2019). Policy analysis and knowledge application for building a healthy health innovation system in developing countries. *Innovation and Development*, *9*(2), 159–168. https://doi.org/10.1080/2157930X.2019.1570627.

Nurhussain, M. (2024). Building a state-of-the-art Healthcare-Medical City Center: Case study in Ethiopia. In J. Kuriansky, & P. Kakkattil (Eds.), *Resilient health: Leveraging technology and social innovations to transform healthcare for COVID-19 recovery and beyond* (pp. 1131–1136). Elsevier.

Odero, P., Sable, S., Cook, J., & Udayakumar, K. (2016). *Healthcare innovation in East Africa, navigating the ecosystem*. Retrieved from: https://www.innovationsinhealthcare.org/East%20Africa%20Landscape%20Report%20-%20SEAD.pdf.

OECD. (2020). *Global outlook on financing for sustainable development 2021: A new way to invest for people and planet*. Paris: OECD Publishing. https://doi.org/10.1787/e3c30a9a-en.

Pain, A., Crouch, J., & Tillman, K. (2024). VillageReach: The COVID 411 initiative. In J. Kuriansky, & P. Kakkattil (Eds.), *Resilient health: Leveraging technology and social innovations to transform healthcare for COVID-19 recovery and beyond* (pp. 1005–1010). Elsevier.

Santra, S., & Mandal, T. S. (2024). Innovation in health delivery using technology: Addressing healthcare disparity in India. In J. Kuriansky, & P. Kakkattil (Eds.), *Resilient health: Leveraging technology and social innovations to transform healthcare for COVID-19 recovery and beyond* (pp. 1019–1027). Elsevier.

Simpkin, V., Namubiru-Mwaura, E., Clarke, L., & Mossialos, E. (2019). Investing in health R&D: Where we are, what limits us, and how to make progress in Africa. *BMJ Global Health*, *4*(2), e001047. https://doi.org/10.1136/bmjgh-2018-001047.

UNCTAD. (2014). *World Investment Report*. United Nations Publication. ISBN 978-92-1-112873-4 eISBN 978-92-1-056696-4.

van Engen, J. (2024). Healthy Entrepreneurs: Transforming access to healthcare through an innovative model of community health entrepreneurship. In J. Kuriansky, & P. Kakkattil (Eds.), *Resilient health: Leveraging technology and social innovations to transform healthcare for COVID-19 recovery and beyond* (pp. 1077–1083). Elsevier.

Wai, S., & Kuriansky, J. (2024). Health innovation and advances in the Republic of Sierra Leone. In J. Kuriansky, & P. Kakkattil (Eds.), *Resilient health: Leveraging technology and social innovations to transform healthcare for COVID-19 recovery and beyond* (pp. 445–462). Elsevier.

Wanjau, W., Rindra-Ralidera, O., & Walumbe, R. (2024). Digital health as an enabler of Universal Health Coverage in the African Region: Contributions of Amref Health Africa. In J. Kuriansky, & P. Kakkattil (Eds.), *Resilient health: Leveraging technology and social innovations to transform healthcare for COVID-19 recovery and beyond* (pp. 45–51). Elsevier.

World Health Organization. (2016). *Framework of engagement with non-state actors*. https://apps.who.int/gb/ebwha/pdf_files/wha69/a69_r10-en.pdf.

Further reading

Ansell, L. (2024). Redesigning healthcare to reach the 52% without access: reach52, a health-tech social enterprise. In J. Kuriansky, & P. Kakkattil (Eds.), *Resilient health: Leveraging technology and social innovations to transform healthcare for COVID-19 recovery and beyond* (pp. 1051–1055). Elsevier.

Clarke, D., et al. (2019). The private sector and universal health coverage. *Bulletin of the World Health Organization*, *97*, 434–435.

O'Hanlon, B., & Raman, V. (2020). *Options how to engage the private health sector to tackle COVID-19. Geneva.*
Sturchio, J. L., Kickbusch, I., Galambos, L., & Franz, C. (Eds.). (2018). *The road to universal health coverage: Innovation, equity and the new health economy* Johns Hopkins University Press. ISBN 9781421429654.
World Health Organization. (2020a). *Strategy report: Engaging the private health service delivery sector through governance in mixed health systems.* Geneva: World Health Organization. Licence: CC BY-NC-SA 3.0 IGO.
World Health Organization. (2020b). *Private sector landscape in mixed health systems.* Geneva: World Health Organization. Licence: CC BY-NC-SA 3.0 IGO.
World Health Organization. (2022a). *Equitable access to COVID-19 tools: Aligning the private sector with national response efforts.* Geneva: World Health Organization. Licence: CC BY-NC-SA 3.0 IGO.
World Health Organization. (2022b). *Towards better engagement of the private sector in health service delivery: A review of approaches to private sector engagement in Africa.* Geneva: World Health Organization. Licence: CC BY-NC-SA 3.0 IGO.

CHAPTER 45

Emerging role of NGOs in social innovations: A paradigm shift

Madhu Deshmukh[a], William C. Philbrick[b], and Patricia N. Mechael[c]

[a]Program Strategies & Impact CARE, Atlanta, GA, United States, [b]Sitara International, Atlanta, GA, United States, [c]HealthEnabled, Washington, DC, United States

Introduction: The rapidly evolving and changing context globally

The COVID-19 pandemic has dramatically impacted the lives of all people throughout the world. The pandemic has exacerbated and further deepened the already existing inequities particularly in the most marginalized communities (Bong et al., 2020). These inequities have sustained and increased challenges of accessing quality and timely healthcare in the context of COVID-19 pandemic especially given how overburdened, weak, and fragile existing health systems and healthcare delivery systems are globally. The consequences of COVID-19 transcend the health sector, impacting economic livelihoods, food security, education, child protection, and all sectors that constitute the global social and economic development ecosystem. It is impossible to isolate the effects to just health, as each sector of the global ecosystem impacts the others.

For example, the global response to preventing the spread of the pandemic has resulted in sweeping school closures, which, in turn, cut off millions of children's and adolescents' only meal during the day (Bong et al., 2020). Without access to food, cases of acute malnutrition and other detrimental health outcomes exponentially increased. Illness and lockdowns have also led to loss of livelihoods, contributing to the vicious cycle of poverty, curtailing access to education especially in settings which were already unable to provide and/or access remote learning (Zar et al., 2020).

Lockdowns, social distancing, quarantine, and isolating exposed or COVID-19 positive diagnosed population have been the prevention and control responses adopted globally. These responses, while much needed to prevent and address the pandemic, have resulted in further marginalization of populations, exacerbating the existing glaring gap between healthcare services and the communities (Okereke et al., 2021).

Resilient Health. https://doi.org/10.1016/B978-0-443-18529-8.00045-7

Simultaneously, and in addition to the impact of the COVID-19 pandemic, the world has witnessed several shifting contexts, including rise in long and complex conflicts, natural disasters, and widespread involuntary migration (National Intelligence Council, 2021). Given the rising racial injustices in the world, the global community has understandably prioritized and put its commitment behind decolonization of aid, localization, community-led and community-driven solutions. These shifts are challenging the traditional approaches, and how business is conducted in the development sector. Stakeholders must increasingly be more inclusive and consider their evolving roles by transcending traditional direct implementation of health interventions, enabling communities to design and lead implementation, applying a systems lens, and supporting strengthening health systems.

These shifts have also catalyzed rethinking of healthcare service delivery, especially to poor and marginalized communities, the often left-out and left-behind populations, in favor of a greater focus on equity. Traditional actors are pivoting and adapting, innovating and finding solutions that build resilience to these external and internal changes. This is particularly necessary especially in the face of the new realities the pandemic brings, including restricted movement, lockdowns, increasing spread of misinformation and disinformation.

Concurrently, and over the last couple of decades, technological and social innovations are playing an increasingly omnipresent role in the delivery of health services and information, resulting in improved health outcomes—especially improved access, quality, and reach, building trust between service providers and end users, strengthening resilience, thereby increasing the effectiveness of the health ecosystem.

The functions of technology, particularly digital technology, are leading to the overall increase in the ability to reach members of the community who may have previously been unable to access quality healthcare even in the most remote places. This improved situation is attributable to: more effective data collection (and therefore better informed decision-making and policy formulation), better follow-up and monitoring of cases, the dissemination of essential health information to both patients and healthcare workers, the delivery of training to frontline health workers, and improving supply chain logistics (Labrique et al., 2013). It is therefore not surprising that digital technologies are being universally and strategically leveraged in the global response to the COVID-19 pandemic along with provisions of other health services. These digital technologies and tools include billions of mobile phones, large online datasets, connected devices, and "relatively low-cost computing resources and advances in machine learning and natural language processing" (Budd et al., 2020; Uohara et al., 2020). Digital technologies are being used for:

- Epidemiological surveillance
- Rapid case identification
- Interruption of community transmission (of disease)
- Public communication
- Clinical care (Budd et al., 2020).

While ultimately the responsibility falls upon the governments for providing and ensuring access to essential quality health services, nongovernment organizations (NGOs) and other civil society stakeholders play a vital and indispensable role in the health system ecosystem. Although different definitions have been ascribed to the term, "civil society," the general definitions that are used include the "(1) the aggregate of NGOs and institutions that manifest interests and will of citizens, or (2) individuals and organizations in a society which are independent of the government" (Wikipedia, 2022; World Economic Forum, 2018. According to the World Bank: "Civil society ... refers to a wide array of organizations: community-based groups, NGOs, labour unions, indigenous groups, charitable organizations, faith-based organizations, professional associations, and foundations" (Ibrahim, 1998). For purposes of this chapter, our discussion focuses on the NGO component of the civil society.

Why NGOs?

Nongovernmental Organizations (NGOs), defined as most stakeholders comprising civil society, possess unique traits and abilities that position them to play an essential role in introducing, leveraging, and implementing technological and social innovations that contribute to the evolving health ecosystem while addressing COVID-19. This is particularly true in low- and middle-income countries (LMIC) because of the unparalleled reach of NGOs in providing critical health services (Sanadgol et al., 2021).

The authors of the chapter have drawn from the learnings from the interventions of CARE (Cooperative for Assistance and Relief Everywhere), an international NGO, to offer insights, while also acknowledging the expertise generated by other NGOs. These illustrative examples highlight the role of an NGO in facilitating social and digital innovations and set them apart from most other stakeholders in the global health ecosystem.

Through a semisystematic review, the authors identify the unique roles that NGOs are best positioned to play for designing, implementing, and scaling innovations in the global health ecosystem working in tandem with other local stakeholders.

NGOs leverage innovations to facilitate trust building and social accountability for quality services and the harnessing of locally led solutions

The key to being an effective stakeholder in the health ecosystem is the ability to garner trust within the communities and to support access to services and information as trusted partners. Working with communities as trusted partners means also supporting community members in holding service providers accountable—government and private—which, in turn, ensures quality and availability of essential health services.

Garnering trust is partly dependent on long-term presence and familiarity within communities. It is also, however, dependent on a history of listening to communities and being responsive to their needs and perceptions of health services, including how they gauge the quality of those services that exist, and what they need to generate solutions designed and led locally by the communities themselves. Engaging the community members in dialogue with healthcare service providers can be empowering and enabling to achieve desired health outcomes.

COVID-19 pandemic restrictions meant, as an example, that community members and service providers had to adapt and innovate to ensure continuing the dialogue to ensure critical sexual and reproductive health (SRH) services and COVID-19-related services reach the communities.

A case study example is as follows:

> CARE developed the Community Score card approach and tool in Malawi in 2002 to ensure high quality and context-specific services by building awareness, understanding, trust and communication between community members and health service providers. Appropriate tools and support are provided to the community members as citizens to help them understand their right to quality services. Through this approach, groups of communities as citizens meet and dialogue with the service providers and public officials to negotiate for expanded and inclusive spaces, and promote mutual accountability to trust building and dialogue. (CARE, 2020).
>
> A clustered randomized controlled evaluation to assess Community score card (CSC's) impact on reproductive health outcomes in Ntcheu, Malawi, showed: (a) home visits by community health workers during pregnancy went up by 20% and after pregnancy by 6%, compared to control and community women's satisfaction with the reproductive health services (b) a 57% greater use of modern contraception than in the control site; (c) increased level of male involvement and youth involvement in reproductive health issues; (d) increased availability of information, and (e) improved relationship between providers and communities. (Gullo et al., 2017).
>
> During the COVID 19 pandemic, the community members, including women, youth and community leaders, as well as service providers, were able to share real time feedback on SRHR and COVID-19 related services via SMS and WhatsApp due to lockdowns and physical distancing protocols in Malawi. Call-in radio dialogues between community members, frontline service providers and district health officials in Ntcheu were aired across Malawi on family planning and critical SRHR service provision to women. Issues of stock outs of PPE and FP products were highlighted through this dialogue as well as the fears of health workers to contract the infection if working without PPE and other equipment. The Government of Malawi noted the Insights and findings from these dialogues and feedback from the communities and frontline workers and thus increased the domestic budget for family planning in Malawi to sustain access. (CARE, n.d.-a, n.d.-b).

GAVI, the Vaccine Alliance, and UNICEF are currently studying another related and emerging concept of "social listening." Social listening is an activity that has emerged in the context of COVID-19 vaccine promotion, which simply means that stakeholders are systematically monitoring and paying attention to discussions of different population groups in different media forums to gain a sense of the relevant sentiments within those groups (HealthEnabled, 2021). The insights drawn from this social listening can be used to inform communication strategies, policies, service delivery, and quality improvement activities with the intention to inform public opinion and improve community demand for immunization services. Digital tools and computerized approaches can help "find the signal through the noise" by automating the collection of large amounts of data that can be gleaned so to truly understand the local context and history, and use the information for improved immunization programming (HealthEnabled, 2021).

An illustrative example:

> The Digital Literacy Activists Network (Japelidi), a trusted NGO in Indonesia, has engaged in monitoring and public discourse, using digital technology, to understand public perception and create digital literacy and to counter vaccine-related misinformation. (HealthEnabled, 2021) Early learnings seem to indicate that systematic use of digital social listening allows health interventions to be more responsive to communities' needs, and lead to increase in trust and confidence and increased vaccination uptake.

NGOs leverage digital technology to reach the unreached and the most marginalized

Because NGOS can create deep relationships in working with the communities, therefore developing a strong connection to the local community groups and local organizations, these groups are able to reach communities and vulnerable populations often inaccessible by traditional government and other private sector service providers. Groups highly marginalized or stigmatized by society, such as the HIV-positive population, women in sex work, or men who have sex with men, are often invisible and living in the margins of society. Services often do not reach them, either because they are invisible or because they are not able to use services anonymously. Using innovative methods including digital technology can make much needed health services available to these marginalized populations.

> In Cameroon, CARE implements a PEPFAR-supported HIV prevention, treatment, and care program known as "CHAMP" (Continuum of Prevention, Care and Treatment of HIV/AIDS for Most-at-risk Populations) for key populations and other hard-to-reach groups vulnerable to HIV, specifically, men-who have sex with men (MSM), commercial and transactional sex workers, and adolescent girls. A key tool in identifying and reaching these hard-to-reach populations includes using digital technology—i.e. websites and mobile phones with essential information regarding testing sites, reminders about treatment adherence, and with the advent of COVID, essential information about testing and prevention measures. The digital work in CHAMP—related to both HIV and COVID-19—has been bi-directional, namely, direct messaging, live chat, and one-to-one callback features as well as one-way push of

information. With the assistance of digital technology, in the 2021 fiscal year (October 2020–September 2021) CHAMP has been able to reach approximately 86, 263 key population members, including MSM, a group that is generally unreachable because of their status as being illegal under Cameroonian law.

(CARE International in Cameroon, 2021)

NGOs use digital solutions for disseminating accurate health information and dispelling myths

Being able to communicate with the community with accurate and complete information on health issues (maternal health, HIV, COVID-19) is essential. This includes about availability of services, and addressing any myths or doubts and how to stay healthy, particularly in situations in which inaccurate information and pure myths are widespread.

Digital health, particularly SMS messaging, is playing an increasingly important role, particularly with NGOs which work with communities, in making sure all community members have accurate knowledge about health options, adherence to medications, and the safety of different interventions (e.g., vaccinations).

Case study example:

> The Elizabeth Glaser Pediatric AIDS Foundation, a global NGO operating in Tanzania, has been measuring and applying the use of digital technology as part of its prevention-of-mother-to-child transmission (PMTCT) programs, and in doing so, generating evidence of the contributions of digital technology to the healthcare system. As part of its SAFI (Supporting Attendance for Facility Delivery and Infant Health) study with HIV-positive pregnant women, the correlation between mobile phone access for pregnant women and PMTCT adherence was measured. The study's conclusions, presented at the 2016 International AIDS Conference, included that: increased client-provider interaction (e.g., through mobile phone) improves PMTCT adherence through promoting ART-PMTCT knowledge and access to facility delivery; and that, accordingly facilitating provider-client communication (partly for the transmission of information) may enhance the effectiveness of PMTCT models.

NGOs strengthen health systems: Enabling, empowering frontline health workers to deliver and track quality healthcare services using digital health tools

NGOs, as one of the critical stakeholders of a health ecosystem, have been instrumental in suggesting innovative approaches and tools to empower and enable service providers to improve the quality, reach, timeliness, and efficiency of the healthcare services and achieve the desired health outcomes.

Frontline health workers play a critical role to connect communities with the healthcare services, both public and private service providers. However, often they do not have the required knowledge, job skill, or tools to enhance their effectiveness and efficiency to deliver quality health services. The effective use of digital health support tools for frontline health workers is often codesigned and implemented by NGOs in a way that aligns with broader health systems strengthening efforts.

Illustrative examples of frontline health workers using digital health tools:

> In Bihar, one of the poor states in India, the frontline health workers (FLWs) *CARE's Bihar Technical Support Program (BTSP)* worked with cadre of FLWs, who provide critical health and nutrition-related counseling and services to poor and marginalized communities. They were burdened with multiple responsibilities without adequate training or support, and with cumbersome, time consuming tracking and reporting tools, and no access to real time data for critical and timely service provision. Recognizing the role of information, Communication and Technology (ICT), CARE's BTSP with partners developed a transformative and innovative mobile phone and related technology-based solution called *Continuum of Care Services (CCS) application* for FLWs. An end-to-end solution that included a range of modules or suite of applications on the mobile phones were available to the FLWs to support timely identification, registration, and provision of step-by-step services for pregnant mothers and newborns. The CCS application was tested and its scale and success evaluated in an randomized control trial in 2015 and a quasi-experimental in 2016 (Balakrishnan et al., 2016; Borkum et al., 2015). The evaluations noted significant improvements in quality and reporting of service delivery, time efficiency of data collection and equity in delivery of services by FLWs using CCS application including higher confidence levels in FLW's ability to perform their job. 49.8% of pregnant women undertook 3 antenatal care visits visited by FLWs with CCS vs. 28.8% in the control site. The Government of Bihar and Government of India were presented with this transformative innovative application and results. CCS has been adapted and scaled by the Government of India since with approximately 480 million population that have been registered and 57 million mothers and children tracked by their names. (Akam et al., 2020)
>
> Living Goods is an NGO which focuses on leveraging digital technology to enable government partners to enhance the capacity of community health workers (CHWs) and improve health outcomes. Living Goods has equipped CHWs with mobile phone technology to report and track patient data, in places like Uganda, contributing to the reduction of reporting times from 4–6 weeks to near real time. This, in turn, has led to better coordination of care, more follow-up on referrals, reminders for treatment adherence and improving the quality of supportive supervision (Kioko & Peter, 2021). Johnson & Johnson, supporting the Living Good projects, reported that Living Goods supported 11,130 CHWs across Kenya and Uganda, and saved an estimated 17,000–19,000 lives in 2020, during the height of the COVID-19 pandemic, and double that in 2019 (pre-the COVID spread) during a period that witnessed a drop in facility-based care.
>
> ***(Kioko & Peter, 2021)***

NGOs support the effective use of digital supply chain platforms to improve timely access to health commodities

A major dilemma plaguing health programs in many parts of the world, but particularly in low- and middle-income countries (LMIC), is the issue of supply logistics. Issues of stock outs of essential medicines, medical supplies (such as gloves), and even paper forms are inherent at many health centers, particularly in rural areas. Over recent years, digital applications that enable tracking of supplies, and even automatic reordering when supplies fall below a certain level, are being used to solve this problem, and in turn, improve the efficiency of health systems.

Digital health tools for supply logistics:

> John Snow International used the supply chain logistics cStock digital system, scaled nationally to address the needs of CHWs in Malawi. cStock allowed the CHWs to use personal phones to submit SMS with stock information, a means for eliciting community level data that is generally hard to capture (Dimagi website, n.d.). That data is then visible to decision makers. A study determined that the use of cStock in Malawi 'contributed to significant reductions in stockout rates and lead times for resupply, as well as improvements in overall stock supply management. (Shieshia et al., 2014)
>
> Catholic Relief Services (CRS) has begun using ***CommCare,*** a well-known software platform (developed by Dimagi) to strengthen institutional supply chains in Ethiopia. CRS began using CommCare to track procurements and dispatches of essential medical supplies and cleaning materials at Catholic health facilities and for partners to be used at distributions.
>
> Although impact data from this digital supply chain intervention is not yet available, it is expected that using a digital application will contribute to more efficient tracking of essential medical supplies, more stable inventories, and minimize the stock-outs of essential medicines and other medical supplies.

Lessons learned

The illustrative examples documented in this chapter, along with the authors' experiences working for and with NGOs, have informed lessons learned about the roles and functions which NGOs can play with designing, implementing, and promoting innovations in places where community connections are vital. NGOs are best positioned because of their ongoing presence in communities that builds trust with community members and allows them to introduce digital innovations that will cumulatively have a positive health impact and contribute to the evolution of a healthy and robust health system.

A summary of the unique role of NGOs, based on global learnings, includes the ability of NGOs to:

- Engage directly with the community and build trust with ecosystem actors
- Serve as the "connector" and liaise with all stakeholders, including governments

- Create partnerships with other potentially influential health and technology ecosystems stakeholders, including technology companies, faith- and community-based organizations (FBOs and CBOs), and academic institutions
- Make referrals to different types of essential service providers in different technical sectors
- Facilitate institutional strengthening and individual capacity building
- Conduct advocacy and awareness-building activities at various levels
- Collect granular data for evidence generation and decision-making
- Provide essential nonhealth life-sustaining services impacted by COVID-19

As NGOs work with communities and governments in their role as the largest service provider of health services to design and scale innovations and transformative solutions, to increase the quality and reach of critical health services, it is crucial to keep in mind the local context within which the solutions are implemented. Context is key to adapt or scale proven innovations.

Adaptation and scaling should be embedded within the local context, factoring in digital capabilities, local leadership capabilities, and building ownership.

While NGOs are a critical link to reach hard-to-reach populations, and to empower and enable health systems including frontline health workers for the transformative solutions to scale and sustain, it is critical to focus on the enduser—the individuals who will use the solutions—in order to ensure their ownership of the solution, their confidence and ability in the solution to address the challenges within the health systems, and their capabilities to implement or use the solution. Capacitybuilding of the local stakeholders, including local government officials, communities, and local organizations, as the endusers is critical.

Building capacities of the implementing organization, including NGOs themselves, to design and scale effective digital and transformative solutions, adopting the technology alone is not sufficient. To truly transform, it is crucial to bring the technology together with people, processes, and culture within the organization, for example, to leverage data for decision-making through knowing, respecting, and integrating a culture of continual learning and adaptive management.

References

Akam, A., Augustin, C., & Koepsell, J. (2020). *Frontline health workers coalition. Digital Health 2.0: getting tools and services to people who need them most (Blog post, June 18, 2020).* https://www.frontlinehealthworkers.org/blog/digital-health-20-getting-tools-and-services-people-who-need-them-most. (Accessed 1 February 2020).

Balakrishnan, R., Gopichandran, V., Chaturvedi, S., et al. (2016). Continuum of care services for maternal and child health using mobile technology—a health system strengthening strategy in low and middle income countries. *BMC Medical Informatics and Decision Making*, *16*, 84. https://doi.org/10.1186/s12911-016-0326-z.

CHAPTER 46

Unlocking the promise of digital health: A civil society perspective

Carthi Mannikarottu

Civil Society Engagement Mechanism for UHC2030 (CSEM), co-hosted by WACI Health, Nairobi, Kenya; Platform for Access to COVID-19 Tools (ACT)-Accelerator Civil Society and Community Representatives, co-led by WACI Health, Global Fund Advocates Network (GFAN) and STOPAIDS

Introduction

In 2012 the United Nations General Assembly endorsed a resolution on Universal Health Coverage (UHC), urging countries to accelerate the transition to universal access to quality and affordable health services for all. The Political Declaration on UHC signed by Member States in September 2019 further reaffirmed global commitments to achieving UHC by 2030. Yet today, around half the world's population lack access to the health services they need. With less than 8 years left to the finish line, there is much more work to be done and critically, there are best practices to assess and accelerate.

The promise of digital health is one such area deserving of heightened attention and prioritization from all stakeholders. Digital technologies and tools have the potential to accelerate progress toward achieving health for all. However, equity is not a given. These policies and programs must be planned, structured, implemented, and evaluated with an equitable, people-centered framework.

The digital divide—the gap in access to technology due to socioeconomic inequalities—is a threat to the promise of digital health for UHC. Even in the context of the COVID-19 pandemic, when the use of telemedicine and health-related phone applications exponentially increased in response to lockdowns and quarantine measures, we have seen that access to digital health was uneven among all the world's populations. Civil society consultations led by the Civil Society Engagement Mechanism for UHC2030 (CSEM) in the United States and Colombia noted that while COVID-19 catalyzed increased telehealth services, many individuals were not able to access these due to gaps in internet access and digital literacy (Civil Society Engagement Mechanism for UHC2030, 2021).

In fact, over 3.6 billion people globally are not connected to the internet and therefore, unable to access digital health tools (World Economic Forum, 2020). This divide is unacceptable, especially when the solution can involve a political choice. Decision-makers can, and must, act

Resilient Health. https://doi.org/10.1016/B978-0-443-18529-8.00046-9

to unlock the potential of digital health for everyone. Rigorous and purposeful approaches are urgently needed to achieve equity along with innovations in technology.

Civil society and communities play a critical role in addressing the equity gap. To unlock the promise of digital health *for all*, civil society is and must continue to be (1) decision-makers in the digital health ecosystem, (2) staunch advocates for the most vulnerable populations, and (3) coowners of community-based solutions.

The CSEM is the civil society constituent of UHC2030, a global multistakeholder movement to build stronger health systems for UHC. CSEM specifically focuses attention on the needs of the most vulnerable and marginalized populations so that, as called for in the United Nations Agenda 2030 for Sustainable Development, "no one is left behind." To do this, the CSEM serves as a focal point to coordinate and strengthen the voice and actions of civil society and community-led organizations to achieve inclusive and equitable UHC policy and programs.

Civil society refers to a diverse constituency of formal or informal networks, as well as nongovernmental and not-for-profit groups or organizations operating in, and on behalf of, the public, representing their members' interests and values. The interests can be wide, including cultural, political, scientific, religious, ethical, or philanthropic perspectives.

Strong advocacy on equity for strong health systems for all

Just as UHC is a political choice, so is addressing the digital divide and achieving equitable access to new technologies. There is a need for increased political will and investments in digitally enabled primary healthcare (PHC) that can truly accelerate the achievement of health for all.

UHC2030 notes that a "PHC-oriented health system" maximizes equity and solidarity, and is the best approach to achieving UHC. Research shows that strong PHC safeguards marginalized communities and brings quality healthcare services to communities (UHC2030, 2022). Investments in PHC are essential for resilient health systems, and good PHC in turn is supported by innovative models of healthcare delivery such as through digital health tools (World Health Organization & United Nations Children's Fund (UNICEF), 2020).

Governments—with the support of all partners—have an opportunity to close the health equity gap by accelerating and expanding equitable access to digitally enabled primary healthcare. For their side, civil society and communities must highlight the potential of digital health to strengthen health systems, especially delivery systems for PHC, improve quality of care, and reduce costs. They must continue to draw attention to access challenges and the dangers of the digital divide on the road to UHC.

Strong advocacy has the potential to change political prioritization and to bring health equity—including digital health equity—to the forefront. This is evident when you consider how quickly data on COVID-19 incidence, testing rates, and mortality—at the subregional levels—have been made available to the global community since the virus was first identified in December 2019, and coincidentally consider the rise in the development and use of digital tools related to health worldwide. Despite the gaps that still remain on quality of data and regional variation, it is worth noting the considerable progress made especially compared to data availability for the epidemics of HIV, TB, and malaria. The decades of work to tackle these three epidemics became the foundations of the COVID-19 response; and yet, the political momentum and speed of actions seen for the COVID-19 response have not caught up for these epidemics. This illustrates how, even when the technology is available, its adoption and implementation require heightened attention from all stakeholders and collaboration across sectors. COVID-19 received some of this needed attention, but as we maintain, so much more is yet to be done.

All of us have a role to play to bring up the levels of public attention and political will to achieve health for all, enabled by digital technologies for health.

Inclusive governance of digital health systems

Civil society organizations as well as formal and informal networks of communities reflect the lived experiences of the end users of the health system. As such, they are best placed to understand and share the needs of the population—or the problems digital health solutions seek to address.

Institutionalizing participatory health governance mechanisms is key to creating resilient health systems that respond to community needs. Strong structures for participation at every stage of the policy and program development process build systems that are trusted, responsive, and accountable to the people. They put people at the center of health systems, helping to maximize the benefits of innovative technology. This "people-centered" approach is consistent with both UHC and with the UN Agenda.

Digital health plans need a whole-of-society approach to be people centered and equitable. The challenges of health data transparency and privacy are discussed in other chapters in this book. It is clear that in order to ensure that all people can benefit from the innovations in health data collection and use, there must be an inclusive governance mechanism that maximizes the public good while ensuring adequate protections. Mechanisms to share data, for example, must be built on trust, protect the right to privacy, and secure against inappropriate use (Transform Health, 2021). Here, we make the case that this can only be achieved through the active and meaningful involvement of civil society and communities in developing the framework for health governance at the national and sublocal levels.

The "Health Data Governance Principles" launched by Transform Health, a coalition of organizations that have been working together since 2019 to advocate for the equitable digital transformation of health systems, illustrate how inclusive governance can become a reality for digital health (Transform Health, 2022). Governments, international organizations, civil society, research institutions, and private sector actors were convened by Transform Health to create a ground-up framework for health data governance with the goal of protecting people and ensuring equity. It is imperative that frameworks like this continue to be designed and used in ways that will allow everyone to enjoy the benefits of digitally enabled care and realize their right to health.

Invest in community-based health service delivery

Civil society and communities are an essential partner to ensure health services reach all, especially the most marginalized populations. Community-led systems for health include the actions and strategies that improve the health of communities through the knowledge, leadership, and participation of the communities themselves. This includes delivering prevention and disease control solutions, connecting individuals to treatment and care, and ensuring accountability.

A global survey by the CSEM and UHC2030 Social Participation Technical Network (SPTN), conducted in 2020, clearly points to the need to support community-based health service delivery, especially during a crisis (Civil Society Engagement Mechanism for UHC2030, n.d.). The results showed that more than half of the 200 civil society respondents reported minor or no involvement of civil society and communities in their governments' COVID-19 responses. And yet, civil society organizations and networks were delivering lifesaving medications during lockdowns and supply chain disruptions, ensuring food supplies to those experiencing shortages, offering psychosocial support and housing, developing public information campaigns, and more (https://www.civicus.org/index.php/covid-19, n.d.). This leadership and engagement also has potential to scale up digital health tools. During the COVID-19 response, mobile health applications were adopted in many countries with community involvement for COVID-19 contact tracing, delivery of antenatal and postnatal care, and prescription renewals (WHO and UNICEF, 2020).

There is even more that can be done with adequately integrated and supported community-led responses. Evidence has shown from the HIV and TB programs that this approach works, specifically, that community-based monitoring is an effective approach to improve data quality and help create solutions that reflect the reality on the ground. Community-led monitoring—in which actors in a community work together to observe, record, and generate relevant information on their experiences—captures individuals living with HIV or TB who are often missed by health systems. For example, the International Treatment Preparedness Coalition (ITPC) has refined a model for community-led monitoring called the community treatment observatory. In 2019 ITPC implemented this model in 11 countries in West Africa with support from the Global Fund. The model presented new insights on the lived experiences of people living with HIV in these countries to increase government accountability and inform targeted solutions (International Treatment Preparedness Coalition, n.d.).

Importantly, such community-led monitoring can be enabled by mobile applications and other digital technologies that support data collection, processing, and analysis. Thus these efforts must be scaled up, integrated into national health programs, and supported by all partners.

The COVID-19 pandemic has shone a light on how critical civil society and communities are in health service delivery. This progress should be maintained and continued with sustained investments. For example, the World Bank is funding digital technology initiatives through civil society that focus in particular on reaching communities with health emergency and vaccination information (Srinivasan & Robinson, 2021). This is an important start, but we must make sure that the support continues beyond the pandemic response and includes institutionalized roles for civil society in the road to achieving UHC.

Conclusion

The promise of digital health tools and technologies is a world where everyone has access to the healthcare they need. Technology has vast advantages, to reduce the distance to clinics, ensure availability of prevention and disease management information, support health workers for service delivery, empower individuals to manage their conditions, and provide decision-makers with an accurate understanding of the health and well-being of populations. However, advances in technology must be made in service to inclusion and equity.

Civil society and communities have a key role to play in ensuring that the progress in digital health does not exacerbate inequities in access to health. We need to work with governments, donors, the private sector, and civil society across sectors to harness the potential of digital health in order to achieve health for all.

References

Civil Society Engagement Mechanism for UHC2030. (2021, December). *From Commitments to Action: Civil Society Perspectives on Reaching Universal Health Coverage*. Accessed from: https://csemonline.net/civil-society-perspectives-souhcc-2021/.

Civil Society Engagement Mechanism for UHC2030. *Civil society's COVID-19 calls to action*. Accessed from: https://csemonline.net/project/civil-societys-covid-19-calls-to-action/.

For examples of how civil society contributed to the COVID-19 response, see: CIVICUS, 'Solidarity in the Time of COVID-19.' Accessed from: https://www.civicus.org/index.php/covid-19.

International Treatment Preparedness Coalition. *Data for a difference: Key findings, analysis and advocacy opportunities from the regional community treatment observatory in West Africa*. Available from: https://itpcglobal.org/blog/resource/data-for-a-difference/.

Srinivasan, S., & Robinson, D. (2021, June). *Centering equity in vaccine delivery: How digital technologies can combat vaccine hesitancy and misinformation among women and marginalized communities*. World Bank

Blogs. Accessed from: https://blogs.worldbank.org/digital-development/centering-equity-vaccine-delivery-how-digital-technologies-can-combat-vaccine.

Transform Health. (2021). *The case for digital health: Accelerating progress to achieve UHC*. Basel: Transform Health. Accessed from: https://2ja3zj1n4vsz2sq9zh82y3wi-wpengine.netdna-ssl.com/wp-content/uploads/2020/12/Transform-Health_The-Case-for-Digital-Health-TH-Report.pdf.

Transform Health. (2022). *The Health data governance principles*. Basel: Transform Health. Accessed from: https://transformhealthcoalition.org/maximising-the-value-of-health-data-for-public-good-the-health-data-governance-principles/.

UHC2030. (2022). *Action on health systems, for universal health coverage and health security*. Geneva: WHO. Accessed from: https://www.uhc2030.org/fileadmin/uploads/uhc2030/Documents/Key_Issues/Advocacy/UHC2030_Health_systems_narrative__actions_paper.pdf.

WHO and UNICEF. (2020). *Community-based health care, including outreach and campaigns, in the context of the COVID-19 pandemic*. WHO and UNICEF. Accessed from: https://www.who.int/publications/i/item/WHO-2019-nCoV-Comm_health_care-2020.1.

World Economic Forum. (2020, April). *Coronavirus has exposed the digital divide like never before*. Accessed from: https://www.weforum.org/agenda/2020/04/coronavirus-covid-19-pandemic-digital-divide-internet-data-broadband-mobbile/.

World Health Organization & United Nations Children's Fund (UNICEF). (2020). *Operational framework for primary health care: Transforming vision into action*. Geneva: WHO and UNICEF. Accessed from: https://www.who.int/publications/i/item/9789240017832.

Ending stigma about mental health, empowering the voice of persons with lived experience, and other recommendations: The Lancet Commission Report

Ananya Raghuram[a,b] and Judy Kuriansky[b,c,d,e,f,g,h]

[a]*Department of Clinical and Counseling Psychology, Teachers College, Columbia University, New York, NY, United States,* [b]*International Association of Applied Psychology, Indianapolis, IN, United States,* [c]*Global Council SDG3 (Health), "SDGs in Action" Initiative, Dubai, United Arab Emirates,* [d]*United Nations Group of Friends of Mental Health and Well-being, New York, NY, United States,* [e]*Office of Ambassador Sidique Wai, Embassy of the Republic of Sierra Leone to the United States, Washington, DC, United States,* [f]*Mission of the Republic of Sierra Leone to the United Nations, New York, NY, United States,* [g]*World Council for Psychotherapy, Vienna, Austria,* [h]*Department of Clinical Psychology, Teachers College, Columbia University, New York, NY, United States*

Introduction

"Access to mental health services continues to be inefficient with one of the main barriers to help-seeking and care noted as stigma around the conversation of mental health."

This was the opening statement by Dévora Kestel, Director of the Department of Mental Health and Substance Use of the World Health Organisation (WHO) headquartered in Geneva, in her role as host of a virtual event launching the *Lancet* Commission Report on *Ending Stigma and Discrimination in Mental Health*, which took place on October 10, 2022, on World Mental Health Day.

Ms. Kestel elaborated that stigma and discrimination toward mental health conditions continue to act as key barriers for individuals seeking mental health help, which in turn exacerbates social exclusion toward individuals living with mental health conditions. The event pertained to the United Nations Sustainable Development Goal (SDG) 3 calling for "Good Health and Wellbeing," in addition to Goal 10, to "Reduce Inequalities," with specific focus on ending global stigma and discrimination of mental health and on respecting the voice of persons with lived experience in mental health conditions.

Resilient Health. https://doi.org/10.1016/B978-0-443-18529-8.00047-0

Following the opening statement, an introduction to key personnel on the Commission was made by the event chair, Ms. Dristy Gurung, co-investigator and project lead on RESHAPE Indigo, a project of the Transcultural Psychosocial Organization in Nepal (TPO Nepal), the comprehensive work of which is covered in other chapters in this volume (Gautam & Koirala, 2024; Khanal & Kuriansky, 2024). Funded by the National Institute of Mental Health (NIMH), TPO Nepal focuses on training primary healthcare workers within the sphere of mental health to provide quality mental healthcare informed by research in behavioral health (TPO Nepal, n.d.). Further, TPO Nepal mobilized major impressive clinical and research projects in response to the COVID-19 pandemic, as well as contributed consultation to government health officials.

The Lancet is one of the oldest medical journals, founded in 1823, which publishes research from diverse fields of medicine, psychology, and public health in the Americas, Europe, Southeast Asia, and Western Pacific (The Lancet, n.d.). The family of journals, numbering over 30,000, address urgent topics, new research and ideas, stimulate debate, and influence decision makers worldwide. The editors occasionally convene commissions of academic partners to explore issues in science, medicine, and global health of specialty interest, aimed at providing recommendations that change health policy or improve practice, and result in major reports. Since 2003, *The Lancet* journals have published more than 200 global health Series and Commissions on issues including global mental health, AIDS, dementia, and climate change. A new Lancet Commission on the Future of Health and Economic Resiliency in Africa was launched in 2022 to address the convergence of factors blocking progress in development, including the COVID-19 pandemic, poor health, poverty, unemployment, illiteracy, malnutrition, climate change, violence against women and girls, natural disasters and conflict (Pate et al., 2022).

This *Lancet* Commission Report on *Ending Stigma and Discrimination in Mental Health* is the collective effort of a global team of more than 50 researchers with combined research on stigma and discrimination and barriers to accessing quality mental healthcare, as well as reflections of the lived experiences of individuals through poetry, interviews, and direct quotations (Thornicroft et al., 2022). The report was published on October 9, 2022, in collaboration with the World Health Organisation (WHO). Speakers for the virtual launch event included a diverse team of panelists from varied countries, from Argentina to Sierra Leone.

The event was opened by Richard Horton, editor-in-chief of *The Lancet*, by sharing a video of a poem called "Make a Difference," written and recited by an individual, Anne Hoffman, with lived experience of schizophrenia. She describes telling people in a bus about her experience, seeking understanding, but encountering some unpleasant reactions. She also disparages media focus only on tragedy of mental conditons.

Horton then described the background of this particular commission and the history of *The Lancet*'s prior publications devoted to global mental health going back to 2007, with its first series on global mental health at a time when "mental health was not recognised as a development challenge in any way through the development of the United Nation's Millennium

Development Goals." This comment refers to the fact that mental health was not included in the iteration of the United Nations Agenda that covered the years 2000–2015, called the Millennium Development Goals (Votruba et al., 2016). Significantly, mental health and well-being was subsequently included in the new Agenda for the years 2015–2030,which outlined the Sustainable Development Goals (SDGs), due to national advocacy as well as concerted advocacy on the intergovernmental level at the United Nations by mental health advocates (incuding the second author of this chapter) partnering with UN member states (Kuriansky & Zinsou, 2019).

In 2018, in partnership with the WHO, *The Lancet* launched the Commission of Global Mental Health and Sustainable Development, with four main goals, to:

(1) emphasize that mental health is a global health matter;
(2) understand that mental health exists on a continuum and not as binary;
(3) reiterate that mental health places great importance on social and environmental determinants, synonymous with several SDGs; and
(4) affirm that mental health is a fundamental human right.

In closing his remarks, Mr. Horton urged panelists and researchers to consider the conclusions and recommendations provided in the report as urgent, and that "the violation to mental health should end, and end now, to be replaced with policies with dignity." This "violation" to mental health refers to derogatory terminology about mental health, mistreatment of people with lived experiences (PWLEs), and disregard of mental healthcare access within the public health sphere and public policy.

Next to speak was Ms. Charlene Sunkel, founder and Chief Executive Officer at the Global Mental Health Peer Network, based in Johannesburg, South Africa, the first global organization dedicated to strengthen and empower the voices and experiences of people with lived experience of mental health conditions. She outlined the aims of the Commission, which include to characterize what stigma and discrimination are and their impact on society, to review interventions and large-scale programs, and to emphasize that international media platforms and organizations have a responsibility to issue "policy statements and action plans" in order to reduce stigma with regard to mental health.

Ms. Sunkel highlighted that the Commission Report features lived experiences of people with mental health conditions (PWLE) collated from a global survey of a population across 45 countries (half of which were upper- to middle-income nations), in six different languages. The results highlight five areas of concern:

(1) The need to consider equality between people with lived experiences and individuals with health conditions that are physical, which is imperative to reducing stigma and discrimination in mental health;
(2) The recognition that individuals with mental health conditions are more susceptible to having worsened mental health sequelae from existing stigma and discrimination;

(3) The urgency for investments on a national scale for initiatives to reduce stigma and discrimination in mental health to be made by governments in order to forge better mental health outcomes for PWLEs;
(4) Understanding the critical point that increasing rates of stigma and discrimination contribute more harm to individuals than actually having a mental health condition; and
(5) Responsibility of the media, given its influence and position in society, to recognize its considerable role in contributing to higher levels of stigma and discrimination against mental health conditions, and to change this.

Outlining how language can influence perceptions of mental health in a disparaging perspective, Ms. Sunkel underscored that it is vital for governments, media, and all stakeholders to understand appropriate language and terms that must be used regarding mental health.

These issues, she said, inform two main principles from the Report:

(1) Human rights and access to adequate mental healthcare between individuals with and without mental health conditions must remain equitable. Specifically, "provisions for social inclusion of people with disabilities are to be applied equally for physical and mental health conditions", a position which echoes the mission and vision of the field of global mental health whereby, as it is commonly stated, "There is no health without mental health" (Patel, 2014).
(2) Advocacy, impact, and primary contribution and collaboration of PWLEs in the development of national- or state-level initiatives are imperative to reducing stigma and discrimination (Thornicroft et al., 2022). Thus, nations have a duty to eliminate "laws that unfairly discriminate against individuals with mental health conditions or suicidality."

Following the presentation by Ms. Sunkel, Sir Graham Thornicroft, Professor of Community Psychiatry at the Centre for Global Mental Health and the Centre for Implementation Science at the Institute of Psychiatry, Psychology and Neuroscience at King's College London in the United Kingdom, summarized recommendations included in the Report for governments, media organizations, and other sectors. He highlighted selected large-scale programs and initiatives outlined in the Commission from different regions of the world and from low-, middle-, and high-income countries that were created to tackle stigma surrounding mental health. These include *Opening Minds* from Canada, *SMART Mental Health* from India, *The Carter Center Mental Health Program* in Liberia, *Like Minds, Like Mine* from New Zealand, and *On the Level* from Czechia.

The role of the media in the context of stigma and discrimination of mental health, he said, is a "double-edged sword." This is significant in the context of the launch of the Commission, since the Report delineates how the media increasingly sensationalizes mental health conditions and is "reinforcing stereotypes" about individuals with mental health conditions, using labels that are a detriment to the progress a nation can make in eradicating mental health stigma and discrimination. Nations like India would benefit from the recommendations in the Report about how media should change their approach, in light of an analysis of news reports in India (Pathare et al., 2020) which found that the general public were not able to ascertain whether media reporting on the increased levels of suicidality during the SARS-CoV-2 pandemic is due to increased awareness on behalf of journalists and media outlets, or if instead, it reflects long-standing and problematic patterns in reporting. This example leads to the point that media outlets in respective countries could aid in decreasing stigma by adhering to "guidelines on responsible reporting" outlined by the World Health Organisation.

Thus, Dr. Thornicroft called on international organizations, governments, employers, health and social care sectors, the media, PWLEs, local communities, and civil society, to implement the following recommendations from the report in their respective capacities, in efforts to reduce stigma and discrimination of mental health (Thornicroft et al., 2022).

(1) International organizations need to evaluate and change existing mental health policies and laws protecting individuals with mental health conditions, in collaboration with WHO. The Commission Report encourages international organizations to use "an evidence-based practical toolkit" by WHO and global organizations within their respective regions.
(2) Governments and UN Member States need to "amend or repeal legislation" that discriminates against individuals with mental health conditions, "including the decriminalisation of suicide."
(3) Healthcare and social sectors must emphasize the need for better training and for courses to be evaluated so that providers in the health and social sectors would be able to provide better care, if they were better trained with "evidence-based sessions on stigma reduction."
(4) Media should encourage the issuance of policies in place for better reporting standards on mental health, and plans as to how media outlets might be able to "promote mental health and contribute to the reduction of stigma and discrimination."
(5) PWLEs, local communities, and civil society should encourage individuals with and without mental health conditions to collaborate on, and cofacilitate, national- and state-level funded programs that focus on building strong peer support networks and having PWLEs inform better care within primary healthcare systems and social services.

These recommendations for all stakeholders, as well as key findings of the Commission, are shown in the infographic in Fig. 1.

Following this presentation on the guiding principles and recommendations highlighted in the Commission, a panel discussion commenced, with panelists discussing how the Commission Report might impact their work moving forward within their respective organizations. Among the panelists were Martin Agrest, a researcher representing Proyecto Suma, a nonprofit organization based in Argentina; Mr. Joshua Duncan, a researcher representing the Mental Health Coalition based in Sierra Leone; Dr. Dan Chisholm, programme manager for Mental

Fig. 1
Recommendations and key findings from the 2022 Lancet Commission Report on "Ending stigma and discrimination in mental health."

Health at the World Health Organisation (WHO) in Geneva; Mr. Benjamin Perks, head of Campaigns and Advocacy at the United Nations Children's Fund (UNICEF); Dr. Nasser Loza, President of the World Federation for Mental Health, an international, multi-professional non-governmental organization, including citizen volunteers and former patients; and Dr. Javed Afzal, President of the World Psychiatric Association, an international umbrella organization of psychiatric societies.

Martin Agrest, a researcher representing Proyecto Suma, a nonprofit civil society association based in Argentina that does research, teaching, and interventions for persons with severe mental illnesses, began the panel discussion by stating how this *Lancet* Report might generate a change in mental healthcare systems within Latin America. He noted that there has been a sustained effort in the field of mental health to move attention away from psychiatric hospitals as the primary vehicle for mental healthcare to other forms of care available. As another critique, he said efforts to move away from the marginalization of individuals with mental health conditions have been "paternalistic" and not "patient-centred, [thus] depriving a patient's ability to recover." He concluded his comments by stating that the Report's findings would assist researchers and providers to "become more humane in our approach."

Mr. Joshua Duncan, representing the Mental Health Coalition based in Freetown, Sierra Leone, noted that the existing structural stigma and discrimination in his country have made it a "mission impossible" to recruit providers and clinicians with experience about mental health conditions. Distressed about the current mental health act in Sierra Leone, Mr. Duncan stated that existing policies meant to guide mental health providers and researchers are "as old as 120 years." The *Lancet* Commission Report, he said, would contribute to four opportunities for improved mental healthcare in Sierra Leone, to:

(1) mitigate the knowledge gap and help individuals obtain a better understanding of the challenges experienced by PWLEs;
(2) raise awareness around self-stigma since the Report reflects firsthand accounts of PWLEs, thus encouraging others to acknowledge that they are not alone;
(3) serve as a document for the media to refer to for better reporting and to reduce stigma and discrimination; and
(4) encourage advocacy to be optimized, since low- to middle-income countries like Sierra Leone "value internationally approved documents."

In his closing remarks, Duncan urged political figures in his country to utilize and implement the findings from the Report for better financial planning and structural development in reviewing the current and long-standing Mental Health Act.

Next, Dr. Dan Chisholm, programme manager for Mental Health in the WHO regional office in Europe based in Copenhagen, Denmark, highlighted three points with regard to the implementation of the Report within the WHO, namely, that:

(1) "Stigma lies behind discrimination and rights abuses, but is also driving the neglect and low prioritisation that is accorded to mental health in national policy and service development";
(2) The WHO is already in agreement with the identified goals of the Report and is in the process of collaborating with WHO Member States in order to act on the recommendations put forward by the Report;
(3) WHO has an existing package of training materials to "promote and protect rights and services of quality and care" currently being utilized by 18 nations which "showed a 40% improvement in attitudes relating to stigma and discrimination."

Following this presentation, Mr. Benjamin Perks, head of Campaigns and Advocacy at the United Nations Children's Fund (UNICEF) in the New York headquarters office, noted three points in which the Report would inform his work. It would:

(1) assist younger populations and their families in addressing shame, stigma, and judgment around mental health and how it might impact life outcomes;
(2) assist in encouraging "literacy and numeracy in mental health," in line with physical health; and
(3) encourage communities and families to give them autonomy and agency to "focus on solutions embedded in better relationships and support systems."

The next speaker, psychiatrist Dr. Nasser Loza, President of the World Federation for Mental Health, an international, multiprofessional nongovernmental organization which includes professionals and also citizen volunteers and former patients, briefly highlighted their existing work through collaborating with younger populations and students to develop "effective, short-term and inexpensive" campaigns that would reduce stigma and discrimination of mental health.

Lastly, psychiatrist Dr. Javed Afzal, President of the World Psychiatric Association, an international umbrella organization of psychiatric societies, noted that the Report would encourage professionals in the field of mental health to "bridge the gap" over the next few years of the need for, and access to, mental health services. He called attention to two points he said are worth considering for the future development of the findings of the Report.

(1) The existing literature on stigma around mental health is well established; however, "we are still at the same point of publishing about the concept of stigma and discrimination." Thus, he called on WHO and WHO Member States to collaborate and focus their efforts on improving mental healthcare access to "required and desired" service provisions for individuals with mental health conditions.

(2) Pertaining to the mental health of mental health providers, Dr. Afzal urged panelists and researchers to use recommendations from the Report to develop programs, activities, or initiatives that would assist mental health providers so that they might not face stigma and discrimination in their work with the public.

To conclude the event, Ms. Kestel mentioned how the inclusion of firsthand accounts, interviews, and poetry from people with lived experiences truly makes the Commission Report "more useful" than previous reports. She noted her personal confirmation that social contact is key if we want to change the current situation. She also emphasized that the Report has pushed for community-based mental health services and pledged that the WHO will follow up on this Report with initiatives moving forward, as mentioned prior by Dr. Chisholm during the panel discussion.

Conclusion

The *Lancet* Commission Report about the urgency to reduce bias and discrimination about mental health conditions, and the recognition of the valuable contributions of persons with lived experience, make a significant step forward in the field of mental health. Recommendations produced by the Commission, experts say, are straightforward, applicable, and scalable. The work brought forth in this Commission further confirms that mental health is integrally part of physical health; that people's lived experiences need to be heard and honored; and that policies indicative of a dignified, honorable, and fair treatment of individuals in the public health system, media, and workplaces are a fundamental human right.

References

Gautam, K., & Koirala, P. (2024). Adapting and scaling up mental health and psychosocial support response through innovative interventions to transform mental health care in Nepal for COVID-19 recovery and beyond. In J. Kuriansky, & P. Kakkattil (Eds.), *Resilient health: Leveraging technology and social innovations to transform healthcare for COVID-19 recovery and beyond* (pp. 1029–1039). Elsevier.

Khanal, M., & Kuriansky, J. (2024). Promoting resilience in the Nepali community in response to the COVID-19 pandemic: A multistage multistakeholder partnership addressing the United Nations Agenda. 2022. In J. Kuriansky, & P. Kakkattil (Eds.), *Resilient health: Leveraging technology and social innovations to transform healthcare for COVID-19 recovery and beyond* (pp. 601–627). Elsevier.

Kuriansky, J., & Zinsou, J. (2019). Mental health and well-being at the United Nations: Activities, achievements and contributions of the IAAP New York Team. *Applied Psychology Around the World, 1*(3), 16–28. https://iaapsy.org/site/assets/files/1082/apaw_vol1i3.pdf.

Pate, M. A., Coll-Seck, A. M., Lo, S., & Horton, R. (2022). A new *Lancet* Commission on the future of health and economic resiliency in Africa. *The Lancet, 400*(10353), 641–643. https://doi.org/10.1016/S0140-6736(22)01435-0.

Patel, V. (2014). Global Mental Health: An interview with Vikram Patel. *BioMed Central Medicine, 12*(1). https://doi.org/10.1186/1741-7015-12-44.

Pathare, S., Vijayakumar, L., Fernandes, T. N., Shastri, M., Kapoor, A., Pandit, D., Lohumi, I., Ray, S., Kulkarni, A., & Korde, P. (2020). Analysis of news media reports of suicides and attempted suicides during the COVID-19 lockdown in India. *International Journal of Mental Health Systems, 14*(1), 88. https://doi.org/10.1186/s13033-020-00422-2.

The Lancet. *About The Lancet Group.* https://www.thelancet.com/about-us.

Thornicroft, G., Sunkel, C., Alikhon Aliev, A., Baker, S., Brohan, E., el Chammay, R., Davies, K., Demissie, M., Duncan, J., Fekadu, W., Gronholm, P. C., Guerrero, Z., Gurung, D., Habtamu, K., Hanlon, C., Heim, E., Henderson, C., Hijazi, Z., Hoffman, C., … Winkler, P. (2022). The Lancet Commission on ending stigma and discrimination in Mental Health. *The Lancet*, *400*(10361), 1438–1480. https://www.thelancet.com/journals/lancet/article/PIIS0140-6736(22)01470-2/fulltext.

TPO Nepal. (n.d.). *Reshape-TPO Nepal.* Transcultural Psychosocial Organization Nepal (TPO Nepal). Retrieved December 2022, from http://www.tponepal.org/reducing-stigma-among-healthcareproviders-reshape/.

Votruba, N., Thornicroft, G., & Funda MentalSDG Steering Group. (2016). Sustainable development goals and mental health: Learnings from the contribution of the FundaMentalSDG global initiative. *Global Mental Health (Cambridge, England)*, *3*, e26. https://doi.org/10.1017/gmh.2016.20.

CHAPTER 48

Startup nation: Lessons from Israel's innovation ecosystem

Pradeep Kakkattil
Health Innovation and Investment Exchange (HIEx), Geneva, Switzerland; Global Council SDG3 (Health), "SDGs in Action" Initiative, Dubai, United Arab Emirates

As a United Nations (UN) bureaucrat who has served within the UN system for nearly two and a half decades, I thought I knew it all. I was on a mission to change the way healthcare is provided, and how new technologies and innovations emerging from all over the world—from small startups to huge corporations—would bring about that change. I had just recently established an office of innovation within UNAIDS (the Joint United Nations Programme on HIV/AIDS), and with my small team, we scouted some absolutely transformative health technologies that we were convinced would change the world of healthcare. We were convinced (naively, I must add) that it was ignorance among the health decision-makers about all these cool tools we had found that was preventing rapid scale up of access to healthcare—and that we would help leverage these innovations for rapid impact.

It was during this time that the Inter-Parliamentary Union (IPU), a global organization of Parliaments based in Geneva and the Knesset (Israeli Parliament), invited me to a seminar for parliamentarians taking place in Jerusalem with a focus on technological innovation and leveraging innovation to achieve the UN Sustainable Development Goals (SDGs) while "leaving no one behind." Sometimes, divine grace and intervention is all that keeps us from going down the wrong path and this event certainly was an eye-opener for me. Looking back, lessons from Israel have more or less shaped the entire work over the last five years of the "Health Innovation Exchange (HIEx)"—the platform which emerged out of our work at UNAIDS [which is described in detail in another chapter in this volume]. And a long overdue thanks to Aleksandra Blagojevic, the amazing program manager at IPU, who made this possible.

As a team, we were excited at this opportunity to go to Jerusalem to learn about what Israel was creating as a nation known for innovation. From transforming desert lands into green orchards through drip irrigation and precision agriculture; inventing the indispensable USB memory sticks to pillcams (an ingestible capsule device equipped with a miniature video camera to visualize the small intestine) that can help diagnose infections in the intestinal tract; to Rewalk,

Resilient Health. https://doi.org/10.1016/B978-0-443-18529-8.00048-2

a wearable robotic exoskeleton that enables people with disabilities get back on their feet; and security technologies that not only protect millions in the physical world, but do so much more in the virtual world. These technologies and innovation were synonymous with Israel. I was very much looking forward to seeing the latest technologies during my trip to that country, including health technologies upon which we could draw in our efforts to scale access to healthcare in the developing world. Indeed, Israel had so many lessons to teach—here are some which I summarize.

Lesson 1: Innovation is not a product but a mindset

"Innovation is not the cool glitzy technologies that you see all around," said our host, an Israeli Parliamentarian, adding, "These are tools and anyone can buy them. What we are after is changing the mindset, changing the culture, to see how we can find solutions—at all levels of the society. If we don't make that shift, we will fail." That is really what global health needs today: a shift in mindset and culture that embraces change and finds new ways of solving old problems. Health has been one of the most conservative sectors in embracing innovation and new technologies, and although I can't emphasize enough the need for due diligence, it is important to be open to examining and embedding innovation within the health ecosystem—otherwise as we have seen, we may be forced to shift during times of crisis, which is suboptimal, a bigger risk, and often we are forced to roll out solutions that are not sufficiently scrutinized during emergencies.

Lesson 2: You need Chutzpah!

Throughout my visit, a word that kept coming up when discussing innovation was "Chutzpah." I had no clue what that word meant until I met an innovator who was working on a futuristic security service mainly around data protection for healthcare operators (imagine all that personalized data about you floating in etherspace; I think he is on to something). Chutzpah is a Yiddish word for the outsize confidence or courage that allows someone to do or say things that are unconventional and often shocking. Here was an innovator who had the Chutzpah to say that human medical data was a treasure of information that was worth protecting with the most sophisticated security system. Indeed, much of health innovation and breakthroughs have come from people having the Chutzpah to question the status quo and find really unconventional ways that at first seems audacious. It calls for breaking down silos, flattening hierarchies, ending bureaucracies, and most importantly willing to fail fast. Health systems globally could do with more Chutzpah!

Another word that often goes hand in hand with Chutzpah: moonshot—aim or plan to do something that is almost impossible (yes, the word did emerge from the Apollo 11 project and the long shot of getting Neil Armstrong to the moon). We need more Chutzpah and moonshots—to end cancer or break the emerging challenges of antimicrobial resistance or stop rising costs of noncommunicable diseases. We need to truly have Chutzpah.

Lesson 3: Seeing is believing

Miracles are commonplace in Israel. The country's history is full of miracles—from drawing water from a rock to quench the thirst of millions, to deaf ears opening and sight restored. That is not just from past accomplishments, this is what Israeli technology is delivering today. All of us on the visit knew about Israel's reputation as a leader in innovation and technology, and several speakers at the seminar spoke of the amazing things that they were doing. It wasn't until we walked into the Peres Center for Peace and Innovation (Peres Center for Peace and Innovation, 2022) established by the late president of Israel and Nobel Laureate Shimon Peres that the reality of the innovations came to life. The center had not yet opened to the public, so we were privileged to preview the work of the center based in Tel Aviv-Jaffa.

The 14,000 square footmodern building overlooking the Mediterranean provides a space where visitors can see the innovations come to life. In this one space, you can see several innovations from the past to the present and the future. They include the latest in cybersecurity, automotive tech, agritech, health and medtech, and a leap into the future of emerging technologies. It was exciting to actually not just hear about the innovations, but to actually touch, feel, and experience them—that made a world of difference. There was a palpable sense of excitement as the parliamentarians that I accompanied went around the center experimenting with the different innovations and discussing how these would be relevant to their own country setting. I was captivated by the eyewear that could literally give sight to the blind (with an inbuilt camera and an AI narrative) and the virtual 3D image of the heart that a surgeon could examine from all angles before entering the operating theater.

Global health clearly needs a similar venue where ministers and key health planners and implementers can bring their challenges, experience, and experiment with the innovations that have demonstrated impact in similar settings and assess potential to translate to their own settings. Based on this incredible experience at the Peres Center, the HIEx is establishing the "Health Innovation Discovery Center" to showcase innovations that can be transformative for healthcare. This "Health Innovation Discovery Center" will be a venue where Ministers of Health and senior leaders can visit, experience, and experiment with innovations that can be relevant to their settings. It will also support innovators to help scale up and maximize impact.

Lesson 4: Nothing inspires like success

Israel ranks third after the United States and the UK on startup ecosystem ranking for 2022 according to Startup Blink, a global startup ecosystem map and research center (Startup Blink, 2022). Also, Tel Aviv, Israel's capital and startup hub, is in the global top 10 startup ecosystem cities ranking. Israel possibly has the most startups per capita than any other nation (Deloitte Israel, 2017). Further, Israel also has more Nasdaq-listed companies than any country, bar the USA and China (US State Department, 2021).

As a small country, Israel shows that size doesn't matter. As only 22,145 sq.km (8630 sq. miles), with less than 9 million people, this small nation has punched way above its weight. In 2021 Israeli startups raised US$25.6B (IVC Data and Insights, 2022). Where governments play an enabling role and provide the essential environment, people find solutions that not only address their local challenges but often create global goods. Israel has clearly shown what clear and enabling policies, financing and incentives for startups, and letting young people lead can deliver.

We are seeing several countries in the global South focusing on establishing startup ecosystems, including in health. I am a firm believer that development in connectivity, big data, and artificial intelligence actually levels the playing field and there is huge potential for health startups that can transform healthcare. We are seeing, for example, India, a global leader in the low-cost quality pharma sector, positioning itself to also take advantage of developments in digital health and medtech. Ghana, Indonesia, Kenya, Nigeria, Rwanda, Senegal—the list is growing and we are seeing some cutting-edge Fintech, Edutech, and healthtech solutions emerging that respond to not only their local contexts but also scale to global markets.

Global health could do with a miracle. We need more countries that enable and empower innovators to do what they do best: finding solutions for challenges that have dogged healthcare over centuries. It will call for enabling policies and better investments that provide incentives for entrepreneurs and thinking on scale. And also, Chutzpah.

References

Deloitte Israel. (2017). Retrieved from: https://www2.deloitte.com/il/en/pages/innovation/article/the_israeli_technological_eco-system.html.

IVC Data and Insights. (2022). *IVC-MEITAR Israeli tech review 2021*. https://www.ivc-online.com/LinkClick.aspx?fileticket=eWRPYkJvwBA%3D.

Peres Center for Peace and Innovation. (2022). Retrieved from: https://www.peres-center.org/en/the-organization/about-us/.

Startup Blink. (2022). *Startup ecosystem report 2022*. Global https://lp.startupblink.com/report-download/.

US State Department. (2021). *2021 Investment climate statements: Israel*. Retrieved from: https://www.state.gov/reports/2021-investment-climate-statements/israel/#:~:text=Israel%20has%20the%20third%20most,the%20United%20States%20and%20China.

CHAPTER 49

Citizen science: Technology tools to empower consumers for more resilient and sustainable healthcare systems during COVID-19 and beyond

Dari Alhuwail[a,b] and Aisha Alhouti[a]

[a]Information Science Department, College of Life Sciences, Kuwait University, Kuwait City, Kuwait, [b]Health Informatics Unit, Dasman Diabetes Institute, Kuwait City, Kuwait

Introduction

There is no doubt that the Coronavirus Disease 2019 (COVID-19) pandemic has challenged many nations globally, claimed more than 500 million lives, and dramatically changed how we live and go about business (Kantrowitz-Gordon, 2021). Engaging and empowering consumers with the appropriate tools and information can be an effective public health strategy to combat a pandemic (Slack, 1997). Digital tools can support these strategies and have shown preliminary success in curbing the spread of COVID-19 (Abd-Alrazaq et al., 2021; Gabarron et al., 2021), including battling the spread of misinformation (Abd-Alrazaq et al., 2020; Galvão, 2021; Muric et al., 2021). Coupling these digitally supported public health strategies with "citizen science" approaches can form powerful weapons against a pandemic and form a resilient healthcare system (Birkin et al., 2021; Vahidi et al., 2021).

Citizen science—also known as community science, crowd science, crowd-sourced science, civic science, or volunteer monitoring—is scientific research, in whole or in part, conducted by amateur or nonprofessional scientists in the community (Gura, 2013). As such, citizen science "has ample capacity for transdisciplinarity and for integrating natural, physical, and health sciences with the humanities and social sciences." (Roche et al., 2020) This means that such research can advance knowledge in the scientific community especially in settings or situations with limited professional capacity and can increase the participation and knowledge of the public.

Resilient Health. https://doi.org/10.1016/B978-0-443-18529-8.00049-4

The rapid advances in information and communication technologies, as well as the increasing access and reach of the internet, created more opportunities for powerful citizen science applications. One of the great powers of citizen science is crowdsourcing, which is a main distinction between the Spanish flu of 1918 and the current COVID-19 pandemic, which are often compared, except that this approach was not available 100 years ago (Bernardo et al., 2021). The human population in this current time is connected in an unprecedented manner, regardless of their geographical place. This makes it easy for anyone with an internet connection to collect information, analyze data, and develop software and hardware, which enables researchers to reach many volunteers to collect a large sample for their research studies and to help build powerful artificial intelligence based on the collation of such large amounts of data and information.

Advances in technology innovations to adapt to COVID-19: Examples around the world

In the State of Kuwait, and shortly after the first case of COVID-19 was announced in the country on February 24, 2020, informaticians and scientists volunteered their time, skills, and abilities to contribute to the fight against the COVID-19 pandemic. This resulted in CoronaMaps Kuwait, shown in Fig. 1, and in Fig. 2, which is an online platform built on an open spatial hub to create, organize, visualize, and share geo-enabled information about COVID-19 (Tara et al., 2021). Specifically, Fig. 1 shows a geo-enabled dashboard of the most recent official statistics (in April 2020) in the country after feeding the data from the Ministry of

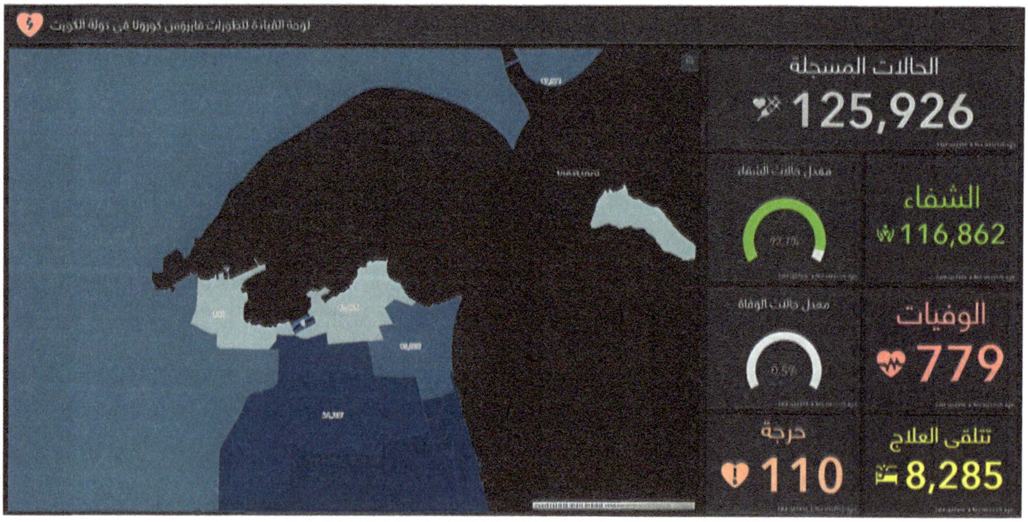

Fig. 1
CoronaMaps Kuwait COVID-19 statistics obtained from the most recent official statistics in the country (April 2020).

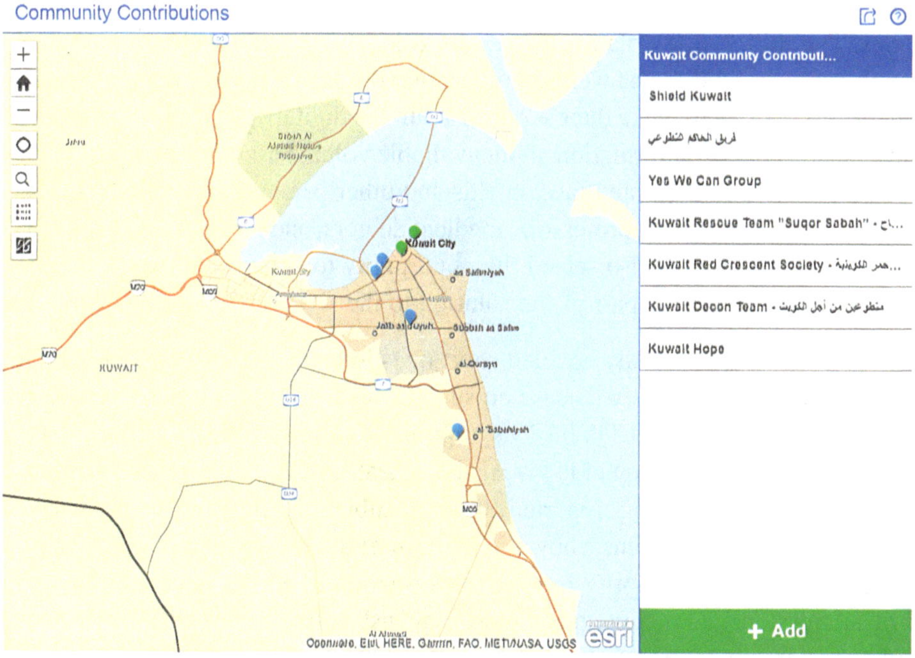

Fig. 2
CoronaMaps Kuwait showcasing available voluntary services to support the community (April 2020).

Health's website, demonstrating an example of a useful dashboard that can help governments connect with citizens and better inform them about the public health crisis. The data in the figure refers to the total number of registered cases at the time (125,926), the running total for recoveries (persons recovered from COVID-19) (116,862), the tally of deaths due to COVID-19, the number of currently active cases (8285), the percentage of recoveries (92.8%), the percentage of deaths due to COVID-19 (0.5%), and the number of critical cases in the ICU being treated at the time (110).

Fig. 2 shows an example of how someone would track available services in the community that are nearby during lockdowns and crisis. It also provides an opportunity for those wishing to contribute or volunteer services in the community during crisis to help, support, and show solidarity. For example, the Shield Kuwait team, comprised of youth in Kuwait, and shown on the list of contributors on the figure, provided free 3D printed face shield to frontline staff. Additionally, also on the list, the Kuwait Red Crescent Society provided food supplies, personal protective equipment, and helped with delivering medications. Such a platform can be useful to other countries wishing to link and connect to local efforts and voluntary services.

With more than 55,000 visits, the platform enabled consumers to locate necessary information about the latest evidence-based COVID-19 guidance as well as to identify the locations and contact information for essential services, like in grocery stores, healthcare clinics, and pharmacies. The utility of the platform came from its ability and function as a hub as most information, services, and forms were on different websites or mobile phone apps, as the government was rushing to make their services available digitally at the time. Consumers were also able to share and post information about available volunteer groups and services available for frontline staff (e.g., 3D printed face shields and other personal protective equipment). Volunteer scientists (university professors, medical doctors, and professional engineers), using the citizen science approach, also seized the opportunity to better empower and engage consumers directly to become part of the solution to the COVID-19 pandemic in Kuwait.

Using an anonymous geospatially enabled short questionnaire, TrackCOVIDKW.com was developed to predict and identify disease hotspots by evaluating how many people in a particular suburb have developed COVID-19 symptoms, as shown in Fig. 3 (Tara et al., 2021). Overall, a visitor to the site can quickly visualize the health status in his or her neighborhood via this heat map. People voluntarily reported (through submissions) if they experienced one or more of the COVID-19 symptoms, shown in the figure; the total population in the suburb was provided from the Public Authority for Civil Information to allow for comparison with the number of submissions. To accommodate more than just Arabic and English, the team also used commonly used emojis to display the results.

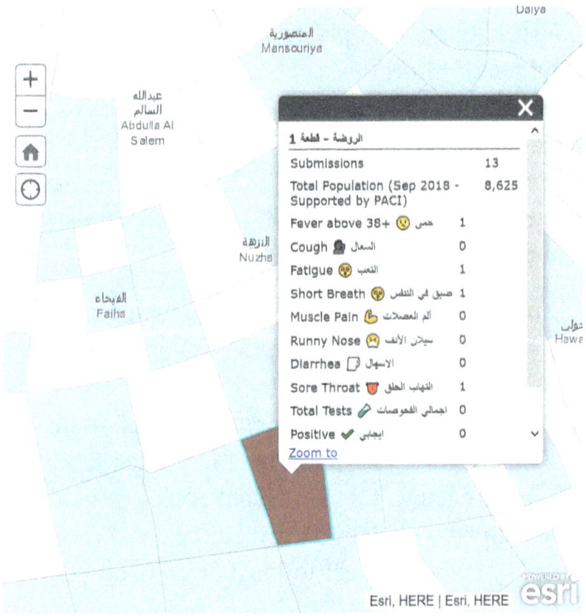

Fig. 3

TrackCOVIDKW.com map showing hotspots based on anonymous responses at a level block.

The authors of this chapter were involved in the development of these technology tools, and note other similar efforts in other countries. In the Islamic Republic of Iran, citizen science-based initiatives were also initiated to combat COVID-19. Examples included creating digital tools for contact tracing, symptom detection via smartphone sensors, and self-assessment of diseases (Vahidi et al., 2021). In the Republic of India, volunteers assisted in visualizing, distributing, and interpreting publicly available unstructured COVID-19 data (Ulahannan et al., 2020). Through engaging the public via social media, the necessary data was gathered and later built into a dataset, which was used to create a web application, as well as associated Application Programming Interfaces (APIs), which help offer services using a common interface, to present outbreak trends, diagnostic testing trends, and hotspot maps.

Additionally, citizen science helped researchers create the COVID Symptom Study (Murray et al., 2021). Through a multinational collaboration consortium, over 360 million self-assessment reports were provided by the public from across the United Kingdom and the United States through mobile applications (Drew et al., 2020). The study gathered all relevant information and analyzed the data to support medical research and the medical community.

Conclusions and recommendations for the way forward

Examples from around the world noted in this chapter highlight the power of citizen science in empowering and engaging consumers in the fight against the spread of pandemics accessing information and networking resources, hence deeming healthcare systems to be more resilient (Katapally, 2020). However, citizen science remains underexplored in healthcare, including public health crises (Ciasullo et al., 2021). Challenges remain in the continued engagement and sustainability of these approaches, as people often experience fatigue after being engaged in one or more initiatives. Additionally, issues around ethics, privacy, and security remain and will continue to challenge integrating citizen science approaches for more resilient healthcare systems (Groot & Abma, 2022; Wiggins & Wilbanks, 2019). Lastly, digitally enabled citizen science approaches need to ensure equal access and opportunity for all, without leaving anyone behind, including the elderly, those with disabilities, or minorities (Fiske et al., 2019; Xie et al., 2020).

The scientific community needs to harness this valuable methodology and identify opportunities to appeal to, and sustain, the engagement of citizens and consumers in such research over in the immediate emergency as well as in the long haul. Certainly, engaging persons in the public, and empowering them with accurate and valuable information and knowledge, can be an effective strategy for building resilient healthcare systems. Collaborative nature and partnership among clinicians and scientists with the general public cannot be put on the back burner; citizen science is a cornerstone to more resilient healthcare systems of the future.

References

Abd-Alrazaq, A., Alhuwail, D., Househ, M., Hamdi, M., & Shah, Z. (2020). Top concerns of tweeters during the COVID-19 pandemic: Infoveillance study. *Journal of Medical Internet Research*, *22*(4), e19016.

Abd-Alrazaq, A., Hassan, A., Abuelezz, I., Ahmed, A., Alzubaidi, M. S., Shah, U., Alhuwail, D., Giannicchi, A., & Househ, M. (2021). Overview of technologies implemented during the first wave of the COVID-19 pandemic: Scoping review. *Journal of Medical Internet Research*, *23*(9), e29136.

Bernardo, T., Sobkowich, K. E., Forrest, R. O., Stewart, L. S., D'Agostino, M., Perez Gutierrez, E., & Gillis, D. (2021). Collaborating in the time of COVID-19: The scope and scale of innovative responses to a global pandemic. *JMIR Public Health and Surveillance*, *7*(2), e25935.

Birkin, L. J., Vasileiou, E., & Stagg, H. R. (2021). Citizen science in the time of COVID-19. *Thorax*, *76*(7), 636–637.

Ciasullo, M. V., Carli, M., Lim, W. M., & Palumbo, R. (2021). An open innovation approach to co-produce scientific knowledge: An examination of citizen science in the healthcare ecosystem. *European Journal of Innovation Management*, *25*(6), 365–392.

Drew, D. A., Nguyen, L. H., Steves, C. J., Menni, C., Freydin, M., Varsavsky, T., Sudre, C. H., Cardoso, M. J., Ourselin, S., Wolf, J., Spector, T. D., Chan, A. T., & COPE Consortium. (2020). Rapid implementation of mobile technology for real-time epidemiology of COVID-19. *Science*, *368*(6497), 1362–1367.

Fiske, A., Prainsack, B., & Buyx, A. (2019). Meeting the needs of underserved populations: Setting the agenda for more inclusive citizen science of medicine. *Journal of Medical Ethics*, *45*(9), 617–622.

Gabarron, E., Rivera-Romero, O., Miron-Shatz, T., Grainger, R., & Denecke, K. (2021). Role of participatory health informatics in detecting and managing pandemics: Literature review. *Yearbook of Medical Informatics*, *30*(1), 200–209.

Galvão, J. (2021). COVID-19: The deadly threat of misinformation. *The Lancet Infectious Diseases*, *21*(5), e114.

Groot, B., & Abma, T. (2022). Ethics framework for citizen science and public and patient participation in research. *BMC Medical Ethics*, *23*(1), 23.

Gura, T. (2013). Citizen science: Amateur experts. *Nature*, *496*(7444), 259–261.

Kantrowitz-Gordon, I. (2021). A new normal after the COVID-19 pandemic. *Journal of Midwifery & Women's Health*, *66*(3), 293–294.

Katapally, T. R. (2020). A global digital citizen science policy to tackle pandemics like COVID-19. *Journal of Medical Internet Research*, *22*(5), e19357.

Muric, G., Wu, Y., & Ferrara, E. (2021). COVID-19 vaccine hesitancy on social media: Building a public twitter data set of antivaccine content, vaccine misinformation, and conspiracies. *JMIR Public Health and Surveillance*, *7*(11). Available from: https://pubmed.ncbi.nlm.nih.gov/34653016/.

Murray, B., Kerfoot, E., Chen, L., Deng, J., Graham, M. S., Sudre, C. H., Molteni, E., Canas, L. S., Antonelli, M., Klaser, K., Visconti, A., Hammers, A., Chan, A. T., Franks, P. W., Davies, R., Wolf, J., Spector, T. D., Steves, C. J., Modat, M., & Ourselin, S. (2021). Accessible data curation and analytics for international-scale citizen science datasets. *Scientific Data*, *8*(1), 297.

Roche, J., Bell, L., Galvão, C., Golumbic, Y. N., Kloetzer, L., Knoben, N., Laakso, M., Lorke, J., Mannion, G., Massetti, L., Mauchline, A., Pata, K., Ruck, A., Taraba, P., & Winter, S. (2020). Citizen science, education, and learning: Challenges and opportunities. *Frontiers in Sociology*, *5*, 613814.

Slack, W. (1997). *Cybermedicine: How computing empowers doctors and patients for better health care*. Jossey-Bass.

Tara, M., Adel, T., Dari, A., Thani, D. A. A., Al-Jafar, E., Hamra, E. A., Ghazal, H., Wahba, H., Bahaadinbeigy, K., AlKhalifa, K., Joseph, L., Zeroug, M., Odeh, M. B., Alhefzi, M. I., Hamdi, M., Househ, M., Al-Shorbaji, N., Hassan, O. E., Al-Salmi, S. A., … Sari, Z. (2021). Middle East and North African Health Informatics Association (MENAHIA): Health informatics response to COVID-19 crisis. *Yearbook of Medical Informatics*, *30*(1), 328–334.

Ulahannan, J. P., Narayanan, N., Thalhath, N., Prabhakaran, P., Chaliyeduth, S., Suresh, S. P., Mohammed, M., Rajeevan, E., Joseph, S., Balakrishnan, A., Uthaman, J., Karingamadathil, M., Thomas, S. T., Sureshkumar, U., Balan, S., & Vellichirammal, N. N. (2020). Collective for Open Data Distribution-Keralam (CODD-K) consortium. A citizen science initiative for open data and visualization of COVID-19 outbreak in Kerala, India. *Journal of the American Medical Informatics Association, 27*(12), 1913–1920.

Vahidi, H., Taleai, M., Yan, W., & Shaw, R. (2021). Digital citizen science for responding to COVID-19 crisis: Experiences from Iran. *International Journal of Environmental Research and Public Health, 18*(18), 9666.

Wiggins, A., & Wilbanks, J. (2019). The rise of citizen science in health and biomedical research. *The American Journal of Bioethics, 19*(8), 3–14.

Xie, B., Charness, N., Fingerman, K., Kaye, J., Kim, M. T., & Khurshid, A. (2020). When going digital becomes a necessity: Ensuring older adults' needs for information, services, and social inclusion during COVID-19. *Journal of Aging & Social Policy, 32*(4–5), 460–470.

CHAPTER 50

Promoting resilience in the Nepali community in response to the COVID-19 pandemic: A multistage multistakeholder partnership addressing the United Nations Agenda

Sadikchhya (Meera) Khanal[a,b,c] **and Judy Kuriansky**[b,c,d,e,f,g]

[a]Russell Reynolds Associates, Dallas, TX, United States, [b]Department of Clinical Psychology, Teachers College, Columbia University, New York, NY, United States, [c]International Association of Applied Psychology, Indianapolis, IN, United States, [d]Office of Ambassador Sidique Wai, Embassy of the Republic of Sierra Leone to the United States, Washington, DC, United States, [e]United Nations Group of Friends of Mental Health and Well-being, New York, NY, United States, [f]Global Council SDG3 (Health), "SDGs in Action" Initiative, Dubai, United Arab Emirates, [g]World Council for Psychotherapy, Vienna, Austria

Introduction

The COVID-19 pandemic, which caused millions of deaths and disrupted economies worldwide, especially impacted Least Developed Countries (LDCs) like Nepal, with an already-fragile health infrastructure and inadequate mental health resources, professionals, and funding (United Nations Conference on Trade and Development, n.d.; Regmi et al., 2004). As the COVID-19 infection spread through the country of Nepal in the Spring of 2020, the "Nepal Resilience-Building Project" was developed to fill the mental health gap caused by the emergency, in order to provide psychosocial support to the Nepali diaspora and in-country population. This was accomplished through events in five phases, which are described in this chapter, that were held on virtual platforms, given government-mandated restrictions on in-person gatherings.

Unique features of the project included creation of multistakeholder partnerships bringing together representatives of government, civil society, and academia; showcasing youth action; involving media for broad outreach; offering advice to questions submitted by the public prior to the events; and intentionally addressing multiple goals of the United Nations 2030 Agenda

Resilient Health. https://doi.org/10.1016/B978-0-443-18529-8.00050-0

for Sustainable Development (United Nations, n.d.). For events in projects such as these, it is always optimal to have cosponsorship of government and nongovernmental organizations (both international and local) as well as a UN agency and voices of people working on the ground—which the Nepal Resilience-building Project did successfully throughout the phases.

Background

The first case of COVID-19 was confirmed in Nepal on January 23, 2020 (Shrestha et al., 2020), but fear spread throughout the population by the time the World Health Organization declared the infection as a Public Health Emergency of International Concern in March 2020 (World Health Organization, 2020), spurring the development and implementation of the first phase of this project. The number of confirmed cases increased to 181 by May 2020 and ballooned to 1834 on June 1, 2020 (World Health Organization, n.d.-a). A year later, a spike occurred in Nepal and neighboring countries of Bangladesh and India, with inadequate vaccine distribution, such that 631,152 people tested positive for COVID-19 and only 3,232,604 vaccine doses had been administered for the population of 30.38 million (Government of Nepal, 2021; World Health Organization, n.d.-b). The spreading infection and repeated government lockdowns seriously impacted the economic and social fabric of the country, including collapse of 'daily wage' jobs (driving rickshaws and taxis) and interruption of the planned "Visit Nepal 2020" tourism promotion campaign (World Bank, n.d.). Furthermore, surveys documented Nepali citizens' stress and anxiety, including fears, anxiety, depression, substance abuse, and increased incidences of suicide and domestic violence (Devkota et al., 2021; Sharma, 2020; Xinnhua, 2020; Yadav et al., 2021).

Nepal is a landlocked country in South Asia situated in the Himalayan mountains and plains bordering India and China. The multicultural, multiethnic, multireligious state, with a history of Hindu and Buddhist influences and Nepali as the official language, is a secular federal democratic republic which has undergone political shifts. A devastating earthquake in 2015 resulted in 8964 people dead, 21,952 injured, and more than 600,000 structures in Kathmandu and nearby towns including centuries-old temples damaged or destroyed, as well as caused anxiety, depression, suicidal ideation, and alcohol-use problems (Sherchan et al., 2017; Shneiderman et al., 2016).

The "Nepal Resilience-building Project" described in this chapter was initiated as the semester project of the first author (Khanal), a Nepali-American national, in order to fulfill requirements for the Columbia University's Teachers College graduate course entitled "Psychology and the United Nations" taught by the second author (Kuriansky), a noted international psychologist, author, professor, who was familiar with Nepalese culture (Kuriansky, 2001). The project evolved as Khanal moved from the New York City college campus to her parents' home in Virginia where she noticed the serious need for mental health awareness and support for the Nepali community. Kuriansky's advocacy at the United Nations, international connections, and emergency psychosocial support projects worldwide provided the foundation for the project phases (Art Saves Lives Foundation, 2019; Kuriansky, 2012, 2013, 2016b, 2019a, 2019b, 2019c, 2019d, 2020a; Kuriansky et al., 2016). Both authors of this chapter reached out to their respective contacts to participate.

Alignment with International agreements and the UN Agenda

Given Kuriansky's successful advocacy to help insure that mental health and well-being was included in the United Nations Agenda 2030 for Sustainable Development and her years-long advocacy at the UN (Kuriansky & Zinsou, 2019), the "Nepal Resilience-building Project" was purposefully designed to align with the framework of the UN 2030 Agenda, specifically, target 3.4 which calls for the promotion of mental health and well-being, Sustainable Development Goal (SDG) 5 about empowering women and girls (particularly at risk during the pandemic), SDG 4 ensuring education for all, SDG 11 about sustainable cities, and SDG 17 encouraging multistakeholder partnerships. The project intentionally consolidated connections between civil society organizations on the ground with government and other stakeholders. Further, the voice of youth was purposefully highlighted. The project further aligned with the WHO Constitution and the Political Declaration of Universal Health Coverage, which assert mental health equity with physical health and as a human right.

Phase I: The first webinar

The first phase of the project was a webinar held on June 13, 2020, entitled, "Coping during COVID-19: Honoring Nepal and Advice about Mental Health and Wellbeing," which purposefully focused on positive psychology principles of resilience, reassurance, and hope. This orientation was consistent with the authors' intentions to increase awareness, provide psychosocial support, reduce stigma, and facilitate policy advocacy about mental health and well-being. For production support and public outreach, a professional media partner, Cyber Headline, founded by Nepali-American communications expert Dr. Ramhari Subedi, was engaged. This leading news outlet, on Facebook with 424,668 followers, promoted the event on social media and broadcast the project webinar on Facebook Live and Zoom. Analytics showed that the webinar reached 25,000 people.

Questions were collected in advance of the event, through a Google Form posted on the Cyber Headline Facebook page, which were answered by psychologist Kuriansky during her presentation. In this presentation, she described and demonstrated specific resilience and strengths-building activities and techniques based on her extensive previous experience providing psychosocial support in emergencies in many settings around the world, including natural disasters as well as epidemics (HIV/AIDS, SARS, and Ebola) which have similarities to the COVID-19 situation.

The three-hours-long webinar was sponsored by the International Association of Applied Psychology (IAAP, an NGO having consultative status with the Economic and Social Council of the United Nations), as well as media outlet Cyber Headline, and a highly respected local NGO, the Transcultural Psychosocial Organization Nepal (TPO Nepal), in collaboration with the Permanent Mission of Nepal to the United Nations. Also, two other Nepali NGOs, Yuwa and ShEmpowered, were invited to be cosponsors and to speak about their work.

The program was purposefully scheduled to start at 10 AM EDT which was 7:45 PM Nepali Standard Time (NST) in order to be a reasonable time to reach both Nepali-Americans and Nepalis in-country.

Fig. 1 shows the flyer for Phase I of the project. Fig. 2 is a screenshot of the speakers on the virtual Zoom platform for the webinar. Their presentations are described later and the concept note, program, and speaker biographies are in Electronic Supplementary Material (ESM) 1.

The program started with an introduction by the first author of this chapter, serving as moderator. She then introduced the first speaker, a high-level representative of the Nepali government, the deputy permanent representative of the Mission of Nepal to the United Nations, Mr. Ghanshyam Bhandari. Mr. Bhandari noted the severe mental health consequences of the pandemic, including the tragic rise in cases of suicide in Nepal, numbering 875 persons in the period of the lockdown alone. This has increased the government's concern for the mental health of the people. In this context, he pointed out legal and constitutional frameworks in which mental health has been recognized in Nepal. Importantly, mental health was recognized as important in the Mental Health Policy of Nepal dating back to the year 1996 and was updated recently in the year 2019 to ensure expansion of mental health services at local, provincial, and federal levels in the country. Also, the Constitution of Nepal 2015 recognizes the fundamental right to health. Further, the government has significantly increased the annual budget for the health sector, and the Fifteenth Plan of the National Planning Commission for Fiscal Year 2019/20–2023/24 assures that "access to mental health services will be expanded at all levels." On the multilateral front, he acknowledged the contributions of frameworks promoting mental health of the WHO Action Plan and the UN Secretary-General policy brief about Mental Health in reaction to the pandemic.

Support for mental health in the current government in Nepal is evident, he emphasized, in the national aspiration of "Prosperous Nepal, Happy Nepali" (Khabarhub, 2021) which has four goals on prosperity and two of six indicators referring to "decent life and wellbeing," and "safe and healthy environment." This aspiration is highly consistent with the theme of resilience building of the project and was acknowledged by psychologist Kuriansky in her presentation later in the program, as a positive example for governments and also to inspire trust in government, since populations' lack of trust in their government has been widely pointed out as a result of the pandemic. His full statement is in ESM 2 in the online version at https://doi.org/10.1016/B978-0-443-18529-8.00050-0.

Fig. 1

Flyer for the Phase I event showing main sponsors and speakers on "Coping during COVID-19: Honoring Nepal and Advice about Mental Health and Wellbeing".

The next speaker, psychiatrist Dr. Kamal Gautam, executive director of Transcultural Psychosocial Organization (TPO) Nepal, reported their survey results showing that after a month of lockdown, the number of individuals suffering from at least one psychosocial problem increased from 42% to 50%. TPO Nepal provided tele-counseling at 76 quarantine centers and for the public, extending hours of their hotline. The extensive activities of TPO Nepal to support the population are reported in a chapter he authored in this volume (Gautam & Koirala, 2024).

Next, psychiatrist Dr. Mahendra Nepal described that he had returned to his native Nepal from Australia during the height of the pandemic and was contacted by former patients about their

Fig. 2

Speakers in the event about "Coping during COVID-19: Honoring Nepal and Advice about Mental Health and Wellbeing" *(top row, left to right)*: youth advocate Ashish Subedi, women's rights advocate Bandana Rana, psychiatrist Dr. Mahendra Nepal, Yuwa youth activist Jenisha Mainali, *(middle row, left to right)*: ShEmpowered founder Amiansu Khanal, TPO Nepal executive director Dr. Kamal Gautam, moderator Meera Khanal, psychologist Dr. Judy Kuriansky *(bottom row, middle)*: DPR of Mission of Nepal to the UN Ghanshyam Bhandari.

depression caused by the pandemic. He also noted an increase in substance abuse, which has been widely reported across cultures as a result of the pandemic.

To respond to extensive reports about an alarming worldwide increase in domestic violence caused by the pandemic (World Bank, n.d.), an influential Nepali advocate for women's rights and gender equality, Ms. Bandana Rana, was invited to speak about help offered by her organization, *Saathi*, a Nepal-based NGO working on violence against women, which provides support and shelters for domestic violence survivors. She described the severe gap between the desperate need for, yet lack of, services and funding for these women, which has been a persistent problem that escalated due to the pandemic lockdowns. According to one study, a shocking 51.9% of women in Nepal have experienced some form of violence in their lifetime (Lamichhane et al., 2011). The rise in domestic violence as a result of the pandemic was being referred to as a "shadow pandemic," including in remarks and tweets by the UN secretary-general (Gutteres, 2020).

Consistent with the UN emphasis on including youth voices, three youth activists were panelists. Nepali native Ms. Jenisha Mainali, speaking from the capital of Kathmandu, described the work of *Yuwa*, a youth organization empowering Nepali youth, and reported the results of their survey, conducted on social media, whereby 42% of respondents stated that the pandemic had strongly affected their mental health.

Nepali-American youth Ms. Amiansu Khanal (the first author's sister) described the nonprofit, ShEmpowered, which she founded to raise awareness about menstruation and girls' rights. She shared her successful grassroots advocacy to end Chaupadi, a practice in rural communities in Nepal, which banishes women and girls from their home during menstruation, exposing them to unsafe housing and abuse (Ranabhat et al., 2015). Her successful efforts involved relentlessly advocating with government officials and police, and counseling parents and youth about the detrimental effects of such "menstrual exile." Her advice to "Push, Probe, and Collaborate!" coincides with SDG 5 about preventing violence and empowering women and girls. Through this experience, she said that she "realized how important the government is, in creating and sustaining change in Nepal."

Also, a recent high school graduate of Nepali heritage, living with his family in Virginia, USA, Mr. Ashish Subedi, shared his challenging personal stress during the COVID-19 pandemic, including disappointment (shared by so many youth) about his high school graduation ceremony being canceled due to the pandemic-imposed lockdowns. He explained his coping by leaning on support of friends and family and ended his presentation with inspiring advice to others "Don't lose hope, we are in it together." This theme of togetherness is supported by much research showing that social connection is a primary factor in positive coping with trauma.

The segment of the event providing specific psychosocial support was conducted by Kuriansky, including a PowerPoint presentation about useful and easily implementable techniques, based on solid psychological principles, to normalize feelings and to build resilience, strength, confidence, hope, a sense of empowerment and silver linings (potential positive outcomes) in the face of the crisis which have been successfully applied worldwide. She also gave answers to questions which were presubmitted online. For cultural relevance, she reminded Nepalese of their inherent resilience and the importance of honoring their cultural traditions and heritage. The submitted questions, examples of which are given in ESM 3 in the online version at https://doi.org/10.1016/B978-0-443-18529-8.00050-0, centered on themes of negative thoughts, anxieties, fear, anger, guilt, and grief. For example, advice in response to a question about negative thoughts, that "I cannot sleep as negative thoughts come into my mind," included the "thought-stopping" technique. Also, her advice about guilt, as in the question, "If I go on long walks and read a book all day, it brings guilt that I am not being productive," or about anger, as in the concern that "I am short-tempered," centered on being compassionate with oneself and permission to have any and all feelings.

The questionnaire, and subsequent advice, given by Kuriansky also included questions about Posttraumatic Growth, namely, positive changes resulting from adversity caused by a trauma like the pandemic (Peterson et al., 2008; Tedeschi et al., 2018). When asked about silver linings, respondents' answers included: spending more quality time with others (family, children, and grandchildren); having time for activities or hobbies previously too busy for (cooking, reading, having fun); self-care and being health conscious (eating well-balanced meals, resting, self-love); realizing a new philosophy of life ("appreciating little things" "You never know what happens, so be happy with whatever you have," "less concern for external things like "property" and richness.") One person referred to climate change, noting a decrease in air pollution, that "Kathmandu valley looks beautiful without a thick layer of dirt in clouds."

Phase II: Webinar on women's rights and mental health

Phase II consisted of a webinar on "Psychosocial Support and Rights of Women and Girls in Nepal: Policies and Programs in the COVID-19 recovery period" held on 20 March 2021, from 9:30AM to 11:00AM EST, during the annual UN Commission on the Status of Women (CSW) Forum convened by the CSW NGO Committee/New York (Global NGO, 2021). This event intentionally focused on the intersection of target 3.4, to promote mental health and well-being, with the mission of SDG 5, to "Achieve gender equality and empower all women and girls."

This webinar in Phase II was a continuation of Phase I, in reconvening the multistakeholder presenters, namely, from government (the Mission of Nepal to the United Nations), civil society (TPO Nepal and the NGO, IAAP), women-led NGOs (Yuwa and ShEmpowered), and the private sector/media (Cyber Headline). The event had the same intent to promote support for mental health and well-being and to review presubmitted questions and survey ratings of pandemic-triggered emotions submitted on the Cyber Headline Facebook page. These main cosponsors were joined by other women-focused grassroots NGOs in Nepal, namely, IPAS Nepal and youth-driven Visible Impact.

The webinar was hosted on the website of the NGO CSW/NY Forum, which publicly posted hundreds of events from organizers all over the world whose submissions were accepted for the program which was made available to over 27,000 registered attendees. The event was broadcast through the NGO CSW Zoom and was later uploaded to Cyber Headline Facebook, with analytics revealing 1200 views.

Fig. 3 shows the flyer for Phase II and Fig. 4 shows a screen shot of the speakers on the virtual Zoom platform for the webinar. A description of the presentations of the speakers is as follows. The concept note, program, webinar link, and speaker biographies are in ESM 1 in the online version at https://doi.org/10.1016/B978-0-443-18529-8.00050-0.

DPR Bhandari of the Mission of Nepal to the UN reprised his participation, this time focused on issues regarding women and girls and reporting about the significant progress of the government with regard to gender empowerment and the rights of women and girls, with a zero-tolerance policy on gender-based violence and criminalization of all forms of sexual discrimination and gender-based violence. Given the rise of domestic violence in Nepal during the COVID-19 pandemic, he relayed the commitment of the Nepali government to promote women's rights, stating, "We have criminalized all forms of sexual discrimination and gender-based violence and have enacted several legislations to eliminate violence against people." He stressed two important pieces of legislations in Nepal: the Domestic Violence Act

Fig. 3
Flyer for Phase II on "Psychosocial Support and Rights of Women and Girls in Nepal: Policies and Programs in the COVID-19 recovery period".

of 2009 (Nepal Law Commission, 2009) and the Crime Victim Protection Act of 2018 (Nepal Law Commission, 2018). These acts, designed and drafted to protect women, represent impressive efforts by the government to support women's rights. However, he noted, challenges arise in their application and cultural relevance to the general public, given the historically patriarchal society of Nepal. Thus he encouraged NGOs and youth to help put these laws into action.

He also pointed out that the government has achieved gender parity at the school level and enacted a Comprehensive School Safety Package, involving distribution of sanitary pads, midday meals, and scholarships that have helped to promote girls' education and to solve some

Fig. 4

Speakers at the event on "Psychosocial Support and Rights of Women and Girls in Nepal: Policies and Programs in the COVID-19 recovery period." *(Top row, left to right)*: IPAS Nepal's Ms. Madhabi Bajracharya, moderator Meera Khanal, Visual Impact Project Manager Ms. Shilpa Lamichhane *(middle row, left to right)*: Nepal DPR Ghanshyam Bhandari, ShEmpowered's Ms. Amiansu Khanal, Yuwa's president Ms. Riu Dhakal, *(bottom row)*: TPO Nepal's Mr. Suraj Koirala, psychologist Dr. Judy Kuriansky, youth Mr. Aashish Subedi.

problems related to mental health. The government has also worked in tandem with civil society and the private sector in programs and services, including hotlines, which, he said, need to continue in efforts on behalf of women and girls in a "whole-of-society" approach.

He also pointed out that this event is being held on March 20, coincident with the United Nations International Day of Happiness and the release of the annual World Happiness Report, which shows Nepal rated as the happiest country in the South Asia region, up from the 3rd ranking previously. He further elaborated that the Fifteenth National Plan, which revolves around the national aspiration of "Prosperous Nepal, Happy Nepali," mentions that globalization, the infodemic, change in lifestyle, and food habits have resulted in an increasing number of communicable diseases and mental health problems, all of which the government is addressing in policy and programs to rebuild and recover long term from the pandemic, including in accessible and affordable basic health services, with prioritizing mental health and including training care providers in psychosocial and social counseling. His full statement is in ESM 2 in the online version at https://doi.org/10.1016/B978-0-443-18529-8.00050-0.

As in Phase I, TPO Nepal was represented, this time by their Executive Manager Mr. Suraj Koirala, who noted their survey findings that 24% of women reported an increase in domestic violence and workload during the lockdowns. Given this startling increase, and the fact that TPO has provided psychosocial counseling support to more than 1000 people where more than 41% were female, Mr. Koirala pledged TPO's commitment to provide psychosocial support to women and girls.

Representatives of organizations in support of the rights of women and girls were purposefully invited given the mission of the CSW Forum. Depictions of their work, taken from photos on their websites, are shown in Fig. 5.

As in Phase I, the voice and programs particularly of young women were showcased. In the first presentation from these groups about their women-focused work, Ms. Madhabi Bajracharya, policy and governance advisor at the nonprofit organization IPAS Nepal, an NGO that advocates for reproductive rights and works (like TPO, in collaboration with the government), shared recent developments in policy promoting Sexual and Reproductive Health and Rights (SRHR). She reported that in September 2018, according to the Constitution, the government

Fig. 5

Examples of Nepal NGOs addressing issues about women in Phase II, as shown in photos from their websites. (*Top row, left to right*): TPO Nepal, (*bottom row, left to right*): ShEmpowered, Saathi, and Visible Impact.

enacted two acts, the Safe Motherhood and Reproductive Health Rights Act, and the Public Health Service Act (apart from the legislation described by DPR Bhandari before). As a result, reproductive services, which include abortion services, are listed as "basic health services," the costs for which are covered by the government. These acts advance the legalization of abortion in Nepal in 2002, allowing more rights for Nepali women to have safe and affordable access to sexual and reproductive medical services.

Next, Ms. Shilpa Lamichhane, program manager of the NGO *Visible Impact*, continued this theme about SHRH and described their activities in support of the SRHR of young people: promoting menstrual health, safe abortion, family planning, comprehensive sexuality education, and marriage equality, including through consultations, awareness-raising using creative drama, innovative social media advocacy, dissemination of menstruation-related products, and research.

Next to speak, Ms. Riju Dhakal, director of the youth-run organization *Yuwa*, echoed the importance of SHRH awareness-raising, through workshops, seminars, and social media.

awareness campaigns. She stressed the importance of raising SRHR awareness to young girls during adolescence, a period of life "closely linked to mental health because of various implications tied to both mental and body changes."

Then, Ms. Amiansu Khanal reprised describing the advocacy of her nonprofit, *ShEmpowered,* on behalf of women and girls (as she had done in Phase I).

In the spirit of CSW to engage males in the empowerment of women and girls, high-school student Ashish Subedi again spoke as he did in Phase I, this time about the South-Asian organization he founded in his school. He also recited a touching poem he wrote, with a message of optimism about women's empowerment, that "misogyny is still alive, but what's livelier is our voices."

As in Phase I, questions were posted for the Nepali diaspora to answer on the Cyber Headline Facebook Page and was reshared on the social media page. The answers were received through a Google form. Questions—and ensuing answers—centered on worries about a "new normal," answered by Kuriansky with reassurance about resilience that you can "bounce back" and "build back better," a phrase increasingly used at the UN.

This advice aspect was deemed important by the event organizers, in order to offer practical psychosocial support to the attendees in the context of the ongoing pandemic, and not just to provide passive information and knowledge during these troubled times. Thus the advice segment of Phase I reprised by Kuriansky consisted of a PowerPoint presentation and answers to presubmitted mental health-related questions from the public. As before, the presentation was interactive, demonstrating dynamic real-time activities, as well as didactic, with messages reinforcing support for women's empowerment—the cornerstone of SDG 5 and CSW—and normalizing feelings caused by the pandemic. She described coping strategies for anxiety,

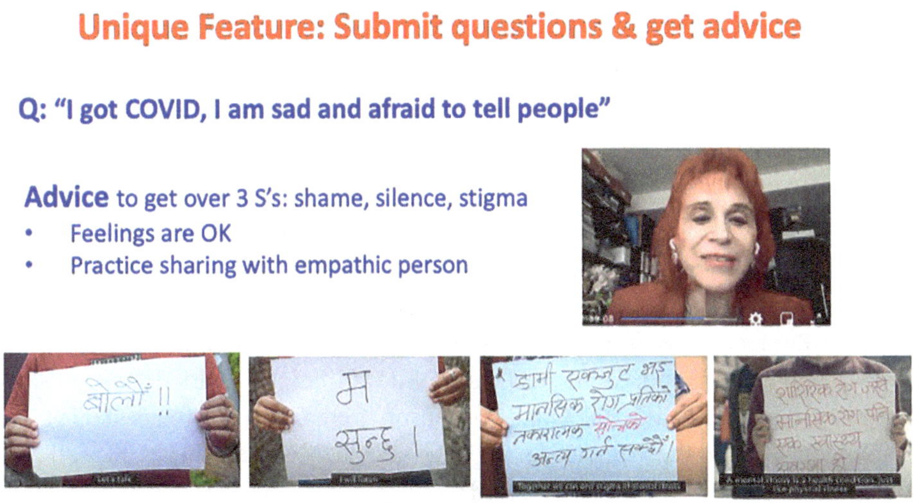

Fig. 6
Screen capture of Dr. Kuriansky's advice to avoid the 3 S's of silence, shame, and stigma of mental health issues, and four signs in Nepali with similar messages, from left to right, "Let's talk," "I will listen," "Mental illness is a health condition just like physical illness," and "Together we can end stigma of mental illness".

stress, and loneliness associated with COVID-19; emphasized self-compassion and self-care; and reported research emphasizing that social support is critical in recovery.

In one example of a question, a woman asked what she can do about being scared to tell her husband about her wishes to do things that are fun in the midst of the tragic pandemic. Kuriansky validated her desire to have fun (as "normal" relief from the pandemic stress) and advised her to feel "no shame" about any emotions (by modeling saying, "My feelings are OK") and to share openly. As shown in Fig. 6, in a slide from the video capture of her PowerPoint presentation, Kuriansky's advice to a question "I got COVID, I am sad an afraid to tell people," included avoidance what she calls "The 3 S's: shame, silence and stigma." These are also demonstrated in the signs in Nepali on the slide which say, "Let's talk," "I will listen," "Mental illness is a health condition just like physical illness," and "Together we can end stigma of mental illness."

Kuriansky further emphasized advice to replace the phrase "social distancing"—commonly used to refer to the recommended safe space between people to avoid transmission of the virus—with the phrase "physical distancing" since considerable research shows that social *connection* helps people cope, thus calling for social *distance* is counterproductive (Kuriansky, 2020b).

As in Phase I, submissions from the Google form included a question about posttraumatic growth from the pandemic. Responses were similar to those in Phase I, mentioning increased quality time for self and family; and positive experiences such as compassion and kindness to and from others, reflecting helping behavior which is shown by research to aid coping with crisis.

Phase III

The project's government and civil society partners, namely, the Permanent Mission of Nepal to the UN and the NGOs IAAP and TPO Nepal, were again brought together in Phase III, this time in a virtual webinar sponsored by the World Health Organization and the UN Mission copresidents of the United Nations Group of Friends of Mental Health and Well-being (governments who are in support of mental health and well-being), entitled "Mental Health and Well-Being in Times of COVID-19." Other cosponsors were UNICEF (focused on youth issues and valuably contributing the voice of another UN agency, besides WHO) as well as the International Federation of Red Cross and Red Crescent and the UN system Workplace Mental Health and Well-being Program (covered in another chapter in this volume) (Kuriansky, Medhizadeh, & Pikkani, 2024). The event was intended to discuss practical tools, resources, and policies that can help navigate through the difficult pandemic times and strengthen mental health and psychosocial well-being. The flyer for the event is shown in Fig. 7.

The event was intentionally held on 30 March 2021, the one-year anniversary since the COVID-19 was declared by WHO as a public health emergency of international concern (PHEIC), to examine the effects of the crisis on mental health and well-being, with reports of country experiences and of experts. In her role as a coorganizer, and with her NGO IAAP being a cosponsor of the event, Kuriansky specifically recommended that Nepal be one of the countries to give a report regarding mental health and well-being in their national COVID-19 response, thus DPR Bhandari was invited. Since a local NGO working on the ground would also be included, she recommended the invitation be extended to Dr. Gautam, which again made for the perfect pairing of representatives from government and from civil society from the same country presenting their perspective. These partners are shown in the webinar Zoom in Fig. 8. Attendees to the webinar included high-level delegates at the UN, all committed to the promotion of mental health and well-being—as is the mission of the Group of Friends—as well as invited guests of WHO and IAAP. A screen shot is shown in Fig. 8 of the speakers, including Project partners DPR Bhandari and Dr. Judy Kuriansky (missing a photo of Dr. Gautam) are shown with the director of the World Health Organization Office at the United Nations, Mr. Werner Obermeyer, director of the WHO Geneva office Department of Mental Health and Substance Abuse Dévora Kestel, and DPR of the Mission of Georgia to the United Nations Elene Agladze who gave her country's perspective of mental health and well-being.

Speakers at the event all examined the effects of the crisis on mental health and well-being.

DPR Bhandari presented the country experience of Nepal, acknowledging the contributions of this project and partnership. He also noted the current challenges by stating that it was not "easy

Fig. 7
Flyer for the event about "Mental Health and Well-being in Times of COVID-19".

for a resource constrained country like Nepal to focus on mental health, on a scale that it really deserved." However, the government continued to focus on mental health as one of the priority areas of the pandemic. He noted, "the government's pandemic response was guided by the legal and policy framework, which recognized mental health and well-being as an important component and provided a trajectory to expand mental health services to the grassroots level." He elaborated that the Nepal emergency response plan for COVID-19 provided for mental health services and psychosocial counseling support for patients and healthcare workers with clinics and a telephone hotline to talk about stress, worries, and anxiety, with the government running noncommunicable disease and mental health services in 51 districts of the country under the epidemic and disease control program. Also, the government adopted a multistakeholder approach with civil society organizations, think tanks, academia, and the private sector. He specifically acknowledged the initiative of this Nepal Resilience-Building Project led by Dr. Kuriansky and her student at Columbia University's Teachers College

Fig. 8

Speakers at the event, cosponsored by WHO, the United Nations Group of Friends of Mental Health and Well-being, and others, on "Mental Health and Well-Being in Times of COVID-19," with sponsors, speakers, and project partners. *(top row, left to right)*: WHO UN office director Werner Obermeyer, WHO Geneva office Department of Mental Health and Substance Abuse director Dévora Kestel, Nepal DPR Ghanshyam Bhandari *(bottom row, left to right)*: Dr. Judy Kuriansky, DPR Mission of Georgia to the Elene Agladze.

(the first author of this chapter) who coorganized the previous events to address the mental health needs of the Nepali diaspora and also recognized the excellent work of civil society organizations that have been partnering with the government of Nepal in the response. He further committed to recognizing mental health as a serious issue in the long term, not just during the pandemic, but beyond; fighting stigma and superstition, misinformation, disinformation, and hatred; and taking a whole-of-society approach, involving government, civil society, community, and the private sector. He concluded with a hopeful invitation to collectively "respond, recover, and rebuild better by leaving no one behind.' His full statement is in ESM 2 in the online version at https://doi.org/10.1016/B978-0-443-18529-8.00050-0.

For her part, panelist Kuriansky spoke on "Projects to Provide Mental Health & Psychosocial Support in COVID-19 times: the Nexus Model and Hybrid Approach." The hybrid approach, she explained, involves a combination of technology tools with in-person visits by community health workers to meet personally with people "where they are," discuss and evaluate their health and mental health needs, and deliver primary services when appropriate. This is a model especially used by several innovations in response to the COVID-19 pandemic, when technological tools became useful given government lockdowns but some in-person contact, when safe, was also critical to serve people, especially in remote villages as reported in other chapters in this volume (Bhavsar & Katara, 2024; Javetski, 2024; van Engen, 2024). These innovations are showcased on platforms, like "Health in Your Hands," which present a valuable space for such projects to network (also described in another chapter in this volume, Kuriansky, 2024).

The Nexus Model of the Sustainable Development Goals highlights the intersection of the sustainable development goals (SDGs) of the UN Agenda 2030 for Sustainable Development, which has been adopted in the "SDGs in Action" initiative of the United Arab Emirates (also described in another chapter in this volume, by Coetzee et al., 2024). This means that achievement and progress in SDG 3 on health and well-being (including mental health) is integrally linked to other goals, like the empowerment of women and girls (SDG 5), education for all (SDG 4), reducing inequalities (SDG 10), and achieving peace and security (SDG 16). In all approaches, though, mental health can be considered "at the heart of the sustainable Development Goals"—the theme of the campaign Kuriansky and H.E. Ambassador Dr. Caleb Otto of the Mission of Palau to the UN used to ensure that mental health and well-being was included in the UN 2030 Agenda (Kuriansky, 2016a).

Further, Kuriansky described the unique features of the Nepali Resilience-building Project that brought together speakers from the government of Nepal with local and international NGOs, as well as presented a section on questions and answers from the public as referred to before. She explained her toolbox of techniques applied in interventions worldwide, including in a holistic program to teach health behaviors, correct myths about the virus, and provide psychosocial support (PSS), which was implemented in the Democratic Republic of the Congo to stave off the COVID-19 spread, reported in another chapter in this volume. The collaboration of a local team and international NGOs in this project aligns with SDG 17 about partnerships. The PSS activities all encourage empowerment, resilience, and social cohesion, which have been applied, she explained, in several other interventions she has conducted during the COVID-19 pandemic for resilience-building, including training volunteer Chinese peer counselors; a special seminar featuring the former UN ambassador of Iraq who founded the UN International Day of Happiness; and a daylong event about positivity and food security, also represented in another example of the nexus model (Parks et al., 2022).

Given that this event was specifically timed to fall on the date a year after COVID-19 was declared by WHO as a PHEIC, she explained the importance of such date selections, that can trigger "anniversary reactions" (Nemeth et al., 2012). In reaction to the question posed to the participants, "How upset have you been related to the COVID-9 pandemic" rated on a scale of "1" (not at all) to "10" (a great deal), their mode score increased from 5 to 7, indicating increased upset a year after the event compared to the previous year. Specifications mentioned included stress, fear, depression, irritability, boredom, and feeling not normal and unsafe.

Phase IV: Training about psychosocial support during another COVID-19 lockdown

In May 2021, while some parts of the world were beginning to get back to normal due to vaccine access, Nepal, like other countries, was experiencing ongoing infection and even spikes, with 1345 COVID-19 related deaths from May 10 to 17, 2021, the highest number in the whole year

(World Health Organization, n.d.-b), forcing another government-mandated lockdown. Given this situation, Khanal and Kuriansky reached out to TPO Nepal's Dr. Gautam to offer any assistance, which led Dr. Gautam to invite Kuriansky to give a training to his staff and frontline workers (Gautam & Koirala, 2024). This constitutes Phase IV of the project.

The training was by held on May 27, 2021, from 9:00am to noon Nepal time (to accommodate their work schedule), for 32 invited psychosocial healthcare workers, including psychologists, psychiatrists, and other mental health care workers from TPO Nepal, Tribhuvan University, and Kanti Hospital. The format, similar to the other phases of the project, consisted of both a didactic component, with PowerPoint slides and videos, as well as experiential exercises. Group discussion was also included.

A week before the training, the participants were sent a Google Form questionnaire about their mental health-related concerns which included personal issues, like burnout and fear of COVID-19 infection, as well as interest in learning professional techniques to help clients. During the training, Kuriansky expressed appreciation to them for their dedication in helping others in these difficult pandemic times and emphasized the importance of self-care—shown by research to be critical for all helpers and frontline workers, to prevent or manage extreme burnout. Techniques she presented derive from varied disciplines, including Cognitive Behavioral Therapy, Gestalt Therapy, Neuropsychology, and Japanese Naikan Therapy (related to mindfulness) (Kuriansky et al., 2017; Kuriansky & Nemeth, 2020). Three lessons learned from her years of advice-giving in diverse media apply: that "your feelings are normal, you are not alone, and there is help and hope." She also demonstrated activities from her toolbox used worldwide for trauma recovery (Kuriansky, 2008, 2019a, 2019c; Kuriansky et al., 2016) and used a culturally relevant metaphor for coping from her experience mountain-climbing including the Annapurna Range in Nepal, (which the trainees appreciated) emphasizing how to meet challenges: with confidence (using self-talk, "I can do this"), having a plan, expecting challenges (rocks, terrain), thinking of past successful experiences, confirming one's motivation, asking for help, and offering help.

The training was planned, coordinated, and moderated by Dr. Gautam, with an introduction and valuable contributions and advice throughout by TPO Nepal chairperson Dr. Mita Rana, who is also professor and clinical psychologist at Tribhuvan University Teaching Hospital in Kathmandu. All participants introduced themselves and shared important professional experiences during the discussion.

To assess the efficacy of the training, participants filled out a 5–10 min mixed-method survey administered through Qualtrics, immediately before and after the training. The quantitative section of the survey consisted of items to assess change in affect (anxiety, depression, anger, confidence) as well as in knowledge about how to cope with that issue. Items were rated on a 10-point Likert scale where "1" indicates extremely low and "10" indicates extremely high. The results of these self-reports are given in ESM 4 in the online version at https://doi.org/10.1016/B978-0-443-18529-8.00050-0. The results support the trainer's (Kuriansky) perception of the staff as a highly competent and resilient group of frontline mental health workers who can benefit from support but who are coping impressively well with the crisis and doing an excellent job serving the population in these troubled times. Accordingly, self-reports of confidence, happiness, and feeling secure were high at the beginning of the training, but positive outcomes from before to after the training were still reflected in the ratings. For example, self-reported confidence, happiness, feelings of safety and security, as well as knowledge about coping techniques (e.g., self-care, managing anxiety and depression) increased, while decreases were noted in stress level and tiredness.

Responses to the question in the posttraining assessment about what participants found helpful included: techniques related to mindfulness, happiness, and gratitude; discussion about Posttraumatic Growth (PTG); activities promoting resilience; and suggestions for self-care. Asked what they enjoyed in the training, replies referred to the interactive activities, the way everyone was included, the "energetic" and "positive" presentation style, and "how Dr. Judy engaged everyone who participated." Also appreciated was Kuriansky's respect for the Nepali culture, tradition, and spirituality—consistent with cultural competence emphasized as essential in professional skills (Sue, 2001) by using examples connecting the group to their roots (e.g., about Hindu deities symbolizing strength and resilience) and her therapeutic orientation integrating East and West techniques. Fig. 9 shows the PowerPoint slide she showed during the training reflecting aspects of Nepal's culture, including: brave Gurkhas (soldiers),

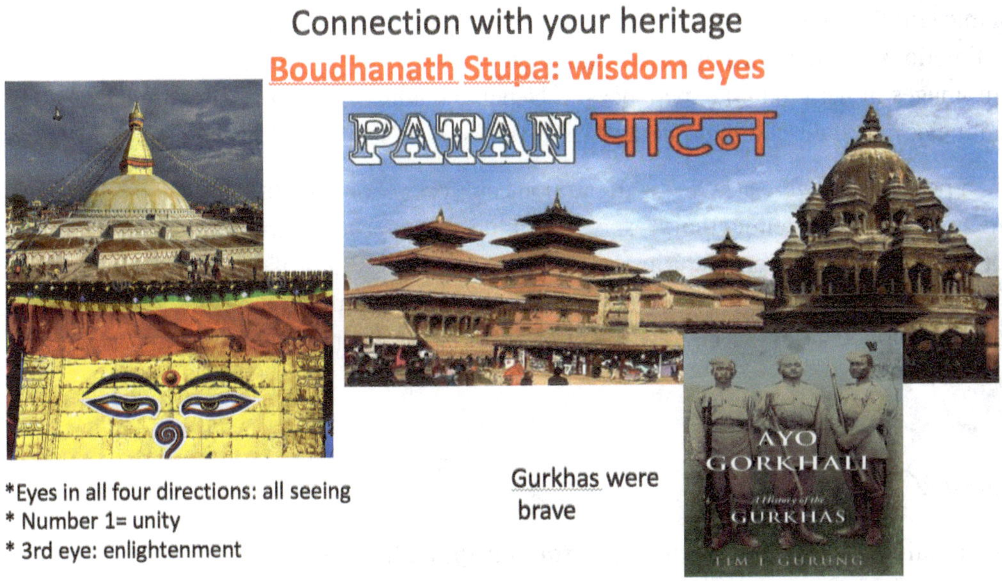

Fig. 9
Slide from presentation by Dr. Judy Kuriansky, honoring Nepal's temples and her experiences in Nepal.

Fig. 10
Slide presentation by Dr. Judy Kuriansky of experiences in Nepal and honoring Nepal's cultural tradition.

notable temples like in Patan, and the internationally known Boudhanath Stupa, with its eyes in all directions and the third eye of enlightenment. Fig. 10 is the PowerPoint slide she showed with images of her cultural experiences in Nepal, including shaman training, thangkas (holy paintings), mountain-climbing, being blessed at the Kali shrine, and participating in a ritual.

When participants were asked—as in previous phases—about whether they experienced silver linings from the pandemic, their responses included: increase in family time, personal growth (better stress management); spiritual growth, and increases in professional skills helpful to provide psychosocial support to their clients (empathy, compassion, technological skills for online therapy and counseling, and inspiration to "work for betterment for my people and myself").

Phase V

In continuing collaboration among the partners in the earlier phases, the Mission of Nepal at the UN and IAAP came together again, this time as cosponsors among many other UN missions, as well as the World Health Organization, and NGOs, for an online event entitled, "Building Back Better' after COVID-19 through Addressing Health Equity, Mental Health, and Well-being:

Contributions of Psychological Science and Practice to Leave No One Behind." The event, reported in another chapter in this volume (Kuriansky, Altungy, et al., 2024), was organized and moderated by Kuriansky and held on July 12, 2021, during the important United Nations High-Level Political Forum, during which governments present their voluntary national reviews about their progress achieving the Sustainable Development Goals (United Nations, n.d.). The goals selected for review that year were SDG 3 on Health and well-being, SDG 10 on reducing inequalities, and SDG 17 on multistakeholder partnerships. In yet another powerful example of the evolving cooperation between the Nepal government on the international level at the UN and civil society working on the ground in response to the pandemic, the DPR of the Mission of Nepal to the UN Ghanshyam Bhandari and TPO Nepal's Dr. Kamal Gautam were again on the program as they had been in Phases I and II, invited specifically by Kuriansky in her role as coorganizer and moderator of the event, and as a perfect example of collective achievements in the three SDGs addressed, namely, SDGs 3, 10, and 17. Fig. 12 is a screenshot of the speakers.

The full program is shown in ESM 5 in the online version at https://doi.org/10.1016/B978-0-443-18529-8.00050-0, but Fig. 11 shows a shortened version to underscore the multistakeholder partnership promoted throughout this project, namely, listing project partners Dr. Judy Kuriansky as moderator, DPR Bhandari under the High-Level speaker segment, and Dr. Gautam's name as a speaker in the segment on "Voices from the Field."

DPR Bhandari made a statement during the high-level segment (among other Member States of the UN including Portugal, Mexico, Ecuador, Qatar, Lebanon, Canada, Georgia, Japan, with the main sponsor of Sierra Leone). Speaking on behalf of the government, he again confirmed his government's support of mental health and their cooperation with civil society.

In the civil society section of the program, on "Voices from the Field: Programs on the Ground,"

Dr. Gautam acknowledged the resilience-building multistakeholder program described in these phases in this chapter and presented the diverse activities of his organization, TPO Nepal, to provide psychosocial support in response to COVID-19 (Gautam & Koirala, 2024). These covered a vastly impressive range of accomplishments in many areas, including *capacity building* (conducting trainings for community psychosocial and health workers, teachers, and police, and webinars for government health workers, service providers, and community people and leaders); *service delivery* (in individual and group in-person sessions quarantine centers, and a free daily tele-mental health helpline); *research*, about an online survey on stress and self-harm during the pandemic); and *advocacy* (with local and provincial governments about mental health, and about the urgent need for funding). Importantly, TPO Nepal had provided technical assistance to the Ministry of Health and Population (MoHP) in drafting a COVID-19 mental health and psychosocial response and preparedness framework for the National Mental Health Strategy Plan to be scaled up in the coming 5 years. The framework includes

July 12, 1:00PM-2:30PM (EDT)

This event is virtual. To REGISTER, COPY AND PASTE LINK: https://bit.ly/3jKUyS8

Moderator: Dr. Judy Kuriansky, Columbia University Teachers College; International Association of Applied Psychology

Welcome: **H.E. Ambassador Alie Kabba**, Permanent Representative, Mission of Sierra Leone to the United Nations

Opening Remarks: H.E. Vice President Epsy Campbell Barr, Government of Costa Rica

Opening Remarks: Dr. Jennifer F. Kelly, President, American Psychological Association

High-level Speakers

H.E. Ambassador Francisco Duarte Lopes, Permanent Representative, Mission of Portugal to the United Nations

H.E. Ambassador Juan Ramón de la Fuente, Permanent Representative, Mission of Mexico to the United Nations; United Nations Security Council 2021-2022

H.E. Ambassador Issa Konfourou, Permanent Representative, Mission of Mali to the United Nations

H.E. Ambassador Jamal Fares Alrowaiei, Permanent Representative, Mission of Kingdom of Bahrain to the United Nations; Co-chair, Group of Friends of Mental Health and Well-being

H.E. Ambassador Cristian Espinosa, Permanent Representative, Mission of Ecuador to the United Nations; Co-chair, Group of Friends of Mental Health and Well-being

H.E. Ambassador Kaha Imnadze, Permanent Representative, Mission of Georgia to the United Nations; Co-facilitator, Political Declaration on Universal Health Coverage (UHC)

H.E. Ambassador Amal Mudallali, Permanent Representative, Mission of Lebanon to the United Nations

H.E. Ambassador Richard Arbeiter, Deputy Permanent Representative, Mission of Canada to United Nations; Co-chair, Group of Friends of Mental Health and Well-being

Mr. Ghanshyam Bhandari, Deputy Permanent Representative, Mission of Nepal to the United Nations

H.E. Ambassador Tetsuya Kimura, Mission of Japan to the United Nations; Co-Chair, Group of Friends of UHC

H.E. Ambassador Sidique Abou-Bakarr Wai, Chief of Mission, Embassy of Sierra Leone to the USA

Mr. Werner Obermeyer, Director, World Health Organization Office at the United Nations

Scientific Experts

Dr. Geoffrey Reed, Department of Psychiatry, Columbia University

Dr. Sharifa Al Emadi, Executive Director, Doha International Family Institute, State of Qatar

Mr. Johan Carlson, Public Health Agency of Sweden

Dr. Willem Hanekom, Africa Health Research Institute

Dr. María Elena Medina-Mora, National Autonomous University of Mexico

Voices from the Field: Programs on the Ground

Sierra Leone: Jordan Levinson, Dimagi

Democratic Republic of the Congo (DRC):

Mr. Gordon Tapper, United African Congress

***Nepal*: Dr. Kamal Gautam**, TPO Nepal

***India*: Mr. Sujay Santra**, iKure

Closing Remarks

Dr. Christine Roland-Lévy, President, International Association of Applied Psychology

Main Sponsor: Permanent Mission of the Republic of Sierra Leone to the United Nations

Canada

MEXICO

República del Ecuador

The Permanent Mission of Lebanon to the United Nations

Permanent Mission of Iceland to the United Nations

Permanent Mission of Sweden to the United Nations

World Health Organization

IAAP

American Psychological Association

United African Congress

Fig. 11

Flyer of the event, "Building Back Better' after COVID-19 through Addressing Health Equity, Mental Health, and Well-being: Contributions of Psychological Science and Practice to Leave No One Behind" at the UN High-level Political Forum, showing the partners of the Nepal Resilience-building Project, with Dr. Kuriansky as the moderator, DPR of Nepal Bhandari in the high-level segment, and Dr. Gautam in the segment about Voices from the Field—projects locally on the ground.

Fig. 12

The multistakeholder partners of the Nepal Resilience-building Project at an event during the UN High-level Political Forum. *(Top row from left)*: Dr. Judy Kuriansky, DPR Bhandari; *(4th row down, 5th from left)* Dr. Kamal Gautam.

components of Psychological First Aid, stress management, and grief counseling for 500 government frontline healthcare workers, to help themselves and the people seeking help. The extent of work in all these dimensions is an exceptional example for the global community of the response of an NGO in response to the COVID-19 pandemic.

Lessons learned, conclusions, and the way forward

Phases of the "Nepal Resilience-building Project" successfully provided psychosocial support during the COVID-19 emergency to the national population, Diaspora, and mental health workers in the country of Nepal, a least-developed country severely affected by the COVID-19 pandemic throughout the period of the project. Several SDGs of the United Nations Agenda for Sustainable Development were addressed, notably target 3.4 about promoting mental health and well-being, as well as SDG 5 about the empowerment of women and girls, and SDG 17 about creating multistakeholder collaborations. Special features of the project's phases included collecting mental health-related questions and providing affirming advice and normalizing feelings, consistent with positive psychology, as well as the value of a media partner to raise awareness. The phases of the project show the importance of cultural competence both in webinars and in training. Training of frontline mental health providers, particularly regarding self-care, was shown to be useful in results of a mixed-methods survey and qualitative feedback about the efficacy of such a staff training, which revealed an increase of knowledge about coping skills. TPO Nepal and the "Nepal Resilience-building Project" partners continue to explore collaborations to address ongoing recovery from the pandemic, especially in light of DPR Bhandari's expression of the Nepal government's strong commitment to continue support of mental health and well-being. The five phases of this project can serve as a model of partnership, community outreach, education, and psychosocial support that can be adapted and applied in other developing countries with fragile mental health infrastructure and limited resources, in order to build resilience of the public, professionals, and health workers during this pandemic crisis and in the ongoing recovery period, in order to bridge the mental health treatment gap. The project further exemplifies the valuable contributions of psychologists, as academicians, clinicians, and researchers; as agents for change; and as partners with multistakeholders, including government. The phases of this project demonstrate the necessity, and value, of nurturing such partnerships, especially to engage government on the national as well as international stage. The project also underscores the importance of ongoing commitment of all stakeholders in the COVID-19 response, to enact policy and design and implement programs which promote mental health and well-being, especially for developing countries, in this time of crisis and moving forward toward recovery and sustainable resilience.

Appendix: Supplementary material

Supplementary material related to this chapter can be found on the accompanying CD or online at https://doi.org/10.1016/B978-0-443-18529-8.00050-0.

References

Art Saves Lives Foundation. (2019). *Mission report: Resilience, empowerment, and connection building training*. Report of training/workshops post-disaster in Sint Maarten https://drive.google.com/open?id=17Gf6uOVZMD3N-Ak4igihhTRewTC5BoE1.

Bhavsar, M., & Katara, K. (2024). Realizing the transformative potential of Community Health Workers with digital innovation | A living goods case study. In J. Kuriansky, & P. Kakkattil (Eds.), *Resilient health: Leveraging technology and social innovations to transform healthcare for COVID-19 recovery and beyond* (pp. 1011–1018). Elsevier.

Coetzee, C., Kuriansky, J., & AlHammadi, R. (2024). The nexus model and the Young Leaders Program: Innovative models for the future of the SDGs in Action Campaign and the Global Councils on Sustainable Development Goals. In J. Kuriansky, & P. Kakkattil (Eds.), *Resilient health: Leveraging technology and social innovations to transform healthcare for COVID-19 recovery and beyond* (pp. 629–646). Elsevier.

Devkota, H. R., Sijali, T. R., Bogati, R., Ahmad, M., Shakya, K. L., & Adhikary, P. (2021). The impact of COVID-19 on mental health outcomes among hospital fever clinic attendants across Nepal: A cross-sectional study. *PLoS One*, *16*(3), e0248684. https://doi.org/10.1371/journal.pone.0248684.

Gautam, K., & Koirala, P. (2024). Adapting and scaling up mental health and psychosocial support response through innovative interventions to transform mental health care in Nepal for COVID-19 recovery and beyond.

In J. Kuriansky, & P. Kakkattil (Eds.), *Resilient health: Leveraging technology and social innovations to transform healthcare for COVID-19 recovery and beyond* (pp. 1029–1039). Elsevier.

Global NGO. (2021). *Youth-led NGO CSW/NY Forum events about Migrant Latinx Women, Women in Middle East Peace, and Rights of Women and Girls in Nepal: Psychosocial issues in COVID-19 times*. http://globalngoexec.org/2021/04/youth-led-ngo-csw-ny-forum-events-about-migrant-latinx-women-women-in-middle-east-peace-and-rights-of-women-and-girls-in-nepal-psychosocial-issues-in-covid-19-times/.

Government of Nepal. (2021). *National Census 2021*. https://censusnepal.cbs.gov.np/Home/Index/EN.

Gutteres, A. (2020). https://twitter.com/antonioguterres/status/1246973397759819776?s=20.

Javetski, G. (2024). Innovation: Exploring Dimagi's Digital Platform CommCare: A widely adopted digital solution used by frontline workers worldwide. In J. Kuriansky, & P. Kakkattil (Eds.), *Resilient health: Leveraging technology and social innovations to transform healthcare for COVID-19 recovery and beyond* (pp. 1073–1075). Elsevier.

Khabarhub. (2021). *'Prosperous Nepal, Happy Nepali' campaign based on martyrs' sacrifice: PM Oli*. https://english.khabarhub.com/2021/29/159540/.

Kuriansky, J. (2001). Trekking the other Nepal: The Kathmandu Valley. *Spirituality & Health*, 57–60. (Spring). https://www.dropbox.com/s/uceyufhd6qsm3lb/Nepal%20Spirituality%20and%20Health%20magazine%20artcile%20DrJudy%20Kuriansky.pdf?dl=0.

Kuriansky, J. (2008). A clinical toolbox for cross-cultural counseling and training. In U. P. Gielen, J. G. Draguns, & J. M. Fish (Eds.), *Principles of multicultural counseling and therapy* (pp. 295–330). New York: Taylor and Francis/Routledge.

Kuriansky, J. (2012). Our communities: Healing after environmental disasters. In D. G. Nemeth, R. B. Hamilton, & J. Kuriansky (Eds.), *Living in an environmentally world: Healing ourselves and our planet* (pp. 141–167). Praeger/ABC-CLIO.

Kuriansky, J. (2013). Helping kids cope with the Oklahoma tornado and other traumas: 7 techniques. *Huffington Post*. http://www.huffingtonpost.com/judy-kuriansky-phd/helping-kids-cope-with-the-oklahoma-tornado_b_3322238.html.

Kuriansky, J. (2016a). Report of the IAAP United Nations Team: Success in the New Global Agenda Advocacy. *Bulletin of the International Association of Applied Psychology*, *28*, 80–88.

Kuriansky, J. (2016b). *The psychosocial aspects of a deadly epidemic: What Ebola has taught us about holistic healing*. Santa Barbara: ABC-Clio Publishing. http://www.abc-clio.com/abc-cliocorporate/product.aspx?pc=a4983c.

Kuriansky, J. (2019a). A model psychosocial support training and workshop during a medical Mission for Syrian refugee children in Jordan: Techniques, lessons learned and recommendations. *Journal of Infant, Child, and Adolescent Psychotherapy*, *18*(4), 352–366. https://doi.org/10.1080/15289168.2019.1690925.

Kuriansky, J. (2019b). Building capacity in low resource settings post-disaster: Empowerment, empathy and esteem in training comforters in the case of Haiti. In W. Jean-Charles (Ed.), *Empathie en Haïti de la Recherche à la Practique à la Croisée de Différents Regards* (pp. 207–230). Coconut Creek, FL: EducaVision. Chapter 11.

Kuriansky, J. (2019c). Expressive arts for helping children heal in crisis and disaster. In I. A. Serlin, S. Krippner, & K. Rockefeller (Eds.), *Integrated health care for the traumatized: A whole person approach*. Lathan, NY: Rowman & Littlefield.

Kuriansky, J. (2019d). Women's issues at the United Nations: Activities, achievements and contributions of the IAAP New York team. *Applied Psychology Around the World*, *1*(3), 37–48. https://iaapsy.org/site/assets/files/1082/apaw_vol1i3.pdf.

Kuriansky, J. (2020a). Chapter 19: Psychosocial recovery after natural disaster: International advocacy, policy, and recommendations. In K. E. Cherry, & A. Gibson (Eds.), *The intersection of trauma and disaster Behavioral health*. New York, NY: Springer Nature.

Kuriansky, J. (2020b). Words matter in combatting covid-19: Continuing to call for "social distancing" can make matters worse. *Black Star News*. https://www.blackstarnews.com/global-politics/others/words-matter-in-combatting-covid-19-continuing-to-call-for-.

Kuriansky, J. (2024). "Health in Your Hands": A platform for innovative healthcare solutions. In J. Kuriansky, & P. Kakkattil (Eds.), *Resilient health: Leveraging technology and social innovations to transform healthcare for COVID-19 recovery and beyond* (pp. 989–1004). Elsevier.

Kuriansky, J., Altungy, P., Twose, G., Liébana, S., Maurice, S. R., Velo, C., Sadia, R., Davcheva, M., & Guimarães, M. A. (2024). Health equity, mental health, and partnerships to build back after COVID-19: Multistakeholder perspectives at a United Nations event. In J. Kuriansky, & P. Kakkattil (Eds.), *Resilient health: Leveraging technology and social innovations to transform healthcare for COVID-19 recovery and beyond* (pp. 115–146). Elsevier.

Kuriansky, J., Margevich, A., Jean-Charles, W., & Daisey, R. (2017). Resilience and recovery in natural disasters and epidemics: Comparisons, challenges, and lessons learned from train-the-trainer projects. In G. Rich, & S. Sirikantraporn (Eds.), *Human strengths and resilience: Developmental, cross-cultural, and international perspectives*. Lanham, MD: Lexington Books.

Kuriansky, J., Medhizadeh, T., & Pikkani, P. (2024). Mental Health Matters at the United Nations: A web and in-person healthy workplace initiative and response during COVID-19 and beyond. In J. Kuriansky, & P. Kakkattil (Eds.), *Resilient health: Leveraging technology and social innovations to transform healthcare for COVID-19 recovery and beyond* (pp. 517–524). Elsevier.

Kuriansky, J., & Nemeth, D. G. (2020). Rebuilding executive functions in environmentally traumatized children and adolescents. In D. G. Nemeth, & J. Glozman (Eds.), *Evaluation and treatment of neuropsychologically compromised children: Understanding clinical applications post Luria and Reitan*. Amsterdam, Netherlands: Elsevier/Academic Press.

Kuriansky, J., Polizer, Y., & Zinsou, J. (2016). Children and Ebola: A model resilience and empowerment training and workshop. In J. Kuriansky (Ed.), *The psychosocial aspects of a deadly epidemic: What Ebola has taught us about holistic healing* (pp. 175–214). Santa Barbara, California: ABC-CLIO/Praeger. Traumatized world: Healing ourselves and our planet. Santa Barbara, California: Praeger Press.

Kuriansky, J., & Zinsou, J. (2019). Mental health and well-being at the United Nations: Activities, achievements and contributions of the IAAP New York team. *Applied Psychology Around the World*, *1*(3), 16–28. https://iaapsy.org/site/assets/files/1082/apaw_vol1i3.pdf.

Lamichhane, P., Puri, M., Tamang, J., & Dulal, B. (2011). Women's status and violence against young married women in rural Nepal. *BMC Women's Health*, *11*(1). https://doi.org/10.1186/1472-6874-11-19.

Nemeth, D. G., Kuriansky, J., Reeder, K. P., Lewis, A., Marceaux, K., Whittington, T., Olivier, T., May, N. E., & Safier, J. A. (2012). Addressing anniversary reactions of trauma through group process: The hurricane Katrina anniversary wellness workshops. *International Journal of Group Psychotherapy*, *62*(1), 129–141.

Nepal Law Commission. (2009). *Domestic Violence (Offence and Punishment) Act, 2066 (2009)*. https://www.lawcommission.gov.np/en/wp-content/uploads/2018/10/domestic-violence-crime-and-punishment-act-2066-2009.pdf.

Nepal Law Commission. (2018). *The Crime Victim Protection Act, 2075 (2018)*. https://www.lawcommission.gov.np/en/archives/category/documents/prevailing-law/statutes-acts/the-crime-victim-protection-act-2075-2018.

Parks, S., Kuriansky, J., & Koch, P. (2022). Confronting the Crisis of COVID-19: Evolution of a Model for Preservation of People and the Planet. *International Perspectives in Psychology Special Issue: Psychology and the COVID-19 Pandemic – A Global Perspective*, *11*(3). https://doi.org/10.1027/2157-3891/a000053.

Peterson, C., Park, N., Pole, N., D'Andrea, W., & Seligman, M. E. (2008). Strengths of character and posttraumatic growth. *Journal of Traumatic Stress*, *21*(2), 214–217. https://doi.org/10.1002/jts.20332.

Ranabhat, C., Kim, C.-B., Choi, E. H., Aryal, A., Park, M. B., & Doh, Y. A. (2015). Chhaupadi culture and reproductive health of women in Nepal. *Asia Pacific Journal of Public Health*, *27*(7), 785–795. https://doi.org/10.1177/1010539515602743.

Regmi, S. K., Pokharel, A., Ojha, S. P., Pradhan, S. N., & Chapagain, G. (2004). Nepal mental health country profile. *International Review of Psychiatry (Abingdon, England)*, *16*(1-2), 142–149. https://doi.org/10.1080/09540260310001635186.

Sharma, J. (2020). *In Nepal, a helpline serves as a lifeline for survivors during COVID-19 lockdown*. https://blogs.worldbank.org/endpovertyinsouthasia/nepal-helpline-serves-lifeline-survivors-during-covid-19-lockdown.

Sherchan, S., Samuel, R., Marahatta, K., Anwar, N., Van Ommeren, M. H., & Ofrin, R. (2017). Post-disaster mental health and psychosocial support: Experience from the 2015 Nepal earthquake. *WHO South-East Asia Journal of Public Health*, *6*(1), 22–29. https://doi.org/10.4103/2224-3151.206160.

Shneiderman, S., Wagner, L., Rinck, J., Johnson, A., & Lord, A. (2016). Nepal's ongoing political transformation: A review of post-2006 literature on conflict, the state, identities, and environments. *Modern Asian Studies*, *50*(6), 2041–2114. https://doi.org/10.1017/S0026749X16000202.

Shrestha, R., Shrestha, S., Khanal, P., & Kc, B. (2020). Nepal's first case of COVID-19 and public health response. *Journal of Travel Medicine*, *27*(3), taaa024. https://doi.org/10.1093/jtm/taaa024.

Sue, D. W. (2001). Multidimensional facets of cultural competence. *The Counseling Psychologist*, *29*(6), 790–821. https://doi.org/10.1177/0011000001296002.

Tedeschi, R. G., Shakespeare-Finch, J., Taku, K., & Calhoun, L. G. (2018). *Posttraumatic Growth: Theory, Research, and Applications* (1st Edition). Routledge.

United Nations (n.d.). Transforming our world: the 2030 Agenda for Sustainable Development. https://sdgs.un.org/2030agenda.

United Nations Conference on Trade and Development (n.d.). UN list of least developed countries. https://unctad.org/topic/least-developed-countries/list.

van Engen, J. (2024). Healthy Entrepreneurs: Transforming access to healthcare through an innovative model of community health entrepreneurship. In J. Kuriansky, & P. Kakkattil (Eds.), *Resilient health: Leveraging technology and social innovations to transform healthcare for COVID-19 recovery and beyond* (pp. 1077–1083). Elsevier.

World Bank (n.d.). Nepal Development Update. https://www.worldbank.org/en/country/nepal/publication/nepaldevelopmentupdate.

World Health Organization. (2020). *COVID-19 public health emergency of international concern (PHEIC) Global research and innovation forum*. https://www.who.int/publications/m/item/covid-19-public-health-emergency-of-international-concern-(pheic)-global-research-and-innovation-forum.

World Health Organization (n.d.-a). India Situation. https://covid19.who.int/region/searo/country/bd.

World Health Organization (n.d.-b). Nepal Situation. https://covid19.who.int/region/searo/country/np.

Xinnhua. (2020). *875 people commit suicide during lockdown period in Nepal*. http://www.xinhuanet.com/english/2020-05/26/c_139089729.htm.

Yadav, R. K., Baral, S., Khatri, E., Pandey, S., Pandeya, P., Neupane, R., Yadav, D. K., Marahatta, S. B., Kaphle, H. P., Poudyal, J. K., & Adhikari, C. (2021). Anxiety and depression among health sciences students in home quarantine during the COVID-19 pandemic in selected provinces of Nepal. *Frontiers in Public Health*, *9*, 580561. https://doi.org/10.3389/fpubh.2021.580561.

CHAPTER 51

The nexus model and the Young Leaders Program: Innovative models for the future of the SDGs in Action Campaign and the Global Councils on Sustainable Development Goals

Christiaan Coetzee[a,b], Judy Kuriansky[c,d,e], and Reem Al Hammadi[a]

[a]Federal Competitiveness and Statistics Centre, Dubai, United Arab Emirates, [b]Etihad Credit Insurance, Abu Dhabi, United Arab Emirates, [c]Global Council SDG3 (Health), "SDGs in Action" Initiative, Dubai, United Arab Emirates, [d]Department of Clinical Psychology, Teachers College, Columbia University, New York, NY, United States, [e]International Association of Applied Psychology, Indianapolis, IN, United States

Introduction

A nexus is defined as a connection or series of connections linking two or more ideas, elements, groups, or things as the central and most important intersection, point, or place. The Nexus Model of the Sustainable Development Goals highlights the intersection of the sustainable development goals of the United Nations 2030 Agenda for Sustainable Development. This model has been adopted in the "SDGs in Action" initiative of the United Arab Emirates (UAE) (Federal Competitiveness and Statistics Centre, 2022a). "SDGs in Action" takes place within the official agenda of the World Governments Summit. The summit is the world's leading platform to shape the future of government. This chapter demonstrates the value of this nexus model and shows how this approach serves as a paradigm of how health innovations in the post-COVID-19 recovery period and beyond can address several sustainable development goals and be creative to introduce innovation and technology into their plan. Two examples are described in this context. The Katsina Eye Care program in Nigeria, which merges SDG 3

Resilient Health. https://doi.org/10.1016/B978-0-443-18529-8.00051-2

(good health and well-being) with SDG 1 (no poverty), SDG 4 (quality education), and SDG 5 (gender equality), was formally presented at the launch of the second term of the UAE's "SDGs in Action" campaign. In another example, a Geospatial Information Systems intervention represents the nexus on education SDG 4 (quality education) and SDG 17 (partnerships for the goals) at the heart, with impact and relevance to all other SDGs.

Leveraging the series of connections and aligned with the UAE's commitment to youth development, the UAE initiated the SDG Young Leaders Program designed to empower youth through knowledge transfer opportunities to building capacity of future sustainable leaders. The program, launched in January 2022 during Expo 2020—Dubai's Global Goals Week, integrated youth representatives into the Global Councils on SDGs initiative. In March 2022, during the World Governments Summit coinciding with the final week of Expo 2020, the program hosted an event themed "WGS Youth for SDGs: Compassion, Passion, and Action for a Better Tomorrow." The event featured internationally renowned personalities, including the first Minister of Youth, the Director-General of the Nuclear Energy Agency, and an international psychologist Dr. Judy Kuriansky, coauthor of this chapter and volume, who interacted with youth from the audience, as described in this chapter.

Background

In September 2015, 193 United Nations (UN) Member States and world leaders gathered in New York at the UN General Assembly to endorse the set of new Sustainable Development Goals (SDGs) as a blueprint for the next 15 years. The scope and ambition of the SDGs individually and as a collective plan are greater than anything that the international community has ever attempted before, and rightly so. The SDGs aim to transform our world and to "leave no one behind," while ensuring the future of all peoples and the planet. The United Arab Emirates (UAE) activated the SDGs with a first-of-its-kind and dedicated event at the World Governments Summit Organization, called "SDGs in Action." The stages of this campaign are shown in Fig. 1.

The inaugural "SDGs in Action" event in February 2016 focused on "*The Role of Monitoring, Reporting and Financing the implementation of the SDGs.*" The follow-up event, held the next year at the World Governments Summit (WGS) in February 2017, discussed "*The Role of Science, Technology and Innovation in achieving the SDGs.*"

The 2018 theme of the "SDGs in Action" focused on "*Partnerships Supporting Innovation for our Common Future,*" in order to break down silos and find innovative means to progress the 2030 Agenda for Sustainable Development. The outcome of the event saw the inception of the *Global Councils on SDGs* 2018–20, with a singular SDG approach at the core of the council's work programs.

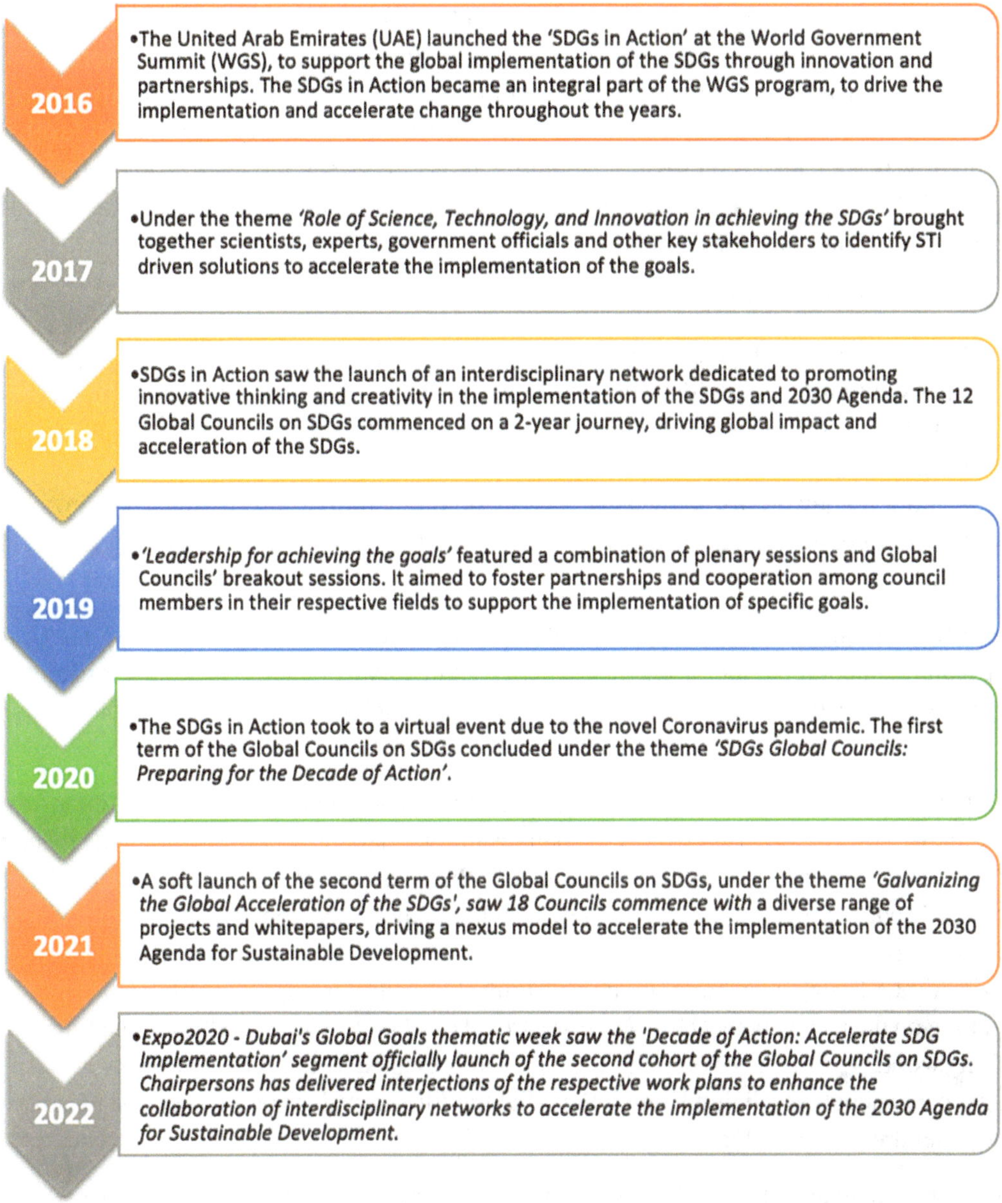

Fig. 1
Stages of the SDGs in Action campaign (Federal Competitiveness and Statistics Centre, 2022a).

Of note, in 2019, several of the innovations which were deemed to have the most potential for advancing the SDGs, including the "Health in Your Hands" platform (presented in another chapter in this volume (Kuriansky, 2024)), were presented at the United Nations at the Science, Technology and Innovation Forum, sponsored by the Mission of the UAE to the United Nations.

The Global Councils on Sustainable Development Goals (Global Councils on SDGs, 2021) is a unique interdisciplinary network of decision makers from governments, international organizations, academia, and the private sector coming together to share innovative practices and discuss the creative implementation of the 17 SDGs at the national and global level. The initiative was launched as a cross collaborative project between the UAE Prime Minister's Office (PMO), the Federal Competitiveness and Statistics Centre (FCSC), and the World Governments Summit Organization.

The mission of the Global Councils is as follows:

(a) Identify factors, which are inhibiting progress toward their particular SDG
(b) Find solutions to issues identified and develop a plan for the implementation of successful policies, programs, technologies, and/or innovative service delivery mechanisms to resolve the issues
(c) Build a framework for knowledge sharing among multiple stakeholders, within and between countries, that will catalyze the council's vision of sustainability well into the future.

The 2019 edition of the "SDGs in Action" focused on "*Leadership for Achieving the Goals*" that featured a combination of plenary sessions and Global Councils' breakout sessions. It aimed to foster partnerships and cooperation among council members in their respective fields to support the implementation of specific goals.

In 2020, amid the novel Coronavirus when travel was restricted, the "SDGs in Action" event took to the virtual world and hosted the councils to celebrate success stories from the inaugural term, under the theme "*SDGs Global Councils: Preparing for the Decade of Action.*" The meeting brought together presenters some of whom have since configured into initial nexus models (described later in this chapter). For example, the Finance Advisory Board, chaired by Marie Aimee Boury, managing director of Impact-Based Finance of Société Generale, speaking on the panel about "The Way Forward... How Do We Get There?" is at the time of this writing working with SDG 3 on Health, which was presented by Dr. Judy Kuriansky, who spoke on the panel about "Realizing Human Potential for All," as well as other councils. Panelists on that panel are shown in Fig. 2.

The event concluded with a Call to Action for the global community to nominate themselves to support and accelerate the implementation of the SDGs through the councils, which has been transformed to form a nexus model approach, especially as the SDGs are interlinked and complex in nature. Two examples of this interlinkage are shown later in this chapter.

Under the theme "*Galvanizing the Global Acceleration of the SDGs*" and on the sidelines of the UN General Assembly 2021, the second cohort of the Global Councils on SDGs met to have general introductions and to set expectations to further progress the implementation of the global goals. The 2021–23 councils are set to drive impact toward the implementation of the

Fig. 2
The panel on "Realizing Human Potential for All" at the "SDGS in Action" program at the WGS, with then-chairs (top, left to right) of SDG 5 on women's empowerment, Vice President, UAE Gender Balance Council, H.E. Mona Al Marri; SDG 1 on poverty, Dr. Mahmoud Mohieldin; SDG 3 on health, Dr. Judy Kuriansky; (bottom, left to right): SDG 4 on education, CEO at Dubai Cares H.E. Dr. Tariq Al Gurg and moderator UN Foundation president and CEO Elizabeth Cousens.

2030 Agenda for Sustainable Development. While continuing chairpersons of the term 2018–20 will focus on their respective singular SDGs, the nexus model approach has been introduced to encourage new council chairpersons to adopt innovative projects, whitepapers, or initiatives that interlink the SDGs and to rebuild sustainably in a post-COVID-19 world.

The SDG in Action meeting in 2022 in Dubai

At the 2022 SDGs in Action, held in Dubai in-person, January 15–17, initiative was presented in its next phase in a special 3-day session. This was held during the Global Goals thematic week coinciding with Expo 2020—Dubai, themed "*Decade of Action: Accelerate SDG Implementation.*" The 17 Global Councils on SDGs 2021–23 officially launched the work programs. This was done in the presence of multiple ministers of the UAE and the United Nations deputy secretary-general, Amina J. Mohammed. Eight of the councils embarked on dedicated work programs to ensure the nexus model remains at the core of the second cohort, while the remainder of councils acknowledged the importance of cross council collaboration and impact across multiple SDGs in their respective work programs. The WGS SDGs in Action,

Fig. 3
Panelists in the "Change Agents Showing Endless Possibilities" during the "SDGs in Action" sessions during the World Governments Summit 2022 with Global Council chairs Dr. Muhammad Pate (Health), Ibrahim al Zu'bi, and Marie Aimée Boury (Finance Advisory Board).

held March 29–30 under the theme "Scaling New Heights for the Future of Humanity," continued its legacy to serve as platform to establish partnerships for the goals.

The focus of the councils incorporated the nexus model, ensuring an impact-driven approach that will support the Global Councils to address different global goals in a single project, whitepaper, or initiative. Each council's nexus model must have a "focus" goal and include a combination of additional goals.

Fig. 3 shows the session "Change Agents Showing Endless Possibilities" with Dr. Mohammad Pate, the cochair of the second term of the GC SDG3 (Health), Julio Frenk, professor of the Practice of Public Health Leadership at the Harvard T. H. Chan School of Public Health (presenting on behalf also of his cochair Dr. Zsuzsanna Jakab, deputy-general of the World Health Organization), with Marie Aimee Boury, chair of the Finance Advisory Board, whose Global Councils are collaborating.

The SDGs in Action meeting in 2023 in Dubai

The SDGs in Action at World Governments Summit 2023 held February 12–15 activated and scaled actions toward the Decade of Action with a theme "Redesigning the Future of Our World: 2030 Principles and beyond." The 2023 version addressed fundamental challenges to implement the global goals and reshaped our common for generations to come. The Global Councils on SDGs commenced with tailored workshops to support and drive impact toward the SDGs and reported their midterm progress and aspirations to conclude work programs in February 2024.

Background of the nexus model

The following are some principles of the nexus model.

(a) Some situations cannot be addressed through single sector actions taken in isolation.
(b) A coherent and integrated nexus approach can help identify synergies and trade-offs, leading to better outcomes for current and future generations.
(c) The 17 SDGs, adopted by UN member states in September 2015, affect practically all aspects of public policy.
(d) Policy Coherence for Sustainable Development (PCSD) assists in understanding the interlinkages between economic, social, and environmental policies when related to access, availability, and sustainability of our planet's natural resources.
(e) The nexus approach will enable the councils to identify linkages, overcome trade-offs, and ensure synergies are utilized.

One council (Federal Competitiveness and Statistics Centre, 2022b) launched a nexus model that drives impact toward SDG 3 (Good Health and Well-being), while several councils introduced work programs interlinking multiple other SDGs.

Examples of the nexus models in action

The nexus model interlinkage is shown in Fig. 4 for two of the examples described in the text, with one set of lines (in the color red) showing the connection between the SDGs for one example, and the other set (in the color green) showing the interconnection of the SDGs for the other example.

The nexus model of SDGs 1, 3, 4, and 5

Dr. Manal Taryam, chief executive officer, and board member Noor Dubai Foundation, serves as chairperson to the council targeting SDG 3 with a nexus model to SDGs 1 (End Poverty), 4 (Quality Education), and 5 (Gender Equality), through cataract and refractive error services, as part of the *Katsina Eye Care Program* in Nigeria. Cataract and refractive error are the two major causes of blindness and visual impairment globally responsible for vision loss in over 2.2 billion people in the world. The interrelationship of these SDGs is shown in Fig. 4 (in red color).

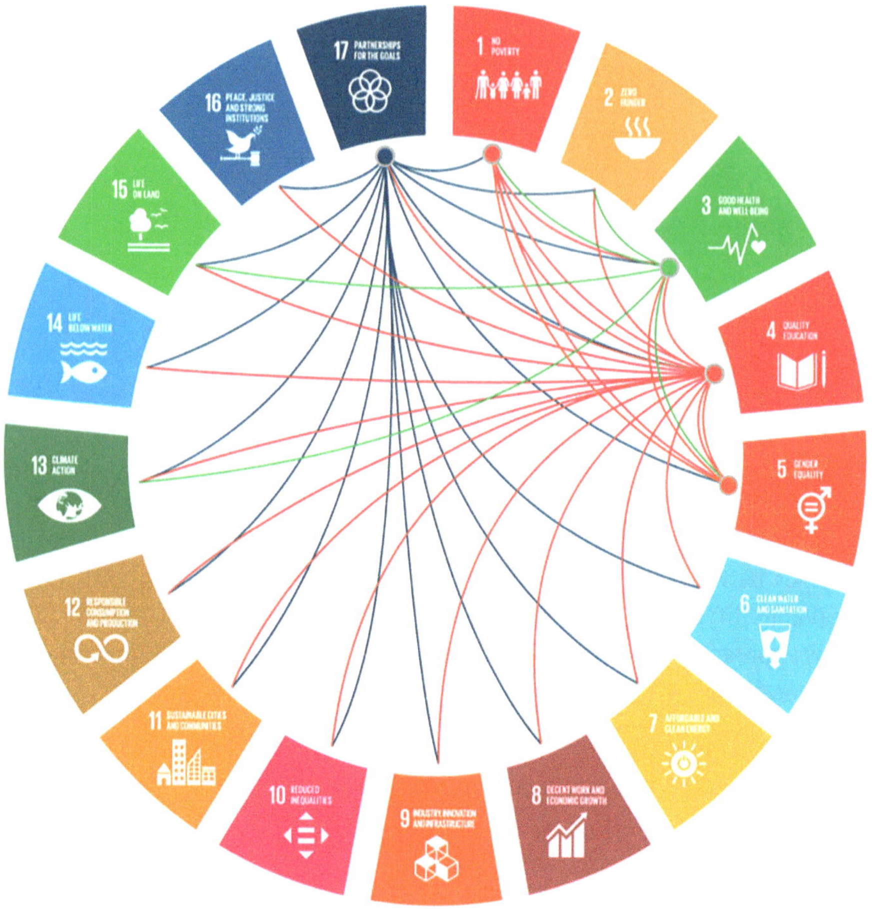

Fig. 4

Visualizing the interlinkages of the nexus model for two examples of the Global Councils on SDGs featured within this chapter, the Katsina Eye Care Project and the Geospatial Information Systems Education nexus (Federal Competitiveness and Statistics Centre, 2022c).

The concept of the Katsina Eye Care Program aims at improving the quality of life and the socioeconomic status of the people of the Katsina state in Nigeria and will contribute to reducing blindness and visual impairment in the state by at least 30%–40%. The objectives of the project include:

(a) Raise the effective cataract surgical coverage in Katsina state by at least 50%, taking into consideration gender equality, within 4 years.
(b) Improve refractive error services among schoolchildren in parts of Katsina state by at least 50%, with an increase in effective refractive error coverage among the students screened in selected schools within 2 years.

The program has identified milestones aligned with the nexus approach and includes:

(a) Conducting 11,000 cataract surgeries in the state in 4 years.
(b) Ensuring more than 80% of operated cases achieve good surgical outcomes.
(c) Ensuring gender equality in the cataract surgery delivery such that at least 50% of recipients are women.
(d) Conducting eye screening of 160,000 elementary school pupils across 160 elementary schools in the state.
(e) Conducting eye screening of 3000 teachers in the 160 schools.
(f) Providing refractive error corrections to at least 1500 pupils and 1500 teachers.

In 2022 the council reported progress in Noor Dubai Foundation's annual report (Noor-Dubai, n.d.). A total of 17,519 screenings took place, complemented with 9273 surgeries. The Katsina Eye Care Program further trained 267 people and provided 250 primary health centers with basic eye screening equipment that benefited 7.5 million people indirectly. In addition, Noor Dubai launched the Katsina School Eye Health Program in 2022 with the main objective of improving access to eye care services to school students and teachers. A total of 161,524 students and 1800 teachers were screened while 88 teachers and ophthalmic nurses were trained. The program actively drives psychological, social, and academic development of children that will have positive long-term impact on society.

The council's expected outcomes address health/medical, social, educational, and economic aspects of life in their respective communities. The council will address multiple aspects including economic (increased employment, productivity, and income), social (reduced extreme poverty and hunger, increased independence, self-esteem and gender equality, and improved social networks), health (improved quality of life measured in DALYs averted and QALYs gained, reduced mortality rates), and education (achieving universal primary education, increased school attendance and academic achievements, and improved teaching by teachers).

Nexus model of SDGs 4 and 17, with relevance to all SDGs

Clint Brown, director at Esri, serves as chairperson of the Geospatial Information Systems Education Council that focuses on SDGs 4 (quality education) and 17 (partnerships for the goals), with relevance to the additional 15 global goals. The council, driven on Esri technology, learned from experiences and is implementing an SDG framework worldwide to promote GIS as an effective tool to address the implementation and report on the progress made toward the 2030 Agenda. The council provides lessons and tutorials for understanding and progressing the SDGs and encourages the global community to adopt technology to accelerate the implementation of the global goals. Resources that are available include websites like SDG

Data Alliance (SDG Data Alliance, n.d.), the UAE's SDGs Data Hub (UAE SDGS Data Hub, n.d.), and the SDG Geospatial Learning Lab (SDG Geospatial Library, n.d.).

The council, with Esri, has achieved key milestones including the launch of the Sustainable Development Goals Solution in November 2022, facilitating government sharing of SDG progress and community engagement. In February 2023, the council released "Storytelling for a Sustainable World" that offers a collection of all SDG stories complemented by data, sample stories, and learning resources. In addition, the SDG Learning Lab was launched in January 2023 and delivers tutorials and lessons for SDG achievement under the name SDG Geospatial Learning Lab. The latter includes unique tutorials and content relevant to each of the 17 global goals, including 14 tutorials and 15 content-related materials on SDG 3 (good health and well-being). The SDG 3 workflows and tutorials utilize problem-solving exercises and hands-on learning experiences through GIS applications, including policy mapping to improve newborn health, monitoring malaria pandemics, and supporting midwives worldwide. As an outcome, the council has reported in the Global Councils on SDGs midterm progress report (Publications (SDGs), n.d.) February 2023 edition that the SDG Geospatial Learning Lab, in February 2023, published a total of 320 tutorials and related content.[a]

The FCSC, in its capacity as Secretariat to the UAE National Committee on SDGs and the Global Councils on SDGs, invites everyone to follow the progress of the Global Councils on SDGs through the UAE SDG Data Hub (UAE SDGS Data Hub, n.d.) (https://sdgsuae-fcsa.opendata.arcgis.com/) and actively engages in the implementation of the 2030 Agenda for Sustainable Development.

In reaction to the "SDGS in Action" campaign, the UAE Minister of State for International Cooperation, Her Excellency Reem bint Ebrahim Al Hashimy, president of the Global Councils on SDGs, has said:

> Today *[16 September 2021]*, we welcome 17 distinguished Councils who will embark on the second term of the Global Councils on SDGs. The term will divert from a singular SDG approach to a nexus approach. Giving the complexity and interlinkages of the SDGs, the Councils will galvanize and accelerate the implementation of the goals through innovative partnerships, inspiring projects, and informative white papers that will advance our understanding of one another as well as the challenges we face as a society today. (World Government Summit, 2021)

[a] The text mentioned before serves as summaries of the two respective Global Councils on SDGs contribution to the Global Councils on SDGs midterm report, published by the Federal Competitiveness and Statistics Centre in February 2023. Full details can be found in the mentioned report, available on the Federal Competitiveness and Statistics Centre official website.

During the official launch of the Global Councils on SDGs 2021/2023, Al Hashimy said:

> The launch of the second cohort of the Global Councils on SDGs during the Expo 2020 – Dubai's Global Goals thematic week, is a reconfirmation of the UAE's commitment to strengthen its role in enhancing the welfare of societies and inclusive sustainable development. The UAE government is driven to support the global implementation of the 2030 Agenda for Sustainable Development, and reiterated its commitment to complement the Decade of Action. (Emirates News Agency, 2022)

During the official launch of the second term of the Global Goals at Expo 2020—Dubai, referred to earlier and described in more detail later, the United Nations deputy secretary-general, Amina J. Mohammed, said:

> Global partnerships are the pathway to achieve global sustainability. The Global Councils for SDGs is a prime example of bringing together decision makers and thought-leaders across multiple disciplines and geographies to tackle, head on, some of the world's most pressing issues, bringing together, world governments, private sector, academia and societies, to deliver decisive action aimed to build a sustainable future that leaves no one behind. (Emirates News Agency, 2022)

The cross-cutting issue of youth as part of the SDGS in Action initiative

Consistent with the guidelines of the UN 2030 Agenda for Sustainable Development, several issues are considered cross-cutting, meaning a topic identified as important that affects most or all aspects of development and therefore should be integrated and mainstreamed throughout all the goals, and also the stages of development from policy to implementation and evaluation. Youth is one of these topics.

Thus the Federal Competitiveness and Statistics Centre, in its capacity as the secretariat of the UAE National Committee on SDGs, has been keen on empowering and enabling youth in the UAE and actively involving them in the UAE journey toward the implementation of the SDGs. The FCSC in partnership with the Federal Youth Authority, UAE CSR Fund, and the UAE Government Leadership Program launched the SDG Young Leaders Program in January 2022 to provide youth with the knowledge, experience, and skills necessary to participate effectively in supporting and leading national and international initiative and programs to achieve the SDGs and become the leaders of the future. The dedicated platform for youth dialogue engages the youth of the UAE in the Decade of Action for the SDGs.

The program includes three different tracks for a new journey toward inclusive youth development:

(1) SDGs Youth Council, where the youth members will be working closely with the Secretariat of the UAE National Committee on SDGs on developing initiatives and

projects that encourage UAE youth involvement in the SDGs and develop SDG-related studies, initiatives, and policy proposals to support the objectives of the UAE National Committee on SDGs.

(2) Youth Media Council for SDGs, this group aims to educate, captivate, and accelerate youth's involvement in the 2030 Agenda for Sustainable Development and highlight the power of media and how youth approach media.

(3) Global Councils on SDGs Youth Representatives, the 2021/23 Councils will each have a youth member to ensure capacity development for future global decision makers.

In fulfillment of these goals, the SDG Young Leaders held a launch event meant to inspire our young leaders with heart-to-heart dialogue with global figures to inspire young leaders in their journey to fulfill the SDGs. The program was held on the margins of the WGS 2022, on the topic of "WGS Youth for SDGs: Passion, Action and Compassion for a Better Tomorrow."

This program, held at the Youth Pavilion at Expo 2020—Dubai, was organized by the Federal Competitiveness and Statistics Centre. Members of the panel and organizers, as well as student attendees, are shown in Fig. 5.

Fig. 5
Organizers and speakers of the "WGS Youth for SDGs: Compassion, Passion, and Action for a Better Tomorrow". (left to right): Her Excellency Hanan Ahli, director of the Federal Competitiveness and Statistics Centre; Mr. William D. Magwood, director-general of the Nuclear Energy Agency; Ms. Reem Al Hammadi, former senior projects manager at FCSC; Dr. Judy Kuriansky, international renowned psychologist and Main United Nations NGO representative; His Excellency Abdulla Lootah, Deputy Minister of Cabinet Affairs for Competitiveness and Knowledge Exchange, vice chair of the UAE National Committee on SDGs, and vice president of the Global Councils on SDGs; Mr. Christiaan Coetzee, freelance consultant on SDGs at FCSC.

The session was moderated by His Excellency Abdulla Nasser Lootah, Deputy Minister of Cabinet Affairs for Competitiveness and Knowledge Exchange, vice president of the Global Councils on SDGs and vice chairman of the UAE National Committee on SDGs, who gracefully and with great dynamism introduced the topic about passion, action, and compassion, embodying these qualities, setting the stage for the speakers.

Opening remarks were made by Mr. William Magwood, noted physicist and director-general of the Nuclear Energy Agency, who described how energy and education is important for youth, highlighted the power of youth in creating change, and encouraged youth to have a sense of urgency and passion.

Ferid Belhaj, vice president of the World Bank in the Middle East and North Africa (MENA) region, who was previously in charge of the banks' programs in Lebanon, Syria, Jordan, Iraq, and Iran, described the Arab region as raising up with its youth, calling youth of the MENA region "The Bright Spot" as he sees the future as a brighter tomorrow with the power of the youth as the real energy, hope, and drivers of change. But it is the responsibility of the leaders in the region to give youth knowledge and hope.

Featured speaker, noted international psychologist Dr. Judy Kuriansky, gave a highly interactive and passionate presentation, entitled a "Heart-to-heart action plan for youth." In this dialogue, she described the positive climate for mental health and well-being, and involvement of youth at the United Nations and globally, reflected in the inspiring approach of "Positive thinking and family & community cohesion" and thematically related books by His Highness Sheikh Mohammed bin Rashid Al Maktoum, vice president and prime minister of the UAE and ruler of Dubai, as well as reflected in the happiness theme incorporated in the UAE government

Fig. 6
Dr. Judy Kuriansky speaking about global leaders' support of mental health and well-being.

policy and plans. An important point, shown in Fig. 6 in her engaging slide presentation, is that mental health is being recognized as a priority by global leaders and critical to achieve inclusive and sustainable development—especially given the emotional issues triggered in populations worldwide trying to cope with the COVID-19 pandemic. Pictured are the World Health Organization Director-General Dr. Tedros Adhanom Ghebreyesus and United Nations Secretary-General His Excellency António Guterres, and also United Nations Deputy Secretary-General Her Excellency Amina J. Mohammed who have all spoken out passionately and definitively about the importance of mental health.

Dr. Judy, as she is fondly called, also presented concrete actions that youth can take to advance the SDGs, within her philosophy of being brave, creative, resilient, and going for your dreams, which inspired youth who were present to give personal examples. In the interactive part of the dialogue, she went into the audience and engaged individuals in selected activities she uses worldwide to promote resilience, much to the assemblage's enjoyment. Several youth volunteered how they appreciated her unique action-oriented end engaging presentation style (unique in contrast to more usual one-way lectures) and how memorable her messages of empowerment were, and how even she had "changed their lives." As shown in Fig. 7, Omar Aldhuhoori, member of the SDG Young Leaders Program, shared the importance to him of her message in the Dream Ball activity about identifying and acknowledging your dream—that he said he would remember for the rest of his life. Following the activity she does worldwide for self-fulfillment, Dr. Judy asked him to announce his dream—to have a happy family and to make a difference in the world—to the group, and for them to repeat it back to him—"Omar, you have a happy family and are making a difference on the world"—so he can feel its strength and power, first, as if it is true in the here and now (not the future) and second, that others are "witnessing" and affirming it as real. The joy and love were palpable in the room.

Closing Remarks were given by Her Excellency Shamma bint Suhail Faris Al Mazrui, UAE Minister of State for Youth. She spoke about the power of passion and mentioned three important points to harness passion: (1) shedding skin, removing the dead cells preventing growth and being open to transformation and change and unleashing one's full capacity of learning; (2) recognizing the metaphor of music, seeing yourself as an orchestra playing a piece of music in harmony, from the inside, and moving from thinking to expressing; and (3) having courage, to act with harmony and choose motion over being static, and to be self-inspired to take action, in the same way as the UAE is a self-inspired nation that seeks to inspire others.

Members of the SDG Young Leaders Program engaged with Dr. Judy Kuriansky after the event to share thoughts and lessons learned, and express appreciation for the lively and memorable presentation. Shown in Fig. 8 are Hamdan Al Khoori, Reem Almusabeh, Omar Aldhuhoori, Dr. Judy Kuriansky, Hamed Al Ali, Abdulla Aljunabi, Mohammed Ahli.

Fig. 7
Dr. Judy Kuriansky and Omar Aldhuhoori, member of the SDG Young Leaders Program, shared the importance **to** him of her message in the Dream Ball activity about identifying and acknowledging your dream.

Fig. 8
Dr. Kuriansky and members of the SDG Young Leaders Program after the event. (left to right): Hamdan Al Khoori, Reem Almusabeh, Omar Aldhuhoori, Dr. Judy Kuriansky, Hamed Al Ali, Abdulla Aljunabi, Mohammed Ahli.

Background: Federal Competitiveness and Statistics Centre

The Federal Competitiveness and Statistics Centre (FCSC) is a government center affiliated with the Ministry of Cabinet Affairs in the United Arab Emirates. FCSC was established according to federal decree number 2/2020, with the aim to develop and enhance the UAE's performance in the areas of global competitiveness, statistics and data, and to support the country's journey in achieving UAE's Centennial Plan 2071. FCSC is actively engaged in building an integrated national statistical system, raising the UAE's competitiveness across various sectors, enhancing the country's position within global competitiveness reports and indicators, and is engaged in achieving the Sustainable Development Goals (SDGs) in its capacity as the Secretariat of the National Committee for the Sustainable Development Goals. FCSC ideates and recommends policies, strategies, legislations, and plans related to competitiveness and statistics to different stakeholders and decision makers in the UAE. FCSC is actively involved in raising awareness on the culture of competitiveness and the importance of providing quality and accurate data and statistical information at the national level.

Background: World Governments Summit Organization

The World Governments Summit Organization is a global, neutral, nonprofit organization dedicated to shaping the future of governments.

The WGS is the largest global platform dedicated to exploring the future of governments driven by technological advances and evolving citizen expectations and highlights future trends in government services, leadership, and innovation. The summit brings together over 4000 leaders, policy makers, academics, and key representatives from international organizations from over 190 countries, thereby creating an exciting opportunity to exchange innovations and experiences and build strong networks for collaboration.

Following the adoption of the SDGs in 2015 and as part of the UAE's commitment to their implementation, the World Governments Summit in its fourth edition in 2016 introduced a special track to focus on the SDGs and their implementation. The inaugural "SDGs in Action" at the World Governments Summit event focused on the role of monitoring and reporting and financing in implementing the SDGs, which later became the platform to launch the Global Councils on SDGs. Subsequent sessions were held at the WGS.

Expo 2020—Dubai

On 27 November 2013, fireworks painted the sky from the world's tallest building, the Burj Khalifa, as thousands of people across the UAE celebrated the bid to host Expo 2020—Dubai, the first World Expo in the Middle East, Africa, and South Asia region. The Expo, held from 1 October 2021 to 31 March 2022 (delayed from 2020 due to the COVID-19 pandemic), under the theme "Connecting Minds, Creating the Future," consisted of a large area of land with pavilions hosted by countries, partners, and organizations, constructed with their own design and content message featuring one of the three subthemes: opportunity, mobility, or sustainability. The exhibits presented visitors with a vast opportunity to experience displays and interactive demonstrations, films, foods, shows, and goods that inspire, empower, and drive positive environmental, social, and economic change about global issues, and as a totality, to feel the unity of the world that is so needed right now.

After the Expo 2020 closed on March 31, 2022, the physical site is being transformed into District 2020, a model global community for the future, and home to world-class innovation, educational, cultural, and entertainment facilities. District 2020 has been designed to support the future of work and living and to facilitate connection and collaboration, serving as home to new ideas and innovations in an inclusive and diverse community thriving in a balanced and sustainable way of life.

The way forward

The FCSC and the WGS remain committed to support the Global Councils on SDGs, with its interdisciplinary network of chairpersons, vice chairs, youth, and members, to continue its journey to support the global implementation of the 2030 Agenda for Sustainable Development and the Decade of Action. The individual councils' work programs focus on innovative projects that follow the nexus model and accelerate the SDGs, and also that are currently, and for the future, working together in the spirit of SDG 17 about partnerships, to accelerate global prosperity for the people and the planet. The Youth Leaders Program continues to expand to represent the importance of the youth role in achievement of the SDGs and in sync with the Global Councils, with a youth member on each council.

References

Emirates News Agency. (2022). *Global Councils for SDGs to strengthen efforts and accelerate implementation.* Retrieved from https://www.wam.ae/en/details/1395303011301.

Federal Competitiveness and Statistics Centre. (2022a). Concept note: SDGs in Action @ World Government Summit 2022. In *World Government Summit. Dubai, UAE.*

Federal Competitiveness and Statistics Centre. (2022b). *Visualizing the interlinkages of the Nexus model.*

Federal Competitiveness and Statistics Centre. (2022c). *Global Councils on SDGs 2021/23—Charter document.* Dubai.

Global Councils on SDGs. (2021). Retrieved 4 March 2022, from https://global-councils-on-sdgs-fcsa.hub.arcgis.com/.

Kuriansky, J. (2024). "Health in Your Hands": A platform for innovative healthcare solutions. In J. Kuriansky, & P. Kakkattil (Eds.), *Resilient health: Leveraging technology and social innovations to transform healthcare for COVID-19 recovery and beyond* (pp. 989–1004). Elsevier.

Noor-Dubai. (n.d.). https://noordubai.ae/MediaCenter/en_AnnualReport.

Publications (SDGs). (n.d.). Federal Competitiveness and Statistics Centre. https://fcsc.gov.ae/en-us/Pages/Media/Publications.aspx#?id1=3.

SDG Data Alliance. (n.d.). SDG Data Alliance. https://www.sdg.org/.

SDG Geospatial Library. (n.d.). https://yoursdgstory-learngis3.hub.arcgis.com/.

UAE SDGS Data Hub. (n.d.). https://sdgsuae-fcsa.opendata.arcgis.com/.

World Government Summit. (2021). *The World Government Summit Organization hosts an international meeting to accelerate the implementation of the SDGs and announces 2nd term of the Global Councils on SDGs initiative with a wide global participation*. Retrieved from https://www.worldgovernmentsummit.org/press/releases/the-world-government-summit-organization-hosts-an-international-meeting.

PART 4

Innovative financing for health

Scaling innovations for health impact requires financing; thus, this Part covers models for financing health and mental health, including for low- and middle-income countries, as well as a personal story about fundraising approaches.

A memorable mantra: money for health and mental health is an investment not an expenditure. Experts in these chapters tell us how and why that's true.

What notables have to say

No one has ever become poor by giving.

—*Anne Frank*

Do what you love. When you love your work, you become the best worker in the world.

—*Uri Geller, Israeli-British illusionist, magician*

When we give cheerfully and accept gratefully, everyone is blessed.

—*Maya Angelou, American memoirist, popular poet, and civil rights activist*

Remember that the happiest people are not those getting more, but those giving more.

—*H. Jackson Brown Jr., American author*

Kindness in giving creates love.

—*Lao Tzu, ancient Chinese philosopher*

It is health that is the real wealth, and not pieces of gold and silver.

—*Mahatma Gandhi, Indian civil rights activist and pacifist*

Our prime purpose in this life is to help others.

—*Dalai Lama, spiritual leader of Tibet*

It's not about giving back if you're successful or a celebrity or how much money you have: it's about your responsibility as an adult to help others.

—Trisha Yearwood, American singer

Every good act is charity. A man's true wealth hereafter is the good that he does in this world to his fellows.

—Moliere, French playwright

CHAPTER 52

Introduction to innovative finance for health

Christoph Benn[a] **and Patrik Silborn**[b]

[a]Joep Lange Institute, Geneva, Switzerland, [b]Theia Advisors, Singapore, Singapore

Since the turn of this century, the world of financing for development has increasingly emphasized innovation with the aim to mobilize more money and achieve improved development outcomes. This is also true in Global Health where new technologies and innovations—if properly funded—can help accelerate the achievement of Universal Health Coverage by 2030.

This chapter presents an introduction to innovative finance and its promise and limitations. It also explores a new emerging approach—Global Public Investments—and how this may help incentivize greater investments in resilient health systems, especially urgent in the period of recovery from the COVID-19 pandemic and beyond.

Background

Innovative finance became a prominent topic of the global discourse in the context of the United Nations Millennium Development Goals (MDGs), eight commitments that all UN member state governments agreed to achieve from 2000 to 2015, and in the subsequent Monterrey Consensus, a declaration of intent about financing for development made in 2002.

Three of the eight MDGs addressed health: MDG 4 (Reduce child morality); MDG5 (Improve maternal health); and MDG6 (Combat HIV/AIDS, malaria, and other diseases).

When the MDGs were first adopted, there was general acknowledgment that traditional sources of development financing—domestic resources and official development assistance (ODA)—would not be enough to meet the ambitious goals.

As a response to the growing gap in development financing between identified needs and available funding through the established sources, the United Nations (UN) organized the International Conference on Financing for Development in Monterrey, Mexico, in 2002. In the resulting Monterrey Consensus document, Heads of States and Governments for the first time "recognize[d] the value of exploring innovative sources of finance provided that those sources do not unduly burden developing countries" (United Nations, 2002). The consensus

Resilient Health. https://doi.org/10.1016/B978-0-443-18529-8.00052-4

also urged banks and financial institutions to "foster innovative development financing approaches" and underscored its support for public-private sector partnerships (United Nations, 2002).

Since 2002, a range of international organizations, governments, and businesses have developed and implemented innovative financing instruments. The UN and its agencies, development banks, and multilateral platforms have also established commissions and studies about the promise—as well as the limitations—of innovative finance (United Nations, 2011).

Definitions of innovative finance

There are many definitions of innovative finance. The World Bank Group defines it as "any financing approach that helps to:

- Generate additional development funds by tapping new funding sources (that is, by looking beyond conventional mechanisms such as budget outlays from established donors and bonds from traditional international financial institutions) or by engaging new partners (such as emerging donors and actors in the private sector).
- Enhance the efficiency of financial flows, by reducing delivery time and/or costs, especially for emergency needs and in crisis situations.
- Make financial flows more results-oriented, by explicitly linking funding flows to measurable performance on the ground" (The World Bank Group, 2022).

Although there are other definitions, these generally accord with the need to mobilize additional funding beyond domestic resources and ODA, increase the efficiency of development financing, and enhance development results.

The early years of innovative finance

During the early years of innovative finance, global leaders were particularly interested in exploring the use of new, earmarked taxes for development. One of the most well-known examples was a tax on air travel launched by France in 2006 and joined by other countries, including Cameroon, Chile, the Republic of the Congo, Madagascar, Mali, Mauritius, Niger, and South Korea (KOICA global disease eradication fund, 2022). This airline levy became the backbone of a new international organization: Unitaid. Another tax that was discussed extensively as a potential source for development finance was a tax on financial transactions (FTT) or the so-called Tobin tax. Despite the support of several European countries, the FTT has not been implemented yet on a multicountry level.

In addition to exploring earmarked taxes, development actors implemented other innovative finance initiatives in the years following the Monterrey Consensus. In the area of Global Health, the Global Fund to Fight AIDS, Tuberculosis and Malaria (the Global Fund), as well as the Gavi Alliance, pioneered several initiatives. These included consumer donations (RED), debt swaps (Debt2Health), capital market solutions (IFFIm), and advanced market commitments to incentivize innovation and production of vaccines (Gavi AMC).

In the wake of the financial crisis in 2007 and 2008, ODA budgets came under increased threat leading to a growing interest in innovative finance as an alternative source of funding. This trend was demonstrated by the commissioning of two seminal reports, both issued in 2011 and still relevant today. The then-secretary-general of the United Nations Ban Ki-moon (UNSG) presented a report to the UN General Assembly on "Innovative mechanisms of financing for development," (United Nations, 2011) and Bill Gates, cochair of the foundation that bears his name, was asked to present a report at the G20 Summit in France, entitled "Innovation with impact: financing 21st century development" (Gates Notes, 2022). Both reports recommended the exploration of new mechanisms but were realistic in terms of expectations. They were adamant that innovative finance should always be additional and complementary to existing channels, in particular ODA. The Gates report called for increased investments from the private sector but concluded: "I am excited by the prospects for private sector investment in development, but I am convinced of the need for countries to keep investing in ODA."

Similarly, the report of the UNSG stated: "While the potential to raise additional resources for development remains significant, appropriate measures need to be taken to make innovative financing mechanisms more durable, predictable and effective."

In 2015 the MDGs were replaced by the Sustainable Development Goals (SDGs), a call for action negotiated over several years and agreed upon as 17 new goals that UN Member State governments committed to achieve by the year 2030. In this new blueprint, SDG3 called for "Good health and well-being" for all. In comparison to the MDGS, the goals for health were subsumed under one heading, and expanded considerably with 13 targets including maternal and child mortality, infectious disease, noncommunicable diseases, mental health, tobacco control, road safety, vaccines, and Universal Health Coverage, measured by 28 indicators. The financing gap to achieve the SDGs in countries in the global south has been estimated at US$ 2.5 trillion per year (OECD, 2020).

The financial gap to achieve the SDGs combined with the human, social, and economic costs of COVID-19 has increased the interest in innovative finance yet again. The world must invest more in Pandemic Preparedness, Prevention and Response, and build stronger and more resilient health systems to reach the goal of Universal Health Coverage by 2030—all at a time of severe budget constraints in every country.

Innovative finance taxonomy

Given the many definitions of what constitutes innovative finance, there are various taxonomies, or descriptions, of the innovative financing landscape and its mechanisms. The taxonomy presented in this chapter, shown in Fig. 1, has been developed by the authors of this

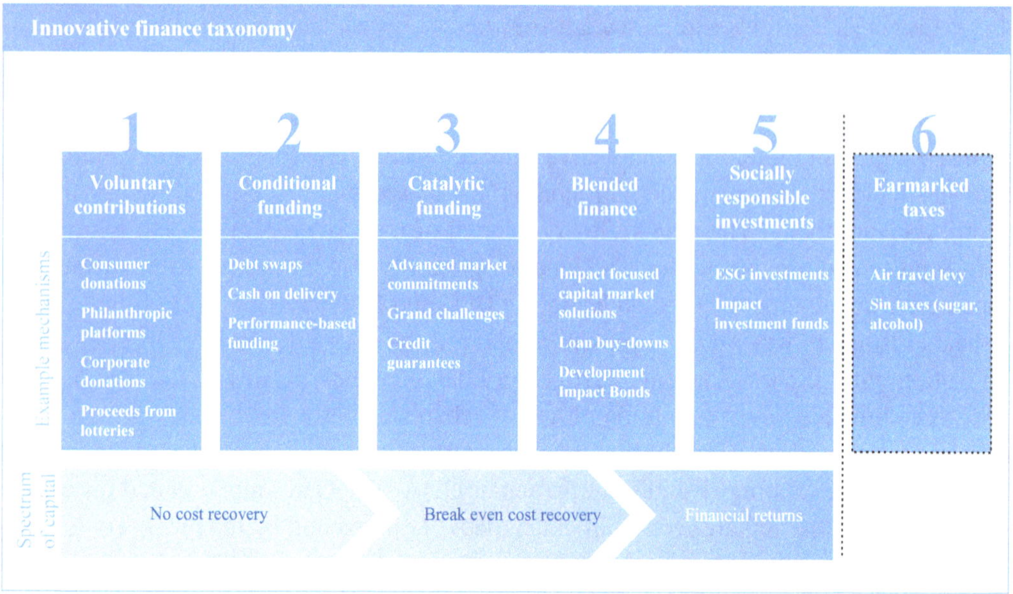

Fig. 1
Innovative finance taxonomy.

chapter, influenced by the work of the Global Fund (The Global Fund, 2019), USAID (USAID, 2019), and other organizations.

The taxonomy introduces six different groups of innovative finance mechanisms, which we define as follows:

(1) *Voluntary contributions*: Grants from the private sector, philanthropists, private foundations, and/or direct fundraising from individuals.
(2) *Conditional funding*: Development funds that are conditional on actions or the achievement of predetermined results by the recipients.
(3) *Catalytic funding*: Development funds that strive to stimulate public/private sector innovations, or market-based solutions that can be delivered at scale.
(4) *Blended Finance*: Development funds that incentivize private sector investment in sustainable development or that facilitate the mobilization of additional development funds in the form of loans or other funding.
(5) *Socially responsible investments*: A broad term often referring to investments that strive to achieve financial (sometimes concessionary), social, and environmental returns for its investors.
(6) *Earmarked taxes*: Specific taxes whose proceeds are earmarked for development purposes. This category stands out for a few different reasons. Rather than being an innovative instrument (it is difficult to argue that taxes, which have been used for millenniums are innovative), it represents a political choice about how resources are being allocated. Contrary to the other categories mentioned, it also requires legislative action

and must be led by governments. The other mechanisms may also be hindered or helped by legislative action. To generalize, however, such mechanisms may still proceed without major political changes and can often be led by organizations or businesses.

The "spectrum of capital" ranges from donors that provide grant funding (no cost recovery) to investors who recover their principal investment (break even cost recovery) to investors who seek social and financial returns. There are no hard lines between these groups of mechanisms, and many of them overlap. The same can be said about the spectrum of capital, where the borders between the categories are fluid.

Examples of innovative finance mechanisms

In the area of Sustainable Development, Global Health organizations stand out as trailblazers in the area of innovative finance. The following examples are drawn mainly from Global Health.

(1) Voluntary contributions examples

- **(RED)**: (RED) is a private sector initiative founded by internationally known Irish musician, activist, and philanthropist—Bono—founder of the music group U2, and Robert Sargent "Bobby" Shriver III, American attorney, activist, and journalist, in 2006, to provide sustainable funding from private corporations exclusively for the Global Fund ((RED) Website, 2022) (RED), which is still operating, partners with well-known companies that produce consumer goods. (RED)-associated companies typically create (RED)-branded products with a portion of the proceeds from the sales going to AIDS programs supported by the Global Fund in Africa. By end of 2021, (RED) had raised close to US$700M of funds through partnerships with more than 160 companies including Apple, SAP, Starbucks, Bugaboo, GAP, and Bank of America.
- **M2030**: M2030 was launched by the Asia Pacific Leaders Malaria Alliance in 2017 to bring together businesses, consumers, and health organizations to fight malaria in South-East Asia. M2030 partnered with businesses headquartered in Myanmar, Indonesia, Thailand, Singapore, and Japan. The initiative leveraged the public platform of participating business leaders to promote more public funding for malaria elimination. Participating businesses also committed to use their platforms to raise consumer awareness of malaria and funds through consumer marketing (e.g., M2030-branded products). M2030 ceased its operations in 2021 due to the disruptions caused by COVID-19.

(2) Conditional funding examples

- **Debt2Health (D2H)**: D2H was launched in 2007, building on the experience of debt swaps in other sectors such as environment and education. D2H is a partnership between a creditor and a debtor country using Global Fund grants and accountability instruments to

BOX 1 Supporting Sri Lanka's health system infrastructure

In June 2021, Debt2Health: the Global Fund approved a new grant to support health systems strengthening in Sri Lanka, which is exclusively funded by the proceeds of a debt swap agreement between the Global Fund, the Government of Sri Lanka (represented by the Ministry of Finance), and the Government of Germany (represented by the KfW Development Bank). Under this agreement, the Ministry of Finance of Sri Lanka is channeling an amount totaling €20M to fund the Health Information and Quality Improvement Project through the Global Fund. In exchange, KfW will cancel Official Development Assistance (ODA) claims owed by Sri Lanka to Germany for the same amount.

invest increased resources into health systems. Under D2H, a creditor country relinquishes its rights to repayment of qualifying outstanding debt on the condition that the debtor country channels all, or part, of this repayment (the "counterpart payments") to health programs in their country. The counter payments are typically channeled through the Global Fund. By early 2022, D2H had converted around US$200 million worth of debt into additional funding for health. There are several examples, one of which is shown in Box 1, of how D2H enabled increased investments in health systems strengthening, providing ministries of health with additional resources from their own ministries of finance through the debt swaps.

- **Cash on delivery (CoD)**: For certain smaller grants, the Global Fund has implemented CoD models to simplify processes and incentivize results. In the Solomon Islands, for example, the country prefinances malaria activities, and based on the achievement of outcome and impact indicators, the Global Fund disburses "reward" funds. The size of the reward is tied to the performance against the predetermined indicators. With the exception of unethical or illegal purposes, the Global Fund reward funds can be spent on any government expenditure that supports achievement of HIV/AIDS, tuberculosis, and malaria goals.

(3) Catalytic funding: examples

- **India Health Fund**: Tata Trusts—the largest endowment trust in India—and the Global Fund established the India Health Fund (IHF) in 2015. IHF, initially funded by US$15M from Tata Trusts, scales innovations to fight infectious diseases in India. Based on its mandate and landscape studies, IHF scouts for innovations to resolve specific challenges. The fund provides catalytic funding, mentorship, and support for market entry. Once the innovations have demonstrated their effectiveness, IHF helps innovators to scale up by identifying other funding sources. Notable investments include molecular diagnostics to improve diagnosis of tuberculosis and COVID-19 at point of care; artificial intelligence to improve diagnosis of lung diseases; and development of accurate, low-cost, rapid tests for malaria (India Health Fund, 2022).

- **Gavi Advanced Market Commitments (AMC)**: Gavi's AMC creates a market for vaccines that are mainly used or needed in low- and middle-income countries. Through the AMC, Gavi commits to procure large quantities of vaccines at established price points. This provides pharmaceutical companies an incentive to develop and produce needed vaccines at a sustainable price. The AMC mechanism has been used for different vaccines, starting with the Pneumococcal AMC in 2009. With the onset of COVID-19, Gavi created COVAX AMC to negotiate and procure vaccines for eligible countries.

(4) Blended Finance examples

- **International Financing Facility for Immunization (IFFIm)**: IFFIm, founded in 2006, leverages long-term, legally binding pledges from donor countries to Gavi and, with the World Bank acting as Treasury Manager, turns these pledges into bonds. The money raised on the capital markets via Vaccine Bonds provides funding for Gavi. Donors' annual payments to IFFIm—or proceeds from new bond issuances—go toward repayment to bondholders. Vaccine Bonds have raised over US$6.4 billion from investors. IFFIm is an accelerator tool. It allows long-term pledges to be converted into health investments faster; but as such, depends on the ability of sovereign donor countries to make long-term commitments, usually through ODA budgets.
- **EU Malaria Fund**: The EU malaria fund is a public-private partnership, mainly funded by the European Union and the European Investment Bank. In total, the EU Malaria Fund is expected to bundle investments of €150 million. As of December 18, 2020, the EU Malaria Fund has selected 17 companies with 30 independent projects, which are eligible to submit formal financing requests. The fund mostly finances companies through venture debt to stimulate innovations in the prevention, detection, and treatment of malaria (EU Malaria Fund, 2022).
- **Development and Social Impact Bonds**: Despite their name, impact bonds are not loan instruments. Rather, these mechanisms represent public-private partnerships that fund social services through performance-based contracts. Impact investors provide the capital to scale the work of service providers. Outcome funders repay those investors if the project achieves preagreed outcomes. The difference between a Development Impact Bond (DIB) and a Social Impact Bond (SIB) depends on the outcome funder. When the outcome funder is a domestic government, it is called a SIB; and when it is an international donor it is called a DIB. Examples where these mechanisms have been used include education projects in India (UBS Website, 2022), and a forthcoming project in South Africa to reduce HIV transmission in young girls (F. Abdullah, 2019).

(5) Socially responsible investments examples

- **The Global Health Investment Fund**: The Global Health Investment Fund (GHIF) is a social impact investment fund that supports the development of drugs, vaccines, and diagnostics against diseases that disproportionally burden low- and middle-income countries. The US$ 100M+ fund seeks opportunities with a high success of commercialization within two to three years. The fund employs a range of investment structures including mezzanine debt, convertible debt, preferred equity, and project financing. Each of the funded projects must both advance social impact objectives and demonstrate potential for commercial success. GHIF was established by the Global Health Investment Corporation—a nonprofit organization—established to provide long-term funding for global health research and development through financing vehicles.

(6) Earmarked taxes: examples

- **Airline levies**: The airline-ticket levy introduced by France in 2006, mentioned before in this chapter, was a milestone in innovative finance. It also led to the creation of *the Leading Group on Innovative Financing for Development*, initiated by France together with Chile, Brazil, and Spain. The idea behind the levy was to introduce a progressive tax, collecting resources only from those able to afford international air travel and to make those revenues directly available to internationally recognized goals. In the case of France, the collection and use of the income from the levy is unique as it is *not* transferred to the Republic's general budget.
  The Republic of Korea (ROK) as another supporter of this levy used the revenue to create a new Global Disease Eradication Fund hosted by the Korean development agency KOICA. Both France and ROK used the airline levy to complement their contributions to the Global Fund and to create a new international organization: Unitaid.
  Unitaid's original aim was to accelerate the availability of drugs for HIV, tuberculosis, and malaria. Funds for its activities would be generated from earmarked taxes, mainly the airline levy. As time passed, however, Unitaid's programmatic aims evolved, and it also started receiving traditional ODA resources from donors and contributions from private foundations.
- **Zimbabwe AIDS levy**: Established by legislation in 1999, the Zimbabwe AIDS Levy entails a 3% income tax for individuals and 3% tax on the profits of employers and trusts. As such, it is one of the earliest examples of earmarked taxes for development, even predating the Monterrey Consensus. Estimates show that the levy has generated significant resources in the order of US$35M per year after 2014. The resources are managed by the parastatal National AIDS Council through a decentralized structure of AIDS Action Committees (Bhat et al., 2016).
- **Other tax-based initiatives**: Countries have also experimented with a range of other taxes to improve health outcomes. Today, high taxes on tobacco are the norm across the world, and the same can increasingly be said about so-called sugar taxes (World Cancer Research Fund International, 2018). Typically, these taxes intend to discourage the use of harmful substances, and thus decrease the burden of noncommunicable diseases. However, these taxes are generally not "earmarked" for health-related projects.

Resource mobilization using innovative finance mechanisms

Although innovative finance mechanisms can have many objectives, including making ODA funding flow more efficient and programs more effective, the original *raison d'être* of innovative finance was to raise new and additional funds for development. Hence, innovative finance is still to a large extent judged, rightly or wrongly, by its ability to raise new money. The authors have used this lens when examining the mechanisms.

The limitation of earmarked taxes for development

Many of the early innovative finance initiatives focused on raising new funds for development. In a twist of history, taxation—which has been used for thousands of years—was reinvented as an innovative concept in the early 2000s.

Although such tax-based initiatives can be useful, there are also inherent challenges. When earmarked taxes are exclusively used for funding a cause or organization, it can create an artificial ceiling limited by the revenues. For example, South Korea provides financing to Global Health causes using the proceeds from an airline levy. When COVID-19 reduced international travel significantly, it had a direct impact on the country's ability to fund such causes. Even before the crisis, the contributions from South Korea to Global Health organizations had not kept up with countries of similar GDP and population size, and it appears that the artificial ceiling created by the airline levy may have been one of the reasons.

Others have argued for the introduction of a global solidarity tax, e.g., a Financial Transaction Tax, to fund development priorities. However, efforts have been stymied by lack of political will, and it is difficult to imagine that this will change in the immediate future given the geopolitical landscape and the differing priorities of leading nations. In 2021, 136 countries agreed to a corporate minimum tax of 15%. OECD estimates that this will generate US$150 billion of new resources annually. These taxes are so far not considered to be earmarked for achieving the SDGs (OECD Website, 2022). These examples highlight the challenge of introducing global or national taxes specifically for funding global goals.

The challenge of replicating successful initiatives

Unlike taxes, many innovative finance mechanisms can be developed and implemented by organizations and donors. Some of these mechanisms have generated significant resources, but they have proven difficult to replicate at scale.

(RED) can be used as an example of this phenomenon. (RED) managed to generate significant buzz because of its charismatic and well-known founders (Bono and Bobby

Shriver) and through generous funding for the (RED) organization by donors. Since the energy, commitment, and network of Bono—one of the world's most famous artists and influencers—is difficult to replicate, it is challenging to reproduce the success of (RED).

D2H, another successful example, is challenging to emulate for other reasons. The initiative is highly dependent on the willingness of political leadership in donor countries to give up debt claims. In addition, the negotiation of tripartite debt swaps is complex and time consuming and involves multiple ministries and departments. At times, this has discouraged debtor nations from pursuing debt-swap deals despite the resources it would generate for health programs.

The question of additionality

Already in 2011, then-UNSG Ban Ki-moon raised the issue of "additionality" of innovative finance mechanisms. In a report, the UNSG explained that "new resources are needed to complement traditional ODA. In a narrow sense, additionality of innovative financing would only apply to resources that are not classified as ODA, whereas in a broader sense all resources raised by innovative financing mechanisms, regardless of their classification as ODA, might be regarded as additional, provided that they do not substitute traditional ODA" (United Nations, 2011).

For some concepts, e.g., voluntary contributions from consumers and the private sector, the level of additionality is clear. However, for other innovative finance mechanisms, it is more challenging to prove the same. Some blended finance mechanisms, for example, use ODA to incentivize private sector investments in commercial projects that benefit the achievement of the SDGs. By focusing on areas that need investments, e.g., water, sanitation, electricity, etc., and by injecting public capital, typical investment barriers (perceived political and financial risks, and low return on investment) can be reduced. Convergence, a global platform that connects private investors and development financiers, states that blended finance has mobilized approximately $160Bn according to their data (Convergence Website, 2022). In the absence of public funding, it is likely that many of those investments would not have occurred. It is, however, difficult to prove that that is the case. This is not a criticism of blended finance platforms, but rather highlights the challenge of proving additionality of innovative finance.

In addition, due to the expansive nature of what can be considered innovative finance, it has proven difficult to objectively assess how much of additional resources that have been generated through these mechanisms.

The distant promise of the capital markets

In recent years, there have been discussions about leveraging the capital markets to raise funds for Global Health and the SDGs. Banks, asset managers, and pension funds manage US$380 trillion of assets. According to OECD, shifting a small percentage of these investments toward sustainable development would go a long way toward meeting the SDGs (OECD, 2020).

Despite the promise, development actors have found it challenging to mobilize significant funds from the commercial capital markets, especially when contrasted with the estimated annual US$ 2.5 trillion financing gap to meet the SDGs (OECD, 2020). Private funds seek high financial return on investments at a minimal risk, which is not the typical investment climate found in countries that are struggling to meet the SDGs. The UN Secretary-General in his Roadmap for Financing 2030 Agenda for Sustainable Development noted that there is a lack of "bankable SDG investment projects," which in turn makes it challenging to attract even well-meaning private investors (The United Nations Secretary-General's Roadmap for Financing 2030 Agenda for Sustainable Development, Geneva, 2020).

Twenty years after the Monterey Consensus, the world has seen a range of innovative finance mechanisms. Although many have proved useful in raising resources for organizations and causes, they have not been able to achieve scale at the levels required to meet the SDGs. It is unlikely that the existing mechanisms, even when combined and at scale, will be able to meet this gap.

Beyond ODA and Innovative Finance: New forms of international cooperation to finance global commons

The world is increasingly facing the challenge of financing global commons that are in the interest of citizens in all countries of all income levels but for which there are no appropriate mechanisms. These global goals include pandemic preparedness, prevention and response, climate mitigation and adaptation, equal opportunities in the rapid digital transformation of the world, and others. These are fundamental problems affecting the economies and societies of all countries. It has become clear that neither domestic resources on their own, traditional ODA, nor innovative finance is sufficient or appropriate for the financing of these global issues. All countries will need national budgets that include international spending for domestic gain.

Therefore considerations about finding new forms of finance need to include concepts that will gradually replace the overstretched and inequitable ODA paradigm, characterized by a small group of high-income countries that finance issues concerning all other countries. The strategic direction and the use of funds are typically determined to a large extent by the same high-income countries, which creates an uneven relationship with recipient countries. The same can be said about most innovative finance mechanisms.

One of the alternatives to the prevailing financing paradigm is a concept called Global Public Investment (GPI). This will require a radical and innovative redesign of how budgets work today.

Global Public Investment

GPI was originally introduced by authors such as Simon Reid-Henry (Reid-Henry, 2019) and Jonathan Glennie and has been developed further in a global consultation process by an international Expert Working Group (Glennie, 2020).

GPI is a proposed new paradigm of international public finance in which governments cooperate to secure international public policy outcomes via fractional contributions from general government revenue. GPI addresses the question of "additionality" directly through the establishment of a non-ODA budget line for global public goods, services, and infrastructure. While not explicitly a funding mechanism for the SDGs, GPI compliant spending overlaps with many of the SDG ambitions and it could go some way to filling the annual financing gap.

GPI stands out from other proposals in three key respects. First, it is a universal funding arrangement. Instead of donor countries providing the capital, all participating countries contribute through a fair share arrangement. In return, the second distinctive feature of GPI establishes the principle that all countries participate equally in the governance of the scheme. Third, rather than being defined as a conditional, blended, or grant-type arrangement, GPI is focused on investments in common needs that generate wider public returns.

GPI is therefore not another new "fund" but a new political process to incentivize and allocate financial contributions from all countries to global issues. It could help harmonize the current financing landscape via fractional commitments to prioritized global needs that are determined in a fair and equitable way and that are presently underfunded. GPI offers an opportunity in that there may yet be ways to address some of the underlying *structural* problems in the financing of international outcomes and developmental needs, including the large financial gaps for global public goods and the high degree of fragmentation and volatility in commitments. As a practical proposition, GPI could be set up through the existing multilateral and international financial institution infrastructure.

Conclusion

Increasing investments in health systems, ensuring equitable access to vital health services, and protecting the health of people in all countries from ongoing and future pandemics have never been more important than now. The current level of domestic and international financing is grossly inadequate, and international health organizations are struggling to mobilize the resources they need to fulfill their mandates and to get the world back on track to achieve the goals of the UN Agenda 2030 for Sustainable Development and of Universal Health Coverage by 2030.

Therefore looking into the best and most innovative ways to achieve these global goals is more important than ever. The last 20 years, since the International Conference on Financing for Development in Monterrey in 2002, have seen multiple initiatives, and what these authors consider as brilliant ideas, in the field of innovative finance. More of those are needed. However, at the same time we are also well advised to be realistic and informed by evidence about what has worked well, what can be brought to scale, and what can support the provision of healthcare, particularly for the most vulnerable and disadvantaged communities around the world at reasonable transaction costs for all implementing partners.

All potential solutions need to be sustainable, additional to current international financing agreements, and responsive to the needs of the people in greatest need. The authors of this chapter have been personally involved in setting up and implementing several of the innovative finance mechanisms described herein. They have provided, as much as possible, an objective assessment of the models that have been implemented with some level of success, as well as their opportunities and challenges. These models deserve to be explored further. However, the authors also argue that to finance broad healthcare needs in a postpandemic world, the international community will need to reexamine the sources of funding, to redesign budget processes to include international spending for domestic gain, and to establish alternatives to the prevailing global governance structures for channeling international contributions to issues of global concern.

The world needs mechanisms that provide the right incentives for all sovereign states to contribute according to their ability and to decide together how scarce resources can be deployed to the best possible effect for the broadest number of people worldwide.

References

(RED) Website. (2022). https://www.red.org/how-red-works.
Bhat, N., et al. (2016). Zimbabwe's national AIDS levy: A case study. *SAHARA Journal*, *13*. https://pubmed.ncbi.nlm.nih.gov/26781215/.
Convergence Website. (2022). https://www.convergence.finance/.
EU Malaria Fund. (2022). https://www.controlmalaria.eu/portfolio.

F. Abdullah, et al. (2019). First social impact bond for the SAMRAC. *South African Medical Journal, 109*(11b), 57–62. https://doi.org/10.7196/SAMJ.2019.v109i11b.14254.
Gates Notes. (2022). https://www.gatesnotes.com/development/g20-report-innovation-with-impact.
Glennie, J. (2020). *The future of aid: Global public investment.* www.globalpublicinvestment.org.
India Health Fund. (2022). http://www.indiahealthfund.org/our-portfolio/.
(2022). *KOICA global disease eradication fund.* http://www.koica.go.kr/koica_en/3458/subview.do.
OECD. (2020). *Global outlook on financing for sustainable development 2021.*
OECD Website. (2022). https://www.oecd.org/tax/international-community-strikes-a-ground-breaking-tax- deal-for-the-digital-age.htm.
Reid-Henry, S. (2019). Global public investment: Redesigning international public finance for social cohesion—A preliminary sketch. *Revue d'Economie du Developpement, 27*(2), 169–201. https://doi.org/10.3917/edd.332.0169.
The Global Fund. (2019). *Bm39 25 innovative finance update.* https://www.theglobalfund.org/media/7435/bm39_25-innovativefinance_update_en.pdf?u=637066574650000000.
The United Nations Secretary-General's Roadmap for Financing 2030 Agenda for Sustainable Development, Geneva. (2020).
The World Bank Group. (2022). *Innovative finance for development solutions.* https://olc.worldbank.org/system/files/Innovative_Finance_for_Development_Solutions.pdf.
UBS Website. (2022). https://www.ubs.com/global/en/ubs-society/philanthropy/optimus-foundation/our-impact/development-impact-bond.html.
United Nations. (2002). *Monterrey Consensus of the international conference on financing for development* (p. 15).
United Nations. (2011). *Innovative mechanisms of financing for development.* Report of the Secretary General.
USAID. (2019). *Investing for impact, Center for Innovation and Impact.* https://www.usaid.gov/sites/default/files/documents/1864/investing-for-impact-may2019-updated.pdf.
World Cancer Research Fund International. (2018). *Building momentum: Lessons on implementing a robust sugar sweetened beverage tax.* Available at: www.wcrf.org/buildingmomentum.

CHAPTER 53

Financing instruments for innovators to scale health innovations in low- and middle-income countries☆

Deepika Devadas[a], Annie Thériault[a], and Anita M. McGahan[b,c,d]
[a]Every Woman Every Child Innovation Marketplace, Grand Challenges Canada, Toronto, ON, Canada, [b]Joseph L. Rotman School of Management, Toronto, ON, Canada, [c]University of Toronto, Toronto, ON, Canada, [d]Institute for Strategy and Competitiveness, Harvard University, Cambridge, MA, United States

Introduction

Health innovations addressing the development agenda in low- and middle-income countries (LMICs) are financed by both public and private funders and investors. The role of each source of funding in the implementation of innovation has not been comprehensively explored. More often, the emphasis in most prior research has been on the role of the public sector in scaling innovations through governmental adoption. Various development initiatives have underscored the potential and the need to mobilize private sector investment to accelerate the pathways to meeting the Sustainable Development Goals (SDGs) (United Nations Global Compact, 2015; USAID, 2019), particularly for the goals related to health (SDG3) and gender equality (SDG5) in resource-limited settings. Lack of financing is often one of the biggest barriers to the scaling of proven solutions.

This chapter presents a guide for innovators seeking funding from both public and private sources to scale innovations in health, as well as for funders that can map their financial offering onto such frameworks to seek complementary funding for their portfolio. Within these broad categories, providers of funding are themselves varied, with governments, donors,

☆ This White Paper was originally published February 16, 2021, in a white paper series by the Every Woman Every Child (EWEC) Innovation Marketplace at Grand Challenges Canada and the University of Toronto, Rotman School of Management. It has been reprinted by permission of Grand Challenges Canada.

Resilient Health. https://doi.org/10.1016/B978-0-443-18529-8.00053-6

universities, multilaterals, foundations, individuals, corporations, and angel investors broadly representing the spectrum of public and private investment sources. This chapter will discuss the types of financing available from these sources—mainly classified as grant, debt, and equity—and their individual characteristics, along with various market insights.

The ideas expressed in this chapter evolved from expertise developed while studying scaling and in working with funders and innovators through the Every Woman Every Child (EWEC) Innovation Marketplace, a strategic alliance of development organizations consisting of Grand Challenges Canada, the Norwegian Agency for Development Cooperation, the U.S. Agency for International Development, and the Bill & Melinda Gates Foundation—an initiative housed at Grand Challenges Canada. The points made should be interpreted as scholarly observations, rather than as an agenda endorsed by the EWEC Innovation Marketplace partners.

Grants

As the most traditional form of development aid, grants dominate the landscape of financing for health innovations that are targeted at low-resource settings. Grant capital can take many forms, depending on the receiving entity, the stage of scaling, and the use of funds. Entities behind innovations include nonprofits, academia, social enterprises, and for-profit companies. Grants are typically the sole source of funding for nonprofit organizations and academic innovators, while still being accessible by others. In the context of scaling, grant funding can be catalytic in helping a social enterprise or product innovation achieve early testing and validation results before the innovation is sufficiently derisked for different and larger types of financing.

Grants, provided by public funders and governments, necessitate a high degree of accountability and can therefore be restrictive in the types of activities they are allowed to fund, requiring highly detailed budgets and, in some cases, the financial agreements can be quite complex and may require a legal review prior to signing. Grants may also include rights to access the intellectual property (sometimes called global access rights) under specified circumstances in specific LMICs, which may or may not be negotiable for the recipient's unique circumstances. While designed to enable the access for public good, these rights are often permanently tied to the intellectual property, which can affect the commercial viability of the venture. Grants provided by private foundations or corporate social responsibility (CSR) groups of large corporations may be similar to those provided by the public sector and governments or they may be more flexible, depending on the funder's philosophy and approach to grant-making. Some private funders are able to allow for more built-in flexibility on the use of funds.

While most grants are not repayable, a recent evolution of this financial instrument is the Repayable Grant, which is technically a loan. These grants are, as the name suggests, repayable to the funder based on agreed-upon milestones, market conditions, and/or growth and impact achievements. Repayable grants are often offered in lieu of debt/loans and offer more flexible

Table 1 Typical characteristics of grants.

BEST USED FOR	• Testing and validating proposed high-risk ideas • Demonstration studies and market-building activities • Nonprofitable impactful activities, such as product refinements to better serve LMIC segments
VARIATIONS	• One-off, tranched, multiyear, repayable, forgivable
MAIN FUNDERS	• Governments, multilateral agencies, foundations, universities, research institutes, or private corporations (often through their CSR arms or Corporate Foundations)
EXPECTATIONS	• Detailed budget • Intellectual property (IP) rights, often called global access rights (for some granting organizations) • Periodic reporting of milestone progress • Ability to measure and translate impact of the innovation
ADVANTAGES	• Nondilutive • Limited or no impact on cash flows • Incentivizes impact • Limited consequences to project failure • High-risk tolerance • Achieve faster growth • Patient capital
CHALLENGES	• Grant reporting may not be proportional to the size of the grant or organization • IP rights can limit ability to attract other forms of capital • Narrow scope (specific project or activities) with little flexibility and often excluding overhead costs needed to build strong organizations • Grant funding cycles may not align with the timetable for project implementation or business needs • Projects that rely on grant funding for long periods can set a growth course and organizational structure that may not be attractive to other forms of capital • Compelling robust impact evidence and expertise in grant writing needed to procure grants in perpetuity

terms to the recipient, including no-interest payments on the loan. In some cases, the repayment feature acts as a punitive deterrent in the event the impact strategy funded by the grant-maker is no longer pursued by the recipient in the future. The typical characteristics of grants are summarized in Table 1.

Insights—Grants

Innovator insights

Innovators often find the grant application process tedious and difficult to navigate, especially in understanding selection criteria, effects of intellectual property rights clauses, flexibility, and reporting requirements. This is especially true for nonacademic teams and/or local innovators who regularly struggle with the format, impact/scientific jargon, expectations, and presentation norms. Innovators frequently underestimate the time and bandwidth associated with the application process, negotiations, and reporting.

Funder insights

Funders expect a detailed, thorough, and well-presented plan at the application stage itself, including the ability to generate detailed performance reports and itemized budgets prior to the start of the project. Many funders have various external, internal, and programmatic limitations on flexibility. That being said, funders, including public funders and governments, increasingly recognize the benefits of flexibility in fueling innovation and creativity; the few funders that do have flexibility on grant terms often pass these on to innovators.

Debt

Debt financing (also called "borrowing") occurs when lenders provide capital to finance a company, the borrower. Loans are usually justified by the borrower's ability to repay the loan in the future. Loans are repayable at a specified interest rate either at regular intervals during the term of the loan—the borrowing period—or at its end. A borrower's ability to service a loan is judged based on evidence of current or future expected stable cash flows, which are often uncertain for early stage organizations. The interest rate charged by the lender to a borrower is greater when the lender's level of confidence that the borrower will have stable future cash flows is lower.

There are different types of debt products available in the market from banks and alternative lending companies. Examples discussed here include senior debt, mezzanine debt, revenue-based lending, asset-backed lending, and convertible debt. There are numerous variations in the interest rates, payment schedules, and other features, so long as the borrower and lender agree to the terms and such terms are legally permissible in the relevant geography.

While impact investors—in the sense defined by the Global Impact Investment Network—seek market rates of return and tend to offer products inline with commercial markets for high-impact organizations, mission-driven investors (such as foundations and government-related entities, with mandates to be catalytic) offer a plethora of potential loan structures, with modifications designed to incentivize impact. When these are offered at or below market terms for high-impact ventures, the loans are called "concessionary debt." The typical characteristics of debt financing are summarized in Table 2.

Types of debt instruments

Senior debt

Senior debt or conventional debt is typically secured by all of the assets of a company and offered to organizations that are more mature and meet stringent financial requirements, such as having stable and growing revenues, margins, cash flows, and profits. Borrowers often need to

Table 2 Typical characteristics of debt financing.

BEST USED FOR	• Growth and expansion financing by organizations that have current or expected predictable cash flows tied to these activities • Working capital (usually to fund short-term cash imbalances created by delayed payments or inventory)
VARIATIONS	• Senior, mezzanine, asset-backed, convertible, revenue-based, concessionary
MAIN FUNDERS	• Foundations, Development Finance Institutions (DFIs), private debt funds and banks, equity investors (convertible notes)
EXPECTATIONS	• Ability to service debt via positive cash flows and profits, conversion to equity during a future equity round (convertible notes)
ADVANTAGES	• Nondilutive (except convertible notes) • Tax-efficient • Typically no loss of control via active positions on boards (though observer positions can be requested; except for convertible notes) • Can increase value for existing equity investors
CHALLENGES	• Often inaccessible or inappropriate for prerevenue or early stages under commercial terms • Can stunt the venture when diverting cash flows away from growth • Available interest rates can be prohibitive • Bankruptcy risk • Refinance risk (risk of refinancing with a new loan or other form of capital, such as equity, under less favorable terms)

make specific commitments to a payment schedule and to financial covenants, which are commitments to maintaining specific metrics of financial performance above specified levels. In the event of a failure to pay back the debt or meet the financial covenants, the borrower may be forced to repay the loan on short notice by (i) refinancing under more strident terms with a new lender, (ii) forced liquidation of assets, or (iii) undergoing a bankruptcy process.

Mezzanine debt

Mezzanine debt is similar to senior debt but the rights of the lender are subordinated to the rights of one or more senior lenders, meaning that in the event of a bankruptcy or liquidation process, the mezzanine lender is repaid after senior lenders receive their capital and interest. For this reason, mezzanine loans have higher interest rates compared to senior debt. Compared to senior lenders, mezzanine lenders have a greater appetite for risk, fewer financial metric targets, and/or weaker or fewer financial covenants.

Revenue-based financing

This type of debt instrument is repaid as a percentage of monthly, quarterly, or annual revenues until the total accumulates to a contractually predetermined amount, which may be either capped or unlimited. Compared to senior and mezzanine debt, this type of financing has the advantage of aligning interests of the lender and company by allowing the borrower to pay more when organizational revenue is higher and less when revenue is lower. The approach is effective as long as the borrower's cash flows are aligned with revenues. Problems arise if the borrower's customers pay for products and services a long time after the borrower books the associated sales revenue, as is often required by standard accounting practices. Lenders in this space are often interested in the potential for higher returns if the company outperforms expectations.

Asset-backed lending

Another related but important category of debt is asset-backed lending. Asset-backed loans are tied to a valuable piece of property or equipment that the lender can take from the borrower if the loan is not paid on schedule. There are a wide variety of assets that can be used by borrowers as collateral, including intellectual property, invoices, receivables, fixed assets, buildings, production equipment, and inventory. The key differentiation between these loans and senior debt is that they tie to just one asset, as opposed to the entire asset base of the company. These loans vary greatly with regard to terms, interest rates, and appropriateness of use.

Convertible note

A convertible note or loan is a debt instrument that combines elements of both conventional debt and equity. This type of debt converts into equity in the company after a period of time, after certain milestones are achieved, and/or at the discretion of the lender. This type of arrangement may be attractive for higher-growth companies requiring significant amounts of growth capital to achieve future milestones. The downside for the borrower is that the interest payments on the debt continue if milestones are not achieved and/or if the lender does not convert its position into equity. This can result in either greater outlay of capital to pay off the debt (if it is paid off) or greater dilution of ownership (if the debt is converted). On conversion to equity, these loans also dilute the ownership position of the borrower and result in some loss of company control, as this is expected by equity investors of a company (see equity section).

Convertible debt is typically used by investors who see the potential of a venture to produce high returns, but are not prepared initially to invest in equity because of uncertainty related to the potential growth and/or valuation of the business. Through convertible notes, investors avoid the complicated negotiations of setting a share price for an early stage company, but ensure they get a "piece of the action" if all goes well (Bussgang, 2014).

Concessionary debt

Concessionary loans are often provided in lieu of a grant to early stage, for-profit social enterprises and companies. These types of loans are typically offered by mission-driven investors (such as foundations and development arms of governments) to provide catalytic financing, with the main or only goal being impact and development. They can be structured like any other form of loan mentioned before, although asset-based structures are less common. Their common feature is that the terms of the loan include some type of concession, such as below-market interest rates, relaxed financial covenants, relaxed milestones, impact-based milestones, longer-term and/or impact-based interest rate reductions. Additionally, should the borrower not succeed or experience unpredictable circumstances, some concessionary lenders may renegotiate the terms of the loan and/or forgive the loan to support the continued impact and sustainability of the entity. Mission investors using the convertible note structure may also be open to renegotiating the conversion rates or converting early in downside scenarios where companies are unable to pay back the debt, both to ensure they do not end up owning too much of the entity and to see impact continue.

Insights—Debt

Innovator insights

The lending market is difficult to navigate for most innovators, regardless of background. However, nonprofits and academics particularly lack the skills or expertise to fully understand lending products available in their market. As a result, favorable options (such as flexible concessionary notes, project financing, or the arrangement of microloans for customers) are often overlooked as an alternative to grant financing. On the other hand, social enterprises often borrow excessively because they are viewed as too profitable for grants but not at the profit level or potential to issue equity. Even when debt is used appropriately, innovators may struggle to negotiate terms that reflect the unique context of their businesses. They also may fail to renegotiate terms early enough with lenders when circumstances or priorities change. In many instances, innovators underestimate the cost of convertible debt (given its unique characteristic to become equity combined with the specialty features of such agreements) or ignore the opportunity that convertible debt may offer when used appropriately. Borrowers frequently choose too much repayable debt, which may stunt their growth by limiting the availability of cash to finance growth. Early stage companies may also be composed of teams that are lacking in financial management expertise and systems to forecast, analyze, and articulate needs and ability to pay back debt effectively.

Lender insights

Many lenders active in global health are able to offer flexible terms and have the ability to renegotiate terms to the benefit of the continued operation of the business, especially when the impact mandate of the borrower is strong. However, lenders do expect portfolio companies to lead the effort by bringing informed proposals, including detailed cash flow projections, to the table. The space benefits from significant activity from new entrants with fintech offerings, new nonbank lenders and microlenders, as well as impact-motivated investors and guarantee facilities that enable local lenders to take on more credit risk. As part of the mission in achieving gender equality, there is also a nascent trend to provide debt capital to female entrepreneurs that female founders should tap into when raising capital.

Equity

In equity financing, the investor provides the company with funds in exchange for a share of ownership in the organization. This arrangement provides equity investors with a claim on the venture's upside value creation. "Because the potential for a large return exists, the equity investor is typically willing to invest in riskier ventures—those developing breakthrough products, working with emerging technologies, or operating with unproven but potentially transformative business models"(Bussgang, 2014). Technology-based start-ups that typically lack early revenue, cash flows, or profits, but hold the promise of significant future growth and value creation are characteristically financed by equity. For these organizations, venture capital (VC) is the predominant source of capital in the early stages.

VC funding for health and impact is available at the angel, seed, Series A and Series B stages, and beyond, depending on the revenue growth, milestones, and/or regulatory stage of the company (Bocken, 2015). Typically, the entrepreneur seeks investors to fill each round based on the company's valuation and financing needs. Equity investors participate in these rounds based on their risk/reward preferences (often dictated by their agreements with their own investors in their funds) and receive a percentage ownership in the company in exchange for their capital. They realize their returns during an "exit," which is a contractually specified event in which the company is sold to or merged with larger industry players or when shares are offered to the public through an initial public offering (IPO), typically on regulated public stock exchanges. Before the exit, investors actively participate and support the company's growth to increase its value. They exercise varying levels of control on the business, depending on the investor group and negotiated terms of the investment. While angel investors typically invest their own capital, it is important that innovators recognize that most VC funds have fiduciary responsibility to their own investors for the capital they invest and that the terms they can offer relate to the terms they negotiated with their own investors. VC funds also have time limits on their funds, which may drive their decisions.

Other key players that invest in private companies are strategic investors such as VC groups that operate from within a larger corporation in the same industry as the innovator. These groups often operate similarly to VC funds, although they may also often request terms specifically related to the strategic nature of the relationship with an investee company. These terms can both help and hurt the company's path toward generating returns for other investors. Another mechanism that can be used to raise capital is crowdfunding, which is a form of financing where companies can raise equity capital or sell their products directly to the public via specialized online platforms. The typical characteristics of equity financing are summarized in Table 3.

Insights—Equity

Innovator insights

In the health and impact sectors, most innovators require much longer to close each round of financing (typically 18–24 months) than companies outside these sectors (typically 6-9 months). First-time CEOs often struggle with pitching to investors and understanding the process. They often underestimate the time required and begin fundraising later than optimal, which can lead to short runways and poor negotiating positions. Another potential issue is that innovators in global health who access grants as early stage financing to achieve proof-of-concept often contractually agree to make their products/services available for the public good (sometimes referred to as "global access"). These clauses can negatively impact their valuation or their ability to raise equity to scale their innovation, if not drafted with sufficient flexibility to enable the crowding in of more commercial investors in the future. Many funders recognize this and will craft reasonable terms at the outset of a grant or negotiate after the grant is in effect to support companies in their subsequent growth by creating win-win scenarios between investors and beneficiaries. This can often be achieved by simple mechanisms (such as the segmentation of rights for different jurisdictions or populations) to ensure the needs of each party are met. Another approach is to enter into distribution and/or pricing commitments (with enforceable consequences if not met), leaving the intellectual property unencumbered. A degree of customization is typically required to ensure the objectives of current and future stakeholders are aligned.

Many equity investors do insist that prior debt either be extended or converted before making their investment. Those companies that carry outstanding debt when equity is raised must

Table 3 Typical characteristics of equity financing.

BEST USED FOR	• Growth and expansion financing by companies that expect high future growth, with varying degrees of uncertainty based on stage of financing
VARIATIONS	• Venture capital, private equity, quasi equity (such as convertible debt)
MAIN FUNDERS	• VC funds, private equity (PE) funds, angel investors, public (publicly traded or crowd-funded), larger companies (strategic investors), DFIs, certain banks
EXPECTATIONS	• Dedication to growing financial value of the company • Percentage of ownership • Control of the company's decisions, typically via board positions and shareholder rights documented in the shareholder agreement
ADVANTAGES	• No repayment obligation • Large amount of capital to facilitate accelerated growth plans and value creation • Access to expertise of the investor group and their networks • Flexible capital approved based on high-level plans for the use of funds
CHALLENGES	• Dilutes ownership • Commercial value creation is a higher priority than impact, even when investors are impact investors • Exit pressures and timing • Loss of control

confront a concern by prospective investors that lenders will not cooperate with the issuance of equity. Cooperation typically requires lenders to extend the payment timeline or convert alongside new investors.

Investor insights

Investors that value impact also expect competitive returns, as they must fulfill their fiduciary duty to generate returns on behalf of their own investors. Other forms of capital such as grants and debt are therefore often essential tools as companies and equity investors seek to create a total capital structure that yields returns for each stakeholder, including the underserved populations that are the focus of impact. Notably, while impact investing has gained increasing traction in the last few years, the sectors that are most often represented are agriculture, livelihoods, and energy. Fewer equity investors in health are available, especially when the targeted populations are in lower-income countries. Within this subsector, fewer still are the investors interested in prerevenue or early revenue stages of health technologies and devices. Those that do invest appear to either be small (thus limiting their financing to initial stages) or very large (thus participating only in the financing of companies that are already scaling). A wide missing middle cohort of investors has left the health sector underfinanced. We note that a number of new efforts are underway to address this problem, but that they face significant hurdles in raising the necessary capital to launch due to the emerging nature of the sector. There are also new technical assistance facilities being raised to support the efforts of these future funds, combining clinical and health expertise, regulatory and IP knowledge, market and local knowledge and local talent.

New mechanisms

Outcomes-based financing

Outcomes-based financing approaches are also increasingly being tested in the development sector. Simply put, the mechanism is structured to involve an outcomes-based funder that agrees to pay an innovator to implement a project if predefined outcomes are achieved, based on independent, third-party verification. The investors who initially finance the project receive a preagreed return, in addition to the capital invested, if the project is successful in achieving the defined outcomes. The investor therefore takes on the risk that the project will fail—a risk that an outcomes funder, typically a government, is not willing to bear, thereby aligning impact goals and risk/reward appetites. The model is gaining popularity, as it aligns impact goals, financial goals, and risk tolerance. Examples include social success notes, conditional cash transfers, advance market commitments, guarantees, and social/development impact bonds. An example is the Kangaroo Mother Care Development Impact Bond (DIB) in Cameroon, which is structured between:

(I) **Outcomes Funder**—Ministry of Public Health of Cameroon (through the Global Financing Facility) and Nutrition International

(II) **Third-Party Verification**—Institute for Research and Behavioural Studies (IRESCO)

(III) **Innovator/Service Provider**—Fondation Kangourou Cameroon

(IV) **Investor**—Grand Challenges Canada

The complex structuring required for its success was facilitated by Social Finance UK and the MaRS Centre for Impact Investing.

The key drawbacks of this financing mechanism are its complexity; high transaction costs related to legal, financial, and technical services for customization; time requirements; and the need for multiple parties from different sectors to agree to common terms. More data is needed to build structures that are replicable and/or restricting these instruments to certain focus areas or minimum funding sizes to achieve better cost efficiency (Itagaki, 2016).

Blended finance

In recent years, blended finance has become the "North Star" in the fields of global health and, more generally, other impact sectors. Blended financing (see Fig. 1) refers to the coordinated structuring process to align risk/reward characteristics of an investment opportunity (including financial and impact characteristics) with risk/reward preferences of various investor types, to include various sources and categories of financial instruments, including a concessionary layer. This is common for larger investment vehicles of ~$20 million or more and has the potential to significantly grow the sector.

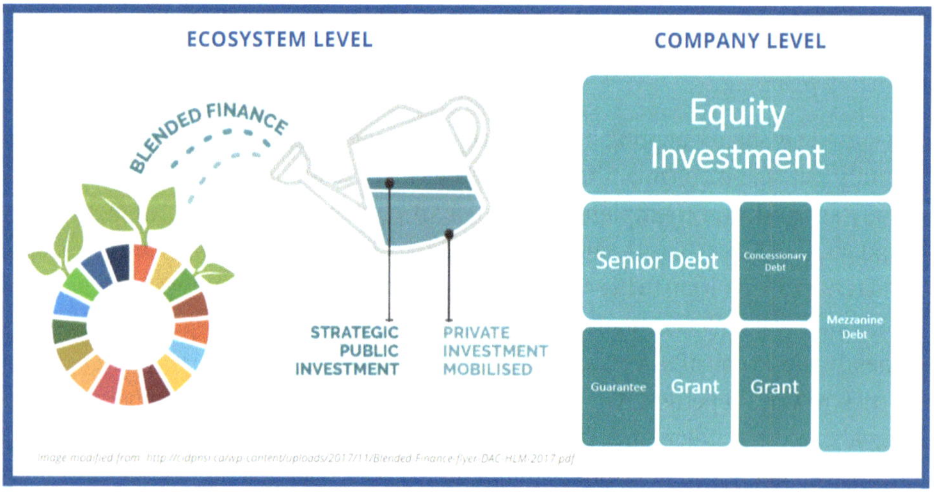

Fig. 1
Blended finance to unlock private sector capital.

At the EWEC Innovation Marketplace, the hands-on support provided to innovators has allowed companies to secure blended finance at the company level for \$1–\$5 million dollars in total capital raised. Because their growth plans encompass both impact and return, the companies supported are able to secure financing from investors and funders with varying risk/return appetites. This includes grants, unique loan structures, equity investment and, in some cases, guarantees and outcomes-based funding. In ideal scenarios, participants in the blended finance structure coordinate transparently to select the risk/reward structure that most closely matches their actual requirements and collaborate to reduce the reporting burden on the innovator. This type of process tends to work better when an advisor (formal or informal) is available to nudge collaboration among investors. More commonly, each investor comes into the structure around the same time with their own structure in an uncoordinated manner and agrees to the hierarchy of lender priority through intercreditor agreements.

Innovators should be aware that blended financing requires thought, advice, and planning to accomplish optimally, including an understanding of future financings required to achieve scale. Three key skillsets innovators should equip themselves with for such a journey include:

(I) Ability to measure and report on impact and outcomes
(II) Ability to forecast, strategize, and manage financials effectively
(III) Ability to identify and negotiate the financial instruments for the growth strategy of an entity

In addition, innovators should seek out formal or informal advisors with a high degree of knowledge and expertise in all potential financing instruments, to assist their journey for the long term.

Leveraging instruments to scale

Hypothetical examples of potentially successful models are provided as follows.

Medical device or technology company

Initial prototyping and testing is conducted using grants from foundations and academia in combination with small angel investments. Clinical trial, regulatory, and product development stages are funded through larger grants from foundations and seed stage investors. Early stage VC supports the preregulatory approval stage where manufacturing and distribution models are being developed. Global access agreements are developed or refined together with the innovator, grant providers, and commercial investors. Funding from strategic investors enables the growth of new distribution channels and/or entry into new markets. Debt financing is leveraged to achieve further growth once the company reaches predictable revenues, with grants potentially available to bring solutions to last-mile or LMIC populations. Long-term agreements are struck between the company, local entrepreneurs/distributors, and governments to embed a solution locally, ensuring the impact survives potential exits.

Operationally intensive service models

Such models include hospital chains, clinics, and other primary care delivery mechanisms that access foundation grants and angel investments to pilot their model in communities, test market acceptability and pricing strategies, begin developing systems for financial management and monitoring and evaluation of impact. As is typical, the organization is revenue-positive earlier on and raises financing rounds composed of concessionary debt to fund working capital to deliver on government contracts and service expansion and raises equity to finance new facilities or new markets. Success usually attracts debt financing from later-stage lenders (such as DFIs and banks), as well as equity from late-stage private equity investors.

Programmatic service models

These are typically led by nonprofit entities or partnerships with service providers. Grants serve as the main instrument for a large period of time, proving aspects of the model especially related to cost-effectiveness, user adoption, and impact. In the long term, public sector adoption may manifest as fee-for-service, cost-sharing, or licensing-in. Debt in the form of working capital can reduce cash flow issues for the nonprofit, created by delayed contract payments from large providers or governments. For large projects, outcomes-based financing may also be suitable in some cases in securing government or donor financing.

Product-based companies

Examples include products focused on nutrition or menstrual health, and commodities that may grow through a combination of early grants followed by debt or equity investment based on their cash flows or equity-return potential. Intellectual property-rich companies are more likely to receive venture capital investment, while companies commercializing commodities are more likely to receive debt. Volume guarantees and advance market commitments may also be feasible for some promising products.

Conclusion

To conclude, innovators can benefit from weighing the costs and benefits of applying different models of financing to determine their best path to growth. The wide range of sources and types of capital described in this note are complex and must be carefully considered to yield an approach that is suited to the unique capabilities and aims of innovators. The relationships between various types of financing tools are important to the long-term growth of organizations, but are infrequently considered or understood by novice investees or their investors or funders who often operate in silos and use different jargon and language to refer to shared aspects. Companies seeking to achieve social impact have unique access to blended finance that attracts both private and public capital, but require assistance to navigate these complex waters. Because the populations served by global health and social sector innovators are vulnerable, the consequences of organizational failure are significant. Investors and funders must understand and embrace the consequences of the terms that they impose through financial instruments and legal requirements for these populations as well as for the future financeability of the investee organization, as innovators cannot scale impact if they cannot raise growth capital. When each stakeholder ensures all aspects of the organization that are financed are well supported throughout scaling, including impact, revenue, growth, and financial return (when investor capital is needed), the approach can indeed be the elusive win-win the sector needs.

References

Bocken, N. (2015). Sustainable venture capital—Catalyst for sustainable start-up success? *J Clean Prod., 108*, 647–658. https://doi.org/10.1016/j.jclepro.2015.05.079.

Bussgang, J. (2014). *Raising startup capital.* Harvard Business School. https://www.hbs.edu/faculty/Pages/item.aspx?num=46528.

Itagaki, W. (2016). *Reducing transaction costs in development impact bonds.* Center for Strategic & International Studies: Prosper. https://csisprosper.com/2016/07/05/reducing-transaction-costs-in-development-impact-bonds/.

United Nations Global Compact. (2015). *Private sector investment and sustainable development.* https://www.unglobalcompact.org/library/1181.

USAID. (2019). *Unleashing private capital for global health innovation: Innovator and investor support opportunities.* https://www.usaid.gov/cii/private-capital.

CHAPTER 54

Financing for mental health: A smart investment, not an expense

Sarah Kline[a], James Sale[a], and Shekhar Saxena[b]

[a]United For Global Mental Health, London, United Kingdom, [b]Department of Global Health and Population Harvard T. H. Chan School of Public Health, Cambridge, MA, United States

Introduction

The statistics about the numbers of people with common mental health conditions worldwide are truly enormous—dwarfing the rates of many common physical health conditions that receive far greater financial investment (Friedli & Niamh, 2007).

According to 2019 data—the most recent we have—an estimated 970 million people around the world have a mental health or substance use condition. That amounts to about 13% of the global population (IHME, 2020). And the numbers are rising, accelerated by the ongoing impact of COVID-19, the global economic crisis, climate change, and conflicts and natural disasters.

At the same time, the level of investment in mental health is dismally low. Funding for mental health typically accounts for between 1% and 4% of a country's spending on health (WHO, 2021b). Even more disappointing, Development Assistance for Mental Health (DAMH) is estimated at around 0.4% of Development Assistance for Health (DAH) (Iemmi, 2020). Moreover, philanthropic funding for mental health is also very limited. While a third of DAMH comes from philanthropic funding, the amount of funding is only equivalent to 0.5% of total philanthropic contributions to DAH (Iemmi, 2020). As things stand, huge swathes of the world have next to no access to mental health services (Patel et al., 2018).

The good news, as United for Global Mental Health (UnitedGMH) sets out in our 2020 report *The Return on the Individual*, is that the potential return on investing in mental health is vast—for our economies, for our societies, and crucially for individual human lives.

In this chapter, we walk you through the key findings of that report, based on data from the partners who make up our Global Mental Health Action Network and the "Speak Your Mind" campaign (Box 1). We explore the growing evidence that mental health can deliver an

Resilient Health. https://doi.org/10.1016/B978-0-443-18529-8.00054-8

BOX 1 United for Global Mental Health

We believe that only adequate, efficient, and sustainable public mental health finance invested primarily at the community level will allow everyone, everywhere, to enjoy their right to good mental health. We work with national advocates, campaigners, and governments to increase and improve government finance for mental health. And, we work with international donors to increase funding, catalyze further mental health finance, integrate mental health into existing programs and emergency response plans, and strengthen coordination among donors.

One example of our work is the development of Countdown Global Mental Health 2030 (UnitedGMH, 2021b): the world's first independent monitoring and accountability mechanism for mental health. Launched in 2021 with our partners the *WHO, UNICEF, GlobalMentalHealth@Harvard, Global Mental Health Peer Network,* and *The Lancet*, the Countdown GMH 2030 provides users with a free and interactive dashboard to search mental health data by country using a range of indicators, combined with an annual monitoring report on what the latest data shows (UnitedGMH, 2021a, 2021b).

overwhelmingly positive return on investment for both businesses and the wider economy (Chisholm et al., 2016). We look at the potential impact of smart mental health investments on everything from the height and weight of infants to rates of TB, from smoking to crime, and from workplace well-being to the wealth of nations. You will see that smart investment has seismic implications for the economy, for society, and for the individual.

The return on investment for the individual

Poor mental health: The cost to the individual

Our growing global mental health crisis is having a devastating impact on the lives of people across the world. Among the 970 million people with a mental health or substance-use condition:

- There are 301 million suffering from anxiety and 280 million from depression (WHO, 2021a)
- An estimated further 50 million people are living with dementia, with this number projected to triple by 2050 (World Health Organization, 2019b)
- Suicide claims the lives of more than 700,000 people every year—with many more attempting to take their own lives (WHO, 2021a).

Behind those staggering statistics are individual human beings who need access to good-quality mental healthcare. Yet, investment in mental health has remained extremely low. Currently, only 2.1% of government health budgets on average are allocated to mental healthcare—a figure that falls to less than 1% in many low- and middle-income countries (WHO, 2021b).

For those people who want to seek treatment, care, or support, there are far too many barriers—from stigma and discrimination to a lack of resources and too few trained health workers. Ironically, while the COVID-19 pandemic increased public awareness of mental health as an critical issue, investment in mental health from a financial standpoint actually fell, in response to more immediate physical health demands.

It is time to put this right.

The smart investment opportunity

Almost a decade ago from the time of this writing, in 2013, the World Health Organisation (WHO) recommended that member states (governments) increase their mental health service coverage by half by the year 2030. Member states agreed, with the result that a 50% increase is a core part of the WHO Comprehensive Mental Health Action Plan 2013–30 (WHO, 2021c). Now, research published by UnitedGMH in 2021/22 shows that reaching this target would involve only a minimal level of investment.

We found that achieving the 50% coverage increase for five common mental health and neurological conditions—anxiety, bipolar, depression, psychosis, and epilepsy—only requires a year-on-year investment increase of only US$0.20 per person globally until the year 2030. For low- and lower-middle-income countries, the cost would be even smaller: from just US$0.02 in lower-middle-income countries to less than US$0.05 in low-income countries (UnitedGMH, 2022).

In stark contrast to this minimal financial investment, reaching this target would make an enormous difference in the lives of people with these five mental health and neurological conditions.

If health coverage were scaled up by 50%, alongside a transition from institutional to primary healthcare and community care, as the WHO Comprehensive Mental Health Action Plan 2013–2030 details, the projected global benefits by 2030 are (UnitedGMH, 2022) as follows:

- nearly 24 million healthy life years gained
- 51.5 million cases of these common disorders prevented
- nearly half a million deaths averted.

If the world went further and managed to achieve Universal Health Coverage (over 90% coverage) for these five common mental health and neurological conditions, the projected gains are even more pronounced, as presented in Table 1. The numbers in Table 1 clearly outline that by the year 2030, 299.7 million cases of mental illness could be averted, 126 million healthy years gained, and 1.6 million deaths avoided (UnitedGMH, 2022). Further, these numbers would be exponentially increased by the year 2050.

Table 1 Approximate projected global health outcomes of achieving UHC (UnitedGMH, 2022).

Achieving universal (90% plus) mental health service coverage	Prevalent cases averted	Healthy life years gained	Deaths avoided
By 2030	299.7 million	126 million	1.6 million
By 2050	1.4 billion	531 million	6.1 million

Test case: The Philippines

The way health coverage would be increased would depend on the country's individual context. Table 2 outlines the case for investment to close the mental health treatment gap (referring to the extent of needs compared to the provisions to meet these needs) for certain clinical packages in the Philippines, a WHO Special Initiative Country (World Health Organisation, 2021). It recommends an investment framework over 10 years and projects the return on investment.

Table 2 presents what would be the impact of increasing the coverage rate of the treatments listed in the table—whether for certain conditions or through certain interventions (curbing alcohol use, curbing access to pesticides and therefore their use in suicide attempts or providing school based mental health support). It calculates the health gains could be huge: 724,195 healthy life years gained, 1,748,020 cases averted, and 5344 deaths would be avoided. Moreover, investing in mental health in the Philippines, as Table 2 presents, will have significant economic benefits, won through increased productivity (PHP 216 billion). That is the area we will move on to discuss now.

The return on investment for the economy

Poor mental health: The economic costs

Mental health and economic performance are closely interlinked. It is estimated that as much as 20% of the world's working population has some form of mental health condition at any given time (Mental Health Commission of Canada, 2014). With an estimated 12 billion productive days lost each year to depression and anxiety alone, poor mental health costs the world economy approximately US$2.5 trillion per year in reduced economic productivity and the direct cost of care (Bloom et al., 2011). This figure is projected to rise to US$6 trillion by 2030.

This projection means that mental illness will account for more than half the economic burden of disease over the next two decades—higher than physical conditions like cancer, diabetes, and chronic respiratory disease combined (Bloom et al., 2011). At the same time, countries with weak and fragile physical and mental health systems will find it harder to achieve sustained growth.

Poor mental health is now one of the main challenges facing the economy.

Table 2 The case for investment in mental health services in the Philippines (WHO, 2021).

<table>
<tr><td colspan="9">Burden of mental health conditions
Total: PHP 68.9 bn (US$1.32 bn)
PHP 2.7 bn direct costs due to health expenditure
PHP 66.2 bn indirect costs due to loss of workforce and reduced productivity
The treatment gap (using basic coverage figures)</td></tr>
<tr><th>Type</th><th>Anxiety (%)</th><th>Depression treatment (%)</th><th>Psychosis treatment (%)</th><th>Epilepsy treatment (%)</th><th>Bipolar disorders treatment (%)</th><th>Alcohol use (%)</th><th>Pesticides access (%)</th><th>School based (%)</th></tr>
<tr><td>Current coverage</td><td>5</td><td>5</td><td>20</td><td>40</td><td>20</td><td>5</td><td>70</td><td>5</td></tr>
<tr><td>Target coverage (by 2040)</td><td>30</td><td>30</td><td>60</td><td>90</td><td>60</td><td>30</td><td>100</td><td>100</td></tr>
<tr><td colspan="9">Recommended investment over 10years
Total Investment: PHP 143 bn (PHP 1306 per capita)
Investment breakdown:</td></tr>
<tr><td colspan="2">Anxiety disorders: PHP 10.4 bn
Epilepsy: PHP 7.4 bn</td><td colspan="2">Depression: PHP 74 bn
Alcohol use: PHP 9.8 bn</td><td colspan="2">Psychoses: PHP 29.1 bn
Pesticide ban: PHP 2.4 bn</td><td colspan="3">Bipolar disorder: PHP 63.2 bn
School-based interventions: PHP 13.6 bn</td></tr>
<tr><td colspan="9">The return on investment
Total return on investment: PHP 217 bn
Healthy life years gained: 724,195
Prevalent cases averted: 1,748,020
Deaths avoided: 5344
Productivity gains: PHP 216 bn</td></tr>
</table>

The smart economic investment opportunity

A study published in the *Lancet* of 36 countries estimated that a linear increase in the treatment of depression and anxiety between the years 2016 and 2030 globally could generate US$230 billion in productivity gains and US$310 billion in the value of extra healthy life years (Chisholm et al., 2016). Low-income groups were expected to gain US$3.3 worth of benefits for every US$1 spent, while for high-income countries, the return was US$4.

National-level analyses have shown the specific benefits that a country can expect to accrue from scaling up investment in mental health services. Here are three examples.

In **Jamaica**, scaling up treatment for depression, anxiety, and psychosis between 2015 and 2030, at a cost of over US$115 million, could deliver total benefits of over US$434 million for the Jamaican economy (United Nations Interagency Task Force on Noncommunicable Diseases, 2020).

In **Lebanon,** a modest 5% increase in making people able to work through mental health treatment could add US$1.7 million to the Lebanese economy, delivering a total of US$6.62 in financial value for every US$1 spent.

In **Australia**, an analysis found that an investment of approximately US$4.4 billion would generate economic returns of between US$8.2 and US$12.7 billion (KPMG, 2018). Most of this would be generated through savings to employers as a result of workplace interventions, likely to be worth around US$4.5 billion. The study emphasizes the advantages of targeted early interventions, which generate the highest cost savings.

The power of workplace investment

As the Australian example illustrates, the workplace offers an opportunity to address mental health issues.

Many leading companies across the globe are seizing this opportunity by prioritizing their employees' mental health and using their public platform to encourage others to follow. This is not simply a question of altruism. These companies now recognize the connection between mental wellbeing, productivity, and company performance (Harnois et al., 2000). They know that protecting the mental health of their staff is essential to build a successful business—something that is more important than ever with the shift toward home working accelerated by the COVID-19 pandemic and the financial stress caused by widespread economic uncertainty.

Companies that invest in employee mental health are making a smart business move (Hilton et al., 2008). These companies have four times the staff retention of companies that do not effectively manage employee wellbeing (Ang, 2009). On average, companies receive an impressive $5 return for every $1 invested in employee mental health and well-being (Deloitte, 2021).

In 2019 leading companies from around the world, supported by United GMH, launched the Global Business Collaboration for Better Workplace Mental Health (GBC, 2022). Around 100 organizations comprising 1.6 million employees have signed up as part of the collaboration, and hundreds of senior leaders have worked together to share best practices for the creation of mentally healthy workplaces. With their support, the collaboration has produced online webinars and resources enabling leaders to discuss challenges and share their own approaches to improve employee well-being.

The return on investment for society

Mental health is a public good and is interwoven into society. With the right investment, we can create a virtuous circle, enhancing mental and societal health. By promoting good mental health, we promote the ability of everyone to fully participate in society, whether in their social networks, workplaces, schools, communities, or families. And by participating in society, people's mental health tends to improve.

The potential impact of either good or poor mental health on society is enormous, yet not fully understood or acted upon. We explore it here in three key areas: physical health, the family unit, and social cohesion.

Physical health

The cost of poor mental health

The link between mental health and physical health is well established—people with poor mental health tend to have significantly worse physical health, and vice versa (Burns et al., 2015). Poor mental health tends to lead to increased

- tobacco, alcohol, and substance abuse (Delgadillo et al., 2013; National Institute on Drug Abuse, 2019; Weaver et al., 2003)
- vulnerability to infectious diseases such as HIV and tuberculosis (TB) (Liu et al., 2017)
- obesity and chronic health conditions (Ashworth et al., 2017; Naylor et al., 2016).

Furthermore, each of these physical conditions has an impact on mental health. For example, around half of people with TB have depression, making them ten times more likely to stop treatment, which puts their lives at greater risk (Ambaw et al., 2018).

The smart investment opportunity

The proven connection between physical and mental health means it's crucial that we integrate mental healthcare into physical health programs. Recent United GMH research shows that integrating mental healthcare in HIV and TB health services would accelerate the reduction of HIV infections by 10%–17%, and similarly reduce TB infections by 13%–20%. This would mean almost 1 million people avoiding contracting HIV and averting as many as 14 million TB cases. Our research shows that integrating mental healthcare into physical health treatment does not need to be expensive, coincident with evidence that the benefits outweigh the costs (UnitedGMH, 2021a, 2021b).

The family unit

The cost of poor mental health

Poor mental health affects entire families. Parental stress and depression, such as postpartum depression, impact maternal-infant bonding. Such discord leads to psychological, developmental, and educational risks for the child. Globally, 10%–20% of children and adolescents experience mental disorders (World Health Organization, 2019a), while 15%–23% of children live with a parent with a mental illness, making them more likely to have mental health problems themselves (Leijdesdorff et al., 2017).

The smart investment opportunities

The "*Thinking Healthy Program*," developed in Pakistan, is the first low-intensity psychological intervention endorsed by the WHO for the treatment of perinatal depression in low- and middle-income countries. By supporting the mental health of perinatal women, the program led to improvements in the physical development of their infants and an increase in immunization uptake, as well as reducing depression in the new mothers.

This program is just one illustration of how mental healthcare in family settings can lead to crucial long-term improvements in physical health. The scope for integrating mental healthcare to improve physical health in other areas of society is vast.

Social cohesion

The cost of poor mental health

The influence of social relationships on health and well-being is well recognized (Berkman et al., 2000). In a vicious cycle exacerbated by the COVID-19 pandemic, evidence is building that mental illness is a barrier to social connections—and poor social connections can result in poor mental health.

People with weak social connections are more likely to develop mental health illnesses (Yu et al., 2015) while those who join social groups are less likely to have depression (Cruwys et al., 2013). Individuals with good-quality relationships are consistently seen to have better mental health (De Silva et al., 2005; Ehsan & De Silva, 2015; Rocco & Suhrcke, 2012). Clinically, social isolation is recognized as a symptom of depression and a precursor for other mental health diagnoses.

The opportunity for smart investment

Treating depression results in an improvement in social connectedness (Renner et al., 2014) and social functioning is proposed as a marker for successful treatment of depression (Zimmerman et al., 2006).

Community engagement approaches, small group activities, individual talk therapy, and multipronged approaches have all been shown to improve social connectedness (Webber & Fendt-Newlin, 2017). These techniques all serve to improve familial and social cohesiveness, with the result that well integrated societies have less crime, lower mortality, and better physical and mental health (Wilkinson & Pickett, 2009). Indeed, the potential benefits of improving social connectedness are far reaching.

Conclusion

The case for investing in mental healthcare is clear. A wealth of evidence shows that investment in mental health and well-being comes with economic, social, and individual health benefits that far outweigh the financial outlay. Therefore such financial outlay must be seen as an investment, not an expense.

All stakeholders, from governments and international funders to community leaders and employers, have a role to play in promoting mental well-being. But, simply increasing funding for mental health will not be enough. Investment in mental health needs to be smart: founded on the kind of evidence-based policies and services outlined in this chapter, and developed in line with human rights frameworks and the given individual cultural context.

As the public and private sector gain a greater understanding of the impact of mental well-being on the economy, society, and individual lives, support for such investment will only grow globally and lead to the healthier world we all seek.

References

Ambaw, F., Mayston, R., Hanlon, C., Medhin, G., & Alem, A. (2018). Untreated depression and tuberculosis treatment outcomes, quality of life and disability, Ethiopia. *Bulletin of the World Health Organisation*, *96*(4), 243–255.

Ang, H. S. K. (2009). *Wellness and productivity management: A new approach to increasing performance*. Philadelphia, PA: Right Manag. Inc.

Ashworth, M., Schofield, P., & Das-Munshi, J. (2017). Physical health in severe mental illness. *British Journal of General Practice*, *67*(663), 436–437.

Berkman, L. F., Glass, T., Brissette, I., & Seeman, T. E. (2000). From social integration to health: Durkheim in the new millennium. *Social Science and Medicine*, *51*(6), 843–857.

Bloom, D. E., Cafiero, E. T., Jane-Llopis, E., Abrahams-Gessel, S., Bloom, L. R., Fathima, S., Feigl, A. B., Gaziano, T., Mowafi, M., Pandya, A., Prettner, K., Rosenberg, L., Seligman, B., Stein, A., & Weinstein, C. (2011). *The global economic burden of noncommunicable diseases*. Geneva: World Economic Forum.

Burns, R. A., Butterworth, P., Browning, C., Byles, J., Luszcz, M., Mitchell, P., et al. (2015). Examination of the association between mental health, morbidity, and mortality in late life: Findings from longitudinal community surveys. *International Psychogeriatrics*, *27*(5), 739–746.

Chisholm, D., Sweeny, K., Sheehan, P., Rasmussen, B., Smit, F., Cuijpers, P., et al. (2016). Scaling-up treatment of depression and anxiety: A global return on investment analysis. *Lancet Psychiatry*, *3*(5), 415–424.

Cruwys, T., Dingle, G. A., Haslam, C., Haslam, S. A., Jetten, J., & Morton, T. A. (2013). Social group memberships protect against future depression, alleviate depression symptoms and prevent depression relapse. *Social Science and Medicine*, *98*, 179–186.

De Silva, M. J., McKenzie, K., Harpham, T., & Huttly, S. (2005). Social capital and mental illness: A systematic review. *Journal of Epidemiology and Community Health*, *59*(8), 619–627.

Delgadillo, J., Godfrey, C., Gilbody, S., & Payne, S. (2013). Depression, anxiety and comorbid substance use: Association patterns in outpatient addictions treatment. *Mental Health and Substance Use*, *6*(1).

Deloitte. (2021). *Mental health and employers: The case for investment—Pandemic and beyond*. Deloitte. Available from: https://www2.deloitte.com/content/dam/Deloitte/uk/Documents/consultancy/deloitte-uk-mental-health-report-2022.pdf.

Ehsan, A. M., & De Silva, M. J. (2015). Social capital and common mental disorder: A systematic review. *Journal of Epidemiology and Community Health*, *69*(10), 1021–1028.

Friedli, L., & Niamh, M. P. (2007). *Mental health promotion: Building an economic case*. Belfast: Northern Ireland Association for Mental Health.

GBC. (2022). *Global business collaboration for better workplace mental health*. GBC. Available here: https://betterworkplacemh.com/.

Harnois, G., Gabriel, P., World Health Organization, & International Labour Organisation. (2000). *Mental health and work: Impact, issues and good practices*. World Health Organization.

Hilton, M., Scuffham, P. A., Sheridan, J. M., Cleary, C. M., & Whiteford, H. A. (2008). Mental ill-health and the differential effect of employee type on absenteeism and presenteeism. *Journal of Occupational and Environmental Medicine*, *50*(11), 1228–1243.

Iemmi, V. (2020). *Philanthropy for global mental health 2000–2015, Global mental health. Vol. 7*. Cambridge University Press.

IHME. (2020). The global burden of disease. *The Lancet*, *396*, 1129–1306.

Klynveld Peat Marwick Goerdeler (KPMG). (2018). *Investing to save: The economic benefits for Australia of investment in mental health reform*. Australia: KPMG.

Leijdesdorff, S., van Doesum, K., Popma, A., Klaassen, R., & van Amelsvoort, T. (2017). Prevalence of psychopathology in children of parents with mental illness and/or addiction: An up to date narrative review. *Current Opinion in Psychiatry*, *30*, 312–317. https://doi.org/10.1097/YCO.0000000000000341.

Liu, N. H., Daumit, G. L., Dua, T., Aquila, R., Charlson, F., Cuijpers, P., et al. (2017). Excess mortality in persons with severe mental disorders: A multilevel intervention framework and priorities for clinical practice, policy and research agendas. *World Psychiatry*, *16*(1), 30–40.

Mental Health Commission of Canada. (2014). *Why investing in mental health will contribute to Canada's economic prosperity and to the sustainability of our health care system*. Mental Health Commission of Canada. Available from: https://www.mentalhealthcommission.ca/sites/default/files/MHStrategy_CaseForInvestment_ENG_0_1.pdf.

National Institute on Drug Abuse. (2019). *Comorbidity: Substance use disorders and other mental illnesses*. NIDA. drugabuse.gov/publications/drugfacts/comorbidity-substance-use-disorders-other-mental-illnesses.

Naylor, C., Das, P., Ross, S., Honeyman, M., Thompson, J., & Gilburt, H. (2016). Bringing together physical and mental health: A new frontier for integrated care. London: The King's Fund.

Patel, V., Saxena, S., Lund, C., Thornicroft, G., Baingana, F., Bolton, P., et al. (2018 Oct 27). The Lancet Commission on global mental health and sustainable development. *Lancet*, *392*(10157), 1553–1598.

Renner, F., Cuijpers, P., & Huibers, M. J. (2014). The effect of psychotherapy for depression on improvements in social functioning: A meta-analysis. *Psychological Medicine*, *44*(14), 2913–2926.

Rocco, L., & Suhrcke, M. (2012). *Is social capital good for health? A European perspective*. Copenhagen: World Health Organization Regional Office for Europe.
UnitedGMH. (2021a). *Bending the curve: The impact of integrating mental health services on HIV and TB outcomes*. June.
UnitedGMH. (2021b). *Countdown global mental health 2030: Using data to inform action*. September.
UnitedGMH. (2022). *Mental health for all: What can achieve if we meaningfully integrate mental health into UHC*. United GMH. Available at: https://unitedgmh.org/UHC-mental-health-integration.
United Nations Interagency Task Force on Noncommunicable Diseases, United Nations Development Programme, Pan American Health Organization. (2020). *Care for Mental Health Conditions in Jamaica: The case for investment. Evaluating the return on investment of scaling up treatment for depression, anxiety, and psychosis*. https://iris.paho.org/handle/10665.2/51834.
Weaver, T., Madden, P., Charles, V., Stimson, G., Renton, A., Tyrer, P., et al. (2003). Comorbidity of substance misuse and mental illness in community mental health and substance misuse services. *The British Journal of Psychiatry*, *183*, 304–313.
Webber, M., & Fendt-Newlin, M. (2017). A review of social participation interventions for people with mental health problems. *Social Psychiatry and Psychiatric Epidemiology*, *52*(4), 369–380.
Wilkinson, R., & Pickett, K. (2009). *The spirit level, why equal societies almost always do better*. London: Penguin Books.
World Health Organisation. (2021). *Prevention and management of mental health conditions in the Philippines. The case for investment*. Manila: World Health Organisation Regional Office for the Western Pacific.
World Health Organization. (2019a, Nov 27). *Child and adolescent mental health [online]*. WHO. Available from: https://www.who.int/activities/improving-the-mental-and-brain-health-of-children-and-adolescents.
World Health Organization. (2019b). *Dementia [online]*. WHO. Available from: https://www.who.int/news-room/fact-sheets/detail/dementia.
World Health Organization. (2021a). *Suicide [online]*. WHO. Available from: https://www.who.int/news-room/fact-sheets/detail/suicide.
World Health Organization. (2021b). *Mental health atlas 2020*. Geneva: WHO. Available from: https://www.who.int/publications/i/item/9789240036703.
World Health Organization. (2021c). *WHO comprehensive mental health action plan (2013-2030)*. Geneva: WHO.
Yu, G., Sessions, J. G., Fu, Y., & Wall, M. (2015). A multilevel cross-lagged structural equation analysis for reciprocal relationship between social capital and health. *Social Science and Medicine*, *142*, 1–8.
Zimmerman, M., McGlinchey, J. B., Posternak, M. A., Friedman, M., Boerescu, D., & Attiullah, N. (2006). Discordance between self-reported symptom severity and psychosocial functioning ratings in depressed outpatients; implications for how remission from depression should be defined. *Psychiatry Research*, *141*(2), 185–191.

Further reading

Institute for Health Metrics and Evaluation. (2018). *Findings from the global burden of disease study 2017*. Seattle, WA: IHME.
Ohrnberger, J., Fichera, E., & Sutton, M. (2017). The dynamics of physical and mental health in the older population. *Journal of the Economics of Ageing*, *9*, 52–62.
UNITNCDs. (2019). *Making the investment case for mental health: A WHO/UNDP methodological guidance note*. UN Interagency Task Force on NCDs. Available at: https://apps.who.int/iris/bitstream/handle/10665/325116/WHO-UHC-CD-NCD-19.97-eng.pdf?sequence=1&isAllowed=y.
Vigo, D. V., Kestel, D., Pendakur, K., Thoricroft, G., & Atun, R. (2019 Feb). Disease burden and government spending on mental, neurological, and substance use disorders, and self-harm: Cross-sectional, ecological study of health system response in the Americas. *The Lancet Public Health*, *4*(2), e89–e96.

World Health Organization. (2017). *Depression and other common mental disorders: Global health estimates.* Geneva: WHO. License: CC BY-NC-SA 3.D IGO. Available from: https://apps.who.int/iris/bitstream/handle/10665/254610/WHO-MSD-MER-2017.2-eng.pdf?sequence=1.

World Health Organization. (2018). *Mental health atlas 2017.* Geneva: WHO. Available from: https://apps.who.int/iris/bitstream/handle/10665/272735/9789241514019-eng.pdf?ua=1.

World Health Organization. (2019c). *The WHO special initiative for mental health (2019-2023): Universal health coverage for mental health.* Geneva: WHO.

World Health Organization. (2022). *Mental disorders [online].* Geneva: WHO. Available from: https://www.who.int/news-room/fact-sheets/detail/mental-disorders.

CHAPTER 55

Financing an NGO addressing mental health and poverty during and after a pandemic: My personal experience

Nicole K. Bulanchuk[a,b,c] and Judy Kuriansky[a,c]

[a]*Department of Clinical Psychology, Teachers College, Columbia University, New York, NY, United States,* [b]*Believe New York Philanthropies, Inc., New York, NY, United States,* [c]*International Association of Applied Psychology, Indianapolis, IN, United States*

Introduction

"How are we going to pay for it?" "Who's going to pay for it?" Those are key, and arguably the most overwhelming, questions facing brave humanitarians who want to start, or sustain, a non-for-profit organization or nongovernmental organization (NGO) in mental health, which is traditionally underfunded since it is hard to assess and monetize the outcome of mental well-being. Many worthy projects fold because of lack of funds, particularly in the COVID-19 era, when world economies collapsed. In this chapter, I describe my personal experiences and strategies raising funds for NGOs focused on mental health and poverty which can inspire aspiring NGOs and start-up not-for-profits.

Background

My experience starts at 18-years old when I traveled door-to-door to raise money for a local NGO. Because of my upbringing and experiences with abuse, neglect, and economic hardship, I took on this opportunity to help others, especially those with far worse situations than my own. Throughout time, this goal became a central theme in every opportunity I pursued—frequently asking myself, "Will this work benefit the lives of others?"

Later, I became the WalksManager (colloquilly known as Fundraising Manager) for the *National Alliance on Mental Illness (NAMI-NYS)*, a United States-based advocacy group founded as a grassroots group by family members of people diagnosed with mental illness,

Resilient Health. https://doi.org/10.1016/B978-0-443-18529-8.00055-X

for which I successfully raised about USD$50,000 for their annual charity walk, *NAMIWalk*. Afterward, for two years, I was a Program Manager with an integrated fundraising company, *Up-Fundraising*, where I managed face-to-face fundraising campaigns for multimillion-dollar NGOs, primarily, *Save the Children Federation* and *Care International*, which focus on reducing global poverty and improving health. During that time, I further exceled at face-to-face fundraising strategies and learned the importance of donor acquisition and diversified donor programs.

In October 2018, I wanted to try my hand at my own projects, so I cofounded a start-up NGO called *Believe New York* (BNY) with limited funding and insufficient manpower. In July 2020, we submitted for nonprofit status, restructed our leadership, and fundraising status. Headquartered in my home city of New York, we connect vulnerable individuals and families to resources through our 24/7 *GetHelp* email and phone service. We also offer a food donation delivery service through our *Food First Program*, as well as free mental health support through *The Open Space*. Occasionally, we host events to service low-income and underserved communities. For example, for Thanksgiving 2021, we hosted Hot Meal and Warm Clothing Handout in the Lower East Side, a neighborhood in the borough of Manhattan with high rates of homelessness and food insecurity. During that event, we served over 80 hot meals to underserved individuals and families, shown in Fig. 1. We mainly rely on donations from family and friends and our local canvassing efforts, but we are expanding our network and approach.

I also serve as the Fundraising Chair on the Alumni Board for the University at Albany student organization, called the *Middle Earth Peer-Assistance Program*, a mental health outreach organization with a student-run peer-support hotline and peer-coaching service. Despite the challenges of COVID-19, we raised tens of thousands of dollars for our 50th Anniversary of the organization in 2020.

Fig. 1
Nicole Bulanchuk *(right)*, packaging hot food during BNY's *Thanksgiving Hot Meal and Warm Clothing Hand Out* in Manhattan, New York, on November 28, 2021.

Sustainable development goals 1 and 3 and COVID-19

My work is grounded in achievement of the United Nations 2030 Agenda for Sustainable Development (United Nations, Department of Economic and Social Affairs, 2020), specifically Sustainable Development Goal (SDG) 1 to "End poverty in all its forms everywhere" and SDG 3 to "Ensure healthy lives and promote well-being for all at all ages" which includes improving access to mental health and physical health, as well as the World Health Organization (WHO) Framework for Action and health systems, and Universal Health Coverage. Unfortunately, the COVID-19 pandemic interrupted and set back progress in achieving these SDGs, for example, over 71 million people were pushed into extreme poverty, exacerbating mental health problems and health disparities, and decreasing peoples' well-being (Kesner, 2020). Aligning with the SDGs was consolidated when I took a very valuable course entitled "Psychology and the United Nations" at Teachers College, Columbia University in New York City, taught by Dr. Judy Kuriansky who has nearly two decades of experience with governments and civil society NGOs and the UN system.

Lack of funding for mental health and NGOs

Helping NGOs be sustainable is critical, according to Dr. Kuriansky, since governments draft and commit to mental-health-positive policy, but it is NGOs who do the work "on the ground." Dr. Kuriansky notably leads, plans, and executes humanitarian projects and programs that firsthand impact communities in need and directly works to achieve the goals outlined in the UN 2030 Agenda (Kuriansky, 2020). Funding is direly needed, but sorely inadequate, for all these interventions.

Although the WHO widely recognizes that mental health is critical to overall health, countries spend on average less than 2% of their health budget on mental health access and care and less than 1% is spent by low-income countries (Kesner, 2020). Yet, mental health problems place a huge burden on global health, where 14% of the global burden of disease is attributed to mental health problems (Iemmi, 2019).

However, we must not be discouraged and must instead develop constantly renewed passion for contributing valuable NGO work and finding funding to complete it.

The art of fundraising

Fundraising is the art of *storytelling*. I am always thinking that I am not simply asking people for money, but I am connecting on a personal level to tell an inspiring story about me and my cause (Phoofolo et al., 2011). At a 1-day Communications Workshop entitled: "Finding Funding for NGOs in a Challenging Global Economy" hosted by the United Nations

Department of Public Information Non-Governmental Organizations Relations Cluster, Mr. David Andrews, of *David J. Andrews Associates LLC*, said, "the best fundraisers are good storytellers."

When I fundraised for *Save the Children*'s programs at the US-Mexico Border detention centers, I told interesting stories of experiences and struggles of my great grandfather who came to the United States from Colombia and opened a small bodega in Harlem, New York City.

Also, I always refine my *elevator pitch*, a 30-second introductory version of who I am and what my NGO does that is concise, honest, passionate, inspiring, and interesting.

Overcoming any resistance, intimidation, or fear of getting a "NO," I make myself ask for the money that I need for my NGO, with a strong sense of urgency. I have said, "We need your donation today, to ensure these children can have food *tonight*" and "Your contribution makes sure that teenagers don't consider killing themselves." When fundraising for *The Nature Conservancy*, I said, "Helping us plant trees today with your donation protects our future 20 years from now when they fully grow."

I always ask for a specific amount that's symbolic. At *Up-Fundraising*, we found that donors who gave money when we provided *symbolic numbers* were more likely to donate again. For example, asking for USD$25 symbolizes the roughly 1 in 4 (or 25%) of Americans who have a mental health disorder. A request for USD$69,000 or USD$6,900 symbolizes the 690 million people who go to bed on an empty stomach every night.

Creating donation tiers which are labeled is lucrative, too, as we did at *NAMI-NYS*; for example, diamond, gold, silver, and bronze levels correspond to a donation amount and specific benefits. At *BNY*, the highest donors for our 2021 Rooftop Fundraiser received free merchandise with our logo and name recognition in our newsletters.

I tell my team to never get discouraged and to keep asking, especially in pandemic recovery times when people might feel sympathetic to others' needs and aspire to help others.

Building a funding model and plan

I have learned the importance of building a Funding Model and a solid business plan. Donor groups do not like to fund "needy charities" but seek *viable business partners*. A business model incorporates choices and structures about costs and values, while a funding model focuses only on how funds move their program and services forward (Bray, 2019). Donor groups also like to see innovative solutions, concrete plans, and worthy causes relevant to current events or crises.

A model I have followed is a *5-year fundraising plan* (Bray, 2019) with these aspects

(1) Determine a reasonable dollar amount my NGO needs.
(2) Evaluate my NGO's fundraising assets (e.g., staff, volunteers, experiences).

Fig. 2
Nicole Bulanchuk *(left)* using *Wasyact* to collect donations from *Save the Children* supporters in Brooklyn, New York, September 2019.

(3) Create a timeline.
(4) Write my strategy in a short, easy-to-follow document or spreadsheet.
(5) Identify sources for funding: individuals, businesses, and foundations.
(6) Find ways to cut time and energy, by using software, for example, a platform like *DonorDrive, or Waysact* by *Evergiving,* for use when doing face-to-face fundraising, as shown in Fig. 2.

Sources of funding

I always brainstorm where money can come from, both from traditional and out-of-the-box sources.

Small businesses. Small business owners used to donate a percentage (6%) of their profits to NGOs (Preston, 2008) partly due to tax benefits, but many of these businesses were hit hard during COVID-19 and tragically forced to close. I learned the importance of offering incentives to these, like promotions on social media, and displaying their business on my website. Also, as business owner myself, I know the value of appealing to the silver lining of the COVID-19 pandemic to join my efforts since "we are all in this together."

Corporate giving. Corporate gifts can come from sponsorships, grants, and/or matching gift programs. During the pandemic, the CARES Act (the two-million-dollar federal stimulus package response to COVID-19) presented an opportunity for foundations and corporations to support NGOs, since the annual cash gift limit was increased from 10% to 25% of corporate taxable income (Simon, 2020). Since corporations care about how many "eyes" are on your campaign—equivalent to advertising dollars they would spend on a traditional campaign or "ad buy"—I always try to be unique in my appeal for their support and match their business goals and values.

Private philanthropists and foundations contributions. There are three main types of foundations: corporate, family, or private foundations, and community foundations. Community foundations, also known as public foundations, give funds through grants and scholarships, usually through an intense application process. Private foundations and family foundations are usually created from endowments from wealthy families, corporate board members, or private philanthropists like *The Rockefeller Foundation* and the well-known *Bill and Melinda Gates Foundation,* which funds health, mental health, and reproductive healthcare. It is in my favor that health is a primary sector funded from private foundations and philanthropists, giving 53% of funds from 2013 to 2015 (Lion's Head, 2018).

I always remember that individual, smaller, contributions from people in the public are an important funding group. MacKenzie Scott, ex-wife of Jeff Bezos, CEO and founder of *Amazon,* the second-wealthiest woman in the world, donated over USD$6 billion in 2020 alone, mostly to small charities and NGOs working on ending poverty and improving health (Kulish, 2020), including a multimillion-dollar grant to the New York City-based Child Mind Institute which focuses on children's mental health. I always think about their tax benefits from donating.

To find funders, I use search engines, like *Candid* (www.candid.org) in the USA (there is also the *European Foundation Centre* in Europe). In doing this research, I'm prepared for a tedious time-consuming process, and I consult experts I find on *LinkedIn*, or networking at NGO workshops and community events.

Reoccurring giving and memberships

A challenge that I and all my colleagues face is retaining supporters and ensuring recurring donations and memberships, parallel to a subscription to *Netflix*. Research confirms that recurring giving programs score higher and stronger retention rates over time (Network for Good, 2020). As good news, according to a survey, 52% of millennials are more interested in giving monthly to causes they care about to stay engaged, rather than giving only one time.

Social media, online fundraising, virtual events, and crowdfunding campaigns

During the COVID-19 pandemic, when we got used to being online, social media and online fundraising became even more vital. As a result, I am now very keenly aware about ensuring high visibility and easy access to my platform's donation button. Also, I create emotional connection with donors on my social media using human interest stories in posts. I also consider options of fundraising 24/7 with online crowdfunding sites like *GoFundMe* or *Patreon*.

One of the strategies with which I've been successful is hosting virtual fundraisers through events over *Zoom* or *Google Meet*. Virtual events are fun and engaging, especially as people emerge from the pandemic being more familiar with video-calling platforms. The events can be interesting workshops, wellness sessions, games, activities, or discussion groups. A virtual event I personally enjoyed was hosted by the NGO *Hope for Haiti*, entitled, "Hike for Haiti Challenge" where supporters paid USD$30 to participate virtually in a livestream hike alongside real people you see in the screen who are hiking in a rural community in Haiti. The virtual hike created a sense of solidarity among the participants, supporting the students, teachers, and healthcare workers who were fighting to keep Haitians alive and healthy during the pandemic.

Face-to-face fundraising

I have a lot of experience in face-to-face fundraising, colloquially known as canvassing, which I and my colleagues have done on public streets, in parks and retail settings, and at events, as shown in Fig. 3. This is one of my favorite ways to fundraise because it is fun and directly engages donors you might have a difficult time reaching.

Peer-to-peer and pledge fundraising

I also do a lot of peer-to-peer and pledge fundraising, in conjunction with an event, which is popular when combined with a walk-a-thon or charity run. In peer-to-peer approach, I ask friends, coworkers, and family members to donate. In pledge fundraising, I ask supporters to "pledge", or promise, to raise a certain amount in a specific period. For example, in the "Virtual Step Challenge"—held in place of a live annual Breast Cancer Awareness Walk and hosted by the NGO *Pink Forward*, to raise money for the *American Cancer Society*—I paid the USD$25 registration fee, pledged how much I would raise, and then simply walked in my community, keeping track on the society's website of how many steps I took by myself or with others. Fundraising incentives were built in, with prizes for how much I raised. Participants walked over 50 million steps for breast cancer awareness during this event, raising decent money for the NGO.

Fig. 3
Nicole Bulanchuk face-to-face fundraising during *Save the Children*'s inaugural "Bridge the Gap Challenge" to advocate for girls who are disproportionally impacted by violence, health issues, and poverty on October 12, 2019, Brooklyn, New York.

Writing grant proposals

I don't love the tedious process of grant writing, but I have to do it. I follow guidelines from Tiedan Haung, grant writer for the *Center for Developing Urban Educational Leaders* (Phoofolo et al., 2011), by including an introduction: my NGO mission and needs, capabilities of my work, and plans for implementing and evaluating my project.

Sustaining existing donors and networks

Knowing the value of appreciation, as a budding psychologist, I always thank donors for their generosity regardless of the size of a donation—even only USD$5—to build our relationship and make supporters feel good and special, especially important in postpandemic times, when people still feel the ongoing sting of being isolated due to physical distance and lockdown restrictions. I retain donors by giving them something in return for their giving, such as a shirt, prize, and/or an award, with my NGO logo. When I worked with *NAMI-NYS*, we hosted an annual "Recognition Dinner" with our supporters, giving out awards, certificates, and prizes. I also send handwritten thank-you notes and think of activities to share, since we all know that deals are made during personal conversations, or even on the golf course or during a shared enjoyable experience (Phoofolo et al., 2011).

I also know that donors always want to hear results and what my organization is achieving with their contribution (Phoofolo et al., 2011), so I communicate my progress often, through phone calls, emails, newsletters, and events, and specify how they helped.

Lessons learned and the way ahead

While fundraising is a challenge in the ongoing economically challenging times, I am still intrepidly driven and inspired to help people through my NGO efforts to overcome poverty and to ensure mental wellness. The pandemic has forced NGOs to become creative with fundraising that will continue in the ongoing recovery period and beyond. I enjoy using traditional methods of raising money but also exploring new ideas. My psychological training has reinforced the value of forging relationships with donors, which is even more crucial in postpandemic times when people are emerging from long periods of feeling isolated and desolate. Now, when times are difficult economically and emotionally for everyone, I am patient but remain hopeful. I am inspired by the story of the brother and sister who founded a project in 1919 to feed starving children in Germany and Austria-Hungary during the allied blockade during World War I (Watson, 2017), that, with innovative marketing and fundraising, has now grown into the multimillion-dollar NGO *Save the Children* to now service over 120 countries. I am resolved to keep up my devotion to improve the lives of the underserved and ask people to help me make the world a better place.

References

Bray, I. (2019). *Effective fundraising for nonprofits: Real-world strategies that work* (6th ed.). NOLO.

Iemmi, V. (2019). Sustainable development for global mental health: a typology and systematic evidence mapping of external actors in low-income and middle-income countries. *BMJ Global Health*, *4*(6), e001826. https://doi.org/10.1136/bmjgh-2019-001826.

Kesner, C. (2020). *The global COVID-19 financial response is leaving mental health behind*. United for Global Mental Health.

Kulish, N. (2020). *Giving billions fast, MacKenzie Scott upends philanthropy*. The New York Times.

Kuriansky, J. (2020). *Psychology at the United Nations*. Retrieved from Teachers College, Columbia University, CCPX 4075.

Lion's Head. (2018). *Financing global mental health*. Loin's Head Global Partners.

Network for Good. (2020). https://www.networkforgood.com/resource/why-recurring-giving-matters/.

Phoofolo, R., Kokoris, C., & Kuriansky, J. (2011). Finding funding for NGOs in today's challenging global economy: A United Nations DPI/NGO Communications Workshop. *IAAP Bulletin*, *23*(3), 26–28.
Preston, C. (2008). *Most small companies make charitable donations survey finds*. The Chronicle of Philanthropy.
Simon, M. (2020). *Now is a great time to give: New charitable rules incentivize generosity during COVID-19*. Forbes.
United Nations, Department of Economic and Social Affairs. (2020). *The 17 sustainable development goals*. United Nations, Department of Economic and Social Affairs.
Watson, N. (2017). *Save the children, an International Organization*. Encyclopedia Britannica.

PART 5

The innovation toolbox: Special populations and interest groups

According to the World Health Organization (WHO), "Health is a human right for all." This declaration is confirmed and underscored by these chapters covering many populations.

Thus, this Part explains how innovations in the pandemic response have been applied to specific populations, especially using digital tools, techniques, and approaches. These populations include women, children, youth, aged persons, families, racial groups, persons with disabilities, indigenous persons, severely mental ill persons, persons with addictions and autism spectrum disorders, persons in the LGBTQI+ community, and indigenous peoples as well as migrants, refugees, and asylum seekers.

Chapters also cover the importance of insuring mental health in the workplace, and innovative technologies relevant to specific fields of health workers, including dentists, nurses, female doctors, and medical students.

What notables have to say

The disproportionate levels of digital abuse faced by women is considered an extension of gender-based violence, which is characterized by the physical, sexual, mental, or economic harm forced on a person, usually a woman or girl, because of their gender identity. Social media platforms that have grown exponentially during the digital age have led to a breeding ground for people to abuse and harass women, often without consequences.

—Michelle Bachelet, former president of Chile and former president of UN Women, currently United Nations High Commissioner for Human Rights, at the Generation Equality Forum (GEF), June-July 2021 on signing a letter with 200 distinguished women, urging tech companies to fight digital abuse

I am not what happened to me. I am what I choose to become.

—Carl Jung, Swiss psychiatrist and psychoanalyst

Healthy citizens are the greatest asset any country can have.

—Winston Churchill, former Prime Minister of the UK

He who has health has hope and he who has hope has everything.

—Arabian Proverb

How wonderful it is that nobody need wait a single moment before starting to improve the world.

—Anne Frank

CHAPTER 56

Unlocking potential: Women's leadership and innovation for health

Roopa Dhatt[a,b,c], Preethi John[d,e], and Rachel Thompson[a]

[a]Women in Global Health, Hayward, CA, United States, [b]Global Council on SDG 3 - Good Health and Wellbeing, "SDGs in Action" Initiative, Dubai, United Arab Emirates, [c]Georgetown University, Washington, DC, United States, [d]Global Business School for Health, Faculty of Population Health Sciences, University College London (UCL), London, United Kingdom, [e]Women in Global Health India Chapter, Delhi, India

Introduction

In 2015 when the Sustainable Development Goals outlined in the United Nations (UN) 2030 Agenda for Sustainable Development were adopted by UN member states, a new era in development was charted—a universal agenda which was inclusive, country-driven, with stakeholders committed to "leave no one behind" (United Nations, 2015). In Sustainable Development Goal 3 (SDG3) regarding health and good well-being, the call for digital solutions and innovation was identified as an accelerator to achieving health targets. Yet, amid the exciting new promise and potential of health tech, digital solutions, and artificial intelligence (AI), long-standing issues of gender and broader equity were not prioritized.

During the COVID-19 pandemic, the pace of innovation in the health sector has been unprecedented. Rapid innovation has occurred in diagnostics, pharmaceutical development and vaccines, home healthcare, telemedicine, and other medical technologies. However, these innovations did not always incorporate sex and gender perspectives (Bali et al., 2020). For example, during the development of the COVID-19 vaccine, pregnant and breastfeeding women were left out of clinical trials in the initial stages, with serious consequences (Costantine, Landon and Saade, 2020; Engjom et al., 2022).

Gender inequality is widespread across the health, innovation, and tech sectors. Women make up 70% of the health workforce but hold only 25% of senior leadership roles in health (World Health Organization, 2019). Technology giants employ less than 15% women on AI projects (Nesta, 2019). A literature review looking at the gender of patent authors found

Resilient Health. https://doi.org/10.1016/B978-0-443-18529-8.00056-1

only 7% of all patents listed a woman as the primary inventor; at the current rate of progress, women are not expected to achieve parity in patenting until 2092 (Milli et al., 2016). This is important because the ability for women to attract funding through venture capital mechanisms is affected by low rates of patenting among women, which in turn impacts the resources and type of products innovators work on, creating a vicious cycle that excludes women from the design and benefits of innovation.

Gender inequality in innovation contributes to the status quo of a world designed primarily for men (Criado-Perez, 2019). Excluding women from innovation excludes their different perspective and experiences, which could result in the development of more gender-appropriate innovations (Scaffidi et al., 2017). Correcting this imbalance, by increasing the proportion of women involved, can shift innovation toward the needs of women. This is evidenced by research on United States biomedical patents, which shows that patents with women inventors are significantly more likely to focus on women's health, and women-led research teams are 26% more likely to focus on women's health outcomes (Koning et al., 2020).

This chapter demonstrates where gender responsive innovation is urgently needed in global health, and the role women can, and must, play in it. Women need better access to medical technologies, for example, to diagnostics such as pregnancy tests, as well as for innovation to improve safe and decent working conditions for women health workers, for example, in the design and provision of Personal Protective Equipment (PPE).

Two case studies are presented in this chapter. The first case study summarizes research into women's perspectives and needs around diagnostics, an orphan of global health in terms of investment and innovation. The work of our organization, Women in Global Health (WGH), a global movement demanding gender equity in health systems and health leadership, with The Foundation for Innovative New Diagnostics (FIND), a global health nonprofit based in Geneva, Switzerland, shows how women's empowerment can lead to more testing and better health for all.

The second case study presents an overview of the issues that women health workers face around PPE, highlighting the impacts of excluding women in product design processes that are still based on a 'reference man,' despite the majority of the workforce being female.

Underlying the case studies, the importance of women's leadership comes through as, arguably, the most important innovation. Women in leadership can expand the agenda, prioritizing neglected health issues and advancing innovation in the areas where women most need it (Dhatt & Keeling, 2021). Recent evidence in the initial stages of the pandemic showed that women-led countries had six times lower COVID-19 death rates, as well as lower peaks in daily deaths per capita, and lower excess mortality (Coscieme et al., 2020). There is a growing body of research showing that women parliamentarians, when present in sufficient numbers, have the power to change the political agenda and prioritize health, and especially women's health (Funk & Philips, 2019). Yet despite this, new reesearch from WGH finds women, and espeically those from low income countries, remain marginalized in health leadership (Women in Global Health, 2023).

Barriers women face

The ecological model illustrated in Fig. 1 helps explain the multiple barriers women face in entering into health leadership. The factors highlighted are also relevant to the innovation sector, where women may be disadvantaged by issues including harassment, historical norms that exclude women from science and technology committees and management positions, fear of criticism, and insecurities. Women make up 70% of the workforce and there is no shortage of women in the pipeline for health leadership; rather, a blockage in the pipeline stops women from entering leadership in equal numbers to their male counterparts (World Health Organisation, 2021). The model in Fig. 1 highlights how action is needed at all levels to drive sustainable change by showing the barriers across several strata of society (shown in concentric circles) from individual factors such as 'imposter syndrome,' to the wider sociopolitical context where, e.g. legal frameworks may be a barrier.

Gender transformative leadership

The chapter ends by considering how to create a more enabling ecosystem for women's leadership in health, using the principles of gender transformative leadership, and the dividends this approach will generate for us all. To improve gender equality in health and innovation, leaders of all genders must play their part. Gender transformative policies are defined as those

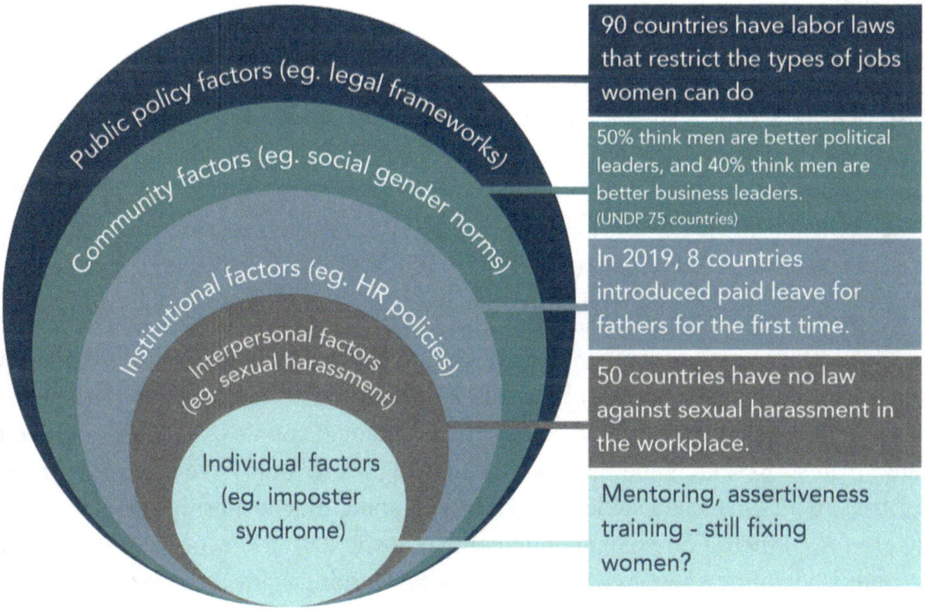

Fig. 1
The ecological model: multi-layered factors impacting women's journey into leadership. *Women in Global Health (2021). Ecological model. http://bit.ly/ECOMODEL.*

that "seek to transform gender relations to promote equality" (Women in Global Health, 2018). Gender transformative leadership approaches are needed to:

- Challenge privilege and imbalances in power that undermine progress in global health innovation.
- Confront intersectionality—the social and personal characteristics that intersect with gender (race, ethnicity, etc.) to create multiple disadvantages for women.
- Recognize different forms of leadership, such as thought leadership, which are not based on simple management hierarchy.

Definition: Gender transformative policies and programming include policies and programmes that seek to transform gender relations to promote equality and achieve programme objectives. This approach attempts to promote gender equality by (a) fostering critical examination of inequalities and gender roles, norms, and dynamics; (b) recognizing and strengthening positive norms that support equality and an enabling environment; (c) promoting the relative position of women, girls, and marginalized groups; and (d) transforming the underlying social structures, policies, and broadly held social norms that perpetuate gender inequalities (Inter-Agency Gender Working Group, 2017).

Case study 1: Innovation in diagnostics

*With permission, this case study is drawn from WGH's report 'Health in their hands: Testing and women's empowerment means better health for all' (Women in Global Health & FIND).

Diagnostic testing has often been an afterthought in global health (Pai & Kohli, 2019). The historical lack of investment and innovation in testing means that for every person diagnosed with a disease, another goes undetected; it is estimated that 50% of patients get no care at all or get care too late (World Health Organization, 2017). Testing is crucial to fulfilling the Right to Health and to achieving SDG 3. Women require tests for conditions and diseases common to both sexes (e.g. HIV/AIDS, COVID-19) but also for conditions unique to women (e.g. antenatal tests related to pregnancy, mammograms).

In part, due to the dearth of women's leadership in health, there has been little investment in women's health diagnostics, as well as a lack of data and research on the issues women face in accessing testing especially in low- and middle-income countries (LMICs), where testing gaps are greatest. Research and development of diagnostic tools have often been based on clinical trials with only male subjects, resulting in some diagnostics that do not work as effectively for women. For example, clinical diagnostic definitions of heart disease have been based on symptoms reported in men, meaning warning signs that are different in women have been ignored, unrecognized, or misdiagnosed (Criado-Perez, 2019).

More positively, self-testing diagnostic innovation as well as point-of-care testing has the potential to expand testing to more women, as well as men. Research suggests that women prefer to self-test at home for pregnancy and cervical cancer. A human papillomavirus (HPV) screening study in Uganda found 93% of women chose to provide a self-collected sample rather than attend a local clinic to have a sample taken by a health provider (White, Meglioli, Chowdhury, & Nuccio, 2017). Learning from HIV self-testing programs in southern Africa has demonstrated how distribution of self-test kits to women can benefit the wider community; for example, female sex workers can pass on kits to male clients to increase testing rates (Napierala et al., 2019). However, many self-testing innovations—even simple pregnancy urine tests—are not widely available in LMICs.

The WGH call to action for innovation in diagnostics

WGH believe that women's leadership and influence at community, health system, and political levels has the potential to drive access to testing. Women political leaders at national and local government levels can drive inclusion of testing in health budgets and gender responsive health services that prioritize diagnostic services needed by all genders. As a result of the earlier mentioned challenges, WGH has called for action for innovation in diagnostics by policymakers, to:

- Invest in innovation for low cost, quality self-testing methods and point-of-care testing devices.
- Enable women community health workers, nurses, and midwives at primary health level through training and resourcing to deliver testing in homes and communities.
- Engage women community leaders and women-led community-based organizations to promote health literacy on testing and support women to attend.
- Enable women political leaders to be testing champions within their countries and communities and to champion investment in health to ensure all women can access testing and treatment.

Case study 2: Innovation in PPE

*With permission, this case study is drawn from WGH's report 'Fit for Women: Safe and decent PPE for women health workers' (Women in Global Health, 2021).

Although women are 70% of the health workforce and have been 90% of health workers in patient-facing roles during the pandemic, most medical PPE is designed for men's bodies. Research by WGH found ill-fitting PPE has exposed women health workers on the pandemic frontlines to infection and left them working with indignity and discomfort, especially during menstruation, pregnancy, and menopause. Gender inequity in health systems limits protection for women who are clustered in roles accorded lower status, making them low priority for PPE when they are least able to buy their own. Global inequity means women in low-income countries are most impacted by shortages of PPE.

Despite the clear need for PPE that fits and protects women health workers, innovation has not been forthcoming. A 2021 survey by WGH of nearly 1000 women health workers in over 50 countries found only 25% of women had adequate access to PPE, only 14% used fitted PPE, and only 10% could use the toilet as often as needed (Women in Global Health, 2021). Average sizes of PPE are too big for most women; masks are too big, leaving gaps in the sides, increasing stress and infection risk for women health workers. Gowns and other body coverings are not modeled for women's bodies and although PPE may be manufactured in a range of sizes, only large sizes may be procured on the assumption that large will fit everyone.

Thus innovation is required in design but also in making procurement policies gender responsive, and to promote equity so that the women who need PPE most (for example, unvaccinated community health workers) have access to appropriate and adequate protection. Good governance and accountability are needed to make sure that the benefits of improved gender responsive PPE reach women on the frontlines in LMICs.

Women in Global Health's research on PPE confirmed that women remain marginalized in health leadership. Women health workers have considered some workplace policies and practices in the pandemic unsafe, for example around reuse of PPE, but have felt powerless to do anything, or when they do speak out, concerns are silenced or ignored. Challenges with PPE have caused women mental distress, worried that they might catch COVID-19 and infect their families. Inadequate PPE has made women health workers feel expendable and lose trust in their employers. Significant numbers of women plan to leave the profession. Designing PPE fit for women is feasible and women know what they need: women want to be protected and to propose solutions for more appropriate PPE, designed by and for women, suitable and comfortable for long shifts, and for hot temperatures and that enable women to use the bathroom.

The WGH Call to action for innovation in PPE

WGH is calling for action from governments to improve PPE for women health workers by:

- Promoting gender equity in leadership in the health sector to enable women to influence decisions on the protection of health workers, including the provision of gender responsive PPE.
- Enabling women health workers to lead and deliver change in PPE, for example by including them in standards-setting committees.

- Engaging professional associations, in nursing and midwifery especially, in setting standards for the design of PPE.
- Enabling women to work collectively through trade unions, professional associations, networks, and women's movements to advoacte for safe and decent working conditions.

More women in health leadership could have many positive impacts to improve the current situation around PPE. In design, it would end the current situation where PPE is designed on a 'reference man,' as considerations such as menstruation and pregnancy and menopause would be proactively addressed, and diversity among women embraced. In manufacturing, more women would be encouraged to enter the industry, breaking glass ceilings in a traditionally male sector. In procurement, having more women in charge would influence decisions to ensure procurement of sizes that fit women, the majority of health workers. In policy, women leaders will help ensure guidelines and policies around PPE are not gender blind. Strengthening women's leadership in the health sector across all parts of the PPE ecosystem will be essential to help prevent a 'Great Resignation' of health workers.

Conclusion

From school classrooms to boardrooms, as highlighted by the ecological model, action is needed at all levels to address challenges for women and gender equality in health innovation. Building the foundations for equality will require a renewed effort in ensuring access of girls to education, especially secondary education. It is critical girls are not deterred from taking Science, Technology, Engineering and Math (STEM) subjects by stereotypes that signify them as "male subjects." In the workplace, cultures and norms that discourage women, for example, from authoring their own patents must be exposed and confronted, using the principles of gender transformative leadership.

The innovation ecosystem must better address women's needs and become more gender responsive. Women's leadership can help drive this change. The case studies presented in this chapter illustrate how gender responsive innovation is required, whether designing diagnostic tools or PPE. Women's needs, and diversity among women, should be the starting point for innovation, not an afterthought.

Addressing the issues highlighted in this chapter is vital to achieving SDG 3 related to good health and well-being. Promoting women's leadership and more diverse innovation will also pay dividends toward progress in several other SDGs (e.g., SDG5 related to gender equality, SDG6 related to innovation). The dual strategy of implementing gender transformative leadership in health (see definition before), as well as enabling women's contribution across the innovation spectrum, is key to the achievement of the UN Agenda 2030 and specifically SDG 3. Global health cannot afford to lose out on women's talent, perspectives, and knowledge.

Recommendations

Given the above, WGH recommends:

For researchers: Adopt a gender responsive approach to global health research agendas. Research and development must include women. A gender lens is needed from ideation, to design, testing, distribution, and analyzing use. It is important that all groups of women and their needs are considered. Pregnant and breast-feeding women must be safely included in research and clinical trials rather than be excluded by default. Research to address biological and behavioral risk factors which are different among men and women must be prioritized. Collecting and publishing sex-disaggregated data must be mandatory as appropriate to the context.

For industry: Recalibrate innovation metrics and standard setting to ensure gender equity. The metrics by which innovation is measured, as well as the standards used to regulate design, need to incorporate a gender dimension. A gender lens in research and sex-disaggregated data is required to understand the problems, as well as how the potential benefits of innovation impact women.

For donors: Fund women's movements including those spearheaded by women entrepreneurs and innovators. There is sufficient evidence to show that women-headed organizations are underfunded particularly those based in LMICs. Inclusive innovation systems call for the involvement of women to be involved in all parts of the innovation process. Capacity building on innovation needs to be incorporated into curricula and policy frameworks. Mentoring and elevating women innovators as role models can help inspire others.

For governments: Commitments on equality in leadership have been made in the SDGs and other global agreements and now need to be delivered. Quotas to promote gender parity on decision-making bodies can help enable women to take the first step into leadership. Enabling women to achieve success in innovation will also require financial and practical support, for example, implementing family-friendly policies that enable women to achieve both personal and professional goals.

For leaders: Improve health for all through the catalytic effect of gender transformative leadership. All leaders must be sensitized to and practice gender transformative leadership. The dividends of gender transformative leadership in the realm of innovation will result in more investment for research and development into issues that affect women, beyond the confines of just reproductive health. Evidence has shown that women in health leadership can prioritize neglected health issues, ultimately improving health for all.

Fix systems, not women. Interventions to promote women's leadership in the past have focused on training for women in areas such as self-esteem and self-presentation, on the assumption that women needed to change to compete in systems and cultures designed for men. This ignored the systemic inequality, bias, and exercise of power that favors men for leadership roles—what we call 'male bonus syndrome' (Women in Global Health, 2023). Change is needed at all levels, from individual to national frameworks to global conventions. Closing the gender gap in health and innovation leadership is a critical foundation for strong health systems and health for all.

References

Bali, S., Dhatt, R., Lal, A., Jama, A., Van Daalen, K., & Sridhar, D. (2020). Off the back burner: diverse and gender-inclusive decision-making for COVID-19 response and recovery. *BMJ Global Health*, *5*(5). https://doi.org/10.1136/bmjgh-2020-002595.

Coscieme, L., Fioramonti, L., Mortensen, L. F., Pickett, K. E., Kubiszewski, I., Lovins, H., ... De Vogli, R. (2020). Women in power: female leadership and public health outcomes during the COVID-19 pandemic. *MedRxiv*. https://doi.org/10.1101/2020.07.13.20152397.

Costantine, M. M., Landon, M. B., & Saade, G. R. (2020). Protection by exclusion: Another missed opportunity to include pregnant women in research during the coronavirus disease 2019 (COVID-19) pandemic. *Obstetrics and Gynecology*, *136*(1), 26–28. https://doi.org/10.1097/AOG.0000000000003924.

Dhatt, R., & Keeling, A. (2021). Delivered by women, led by women: the future of global health leadership. *Leader to Leader*, *2021*(100), 43–47. https://doi.org/10.1002/ltl.20568.

Engjom, H., van den Akker, T., Aabakke, A., Ayras, O., Bloemenkamp, K., Donati, S., & Knight, M. (2022). Severe COVID-19 in pregnancy is almost exclusively limited to unvaccinated women–time for policies to change. *The Lancet Regional Health-Europe*, *13*, 100313. https://doi.org/10.1016/j.lanepe.2022.100313.

Funk, K. D., & Philips, A. Q. (2019). Representative budgeting: Women mayors and the composition of spending in local governments. *Political Research Quarterly*, *72*(1), 19–33. https://doi.org/10.1177/1065912918775237.

Inter-Agency Gender Working Group. (2017). *Gender integration continuum training session user's guide*. https://www.igwg.org/wp-content/uploads/2017/05/FG_GendrIntegrContinuum.pdf.

Koning, R., Samila, S., & Ferguson, J. (2020). Inventor gender and the direction of invention. *AEA Papers and Proceedings*, *110*, 250–254. https://doi.org/10.1257/pandp.20201045.

Milli, J., Gault, B., Williams-Baron, E., Xia, J., & Berlan, M. (2016). *The gender patenting gap* (p. C440). Institute for Women's Policy Research, Briefing Paper. https://www.research.uky.edu/uploads/iwpr-gender-patenting-gap; (2016).

Napierala, S., Desmond, N., Kumwenda, M., Tumushime, M., Sibanda, E., Indravudh, P., ... Cowan, F. (2019). HIV self-testing services for female sex workers, Malawi And Zimbabwe. *Bulletin of the World Health Organization*, *9*, 764–776. https://doi.org/10.2471/BLT.18.223560. https://www.who.int/bulletin/volumes/97/11/18-223560/en/.

Nesta (2019). Gender diversity in AI research. Policy Paper. Available at: https://www.nesta.org.uk/report/gender-diversity-ai/.

Pai, M., & Kohli, M. (2019). *Essential diagnostics: A key element of universal health coverage*. Dr. Sulaiman Al Habib Medical Journal. https://www.atlantis-press.com/journals/dsahmj/125906081.

Criado-Perez, C. (2019). *Invisible women: Data bias in a world designed for men*. New York: Abrams Press.

Scaffidi, J., Mol, B. W., & Keelan, J. A. (2017). The pregnant women as a drug orphan: a global survey of registered clinical trials of pharmacological interventions in pregnancy. *BJOG*, *124*, 132–140. https://doi.org/10.1111/1471-0528.14151.

White, H., Meglioli, A., Chowdhury, R., & Nuccio, O. (2017). Integrating cervical cancer screening and preventive treatment with family planning and HIV-related services. *International Journal of Gynecology & Obstetrics*, 41–46. https://doi.org/10.1002/ijgo.12194.

Women in Global Health. (2018). *Gender Transformative Leadership: A New Vision for Leadership in Global Health*. https://www.womeningh.org/single-post/2018/11/17/gender-transformative-leadership-a-new-vision-for-leadership-in-global-health.

Women in Global Health. (2021). *Fit for women? Safe and decent PPE for Women Health and care workers*. https://www.womeningh.org/fitforwomen.

Women in Global Health. (2023). *The state of women and leadership in global health.* https://womeningh.org/sheshapes/.

Women in Global Health & FIND. (2020). *Health in their hands: Testing and women's Empowerment means Better Health for all.* https://www.womeningh.org/_files/ugd/ffa4bc_48abbc49eea14129ab2f9331382b9f22.pdf.

World Health Organization. (2017). *Strategic purchasing for universal health coverage: Unlocking the potential.* Global Meeting Summary and key messages. Geneva. Available at: https://www.who.int/health_financing/events/strategic-purchasing-meeting-2017-report.pdf?ua=1.

World Health Organisation. (2021). *Closing the leadership gap: Gender equity and leadership in the global health and care workforce. Policy action paper.* World Health Organisation. https://apps.who.int/iris/bitstream/handle/10665/341636/9789240025905-eng.pdf.

World Health Organization. (2019). Delivered by women, led by men: A gender and equity analysis of the global health and social workforce. *Human Resources for Health Observer Series No. 24.* Geneva: World Health Organization.

United Nations. (2015). The 17 goals. Sustainable Development. Retrieved from https://sdgs.un.org/goals.

CHAPTER 57

Innovations and well-being of persons with disabilities

Daniela Bas
SustainABILITY Advisory Boutique, Inc., Gredo, Italy

Introduction

Millions of people in the world have and acquire physical, mental, intellectual, or sensory impairments which, coupled with diverse attitudinal and environmental barriers, too often hinder their full participation in society on an equal basis with others. The well-being of persons with disabilities has been more compromised during the COVID-19 pandemic. The silver lining to this situation is that more awareness has been raised on the plight, and solutions, for and with persons with disabilities.

In this chapter, I explore innovations in the "space" (as I like to call it) where healthcare and well-being of persons with disabilities is present, as well as policies that are agreed upon globally. Old and new existing technologies are examined with specific emphasis on assistive technologies that focus on inclusion and participation which improves quality of life by connecting and empowering individuals of all ages, abilities, ethnicities, and genders, in all dimensions of life. Finally, the way forward for promoting and implementing equitable technological innovations is proposed.

Yes, there is a space!

One may wonder, "what" is this "space" where needs are discussed and where solutions like innovations and best practices for the healthcare and well-being of persons with disabilities are found. Indeed, innovations are emerging and expanding to address the needs of many persons with disabilities.

So, "where" is this space? The answer, for me, is that the United Nations Organization (UNO) has the convening power to put this space at the disposal to multistakeholder entities and

Resilient Health. https://doi.org/10.1016/B978-0-443-18529-8.00057-3

individuals. In fact, the UN offers a setting in which, at a multilateral level, stakeholders meet, in person or virtually, to discuss policies and programs for the well-being of all people and the planet, to "leave no one behind"—the commonly used phrase to refer to equity rights of all and develop policies, which include persons with disabilities, who are all too often stigmatized, marginalized, and left behind.

Terms

The term, once commonly used to refer to persons with disabilities, has been "handicapped," but this term has evolved over the years to a term derived from the medical model which focuses on the disability, hence the word "disabled," which then evolved into the social model, focusing on the "person," which is the term presently adopted by the UN, that is, "person with disability." It is important to be people centered, as this is one of the main purposes of the UN and the World Social Development Summit of 1995, which leads the activities and mandates of the Division for Inclusive of Social Development in the UN Department of Economic and Social Affairs (DISD/UNDESA), which I guided as director till March 2023.

In thinking about persons with disabilities, it is crucial to think of the term "inclusive." This is reflected in the name of the Division for Inclusive Social Development (DISD) of the UNDESA.

The Division researches and executes policy analysis of national and global socioeconomic trends, identifies emerging issues, and assesses their implications for social policy at the national and international levels and makes recommendations for consideration to Member States part of the UN. We cover vulnerable social groups globally, including older persons, youth, persons with disabilities, and Indigenous peoples as well as topics like sport for development and peace, cooperatives and themes that impact these groups, like reduction of poverty and of social inequalities, promote decent work, and their participation and inclusion in all dimensions of life. As of 2015, we also support the acceleration of the 2030 Agenda for Sustainable Development and its 17 Sustainable Development Goals (SDGs). DISD serves also as secretariat to the Commission for Social Development, the Conference of State Parties to the Convention on the Rights of Persons with Disabilities; the Permanent Forum of Indigenous Issues, the Open-Ended Working Group on Ageing, and supports Member States in their endeavors including through capacity development and technical cooperation projects.

The United Nations Conference of State Parties to the Convention on the Rights of Persons with Disabilities (COSP CRPD) is held annually, usually in June at the UN where participants discuss the implementation of chosen articles of the CRPD and hold many events at its margins. One of the side events in a previous COSP CRPD was devoted to technology and disability, entitled "Buying Social Support Through Online Platforms: What Way Forward?", which discussed how online platforms can support independent living, yet also confront barriers, limitations, and threats (Chatelin, 2021). Matthieu Chatelin, member of the youth Committee of the European Disability Forum, asserted that while to date most support to persons with disabilities does not have online alternatives and therefore must be provided in person, online platforms can be developed to put persons with disabilities in contact with providers and match support workers with special skill sets or interests with persons with disabilities, whether for work or hobbies, logistically organize support networks, and bring persons with disabilities together for peer support or mentoring (Chatelin, 2021).

Side events at COSP have also covered many related issues, including those on mental health. For example, an event of COSP on 12 June 2019 was on "Reaching Higher Standards of Health and Well-Being through Cultural Engagement, Empowerment, and Inclusion," and included a young man sharing his challenges with severe mental stress, causing him to attempt to end his life (Kuriansky, 2019a).

Another mechanism in support of the rights of persons with disabilities is the United Nations Committee on the Rights of Persons with Disabilities, a body of independent experts which monitors the implementation of Convention by the State Parties, which is under the Office of the High Commissioner for Human Rights (OHCHR), to promote inclusion and advocate for the human rights of all persons with disabilities.

Background

Over 1 billion people, or approximately 15% of the world population, and 46% of older persons (Department of Economic and Social Affairs, 2021a) have some form of sensory, physical, cognitive, or invisible disability (ILO, 2014). All persons with disabilities or with a disabling temporary or permanent condition are incredibly diverse in their identities. With increasing climate change that has caused natural disasters, and with manmade disasters like violence and wars, the number of individuals with some form of disability is only growing.

Also, in light of an aging world, with an increasing population, the number of persons who may acquire a disabling condition is going to increase. The 2020 World Population Data Sheet indicates that the world population is projected to increase from 7.8 billion in 2020 to 9.9 billion by 2050 (SDG Knowledge Hub, 2020).

Persons with disabilities comprise the largest minority group in the world. To protect and empower persons with disabilities, there are international agreed-upon tools, notably, the United Nations Convention on the Rights of Persons with Disabilities (UNCRPD) and the United Nations 2030 Agenda with its 17 Sustainable Development Goals (SDGs). The 17 SDGs were based on a multilateral consensus of governments and other stakeholders adopted in 2015 by the United Nations General Assembly, intended to be achieved by the year 2030 (Tarigan, 2020). This 2030 Agenda is the most comprehensive global blueprint at the present time for the well-being of people, with overarching universal goals, *inter alia*, of reducing poverty and the gap of socioeconomic and environmental inequality, as well as to protect our planet (United Nations, 2021).

In these documents, persons with disabilities are acknowledged as equal citizens on an equal basis with others, with dignity, respect, and a right to appropriate healthcare, education, employment, fun, love, and freedom. I myself have acquired a disability since when I was 6 years old. In fact, I had a cancer pressing on the spine, which caused me to become paraplegic. Doctors performed a delicate spinal surgery to save my life and since then I regained my autonomy and independence using a wheelchair.

As persons with disabilities, it is important to engage and play in active roles as actors of change to achieve these Goals. Inclusion, equity, and well-being is a two-way process, and I encourage all to apply the values for which we champion.

How can we achieve these results in a sustainable manner, finding common denominators that allow us to work in global partnership, and unite in multilateralism with stakeholders? As I mentioned, we use the road map of the 17 SDGs of the UN Agenda 2030 for Sustainable Development, to ensure the well-being and prosperity of all countries, for people of all ages and abilities (United Nations Social Development Network, 2021a). Note that "ALL" people include persons with disabilities. In particular, reference to persons with disabilities is made in the SDG 4 (education), SDG 8 (employment), SDG 10 (inequalities), SDG 11 (inclusive, safe, resilient, and sustainable cities), and SDG 17 (international cooperation) (Department of Economic and Social Affairs, 2021b).

Let us look at some of these 17 milestones that are the SDGs and how they are interlinked for persons with disabilities. SDG 1 focuses on eliminating extreme poverty, and as said, SDG 8 on decent employment and SDG 10 on reducing the gap inequalities. In my view, these SDGs could be more easily achieved if the SDG 3 on "health and well-being for all" and SDG 4 about "education for all" (including for children and youth as well as lifelong learning for adults with disabilities) are accelerated in their implementation (United Nations, 2021).

Multiple intersecting factors impede persons with disabilities from gaining access to decent work and employment. This impacts their mental, emotional, and physical well-being and their environment in all settings. By taking care of our health (SDG 3) and with education (SDG 4) and/or vocational training, persons with disabilities are less likely to be in vulnerable types of employment (SDG 8), whether self-employed or in part-time jobs (United Nations Department of Economic and Social Affairs, 2020a).

I believe that if we can find innovative ideas and put them into action, we can solve the challenge of SDG 10 (to reduce the gap of inequalities); strengthen resilience and adaptive capacity of people to climate-related hazards and natural disasters (SDG 13), particularly persons with disabilities; provide healthcare and mental healthcare services (SDG 3 and target 3.4); achieve more just, peaceful and inclusive societies (SDG 16); and work united in partnership (SDG 17) (Department of Economic and Social Affairs, 2021b). Putting SDGs into action is the turbo for a new world based on trust, social and economic justice, and equity.

As a human being, and not just as a person with a disability myself, I see most of the SDGs apply to my life. For example, sport is an enabler of development and peace, as stated in the preamble of the UN 2030 Agenda. I love swimming and archery as well as watching classic ballet or rugby. The world has come to embrace sport competitions where athletes, depending on their abilities, compete in the Olympics, Paralympics, or Special Olympics.

SDG 3 (about good health and well-being) applies to me, recalling the time when I was diagnosed with cancer and fortunately healed. I am particularly passionate about SDG 10 as it refers to reducing the gap of socioeconomic inequalities. All too often, persons with disabilities due to prejudice, discrimination, or lack of education mostly because there are architectural barriers and cultural obstacles to overcome are offered lower income jobs and fewer opportunities to utilize their real potential and skills. I am particularly in harmony with SDG 4 (about education) which has allowed me to find my career path and employment. Also, SDG 16 is relevant to me, as I have become personally and professionally engaged in achieving more just societies. And regarding SDG 17, my work involves promoting international cooperation through the multilateral convening power of the UN.

The advent of new technologies

I was born before the current era when new technological devices that facilitate independent living are available.

When I was growing up, there was no internet. Instead, for example, we worked with books and pens, such that math was done without calculators. Nowadays, technologies, particularly digital ones, offer considerable independence, inclusion, and equity to those who face disabling conditions, as far ranging as like having dyslexia or visual or hearing impairments.

In my case, I was very blessed to have parents who empowered me and taught me to develop my own resilience and rely on my own abilities. My parents were influential. As I told journalist Robert Golomb, who interviewed me a few years ago: "My parents never saw me as a child with a disability but rather as a person who could overcome any challenge…As I grew up, I felt very lucky to have survived rather than feeling bitter about my physical restrictions…My belief about why I survived eventually became that life knew I had a mission. And, of course, my parents always encouraged me to see my life that way."

This experience and encouraging message formed my philosophy of life, which guides me in my work, that is, to promote that all persons have the same dignity, rights, and opportunities. This equality for all is affirmed in the principles of the UN which I love.

The role of innovative, digital, and assistive technologies for all

Digital innovation empowers people to connect with one another; access knowledge, information, and tools; and actively participate in their community. Digital innovation through connectivity has transformed our lives by strengthening our relationships and fostering active participation to achieve poverty eradication, social integration and reduction of inequalities, and productive and decent work for all, which are the three cores of social sustainable development (Bas, 2013).

Digital technologies, artificial intelligence, and assistive technologies play an important role as tools to benefit the mental and physical well-being of people. These innovative technologies keep our planet healthy, raise awareness, reduce the digital divide that is causing more poverty and inequalities, and build resilience to shocks like the COVID-19 pandemic that first challenged the world starting in late 2019 and continuing for years with variants (United Nations Social Development Network, 2020).

Assistive technologies can facilitate inclusion and participation that in turn improve quality of life by connecting and empowering people of all ages, abilities, ethnicities, and genders in all dimensions of life: social, economic, cultural, environmental, and political. They help to change behaviors of people and mind-sets of the public and private sector in positive ways and make invisible voices visible and heard globally (WIPO, 2021).

For example, digital technologies can foster social inclusion by facilitating the full participation of all people in society and by ensuring their access to quality education, healthcare, decent work, affordable housing, and social protection. No one should be left behind in the digital revolution (WIPO, 2021).

An example is DO-IT (Disabilities, Opportunities, Internetworking, and Technology), a project that provides resources and content to increase the successful participation of individuals with disabilities in challenging academic programs such as those in science, engineering, mathematics, and technology (Burgstahler, 1992).

The gap in the need and access to technology for persons with disabilities

About 80% of persons with disabilities live in developing countries and according to the World Health Organization (WHO), globally, only 10% of persons with disabilities who need assistive technologies have access to and can afford them (WIPO, 2021).

The COVID-19 pandemic intensified the need for technology to help provide healthcare services and products. Yet in the new wave of rapid digital innovations, the risk that persons with disabilities could be "left behind or out" is very high (Department of Economic and Social Affairs Social Inclusion, 2021a).

Persons with disabilities are included in the group of those who can suffer most from the "digital divide." The United Nations Secretary-General (UNSG), in his report of 2021 to the United Nations Commission for Social Development on "Socially just transition towards sustainable development: The role of digital technologies on social development and well-being of all," underlines how the COVID-19 crisis accelerated the pace of digital transformation (Department of Economic and Social Affairs Social Inclusion, 2021a). The digital revolution has already brought tremendous benefits to socioeconomic development and improved the quality of life of many. Nevertheless, it risks widening the digital divide, further exacerbating existing inequalities and polarizing the labor market.

There is an urgent need to promote a people-centered approach to maximize the benefits and minimize the risks brought on by digital technologies.

Recovery from the COVID-19 pandemic crisis is an opportunity to reset socioeconomic policies to restart economic growth. This means sustainably improving the living standards and well-being of all people, as an integral part of efforts to promote the "socially just transition to sustainable development" envisioned in the 2030 Agenda. Digital technologies can facilitate that transition and create a more inclusive, equitable, resilient, and sustainable society for all (Department of Economic and Social Affairs Social Inclusion, 2021a).

In his report, the UNSG highlights that the role of digital technologies on social development and well-being of all has its benefits and opportunities, but also potential risks. Digital technologies offer new opportunities for achieving the three core objectives of social development, as set out at the World Social Development Summit of 1995 that in those years was the largest gathering of world leaders reaching the consensus that people had to be put at the center of development to namely: (1) eradicating poverty, (2) promoting full and productive employment, and (3) fostering social inclusion.

Regarding (2) promoting full and productive employment (SDG 8), digital technologies can enhance productivity in the world of work, expand the way individuals search for work, assist in companies' recruitment of talent, and enhance the goods and services that are produced and distributed. They can enable employees to work remotely and promote improved work-life balance, while expanding income-generating opportunities (Department of Economic and Social Affairs Social Inclusion, 2021a). For example, e-commerce has helped some rural communities move out of poverty (Department of Economic and Social Affairs Social Inclusion, 2021a).

With regard to (3) digital technologies can foster social inclusion by facilitating, for example, healthcare, e-health is increasingly considered to be an essential component in the achievement of Universal Health Coverage as laid out in the Political Declaration, which was agreed upon by governments in 2019. More than half of Member States of the World Health Organization now have e-health strategies in place. Mobile health (m-health) supports the dissemination of information, including through targeted health campaigns (World Health Organization, 2011). Tele-health offers the direct provision of healthcare through mobile devices, thereby overcoming infrastructure constraints and healthcare worker shortages in remote populations and underserved communities. Tele-health services have proved to be critical during COVID-19 lockdowns. In addition, e-learning facilitates the training of health workers; electronic health records can provide accurate and timely patient information, thus improving patient diagnosis and treatment; and artificial intelligence can provide valuable assistance in diagnosis (United Nations Social Development Network, 2021b).

In this report, the UNSG explains that "the lack of digital skills is increasingly a leading barrier to digital inclusion" (Department of Economic and Social Affairs Social Inclusion, 2021b). Urgent investment in education, literacy and digital skills is required. Schools are at the heart of such efforts, as the level of educational attainment is one of the strongest indicators of digital skills and proficiency. Promoting greater secondary school enrollment remains an important step in tackling the digital divide. Countries should develop national digital skill strategies that mainstream digital skills within national curricula, promote school connectivity, and support lifelong digital learning (Department of Economic and Social Affairs Social Inclusion, 2021b). For instance, "Giga," a global initiative of the United Nations Children's Fund (UNICEF) and ITU, is aimed at connecting every school to the internet. Such measures can also address the skills mismatch that exists between young people's education and the digital skills demanded by the labor market.

The report's conclusion brings the attention of Member States and other actors to the fact that the world is at a crossroads, that the COVID-19 pandemic is posing unprecedented and multifaceted challenges to social development and the well-being of people worldwide, hitting those in vulnerable situations the hardest. Those challenges intertwine with high and rising inequalities, extreme poverty, unemployment, exclusion, overconsumption, environmental degradation, food insecurity, and climate change. They amplify each other, further exposing the fragility of the current systems and highlighting the need for a socially just transition to sustainable development (Department of Economic and Social Affairs Social Inclusion, 2021a).

The value of information and communication technology (ICT)

Digital technologies can empower marginalized or disadvantaged social groups and foster to improve their quality of life, enhance their inclusion in the community, and make independent living possible (ILO, 2014). For example, with regard to three groups of persons with disabilities, online platforms enable persons with mobility difficulties to access both public services and jobs; e-learning materials adapted to the needs of students with cognitive disabilities remove barriers to learning; and digital devices and assistive technologies promote the inclusion, participation, and engagement of older persons and persons with disabilities in their communities (United Nations Social Development Network, 2021b).

Information and Communication Technology (ICT) is increasingly enabling inclusion of persons with disabilities to access lifelong education, human rights programming, skills development, connection to markets for income generation and employment, thereby reducing poverty, according to a World Bank report (Raja, 2016). This report also points out how mobile devices and desktop computers facilitate communication and information access for persons with disabilities in ways they prefer and need, with features like text-to-speech and voice recognition, ability to change contrast and color schemes, touch and gesture input, and screen magnification. For example, visually impaired persons can use speech-to-text functionality, those with hearing impairments can communicate with SMS or instant text messaging, and those with mobility impairments can use voice recognition to operate their digital device.

Issues about technologies and indigenous peoples who have disabilities

While there is no data regarding indigenous peoples with disabilities globally, available statistics show that indigenous peoples are disproportionately likely to experience disability in comparison to the general population. Indigenous peoples often experience multiple forms of discrimination and face barriers to fully enjoy their rights (Department of Economic and Social Affairs Disability, 2021).

DISD at DESA holds an annual conference, the Permanent Forum on Indigenous Issues (PFII), mandated to deal with indigenous issues related to economic and social development, culture, the environment, education, health, and human rights. In past sessions, it also discussed topics on health systems and technology innovations specifically on disability (Department of Economic and Social Affairs Disability, 2021).

Digital technologies and the internet can provide opportunities for indigenous peoples, including those with a disability, to preserve and share their culture. ICT can be a powerful platform for making the voices and interests of indigenous peoples known, while also helping them overcome geographical remoteness, and thereby ensure greater participation in all aspects of society (Raja, 2016). However, one must be cautious, as technologies can also pose a potential threat by diluting indigenous peoples' traditions.

A solution involves forging close partnerships between indigenous leaders and communities with public sector organizations and private companies to improve inclusion. Also, the media can cover more issues relevant to indigenous peoples (Department of Economic and Social Affairs Disability, 2021).

Dangers of technology to persons with disabilities

Although digital technologies have the potential to help address challenges and accelerate the realization of the UN 2030 Agenda, they also give rise to new risks. The accelerated pace of digital transformation is expected to further deepen the digital divide in many parts of the world, which could, in turn, exacerbate existing socioeconomic inequalities (WIPO, 2021). Since persons with disabilities often suffer from being in lower socioeconomic conditions, they are at risk.

Also, there has been little alignment between science, technology, and innovation on the one hand, and socioeconomic policies on the other. As a result, new technologies may not necessarily advance social development or improve the well-being of all people, including those with disabilities. There is some hope, though, since recovery from the COVID-19 pandemic provides a window of opportunity for promoting a socially just transition to sustainable development, where economic growth is used to advance human well-being and capabilities, while protecting the planet. But enabling a socially just transition requires a shift in mind-set and approaches, from pursuing short-term economic and material gains to rebalancing economic, social, and environmental objectives within the framework of the 2030 Agenda, to build a common sustainable future for all (WIPO, 2021).

Promoting the well-being of persons with disabilities through sport

While sport may not initially come to mind as related to technology for persons with disabilities, the opposite is true. In fact, major assistive technologies in sports have vastly improved the lives of persons with disabilities. Sport is a critical way to promote healthcare and well-being for all, including persons with disabilities. Appropriately, sport is recognized by the UNSG as "a pioneer and promoter of inclusive and sustainable development in an unstable and unequal world" (Department of Economic and Social Affairs, 2021c).

At the Multi-Stakeholder Dialogue on Sport and Digital Technology organized by UNDESA/DISD on 15 December 2020, it emerged that "the practice of sport is deeply entwined with the use of technology, which influences how it is carried out, the equipment used, how athletes and fans travel to matches, and how it is covered by the media and enjoyed by sports fans from afar. Similarly, technology, and recently digital technology, play an increasing central role in facilitating the use of sport to advance development and peace" (United Nations Department of Economic and Social Affairs, 2020b).

For example, technology allows organizations across the globe to connect and exchange best practices in networks such as StreetFootball World (Social Change on a Global Scale, 2021) or Laureus Sport for Good (Laureus Sport for Good, 2022). It contributes to the creation of affordable and durable sport equipment, such as the One World Futbol (One World Play Project, 2020). It can increase access to sport for all, including through the development of assistive technologies, such as those promoted, utilized, and provided by Motivation in the UK (One World Play Project, 2020) and Jumping Kids in South Africa (Jumping Kids in Ability, 2020). In 2020 the world had to adapt to a new reality in which the usual means of practicing sports and physical activities with others was no longer possible. Global events such as the

Olympic Games or regional events such as the UEFA Euro 2020 were postponed, and many community-based organizations stopped their activities. In the absence of in-person opportunities for sport and physical activity, the use of technology has become even more central in the world of sport, playing a critical role in enabling many to stay active, for example, by participating in remote online classes or utilizing dedicated fitness apps and wearable devices. Access to such opportunities, however, has been far from universal, as 3.7 billion people, many from poor communities and in developing countries, remain offline. In some cases, other forms of technology, such as radio and television, have been used to encourage physical activity for those who lack online access (Deputy Secretary-General Warns General Assembly, 2021).

A two-way street: Discoveries for the well-being of all

The World Intellectual Property Organization (WIPO) in its 2021 publication on trends of technology entitled Assistive Technology states that: "Over 1 billion users currently need assistive technology. This figure is expected to reach two billion by 2050 as the population ages and consumer electronics and assistive products converge. The market is shaped not only by demographics and the demand for consumer electronics, and the investment this attracts, but by legislation and policies. The Convention on the Rights of Persons with Disabilities (CRPD) recognizes access to assistive technology as a human right, bringing with it state obligations and expected market influence" (WIPO, 2021).

Persons with disabilities are not only beneficiaries of change, but rather they are actors of change and have led or triggered inventors to make many interesting discoveries and innovations to provide healthcare and well-being. There are many positive practices and successful technologies we can take example from, thought initially for persons with a disability, which have evolved in services and products that have improved the healthcare and well-being of the public at large throughout the centuries. Many innovations have been adapted for use by a wider consumer base (Steenhout, 2022).

The importance of community to offset risks of technology

People-to-people contact is crucial for the socialization of persons with disabilities, essential in modern times when overuse of technological tools can exacerbate the isolation from which persons with disabilities can suffer. A community-based model applied in the country of Georgia was featured in a side event during the UN Conference on the Rights of Persons with Disabilities on "Reaching Higher Standards of Health and Well-Being through Cultural Engagement, Empowerment, and Inclusion." Additionally in Georgia, in an innovative "Mobile Crisis Team" a team of health service providers went to visit people at home for care (Kuriansky, 2019a).

Another community-based model of note, The Arche, was founded by the Canadian humanitarian Jean Vanier, in 1964 as an international federation active in 35 countries dedicated to support people with mental disabilities. A discussion and film screening on Mental Health at the UN on *Jean Vanier: The Sacrament of Tenderness* was held at the UN cosponsored by the Permanent Missions of Canada, Belgium, and Bahrain; the World Health Organization; the City of New York; the Francophonie, moderated by mental health advocate and this book's coeditor and my friend, Dr. Judy Kuriansky, who is a strong advocate at the UN for the human rights of all peoples (Vanier, 2021).

The history of assistive technologies

History from thousands of years ago to the present shows us examples of innovative products and services which were initially for persons with disabilities, and which ended up benefitting billions of people today, and have even become status symbols, like fashionable sunglasses (Steenhout, 2022) and luxury jacuzzies! (Jacuzzi, 2021).

In the AALOS brief history article about the development of assistive technologies for the well-being and healthcare of people, archeologists discovered a 3000-year-old mummy with a wooden, prosthetic flex toe attached to the foot and leg! In 1910 the "Capua Leg" was discovered north of Naples, Italy, dating back to 300 BC. Up to the early 2000s, high-performance prosthetics were made with carbon fiber and a wealth of advanced technology that were motorized and adaptable to any terrain and activity (Alabama Artificial Limb & Orthopedic Service, 2021).

Eyeglasses are a good example. The Roman tragedian Seneca (4 BC–AD 65) used "spectacle lenses" consisting of a glass globe of water as a magnifier to read "all the books of Rome." (Glasses History, 2022) Later in the 13th century in Venice, Italy, glass blowers produced reading stones made of glass in single lens-type frames, similar to the magnifying lenses we have in our homes and offices. It took until 1284 in Italy, when the first wearable eyeglasses were invented by Salvino D'Armate (Vanier, 2021). Wearing glasses, whether they are for nearsightedness or farsightedness, or sunglasses, has become an object of beauty not considered an "assistive technology" anymore only for people who have vision impairments. In reality, most of the world is therefore a person with a disability since sooner or later we shall all wear eyeglasses (Glasses History, 2022).

Another example is the typewriter that throughout decades has become the keyboard we all use on our computers, laptops, or smartphones (Steenhout, 2022). Many inventors tried to design and create machines that allowed people to write for a long time. The early typewriter design was aimed at helping persons with hearing impairments. In fact, the principal at the Danish Royal Institute for the Deaf came up with the right design to help deaf students "to write as fast as they fingerspelled" (Steenhout, 2022).

Also, the most unexpected assistive technology created for the healthcare and well-being of persons with disabilities is nowadays considered a luxury high tech product: the Jacuzzi (Jacuzzi, 2021).

The Jacuzzi brothers—who ultimately designed the well-known whirlpool bath called the Jacuzzi—moved from the town of Casarsa della Delizia in northeast Italy (in the Region of Friuli where I am originally from, at the border of Austria and Slovenia) to California early in the last century (Jacuzzi, 2021). In 1915 the brothers founded their company producing hydraulics and thermodynamics, and eventually moved into agriculture and created a special agricultural pump. In the late 1950s, a toddler in the family, named Ken, started suffering from rheumatoid arthritis. So, the brothers adapted their special agricultural pump that they created for agricultural purposes for the bathtub for hydrotherapy treatments for the child (Jacuzzi, 2021). Thus an assistive technology created to provide healthcare and well-being—a Jacuzzi creation—became the first jet whirlpool bath, that later became a luxury in the world, that famously accommodates large numbers of people relaxing in warm swirling water or sipping a cocktail during happy hour.

The way forward

Technologies and assistive technologies are an important ingredient to build resilience and support our well-being despite whatever challenges we face in life. They also accelerate the implementation of the United Nations 2030 Agenda and its 17 SDGs with a holistic approach that goes beyond disability-related matters. By implementing the UN 2030 Agenda of Sustainable Development and promoting its SDG 17 (calling for international cooperation and partnerships), digital technologies can create a virtual space where multilateralism happens among multistakeholders "united in a common goal for the health and well-being of all people: social justice, human rights, and peace on heart" (Department of Economic and Social Affairs, 2021b).

As an overall theme to the application of technology to the health and well-being of all persons, including those with disabilities, and as an Italian neighbor to France, I like to invoke as a summary the French motto, "Liberté, Égalité, Fraternité" that is "Freedom, Equality, and Togetherness."

As a professional in the field of social development and inclusion, and as a person with a disability, I would like to see this motto define digital innovations also for those who have disabilities. We, all people, have to be able to experience freedom to express ourselves, and be equal in our diversities together, advancing our personal and collective goals while contributing to creating and developing, as well as benefitting from, digital and assistive technologies. I therefore call upon more innovators to devote their vision and resources to making this a reality, where no one is left behind.

Acknowledgments

I wish to express great appreciation to Brianna Fitapelli for her exceptional assistance in the research and locating references for this chapter. Ms. Fitapelli earned her master's degree in clinical psychology from Columbia University Teachers College. Also, I thank Dr. Judy Kuriansky of Columbia University Teachers College for her outstanding help in the editing process.

References

Alabama Artificial Limb & Orthopedic Service. (2021). *A brief history of the development and evolution of prosthetic limbs*. https://www.aalos.com/brief-history-development-evolution-prosthetic-limbs/. (Accessed 22 May 2022).

Bas, D. (2013). How to promote development through the global platform of digital services for development. In *Division for inclusive social development. Talk presented at: Infopoverty world conference 2013: Innovations for the power of people*.

Burgstahler, S. (1992). *Working together: People with disabilities and computer technology. DO IT*. https://www.washington.edu/doit/working-together-people-disabilities-and-computer-technology. (Accessed 22 May 2022).

Chatelin, M. (2021). Buying social support through online platforms: What way forward? In *Oral presentation at the 14th session of the conference of state parties to the UN CRPD*Virtual.

Department of Economic and Social Affairs. (2021a). *Ageing and disability*. https://www.un.org/development/desa/disabilities/disability-and-ageing.html. (Accessed 3 June 2022).

Department of Economic and Social Affairs. (2021b). *The 17 goals | sustainable development*. https://sdgs.un.org/goals. (Accessed 22 May 2022).

Department of Economic and Social Affairs. (2021c). *More than a game—Recovering better with sports*. https://www.un.org/en/desa/more-game-%E2%80%93-recovering-better-sports. (Accessed 22 May 2022).

Department of Economic and Social Affairs Disability. (2021). *The United Nations and indigenous persons with disabilities*. https://www.un.org/development/desa/disabilities/issues/the-united-nations-and-indigenous-persons-with-disabilities.html. (Accessed 22 May 2022).

Department of Economic and Social Affairs Social Inclusion. (2021a). *Socially just transition towards sustainable development: The role of Digital Technologies on social development and well-being of all | DISD*. https://www.un.org/development/desa/dspd/2020-meetings/socially-just-transition-digital-technologies.html#:~:text=being%20of%20all%E2%80%9D-,Socially%20just%20transition%20towards%20sustainable%20development%3A%20The%20role%20of%20digital,for%20all%20segments%20of%20society. (Accessed 22 May 2022).

Department of Economic and Social Affairs Social Inclusion. (2021b). *Leveraging digital technologies for social inclusion*. https://www.un.org/development/desa/dspd/2021/02/digital-technologies-for-social-inclusion/. (Accessed 22 May 2022).

Deputy Secretary-General Warns General Assembly. (2021). *With almost half of world's population still offline, Digital divide risks becoming 'new face of inequality'*. https://www.un.org/press/en/2021/dsgsm1579.doc.htm. (Accessed 22 May 2022).

Glasses History. (2022). *History of eyeglasses and sunglasses. Eyeglasses history—Origins of glasses and sunglasses*. http://www.glasseshistory.com/. (Accessed 22 May 2022).

ILO. (2014). *Sustainable development: The promise of technology. International day of persons with disabilities—3 December*. https://www.ilo.org/skills/events/WCMS_322480/lang- -en/index.htm. (Accessed 22 May 2022).

Jacuzzi. (2021). *Wikimedia Foundation*. https://en.wikipedia.org/wiki/Jacuzzi. (Accessed 22 May 2022).

Jumping Kids in Ability. (2020). *The gift of mobility*. http://www.jumpingkids.org.za/about/. (Accessed 22 May 2022).

Kuriansky, J. (2019a). *Disability rights promoted at the United Nations*. Black Star News. http://www.blackstarnews.com/global-politics/others/disability-rights-promoted-at-the-united-nations.html.

Laureus Sport for Good. (2022). *Using the power of sport to end violence, discrimination and disadvantage.* https://www.laureus.com/sport-for-good. (Accessed 23 May 2022).

One World Play Project. (2020). *The one world Futbol.* https://www.oneworldplayproject.com/. (Accessed 22 May 2022).

Raja, D. S. (2016). *Bridging the disability divide through digital technologies.* World Bank Group. https://pubdocs.worldbank.org/en/123481461249337484/WDR16-BP-Bridging-the-Disability-Divide-through-Digital-Technology-RAJA.pdf. (Accessed 1 June 2022).

SDG Knowledge Hub. (2020). *World population to reach 9.9 billion by 2050: News.* https://sdg.iisd.org/news/world-population-to-reach-9-9-billion-by-2050/. (Accessed 22 May 2022).

Social Change on a Global Scale. (2021). *Our impact.* https://www.streetfootballworld.org/who-we-are/our-impact. (Accessed 22 May 2022).

Steenhout, N. (2022). *The evolution of assistive technology into everyday products.* https://incl.ca/the-evolution-of-assistive-technology-into-everyday-products/. (Accessed 22 May 2022).

Tarigan, N. P. (2020). Online festival as method for a network for action and well-being for disability and all based on CRPD and SDG2030. *International Journal of Social, Policy and Law, 1*(1), 17–22. https://doi.org/10.8888/ijospl.v1i1.12.

United Nations. (2021). *59th Session of the commission for social development—CSOCD59 | DISD.* https://www.un.org/development/desa/dspd/united-nations-commission-for-social-development-csocd-social-policy-and-development-division/csocd59.html. (Accessed 22 May 2022).

United Nations Department of Economic and Social Affairs. (2020a). *Sustainable development goals (SDGs) and disability.* https://www.un.org/development/desa/disabilities/about-us/sustainable-development-goals-sdgs-and-disability.html. (Accessed 31 May 2022).

United Nations Department of Economic and Social Affairs. (2020b). *Multi-stakeholder dialogue on the role of technology in advancing sport for development and peace.* https://www.un.org/development/desa/dspd/wp-content/uploads/sites/22/2021/03/DraftSummary-SPORTMSD25Feb.pdf. (Accessed 1 June 2022).

United Nations Social Development Network. (2020). *Reimagining cooperation on sport for persons with disabilities.* https://unsdn.org/2020/11/11/reimagining-cooperation-on-sport-for-persons-with-disabilities-leaving-no-one-behind-2/.

United Nations Social Development Network. (2021a). *Digital transformation and its impact on education & health.* https://unsdn.org/2021/04/22/digital-transformation-and-its-impact-on-education-health/. (Accessed 22 May 2022).

United Nations Social Development Network. (2021b). *Digital technology and its impact on health for persons with disabilities.* https://unsdn.org/2021/04/29/digital-technology-and-its-impact-on-the-accessibility-to-health-and-srhr-services-for-persons-with-disabilities/. (Accessed 22 May 2022).

Vanier, J. (2021). *Jean Vanier, the sacrament of tenderness.* Vimeo. https://vimeo.com/324929286. (Accessed 1 June 2022).

WIPO. (2021). *WIPO technology trends 2021: Assistive Technology.* Geneva, Switzerland: World Intellectual Property.

World Health Organization. (2011). *MHealth new horizons for health through mobile technologies.* https://www.who.int/goe/publications/goe_mhealth_web.pdf?ua=1>. (Accessed 23 May 2022).

Further reading

Giga. (2021). *UNICEF Office of innovation.* https://www.unicef.org/innovation/giga#:~:text=Giga%2C%20an%20initiative%20launched,and%20opportunity%20in%20the%20%22time. (Accessed 22 May 2022).

Kuriansky, J. (2019b). Disabled persons rights at the United Nations: Activities, achievements and contributions of the IAAP New York team. *Applied Psychology Around the World, 1*(3), 71–78. Accessed at: https://iaapsy.org/site/assets/files/1082/apaw_vol1i3.pdf.

WeAreMotivation. (2022). *UK aid match and motivation's all stars project—One year on.* https://www.motivation.org.uk/Blog/uk-aid-match-and-motivations-all-stars-project-one-year-on. (Accessed 22 May 2022).

CHAPTER 58

Innovative approaches in meeting the health needs of migrants, refugees, and displaced persons

Ian O'Donnell and Lasha Goguadze

International Federation of Red Cross and Red Crescent Societies (IFRC), Chemin des Crêts 17, Geneva, Switzerland

Introduction: The big numbers

In recent years, the world has witnessed record levels of migration and displacement, creating significant humanitarian needs. Worldwide, there were an estimated 281 million international migrants as of 2020 (UNDESA, 2020). As of 2022, estimates by the UNHCR show that there were 40.7 million refugees and asylum seekers globally and a further 5.2 million other people in need of international protection, including for example Venezuelans displaced abroad but not included in the formal total of refugees and asylum seekers (UNHCR, 2023). The factors influencing these human mobility dynamics are diverse, including that people are compelled to flee across borders by conflict, violence, and persecution, while others are driven by poverty, inadequate livelihoods, changes in their environment, or the search for better opportunities. People also migrate within their own countries for similar reasons or become displaced internally. In 2022 conflict and disasters triggered an estimated 60.9 million internal displacements within 151 countries and territories (IDMC, 2023).

One of the main concerns for people on the move—and at all stages of their journey—is the ability to access essential and lifesaving assistance and protection, including access to health services. Migrants and displaced persons, whether they move across borders or within their own country, are exposed to a variety of intersecting risk factors that reduce and complicate their ability to meet basic needs and access services. Migrants and displaced persons may also "have specific physical and mental health needs and vulnerabilities, which often go unrecognized and unaddressed, and which have been exacerbated during the [COVID-19] pandemic" (WHO, 2021). Research conducted by the International Federation of Red Cross and Red Crescent

Resilient Health. https://doi.org/10.1016/B978-0-443-18529-8.00058-5

Societies (IFRC) and ACAPS (started as the "Assessment Capacities Project") on the socioeconomic impacts of the COVID-19 pandemic indicated that having a migrant background was the second biggest vulnerability factor in terms of the socioeconomic impacts, after being homeless (IFRC, 2021a).

In at least 155 countries, Red Cross and Red Crescent National Societies provide assistance and protection to migrants and displaced persons, regardless of their status, based on needs and vulnerabilities. Extending these services to maintain a continuity of health service and treatment for migrants and displaced persons within and between country contexts is increasingly recognized as a priority. This chapter draws insights from this experience in National Societies.

Addressing access to basic services

The IFRC, including its network of Red Cross and Red Crescent National Societies, recognizes the need to provide humanitarian assistance to migrants and displaced persons at diverse points throughout their journey. Humanitarian Service Points (HSPs), positioned at key points along migratory routes, have emerged as key element of the IFRC's efforts to address the protection and assistance of people on the move. An HSP is a safe and neutral space operated by the IFRC or a National Society where migrants and displaced persons, irrespective of their status, can access essential humanitarian services and specialized referrals that might otherwise be inaccessible without fear of arrest or being reported to the authorities. HSPs can be located at strategic points along migratory routes, in countries of transit, destination, and return. They may be fixed or mobile and can include kiosks, information desks, or equipped vans that provide a range of services along migration routes. HSPs help to ensure that people on the move have access to targeted and reliable information; receive appropriate referrals; can communicate or connect with their families through restoring family links services; and have access to critical humanitarian assistance such as emergency and maternal healthcare, shelter, food, water and relief items, and psychosocial support.

According to a global survey and analysis conducted by the IFRC in 2021, there were over 275 HSPs operated by National Societies in 42 countries (IFRC, 2021b). The most popular services provided at HSPs globally were information provision, restoring family links services, safe referrals, first aid, basic health services, and mental health support.

The tech solution

As for most organizations and countries in this increasingly technological age, and given the urgencies of communication created by the COVID-19 pandemic, the IFRC has had to adapt its approaches and incorporate innovative methods as well as technology into its services. For example, mobile health clinics, equipped with modern technology, are increasingly a means to provide integrated, multipurpose services (including basic health services, mental health support, COVID-19 testing, treatment, and vaccination), adopted flexibly in different locations as needed. As this chapter is being written, the IFRC is working with National Societies in Ukraine and neighboring countries to expand mobile health clinics to serve the needs of people displaced in the Ukraine crisis.

National Societies are also extending access to marginalized communities living on the edges or outside of formal services and who may be at increased risk to endemic health threats. While these programs do not focus exclusively on migrants and displaced persons, they expand access to health services in an inclusive manner that increases the availability of services for migrants and displaced persons. Examples of these types of programs which have a focus on technology, and are implemented in countries worldwide, include:

- The AIDS Research Center in Thailand (en.trcarc.org), which is run by the Thai Red Cross and is part of a collaborative network linking HIV Research Centers in Bangkok, Amsterdam, and Sydney, is strengthening the capacity of community-based organizations (CBOs) to provide Key Population-led Health Services (KPLHS), working with lay providers to provide quality HIV services which were previously lacking in their communities and especially targeting displaced and marginalized persons. The KPLHS approach supports Thailand's National AIDS Strategies to enhance uptake of HIV services along the Reach-Recruit-Test-Treat-Prevent-Retain cascade (USAID, n.d.). All services are delivered in community settings by trained lay providers, tailored to the needs of each community and linked closely with the public health sector. The program makes use of the ENGAGE TA Platform, an E-registry database to enable lay providers to plan for training, coaching, assessment, and renewal of competencies toward certification. The center's online services also include an online medicine service, an online counseling service, and an HIV self-test service. The system also supports community-based monitoring, providing enhanced data in the provinces and in each location to both the lay providers and the program managers.
- The Villa Maraini Foundation in Italy (villamaraini.it), a technical hub of Red Cross Red Cross Partnership on Substance Abuse hosted by the Italian Red Cross, has pioneered compassionate harm reduction strategies with marginalized persons, focusing on persons with substance use issues in Italy using a Street Unit, Emergency Unit, and a Prison Project to reach people excluded from local support networks, including migrants and other displaced populations. The Villa Maraini outreach program provides immediate medical assistance, preventive information, and referrals to drop-in centers, night shelters, and outpatient clinics. The foundation has a 24-hour online center providing assistance and connection to their services, such as emergency units, an HIV unit, the street unit, a night center, the first reception center, and an alternative detention center. Specific attention is paid to the diagnosis of blood and/or sexually transmitted diseases, in particular HIV infection and hepatitis C, and to pre/posttest monitoring to assess the patient's immune status.

- The Kyrgyzstan Red Crescent, in Kyrgyzstan, is increasing treatment adherence among people with drug-resistant TB (DR-TB), using a combination of traditional (public health supporters) and digital technological methods. Electronic Dose Monitors for required treatment reduce the necessity for frequent home visits by the Red Crescent staff and volunteers, allowing for greater responsiveness by mobile test brigades. In addition, the program gives more options for patients to choose which method is preferable for them to maintain the treatment course and also facilitates enhanced data collection on treatment adherence, which enables Red Crescent staff and volunteers to target further follow-up and support where needed (Tropical Medicine and Infectious Disease, 2022; USAID, 2021).

"Information as aid"—Digital accompaniment in various National Societies

The access of migrant and displaced persons to reliable and quality information is always a challenge, which increases the generation and dissemination of rumors and misinformation, which in turn raises stress levels, uncertainty, and risk. Promoting access to relevant, timely, and reliable information that responds to the needs of migrants and displaced persons and facilitates their access to local services helps people in situations of vulnerability to make decisions for safe transit and settlement, which mitigates risk factors in their journey and engages migrants and displaced persons themselves as part of the solution. This approach also relies on the creation of mechanisms for feedback and complaints for migrants, displaced persons, and host communities, guiding improvement in the relevance and effectiveness of local humanitarian services (IFRC, 2022).

Recognizing the importance of "information as aid" added into this system, the IFRC and other humanitarian actors are increasingly using technological tools. These include offering referral information via web apps and social media as a form of direct digital aid. Such options increase the availability and use of mobile devices by migrants and displaced persons, permitting them to stay in touch with family in other locations and to exchange information and experience during their journey. Since support for people on the move can be fragmented over many localized organizations and initiatives, a web application with referral information helps them to find their way to the support available to them, for instance related to medical services, COVID-19, shelter, legal advice, and protection. Two-way communication is also essential. Helpdesk services, for example, using WhatsApp, can ensure that people on the move are able to reach out to the local HSPs for questions and support at any time and from any location. This makes them more autonomous and less dependent on aid organizations reaching out to them.

Relevant technological applications have been implemented worldwide. These include the following examples from countries in different regions and cultures.

To reach migrants in irregular situations as a specific target group in their COVID-19 response program, the Netherlands Red Cross has developed new approaches to digital outreach engagement combining a web referral app, WhatsApp helpdesk, and digital voucher aid

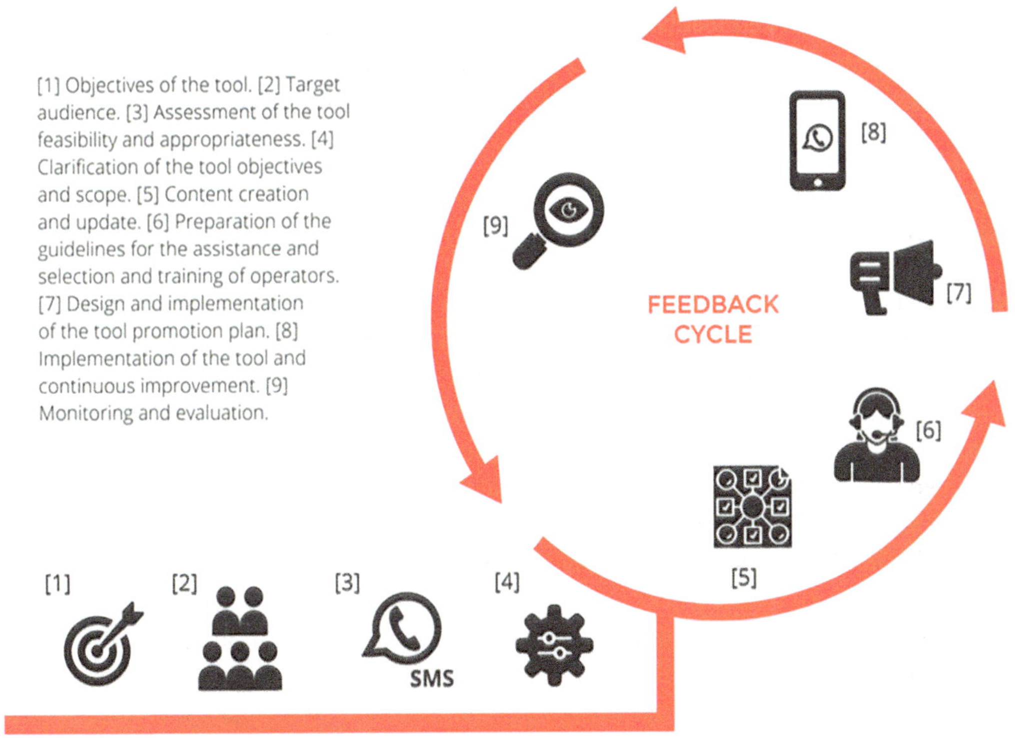

Fig. 1

Configuration and setup of the WhatsApp COVID-19 line in Peru. *IFRC (2020), COVID-19/migration Whatsapp business line: case study, https://communityengagementhub.org/resource/covid-19-migration-whatsapp-business-line.*

program as a complementary set of services to increase the autonomy and self-reliance of migrants (Red Social Innovation, 2022). Similarly, the Ecuadorian Red Cross and the Peruvian Red Cross have used a WhatsApp for Business line to provide remote assistance to migrants and others affected by the COVID-19 pandemic, whereby the service answers questions or doubts about the pandemic; provides key information to promote protective and preventive behavior; and identifies, monitors, and addresses rumors or fake news related to the pandemic (IFRC, 2020). Fig. 1 shows the enhanced feedback cycle with vulnerable communities, including migrants and displaced persons, that the WhatsApp for Business line has enabled in Peru. The German Red Cross has been collaborating with the government to provide a wide range of information to facilitate settling in Germany through an app where migrants can send texts and voice messages in the language with which they are more comfortable. Information includes how to access basic healthcare and mental health support in the country with useful guidance for newborns, children, and adults. The IFRC regional office for the Americas is now creating a strategy to digitally link HSPs in Latin America as an extended access network that will provide increased access and continuity of information services for migrants and displaced persons (IFRC, 2022).

Fig. 1 shows the enhanced feedback cycle with vulnerable communities, including migrants and displaced persons, that the COVID-19 WhatsApp for Business line has enabled in Peru and how the communities' feedback has been used to iteratively improve the communication service.

Cross-border cooperation for continuity of treatment and services

Cross-border cooperation on migrants and displaced persons living with HIV/AIDS in Cambodia, Lao People's Democratic Republic, and Thailand is not a new need, but represents an area where "innovation" is taking place in terms of steady effort to advance progress to provide a continuity of treatment between country contexts. The key change has been in the strengthening of community engagement and accountability. The initial step has been ensuring that the needs, capacities, and vulnerabilities of people of different gender and diversity (age, disability, socioeconomic status, religion, ethnic origin, nationality, sexual orientation) are addressed in developing training curricula and delivery of services to migrants and displaced persons living with HIV/AIDS. This has been complemented with the establishment of referral mechanisms for migrants and displaced persons living with HIV that provide necessary information consistently among the three country contexts, engage migrants and displaced persons directly through digital community engagement and accountability tools, and leverage the National Societies' networks of local branches in these three countries to further extend outreach and accompaniment.

In Burkina Faso, the Gambia, Mali, Niger, and Senegal, a consortium of Red Cross and Red Crescent partners are collaborating in the establishment and implementation of HSPs and "listening centers." In these spaces, migrants and displaced persons are offered psychosocial support among other services. Staff and volunteers use Open Data Kit (ODK) questionnaires to collect disaggregated data and information on vulnerabilities from migrants in order to provide personalized assistance. At the first registration, migrants are provided with a QR code. While guaranteeing data protection and confidentiality, the QR code also enables Red Cross and Red Crescent staff to retrace the services and medical treatment that the person has received and their vulnerabilities. Through ODK questionnaires, migrants can also provide feedback on the services received.

Effective continuity of treatment also involves recognizing and addressing ongoing and chronic needs in addition to looking beyond assistance for immediate needs.

Recent research by the World Health Organization (WHO) on caring noncommunicable disease (NCD) management during the migration cycle highlights that NCD risk factors can be exacerbated by social determinants of health surrounding the migration process. Migrants and displaced persons may face barriers in accessing healthcare due to limited availability of services, inconsistent application of relevant laws and policy, distrust, language differences, and discrimination, often leading them to adopt suboptimal health behaviors that further exacerbate their vulnerability (IFRC, 2021a; WHO, 2022). In addition, digital exclusion has also been identified as a barrier to access basic services during the COVID-19 pandemic for migrants, especially the undocumented ones (Global Migration Lab, 2021).

Advocacy

Policy-making also remains an important domain for innovation. Additional research by WHO indicates the importance of aligning and integrating migration and public health policies. These include border policies to ensure that protection-sensitive access to asylum procedures is not blocked by border lockdowns; migration policies to ensure safe and lawful access for foreigners within national borders to health services; and policies on access to healthcare that include refugees, migrants, and asylum seekers to ensure nondiscriminatory access to healthcare. The research identified that numerous countries adopted flexible administrative procedures in the COVID-19 response while several "also suspended forced returns and prioritized alternatives to immigration detention" (WHO, 2021).

Ensuring security and privacy

Data protection of health status and related demographic information is a critical issue for migrants and displaced persons. Safe management and self-management of personal data of the people supported is crucial to ensure a do-no-harm approach. Therefore data security and data responsibility are essential criteria in the design and setup of services for migrant and displaced persons.

The IFRC is committed to using human-centered design (HCD) approaches to codesign and user test migration services with both people on the move and aid workers. Applying such an HCD approach throughout the various steps of the service creation process ensures that the services that are built reflect the needs, priorities, and user preferences of the end user. At the same time, the approach guarantees that the tools/services are within the capacity and preferences of the National Societies to implement and maintain.

The IFRC is also testing use of Digital ID to provide increased anonymity and control of personal data by people in vulnerable contexts and outside of existing digital financial systems. Migrants and displaced persons may be less willing to share their personal information and prefer "to stay invisible to continue on [their] way" with minimal engagement with local authorities, reinforcing the need for minimized collection of personal data in any digital ID solution. A pilot project, called the Community Inclusion Currencies, by the Kenyan Red Cross in 2021, supported by Innovation Norway, is currently being extended to migrant and displaced persons and provides a foundation for extending additional cash-based and insurance systems to address health, disaster, and socioeconomic risks, building on the increased access to financial services that have been provided in local communities through the project (IFRC, 2021c).

Shared learning

To enhance research and learning on socioeconomic needs among diverse segments of the population, the Argentinian Red Cross has created a new Humanitarian Observatory in collaboration with several partners, namely, the Vatican Observatory (Specola Vaticana), the Consejo Interuniversitario Nacional (CIN), the Consejo de Rectores de Universidades Privadas (CRUP), the Observatorio Social de Infancias y Juventudes de Scouts de Argentina. The observatory conducts public surveying and social media analysis across the different Argentine regions to measure the social, economic, and psychoaffective state in a postpandemic context in relation to older adults; quality of life; human mobility; livelihoods and labor force; youth and digitization; climate change; risk and environment; protection, gender, and inclusion (PGI); and homeless people. In 2022 the Humanitarian Observatory is planning specific research on socioeconomic and health impacts on migrants and displaced persons (Argentine Red Cross, 2021).

The Global Migration Lab, currently hosted by the Australian Red Cross and planned as a rotating IFRC research center, has undertaken a variety of research studies exploring how COVID-19-related policy measures, including lockdowns and movement restrictions, have reduced the ability of migrants and displaced persons to meet their basic needs and access services. The lab ensures that the voices and experience of migrants, displaced persons, and the communities in which they live, work, and transit are well understood and amplified to guide migration policy and operations (Global Migration Lab, 2021).

To share insights and increase capacities among humanitarian actors, the IFRC has also developed a wide range of learning resources on health and migration that are publicly accessible. As presented in Table 1, these include names of courses, intended audience, and platforms on which they were delivered. These courses provide a diverse set of learning

Table 1 IFRC learning resources on health and migration.

Course	Audience	Platform
Online course: Things You Should Know about Health and Migration	Volunteers	IFRC Learning Platform
Practical Guidance for Red Cross and Red Crescent on Health and Care for Migrants	Volunteers	IFRC Learning Platform
Mental health and migration resources from the IFRC Psychosocial Support Centre	Humanitarian service providers	Psychosocial Centre website
Online course: Caring for children moving alone	Public	FutureLearn *(periodic offering)*

opportunities to volunteers and staff of the Red Cross Red Crescent and of other local organizations and are also open to the public and accessible to members of both migrant and host communities.

Outstanding challenges to bridge the digital divide

It is widely understood that investing in connectivity and digital literacy will help bridge the digital divide and aid localized approaches and that collaboration with other actors will boost synergies between areas of expertise and save resources. However, there are still barriers hindering the widespread adoption of digital solutions by humanitarian organizations. This includes limitations on access and inclusion and challenges pertaining to the ethical use of digital tools and data collection by humanitarian organizations. While the past few years, spurred by the urgencies and the global reach of the COVID-19 pandemic, saw an increase in the use of data and digital technologies in the humanitarian sector, the digital divide remains a significant and persistent challenge. Many humanitarian actors are deprived of the benefits of modern technology. National Societies are struggling to use the technology they have, find the technology they need, and plan for the right technology that will help them in the future. This digital divide is further exacerbated by the inability of actors to acquire, develop, or retain the right skills. Data and digital technologies certainly offer a tremendous potential to improve the speed, quality, and cost-effectiveness of humanitarian aid, but the sector also needs to make sure that it does not inadvertently cause harm when using these new technologies. The principles and frameworks to guide the design and use of technologies in ways that are ethically, socially, and environmentally responsible are only effective to the extent that the organizations can operationalize them across their whole systems and into the teams they hire, the problems they want to solve, the processes they use to incentivize the use of those principles and frameworks, with whom they partner, and other facets of their decision-making and management processes. Ultimately, digital enabling and transformation of humanitarian services is as much about people and culture as about leveraging data and technology.

Conclusion and steps for the future

The two years during the height of the COVID-19 pandemic (2020-2022) have been spent meeting the tremendous set of challenges to provide innovative and expanded health services to address the urgencies of the pandemic while still coping with the wider set of health needs and services needed by the world's growing population. On top of these challenges, meeting the needs of migrants and displaced persons has often required distinct and dedicated solutions to address the additional risk factors, lack of access to services, and lack of security and trust faced by migrants and displaced persons. Innovative approaches have been essential to the efforts of the IFRC, Red Cross Red Crescent National Societies, and their humanitarian partners to adapt and expand existing health services and to meet the needs of people on the move throughout the stages of their journey. These approaches include: to improve access to basic services, expand the provision of "information as aid," strengthen cross-border cooperation for continuity of treatment and services, ensure security and privacy, and share learning among humanitarian service providers and with migrants and displaced communities.

Many of the approaches discussed in this chapter are new initiatives, based on emerging technologies and new expectations for cooperation. Locking in these advances in a postpandemic world, and expanding them further to meet migration and displacement challenges that continue to grow, will require further investment, consistent practice, and continued innovation and learning. The IFRC's strategy is rooted in increased engagement of the people and resources within its own humanitarian network, of partner organizations and local agencies, and most importantly of migrants and displaced persons themselves, to provide health services and related support in a distributed and flexible manner to migrants and displaced persons throughout their journey.

Acknowledgments

The authors thank the following IFRC staff and consultants for their valuable contributions to this chapter: Sanjula Weerasinghe (Migration and Displacement Coordinator, Geneva), Stefania Tranfo (Migration and Community Engagement Accountability Officer, Geneva), and Amalia Gatopoulou (Community Health consultant).

References

Argentine Red Cross. (2021). *Humanitarian observatory of the Argentine Red Cross.* https://cruzroja.org.ar/observatorio-humanitario/wp-content/uploads/2022/02/HUMANITARIAN-OBSERVATORY-OF-ARGENTINE-RED-CROSS.pdf.

IFRC. (2020). *COVID-19/migration whatsapp business line: case study.* https://communityengagementhub.org/resource/covid-19-migration-whatsapp-business-line.

IFRC. (2021a). *Drowning just below the surface: the socioeconomic consequences of the COVID-19 pandemic.* https://www.ifrc.org/sites/default/files/2021-11/IFRC-2021-SocioeconomicReport-EN.pdf.

IDMC. (2023). *Global report on internal displacement.* https://www.internal-displacement.org/global-report/grid2023.

IFRC. (2021b). *'IFRC Migration and Displacement: Results from the 2021 Global Survey', results from unpublished report, available upon request.*

IFRC. (2021c). *Digital identity: enabling dignified access to humanitarian services in migration.* https://preparecenter.org/wp-content/uploads/2021/06/Digital-Identity-Enabling-dignified-access-to-humanitarian-services-in-Migration-Final.pdf.

IFRC. (2022). *'RED Data Concept Note' an unpublished note outlining a digital Community Engagement and Accountability (CEA) approach for engaging and supporting migrant and displaced persons.* Available on request from chapter authors.

Global Migration Lab. (2021). *Locked down and left out? Why access to basic services for migrants is critical to our COVID-19 response and recovery.* https://www.redcross.org.au/globalassets/cms-migration/documents/migration-services/en-rcrc-global-migration-lab-locked-down-left-out-covid19.pdf.

Red Social Innovation. (2022). *Helpful Information Web-app: a Netherlands Red Cross initiative with referral information for undocumented migrants.* https://red-social-innovation.com/en/solution/helpful-information-web-app-a-netherlands-red-cross-initiative-with-referral-information-for-undocumented-migrants.

Tropical Medicine and Infectious Disease. (2022). *Time trend analysis of tuberculosis treatment while using digital adherence technologies—an individual patient data meta-analysis of eleven projects across ten high tuberculosis-burden countries*. https://www.mdpi.com/2414-6366/7/5/65.

UNDESA. (2020). *International migrant stock 2020*. https://www.un.org/development/desa/pd/content/international-migrant-stock.

UNHCR. (2023). *Global trends forced displacement in 2022*. https://www.unhcr.org/sites/default/files/2023-06/global-trends-report-2022.pdf.

USAID. (2021). *Cure tuberculosis project: Year 2 Annual Report, October 1, 2020—September 30, 2021*. https://pdf.usaid.gov/pdf_docs/PA00XWN7.pdf.

USAID (n.d.). Differentiated HIV-Service Delivery along the cascade for men who have sex with men and transgender women in Thailand: lessons learned from the LINKAGES project. https://www.aidsdatahub.org/resource/differentiated-hiv-service-delivery-along-cascade-men-who-have-sex-men-and-transgender.

WHO. (2021). *Refugees and migrants in times of COVID-19: mapping trends of public health and migration policies and practices*. https://www.who.int/publications/i/item/9789240028906.

WHO. (2022). *Continuum of care for noncommunicable disease management during the migration cycle*. https://www.who.int/publications/i/item/9789240044401.

CHAPTER 59

Technologically mediated resilience for and by refugees, migrants, and asylum seekers

Reem Talhouk

Northumbria University, Newcastle upon Tyne, United Kingdom

Introduction

The COVID-19 pandemic and the subsequent governmental responses, such as prolonged lockdowns and social distancing measures, disproportionately impacted refugees, migrants, asylum seekers' livelihoods and health (Orcutt et al., 2020). This population already experiences challenges in accessing healthcare in their host countries due to legal (Norredam et al., 2006), cultural (Almohamed et al., 2022), transport (Reese Masterson et al., 2014; Talhouk et al., 2016), and language-related (Mangrio & Sjögren, 2017) barriers that were further exacerbated during the pandemic. Indeed, health services provided by governments, local solidarity networks, charities, and nongovernmental organizations (NGOs) were stalled as the world grappled with the need to rapidly digitalize their services. The transition to digital services during the pandemic, such as the increased use of telehealth (Benjamen et al., 2021; O'Mara & Monani, 2021), further amplified the marginalization of some refugees, asylum seekers, and migrants within the healthcare system due to limited access to technologies (Benjamen et al., 2021; Bryant & Holloway, 2020; McMullin, 2021), low digital literacy (Grieco-Page et al., 2021), privacy concerns (Benjamen et al., 2021), and health policies that did not account for the complexity of refugee experiences (Mukumbang et al., 2020; O'Mara & Monani, 2021). Furthermore, the pandemic negatively impacted these populations' livelihoods which is intimately tied to their health and well-being (Bukuluki et al., 2020). It is within the pockets of digital, political, and socioeconomic marginalization of refugees, migrants, and asylum seekers that we witnessed the technologically mediated resilience of these communities and the humanitarian and health systems with which they engage. Technologically mediated resilience encompasses the design, use, and/or adaptation of digital technologies in the preparedness for, response to, and/or recovery from crises.

Resilient Health. https://doi.org/10.1016/B978-0-443-18529-8.00059-7

As the ubiquity of technologies both within healthcare/humanitarian and among refugees, asylum seekers, and migrants has increased, digital technologies have begun playing a greater role in mediating resilience in the face of shocks and stressors. Prior to the COVID-19 pandemic, humanitarians had begun embedding technologies within their services and interventions in order to enhance refugee self-reliance and resilience, as called for in the Global Compact on Refugees (United Nations, 2018). More specific to health, Mesmar et al. (Mesmar et al., 2016) identified how digital technologies are increasingly playing a role in supporting humanitarian preparedness, response, and recovery. This is especially true within interventions that address the determinants of refugee health, including employment, education, personal health practices, and accessibility of health services (Mesmar et al., 2016).

In this chapter, I provide key examples of technologically mediated resilience enacted by humanitarian/health workers and refugees, asylum seekers, and migrants in the context of the COVID-19 pandemic. Through these examples, we see how resilience was mediated by a multitude of digital technologies to address the gaps in services and information experienced by the population. I then outline the challenges that arise when digitalizing health and humanitarian services for this population. I argue that these challenges are key sociotechnical factors that need to be accounted for when formulating new pathways for resilience. Lastly, the chapter concludes with key considerations for the future of technologically mediated transformative resilience that challenges the status quo at the root of the population's marginalization.

Technologically mediated resilience within the health and humanitarian system

Programs and interventions were able to respond to the shock of the pandemic through using technologies to maintain the services they were providing. Health assessments were conducted through the deployment of remote data collection protocols that leveraged popular social media platforms such as WhatsApp (Chen et al., 2020; Sarr et al., 2022; Walker et al., 2021). Digital cash was transferred through mobile money to ensure that humanitarian aid that refugees rely on for their health and well-being was still accessible during lockdowns (Bryant & Holloway, 2020; UN Women, 2020). Mental health and trauma programs transitioned to online delivery during lockdowns through the provision of teletherapy and online psychosocial support (Benjamen et al., 2021; Endale et al., 2020). The United Nations Relief and Works Agency for Palestinian Refugees used their already existing Electronic and Child Health and Noncommunicable Disease mobile applications to communicate to beneficiaries COVID-19 related guidance and information (UNIC Tokyo, n.d.). The mobile applications were also used to send reminders to beneficiaries to take medicine, check blood pressure, and exercise in lieu of the close monitoring and follow-up usually provided in healthcare clinics (UNIC Tokyo, n. d.). Furthermore, mobile phone applications were specifically developed and deployed to provide refugees with COVID-19 related information in multiple languages using audio and visual media (Logie et al., 2021; Narla et al., 2020). In response to COVID-19, NGOs launched refugee-focused digital communication campaigns, through their websites and WhatsApp (Golesorkhi et al., 2020; Mancini et al., 2021), that were found to increase refugees' health-related risk perception (Mancini et al., 2021). Furthermore, NGOs were able to maintain interactive contact with their beneficiaries through online chat applications (Mancini et al., 2021). The transition to online delivery of health services and information was partially supported by technology literacy programs, run prior to the pandemic, that prepared refugees, asylum seekers, and migrants for technologically mediated communications and engagements (Golesorkhi et al., 2020). These examples highlight the resilience of health and humanitarian researchers, workers, and organizations that rapidly digitalized services so that the health of this vulnerablized population that they serve does not deteriorate.

Technologically mediated resilience within refugee, asylum-seeking, and migrant communities

Resilience within the system, however, is only one side of the story. Refugees, migrants, and asylum seekers themselves partnered with actors within the humanitarian and health system (Bezbaruah et al., 2021) and with one another to work toward healthy and knowledgeable communities that is a cornerstone of a resilient community (IFRC, 2014). Indeed, in Germany, refugee women partnered with local community actors to produce short videos in Farsi and Arabic that communicated health-related information (Mohammadi et al., 2021). These videos were shared through social media and the women initiated online chats through which they collated questions from refugees that were then answered by medical doctors (Mohammadi et al., 2021). Similarly, in response to the lack of health information in the Rohingya Language, refugees in Cox's Bazaar, a town on the southeast coast of Bangladesh, created and digitally disseminated videos that raised awareness about the pandemic in the Rohingya language (UNHCR, 2021). The Refugee Youth Education Hub in Dadaab Refugee camp in Kenya hosted radio shows in Somali, Oromo, Swahili, and English that debunked COVID-19 misinformation (UNHCR, 2021). Ipas African Alliance partnered with a refugee-led organization in Kenya to deliver sexual and reproductive health education to refugees through WhatsApp and SMS (UNHCR, 2021).

Additionally, in response to the pandemic, refugees who can no longer access traditional labor markets turned to technologies as a means of maintaining livelihoods conducive to good health and well-being. Jumpstart Refugee Talent in Canada, a refugee-led nonprofit organization serving refugees seeking meaningful economic inclusion, provided online mentoring and developed a talent hiring portal that enabled refugees to learn about employment opportunities and to connect to potential employers (Jumpstart Refugee Talent, 2022). Similarly, in the Bidibidi refugee settlement in Uganda, online entrepreneurship training was provided to support women in farming endeavors that would support their livelihoods (UNHCR, 2021). A refugee-led initiative in South Africa that aims to support LGBTQI+ refugees ran a crowd-funding campaign to raise money to support their community in meeting their daily needs (UNHCR, 2021). Refugee-led initiatives also rose to the challenge of lockdowns which made access to education difficult. For example, several initiatives, including Latin Vox Machine, Roshan Learning Center, and the Refugee Youth Education hub, designed online learning programs that were delivered through dedicated platforms and/or learning packets that can be accessed via smartphones (UNHCR, 2021).

Challenges that remain

Despite the aforementioned initiatives and online provision of services and information that came about in response to the pandemic, challenges for technologically mediated resilience remain. While the refugee, migrant, and asylum-seeking population is increasingly digitally connected (UNHCR Innovation Service, 2019), the costs of remaining connected often become untenable, facing shocks and stressors that negatively impact their livelihoods. Indeed, Syrian refugees in Lebanon indicated that when income is low, they often forgo paying for their household internet connection as they prioritize expenditure on food and rent (Talhouk et al., 2022). Furthermore, while the population often has access to a smartphone, this is often shared within the household, thus making the delivery of personalized and often sensitive healthcare services and education difficult. This is further compounded by the population often residing in compact spaces that lack privacy.

In their examination of mental health mobile applications through a socioecological model for mental health resilience (Ungar et al., 2007), Tachtler et al. (Tachtler et al., 2021) found that often young unaccompanied asylum seekers in Austria are unable to follow the exercises offered by the applications due to sharing their accommodation with several others. Research has also shown that while refugees are technologically literate in using social media applications on their smartphones (Dekker et al., 2018; Udwan et al., 2020), some face challenges in using social media apps for new purposes, such as generating income through social commerce (Talhouk et al., 2022). These challenges highlight the need to address the digital poverty of this population in order to ensure that they are not excluded from the digitalization of services (IOM UN Migration, 2020) and the resilience that technologies may afford. Furthermore, the pandemic highlighted that in order to enable inclusive technologically mediated resilience, restrictions experienced by this marginalized population in their lived environments (e.g., lack of privacy within living accommodation) need to be accounted for.

Moving toward transformative resilience

The concept of resilience has been critiqued for its contribution to a narrative of abandonment (Duffield, 2012; Duffield, 2016; Evans & Reid, 2014) where marginalized communities are left to cope for themselves and fill in the cracks within humanitarian and healthcare systems.

Indeed, we can view the refugee-led technological initiatives presented before as exactly that. These initiatives not only responded to the pandemic but also to the marginalization of the population that excluded them from national responses (e.g., health information not being communicated in multiple languages). Furthermore, the initiatives presented are primarily focused on responding to the challenges that arose during the pandemic, and there has yet to be an exploration of how refugee-led initiatives may support in the recovery of this marginalized population. Therefore, as the world learns to live with COVID-19 and prepare for future pandemics, we should consider how technologies may play a role in supporting refugee, migrant, and asylum-seeking communities in bouncing back better. Evans and Reid (Evans & Reid, 2014) posit that for resilience to be conducive of utopian futures, in which vulnerablized communities bounce back better, we need to consider how mechanisms of resilience can and should challenge the status quo that is at the root of marginalization. This is very much in line with the UN's Research Roadmap for the COVID-19 Recovery (United Nations, 2020) where an emphasis has been placed on the need to work toward transformative changes that enable an equitable, resilient, and sustainable future for all; and how we configure technologically mediated resilience is integral to these transformative changes.

Through their scoping work for the aforementioned roadmap, Jewett et al. (Jewett et al., 2021) identify the importance of technologies in building social capital and in mediating trusted community engagements. As such, we should consider how technologies deployed in and by the health and humanitarian system can be configured so that refugees, asylum seekers, and migrants are not just beneficiaries of digitalized health services but rather are agents within the system who are informing and shaping system preparedness, response, and recovery to crisis. Indeed, research has identified how health technologies may play a dual role in improving refugee health, and in turn their resilience, by the provision of health services and by creating a space in which refugees may increase their agency within their relationship with the health system (Talhouk et al., 2017, 2020). Similarly, in regard to refugee-led technological initiatives that emerged from the pandemic, we need to consider how such initiatives may be placed in dialogue with the healthcare and humanitarian system so that they may synergistically complement one another. This requires us to view these technological initiatives not just as that of crisis response but rather as initiatives that may morph into sociotechnical infrastructures of community engagement akin to civic technologies designed for trustful engagements between community members, governments, and organizations (Corbett & Dantec, 2021; Le Dantec et al., 2021; Vlachokyriakos et al., 2016).

As such, configuring technologically mediated resilience for transformative change requires a view of resilience and the technologies that mediate it as a thread that weaves, knits, and cuts across systems, programs, and communities (Talhouk, 2020). In nonrefugee contexts, research has already shown the challenges of working across system-level resilience and community resilience. Fransen et al. (Fransen et al., 2021) highlighted that during the COVID-19 pandemic, urban community resilience initiatives struggled to connect to, and coordinate their activities with, governmental agencies considered by the initiatives to be unsupportive.

Additionally, top-down paternalistic approaches to building resilience that undermine communities' existing social networks have been indicated as contributors to diminishing trust between systems and communities during crisis (Jewett et al., 2021). Therefore it is imperative that extensive research is done to map how refugee-led technological initiatives functioned in relation to wider humanitarian and health system resilience and how in preparation for future crises new pathways and policies for resilience can be codesigned (Mullagh, Cooper, et al., 2021; Mullagh, Jacobs, et al., 2021) with all stakeholders including refugee, migrant, and asylum-seeking communities. Furthermore, this requires actively working with actors within the humanitarian and healthcare system to embrace this marginalized population's own understanding of resilience and the ways in which they want to configure and deploy technologically mediated resilience that may challenge the status quo (Talhouk, 2020).

References

Almohamed, A., Talhouk, R., & Vyas, D. (2022). Towards a conceptual framework for understanding the challenges in refugee re-settlement. *Proceedings of the ACM Human-Computer Interaction*, 6(GROUP). https://doi.org/10.1145/3492856.

Benjamen, J., Girard, V., Jamani, S., et al. (2021). Access to refugee and migrant mental health care services during the first six months of the covid-19 pandemic: a canadian refugee clinician survey. *International Journal of Environmental Research and Public Health*, *18*(10), 1–11. https://doi.org/10.3390/ijerph18105266.

Bezbaruah, S., Wallace, P., Zakoji, M., Padmini Perera, W. S., & Kato, M. (2021). Roles of community health workers in advancing health security and resilient health systems: emerging lessons from the COVID-19 response in the South-East Asia Region. *WHO South-East Asia Journal of Public Health*, *10*(3), 41. https://doi.org/10.4103/2224-3151.309872.

Bryant, J. H. P. G., & Holloway, K. (2020). *Bridging humanitarian digital divides during Covid-19 key messages. 44*. https://www.odi.org/sites/odi.org.uk/files/resource-documents/digital_covid_briefing_note_web.pdf.

Bukuluki, P., Mwenyango, H., Katongole, S. P., Sidhva, D., & Palattiyil, G. (2020). The socio-economic and psychosocial impact of Covid-19 pandemic on urban refugees in Uganda. *Social Sciences & Humanities Open*, *2*(1), 100045. https://doi.org/10.1016/j.ssaho.2020.100045.

Chen, A., Tossyeh, F., Arnous, M., et al. (2020). Phone-based data collection in a refugee community under COVID-19 lockdown. *The Lancet Psychiatry*, *7*(6), e31. https://doi.org/10.1016/S2215-0366(20)30189-9.

Corbett, E., & Dantec, C. A. L. (2021). Designing civic technology with trust. *Conference on Human Factors in Computing Systems*. https://doi.org/10.1145/3411764.3445341.

Dekker, R., Engbersen, G., Klaver, J., & Vonk, H. (2018). Smart Refugees: How Syrian Asylum Migrants Use Social Media Information in Migration Decision-Making. *Social Media on Society*, *4*(1). https://doi.org/10.1177/2056305118764439.

Duffield, M. (2012). Challenging environments: Danger, resilience and the aid industry. *Security Dialogue*, *58*, 21–36. https://doi.org/10.1177/0967010612457975.

Duffield M. The resilience of the ruins: towards a critique of digital humanitarianism. Resilience 2016;4(3):147-165. https://doi.org/10.1080/21693293.2016.1153772.

Endale, T., St. Jean, N., & Birman, D. (2020). COVID-19 and refugee and immigrant youth: a community-based mental health perspective. *Psychological Trauma: Theory, Research, Practice, and Policy*, *12*, S225–S227. https://doi.org/10.1037/tra0000875.

Evans, B., & Reid, J. (2014). *Resilient life: the art of living dangerously*. Polity Press.

Fransen, J., Peralta, D. O., Vanelli, F., Edelenbos, J., & Olvera, B. C. (2021). The emergence of urban community resilience initiatives during the COVID-19 pandemic: an international exploratory study. *European Journal of Development Research*, (1). https://doi.org/10.1057/s41287-020-00348-y.

Golesorkhi, L. Z., Fortson, G., Harder, K., & Riedmann, T. (2020). Centering community in COVID-19 responses: refugee women's livelihoods in a global pandemic. *International Journal of Sociology and Social Policy*. https://doi.org/10.1108/IJSSP-07-2020-0332.

Grieco-Page, H., Black, C. J., Berent, J. M., Gautam, B., & Betancourt, T. S. (2021). Beyond the pandemic: leveraging rapid expansions in U.S. telemental health and digital platforms to address disparities and resolve the digital divide. *Frontiers in Psychiatry*, *12*. https://doi.org/10.3389/FPSYT.2021.671502/FULL.

IFRC. IFRC Framework for Community Resilience. 2014. Accessed March 17, 2022. http://www.ifrc.org/Global/Documents/Secretariat/201501/1284000-Framework for Community Resilience-EN-LR.pdf.

IOM UN Migration. The power of digitalization in the age of physical distancing : table of contents. Published 2020. Accessed March 17, 2022. https://www.iom.int/sites/g/files/tmzbdl486/files/documents/disc_digest_4th_edition_digitalization_and_migrant_inclusion_final.pdf.

Jewett, R. L., Mah, S. M., Howell, N., & Larsen, M. M. (2021). Social cohesion and community resilience during COVID-19 and pandemics: a rapid scoping review to inform the united nations research roadmap for COVID-19 recovery. *International Journal of Health Services*, *51*(3), 325–336. https://doi.org/10.1177/0020731421997092.

Jumpstart Refugee Talent. (2022). *Refugee Talent Hub*. https://jumpstartrefugee.ca/programs/talent/. (Accessed 11 March 2022).

Le Dantec, C. A., Garcia, A. A., Connelly, C., & Meng, A. (2021). Resisting resolution: enterprise civic systems meet community organizing. *Multimodal Technologies and Interaction*, *5*(4). https://doi.org/10.3390/MTI5040020.

Logie, C. H., Okumu, M., Berry, I., et al. (2021). Kukaa Salama (Staying Safe): study protocol for a pre/post-trial of an interactive mHealth intervention for increasing COVID-19 prevention practices with urban refugee youth in Kampala, Uganda. *BMJ Open*, *11*(11), 1–8. https://doi.org/10.1136/bmjopen-2021-055530.

Mancini, T., Imperato, C., Vesco, S., & Rossi, M. (2021). Digital communication at the time of COVID-19: relieve the refugees' psychosocial burden and protect their wellbeing. *Journal of Refugee Studies*. https://doi.org/10.1093/jrs/feab082.

Mangrio, E., & Sjögren, F. K. (2017). Refugees' experiences of healthcare in the host country: a scoping review. *BMC Health Services Research*, *17*(1), 814. https://doi.org/10.1186/s12913-017-2731-0.

McMullin, C. (2021). Migrant integration services and coping with the digital divide: challenges and opportunities of the COVID-19 pandemic. *Voluntary Sector Review*, *12*(1), 129–136. https://doi.org/10.1332/204080520X16076177287917.

Mesmar, S., Talhouk, R., Akik, C., et al. (2016). The impact of digital technology on health of populations affected by humanitarian crises: recent innovations and current gaps. *Journal of Public Health Policy*, *37*(S2), 167–200. https://doi.org/10.1057/s41271-016-0040-1.

Mohammadi, M., Wächter-Raquet, M., & Odukoya, D. (2021). Health promotion through digital means: Refugee women using video and social media in times of COVID-19. *European Journal of Public Health*, *31*(Suppl 3), 2021. https://doi.org/10.1093/eurpub/ckab164.396.

Mukumbang, F. C., Ambe, A. N., & Adebiyi, B. O. (2020). Unspoken inequality: how COVID-19 has exacerbated existing vulnerabilities of asylum-seekers, refugees, and undocumented migrants in South Africa. *International Journal for Equity in Health*, *19*, 141. https://doi.org/10.1186/s12939-020-01259-4.

Mullagh, L., Cooper, R., Thomas, L., Sacks, J., Lloyd Jones, P., & Jacobs, N. (2021). Designing for a pandemic: towards recovery and resilience. *Strategic Design Research Journal*, *14*(1), 161–174. https://doi.org/10.4013/sdrj.2021.141.14.

Mullagh L., Jacobs N., Cooper R., Kwon N. Design strategies for future governance in emergencies: a case study. International Research Society for Public Management. Published online 2021.

Narla, N. P., Surmeli, A., & Kivlehan, S. M. (2020). Agile application of digital health interventions during the covid-19 refugee response. *Annals of Global Health*, *86*(1), 1–5. https://doi.org/10.5334/aogh.2995.

Norredam, M., Mygind, A., & Krasnik, A. (2006). Access to health care for asylum seekers in the European Union—A comparative study of country policies. *European Journal of Public Health*, *16*(3), 285–289. https://doi.org/10.1093/eurpub/cki191.

O'Mara B., Monani D., Carey G. Telehealth, COVID-19 and refugees and migrants in Australia: policy and related barriers and opportunities for more inclusive health and technology systems. International Journal of Health Policy and Management 2021. http://doi.org/10.34172/ijhpm.2021.31.

Orcutt, M., Patel, P., Burns, R., et al. (2020). Global call to action for inclusion of migrants and refugees in the COVID-19 response. *Lancet*, *395*(10235), 1482–1483. https://doi.org/10.1016/S0140-6736(20)30971-5.

Reese Masterson, A., Usta, J., Gupta, J., & Ettinger, A. S. (2014). Assessment of reproductive health and violence against women among displaced Syrians in Lebanon. *BMC Women's Health*, *14*(1), 25. https://doi.org/10.1186/1472-6874-14-25.

Sarr I., Beltramo T., Loschmann C. Rapid phone surveys in Kenya show disproportionately negative impact on refugees. Published 2022. Accessed March 7, 2022. https://www.unhcr.org/blogs/rapid-phone-surveys-in-kenya-show-disproportionately-negative-impact-on-refugees/.

Tachtler, F., Talhouk, R., Michel, T., Slovak, P., & Fitzpatrick, G. (2021). Unaccompanied migrant youth and mental health technologies: a social-ecological approach to understanding and designing. In *Proceedigns of CHI 2021 Conference on Human Factors in Computing Systems* (pp. 1–19). https://doi.org/10.1145/3411764.3445470.

Talhouk, R. (2020). *Exploring the role of technologies in building syrian refugee community resilience*. PhD Thesis Newcastle upon Tyne, United Kingdom: Newcastle University.

Talhouk, R., Akik, C., Araujo-Soares, V., et al. (2020). Integrating health technologies in health services for Syrian refugees in Lebanon: qualitative study. *Journal of Medical Internet Research*, *22*(7). https://doi.org/10.2196/14283.

Talhouk, R., Bartindale, T., Montague, K., et al. (2017). Implications of synchronous IVR radio on Syrian refugee health and community dynamics. *Proceedings of the 8th international conference on communities and technologies (c&t'17)* (pp. 1–10). https://doi.org/10.1145/3083671.3083690.

Talhouk, R., Mesmar, S., Thieme, A., et al. (2016). Syrian refugees and digital health in lebanon: opportunities for improving antenatal health. *Proceedings of the chi conference on human factors in computing systems (CHI'16). ACM* (pp. 331–342). https://doi.org/10.1145/2858036.2858331.

Talhouk R., Montague K., Ghattas H., Araujo-Soares V., Ahmad B., Balaam M.. Refugee food insecurity & technology: surfacing experiences of adaptation, navigation, negotiation and sharing. Computer Supported Cooperative Work March 2, 2022. https://doi.org/10.1007/s10606-022-09423-w.

Udwan, G., Leurs, K., & Alencar, A. (2020). Digital resilience tactics of syrian refugees in the netherlands: social media for social support, health, and identity. *Social Media*, *6*(2). https://doi.org/10.1177/2056305120915587.

UN Women. Using technology to assist vulnerable Syrian refugees during COVID-19 lockdown. Published 2020. Accessed March 7, 2022. https://www.unwomen.org/en/news/stories/2020/4/feature-technology-to-assist-refugees-during-covid-19-lockdown.

Ungar, M., Brown, M., Liebenberg, L., Man Kwong, W., Amrstrong, M., & Gilgun, J. (2007). Unique pathways to resilience across cultures adolescence. *Adolescence*, *42*(166), 1–24. 10.1111/1540-4560.t01-1-00005. \n10.1177/0044118x03257030.

UNHCR. Refugee-Led Innovations During Covid-19: Best Practices from the 2020 NGO Innovation Award. Published December 10, 2021. Accessed March 17, 2022. https://reliefweb.int/sites/reliefweb.int/files/resources/61b312f24.pdf.

UNHCR Innovation Service. Connectivity for Refugees—UNHCR Innovation. Published online 2019. Accessed March 17, 2022. http://www.unhcr.org/innovation/connectivity-for-refugees/.

UNIC Tokyo. *Smartphone apps help Palestine refugees seek healthcare during COVID-19*. https://www.un.org/en/coronavirus/smartphone-apps-help-palestine-refugees-seek-healthcare-during-covid-19. (Accessed 7 March 2022).

United Nations. Global compact on refugees. Published 2018. Accessed March 17, 2022. https://www.unhcr.org/5c658aed4.

United Nations. UN research roadmap for the COVID-19 recovery. Published 2020. Accessed March 17, 2022. https://www.science.org.au/files/userfiles/support/submissions/2020/consultation-un-research-roadmap-for-covid-recovery.pdf%0Ahttps://www.science.org.au/supporting-science/science-policy-and-analysis/submissions-government/submission-un-research-roadmap.

Vlachokyriakos, V., Crivallero, C., Le Dantec, C. A., Gordon, E., Wright, P., & Olivier, P. (2016). Digital civics : citizen empowerment. *Proceedings of the CHI extended abstracts on human factors in computing systems (CHI EA'17)* (pp. 1096–1099). https://doi.org/10.1145/2851581.2886436.

Walker, L., Bailey, D., Churchill, R., & Peckham, E. (2021). Remote data collection during COVID-19 restrictions: an example from a refugee and asylum-seeker participant group in the UK. *Trials*, *22*(1), 21–23. https://doi.org/10.1186/s13063-021-05058-2.

The rise of racial minority health apps☆

Udani Samarasekera

London, England, United Kingdom

"As a Black woman with a chronic mental health illness, racism and lack of cultural competency in health care scares me. Sometimes I am faced with huge barriers when trying to get the support that I need, and I am tired of it," tweeted Jasmin Pierre, an American mental health advocate and app developer in early February 2022. Her sentiments are echoed by many Black Americans, according to recent data. Results from the 2020 Survey On Race And Health, a nationally representative survey of US adults by American nonprofit Kaiser Family Foundation and digital platform The Undefeated, revealed that one in five Black adults said that they had personally experienced race-based discrimination in the past year.

Additionally, two-thirds of Black adults said it was difficult to find a healthcare provider who shared their background and experiences, and a quarter said they have never received care from a Black doctor. Black adults were also more likely than White adults to report negative experience with healthcare providers, including providers not believing they were telling the truth or refusing to provide pain medication.

Such responses from Black individuals are unsurprising given the long-standing and well-documented racial disparities in health and healthcare in the United States that have arisen from structural and systemic inequities. But while the United States and other countries continue to grapple with this issue, particularly in the wake of the global racial reckoning that occurred in 2020, could digital health be part of the solution? Several Black technology entrepreneurs think so. They have stepped in to try to help address racial health inequities in their communities by designing culturally relevant consumer health apps. The past 5 years have seen a proliferation of such apps in the United States, specifically for Black people and other people of color. Areas that the apps focus on include increasing access to Black healthcare providers, providing mental health support, and improving reproductive care.

☆ Reprinted with permission from the publisher, Elsevier. Original "The rise of racial minority health apps" available at: *https://www.thelancet.com/journals/landig/article/PIIS2589-7500(22)00043-7/fulltext*

Resilient Health. https://doi.org/10.1016/B978-0-443-18529-8.00060-3

An example of these apps is Health In Her HUE, launched in 2018 by two Black female entrepreneurs Ashlee Wisdom and Eddwina Bright. Wisdom has expertise in public health and Bright in business. The app connects Black women and women of color to 1000 culturally competent and sensitive healthcare providers across the United States, including doctors, midwives, and nutritionists. In August 2021, the app attracted US$1 million in funding to build the platform, mostly from Black-owned companies.

In the mental health arena, the Ayana Therapy app founded by Eric Coly, a Senegalese American and mental health advocate, links marginalized and intersectional communities to 200 licensed mental health professionals who might share their race, ethnicity, culture, gender, sexual orientation, or religion. It uses a detailed and culturally sensitive questionnaire to match people with a counselor. Users are then able to use the app to communicate with the therapist via text messages, phone calls, or video sessions. Another mental health platform is The Safe Place, which is a free app created by Pierre in 2018. It is geared toward the Black community and features mental health statistics and self-care tips, including on coping after police brutality and talking to family members about mental illness. "All races go through mental illness, but we also can experience it differently because of our race and social backgrounds," Pierre writes in her app. "African Americans are 20% more likely to experience serious mental health problems than the rest of the general population. However, many Black people still do not wish to seek professional help for their mental illnesses." The purpose of the app "is to bring more awareness, education, and hope to this serious issue," Pierre states in her app.

Black doctors and researchers working in racism and health have largely welcomed the rise of health apps for minorities by Black people. Ayana Jordan, assistant professor of Psychiatry, Yale School of Medicine, CT, USA, whose work focuses on substance use and mental health in minority communities, tells *The Lancet Digital Health* that the increase in such apps "demonstrates that Black people are attempting to find solutions to the very real need to be seen and treated with humanity in a very racist mental health and health-care system." Using technology makes sense too. Eighty-three percent of Black Americans now own a smartphone, according to a Pew Research Center survey in 2021. With the rise of smartphones and social media, Jordan tells *The Lancet Digital Health* that "culturally based apps allow people to readily identify Black and other treaters of colour" and find "which providers are safe and can understand and empathise with the Black experience …or minimise the impact of real consequences of being Black and othered in a system not created or meant to integrate your cultural values."

Joia Crear-Perry is president of the US nonprofit, the National Birth Equity Collaborative, and a thought leader around racism and inequity. She thinks the proliferation of health apps directly marketed to Black people in the past 2–5 years is not surprising. "We spend a lot of time and money on technology that we find value in and I am happy to see companies understanding that and creating products that reflect our unique experiences while also making quality health care, particularly for mental health, more accessible to our community," she comments.

Only 5% of active physicians in the United States identify as Black or African American, according to the Association of American Medical Colleges. In view of such statistics, Jordan is overall in support of Black entrepreneurs "responding to the real need for representation in a global milieu where there are so few Black physicians" but she notes some important considerations for these initiatives. She questions whether these health apps utilize evidence-based treatments. Although this is "not mandatory" she says, Black people should be offered options to evidence-based practices and "given the opportunity to learn about the latest treatment options, including medication."

Jordan also thinks that there needs to be a way to distinguish self-help support apps from treatment apps where an intervention is occurring or where you can connect to doctors, psychiatrists, or other mental health providers. Importantly, she notes that "if people are feeling unsafe (acutely wanting to hurt someone or themselves), how will the app address that urgency or interface to connect a person to more immediate care?" Although she does not consider these insurmountable concerns, she thinks that they need to be carefully considered by developers.

Meanwhile, Crear-Perry thinks that although developing health apps specifically for the Black population is useful and necessary, "it is important that we not reinforce or present harmful narratives that suggest that our bodies are somehow biologically different than other races. They are not. However, our experiences and how the effects of those experiences play out in terms of our health, mental and physical, absolutely need to be reflected in the type of care we receive from our provider or via an app in order for us to feel safe." She is also concerned about potential exploitation. "Now that Black companies have proven there is a market for this technology…you can see how others are attempting to lean into this space as well." She is "instantly wary when those involved—and profiting—are not Black people ourselves."

Noncommercial health apps for African Americans have also grown in recent years as researchers try to tackle racial health disparities. These apps focus on health promotion and health behavior change, are evidence based, and have a longer history than the newer consumer health apps for the Black community. Areas of health covered include cardiovascular health, reproductive health, HIV prevention, and physical activity and many include or are led by Black researchers. Some, such as the Fostering African American Improvement in Total Health (FAITH!) app, are reporting promising results. Researchers led by LaPrincess Brewer at the Mayo Clinic, Rochester, MN, USA, codesigned the culturally tailored FAITH! app to promote cardiovascular health within the African American church community using an iterative participatory approach between academia and the community. In the pilot study, participants achieved improvements in key cardiovascular health behaviors and factors and in a formative evaluation reported overall high acceptability and satisfaction of the intervention.

Such codesigning of apps is key, notes Sherry Pagoto, professor in the Department of Allied Health Sciences at the University of Connecticut and director of the UConn Center for mHealth and Social Media, Storrs, CT, USA.

"Experts from the target population should be at the design table and developers should heavily engage potential users who are members of the target population throughout the design process. If you aren't co-designing in this way, there is a good chance your app won't be of interest to an entire population of people," she says.

An issue with any app in the health promotion realm, however, is getting users to keep returning. "A major challenge with mobile apps is that they often get downloaded and then ignored. A significant tension exists between in-the-pocket convenience and adherence. Mobile apps are incredibly convenient but getting people to use them enough to produce clinical outcomes has been an elusive goal. All of these factors require a team that has the proper expertise represented," Pagoto notes.

Whether it is an app to link the Black community with Black healthcare providers, culturally relevant information, or to encourage health-promoting behaviors, the common goal is to improve the Black healthcare experience and reduce racial health disparities. Will these Black entrepreneurs and researchers succeed? Only time will tell. But that the development of these apps is being led by the Black community looks to be a step in the right direction.

CHAPTER 61

The imperative for mental health in the workplace post-COVID-19

Patricio V. Marquez

World Bank Group, Washington, DC, United States; Centre for Health Economics, The University of the West Indies, Port of Spain, Trinidad and Tobago; Johns Hopkins University Bloomberg School of Public Health, Baltimore, MD, United States; Charles Darwin Foundation for the Conservation of the Galapagos Islands, Puerto Ayora, Ecuador

Introduction

The COVID-19 pandemic disrupted the workplace, as illustrated by the commonly mandated use of masks (Fig. 1). This chapter focuses on changes in work arrangements during the pandemic, their implications going forward, and elaborates on options for contributing to the development of healthy workplaces, with particular attention to the need to tackle unmet mental health needs of employees.

Workplace disruptions during the pandemic

The workplace has been drastically disrupted by the pandemic. One dramatic manifestation has been the "Great Resignation," a trend in which a large number of people voluntarily resigned from their jobs, beginning in early 2021, primarily in the United States and other rich countries (Cohen, 2021). This trend has been acutely felt in the service industries, such as retail, hospitality, food service, and healthcare (Chugh, 2021).

The experience of US workers across various ages, industries, and income brackets shows that they have experienced heightened levels of fatigue, burnout, and general dissatisfaction toward their jobs since the pandemic's start, with more young people airing these indignations and jaded attitudes on the internet (North, 2022). The findings of a new Pew Research Center survey of the US labor market found that low pay, a lack of opportunities for advancement, and feeling disrespected at work are the top reasons why Americans quit their jobs in 2021

Resilient Health. https://doi.org/10.1016/B978-0-443-18529-8.00061-5

Fig. 1
A common feature of the COVID-19 pandemic: mandate use of masks in the workplace. *Source: Stock Photo ID: 1866510874. Free download from author's personal subscription.*

(Parker & Menasce Horowitz, 2022). The survey also finds that those who quit and are now employed elsewhere are more likely than not to say their current job has better pay, more opportunities for advancement, and more work-life balance and flexibility. These findings are reinforced by data from a nationally representative survey conducted by the Harris Poll for Bloomberg News that suggest that about 55% of respondents are likely to seek out job offers from other companies to get raises at their current firms (Wells et al., 2022). Millennials are the most likely to jump ship, followed by Gen Z, Gen X, and Boomers.[a]

In low- and middle-income countries (LMICs), while people in the formal sector of the economy have had flexible work arrangements or remained at their employment to save their jobs, large numbers of informal workers (in Latin America, for example, 55% of total workers) have not had the same choice (Marquez et al., 2020). They continued to work in their regular settings to earn a living since they were unlikely to have much savings or to receive support from social protection programs at a level that would allow them to feed their families.

Although experiences are different across countries, there is growing consensus globally that how work is organized and how we work together, interact, and communicate will continue to change and be shaped by technological advances, especially information and communication technology. This should not surprise us. Even before this COVID-19 pandemic, there was much

[a] Millennials refer to a person born between the early 1980s and the late 1990s. Gen Z or zoomers refer to the generation born between 1997 and 2012, who is the first social generation to have grown up with access to the internet and portable digital technology from a young age, and who tend to be more concerned than older generations with academic performance and job prospects. Gen X is the name given to the generation of born between the mid-1960s and the early 1980s. Boomers refer to people born from 1946 to 1964, during the post–World War II baby boom.

discussion about how technological innovations were fueling momentous change in our daily lives and in the workplace. Indeed, as Professor Klaus Schwab, chairman of the World Economic Forum, claimed, a range of new technologies are fusing the physical, digital, and biological worlds, impacting all disciplines, economies, and industries, and unleashing changes "unlike anything humankind has experienced before." They underpin what he calls "the fourth industrial revolution" (Schwab, 2017).

In workplace arrangements, the "ripple effects" from the pandemic have compounded, deepened, and expanded the impact of rapid and disruptive economic and psychosocial changes that were already underway, fostered by rapid advances and the increased reliance on technology. The introduction of new information technology platforms and tools, for example, has had a significant impact on the structure, management, and function of most organizations, requiring the development of new patterns of work organization, as well as the redefinition of individual jobs and responsibilities, the structure of groups and teams, and the nature of supervision and managerial roles (UKEssays, 2018). These changes have dislocated or displaced workers, causing concerns and fears among employees and managers alike, and give rise to the question: What can be done to build social resilience, address mental health challenges, and protect and keep people front and center of the development process?

Redefining working arrangements postpandemic

The adoption and widespread use of new information technologies have been disrupting and changing significantly the nature of work and employment conditions across the world over the past decades. Indeed, cloud and mobile computing, big data and machine learning, sensors and intelligent manufacturing, advanced robotics and drones, and clean energy technologies are transforming the very foundations of global business and the organizations that drive it (Cascio & Montealegre, 2016).

These technologies are not just helping people to do things better and faster, but they are enabling profound changes in the ways that work is done in organizations. For example, new forms of work have altered the conditions of work and employment, especially telecommuting, outsourcing, and off-shoring activities (UKEssays, 2018). Technological advances, particularly in electronic communication, continued to decrease the need for traditional office settings while increasing the number of telecommuters. New digital technology has decreased direct face-to-face communication between organizations, their suppliers, business partners, and their customers, while employees have been benefitting from becoming mobile, conducting work-related activities teleconferencing, using cellular phones and laptop computers, either from their homes or from some other remote location. This trend has been facilitated by easier access to the internet, which has allowed more employees to become telecommuters.

During the pandemic, remote work, carried out on a virtual platform, became part of a "new normal." This arrangement is attractive to employees who value the independence of working remotely from anywhere and who flourish amid flexible arrangements, and to younger cohorts of workers who are technologically savvy. Employers also glean value, as in lower costs of office space, equipment, and access to otherwise unavailable labor.

Working from home is a relatively new experience for a majority of workers with jobs that can be done remotely. Recent findings from another Pew Research Center survey of the US labor market provide evidence that the impetus for working from home has shifted considerably since 2020 (Parker et al., 2022). Today, more workers say they are doing this by choice rather than necessity. Among those who have a workplace outside of their home, 61% now say they are choosing not to go into their workplace, while 38% say they're working from home because their workplace is closed or unavailable to them. Earlier in the pandemic, just the opposite was true: 64% said they were working from home because their office was closed, and 36% said they were choosing to work from home. While more say a preference for working from home is a major reason they are doing so (76% now vs. 60% in 2020), there has also been a significant increase since 2020 (from 9% to 17%) in the share saying the fact that they have relocated away from the area where they work is a major reason why they are currently teleworking.

Indeed, as observed by Dr. Isabell Welpe, professor of Strategy and Organization at the Technical University of Munich, Germany, virtual and remote work are mostly here to stay postpandemic (Chugh, 2021). Many companies have already announced that their employees will never have to return to the office full time, she notes. New postpandemic working arrangements are moving away from a one-size-fits-all approach to allow individual and asynchronous (not synchronized or coordinated with time) organization of work and work settings, e.g., a part of the workforce works fully remotely most of the time and another part comes to the office only 1–2 days a week. In Welpe's view, office buildings will cease being a place for daily work and evolve into places for meetings, recruiting, meeting customers, and boot camps to facilitate interpersonal exchange.

However, as Welpe also predicts, the degrees of remote work will also depend upon how well firms manage the challenges that come with such remote work, which includes overcoming communication silos (separation between workers and across departments which was already pervasive prepandemic) as well as sharing noncodified information and knowledge. A different leadership style and new skills from workers will also evolve, for example, that management through leaders will be gradually replaced by self-leadership. Also, workers who are healthy, able, and motivated to take on responsibility will thrive and foster greater agility in their organizations. Others, who could work from home but are opting not to, stated that a major reason is that they feel more productive at their workplace, or indicate that not having the space or resources at home to work effectively is a major reason why they rarely or never work from home (Parker et al., 2022).

Ways to support the establishment of new work arrangements

Technological changes and new forms of workplace organization over decades have led to workers assuming increased responsibilities and more autonomy than ever before (Appelbaum et al., 2000; Hafner et al., 2015). Although this trend has increased overall productivity, downsides include that flexible working practices have also increased daily job demands, requiring employees to multitask, which leads to increased levels of workplace stress and unrealistic time pressures that contribute to increased absences due to sickness, and also to the emergence of "presenteeism," when employees attend work while in suboptimal health (Bevan, 2021).

A healthy workforce is vital to the well-being of workers and that of their families, and key for the continuous development of human and social capital in the workplace. It enhances the productivity, competitiveness, and bottom line of firms. The cost of ill health in the workplace is easily grasped when one considers information from the UK that shows that 140 million working days are lost to sickness absence and 300,000 individuals leave the workplace each year due to ill health (Hafner et al., 2015), with a cost of sickness absence to British business estimated at approximately £15bn or about US$19bn (Black & Frost, 2011), and another £15bn or US$19bn due to presenteeism (employees attending work while sick) driven by mental ill health (Centre for Mental Health, 2011).

One way to address these health challenges is to heed the advice of Professor Michael Porter of the Harvard Business School, an expert in management and competitiveness, that the next transformation of business thinking lies in the principle of shared value: creating economic value in a way that also creates value for society by addressing its needs and challenges (Porter & Kramer, 2011). At the workplace, the adoption of the shared value proposition means that organizations and businesses have in their role as employers a great responsibility to nurture resilience of employees (and their families), particularly in these changing times. While money is an important thing people want from work, requiring better compensation and benefits to attract and retain workers, people also place value in finding meaningful work and deriving a sense of well-being and fulfillment (UC Press Blog, 2021). If we accept the previously mentioned considerations, then, it becomes clear that disease prevention and promotion of good health are vital investments to accompany the new working arrangements postpandemic.

Indeed, while employers have a responsibility to provide a safe and hazard-free workplace, they also have abundant opportunities to promote individual physical and mental health, fostering a healthy work environment to reduce health risks, and improve the quality of life for workers and their families. This can be achieved in several ways, including by introducing an employee-centered wellness culture into the workplace and providing supportive environments where safety, respect, and equal treatment of all are ensured. Also, employers can facilitate a more optimal balance between work and life for their employees, by adopting institutional norms and nudges to encourage workers and managers to take time away from email and internet communications after office hours, on weekends, and on holidays, and by facilitating access and opportunities for employees to engage in a variety of workplace health and wellness programs.

There is growing evidence of the significant impact of such workplace health and wellness programs. In the United States alone, these programs are now a US$6 billion industry, with more than half of firms with at least 50 employees offering these programs. Results of a national survey showed measurable improvements among program participants in their physical activity, smoking cessation, and weight control programs over a four-year period. The study also found that participation in a wellness program over five years is associated with a trend toward lower healthcare costs (Mattke et al., 2015). The return-on-investment is noteworthy when comparing the ratio of reductions in the cost of healthcare (e.g., keeping people healthy and out of hospital) to the costs of the programs: wellness programs generated a return of $1.50 for every $1 invested and a return of $3.80 for every $1 spent on disease management.

Another study covering 25 firms with 2 million employees in 125 countries around the world similarly found that firms that champion workplace wellness reap significant benefits in terms of increased productivity, reduced cost of employee healthcare, and increased employee engagement leading to reduced turnover (World Economic Forum, in Collaboration With FTI Consulting, 2013).

The growing burden of mental and substance use disorders

Mental and addictive disorders affect more than 1 billion people globally, causing 7% of all global burden of disease as measured in DALYs and 19% of all years lived with disability—depression was associated with most DALYs for both sexes, with higher rates in women as all other internalizing disorders, whereas other disorders such as substance use disorders had higher rates in men (Rehm & Shield, 2019).

Mental and physical health are interrelated: people who suffer from a mental disorder also have a higher lifetime risk for serious physical health problems (Matheson et al., 2014). Mental disorders can result in worse treatment adherence and outcomes for commonly cooccurring diseases, such as tuberculosis, diabetes, cardiovascular disease, and cancer (Prince et al., 2007).

After adjustment for the COVID-19 pandemic in 2020, the estimated prevalence of major depressive disorder was 3152·9 cases per 100,000 population, equivalent to 246 million people, and the global prevalence of anxiety disorders was 4802·4 per 100,000 population, equivalent to 374 million people (GBD 2019 Diseases and Injuries Collaborators, 2021). This burden was high across the entire lifespan, for both sexes, and across many locations. Increases in the prevalence of major depressive disorder and anxiety disorders during 2020 were both associated with increasing SARS-CoV-2 infection rates and decreasing human mobility. Social restrictions, unemployment, financial instability, and school closures are among COVID-19-related factors that contributed to worsened mental health outcomes (Whiteford et al., 2013).

Globally, 107 million people or 1.8% of the total population are estimated to have an alcohol use disorder (Ritchie & Roser, 2018a). The results of a study on mental health and alcohol use in developing countries showed that the consumption of alcohol is heavily gendered, characterized by a high proportion of hazardous drinking among men compared to women. Hazardous drinkers not only consume large amounts of alcohol, but also do so in high-risk patterns, such as drinking alone and binging, both of which are associated with depressive and anxiety disorders as well as suicide and domestic violence (Patel, 2007). In terms of drug use, it is estimated that globally in 2017 around 71 million people had a drug use disorder. Most of these people have an addiction to opioids, which accounts for around 55% of drug use disorders globally (Ritchie & Roser, 2018b).

Excessive alcohol consumption can influence COVID-19 susceptibility and severity, as it both activates the immune system, causing inflammation, and interferes with the body's immune response to viral and bacterial infections. And, in a vicious cycle, the broad effects of the pandemic are also likely to lead to excessive alcohol consumption. As documented in a previous post, people are more likely to drink more as a coping mechanism "during times of uncertainty and duress," particularly in the face of raising unemployment and loss of income (Marquez, 2020c). For example, with bars and restaurants closed in the United States, sales of alcohol in the United States have spiked nearly 27% since the start of the pandemic, with consumers gravitating to larger pack sizes.

The costs to society of alcohol and drug in terms of lost productivity, healthcare costs, traffic accidents, and personal tragedies are staggering (Goplerud et al., 2017). For example, numerous studies and reports in the US have been issued on the workplace costs of alcoholism and alcohol abuse, and they report costs that range from US$33 bn to US$68bn per year (US Office of Personnel Management, 2022). Likewise, data from the US show drug use costs the economy more than US$700bn annually in increased healthcare costs, crime, and lost productivity (National Institute on Drug Abuse (NIDA) Site, 2020).

Tackling mental and substance use disorders in the workplace

As the world entered into the postpandemic period in 2023, there is growing recognition that the mental health "tail" of this global crisis will likely prove to be quite a "long haul" for many people, compounding and deepening the impacts of social and economic disruptions already underway prepandemic. As documented in different articles and reports, the pandemic has changed our lives, overwhelming people with feelings of instability (Marquez, 2020b). Anxiety about the future adds to this burden. While shutdowns in the spring of 2020 extracted a heavy toll, the Delta and Omicron variants of the COVID-19 virus onslaught in the second half of 2021 not only spiked the coronavirus case numbers but the disruptions that followed brought a sense that the pandemic would never end despite growing availability of vaccines and new therapies (Marquez, 2021).

While physical health-related metrics associated with workplace health and wellness programs are promising, tackling mental illness in the workplace is lagging. This is a major challenge that needs to be addressed head on. The enormous burden of mental ill health at home and at the workplace is aggravated by widespread stigma and discrimination of affected people.

It is imperative, therefore, to recognize that mental health is an integral and essential component of health demanding the same priority attention as physical health. Mental health is a form of human capital for individuals, families, communities, and society, which can be looked after and invested in. It is a state of well-being in which individuals realize their own abilities, can cope with the normal stresses of life, can work productively, and are able to contribute to the community (World Health Organization Site, 2018). Good mental health improves opportunities for positive interpersonal relationships and success in studies and work, and it strengthens trust, reciprocity, engagement, and a sense of belonging in a society. It manifests as general well-being and reduced levels of strain on families, in communities, at the workplace, and across the entire society (Ministry of Social Affairs and Health, Finland, 2020).

Poor mental health is a major issue for employers and employees alike (Business Group on Health Site, 2021) and costs organizations billions of dollars every year and worsens the lives of millions of people, in the workplace and beyond. For example, in wealthy OECD countries, which spend on average 9% of GDP on healthcare (OECD, 2021), the high economic cost associated with mental conditions is largely driven by lost productivity in the working-age population rather than by mental healthcare expenditure (OECD, 2012). People suffering from mental ill health are not only less productive at work but are more likely to be out sick from work, and when they are out sick are more likely to be absent for a longer period—around 30% to 40% of all sickness and disability caseloads in OECD countries are related to mental health problems.

Despite very high costs to individuals and the economy, there is limited awareness about the connection between work and mental health, and of the drivers behind the labor market outcomes and the level of inactivity of people with ill mental health (OECD, 2021). Work, as a social arrangement, links and mediates countless relationships, both casual and intimate (Malesic, 2022). The disruptions in workplace arrangements, which were further accentuated during the pandemic, have severed those relationships, and workers have become more isolated and despondent. Understanding these drivers is critical for developing more effective policies to deal with workplace mental health issues during the pandemic and its aftermath.

Addressing mental ill-health risks in the workplace could generate significant benefits for workers and firms alike. A study prepared for the 2016 World Bank Group—World Health Organization Global Health Conference estimated that the returns on this investment in a country can be substantial as measured by a favorable benefit-to-cost ratio, ranging between 2.3–3.0 and 1 when economic benefits only are considered and 3.3–5.7 and 1 when social returns are also included (Kleinman et al., 2016).

Workplace mental health is a good corporate investment

The high prevalence of stress-related and mental health conditions and substance use disorders in the workplace, and the common cooccurrence with other medical conditions, has a significant impact on work performance and organizational effectiveness. To deal with this challenge, different private and public employers have expanded the focus of their health benefits programs beyond physical health to include workplace mental health and well-being interventions.

A number of organizations such as Bell Canada, British Telecom, American Express, Johnson & Johnson, Sprint Corporation, and Bank of England (Health Enhancement Research Organization (HERO), 2020), as well as multilateral organizations such as the World Bank Group (World Bank Group, 2022a) that launched its Mental Health and Well-being Strategy in May 2022, covering staff working at headquarters and more than 120 offices worldwide (World Bank Group, 2022b), offer examples of emerging good practices for supporting mental health and well-being in the workplace.

The experience of Bell Canada (Bell), with a workforce of more than 52,000 employees, is instructive of what investing in employee mental health looks like in practice (Kangasniemi et al., 2019). It combines workplace programs, practices, and policies to raise awareness about mental health, including leadership training and buy-in, and employee awareness and educational initiatives, and to offer employee and family assistance programs, psychological care benefits, appropriate processes related to taking leave and returning to work, and policies that promote mental health and well-being.

"Bell Let's Talk" is the largest-ever corporate initiative supporting mental health in Canada. Launched in September 2010, it promotes awareness of mental health issues and drives action with a strategy built on four pillars: fighting the stigma of mental health, improving access to care, supporting research, and leading by example in workplace mental health.

On "Bell Let's Talk" Day 2022, more than 164 million "messages of support for mental health" were logged (Bell Canada Site, 2022). Those messages took the form of voice calls and text messages by Bell customers, using the #BellLetsTalk hashtag on Twitter and TikTok, and views of the Bell Let's Talk Day video on Facebook, Instagram, LinkedIn, Pinterest, Snapchat, TikTok, Twitter, and YouTube. For every "interaction" on "Bell Let's Talk" Day, the company pledged to donate five cents toward Canadian mental health, which is then divided among various community projects and major institutions across Canada. The donation in 2022 amounted to more than $8.2M for mental health initiatives.

Bell's experience also illustrates that investment in mental health programs is more likely to achieve positive returns on investment (ROI) when they support employees along the entire spectrum of mental health, from promotion of well-being to intervention and care, as well as the elimination or reduction of workplace hazards that could psychologically harm an employee. Bell's ROI in this area over the past decade was estimated at $4.10 for every Canadian dollar invested (Kangasniemi et al., 2019).

Leveraging new technologies to expand access to mental health services

Technology has opened a new frontier in mental health support and data collection, offering new ways to expand access to mental health services across society.

Mobile devices like cell phones, smartphones, and tablets are offering users, doctors, and researchers new ways to access help, monitor progress, and increase understanding of mental well-being (National Institute of Mental Health (NIMH) Site, 2022). There are up to 20,000 apps for mental health on the market, and many of the most popular have millions of users (Pappas, 2020).

Digital care options through teletherapy and new apps have seen explosive growth during the COVID-19 pandemic (Doraiswamy et al., 2021). They offer alternative service delivery models that help overcome obstacles that hinder access to care, such as transportation barriers, stigma associated with visiting mental health clinics, personnel shortages, and high costs. These platforms, especially in mobile formats, can offer remote screening, diagnosis, monitoring, and treatment; facilitate remote training for nonspecialist healthcare workers; and enhance online peer-to-peer support and self-care (Staglin & Herrman, 2021).

Other tech innovations benefiting families include prescription video games that challenge children ages 8 to 12 with Attention-Deficit/Hyperactivity Disorder to focus on multiple tasks simultaneously within the game environment (Asar, 2020). Also, artificial intelligence (AI) chatbots that can help patients practice cognitive behavioral therapy (CBT) to become aware of inaccurate or negative thinking so that they can view challenging situations clearly and respond more effectively (Makadia, 2021).

Online symptom tracking apps allow users to have a greater control over symptoms and an opportunity to learn how to self-manage them, but can be used to prompt patients to share data daily which are analyzed with an AI algorithm to identify patterns and alert providers in real time of any warning signs among their patients (Weinberg, 2020).

Public and private sectors in several countries have developed online mental health support and services such as Togetherall, ReachOut, 7 Cups, and UCLA STAND, often codesigned and advised by community peers, that have become trusted sources of help and referral for young people, their families, and others in distress, to cope with mental ill health (Fox, 2021).

Several initiatives have been launched to deal with scarcity or lack of providers needed to reach people in need (Hijazi & Gornitzka, 2021). For example, the World Health Organization Academy aims to provide digital learning to health workers across the world on a range of topics, including mental health; EMPOWER, a Harvard Medical School initiative, provides digital training to support delivery of psychosocial interventions; and UNICEF's Caring for the Caregiver program targets community workers using online and in-person training.

Going forward

Integrated health and wellness programs in the workplace that focus on the physical and mental health of workers and their dependents are a "good buy" for any organization and business. They generate significant returns in terms of health and economic gains. Investing in these programs can contribute to developing and strengthening new work arrangements postpandemic and contribute to accelerate the progressive realization of universal health coverage by engaging and leveraging resources and know-how from firms and businesses for the benefit of workers and families alike.

As evidenced during COVID-19, leveraging new technologies can also help businesses expand telehealth options to improve provision of mental health services for employees and managers alike by the use of online tools for consultation and counseling appointments, covering an entire course of therapy or multiple sessions to ensure continuity of care. Also, online learning courses can help staff and those who manage staff understand the implications of mental health issues and how best to support one another in times of need.

Nurturing the development of healthy work environments that promote the physical and mental well-being of employees is not only the right thing to do, but it is a smart economic strategy to enhance human and social capital and to improve productivity and competitiveness of firms, and help national economies combat poverty and achieve sustainable development.

Indeed, business leaders and senior managers need to understand that the adoption of this "shared value" course of action, reflected in corporate decisions to support and sustainable fund health and wellness programs, integrating physical and mental health activities, is "not corporate social responsibility, philanthropy, or even sustainability, but a new way to achieve economic success" in the "new normal" postpandemic (Harvard Business Review Site, 2022).

References

Appelbaum, E., Bailey, T., Berg, P., & Kalleberg, A. L. (2000). *Manufacturing advantage: Why high-performance work systems pay off.* Ithaca, NY: Cornell University Press. Accessed at: https://eric.ed.gov/?id=ED456338.

Asar, A. (2020). *Five tech innovations that changed mental health in 2020.* Forbes. Accessed at: https://www.forbes.com/sites/forbestechcouncil/2020/11/25/five-tech-innovations-that-changed-mental-health-in-2020/?sh=6b0260b81e9c.

Bell Canada site. (2022). *Bell's let talk.* Accessed at: https://letstalk.bell.ca/en/.

Bevan, S. (2021). *Good work, high performance and productivity.* European HRD Circle Post. Accessed at: https://european-hrd-circle.org/stephen-bevan-good-work-high-performance-and-productivity/.

Black, C. D., & Frost, D. (2011). *Health at work—An independent review of sickness absence.* The Stationery.

Business Group on Health site. (2021). *Mental health and emotional well-being: Key insights from May 2021 industry benchmarking calls.* Accessed at: https://www.businessgrouphealth.org/en/resources/key-insights-from-may-2021-industry-benchmarking-calls.

Cascio, W., & Montealegre, R. (2016). How technology is changing work and organizations. *Annual Review of Organizational Psychology and Organizational Behavior*, *3*, 349–375. https://doi.org/10.1146/annurev-orgpsych-041015-062352.

Centre for Mental Health. (2011). *Managing presenteeism: A discussion paper.* Cited at Hafner, M, Van Stolk, C, Saunders, CL, Krapels, J, Baruch, B. 2015. "Health, wellbeing and productivity in the workplace: A Britain's Healthiest Company summary report". Santa Monica, CA: RAND Corporation. Accessed at: https://www.rand.org/pubs/research_reports/RR1084.html.

Chugh, A. (2021). *What is 'The Great Resignation'? An expert explains.* World Economic Forum Blog. Accessed at: https://www.weforum.org/agenda/2021/11/what-is-the-great-resignation-and-what-can-we-learn-from-it/.

Cohen, A. (2021). *How to quit your job in the great post-pandemic resignation boom.* Bloomberg Businessweek. Accessed at: https://www.bloomberg.com/news/articles/2021-05-10/quit-your-job-how-to-resign-after-covid-pandemic.

Doraiswamy, P. M., Fox, C., & Gordon, J. (2021). *How the pandemic is changing mental health.* Scientific American. Accessed at: https://www.scientificamerican.com/article/how-the-pandemic-is-changing-mental-health/.

Fox, C. (2021). *These are the top 10 innovations in mental health.* World Economic Forum. Accessed at: https://www.weforum.org/agenda/2021/09/these-are-the-top-10-innovations-tackling-mental-ill-health.

GBD 2019 Diseases and Injuries Collaborators. (2021). Global prevalence and burden of depressive and anxiety disorders in 204 countries and territories in 2020 due to the COVID-19 pandemic. *The Lancet*, *398*(10312), P1700–P1712. Accessed at: https://www.thelancet.com/journals/lancet/article/PIIS0140-6736(21)02143-7/fulltext.

Goplerud, E., Hodge, S., & Benham, T. A. (2017). A substance use cost calculator for US employers with an emphasis on prescription pain medication misuse. *Journal of Occupational and Environmental Medicine*, *59*(11), 1063–1071. https://doi.org/10.1097/JOM.0000000000001157.

Hafner, M., Van Stolk, C., Saunders, C. L., Krapels, J., & Baruch, B. (2015). *Health, wellbeing and productivity in the workplace: A Britain's Healthiest Company summary report.* Santa Monica, CA: RAND Corporation. Accessed at: https://www.rand.org/pubs/research_reports/RR1084.html.

Harvard Business Review site. (2022). *Creating shared value.* Accessed at: https://www.isc.hbs.edu/creating-shared-value/Pages/default.aspx.

Health Enhancement Research Organization (HERO). (2020). *Employee mental health and well-being: Emerging best practices and case study examples.* Report prepared by the Workplace Performance Study Committee (WP) and their Workplace Mental Health and Well-Being Workgroup.

Hijazi, Z., & Gornitzka, C. (2021). *Digital tools help train providers to fill gaps in mental health care.* Scientific American. Accessed at: https://www.scientificamerican.com/article/digital-tools-help-train-providers-to-fill-gaps-in-mental.

Kangasniemi, A., Maxwell, L., & Sereneo, M. (2019). *The ROI in workplace mental health programs: Good for people, good for business.* Deloitte Insights. Accessed at: https://www2.deloitte.com/us/en/insights/topics/talent/workplace-mental-health-programs-worker-productivity.html.

Kleinman, A., Lockwood Estrin, G., Usmani, S., Chisholm, D., Marquez, P. V., Evans, T. G., et al. (2016). Time for mental health to come out of the shadows. *The Lancet, 387*(10035), P2274–P2275. https://doi.org/10.1016/S0140-6736(16)30655-9. Accessed at: https://www.thelancet.com/journals/lancet/article/PIIS0140-6736(16)30655-9/fulltext.

Makadia, M. (2021). *How AI Chatbots can help you beat the competition.* Business 2 Community site. Accessed at: https://www.business2community.com/marketing-automation/how-ai-chatbots-can-help-you-beat-the-competition-02410881.

Malesic, J. (2022). *What it would take to make us love our jobs again.* Vox.

Marquez, P. V. (2020b). *Let's not forget about Mental Health during COVID-19!.* Blog. Accessed at: http://www.pvmarquez.com/digital%20media_impacts.

Marquez, P. V. (2020c). *Addressing alcohol use disorders through taxation: A public health and economic win-win in the time of COVID-19.* Blog. Accessed at: http://www.pvmarquez.com/alcoholuse.

Marquez, P. V. (2021). *Revisiting the mental health impact of the COVID-19 pandemic and a way forward.* Blog. Accessed at: http://www.pvmarquez.com/revisitingmentalhealth.

Marquez, P. V., Herrera, S., & Butron Calderon, L. (2020). *Have South and Central America become the new coronavirus (COVID-19) epicenter?.* World Bank Group Investing in Health Blog. Accessed at: https://blogs.worldbank.org/health/have-south-and-central-america-become-new-coronavirus-covid-19-epicenter.

Matheson, F. I., Smith, K. L., Moineddin, R., Dunn, J. R., & Glazier, R. H. (2014). Mental health status and gender as risk factors for onset of physical illness over 10 years. *Journal of Epidemiology and Community Health, 68*(1), 64–70. https://doi.org/10.1136/jech-2013-202838. PMID: 24014840; PMCID: PMC3888625. Accessed at: https://europepmc.org/article/PMC/3888625.

Mattke, S., Kapinos, K. A., Caloyeras, J. P., Taylor, E. A., Batorsky, B. S., Liu, H. H., Van Busum, K. R., & Newberry, S. (2015). Workplace wellness programs: Services offered, participation, and incentives. *RAND Health Quarterly, 5*(2), 7. Accessed at: https://www.rand.org/pubs/periodicals/health-quarterly/issues/v5/n2/07.html.

Ministry of Social Affairs and Health, Finland. (2020). *National mental health strategy and program for suicide prevention 2020–2030.* Helsinki. Accessed at: http://urn.fi/URN:ISBN:978-952-00-5401-4.

National Institute of Mental Health (NIMH) Site. (2022). *Technology and the future of mental health treatment.* Accessed at: https://www.nimh.nih.gov/health/topics/technology-and-the-future-of-mental-health-treatment#part_2437.

National Institute on Drug Abuse (NIDA) Site. (2020). Accessed at: https://nida.nih.gov/about-nida/frequently-asked-questions#:~:text=The%20National%20Clearinghouse%20for%20Alcohol%20and%20Drug%20Information,information%20and%20materials%20about%20alcohol%20and%20other%20drugs.?msclkid=113e166eb8fe11ec9f4811648300ca50.

North, A. (2022). *How to heal our national exhaustion. Americans have burnout on top of burnout on top of burnout. Now what?.* Vox. Accessed at: https://www.vox.com/the-goods/22902543/covid-omicron-pandemic-tired-burnout.

OECD. (2012). *Sick on the Job? Myths and realities about mental health and work, mental health and work.* Paris: OECD Publishing. https://doi.org/10.1787/9789264124523-en.

OECD. (2021). *Fitter minds, fitter jobs: From awareness to change in integrated mental health, skills and work policies, mental health and work.* Paris: OECD Publishing. https://doi.org/10.1787/a0815d0f-en.

Pappas, S. (2020). *Providing care in innovative ways.* American Psychological Association. Accessed at: https://www.apa.org/monitor/2020/01/cover-trends-innovative-ways.

Parker, K., & Menasce Horowitz, J. (2022). *Majority of workers who quit a job in 2021 cite low pay, no opportunities for advancement, feeling disrespected.* Pew Research Center. Accessed at: https://www.pewresearch.org/fact-tank/2022/03/09/majority-of-workers-who-quit-a-job-in-2021-cite-low-pay-no-opportunities-for-advancement-feeling-disrespected/?msclkid=df67631 9b8e411ec8682ff173b8eb842.

Parker, K., Menasce Horowitz, J., & Minkin, R. (2022). *COVID-19 pandemic continues to reshape work in America.* Pew Research Center. Accessed at: https://www.pewresearch.org/social-trends/2022/02/16/covid-19-pandemic-continues-to-reshape-work-in-america/.

Patel, V. (2007). Alcohol use and mental health in developing countries. *Annals of Epidemiology, 17*(e5, Supplement), S87–S92. Accessed at: https://www.sciencedirect.com/science/article/abs/pii/S104727970700021X.

Porter, M. E., & Kramer, M. R. (2011). *Creating Shared Value. How to reinvent capitalism—And unleash a wave of innovation and growth.* Harvard Business Review. Accessed at: https://hbr.org/2011/01/the-big-idea-creating-shared-value.

Prince, M., Patel, V., Saxena, S., Maj, M., Maselko, J., Phillips, M. R., & Rahman, A. (2007). No health without mental health. *The Lancet, 370*(9590), 859–877. https://doi.org/10.1016/S0140-6736(07)61238-0. PMID: 17804063. Accessed at: https://pubmed.ncbi.nlm.nih.gov/17804063/.

Rehm, J., & Shield, K. D. (2019). Global burden of disease and the impact of mental and addictive disorders. *Current Psychiatry Reports, 21*(2), 10. https://doi.org/10.1007/s11920-019-0997-0. PMID: 30729322.

Ritchie, H., & Roser, M. (2018a). *Alcohol consumption.* OurWorldInData.org. Retrieved from: https://ourworldindata.org/alcohol-consumption.

Ritchie, H., & Roser, M. (2018b). *Opioids, cocaine, cannabis and illicit drugs.* OurWorldInData.org. Retrieved from: https://ourworldindata.org/illicit-drug-use.

Schwab, K. (2017). *The fourth industrial revolution.* Geneva: World Economic Forum. Accessed at: https://www.weforum.org/about/the-fourth-industrial-revolution-by-klaus-schwab.

Staglin, B., & Herrman, H. (2021). *Inventive ways of delivering mental health care thrive during the pandemic.* Scientific American. Accessed at: https://www.scientificamerican.com/article/inventive-ways-of-delivering-mental-health-care-thrive-during-the-pandemic/.

UC Press Blog. (2021). *Q&A with Erin Cech, author of The Trouble With Passion.* University of California Press. Accessed at: https://www.ucpress.edu/blog/57580/qa-with-erin-cech-author-of-the-trouble-with-passion/.

UKEssays. (2018). *Impact of information technology on work organizations.* Accessed at: https://www.ukessays.com/essays/management/the-impact-of-information-technology-on-work-organisations-management-essay.php?vref=1.

US Office of Personnel Management. (2022). *Alcoholism in the workplace.* Accessed at: https://www.opm.gov/policy-data-oversight/worklife/reference-materials/alcoholism-in-the-workplace-a-handbook-for-supervisors/.

Weinberg, J. (2020). *Are symptom tracker apps the future of healthcare?.* Raconteur.net. Accessed at: https://www.raconteur.net/healthcare/symptom-tracker-apps/.

Wells, C., Ballentine, C., & Cachero, P. (2022). *Pay me more or I quit: Workers play risky game with their bosses.* Bloomberg. Accessed at: https://www.bloomberg.com/news/articles/2022-02-04/should-i-quit-my-job-for-more-money-great-resignation-has-workers-taking-risks?utm_campaign=news&utm_medium=bd&utm_source=applenews.

Whiteford, H. A., Degenhardt, L., Rehm, J., Baxter, A. J., Ferrari, A. J., Erskine, H. E., Charlson, F. J., Norman, R. E., Flaxman, A. D., Johns, N., Burstein, R., Murray, C. J., & Vos, T. (2013). Global burden of disease attributable to mental and substance use disorders: Findings from the Global Burden of Disease Study 2010. *The Lancet, 382*(9904), 1575–1586. https://doi.org/10.1016/S0140-6736(13)61611-6. Epub 2013 Aug 29. PMID: 23993280.

World Bank Group. (2022a). *Open learning campus. Series mental health at the World Bank Group.* Accessed at: https://olc.worldbank.org/content/mental-health-world-bank-group-0.

World Bank Group. (2022b). *Better together: A mental health and well-being strategy for the World Bank Group.* Washington, DC: WBG.

World Economic Forum, in Collaboration With FTI Consulting. (2013). *The workplace wellness alliance making the right investment: Employee health and the power of metrics.* Geneva: WEF. Accessed at: https://www3.weforum.org/docs/WEF_HE_WorkplaceWellnessAlliance_Report_2013.pdf.

World Health Organization site. (2018). *Mental health: Strengthening our response*. Accessed at: https://www.who.int/news-room/fact-sheets/detail/mental-health-strengthening-our-response.

Further reading

Bloom, D. E., Cafiero, E. T., Jané-Llopis, E., Abrahams-Gessel, S., Bloom, L. R., Fathima, S., Feigl, A. B., Gaziano, T., Mowafi, M., Pandya, A., Prettner, K., Rosenberg, L., Seligman, B., Stein, A. Z., & Weinstein, C. (2011). *The global economic burden of noncommunicable diseases*. Geneva: World Economic Forum. Accessed at: https://www3.weforum.org/docs/WEF_Harvard_HE_GlobalEconomicBurdenNonCommunicableDiseases_2011.pdf.

CHAPTER 62

The crucial role of nurses in telehealth

Jennifer Knox
Woodward Academy, College Park, GA, United States

Introduction

Nurses are the largest and most intimately engaged population within the healthcare field and their inclusion to develop emerging telehealth models provides more direct and informed routes to support individuals' needs who are being served. Nurses play an especially important role during emergencies such as the COVID-19 pandemic, given the low resources for qualified medical personnel worldwide and particularly in low-resource settings. As the globe grapples with how to best meet the growing health and mental health-related needs made more urgent by the devastating COVID-19 pandemic, remote access to nurses and telehealth solutions have the potential to reach beyond the sphere of health to meet the growing needs related to resilience cultivation.

To address these issues, the International Society for Telemedicine and eHealth's Telenursing (ISfTeH) Working Group hosted a webinar on May 17, 2022, featuring four speakers who highlighted the importance and diversity of nurse involvement in shaping the effectiveness of telehealth solutions to healthcare worldwide, including Finland; Nigeria; and the collective region of Europe, Middle East, and Africa. An attendee question section and a moderated discussion followed the presentations (International Society for Telemedicine and eHealth, 2022).

The four speakers were the vice president of the ISfTeH and secretary of the Finnish Society of Telemedicine and eHealth, Pirkko Kouri, Finland's Telehealth Strategy Built for Nurses. The director of Nursing EMEA for International SOS, Kerryn McGowran, described examples of worldwide applications of digital health by her organization's Global Medical Assistance program. Two speakers spoke about the opportunities and challenges of building Nigeria's telenursing ecosystem, namely, the CEO of the Institute for Nursing Research in Nigeria, Oluwadamilare Akingbade, and the CEO-founder of Care Code Digital Health Hub, Daniel Ayinla. The event was moderated by Claudia Bartz, cochair of the ISfTeH Telenursing Working Group.

Resilient Health. https://doi.org/10.1016/B978-0-443-18529-8.00062-7

Finland's telehealth strategy built for nurses

The opening presentation was given by Pirkko Kouri, vice president of the ISfTeH and secretary of the Finnish Society of Telemedicine and eHealth, on "Finnish Nurses Association's Digital Social and Health Services Strategy" (Ahonen et al., 2021). She stressed the importance of nurses being part of the national development process since they represent the largest group among healthcare professionals, and how this strategy is evolving as society changes. As she said, "It's crucial that [the nurse voice] is heard when we are developing new systems and we have to actively follow what is happening around and participate in issues concerning healthcare."

Statistics in Finland reveal that the digital divide is narrowing due to increased internet use, the widespread use of coronavirus app, and significant growth in real-time engagement and transactions in healthcare centers. Also, citizens are expecting digital services.

Kouri described that historically, in 2015, Finland engaged in a national debate on the role of nurses in developing and running digital services. After 5 years, the Finnish Nursing Association, along with experts from education and healthcare development, created a national strategy under the name "Digital, Social and Health Care Services Strategy in Finland" (Ahonen et al., 2021). The new strategy consists of six areas and supports the participation of clients with joint decision-making between a nurse and client. The areas are as follows:

1. Digitalization supports client participation—emphasizes the role of the client as an acting part of their care
2. Digital health and social services as part of nurses' work—nurses use and support client's use of an expanding range of digital services
3. Security and ethics in the digital environment—nurses cultivate ethical competencies which inform decision-making to support action on behalf of client's best interest
4. Digital health services and skills—nurse education includes postgraduate level development and digital health education, a nationwide project called Sote Peda 24/7 defined 12 digital competencies for nurse education (Ahonen et al., 2021)
5. Digital social and health services and management—nurse managers play a key role in developing healthcare services, supporting skills building and inform decision-making
6. Research and development of digitalization of health services—nurses provide research and evaluation data, participate as experts in development of information systems and participate in national and international networks to advance the work

Fig. 1 shows the 12 digital competencies of Sote Peda 24/7 for nurse education.

Kouri underscored the important role of research and development, saying, "I think [research] is important because the digitalization is a process of transforming practices, internal processes and services through the use of technology and research in all phases is important and helps develop and understand new solutions when new technologies emerge." She noted the

Fig. 1

Finland's national Digital Health Care and Social Welfare Competencies, Sote Peda 24/7. *From: Sote Peda 24/7. Ahonen, O., Id-Korhonen, A., Juvonen, S., Koivisto, J. M., Kuosa, P., Pekkarinen, V., … & Vuolas, M. (2021). Improve your digital health care and social welfare competencies 24/7!.*

importance of publishing nurse-led studies and follow-up studies related to telenursing and digital nursing as promoted by Claudia C. Bartz, a telehealth nursing advocate who was the coordinator for eHealth programs for the International Council of Nurses for 10 years. The digital tool-rich landscape makes it is easier for nurses to share information and best practices as participants in both national and international networks, a good example of this being the Finnish Nursing Association webpage which gives nurses the opportunity to provide feedback that shapes the strategy woven into nurse education as well as the promotion of digital health services.

Actualization of a telehealth strategy

The next speaker, Kerryn McGowran, director of Nursing EMEA (Europe, Middle East, and Africa) for International SOS, described her organization's global medical assistance program, presented on the actualization of a comprehensive global telehealth strategy. The model integrates medical and security services. The international organization works on behalf of large multinational clients to offer medical assistance and advice to their members worldwide.

The organization has 12,000 employees, about half are other medical professionals including nurses, who make up medical teams. Medical teams collaborate to provide telehealth solutions to subscribed members affiliated with client organizations. When members fall ill by accident or injury, they call for medical assistance, and nurses respond, working in collaboration with doctors and logistics experts to solve a variety of the presented complex problems with telehealth and security solutions. There is an annual rate of over four million calls to clients and members. Communications and outreach are paramount to the process.

The structure of the organization involves 27 regional assistant centers as hubs located around the world to support a region. For example, the London assistant center supports all of Scandinavia, the United Kingdom, and the Netherlands; the Paris hub supports West Africa; and the Beijing center looks after China. Hubs are staffed with full medical security and operational teams to triage client and member calls and provide relevant services. McGowan stated, "You see, all these different centers then work closely together as a kind of hive mind that have different levels of expertise and knowledge about the medical standards of care and the healthcare needs of those regions they cover."

Additionally, 64 medical clinics were built, and staffed by local and international personnel in areas where healthcare is more variable, making on-the-ground emergency services available to larger client populations. These clinics collaborate with the regional hub to determine the best course of action. Linguistic needs are met in that about 99 languages are spoken by team members; for example, in the London regional assistant center, 56 different languages are spoken which makes up a unique multicultural network. Security consultants are also provided to members who move to different locations.

The nurses' role includes a broad scope of services, but are hired with particular experience in emergency and intensive unit care, familiar with triage skills, and corporate levels of service. Driven by an international mindset, most nurses manage patient care over the phone from an office space. The nurses who continue to work on the ground in company-run clinical settings are up to date with hands-on clinical skills. Nurse escorts are also made available to evacuate and accompany a patient if needed, on commercial airlines.

Day-to-day roles include telediagnostics, teleassistance, teleconsulting, and specialist telemedical consultations. Medical and risk assessment skills are essential, due to the variety and complexity of cases. Referrals are evidence based with quality assurance. Local care is benchmarked against international standards and the nurses can access internal resources, maps, and databases to ensure the best care and referral for each case.

Fig. 2 shows the roles included in nurse assistance with medical advice and collaboration with doctors being most common.

During the pandemic, the organization adopted the model due to the COVID-19 spread and increased need for specialists to mitigate issues. In a pertinent case study, McGowan described a patient living on a remote island who had been given a diagnosis of end-stage lung fibrosis needing a lung transplant which would require a high-risk evacuation. The medical team

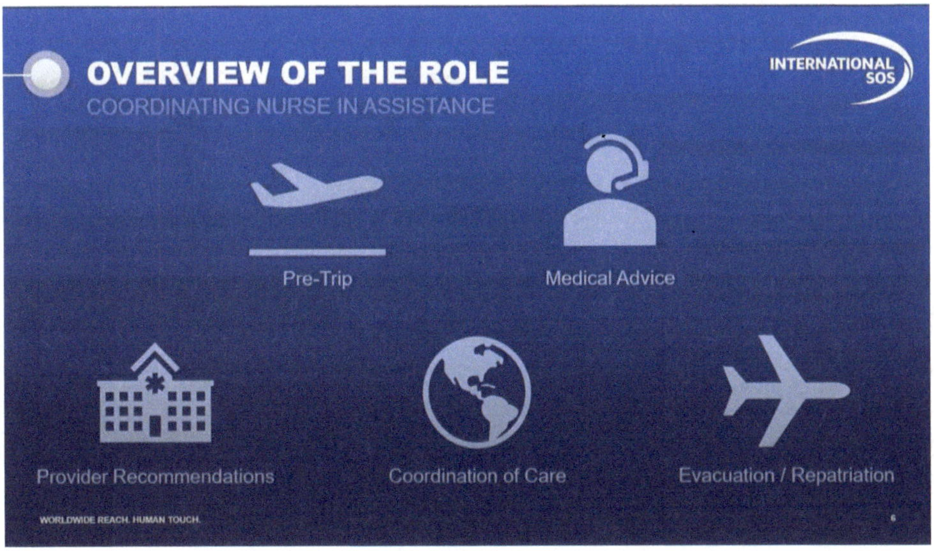

Fig. 2
The overview of the role of nurse assistance at International SOS.

received a second opinion from a respiratory specialist in the United Kingdom and thoracic radiologists, who confirmed a new diagnosis that did not require a lung transplant. Instead, follow-up health guidance was provided remotely to the patient. McGowan concluded that the larger telehealth community has much to learn from the experiences of this large-scale multinational model.

Nigeria's telehealth opportunities

The final two speakers, Oluwadamilare Akingbade and Daniel Ayinla, provided an overview of the current state of Nigeria's nursing education and offered compelling strategies to build Nigeria's telenursing ecosystem based on existing opportunities and challenges.

Fig. 3 depicts Nigeria's current telenursing activities, including interactive activities, training, and administration.

Daniel Ayinla, CEO-founder of Care Code Digital Health Hub, spoke about the present state of nursing in Nigeria. He said that more nurses and healthcare professionals are understanding what telehealth is, and using mobile phones and web platforms to provide care to patients in health education and health advice in the areas of maternal health, chronic disease management, diabetes, and hypertension. This is complemented by the uptick in Nigerians themselves using mobile devices, according to news reports. The crisis of COVID-19 popularized telehealth and telemedicine, which opened doors for start-up tech companies. Due to this current context,

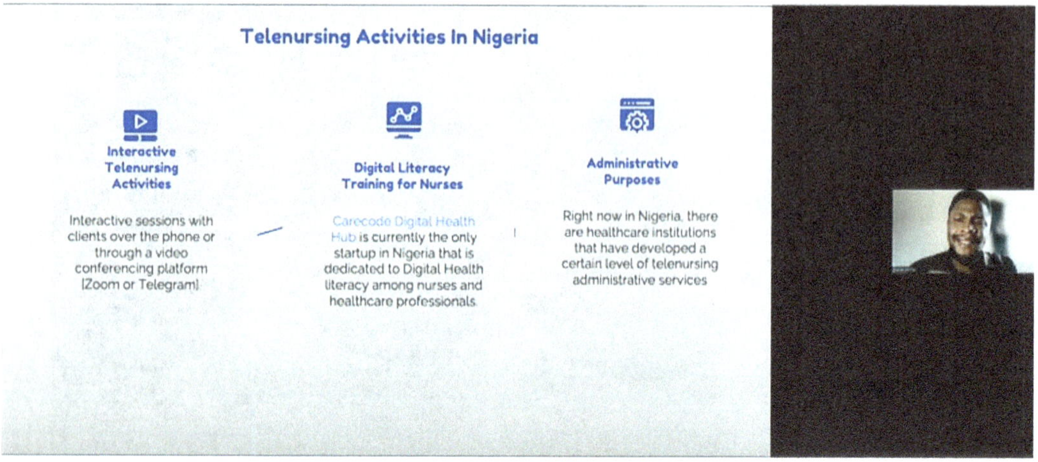

Fig. 3
Daniel Ayinla presented an overview of the telenursing activities in Nigeria.

Akingbade strongly encouraged more investment in the area of telehealth, stating, "We [in Nigeria] have about 200 million people, that's a very low population, and about 60% of them are young. And that means they are connected to the internet, that's a very good opportunity for the system in Nigeria." The area of nursing, he said, is divided into three parts:

1. Interactive diagnosing activities. Clients call numbers and require medical advice be sent to them over the phone.
2. Digital literacy training for nurses and healthcare professionals. This is in development phases, led by the start-up Care Code Digital Health Hub since 2019, aiming to build a community of nurses and healthcare professionals with advanced understanding of digital solutions to create effective telehealth strategies.
3. Administrative opportunities. Monthly meetings are held for patient reviews and planning the launch of a data health conference for experts in the areas of data health to share entrepreneurial and business ideas, with active engagement of nurses. As a follow-up to the conference, a 90-day career training provided health professionals with the core skills necessary to build telehealth systems.

Fig. 4 depicts the final presentation by Oluwadamilare Akingbade, CEO of the Institute for Nursing Research in Nigeria, including lack of access, trust, and skills.

The final presenter, Oluwadamilare Akingbade, CEO of the Institute for Nursing Research in Nigeria and PhD candidate at the Chinese University of Hong Kong, described telenursing research about telenursing in rural populations conducted in Nigeria in 2018 and published in the *Journal of the International Society for Telemedicine and eHealth*. Findings showed that women respondents, 91% of whom possessed mobile phones, embraced mobile phones as a potentially effective vehicle for maternal care and that the percentages of users increased over time (Odetola et al., 2018).

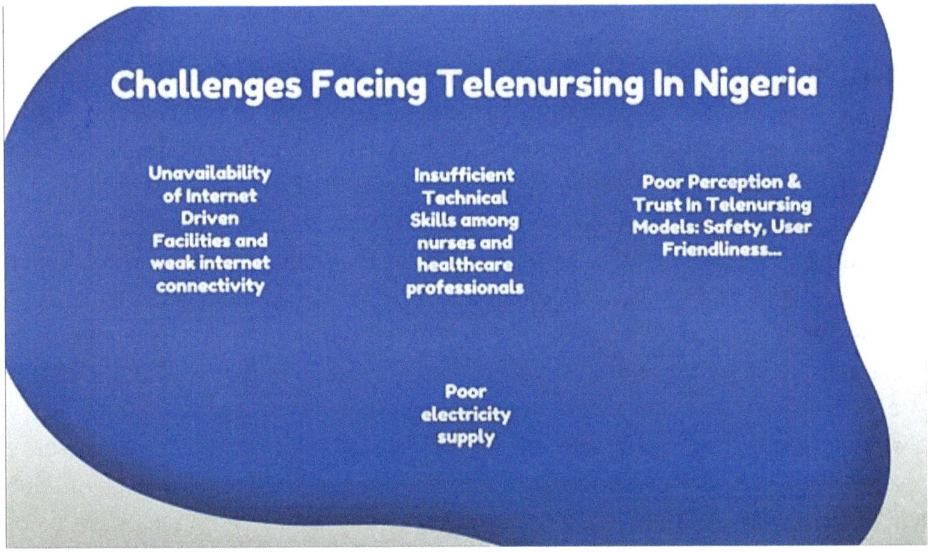

Fig. 4
Oluwadamilare Akingbade shares the challenges facing telenursing in Nigeria.

Akingbade reported about another study, conducted in 2017 and published in the *Texila International Journal of Nursing*, which found that 84.8% of nurse participants in Plateau State, Nigeria, were interested in using telenursing for improving communication and collaboration (Dangyang, 2017). Oluwadamilare Akingbade has developed a mobile application called CENBA to deliver psychoeducational support for Nigerian women receiving breast cancer chemotherapy and plans an upcoming randomized control trial to test the intervention's effectiveness (Akingbade, Adediran, et al., 2022; Akingbade, Nguyen, & Chow, 2022).

Challenges facing telenursing in Nigeria, outlined by Akingbade, include: availability to internet facilities, weak connectivity, insufficient technical skills among nurses and healthcare providers, poor perceptions and trust, poor electricity supply, and a nursing shortage plaguing the country.

Yet, opportunities for telenursing in Nigeria include: leveraging the young population of Nigeria who are concerned and innovative population, the collective tribe intelligence system" with an increasing number of start-ups being founded yearly, widespread use of mobile and internet devices, and a robust tech ecosystem.

Offering recommendations, Akingbade said that "telenursing startups should be developed in Nigeria, healthcare facilities should start teaching nurses and healthcare professionals about digitalis, more investment is needed in telenursing, and leaders [should encourage] members to do the same."

The questions and answer section provided attendees with an opportunity to expand the conversation to include their mental health concerns, invite interdisciplinary collaboration at conferences and upcoming webinars, and invite everyone to engage with the telenursing working group hosted by ISfTeH.

Conclusion and future directions

In a webinar as part of their ongoing series on telemedicine and eHealth and services to members, the International Society for Telemedicine and eHealth (ISfTeH) provided diverse perspectives on the topic of nurse involvement in global telehealth solutions worldwide. The three presentations highlighted Finland's comprehensive nationwide strategy which evolved with nurse feedback and societal changes, an application of telehealth solutions by medical teams including nurses who work within a known international organization (SOS) committed to providing security and medical solutions for clients and members, and challenges and opportunities in Nigeria to create evidence-based pathways to expand medical information access via mobile phone and nurse education programs including digital solutions. The future of cost-effective, efficient, and accessible healthcare must highlight and expand the voice of nurses who constitute the largest community in the field of medicine and honor nurses' capacity to access and fulfill the needs of people whom they serve. Models serving populations in Finland, Nigeria, and other world regions can meet people's essential needs in this ongoing period of recovery from the COVID-19 pandemic and its evolving variants plaguing nations.

References

Ahonen, O., Id-Korhonen, A., Juvonen, S., Koivisto, J. M., Kuosa, P., Pekkarinen, V., … Vuolas, M. (2021). *Improve your digital health care and social welfare competencies 24/7!*.

Akingbade, O., Adediran, V., Somoye, I. E., Alade, A. S., & Chow, K. M. (2022). *Perception of mhealth intervention for psychoeducational support among Nigerian women receiving chemotherapy for breast cancer: A qualitative study. ORCID*. https://doi.org/10.21203/rs.3.rs-1237217/v1.

Akingbade, O., Nguyen, K. T., & Chow, K. M. (2022). *Effect of mHealth interventions on psychological issues experienced by women diagnosed with breast cancer receiving chemotherapy: A systematic review and meta-analysis*. https://doi.org/10.1101/2022.02.17.22271140.

Dangyang, S. Y. (2017). Knowledge and practice of telenursing among nurses in Plateau State, Nigeria. *Texila International Journal of Nursing*, *3*(2), 1–12. https://doi.org/10.21522/tijnr.2015.03.02.art001.

International Society for Telemedicine & eHealth. (2022). *Webinar—ISfTeH Telenursing Working Group—18 May 2022—Nurse involvement in telehealth*. Retrieved March 18, 2023, from https://youtube.com/watch?v=OiVYGc_2WoA&si=EnSIkaIECMiOmarE.

Odetola, T. D., Ayamolowo, L. B., & Ayamolowo, S. J. (2018). Childbearing women's perception about the use of mhealth for maternal health information in rural communities, Ile-Ife, Nigeria. *Journal of the International Society for Telemedicine and eHealth*, *6*. e9-1.

Further reading

Bartz, C. C., Kouri, P., Thouvenot, V., McGowan, K., Griffith, M. Y., & Kwankam, Y. (2017). International society for telemedicine & ehealth. *Journal of Nursing Scholarship*, *2*(15), 4–6.

Värri, A. O., Kinnunen, U. M., Pöyry-Lassila, P., & Ahonen, O. (2019). The national SotePeda 24/7 project develops future professional competencies for the digital health and social care sector in Finland. *Finnish Journal of eHealth and eWelfare*, *11*(3), 232–235.

CHAPTER 63

The top 10 new technologies in dentistry☆

123Dentist
Headquarters of 123Dentist, Burnaby, BC, Canada

The field of dentistry is constantly evolving and changing to best suit patient needs. 123Dentist is always on the lookout for new and emerging technology to make our patients' experience as comfortable as possible. Here is just a few of the different technologies that 123Dentist practices use to ensure we provide world-class service to all our patients:

1. *VELscope*—VELscope is a special type of light that a dentist will shine in a patient's mouth to detect any abnormalities. This new form of technology is effectively used to detect early forms of cancer or disease during an oral cancer screening.
2. *Digital X-rays*—Digital X-rays are faster and contain less radiation than traditional X-rays. When a patient has a digital X-ray done, the image appears on the computer in a matter of seconds. Your dentist can then zoom into the image to better assess and educate the patient regarding their oral health. Digital X-rays are also less harmful for patients as they contain up to 90% less radiation compared to traditional X-rays.
3. *Invisalign*—Invisalign are clear, practically invisible braces that can gently straighten your teeth. They provide an effective and comfortable way to straighten your smile without the inconvenience of wearing heavy, metal braces. Invisalign braces are easy to take out for cleaning and don't require a restriction on what type of foods to eat. They get the job done in less time with less hassle.
4. *Laser dentistry*—Lasers are used to improve efficiency and eliminate discomfort in a number of dental procedures, including filling cavities, reducing tooth sensitivity, getting rid of tumors, and whitening. Laser dentistry is fast, painless, and can effectively eliminate any form of bacteria during the procedure to avoid any further complications or problems.
5. *Dental implants*—Dental implants are screw replacements for the root portion of missing teeth. Implants are used to restore healthy smiles when patients have missing teeth. Implants are effective because they can replace missing teeth, while still giving the patient the feel and look of a natural tooth.

☆ This article is reprinted from 123dentist, with permission; original publication date of 2012-08-14.

Resilient Health. https://doi.org/10.1016/B978-0-443-18529-8.00063-9

6. *HealOzone*—HealOzone is a fast, easy, and painless way to eliminate tooth decay. HealOzone is effective because it contains ozone (O3), which is a common, natural gas that effectively kills bacteria and fungus. HealOzone is a great tool to detect and get rid of any early signs of tooth decay before it progresses to a more advanced stage.
7. *DIAGNOdent*—DIAGNOdent is a new, state-of-the-art device that dentists use to detect cavities that are hiding in places regular X-rays can't find. It ensures that your mouth has been thoroughly checked for early signs of cavities to avoid you having to spend more in the future if the cavities progress and expand.
8. *Intraoral camera*—The intraoral camera is a tool that dentists use to gain precise and well-defined pictures of hard to see places within a patient's mouth. The camera also allows dentist to show these images to patients while assessing and educating the patient's needs. This new form of technology allows dentist to conduct a thorough checkup of your mouth and better assess their patient's oral care needs.
9. *Zoom! Whitening*—Zoom! Whitening is a new, state-of-the-art whitening treatment that gives our patients fast and easy results. In just one appointment, Zoom! Whitening can make a significant change in a person's smile and can make their teeth up to eight shades whiter.
10. *Nitrous oxide and (intravenous) IV sedation*—Nitrous oxide, also known as laughing gas, can calm a patient down to a point that they are relaxed but can still interact with their dentist. On the other hand, IV sedation puts a patient to sleep completely so that they are unaware of what happened during their dental session. This is usually recommended for patients who are fearful of going to the dentist or for procedures that are painful such as wisdom teeth extraction.

CHAPTER 64

Gender disparities faced by female doctors in Pakistan: Training but turning into trophy wives

Fatima Rizwan

Department of Clinical Psychology, Teachers College, Columbia University, New York, NY, United States; International Association of Applied Psychology, Indianapolis, IN, United States

Introduction

As of 2020, women represented an overwhelming majority—approximately 80 percent—of the student body in medical schools of Pakistan. However, less than 50% of these female students reportedly continued to practice after graduation (Mohsin & Syed, 2020; Junaidi, 2014). Another report states that over 75% of female students did not practice postgraduation (Kazim, 2013). The ratio of female to male students—75% to 25%—in medical schools and colleges predicts the expansion of female representation in the medical field; however, the female workforce in medicine is largely absent due to many intersecting sociocultural, organizational, and individual factors (Mohsin & Syed, 2020; Zakaria, 2013).

According to the president of the Pakistan Medical and Dental Council (PMDC), "Pakistan is producing around 14,000 doctors per year, out of whom 70% are women. There are around 50,000 to 60,000 medical practitioners against the demand of 600,000 in the country. It is feared that almost 50% of the medical students would never work. The government spends almost Rs 2.5 million on each MBBS student. Unfortunately, after completion of the course, most of the female medical graduates never work." According to a report in *Al Jazeera*, "Pakistan's public health system is overwhelmed and underfunded" (Syed, 2016). Since nearly half of female students withdraw or never pursue their practice, the PMDC was forced to amend their regulations and were directed to increase their male student intake and limit the placements for female students (Abid, 2014).

This chapter addresses the importance for trained Pakistani female doctors to continue practice in the field (Shekhani & Moazam, 2019) and challenges faced by Pakistani female doctors in

Resilient Health. https://doi.org/10.1016/B978-0-443-18529-8.00064-0

the cultural and societal context, including research pertaining to gender disparities. Solutions are also posed, for example, a women-run, technology-based community initiative to retain female doctors in the workforce, while still maintaining their family responsibilities, allowing them to fulfill their professional obligations, reach their potential, serve as a positive role modelof female empowerment in the healthcare community, and tend to the urgent medical needs of the Pakistani population. Two impressive and innovative solutions are described—*DoctHERS* and *Sehat Kahani*—to allow female doctors to secure their place in the health profession, while also upholding responsibilities toward their family to meet the expectations of a patriarchal society.

Background

Gender stereotypes, disparities, and discrimination deeply woven into the fabric of culture and society have serious consequences prevalent in some cultures like Pakistan (Ali et al., 2011). As a patriarchal society within a collectivistic culture, the opportunity for women in Pakistan to exercise personal autonomy is restricted (Hofstede, 1980; Merkin, 2016). According to the Global Gender Gap Index, which measures a country's progress toward gender equality across education, economic participation, health, and political involvement, Pakistan ranked as second to last for five consecutive years (2012–17) (Ahmed, 2018). In Western societies, couples make collaborative effort to run dual-income households, while sustaining a healthy family life where some wives earn higher wages than their husbands and even allow men to opt for the role of stay-at-home dads. In contrast, in Pakistan, men may remain at low-paying jobs while women who are highly qualified, including those in the medical field, are usually forced to eventually stay at home and raise families.

An article published in a Pakistan-based newspaper, *The Express Tribune,* reports that "Pakistan has a shortage of qualified people. We need more doctors, more engineers, more teachers, more scientists—more of everything. What we don't need are more stay-at-home mothers" (Kazim, 2013).

Women's independence in a collectivistic, patriarchal society

Social psychologist, anthropologist, and pioneer of a cross-cultural research Geert Hofstede (1980) developed a four-dimensional model of studying national culture, including individualism, masculinity, uncertainty avoidance, and power distance. The latter, power distance, is defined as "the extent to which the less powerful members of institutions and organizations within a country expect and accept that power is distributed unequally."

Power distance and masculinity have been shown in cross-cultural research to score high in Pakistan, while individualism, defined as the degree of interdependence in a society maintained among its members (Hofstede, 1991, p. 23–158), scores low (Routamaa & Hautala, 2008). In a collectivistic culture like Pakistan, family structures and sociocultural norms are salient parts of the decision-making process, limiting individuals from making autonomous decisions (Mohsin & Syed, 2020). The hierarchy of decision-making begins with the wider community that defines cultural norms, and trickles down via patriarchal gender-based norms, to elder members of a family and to the husband who maintains authoritative control in the household (Khatwani, 2017; Qazi et al., 2021). Cross-cultural research has also revealed differing meanings of autonomy and independence in individualistic and collectivistic cultures. Western cultures supported individuality, autonomy, and equality, whereas collectivistic cultures emphasized in-group harmony, interdependence, and avoidance of conflict (Mines, 1994; Rhee et al., 1996). Research with a group of Pakistani women revealed that words like "free" and "independent" held negative connotations, implying that women who were free or independent were "immodest" or "shameless." Hence, women refrained from such terminologies and preferred the word "responsible" instead of "independent" to avoid the attachment of negative labels with themselves (Khatwani, 2017).

Moreover, family systems and dynamics present an additional impediment to women's decision-making, as women living in extended families have lesser control over their choices compared to women living in nuclear or seminuclear settings (Khatwani, 2017; Qazi et al., 2021). Extended family members were found to add to parents' opposition to coeducation of genders, fearing it would morally 'corrupt' daughters (Mehmood et al., 2018). While some females may be allowed to express opinions in family matters, the final verdict is up to the 'man of the house.' In an interview, one female respondent said, "*I can suggest but I cannot decide. My husband asks me for the suggestion, sometimes he appreciates my suggestion and sometimes does not, but finally he makes all the decisions*" (Khatwani, 2017).

Females in the medical field: Challenges and barriers

An analysis of the absence of female doctors in the medical field found that marriage was considered a woman's primary duty, whereas a professional career was a privilege requiring permission to pursue (Mohsin & Syed, 2020). Despite a professional woman's assertion of her independence, she still adheres to deeply rooted cultural norms, preventing the exercise of complete autonomy (Khatwani, 2017). Women who challenge male authority within their household by attempting to make their own choices regarding academics and careers usually face opposition instead of favorable, supportive responses (Merkin, 2016). In a qualitative study based in urban Pakistan, addressing the gender inequity in the medical workforce, one of the respondents is quoted as saying, "My sister is also a medical doctor, and she is often beaten up by her husband whenever she tries to decide things for her career. Her husband openly says that too much education has gotten to her head and that is why she will never be a good wife." Another respondent stated, "My husband left me just because I did not leave my job and I was more competent than him. Everyone in my family blames me for being too competitive, no one ever saw how insecure and aggressive he was" (Anjum et al., 2018).

The decision for women to pursue a medical practice after training is not only not in their own hands but the result of many intersecting factors. A main factor is responsibilities of family life and work–family conflicts (Khursheed et al., 2019). In a survey administered to final year medical students in Karachi, 92.2% of female students expressed wanting to continue or complete their residency, whereas 87.9% of respondents showed ambition to pursue postgraduate qualification. However, 76.6% of female students reported that their husband or future husband's opinion on their professional career plans would be "important" or "very important" to their decision-making (Shaikh et al., 2018). In addition, the demands and needs of spouses, in-laws, childbearing and rearing, household duties, inflexible working hours, and unsatisfactory working conditions create hindrances in women's ability and capacity to work in their chosen professions (Shekhani & Moazam, 2019).

Not all pressures are external, as women also experience internal pressures when feeling conflicted between their professional and personal lives. Female doctors found difficulty equally managing their roles as a doctor and wife and often are only able to fully commit to one role at the expense of the other. Medical practice responsibilities compromised their family life due to stringent working hours, and unavailability or lack of daycare centers, making it difficult to spend time with their children, while also managing a demanding medical career. As a result, some women who remain in the workforce resort to administrative or teaching positions, which are less time consuming and allow for more flexibility and freedom in their schedule to spend time with their families (Shekhani & Moazam, 2019).

Socio-cultural norms

Gender role stereotypes pervasive within individuals, societies, and countries become an accepted way of life and can lead to stigmatization, if opposed. In a qualitative study, experts from Aga Khan University in Pakistan and the Karolinska Institute in Sweden held five focus group discussions to understand the public's perception of gender roles, observe their fixity in culture, and evaluate how they facilitate or impede life circumstances for men and women (Ali et al., 2011). Findings revealed serious gender inequalities and human rights violations against women in the Pakistani society, specifically, that women themselves believed that 'a good woman' is selfless, compromising, empathic, completes her household chores, fulfills family obligations, tolerates any mistreatment, sacrifices her dreams, and never shares her opinions. In contrast, a 'good man' is sensible, strong, a good leader and advisor, sincere, financially stable, and a fair decision-maker. While women adhered to traditional gender roles, they voiced struggles about their desire to have a healthy balance of a professional life with the expected personal life. Women who worked in the corporate sector in urban Karachi exerted equal effort as men in the workplace, but men did not reciprocate the same equal effort in household responsibilities. In another study with similar findings, women complained about having to do a 'second shift' of work after coming home, without any voluntary assistance from their husbands (Hochschild & Machung, 2012). Consequently, women were burdened with additional household chores from which their husbands were invariably exempt. In addition to physically relaxing after returning home from work, men also demanded additional attention from their already exhausted wives to satisfy their sexual needs. In one of the focus groups, women expressed that even if they had the same office hours as their husbands, they were still obligated to complete household chores, and felt obliged to comply without frowning or complaining (Ali et al., 2011). Interestingly, cross-cultural comparisons reflect similar, traditional gender norms. In the UK, an individualistic culture, studies revealed that among married couples where both spouses were physicians, only 2% of male physicians contributed to household tasks, leaving the female physicians to manage household tasks, even without hired help (Elston, 1980). Comparably, in Australia, only 6% of male physicians with young children assisted in childcare. These results indicate that despite female doctors having professional careers, they did not relinquish domestic workload and exerted just as much effort into homemaking duties as full-time housewives (Fett, 1976).

Gender gaps in education

Higher education has been a neglected sector in Pakistan, with a Gross Enrollment Ratio (GER) of only 10% compared to its neighboring states, namely, 24% in India, 21% in Sri Lanka, 16% in Nepal, and 13% in Bangladesh (MFEPT, 2017). Despite these statistics, gender parity in education in Pakistan has improved over the years, with the enrollment rate of females in higher education increasing from 36.8% to 47% between 2001 and 2014. However, a combination of economic and sociocultural factors creates and enforces barriers that contribute to women's low literacy rates in Pakistan (Mehmood et al., 2018). Logistical issues present a major roadblock, in the lack or unavailability of safe public transportation, as well as long travel distance, costs, and timings which can compromise women's safety, leading to families' reluctance to allow young girls access to education (Mehmood et al., 2018). Also, an increase in sexual harassment at the hands of male students and faculty members in public and private institutes has posed danger for girls attending school, and consequently raised safety concerns among families (Ashraf et al., 2015).

An analysis of challenges faced by women in higher education in Pakistan found alarming rates of sexual harassment, with 62.2% of females reporting being harassed during their education by male students (42.6%) or faculty (17.6%), and 68.4% reporting incidents of sexual harassment when using public transport (Mehmood et al., 2018). Additionally, prevailing conservative norms and gender-role attitudes in society favor education for males over females, and as a result, parents invest their expenses on their son's education accordingly (Mehmood et al., 2018).

An article published in World Bank Blogs highlighted that women from poor educational backgrounds in lower socioeconomic classes struggled harder to be liberated from strictly enforced cultural norms, safety concerns, and financial incapacities to pursue education; therefore, they received fewer opportunities to advance in their chosen fields (Redaelli &

Rahman, 2021). In more remote areas and rural populations, girls are especially discouraged from leaving their homes and going to school, with men using religious decrees and family honor to maintain social control. Especially in these settings, the priority for females is to get married at a 'timely' age and embrace domestic responsibilities, including taking care of their husband, in-laws, children, and household chores (Maqsood et al., 2012).

Work–family conflicts

Gender discrepancy in education levels has been a leading cause of the underrepresentation of women in the workforce in Pakistan, despite some advances (Redaelli & Rahman, 2021). Even women who have successfully made professional achievements struggle to maintain or progress in their career development. In the male dominated collectivistic society, women battle with work–family conflicts and leading a balanced work–family life (Frone, 2000). According to researchers, "work-family conflict is a form of inter-role conflict in which the role pressure from work and family domain are mutually incompatible, in that participation in a work (or family) role is made more difficult by virtue of participation in the family (or work) role" (Bethge & Borngräber, 2015). A literature review identified and categorized work–family conflicts into three distinctive groups: work related (employee turnover, absenteeism, job satisfaction); nonwork related (marital, family, and life satisfaction); and stress-related (burnout, depression, psychosomatic symptoms) (Allen et al., 2000). In a study of the impact of work–family conflict among married, female doctors in Pakistan, findings revealed a positive correlation between work–family conflicts and turnover intentions (Tariq et al., 2021), which are consistent with existing research (Malik et al., 2010; Syed et al., 2018). Female doctors reported complaints from their family and in-laws for neglecting family responsibilities and thus, felt unable to perform adequately at their workplace. Female doctors with children especially reported being unable to make time for their family, as work commitments monopolized majority of their time. Although female doctors reported trying their best, they felt guilty for not fulfilling their role as an ideal mother or wife, which, in turn, negatively affected their career aspirations and decisions (Yamazaki et al., 2017). In another study comparing career and family life, female doctors burdened with increased domestic responsibilities dedicated fewer hours than their male colleagues to professional practice, resulting in lesser earnings and more frequent premature resignations (Uhlenberg & Cooney, 1990).

Unfavorable work environment

In addition to work–life conflicts, certain push factors related to the workplace environment make it difficult for women to continue working. These factors include inflexible working hours, sexual harassment, conveyance issues, and threats to personal safety (Khursheed, Mustafa, Arshad, & Gill, 2019). Research using interviews and focus group discussions with female healthcare providers in Pakistan, to identify and document potential reasons for their scarcity in the medical field, found a common theme of abuse of power by authority figures (Mumtaz et al., 2003). Male colleagues and supervisors showed disrespect, excluded women from decision-making, demanded sexual favors, and threatened to file a complaint or cutting off wages if compliance was denied. Males were also insensitive to women's boundaries, often forcing them to manage culturally inappropriate tasks. Further, inflexible hours, coupled with commuting long distances, disabled women from full participation at work. Since relocating the home for the sake of a wife's workplace convenience was deemed "socially degrading" for men, women relegated to live far from their workplace were often compelled to skip work due to competing home responsibilities, leading to higher rates of absenteeism and eventually higher rates of female staff attrition.

The appeal of the trophy wife

The surge of female students in medical schools of Pakistan is partly explained by the fact that such professional qualification of young female doctors makes them more attractive for marriage and elevates the status of the men they marry, since "doctor brides" have become a status symbol in society (Das, 2020; Kazim, 2013). Unfortunately, these doctor brides are ultimately persuaded, or even forced, to abandon their career ambitions and assume the role of a traditional housewife. After marriage, their husband or extended families often discontinue support of their practice, expecting them to devote their time and energy towards familial obligations, including responsibilities of looking after their husband's entire family.

The prospect of a good wife was once predicated on the physical appearance and personality of a woman (i.e., fair skin, beauty, charm, and an agreeable nature). In more modern times, the criteria have evolved to include preference for a woman in the medical field, where her medical degree is an attractive status symbol, while holding no value in terms of her potential for actual practice in the professional field (Zakaria, 2013).

In a moving article about her personal experience, published on *The Tempest* (a storytelling platform for women, including many from Pakistan), Imtiaz (2016) muses about her desire to become a psychiatrist, which requires years of medical school training; yet she worries about how marriage will affect her medical career. "The demand for a wife that's a doctor is very high—top of the list for every rishta aunty," she writes, referring to a Hindi/Urdu matchmaker who has a 'database' of profiles of men and women of her community. "Everyone wants a doctor wife, but nobody wants us to work as a doctor," she adds, expecting that the medical degree she strives to earn and that takes years to achieve will sit on the shelf collecting dust, because the family expects her to "cook, clean and stay at home." She despairs at the case of a friend's cousin, a female cardiologist, who heeded her husband's request and dropped out of practicing medicine after over twelve years of education, to prioritize taking care of her family and home.

DoctHERS—The e-doctor initiative

One notable initiative established in Pakistan to help female doctors tackle the barriers to practicing medicine is DoctHERS. This initiative, whereby the name specifically implying services for women, is a digital health platform that connects the underutilized female doctors with healthcare consumers, especially in low-income communities (https://docthers.com/) allowing women in the medical profession to practice from home, thus, counteracting the gender disparity and absence of women in the medical field. The goal of this e-platform is to extend cost-effective and quality healthcare services to marginalized women in rural and impoverished regions of Pakistan, where access to healthcare is difficult to attain (Syed, 2016). The platform predominantly provides online sessions, as well as follow-up face-to-face appointments, where required.

The DoctHERS initiative, founded in 2015, is currently led by the Executive Chairman Asher Hasan, with a team of 700 female doctors, and a target to bring 35,000 female doctors, who were unable to practice due to family obligations, back into the medical field (Yusufzai, 2019). DoctHERS has received international recognition by news channels such as *Al Jazeera*, which have reported about major achievements, struggles, and outcomes of the project. Doctors work from "teleclinics," where they provide medical consultations from their homes to patients, with nurses serving as intermediaries to facilitate the doctor–patient communication and deliver any required treatment in person. The doctor has access to the clinic's diagnostic equipment and can monitor the patients' vital signs remotely with the help of in-person staff (Syed, 2016).

Challenges of the initiative included issues with poor internet connectivity, power shortages, and unfamiliarity and mistrust of people in rural areas toward video-calling with a "stranger" on a screen, whom they initially refused to believe was a "real" doctor.

On the other hand, some patients responded positively. A thirty-year-old woman said about her experience with DoctHERS, "I have found this system to be better. They have checked me. [At the government hospital] the male doctor was assigned to see me. In fact, they don't even check the patients in big government hospitals. There is no care. People will arrive and then leave. They don't take care of visiting patients despite charging a fee."

Given the scarcity of women in the medical field, the DoctHERS initiative lessens the gap between the demand and supply of female doctors in Pakistan. After the sign-up process, female doctors are enrolled in a six-month training program that includes family medicine, followed by training in basic life support, obstetrics and gynecology, and the use of electronic medical records and video-assisted mobile consultations. The success is impressive, with over 3000 villages introduced to healthcare professionals through this system, an average of 1 million patients who received treatment, and at least 10 appointments scheduled a day per doctor with women belonging to underprivileged areas in Pakistan.

DoctHERS is supported by the Pakistan Ministry of Health and sponsored by various organizations such as UNICEF, UK Aid, MIT, Unilever. The platform shows promising potential to revitalize the medical industry, by enabling and empowering female doctors in the workforce. With the online provision of resources, consultations, and guidance, the female doctors themselves gain self-esteem and fulfill their potential in their career. Through their entry or reentry into the workforce, there is hope that the healthcare system in Pakistan can reach and improve the lives of so many individuals and communities direly in need.

Sehat Kahani: The story of health, another med-tech innovation

A similar e-health platform, named Sehat Kahani, translating to the "Story of Health" in Urdu, was designed and implemented by Dr. Sara Saeed Khurram (former cofounder of DoctHERS in 2015) and Dr. Iffat Zafar Aga, two female entrepreneurs who created a network of e-clinics that connect female doctors with underserved communities in Pakistan, mediated by qualified nursing and community staff. In addition to e-clinics, Sehat Kahani has its own mobile-based application, through which users can access audio and video resources for their healthcare needs and set up virtual appointments with a preferred doctor in three simple steps: logging in, choosing a doctor, and scheduling an appointment. This start-up was founded in 2017 and has grown considerably, with 5000 actively working female doctors, 200 community staff members, and 35 e-health clinics across Pakistan. On 19 January 2022, Sehat Kahani, along with other start-ups that secured national and international funding, was invited for a meeting with prime minister of Pakistan Imran Khan, to be appreciated for their efforts toward a healthier Pakistan. The CEO of Sehat Kahani, Dr. Khurram, received an honorable mention from the prime minister, especially considering she was the only female-led start-up in the group (Kahani, 2022). In an interview shared in Health Management's issue on Human Matters, the two founders of Sehat Kahani (Khurram & Aga, 2021) shared the inspiration that led to its initiation, and their three goals behind the project:

1. Reintegration of the female health workforce, through providing a gender-inclusive employment platform, using technology to provide capacity building and skillset training.
2. Improving healthcare efficiency by providing access to affordable and equitable healthcare while reducing travelling expenses for both patients and clinical staff, and increasing the female health workforce productivity.
3. Enhancing healthcare utilization by improving early diagnosis and fostering access to new technologies and interventional strategies through telemedicine.

When asked to share about the values of technology in providing healthcare by and for women, especially in the age of COVID-19, and advice for "doctor brides," who, for any reason, were unable to pursue their ambitions in the medical field, Dr. Khurram said:

> In a country where half of the population lacks access to a healthcare professional, I would say that you are the greatest asset to the nation, especially in these troubling and uncertain times. With the world going virtual, doctors—especially female doctors, do not need to go out in the field or hospitals to practice. Telemedicine has emerged as a beacon of hope for both patients and doctors to ensure that the help is provided to those in need. Hence, amidst this resurgence of COVID-19, this nation needs more and more doctors willing to serve through any means available to them. Involving the healthcare community will be instrumental in ensuring adherence, monitoring, and liability regarding telemedicine integration in our country. If you were not able to continue your medical practice due to any reason, now is the perfect time to gear up and get back in the field. Faced with a common threat that is blind to wealth, gender, and social status, we can create the equity in healthcare we have craved. This "once-in-a-century pandemic" is our chance to change the healthcare delivery model in Pakistan. This is the time when we, as healthcare professionals, can face our vulnerabilities head-on and choose to be compassionate catalysts for a better tomorrow. (Khurram & Aga, 2021, *p.165)*

Looking to the future

Innovative interventions like DoctHERS and Sehat Kahani noted in this chapter, in addition to more modernization of attitudes in Pakistani society, and increased assertion of independence and empowerment by Pakistani women, offer more hope for female Pakistani doctors to continue their professional career, and therefore to benefit themselves, serve as examples for future female as well as male generations, and improve the availability of medical care for the Pakistani society.

References

Abid, A. M. (2014, October 22). Female doctors becoming trophy wives: Is quota the right move? In *DAWN.COM*. https://www.dawn.com/news/1139639.

Ahmed, I. (2018, March 10). No country for working women. In *The Express Tribune*. https://tribune.com.pk/story/1655699/no-country-working-women.

Ali, T. S., Krantz, G., Gul, R., Asad, N., Johansson, E., & Mogren, I. (2011). Gender roles and their influence on life prospects for women in urban Karachi, Pakistan: A qualitative study. *Global Health Action*, *4*(1), 7448. https://doi.org/10.3402/gha.v4i0.7448.

Allen, T. D., Herst, D. E. L., Bruck, C. S., & Sutton, M. (2000). Consequences associated with work-to-family conflict: A review and agenda for future research. *Journal of Occupational Health Psychology*, *5*(2), 278–308. https://doi.org/10.1037/1076-8998.5.2.278.

Anjum, G., Kamal, A., & Bilwani, S. (2018). Antecedents of gender gap in workforce participation: A phenomenology of psychologists and medical doctors in urban Pakistan. *Journal of Human Behavior in the Social Environment*, *29*, 282–299.

Ashraf, E., Younis, M., & Shurgeel, H. K. (2015). A review, of rural women's education in Pakistan. *Science International (Lahore)*, *27*(1), 555–559.

Bethge, M., & Borngräber, Y. (2015). Work-family conflicts and self-reported work ability: cross-sectional findings in women with chronic musculoskeletal disorders. *BMC Musculoskeletal Disorders*, *16*, 58. https://doi.org/10.1186/s12891-015-0515-4.

Das, A. (2020, December 21). *Why The Term 'Trophy wife' is Nothing But Derogatory*. SheThePeople TV https://www.shethepeople.tv/top-stories/opinion/why-the-term-trophy-wife-is-nothing-but-derogatory/.

Elston, M. A. (1980). Medicine: Half our future doctors? In R. Silverstone, & A. Ward (Eds.), *The careers of professional women* (pp. 99–139). London: Croom-Helm.

Fett, I. (1976). The future of women in Australian medicine. *The Medical Journal of Australia, Supplement*, *2*, 33–39.

Frone, M. R. (2000). Work–family conflict and employee psychiatric disorders: The national comorbidity survey. *Journal of Applied Psychology*, *85*(6), 888–895. https://doi.org/10.1037/0021-9010.85.6.888.

Hochschild, A. R., & Machung, A. (2012). *The second shift*. Penguin.

Hofstede, G. (1980). *Culture's consequences: International differences in work-related values*. Beverly Hills, CA: Sage.

Hofstede, G. (1991). *Cultures and organizations: Software of the mind*. London: Macmillan.

Imtiaz, M. (2016). *I refuse to be your trophy doctor wife. The Tempest*. https://thetempest.co/2016/08/14/culture-taste/31309/.

Junaidi, I. (2014). *50pc of female doctors never work after graduation. DAWN.COM*. https://www.dawn.com/news/1139557.

Kahani, S. (2022). *Sehat Kahani's Co-Founders awarded as the Global Heroines of Health 2022*. https://sehatkahani.com/sehat-kahanis-co-founders-awarded-as-the-global-heroines-of-health-2022/.

Kazim, J. (2013, November 10). Trophy wives. In *The Express Tribune*. https://tribune.com.pk/story/630041/trophy-wives.

Khatwani, M. (2017). Professional women's experience of autonomy and independence in Sindh-Pakistan. In *Gender differences in different contexts* IntechOpen. https://doi.org/10.5772/66093.

Khurram, S., & Aga, I. (2021). Sehat Kahani: Addressing the social phenomenon of doctor brides. *HealthManagement.org the Journal*, *21*(3), 161–165. https://healthmanagement.org/uploads/article_attachment/hm3-v21-sehat-kahani-addressing-the-social-phenomenon.pdf.

Khursheed, A., Mustafa, F., Arshad, I., & Gill, S. (2019). Work-family conflict among married female professionals in Pakistan. *Management Studies and Economic Systems (MSES)*, *4*(2), 123–130. http://www.msaes.org/article_85823_4340b8afdd31f7ff8dbe020c98006a7d.pdf.

Malik, M. I., Zaheer, A., Khan, M. A., & Ahmed, M. (2010). Developing and testing a model of burnout at work and turnover intentions among doctors in Pakistan. *IJBM*, *5*(10), 234–247. https://doi.org/10.5539/ijbm.v5n10p234.

Maqsood, F., Maqsood, S., & Raza, S. (2012). Getting higher education: Is it really challenge for females in Pakistan. *Academic Research Journal*, *2*(3).

Mehmood, S., Chong, L., & Hussain, M. (2018). Females higher education in Pakistan: An analysis of socio-economic and cultural challenges. *Advances in Social Sciences Research Journal*, *5*(6), 379–397.

Merkin, R. (2016). Pakistani cultural characteristics: Updated VSM scores and facework geared towards increasing women's access to education. In S. Roy, & I. S. Shaw (Eds.), *Communicating differences*. London: Palgrave Macmillan. https://doi.org/10.1057/9781137499264_11.

MFEPT. (2017). *National Education Policy 2017–2025*. Islamabad: Ministry of Federal Education and Professional Training, Government of Pakistan.

Mines, M. (1994). *Public faces, private voices: Community and individuality in South India*. California: University of California.

Mohsin, M., & Syed, J. (2020). The missing doctors — An analysis of educated women and female domesticity in Pakistan. *Gender, Work and Organization*, *27*, 1077–1102.

Mumtaz, Z., Salway, S., Waseem, M., & Umer, N. (2003). Gender-based barriers to primary health care provision in Pakistan: The experience of female providers. *Health Policy and Planning*, *18*(3), 261–269. https://doi.org/10.1093/heapol/czg032.

Qazi, M. A., Schofield, S., & Kennedy, C. (2021). 'Doctor Brides': A narrative review of the barriers and enablers to women practising medicine in Pakistan. *JPMA. The Journal of the Pakistan Medical Association*, *71*(9), 2237–2243. https://doi.org/10.47391/JPMA.119.

Redaelli, S., & Rahman, N. (2021). In Pakistan, women''s representation in the workforce remains low. *Blogs.worldbank.org*. https://blogs.worldbank.org/endpovertyinsouthasia/pakistan-womens-representation-workforce-remains-low.

Rhee, E., Uleman, J. S., & Lee, H. K. (1996). Variations in collectivism and individualism by ingroup and culture: Confirmatory factor analysis. *Journal of Personality and Social Psychology*, *71*(5), 1037–1054. https://doi.org/10.1037/0022-3514.71.5.1037.

Routamaa, V., & Hautala, T. (2008). Understanding cultural differences, the values in a cross-cultural context. *International Review of Business Research Papers*, *4*(5), 129–137.

Shaikh, M. A., Ikram, S., & Zaheer, R. (2018). Influences on medical career choice and future medical practice plans among women: Perspective from final year students and house officers. *JPMA. The Journal of the Pakistan Medical Association*, *68*(2), 272–275.

Shekhani, S. S., & Moazam, F. (2019, March 12). The myth of doctor brides. In *DAWN.COM*. https://www.dawn.com/news/1460104.

Syed, A., Ahmad, M., Ali, H., Arif, M., & Gohar, A. (2018). Work-family conflict and turnover intentions: Moderated mediation model. *Human Resource Research*, *2*(1), 95–98. https://doi.org/10.5296/hrr.v2i1.13925.

Syed, R. (2016, April 7). DoctHERs in Pakistan empowering female doctors. In *Women's Rights | Al Jazeera*. https://www.aljazeera.com/features/2016/4/7/docthers-in-pakistan-empowering-female-doctors.

Tariq, I., Asad, S., Majeed, A., & Fahim, U. (2021). Work-family conflict, psychological empowerment, and turnover intentions among married female doctors. *Bangladesh Journal of Medical Science*, *20*(4), 855–863. https://doi.org/10.3329/bjms.v20i4.54145.

Uhlenberg, P., & Cooney, T. M. (1990). Male and female physicians: family and career comparisons. *Social Science & Medicine (1982)*, *30*(3), 373–378. https://doi.org/10.1016/0277-9536(90)90192-u.

Yamazaki, Y., Uka, T., & Marui, E. (2017). Professional fulfillment and parenting work-life balance in female physicians in Basic Sciences and medical research: a nationwide cross-sectional survey of all 80 medical schools in Japan. *Human Resources for Health*, *15*(65). https://doi.org/10.1186/s12960-017-0241-0.

Yusufzai, A. (2019). Pakistan attracts 700 female doctors back into practice through online service. *BMJ*, *l6752*. https://doi.org/10.1136/bmj.l6752.

Zakaria, R. (2013, July 26). The doctor brides. In *DAWN.COM*. https://www.dawn.com/news/1032070.

CHAPTER 65

Medical students' efforts toward the digital transformation of healthcare: research and policy

Maria Inês Francisco Viva
International Federation of Medical Students' Associations (IFMSA), Copenhagen, Denmark

Introduction

Digital health and technological innovations are shaping the future of health and well-being of populations, particularly in reaction to the COVID-19 pandemic. It is evident that healthcare providers had to adapt to the needs of populations when government lockdowns were imposed, motivating many services to develop technological adaptations. These will only increase in the future. Medical professionals will therefore have a prominent role to play in the sensitization of communities to concepts of digital health and the development and implementation of digital health services. Therefore, in recent times, the International Federation of Medical Students' Associations (IFMSA) has intensified its efforts to build the capacity of medical students worldwide regarding digital health.

Our Medical Students Association's attitude to digital health

IFMSA believes that technological innovation can lead to better health outcomes and more efficient healthcare delivery. We believe that more investment is needed in digital health technologies and that these innovations should be equitably accessible by everyone. Implementing an ethical and regulated approach to innovation in healthcare and ensuring its continued evaluation is crucial to ensure these advances are monitored and improved upon over time. Developing safe and reliable electronic health records and taking advantage of mobile digital innovations can significantly improve health literacy and empower patients.

Resilient Health. https://doi.org/10.1016/B978-0-443-18529-8.00065-2

Our policy work

In 2019 the IFMSA General Assembly adopted its first-ever Policy Document on Digital Health (IFMSA, 2019) entitled "Technological Innovation in Healthcare." In this Policy Document, we stressed that even though technological innovation has had an enormous expansion over the last few decades, robust regulation and monitoring to ensure the safety and quality of health technologies have not been implemented in many parts of the world. As I write this chapter, we are updating this Policy Document, focusing on the digital health transformation of healthcare. Hopefully, by the beginning of August 2022, our General Assembly, composed of Medical Student representatives from 129 countries, will adopt the updated IFMSA Policy Document Digital Transformation in Healthcare.

Collaborations

IFMSA efforts related to digital health positively impacted our collaborations with partners, translating into several joint initiatives and contributions. Such efforts include IFMSA's contribution to the Youth Statement+Call for Action of *The Lancet and Financial Times Commission on Governing Health Futures 2030: Growing up in a Digital World* (The Lancet and Financial Times Commission on Governing health, 2021). Recognizing that the Coronavirus increased inequities, upended education and job prospects, and escalated mental and physical health problems for youth, and reacting to the expanded use of digital technology by youth yet widened digital divides, the report emphasized that digitally enabled health systems and digital health technologies must be accessible and are available to all, of good quality, and not cause financial hardship. Recommendations to all stakeholders were to: protect youth from violence like cyberbullying; support youth in creating innovative digital solutions; develop regulatory bodies for AI-based health technologies; facilitate dialogues for developing, implementing, and evaluating digital health policies, programs, and services; and invest in digital education.

The federation also participated in the first-ever Digital Health Week in 2021, organizing an event under the topic of "Digital Health in Medical Education" (Digital Health Week, 2021).

Research findings on medical students' attitudes to digital health

Medical students worldwide contributed to the efforts of IFMSA concerning digital health by contributing their perspectives on the current utilization of technologies in healthcare and its essential place during and in settings. These perspectives were materialized in IFMSA's Medical Student International (MSI) August 2021 edition, a biannual magazine focusing on pressing global health topics (ISSUU, 2021).

IFMSA developed a global digital health survey to understand the current status of digital health inclusion in the medical curriculum. More than 85% of respondents agreed or strongly agreed that digital health can contribute to the efficiency, access, and quality of healthcare. Additionally, 87.7% of respondents agreed or strongly agreed that future health professionals should receive education on digital health as part of the core medical curriculum to prepare them for their future work as health professionals ($n=626$). In conclusion, the results of this survey showcased that the current status of medical education does not provide sufficient exposure to digital health within medical schools globally and remains a neglected area.

What's in the future?

A key ask from young people is the need for improved digital health literacy. In line with this key ask, we are developing capacity-building initiatives to bridge this gap, which shall be made available to medical students. Furthermore, with the adoption of our Policy Documents, we will continue to advocate for digital health inclusion in the medical curricula, aiming to partner with key stakeholders in the field of medical education and digital health.

References

Digital Health Week. (2021). *Program*. https://digitalhealthweek.co/events.

IFMSA. (2019). *IFMSA policy document technological innovation in healthcare*. https://ifmsa.org/wp-content/uploads/2021/02/AM19-Technological-Innovation-in-Healthcare.pdf.

ISSUU. (2021). *Digital Health l Medical Students International*. https://issuu.com/ifmsa/docs/msi_final.

The Lancet and Financial Times Commission on Governing health. (2021). *futures 2030: Growing up in a digital world*. https://www.governinghealthfutures2030.org/wp-content/uploads/2021/12/Youth-Statement-and-CfA.pdf.

CHAPTER 66

citiesRISE: A transdisciplinary, youth-driven approach to scale social and technological innovations in mental health and well-being locally and globally

Matthew Hughsam[a], Mariam Claeson[b], Lian Zeitz[a], Suresh Kumar[c], Nathaniel Foote[a,d], and Moitreyee Sinha[a]

[a]citiesRISE, New York, NY, United States, [b]Global Public Health, Karolinska Institutet, Solna, Sweden, [c]citiesRISE, Chennai, India, [d]TruePoint, Lincoln, MA, United States

Introduction: Challenge and opportunity

In recent years, the world has seen immense challenges, from the effects of climate change to the COVID-19 pandemic. We are a global community under enormous stress and our ability to adapt depends on how well we support the mental health and well-being of emerging generations. Comprising a population of over 1.8 billion, there are more youth today than ever before (United Nations Population Fund, 2022). Mental ill health causes significant mortality and morbidity among youth, representing 13% of the global burden of disease among 10–19-year-olds (World Health Organization, 2021). The estimated annual loss in human capital arising from mental ill health in this age group is US$387.2 billion (United Nations Children's Fund, 2021). Mental health is not just the absence of illness—it is central to human flourishing and an essential building block for human capital and sustainable development. While some progress has been made toward achieving the United Nations Sustainable Development Goals (SDGs) by the year 2030, unprecedented threats have emerged that impede efforts to accelerate progress on the mental health target and other interrelated targets (Clark et al., 2020; Patel et al., 2018).

The launch of the SDGs in 2015 marked a shift in the importance given to mental health in global development, evident in Target 3.4 calling for the promotion of mental health and well-being (United Nations, 2022a). The relationship between youth mental health and progress on other SDGs is interdependent and reciprocal: strengthening youth mental health

Resilient Health. https://doi.org/10.1016/B978-0-443-18529-8.00066-4

requires progress in education, economic development, peacebuilding, and other areas, and vice versa (Patton et al., 2016). Further, youth mental health is heavily impacted by the social and structural determinants of health—addressing social, economic, and other inequities within and between countries improves mental health (Lo & Cheng, 2012). Youth have a prominent role to play in the social movements needed to address such systemic inequities, evident, for example, by efforts of youth like Greta Thunberg, Vanessa Nakate, and others leading the *#FridaysForFuture* movement demanding global climate justice (Time, 2021). There is a related significant need for more youth-driven multisectoral action and movements to drive mental health transformation.

Despite growing recognition of the importance of mental health, awareness has not been accompanied by significant investments of resources. The current paradigm in the mental health field focuses narrowly on expanding access to traditional approaches, neglecting central concepts such as inner-led development. Traditional approaches overly rely on clinical models and treat youth as passive beneficiaries. They also largely fail to target the intersecting inner, social, and environmental factors critical to youth mental health (Sinha et al., 2020). While social and technology-enabled innovations that address some of these needs are designed and piloted, these are often not brought to scale. The need for a fundamental paradigm shift is clear.

The COVID-19 pandemic has heightened the urgency of this paradigm shift. COVID-19-related shocks have created significant challenges to youth mental health and well-being, leading to a 'shadow pandemic' driven by factors including trauma, social isolation, and disruption of essential services such as education and child protection spaces. The pandemic has also exacerbated preexisting systemic inequities, further impacting the most vulnerable. Globally, we can expect a level of unparalleled need at a time of systemic depletion and fragility (de Miranda et al., 2020; Sharma et al., 2021). While acknowledging the distress and harm experienced by youth, unique opportunities have also emerged for rapid disruption and acceleration of innovations in partnership with youth, for example, in the digital environment. A greater focus on technology-enabled innovations that address the mental health continuum (promotion, prevention, treatment, care) and consider youth's broader life journeys and environments is needed to address pressing challenges posed by the pandemic.

citiesRISE was founded in 2017 in response to the immense scale and complexity of challenges in youth mental health. Recognizing the need to not just increase access to care, but also drive a paradigm shift in how mental health is conceptualized and addressed, *citiesRISE* itself is constantly innovating in ways to catalyze social and technology-enabled innovation. *citiesRISE* has evolved to take a transdisciplinary, youth-driven approach to scale innovations in mental health and well-being, linking local and global action. In this chapter, we share *citiesRISE*'s journey—what shaped the initial approach, lessons learned along the way, how these lessons have led to a sharpened strategy, and the way forward.

citiesRISE's initial approach

citiesRISE is a global platform committed to transforming mental health through local innovation, coalition building, and youth-driven action. Starting out with a vision of mental health as the potential to become one of humanity's greatest assets, *citiesRISE*'s initial focus was on a paradigm shift from solely clinical care to integrating mental health promotion and prevention, and centered on flourishing (VanderWeele, 2017). Recognizing that this ambitious mission would require collaborative action between stakeholders across disciplinary, sectoral, generational, cultural, geographical, and other boundaries, *citiesRISE* was designed as a multistakeholder platform capable of facilitating cross-boundary collaboration (Sinha et al., 2019). The following design principles, cocreated with youth, communities, and other experts from around the world, guided *citiesRISE*'s early work: experimentation, trying things out with a close connection to innovators in communities; using rigor and research to distill what is working; increasing capacity to accelerate and scale; and embracing movement building.

citiesRISE initiated a global listening tour engaging youth, innovators, private sector partners, research experts, system leaders, and other stakeholders to identify key levers for turning the tide on mental health. Two levers quickly emerged to drive *citiesRISE*'s strategy: youth and cities. Demographic trends suggest that 60% of all city dwellers will be under the age of 18 by 2030 (United Nations, 2022b). This presents an opportunity to harness the ingenuity of the young, who have advantages of unique insights into how mental health affects them, and also the capacity for driving inner-led change. Youth have been at the forefront of many key technological innovations and social movements. Further, cities have long demonstrated a willingness to pursue innovative practices and policies that can be subsequently scaled more widely (McAllister, 2008).

Early progress

The following examples illustrate how *citiesRISE* has supported incubation and scaling up for social and technology-enabled innovations anchored in youth and cities as the key levers.

City innovation hubs

Rooted in a commitment to cross-cultural collaboration, as well as the importance of adaptive capacity to address evolving challenges, *citiesRISE* established a linked network of city innovation hubs spanning South and North America, Africa, and India, specifically: Bogotá in Colombia, Chennai in India, Nairobi in Kenya, and Sacramento and Seattle in the United States, as shown in Fig. 1. These cities were selected based on criteria such as their regional influence and diversity of contexts across the network. In each city, rapid assessments were conducted to map local landscapes, identify system gaps, and lay the foundation for youth-driven

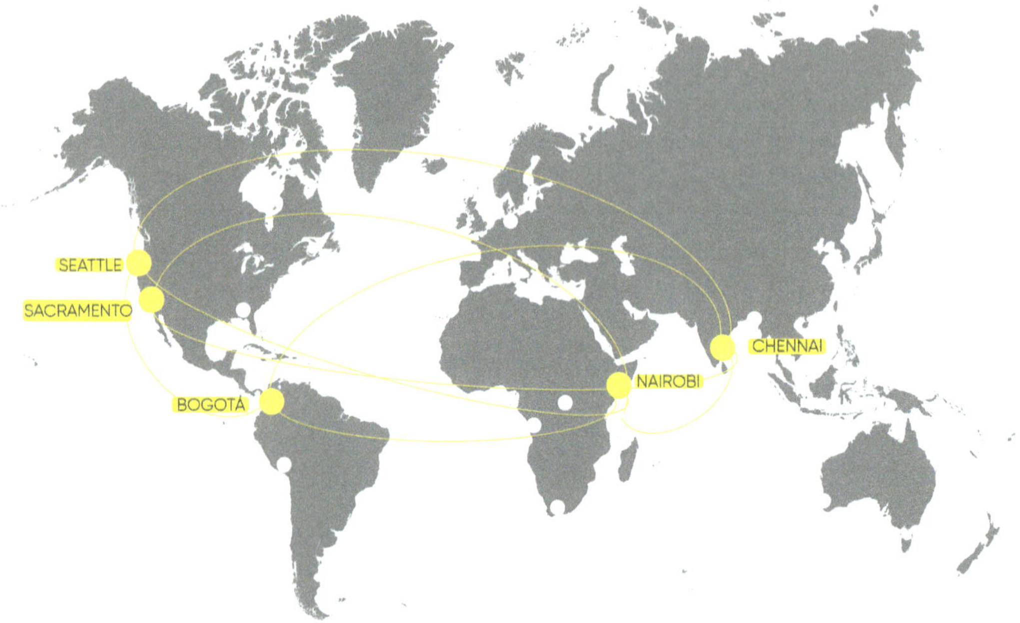

Fig. 1
Network of city innovation hubs underpinning the ability to innovate in local contexts while being part of a shared journey toward mental health transformation.

transformation (Kumar, Mohanraj, et al., 2021). Grounding all of *citiesRISE*'s programs in this network approach has enabled highly contextualized local work, linked via a global network and a common results framework.

citiesRISE initiated work in each city through a *Youth Challenge Award*, supporting a total of 17 social innovations, approximately half of which were led by youth themselves. One example is *Amazing Minds Africa*, a youth-led organization using theater and youth networks to create awareness and provide mental health support on college campuses in Kenya. Fig. 2 shows the theater-based model in action. *citiesRISE* partnered with *Amazing Minds Africa* to expand from one to four colleges in Kenya and adapt the model to colleges in Chennai, India.

Chennai, India: Whole school approach

Schools were identified in the Chennai rapid assessment, outlined before, as the top priority setting for innovation. Based on a review of the global evidence base, and listening to local stakeholders to ensure cultural adaptability, *citiesRISE* approached transformation of the school system from two angles: the curriculum and the broader school environment. Two local partners, Schizophrenia Research Foundation and NalandaWay Foundation, were crucial to this work, including to leverage technology when COVID-19 hit, using interactive voice response technology to adapt content and to support parents when children had to stay at home due to

Fig. 2
Amazing Minds Africa's use of youth-led theater and youth networks creates accessible and transformative spaces for college students to engage with mental health.

lockdowns and quarantines. The Teen Mental Health curriculum, developed in Canada to teach mental health literacy to students grades 9–12, was augmented with an additional component involving people with lived experience and applied in secondary schools (Kutcher & Wei, 2018). Results showed that while changes in mental health knowledge and stigma were similar among students who received intervention with or without augmentation, a statistically significantly greater improvement in attitude toward mental health was observed in the augmented intervention group, indicating a strength of involving people with lived experience (Kumar, Hanisha, et al., 2021). In addition, the Hopeful Minds curriculum, originally developed by iFred based on work in Ireland to teach hope to students ages 14–18 years in schools, was adapted for India for students in primary school (ages 5–10 years) and integrated with an arts-based delivery model and implemented to reach 4000 students (Hopeful Minds, 2022).

citiesRISE and partners also worked to design a *Whole School* transformation-focused intervention including multiple components. One such component is building the capacities of resilience teams consisting of teachers, parents, and peers to create a more supportive school environment with the engagement of mental health professionals trained using a practice transformation curriculum on youth-friendly mental health support, developed in partnership with the World Psychiatric Association.

Nairobi, Kenya: Youth gathering spaces

In Nairobi, the *Youth Gathering Spaces* model is a social innovation that emerged from collaboration between youth, communities, and professionals. *citiesRISE,* in partnership with

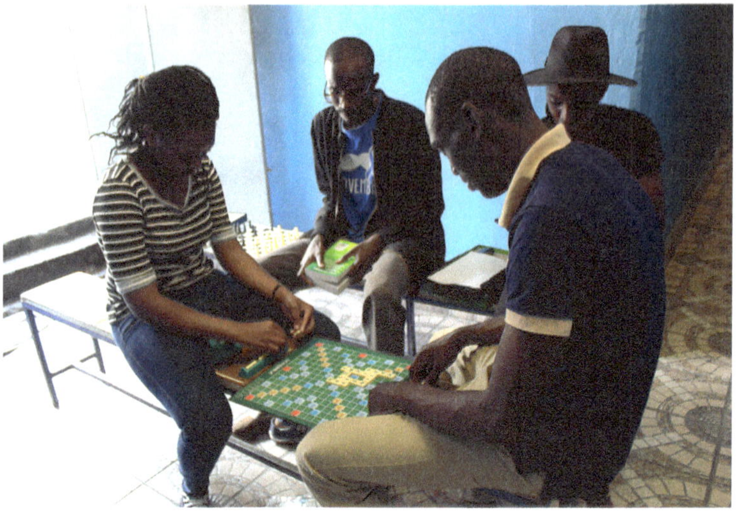

Fig. 3
Pro-social activities in the *Youth Gathering Space* in the informal settlement of Huruma in Nairobi, Kenya.

Nzumari Africa (a local youth-led advocacy organization) and Salama Youth Hub (an existing youth-focused space), as well as mental health experts from organizations including the World Psychiatric Association and the Kenya Psychiatric Association, developed an acceptable and feasible model for providing safe, supportive gathering spaces for youth ages 18–30 living in informal settlements. These spaces provide pro-social environments where youth can congregate with peers and participate in creative arts-based activities that weave in mental health literacy, link youth to peer support, and act as points of referral for those who need more formal services. Fig. 3 shows such pro-social activities in the *Youth Gathering Space* in the informal settlement of Huruma in Nairobi, Kenya. Qualitative results from informal evaluation showed significant uptake in youth populations in informal settlements, reduction in stigma, and an increase in awareness among users, as well as stronger cohesion with families and other community members (study underway to formally evaluate effectiveness of this approach). In response to the national lockdown due to the COVID-19 pandemic, a range of low-threshold social media tools were used to maintain connections, and digital care tools were also prototyped. *citiesRISE* is currently adapting this *Youth Gathering Spaces* model for implementation in Chennai, Sacramento, and Seattle.

India, Kenya, Rwanda, South Africa: Youth and mental health science

There is a significant time lag, estimated at approximately 17 years, between the generation of science and research results, and translation of these findings into programs with tangible impacts on youth's lives (Morris et al., 2011). In response, *citiesRISE* conducted a youth-driven mixed-methods study in partnership with Wellcome Trust and research partners in India, Kenya, Rwanda, and South Africa to understand youth's journeys when engaging with mental health science, information, and support. Together with collaborating youth researchers and senior researchers, initial research findings were integrated with insights from the consumer marketing field to create a new model mapping the dynamic and longitudinal natures of youth's journeys (Kumar, Sinha, et al., 2021). This study also elucidated key drivers of mental health intervention uptake and sustainability; for example, the importance of trust in decisions about whether to access an intervention, or the weight that someone may place on endorsement of an intervention by a peer over evidence of effectiveness. This model, and lessons learned from this study, can serve as the foundation for action research to get evidence-based science and research results into the hands of youth quicker, and in more empowering and respectful ways.

Global: Accelerator and learning platform

Building on the experience of running the *Youth Challenge Award* and of facilitating learning and collaboration across city innovation hubs, *citiesRISE* partnered with the *Saving Brains* platform hosted by TruePoint (Radner et al., 2018) to develop an *Accelerator and Learning Platform*, which has provided support to over 60 innovators in mental health across over 20 countries. Innovators range from grassroots youth-led collectives to multinational organizations reaching hundreds of thousands of people. Working with partners including Grand Challenges Canada, Rural India Supporting Trust, and Pivotal Ventures, *citiesRISE* built the platform to facilitate learning communities and provide technical assistance, and also to drive field-level paradigm shifts. The platform draws on *citiesRISE*'s own research and innovation experience, and convenes leading experts in key areas including path-to-market, monitoring and evaluation, organization development, and youth engagement. Through this work, *citiesRISE* has supported a range of technology-enabled innovations. One example is the *#ShareWeCare* online mental health peer-counseling platform developed by Teenergizer, a youth-led movement in Eastern Europe (Teenergizer, 2023).

Key lessons learned

The combination of *citiesRISE*'s evidence generated bottom-up from the network of city innovation hubs, insights from working with innovators in over 20 countries, and overall experience supporting youth- and city-led action globally has led to several key lessons learned:

Mental health friendly cities: A multidimensional framework

The global listening tour that informed *citiesRISE*'s initial strategy revealed significant interest from various stakeholders in a shared journey between different communities toward transforming mental health. Bottom-up insights from *citiesRISE*'s city innovation hubs,

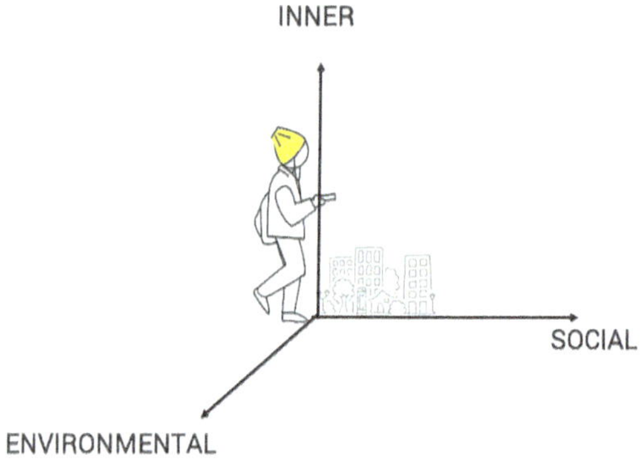

Fig. 4
The three-dimensional (inner, social, environmental) model of the life journey of a young person toward wholeness and connection.

merged with perspectives from a range of global stakeholders and supported by findings from a global research Delphi synthesizing inputs from over 500 informants in over 50 countries, led to the development of a multidimensional framework mapping youth's mental health and broader life journeys. The resulting *Mental Health Friendly Cities (MHFC) Framework* is grounded in the three-dimensional model of the intersecting and interacting dimensions of youth's life journeys toward wholeness and connection, which are shown in Fig. 4. These are as follows:

- Inner dimension, which links a young person's emotions, thoughts, behaviors, and spirituality to a vast constellation of past and present experiences, and future aspirations, and includes key adaptive capacities and skills for navigating life.
- Social dimension, which reflects a young person's interactions and relationships with the people around them, from immediate family and peers to the broader community and societal levels.
- Environmental dimension, which includes a young person's interactions and relationships with the natural and built environments around them, from the immediate space they inhabit to the places they grow up in and broader environments, shaped by structural determinants such as conflict and climate change.

As youth navigate through their life journeys, they encounter crucial transition points where some paths lead to flourishing, compared dramatically to the opposite, whereby other paths lead to crisis, mental ill health, and other negative outcomes. Importantly, the overall mental health and well-being of the young person are impacted by factors in each of the inner, social, and environmental dimensions. Given that a young person's life journey consists of the intersections and interactions across these dimensions, the overall impact of changes in these dimensions cannot be decoupled. Change in one dimension can have amplifying or reinforcing effects across the others. For example, a student who undergoes inner growth may be better equipped to build supportive relationships and to work to create more supportive spaces in their school. Further, more supportive relationships and a more supportive school environment may help unlock further inner growth for the student. This framework incorporates and expands beyond a narrow biomedical model of mental ill health by addressing core dimensions of the human experience, such as spirituality, belonging, hopefulness, and relationship to nature. Rather than focusing on alleviation of symptoms in acute cases of anxiety or depression alone, this framework holistically considers youth's life journeys, which can be significantly shaped by acute experiences of mental ill health. This evolving framework is the conceptual grounding for all *citiesRISE*'s work.

Youth-driven transformation

citiesRISE's early progress has confirmed that the creativity, wisdom, passion, and lived experience of youth are crucial to driving progress. *citiesRISE*'s approach to mental health engages youth as the center of transformation, working in partnership with other stakeholders. This approach is rooted in a strengths-based philosophy and commitment to leveraging the unique insights and skills of youth, mental health professionals, researchers, system leaders, community leaders, and other partners. In this partnership, each partner brings unique skills to the table. For example, youth have not only the insights to know what types of interventions and delivery methods will resonate with them, but they also offer a range of needed knowledge and skills. Mental health professionals bring technical expertise and knowledge of the evidence for, and strengths and limitations of, different interventions to help maximize effectiveness and appropriateness. And, system and community leaders bring the political capital and access to resources required for systemic change. *citiesRISE*'s approach brings these different stakeholders together to collaborate as full, respected partners, laying the foundation for more robust and equitable mental health systems that promote well-being and are more likely to be sustained.

citiesRISE's approach to youth-driven transformation differs significantly from the status quo of designing and delivering interventions top-down with little to no meaningful engagement of youth. It also differs from approaches where youth are engaged but only in a single, limited advisory capacity. The *citiesRISE* approach matches youth's unique insights and skills to specific types of work, where youth work in partnership with others as innovators, researchers, advocates, and in service of others, based on their interests and capacities.

Core ingredients of effective interventions

A significant feature in *citiesRISE*'s strategy has been a shift away from an initial focus on traditional scaling pathways for innovations toward a focus on distilling and spreading the core ingredients of effective interventions that drive significant, durable mental health improvements. The range of research and innovation efforts of *citiesRISE*, together with evidence generated by other innovators, has resulted in an initial set of core ingredients: content that supports youth's inner growth, such as cultivating hope; development of key protective factors, such as emotional regulation and social connectedness; affective learning facilitated through creative and contemplative practices; intervention components that target social and environmental factors in youth's lives; and meaningful youth engagement.

Leveraging technology and technology-enabled innovations

Across *citiesRISE*'s areas of work, technology and technology-enabled innovations present significant opportunities. Most global resources are currently invested in digital solutions that provide individualized support. These solutions are important since applications of technology enable customization while deploying at scale, aggregation of data, and targeting of resources where most needed. However, these solutions largely aim to increase access to services provided by current systems, without addressing issues that contribute to youth falling through the cracks. *citiesRISE* has identified several ways that technology-enabled innovations can incorporate the paradigm shift in how mental health is conceptualized and addressed, including: innovations leveraging technology to address factors in youth's broader social and environmental contexts, technology-enabled innovations that support youth engagement and accountability in mental health services and systems, applications of technology to support innovators in mental health to learn and collaborate with each other, and applications of technology to support youth mobilization and movement building.

The result: Sharpened citiesRISE strategy

In 2021 *citiesRISE* undertook a strategy refresh process to chart the path forward. The sharpened strategy reflects the input and wisdom of countless youth and other experts, with three pillars:

(1) Research and Innovation—Conduct research and innovation projects, rooted in a network of city innovation hubs, that build the evidence base for and grow the set of core ingredients of effective interventions.
(2) Learning and Acceleration—Strengthen and spread the core ingredients through global communities of social innovators, cities, and youth, supported by the *Accelerator and Learning Platform.*
(3) Mental Health Friendly Cities (MHFC) Movement—Grow a global movement of MHFCs driving collective action among youth, system leaders, and innovators in cities around the world toward system changes centering on the core ingredients and driving the paradigm shift.

The way forward

We are at a unique moment in history where the importance of youth mental health is recognized more than ever before. However, median country spending on mental health remains at less than 2% of total government health spending globally, with most spending going toward treatment of acute cases of mental ill health (Vigo et al., 2019). The *citiesRISE* experience has confirmed the urgent need for not just increased access to care but also a fundamental paradigm shift. The scale and complexity of current challenges in youth mental health require massively scaled up investment and action.

Our hope is that *citiesRISE*'s experience, shared in this chapter, demonstrates the importance of transdisciplinary and youth-driven approaches, and highlights the readiness of youth, communities, experts, and other stakeholders in cities around the world to drive change. As countries continue to address the COVID-19 pandemic, through recovery and beyond, we have an opportunity to build forward better. We need collaboration across disciplinary, sectoral, generational, cultural, geographical, and other boundaries as part of the solution to one of the most significant global challenges and opportunities of our time.

Acknowledgments

We wish to acknowledge the inaugural cities and local teams and partners, as well as the Steering Group that supported the early thinking of the citiesRISE platform: Helen Herrman, Chris Underhill, Graham Thornicroft, Tom Insel, Vikram Patel, Karlee Silver, Andrew Stern, Pamela Collins, Lukoye Atwoli, Ada Williams Prince, John Boyd, Shweta Rawat, Patricio Marquez, Yvonne Ochieng, Charlene Sunkel, R. Thara, Craig Kramer, Pete McDermott, and Norman Sartorius. We also thank Co-Impact, Pivotal Ventures, Rural India Supporting Trust, Sutter Health, Grand Challenges Canada, and Johnson & Johnson for their support in establishing the formative phase of the platform. We would also like to thank Olivia Biermann and Stefan Swartling-Peterson at Karolinska Institute for their partnership to deepen the evidence base for transformative change on adolescent mental health.

References

Clark, H., Coll-Seck, A. M., Banerjee, A., Peterson, S., Dalglish, S. L., Ameratunga, S., … Costello, A. (2020). A future for the world's children? A WHO-UNICEF-Lancet Commission. *The Lancet*, *395*(10224), 605–658.

de Miranda, D. M., da Silva Athanasio, B., Oliveira, A. C. S., & Simoes-e-Silva, A. C. (2020). How is COVID-19 pandemic impacting mental health of children and adolescents? *International Journal of Disaster Risk Reduction*, *51*, 101845.

Hopeful Minds. (2022). *Hopeful Minds Curriculum*. Retrieved from: https://hopefulminds.org/hopeful-minds-curriculums/.

Kumar, S., Hanisha, A., Goutham, S., Priya, D., Chiranjeevi, Azariah, F., Sinha, M., Raghavan, V., & Thara, R. (2021). *Effectiveness of mental health literacy interventions on knowledge, attitude, and stigma towards mental health among secondary school students in Chennai, South India: A cluster randomised controlled trial*. Trial registration ID: CTRI/2019/10/021488.

Kumar, S., Mohanraj, R., Lidiya, A., Karthikeyan, D., Kannan, L., Azariah, F., … Thara, R. (2021). Exploring perspectives on mental well-being of urban youth from a City in South India. *World Social Psychiatry*, *3*(2), 87.

Kumar, S., Sinha, M., Skeen, S., Laurenzi, C. A., Tomlinson, M., et al. (2021). *Youth and mental health science: Developing a model for young people's journeys engaging with mental health information and support through youth-driven research spanning India, Kenya, Rwanda, and South Africa*. Report commissioned by Wellcome Trust.

Kutcher, S., & Wei, Y. (2018). *Mental health & high school curriculum guide: Understanding mental health and mental illness*. Retrieved from: http://mentalhealthliteracy.org/wp-content/uploads/2018/05/FINAL-Canada-Guide-2018-Full-Online.pdf.

Lo, C. C., & Cheng, T. C. (2012). The role of social structural factors in treatment of mental health disorder. *Journal of Mental Health, 21*(5), 430–438.

McAllister, C. (2008). Child friendly cities and land use planning: Implications for children's health. *Environments, 35*(3), 45.

Morris, Z. S., Wooding, S., & Grant, J. (2011). The answer is 17 years, what is the question: Understanding time lags in translational research. *Journal of the Royal Society of Medicine, 104*(12), 510–520.

Patel, V., Saxena, S., Lund, C., Thornicroft, G., Baingana, F., Bolton, P., … UnÜtzer, J. (2018). The lancet commission on global mental health and sustainable development. *The Lancet, 392*(10157), 1553–1598.

Patton, G. C., Sawyer, S. M., Santelli, J. S., Ross, D. A., Afifi, R., Allen, N. B., … Viner, R. M. (2016). Our future: A lancet commission on adolescent health and wellbeing. *The Lancet, 387*(10036), 2423–2478.

Radner, J. M., Ferrer, M. J., McMahon, D., Shankar, A. H., & Silver, K. L. (2018). Practical considerations for transitioning early childhood interventions to scale: Lessons from the saving brains portfolio. *Annals of the New York Academy of Sciences, 1419*(1), 230–248.

Sharma, M., Idele, P., Manzini, A., Aladro, C. P., Ipince, A., Olsson, G., … Anthony, D. (2021). *Life in lockdown: Child and adolescent mental health and well-being in the time of COVID-19*. UNICEF.

Sinha, M., Collins, P., & Herrman, H. (2019). Collective action for young people's mental health: The citiesRISE experience. *World Psychiatry, 18*(1), 114.

Sinha, M., Kumar, M., Zeitz, L., Collins, P. Y., Kumar, S., Fisher, S., … Atwoli, L. (2020). Towards mental health friendly cities during and after COVID-19. *Cities & Health*, 1–4.

Teenergizer. (2023). *Website*. Retrieved from: https://teenergizer.org/en/.

Time. (2021). *An open letter to the global media by Greta Thunberg and Vanessa Nakate*. Retrieved from: https://time.com/6111851/greta-thunberg-vanessa-nakate-open-letter-media/.

United Nations. (2022a). *Transforming our world: the 2030 Agenda for Sustainable Development*. Retrieved from: https://sdgs.un.org/2030agenda.

United Nations. (2022b). *UN-Habitat and youth*. Retrieved from: https://www.un.org/youthenvoy/2013/08/un-habitat-and-youth/#:~:text=60%25%20of%20this%20urban%20population,young%20people%20and%20their%20societies.

United Nations Children's Fund. (2021). *The state of the World's children 2021: On my mind—Promoting, protecting and caring for children's mental health*. New York, NY: UNICEF.

United Nations Population Fund. (2022). *State of World Population 2022: Seeing the Unseen*.

VanderWeele, T. J. (2017). On the promotion of human flourishing. *Proceedings of the National Academy of Sciences, 114*(31), 8148–8156.

Vigo, D. V., Kestel, D., Pendakur, K., Thornicroft, G., & Atun, R. (2019). Disease burden and government spending on mental, neurological, and substance use disorders, and self-harm: Cross-sectional, ecological study of health system response in the Americas. *The Lancet Public Health, 4*(2), e89–e96.

World Health Organization. (2021). *Adolescent mental health*. World Health Organization. Retrieved May 27, 2022, from: https://www.who.int/news-room/fact-sheets/detail/adolescent-mental-health.

CHAPTER 67

Investigating children's health and wellbeing in a digital world: A rapid review of the evidence on children's internet use and outcomes☆

Mariya Stoilova[a], Sonia Livingstone[a], and Rana Khazbak[b]
[a]Department of Media and Communications, London School of Economics and Political Science (LSE), London, United Kingdom, [b]School of Education, Communication and Society, King's College London, London, United Kingdom

Editor's introduction

With youth using technology tools at an increasing rate, especially due to the COVID-19 epidemic, it is important to assess related risks and opportunities. This chapter is an abstract of key points specifically on the topic of health and mental health, in a much more extensive Innocenti Discussion Paper, 2021.

Executive summary

Children's lives are increasingly mediated by digital technologies, yet our knowledge of how this affects their well-being is far from comprehensive. We know, for example, that the online environment exposes children to new ideas and more diverse sources of information. The use of digital technologies can expand their opportunities, reduce inequalities, and contribute to the

☆ **Note:** Printed with permission, an abstracted from a report: UNICEF Innocenti Discussion Paper 2020-03 February 2021. References are accessible in that original report. The material is available free of charge on the UNICEF Office of Research—Innocenti website and here: https://www.unicef-irc.org/publications/1183-investigating-risks-and-opportunities-for-children-in-a-digital-world.html
Citation: Stoilova, Mariya; Livingstone, Sonia; Khazbak, Rana (2021). Investigating Risks and Opportunities for Children in a Digital World: A rapid review of the evidence on children's internet use and outcomes, Innocenti Discussion Papers no. 2021-03.

Resilient Health. https://doi.org/10.1016/B978-0-443-18529-8.00067-6

realization of children's rights. We also know that when children seek information online and want to learn, they risk being exposed to inappropriate or potentially harmful content. Yet, when it comes to determining the long-term effects of internet use and online experiences on children's well-being, mental health, or resilience, the best we can do is make an educated guess.

This is just one evidence gap among many that need to be filled in order that society can support children's positive use of digital technologies, develop children's skills, and protect those who are vulnerable. Filling these gaps would benefit children. Their education, relationships, entertainment, and participation in a connected world increasingly depend on digital technologies. Filling these gaps would also help to guide policy and program responses and maximize the potential of technological advancements. Our need for this knowledge has become even more acute as internet use rises during the global COVID-19 pandemic.

This chapter identifies, evaluates, and synthesizes what has been learned from the most recent research about children's experiences and outcomes relating to the internet and digital technologies. It aims to inform policymakers, educators, child protection specialists, industry, and parents on the latest and best evidence, and it proposes a future research agenda.

The conceptual framework for the chapter is based on the well-established models of children's internet use that underpin research by three international projects: EU Kids Online, Global Kids Online, and Disrupting Harm (Fig. 1).

What do we know about how children's internet use can contribute positively to child outcomes and well-being?

Children around the world primarily use mobile devices to access the internet for information and entertainment, to connect with their friends, and to engage with their community. Increased internet and device uptake by society diminishes the gaps in internet access experienced by different groups of children. In low-access contexts, children lead the way in the adoption of new technologies and are online more than adults.

Some children experience the benefits more strongly than others. Children from global South countries and poorer backgrounds face significant inequalities in relation to access to devices and the internet, the quality and cost of connectivity, and the availability of competent support from those around them. Such inequalities create a domino effect in other areas, resulting in cross-country inequalities in online activities and digital skills.

The online environment exposes children to new ideas and more diverse sources of information. This helps children to engage with others and become aware of different views. The internet is also an important source of physical and mental health knowledge for children. They value that it offers anonymity, access to reliable information, peer experiences, and support-seeking options.

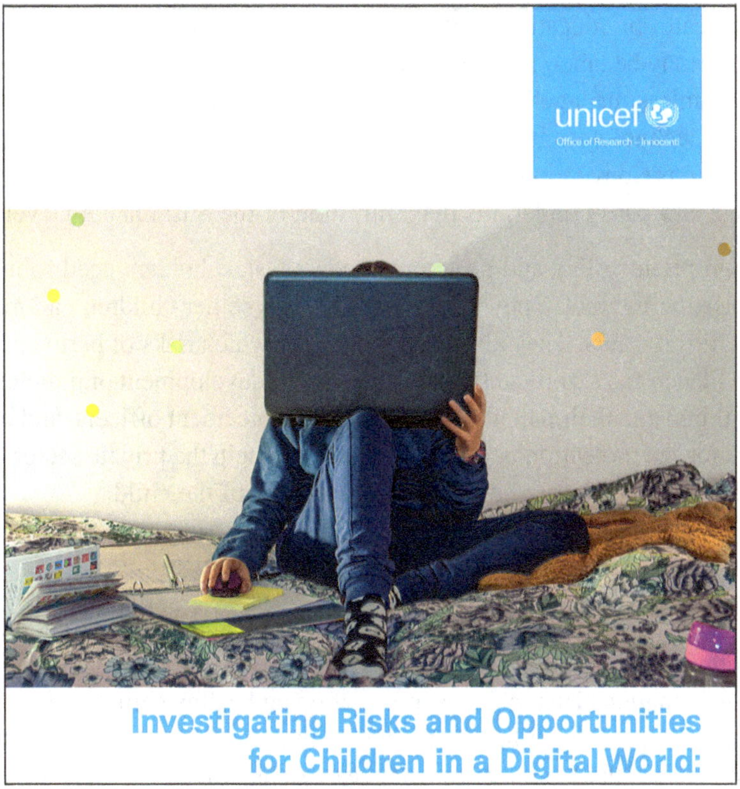

Fig. 1

Cover of the full report. *Credit: Cover Photo: © UNICEF Innocenti/Mannocchi. A lower secondary school student does her homework online, Rome, Italy, January 2021.*

The internet can have a positive effect on mental health and well-being. Children who receive mental health support online or who can talk about their problems with friends online have improved outcomes.

What happens to children offline affects their experiences and outcomes online. Their age, gender, home context, and relationships with family, peers, and school are all important factors. So too are such factors as the safety of their neighborhood or whether they are overweight, engage in substance abuse, take sexual risks, or are a survivor of violence or sexual abuse.

How can children's internet use amplify the risk of harm and potentially undermine their well-being?

Introduction

An estimated one in three internet users globally is a child. This number is likely to be even higher in the global South. Access to digital technology can help children to develop the skills and literacies they need to learn and prepare for life in an increasingly digital society. It can give them access to health or education services or information on topics that are important for young people but may be taboo in their societies. Yet internet access may also compound risks to children that, unless managed, could undermine their well-being. Most importantly, engagement with technologies is increasingly a means of realizing children's rights, as set out in the United Nations Convention on the Rights of the Child. Use of digital technology is a key driver of change and partly underpins the realization of the Sustainable Development Goals.

To develop appropriate policy and program responses, stakeholders need robust evidence on how children use the internet. They need evidence on whether children can make use of the many opportunities it offers. They also need to identify which risks of harm children may face online. Such evidence can significantly contribute to the development of policies and programs that support children and their parents, teachers, law enforcement officers, and others who bear responsibilities for the protection of children. It can also help the private sector to protect child rights and make decisions informed by the best interests of the child.

Digital health

On an individual level, digital health refers to the use of the internet and digital technologies to maintain or improve health and lifestyle. It may involve searching for a broad range of health-related information. This includes information on healthy eating and nutrition, exercise or sports, sexual or reproductive health, mental health, disease or injury, and treatment or medication. Digital health also refers to using specific applications, software programs, or devices to monitor, plan, share, or discuss issues related to health or lifestyle. Such issues include weight loss, exercise, food intake, and mood.

In an evidence review of 359 empirical studies of children's Internet use published since 2016, 34 concerned mental health and well-being, of which six concerned digital health. The six studies reviewed mainly cover countries in Europe and North America, including Germany, the Netherlands, Northern Ireland (in the United Kingdom), and the United States. One study was conducted in Nigeria.

Key findings

- The digital environment is an important source of health information and support for children, both in relation to physical and to mental health.
- Children value that the internet provides anonymity, access to reliable information, glimpses of the experiences of peers, and support-seeking options.
- Children use a variety of online channels to access digital health content. They usually try to judge the reliability of the source and the particular benefits it provides, but this can sometimes be difficult for them.
- Not everyone is equally able to benefit. Looking for health information online is positively associated with having a higher level of digital skills.
- Digital health outcomes are rarely discussed and the evidence on effectiveness is somewhat mixed and insufficient.

Several studies explore how children seek health information online. They show that trusted and high-quality online health information is important to young people. A survey in Germany of 1,162 students aged 13–16 years shows that the aspects most frequently cited as important for digital health information are easily comprehensible wording (cited by 88 per cent of respondents), clear layout (81 per cent), and perceived reliability of the website's publisher (79 per cent). The visual style of a website and whether it is explicitly aimed at students are deemed less important. Girls are nearly twice as likely as boys to attach importance to the publisher's reliability, especially those girls who do not have a migrant background or who attend a school in the highest academic tier.

Children also evaluate the reliability of online information by determining whether the same information appears on multiple websites, and if it corresponds with what they know from offline sources such as family, friends, or school. Websites associated with medical institutions, such as pharmacies or healthcare providers, are generally considered reliable, as are websites explicitly aimed at educating youth about sexual and reproductive health. 50 Forums with user-generated content are seen as beneficial for offering glimpses of the experiences or opinions of others, and for providing opportunities for informal support through online peer networks.

A survey in Nigeria of 1,186 students aged 11–19 years found that 63 per cent of children and young people use the internet to seek health information on a weekly basis. The most popular search topics include: healthy diet and nutrition; body size, shape, and image; sexually transmitted infections; and other sexual issues. The availability of easily accessible online information is particularly valuable for the hard-to-reach children who struggle to seek help face to face. A study of 527 boys from Northern Ireland aged 11–16 years found that one quarter of those who seek information online will not tell anyone about a mental health issue. They place particular value on the anonymity of digital health information, resources, and support services.

The internet also provides opportunities for children to access information on sexual health and relationships, mental health support, and general health services. A study in the Netherlands shows factors related to children's identity and circumstances may influence how they engage with online information. For example, LGBTQI children and those with more knowledge about sex and sexuality are more likely to consult sexual information online. Girls, children who know more about sex and sexuality, and those who communicate with friends (rather than parents) about sex tend to seek sexual information from professional websites. Boys and children with low sexual self-esteem and high sexual curiosity are more likely to use interactive, user-generated online content.

The study in Nigeria revealed that children who have more advanced digital skills are more likely to use the internet for health information. The study in Northern Ireland found no significant impact of economic status, geographical location, or educational attainment on the online help-seeking habits of boys in relation to health. The study suggests that boys who use online help-seeking to supplement offline support may gain more mental health benefits. Boys who have larger online friend networks, and those who have confided in online friends regarding personal problems, also have better mental health outcomes. This suggests that a lack of online social support may create inequalities among children. Long-term digital health outcomes are rarely discussed. The evidence on effectiveness is both insufficient and somewhat inconsistent.

A review of the evidence on changes in children's health behavior found that mobile health interventions can have a positive effect on children's health outcomes.

Furthermore, interventions supported by caregivers produce larger changes in health outcomes, on average, than those that exclusively target children. Yet, another review of the evidence on the effects of mobile health, including use of wearable activity trackers to increase physical activity outcomes among healthy children, found no evidence of an effect among all interventions.

Limitations and suggestions for further research

- More evidence that is methodologically robust and comparable is needed to assess how children benefit from digital health information, resources, and support services. The conclusions from current findings are limited by the small number of studies, differences in measurement, dissimilarities in the digital health content assessed, and the methodological challenges of the study designs.
- More evaluations are needed of digital health interventions. Another area for future research is to explore how the relationship between youth health behaviors and outcomes is related to factors at the individual, family, and community level, and influenced by healthcare systems.
- Further analyses of the relationship between seeking mental health information and support and children's exposure to online harm could be helpful, especially in relation to targeting online harm interventions.

Mental health and well-being

The studies we reviewed typically offer a narrow measure of children's psychological well-being. A more comprehensive model is needed. A promising solution can be offered by the work of Ryff, an American academic and psychologist, which is among the most popular conceptualizations of psychological well-being. Her model of well-being incorporates measures of six dimensions: self-acceptance, positive relations with others, personal growth, purpose in life, environmental mastery, and autonomy. This model has been successfully applied by the Children's Society in its international survey of children's well-being (Children's Worlds), which in wave 3 involves child participants aged 8–12 years from

40 countries around the world. This research makes connections between well-being and children's lives and daily activities, their time use, and their own perceptions and evaluations of their well-being. Ryff's model can be supplemented by measures of depression, self-harm, and life satisfaction.

It is worth noting that well-being may be positioned as an outcome or as an input, for example, as a measure of child vulnerability.

As already noted, we identified 34 studies on digital technology and its relation to mental health and well-being, most of which are based in Europe (Belgium, Estonia, Hungary, Italy, Lithuania, the Netherlands, Northern Ireland [United Kingdom], Norway, Spain, Sweden, and Wales [United Kingdom]) and North America (Canada and the United States). A smaller number of studies address the context in parts of Asia and the Middle East (China, the Islamic Republic of Iran, Israel, the Republic of Korea, Taiwan Province of China, and Thailand).

Key findings

- Long-term outcomes related to children's well-being, mental health, or resilience are rarely explored. The evidence on how factors related to digital device and internet use may intersect with offline factors and predict child well-being requires development.
- The evidence on the impact of children's internet use on mental health and well-being is mixed and inconclusive. Some studies show a positive association between internet use and negative outcomes, such as anxiety, depression, suicidal thoughts, and panic disorder, while others show a negative association, and some studies show no relationship at all.
- In most cases, there is only evidence about an association, not causality. A child who uses the internet excessively and experiences poor mental health may be using digital technologies as a way of coping with preexisting issues.
- The internet can have a positive effect on mental health and well-being. Children who receive mental health support online or who can talk about their problems with friends online have reduced depression levels.

A recent UNICEF review on the impact of digital technology use on children's mental health and well-being found that the evidence to date is largely inconclusive. It concludes that children's use of digital technology is more often assessed in terms of the time spent online rather than by measuring such factors as the type of online activities children engage in or the context of internet use. This evidence review found similarly mixed results. While some studies show a positive association between internet use and mental health problems, others find a negative association or no association at all. For example, findings from cross-sectional and longitudinal studies conducted in China, the Islamic Republic of Iran, the Republic of Korea, and the United States show a significant association between increased digital, social media, or internet use and higher levels of anxiety, depressive symptoms, suicidal ideation, panic disorder, and attention deficit hyperactivity disorder (ADHD). This association is stronger among children who do not live with either of their parents.

In contrast, three longitudinal studies from Canada, European countries, and the United States found no significant direct impact of the time children spend using digital technologies, such as social media, the internet, and texting, on their mental health in terms of depression and anxiety. Another study found that moderate social media use was associated with the highest levels of mental well-being, compared with either too much or too little social media use.

It is plausible that these conflicting results stem from the different ways in which digital media use is measured. Comparing longitudinal studies, it seems that those that find no relationship with mental health measure use in terms of hours spent online per day or per week. Studies that do find an association measure the number of times per day or per week that children use certain digital platforms or do specific activities such as sending messages or using social media.

Several studies show that what matters for children's mental health is not merely the time spent using the internet and social media, but rather how children use this time, what they experience online, and the consequences of that use. More specifically, it is the negative perceived impact of high levels of internet use or social media stress on sleep loss and withdrawal (negative mood when the internet is inaccessible) that indirectly mediates the adverse effect on mental health over time. Another study found that children who report frequent, daily use across all social media platforms suffer worse mental health problems compared with low social media use subgroups.

The possible causal relationship between intense internet use and decreased mental health and well-being is rarely studied. Hence, it is often difficult to say whether excessive use of digital technologies is the cause or consequence of poor mental health. In fact, one study found evidence of reverse causation, where greater depressive symptoms predicted more frequent social media use, although only among girls. Most studies do not show significant gender differences in these associations.

A study in the United States of 5,593 students aged 12–17 years examined digital self-harm, defined as the "anonymous online posting, sending, or otherwise sharing of hurtful content about oneself." The authors found that it was significantly correlated with experiences of depressive symptoms. A qualitative study with children with a previous history of self-harm revealed that the internet can act as a catalyst for children's self-harm behaviors, normalizing them through their participation in online communities where others share self-harm practices and techniques, and through easy access to images that invoke a physical reaction and inspire behavioral enactment.

Another group of studies demonstrates the positive impact of internet use on mental health and well-being. Higher levels of children's well-being are positively associated with receiving mental health support online and with speaking to friends online regarding personal problems.

Similarly, decreased levels of depression are associated with freely revealing negative emotions online. Social media use can be a positive coping mechanism for socially rejected children because they tend to use it to share their personal difficulties. A positive effect on mental health is also observed when internet use is perceived to make life more meaningful, improve school grades, or enhance performance. Thus, besides time spent online, it is important to understand the ways in which children use the internet and social media, what they are exposed to while online, and the mediating perceived consequences of that use, to have a better idea of the impact on their mental health.

Digital technologies have profoundly changed childhood and adolescence. The internet and the means to access it, such as tablets and smartphones, along with social media platforms and messaging apps, have become integral to the lives of youth around the world. They have transformed their education and learning, the way they make and maintain friendships, how they spend their leisure time, and their engagement with wider society. UNICEF's *State of the World's Children 2017: Children in a Digital World* report reveals that one in three internet users is younger than 18 years and 71% of 15–24-year-olds are online, making them the most connected age group worldwide.

However, the so-called digital divide is substantial: 346 million youth are not online, with African adolescents the most affected (60% are not connected compared with 4% in Europe). Young people who lack digital skills, live in remote regions, or speak a minority language are also being left behind in harnessing the opportunities that digital technologies can provide. These benefits include access to education, training, and jobs, which can help break intergenerational cycles of poverty, and access to news and information sources that can help protect their health, safety, and rights.

Along with the substantial opportunities the digital age brings, come a diverse range of risks and harms. Digital technologies have increased the scale of child sexual abuse and exploitation. Child sex offenders have increased access to children through unprotected social media profiles and online gaming forums.

Technological advances have allowed individual offenders and trafficking rings to evade detection through encrypted platforms and the creation of false identities and have enabled them to pursue multiple victims at the same time.

Young people are not only at risk from adult exploitation; children can also obtain sexual or intimate photos of a peer and share them with or without the individual's consent. Digital advances have meant that bullying is no longer left at the school gates; *cyberbullying* is a new method for bullies to hurt and humiliate their victims with the click of a button. Words and images posted online that are designed to cause harm are also difficult to delete, increasing the risk of revictimization.

Concerns have also been raised about the time children spend using digital technology and its effect on their physical activity and mental health. Evidence suggests moderate use of digital technology can be beneficial to children's mental well-being, whereas excessive use can be detrimental. Children's social relationships seem to be enhanced by digital technology, especially since most of their social circle is now online. Evidence on physical activity is mixed, and better research is needed in this area.

The effect of the content children encounter online should be a focus of future studies. Some websites and apps present a risk to the health of vulnerable youth, such as those that promote self-harm, suicide, or anorexia. Others support underage activities, such as gambling. A 2017 survey found that 3% of 11-16-year-olds in the UK have spent some of their own money on online gambling, and 7% have used their parents' accounts, with or without their permission.

Despite these issues, the popularity and use of digital technologies will continue to grow. However, an intergenerational gap exists in digital knowledge and literacy. Teachers and parents need training to teach digital skills and online safety to children. They are also crucial to helping young people assess reliable news and information sources and navigate the pressures social media, as highlighted in a report released on January 4 by England's Children's Commissioner. Schools are also an important forum for discussing cyberbullying, and antibullying programs need to include online harassment.

Some children have never been online or have little access and are missing out on the benefits of connectivity. UNICEF recommends that market strategies to foster competition among providers, and tax and other incentives for technology companies could help reduce the cost of access. Digital companies, meanwhile, have a responsibility to protect young users from harm. The WePROTECT Global Alliance to End Child Sexual Exploitation Online, involving technology companies, international organizations, and countries, is one good example of a multistakeholder, coordinated approach to tackle this threat that has no borders.

Children and adolescents have the most to gain and are most at risk from digital technologies. They must be at the forefront of national and global digital policies, not only to protect them from online harm but also to allow technology to help them fulfill their full potential.

CHAPTER 68

Impact of technology on health and well-being of the aging population in the COVID-19 era

Florence L. Denmark[a], Taylor Mulligan-Stark[a,b,c], Anna Stauber[a,c,d], and Judy Kuriansky[e,f,g,h,i]

[a]Department of Psychology, Pace University, New York, NY, United States, [b]McShane Center for Psychological Services, Pace University, New York, NY, United States, [c]The NGO Committee on Ageing, New York, NY, United States, [d]Duluth Aging Support, Duluth, MN, United States, [e]Department of Clinical Psychology, Teachers College, Columbia University, New York, NY, United States, [f]International Association of Applied Psychology, Indianapolis, IN, United States, [g]World Council for Psychotherapy, Vienna, Austria, [h]United Nations Group of Friends of Mental Health and Well-being, New York, NY, United States, [i]Global Council SDG3 (Health), "SDGs in Action" Initiative, Dubai, United Arab Emirates

Technology has become a valuable resource for managing everyday life, especially given conditions of the COVID-19 pandemic, in order to maintain social interactions and relationships but also to survive by being able to access healthcare and even order necessities while sheltering or quarantining at home. Electronic devices, tablets, computers, internet access, and an array of applications for making life more convenient and efficient, but constant advancements require new learning. Adjustments to this evolving world with technology are challenging for older generations compared to younger people. But, supporting older generations with technology is crucial for their thriving and also for an inclusive society (Chen et al., 2021). A growing field of interest involves the need to take action to protect older adults and contribute to their physical and mental wellness especially in the ongoing COVID-19 recovery period (Osama et al., 2020). This chapter explores issues regarding the aging population and technology, made more urgent due to lifestyle adjustments required by the COVID-19 pandemic.

Ubiquitous tools, such as smartphones, computers, and tablets, have become increasingly more advanced and complicated, accompanied by multiple choices of wireless services and data plans, online payments, applications, security measures, and setting passwords. Adapting to this new age can feel intimidating and unattainable by older persons, especially upon first interaction during the pandemic. In fact, older adults were disproportionately affected by the COVID-19 pandemic with higher rates reported of environmental, social, economic, and healthcare inequalities (Barth et al., 2021).

Resilient Health. https://doi.org/10.1016/B978-0-443-18529-8.00068-8

Background

One study, as early as 1992, found that both younger and older participants reported their belief that it is atypical for older persons to engage capably with technology (Ryan et al., 1992). Added to this pervasive belief and rapid advances in technology, a multitude of factors contribute to technology apprehension among the aging population. Prepandemic research has shown that when compared to younger generations, older generations are less flexible and able to change and adapt to new technologies, since such devices are user-friendly primarily for younger cohorts, due to features like small buttons, small icons, and only online instruction manuals (Hunsaker & Hargittai, 2018). According to the Pew Research Center (2019), nearly three out of four (73%) Americans over age 65 use the internet, with over half (59%) having broadband services at home; however, when compared to younger cohorts, 100% of adults between the ages of 18 and 29 use the internet and three out of four (75%) in this age group use broadband services. These figures are drastically lower in less developed countries. Many factors contribute to lower usage of technology by older persons, including infrastructure-based limitations, inability to access instructions or manuals on how to efficiently use technology, and limited access to support groups (Pew Research Center, 2019). The pandemic forced a change to everyday lifestyles, whereby using technology became essential for older adults to remain connected.

While all age groups were affected by COVID-19, older adults were more likely to experience severe physical complications, with mortality increasing by 8% in patients between 70 and 79 years old and up to 14.8% in persons 80 years and older, compared to younger adults with similar health status before the COVID-19 epidemic (Sanchez-Rodriguez et al., 2020). Yet, government-imposed lockdowns in response to the pandemic led to more people being sequestered, which created the possibility of long-term isolation experienced by all ages. Neglect and isolation have been acknowledged as being particularly difficult for older adults; thus, staying connected through virtual means became imperative to support healthy aging (Batsis et al., 2021). For example, older persons who had limited access to technology during the COVID-19 pandemic were faced with making tough choices when following emerging CDC guidelines discouraging in-person contact, about whom to contact and rely on to meet their needs and to stay connected. Instructions to maintain physical distance from human contact and to disinfect everything before it entered their homes, led older individuals to feel further secluded and detached. Those living with family members were able to fare better, especially if younger family members were present or close by who were knowledgeable to support or guide older persons through using the internet for online services while sheltering at home, for example, ordering groceries and using delivery services for medical prescriptions or other needs.

The restrictions imposed during the pandemic led to pressure for older adults to transition quickly into connecting through virtual means (Weil et al., 2021). While video applications such as Zoom, Skype, and FaceTime had been introduced and used prior to the pandemic, these methodologies were largely only used by those who had the infrastructure in place (e.g., a smart phone, tablet, video-capable computer, and internet access) and working knowledge of how to use such tools. Many older persons globally had limited prepandemic exposure to these technological devices, compounding their worries about the pandemic itself, which altogether deterred older adults from approaching technology with a positive or optimistic outlook (Fang et al., 2019). The pandemic's widespread shutdowns and stay-at-home orders created a gap between the technological "haves" and "have-nots" which widened significantly, causing unprepared populations, such as older and more isolated people, to become even more neglected (Weil et al., 2021). Aside from concerns over physical wellness, throughout the pandemic older people experienced a lack of social support as well as financial constraints barring access to some technology. Such barriers to access made older adults less likely to receive necessary and timely care, which was compounded by cognitive and physical decline, including memory loss, impaired vision, and hearing loss (Wu et al., 2019). Further compounding the problem, these symptoms created additional hurdles to effectively interface with technology (Wu et al., 2019). This complexity of factors led to whether or not an older person was able to become tech savvy in the emergency, as well as whether they were willing to adapt.

Impact of the pandemic: Older persons at risk and technology hesitancy

During the early months of the pandemic, caregivers were required to maintain a 6-foot distance, and to limit contact with vulnerable others, especially older persons, many of whom had preexisting health conditions, requiring extra care (Sanchez-Rodriguez et al., 2020). Older persons in nursing homes were particularly cut off from loved ones during the pandemic, evident in many reports of incidences of aged persons being totally alone or seeing relatives only through windows. Research shows that high levels of social support from family and friends, as well as participation in leisure activities, predict lower levels of loneliness among nursing home residents 65 years or older (Chang et al., 2021). In fact, extensive research has shown that social support improves one's ability to cope with difficult experiences (Rodrigues et al., 2021); however, many older people were particularly secluded during the pandemic and, due to technological limitations, could not reap the psychological benefits social support offers to mental wellness (Chang et al., 2021). To avoid the negative repercussions of the association between social connection and COVID-imposed isolation, one psychologist early on strongly recommended replacing the all-too-comonly used warning to keep "social distance" with a more neutral yet still accurate wording of "physical distance" (Kuriansky, 2020).

The impact of the pandemic on the mental health of older persons is of considerable concern. High levels of perceived loneliness have been linked to negative outcomes, such as anxiety, depression, cognitive decline, and even mortality (Gorenko et al., 2021). Further, the psychosocial effects of the digital divide between those who use technology properly and those who do not use technology appropriately were thrown into sharp focus (Cosco et al., 2021). While technology has the advantage to create a more cohesive and connected lifestyle, older adults have been more likely to find using such tools frustrating, leading to the opposite effect, namely, of increased feelings of alienation. In fact, research has shown that older persons were found to have increased feelings of isolation and digital frustration during the pandemic, which further escalated emotional, social, and psychological distress (Losada-Baltar et al., 2020; Weil et al., 2021).

Research also shows that technology hesitancy exists in greater levels among those with cognitive, auditory, or visual impairments (Fang et al., 2019; Fisk et al., 2009), which are likely to occur in older persons. Yet, some findings support a more positive view that older adults' interest in gaining access to technology accelerated with the onset of the COVID-19 pandemic (National Telecommunication and Information Administration, 1994; Weil et al., 2021). A study by Lee and Coughlin (2014) found that technology is an essential tool for older adults to attain social and physical independence, and when used to its fullest capacity, can increase participation in social activities, predict positive health outcomes, and promote independence. Hesitancy is rooted in fears of the unknown; thus highlighting advantages of adopting a digital mindset is imperative to pave the way forward for a more supportive model of growth surrounding technology and older persons.

Advantages and services for older persons

Features of digital tools can mitigate limitations in older persons' experience. Tools like text enlargement and read-out-loud capabilities can help older adults gain access to services offered online, such as Telehealth and virtual medical visits, and can serve older populations where they are, without requiring time, effort, or money for travel (Weil et al., 2021). Recent research has also provided insight into how older persons are actually using technology, given that the appropriate use of technology can have a direct positive impact on older persons' physical health, for example, when such tools are used to monitor health conditions (Halpert, 2019). Cell phone apps can be used for tracking heart rate, medications levels and schedules, immediate notification in medical emergencies, alerts for accidents or falls, as well as health-related patches, implants, and medication assistance call systems. Technological advances have also offered aid for older persons with mobility problems to access more services at home, especially during emergencies like COVID-19, when many business models pivoted to primarily online services. The lingering effects of the pandemic have also led to intensified development and implementation of e-mental health services and telepsychiatry, aimed to combat feelings of loneliness and alleviate symptoms (Di Carlo et al., 2021).

Although these services are not always within reach of older populations, if accessed, they could be very helpful for older persons' overall physical and mental health and well-being. A prepandemic study to integrate telehealth literacy into the lives of 1052 older patients of a medical clinic in New York City found that once given the necessary skills to utilize a telehealth system, older persons used the service with similar frequency as younger age participants (Greenwald et al., 2018). This and other research suggest that older adults would be amenable to try new technologies and can indeed perceive value in services they provide, as long as access to these tools are not restricted by limitations of affordability, accessibility, and availability

(Lee & Coughlin, 2014). Overall, there are varied barriers to technology adoption than just unwillingness to try (Jiménez-Etxebarria et al., 2021; Lee & Coughlin, 2014).

A literature review in the year 2021 identified 14 research papers from 2010 to 2014 that examined the effectiveness of telehealth technology in coaching health behaviors for older adult samples and found that online platforms of telehealth, when used over a period of time and with the availability of technical support, can be especially suited to older populations (Markert et al., 2021). The findings indicated that with training, older adults may benefit greatly from utilizing virtual healthcare options, paving the way for a promising future for adults even before the pandemic became a global challenge. In fact, some research found that older adults displayed considerable resilience in the face of COVID-19, especially related to telehealth services (Jiménez-Etxebarria et al., 2021), leading to the suggestion that perseverance, in combination with education, is a way forward to help older adults become more fluid with the digital age. Technology access and literacy is imperative for social connectedness and maintenance of physical and mental health through telehealth services and mental health apps, such as the COVID Coach app from the US Department of Veteran Affairs; therefore, it is important for clinicians and patients to communicate about the benefits of online services for mental health to reduce technology hesitancy and encourage technology literacy and patient buy-in of technology-based services (Gould & Hantke, 2020).

In a relevant study with important results, Tech Allies in San Francisco, California, an organization whose mission is to provide companionship and support to isolated seniors, devised an 8-week in-home technology training program to combat technological hesitancy in older adults, providing participants with a tablet, access to broadband, and use lessons (Fields et al., 2021). Self-reports showed significant increases in the levels of technology confidence and usage in the participants. While the provision of the technology tools presented limitations of the study's methodology being replicated in other settings, the results showed that technology literacy training offers promise for older adults to decrease social isolation and improve health outcomes (Fields et al., 2021). Additionally, a study by Older Adults Technology Services (OATS) suggests how COVID-19 accelerated digital dependency and highlighted the importance of digital inclusion of older persons (Weil et al., 2021). In summary, training, willingness, and policy all play a role in how older persons perceive technology and how a digital shift in older adults can lead to a more inclusive society for all.

Action by the United Nations on behalf of older persons

Education is essential, as called for in Goal 4 of the United Nations Agenda 2030 for Sustainable Development, to "Ensure inclusive and equitable quality education and promote lifelong learning opportunities for all." The entire Agenda itself is focused on the importance of "leaving no one behind" (United Nations, 2015). Older persons are among the vulnerable populations who are at risk to be left behind, especially triggered by the COVID-19 pandemic. While older persons are responsible themselves, for example, to overcome technology hesitancy, other bodies—local, national, and international—need to take action in support of older persons. Established in 2010, the Open-Ended Working Group on Ageing (OEWGA) has worked to strengthen the protection of the human rights of older persons. In 2019, the OEWGA turned its focus toward older adults' access to education, training, life-long learning, and capacity building as well as social protection and security (Kuriansky, 2019). In December 2020, the United Nations endorsed "The Decade of Healthy Aging," an initiative created as a catalyst for nations around the globe to work within their own bounds to reframe negative attitudes toward age, foster abilities for older people, adopt universal health coverage, and ensure long-term care for older persons (World Health Organization, 2020). This initiative was created as a partnership and a continuance of the 15-year project surrounding the Sustainable Development Goals (SDGs), guiding the globe toward a more equitable and sustainable future for everyone. The Decade of Healthy Ageing brings together governments, civil society, international agencies, professionals, academia, and individuals and create awareness and accountability around improving the lives of older people, their families, and the communities in which they live (World Health Organization, 2020). Many projects are ongoing under this umbrella to advance research, policy, and programs for active and healthy aging, in countries worldwide including Chile, Japan, Qatar, and the Philippines (Khallouf, 2023).

In 2012, the World Health Organization launched a global initiative to help cities, states, countries, and communities around the world create "age-friendly states and communities." Communities that sign on to this project use resources to improve housing, transportation, the built-environment, city planning connectedness, and other aspects of community living for people of all ages, with particular focus on improving long-term livability factors for older persons. Many countries or regions have a national organization leading age-friendly initiatives in their respective location, including the Age Friendly Ireland programme in Ireland, the Servicio Nacional del Adulto Mayor (SENAMA) in Chile, AGE Platform Europe that helps lead projects in varied European countries, and AARP (formerly called the American Association of Retired Persons) in the United States. As of Spring 2022, twelve of the states in the United States of America have signed on to the list of affiliates of AARP, and 663 individual communities have joined the network, with more joining monthly (AARP, 2022). Individual regions of larger countries are also responding to this need, for example, two Australian governing bodies (the Governments of Queensland and of Western Australia); the Municipalité amie des aînés (MADA) in Quebec, Canada; and the Kanagawa Prefectural Government in Japan (United Nations, 2022a, 2022b).

NGOs helping older persons

The authors of this chapter have all participated in the Nongovernmental Committee on Ageing in New York (NGO COA-NY), dedicated to advancing the rights of older persons and eliminating the stigma of ageism. The committee hosts numerous events to generate awareness and support for older persons. The committee also collaborates with other NGO committees and entities at the UN, including three NGO committees (on Mental Health, the Family, and Migration), as well as the Commission on the Status of Women, the International Federation on Ageing, and others. These entities have taken advantage of the opportunity of increased access that the pandemic presented for wide-scale education and advocacy work to take place on a virtual platform. Coauthors have hosted many virtual events to raise awareness about topics surrounding the digital divide disparities faced by older women, elder abuse, ageism, and challenges faced by older indigenous women and older adults in the workforce (United Nations, 2022a).

In another positive advance, June 15 has been declared as World Elder Abuse Awareness Day (WEADD), introduced by the International Network for the Prevention of Elder Abuse (INPEA) in 2006 and recognized as a United Nations Day by the General Assembly in its resolution A/RES/66/127 adopted in 2011 (Penhale, 2006). This day builds awareness on the importance of strong support for older persons and highlights situations of mistreatment of older adults, particularly older women and older disabled persons, in efforts to protect the rights of older persons. With regard to gender, older women are especially vulnerable, being more frail physically and having less access than men to financial resources (Tuohy & Cooney, 2019). Since individuals 60 years and older are projected to steadily become the majority of the total worldwide population, the events and advocacy by the Committee on Ageing are essential in making a difference in global views of the older population.

Additionally, the United Nations has recognized October 1 to celebrate the International Day of Older Persons (UNIDOP). Officially recognized in 1991, the UNIDOP is jointly hosted by the UN Department of Economic and Social Affairs (ECOSOC) and the NGO Committee on Ageing. This event traditionally took place in person at the United Nations headquarters in New York in the ECOSOC chamber, and pivoted to a virtual platform for 2020 and 2021 during the pandemic, but transitioned in 2022 to a hybrid model, which still allowed inclusion for those who may not have the ability or the access to travel to New York City.

Furthermore, the Open-Ended Working Group on Ageing (OEWGA) was established in 2010 as a UN body dedicated to discussing ways to strengthen the protection of human rights as we age. In April 2023, OWEGA will host a number of events for its 13th Session at the UN Headquarters in New York with the aim to bring awareness to the purpose of strengthening and protecting the rights of older persons. Their current work is grounded in the creation of a UN Convention to protect the Rights of Older Persons similar to protections provided by the Convention on the Rights of the Child (1990) and the Convention on the Elimination of All Forms of Discrimination Against Women (1979). In 2021, a new aging advocacy group was formed out of the executive committees of the International Federation on Ageing, the NGO Committee on Ageing, the International Longevity Centre Canada, The Global Alliance for the Rights of Older People (GAROP), and the International Association of Gerontology and

Geriatrics (IAGG). This new group, called Stand Up For The Rights of Older Persons, was formed as a support system for the OEWGA, in an attempt to move the needle on this prospective convention (International Federation on Ageing, 2021). Aging advocacy organizations throughout the world are pushing this agenda item forward, including organizations in the Philippines (Pasali Philippines), South Africa (Age In Action), Egypt, Jordan, Saudi Arabia, Syria, Tunisia, and the United Arab Emirates (UNFPA Arab States), as well as HelpAge International which is present across the globe from Nicaragua and Chile to Kenya and Laos (United Nations, 2022a, 2022b).

Conclusions and the way forward

The COVID-19 pandemic presented an unprecedented opportunity and requirement for change in the lives of the global community, especially impacting older adults as well as other vulnerable populations. While some older adults show resilience and an ability to adapt to a rapidly changing society with technological advances, others have reservations, and still others have physical conditions limiting their abilities, including to integrate technology into their lives. Programs must be developed and policies put into place that support older persons in the journey toward equal access and adaptation to technology. Comfort around technology must be curated to help people maintain their physical and mental health and remain self-sufficient into older age. Some evidence reveals a growing population of older persons, 65 years and older, that feels positively about the use of technology, faced with the COVID-19 emergency that forced a change in communication and connection as well as access to food, medicines, and basic needs, proving that people can learn and adapt to new things with the right access and support. It is important to note, however, that while evidence shows that technology use can greatly improve the lives of older persons—even allowing connection through a phone or computer screen—in-person contact is still essential for well-being (Rodrigues et al., 2021). As the future is filled with many unknowns, it is imperative that health experts and all stakeholders work together to support older adults to thrive during the new age of technology.

References

AARP. (2022). *AARP network of age-friendly states and communities*. https://www.aarp.org/livable-communities/network-age-friendly-communities/info-2014/member-list.html.

Barth, N., Guyot, J., Fraser, S. A., Lagacé, M., Adam, S., Gouttefarde, P., … Célarier, T. (2021). COVID-19 and quarantine, a catalyst for ageism. *Frontiers in Public Health*, *9*, 589244. https://doi.org/10.3389/fpubh.2021.589244.

Batsis, J. A., Daniel, K., Eckstrom, E., Goldlist, K., Kusz, H., Lane, D., Loewenthal, J., Coll, P. P., & Friedman, S. M. (2021). Promoting healthy aging during COVID-19. *Journal of the American Geriatrics Society*, *69*(3), 572–580. https://doi.org/10.1111/jgs.17035.

Chang, L.-C., Dattilo, J., Hsieh, P.-C., & Huang, F.-H. (2021). Relationships of leisure social support and flow with loneliness among nursing home residents during the COVID-19 pandemic: An age-based moderating model. *Geriatric Nursing (New York, NY)*, *42*(6), 1454–1460. https://doi.org/10.1016/j.gerinurse.2021.08.014.

Chen, A. T., Ge, S., Cho, S., Teng, A. K., Chu, F., Demiris, G., & Zaslavsky, O. (2021). Reactions to COVID-19, information and technology use, and social connectedness among older adults with pre-frailty and frailty. *Geriatric Nursing*, *42*(1), 188–195. https://doi.org/10.1016/j.gerinurse.2020.08.001.

Cosco, T. D., Fortuna, K., Wister, A., Riadi, I., Wagner, K., & Sixsmith, A. (2021). COVID-19, social isolation, and mental health among older adults: A digital catch-22. *Journal of Medical Internet Research*, *23*(5), e21864. https://doi.org/10.2196/21864.

Di Carlo, F., Sociali, A., Picutti, E., Pettorruso, M., Vellante, F., Verrastro, V., Martinotti, G., & di Giannantonio, M. (2021). Telepsychiatry and other cutting-edge technologies in COVID-19 pandemic: Bridging the distance in mental health assistance. *International Journal of Clinical Practice*, *75*(1), e13716. https://doi.org/10.1111/ijcp.13716.

Fang, M. L., Canham, S. L., Battersby, L., Sixsmith, J., Wada, M., & Sixsmith, A. (2019). Exploring privilege in the digital divide: Implications for theory, policy, and practice. *The Gerontologist*, *59*(1), e1–e15.

Fields, J., Cemballi, A. G., Michalec, C., Uchida, D., Griffiths, K., Cardes, H., … Lyles, C. R. (2021). In-home technology training among socially isolated older adults: Findings from the Tech Allies Program. *Journal of Applied Gerontology*, *40*(5), 489–499.

Fisk, A. D., Rogers, W. A., Charness, N., Czaja, S. J., & Sharit, J. (2009). *Designing for older adults: Principles and creative human factors approaches* (2nd ed.). Boca Raton, FL: CRC Press.

Gorenko, J. A., Moran, C., Flynn, M., Dobson, K., & Konnert, C. (2021). Social isolation and psychological distress among older adults related to COVID-19: A narrative review of remotely-delivered interventions and recommendations. *Journal of Applied Gerontology*, *40*(1), 3–13. https://doi.org/10.1177/0733464820958550.

Gould, C. E., & Hantke, N. C. (2020). Promoting technology and virtual visits to improve older adult mental health in the face of COVID-19. *The American Journal of Geriatric Psychiatry*, *28*(8), 889–890.

Greenwald, P., Stern, M. E., Clark, S., & Sharma, R. (2018). Older adults and technology: In telehealth, they may not be who you think they are. *International Journal of Emergency Medicine*, *11*(1), 2. https://doi.org/10.1186/s12245-017-0162-7.

Halpert, J. (2019). *Seven new tech devices for elder care that help seniors live happier, healthier lives.* Retrieved from https://www.cnbc.com/2019/09/12/7-new-tech-devices-for-that-help-seniors-live-happier-healthier-lives.html.

Hunsaker, A., & Hargittai, E. (2018). A review of Internet use among older adults. *New Media & Society*, *20*(10), 3937–3954. https://doi.org/10.1177/1461444818787348.

International Federation on Ageing. (2021). *Stand up for rights of older persons: A call to action focused on Canada and the United States.*

Jiménez-Etxebarria, E., Bernaras Iturrioz, E., & Jaureguizar, J. (2021). Impact of the COVID-19 Pandemic as perceived by older people in Northern Spain. *Psychology Research and Behavior Management*, *14*, 1789–1803. https://doi.org/10.2147/PRBM.S308537.

Khallouf, C. (2023). Advancing health and well-being of older persons at the United Nations and beyond: An integrated and multi-stakeholder perspective. *Psychology Around the World*, *5*(1), 95–101. https://iaapsy.org/site/assets/files/1082/apaw_v5i1_feb_2023.pdf.

Kuriansky, J. (2019). Youth and ageing at the United Nations: Activities, achievements and contributions of the IAAP New York Team. *Applied Psychology Around the World*, *1*(3), 88–94. https://iaapsy.org/site/assets/files/1082/apaw_vol1i3.pdf.

Kuriansky, J. (2020). Words matter in combatting COVID-19: Continuing to call for "Social distancing" can make matters worse. In *Black Star News*. https://www.blackstarnews.com/global-politics/others/words-matter-in-combatting-covid-19-continuing-to-call-for.

Lee, C., & Coughlin, J. F. (2014). PERSPECTIVE: Older Adults' adoption of technology: An integrated approach to identifying determinants and barriers. *Journal of Product Innovation Management*, *32*(5), 747–759. https://doi.org/10.1111/jpim.12176.

Losada-Baltar, A., Jiménez-Gonzalo, L., Gallego-Alberto, L., Pedroso-Chaparro, M., Fernandes-Pires, J., & Márquez-González, M. (2020). "We're staying at home": Association of self-perceptions of aging, personal and family resources and loneliness with psychological distress during the lock-down period of COVID-19. *The Journals of Gerontology: Series B*, *76*(2), e10–e16. https://doi.org/10.1093/geronb/gbaa048.

Markert, C., Sasangohar, F., Mortazavi, B. J., & Fields, S. (2021). The use of telehealth technology to support health coaching for older adults: Literature review. *JMIR Human Factors*, *8*(1), e23796.

National Telecommunication and Information Administration. (1994). *Falling through the net*. Census Bureau, Population Division. Education and Social Stratification Branch.

Osama, T., Pankhania, B., & Majeed, A. (2020). Protecting older people from COVID-19: Should the United Kingdom start at age 60? *Journal of the Royal Society of Medicine*, *113*(5), 169–170. https://doi.org/10.1177/0141076820921107.

Penhale, B. (2006). World elder abuse awareness day. *The Journal of Adult Protection*, *8*(3), 33.

Pew Research Center. (2019). *Internet/broadband fact sheet*. https://www.pewresearch.org/internet/fact-sheet/internet-broadband/.

Rodrigues, N. G., Han, C. Q. Y., Su, Y., Klainin-Yobas, P., & Wu, X. V. (2021). Psychological impacts and online interventions of social isolation amongst older adults during COVID-19 pandemic: A scoping review. *Journal of Advanced Nursing*. https://doi.org/10.1111/jan.15063.

Ryan, E. B., Szechtman, B., & Bodkin, J. (1992). Attitudes toward younger and older adults learning to use computers. *Journal of Gerontology*, *47*(2), P96–P101.

Sanchez-Rodriguez, D., Annweiler, C., Gillain, S., & Vellas, B. (2020). Implementation of the integrated care of older people (ICOPE) app in primary care: New technologies in geriatric care during quarantine of COVID-19 and beyond. *The Journal of Frailty & Aging*, *1–2*. https://doi.org/10.14283/jfa.2020.24.

Tuohy, D., & Cooney, A. (2019). Older women's experiences of aging and health: An interpretive phenomenological study. *Gerontology and Geriatric Medicine*, *5*. https://doi.org/10.1177/2333721419834308.

United Nations. (2015). *Transforming our world: The 2030 agenda for sustainable development*. http://www.un.org/ga/search/view_doc.asp?symbol=A/RES/70/1&Lang=E.

United Nations. (2022a). *UN decade of healthy ageing (2021–2030)*. https://www.who.int/initiatives/decade-of-healthy-ageing.

United Nations. (2022b). *List of NGOs accredited to the open ended working group on ageing*. https://social.un.org/ageing-working-group/accreditedngos.shtml.

Weil, J., Kamber, T., Glazebrook, A., Giorgi, M., & Ziegler, K. (2021). Digital inclusion of older adults during COVID-19: Lessons from a case study of older adults technology services (OATS). *Journal of Gerontological Social Work*, 1–13.

World Health Organization. (2020). Decade of healthy ageing (2021–2030). In *World Health Organization*. https://www.who.int/initiatives/decade-of-healthy-ageing.

Wu, Y.-H., Lewis, M., & Rigaud, A.-S. (2019). Cognitive function and digital device use in older adults attending a memory clinic. *Gerontology and Geriatric Medicine*, *5*. https://doi.org/10.1177/2333721419844886.

Families and new technologies: Research, programs, and launch of event series celebrating the International Day of the family

Dasoo Milton Yoon

Human Development and Family Studies, University of Wisconsin-Madison, Madison, WI, United States; Department of Clinical Psychology, Teachers College, Columbia University, New York, NY, United States; International Association of Applied Psychology, Indianapolis, IN, United States

Introduction

New technologies are shaping our world today for all populations, especially in the wake of the COVID-19 pandemic, which includes not only individuals and societies, but also the ecosystem of the family.

Recognizing the importance of the impact of technology on the family and the subsequent adaptations families have had to make in response to the pandemic, the Division for Inclusive Social Development (DISD) of the United Nations Department of Economic and Social Affairs (UN DESA) intentionally raised awareness of megatrends that have emerged in recent times and presented research on how these trends impact families. Accordingly, the observations of the International Day of Families from 2021 to 2024 are planned to focus on these identified trends, in preparation for the 30th anniversary of the International Year of the Family expected in 2024.

The themes for the observance of the International Day of Families kicked off in 2021 with an event on May 14 focused on new technologies, entitled: "Families and New Technologies." Due to the ongoing restrictions of the COVID-19 pandemic, the event was conducted online over Zoom, allowing attendees to join by following the presentations and discussion through Facebook Live. The invite for the webinar event is shown in Fig. 1.

Resilient Health. https://doi.org/10.1016/B978-0-443-18529-8.00069-X

Fig. 1
Invite for the event about "Families and New Technologies".

The UN DESA is responsible for the follow-up to major UN summits and conferences, assisting countries around the world in agenda setting and decision-making with the goal of meeting their economic, social, and environmental challenges, and achieving the goals as set out in the UN Agenda 2030 for Sustainable Development. The DISD aims to strengthen international cooperation for achieving social inclusion and the reduction of inequalities through fostering effective policy impact and intensified global dialogue on social development issues.

The decision to focus on megatrends stems from a growing need to facilitate their impact on family lives and respond by recommending family-oriented policies to harness the positive aspect of such trends while counteracting the potentially negative facets. The trends that the DISD has chosen as drastically shaping the modern world include new technologies, demographic shifts, rapid urbanization, and migration trends.

The focus on new technologies is brought to attention at an ideal time—when the prolonged COVID-19 pandemic has firmly demonstrated the sheer importance of digital technologies in all aspects of life, including work, education, and communication. Although on the positive side, the pandemic has accelerated technological changes to improve societal functioning, concerns about the negative aspects of technology have also been raised. These concerns include challenges for parents adjusting to working from home, juggling work responsibilities with supervising children's internet learning, and caring for children forced to stay at home due to school closures. These problems, compounded by issues like cyberbullying and screen fatigue, have exacerbated already rising levels of parental burnout, a strong predictor for negative long-term effects on child well-being. As such, families and parents must develop a strong understanding of new digital technologies to empower their abilities to successfully manage their parental and family obligations in a healthy manner.

The importance of parenting education

Parenting education has been identified as the key pathway to allowing parents to ensure their children's and family's well-being, as well as being fundamental to ensuring children's healthy development. By definition, such education involves intervention or services aimed at improving parental interactions, behaviors, knowledge, beliefs, attitudes, and practices, and encompasses strategies of positive parenting, psychosocial stimulation, and maltreatment prevention.

As such, the 2021 International Day of Families aimed to raise awareness of the importance of such education programs for healthy child and parental functioning and well-being during and after the pandemic, with the specific focus on recognizing technologies as having considerable potential to advance parental education and empower parents for the well-being of their families. Along with a panel discussion composed of speakers from academia, civil society, and the private sector, the observance explored the launch of a major research paper on the impact of digital technologies on children at different stages in their lives as well as on parents and parenting.

Overview of the event

The flyer for the event is shown in Fig. 2, giving the particular logistics of the webinar event and program showing the speakers and their affiliations.

The identified objectives of this 2021 observance of the International Day of Families were to:

- Demonstrate the importance of family-oriented policies and programs to effectively respond to the challenges posed by the rapid expansion of new technologies
- Present current research on the growing potential of information and communication technologies to empower parents through parenting education tools to benefit families and society at large
- Warn of negative impacts of new technologies on children and families
- Share good practices in harnessing digital technologies for parenting education and overall family well-being.

Key speakers during the webinar observance, as shown in the screenshot in Fig. 3, were as follows:

- Opening Speaker: Daniela Bas, Director, of the DISD of UN DESA
- Keynote Speaker: Susan Walker, PhD, Associate Professor of Family Social Science, University of Minnesota, USA

Fig. 2
Flyer for the event showing the logistics and program of the webinar.

Fig. 3
Screenshot of the speakers at the International Day of Families on "Families and New Technologies".

- Moderator: Mr. Oliver Yao, President of the International Federation for Family Development (IFFD)
- Panelists:
  - Mr. Zach Sims, Cofounder & CEO, Codeacademy
  - Ms. Amina Fazlullah, Directory of Equity Policy, Common Sense Media
  - Ms. Mary Dozier, PhD, Professor of Psychological and Brain Sciences, University of Delaware, USA
  - Ms. Jessica Navarro, PhD Candidate in Human Development and Family Studies, University of North Carolina at Greensboro, USA

The event started with opening remarks from Ms. Daniela Bas, director of DISD. She highlighted the importance of technology in the 21st century, emphasizing that the importance of the present topic is evident by the fact that it is due to technology that people from all over the globe could attend this online observance. She added that this year's theme came at an appropriate time, due to the increasing need for technology caused by the limitations caused by the COVID-19 pandemic. She highlighted that, unfortunately, not everyone is included in this new digital world, as there are families without access to the internet or new technologies. Statistics show that almost 50% of the world's population is still offline, and thus faces more challenges to provide for their children during emergency or even nonemergency times. The pandemic has presented the world with a critical opportunity to build more effective, inclusive, and resilient systems to protect vulnerable families, especially those who live in extreme poverty. To build an inclusive society—which is the goal of the UN DISD and inherent in the UN Agenda 2030 for Sustainable Development which her division supports—the digital divide cannot continue. Access to the internet must be considered a universal right and needs to be put into the heart of global and national agendas. She concluded by emphasizing that the potential of new technologies is immense and should be harnessed to benefit families worldwide.

Following this inspiring opening speech, Mr. Oliver Yao introduced himself as the moderator of the observance, a role given to him due to his position as the president of the International Federation for Family Development (IFFD). The IFFD is a nongovernmental, independent, and nonprofit organization that supports families across the globe through training (International Federation for Family Development, 2021). He provided a brief but succinct overview of the event before handing the floor to the keynote speaker, Dr. Susan Walker, PhD, associate professor of Family Social Science at the University of Minnesota in the USA, to launch her comprehensive background paper entitled "Technology Use and Families: Implications for Work-Family Balance and Parenting Education."

Dr. Walker described her research work, documented in this paper, on integrating technology in family education, early childhood parenting, childcare policy, and parent learning. In practice, she has designed technology to supplement the face-to-face delivery of parenting education. Her keynote speech discussed the main points from her paper, connecting them to policies that the UN could adopt and pursue in alignment with the Sustainable Development Goals (SDGs). The four policies that she set forth during this presentation were as follows:

1. Ensure internet access (equity), safety, and promote digital rights
2. Include parenting education as a family support strategy and when addressing the range of children's outcomes
3. Integrate technology in parenting education—as content, pedagogy, and in maintaining relationships
4. Provide a system of support and standards that empower parenting

She proposed that by working toward these four proposed policies that effectively maximize the overlap between technology and families, the UN will be making significant progress toward multiple SDGs, including SDG 1 (end poverty), SDG2 (end hunger), SDG3 (good health and wellbeing for all), SDG4 (education for all), SDG5 (empowerment of women and girls), 8 (decent work), SDG9 (industry, innovation, and infrastructure), and SDG10 (reduce inequalities).

She also emphasized three points:

1. the importance of understanding the past in planning for the future,
2. the duality of technology that has both a positive and negative impact on human development, and
3. current systematic challenges for educators focused on helping parents and families in need.

The next panelist was Mr. Zach Sims, cofounder and CEO of Codecademy, an American online interactive platform for learning technical skills. Codecademy's central mission is to teach people modern skills online (programming, web development, data science) in a way that is engaging, flexible, and accessible, and that is necessary in the job market—either to start or advance a career—which is especially relevant in the difficult economic times of the pandemic and recovery (Codecademy, n.d.). He discussed the rise of online education, triggered by the COVID-19 pandemic, and its ultimate impact on parents and families. He noted three pillars of online education: engagement, accessibility, and flexibility. To summarize his main points, he commented on the unique impact of online education tools to keep parents in the workforce through technology, which will ultimately help families stay together and thrive.

Ms. Amina Fazlullah, the director of Equity Policy in the Washington DC office of Common Sense Media, a nonprofit organization that helps kids thrive in a world of media and technology, described her organization's work on a range of issues, including expanding access to technology, digital well-being, and privacy. The goals of the organization are to improve the lives of children and families by providing them with information, education, and an independent voice (Common Sense Media, n.d.). Her presentation included the logistical and pragmatic side of utilizing technology to positively impact families, with a central focus on how society can get access to the underlying networks that drive all the benefits gained through technologies for families. She emphasized the importance of collaboration among stakeholders in the various different fields of influence, including policymakers, politicians, and the private sector. A key example of how technology can be used to benefit families in need is through the delivery of essential services, such as healthcare, education, and workforce, by the government. Throughout her speech, she emphasized the need for proper infrastructure to support these policies and successfully transition plans into real life in order to impact real people.

The next panelist was Ms. Mary Dozier, PhD, professor of Psychological and Brain Sciences at the University of Delaware, USA. Her work revolves around the development of young children in foster care and children living with neglecting parents, examining challenges in attachment and regulatory capabilities. She introduced a parenting educational program she developed over several decades, called "Attachment and Biobehavioral Catch-up." This program focuses on three parenting behaviors: (1) nurturing when the child is distressed, (2) following the child's lead, and (3) avoiding frightening behaviors. When parents use these three behaviors combined, they positively benefit the child's behavioral and brain development. Her original program was designed to have 10 at-home sessions, with parent coaches providing "in-the-moment" comments as a key contributor to effecting changes in the parents' behavior. However, when the COVID-19 pandemic hit, coaches were unable to physically be at the families' homes to provide this coaching and feedback to the families. Since the target of the program was to help vulnerable parents and families, and since the administrators wanted the same services to be available to everyone, the program was pivoted to an online format, through online teleconferencing where the coaches could provide valuable comments. She and her colleagues were enthused to find that the remote setting not only was successful, but that the parent coaches were able to surpass their target comment hit rates, and positive changes were even found in essential measures such as sensibility and intrusiveness.

The last panelist, Ms. Jessica Navarro, a PhD candidate in the Department of Human Development and Family Studies at the University of North Carolina at Greensboro in the USA, presented preliminary findings from an international focus group organized by the IFFD that brought together experts on technology and family across the globe to address how technology is shaping the lives of children and families in the 21st century. The specific questions posed to these experts, as well as a brief overview of the experts' collective responses (noted as Q & A in the program), were:

- Q: Should the universal coverage of the internet be considered a human right?
  A: Yes.
- Q: What are the best indicators to measure households' access to new technologies?
  A: Broadband internet access, functional equipment to access the internet, devices per person, demographics of household related to digital needs, digital skills and knowledge, bandwidth, global perspective (social norms related to access).

- Q: Does teleworking contribute to work–family balance or is it just a way to work more?
  A: Both positive (flexibility, reduces commuting, multitasking, efficiency, opportunities for individuals with impairments) and negative (lack of social interaction, juggling multiple roles, boundary permeability, extended schedules, problems "logging out").
- Q: Can the remote learning experiences of COVID-19 help to improve education?
  A: Yes, but innovation must be based on pedagogy and these approaches need to be pushed forward in the right way that ensures their acceptance and integration into educational systems.
- Q: Which specific policies can help to bridge the intergenerational digital divide?
  A: Access and training for older generations, intergenerational programming, and education for family professionals; albeit more data is needed to identify the exact statistics to support these targets.
- Q: Should coding be introduced as a classroom subject, like other languages?
  A: It should not be promoted as a "language" per se, but should be considered a useful skill and requires the necessary access and devices to support its place in education.
- Q: What instruments should be developed to ban child abuse materials online? What tools can parents use?
  A: Strong parent–child relationships are central, and methods whereby parents can better monitor what information their children have access to—while teachers, educators, industry, and policymakers are all responsible as well.

Following this panel discussion, the event was open to another Q&A session, this time with the attendees, where the following themes were identified and discussed:

- Bridging the digital divide: providing equity, access, and accommodations to all
- New technologies: the impact on family well-being, family relations, children, parents, and other caregivers
- The impact of digital technologies on families during the COVID-19 pandemic
- The importance of parenting education to adjust to digital technologies
- Parenting education as a global investment in family support and women's empowerment
- Preparing family professionals to integrate technology into their practice
- The rise of online learning and its role in providing a flexible, accessible way for parents to advance their careers or pursue new opportunities
- The benefits of online learning for the entire family
- The ways in which public and private sectors can work together to bridge the digital divide, support parents in the workforce, and help families prosper through technology
- Good practices in parenting education and other areas under discussion in academic, civil society, and private sector

Conclusions and personal reflections

The topic of this event presented an exceptionally unique perspective on the role of technology to contribute to the flourishing of families in this age of COVID-19 recovery and beyond. Much is said about individuals and societies, but this focus on the family is critical, as the family unit is a crucial system within the infrastructure of society and a link between individuals and their communities. The panelists' approach that technology is a relevant topic to cover within the context of the family is especially timely, due to the abrupt changes that the COVID-19 pandemic forced on the world's populations. The experts on the panel from different sectors presented provided useful overviews as well as practical interventions—for example, on parent education—that are working and can potentially be applied in other contexts.

From a personal perspective, this topic aligns with my research interest and career goals to focus on family dynamics and relationships in the family system. Thus this four-year program by the Division of Inclusive Social Development of the United Nations Department of Economic and Social Affairs, including observations of the International Day of Families from the years 2021 to 2024, with their focus on important megatrends revolving around the family leading up to the 30th anniversary of the International Year of the Family expected in 2024, is a perfect match to my work. It is also a critical view that should be recognized by all stakeholders in government and civil society, and integrated into all policy and programming, to positively impact achievement in all areas of the UN Agenda.

References

Codecademy. (n.d.). About. https://www.codecademy.com/about.
Common Sense Media. (n.d.). Our mission. https://www.commonsensemedia.org/about-us/our-mission.
International Federation for Family Development. (2021, December 17). About us. https://iffd.org/about-us/.

Further reading

Bas, D., Walker, S., Yao, O., Sims, Z., Fazlullah, A., Dozier, M., & Navarro, J. (2021, May 14). Families and New Technologies [Online Conference]. 2021 Observance of the International Day of Families, Online. https://www.youtube.com/watch?v=Y71_dHazvVA.

United Nations. (n.d.-a). In 2021 International day of families on "Families and New Technologies". United Nations Department of Economic and Social Affairs. https://www.un.org/development/desa/family/international-day-of-families/2021-2.html.

United Nations. (n.d.-b). Division for inclusive social development. United Nations Department of Economic and Social Affairs. https://www.un.org/development/desa/en/about/desa-divisions/social-policy.html#:%7E:text=The%20Division%20seeks%20to%20strengthen,dialogue%20on%20social%20development%20issues.

United Nations. (n.d.-c). About us. United Nations Department of Economic and Social Affairs. https://www.un.org/en/desa/about-us.

CHAPTER 70

Addiction treatment and recovery using technology innovation to transform healthcare during the COVID-19 pandemic and beyond: The case of St. Joseph's Addiction Treatment & Recovery Centers

Robert Ross*

St. Joseph's Addiction Treatment & Recovery Centers, Saranac Lake, NY, United States

Background

As the COVID-19 pandemic raged, so did the addiction epidemic, as shown in shockingly increasing rates of alcohol and substance abuse, putting greater strain on the need for addiction and related mental health services.(O'Gorman & Morgan, 2022)

St. Joseph's Addiction Treatment & Recovery Centers, a 52-year-old nonprofit organization, rapidly implemented a series of technological advances and staffing adjustments at the start of the COVID-19 pandemic permitting us to stay open and save lives.

These adjustments allowed us to provide remote addiction and mental health services to over 2000 individuals and families across 4000 miles in mostly rural, economically disadvantaged northern New York State.

Located in the heart of the Adirondack Mountains in northern New York State where it snows easily from October through May, as shown in Fig. 1, the center has been called the "Magic Mountain" by generations of residents.

* Recently retired as CEO of St. Joseph's Addiction and Recovery Treatment Centers, and now operates a healthcare and organizational development consultancy.

Resilient Health. https://doi.org/10.1016/B978-0-443-18529-8.00070-6

Fig. 1
Exterior of St. Joseph's Addiction Treatment & Recovery Centers, main facility.

Fig. 2
Dedication to veteran health.

To address the urgent and dire needs of those suffering from addiction, St. Joseph's:

- expanded crisis services for individuals and families in acute addiction and mental health distress.
- dramatically increased telehealth outpatient services in all locations.
- kept residential programs open and staffed, providing 144 beds in
  - five long-term residential addiction, health, and mental health services for youth and adults;
  - two programs for veterans and their families, given our dedication to veterans, as shown in Fig. 2;(O'Gorman et al., 2020)

    - four supportive housing programs for 26 families/individuals who have addiction and are homeless.
- expanded programming by identifying a major need for a detox center for teenagers and adults, as shown in Fig. 2, as there were no such centers within a 2½-hour drive.
    - We opened this facility in March 2022, joining this to our outpatient, inpatient, and supportive housing continuum of care.

St. Joseph's is committed to constantly expanding access to needed addiction and mental health services for individuals, families, and communities in the geographic areas of New York State.

We accomplished this goal by broadening its continuum of care through:

- adopting new technologies
- expanding strategic partnerships
- identifying and securing additional funding sources
- identifying underserved populations, both geographic and socioeconomic
- adopting new best-treatment practices and evidence-based programs.

Problem/challenge this innovation addresses

The COVID-19 pandemic brought an unprecedented surge in addiction of more than 50%, which created a huge demand for addiction/mental health services.(O'Gorman & Morgan, 2022) We could not let COVID-19 close us down. People suffering from addiction were going to die(O'Gorman, 2021).

Our challenge was how to keep offering services while following strict regulations from New York State and local county health departments, which required a moratorium on inpatient/residential admissions if either staff or residents tested positive.

To ensure we could continue to serve those with addiction, we

- delivered outpatient services remotely;
- admitted individuals in need and provided meaningful and robust services, remotely if necessary;
- kept the residents and staff in our inpatient/residential programs safe by
    - testing all residents upon admission and keeping them quarantined for five days (prior to vaccine availability)
    - if a resident tested positive, we kept them in isolation for a total of ten days; now that vaccines are available, this protocol is still in effect.

- kept facilities adequately staffed while requiring staff to isolate at home if they had been exposed to or had COVID-19. To ensure that all outpatient and inpatient/residential clients received the critical services they needed, we employed technology and significant altering of staffing patterns to provide lifesaving services.

How and where is innovation being implemented, what technology is used, how is it targeted and to whom, and how is the service being delivered?

Our agency is comprised of 18 sites and 275 staff. We utilize many technology tools, including:

- healthcare via Zoom to provide patient confidentially and secure data transmission for individual, family, and group treatment
- electronic health records (EHR) to facilitate communication between providers
- office chairs, desks, and dual computer monitors to assist providers working remotely with seeing clients and documenting into EHR
- cell phones issued to staff equipped with hotspots to provide internet to providers in rural, isolated locations
- utilized the software NetExtender to provide secure internet connection for staff working remotely.
- large screens to enable large group meetings easily on Zoom

To support our clients during the pandemic, we:

- provided paid phone time for telehealth
- offered and delivered food when clients had no access to food during COVID-19
- utilized iPads for clients in all residential programs while they are in isolation or quarantined, to connect clients to medical staff and allow participation in individual, family, and group counseling
- dispatched mobile vans to provide internet to clients in rural locations to enable them to have treatment with their providers
- provided vans with protective shielding to take clients to appointments

With regard to helping our veteran population, a good example of an innovative pivot using technology that we instituted due to COVID-19 relates to our comedic improv group which had been running for two years in person with one of our psychologists and our artistic director. In order to keep the program going during COVID-19, given proof of its effectiveness(O'Gorman et al., 2020), the staff ran the comedic program on Zoom. After the dangers of the pandemic subsided, the program was resumed in person.

Continuum of care using technology was also accomplished by facilitating former patients to continue at home their lessons learned at St. Joe's using their personal technology devices (Fig. 3).

Fig. 3
Using technology for continued care learned at St. Joseph's

Impact, successes

The COVID-19 pandemic, while tragic, provided an opportunity for us to dig deep into our agency philosophy, with staff and management rising to the crisis. The result: a stronger agency culture, as seen by St. Joe's being recognized in several ways. These include being:

- selected as a "Best Company to Work for in New York" for the 10th year(Best Companies Group, 2022)
- designated by *Newsweek* magazine as one of "America's Best Addiction Treatment Centers 2021"(Cooper, 2022)
- certified by the Commission on Accreditation of Rehabilitation Facilities (CARF) with its highest ranking(Office of Addiction Services and Supports, n.d.)
- chosen by *Modern Healthcare* magazine as a best place to work in healthcare in the United States, and one of the top five places to work in terms of a family-friendly environment (Modern Healthcare, 2022).

St. Joe's promotes a positive team culture where staff's needs are heard. Our meetings begin with a check-in, where each participant answers questions about:

- how they feel
- what their goals are for the day
- "Who can help you?"

Our staff have demonstrated their commitment to St. Joe's through strict adherence to COVID-19 protocols, high rates of vaccination, attendance, and volunteering to cover shifts in their departments and in others. Heavy reliance on technology has brought our staff into more contact with each other through the immediacy of zoom despite our wide geographic area, creating closer personal ties.

Program development continued with a new lifesaving, integrated detox program for 16- and 17-year-olds as well as for adults. This program is unique in New York State and is attached to our Open Access Center, creating a physical continuum of care within one building and ensuring a reduction in gaps of service.

Challenges

The COVID-19 pandemic presented very significant challenges, in several areas:

Policy: Most of our agency policies required updating to be compliant with state regulations for operating licenses, against which services as ours are measured. This is a time-consuming process that needed to be repeated as directives changed.

Financial: We had to identify eligibility for all federal and state grants related to COVID-19 and apply for them at the earliest potential opportunity to secure needed funds to continue operating.

Revenue: Significant revenue was lost because of diminished residential occupancy and outpatient visits. Also, the government grants that helped stabilize financial conditions in 2020 and 2021 during the crisis were not available to a similar degree the next year, in 2022.

Technology: Technology had to be secured and distributed in a timely manner to staff, including those working remotely, and necessary training had to be provided.

Health agency guidance: Federal, state, and local health agencies produced inconsistent and, at times, contradictory guidance and regulations, which individual providers like us needed to respond to, while maintaining compliance and consistently providing access of quality services to clients.

Strategic partners

- Local hospitals for effective care coordination
- Small local agencies struggling with COVID-19 protocols
- Federally qualified health centers providing psychiatric services and primary care
- Private-sector vendors who coordinated access to PPE, tests, and sanitizing materials
- County and state government health agencies, as partners collaborating on COVID-19 safety
- State regulatory agencies
- Integrated provider associations pursuing value-based payment (VBP) opportunities with payors

- Public–private partnerships to secure more diversified funding streams
- Federal, statewide, and regional behavioral health advocacy groups
- State legislative leaders of health and behavioral health committees and local elected officials
- Universities, think tanks, and foundations involved in healthcare policy and research

Business plan

We are focused on essential funding sources to continue our exemplary service delivery. St. Joe's receives most of its revenue from Medicaid and insurance reimbursement and considerable grant funding from state, federal, and foundation sources for both operational and capital support. During COVID-19, there was a major increase in federal grant funding, especially targeted at replacing lost operational revenues caused by the pandemic. New sources need to be secured.

Sustainability plans

To continue or work, we plan to:

- Maximize revenue from residential, inpatient, and housing facilities by increasing occupancy above 90% levels.
- Increase outpatient services by continuing to expand integrated mental health services to individuals with this primary diagnosis.
- Build upon the experience of daily leadership meetings over Zoom, for example, staff working across programs are now being encouraged to meet over Zoom when appropriate rather than be required to travel to all meetings.
- Continue our new and successful initiative of holding staff retreats that support our workforce by providing a time of relaxation and reflection, which led to offering retreats to other organizations as a service, and as a new source of revenue.

How has COVID-19 affected your work and pivoting to respond to the pandemic

In response to the pandemic, we have implemented four major changes:

1. Having more staff work from home has allowed greater productivity and job satisfaction, as well as improving retention and recruitment.
2. Unfortunately, we had to reduce census in residential and inpatient programs to ensure proper physical distancing and risk mitigation. But, because of this, we learned to be more creative in residential program scheduling and space utilization.

3. We initiated the establishment of a new position for a director of Foundation and Development, whose role is to secure sustained private-sector funding and increase Foundation support.
4. We developed robust participation in value-based payment arrangements with managed care and insurance companies, to benefit from the competitive advantage of improved patient outcomes and financial efficiency.
5. In response to a 2022 significant operational deficit, largely attributed to circumstances related to the COVID-19 Pandemic, St. Joseph's initiated efforts to affiliate with a larger healthcare organization. This will advance St. Joseph's dual goals of continuing to improve the quality of services and promoting greater fiscal sustainability in meeting the challenges of a rapidly changing healthcare environment.

Sources of additional information

www.stjoestreatment.org
facebook.com@sjartc
instagram.com/stjoestreatment

References

Best Companies Group. (2022). *Do you have what it takes to be considered best companies to work for in New York?*. Retrieved May 31, 2022 from https://bestcompaniesgroup.com/best-companies-to-work-for-in-new-york/.

Cooper, N. (Ed.). (2022). *America's best addiction treatment centers 2021* Newsweek. Retrieved June 7, 2022, from https://www.newsweek.com/americas-best-addiction-treatment-centers-2021.

Modern Healthcare. (2022). *Best places to work in healthcare—2022*. Modern Healthcare. Retrieved May 31, 2022 from ModernHealthCare.com/awards/best-places-to-work-2022.

O'Gorman, P. (2021). Suicide by intentional overdose: the missing variable in tackling teen suicide is the addiction epidemic. *Psychology Today*. Retrieved May 31, 2022 from https://www.psychologytoday.com/us/blog/psychological-trauma-coping-and-resilience/202102/suicide-intentional-overdose-in-children.

O'Gorman, P., Hall, S. M., & Merkel, C. D. (2020). Can Zoom increase group therapeutic outcomes? A case study involving therapeutic improvisational theater for veterans with addiction and PTSD. *Trauma Psychology News*. Retrieved May 31, 2022 from https://traumapsychnews.com/2020/08/can-zoom-increase-group-therapeutic-outcomes/.

O'Gorman, P., & Morgan, C. W. (2022). Addiction and COVID: A toxic cocktail. *Psychology Today*. Retrieved May 31, 2022 from https://www.psychologytoday.com/us/blog/psychological-trauma-coping-and-resilience/202205/addiction-and-covid-toxic-cocktail.

Office of Addiction Services and Supports. (n.d.). Strengthening addiction prevention services. New York State. Retrieved May 31, 2022 from http://www.carf.org/providerProfile.aspx?cid=38397.

CHAPTER 71

Autism spectrum disorder and technological interventions

Sripriya Bikkani

Department of Clinical and Counseling Psychology, Teachers College, Columbia University, New York, NY, United States; International Association of Applied Psychology, Indianapolis, IN, United States

Introduction

Autism Spectrum Disorder (ASD) is a broad group of neurodevelopmental disorders that result in deficits in social, communication, and restrictive repetitive patterns of behavior, interests, and activities (Harris & Coyle, 2023). The term "spectrum" is an umbrella term used to loosely describe individuals with a DSM-IV diagnosis of autistic disorder, Pervasive Developmental Disorder (PDD), and Asperger's Disorder (American Psychiatric Association, 2022 The *Diagnostic and Statistical Manual of Mental Disorders, Fifth Edition, Text Revision* (5th ed.; DSM-5-TR)).

The diagnostic criteria for ASD include deficits in social communication, social interaction, restricted and repetitive movements, fixated interests, abnormal reactivity to sensations, and rigidity of routine among others (American Psychiatric Association, 2022).

Symptoms are usually present in early childhood and developmental phases although may not be fully manifested. The symptoms can affect an individual in social, occupational spheres, as well as impair other important areas of functioning. These symptoms are also not due to an intellectual disability; however, ASD can be accompanied by intellectual impairment, language impairment, and/or with a genetic or medical condition (American Psychiatric Association, 2022).

This chapter reviews the prevalence, diagnosis, assessment, and remediation approaches to ASD and the technological interventions to help persons with ASD function better in society.

Diagnosis and assessment

According to the *DSM-5-TR* (American Psychiatric Association, 2022), the two main characteristics for diagnosis of ASD are (1) impairment in social communication and interaction and (2) restricted behavior, interests, and activities.

Resilient Health. https://doi.org/10.1016/B978-0-443-18529-8.00071-8

With respect to social communication impairments, individuals must present with difficulties with social reciprocity, which is the inability to participate in dialogue and discussion; difficulties with nonverbal communication, which is typically a lack of understanding of nonverbal communication such as eye contact or facial expressions; and difficulties in creating or maintaining relationships with others, which is manifested as disinterest in fostering relationships or the inability to navigate social situations.

With respect to patterns of behavior, individuals with ASD must present with at least two of the following: repetitive speech, motor movements, or use of objects; following routines or patterns of behavior excessively or exhibiting a strong resistance to changes in schedule; intense fixations on certain objects or topics of interest; and atypical response to sensory stimuli such as light, texture, smell, and taste.

The *DSM-5* also categorizes ASD based on three levels of severity: Level 1, "Requiring support"; Level 2, "Requiring substantial support"; and Level 3, "Requiring very substantial support."

Clinicians including pediatricians may conduct screening tests for ASD during routine checkups and visits. In the case of youngsters, the clinician may ask parents questions regarding any of the aforementioned symptomatic features to identify early signs of ASD.

In addition to the informal assessments, clinicians may also opt to use more formal Autism screening tools such as The Ages and Stages Questionnaire (ASQ), to investigate developmental delays or challenges at specific ages; the Screening Tool for Autism in Toddlers and Young Children (STAT), an interactive and observational tool that consist of activities which assess play, communication, and imitation skills; and/or the Modified Checklist for Autism in Toddlers-R/F (M-CHAT-R/F), a 20-item 2 part screen in which the first item is a checklist to be completed by parents and the second item is to be completed by staff or clinicians. Many other screening tools are available and some have been found to be more effective for screening females, minorities, or individuals from low-socioeconomic backgrounds (Autism Research Institute, 2022). In general, assessments for young children focus on obtaining a diagnosis, whereas for older individuals, the focus is on assessing skills and understanding what interventions they may require to increase functioning in the impaired areas (Goldstein & Ozonoff, 2018).

Epidemiology

The data for the reported prevalence of ASD is restricted to high-income countries due to differences in access to research, data collection, as well as different methods of measurement (Elsabbagh et al., 2012). However, a 2012 systematic review of epidemiological surveys found that the median prevalence of ASD was 62 cases per 10,000 people (Elsabbagh et al., 2012). The reported estimated prevalence of ASD has increased over time, although this may be attributed to increased diagnostic criteria, availability of diagnostic services, and increased awareness of ASD in the public and professional spheres (Elsabbagh et al., 2012).

A 2018 report on the prevalence of ASD among children aged 8years in 11 specific states in the continental United States found that one in 44 children was diagnosed with ASD. They also reported that ASD was 4.2 times more prevalent among boys than girls, and that ASD was prevalent in all racial, ethnic, and socioeconomic groups; however, estimated prevalence was slightly higher among non-Hispanic white children compared to non-Hispanic black children and Hispanic children aged 8 years (Christensen et al., 2018).

Currently there are a number of available treatment interventions for individuals with ASD. The National Institute of Child Health and Human Development (NICHD) advises that the most effective therapies and interventions should be tailored to suit the individual; however, early diagnosis and interventions are most beneficial in remediating symptoms and improving skills. Some treatments include behavioral management therapy, cognitive behavior therapy, educational and school-based therapies, medication treatment, occupational therapy, social skills training, speech-language therapy among many others (NICHD, 2022).

Technological interventions

Despite advancements in assessment and treatment for ASD, there are many barriers in the accessibility to these services to individuals in need. According to Buescher et al. (2014), the cost of evaluation and ongoing treatment can be exorbitant for families, although there is no significant correlation with respect to socioeconomic strata and prevalence of ASD. Furthermore, access to high quality services can be restricted to certain geographical locations, and individuals in rural areas usually have to travel further and for more time to obtain services (Mello et al., 2016).

Technological-based interventions promise to fill this gap, whereby the use of technology-mediated devices can help deliver quality care and services to individuals with ASD. The 21st century has seen a large swathe of innovation in technology-based interventions being delivered in a variety of spheres, especially in healthcare and education (Ramsey & Montgomery, 2014). With technological devices becoming more common and affordable, we can incorporate technology into delivering care for individuals with ASD in an efficacious and promising manner (Goldsmith & LeBlanc, 2004).

In general, the types of technological interventions that exist to serve individuals with ASD can be categorized into: communication-based aids, video-based aids, algorithmic aids, interaction-based aids, and, to a lesser extent, script based or storyline formats, games, and artificial intelligence. Devices used are primarily mobile devices, computers, televisions, DVDs or recorded content, other internet-based devices, and also occasionally robotic devices. The interventions mentioned in this chapter are primarily for children and young adults as opposed to adults because intervention is generally most effective for individuals with ASD if it takes place early in age (Orinstein et al., 2014).

Although ASD can be diagnosed in individuals from 18 months of age, there are still some delays in diagnosis reported in the United States (Gordon-Lipkin et al., 2016). This diagnostic delay can be attributed to time-intensive assessments, lack of available trained clinicians, as well as long wait times in clinical centers. As such, a diagnosis of ASD can manifest nearly 2 years after the earliest signs of developmental concerns by parents.

Additionally, since intervention is most effective when introduced earlier, it can be beneficial to use technological aids to close this gap (Orinstein et al., 2014).

Megerian et al. (2022) have investigated the efficacy of an artificial intelligence-based device in augmenting the necessary assessments provided by clinicians to diagnose for ASD. The device is a Software as a Medical Device (SaMD) which uses an algorithm based on preexisting data on children diagnosed with ASD, to aid the evaluation of ASD. The SaMD receives input from three sources: a caregiver questionnaire, an analysis of two short home videos, and a questionnaire from a healthcare provider. Using the algorithm, it deduces the individual to either be ASD positive, ASD negative, or indeterminate. The device was found to be 80.8% accurate in diagnosing positive diagnoses and 98.3% accurate for negative diagnoses, signifying high degree of accuracy. In 91% of the cases, the device resulted in an indeterminate output which was later identified as having one or more complex neurodevelopmental disorders by specialists. Thus, this type of artificial intelligence is a promising tool to aid in the diagnostic evaluation of ASD. It can also facilitate greater efficiency in terms of allocating specialist resources and expedite early intervention.

Many advancements in technology provide language support and facilitate communication in individuals with ASD. For example, augmentative and alternative communication (AAC) technology refers to strategies and tools that aid in language and communication (Shane et al., 2012). AAC approaches can be through the implementation of hardware platforms and/or software programs. According to Shane et al. (2012), there are many such digital applications or "apps" designed to serve this very purpose and with the ubiquitous *TM* nature of portable hardware like the Apple *iPad,* individuals can utilize these services frequently and with ease. Shane et al. (2012) elucidate some AAC apps designed to be used on such devices are *Proloquo2go and MyTalk*, which are speech-generating applications (SGAs) that produce digitized speech.

In addition to applications, there are also speech-generating devices (SGDs) that are specialized AAC hardware which do the same. Some examples of SGDs are *GoTalk* and *DynaVox V*.

SGDs and SGAs can be adapted to suit the skill level of the user and can also incorporate text, symbol, or multimedia content. In general, application-based software that can be used on personal, portable devices is more accessible, cost effective, user friendly, and customizable, compared to AAC hardware which is more expensive and more specialized (Shane et al., 2012).

Video modeling interventions are innovations in technology that help address social-communication skills, functional skills, and behavioral functioning in young adults and children with ASD (Bellini & Akullian, 2007). The concept of video modeling is based on the work of social psychologist Albert Bandura on social learning theory and observational learning. According to Bandura (1977, 1997), modeling, observational learning (including self-observation), and imitation have a profound impact on a child's development. In a video modeling intervention, an individual can observe a video demonstration of a certain behavior and can then mimic the behavior. Studies have shown that video modeling can be an effective intervention to teach individuals with ASD relevant life skills as well as behaviors (Bellini & Akullian, 2007; Gardner & Wolfe, 2013).

In terms of social interaction in children with ASD, research has shown that these children can exhibit certain social behaviors with robots but not with peers. These behaviors include, but are not limited to, imitation, maintaining eye contact, and attention (Robins et al., 2004). Ricks and Colton (2010) explored the use of robots in autism therapy and found numerous applications such as: diagnosing ASD by way of eye tracking, facilitating interactions in children, involving them in turn-taking activities, and incorporating imitation-based activities to teach children emotional recognition. Robot-based therapies are promising interventions in the treatment of ASD; however, there are several limitations, including excessive cost, variability of effectiveness depending upon the depiction of the robot (humanoid or not), and the application of skills learned with the robot to the outside world. Engineers at Brigham University, Ricks and Colton (2010), argue that the most effective use of robot-mediated therapy is by way of 'Triadic Interactions' where there is interaction between the child, the robot, and a third human who may be another child or adult. This unique relationship can teach children with ASD social interaction with another human and have the robot serve as a kind of "social pivot."

Virtual Reality (VR)-based interventions are other adaptations in which artificial environments are created to simulate real-world environments for individuals with ASD, including images, sensations, and products (Nordahl-Hansen et al., 2020). VR Head-Mounted Displays or HMDs are portable hardware equipment containing goggles which allow the user to experience a virtual environment and interact with the surroundings. This technology allows individuals to practice real-world and behavioral skills in a realistic environment which also allows for control, safety, monitoring, and the ability to repeat scenarios (Bradley & Newbutt, 2018). Dechsling et al. (2021) reviewed that VR as an intervention tool can increase accessibility to therapy and advance the clinical practice in remediating ASD. However, there is limited research on VR-induced therapy and while promising, more research is warranted in this particular field.

Other technological interventions for children with ASD are script or story-based interventions in which the child reads, listens to, or watches a story by way of a technological device and the use of algorithmic games that deliver animations based on the child's choice of action.

However, there has been limited research on these types of interventions and future research should aim to design and replicate studies examining these interventions (Heng et al., 2021).

Conclusions

Technology-assisted interventions for children with ASD are increasing and are promising, to deliver effective remediation and therapy to help increase skills in communicating and relating, boost self-esteem, and improve daily functioning. In general, such interventions can increase accessibility of services, be personalized to suit the needs of the child, offer supplemental support when human resources are lacking, and simulate repeatable real-world scenarios where applicable. Technology-mediated interventions are more cost effective when using electronic devices that are widely available, while other interventions such as robots which are promising are expensive and therefore wide use and availability are restricted. Technological interventions should not replace human-assisted therapy, since it is important for persons with ASD to build skills in relating to others, but instead, be an additional tool where necessary, for example, when human resources are limited. While studies have been done about video modeling and communication-based interventions, the application of other tools, like VR and games, needs to be researched. Given advancements in technology in this digital age of the 21st century, the potential for technology-assisted therapy for ASD is encouraging and uncharted.

Personal thoughts

I have personal and professional experience working with children with ASD and other neurodevelopmental disorders and have firsthand experience with some of the devices referred to in this chapter, such as video-based communication aids. This experience has enhanced my interest in exploring the potential of digital health especially for neurodevelopmental disorders. In addition, my experience working with individuals with learning disabilities has piqued my curiosity to study the efficacy of novel educational technology.

References

American Psychiatric Association. (2022). Neurodevelopmental disorders. In *Diagnostic and statistical manual of mental disorders* (5th ed.). American Psychiatric Association Publishing. text revision.

Autism Research Institute. (2022). *Screening & assessment*. https://www.autism.org/screening-assessment/. (Accessed 8 May 2022).

Bandura, A. (1977). *Social learning theory*. Prentice-Hall.

Bandura, A. (1997). *Self-efficacy: The exercise of control*. W.H. Freeman.

Bellini, S., & Akullian, J. (2007). A meta-analysis of video modeling and video self-modeling interventions for children and adolescents with autism spectrum disorders. *Exceptional Children*, *73*(3), 264–287. https://doi.org/10.1177/001440290707300301.

Bradley, R., & Newbutt, N. (2018). Autism and virtual reality head-mounted displays: A state of the art systematic review. *Journal of Enabling Technologies, 12*(3), 101–113. https://doi.org/10.1108/JET-01-2018-0004.

Buescher, A. V. S., Cidav, Z., Knapp, M., & Mandell, D. S. (2014). Costs of autism spectrum disorders in the United Kingdom and the United States. *JAMA Pediatrics, 168*(8), 721–728. https://doi.org/10.1001/jamapediatrics.2014.210.

Christensen, D. L., Braun, K. V. N., Baio, J., Bilder, D., Charles, J., Constantino, J. N., Daniels, J., Durkin, M. S., Fitzgerald, R. T., Kurzius-Spencer, M., Lee, L.-C., Pettygrove, S., Robinson, C., Schulz, E., Wells, C., Wingate, M. S., Zahorodny, W., & Yeargin-Allsopp, M. (2018). Prevalence and characteristics of autism Spectrum disorder among children aged 8 years—Autism and developmental disabilities monitoring network, 11 sites, United States, 2012. *MMWR Surveillance Summaries, 65*(13), 1–23. https://doi.org/10.15585/mmwr.ss6513a1.

Dechsling, A., Shic, F., Zhang, D., Marschik, P. B., Esposito, G., Orm, S., Sütterlin, S., Kalandadze, T., Øien, R. A., & Nordahl-Hansen, A. (2021). Virtual reality and naturalistic developmental behavioral interventions for children with autism spectrum disorder. *Research in Developmental Disabilities, 111*, 103885. https://doi.org/10.1016/j.ridd.2021.103885.

Elsabbagh, M., Divan, G., Koh, Y.-J., Kim, Y. S., Kauchali, S., Marcín, C., Montiel-Nava, C., Patel, V., Paula, C. S., Wang, C., Yasamy, M. T., & Fombonne, E. (2012). Global prevalence of autism and other pervasive developmental disorders: Global epidemiology of autism. *Autism Research, 5*(3), 160–179. https://doi.org/10.1002/aur.239.

Gardner, S., & Wolfe, P. (2013). Use of video modeling and video prompting interventions for teaching daily living skills to individuals with autism spectrum disorders: A review. *Research and Practice for Persons with Severe Disabilities, 38*(2), 73–87. https://doi.org/10.2511/027494813807714555.

Goldsmith, T. R., & LeBlanc, L. A. (2004). Use of technology in interventions for children with autism. *Journal of Early and Intensive Behavior Intervention, 1*(2), 166–178. https://eric.ed.gov/?id=EJ848688.

Goldstein, S., & Ozonoff, S. (Eds.). (2018). *Assessment of autism spectrum disorder* (2nd ed.). The Guilford Press.

Gordon-Lipkin, E., Foster, J., & Peacock, G. (2016). Whittling down the wait time: Exploring models to minimize the delay from initial concern to diagnosis and treatment of autism spectrum disorder. *Pediatric Clinics of North America, 63*(5), 851–859. https://doi.org/10.1016/j.pcl.2016.06.007.

Harris, J. C., & Coyle, J. T. (2023). Autism spectrum disorder. In *Neurobiology of brain disorders* (pp. 69–88). Academic Press.

Heng, E., Lanovaz, M. J., & Beauregard, A. (2021). Research on technological interventions for young children with autism spectrum disorders: A scoping review. *Review Journal of Autism and Developmental Disorders, 8*(2), 253–263. https://doi.org/10.1007/s40489-020-00215-w.

Megerian, J. T., Dey, S., Melmed, R. D., Coury, D. L., Lerner, M., Nicholls, C. J., Sohl, K., Rouhbakhsh, R., Narasimhan, A., Romain, J., Golla, S., Shareef, S., Ostrovsky, A., Shannon, J., Kraft, C., Liu-Mayo, S., Abbas, H., Gal-Szabo, D. E., Wall, D. P., & Taraman, S. (2022). Evaluation of an artificial intelligence-based medical device for diagnosis of autism spectrum disorder. *NPJ Digital Medicine, 5*(1), 57. https://doi.org/10.1038/s41746-022-00598-6.

Mello, M. P., Goldman, S. E., Urbano, R. C., & Hodapp, R. M. (2016). Services for children with autism spectrum disorder: Comparing rural and non-rural communities. *Education and Training in Autism and Developmental Disabilities, 51*, 355–365.

NICHD. (2022). *What are the treatments for autism?*. https://www.nichd.nih.gov/health/topics/autism/conditioninfo/treatments. (Accessed 9 May 2022).

Nordahl-Hansen, A., Dechsling, A., Sütterlin, S., Børtveit, L., Zhang, D., Øien, R. A., & Marschik, P. B. (2020). An overview of virtual reality interventions for two neurodevelopmental disorders: Intellectual disabilities and autism. In D. D. Schmorrow, & C. M. Fidopiastis (Eds.), *Augmented cognition. Human cognition and behavior* (pp. 257–267). Springer International Publishing. https://doi.org/10.1007/978-3-030-50439-7_17.

Orinstein, A. J., Helt, M., Troyb, E., Tyson, K. E., Barton, M. L., Eigsti, I.-M., Naigles, L., & Fein, D. A. (2014). Intervention for optimal outcome in children and adolescents with a history of autism. *Journal of Developmental & Behavioral Pediatrics, 35*(4), 247–256. https://doi.org/10.1097/DBP.0000000000000037.

Ramsey, A. T., & Montgomery, K. (2014). Technology-based interventions in social work practice: A systematic review of mental health interventions. *Social Work in Health Care*, *53*(9), 883–899. https://doi.org/10.1080/00981389.2014.925531.

Ricks, D. J., & Colton, M. B. (2010, May). Trends and considerations in robot-assisted autism therapy. In *Robotics and automation (IcrA), 2010 IEEE international conference on* (pp. 4354–4359). IEEE.

Robins, B., Dickerson, P., Stribling, P., & Dautenhahn, K. (2004). Robot-mediated joint attention in children with autism: A case study in robot-human interaction. *Interaction Studies*, *5*(2), 161–198. https://doi.org/10.1075/is.5.2.02rob. Social Behavior and Communication in Biological and Artificial Systems.

Shane, H. C., Laubscher, E. H., Schlosser, R. W., Flynn, S., Sorce, J. F., & Abramson, J. (2012). Applying technology to visually support language and communication in individuals with autism spectrum disorders. *Journal of Autism and Developmental Disorders*, *42*(6), 1228–1235. https://doi.org/10.1007/s10803-011-1304-z.

Further reading

Goldstein, S., & Naglieri, J. A. (Eds.). (2013). *Interventions for autism spectrum disorders: Translating science into practice* Springer.

National Collaborating Centre for Mental Health (UK). (2012). *Case identification, assessment and diagnosis.* British Psychological Society. https://www.ncbi.nlm.nih.gov/books/NBK327413/.

CHAPTER 72

Technology advances addressing the mental health needs of the LGBTQI+ community

Taylor Hirschberg

Hearst Endowed Scholar and Visiting Research at Columbia University Global Centers in Istanbul Program on Forced Migration and Health, Heilbrunn Department of Population and Family Health, Columbia University Mailman School of Public Health, New York, NY, United States

The LGBTQI+ community is a demographically significant group that over the past decade has grown globally in size and visibility. While visibility has increased, there remains intersecting systems of oppression, as well as violence, toward the LGBTQI+ community, that limit access to healthcare, safety, and basic needs including food and housing. However, alongside this growing visibility, as well as intersections of structural violence, exist tremendous advances in the quest of LGBTQI+ equality worldwide through the use of digital technologies, which have led to a profound impact on the health and well-being of the LGBTQI+ community. Perhaps the most notable impact has been in new ways that this group is now able to identify and build community through the adoption of technologies such as social media and phone-based applications.

Arguably, it is through technological advances in the use of cellular phones and smart devices, and social adaptation to these tools, that you see a magnified change in how innovation and technology impact the well-being of the LGBTQI+ community. In many cases, these devices act as a new portal through which a person can connect to needed services in health and other life areas such as housing, as well as to a community of friends and even romantic partnerships. Moreover, and more unexpected, these devices act as unsung heroes to the LGBTQI+ community across disparate geographies, especially for those who come from the 71 countries that criminalize homosexuality, 11 by capital punishment, and put the members of the community in grave danger, in that the proliferation of smart devices allowed these members of the community to create networks of a secret society, asylum, and safety (Human Dignity Trust, 2021).

However, as historically seen, innovation is not always distributed equally and, in some contradictory ways, act against the ethos of its own existence. During the COVID-19 pandemic, technology took on a new meaning for everyone, whereby you really began to see contrasting experiences in the use of technology. While many advantages emerged, in other cases, not having access to technology widened the gap in access to mental health and psychosocial support among vulnerable populations that include the forcibly displaced LGBTQI+ community.

Resilient Health. https://doi.org/10.1016/B978-0-443-18529-8.00072-X

Positive applications of technology and innovation

Limited literature is available on the outcomes of technology-based interventions aimed at the mental health wellness of the LGBTQI+ community. Given this paucity of data, and to better understand the individual experiences and perhaps give some insight on how the community at large could be served using technology, I conducted over 20 interviews of members who represent every letter of the LGBTQI+ across gender, race, socioeconomics, and geographies.

When those interviewed were asked—"How did technology help you with your mental health and well-being during COVID-19?"—various themes arose. Some themes were similar as for peoples other than those in the LGBTQI+ community, such as minimizing isolation, improving technology skills, and expanding new otherwise unknown connections.

One particularly notable theme was that participating in zoom calls was awareness-raising about one's identify and affiliation. For example, one woman spoke to an online group hosted by her local LGBTQI+ community center about her experience of coming out to her loved ones. She reflected to me, "I would have never known I was 'queer.' I thought I had to be a lesbian or bisexual, but I learned what 'queer' was by attending a zoom meeting I found online. I joined this meeting because I could have some level of anonymity."

While "queer" is sometimes used as a hurtful word that has caused pain in the past, for this woman as for others, the word has become more acceptable, and in some cases, well expresses constant fluidity in a person's sexual orientation and gender identity. When asked how her mental health was now, she mentioned that she has continued to go to this online group, exploring her identity, and that she was much happier than she was prior to the COVID-19 pandemic.

This example also shows that during the COVID-19 pandemic, and even online, LGBTQI+ community centers served as a safe space for members of the LGBTQI+ community worldwide to share about themselves safely.

Another theme from my interviews emerged from the fact that as COVID-19's first deadly assault hit the world hard in March of 2020, many LGBTQI+ communities in postindustrial countries adapted by quickly moving many of their services and programming online, thus giving access to these hard-to-reach populations within the communities that they serve. In some places, like Denver, Colorado, in the United States, Pride Celebrations took place online providing opportunities for individuals who were less mobile or in high-risk health categories a chance to enjoy pride from the safety of their own homes. "It is hard to get in and out of Pride celebrations when you have a disability," a local Denver resident with multiple sclerosis told me, adding, "It was quite a treat to experience something that has not always been friendly to people in wheelchairs."

Unsurprisingly, the interviews revealed that members of the LGBTQI+ community had similar experiences to what many others did who had access to a computer and the internet. Every person interviewed spoke about changes to their interpersonal relationships. Many spoke about their experiences of watching their partners work online instead of at an office. One commented on how watching their partner work at home helped to build a new respect for their partner's work style. "I learned a few things about how to manage people more effectively from watching [my partner]" said a gay man from Lexington, Kentucky. Several people interviewed spoke about how technology brought laughter into their homes when teaching their older parents how to use Zoom and FaceTime, or when organizing new ways to socialize by having Zoom happy hours, game nights, and costume parties.

Social media was a popular theme. Every person I interviewed spoke about how social media became a central theme of conversation in their homes and often brought emotions of sadness or joy. Each saw social media as entertainment, a way of coping with the realities of the world, and as a place to get information about how to prevent the spread of COVID-19. Every person interviewed spoke about taking time away from social media and deleting apps on their smart phones in order to prevent the temptation of access to their social media accounts. "I would get depressed," said a lesbian living in Washington State when speaking about limiting her social media use during COVID-19 lockdowns. "Apparently, everyone is an expert on Facebook."

Applications of technology for international LGBTQI+ community

The international community during COVID-19 imposed varied levels of restrictions, that included lockdowns and quarantines, that in some cases did not allow people to leave their home at all. These measures especially impacted members of the Trans community who often experience higher rates of suicide in comparison to other members of the LGBTQI+ community. To combat this, many organizations rallied to find unique ways to support the Trans community with potential depression from losses of work and relationships. One member of the Trans community living in Bangladesh spoke about his experience, "I was sent a computer that helped me stay connected," he said. "I was also able to get a job online doing customer service because everything moved online." Fortunately, the organization he worked with, Pawan Dhall, shifted funding used for their advocacy work into service, such as housing support and purchasing computers. "We were surprised of the outcome," said the organization's executive director Ryan Figerou, referring to the successful outcome of the purchase of computers for members of the Trans community. "We never expected that our community members would also have access to new jobs. It certainly was an unexpected and welcomed surprise."

As we all have experienced, innovation and technology move rapidly, changes which are often unseen unless you work in those fields, but which directly impact the community to which you belong. Media professionals uniquely find themselves at the nexus of invention and at the pulse of its application, as a requirement of their job and sometimes as a result of their own curiosity. Michael Lavers has a unique perspective on how global innovations have impacted the LGBTQI+ community, specifically in Latin America during the COVID-19 lockdowns. Michael is an award-winning international news editor for the Washington Blade, the oldest LGBTQI+ newspaper in the United States, who has been reporting on international LGBTQI+ issues for more than a decade.

In July 2021, Lavers was on assignment in Honduras, a country notoriously known for dangerous gang activity limiting freedom of movement for many LGBTQI+ community members. In Honduras, as in other parts of the world, gay men and trans women who have access to smartphones use applications such as Grindr and Scruff to be connected for friendship, romance, and sexual partnership to other gay men and trans women.

The International Labour Organization (ILO), funded by USAID, leveraged the use of these applications and targeted vulnerable and in-hiding populations who lived in neighborhoods where gang violence happened frequently. One of ILO's missions is to inform gay men and trans women, who used these applications, about sexual health and where they can access HIV testing and treatment. The important point was that attention to HIV/AIDS should not be overshadowed entirely by the risks of COVID-19. Consequently, during the COVID-19 pandemic, these organizations ramped up their outreach programs and expanded their services to help the community at large.

Data and the LGBTQI+ community

While technology has been paramount to create outreach programs, such tools are equally instrumental in collecting valuable information that could help advance LGBTQI+ causes. According to Fabrice Houdart, a human rights specialist and LGBTQI+ activist who fights for economic inclusion for the LGBTQI+ community, technological advancement has assisted in extracting and collecting important data that once was nearly impossible to get about the LGBTQI+ community members. He explained, "Technology had been one of the few ways that you could collect data on the LGBTQI+ population that is self-reported. This information is important because data and public policy have a relationship, and you cannot make a claim to change or create policy without it."

Overall, this increase in data collection and the information it reveals are largely due to the willingness of LGBTQ+ people once hesitant to disclose information or self-identify in face-to-face or phone surveys, to now reveal such information given the anonymity that a computer screen provides. Houdart anecdotally shared that this kind of data collection conclusively dispelled the previously held bias by the general public that LGBTQI+ community members are financially well off when they are actually more often economically vulnerable than other groups. Such evidence is especially important for those who support health communities since research shows that economically disadvantaged communities, on average, experience higher incidences of poor health outcomes, including poor mental health outcomes.

Houdart further told me that data collection using app- and chat-based applications, such as Trevor Lifeline, TrevorText, and TrevorChat, as well as the world's largest safe space social networking site for LGBTQ youth, TrevorSpace, was able to detect early on the influences of COVID-19 and suicidality among queer youth. With this information, they were able to tailor their organization's programming and aid in helping other organizations while assisting in building policy that served both the LGBTQI+ community and the teen community at large.

Innovation has not just been exclusive to technology but also in programming and public health interventions. In the summer of 2020, I was a part of a research team that designed a nature-based wellness program for gay men who are over the age of 50 and living with HIV. The research aimed to understand if microbiome exposure through nature walks in a local botanical garden would help people living with HIV aged 50 and older with medication adherence and overall well-being during the COVID-19 pandemic.

The long-term effects of the program have yet to be reported; however, anecdotally, I can say that many in the group have spoken about how the isolation imposed by COVID-19 restrictions triggered memories of the AIDS epidemic. As in many instances, a participant shared that the deaths caused by COVID-19 triggered memories of the trauma and loss he experienced when 10 of his friends died in 1 year from HIV/AIDS. Many people in the LGBTQI+ community reported to me that having a group to talk to about the anxiety of being a part of an especially vulnerable health population during COVID-19 made them feel less anxious and depressed.

Regrettably, access to new technological approaches to mental healthcare and psychosocial support have not been universal for all members of the LGBTQI+ community, whereby some experienced an even greater gap in access to care. These include those based on their sexual orientation and gender identity alone due to stigma or oppressive laws, or those forcibly displaced from their homes due to conflicts or war in their country, or those in developing countries where displaced populations have settled worldwide and where a silent epidemic of mental disease has crept into the LGBTQI+ community.

Opportunity for technology and innovation

Mental illnesses and substance abuse disorders are now the leading causes of disability worldwide, especially impacting marginalized populations, including the LGBTQI+ community. This has added to an already ever-increasing mental health treatment gap, which has only compounded a growing global burden of disease. Tragically, more than 70% of persons worldwide who need mental health services lack access to care, particularly marginalized communities including the LGBTQI+ community (Henderson et al., 2013). This disturbing trend grew even faster as the COVID-19 pandemic continued. This gap, between those who need care yet don't have access, continues to grow, even when evidence-based mental health interventions have been found effective in low- and middle-income settings.

It is important to note the situation of mental health in the LGBTQI+ community, which suffers higher rates of mental health diagnoses, such as depression, anxiety, substance abuse, body image and eating disorders, as well as borderline personality disorder. Tragically, internationally, these mental health needs of the LGBTQI+ community have been largely overlooked. Much of the international literature that exists on LGBTQI+ populations is primarily focused on sexual health issues and on human immunodeficiency virus (HIV), which is assuredly critical work, but overshadows other important mental health needs.

In this context, LGBTQI+ people globally are subjected to institutionalized prejudice, social stress, social and family exclusion, hatred, and violence. And when you intersect the identity of being LGBTQI+ to that of being forcibly displaced and/or stateless, a more severe disparity arises in access to quality care and the availability of mental health treatment. In fact, this group has historically experienced—on a sometimes-daily basis—stigma, discrimination, and xenophobia while seeking health services, while experiencing higher levels of mental health distress and disease state.

In response to the treatment gap—that exists for all populations—the World Health Organization has developed guidelines, in their Mental Health Gap Action Programme Intervention Guide, meant to be used by specialized and nonspecialized healthcare providers adapting to the needs of the population with which they are working (World Health Organization (WHO), 2016). This is a valuable guide whose implementation and integration into health policy are still lacking. I argue that technology and social innovation can find a place to grow and be applied, specifically when building upon this Mental Health Gap Action Programme Intervention Guide. Now more than ever, capacity building and implementation of mental health services are urgently needed, as nearly 84 million people by the end of 2021 were forcibly displaced from their homes as a result of persecution, conflict, violence, human rights violations, and climate change. (United Nations High Commissioner for Refugees (UNHCR), 2021).

Conclusion and suggestions for the future

The advent of technology into today's world has changed everything about the way we work, live, and even love, as well as ways we do research, teach, and even provide care. In today's technological era, we can transfer and receive information in an instant, create virtual global communities that have the power to alter policy and create lasting change, and give people access to opportunities once out of reach. While technology is a gateway to once-insurmountable opportunities that can change the world, we mustn't forget to include access for those persons, including in the LGBTQI+ community, who are often forgotten. As we move into the future, I challenge each of you to apply your creativity and find solutions to serve the hardest-to-reach communities among us.

References

Henderson, C., Evans-Lacko, S., & Thornicroft, G. (2013). Mental illness stigma, help seeking, and public health programs. *American Journal of Public Health*, *103*(5), 777–780.

Human Dignity Trust. (2021). *Map of countries that criminalize LGBT people*. Human Dignity Trust. Available from https://www.humandignitytrust.org/lgbt-the-law/map-of-criminalisation/ (Retrieved 27 May 2022).

United Nations High Commissioner for Refugees. (2021). *Refugee statistics*. UNHCR. Available from https://www.unhcr.org/refugee-statistics/ (Retrieved 27 May 2022).

World Health Organization. (2016). MhGAP intervention guide (mhGAP-Ig). In *mhGAP Intervention Guide—Version 2.0* World Health Organization. Available from https://www.who.int/publications/i/item/9789241549790 (Retrieved 27 May 2022).

CHAPTER 73

Immersive technology tools and new media for Indigenous culture and wellness

Tahila Mintz
OJI:SDA' Sustainable Indigenous Futures, Haudenosaunee Territory, Spencer, NY, United States

Our collective human experience has been one of resilience through challenging environmental, social, and political conditions. The world can be a hard place to live, threatening our health and well-being, and this has certainly been made more apparent during the COVID-19 pandemic. The pandemic has been like a magnifying glass focused on a piece of paper; when the sun hits the spot just right, it lights a fire. Our lives and our souls have similarly been set aflame.

All peoples have struggled with the destructive and disruptive blaze of the pandemic. For the tragedies that have befallen us due to the pandemic, I mourn. But the pandemic has also brought an expansion, for which I am grateful. The pandemic has forced us to expand to new ways of looking at, and living, our life, shining new light on creating and reconnecting. These new lights have come in the form of including new technology in our perspective and ancient practices in our daily lives.

It is important to recognize that the life issues we are addressing during this pandemic are not new. Indigenous communities trust in wisdom from the elders, thus innovation now builds on what we have learned before, illuminating what has worked and also not worked. This is especially poignant to support the mental health of Indigenous youth who are currently at high risk of mental distress and suicide.

In this chapter, I outline some of the mental health challenges faced by Indigenous people, particularly in the realm of youth suicide, and then I describe a 360 Virtual Reality (VR) project, the *Ancestral Gratitude Bridge*, which facilitates a potent empowering immersive experience in Indigenous culture and heritage.

A growing number of Indigenous people, including myself, are working with new media and technology tools in creative and compelling ways to share our traditional stories and contemporary narratives, while promoting authentic visibility. The rich content from these approaches is changing the identifiable landscape especially for youth in our community, to validate their understandings, worldview, and culture.

Resilient Health. https://doi.org/10.1016/B978-0-443-18529-8.00073-1

Before the pandemic, Indigenous youth suicide rates were already much higher than national averages globally—up to 11 times greater in some countries. During the pandemic, youth suicides skyrocketed around the world, with Indigenous communities continuing to be severely impacted. Given these grim statistics, it is urgent to support the mental health of Indigenous youth through means that connect them to a sense of support and hope through their communities' cultural traditions (Cultural Survival, 2020; Raypole, 2021; Toomey, n.d.; United Nations, 2016).

Virtual reality for Indigenous well-being

In seeking to mitigate these horrific high rates of Indigenous youth suicide, specifically in Haudenosaunee communities, I have been creating an immersive Virtual Reality experience entitled the "Ancestral Gratitude Bridge." I use this tool as an example to demonstrate important considerations and best practices when working with Indigenous peoples for mental healthcare support. The idea about this technique came to me before the pandemic, but was put into action during the first year of the pandemic, as the number of youth suicides alarmingly climbed.

The *Ancestral Gratitude Bridge* has shown the potential for supporting a healthy sense of cultural identity and belonging. This tool uses 360 virtual reality (VR), which is an immersive experience that presents a guided audiovisual simulation that surrounds the user, allowing them to look around in all directions, just as in real life.

The 360 VR is useful for the young population since they are already familiar with technology tools like those in video games. Applying technology with intention for growth and not just entertainment provides an excellent opportunity for connecting youth to a greater sense of themselves, their culture, and the natural world.

Our relationship to the earth and cultural teachings—Mental wellness and balance

I was taught that my primary responsibility was as a guardian of the earth and all that is living on her. I was also taught that it is through our relationship with the natural world that our mental and physical equilibrium can be restored when we become imbalanced. These teachings were not just taught to me, and are not only teachings of my tribe, but rather are core understandings passed down for generations for many Indigenous Peoples. Here I speak from my knowledge base, from what has been shared with me.

Our relationship to the natural world is that of community, family, and a connected support system, always with respect. The natural world is inclusive of the trees, the waters, the four-leggeds, the winged ones, the crawlers, and the swimmers, and all living organisms existing in and on this planet. Scientists and healthcare professionals are beginning to "discover" that trees communicate with and support one another; we are all community. Research shows that people feel calmer and less depressed after spending time in nature, as we are nurtured and healed by being in balance with the natural world. These understandings are ancient.

The elders who give us teachings can be embodied in a person, rich with wisdom from years of life and what was shared with them, or can be a seed passed down for generations, whose stories are shared with us and teachings passed on to us as we learn how to plant, and care for and prepare foods and medicine. Our teachers are members of our community in the natural world, of which humans and all beings and elements are a part. As humans, the act of listening to and caring for those around us is an experience of reciprocity in which we are providers and also provided for. Since we include all beings, our Indigenous community framework is more expansive than that of the Western model. The health and well-being of our whole community is acutely essential for each of us. The state of the natural world is a fundamental part of our health and well-being, physically, emotionally, cognitively, and spiritually, and affects us deeply on all these levels.

Connection disruption—Historical and contemporary colonization

Our societal model is not that of individualism but rather of interdependence. It is not of capitalism but of community. The Western societal model clashes with our model of balance, community, and life purpose. In our communities, we thrive in interwoven relationships with one another, with humans and with the natural world.

Throughout the world, there have been violently aggressive campaigns of assimilation and annihilation of Indigenous Peoples. This does not just refer to times long ago, but is a contemporary issue, evident in the erasures of our heritage in the education system and in the destruction of people's natural habitats. The "modern world" disregards our essential truth that we are part of something bigger. We are a piece of a world balance that does not revolve around humans alone. Indigenous peoples' devotion to the protection of the natural world positions them as a threat to those who want to "own" the land and all that is under and above the soil. Tactics of devaluing and denying the existence of Indigenous peoples continue.

It is not unusual for a stranger to tell a Native person that they "didn't realize Native people still exist." It is common to be in a school classroom and hear rhetoric of the standard curriculum that only refers to Indigenous people in past tense. This is not an isolated experience in only one country. There have been many intentional campaigns to eliminate Indigenous visibility and to sever the lines of traditional life pathways that should be shared down generationally. Indigenous people are deemed as second-class citizens still by those who do not know better than to put themselves above others.

I will speak more specifically about the Haudenosaunee. Their history is shared by, or holds, many similarities to various tribes of North America and some around the globe. The generational genocide of the Haudenosaunee people includes historic land grabs, the Sullivan-Clinton scorched earth campaign, the introduction of alcohol, broken treaties, and various other means which continue in contemporary times. Only recently did residential schools for Indigenous peoples close, which had furthered the systematic separation of people from their culture and their community support systems through abuse and terrors found in these institutions. "Kill the Indian, and save the man" was the motto. Residential boarding schools promoted extermination or conversion. This system broke down traditional family structures of nurturing, loving, support, and community. The original cosmological teachings, traditional environmental relationships, language, food systems, ceremonies, and oral traditions were robbed from many children's life experiences. Instead, the absence of cultural identity was passed down.

Colonization of Indigenous peoples runs deep and tears wounds in the family, community, and hearts. As a result, many youth encounter challenges in accessing traditional teachings. Few healthy elders are alive today and many youth go to Western schools with racist curricula, creating further detachment from their cultural education and disempowerment of their cultural identity. America and Canada continue to disempower Native Americans through systematic and societal racism. Community members' lives are claimed every day from drug overdoses, alcohol abuse, suicide, and violent crimes. As the disconnection from healthy traditional ways is widespread, abuse comes also from within, running rampant in some homes.

Intervention innovation—Cultural healing, affirmation, and media technology

Immersive technology offers an opportunity to strengthen the physical, mental, and spiritual health of Indigenous people. The pandemic deepened feelings of isolation and vulnerability of Indigenous youth. Therefore the goal of innovations using technology for Indigenous youth is to create something that could be used by them when they are having a hard time—or even before bad times happen—to interrupt negative cycles. Through new media, there is an opportunity to meet youth in the conversational space with which they are accustomed, in a whole and safe way, and to reintroduce teachings that are a part of them but which have been denied from them for all the circumstances I have outlined before.

As an Indigenous media maker and a member of the Onondaga Nation suicide prevention group, I wanted to create a tool for first responders (those of us who reach out to troubled persons) to use in the community when they go out on calls for youth emergencies and also for youth to engage with on their own. Since Indigenous youth are plugged into "new" media technology, such as gaming, social media, television, and the internet, I created this tool in 360 Virtual Reality, a media format that I know would be exciting to young people and more likely for them to engage with. Utilizing new media to present material that would focus them on their culture, teachings, and territory, let them know that they are seen and valued. Furthermore and importantly, they would remember the community connection with the natural world that is always there to nurture them.

Immersion into Haudenosaunee cosmology

The Haudenosaunee homelands are in what is today New York State and southern Ontario. The Haudenosaunee Confederacy, also known as the Iroquois Confederacy, includes: the Mohawk (Kanien'kehá:ka), the Oneida (Onyota'a:ka), the Onondaga (Onoñda'gegá'), the Cayuga Cayuga (Gayogo̱hó:nǫ'), the Seneca (Onödowa'ga:'), and the Tuscarora (Skaru:rę̨). The words in parentheses are how each nation calls themselves in their language.

Two quotes express well the Haudenosaunee worldview.

> The Haudenosaunee give thanks daily, not just once a year. They give thanks for all things, from the water and sun to the insects and animals. Their thanksgiving address, called the Ganö:nyök (ga-NOOHN-nyonk), is a very important part of ceremonial and social gatherings. All social and ceremonial gatherings start and end with the Ganö:nyök, which is sometimes called "the words that come before all else." "The Ganö:nyök serves as a reminder to appreciate and acknowledge all things. The words express thanks for fellow human beings, Mother Earth, the moon, stars, sun, water, air, winds, animals, and more." (National Museum of the American Indian, 2009)

> It lays our relationship to the Universe…A summation of our cosmology, takes us to an orientation of gratitude…that we are not alone in the world but that we are all in community with all the beings that we share the planet with… Let us bring our minds together as one… The words before all else…to say thank you for the gifts of life everyday…renewal continuation, great authority, great instructions, good values…takes you to that place of feeling gratitude for what you have. (Skä•noñh Great Law of Peace Center, 2018)

The 360 virtual reality experience—The "Ancestral Gratitude Bridge"

In this Virtual Reality (VR) experience, which I call the "Ancestral Gratitude Bridge," people are provided goggle headsets that have a prerecorded video embedded, such that you are immersed 360 degrees in the scene audio-visually. One can turn and see in all directions, including side to side and above and below, as well as experience surround sound. The scenes are presented sequentially in a slow enough time to absorb the variety of landscapes and topical content that is happening.

Fig. 1
The two honored speakers in the Virtual Reality experience.

With the headset on your head, you open your eyes to first find yourself in a field of high grasses and plants, with tall trees in the distance, and warm morning sunlight illuminating the landscape. A melodic resonant voice starts talking. You can turn to find the source of the voice you hear, speaking in a language you most likely don't know, rotating your head, to encounter a man and a woman standing together, as shown in Fig. 1. The man speaking first is Clayton Logan, an internationally known traditional Smoke Dancer, a carrier of ceremonial responsibilities and one of the last few Senecas fluent in the language. Standing next to him is Nicky Thompson, a Water Protector, the Minister for Native Americans in New York Correctional facilities and a Knowledge Carrier, who will speak later.

In our cultures, it is an honor to sit and listen. The time held in that space may seem long to people accustomed to the Western model of communication and time, but in this experience, we are reminded to slow down, witness, and breathe. When he stops talking, the visual begins to change. A forest now surrounds you as you begin to hear a female voice speaking in English. It is now the voice of Nicky Thompson. She shares teachings of connections to the natural world and of ancestral stories, self-acknowledgment and care, and recognition of the many elements we give gratitude to in the Ganö:nyök.

The scene with Nicky and Clayton changes to trees whose tips are blowing in the wind and the image changes to grasses by the ocean shore, blowing and covering you over and over. Next, we find ourselves by the gentle flowing waters of the Ohi:yo' river, as clouds come in for the start of rain. Through the clouds, and on the water, people are seen appearing. Nicky has paused in her speaking, and a drumbeat is getting loud and a voice is strongly calling. In each direction you turn, there is a different person, dressed in traditional regalia, dancing the Smoke Dance, while Bill Crouse, a cultural performer and ambassador, and an artist, is drumming. They are of different ages and genders, including a young female child.

These people fade away, and the sun changes, but we stay in the same field. Nicky continues speaking and a herd of buffalo comes to join us. They wander up close to us, smell us, and investigate until we are in the center of the herd. Then, the scene transitions to a table, laid out with traditional foods, corn, berries, maple syrup, and strawberry drink. People around us now are speaking in Seneca and a small child is asking questions about the moon she sees in the distance. We enter the landscape of medicine plants again. Nicky returns to speaking to us, in her calming clear voice, as we are standing in flowers on the Great Hill, the hill of origin for the Seneca people. The video presents a perspective as if we were a small being watching the sun moving out to shine on the day. Nicky is speaking of possibilities, oneness, gratitude, and love.

The scenes making up this story, which I have created in this video, were chosen with great intention. Many elements to which we give thanks overlap in different parts of the experience. This is meant to demonstrate that our lives are not in isolation but are continually interwoven in the larger community with relationships to the natural world as a whole. Though all of the people featured in the Ancestral Gratitude Bridge are of the Seneca Nation, the cultural content is relevant to all of the six nations of the Haudenosaunee Confederacy.

Throughout the experience, you have ample time to view the environment that surrounds you in each scene, while listening meditatively to the messages of the voices. Scenes and moments are intended to be exciting and inspiring, and for you to have the feeling of shared experiences immersed in the environment and with the people, animals, and nature, in spaces you would not normally have access to, or know how to go to.

For us, by us: Work from the center out

Making work for, and within, Indigenous communities begins with the community, through community participation, community advice, community focus, and community support.

My project of the Ancestral Gratitude Bridge can serve more populations than only a specific community. But, the experience must be defined by each community on their own. The content and rooted stories are deeply intimate to a specific community and therefore, it is theirs, existing to resonate and support them primarily. Then, it can be shifted, or shared with other communities, if permission is received from the original community whose story is being shared.

I create work that is primarily for the community to whom the stories belong. My projects are not made in isolation; instead, I cocreate with the community. After that, the project can be shared with other Indigenous communities. And then, it can go out into the world, but only with the permission of the original community. Following these guidelines, my project of the Ancestral Gratitude Bridge has been created in collaboration with the community members, through sharing teachings and sharing needs of the youth, with guidance being facilitated, for example, by a safe visit with the Buffalo.

The project I have described in this chapter was also made possible by the support of Insta360 and Cornell Virtual Embodiment Lab with the loan of VR equipment and support from Goethe-Institut Chicago to help with production costs.

The media tech camp

To support the shift toward involving more Indigenous creators in such technology-based innovations for Indigenous populations, OJI:SDA..... Sustainable Indigenous Futures has started an Indigenous youth summer camp, called Mochik Ania—Camp Turtle. The camp provides youth with opportunities for self-discovery and expression grounded in the land, Indigenous knowledge, and cultural traditions. Media Arts and Traditional Arts are a main part of the camp experience, as extensions of the storytelling we do in relationship to the lands from which each camper comes, and where the camp is held. The campers come from diverse nations, allowing them to nurture connections across Indian Country while feeling "seen" as they share their experiences with each other. Additional digital media intensive trainings are held at other times in the year. Also, the youth can nurture their continued articulation of these skills and empowerment of their voices and connections to the land by returning annually to the camp.

Sustainability, replicability, and access: Reaching more people

This project's usefulness extends beyond pandemic times, to support the mental health and well-being of Indigenous youth. Also, it can positively impact people of other cultural traditions and ages. Such a reconnection to the earth and teachings of respect and gratitude crosses cultural boundaries, to boost mental health support and mitigate PTSD, suicide, and depression. The 360 VR, along with other experiences like storytelling, art, and lessons of the natural world, becomes potent tools for good mental health support programming.

For the example given before, there is a selected location where people can go to have the experience and a number of headsets can be taken home on loan to a family or youth for a time period. Ideally, the application of this VR experience can reach everyone in the community. Unfortunately, economic realities limit both the technology access and the requisite training for content creation. In order to facilitate broader access, sustainable partnerships must be formed internally and externally, particularly with the manufacturers of these tools.

Guardianship of ancestral teachings is important for our communities, along with sharing contemporary perspectives and realities. This is essential to preserve our cultural knowledge. In a positive step, a lot of work is being done to record Indigenous languages, since many communities in the North are experiencing the loss of their first language speakers who are transitioning out of this life. Sharing is also necessary; some of us work only in our own community, while others, like me, work in many different Indigenous communities so we can reach a broader audience.

Conclusion

Utilizing new media tools to share the Ganö:nyök message to Haudenosaunee youth shows great promise to nurture their mental health and to lower susceptibility to depression, suicide, and substance abuse, by boosting their connection to the wisdom of their heritage. The Ancestral Gratitude Bridge project I have developed, using technological tools, facilitates an engagement with powerful messages, connecting viewers with balance, honor, beauty, dignity, and respect within ourselves and the world surrounding us. It serves as a reminder of how we are held within the universe and never alone or isolated. This message is important for all Indigenous youth, including those from non-Haudenosaunee traditions, as it resonates with many Indigenous peoples' worldviews. Application of technology has been a gift in this process, acting as a medium to help youth feel reconnected to their roots and feel unity with their community, which includes all of the natural world. Digital media youth camps continue these positive lessons.

The programmatic media content must be facilitated by taking into account the needs of a community cocreating it, with deep and real understanding of the culture of the community. New technology by itself is not the key to designing transformative innovations. Media tools can help convey the teachings, stories, and worldview of a community, especially the wisdom-filled narratives of the elders that are at the root of their culture. While a given community may not be familiar with the latest technology, many people, especially youth, are willing to learn. A good model for creating and applying new technology is to do so collaboratively. In these ways, introducing technology to Indigenous peoples lives can help make our futures be in greater balance, with improved mental well-being and strengthened social ties.

References

Cultural Survival. (2020). Combatting suicide amongst indigenous peoples: An interview with Pat Dudgeon. In *Cultural Survival Quarterly Magazine*. https://www.culturalsurvival.org/publications/cultural-survival-quarterly/combatting-suicide-amongst-indigenous-peoples-interview.

National Museum of the American Indian. (2009). *Smithsonian*. George Gustav Haye Center. www.americanindian.si.edu.

Raypole, C. (2021). Depression in indigenous communities: Cultural understanding matters. In *Psych Central*. https://psychcentral.com/depression/depression-in-indigenous-communities.

Skä•noñh Great Law of Peace Center. (2018). *Thanksgiving Address [video file]*. http://vimeo.com/271474122.

Toomey J., n.d. Mission Australia. Can we talk? Seven-year youth mental health report—2012–2018. Mission Australia. https://www.missionaustralia.com.au/media-centre/media-releases/persistent-rise-in-youth-psychological-distress-spurs-urgent-call-for-action.

United Nations. (2016). Discriminatory pressures confronted by indigenous youth, traditional land rights of peoples, among issues highlighted as permanent forum continues session. In *Meetings Coverage and Press Releases. Permanent Forum on Indigenous Issues*. https://www.un.org/press/en/2016/hr5300.doc.htm.

CHAPTER 74

Innovative technologies for mental healthcare: An application for severely mentally ill persons

Brianna Fitapelli

Department of Clinical and Counseling Psychology, Teachers College, Columbia University, New York, NY, United States; International Association of Applied Psychology, Indianapolis, IN, United States

Introduction

Amid the fourth industrial age of the digital revolution, innovative technologies have emerged for the diagnosis and treatment of mental health conditions (Bucci et al., 2019). Not only have these technologies allowed for more precise and reliable practices, but they can also offer access and scalability that was limited with traditional modalities (World Economic Forum, 2019). The use of Artificial Intelligence (AI) and Virtual Reality (VR) has gained much attention in recent years (World Economic Forum, 2019), with the COVID-19 pandemic ramping up technological production, research, and use even further in the two years since the pandemic hit a peak in early 2020 (Renu, 2021).

For individuals with severe mental illness (SMI) and other vulnerable populations, it is often a challenge to interact with the rapid growth of digital technologies (Harvey & Keefe, 2012). SMI is an emotional, mental, or behavioral disorder that results in severe impairment in functioning, substantially interfering with or limiting at least one major life activity (U.S. Department of Health and Human Services, 2022). The most common forms of SMI include bipolar disorder, major depressive disorder (MDD), and schizophrenia (Substance Abuse and Mental Health Services Administration (SAMHSA), 2022). Data from 2020 showed that 14.2 million adults in the United States have a SMI, representing 5.6% of the adult population. Further, prevalence rates of SMI are higher for females (7.0%) compared to males (4.2%), and highest among individuals that report two or more races (9.9%), followed by American Indian/Alaskan Natives with an adult prevalence of 6.6%. Young people in the U.S. population aged 18–25 hold the highest prevalence of SMI at 9.7% versus a 6.9% prevalence of adults aged 26–49, and a

Resilient Health. https://doi.org/10.1016/B978-0-443-18529-8.00074-3

3.4% prevalence of individuals 50 years of age and older (U.S. Department of Health and Human Services, 2022).

With the high prevalence rates of individuals living with SMI and the increased acknowledgment of the critical role mental health plays in achieving both personal and global development goals, quality mental health services and treatments are fundamental (World Health Organization (WHO), 2022). Inpatient hospitalization is one form of mental health treatment for individuals in need of supervised, in-resident care. There are over one million hospitalizations in a given year for mental health-related conditions, with schizophrenia and mood disorders as the two leading causes (Agency for Healthcare Research and Quality, 2008). Schizophrenia is a serious psychiatric disorder affecting nearly 1% of individuals worldwide (Green et al., 2015). Among all mental illnesses, schizophrenia makes up approximately 50% of all hospitalizations in the New York City area (New York City Department of Health and Mental Hygiene, 2016) and is one of the most disabling mental health disorders (Green et al., 2015). Schizophrenia presents as a wide range of symptoms, including delusions, hallucinations, disorganized thinking and speech, disorganized or abnormal motor behavior (i.e., positive symptoms), social withdrawal, inability to experience pleasure, and blunted affect (i.e., negative symptoms; Green et al., 2015). These symptoms can severely impact the ability to live a long functional life, evidenced by the 15 to 20 year reducation in life expectancy for individuals with schizophrenia compared to the general population (Fleischhacker et al., 2014). To reduce and manage symptoms, inpatient hospitalization is often needed and although the quality of care received within these facilities has improved in recent years, further advancement is necessary to help individuals fulfill a meaningful, healthy life (Fleischhacker et al., 2014).

Throughout an inpatient hospital stay for individuals with schizophrenia, antipsychotic medication is most frequently prescribed and often effective to alleviate the distressing symptoms of hallucinations and delusions but less effective for other symptoms like social withdrawal and the inability to experience pleasure (Fleischhacker et al., 2014). To target these symptoms, psychosocial interventions are utilized concurrently with medication. Psychosocial treatments are cost effective, promote recovery, and reduce relapses and future hospitalizations. However, the funding provided from public healthcare systems varies greatly between countries; therefore, many individuals have no access to, or are denied such treatments (Fleischhacker et al., 2014). Furthermore, lack of staff to provide such treatments causes another major obstacle for individuals in need of further mental healthcare. With only 27% of America's community hospitals having separately organized and adequately staffed psychiatric units, there is a clear need for advancement (Weil, 2015).

One area of focus that has advanced rapidly in recent years, especially throughout the COVID-19 pandemic, is the emergence of innovative, digital technologies for the treatment of mental health (Renu, 2021). All societal actors were forced to implement technological tools that were available but not extensively used, and even more, to think outside the box and develop innovative interventions and practices. Technological strategies and methods of communication are now at the forefront of daily life and started the monumental movement toward more advanced and precise modalities for the treatment, prevention, and diagnosis of mental health conditions. In this review, two innovative technology programs are discussed that were implemented over the course of the past 2 years, throughout the COVID-19 pandemic at Manhattan Psychiatric Center's (MPC's) inpatient unit.

Innovation 1

The inpatient unit at Manhattan Psychiatric Center (MPC) primarily houses individuals with chronic and acute schizophrenia who are in need of long-term mental healthcare. The research conducted at MPC aims to better understand the progression and treatment of symptoms of schizophrenia and related disorders. One ongoing project uses an artificial intelligence (AI) software, Modality, which is the first clinically validated, multimodal, automated system for assessing psychiatric and neurological states, both remotely and in the clinic (Suendermann-Oeft et al., 2022). The software has a conversational AI system that can monitor psychiatric and neurological disorders with precision, consistency, and accuracy for more cost-effective and efficient treatment responses (Suendermann-Oeft et al., 2022). The software also includes a HIPAA-compliant system, allowing users to be safely and securely monitored in the comfort of their own homes (Suendermann-Oeft et al., 2022).

At MPC, Modality is being implemented to assess symptom severity in schizophrenia, particularly for disorganized thinking and speech, disorganized or abnormal motor behavior, social withdrawal, inability to experience pleasure, and blunted affect. Preliminary results showed that (1) patients with greater symptom severity are louder and exhibit slower articulatory kinematics (i.e., jaw and lower lip movements) during repeated speech; (2) those with greater negative symptom severity are louder and exhibit greater eyebrow displacement; (3) reading scores correlate with the temporal dynamics of speech production during reading and repeating tasks; and (4) correlations were found with standard scores used to track depression, treatment responses, parkinsonism, and tardive dyskinesia.

Twenty-one inpatients were included in a sample for study who were diagnosed by their treating inpatient psychiatrists as having schizophrenia according to the criteria of DSM-5. The mean age of the sample was 40years old, including 16 males and 5 females, with 66.7% of the sample identifying as Black and 33.3% of the sample identifying as White. The results of the study were presented at international conferences such as the Society for Neuroscience (SfN) Annual Meeting in 2021 (Kothare et al., 2021). In this study, Modality was presented to inpatients remotely (making this at a low cost to researchers and users), for 20min, and included a virtual conversational agent. The software captures audiovisual data, automatically extracts speech and facial metrics, and provides a user-friendly dashboard which includes the metrics of each session. The conversational agent asks participants to discuss their difficulties with speech, salivation, and swallowing and while the participant is doing so, the AI software assesses phonation, the repetition of syllables (oral diadochokinesis), and sentences (sentence intelligibility test). Then, the conversational agent asks the participant to read specific passages out loud and finally, asks them to respond spontaneously to vignettes.

The researchers concluded that the multimodal remote assessment approach to assess symptom severity in schizophrenia, which the Modality software uses, may help to combat the shortage of mental health services in many countries which has been exacerbated by the COVID-19 pandemic. Further, the Modality software has the potential to be an effective remote tool for monitoring both positive (i.e., hallucinations and delusions) and negative symptoms (i.e., social withdrawal and inability to experience pleasure) and to improve the efficacy of diagnostic and treatment strategies of schizophrenia. Although larger sample sizes are needed, with testing replicability and generalizability, this groundbreaking innovation holds much promise for the future.

Innovation 2

Currently, there is a short supply of healthcare workers and major disruptions to essential health services resulting from the COVID-19 pandemic (United Nations (UN), 2021). These factors, on top of the nuances and complexities of a schizophrenia diagnosis, clearly warrant a need for advancements in the strategies we use to lessen the burdens of cost, and lack of adequate staff and access we have suffered with for far too long.

Along with the Modality study at MPC, the research team also has an ongoing feasibility and preliminary effectiveness study using an immersive virtual reality (VR) mindfulness-based program, TRIPP, to improve aggression and violence in individuals diagnosed with schizophrenia and schizoaffective disorders (Khan et al., 2021). In this randomized-controlled trial, individuals are randomly assigned to a group using either TRIPP VR or a mindfulness-based digital application, Headspace, via an iPad. Twenty-eight inpatients were included in the preliminary data, which was presented at the Technology, Mind, & Society Conference in 2021. The mean age of participants was 35.1years old, including 12 males and 3 females with 17.9% of participants identifying as Black, 17.9% White, and 17.9% Other. The mean age of first psychiatric hospitalization was 22 years old. Preliminary results showed decreases in aggressive episodes and hostility for participants in the TRIPP VR group compared to those in the Headspace group. Regarding the feasibility of implementing a VR program in an inpatient facility, results showed that individuals preferred mindfulness-based TRIPP VR over the mindfulness-based digital app, Headspace, in terms of acceptance, engagement, functionality, and aesthetics (Khan et al., 2021).

Participants in each group received 18 10-minute sessions over the course of 4–6 weeks (Khan et al., 2021). Individuals randomized to TRIPP VR were given goggles with head-and eye-tracking technology that includes augmented and mixed reality. The software includes geometric images that are presented to the participant in a 360-degree view, while the person listens to self-guided mindfulness practices (TRIPP INC, 2021). The virtual world changes in real time, to redirect the person's attention to specific stimuli, sounds, and visuals intended to target the active control of attention. While generating realistic images, sounds, and other sensations, the participant's physical presence is stimulated in the virtual environment. Individuals randomized to the Headspace group engaged in single sessions that guided them through a mindfulness and meditation practice through audio-only modules. Participants are taught how to handle positive and negative emotions through topics such as "personal growth" and "stress and anxiety" (Khan et al., 2021). The audio-only modules are presented to the participant on an iPad, and they are given headphones to minimize external distractions. Although results indicated that participants found both groups, TRIPP VR and Headspace, to be engaging, reports showed TRIPP VR to have significantly higher levels of acceptability and aesthetics, compared to Headspace.

Overall, mindfulness practices have been shown to be beneficial for mental health treatment by helping individuals strengthen their attention flexibility and adopt an orientation toward experiences that reduce the reliance on automatic thoughts or maladaptive emotion regulation strategies (Aldao et al., 2010; Blanck et al., 2018). However, mindfulness practice requires a conscious effort and can be difficult to sustain, and there are several experiential challenges that can arise during mindfulness therapies and interventions (Anderson et al., 2019). These include affective demands such as cognitive effort and frustration; task demands presented by the physical environment (e.g., noisy settings and people); and negative emotional or psychological outcomes, such as boredom, upsetting thoughts, and emotions. Adapting mindfulness practice via VR to reduce the likelihood or impact of these challenges may be an important consideration for improving the success of mindfulness-based interventions for patients with schizophrenia spectrum disorders (Anderson et al., 2019).

Conclusion

Research projects such as these that support the benefits of innovative technology can not only improve treatment practices for severely mentally ill patients but also enhance objectivity in data collection and increase access to mental healthcare for the broader population. Although this research is still in its preliminary phase and larger sample sizes are needed, results thus far are noteworthy. The work at MPC is currently expanding, with future goals to implement VR to help reduce social cognitive, neurocognitive, and vocational deficits seen in schizophrenia spectrum disorders.

Not only have research departments across the world made strides toward the advancement of technology for mental healthcare, but the United Nations (UN) has promoted the use of digital technologies to advance the 17 Sustainable Development Goals (SDGs) for the 2030 Agenda (UN, 2021). Furthermore, the World Health Organization (WHO) recognizes that technology, specifically AI, has been shown to improve research, facilitate drug development, support diversity in health interventions, promote autonomy of one's own healthcare, and act as a bridge in underresourced countries to increase access to health services (World Health Organization (WHO), 2021). The potential is vast for improvements in not only SMI, but common mental health issues, as well as for widened and strengthened social support networks, access to information and quality education, and relapse prevention (United Nations (UN), 2016).

AI and VR are still underresearched and not easily accessed by the masses, and the full potential of these innovations is yet to be unlocked. However, evidence shows the potential for these modalities in ameliorating both common and SMIs in populations who have suffered the downfalls of mental health services. With increased attention and research on innovative technologies in mental healthcare, the cost of treatment can potentially decrease, and individuals of all backgrounds and mental health status can begin to have access to the care they need and are entitled to.

References

Agency for Healthcare Research and Quality. (2008). *Statistical brief #62*. Healthcare Cost and Utilization Project. Available from https://www.hcup-us.ahrq.gov/reports/statbriefs/sb62.jsp (Retrieved 15 March 2022).

Aldao, A., Nolen-Hoeksema, S., & Schweizer, S. (2010). Emotion-regulation strategies across psychopathology: A meta-analytic review. *Clinical Psychology Review*, *30*(2), 217–237. https://doi.org/10.1016/j.cpr.2009.11.004 122.

Anderson, T., Suresh, M., & Farb, N. A. (2019). Meditation benefits and drawbacks: Empirical codebook and implications for teaching. *Journal of Cognitive Enhancement*, *3*(2), 207–220. https://doi.org/10.1007/s41465-018-00119-y.

Blanck, P., Perleth, S., Heidenreich, T., Kröger, P., Ditzen, B., Bents, H., & Mander, J. (2018). Effects of mindfulness exercises as stand-alone intervention on symptoms of anxiety and depression: Systematic review and meta-analysis. *Behaviour Research and Therapy*, *102*, 25–35. https://doi.org/10.1016/j.brat.2017.12.002.

Bucci, S., Schwannauer, M., & Berry, N. (2019). The digital revolution and its impact on mental health care. *Psychology and Psychotherapy: Theory, Research and Practice*, *92*(2), 277–297. https://doi.org/10.1111/papt.12222.

Fleischhacker, W. W., Arango, C., Arteel, P., Barnes, T. R., Carpenter, W., Duckworth, K., Galderisi, S., Halpern, L., Knapp, M., Marder, S. R., Moller, M., Sartorius, N., & Woodruff, P. (2014). Schizophrenia—Time to commit to policy change. *Schizophrenia Bulletin*, *40*(Suppl. 3). https://doi.org/10.1093/schbul/sbu006.

Green, M. F., Horan, W. P., & Lee, J. (2015). Social cognition in schizophrenia. *Nature Reviews Neuroscience*, *16*(10), 620–631. https://doi.org/10.1038/nrn4005.

Harvey, P. D., & Keefe, R. S. E. (2012). Technology, society, and mental illness. *Innovations in Clinical Neuroscience*, *9*(11–12), 47–50.

Khan, A., Fitapelli, B., Lindenmayer, J.-P., Ahmad, S., Insel, B., Prokop, S., Seddo, M., & Padua, T. (2021, November 12). Immersive, mindfulness-based virtual reality (VR) to improve aggressive behaviors in an inpatient setting: Pilot data (Poster Presentation). In *Technology, mind, & society conference, virtual*. https://assets.pubpub.org/etvo842c/31634852733528.pdf.

Kothare, H., Neumann, M., Roesler, O., Liscombe, J., Pautler, D., Suendermann-Oeft, D., Prokop, S., Yavorsky, C., Khan, A., Lindenmayer, J.-P., & Ramanarayanan, V. (2021, November 8). Multimodal conversational technology for remote assessment of symptom severity in people with schizophrenia (Poster Presentation). In *Society for neuroscience (SfN) annual meeting, virtual*. https://drive.google.com/file/d/1yCuCgH9EUDcJFlhGExr4x_A0PCTcjSyJ/view.

New York City Department of Health and Mental Hygiene. (2016, June). *EPI data brief—Welcome to nyc.gov | city of New York*. Available from https://www1.nyc.gov/assets/doh/downloads/pdf/epi/databrief71.pdf (Retrieved 12 December 2021).

Renu, N. (2021). Technological advancement in the era of covid-19. *SAGE Open Medicine, 9*. https://doi.org/10.1177/20503121211000912. 205031212110009.

Substance Abuse and Mental Health Services Administration. (2022, October 2). *Living well with serious mental illness*. SAMHSA. Available from https://www.samhsa.gov/serious-mental-illness (Retrieved 15 March 2022).

Suendermann-Oeft, D., Pautler, D., Fox, D., & Ramanarayanan, V. (2022). *Modality.ai*. Available from https://modality.ai/ (Retrieved 5 March 2022).

TRIPP INC. (2021, September 10). *Fitness for your inner self. TRIPP®: Fitness for your inner self*. Available from https://www.tripp.com/ (Retrieved 5 March 2022).

U.S. Department of Health and Human Services. (2022, January). *Mental illness*. National Institute of Mental Health. Available from https://www.nimh.nih.gov/health/statistics/mental-illness (Retrieved 10 March 2022).

United Nations. (2016). *A healthy workforce for a better world: United nations systemmental health and well-being strategy*. Available from https://www.un.org/en/healthy-workforce/files/Implementation%20Board%20ToR.pdf (Retrieved 6 March 2022).

United Nations. (2021). *The 17 goals | sustainable development*. United Nations. Available from https://sdgs.un.org/goals (Retrieved 5 March 2022).

Weil, T. P. (2015). Insufficient dollars and qualified personnel to meet United States mental health needs. *The Journal of Nervous and Mental Disease, 203*(4), 233–240. https://doi.org/10.1097/nmd.0000000000000271.

World Economic Forum. (2019, April). *Health and healthcare in the fourth industrial revolution global future council on the future of health and healthcare 2016-2018*. Available from https://www3.weforum.org/docs/WEF__Shaping_the_Future_of_Health_Council_Report.pdf (Retrieved 11 March 2022).

World Health Organization. (2021). *Ethics and governance of artificial intelligence for health*. World Health Organization. Available from https://www.who.int/publications-detail-redirect/9789240029200 (Retrieved 27 February 2022).

World Health Organization. (2022). *Mental health*. World Health Organization. Available from https://www.who.int/health-topics/mental-health#tab=tab_1 (Retrieved 15 March 2022).

PART 6

Reports, forums, conferences, and networks shaping innovations and expanding partnerships

Major international conferences, multiday events, and collaborative networks provide an exceptional platform for varied stakeholders in the health ecosystem, from governments to civil society, innovators and investors, to present their work and network.

Chapters in this Part are an exceedingly rich resource about the diverse and exciting projects and innovations in policy, research, and groundbreaking initiatives and programs. These are reimaging healthcare and forging a new world of digital health and e-mental health. Techniques to achieve a major goal—to close the treatment gap—are described.

What's exceptionally exciting is that chapters in this Part cover work being done all over the world, with attention given to especially underrepresented and less well-known regions that have not been previously highlighted as centers of telehealth. For example, chapters review presentations at conferences focusing on the MENA and GCC region, as well as the RECAINSA conference that highlights telehealth in the Central and South American regions. These presentations from government and civil society representatives are so rich, they can be entire chapters on their own!

Also, with e-health being the wave of the future, a valuable mental health collaborative is presented. In addition, a review is included of the important new World Mental Health Report from the World Health Organization, updated from two decades ago, that provides governments, academics, health professionals, civil society, and all stakeholders with an essential blueprint about transforming mental health worldwide.

Valuable opportunities to join networks and coalitions about health and mental health professional affiliations are presented in these chapters.

What notables have to say

All our dreams can come true—if we have the courage to pursue them.

—Walt Disney

The people who are crazy enough to think they can change the world are the ones who do.

—Steve Jobs, Apple founder

Alone we can do so little; together, we can do so much.

—Helen Keller, American disability rights advocate

We can't help everyone, but everyone can help someone.

—Ronald Reagan, former American president

Digital transformation of health featured during Digital Health Week 2021: Profiles of partnerships and health innovations using technology in countries around the world

Hanna Lissinna

Department of Clinical Psychology, Teachers College Columbia University, New York, NY, United States; International Association of Applied Psychology, Indianapolis, IN, United States

Introduction

Only 53.6% of the population around the world use the internet, leaving 3.6 billion people unconnected, thereby widening the digital divide (Digital Health Week, n.d.-a, n.d.-c). This statistic is even more concerning in the context of the fact that "93% of the world's population live within physical reach of internet or mobile broadband services" (Digital Health Week, n.d.-b, n.d.-c).

Given these statistics, the increase in health innovations using digital tools provides solutions in offering access to mental and physical healthcare for a broad population, especially those in low-resource, fragile, and remote settings. The rise in such digitalization offers promise to close the general health gap, namely, the distance between the need for and availability of services. Currently, countries vary greatly in their health infrastructure, as well as in access and cost of care in general, much less in their ability to apply technological solutions, leading to vast inequities. Stigma surrounding mental health exacerbates this inequality. Yet, many of these inequities can be mitigated by digital health interventions. Hence, the development, implementation, and evaluation of such approaches is increasing and crucial.

To highlight, showcase, and upscale such innovations, an increasing number of exhibitions, showcases, and competitions have been produced. Selected events during one such effort, called "*Digital Health Week*," are profiled in this chapter (Digital Health Week, n.d.-a, n.d.-b, n.d.-c).

Resilient Health. https://doi.org/10.1016/B978-0-443-18529-8.00075-5

Overview of projects

The selected projects presented in this chapter are the following, listed by the country/countries they predominantly serve, and including their main objective:

Ghana: *BISA*, a digital platform to manage the COVID-19 pandemic.
Uzbekistan: *ELSA*, development loan from KfW, a German state-owned investment bank.
Nigeria (as well as Tanzania, sub-Saharan Africa): *Take Care Africa*, a pan-African healthcare information platform.
India: *Lighthouse Initiative India*, digital solutions across the Indian health system.
Nicaragua: *Estación Vital*, access to preventive checkups and personalized plans to change health behavior.
Africa: *AfyaPro*, a digital hospital and patient data system, for better health management.
Kenya (Nairobi): *Penda Health*, a model of medical centers in low-income areas.
International: *Solvoz*, a platform to facilitate efficient humanitarian tendering/procurement.
Africa: *Essential Medicine Guidance* (EMG), a mobile reference and learning platform, which provides free healthcare information.

Background

Justification for developing technological innovations for healthcare delivery is grounded in several international documents. The preamble of the Constitution of the World Health Organization (1946) defines health as the "state of complete physical, mental and social well-being and not merely the absence of disease or infirmity." The United Nations Human Rights Council further asserts that the "right to the highest attainable standard of physical and mental health is a fundamental human right" (Human Rights Council, 2017, p. 3). Sustainable Development Goal (SDG) 3 of the United Nations Agenda 2030 for Sustainable Development affirms the aspiration to "ensure healthy lives and promote well-being for all at all ages" (United Nations, n.d.). An important feature of these documents is the inclusion of mental health, with Goal 3.4 including to "promote mental health and well-being," and to reduce the *suicide mortality rate,* as an indicator for success (United Nations, n.d.).

Additionally, SDG 3.8 notes the aspiration to "achieve universal health coverage" as proposed in the Political Declaration for Universal Health Coverage (UHC) (United Nations, n.d.). The Committee on Economic, Social and Cultural Rights defined such health coverage as "health facilities, goods and services be located within safe physical reach for all sections of the population" (Human Rights Council, 2017, p. 15). Information and Communications Technology plays a crucial role in achieving the SDGs (ADEC Innovations, n.d.).

Digital Health Week

The five-day *Digital Health Week,* held November 29 to December 3, 2021, organized by a global coalition called *Transform Health,* presented over 50 events by organizations, individuals, and institutions committed to achieving UHC by using digital technologies and data, providing a "global week of action to celebrate, promote, and advocate for digital health" (Transform Health, n.d.-a, n.d.-b). The event provided a platform to exchange ideas and experiences as well as discuss challenges and opportunities, addressing what needs to happen in order to "ensure digital transformation of health to achieve universal health coverage" (Digital Health Week, n.d.-a, n.d.-b, n.d.-c).

Collaborating for digital health and better care

The opening session on November 29, organized by the Digital Connected Care Coalition (DCCC) and the UHC 2030 Private Sector Constituency, provided an initial understanding of challenges and opportunities in scaling up inclusive digital health tools with a focus on finding common elements of public and private incentives toward digital innovation (Lakew et al., 2021).

Panelists from the private and public sector reflected on digital health tools which have worked and what can be improved for a united effort in working toward UHC (Fig. 1). The moderator, Ms. Desta Lakew, the group director of Partnerships and External Affairs at Amref Health Africa, the largest health development nongovernmental international organization based in Africa, confirmed the idea of using technology as a tool to spread healthcare with the eventual goal of achieving UHC. She stressed that the biggest challenge in achieving this goal is creating cooperation between governments and the private sector to achieve access and reasonable cost, as these issues cannot be tackled by either side alone.

Four panelists gave presentations. The first speaker, Ms. Marelize Gorgens Prestidge, health innovation lead at the World Bank, noted that (1) the pandemic led to the implementation of digital health interventions, and (2) this has brought value to both patients as well as providers. While digital health holds tremendous potential and promise, she emphasized the need to go "from hypes and hopes to hands-on solutions," which requires new partnerships.

The next speaker, Dr. Anthony Adofo Ofosu, deputy director general of Ghana Health Service in the Ghana Ministry of Health, built on this point, emphasizing that the roles of the private and public sectors need to be better aligned for united efforts in digital innovations, and that their collective skills and strengths need to be leveraged to make the most of the potential of technologies.

Next, Mr. Jan-Willem Scheijgrond, global head of Government and Public Affairs at Royal Philips, a global network which leads large-scale healthcare transformation, acknowledged the

Fig. 1
Speakers (from left to right): Desta Lakew, Dr. Anthony Adofo Ofosu, Marelize Gorgens Prestidge, Jan-Willem Scheijgrond, and Niti Pall.

current pivotal point in transformation of digital healthcare (World Health Organization, 2022). While this transformation has been talked about for a decade, he said, the health sector is the last to go digital, despite the fact that there is a pressing and obvious need to inject digital health in all healthcare systems. He added that the digital divide is deepening between developing and developed countries, regardless of the pressing need to close this gap.

The last speaker, Ms. Niti Pall, senior advisor Digital Health at AXA, a multinational insurance company, speaking on behalf of consumers, stressed how consumers look for empathy and trust just as much as expertise and information (AXA, n.d.). Thus consumers struggle to find the best information for their families, but do not know how to navigate finding the best care, and even if they do, they may lack the means to pay. For her, financial inclusion, therefore, goes hand in hand with health inclusion and adoption of technologies.

The moderator asked what has been learned from the COVID-19 pandemic, and if it has brought accelerated acceptance of technologies in healthcare. Ms. Prestidge answered that digital solutions are valuable in contact tracing of infections. She noted the failure of following through and maintaining momentum on the use of technological tools during the Ebola crisis in West Africa in 2014–15, despite the fact that these approaches were developed at that time, but said that the circumstances differ with COVID-19, due to the global level of the COVID-19 pandemic and the accompanying larger impact, which has changed the situation for providers as well as expectations of clients. She further mentioned a study which found that about 40% of visits to primary care doctors are for basic health promotion which could have been provided digitally, and therefore less expensively and more efficiently. Digital literacy must be expanded, for example by using WhatsApp, a platform which many international people are familiar with and trust. At the same time, challenges exist in areas such as data governance, especially in providing access cross-border, and in the lack of physical infrastructure in developing countries.

In closing remarks, the moderator Dr. Ofosu mentioned that the responsibility of tackling these issues belongs to all stakeholders, but that government needs to protect the privacy and rights of patients, since the private sector is always concerned with profitability. The connection of all sectors was also emphasized by Mr. Scheijgrond, that it takes everyone to transform and digitalize the system. He added the point that issues like reducing long wait times and deciding which patient should be treated first could be tackled by using Artificial Intelligence (AI) to sort patients by priority.

Digital health projects in the German Development Cooperation Agency

An event held on November 30 was organized by the Deutsche Gesellschaft für Internationale Zusammenarbeit GmbH (GIZ), the German Development Cooperation Agency involved in development and international education, which has innumerable technology-related projects around the world (Lembcke et al., 2021). This agency follows the "principal of digital by default" meaning that projects systematically explore digital solutions and decide on project-specific digital components to achieve their goals. This principle was created by the Federal Ministry for Economic Cooperation and Development (BMZ) in 2018 and made health an area of priority for the implementation of digitalization (Federal Ministry for Economic Cooperation and Development, 2018).

The moderator, Mr. Ernesto Lembcke, advisor for Digital Health at GIZ, introduced their multiple digital health projects around the world. Using a mixed methods approach, GIZ created a map showing where and to what extent digital health projects have been implemented, as shown in Fig. 2. The figure shows the project implemented by GIZ in red, the ones from KfW in blue, and those coordinated by both are stripped with most of the projects concentrating on the African continent and countries in the Middle East as well as some in South East Asia. Their assessment shows a differentiation between projects in the areas of information technology infrastructure (IT infrastructure), mobile health (mHealth), electronic health (eHealth), health management information system (HMIS), telemedicine, and electronic learning (eLearning). Among the 26 presented projects, all except five projects include either an mHealth or eLearning intervention and thus implement smartphones or tablets in their digital intervention.

Two projects were described in greater detail. One project is a partnership with the government of Ghana. The project representative and technical advisor for GIZ in Ghana, Ms. Christine Esenam Afewu, gave an overview of the health sector in Ghana and introduced project BISA, a digital platform to manage the COVID-19 pandemic in Ghana. While the Ministry of Health of Ghana holds the mandate to ensure the attainment of UHC and striving for a healthy population, Ms. Afewu pointed out many challenges in this pursuit. All these challenges are in the context of the uneven distribution of doctors, concentrated in urban areas, with a lack of enough health workers in rural areas. Then, Health Management Information Systems (HMIS) are not very

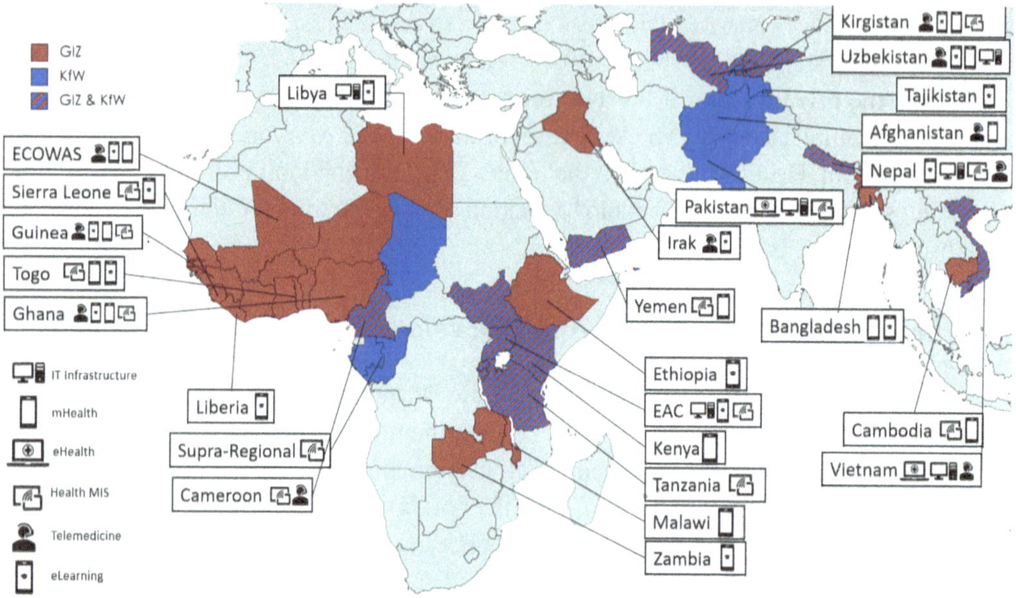

Fig. 2
GIZ overview of digital health projects.

widespread, which causes most health professionals in healthcare facilities to continue to rely on physical health records for their patients. This use of physical records instead of electronic records (eRecords) creates longer than necessary lines in healthcare facilities. Further, many health professionals, as well as users, lack Information and Communications Technology (ICT) skills, such as knowing how to use specific software as well as hardware such as tablets and computers. Thus, despite any political will to implement digital tools, health professionals and patients need to be educated on ICT skills and use of emerging digital health innovations. Ms. Afewu stressed that the current eHealth strategy in Ghana is therefore outdated, since these skills are not considered.

The next innovation described by Ms. Afewu is the telemedicine app called BISA, which reduces face-to-face doctor consultation at health facilities by giving patients direct access to healthcare providers through texting each other. This texting frees up health facilities, thereby allowing doctors to have time to be more available for in-person emergencies. So far, the app has enabled daily follow-ups at home and the provision of reliable COVID-19 information to the public. The app has been promoted in social media campaigns, leading to it being downloaded 10,000 times, with about 7000 questions asked.

Apart from these successes, Ms. Afewu pointed out challenges in using BISA. The use of the app highlights the digital divide between both rural and urban areas, stemming from availability of internet connectivity, cost of smartphones, lack of ICT skills, and language barriers. While 56% of males and 52.4% of females in their assessments were shown to have access to a smartphone in Ghana, the project needs to reach nonsmartphone users by implementing bulk messaging. Other means to bridge the digital divide and to build capacity in rural communities, she suggested, are: to include local languages, and to collaborate with projects such as eSkills4Girls, an initiative with the goal to increase access to the digital world for women and girls (eSkills4Girls, n.d.). The latter shows great promise in gender empowerment and equity in that the app has also shifted social and cultural norms by answering questions about stigmatized topics, such as birth control and HIV, which are often also asked by males. Another challenge of the app concerns patient confidentiality and data protection, but these were tackled by using anonymized data and adhering to rules according to the Ghana Data Protection Act. In summary, through this partnership of government and the private sector, BISA has provided the government with real-time data for decision-making and risk communication for outbreaks, and supported emergency response.

The second project from the German Development Cooperation Agency, called ELSA, was started in Uzbekistan, and presented by its representative Ms. Anna Berezovskaja, portfoliomanger at Kreditanstalt für Wiederaufbau (KfW), a German state-owned investment and development bank. In first giving some background information to understand the health sector of the country, she described that Uzbekistan is a former dictatorship, which has seen GDP growth, even during the pandemic, since the liberalization in 2016. The health sector faces challenges such as double disease burden and a fragmented system with parallel structures which make referrals difficult and impede sharing of resources. Further restrictions are the weak sector coordination and low management capacity on a national level. To address these challenges, Ms. Berezovskaja described the ambitious reform agenda of the government to digitalize the health sector on all levels, to introduce a mandatory health insurance, and to improve the general health service delivery.

A recent feasibility study found no systematic use of digital applications in Uzbekistan, as although they are available, the IT literacy among health personnel is low, with only about 22% of personnel reporting experience in using a computer. Other limitations are that little IT infrastructure exists and hardware is outdated. With a development loan of 45 million Euros, the Digital Health Masterplan is being supported by the KfW to assist the government in its implementation. Projects include software solutions like a digital health platform and medical information systems as well as acquisition of hardware and building of networks. Another grant of 5.5 million Euros supports IT literacy training for different user groups and support for change management.

Closing remarks of this session highlighted the insight and value of these projects in digital healthcare. The project representatives from Ghana and Uzbekistan emphasized how the German government advocates for digital health through the BMZ and promotes insight in data protection and safety. There is also a strong focus on "leaving no one behind," the aspiration of the United Nations UN Agenda, and closing the digital gap, especially in the global South (UN Sustainable Development Group, n.d.).

The DCCC projects in collaboration with the private sector

The next event on December 2 was again hosted by Digital Connected Care Coalition (DCCC) and the UHC 2030 Private Sector Constituency, this time to discuss the role of the private sector in promoting digital healthcare (Morris et al., 2021). The presentations elaborated challenges as well as successes in setting up new digital health projects, focusing on two sample cases.

Claire Morris, director of Social Impact at *Babylon Health,* a company that uses artificial intelligence (AI) to make healthcare more affordable, moderated the event and stressed the importance of upgrading the quality and consistency of digital health tools, since half of the world population does not have access to essential healthcare (Babylon Health, n.d.).

The first case example, "*Take Care Africa,*" was introduced by Dr. Ademola Serrano, innovation manager at PharmAccess Foundation, Nigeria. This pan-African healthcare information platform was started in Nigeria to empower health consumers with accurate, reliable, and relatable health information. Given the growing need for this information throughout sub-Saharan Africa, the developers aim to expand to other nearby countries, starting in Tanzania. Dr. Serrano emphasized the importance of providing locally relevant information in an easy-to-use platform. To design the project, they conducted local, qualitative research, administering local surveys with potential users. Initial findings showed that participants first seek health information from local experts and then international experts.

Dr. Serrano pointed out that digital penetration has been much slower in sub-Saharan Africa compared to the European Union. This is partially due to the fact that only 66% of the population has access to a smartphone and only half have daily access to the internet in Nigeria and Tanzania specifically. In order to be able to expand the platform, Dr. Serrano noted the need to bring in more users which leads to more investment and simultaneously more content creation, eventually making this a self-sustaining project. While the project works together with the National Medical Association of both Nigeria and Tanzania, more partnerships are needed. As future directions of the project, Dr. Serrano highlighted the potential of the project to expand into new geographic regions and to include more value-added services such as mental health services.

The second example, "*Lighthouse Initiative India,*" was presented by Dr. Neeraj Jain, country director of Path India, a nonprofit organization with the goal to advance health equity (PATH, n.d.). This project aims to advance Health & Wellness Centers by finding digital solutions for improving the standard of care across the Indian health system. In a country of 1.4 billion people, managing primary healthcare is difficult, causing the necessity to find evidence to present to the government in order to justify expanding the project across the country, Dr. Jain noted. Demonstrating feasibility in a small-scale intervention in one district, the slums of Mumbai, would then demonstrate the scalability to the other 700 districts, and eventually other countries. She also pointed out how one organization cannot tackle the digitalization of the health system on their own, suggesting the need for collaboration and partnerships. Presenting their model to the government, she said, offers the possibility of receiving funding.

In subsequent discussion among the speakers, Dr. Serrano was asked about the relevance of finding partners in order to be able to scale up his project. He responded that they reached out to local healthcare workers, since research has indicated how users trust local workers and that it is important to provide what users want. On the question of building trust, Dr. Jain pointed out how, for example, the initiative's work in providing the COVID-19 vaccine has helped to build trust, while working with the Indian government also automatically increases trust among the population. Dr. Serrano agreed with the importance of partnering with governments, since, he said, they are the biggest healthcare providers in most African countries.

Scalable digital health models to increase access to quality care

Another event in the Digital Health Week, held on December 3, addressed the importance of scalable digital health models, focused specifically on entrepreneurial models to increase access to quality digital healthcare (Bosche et al., 2021). The Philips Foundation organized this webinar to collaborate with social enterprises who combine societal impact with scalable, self-supporting business models. Speakers focused on two different approaches to provide solutions: based on either care improvement or information sharing (Fig. 3).

The first speaker on improving care was Mr. Marcos Lacayo Bosche, founder of the healthcare platform *Estación Vital,* which provides access to preventive checkups and personalized plans to change health behavior in underserved communities in Nicaragua (Estación Vital, n.d.). With 60% of people in Latin America being obese, the app aims to improve the internal dialogue to lose weight. The disruptive aspect of this app is how money is generated through the use of the app to subsidize free kiosks in the community.

Next, Mr. Nicolaas Moens, director of *AfyaPro,* a social impact organization, talked about the fragmentation of services and data across the healthcare system in African countries, ranging from care regarding diabetes to domestic violence (AfyaPro, n.d.). Therefore *AfyaPro* was created to provide a digital hospital and patient data system that aims to provide one comprehensive platform. The main goals he emphasized were to provide better practices, better health financing, and speeding up the healthcare process. Having these clear objectives help in the continuous improvement cycles the project undergoes.

The third speaker on improving care was Ms. Rama Siva, chief product officer at *Penda Health,* who works on a model of medical centers in low-income areas of urban Nairobi. The *Penda*

Fig. 3
Speakers (from top left to bottom right): Mohammed Dalwai, Dr. ck Andrade, Claire Barnhoorn, Marcos Lacayo Bosche, Rama Siva, Nicolaas Moens.

Health solution is a citywide initiative in Nairobi which includes 21 primary care clinics and serves about 30,000 patients a month with about 1 million patients overall. The project intends to develop a blended digital health system to provide low-cost, high-quality care to families.

The first solution based on information sharing was introduced by Ms. Claire Barnhoorn, founder of Solvoz, an open international platform to facilitate efficient humanitarian tendering and procurement. With about 10 million NGOs having their own procurement departments, they represent the 5th largest economy in the world and have the potential to operate more efficiently. Ms. Barnhoorn explained how the disruptive idea of providing procurement as a service can empower those who cannot get funding, manage tenders, provide decision-making tools, and bridge the gap between providers and NGOs.

Another approach to information sharing was presented by Mr. Mohammed Dalwai, cofounder of *EMGuidance*, which created a mobile reference and learning platform called *Essential Medicine Guidance* (EMG). The platform provides free healthcare information to more than 30,000 African medical professionals and pharmaceutical companies. Mr. Dalwai stressed that generic information is not enough, since that healthcare providers require local information to make informed decisions which are sustainable and impactful for their patients. With a million searches per month, EMG is the largest medical platform for South Africa.

A subsequent panel discussion was held for the two approaches of care improvement and information sharing. Panelists who presented the information sharing approach summarized their goal as being "humanitarians with a business model," using their profit to be sustainable and have a continuous impact. Access to knowledge is seen as the underlying mechanism to achieve access to healthcare, requiring a focus on the ability of patients as well as providers to access the right information at the right time. Panelists with the approach of improving care emphasized the need to be willing to stick to a problem until it is solved regardless of how long it takes.

One question that stood out in the discussion was what high-income countries can learn from the delivery of healthcare in low-income countries. The answer was that it is possible to reduce cost while preserving the quality of care by shifting tasks from doctors to nurses, reducing the complexity of structures, training specific personnel, and thinking creatively. While the market in low-income countries may be smaller, the opportunities are larger, and there is more space to develop with the constant challenge of needing to adapt to change.

Conclusion

A wide range of projects in different countries exist and are presented in special conferences as well as weeklong showcases of digital innovations, like in the Digital Health Week reported in this chapter. These presentations show promising insight into the digitalization of the healthcare system around the world. In sessions during this week, many experts advocated for assessing the specific needs and problems of a community or region before implementing a project. Knowing what people want is essential in order to ensure that they profit from the digital health interventions and are empowered by the innovations; therefore, the digital transformation "must be systematic, person-centered, and equitable" (Digital Health Week, n. d.-a, n.d.-b, n.d.-c).

Another prominent aspect is the importance of ensuring that projects will be self-sustaining to have a lasting, positive impact on the community. Also, there has to be a willingness on the part of users to adapt to changes required by technological approaches; fortunately, many users as well as healthcare providers seem in current times to be open to the idea and programs offering digitalization of the healthcare system.

Advancements in technology have offered a unique opportunity to transform healthcare systems globally, with the goal to "enhance the availability, accessibility, affordability and quality of health services" (Digital Health Week, n.d.-a, n.d.-b, n.d.-c). Much potential exists to create meaningful interventions with a sustainable impact on a large scale, thus progressing the goals of UHC, the UN Agenda 2030 for Sustainable Development, and the World Health Organization, as well as the aims of many other international and national agendas.

References

ADEC Innovations. (n.d.) Resources. https://www.adecesg.com/resources/blog/.

AfyaPro. (n.d.). About us. https://afyapro.net/about-us/.

AXA. (n.d.). About us. https://www.axa.com/en/about-us.

Babylon Health. (n.d.). About our mission. https://www.babylonhealth.com/en-us/about.

Bosche, M., Moens, N., Siva, R., Barnhoorn, C., & Dalwai, M. (2021 December 3). Entrepreneurial digital health models to increase access to quality care [Conference presentation]. In *Digital Health Week 2021*. https://youtube.com/watch?v=qNC0V5ym6nQ.

Digital Health Week (n.d.-a). Events. https://digitalhealthweek.co/events.

Digital Health Week. (n.d.-b). FAQs. https://digitalhealthweek.co/faq.

Digital Health Week. (n.d.-c). Health for all in a digital age: Key messages for Digital Health Week. https://digitalhealthweek.co/image/Key_Messages/Keymessages_DHW.pdf.

eSkills4Girls. (n.d.). About. https://www.eskills4girls.org.

Estación Vital. (n.d.). About us. https://www.estacionvital.com/website/nosotros.

Federal Ministry for Economic Cooperation and Development. (2018). *Development policy 2030. Division for political analysis and planning*. https://www.bmz.de/resource/blob/23770/71cf4bb9fee375d369a42c1abf29b64d/strategiepapier452-10-2018-data.pdf.

Human Rights Council. (2017). Mental health and human rights: Report of the United Nations High Commissioner for Human Rights. United Nations High Commissioner for Human Rights and reports of the Office of the High Commissioner and the Secretary-General. https://documents-dds-ny.un.org/doc/UNDOC/GEN/G20/109/10/PDF/G2010910.pdf?OpenElement.

Lakew, D., Prestidge, M. G., Ofosu, A. A., Scheijgrond, J.-W., & Pall, N. (2021, November 29). UHC2030 & DCCC Digital Health Series #1: Collaborating for Digital Health and Better Care [Conference presentation]. In *Digital Health Week 2021*. https://us06web.zoom.us/rec/play/KGHMNzowVZvTSw1qLGcLHtjEKw5D10Ad1QhW5N7tC3TpqFFGVUNtk1k6wB1PITxC9bh87hn2OYzH4x1i.3IZVZYHFtDhXfaMh.

Lembcke, E., Afewu, C. E., & Berezovskaja, A. (2021, November 30). GIZ (Deutsche Gesellschaft für internationale Zusammenarbeit GmbH): Digital by Default - Digital Health Projects in German Development Cooperation [Conference presentation]. In *Digital Health Week 2021*.

Morris, C., Serrano, A., & Jain, N. (2021 December 2). UHC2030 & DCCC Digital Health Series 3#3: The DCCC projects: Learnings on collaborating with private sector in digital health [Conference presentation]. In *Digital Health Week 2021*. https://us06web.zoom.us/rec/play/khEj-Z64dEstaQKgJ13bladwoCSLTjHfZ8NGiaDeFlI1wvFTwU6aYvE8lNDQh52cLqrSwr-h-GSVOvm9.TG7W9KJ551nWK7t.

PATH. (n.d.). About us. https://www.path.org/about/.

Transform Health. (n.d.-a). Participate in digital health week: a global moment for digital health. https://transformhealthcoalition.org/blogs/th-news/participate-digital-health-week/.

Transform Health. (n.d.-b). Who we are. https://transformhealthcoalition.org.

UN Sustainable Development Group (n.d.). 2030 Agenda. https://unsdg.un.org/2030-agenda/universal-values/leave-no-one-behind.

United Nations. (n.d.). Department of economic and social affairs: Sustainable Development Goal 3. https://sdgs.un.org/goals/goal3.

World Health Organization. (1946). *Constitution of the World Health Organization*. https://apps.who.int/gb/bd/PDF/bd47/EN/constitution-en.pdf?ua=1.

World Health Organization. (2022). *Royal Philips*. https://www.who.int/pmnch/about/governance/board/philips/en.

Further reading

Carter, H. (2021). The emergence of digital mental health in low-income and middle-income countries: A review of recent advances and implications for the treatment and prevention of mental disorders. *Journal of Psychiatric Research*, *24*(133), 223–246. https://doi.org/10.1016/j.jpsychires.2020.12.016.

Naslund, J., et al. (2021). Identifying challenges and recommendations for advancing global mental health implementation research: A key informant study of the National Institute of Mental Health Scale-Up Hubs. *Asian Journal of Psychiatry*. https://doi.org/10.1016/j.ajp.2021.102557.

World Health Organization. (2021, April). *Universal health coverage (UHC)*. https://www.who.int/news-room/fact-sheets/detail/universal-health-coverage-%28uhc%29.

CHAPTER 76

Raising the profile of digital health in Latin America: The RECAINSA digital health forum and network

Daniel Otzoy-García[a], Maria Isabel Iñigo Petralanda[b], and Paula Aljovin,[c,d,e]

[a]Central American Network for Health Informatics (RECAINSA), [b]Ethics, Privacy and Security Chapter of the Central American Network for Health Informatics (RECAINSA), City of Buenos Aires, Argentina, [c]Mental Health Team, Vine Trust Amazon Hope Medical Programme, Loreto, Peru, [d]Department of Clinical Psychology, Teachers College, Columbia University, New York, NY, United States, [e]International Association of Applied Psychology, Indianapolis, IN, United States

Introduction

The Central American Health Informatics Network (Red Centroamericana de Informática en Salud, RECAINSA, in Spanish) is a digital health network formed by Latin American multidisciplinary professionals working in the health sector and digital technologies. Founded in 2013, the mission of RECAINSA is to support national and subnational strategies about digital health and to establish a public forum space—mostly virtual—for sharing experiences and best practices in the public and private sectors. To accomplish this, RECAINSA collaborates with multilateral and bilateral agencies, private foundations, nonprofits, and other key actors in the health sector both locally and regionally, leveraging partner capabilities to implement its strategic objectives. The organization joins the global effort led by the World Health Organization (WHO) to promote universal access and coverage in health through digital transformation. This allows implementation of national strategies that have a significant impact on efficiency and quality of health systems, contributing to reducing inequities affecting countries in Latin America.

The digital health forum

The need to maintain essential health services throughout the COVID-19 pandemic made the implementation of digital technologies an urgent matter (Banco Bilbao Vizcaya Argentaria [BBVA], 2021). The new, broadened use of teleworking, eLearning, and telehealth was

Resilient Health. https://doi.org/10.1016/B978-0-443-18529-8.00076-7

unthinkable before the health emergency (Robles, 2020). However, the question that the health sector experts asked themselves is whether this transformation is inclusive, or if it has further deepened the digital divide that was already quite marked before the pandemic (Corporación Andina de Fomento—Banco de Desarrollo de América Latina [CAF], 2021).

As suggested by the United Nations High-Level Panel on Digital Cooperation (2019), we are in urgent need of improved digital cooperation in order to support achievement of the Sustainable Development Goals (SDGs). The digital transformation must be grounded in common human values such as inclusiveness, respect, human centeredness, human rights, international law, transparency, and sustainability. Following the UN's suggestions, international organizations have made a 'call for action' so that *no one is left behind in the era of digital interdependence*. One of the requirements for effective digital cooperation is to strengthen multilateralism and multistakeholderism.

In order to carry out our strategic goals of Red Centroamericana de Informática en Salud [RECAINSA] (2022)—to create spaces for the exchange of experiences and good practices—activities at the core of our work include monthly webinars and blog posts with guest writers. More recently, with the proliferation of virtual workshops necessitated by the COVID-19 pandemic, we have created one of the most prestigious platforms to allow for the generation and exchange of digital health knowledge. This has been one of our dreams come true. Organizations that trust our work as a valuable network in Latin America and with whom we have partnered in various initiatives include UNICEF, PAHO/WHO, The Broadband Commission, Transform Health, The Patrick J. McGovern Foundation, and PATH/Digital Square.

We at RECAINSA fulfilled another of our dreams by launching the *First RECAINSA Digital Health Forum*, within the framework of the celebration of Digital Health Week worldwide—an initiative led by Transform Health (2021). The forum (Corporación Andina de Fomento—Banco de Desarrollo de América Latina [CAF], 2021) was the perfect occasion for us to celebrate an extraordinary year and expand our "social mass," both of individuals and organizations. We have also continued our evolution toward having more impact, as a promoter of digital transformation in health to boost universal health coverage equitably and inclusively.

The slogan chosen for our 2021 Forum was "Digital Health for All." This means that with digital health we can make significant contributions to achieve universal health coverage and ensure that health services are accessible, efficient, and reliable. This effort contributes decisively to reducing the inequities that affect the inhabitants of Latin American countries and the world in general.

The event was broadcast live between November 29 and December 3, 2021. There were 10 keynote conferences, 15 discussion panels, and five workshops, with more than 75 speakers from four regions over the world, presenting in two official languages (Spanish and English) with live interpretation, and more than 2000 participants virtually, both synchronous and asynchronous.

The agenda was based on thematic areas considered of interest to enhance the digital transformation of the health sector in the Latin American region. Each day consisted of an opening keynote conference, three discussion panels, and a closing keynote conference.

The following is a review of each of the days of the forum and the most relevant aspects of them.

Digital transformation principles, governance, and regulation

The first day started with a keynote address on "8 Principles to Digitally Transform the Region of the Americas" by Mr. Marcelo D'Agostino, senior regional advisor for Digital Health and Information Systems for Health of the Pan American Health Organization (PAHO). Following the UN's suggestions, the PAHO has proposed eight guiding principles for the digital transformation of the health sector in the region of the Americas, defining a roadmap to support countries to make health services more accessible to the most vulnerable and underserved populations. As D'Agostino explained, these principles include.

(1) universal connectivity in the health sector.
(2) create digital public health goods for a more equitable world.
(3) inclusive digital health.
(4) open, sustainable, interoperable digital information and health systems.
(5) human rights across all areas of digital transformation in health.
(6) cooperation on artificial intelligence and any emerging technology.
(7) confidentiality and security of information in the digital public health setting.
(8) renewed public health architecture for the age of digital interdependence.

Main takeaways from D'Agostino's presentation included the importance of a multisectoral and interdisciplinary approach to achieve an inclusive digital transformation that can reach marginalized populations. He recommended that actions must pay close attention to local needs and contexts, encouraging countries to adapt PAHO's guiding principles to their particular necessities. Further, he highlighted the importance of education in digital health and called for institutions to include such training at the undergraduate level. Lastly, he emphasized that "data can save lives" and insisted that public health must ensure open access to data, software, algorithms among other public health goods.

The next discussion panel addressed three topics relevant to the central theme of the day. The first panel, "Digital Governance and Health: How do we build it?", dealt with experiences and lessons learned in Uruguay, Southeast Asia, and Costa Rica through speakers from the Agencia de Gobierno Electrónico y Sociedad de la Información y del Conocimiento (AGESIC) Salud.Y, Asia *E*-Health Information Network (AeHIN), and RECAINSA, taking into consideration the challenges faced when building strong Digital Health Governance in low- and middle-income countries (LMICs). Identified barriers and challenges included the lack of required infrastructure, including poor or inexistent connectivity and hardware in certain regions; fragmentation of systems; lack of training of health professionals; and low digital literacy among certain sectors of the population.

The second panel, on "Digital Transformation of Primary Health Care,''included the participation of three Central American NGOs: TELESAN,[a] TulaSalud,[b] and Wuqu' Kawoq.[c] These organizations are incorporating digital innovations at the first level of healthcare in Honduras and Guatemala, with the ultimate goal of closing the health gap and reaching underserved communities who have been traditionally neglected by healthcare systems.

The third panel involved representatives from the Argentinian and Peruvian Ministries of Health, the National Center for Information Systems (CENS[d]) of Chile, and the international NGO PATH[e] working mainly in Africa. Panelists discussed the topic "National Digital Health Strategies: Lessons learned," commenting on the challenges of developing national Digital Health strategies in complex contexts where multiple environmental elements can condition their success.

For example, the Argentinian representative described how the geographic heterogeneity in her country had become a barrier to the implementation of a digital health ecosystem. Regions tend to work fragmentarily, leading to segmented coverage and disintegrated services. In turn, this makes it difficult to use electronic medical records at a national level. Further, the Health Ministry in Argentina is currently trying to foster responsible stewardship and governance, strengthen data security, engage multiple stakeholders to define interoperability standards, and to promote the importance of mental health and sustainability. In the case of Peru, the pandemic brought about a heightened interest toward telemedicine that led to the modification of legislation. Challenges have included efforts to decrease resistance toward telemedicine, both from the general public and health professionals themselves. Also, due to low digital literacy among health professionals, experts have offered training on the use of electronic medical records, videoconferences, and Zoom calls. Moreover, the Health Ministry designed a software adjusted to local needs where patients could request free medical and psychological services remotely. The telehealth platform allowed Peruvians to register for consultations and receive remote monitoring from health professionals. Likewise, the Health Ministry also put forward a contact/call center that patients could reach out to.

---

[a] https://telesanhn.org/

[b] https://www.tulasalud.org/

[c] https://www.wuqukawoq.org/

[d] https://cens.cl/

[e] https://www.path.org/

The first day of conferences ended with the keynote speech entitled "Legislation, regulation and compliance for a digital transformation in health with a human rights approach" by Marisa Aizemberg, director of the Health Observatory of the Law School of the University of Buenos Aires.[f] She emphasized how regulatory frameworks that enable each country's digital transformation of health sectors must be strengthened in such a way that equity and inclusion are always taken into account. It is thus crucial to "leave no one behind" and to protect the most vulnerable populations, she said, while ensuring an ethical and secure use of health data by ministries and other key actors.

Digital transformation of the health information systems in Latin America

The second day began with the keynote conference entitled "Health Data Security in Latin America" given by Daniel Otzoy, Digital Health International consultant and executive director of RECAINSA, who shared information regarding cybersecurity strategies in Latin America. He highlighted the principal gaps and the main recommendations to close them, taking into account the different contexts and stage of maturity of the countries in the Latin American region.

These recommendations included the importance of promoting research to better understand the gaps in health data security across the Latin American region. He emphasized that countries must have national health strategies with carefully outlined plans to protect the digital information systems. Lastly, he called attention to the need of training professionals in the matter, as well as the importance of instigating a cybersecurity culture; as the virtual world multiplies efficiency and possibilities, it is important to understand its risks and take precautions to mitigate such threats.

The first discussion panel was on the topic of "Electronic Clinical History: A citizen, a record" with participation of representatives of the Ministry of Health of the Autonomous City of Buenos Aires, the Costa Rican Social Security Fund,[g] and HIMSS[h] for Latin America. During this session, panelists presented the experience of adopting electronic medical records in Buenos Aires and Costa Rica, discussing the way in which this initiative was promoted, challenges faced to date, and perspectives for the future. For example, the Costa Rican and Argentinian representatives highlighted the importance of promoting synergy between health and IT professionals. They both recommended starting at the least complex health centers (first level of care) in order to gain experience and refine the digitalizing strategy before reaching the tertiary level services, i.e. hospitals. The Costa Rican representative stressed how helpful it had been to have one sole system of specialized public health providers that covered 92% of the

[f] http://www.derecho.uba.ar/institucional/observatorio-salud/

[g] https://www.ccss.sa.cr/edus/

[h] https://www.himss.org/

country's demand. Lastly, all speakers agreed on the importance of framing the health digitization strategies within the government's larger agenda, especially in countries where provinces work independently from one another (federal systems).

The next panel on the topic of "Open Source and Digital Health: Myths and Realities" included representatives from the University of Oslo,[i] RECAINSA, and OpenHIE.[j] Speakers discussed the status of the adoption of Open Source and Open Innovation in Latin America, as well as the main challenges faced and the perspective about the future given the acceleration in the adoption of this kind of technology during the COVID-19 pandemic.

Speakers discussed the importance of using open source as a way to promote local empowerment and sustainability, given that open communities of practice change power dynamics and enable those with ability and interest within a local ecosystem to thrive. An open system provides the opportunity for a country to build the capacity locally and with the support of whoever wants to learn and collaborate. Further, emphasis was placed on the importance of developing flexible software that can quickly adapt to changes in the care delivery system.

The closing panel was about "The role of the Pharma Industry in the Digital Transformation of the Health Sector" with the participation of CEOs of two of the most important international pharmaceutical companies, Pfizer and Roche. Representatives talked about the industry's prioritization and investment in health innovation, including digital health and data management. Each speaker discussed the way in which they were investing in health innovation, both within their own companies and in the countries where they collaborate with governments. Mr. Álvaro Soto, the CEO of Roche for the Central American region, emphasized the importance of building trust and collaboration between the private and public sectors, entrepreneurs, and the academic world, with the purpose of cocreating solutions. He urged the industry to tackle the problem of lack of transparency and to demystify the prejudice surrounding data use. Lastly, he stressed that we must all act with a sense of urgency given that key actors can make a country stop if they do not keep up to the speed of scientific advances.

Bradley Silcox, the CEO of Pfizer in Central America, described the company's aim to constantly innovate and experiment. Throughout the pandemic, their clinical trials were in danger because they could not work as usual. Hence, Pfizer and BioNTech had to quickly adapt to enable clinical trials digitally. Similarly, digital tools had to be implemented to track the distribution of medicines and vaccines to all locations around the world, guaranteeing low temperatures. Moreover, Silcox mentioned that Pfizer is supporting accelerated diagnostic solutions, disease modeling, big data, and real-world evidence, with the ultimate purpose of developing the most intelligent platforms to connect patients, doctors, and diagnostic results.

---

[i] https://www.mn.uio.no/hisp/english/

[j] https://ohie.org/

Likewise, he mentioned the importance of investing in artificial intelligence and machine learning to accelerate the discovery of new medicines, treatments, and vaccines.

The second day concluded with a keynote conference on the topic of "Health Information Systems and the Use of Health Data: CHISU and its Global Impact" by Steve Ollis, director of the Country Health Information Systems and Data Use project (CHISU),[k] a global project funded by USAID. Ollis spoke about CHISU's aim of strengthening country capacity, leadership, and self-reliance to manage and use high-quality health information systems in order to improve evidence-based decision-making. More specifically, he focused on the importance of strengthening governance, increasing availability and interoperability of quality health data and information systems, increasing use of health data to address the existing gaps, and strengthening the organizational development of local partners for sustained health data use. He illustrated these actions through a description of CHISU's work in Eastern Europe and African countries.

Workforce, digital health competences, and academia's role

The third day began with a keynote conference on "Data Governance for Universal Health Coverage" by Ms. Kirsten Mathieson, policy lead in the Transform Health[l] Coalition. She spoke about the health data governance principles promoted by the coalition, which include protecting individuals and communities, ensuring data security, building trust in data systems, enhancing health systems and services, promoting data sharing and interoperability, facilitating innovation using health data, promoting equitable benefit from health data, and establishing data rights and ownership. She also described the origins of this initiative, its approval in the World Health Assembly in 2022, and the subsequent global launch.

Subsequent panels began with a discussion on "The Power of Digital Health Networks for Sharing Knowledge, Experiences and Best Practices,''with participation of the most important digital health networks globally: the African Center for eHealth Excellence (ACFEE[m]), the Asian eHealth Information Network (AeHIN[n]), the Global Digital Health Network (GDHN[o]), and RECAINSA. During the conversation, representatives talked about the initiative of the "Network of Networks," which aims to enhance the relationship between regional networks and to generate spaces to exchange experiences and best practices. The initiative is closely aligned to the Digital Health Strategy of the WHO 2020–2025 (World Health Organization, 2021).

[k] https://www.jsi.com/project/country-health-information-systems-and-data-use-chisu/

[l] https://transformhealthcoalition.org/

[m] https://www.acfee.org/

[n] https://www.asiaehealthinformationnetwork.org/

[o] https://www.globaldigitalhealthnetwork.org/

The second panel on the topic of "Knowledge Management and Research in Digital Health" included academics and researchers from the National Institute of Public Health of Mexico (INSP[p]), the National Education and Research Network/University Telemedicine Network of Brazil (RNP[q]/RUTE[r]), and the Continental University of Peru.[s] Speakers presented how their institutions were promoting digital health research and generating evidence with the purpose of informing policy and decision-making. They shared their views on how to strengthen this kind of research in the Latin American region, where Science and Technology policies do not tend to prioritize the topic. All the speakers deemed it crucial to establish alliances between the private and public sector.

The third panel on "Training Offer in Digital Health: Advances and Challenges in Latin America" involved representatives of important universities and training centers in the region, such as the University Institute of the Italian Hospital of Buenos Aires,[t] Argentina; the University of Chile[u]; and the Universidad El Bosque de Colombia.[v] An interesting discussion was held on the experience of generating pioneering academic opportunities on digital health training, both nationally and abroad. Speakers highlighted the important gap that exists in health professionals' training on digital health, and the joint work that must be carried out with all sectors to include and strengthen this area of study for undergraduate degrees. Further, they pinpointed the importance of generating specific jobs where these competencies are valued by employers.

The closing conference of the day, on "Human Resources in Health 4.0: New competencies for the new work paradigm," was given by Jose Francisco Garcia Gutierrez, senior regional advisor in Human Resources for Health at the PAHO. He spoke about the evolution in the training of human resources according to the challenges posed by the acceleration in the adoption of digital technologies in society, including public health, sharing his extensive experience both in Spain and in the Americas.

Culture, environment, and digital health enablers

The fourth day opened with a keynote address by Tina Purnat on "Infodemic in Public Health: Its Global Impact and How To Manage It." Based on her experience as team lead of Infodemic Management at the WHO, she described PAHO's strategies to counteract something that we all

[p] https://www.insp.mx/

[q] https://www.rnp.br/en

[r] https://rute.rnp.br/

[s] https://ucontinental.edu.pe/

[t] https://www.hospitalitaliano.org.ar/#!/home/educacion/nuevo

[u] https://www.uchile.cl/

[v] https://www.unbosque.edu.co/

witnessed throughout the pandemic: the "infodemic," which refers to the abundance of information—some accurate and some not—that occurs during an epidemic. She provided recommendations on how to address this problem and described a framework now being used globally through the training of specialized personnel.

Specifically, to counteract misinformation and the infodemic, Purnat recommended the need to address information voids by proactively anticipating what people's concerns and questions are, so that authorities as well as communication and infodemic response efforts can meet the public's needs at all times. For this purpose, in a pandemic context, real-time tools and their progressive implementation can be a solution to avoid these issues. Finally, Purnat stressed the importance of educating people on how to handle data and information wisely and virtuously.

The first panel of the day was on "The role of international cooperation in the digital transformation in health" which included participation of representatives of the Bill & Melinda Gates Foundation (Mr. Tim Wood),[w] the Carlos Slim Foundation (Mr. Carlos Tapia),[x] the Patrick J. McGovern Foundation (Mrs. Rebecca Distler),[y] and USAID (Mrs. Adele Waugaman),[z] which are among the most important investors globally in the digital transformation of public health. Speakers stressed the importance of coordinating among donors in order to optimize resources and avoid the duplication of efforts. They also shared their views for the future few years, given the acceleration in the adoption of digital health as an important driver to achieve Universal Health Coverage and the United Nations 2030 agenda.

The representatives agreed on the importance of international cooperation as a requirement to achieve a global digital health transformation. There was unanimous consensus that connectivity, health literacy, interoperability, and digital health standards are the foundation of any digital health plan. They collectively concluded that success will depend on the digital moving from margin to mainstream. In order to do so, a dynamic collaboration between stakeholders across regions needs to take place. Private and public sectors interacting to understand digital health are a key element to transform health provision and community outreach. Lastly, privacy and security were highlighted as grounding determinants for duly respect and good practices that will guarantee the success in the digital health transformation.

The second panel, entitled "Digital Transformation in Health with Equity," included women leaders promoting digital health, representing the private sector, public sector, and international cooperation. Speakers discussed the existing gender gaps in digital health strategies that Latin American countries are undertaking, highlighting the low participation of women in STEM[aa]

[w] https://www.gatesfoundation.org/

[x] https://fundacioncarlosslim.org/english/

[y] https://www.mcgovern.org/

[z] https://www.usaid.gov/digital-health-vision

[aa] https://www.britannica.com/topic/STEM-education

degrees and the difficulties that women unjustly face when they seek participation in decision-making processes. These difficulties are mainly the lack of metrics or statistics as evidence of women working in the digital health sector. According to UN Women, only three out of 10 students in STEM majors are women. When it comes to AI, even fewer, namely, two of every 10 employees are female, and, in the world of research, only one in 10 are women. Furthermore, only 3% of the Nobel Prizes are awarded to women. All speakers agreed that gender bias should be considered in public policies so that the role of women in the digital health sector is guaranteed. Lastly, the situation for women is even worse when ethnic or racial bias occurs together with gender bias.

The closing panel featured a discussion on "Interoperability in Health in Latin America: Use Cases." Speakers included experts on health data interoperability standards, which aim to facilitate the exchange and sharing of health information between different systems or organizations. Other speakers included professionals working on projects financed by PAHO, the Inter-American Development Bank, and USAID, with the aim of implementing interoperability standards in the surveillance of possible adverse effects of vaccination and immunization (ESAVI), as well as the use of COVID-19 vaccination passports, and data management for public health programs.

Fernando Portillo, digital health senior advisor and health interoperability standards specialist, talked about his work experience in Uruguay and Colombia, where he helped push the national digital health reforms. He stressed the importance of having the governments' full commitment to the process of implementing interoperability standards, as well as the setting up of an independent, specific technical team to be in charge of the project. Further, he highlighted the importance of fostering trust among all stakeholders as a foundational requirement for later strategies to succeed. A second panelist, Daniel Rizzato, who was the director of the Health Information Systems Development at the Argentinian Ministry of Health between 2018 and 2020, confirmed Portillo's views and placed emphasis on the importance of having actors collaborate, share their expectations, and build an environment of trust. Once stakeholders have decided what kind of data they wish to exchange among each other, it is crucial to consult experts and follow the paths of countries that have already been successful in the task. In Argentina's case, Rizzato mentioned that Canada, Uruguay, and Australia were used as examples of success. Lastly, both Portillo and Rizzato agreed on the need of catering strategies to the specific needs and characteristics of each nation.

The day's agenda ended with a keynote address from Victor Medina, president of HL7 Mexico,[ab] who spoke about the "Stage of the HL7/FHIR adoption in Mexico." HL7 is an interoperability standards protocol that was created around 40 years ago. Over time, it has evolved to allow quick communication between applications and web platforms. Fast

[ab] https://www.linkedin.com/company/hl7mx/

Healthcare Interoperability Resources (FHIR) connects applications in a simple, efficient, and transparent way (Bello, 2021).

Medina described the process through which the HL7/FHIR standard was adopted in his country, Mexico, in August 2021, after the birth of the HealthTech México association, which brings together software developers, vendors, and academic organizations with the common aim of building a digital health ecosystem (Bello, 2021). One of the problems faced is the rapid speed at which digitalization is moving in the health industry in contrast to the slower pace of regulators to establish clear and sufficient norms. It has been difficult for Mexico to keep up to date with the fast speed of innovation and the constant changes in regulations. He stressed the importance of bringing together actors from the private and public sectors, and offered the example of how the public health system is in great need of these standards in order to allow for interoperation within and between their health institutions. This kind of recommendation was particularly useful considering our current interest in RECAINSA to create a chapter of HL7 Central America.

Cutting-edge innovation and sustainable scalability

The last day of the forum's activities began with a keynote address on "International medicine in Latin America: Motivations, uses, results, strategies and policies." The speaker was Francesc Saigí Rubió, associate professor of the Faculty of Health Sciences and head of the WHO Collaborating Center in eHealth at the Open University of Catalonia (UOC[ac]). He shared the results of a study conducted in the Americas with the support of the Inter-American Development Bank, whose main objective was to determine the feasibility of adopting strategies of international telemedicine in specific countries. He highlighted the importance of taking into account local limitations, such as the available human resources in the country's health sector, the legal frameworks, and the national policies related to the provision of health services and data management.

The last day's discussion panel opened with the following inquiry: "How do we stimulate innovation and entrepreneurship in digital health in Latin America?" Participants included several women entrepreneurs and innovators who had designed and implemented digital health initiatives in European and South American countries. Speakers shared the way in which they made use of available funding opportunities for innovation and entrepreneurship to create and launch their start-ups. One of the main takeaways was the importance of stimulating entrepreneurship hubs in digital health, given the economic potential and impact that they have in the region.

The second panel, entitled "Artificial Intelligence in Health: Myths and realities," was one of the most crucial discussions of the forum because of the ever-increasing interest in AI. As a

[ac] https://www.uoc.edu/portal/en/index.html

trending topic, people have many expectations surrounding the implementation of AI to support clinical decisions, improve diagnostics, and provide predictive models. The pandemic has revealed the real stage of maturity of AI applied to health and has also unveiled unethical businesses where companies have sold invalid and phony AI solutions.

The third and final panel of the event was on the topic of "Acceleration in the use of telemedicine in Latin America," with speakers who have a long history in telemedicine and represented the Hospital Italiano de Buenos Aires, the University Network of Telemedicine of Brazil (RUTE[ad]), and the Ibero-American Network of Digital Health (RISAD[ae]). Speakers shared advances in the application of telemedicine in recent years and how the pandemic has accelerated the use of strategies such as teleconsultation. They all highlighted that the implementation of such tools has not always been based on good and ethical practices, identifying that some initiatives have failed to guarantee safety in case management, as well as protection and confidentiality of patient information.

The RECAINSA 2021 Digital Health Forum ended with one of the most anticipated discussions, on the topic of "Digital Public Goods (DPG): An opportunity to make digital health more inclusive." Ms. Karin Källander, senior digital health advisor at UNICEF,[af] reviewed the main advances in the generation of digital public goods in health that can be used openly by countries without incurring licensing costs or long and bureaucratic processes for use. Open source software, open innovation, open knowledge, open data, open standards, open content, open AI models, evaluation tools, and other digital assets are today part of the existing offerings of Digital Public Goods (DPGs) which are meant to adhere to privacy and other applicable laws and best practices, do no harm by design, and help attain the Sustainable Development Goals (SDGs). However, as Källander highlighted, these tools have not been as widely adopted in the Latin American region due to several local barriers. She therefore provided some strategies to expand the use of Digital Public Goods (DGPs) and allow countries to take advantage of them, especially in the current pandemic situation.

The DPG Strategic Objectives are based on four assumptions: (1) DPGs are discoverable, sustainably managed and accessible; (2) the UN, and multilateral and other institutions, promote and support the adoption of DPG; (3) LMICs have the capacity to adopt DPGs; and (4) LMICs have vibrant ecosystems and capacity for creation.

According to Källander, it is important to maintain the Registry of DPGs in order to increase their discoverability, as well as to convene communities of practice to further identify high potential DPGs within specific priority areas. She also highlighted the importance of leading pathfinding pilots, that is, entities in LMICs piloting new ways to change the power balance

[ad] https://rute.rnp.br/

[ae] https://risadsaluddigital.org/

[af] https://www.unicef.org/innovation/digitalhealth

around technology solutions. Lastly, she expressed that these strategic actions will contribute to maintain access to quality, openly licensed and quickly deployable DPGs; transform the complex systems, structures, and practices in which DPGs are embedded; and mobilize the DPG ecosystem to sustainably fund and uplift DPGs.

Reflections for the future

The RECAINSA 2021 Digital Health Forum (Red Centroamericana de Informática en Salud [RECAINSA], 2021) was considered one of the most relevant events of the year in Latin America in terms of digital health. It has enjoyed wide dissemination and received excellent feedback by attendees, RECAINSA partners, and prestigious organizations that promote the digital transformation of health as an important driver toward universal health coverage. The success is due in part to the methodology used to structure the forum, by including thematic areas, captivating and timely content in the different sessions, and high quality of speakers. These features are in contrast to other events where the focus is placed on presentation of individual papers, which are even not necessarily published in peer-reviewed journals.

The success in the 2021 Forum is aligned with RECAINSA's motto: *Together we are stronger*.

Looking to the future, we at RECAINSA intend to organize this forum annually, creating a *strong forum area* for the exchange of experiences, knowledge, and good practices, while guaranteeing its quality, and engaging all the stakeholders who make up the digital health ecosystem in Latin America and worldwide.

References

Banco Bilbao Vizcaya Argentaria [BBVA]. (2021). *The digital transformation in Latin America is accelerating with the pandemic*. BBVA News. https://www.bbva.com/es/the-digital-transformation-in-latin-america-accelerates-with-the-pandemic/.

Bello, M. (2021). *Mexico's path to interoperability standards*. https://mexicobusiness.news/health/news/mexicos-path-interoperability-standards.

Corporación Andina de Fomento—Banco de Desarrollo de América Latina [CAF]. (2021). *Inequality 4.0: To close the digital divide*. https://www.caf.com/es/actualidad/noticias/2021/05/desigualdad-40-a-cerrar-la-brecha-digital/.

Red Centroamericana de Informática en Salud [RECAINSA]. (2021). *Foro de Salud Digital RECAINSA*. https://recainsa.org/foro-de-salud-digital-recainsa-2021/.

Red Centroamericana de Informática en Salud [RECAINSA]. (2022). *Strategic Framework*. https://recainsa.org/marco-estrategico/.

Robles, J. M. (2020). *The great leap forward of technology*. The World. https://lab.elmundo.es/coronavirus/tecnologia.html.

Transform Health. (2021). *Digital Health week | A global moment to champion health for all*. Digital Health Week. https://digitalhealthweek.co.

United Nations High-Level Panel on Digital Cooperation. (2019). *The age of digital interdependence: A report of the UN Secretary-General's High-Panel Level on Digital Cooperation*. New York: UN.

World Health Organization. (2021). *Global strategy on digital health 2020–2025*. https://apps.who.int/iris/bitstream/handle/10665/344249/9789240027633-chi.pdf. Chicago.

CHAPTER 77

Raising the profile of health technology in the Middle East and North Africa (MENA) region

Kiyan Irani[a,b], **Megha Siddhanta**[a,b,c], **and Judy Kuriansky**[b,d,e,f,g]

[a]*Department of Clinical and Counseling Psychology, Teachers College, Columbia University, New York, NY, United States,* [b]*International Association of Applied Psychology, Indianapolis, IN, United States,* [c]*Harpaz-Rotem PTSD Treatment & Research Lab, Yale School of Medicine, New Haven, CT, United States,* [d]*Global Council SDG3 (Health), "SDGS in Action" Initiative, Dubai, United Arab Emirates,* [e]*Office of Ambassador Sidique Wai, Embassy of the Republic of Sierra Leone to the United States, Washington, DC, United States,* [f]*Mission of the Republic of Sierra Leone to the United Nations, New York, NY, United States,* [g]*Department of Clinical Psychology, Teachers College, Columbia University, New York, NY, United States*

Introduction

Given the unprecedented transmissibility of COVID-19, the world's healthcare systems faced the burden of a tragic global pandemic leading to excessive unmet needs, the magnitude of which proved overwhelming in response to the influx of seriously ill patients and inadequate infrastructure and resources. As a result, the demand for digital solutions and services grew exponentially. Countries pivoted as much as possible to incorporate technological innovations and tools to meet pressing needs. Recent years have witnessed progress in telehealth in the regions of the Middle East and North Africa (MENA) and the Gulf Cooperation Council (GCC, a political and economic alliance of six Middle Eastern countries of Saudi Arabia, Kuwait, the United Arab Emirates, Qatar, Bahrain, and Oman), but these have received less international attention.

To address this progress and imbalance, region-specific conferences about telehealth were organized to focus on advances in the MENA/GCC region. The first region-specific Telehealth Conference for the MENA region was held in Dubai in January 2020, mere months after the COVID-19 pandemic spread wildly throughout the world. The conference focused on establishing strategies and partnerships in health, with presentations on various topics, including the global development of telehealth, patients' and professionals' experience with telehealth, opportunities and challenges of telehealth and telemedicine techniques, innovations

Resilient Health. https://doi.org/10.1016/B978-0-443-18529-8.00077-9

with artificial intelligence and other methodologies, and telehealth surgical applications. Regional voices were highlighted, from the United Arab Emirates (UAE), the Kingdoms of Saudi Arabia and Bahrain, Jordan, Egypt, Algeria, and Sudan, as well as contributions from experts in other parts of the world. Formats included roundtable discussions and panels (MENA Telehealth Conference, 2020).

The second MENA Telehealth Conference was held in February 2021 to continue this focus (MENA Telehealth Conference, 2021). Over 2 days, sessions similarly covered a variety of relevant topics and formats. These included an overview of the present telehealth landscape to build solid and sustainable cases; a keynote on the future, entitled "The Cloudy Post-Pandemic Future of TeleHealth: Where to go from here?"; successes in the region; data protection; telehealth accreditation; elements of sustainable and safe platforms; reimbursing strategies; and ways to reduce costs of technology approaches. Regional voices were present again from the UAE, the Kingdoms of Bahrain and Saudi Arabia, Jordan, and Algeria, with added experts from Egypt, the Islamic Republic of Iran, Kuwait, and Morocco as well as from neighboring countries of Pakistan, India, and Turkey.

In the third year, in an attempt to leverage the year-over-year growth in attendance at this conference and to create synergistic influence, the event organizers decided to merge the MENA Telehealth Conference with the International Society for Telemedicine and eHealth Conference (ISfTeH). The resulting third MENA Telehealth Conference was held over 2 days from March 7–8, 2022, at the Emirates Towers Hotel in Dubai. Considering the ongoing travel restrictions and cautions of the persisting COVID-19 pandemic—despite some degree of recovery in parts of the world—this event followed a hybrid format, namely, in person but also virtually. The location of Dubai was deliberate, with the Emirates being ranked first in the Arab region and sixth in the world for health tourism (travel for medical purposes) by the International Centre for Health Care Research, attributable to its world-class patient care, cutting-edge treatment methodologies, modern medical facilities, advanced healthcare infrastructure and technology, and skilled healthcare professionals (Haziq, 2021). More than 350 people attended the conference. The event presentations, sessions, and panels offered speakers and delegates a valuable opportunity to exchange international work about telehealth, with the specific intention of exploring what such progress means for the MENA and GCC regions. The subject matter of the presentations was wide-ranging but can be generally categorized into these major themes: (1) game-changing models to improve healthcare outcomes; (2) groundbreaking technologies, regulation reform, and business modeling; and (3) implications of telehealth implementation on healthcare access and equity. Notably, the conference coincided with International Women's Day on March 8 and thus included a special session on women's health (Mena telehealth and ISFTEH conference 2022- highlights 2022).

The conference was purposefully held jointly with the 26th International Conference of the International Society for Telemedicine and eHealth (ISfTeH), given that organization is the world's largest digital health network with members in regions all over the globe. The MENA meeting was a pivotal event for the ISfTeH to fulfill its purpose of sharing best practices in developing, implementing, consolidating, and sustaining telehealth regulations and guidelines and to "facilitating the international transmission of knowledge and expertise in telemedicine and eHealth, as well as offering access to acknowledged specialists in the field around the world" (Mena telehealth and ISFTEH conference 2022- highlights 2022).

This chapter covers events on both of the 2 days of the conference. The meeting is significant for the world of digital technology and health, considering that the ISfTeH holds its annual international conference in a different country each year, but this was the first time the ISfTeH International Conference was hosted in the Middle East, where the ISfTeH has seen a considerable increase in membership in recent years. While some presenters came from the Americas and European region, many others were from countries in the MENA region and covered projects in those countries which have previously received little attention (Mena telehealth and ISFTEH conference 2022- highlights 2022).

This report gives invaluable insight into innovations in this region which complements the efforts of another conference similarly focusing on a specific region, namely, the RECAINSA Conference, which highlights projects and innovations in Central and South America, covered in a chapter in this volume (Otzoy-García et al., 2024).

Background

The ISfTeH event has been held around the world over the past three decades. The most recent editions were in Japan (in collaboration with the Japanese Telemedicine and Telecare Association—JTTA), Portugal (in collaboration with the Portuguese National Center for Telehealth—CNTS), Finland (in collaboration with the Finnish Society of Telemedicine and eHealth—FSTeH/STeHS), and Morocco (in collaboration with the Moroccan Society for Telemedicine and eHealth—MfSTeH/AMTe). Partnering with one of its national members each year helps the ISfTeH showcase local/regional initiatives to the international community and gives members of its worldwide network the opportunity to present their activities and experience and build connections by networking with those local and regional stakeholders.

The 26th ISfTeH International Conference, held in conjunction with the third annual MENA Telehealth conference in 2022, was also in collaboration with the Emirates Health Informatics Society (EHIS), its national member in the United Arab Emirates. Fig. 1 shows the conference logo found on the organization's website and promotional materials (MENA Telehealth Conference, 2022).

Over half of the 17 Scientific and Advisory committee members came from the MENA region. The conference was managed by the founder and managing director of MWAN Events, Svetlana Efimova, based in the United Arab Emirates. MWAN Events is a conference and training business focused on healthcare and life sciences that organizes high-level scientific events with core themes of digital health, health technologies, medicines safety, and health regulation.

Fig. 1
MENA Telehealth Conference logo (MENA Telehealth Conference, 2022).

The pressing need for telehealth

Albert Einstein once stated, "In the middle of every crisis, there lies immense potential" (HIVE Media, 2021). This statement reflects the severity of the COVID-19 pandemic. Despite the worldwide calamity with massive loss of life and catastrophic impacts on economies, social structures, enterprises, families, relationships, and health, prospects for progress and innovation emerged in healthcare.

In the tragic and unpredictable times of the pandemic, healthcare systems worldwide were compelled to discover a way to deliver healthcare to the masses safely, leading to the invention and advancement of various technologies. Virtual care has been at the forefront of these new healthcare changes, for both the delivery of healthcare services (via telemedicine and techniques like online appointment booking) and the prevalent use of digital-assisted methodologies (head-mounted displays, augmented reality, virtual reality, mixed reality, Artificial Intelligence). As in all parts of the world, the COVID-19 pandemic prompted the emergence of healthcare innovations in countries in the MENA and GCC regions, which had to redesign healthcare delivery systems given that the pandemic exposed several flaws in their current paradigm which uncovered opportunities for better virtual care delivery.

The current healthcare landscape in MENA

MENA is an acronym in the English language popularly used by academic, economic, business, social, international, and media organizations to refer to a grouping of countries in and around the Middle East and North Africa. This region generally encompasses 20 countries (although, at times, the number varies), including some of the world's wealthiest and poorest countries:

Algeria, Bahrain, Egypt, Iraq, Iran, Israel, Jordan, Kuwait, Lebanon, Libya, Morocco, Oman, Palestine, Qatar, Saudi Arabia, Sudan, Syria, Tunisia, United Arab Emirates, and Yemen.

The wealth disparity in this region creates a wide-range of needs for each country's population and some variation between countries. For example, infectious diseases (malaria, cholera, TB) more often afflict those living in poverty-stricken countries, while chronic diseases (diabetes, heart conditions) are more common among those living in more developed societies. As a result of the COVID-19 pandemic, countries in the MENA region—as in other parts of the world—suffered significant economic losses and social disruption. According to a recent essay by the Wilson Center, a nonpartisan global policy forum:

> In its latest economic outlook, the International Monetary Fund estimated that the MENA region's recovery rate of 4% of GDP [gross domestic product] would lag behind global growth of 6% for 2021. Due to an economic contraction of 3.4% last year, the Arab States in the region, for example, lost the equivalent of 5 million full-time jobs, according to the International Labor Organization. The region witnessed a sharp increase in poverty rates due to COVID-19. The World Bank estimated in January that the number of new poor at the middle-income poverty line of living on $5.50 per day increased by 13 million people in 2020—previous estimates said this number could reach 23 million by the end of this year. (Langendorf & Farley, 2021).

The need for telehealth care in the MENA region

Some countries in the MENA region have been lagging behind nations in other parts of the world in terms of capacity and capability of healthcare services and delivery; thus, virtual care has the potential to relieve the strain on existing traditional health systems (Langendorf & Farley, 2021). According to Jalil Allabadi, founder and CEO of Altibbi, a leading digital health platform in the MENA region, there are multiple reasons for the gaps in healthcare systems across the region, including a weak primary care infrastructure, overreliance on larger secondary and tertiary hospitals, cultural taboos, and the emergence of lifestyle diseases (HIVE Media, 2021).

Access to immediate and affordable primary care is critical to addressing people's health needs. Telehealth and virtual care enable individuals to get attention for basic health concerns without difficulties like excessive time, dangerous travel, queues at clinics, or huge costs at some facilities or private care. Virtual care rose to prominence during the COVID-19 pandemic, allowing healthcare providers and caregivers to better respond to the needs of people isolating at home. Increasingly, both consumers and providers in the region—and globally—perceive telemedicine and related services as a convenient, accessible, and safe alternative to traditional methods of care.

Event overview: 2022 MENA telehealth conference, Day 1

Day 1 of the MENA conference consisted of presentations by speakers from across various disciplines, addressing issues including government regulation, policy, telehealth, business models, cybersecurity, and technologies like artificial intelligence.

The event was kickstarted in an opening address by H.E. Dr. Amin Hussain Al Ameeri, Assistant Undersecretary of the Health Regulation Sector in the Ministry of Health and Prevention (MoHAP) of the United Arab Emirates. He presented a brief history of telemedicine that started in the region in 1992 and has since greatly expanded to different platforms to reach a wider audience. Dr. Al Ameeri emphasized the role of ICT for the exchange of information for the diagnosis and treatment of diseases and injuries, as well as for research and evaluation, and continuing education of health professionals. In his presentation on the topic of "*Main elements of telehealth*," he outlined the main elements of telehealth, as shown in Fig. 2, identifying these as telemedicine, electronic health record systems, big data, real-world evidence (RWE), mobile applications, wearable and implantable devices, and continuing professional education and interdisciplinary care.

The keynote address of the conference was delivered by Mona Sahman Al-Subaie, digital health advisor to the Assistant Minister in the Ministry of Health of the Kingdom of Saudi Arabia (KSA), who is also CEO of the Digital Health Center of Excellence and acting executive director of the newly launched SEHA Virtual Hospital. The SEHA Virtual Hospital is a network of 130 affiliated hospitals, the largest of its kind in the world, only rivaled in the US which has 43 affiliated hospitals. Al-Subaie had been involved in the successful development of a treatment system that electronically links Saudi medical commissions with health attachés abroad.

Her presentation showcased the rapid digital revolution in Saudi Arabia and efforts to replace traditional processes with digital innovations to better facilitate services. Notably, the KSA was ranked first as the most developed country in digital competitiveness, according to the Digital

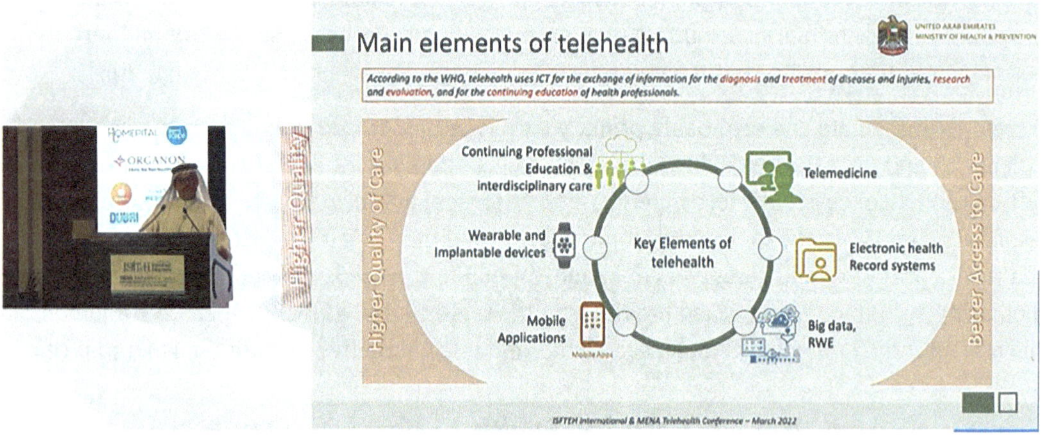

Fig. 2
Dr. Amin Hussain Al Ameeri addressing the main elements of telehealth (Al Ameeri, 2022; conference recording).

Riser Report of 2020 released by the European Center for Digital Competitiveness of the Business School of ESCP (École Supérieure de Commerce de Paris) and ranked second in corporate cybersecurity worldwide. Al-Subaie explained that digital health is transforming the KSA, emphasizing the urgent need in the MENA region to implement telehealth systems for treatment and care. This need is driven by the fact that the KSA has seen an increase in the disease rate in recent years, in addition to geographical barriers to receiving proper care and a significant human resource gap. As a result, there have been increases in treatment costs, treatment delays, and the need for health security. Hence, she said, there are growing needs to take advantage of data; continue to employ modern technologies, IOT, and artificial intelligence; and increase research and innovation in this sector.

The next presentation, "*The How-To Guide for Integrating Telehealth into the National Digital Health Strategy*," highlighted the importance of capacity-building and collaboration. It was delivered by two experts: Mr. Derrick Muneene, the unit head of the Capacity Building and Collaboration Digital Health and Innovation Department of the World Health Organization (WHO) in its headquarters in Geneva, Switzerland, and Dr. Henrique Martins, a Portuguese internist who is an expert in digital health and Associate Professor in Health Management and Leadership at the University of Beira Interior and University Institute of Lisbon. Mr. Muneene explained how telepsychiatry has emerged as a significant trend among digital health interventions. He described the results of a WHO survey in 2015 that telepsychiatry was one of the most impacted domains when technology platforms were being applied to disseminate services, resulting in a manual of telemedicine developed by the WHO to be implemented by their member states (World Health Organization, 2016). He emphasized that tele-mental health services must be implemented in low-care settings and integrated into primary healthcare services. However, barriers need to be addressed, including (1) lack of funding to develop and support telemedicine programs, (2) competing health systems priorities, and (3) the need for legislation or regulations covering telehealth/telemedicine programs. Only after resolving these issues, he said, will we be able to see a better healthcare system, especially in primary care facilities and in least developed settings.

As shown in Fig. 3, Mr. Muneene illustrated the significant global progress in telemedicine as reported by the WHO Global Observatory on eHealth in 2016 (World Health Organization, n. d.). Notably, Mr. Muneene revealed that telepsychiatry emerged as the highest utilized digital health intervention globally.

The next presenter was Mr. Willo Brock, Vice President of External Relations & Strategy of the Foundation for Innovative New Diagnostics (FIND), a global health nonprofit that collaborates with over 150 partners to facilitate the development, evaluation, and implementation of diagnostic tests for poverty-related diseases, who spoke on the "*Rules of Engagement for Establishing a Telehealth Ecosystem: An NGO's Perspective*." He explained that digital health is an excellent equalizer since a greater population can be reached in low- and middle-income

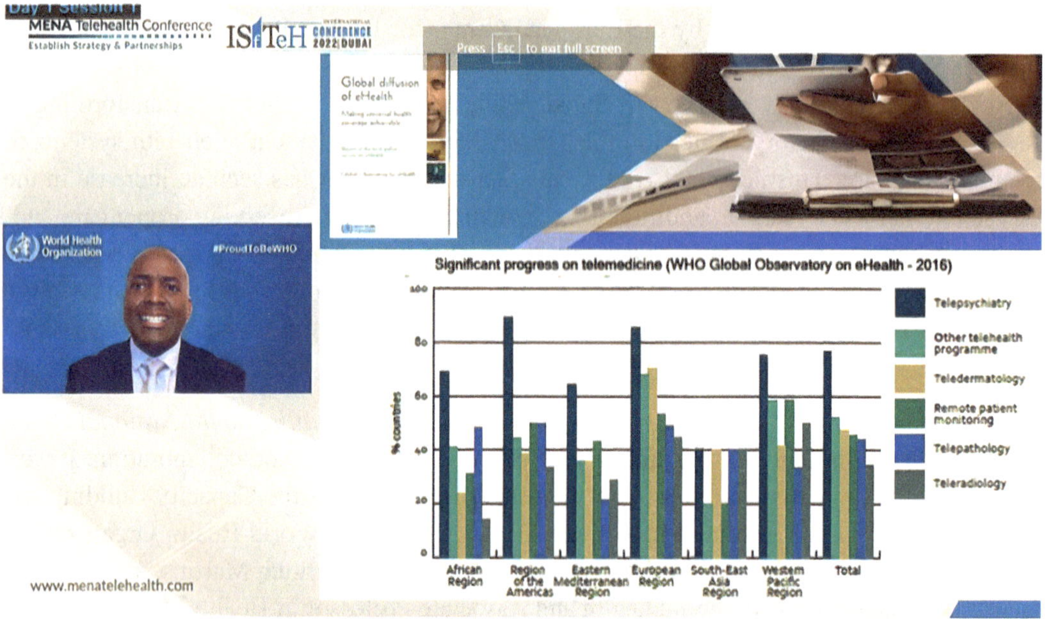

Fig. 3

Mr. Derrick Muneene from WHO presenting the significant progress in telemedicine (Muneene, 2022; conference recording).

countries. Thus, eHealth is especially important for countries like Afghanistan and Malawi where there are currently no diagnostic capabilities. FIND collaborates with countries, communities, decision-makers, and healthcare providers to spur diagnostic innovation and make testing an integral part of sustainable, resilient health programs. He explained that the organization has been catalyzing collective actions to support the development and scaling up of fit-for-purpose digital solutions that can be integrated into existing workflows and address user pain points, which are suitable for healthcare workers while taking into account digital literacy levels and local digital infrastructure.

Mr. Brock described the promise of telehealth to enhance the quality and efficiency of health service delivery. Specifically, he pointed out that telehealth can increase patient access to screening and diagnosis, facilitate decentralized testing and imaging, and improve data collection and management.

After a 20-min break, a series of panel discussions were held on different topics. The first panel was on "*Best International Practices in Developing Scalable TeleHealth Regulatory*." Moderated by Dr. Mazin Gadir, digital health adviser to healthcare authorities and ministries in the Gulf Cooperation Council (GCC), panelists included Mr. Bogdan Tušar, director of the Directorate for Health System Development in the Ministry of Health of the Republic of

Slovenia; Dr. Adam Chee, Chief of the Smart Health Leadership Centre in the Institute of Systems Science at the National University of Singapore; Dr. Caio Soares, President of SDB Saúde Digital Brasil and Medical Director of Teladoc Health Brazil (Brazil); and Mr. Kyle Zebley, vice president of Public Policy of the American Telemedicine Association (ATA) and Executive Director of ATA Action.

In his presentation on "*Best International Practices in Developing Scalable TeleHealth Regulatory Brazilian Experience*," Dr. Soares described how telemedicine in the pre-COVID-19 pandemic period in Brazil had very low utilization with few providers and rudimentary technology. However, in the recovery period, greater acceptance emerged of health interventions that can be provided digitally. Also, the National Congress of Brazil has been discussing needed regulations, a prominent theme in this conference and in the field.

Speaking from the Slovenian point of view, Mr. Tušar addressed how different European countries have advanced their telehealth solutions to reach the masses and that changes in policy and regulations have been instrumental in building capacity and increasing access to delivering interventions.

The Asian context was presented by Dr. Chee of the National University of Singapore (NUS), who stated that it was important to remember that telehealth is not a new concept in Asia but that it has been around since pre-COVID-19. In fact, teleheatlh has been very successful in the past, but the pandemic has removed even more barriers to digital health tools and has changed patient behavior to be more receptive to these techniques.

Dr. Chee explained that the outbreak of COVID-19 helped promote broad-scale support of telemedicine in the Asia region. For example, South Korea eased telehealth restrictions to allow for remote treatment of COVID-19 patients, and the Ministry of Health in Indonesia formed vital partnerships to enable timely COVID-19 diagnostics in remote regions.

Another panel followed on the topic of "*Revealing the Secrets of the Pioneers of Virtual Hospital Care Models*." The moderator was Dr. Erik Koornneef, Director of Research and Innovation for Pure Health, an integrated healthcare solutions platform of hospital management, laboratories, and medical services in the United Arab Emirates.

The first speaker, from Finland in the north of Europe, was Mr. Aki Puustjärvi, Development Manager of IT Management of the Helsinki University Hospital, who described the Health Village (also called the Digital Health Village) in Helsinki. The Health Village is a comprehensive digital service platform for healthcare, developed in cooperation with patients, healthcare professionals, IT specialists, and patient organizations. Putting patients at the center of care, it intends to provide quality, safe, and efficient public health services for the general public, digital pathways for patients, and electronic training for professionals, as well as online peer support.

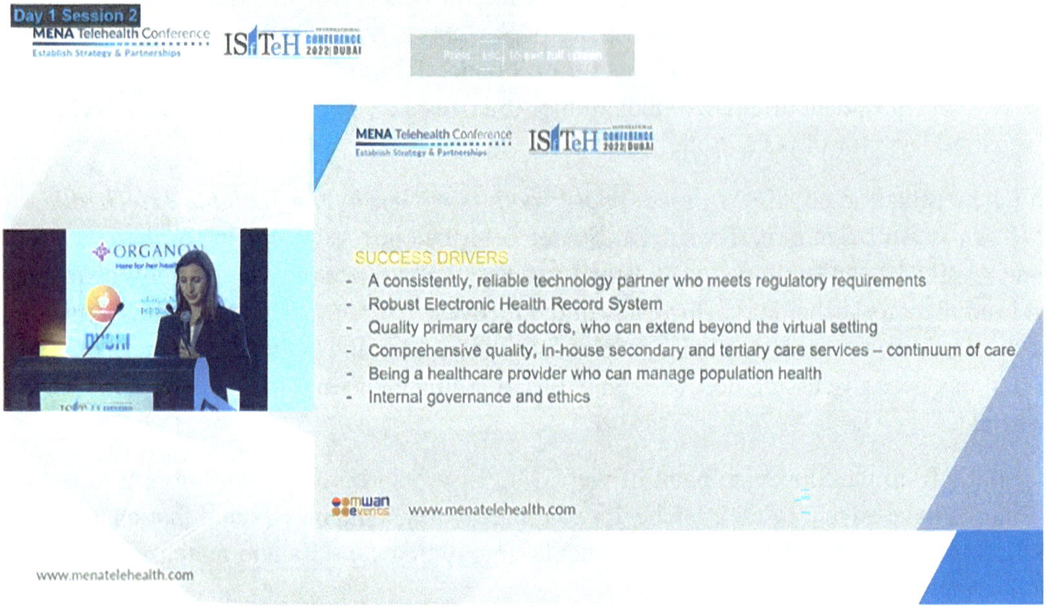

Fig. 4
Dr. Micaela Seemann Monteiro pointing out the success drivers of telehealth in Portugal (Monteiro, 2022; conference recording).

The next panelist, also from Europe, was Portuguese physician, Dr. Micaela Seemann Monteiro, who is Chief Medical Officer for digital transformation at CUF (a brand initially named Companhia União Fabril in Portuguese), which is now the largest network of private healthcare services in Lisbon. She highlighted the Portuguese National Health Service approach built on a hybrid health model, with a high presence of telehealth-enabling technology in its ICT systems. The model allows mobile access to services for citizens and healthcare professionals that is safe and easily used. As shown in Fig. 4, Dr. Monteiro outlined the key drivers of telehealth success in Portugal, including a reliable technology partner, robust electronic health record system, quality healthcare providers, comprehensive continuum of care, and strong internal governance and ethics of care.

Representing an innovation from Saudi Arabia, Dr. Safaa Almajthoub described the Digital Health Center of Excellence (DHCOE) of the Ministry of Health in the Kingdom of Saudi Arabia, where she serves as the Director of Strategic Planning and Governance. The center aims to transform the country's healthcare system by developing regulations, exchanging expertise, empowering investors and entrepreneurs, and conducting research to apply digital health solutions in the Kingdom as well as to define and enact the National Digital Health Strategy (Almajthoub, 2022).

The ultimate challenge, she noted, is to "strike the balance between innovation and regulation" which the center aims to do. This mission involves aligning stakeholders in order to safely and successfully facilitate the adoption of novel technologies in healthcare. She named four major stakeholders: regulators, entrepreneurs, patrons, and investors. She noted that the center is a valuable place to bring together these stakeholders, with their diverse expertise in digital health, that can then be applied nationally. The benefits include empowering investors, reducing barriers to market entry, reducing the gap between regulation and development, and finding alignment between relevant stakeholders by harmonizing the flow of information and optimizing processes to accelerate digital health adoption.

Another solution to the challenges, Dr. Almajthoub said, lies in the concept of a Digital Health Regulatory Sandbox implemented in countries including the United States and the United Kingdom. A Regulatory Sandbox provides a secure environment where digital health solutions can be safely tested for proof of concept and regulatory compliance in order to test and adapt regulations for efficiency and safety, and to open the field of digital health to investors. Dr. Almajthoub detailed the steps taken by the DHCOE in implementing the Regulatory Sandbox framework, which provides valuable insights that can aid in adopting the concept to other countries or contexts in the future in the continued quest to revolutionize healthcare globally (Almajthoub, 2022).

The last speaker of this panel was Ms. Nicolene Kruger, business process officer of Mediclinic Middle East, a private hospital group in South Africa, Namibia, Switzerland, and the UAE, who gave an overview of the local telehealth scenario in the UAE and the wider MENA region. She focused on the virtual primary care model that has been successful for the past few years but that also faces several challenges. These challenges include, (1) that technology adoption for healthcare services is slow, compared to other commercial services; (2) lack of community trust in the use of technology to provide quality healthcare; and (3) that patients are more familiar with, and rely more on, in-person healthcare services in the UA. These challenges can be overcome, she said, with "an acute focus on change management to facilitate mentality change." Also, more attention must be paid to making telemedicine a regulated and evidence-based healthcare delivery service.

Day 2 of the conference

The second day of the conference focused on various themes, including emerging telemonitoring, digital healthcare ecosystems, telehealth medical tourism, digital transformation of primary healthcare, and funding telemedicine practices. The final session of the day, which happened to coincide with March 8 as International Women's Day, consisted of a three-part event focused on women's health issues and innovations as growing areas of interest.

Emerging telemonitoring

The first panel of the day introduced telemonitoring as one avenue of telehealth showing promise for the future goals of telehealth policies. This technology emphasizes patients taking more of an active role in the healthcare process. Panelists on this topic included Dr. Tufia Haddad, MD, an oncologist and Medical Director for Digital Strategy at the internationally recognized Mayo Clinic Cancer Center in Minneapolis, Minnesota in the United States; Meshari F. Alwashmi, PhD, who is Chief Scientific Officer at BreatheSuite, as well as an associate member at Michigan University's Mixed Methods Program, and Director at Tech NL which is a not-for-profit industry association representing Canada's Newfoundland province's technology and innovation sector, with vast experience in new technologies like wearables and mobile health; Ram Fish, CEO of 19Labs Inc., a telemedicine platform for remote care in rural communities; and Dr. Harold Arkoff, MD, CEO of OneSource Solutions International (OSSI), a health information technology company.

To start the conversation, Dr. Tufia presented her research about how web-mediated follow-up in the form of electronic patient report outcomes (ePROS) significantly increases the likelihood of survival among patients previously diagnosed with lung cancer. The significance of these effects also held true when compared to the benefits of routine surveillance. Data also shows that patients participating in telemonitoring reported improved quality of life (QOL) and were hospitalized less frequently than their usual-care counterparts. She said that the results show promise in decreasing patient health complications, hospital administrative demands, and financial burdens for ongoing surveillance.

The panelists acknowledged the soundness of her research methodology and shared insights into potential applications of this technology to expand beyond the field of oncology. For example, Dr. Meshari F. Alwashmi mentioned that many chronically ill patients could benefit from intelligence-based medical devices such as telemonitoring instruments for cardiopulmonary disease, cystic fibrosis, and heart disease.

Digital healthcare ecosystems

Following the telemonitoring panelist discussion, Dr. Waqas Ahmed, MD, CEO and founder of American TelePhysicians, a physician-led digital healthcare and telemedicine company with expertise in building 360° healthcare ecosystems, gave a presentation on the importance of digital health ecosystems and the impressive growth trajectory of telehealth infrastructures. He gave particular credence to the notion that organization-centered healthcare models have the potential to be transformed into universal standards in the near term. He emphasized that the primary purpose of these systems is to deliver multidisciplinary and collaborative health services, which would decrease patient hospital readmissions and clinic overhead.

Telehealth medical tourism

A panel discussion followed on the topic of telehealth and medical tourism, moderated by Ziad Tabe, an independent consultant for best practices in global telehealth and medical tourism. Medical tourism is a global practice wherein a person travels abroad for medical care that they could not acquire or afford, or do not prefer, in their home country. In recent years, there has been a surge of people from developed countries traveling to developing countries for discounted medical treatments (Horowitz et al., 2007).

Panelists included Mohamed Al Mheiri, Director of the Health Tourism Department at the Dubai Health Authority in the UAE; Professor Piotr H. Skarzynski, MD, PhD, and MSc who is the director of the World Hearing Center in Poland, where he coordinates and manages national and international projects related to telemedicine and eHealth across Europe, Asia, and Africa, which includes teleconsultations, tele-rehabilitation, and remote hearing implant systems; and Dr. Dania Alsammarae, COO of King's College Hospital London and cofounder of Anglo Arabian Healthcare, an integrated healthcare provider in the UAE.

In an interchange, Professor Skarzynski raised the general benefits of applying telehealth across national borders to provide medical expertise to countries lacking technical resources or proper medical training. In reaction, Dr. Alsammarae introduced the idea that patients with rare conditions could benefit tremendously from outsourcing medical acumen from countries with staff trained in relatively niche patient-care protocols. However, she warned that some countries might experience a rebounding effect due to medical tourists seeking elective procedures performed in countries with less stringent healthcare regulations. Thus, federal officials and medical professionals should be made aware of the potential harm this practice can have on medical tourists. Consumers themselves also need to be aware of these dangers.

Extending ICU and stroke management beyond tertiary hospitals

After a short intermission, the conference resumed with another panel discussion moderated by Dr. Taha K. Al Hazarmerdi, UAE country General Manager of the Dr. Sulaiman Al-Habib Medical Group, and Dr. Sulaiman, Chairman of the Board of Directors at the Al Habib Hospital-Dubai. Panelists included Dr. Zohair Al Aseri, MD, consultant and associate professor of Emergency Medicine at King Abdul-Aziz University College of Medicine in Dar Al Uloom University, a private university in Riyadh, Saudi Arabia; Dr. Mohammed Aljohani, MD, Director of the Ministry of Health in Saudi Arabia; and Dr. Maike Schumacher, consultant for the Pediatric Intensive Care Unit and Chair of Critical Care at Sheikh Khalifa Medical City in Abu Dhabi, UAE.

As the first panelist to speak, Dr. Al-Aseri shared an anecdote that demonstrated the value of an innovative technology tool in the management—and lifesaving—of a patient. The patient, whom he treated last year, was initially not allowed to leave the hospital given the recurrence of transient ischemic attacks (TIAs), characterized by the temporary expression of mild-to-moderate stroke symptoms. These incidents typically last only a few minutes and usually do not cause permanent damage; however, given that TIAs are often precursors to actual strokes, the hospital staff were not comfortable discharging the patient until he had a permanent caretaker. For over 6 months, this man was forced to live in the hospital, away from his wife, who had a moderate physical impairment. However, after introducing a stroke symptom activity tracker (SSAT)—which helps the patient monitor their symptoms, nutritional requirements, and medication—into the patient's care regimen, the treating physicians found that the patient, when outfitted with an SSAT, could monitor his condition just as thoroughly as the medical staff and with greater accuracy. Thus, he was discharged from the hospital to his home. Three months after discharg, the patient reported an increase in specific prestroke indicators on his SSAT and was therefore admitted to the hospital. Tests concluded that he had a minor blood clot that, if left untreated, could have caused a life-threatening stroke. Given the accuracy and ease of using the SSAT, the patient caught this potential disaster by himself at perceivably the same time it would have taken to catch with round-the-clock medical monitoring, and cerainly earlier than if he had to schedule a medical appointment.
Dr. Schumacher concluded by noting how this process represents a major contribution of a technology tool to medical care, that in many cases can be life-saving. He punctuated his conclusion by emphasizing the importance of collecting and evaluating tertiary hospital performance data before and after implementing telehealth monitoring technology similar to SSAT.

Digital transformation of primary care

Digital health is in the midst of a technology boom. As such, the industry continues to outdate technology almost as fast as any tool is implemented. Therefore, hospitals and healthcare systems alike are faced with making many decisons, including about long-term investment choices among innovative technological solutions. In a brief presentation, Dr. Minal Bakhai, Clinical Director for General Practice Transformation and Digital First Primary Care at the National Health Service (NHS) in the United Kingdom, discussed telehealth solutions implemented in the United Kingdom during the COVID-19 pandemic. She gave special emphasis to digital systems that lessened the burden of the pandemic on the United Kingdom and other nations that had already implemented similar technology.

Following this presentation, Dr. Alia Mohammad Rafie, consultant family physician and acting Director of Specialized Program Services Department at Dubai Health Authority in the UAE, spoke about primary healthcare in the Gulf region and telecommunication initiatives in Dubai. In her talk, she made direct parallels between Western European telehealth initiatives with the framework outlined in Dubai's digital healthcare goals. She shared that the Dubai Health Authority is implementing numerous strategic programs to reduce unnecessary costs of care to patients and healthcare providers.

Funding telemedicine practices

After a brief intermission, another panel discussion commenced. The moderator was Abdullah Alsharqi, the Chief Technology and Digital Transformation Officer at the Council for Cooperative Health Insurance, an independent government body that promotes transparency, equity, and value-based care. Speakers included Dr. Sara Al Dallal, President of Emirates Medical Association's Emirates Health Economics Society in the UAE; Milo Wicinski, business development executive at Infermedica, a medical knowledge platform; Hein Van Eck, Chief Strategy Officer at Mediclinic Middle East which operates hospitals and clinics in the UAE; Heba Al Hasan, head of Medical Network and Vendor Management Gulf at MetLife, the well-known international health insurance company; and Othman AlKassabi, CEO of the Health Sector at Tawuniya, the first national insurance company licensed in the Kingdom of Saudi Arabia to practice all types of insurance business in accordance with the cooperative insurance principle accepted by Islamic Sharyia.

The discussion focused on how payers and healthcare providers can sustainably fund telemedicine practices. Mr. Wicinski proposed that all hospital administrations should have an in-house financier on staff, separate from the accounting department, to focus on maintaining long-term wealth to account for unavoidable malpractice lawsuit settlements. The conversation segued into discussions of healthcare payment models (deferred payments, interest deferral, and mutual bonds) and digital billing applications on mobile devices. Mr. Wicinski also explained the potential benefits of reimbursement incentives for virtual appointments and preventative healthcare through methods like annual checkups, plasma monitoring, and frequent hormonal testing. Hospital auditing practices, he said, should be considered in the context of preventing insurance fraud and other nonessential patient testing procedures.

The future of women's health: A 3-part special session

Women's physical and mental health continues to be a major topic of concern in both developing and developed countries alike. In fact, all issues related to women's rights and empowerment are enumerated in the United Nations Agenda 2030 for sustainable Development, which serves as a blueprint for member states (governments) of the UN and all stakeholders for progress of the people and protection of the environment. The UN Agency known as UN Women is the global champion for women and girls. Certain days of the year are identified by the UN as specifically focusing on women's issues. For example, March 8 was officially declared by the UN General Assembly as International Women's Day, which kicks off a two-week conference at the UN and is celebrated by stakeholders around the world in their own way. Thus, the third MENA Telehealth Conference ended with a three-part session held on March 8 to purposefully focus on women's health issues prevalent in all corners of the world.

A roundtable on "Women in eHealth—Partnering Globally in Digital Health" was organized by the ISfTeH Working Group on Women (WoW) and the Telenursing Working Group. The WoW works to support and promote the critical role of women in telemedicine and eHealth globally by forming collaborations and international partnerships geared toward increasing female visibility in international conferences and scientific publications. The session educated participants about the broad scope of digital health solutions relevant to women across several specialties and settings. Participants were encouraged to consider how they could advance themselves in the field while also contributing to healthcare delivery systems and positive patient outcomes.

The sessions were moderated in succession by Dr. Louiza Chitour, Senior Program Manager at HealthTech Plug and Play, a technology innovation platform headquartered in California, USA; Sarah Hariz, lead of Policy and Government Affairs at Organon, the global healthcare company catering to women; and Sawsan Al Madhi, Director General of Friends of Cancer Patients (FoCP) in the UAE, which is part of consolidated efforts in the fight against cancer of the Gulf Cooperation Council, a regional coalition of academic institutions with initiatives to expand research activities in cancer prevention, screening, and treatment.

As part of its activities among the UAE community, in collaboration with the government and private sector, the FoCP's "Pink Caravan" initiative raises awareness about breast cancer and the importance of early detection. The field of psychology has been a key component of the FoCP strategy, evident in initiatives like "Colour My World" that extends emotional support to cancer patients and the "Joy Cart" initiative that caters to the psychological needs of children in hospitals.

Speakers included Dr. Rasheed Al Hammadi, Director of Medical Research and Development at the Department of Health in Abu Dhabi in the UAE; Sophie Smith, founder and CEO of Nabta Health UAE, a disruptive technology company building a decentralized, patient-centric healthcare platform for women in emerging markets; Farouk Shamas Jiwa, lead of Policy, Communications, Market Access and Pricing at Organon, a well-known American global healthcare company dedicated to the healthcare needs of women; and Joe Hawayek, head of Virtual Health at Aetna International, the well-known global insurance company.

Topics ranged from in vitro technological enhancements assessing fetal heart rate to preventative telehealth birthing trackers to prevent mother and fetal mortality. Ms. Smith explained numerous benefits of utilizing a systematic and multifaceted approach to provide equitable health and security resources for women. Innovations in fertility and birth planning technology were also discussed, as were birthing complications, like those associated with scar tissue from female genital mutilation.

Female genital mutilation was described as one of many practices that women-centered medical, political, and social-justice professionals aim to eradicate, despite the controversy, given that some societies consider such cutting a cultural practice. Noting dangers to a woman's health from the practice, Mr. Hawayek brought up an evocative point that "women's health issues are a man's problem," meaning that such procedures of genital mutilation derive from a patriarchal society and that the fastest and most enduring way to dismantle such harmful patriarchal structures is from within the cultural norms. Not surprisingly, tech-mavens offer a digital solution to mitigate negative outcomes, she said. For example, telemonitoring following genital mutilation can significantly help reduce the risk of developing life-threatening complications during pregnancy and birth associated with the resulting overdevelopment of scar tissue. Other technologies to assess biomarkers related to the development of certain cancers were discussed.

Conclusion

Telehealth has burgeoned during the COVID-19 pandemic to become a major solution to healthcare delivery worldwide and has remained pivotal in maintaining adequate healthcare systems during the pandemic's recovery period. As this field progresses, so does its ability to increase access to safe and effective treatment and expand physical, mental, and behavioral health services. Comprehensive events and expos designed to highlight regional developments, like the MENA Telehealth Conference aimed at the MENA and Gulf region, offer researchers, professionals, and academics unique opportunities to expand professional networks both regionally and internationally. These events also provide a space for industry leaders to collaborate on innovative projects and showcase pioneering scholarship. The 2-day MENA conference, which has been held for 3 years to date and was expanded in the year 2022 to coincide with a large international eHealth network, offered experts from the region and around the world, including from Europe, Southeast Asia, and North and South America—a hub to exchange ideas and information to progress telehealth innovations, implementations, and applications as well as to discuss important issues like regulation. Holding a conference of this nature in the United Arab Emirates opened the door for this and other countries in the MENA and GCC region to attract global scholarly attention that will boost the profile and networking of experts in the region as well as enhance the lives of all people in this region and beyond.

References

Al Ameeri, A. H. (2022). *Main elements of telehealth [screenshot]*. United Arab Emirates: Ministry of Health and Prevention.

Almajthoub, S. (2022). *Digital health center of excellence [PowerPoint slides]*. Kingdom of Saudi Arabia: Ministry of Health. https://events.mwancloud.com/company/136/events/11/presentations/Safaa%20Almajthoub164740349173.pdf.

Haziq, S. (2021). *Dubai is a top health tourism hub despite the pandemic*. Khaleej Times. https://www.zawya.com/en/business/dubai-is-a-top-health-tourism-hub-despite-the-pandemic-d2y9kc02.

HIVE Media. (2021). *The state of virtual care in MENA*. Hippo HIVE. https://hippohive.org/wp-content/uploads/2021/10/Hippo-Virtual-Care-MENA-092621-04.pdf.

Horowitz, M. D., Rosensweig, J. A., & Jones, C. A. (2007). *Medical tourism: globalization of the healthcare marketplace*. Medscape General Medicine.

Langendorf, M., & Farley, A. (2021). *Digital transformation and COVID-19 in MENA: turning challenge into opportunity*. Wilson Center. https://www.wilsoncenter.org/article/digital-transformation-and-covid-19-mena-turning-challenge-opportunity.

MENA Telehealth Conference. (2020). *MWAN*. http://www.swissbcuae.com/wp-content/uploads/2020/01/Final-Program-TeleMedicineConference-2020-2.pdf.

MENA Telehealth Conference. (2021). *MWAN*. https://www.menatelehealth.com/mena-telehealth-conference/agenda.

MENA Telehealth Conference. (2022). *About event & scientific and advisory committee*. Retrieved July 18, 2022, from https://www.menatelehealth.com/mena-telehealth-conference/page/about-us/27.

Monteiro, M. S. (2022). *Success drivers [screenshot]*. CUF.

Muneene, D. (2022). *The how-to guide for integrating telehealth into the national digital health strategy [screenshot]*. World Health Organization.

Otzoy-García, D., Iñigo Petralanda, M. I., & Aljovín. (2024). Raising the profile of digital health in Latin America: The RECAINSA digital health forum and network. In J. Kuriansky, & P. Kakkattil (Eds.), *Resilient health: leveraging technology and social innovations to transform healthcare for COVID-19 recovery and beyond* (pp. 897–909). Elsevier.

World Health Organization. (2016). *From innovation to implementation: eHealth in the WHO European region*. https://www.euro.who.int/__data/assets/pdf_file/0012/302331/From-Innovation-to-Implementation-eHealth-Report-EU.pdf.

World Health Organization. (n.d.). Global observatory for eHealth. https://www.who.int/observatories/global-observatory-for-ehealth.

YouTube. (2022, March 13). *Mena telehealth and ISFTEH conference 2022- highlights*. YouTube. https://www.youtube.com/watch?v=tEN7zPczukY.

CHAPTER 78

eMental Health International Collaborative: The global peak-body for digital mental health implementation

Anil Thapliyal
eMental Health International Collaborative, Auckland, New Zealand

Background

Available international mental health data trends indicate that access to timely care, information, and treatment at a time and place of person's choosing will continue to be the biggest public health challenge in the post-COVID-19 era.

At the same time, governments and public health services are contending with continued conflict around the world, economic disparity, forced migration, and numerous remote and rural issues. The way we communicate, engage with each other, and access services has come a long way in just under a decade but we still have a long way to go to meet the mental health needs of a global population.

While the scale of challenges in front of us can seem overwhelming, there is also opportunity for us to collectively shape the future together. An international collaborative of mental health professionals, called eMental Health International Collaborative (eMHIC), was formed in 2015 in response to professionals working in the emerging field of digital mental health expressing an urgent need for cross-border collaboration. Its international board was comprised of experts and government officials from Sweden, USA, Canada, Australia, and New Zealand, committed to sharing learnings to support the development of international best practice.

Since its inception, eMHIC has been consistently working to enable collaboration across borders. eMHIC's goal is to improve global mental health outcomes by promoting and enabling the use of safe and effective digital mental health solutions for all populations.

eMental Health (which stands for "electronic mental health")—otherwise known as digital mental health—is defined as "the use of the internet and related technologies to deliver mental

Resilient Health. https://doi.org/10.1016/B978-0-443-18529-8.00078-0

health information, services and care" (The Royal Australian College of General Practitioners, 2018). Examples of digital mental health solutions include self-help apps, chatbots, virtual reality interventions, website-based information and resources, and online peer support communities. Other solutions are rapidly emerging. One example is digital phenotyping where digital devices monitor a person's mental health biomarkers and use that data to send them personalized mental health support. Another is healthcare interoperability: the sharing of patient information across health organizations to provide patients with a more streamlined experience and a higher level of care.

The need for digital solutions is clear, with the majority of eMHIC member countries not equipped to deliver traditional in-person mental health services. Global demand for mental health services is continuing to outpace service capacity, with patients left unattended on waitlists for weeks to months, or not being able to access services at all, especially in remote, low-resource settings. Digital solutions implemented as alternatives or adjuncts to existing services can significantly reduce this burden on overstretched healthcare systems. At the same time, service users benefit from instant access to evidence-based support from their device at a time and place of their choosing.

Evolution of the eMental Health International Collaborative

In 2011 a group of senior mental health leaders—Geoff Huggins (Scotland), Dr. Ian McPherson (England), Dr. Dennis Morrison (USA), Martin Rogan (Ireland), and Anil Thapliyal (New Zealand)—came together to write a white paper entitled "eTherapies." This was at the request of the Board of the International Initiative for Mental Health Leadership (IIMHL), an international collaboration of mental health professionals in nine countries (of which the group were all members), to help spread innovation and best practices. That paper elicited considerable interest from the IIMHL Board and gave birth to an eMental Health working group within IIMHL.

With an increasing number of experts from other countries wanting to be part of developing the Mental Health domain, this working group developed a life force of its own, and soon formed their own organization: eMental Health International Collaborative (eMHIC). Since then, senior country leaders from Sweden, USA, Canada, Australia, and New Zealand have provided eMHIC with invaluable foundational support.

The eMHIC International Board has worked tirelessly over the years to establish networks, engage a number of international experts as advisors, and build pathways for knowledge exchange and innovation. eMHIC has since expanded its global network, making it possible to achieve the widespread level of collaboration essential for ensuring the best mental health outcomes for all populations.

It is important to acknowledge the early stage support of eMHIC in 2015 by Auckland University of Technology, New Zealand, under their then newly established Centre for eHealth. Their support provided a platform for establishing eMHIC, which was shortly followed by exceptional endorsements and formal backing of the Mental Health Commission of Canada, Health Promotion Agency of New Zealand, National Mental Health Commission of Australia, Swedish Association of Local Authorities and Regions (SALAR) in Sweden, and IIMHL.

More recently, there has been a dramatic shift in the way mental healthcare is delivered. Since the start of the COVID-19 pandemic, ongoing public health restrictions have resulted in digital solutions becoming a vital part of the care continuum. There is now an unprecedented amount of interest in digital mental health from mental health policymakers, funders, planners, administrators, researchers, people with lived experience, practitioners, and the wider industry. This interest has also led to greater acceptance of how digital mental health interventions have been integrated into care pathways. Examples of the range of international developments in the eMental Health domain include:

1. Canada publishing its e-Mental Health briefing paper in 2014 (Mental Health Commission of Canada, 2014)
2. Australia publishing its Digital Mental Health Standards in 2020 (Australian Commission on Safety and Quality in Health Care, 2020)
3. United States launching its Digital Health Software Precertification (Pre-Cert) Program in 2020 (Digital Health Software Precertification (Pre-Cert) Program, 2021)
4. Australia updating its 2012 Digital Mental Health Strategy in 2021 (Australian Government Department of Health, 2021)
5. New Zealand publishing its eMental Health Framework in 2021 (New Zealand Ministry of Health, Health Navigator NZ, and eMental Health International Collaborative, 2021)
6. The Mental Health Commission of Canada has commenced its journey toward developing an e-Mental Health Strategy for Canada in 2022

Once viewed as a fringe innovation in the mental health domain, digital mental health has become a core and integral part of care pathways and a conduit of positive change.

About eMHIC

eMHIC is comprised of a group of international subject matter experts, collaborating and sharing knowledge on all topics related to eMental Health.

Their global community is working toward a future where all populations have access to a safe and effective digitally-enabled mental health promotion, prevention, screening, treatment, and social support. Through enabling support and recovery at a time and place of a person's choosing, such initiatives are a vehicle to promote and achieve population well-being and prevention of mental ill health or distress.

The mission of eMHIC is to support and enable collaboration and knowledge exchange to foster global best practice in digital mental health. As a single international point of reference, eMHIC's collaborative platform enables key mental health country leaders, researchers, clinicians and digital product innovators to learn from each other's successes and challenges in digital mental health implementation. Shared learning creates tangible, compounding benefits, helping to rapidly transform the current level of digital mental health care into one befitting all populations.

The founding board members, as shown left to right in Fig. 1, included: Christine Morgan (CEO, National Mental Health Commission of Australia), Derek McCormack (former vice-chancellor of the Auckland University of Technology), Fredrik Lindencrona (International Mental Health Policy and Strategy Expert), Louise Bradley (former CEO, Mental Health Commission of Canada), Anil Thapliyal (CEO, eMHIC), Laurianne Reinsborough (Director Health Promotion, Wellbeing SA, South Australia), and Fran Silvestri (President and CEO of the International Initiative for Mental Health Leadership).

eMHIC has become the main trusted organization and credible source of information for decision-makers internationally when considering effective implementation of digital mental health solutions.

eMHIC membership is available as a monthly or yearly subscription at three levels: individual, group, and country. Individual and group memberships provide full access to all of eMHIC's membership benefits including opportunities to interact with global leaders and subject matters in their fields of interest, invitations to prestigious eMHIC events, and staying ahead of the curve and maintaining best practice through the eMHIC Knowledge Bank.

Fig. 1
Founding board members of the eMHIC Board, 2019.

Country partners are a collaborative membership type, inviting country-level organizations to contribute to eMHIC and leverage the eMHIC network. This allows eMHIC to foster global connections among industry giants, tailor opportunities to harness the strengths of different nations, and provide platforms for world leaders to showcase their work and drive meaningful change in global mental health.

Key value propositions

- *Collaborate internationally:* Work with other international organizations to encourage the proliferation and implementation of safe and effective eMental health solutions.
- *Promote self-help:* Promote eMental health's role in strengthening population mental health, through preventative self-help and collective-help approaches.
- *Advance the state of the art:* Improve population mental health planning using technology advances in analytics and AI.
- *Foster innovation:* Create an international network of innovators to exchange knowledge, identify current gaps, and explore opportunities for the development of more effective digital solutions.
- *Raise awareness:* Increase public understanding of the effectiveness of eMental health, and the commonality of mental health issues to reduce discrimination and stigma.
- *Network building:* Facilitate and enable leadership exchanges to identify and share information about successful implementation of eMental health, workforce development, and change management. Explore scalable approaches to implementation and mutually beneficial partnership strategies.
- *Support quality and improvement:* Endorse quality standards in eMental health and encourage the development and adoption of international best practice based on evidence and a culture of improvement.

International collaboration

All eMHIC members in the previously mentioned three categories, e.g., government organizations representing various countries, mental health provider organizations, and also individuals, have expressed their desire to share knowledge and experiences, to learn and support each other, and to explore together the effectiveness of digital mental health tools and services. Also important to members is the importance of learning the best ways to incorporate digital approaches in an integrated way within patient care pathways.

The networks and relationships forged through eMHIC's global collaborative platform are proving to be a crucial component for the sharing of this ongoing work. Through eMHIC's webinars, meetings, Knowledge Bank, newsletters, and the eMental Health International

Congress, eMHIC has witnessed a strong appetite for this knowledge, and for further in-depth discussion on important areas.

These include the creation of effective and sustainable digital mental health policies, strategies, and frameworks, along with guidelines for effective implementation. Equally important are the quality of digital solutions and concerns around equity and access, acknowledging that various international jurisdictions often have different approaches and needs.

eMHIC's foundational imperatives

The successful formation of eMHIC can be attributed to its focus on achieving balanced collaboration across the following five significant areas recognizing the immense unique value each brings:

1. Lived Experience inclusion
2. Policy, Strategy, Standards and Frameworks
3. Evidence Creation
4. Workforce Development
5. Health IT Industry

With regard to (1) mentioned before, it is important to incorporate authentic participation of people with Lived Experience in the design, development, and implementation of digital mental health initiatives. This is to ensure that solutions focus on the service user, their families, and carers. If these solutions do not work for them, then they do not work at all; a crucial sentiment illustrated by the Māori proverb in Box 1 that expresses the priority of putting the people at the center of everything, by which we abide.

The eMental Health International Congress

Since 2015, the eMental Health International Congress has brought together global thought leaders dedicated to eMental Health promotion, prevention, screening, self-help, self-management, treatment, and social support to collaborate and share learnings from their latest initiatives.

BOX 1 Maori proverb, New Zealand

"He aha te mea nui o te ao
What is the most important thing in the world?
He tangata, he tangata, he tangata
It is the people, it is the people, it is the people"

The Congress is the only one of its kind, focused on implementation science. At each Congress event, local and international speakers showcase groundbreaking work taking place around the world, highlight trends and challenges, and share their experiences around what is working in digital mental health implementation.

The 2020 Congress, held in Auckland, New Zealand, as shown in Fig. 2, featured an agenda tightly packed with 27 speakers from eight countries representing government policy and strategy, the health IT industry, healthcare professionals, and people with lived experience.

The Website

The eMHIC website is a hub of knowledge and collaboration, as shown in Fig. 3, which shows the homepage of the website. The core services provided through the website are the Knowledge Bank, discussion forums, newsletters, and events including eMHIC's Global Webinar Series.

Members can expect to find access to the latest digital mental health news and thought leadership in eMHIC's Knowledge Bank. The eMHIC team keep their finger on the pulse of all things digital mental health and ensure that the Knowledge Bank is kept up to date with the latest research, new tools, and innovations from the sector; international policy and strategy developments; critical guidelines and frameworks; insightful case studies; examples of implementation; and thought leadership pieces from industry leaders.

Fig. 2
eMental Health International Congress 2020, held in Auckland, New Zealand; one of the prestigious events hosted by eMHIC.

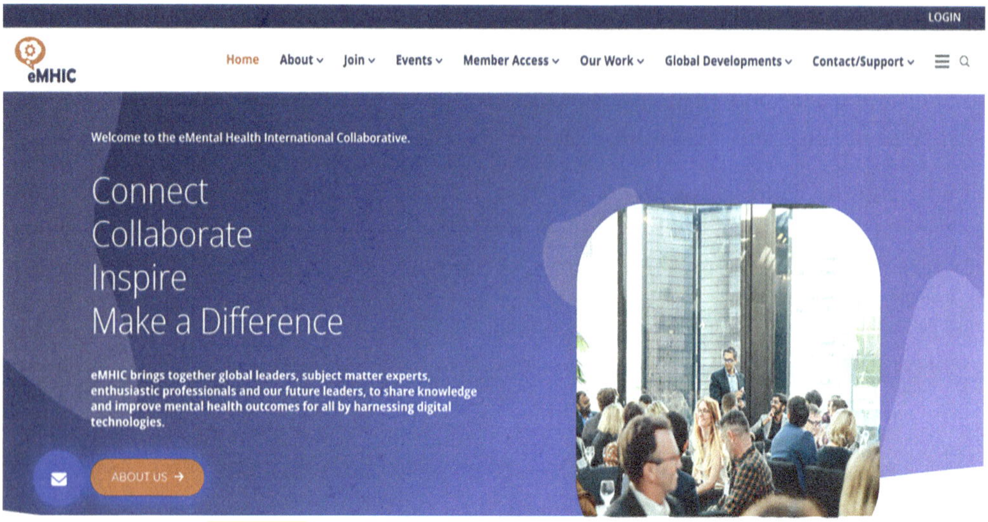

Fig. 3
Homepage of the eMHIC website: www.emhiglobal.com.

Examples of articles on the website that cover technological advances in mental health are as follows:

- *Four ways artificial intelligence is improving mental health therapy:* How are mental health professionals using artificial intelligence to improve the accuracy of diagnosis and treatments?
- *Transforming mental health care with prescription digital therapeutics:* Mental health professionals can now prescribe evidence-based software and digital tools as treatments for mental health disorders.
- *Solving our mental health crisis with technology:* How the app Woebot Health delivers scalable and effective solutions to the growing mental health crisis.
- *The growing field of digital psychiatry:* current evidence and the future of apps, social media, chatbots, and virtual reality: A comprehensive update on the rise of digital mental health tools and their explosive growth in the midst of the pandemic.

In addition, eMHIC provides easy access to important documents from around the globe including national frameworks and domain-advancing position statements. The Knowledge Bank is an invaluable resource for anyone wanting to stay up to date with the latest in global eMental health news and information.

Another highlight of the eMHIC website is the ability to connect and collaborate with the global eMental health community. Members can open or contribute to discussions through eMHIC's online forums, enabling them to connect with digital mental health experts around the world.

Further, eMHIC hosts regular webinars drawing together panels of leaders in their respective fields for inspiring and informative discussions about a range of topics. These include ethics and law, digital mental health in the workplace, eMental health accreditation, the role of artificial intelligence in mental health, and global policy and strategy. The eMHIC webinars are known for the caliber of their panelists, their range of international perspectives and the robust discussions and enduring collaborations they inspire between webinar participants.

eMHIC members also have unique access to the webinar archive which sits on the website. In doing so, they are able to revisit all webinars at a time of their choosing to glean new insights or consolidate their learning.

Position statement on the ethics and law of eMental health

According to a recent report by market intelligence firm CB Insights, digital health investment around the world hit an all-time high of $57.2 billion in funding in 2021. This 79% jump (from the $32 billion raised globally in 2020) fueled by health consumer demand for digital mental health solutions has caught policymakers and service providers by surprise (CB Insights, 2022).

It is a well-documented fact now that technology has reduced barriers to high-quality mental healthcare (World Economic Forum & Deloitte, 2021). But with less than 30% of the available solutions found to be safe and effective, there is an urgent need to focus on responsible governance (What is the health of mental health apps?, 2020). Mental health system administrators and commissioning entities want to know which digital mental health solutions are safe and effective, where they can be found, and how trustworthy they are in terms of data safety and security. Governance frameworks are still catching up with the rapid evolution of clinical practices but in a fragmented manner within siloed commissioning structures.

In response to this need, the eMHIC Special Interest Group on Ethics and Law in eMental Health released in 2021 its first position statement on Ethics and Law in e-Mental Health (Gooding & Wee, 2022) as shown in Fig. 4.

The development of the first version of this Position Statement has been an international effort with high-quality input and advice from experts in Australia, New Zealand, Singapore, Norway, UK, USA, and the APEC Digital Mental Health Hub.

The Position Statement articulates ethical and legal issues raised by eMental health practices and aims to help ensure that such practices are guided by ethical principles, professional rules of conduct, and established norms of law. It takes into account the diversity of views and experiences among people who have used eMental health services, mental health professionals,

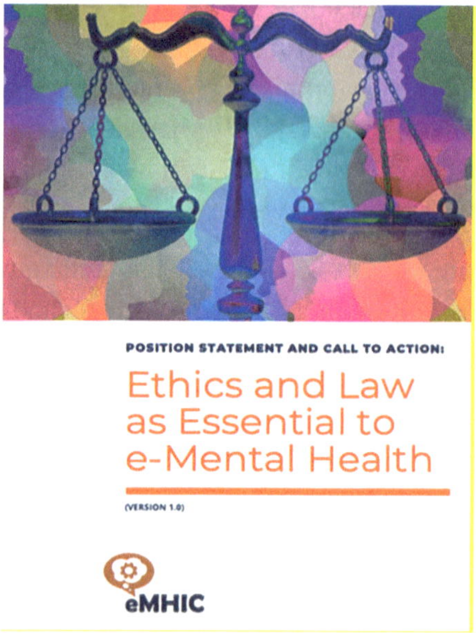

Fig. 4
eMHIC position statement and call to action: ethics and law as essential to e-Mental health.

technologists, and service providers, and in particular, people with lived experience, and their families and carers.

The Position Statement is a useful resource for mental health system administrators looking at how to implement safe and effective eMental health solutions at scale. It provides guidance about key principles relating to law and ethics, and practical calls to action.

Members of eMHIC who represent various countries have endorsed this statement. They include

Canada

"The Mental Health Commission of Canada (MHCC) applauds eMHIC on the release of the Ethics and Law in eMental Health position statement. This work underscores the tremendous value eMHIC holds within the global eMental health community."

Nicholas Watters, Mental Health Commission of Canada

Sweden

"This new position statement and call to action is truly an important breakthrough for progressing safe, ethical, inclusive support. In Sweden, we will start using this asset as a guiding set of principles to review and improve existing core services and use it to design the next generation."

Fredrik Lindencrona, Sveriges Kommuner och Landsting, Sweden

UNICEF

"Ethical principles, professional codes of conduct, regulations and norms are critical for ensuring quality of care & support services. eMHIC's position statement provides an excellent basis for further development, and inclusion of all key players including children, adolescents and youth with personal lived experiences."

Dr. Nina Ferencic, UNICEF, Switzerland

Position statement summary

Digital mental health initiatives have the potential to improve mental healthcare services comprehensively, yet there are ethical concerns and legal issues that will need to be addressed as digital technology develops. The mental health and well-being of service users, patients, and consumers must remain central to these considerations. Organizations, groups, or individuals who are developing and implementing eMental health practices need to be in partnership with service users, patients, and consumers, and to consider key ethical issues.

The purpose of the Position Statement is (1) to recognize the important role eMHIC can play in articulating the ethical and legal issues raised by eMental health practices, and (2) to support action to ensure such practices are guided by ethical principles, professional codes of conduct, and established norms of law.

The Position Statement was developed to set direction and a practical basis for action. Of central concern is promoting practices that benefit individuals, communities, and populations, and preventing and minimizing harms. It aims to encourage and support adherence to the highest ethical standards in the design, development, and use of digital mental health technologies. Typical bioethics principles are likely to be commonly understood in mental health systems; however, this approach may not be the most appropriate lens through which to evaluate eMental health initiatives.

Seven principles are proposed to help guide emerging practices in relation to ethical frameworks. These principles should be considered "living principles," insofar that they are likely to evolve as the eMental health field evolves. They are as follows:

1. Partnership and active involvement of people with "lived experience"
2. Privacy
3. Accountability
4. Safety and security
5. Transparency and explainability
6. Fairness, nondiscrimination, and equity
7. Professional responsibility and evidence-based practice

The seven principles offer a starting point to establish foundational requirements for eMental health initiatives that are ethical and respectful of international law. The true challenge arises in applying these high-level concepts in the real world. eMHIC takes the lead in helping generate awareness and discussions in this area for all to be involved.

Conclusion

eMHIC is dedicated to the promotion of best practices in digital mental health by connecting global leaders, subject matter experts, and enthusiastic professionals to exchange knowledge and collaborate through a diverse range of channels. Working together as a collaborative, eMHIC's global community is inspiring positive change, building on what is working around the world to get the implementation of eMental health solutions right by design. In doing so, eMHIC's community is making an important difference to global mental health outcomes, ensuring safe and effective mental health support is accessible to all. For more information, visit the eMHIC website: www.emhicglobal.com

References

Australian Commission on Safety and Quality in Health Care. (2020). *National safety and quality digital mental health standards*. Australian Commission on Safety and Quality in Health Care. https://www.safetyandquality.gov.au/.

Australian Government Department of Health. (2021). *National digital mental health framework. Australian Government Department of Health*. https://www.health.gov.au/.

CB Insights. (2022). State of digital health. *CB Insights*. https://www.cbinsights.com/.

Digital Health Software Precertification (Pre-Cert) Program. (2021). *US Food and Drug Administration*. Retrieved May 17, 2022, from https://www.fda.gov/medical-devices/digital-health-center-excellence/digital-health-software-precertification-pre-cert-program.

Gooding, P., & Wee, R. (2022). eMHIC position statement and call to action: ethics and law as essential to e-Mental Health (1). *eMHIC Global*. https://emhicglobal.com/ethics-law-in-emental-health-position-statement/.

Mental Health Commission of Canada. (2014). E-mental health in Canada: Transforming the mental health system using technology. *Mental Health Commission of Canada*. https://www.mentalhealthcommission.ca/.

New Zealand Ministry of Health, Health Navigator NZ, & eMental Health International Collaborative. (2021). Digital mental health and addiction tool (DMHAT). *eMental Health Framework*. https://emhf.emhicglobal.com/.

The Royal Australian College of General Practitioners. (2018). e-Mental health: A guide for GPs. *RACGP*. https://www.racgp.org.au/.

What is the health of mental health apps? (2020). ORCHA. Retrieved May 16, 2022, from https://orchahealth.com/health-of-mental-health-apps/.

World Economic Forum & Deloitte. (2021). Global governance toolkit for digital mental health: building trust in disruptive technology for mental health. *World Economic Forum*. https://www.weforum.org/.

CHAPTER 79

Reimagining healthcare: Technological solutions to close the treatment gap and achieve equity in the wake of COVID-19

Daniella Diaz[a,b] **and Judy Kuriansky**[a,b,c,d,e,f]

[a]*Department of Clinical Psychology, Teachers College, Columbia University, New York, NY, United States,* [b]*International Association of Applied Psychology, Indianapolis, IN, United States,* [c]*World Council for Psychotherapy, Vienna, Austria,* [d]*Office of Ambassador Sidique Wai, Embassy of the Republic of Sierra Leone to the United States, Washington, DC, United States,* [e]*United Nations Group of Friends of Mental Health and Well-being, New York, NY, United States,* [f]*Global Council SDG3 (Health), "SDGs in Action" Initiative, Dubai, United Arab Emirates*

Introduction

The COVID-19 pandemic had a dramatic global impact, with over five million deaths worldwide, more millions of people ill and socially isolated, a surge of unemployment, burned-out frontline workers, and upended lives which led to dangerous spikes in psychological distress combined with a lack of resources and personnel to deliver care, resulting in unmet needs for psychosocial support. The resulting dramatic treatment gap stimulated efforts to create innovative ways to deliver care, especially to populations with few resources in countries with poor health infrastructures. To address this problem, innovation through the use of technological approaches is increasingly being incorporated into health systems.

This chapter gives an overview of the treatment gap and technological solutions presented in an online event by experts involved in innovative approaches to mental healthcare, including the use of technology with some preliminary research observations and recommended policies to create a more effective, sustainable, and equitable mental healthcare system.

The treatment gap

The treatment gap refers to "the difference in the proportion of people who have disorders or a particular disorder (prevalence) and the proportion of those individuals who actually receive care" (Kazdin, 2017). Globally, international studies have estimated that the treatment gap is

Resilient Health. https://doi.org/10.1016/B978-0-443-18529-8.00079-2

currently around 50% or higher; for example, the European Study of the Epidemiology of Mental Disorders (ESEMeD) found only "48% of people with mood, anxiety and substance use disorders severe enough to significantly interfere with everyday life received any formal healthcare" (Werlen et al., 2020). In another report, approximately 25% of the United States population experiences a psychological disorder every year and 50% of Americans experience a disorder in their lifetime, yet about 70% of those who need mental health services are not able to receive them, with ethnic minority groups (e.g., African, Hispanic, and Native Americans) at greater risk of not having access compared to European Americans (Kazdin, 2017).

Many factors contribute to this treatment gap. These factors include lack of access and affordability due to cost, inadequate insurance, degree of proximity to services, inadequate staff, cultural training, and stigma (Kazdin, 2017). This situation is poised to worsen, given that the Health Resource and Services Administration (HRSA) estimates a significant decline in the number of psychiatrists in the United States by 2030, reflecting a lack of providers which will not even be alleviated by task-shifting, meaning adding varied level practitioners (physicians assistants, nurse practitioners, and others) into the workforce (Beyer & Boazak, 2021). Additionally, fewer practitioners reside and/or practice in rural small towns compared to qualified professionals more often found in highly populated affluent urban areas (Kazdin, 2017). All these factors contribute to inequities in health and mental health service delivery.

Context of the COVID-19 pandemic

The prevalence of mental health needs in the wake of the COVID-19 pandemic from early 2020 led to increases in the numbers of people suffering from elevated anxiety, anger, confusion, and posttraumatic symptoms. Contributing factors in this pandemic situation include "spatial distancing, self-isolation, quarantine, social and economic discord, and misinformation," which in turn led people to feel more fearful, helpless, lonely, nervous, and sadder than previously (Khan et al., 2022).

Vulnerable populations include frontline workers under constant exposure to the virus and fearing transmitting the virus to family members at home (Khan et al., 2022). Symptoms were also reported by people who have previously never experienced mental illness. The extent of the problem was exacerbated by lack of resources and funding of institutions tasked to serve. These combined factors led to considerable healthcare insecurity.

An important recommendation to address this problem, according to international organizations, including the World Health Organization (WHO), is that mental health and psychosocial support should be integrated into government responses to COVID-19 (Moreno et al., 2020). According to a United Nations policy brief, this integration would allow investment in mental health now, which would reduce effects in the future (United Nations, 2020); however, economic breakdowns in societies caused by the pandemic delayed adequate government mental health response (Moreno et al., 2020).

Despite extreme distress on multiple fronts, many experts agree that the pandemic offered an opportunity for transformation in mental healthcare systems, including escalating the use of alternate, technological methodologies.

E-Mental health

E-Mental healthcare can be defined as "the use of information and communication technologies to support and improve mental health, including the use of online resources, social media, and smartphone applications" (Moock, 2014). This form of mental healthcare delivery has the opportunity to deliver care with considerable advantages, including low cost, easy access, and anonymity for users. However, as a relatively new practice in the mental health field, such technological approaches entail barriers that still need to be addressed, including ethical issues of privacy and confidentiality.

Despite limitations, technology provides clinicians unprecedented tools for outreach. One study reported that the mental healthcare sector is developing web-based solutions for individuals suffering from mental illness (Moock, 2014). Examples of web-based solutions for mental health include: a website for clients that provides information about mental health in everyday language, peer support on forums for individuals to have a safe space to share experiences and support one another, and treatment programs that utilize computer-administered therapy.

Other innovative and useful approaches have been identified (Insel, 2019). One approach detects mental health issues through digital phenotyping—collecting data from sensors, speech analytics, and keyboard use, which provides biomarkers that predict relapse or recovery. Also, smartphone applications have been developed for interventions ranging from crisis services to evidence-based psychotherapies. Both of these approaches were found to produce similar outcomes compared to conventional therapy sessions (Insel, 2019).

The World Health Organization (2018) has maintained that mental health is "fundamental to our collective and individual ability as humans to think, emote, interact with each other, earn a living and enjoy life." Given how immersive technology is in contemporary everyday life, this statement can usefully be extended to refer to mental health being fundamental to individuals' ability to interact with others online and to enjoy life when engaging in online activities.

Conference event overview

During the COVID-19 period, with increased awareness about mental health consequences of the pandemic and increased use of online communications due to infection spread prevention methods such as lockdowns, many webinars were held about technological approaches to health and mental healthcare. One of these is elaborated in this section.

The webinar, held virtually on October 29, 2021, was on the topic of "Mental Health Care in Flux: An Overview of Innovative Approaches being Implemented to Deliver Care in the Wake of the Covid-19 Pandemic" (Grande et al., 2021). The virtual conference was sponsored by the University of Pennsylvania Center for Mental Health in the Department of Psychiatry and the Leonard Davis Institute of Health Economics, with its Director of Policy and Associate Professor of Medicine, Dr. David Grande, serving as host. Panelists addressed providing more effective, sustainable, and equitable mental healthcare in the United States in the wake of the pandemic, using technological innovations.

The conference was in three sections: a keynote address, an innovator session, and a panel discussing research and policy priorities for future mental healthcare.

Keynote address

The opening keynote address was given by Benjamin Miller, Psy.D., President of the Well Being Trust, who discussed the broad theme of mental health, challenges in mental healthcare both generally and in response to the pandemic, and opportunities these challenges bring to improving mental healthcare delivery. The Well Being Trust is a national foundation headquartered in California in the United States, dedicated to advancing mental, social, and spiritual health in the USA (Well Being Trust, 2022).

"Our nation is in pain and hurting," Miller said, offering that solutions lie in transforming how we think about mental health, and rethinking the structures in our culture that see mind and body as separate. An impediment to this change, he explained, is lack of data to validate new programs, leading to a "dysfunctional web" between program and policy.

Three ideas to help reimagine the mental health system that he posed are as follows:

(1) Bring mental healthcare to where people are located. This requires reassessing where a person can access mental healthcare, shifting to community-based care, and adding use of technology that can reach them.
(2) Rethink our work force with an eye on who does what, where, and for whom. A third (33%) of people in the United States who seek mental healthcare have to wait more than a week or 6 weeks or more to access mental healthcare, and also travel great distances. These conditions increase disparities. Further, many towns and counties in the US don't have a psychiatrist at all, making us rethink where care is delivered, such as making clinicians available in common community locations (e.g., community centers, schools, barber shops). This requires reimagining the workforce either "in community, by community, and for community," or, remotely. Task-shifting is essential, delegating tasks done by specialists to less specialized healthcare workers, thus, increasing capacity.
(3) Gather people together from different organizations and policy positions with a common vision and framework to come up with new ideas for a newly designed healthcare system,

while consulting the community. Doing this requires changing current policy to focus on long-term commitment to the new design, and creating relationships with agencies and constituents including from the next generation who will continue the work.

Innovator session

The ensuing 45-min innovator session, moderated by Dr. Courtney Benjamin Wolk, PhD, Assistant Professor at the Penn Center for Mental Health in the Perelman School of Medicine at the University of Pennsylvania, highlighted different innovations in mental healthcare, using technology, which have been accelerated by the pandemic.

The first speaker, Jonathan S. Comer, PhD, Director of the Mental Health Interventions and Technology Program, an interdisciplinary clinical research program at Florida International University, presented his innovative work using telehealth in mental health practice. He defined telehealth as the "electronic ease of the electronic media that facilitates real time interactions between patients and providers for the delivery of care that's traditionally delivered in person." Despite being used prior to the pandemic to overcome geographical workforce shortages, stigma, and transportation issues, it wasn't until the virus spread, imposing isolation and lockdown limitations, that tele-mental health services accelerated into clinical mainstream, especially reaching needy and inaccessible patients. In Dr. Comer's method of telehealth, called "Internet Delivered Parent-Child Interaction Therapy," patients wear a Bluetooth earpiece in their home while the therapist provides live coaching. He claimed that research finds treatment outcomes for this approach are as good or better than in-person sessions, especially for lower income families and persons of color.

"Our thinking was not only are we expanding the accessibility of treatment, but with tele-mental health you're also potentially expanding the ethological validity of care," Dr. Comer explained. Further, treating families in their natural setting allows the therapist to better understand them because they get a look into the patient's real world.

Next, psychologist Thomas Derrick Hull, PhD, Project Director at Silver Hill Hospital and Research Director at *Talkspace* discussed the innovative aspects of their private sector endeavor using technology. *Talkspace* is an online and mobile therapy company based in New York City, giving users access to licensed therapists through their website or mobile app. The business has two innovative foci:

(1) A "boutique" network of providers offers online mental health support for the public meant as a "one-stop shop" with various types of services.
(2) Asynchronous message-based care. The service expands the simplified hybrid approach of combining in-person visits with commonly used technology (video chat and phone calls) whereby clients and "therapists" answer chat messages in their own time. Advantages

> include reducing labor shortage since more mental health workers with restricted schedules can enter the workforce and others can conveniently schedule their work time, as well as serving more people. Critics, however, are skeptical about this approach, because it eliminates valuable dynamics through an exchange in the same time frame.

Studies on asynchronous message-based care date back about six years, he said, with evaluations following patients with conditions of obsessive-compulsive disorder (OCD), anxiety, depression, and trauma, showing that the outcome is similar to what is expected in traditional in-person settings of patients working with a licensed professional. Dr. Hull reported a recently completed phase one randomized controlled trail (RCT) study, supported by an NIMH (National Institute of Mental Health) grant, which found no difference in the treatment outcome of 300 participants each in message-based care or weekly video sessions (a more traditional telehealth model). The outcome was tracked with various measurements, i.e., the PHQ-9, the GAD-7, and the Sheehan disability scale. The data further showed more engagement in the message-based care condition, likely due to factors such as time convenience of the contact. Results also showed that participants in the video session group were more likely to miss a session than those in the messaging condition were likely to stop sending messages. These results, he said, emphasize the importance of a more equitable form of care delivery, although, he admitted, more research needs to be done.

An added value, Hull said, is reduction in stigma due to establishing a "consumer brand" whereby mental health is not seen as intimidating and/or frightening. He added, "people like to do business with brands that they know and trust."

The final panelist was Maninder "Mini" Kahlon, PhD, Vice Dean for the Health Ecosystem and Associate Professor of Population Health at Dell Medical School of the University of Texas in Austin, Texas, whose innovative mental health clinical model involves "sunshine calls," a program that provides "empathetic relationship-building phone calls" with the added value of engaging lay community members in the workforce and reaching young people. Lay community workers are assigned to call six to nine recruited 17- to 23-year-olds. Participants were homebound and all reported having at least one chronic health condition. This approach, similar to that described earlier by Dr. Miller, increases capacity of the workforce by recruiting lay community members.

A unique aspect about this intervention is that participants define the parameters of the contact (when to be called, if they want to talk when called) which Dr. Kahlon describes as the "north star" of the program. An assessment of the impact over the four-week program revealed reduction in the levels of depression and anxiety, and an average improvement in participants' feelings of loneliness of more than one point on a seven-point scale (The University of Texas at Austin: Dell Medical School, 2021).

Similar to Dr. Miller's earlier point, Dr. Kahlon claimed that the program overcomes disparity within the mental healthcare system and closes the treatment gap, by expanding the workforce to include lay people and those in the community. "I think there's a great potential for radical change, for a different kind of workforce where people are recruited for empathy," she said. The importance of establishing rapport, authenticity, and trust in the relationship between the healthcare provider and patient in these sunshine calls was emphasized.

While establishing the evidence-based gold standard for these technological applications, and identifying their long-term effects and generalizability, has yet to be established, much promise has already been shown in their bridging of the treatment gap and providing access and affordability to many persons in need.

Research and policy priorities in mental healthcare

The final hour-long segment of the conference was dedicated to discussion with a panel about how to create the evidence, research, and policy agenda needed to improve the current mental healthcare system and create a better future in mental healthcare.

The moderator, David Mandell, ScD, professor of Psychiatry at the University of Pennsylvania Perelman School of Medicine, posed a question: "Do you think there are policy, practice or research steps that we should be taking next?"

First to answer, Lynn Bufka, PhD, Assistant Executive Director of Practice Research and Policy of the American Psychological Association, acknowledged the current technological-based innovations for closing the mental healthcare treatment gap, but said they are not systematized to be accessible to people across the United States. She said, "We really need to look at how do you leverage these different kinds of innovations so that people can access the level of care that they need at the time that they need care, so that we're not having everyone get care at the crisis point or the point of diagnosis in extreme symptoms, but [instead] get care that might be appropriate for them at an earlier point in their process."

She added that the issue of payment must be addressed. The current mental healthcare system is largely a fee-for-service model, and insurance models may not cover the kinds of innovations discussed in this session.

Dr. Glenda Wrenn Gordon, MD, MSHP, President of Mindoula Clinical Services, a Maryland-based behavioral health management company, pointed out the importance of equity when discussing such healthcare services, with everyone having access to the same level of services.

"Over 1.5 billion dollars was invested in mental health in 2020 alone, so there is a hunger for solutions in this space," she said, but there has been a great divide between what is considered usual care versus innovations in providing care. Closing that gap in an equitable way, she argued, requires "looking at outcomes but [also…] trying to figure how we really come up with solutions that are sustainable…in terms of having our fundamental system evolve in payment reform."

"About half of our behavioral health workforce is completely opting out of the insurance frame," she stated, which complicates the problem of access and insists that we find a way to make a trained workforce more equitably accessible to people who need care.

Victoria Soltis-Jarrett, PhD, PMHNP-BC, FAANP, FAAN, who is the Carol Morde Ross Distinguished Professor of Psychiatric Mental Health Nursing at the University of North Carolina at Chapel Hill School of Nursing in North Carolina, a private group practice, echoed the points made by Dr. Bufka and Dr. Wrenn Gordon about accessibility being key.

"Making sure that federally-qualified health centers are not left out because of what's happened previously with the Cares Act is important," she said. "We need to make sure that all legislation moving forward considers those that need it the most."

"Accountability matters", she added. "That focuses on who's holding people accountable for monies that are apportioned." There is also a need to focus on reimbursement for these innovations.

Cecilia Livesey, PhD, Chief Mental Health Officer at UnitedHealth Group and Assistant Professor of Psychiatry at the University of Pennsylvania, echoed that such new technology tools can provide equity, empathy, accessibility, and "a new workforce of volunteers," which then "allows clinicians to work at the top of their license." New tech applications, she said, provide "care personalization" by identifying the best type of care, provider, and medium for each patient.

Dr. Mandell posed the question: "If we are very excited by the promise of remote care telehealth—either synchronous or asynchronous—and we needed to make changes right now to initiate that, to make sure gains that happened during the pandemic because of the relaxation of various regulations weren't lost, what would you want to see put in place?"

Dr. Bufka answered, "Number one, you have to have payment equity. You cannot have differential payment for remote versus in-person care. Most clinicians providing mental health services don't see a difference in the cost of providing care in-person versus online." Thus, there should be equitable payment across the kinds of care provided.

Another payment problem is that "We don't currently have payment models for other kinds of technology care, and it often falls on the burden of the person seeking it whether it's the asynchronous apps or different kinds of providers who aren't within the reimbursable network."

As a result, she would like to get rid of the fee-for-service mental healthcare model. As long as that system is in place, more codes will be needed to reflect the range of care that different providers can provide. "We need to have codes that reflect better the kinds of services and the depth of services that people are providing as well," she said, including tele-mental health. She explained that the American Psychological Association (APA) is trying to expand the codes to be available, appropriate, and accessible for psychologists and other providers who aren't physicians. However, all this work is happening within the fee-for-service model; but, to do the kinds of innovative care discussed in this session, she said, we need to have access to models beyond that.

Dr. Livesey added that we're moving in the right direction with collaborative care codes which pay for team-based services and allow involvement of different types of providers.

Another question was posed to the panel: "What exactly is the right type of mental healthcare?"

Dr. Livesey answered, "Getting people the right care at the right time with the right provider at the right intensity."

Dr. Soltis-Jarrett answered that not-for-profit healthcare systems are important, clarifying that, "We really have to look at healthcare as not being a profit-based entity. When it becomes for-profit, then what we end up finding out is that providers are trying to do it smarter and see more patients." But, quantity, so often prioritized in profit services, she stressed, cannot compromise quality.

"I think there's a balance between non-profit and for-profit," Dr. Wrenn Gordon was quick to reply. "Some benefits are that they [profit] can sort of make these investments to bring things to scale; you just have to know where to look and you have to know who to have a seat at the table thinking about gender differences, access to care and acceptability to care."

To achieve equity, she said, requires to "include people [in the population you're interested in] in the design process, informing how you're developing it." Even while these technological innovations need to engage people in the design so that they are acceptable, she pointed out, unfortunately, "In the US, we are not able to get self-reported racial, ethnicity, sexual orientation, and gender identity data."

The next question posed to the panel was: "What policies and practices do we need to put in place to do a better job of speeding up the use of these interventions in practice?"

Dr. Livesey responded, "First and foremost is efficacy." Data needs to be transparent, with practices put in an ethical framework, conscious of bias when creating "machine learning loops" in these new innovative technologies that could potentially strengthen bias. "Data touches on all facets of healthcare in terms of research and policy and payment, being efficacy data and value data... especially from an equity standpoint."

Dr. Bufka answered by pointing out challenges of research and the importance of not being paralyzed by not having perfect data. "On the academic side, we struggle so much. We want to have the perfect data, we want to have all the questions answered before we move forward and we can't." This is especially true with technology, she explained, because technology is going to change faster than research is going to get done and out there. To move ahead with tech solutions, we need to be sure that our data is sufficient and has enough rigor and acceptability, while at the same time, we need to acknowledge where we need to build.

Instead of focusing on all the limitations, she said, we need to take important findings and put them into action without all the qualifications, which prevent us from making steps toward change. She concluded, "We can't let the notion of searching for perfect data get us to the point that we don't do anything policy wise."

Dr. Wrenn Gordon responded that "We've come so far in a short time of adopting measurement-based care and behavioral health." She elaborated that when accountable care organizations (ACOs) first emerged in health, behavioral health was often left out of the equation because our measures were process measures. Referring to a short depression assessment, she said, "Now, at least we have the PHQ-9. Everyone sort of knows what that is now and it's a good leading indicator. It's not perfect, but I'm glad that we're actually using these measures…which will be 'the great equalizer' to answer the question of whether or not you can get effective outcomes with the lay provider."

The next question posed to the panel was about lay workers being used in tech innovation. The panelists agreed that there's much need for personnel. Lay personnel can be recruited in some tech applications, Dr. Wrenn Gordon said, imploring lay personnel to "Please, take the piece that you can do and do your part because our society needs mental healthcare across the full spectrum."

Reflecting on the overall topic, Dr. Wrenn Gordon added that while the innovations presented in this session may not be easy to implement, quality measures that are actual, functional and clinical outcomes can continue to show the value of these innovative approaches, which will in turn help set policies to see what should be amplified in scale.

Dr. Soltis-Jarrett emphasized again the importance of participatory research—which is engagement of communities—to empower people to be part of the discussion, despite differences between communities.

The next question to the panel addressed the lack of a diagnostic system in some of these models. Some panelists felt a diagnosis may not be necessary. While getting a traditional diagnostic label can "open doors and opportunities for people to normalize the experience of talking to a professional," the flipside of the coin is that without a diagnosis, you could get support and develop strategies and skills to better handle different adversities of life before issues become labeled as disorders.

To this point, Dr. Wrenn Gordon advised a "population health" lens, so that some interventions mentioned using technology would happen without going through the traditional multilevel complicated intake process.

Also, she mentioned that payment through tech tools can follow different models than in-person. In-person payment needs to be streamlined, with less time spent on filling out detailed paperwork and with reimbursement understood. Any system needs infrastructure through ongoing technical support, especially for complicated billing codes.

Further, the recommendation was made to use the practice of "precision engagement," whereby types and frequency of engagement are not generalized [activity] but tailored to different populations, which would make healthcare more effective and efficient.

The moderator asked the panelists about what regulations and laws are needed.

Dr. Livesey recommended keeping in place the telehealth emergency order so that there can be telehealth across state lines, saying, "That alone would be a huge regulatory win, I think."

In bringing up barriers, Dr. Wrenn Gordon said a lot of engagement is required at the federal, state, and micro policy level because everything impacts barriers to access quality mental healthcare.

Medications could be a further problem in tele-mental health. Given her experience serving on her state's Medicaid Drug Utilization Review Board, Dr. Wrenn Gordon said she often finds herself questioning so many prior authorizations required for first-line medication treatments for serious mental illness, leading to her efforts to educate other board members about why this needs to change.

Age also applies to tele-mental health. Dr. Soltis-Jarrett pointed out that many older adults (and others) do not have access to video conferencing or a smartphone for video access. Thus, audio has value, consistent with the fact that older people tend to use a regular older-style simple phone device anyway. Thus, the "Sunshine" innovation that connects via a simple phone call is valuable to improve access to this older population, and potentially to others. Dr. Bufka added that audio-only sessions (versus video) are also important for privacy reasons. "I have several patients who have access to video conferencing but no privacy for it, so that's another reason why audio only is very essential." The data supports that audio support by phone is effective, she said.

But training and needs of providers in order to effectively deliver care using different modalities like audio or video are important, she added.

The panelists responded to questions in the chat box. One question asked about innovative reforms and how to pay for delivery of care. Dr. Wrenn Gordon spoke about her work in the mental health program of the First Lady of New York City, Chirlane McCray. She said this program demonstrates the power of mayors, evident in the "Cities Thrive" network of elected officials who are trying to think across multisectoral systems, and to identify how healthcare intersects with education, labor, and transportation in a population perspective.

Technology solutions have to fit into the system of mental health first aid, Dr. Wrenn Gordon responded, namely, within "that continuum of care from just living your life to 'oh I have a problem' to 'what do I do now?' to making sure that there's resources within the community to help people access care."

States have been looking at how to implement collaborative care across the state, and how to be innovative within Medicaid and Medicaid expansion to try to make it more accessible, she said.

Working with the Veterans Affairs (VA), Dr. Bufka said, has been a good model of continuity of care with providers including mental health specialists who can communicate and engage with each other in a way that still protects patient privacy and confidentiality. Other key elements, she mentioned, are making any technology to be used affordable and also for people to be able to use it effectively. In summary, she suggested finding ways to capitalize on the promise that electronic health records would bring to all providers, so they can all communicate with one another about anything related to the patient's health.

The final question posed to the panelists was, "If you had to take one lesson that we learned from this pandemic about issues related to mental health and mental healthcare and carry it forward in policy post pandemic, what would that lesson be?"

Dr. Livesey answered, "We were constantly moving and evolving based on context, kind of flexing to individual needs, being part of a fluid system and sort of a force for change in healthcare and society." She also briefly mentioned the speed with which places were able to test, iterate, and change the system during COVID-19, yet speed can prevent listening to all voices, especially those who are marginalized. It has been, and will continue to be, important, she said, "to remain patient, focused, and meeting people where they are rather than use a one-size-fits-all solution."

Dr. Bufka said that an important lesson learned from COVID is "the fragility of health and crumbling of support in skills and structures that we hoped people had when everybody is impacted by something, like the pandemic."

Dr. Soltis-Jarrett mentioned three lessons learned. They are: (1) We need to give up the phrase 'We've always done it this way,' in order to be more open to flexibility and change; (2) It's important to achieve health literacy and educate patients and clients about their options and explain it in a way that they understand; and, (3) Barriers to care must be removed.

She hopes that everyone from the event leaves thinking about how can we really make the most of this opportunity to reform the mental healthcare system, using the technological approaches discussed in this event or others.

Context of the UN Global Agenda

The problem of mental healthcare, and mental health in general, is linked to the Sustainable Development Goals (SDGs) of the United Nations Agenda 2030 for Sustainable Development. Adopted in 2015, this agenda is a framework agreed upon by all countries—developed and developing—for the people and planet to live in global partnership (The 17 Goals—United Nations Sustainable Development, 2022). Mental health is integrated into SDG 3 which calls for ensuring healthy lives and promoting well-being for all at all ages (Goal 3: Ensure Healthy Lives—United Nations Sustainable Development, 2022). The specific target in SDG 3 that cites mental health is target 3.4, which calls for the promoton of mental health and well-being (United Nations, 2022). An indicator for this target is the reduction in the suicide mortality rate.

Mental health was also included in the Universal Health Coverage Political Declaration, adopted by the governments of the United Nations on 23 September 2019. Emphasizing the need to move together to achieve a healthier world, the declaration reaffirms "the right of every human being, without distinction of any kind, to the enjoyment of the highest attainable standard of physical and mental health" (United Nations, 2019).

Summary reflections

While we still have a long way to go for an efficient and effective mental health system, innovations presented in this event, using different technological tools and unique approaches, are promising. Such innovations show that care for mental health and well-being needs to be, and can be, reimagined and that solutions from researchers, clinicians, academicians and even the private sector are valuable. New models using technology are proving to be effective, despite limitations. Certainly more research about outcomes is needed and approaches need to be refined, but given the extent of people's needs, applications and methodologies using technology offer opportunities to close the treatment gap, and more options for making care accessible to more people around the world.

References

Beyer, J. L., & Boazak, M. (2021). The mental healthcare system: Organization and structure. In J. D. Tenenbaum, & P. A. Ranallo (Eds.), *Mental health informatics* (pp. 81–96). Cham: Springer. https://doi.org/10.1007/978-3-030-70558-9_4.

Goal 3: Ensure Healthy Lives—United Nations Sustainable Development. (2022). https://sdgs.un.org/goals/goal3.

Grande, D., Miller, B. F., Wolk, C. B., Comer, J. S., Hull, T. D., et al. (2021). *Mental health care in flux: Exploring new ways to deliver care* (Virtual Conference). Penn Leonard Davis Institute of Health Economics. https://ldi.upenn.edu/our-work/research-updates/strategies-to-repair-a-broken-system-mental-health-care/.

Insel, T. R. (2019). Bending the curve for mental health: Technology for a public health approach. *American Journal of Public Health*, *109*(53), S168–S170.

Kazdin, A. E. (2017). Addressing the treatment gap: A key challenge for extending evidence-based psychosocial interventions. *Behaviour Research and Therapy*, *88*, 7–18.

Khan, K. S., Mamun, M. A., Griffiths, M. D., & Ullah, I. (2022). The mental health impact of the covid-19 pandemic across different cohorts. *International Journal of Mental Health and Addiction*, *20*(1), 380–386. https://doi.org/10.1007/s11469-020-00367-0.

Moock, J. (2014). Support from the internet for individuals with mental disorders: Advantages and disadvantages of e-mental health service delivery. *Frontiers in Public Health*, *2*, 65. https://doi.org/10.3389/fpubh.2014.00065.

Moreno, C., Wykes, T., Galderisi, S., Nordentoft, M., Crossley, N., Jones, N., et al. (2020). How mental health care should change as a consequence of the covid-19 pandemic. *The Lancet Psychiatry*, *7*(9), 813–824. https://doi.org/10.1016/S2215-0366(20)30307-2.

The 17 Goals—United Nations Sustainable Development. (2022). https://sdgs.un.org/goals.

The University of Texas at Austin: Dell Medical School. (2021, February 23). *Empathy-focused phone calls alleviate loneliness, depression, anxiety during covid-19 pandemic*. https://dellmed.utexas.edu/news/empathy-focused-phone-calls-alleviate-loneliness-depression-anxiety-during-covid-19-pandemic.

United Nations. (2019). *Political declaration of the high-level meeting on universal health coverage*. https://undocs.org/en/A/74/L.4.

United Nations. (2020). *Policy brief: Covid-19 and the need for action on mental health*. https://www.un.org/sites/un2.un.org/files/un_policy_brief-covid_and_mental_health_final.pdf.

United Nations. (2022). *Goal 3: Ensure healthy lives and promote well-being for all at all ages*. https://unstats.un.org/sdgs/metadata/Text1⁄4&Goal143&Target1⁄43.4.

Well Being Trust. (2022). *Healing the nation: Advancing mental health and addiction policy*. https://wellbeingtrust.org/areas-of-focus/policy-and-advocacy/.

Werlen, L., Puhan, M. A., Landolt, M. A., & Mohler-Kuo, M. (2020). Mind the treatment gap: The prevalence of common mental disorder symptoms, risky substance use and service utilization among young Swiss adults. *BMC Public Health*, *20*(1), 1–10. https://doi.org/10.1186/s12889-020-09577-6.

World Health Organization. (2018). *Mental health: Strengthening our response*. https://www.who.int/en/news-room/fact-sheets/detail/mental-health-strengthening-our-response.

CHAPTER 80

Transforming mental health for all: Launch of the 2022 World Mental Health Report

Dasoo Milton Yoon[a,b,c] **and Judy Kuriansky**[b,c,d,e]

[a]Human Development and Family Studies, University of Wisconsin-Madison, Madison, WI, United States, [b]Department of Clinical Psychology, Teachers College, Columbia University, New York, NY, United States, [c]International Association of Applied Psychology, Indianapolis, IN, United States, [d]United Nations Group of Friends of Mental Health and Well-being, New York, NY, United States, [e]Global Council SDG3 (Health), "SDGs in Action" Initiative, Dubai, United Arab Emirates

Background

the COVID-19 pandemic on modern life worldwide is the huge burden the disease has inflicted, and continues to plague, on the mental health of people of all ages universally. Rates of already widespread conditions, such as anxiety and depression, rose by more than 25% in the first year of the global pandemic, reaching a point where about one in eight people in the world live with a mental disorder (World Health Organization, 2022b). Yet, these mental health conditions are severely underserved as health systems all over the world are marked by major gaps and inequities. On average, countries allocate less than 2% of healthcare budgets to mental health, and around half of the global population lives where just one psychiatrist serves 200,000 or more people (World Health Organization, 2022b). Thus, most mental health conditions go completely untreated and people are suffering.

To address this serious issue, and to inspire and inform better mental health for all, the World Health Organization (WHO) published the World Mental Health Report on 17 June 2022. This chapter presents an overview of the report and its launch event with high-level speakers along with voices of experts and those with lived experience of mental health issues. The launch of the WHO report supports the WHO's goal to transform attitudes, to motivate actions and approaches to promote and protect mental health, and to provide care for those in need, in alignment with the global objectives set out in the WHO's Comprehensive Mental Health

Resilient Health. https://doi.org/10.1016/B978-0-443-18529-8.00080-9

Action Plan (2013–2030) and the United Nations Sustainable Development Goals as outlined in the UN Agenda 2030 for Sustainable Development.

Introduction

Despite the fact that mental health is an intrinsic part of individual and collective health and well-being, counterproductive negative attitudes and stigma persist, and services are severely insufficient, inaccessible, and inadequate. Highlighting why change is urgently needed and how it can best be achieved, the newly released World Mental Health Report draws on the latest evidence and data, raises the voices of those with lived experience, and showcases examples of best practices from nations in different parts of the world. The report calls on all stakeholders to work together to deepen the commitment and value given to mental health, reshape environments that influence mental health, and strengthen the infrastructure and systems that provide help. The front cover of the report is shown in Fig. 1.

Fig. 1
The front cover of the World Mental Health Report.

A momentous step forward in the efforts to improve mental health for all, the World Mental Health Report is the most extensive review of world mental health since the turn of the century. The detailed document provides a blueprint for governments, academics, health professionals, civil society, and others that support the transformation of mental health worldwide. The ultimate vision of the report is to create a world where mental health is valued, promoted, and protected; mental health conditions are prevented; anyone can exercise their human rights and access affordable, quality mental healthcare; and everyone can participate fully in a society free from stigma and discrimination.

The launch of the report took place on 17 June 2022 (World Health Organization, 2022a). Due to the ongoing restrictions of the COVID-19 pandemic, the event was conducted online over Zoom, with registered viewers able to follow the presentations through Zoom and YouTube, and enter reactions, comments, and questions. Simultaneous translation was available in all five official UN languages.

Overview of the report

The centerpiece of the event was the overview and discussion of the report, led by the Director of the Department of Mental Health and Substance Use at the WHO, Ms. Dvora Kestel.

The Report was designed to support the global transformation of mental health that has long been called for. It aims to strengthen the value and commitment to mental health as a core contributor to not only health but also to social well-being and economic development. Through such efforts, the WHO hopes to inspire a change in attitudes, approaches, and actions toward better mental health. The report draws on the latest available evidence to highlight why and where change is needed and how it can be achieved on the ground. While intended to make a positive impact on multiple sectors, the report was especially designed for decision-makers in the health sector, i.e., those tasked with developing mental health policy and delivering mental health systems and services (World Health Organization, 2022b).

The precedent for this Report dates back to 2001, Ms. Kestel described, which marked the first time that the WHO produced a global report on mental health, which—at the time—generated momentum to increase attention and awareness of the dimensions of the problems and challenges, and also included recommendations to push forward on a global scale. Fast forward two decades to 2022, she said, and the same recommendations are still valid. While some progress has been made, she added, the general consensus—backed by data—is that we still need a change in mental health. In short, despite efforts to raise awareness, tangible progress has been slow. But the good news is that progress is evident in the fact that the UN Sustainable Development Goals includes the promotion of mental health and well-being in target 3.4, and that the WHO Comprehensive Mental Health Action Plan—which was endorsed by all UN member states in 2013—has been renewed over the past few decades, among other targets, actions, and commitments that have been identified. Additionally, mental health has been recognized in other international documents, notably on Universal Health Coverage as well as policy statements by the UN Secretary-General about the importance of mental health, in reaction to the pandemic.

The COVID-19 pandemic tragically increased the number of people suffering from mental health problems, Ms. Kestel pointed out (as did other speakers in the panel), but on a positive note, the worldwide pandemic increased the visibility and awareness of mental health. This situation led the WHO to decide that this was the opportune moment to bring all this information together and publicly present the current status of mental health in the report.

Three key paths to action are set forth by the report, as shown in Fig. 2.

As explained by Ms. Kestel, and shown in Fig. 2, the three paths are as follows:

(1) *To deepen the value and commitment we give to mental health.*
Mental health must be valued as an integral part of health. The process starts with all individuals taking care of ourselves, dedicating time and paying attention to our own mental well-being, and also being aware of those around us. This acceptance must then be extended to the family and the community, until it is accepted as a basic human right. By introducing this value, all stakeholders will commit to taking action, with more resources and investments.

(2) *To reshape environments that influence mental health.*
This path is focused on protecting mental health, as well as preventiing mental health conditions. For people already suffering from mental health conditions, this translates into an issue of equal opportunity—to live, learn, work, have fun, and be equally a part of the community and society. To do this, we need to understand the structure of mental health and need to reduce risks whenever possible, in order to build resilience and eliminate barriers for those with mental health conditions to access care they need. Adequate policies on a larger scale also need to be put into place, and areas outside of healthcare also need to be involved, given that spaces such as the home, workplace, and schools influence mental health, either negatively or positively.

(3) To develop and strengthen community-based mental health services
The World Mental Health Report urges that we need to change where, how, and by whom mental health services are offered. A movement toward community-based mental health services is crucial, grounded in the culture, context, and conditions ofeach country. Also, such services must be open and accessible to all people in need, including for persons with severe mental health conditions, and also, importantly, people with common mental health conditions who frequently are not considered a responsibility of mental health structure today, but should be. The Report lists an extensive network of mental health services that should be established, startingwith the integration of mental health into primary healthcare and the health system in general, since mental health cannot be continued to be seen as an isolated and segregated area. Further, a more comprehensive network must be established through the incorporation of other sectors, such as education, justice, sports, culture,

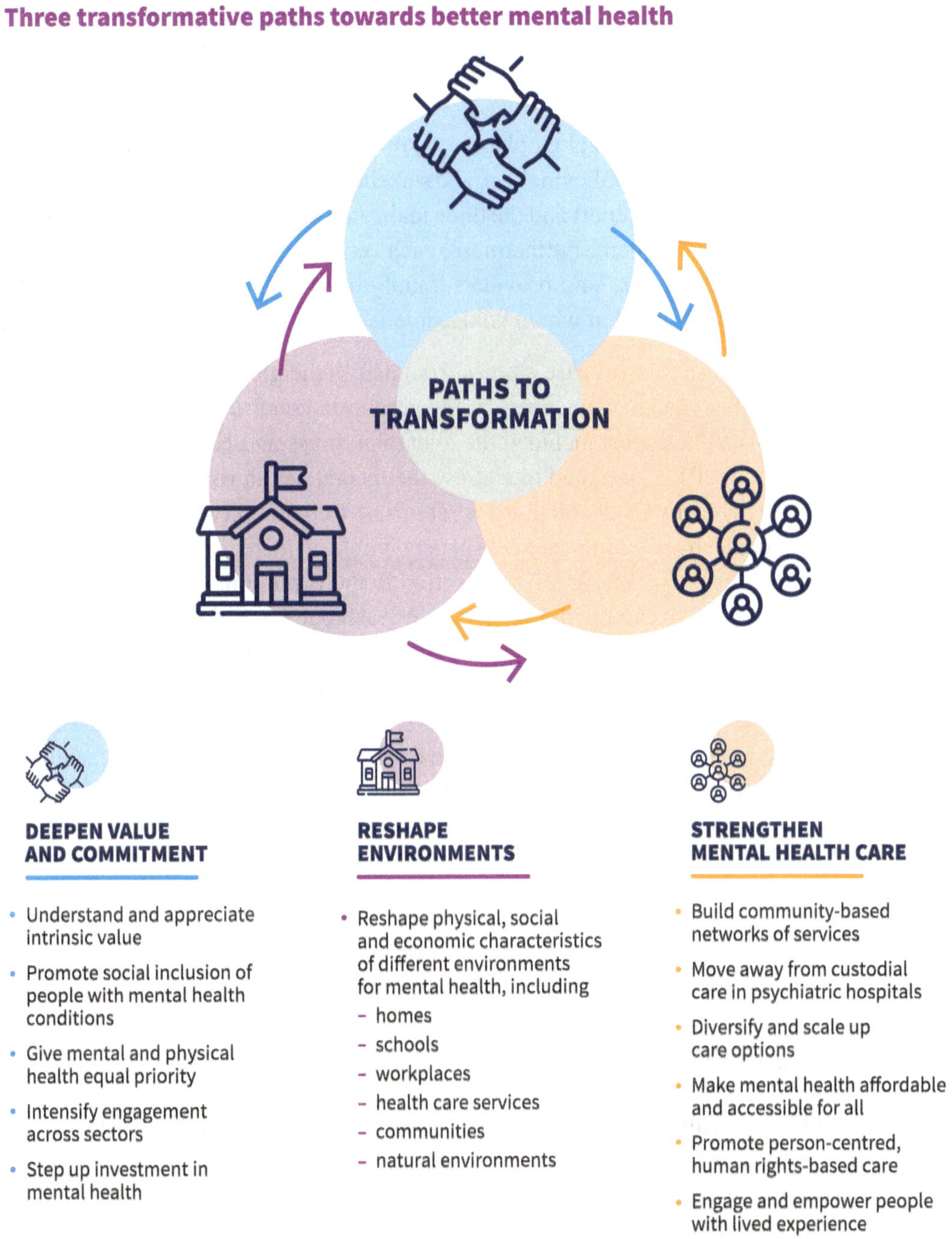

Fig. 2
The report's three proposed paths to transforming mental health.

prison systems, and housing, since all these should, will, and do have a role to play. All such sectors should come together to form a network of services of mental healthcare that each community and country should develop according to its own possibilities and community needs.

In her concluding remarks, Ms. Kestel emphasized how these paths to transformation are all linked together. Once authorities, policymakers, and stakeholders decide it is time to change the situation, they can reference the Report and the three main paths to decide what to prioritize and how to strategize their development. Furthermore, each sector has a crucial role to play in achieving these paths: individuals, care providers, family members, the community, NGOs, academia, and the health sector, all of whom hold a unique role in achieving the paths outlined.

The Report also includes successful case studies from across the globe, which are well summarized to prove that change canhappen on the ground in countries, no matter the conditions or challenges. Examples included the Vidarbha Stress and Health Program (VISHRAM) of India in 2014. Designed to address the mental health risk factors for suicide, VISHRAM utilized a tiered model of collaborative care where community health workers acted as the first point of contact by reaching out to the general public to increase awareness of mental health conditions and services. During this connection to the community, they also helped identify people with symptoms and persuaded them to reach out for professional help. By the end of the 18-month duration of the project, the prevalence of suicidal thoughts among the target population had halved, the prevalence of depression had fallen by 22%, and a sixfold increase emerged in people seeking mental healthcare. Another example was a process of mental health reform in Peru in 2013, which focused on shifting the focus of mental healthcare from specialized psychiatric hospitals to community-based care. Findings showed that the reform supported a gradual and significant expansion in mental health services, to the point where 26% of people who needed care were covered by 2018, compared to 9% 10 years earlier in 2009 (World Health Organization, 2022b).

Ms. Kestel hopes that this section of the Report will encourage others to learn from existing experiences about what is being done successfully in countries, and how to replicate such efforts.

Details of the launch event

The launch event began with opening remarks by Ms. Claudia Hammond, a British broadcaster and writer specializing in the topic of mental health. She reflected on the cultural and societal developments around mental health that she has observed throughout her long career, summarized in two observations: (1) people with mental health conditions are now more open to using their real names when she interviews them for articles, and (2) that these persons believe that they have a genuine and influential voice when telling their story. It is essential for the world to take these issues seriously if we are to flourish, she said, especially now that there is a blueprint, with the release of the World Mental Health Report as the most comprehensive report on mental health since the turn of the century. She hoped that this Report will urge mental health decision-makers and advocates to step up commitment levels and action, while changing attitudes and approaches to mental health. Following her, remarks, she introduced the speakers and panelists.

The flyer for the event is shown in Fig. 3, showing the logistics of the webinar event, the speakers, and their affiliations.

Key speakers were stakeholders representing different constituencies, including:

United Nations high-level representatives:

- H.E. António Guterres, United Nations Secretary-General (video message)
- Dr. Tedros Ghebreyesus, Director-General of the WHO

Government:

- Mr. Alberto Fernández, President of the Republic of Argentina (video message)
- Mrs. Neo Masisi, First Lady of the Republic of Botswana
- Mr. Andrew Little, Minister of Health of New Zealand
- Ms. Bjørg Sandkjær, State Secretary of International Development of Norway

International Organizations

- Ms. Dévora Kestel, Director of the Department of Mental Health and Substance Use at the WHO headquarters in Geneva, Switzerland
- Ms. Anju Joy, youth speaker who is Chief Programs Officer of Generation Mental Health

Persons with lived experiences, a constituency that is increasingly honored at such events:

- Dr. Ahmed Hankir, MD, psychiatrist and person with lived experience
- Ms. Enoch Li, person with lived experience

Performance in Cultural Interlude:

- Ms. Juliana Velásquez, 23-year-old Colombian singer-songwriter

The first speaker was Mr. Alberto Fernández, President of the Republic of Argentina. He shared the development of mental health in his country, ranging from reflecting on experiences in the past to changes expected in the near future. The topic of mental health has increasingly become present in daily life, he said, especially in response to the pandemic which caused uncertainty, loss, stress, suffering, and discomfort both individually and collectively, leading to a steep rise in clinical cases in the country. As a result, the government's goal is to work toward complete recovery of people's psychosocial well-being, as well as transforming mental health for everyone. Given his belief that "there are not any vaccines for mental health, but only actions," the government of Argentina is launching an integrated federal strategy to improve mental health in a comprehensive, collaborative effort, including the Ministry of Health, provinces, ministries, scientists, civilians, and the media. He affirmed that mental healthcare is a right, free of violence, and with respect for diversity.

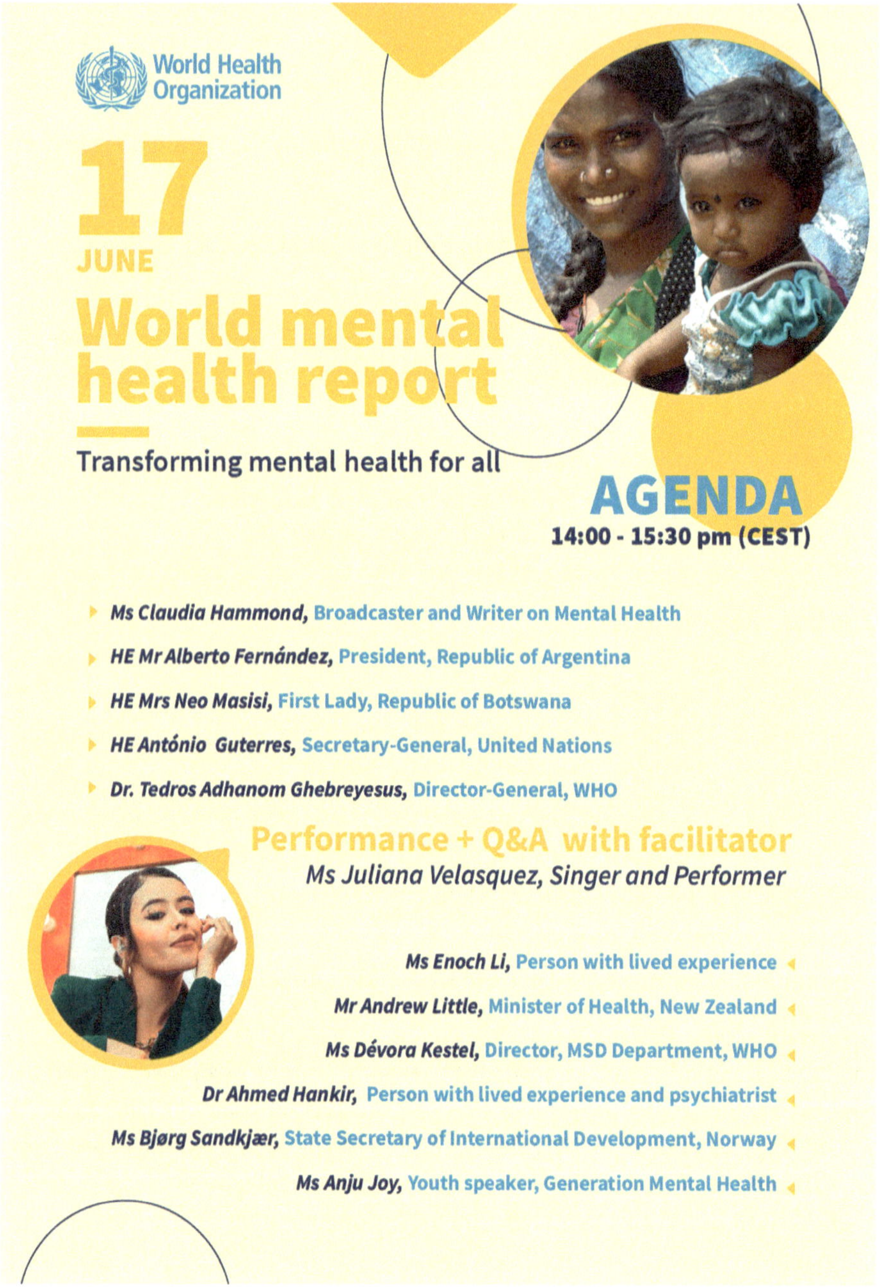

Fig. 3
Agenda and list of speakers of the launching event.

The next speaker was the First Lady of the Republic of Botswana, Mrs. Neo Masisi. She emphasized the importance of fully understanding the presently pervasive values of mental health in order to truly transform them, which range from embedded traditional and cultural norms to peer and societal prejudice. An essential step to combat such stigma and misperceptions is to genuinely appreciate mental health, she said, where everyone accepts mental health as a part of daily struggles that everyone encounters. Coupled with striving to become an accompanying and empathetic community, this approach will transform how mental health is perceived and addressed. This change, in turn, will influence the way the world defines and plans for mental health. She emphasized that an all-of-government and all-of-society approach is the only way to bring this vision into reality. An example of a progressive effort in transforming mental health is the push by the Republic of Botswana for a comprehensive draft law ensuring the quality of mental health services and protecting the rights of people who require mental health services. She ended by emphasizing how happy and passionate she was about using her voice to support efforts that can help transform mental health.

The Secretary-General (SG) of the United Nations, H.E. António Guterres, was the next speaker. He called attention to the fact that the world is currently living through a global mental health crisis, as nearly a billion people currently have a mental health condition, most of them lacking access to treatment. These individuals are also at an increased risk of physical and emotional abuse, denial of education and employment, and other human rights violations. The costs—both human and financial—are immense, as depression and anxiety cost the global economy an estimated $1 trillion each year. But, as he has been saying in many speeches about COVID-19, while the pandemic has increased mental health problems (as well as other socioeconomic devastation), the pandemic has also led to a "silver lining" of a far greater understanding of the importance of positive mental health. Yet, mental health remains the most neglected area of health policy in most nations. As a response, he officially introduced the WHO's World Mental Health Report as a roadmap to guide countries in improving their mental health systems, by highlighting where improvements are and how other advances can be achieved. The Report outlines how to reduce risks, build resilience, and dismantle barriers that prevent people with mental health conditions from participating fully in society. He concluded by recommending the Report to governments and mental health stakeholders everywhere.

The Director-General of the WHO, Dr. Tedros Adhanom Ghebreyesus, began his presentation by agreeing with SG Guterres by calling attention to the huge toll the pandemic has taken on people's mental health, yet noting that from a positive point of view, that this has underscored the integral link between mental and physical health. The pandemic has also highlighted gaps in the capacity of health systems around the world to address mental health conditions. Yet, most unfortunately, investment in mental health remains low despite the fact that the number of cases of mental ill health and the incidence of stigma remain high, and the fact that mental health conditions remain the main contributor to the global burden of disease, In fact, this lack of financial investment is so dramatic, that on average, countries spend on average less than 2% of their health budget on mental health. Exacerbating the serious treatment gap is the fact that despite the need, in some countries, there are only two mental health workers for every 100,000 citizens. This treatment gap can be reduced, he said, through the three key developments presented ealier which are set forth in this Report reiteratng the need to:

(1) deepen the value and commitment we give to mental health
(2) reshape environments that influence mental health
(3) develop and strengthen community-based mental health services

He reinforced the point made by Ms. Kestel that the Report presents examples of positive change from countries in different regions across the globe, showing how every country, no matter its situation, can significantly improve the mental health of its citizens, both adults and children. He pointed out the links between mental health, public health, human rights, and socioeconomic development, and expressed hope that transforming policy and practice in mental health can deliver real, substantial benefits for individuals, communities, and countries everywhere. Ultimately, "investing in mental health is investing in a better future for everyone," he said, ending with the commonly repeated phrase that "There is no health without mental health."

In a cultural interlude, 25-year old Colombian singer-songwriter Juliana Velásquez, who had just won the Latin Grammy award for Best New Singer, performed a song about the difficulties for children to have a peaceful childhood. This message was inspired, she explained, by the tragedy of one of her students from her acting academy committing suicide, which caused her much pain and questioning, transformed her soul, and inspired her to want her music team and fellow musicians to join her in some action. She claims that artists have a voice and platform, as well as a responsibility, to share important messages and resources with the world. She wrote this song to help everyone who has never experienced a depressive episode understand that depression is an illness, and to put listeners in the shoes of those who have suffered through such feelings yet not had anyone to talk to about it. She concluded by saying that changing mental health is not the responsibility of just large organizations like the WHO, but belongs to all humans, as we can all save someone's life if we try through small actions.

Consistent with the personal appeal of Ms. Velásquez in her song, the next speaker, Ms. Enoch Li, spoke about her "lived experience"—personal accounts which have become an increasingly commonly added voice in such meetings as well as by advocacy groups. She described her life journey, from initially being dismissive of mental health problems to ironically getting a reversal "wake-up call" by experiencing depression and suicidality herself. From this experience, she founded a consulting and training company advocating for mental health at work and became a board member for United for Global Mental Health, a consortium dedicated to catalyzing global mental health through their expertise in policy, advocacy, and financing. Being on both sides of the fence – with her personal story and professional role – allowed her to experience the disconnect and misunderstanding surrounding mental health that grows into stigma and discrimination. Recalling her own misunderstandings about mental health and her own reluctance to get treatment due to self-stigma and not understanding what treatment meant, even when she was offered help from others, she now advocates for change. She claims that the world needs to completely change its attitude toward mental health, which requires becoming "kinder to ourselves and to each other," which in turn would make the world a better place.

Next, the Minister of Health of New Zealand, Mr. Andrew Little, spoke about the approach he has been taking to mental health in his country that is closely aligned with the direction championed in the World Mental Health Report. He pointed out the gap that mental health is still often seen as a taboo subject and unacknowledged as a legitimate health condition, in stark contrast to the large portion of the world's population who suffer. " We can, and must, do better," he said, and applauded the milestone effort by the WHO Report. He is encouraged that the direction outlined in the Report strongly aligns with the work currently being done in New Zealand where the government has been hoping to change the underfunded and undersupported mental health services. One of the many considerable efforts being made in the nation is a push to replace mental health legislation—that was seen as forward-thinking 30 years ago but that has not kept pace with modern changes—to better protect human rights. He remains hopeful that while there is plenty of work ahead, change is already imminent, and the commitment and momentum are already here. This WHO Report makes it clear that the world can, and must, do more to improve mental health, a direction directly adopted by New Zealand. He concluded with his strong belief that supporting people's well-being, both physical and mental, is vital to ensuring a strong, resilient, and healthy community.

The moderator, Ms. Hammond, then moved to the section of the event presenting the content of the Mental Health Report, presented by Dévora Kestel, as described earlier in this chapter.

Following this, Dr. Ahmed Hankir similarly spoke from his professional and personal point of view. As a professional, he is a psychiatrist and the Public Engagement and Education lead at the Institute of Mental Health in Nottingham, UK, but he noted that he is also a person with a mental health condition himself, who identifies more with the latter (as a person with "lived experiences" of a mental health condition) than with the former (as a professional in the mental health field). But drawing on both his personal and professional expertise, he emphasized that people with mental health conditions must have a real say in policy, that "we have a "voices, we can speak for ourselves, and we will be heard." In this vein, he stressed that it is crucial that persons with mental health conditions be involved in the design, development, and delivery of mental healthcare services at all levels, as this is the "right way" to transform psychiatric care. While empirical knowledge from data is crucial, he said, the insights and wisdom derived from "lived experience" are equally important. Recovery is possible for the many, rather than just for the few. Amplifying the voices of those with experienced will contribute to a cultural revolution that normalizes mental health conditions and empowers, dignifies, and humanizes living with mental health conditions. In this way, the world will be able to reduce now-prevalent and destructive mental health-related stigma. He called on those currently living with a mental health condition to be more involved in improving the landscape of mental healthcare.

Turning to another government perspective, the next speaker, Ms. Bjørg Sandkjær, State Secretary of International Development of Norway, talked about how mental health has been a neglected aspect of the development agenda. It is imperative, she said, that mental health issues be integrated and prioritized in our work in providing Universal Health Coverage and strong health systems. She highlighted how the World Mental Health Report presents solutions, supporting the fact that both a comprehensive and transformative approach—both within and outside the health sector—are required to provide good mental health for all. To accomplish this, we need to reshape environments that influence mental health and strengthen health systems to combat stigma and exclusion. Another point she emphasized is that mental health must be nurtured in all settings, at home, in school, at work, and in communities.

She also stressed that it is important to have a primary healthcare system for the prevention, diagnosis, and treatment of noncommunicable diseases (NCDs), including mental health (which has been recognized by WHO as an NCD). Unfortunately, the world is far from that comprehensive situation, especially in most low-income countries. In fact, most healthcare systems are unfortunately not equipped to provide good mental health services and psychosocial support, though they should be.

Ms. Sandkjær was proud that Norway supports the WHO's Special Initiative on Mental Health, a program that supports governments to develop strategies, policies, and capacities to provide professional mental health services for the population. Investing to transform mental health means investing in a better life and future for all. "Health is a human right," she emphasized, and her country Norway will continue to support the mental health agenda, including through efforts to promote Universal Health Coverage.

Representing the youth perspective—as is increasingly required in UN meetings—the next presenter, Ms. Anju Joy, spoke about how she spent over half a decade working to reform the public school system in India and is currently working at Generation Mental Health, a youth-based global mental health organization, in the role of Chief Program Officer. A fact that often goes completely under the radar and unacknowledged is the role of mental health in achievement, especially with regard to learning. During her experience teaching in public schools in India, her biggest observation was that students who struggled with emotional and mental health problems were also the ones struggling to access higher levels of academic achievement. Given the bidirectionality of this relationship—along with the amount of time young children and adolescents spend in school—it is important to integrate mental healthcare into education.

"[Improving] mental health is everybody's job," she emphasized; thus, teachers and other stakeholders in the education system all have a critical role in creating an inclusive environment for all. Orientation and training should be required for all stakeholders in education to be more knowledgeable and sensitive toward mental health issues, in order to provide safe learning spaces for children. "Kids need to first feel safe and loved in order to learn," she said. This mindset will transform the way we look at classrooms and education. By doing so, we will create a generation of individuals who feel empowered to make decisions to improve mental health. On a positive note, in her career, she has also seen greater recognition of issues regarding mental health among the younger generation.

Following these presentations, the moderator Ms. Hammond invited Ms. Kestel to offer concluding thoughts. Ms. Kestel mentioned that the main focus of the Report was to help those who were trying to make a change in mental health by supporting them with ways to facilitate and support their efforts. With this Report, she hopes that a future will come where:

(1) Mental health care is offered and available for everybody who needs it at the community level.
(2) Places of isolation, segregation, and punishment due to mental health conditions become nonexistent, and human rights violations are not associated with mental health conditions.
(3) People with mental health conditions speak openly everywhere about their challenges and achievements and will have a voice in every decision.
(4) Anybody, at any place, at any time, is able to go to a healthcare provider to talk about their mental health needs or issues and will be listened to, and that whoever will be on the other side will know what to do.
(5) For all this to become a "reality, and not just slogans."

The launch event ended with Ms. Hammond saying, "We must, and we can, do better for the world and its mental health."

To supplement this launch event and release of this World Mental Health Report, other mediums of promoting mental health have been put in place. One of these is a webpage related to the Report, which includes a brief summary of the Report and resources that supplement its content, including multimedia files, well-organized graphics on some key topics covered during the event, and infographics on country case studies representing successful governmental efforts by nations all over the globe to improve mental health. Examples of these are shown in Figs. 4–6 from Brazil, Singapore, and Türkiye, respectively. The webpage also includes feature stories from all over the globe, from persons living with mental health conditions. Topics covered in these anecdotes include:

- Reshaping work environments to promote and protect mental health
- The impact of COVID-19 on mental health

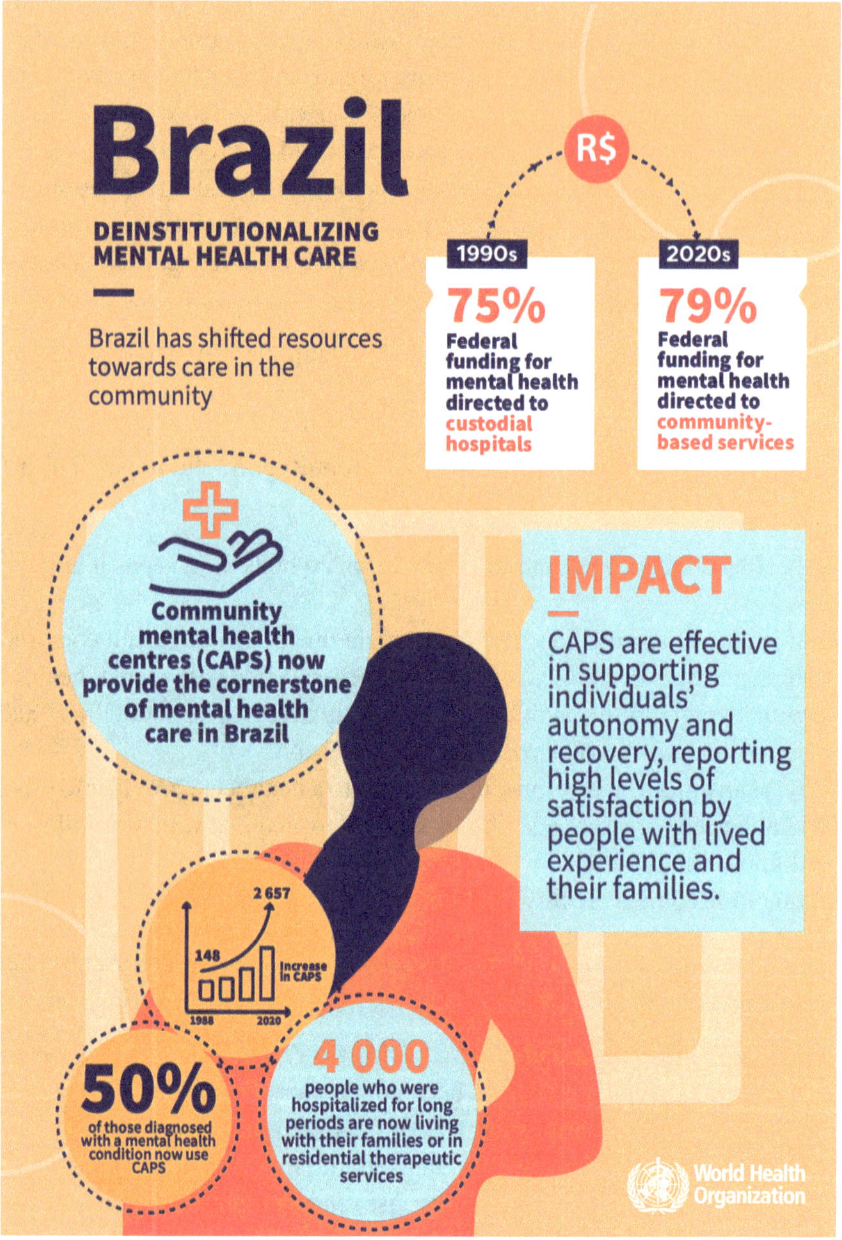

Fig. 4
National efforts to improve mental health in Brazil.

- Autonomy in health decision-making as a key to recovery in mental healthcare
- Sparking a renaissance in mental health through strengthening community-based care for all in need
- Tackling stigma through social contact

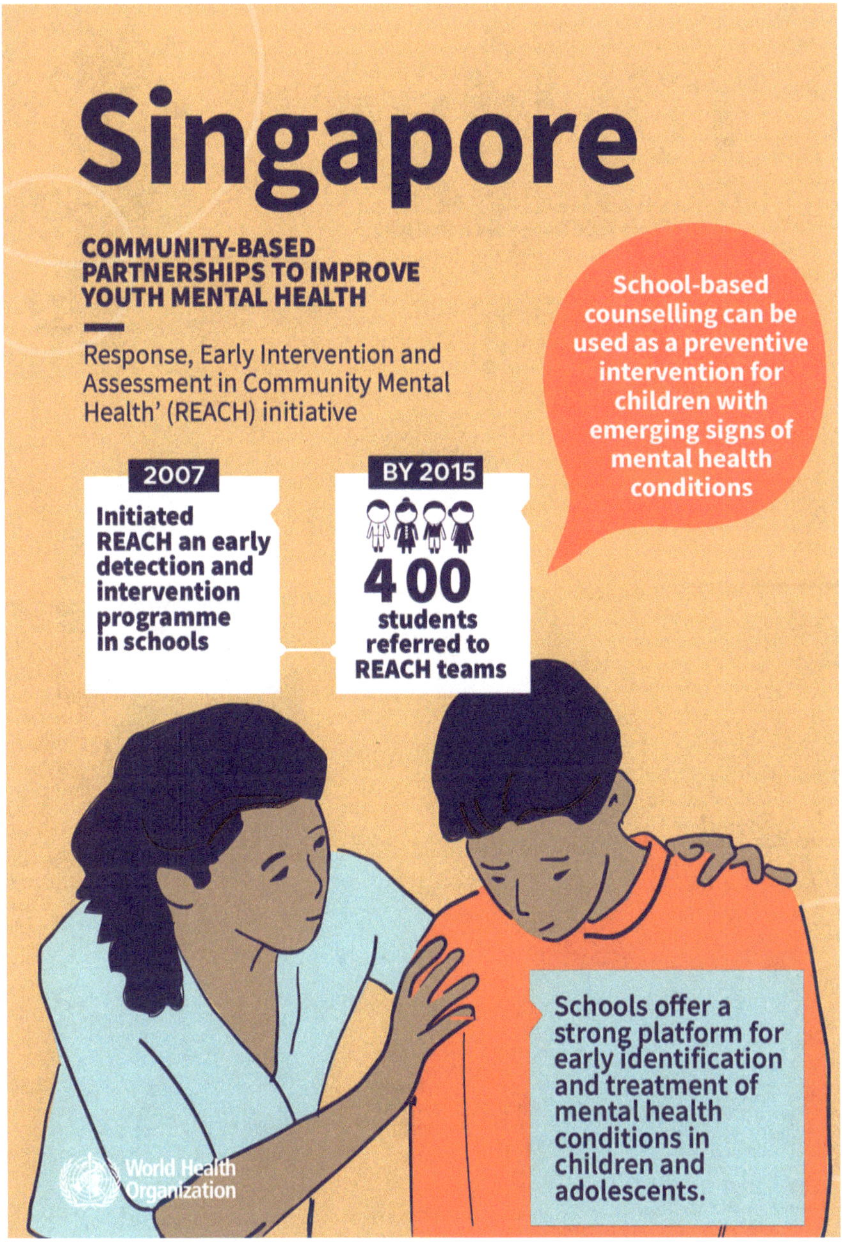

Fig. 5
National efforts to improve mental health in Singapore.

Fig. 6
National efforts to improve mental health in Türkiye.

Summary

The launch of the WHO World Mental Health Report represents a significant step forward in progressive development in the field in the last few decades. Its importance was reflected by the outstanding presentations made at the launch by the impressive lineup of speakers, representing the multilateral contributions of essential stakeholders, including government, UN agency, and civil society, as well as the voice of persons with lived experience. The voices from all these various sectors collectively recognized the importance of advancing mental health, as well as taking action on individual and organizational levels. The Report inspires needed change in the field of mental health that portends a better future for all people worldwide.

References

World Health Organization. (2022a). *Launch of the world mental health report with Dr. Tedros.* https://www.youtube.com/watch?v=LupaPMoU0Gk.

World Health Organization. (2022b). *World mental health report: Transforming mental health for all.* Department of Mental Health and Substance Use. https://www.who.int/publications/i/item/9789240049338.

CHAPTER 81

The UN innovation network

Pradeep Kakkattil

Health Innovation and Investment Exchange (HIEx), Geneva, Switzerland; Global Council SDG3 (Health), "SDGs in Action" Initiative, Dubai, United Arab Emirates

It was June 2018 at a side event on the margins of the UN General Assembly meeting. I sat there listening as one after the other ministers of health and several of the health leaders from various UN entities and civil society organizations spoke of the challenges in providing access to healthcare for the most vulnerable populations. The theme of the session was leaving no one behind, but what was becoming clear was that several, a significant number of people, were being actually left behind when it came to access to essential healthcare. One out of two people globally lack access to essential healthcare according to WHO and every year 100 million people are pushed into extreme poverty due to catastrophic healthcare costs (World Health Organisation, 2021).

I have worked on health—actually HIV—for over three decades and in some ways, while this statistic was not really new and was in the back of my head, it hadn't really touched me until that moment. Why was it that even after significant investments in health over decades, healthcare was practically nonexistent for half the world? The AIDS movement had demonstrated that with a clear vision, a truly community-owned response, and investments, it was possible to reverse one of the biggest threats to humanity and turn it into a manageable disease. It was clear that innovation was the key.

I managed to convince (perhaps it may have bordered on harassing) the UNAIDS executive director into agreeing to establish a small team within UNAIDS to leverage lessons from the AIDS movement for Universal Health Coverage. It was great to have the agreement—but then where does one go to get help within the UN system to understand what resources were available? What has been the experience of other UN entities that had successfully leveraged innovations? What worked and what were the internal challenges? Who were likely partners externally that we could work with? Or did we have to reinvent the proverbial wheel all over?

That is when one of my colleagues pointed me to the UN Innovation Network—UNIN (United Nations, 2023)—an informal, collaborative community of UN innovators interested in sharing their expertise and experience with others to promote and advance innovation within the UN system. It was a great starting point for us to connect with UNIN and its members—from

Resilient Health. https://doi.org/10.1016/B978-0-443-18529-8.00081-0

several UN entities who had managed to push the boundaries within their agencies as well as leverage the power of innovation—from policy to data to technology and financing—within their impact areas. The work of the UNIN is driven by the interests and needs of its members.

What was also helpful was that the UNIN is not just an internal UN closed club—but truly open to innovators from all UN entities as well as external partners—which brought in the much needed diversity and an "outsider" perspective into the UN embracing innovation. Currently the UNIN boasts over 3000 members from over 70 UN entities. Members are spread across over 140 countries.

Apart from learning from some of the brightest minds on development as a starting point, it was great to connect informally in follow-up with some of the agencies who were really working in areas that were relevant to our vision. Colleagues at UNICEF, UNDP, and UNHCR were particularly helpful as we shaped our strategy and some of them gave liberally—their time and networks—to help get us on our feet. The sharing forums went in depth into the work of specific agencies' innovation portfolio and sometimes specific themes—from use of AI to gamification to robotics—and how these could help the UN usher the world toward achieving the UN SDG targets. It also helped highlight the best practices within the UN and often sidebar chats helped look at what didn't work and where we failed. It was a goldmine of information and advice that truly helped us at UNAIDS Innovation fast track our learning process in both internal innovation and our programmatic response.

Apart from promoting cross organizational partnership, the network also helped build common tools including the UN Innovation Toolkit (UN, 2020)—an outstanding resource that is relevant for the UN but beyond for entities committed to SDGs and innovation.

Founded in 2015, the network is jointly chaired by UNDP, UNICEF, and WFP. While it has predominantly been a network for individual innovators, it is also open for organizations to join. The Principles for Digital Development (Principles for Digital Development, 2017) form the shared values of all UNIN members and create a basis for joint innovation efforts.

Conclusion

It is nearly 5 years that I have focused on innovations for health. The UNIN was indispensable in the early days as it not only provided us the tools and lessons from across the UN system but introduced us to a community of like-minded individuals across the UN and beyond. Today as the UN and the broader development sector face some of the biggest challenges ever it is essential to make innovation the new normal. And the UNIN is an indispensable part of that architecture.

References

Principles for Digital Development. (2017). https://digitalprinciples.org/about/.
UN. (2020). *UN innovation toolkit*. https://www.uninnovation.network/un-innovation-toolkit.
United Nations. (2023). *UN innovation network*. https://www.uninnovation.network/. (Accessed 17 April 2023).
World Health Organisation. (2021). https://www.who.int/news/item/12-12-2021-more-than-half-a-billion-people-pushed-or-pushed-further-into-extreme-poverty-due-to-health-care-costs.

PART 7

The innovation toolbox: Innovations driving access to healthcare

How capable are you to come up with a totally new idea, and then see it through, to improve people's lives? Can you visualize and realize something in a new dimension?

People who can do that are designing and implementing highly creative methodologies that impact people's health and mental health for the better. This Part presents specific innovations in health and mental health using technology and community connection, and how project managers, developers, and organizations pivoted to address issues of the pandemic. These provide help and hope to rebuild health systems in the recovery period and beyond.

We call it a toolbox because the projects are exceptionally valuable and creative examples of what is being done "in the field" and "on the ground." Examples come from countries in different regions of the world, from Dubai to Nepal, India, Ghana, Ethiopia, Peru, and others. Many are partnerships between public and private sectors and government—such multistakeholder partnerships that are essential for success. Another successful model involves platforms, shown in examples in these chapters, where innovations are showcased and partners can connect.

To create a new efficient and effective health ecosystem and respond to expanded digital healthcare in this evolving technological age, new data systems are essential, models of which are presented. With education also needing to adapt in the digital age, a graduate school course example is presented using a hybrid (in-person and virtual) approach. Also described is a creative system using Whole Brain Thinking that is exceptionally useful for innovators and partnerships.

What notables have to say

> *Imagination is not only the uniquely human capacity to envision that which is not, and therefore the foundation of all invention and innovation. In its arguably most transformative and revelatory capacity, it is the power to that enables us to empathize with humans whose experiences we have never shared.*
>
> ***—J. K. Rowling, British author of the* Harry Potter *book series***

There are different ways to do innovation. You can plant a lot of seeds, not be committed to any particular one of them, but just see what grows. And this really isn't how we've approached this. We go mission-first, then focus on the pieces we need and go deep on them and be committed to them.

—Mark Zuckerberg, cofounder of Facebook

Do one thing every day that scares you.

—Eleanor Roosevelt, American former First Lady and a drafter of the Declaration of Human Rights

Innovation distinguishes between a leader and a follower.

—Steve Jobs, American entrepreneur, business magnate, and CEO of Apple

Every once in a while, a new technology, an old problem, and a big idea turn into an innovation.

—Dean Kamen, inventor of the powered wheelchair

Innovation, requires the ability to collaborate and share ideas with other people, and to sit down and talk with customers and get their feedback and understand their needs.

—Bill Gates, cofounder of Microsoft

Once you have an innovation culture, even those who are not scientists or engineers - poets, actors, journalists - they, as communities, embrace the meaning of what it is to be scientifically literate. They embrace the concept of an innovation culture. They vote in ways that promote it. They don't fight science and they don't fight technology.

—Neil deGrasse Tyson, American astrophysicist and "futurist"

All our dreams can come true if we have the courage to pursue them.

—Walt Disney, American pioneer of animation, namesake of Disneyworld

Breaking the silos: How the Health Innovation and Investment Exchange (HIEx) helps bridge the health innovation ecosystem

Pradeep Kakkattil[a,b], Anand Tiwari[c], and Biju Jacob[a]

[a]Health Innovation and Investment Exchange (HIEx), Geneva, Switzerland, [b]Global Council SDG3 (Health), "SDGs in Action" Initiative, Dubai, United Arab Emirates, [c]Joint United Nations Programme on HIV/AIDS (UNAIDS), Geneva, Switzerland

Introduction

Since 2019, the Health Innovation Exchange (HIEx.ch) has tackled the challenge of fragmented health innovation ecosystems. HIEx was formally established by UNAIDS and other stakeholders in 2019 as a neutral partnership platform bringing together policymakers, innovators, and investors to leverage health innovations to save and improve lives around the world. UNAIDS, the Joint United Nations Programme on HIV/AIDS headquartered in Geneva Switzerland, leads the global effort to end AIDS as a public health threat by 2030 as part of the Sustainable Development Goals. The launch of HIEx, shown in Fig. 1, took place at the United Nations office in Geneva, during the World Health Assembly, that brought ministers of health, private sector innovators, and community representatives.

While it is critical to continue to invest in building and expanding infrastructure for health—health centers, clinics, labs, hospitals—to scale up access to quality healthcare, the pace at which these are expanding will not be sufficient to get the world to reach the 2030 Universal Health Coverage targets agreed at the first ever United Nations (UN) High-level Meeting (HLM) on Universal Health Coverage (UHC) in September 2019 (United Nations General Assembly, 2019).

Technological developments and innovation can help accelerate the pace of expansion of healthcare access (Fig. 2). Several developments of interest for the health sector that could be leveraged include:

Resilient Health. https://doi.org/10.1016/B978-0-443-18529-8.00082-2

Fig. 1
Launch of HIEx at the UN in Geneva during the World Health Assembly.

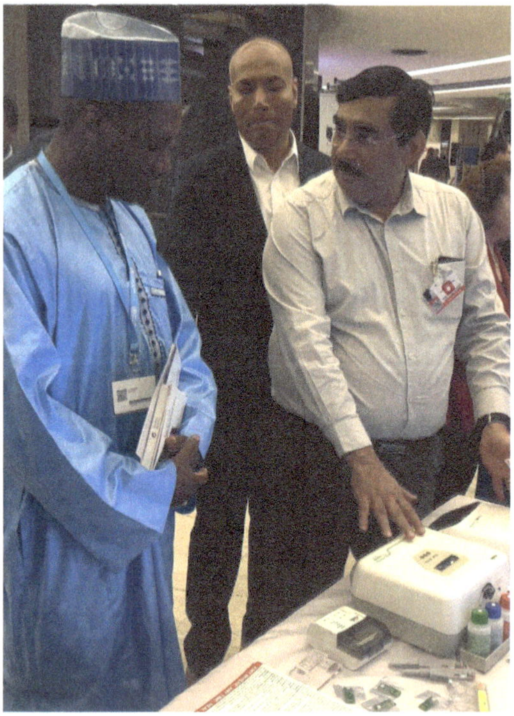

Fig. 2
One of the innovators in HIEx, Molbio, demonstrating a rapid point of care testing technology to a Health Minister.

(1) **Increased connectivity**. Rapid growth of mobile phone usage across the world including in LMICs (low- and middle-income countries) provides an opportunity to leverage digital infrastructure to scale access to healthcare services. From providing real-time information from remote areas on emerging disease outbreaks to providing telemedicine, the mobile data connectivity could help change the way healthcare is conceived and delivered. The COVID-19 pandemic clearly demonstrated the potential of how digital infrastructure could be used to deliver services, including telemedicine and e-health. With 93% of the world having access to a mobile broadband network, including nearly 80% in sub-Saharan Africa (International Telecom Union, 2020) it is possible to expand access on a scale and speed that the traditional brick and mortar model cannot.
(2) **Leverage technology.** One of the challenges that hampers expansion of healthcare services is the lack of human resources. Developments in Artificial Intelligence (AI) and ability to leverage distributed capacities through digital health can help maximize existing limited human capacity and facilitate deployment of technological advances to expand coverage. The launch of ChatGPT in late 2022 has raised significant interest in the potential of AI and large language models that can be leveraged to address human resource shortages—for example, in mental health related counseling or for sexual health for young people. The development of low cost, portable, rapid point-of-care diagnostics provides huge promise of expanding quality diagnosis and access to healthcare.
(3) **Intersectorality**. While expanding mobile phone coverage is a major enabler, health systems across LMICs face several barriers to delivering quality healthcare. In parts of sub-Saharan Africa, for example, only 28% of health facilities are estimated to have access to reliable electricity (Adair-Rohani et al., 2013). The need for close collaboration across sectors to deliver health outcomes is more than evident in the energy sector—from basic lighting to running water to refrigeration—since without electricity, we lose the ability to leverage digital health.
(4) **Financing**. While the share of health in government spending increased over the past two decades, in LMICs, healthcare was financed primarily by out-of-pocket spending (OOPS; 44%) and external aid (29%) (World Health Organisation, 2021). Though Overseas Development Assistance reached nearly US$22.5 Billion in 2019 (Donortracker.org, 2022), it is clear that for scaling healthcare access, there is need to leverage private sector capital and expertise to build sustainable health businesses.

Background

Despite remarkable opportunities offered by new technologies and innovations, access to healthcare has not advanced in line with the potential. While there are a number of innovations that could be transformative in scaling access to healthcare, many of these often do not fully recognize the real "pain points" (problems faced in the marketplace) of the health systems in the countries for which they are designed. A disconnect between policymakers, healthcare providers, innovators, and investors has hindered the opportunities provided by recent technological developments.

In order to strengthen the health innovation ecosystem, HIEx fosters public-private partnerships that can help scale proven technologies and innovations to maximize reach and health impact, especially for the most vulnerable communities. HIEx has an extensive partnership network globally, of the diverse stakeholders, including governments, private sector companies, civil society and nongovernmental organizations, investors, and academia, as well as innovation hubs, incubators, and accelerators.

What does the HIEx do

HIEx leverages innovations for health in a cross-sectoral manner to enable people to live healthier lives, to create economic opportunities, and to fast-track progress toward the SDGs. HIEx works with countries to map key needs of countries and "pain points" in healthcare systems, existing efforts, and enables the broader innovation ecosystem. This includes working with Ministries of Health in countries to identify bottlenecks that can be overcome through innovations in service delivery, policies, and practices.

This cross-sector collaboration, which can either work smoothly, or be disconnected, is shown in a diagram in Fig. 3. In order to achieve the targets set in the UN Sustainable Development Goals (SDGs), particularly SDG3, there is need for seamless collaboration between the public sector, innovators, and investors. The ecosystem needs to incentivize users, markets, investors and lead to sustainable impact (people and planet).

Based on the pain points identified in country, HIEx works with its network of innovation accelerators to identify **high-potential innovative solutions** that can transform access to healthcare on the ground and nurture engagement of communities in translating innovations for impact. These innovations could come from anywhere in the world, but often have been tested in similar contexts at different scales and proven to be effective. As a neutral and trusted "broker," HIEx is able to facilitate and connect implementers, decision-makers, relevant innovators and investors across sectors to build unique partnerships for health impact.

HIEx has over the years also built a network of investors—the Investor Coalition—committed to scaling access to healthcare. This coalition includes development financing institutions, venture capital firms, private equity and philanthropic organizations. HIEx helps guide innovators and start-ups to **access public, private, and community partners** and a variety of **financing options**, including investments, nondilutive grants, and governmental investments. HIEx is also creating new models of financing through use of Web3 and NFTs to finance women-led enterprises for scaling access to healthcare through the Women in Innovation Fund (theWiNFund.org).

Fig. 3
Innovation ecosystem disconnects and opportunities.

Over the last four years, HIEx has seen remarkable growth in innovation ecosystems from the global South. Innovators in China, Ghana, India, Kenya, Nigeria, and Rwanda are all developing health solutions that addresses local challenges but also have huge potential to scale across countries. HIEx works closely with several of these to enable and advocate for greater investment in support of local innovation ecosystem. HIEx also works on building capacities, supports innovators to access markets, advocates for common health technology assessments, and encourages procurement of innovations within public health systems.

Leveraging international convenings

Over time, HIEx has played a major role in leveraging international forums, particularly global health events to advocate on the power of innovations for scaling access to healthcare and the need for investing in the health start-up and innovation ecosystem. It convenes ministers of health, finance, other relevant sectors with innovators and investors at these events to have a constructive dialogue on innovations in health. These events also bring together a marketplace where healthcare providers can experience the utility of the different innovations and learn about the potential in a neutral space. As an example, the <10 cents access to insulin initiative was launched on the margins of the UN General Assembly. A private company, Biocon, is making recombinant human insulin available at less than US$10 cents per day in low- and middle-income countries (LMICs) in stark contrast to the current US list price in retail of over US$5 per day.

Evolution of HIEx

In 2022 HIEx has become an independent nonprofit based in Geneva, Switzerland, and continues to advocate and enable partners to leverage technology and innovation to scale access to healthcare. Priorities of current focus are to:

- Scale investments in women-led health innovations. This is being accomplished through establishment of the "Women in Innovations Fund" (WiNFund). The WiN Fund is established in partnership with Reckitt, Ecobank Foundation, the Kofi Annan Foundation, and Tokenproof and aims to support small and medium enterprises and start-ups founded by women in Africa and Asia through making available much needed capital. Less than 2% of the Venture Capital Funds go toward women-led enterprises (Bittner & Lau, 2021) and HIEx is committed to addressing the inequities in access to capital and empower female health entrepreneurs.
- Showcase innovations that could be transformative in healthcare. This is being accomplished by the establishment of a Health Innovation Discovery Center in India and Dubai with the support of the Government of India. The center works closely with health leaders across countries to familiarize them with potential solutions that have been tried and tested in other settings and enables transfer of technology where appropriate, as well as through investors to enable financing for scale up.
- Convene committed investors. HIEx is convening over 50 investors committed to healthcare and investing in low- and middle-income countries (LMICs). The Investor Coalition consists of development financing institutions, venture capital firms, private equity, and key philanthropic institutions and plays a major role in enabling access to a variety of capital for start-ups and growth stage innovators. Focus of the Investor Coalition has been the regions of Asia and Africa, although investors are based across the world. One of the priorities for the coalition is to incentivize dual market strategies to enable critical innovations translate to LMIC markets.
- Create partnerships. HIEx has partnered with the International Solar Alliance (ISA), an intergovernmental body to scale up access to solar solutions for healthcare. Between 50% and 70% of health infrastructures do not have access to stable energy, which limits the ability of health systems to provide quality care. Through this partnership, HIEx and ISA work closely with health and humanitarian actors to better leverage solar solutions to address health as well as climate change. A "Solar for Health Fund" is also being developed that will attract investors in solar solutions for resilient health systems that also address climate impact on health.

HIEx is founded on the principle that innovation by nature is dependent on leveraging strengths across the ecosystem and the centrality of partnership. HIEx has partnered with Social Alpha—a leader in the innovation ecosystem in India. Similarly, it has entered into a partnership with Health in Your Hands, which was announced at the World Health Assembly in Geneva,

Switzerland, in 2021. No one entity can accomplish the ambitious goals of making health available and accessible to all; thus, joining forces with another platform with similar structure, features, and goals multiplies the reach and impact for health outcomes (Kuriansky, 2024). Further, the lead author of this chapter has been invited to join the Global Council SDG 3 (Health) hosted by the United Arab Emirates "SDGs in Action" campaign, comprised of eminent leaders in the health field from diverse expertise. These including the fields of finance (Devadas et al., 2024) and social innovation (Hughsam et al., 2024) as examples to collectively and comprehensively advance the UN Agenda and achieve the health goals and targets, as well as and maximize visibility, scalability, and financial resources of the innovations. These councils are now seen in a nexus model, interacting with each other to expand their collective reach (Coetzee et al., 2024).

Conclusions and way forward

HIEx has clearly demonstrated the power of a multisectoral approach to healthcare innovation. It has a portfolio of over 100 disruptors with the potential to transform access to health—from Blackfrog Emvolio that played a major role in the COVID-19 response in last mile delivery through its portable backpack medical refrigerators (Blackfrog Technologies, 2022) that reduced wastage of vaccine, to Proximie (Proximie, 2022) that uses Augmented and Virtual reality to enable a surgeon to scrub in to remote surgery while also building capacities across the surgical systems, to Biovitae (Biovitae, 2022), a simple LED bulb that uses a specific nonultraviolet bandwidth to kill bacteria and virus (including demonstrated effectiveness against Corona virus) and could have huge implications for antimicrobial resistance and in preventing hospital acquired infections, to the power of Molbio's (Molbio Diagnostics, 2022) fast portable PCRs for nCoV19 which can now be repurposed for preventive diagnostics, early detection of cancers, drug-resistant TB—these are all inspiring innovations that will truly shape the health systems. The pace with which innovations emerge—for example, the Corona virus Resilience Innovation Map (HIEx and Startup Blink, 2022) has over 1000 innovations—several of them either developed or repurposed for the COVID-19 response—demonstrates the potential of the innovation community in rapidly prototyping and delivering solutions that responds to crisis.

Clearly, it will take disruptive innovations combined with truly innovative financing models to transform access to healthcare and build resilient systems for health that can cope with challenges and the likely pandemics to come. HIEx creates decentralized, interdisciplinary ground for nurturing innovations and works with governments and policymakers to embrace this disruption in an effective and accelerated manner—not only in times of crisis but as a deliberate strategy. The future of healthcare is beyond the four walls of the hospital, a ubiquitous care model embedded in the fabric of community through leveraging technology provides cause for optimism.

However, while technologies including blockchain, web3, artificial intelligence, distributed healthcare including telemedicine, augmented and extended reality, genomics, and personalized medicine will continue to transform healthcare, the divide between the global North and South continues to widen. While several low- and middle-income countries see the life expectancy of their citizens still below 60 years, much of the medical and health innovations in the global North will continue to extend quality life expectancy and likely in the coming years closer to 100 years. This widening gap will have severe ramifications for global health and security (the Goalkeepers 2019 report is an insightful read (Bill and Melinda Gates Foundation, 2019)). It is important to maximize the use of technology and innovation to reduce the inequities in healthcare access.

HIEx will continue to play an advocacy role in highlighting the need for greater focus on the healthcare needs of the developing world; promote greater north-south, south-south, and triangular cooperation; and build a truly public-private-community partnership to leverage technologies and innovation to benefit the most vulnerable.

References

Adair-Rohani, H., Zukor, K., Bonjour, S., Wilburn, S., Kuesel, A., Hebert, R., et al. (2013). Limited electricity access in health facilities of sub-Saharan Africa: A systematic review of data on electricity access, sources, and reliability. *Global Health Science Practice*, *1*(2), 249–261.

Bill and Melinda Gates Foundation. (2019). *Goalkeepers*. Global: Bill and Melinda Gates Foundation. https://www.gatesfoundation.org/goalkeepers/report/2019-report//#ExaminingInequality.

Biovitae. (2022). https://biovitae.de. https://biovitae.de.

Bittner, A., & Lau, B. (2021). Women-led startups received just 2.3% of VC funding in 2020. *Harvard Business Review*. https://hbr.org/2021/02/women-led-startups-received-just-2-3-of-vc-funding-in-2020. In this issue.

Blackfrog Technologies. (2022). https://www.blackfrog.in.

Coetzee, C., Kuriansky, J., & AlHammadi, R. (2024). The nexus model and the Young Leaders Program: Innovative models for the future of the SDGs in Action Campaign and the Global Councils on Sustainable Development Goals. In J. Kuriansky, & P. Kakkattil (Eds.), *Resilient health: Leveraging technology and social innovations to transform healthcare for COVID-19 recovery and beyond* (pp. 629–646). Elsevier.

Devadas, D., Thériault, A., & McGahan, A. (2024). Models to scale healthcare technologies for low- and middle-income countries. In J. Kuriansky, & P. Kakkattil (Eds.), *Resilient health: Leveraging technology and social innovations to transform healthcare for COVID-19 recovery and beyond* (pp. 329–337). Elsevier.

Donortracker.org. (2022). https://donortracker.org/sector/global-health.

HIEx, & Startup Blink. (2022). *Corona virus resilience innovation map*. Global. https://coronavirus.startupblink.com.

Hughsam, M., Claeson, M., Zeitz, L., Kumar, D., Foote, N., & Sinha. (2024). citiesRISE: A transdisciplinary, youth-driven approach to scale social and technological innovations in mental health and well-being locally and globally. In J. Kuriansky, & P. Kakkattil (Eds.), *Resilient health: Leveraging technology and social innovations to transform healthcare for COVID-19 recovery and beyond* (pp. 797–808). Elsevier.

International Telecom Union. (2020). *Measuring digital development facts and figures 2020*. Geneva: ITU.

Kuriansky, J. (2024). "Health in Your Hands": A platform for innovative healthcare solutions. In J. Kuriansky, & P. Kakkattil (Eds.), *Resilient health: Leveraging technology and social innovations to transform healthcare for COVID-19 recovery and beyond* (pp. 989–1004). Elsevier.

Molbio Diagnostics. (2022). https://www.molbiodiagnostics.com.

Proximie. (2022). https://www.proximie.com.

United Nations General Assembly. (2019). *United Nations*. https://www.un.org/pga/73/wp-content/uploads/sites/53/2019/07/FINAL-draft-UHC-Political-Declaration.pdf.

World Health Organisation. (2021). *Global expenditure on health: Public spending on the rise?*. Geneva: World Health Organisation.

"Health in Your Hands": A platform for innovative healthcare solutions

Judy Kuriansky

Global Council SDG3 (Health), "SDGs in Action" Initiative, Dubai, United Arab Emirates; Department of Clinical Psychology, Teachers College, Columbia University, New York, NY, United States; International Association of Applied Psychology, Indianapolis, IN, United States; United Nations Group of Friends of Mental Health and Well-being, New York, NY, United States

Introduction

Innovation in healthcare system delivery is key to healthy lives for all populations, especially those "left furthest behind" in low-resource settings with fragile health infrastructure. Initiatives have increasingly been developed and implemented in countries throughout the world to accomplish this aim, and to achieve the goals outlined in the United Nations Agenda 2030 for Sustainable Development and in the Political Declaration for Universal Health Coverage.

The UN 2030 Agenda with its 17 Sustainable Development Goals and 169 targets was agreed upon by the 193 member state governments of the United Nations in September 2015 (United Nations, n.d.). Ambitious goals were identified, for example, to achieve good health and well-being for all (SDG3), eradicate poverty (SDG1), take urgent climate action (SDG13), empower women and girls (SDG5), achieve universal education (SDG4), and promote peace and security (SDG16), through partnerships (SDG17).

In the ensuing years since 2015, progress was slow and then gravely restricted due in great part to the devastating impact of the worldwide COVID-19 pandemic. To step up progress to meet the agenda commitments, the United Arab Emirates (UAE) initiated a campaign in 2017 called the "SDGs in Action" composed of experts forming Global Councils to address ways to accelerate achievement of selected SDGs. Activities included meetings convened by the UAE, expanding the teams' membership, and working together in collective actions across the councils following a nexus model (Coetzee et al., 2024), given the value of partnerships to achieve the goals.

Resilient Health. https://doi.org/10.1016/B978-0-443-18529-8.00083-4

Fig. 1
Health In Your Hands logo.

This chapter describes the evolution of the "Health In Your Hands" (HIYH) innovation which emerged from the team of the Global Council 3 on health and well-being, tasked to address Sustainable Development Goal 3 of the UN Agenda 2030 (SDG3). Goal 3 states specifically "To ensure healthy lives and promote well-being for all at all ages," with target 3.4 including to "promote mental health and well-being." As the name of the innovation implies, the aim was to put health in the hands of those "left furthest behind" or "at the last mile." The logo of HIYH is shown in Fig. 1. HIYH is now integrated into the Health Information Exchange platform.

Background

Experts from various fields were invited by the UAE to attend the initial "SDGs in Action" Lab, which convened a conference during the World Governments Summit (WGS) in Dubai in 2017. Invitees then self-selected into various teams, called "Global Councils" to address various SDGs. These initial teams were tasked to meet over the course of the ensuing two days of the WGS to come up with an innovation that would accelerate action on their particular SDG. These were presented in a final session with a competition whereby the assembled members voted for the most promising ones to take forward. In a final ceremony, the winning teams were announced and presented with a statue, shaped as a light bulb (signifying an idea or innovation, consistent with the theme of the initiative) upon which were printed all the SDG official logos of the United Nations, as shown in Fig. 2. Of the multistakeholder teams, three winners were selected: "Health in your Hands" (Goal 3); "Game-In" (Goal 5, empowerment of women and girls); and "Goal Hub" (Goal 10, end inequities). The selected teams elaborated their strategies to further their initial ideas into prototype development and implementation phases.

In the next step, the winning three Global Councils were invited to a session sponsored by the UAE at the United Nations Science Technology and Innovative Forum held on May 16, 2017 at the United Nations headquarters in New York City in the Delegates Dining Room, to present their work at a side event entitled "SDGs in Action: From Ideas to Solutions," as shown in Fig. 3. The four most active members of the HIYH team at that point took the stage and described the HIYH innovation platform. Myself and Dr. Shariha Khalid gave an overview of

Fig. 2
Sculpture of the award signifying innovation.

the initiative and the other two members described examples of their innovations featured on the HIYH platform. Mr. Yosef Safi Harb, founder of Happitech, explained that 10% of babies look lifeless at birth and are shockingly put aside as if dead, but their feeble heart rates are more able to be detected with android smartphones used by health workers, whereupon 99% of these babies were resuscitated. The company is now successfully launching heart monitors for adults. Dr. Ritsaart van Montfrans, M.D., who is CEO of INCISION, a Netherlands Essential Surgery business—to date also in more advanced development—explained how his innovation assists surgery procedures in remote locations with Apps that link local medical workers with expert staff elsewhere and with online workflow management by local healthcare workers for diagnosis and follow-up. They also offer visual aids and training.

About the "Health In Your Hands" (HIYH) initiative

In the ensuing years, Global Council 3 continued to meet, with the core membership shifting as members' schedules permitted, since all participation was on a volunteer, nonpaid basis, and all expenses (maintenance of the website, ancillary costs) were absorbed by the members. The chairs, who drove the progress, were the author of this chapter and Shariha Khalid, a former

Fig. 3
Team members presenting the HIYH initiative at the United Nations Science Technology and Innovation Forum at the UN headquarters in New York City (left to right): Dr. Judy Kuriansky, Yosef Safi Harb, Dr. Ritsaart van Montfrans, and Dr. Shariha Khalid.

surgeon and current impact investor and managing partner of Mission & Co. Team members continued to communicate by phone and in person at the annual WGS, to which GC teams were invited. A project manager who served as a portfolio manager for Mission & Co was engaged, Ralph Mbonisi Mpofu, and various assistants, interns, and students participated at various stages.

In the second term, the co-chairship was passed to Dr. Muhammad Pate, Julio Frenk Professor of Public Health Leadership at the Harvard T.H. Chan School of Public Health, who was formerly Global Director for Health, Nutrition and Population at the World Bank Group and Director of the Global Financing Facility for Women, Children and Adolescents, as well as former Minister of State for Health in Nigeria, in the shared chair role with Dr. Zsuzsanna Jakab, Deputy-director of the World Health Organization in Geneva. New members from diverse fields joined the team, including the coeditor of the volume, Pradeep Kakkattil. I stayed on as a member, carrying the tradition of the Council and of the HIYH project which was integrated into the Council's work flow. Dr. Khalid transitioned to join the Finance Advisory Board (FinAB) of the Global Councils, tasked with collaborating with the other councils for fundraising including a focus on the HIYH innovations.

From the start, the development of the HIYH platform was a focus of the GC3 output, to feature game-changing health innovations from organizations with forward-thinking approaches to deliver health services, and particularly to empower patients and health service providers at the "last mile" (most remote settings), using both direct and indirect approaches to improving health outcomes. The platform took an ecosystem approach, intersecting digital health data, techniques to deliver healthcare that address maternal and child health, noncommunicable diseases, food and nutrition, energy, water and sanitation, and supply chain. Platforms like HIYH offer great value—including visibility, partnerships, and leveraging resources—to innovations but also have inherent challenges, in requiring management, accountability, and funding (Khalid et al., 2024).

As a multistakeholder initiative, HIYH was dedicated to enabling the delivery of accessible, affordable, acceptable, and sustainable health solutions to "those left furthest behind" through the platform which showcases health outcomes-focused innovations worldwide, and which places health technologies, tools, and access to finance in the hands of providers working at the frontlines, communities in remote settings with fragile healthcare infrastructure, and other stakeholders. These innovations range from market-driven enterprises which deliver products and services, to initiatives involving mobile-based applications, community-based outreach and training for healthcare providers to deliver care to beneficiaries. The HIYH initiative aligns with the World Health Organization definition of holistic health, which includes both physical and mental health.

The HIYH platform was purposely aligned with Goal 17 of the UN Agenda 2030 for Sustainable Development, by including governments, the private sector, civil society, academia, media, and others, to create shared value and catalyze partnerships to achieve scalable impact to overcome challenges in healthcare provision and prevention.

Importance of the GC3 HIYH initiative

The HIYH initiative offers a solution to the wide treatment gap between the dire need for, yet inadequate access to, healthcare, which has been extensively documented and gravely exacerbated during the recent COVID-19 pandemic (Diaz & Kuriansky, 2024). In doing so, the initiative assists in strengthening countries' health ecosystems to promote good health and well-being outcomes. Importantly, HIYH serves as a unique platform for accelerating partnerships across sectors by leveraging innovative approaches and supporting financing mechanisms through a nexus (interconnected) model for creating impact.

HIYH showcases solutions to the problem of "leaving no one behind" called for in the UN Agenda by showing how the innovations can unlock remote, underserved, at-risk communities with access to basic healthcare, and thereby, serves to connect those at the "last mile" to first-mile healthcare. The projects show that access and distribution of medical care in remote areas is possible, as well as affordable, effective, and sustainable—the criteria for Universal Health Coverage (WHO, n.d.).

Scouting and supporting early stage innovations

To achieve our goal as a multistakeholder platform, we needed to map and understand the existing landscape, identify key stakeholders, and leverage existing networks. We assessed existing relevant platforms and constructed a unique framework for *Health In Your Hands*. For the purpose of finding the best game-changing innovations and service delivery models, we identified key criteria to support the scouting and selection process.

A call for applications was launched in May 2019 at the World Health Assembly, with an extensive outreach and engagement exercise implemented throughout 2019 and 2020. To reach the right audience and to ensure quality applications, we engaged with our wider network of ecosystem partners to support outreach efforts and to nominate the most relevant innovations and solutions. By leveraging the experience and knowledge of ecosystem partners, we were able to streamline our vetting process of potential applications.

The selection process

The selection process was structured into three stages, to ensure that quality innovations were comprehensively reviewed and selected. Over 1500+ innovations were identified and vetted by the *Health In Your Hands* team with the support of select ecosystem partners, which included startup accelerators, incubators, and competition platforms.

Following the first round evaluation by the HIYH team and select partners, a shortlist of 200 innovations was presented to an advisory panel for further deliberation. Concurrently, due diligence calls and meetings were conducted with each shortlisted innovation to verify information, and in some cases to request additional data/information to be reviewed by the advisory panel. Out of that number, 50 game-changing innovations and service delivery models were identified across geographies and settings, meeting selection criteria to deliver needed health prevention and treatment services to underserved, at-risk, and marginalized communities, with potential for replicability, scalability, capacity-building, and sustainability. This final list of 50 innovations to focus on was selected on the basis of their meeting the criteria to deliver needed health prevention and treatment services to underserved, at-risk, and marginalized communities.

Specific selection criteria were impact, scalability, innovation, assisting in universal health coverage, research, use of technology, capacity-building, communications, financial sustainability and profitability, viable management team, contribution to the health ecosystem.

Announcement of the 50 game-changing innovations was delivered in three phases: the first 20 innovations were announced during the Universal Health Coverage Day 2019. The next 10 innovations were announced during World Health Day 2020 and the final 20 innovations were announced during the World Governments Summit 2020. The final selected innovations ranged from market-driven enterprises that deliver products and services, to simple mobile-based applications and hardware innovations delivered directly to healthcare workers and patients. Upon announcement, each selected organization went through an onboarding process to identify key development goals and support needs.

Alignment with international instruments

HIYH is aligned with several important international instruments. These include:

(1) The United Nations 2030 Agenda for Sustainable Development, specifically SDG3 which calls for ensuring "health and well-being for all, at every stage of life" but intersecting with other SDGs as shown in Fig. 4.
(2) The Political Declaration on Universal Health Coverage which emphasizes the human right to high standards of both physical and mental healthcare.
(3) The Astana Declaration on Primary Healthcare, which highlights the importance of primary healthcare in ensuring high standards of healthcare for all.

Fig. 4
Diagram of the HIYH platform at the center of various SDGs which are numbered.

(4) The Sendai Framework for Disaster Risk reduction, which calls for providing health, resilience, and psychosocial support.
(5) The Global Compact on Migration which contains references to providing health, mental health, and psychosocial care, as well as actions for resilience in the face of sudden and slow onset disasters.

Key outcomes

The initiative was adopted as the signature project of the GC3 in the first term and reaffirmed in the second term, as part of the Work Flow. Additionally, the following was accomplished:

- *Visibility:* HIYH has been visible in important global fora and events.
- *Thought leadership*: The initiative and the innovations on the platform served as a model for other innovations to achieve the goals of SDG3 and of Universal Health Coverage and to showcase the nexus model for the SDG Global Councils.
- *Partnerships:* Multistakeholder partnerships were encouraged among governments, civil society, academia, the private sector, and other stakeholders in the spirit of SDG17 and as the model through which SDG3 and the entire SDG Agenda can be advanced and achieved.
- *Sustainability:* The initiative actively explored pathways to harness innovative financing approaches and instruments to unlock new markets and resources for access to quality health solutions and achieve UHC and SDG3.
- *Solutions repository:* The innovations conveyed a body of knowledge about how to bring about successful outcomes in health, particularly for those left furthest behind in low-resource and conflict settings.

Value add

HIYH has supported the innovations in various ways, including opportunities for visibility not only on the platform itself, but also in major conferences and global events. For example, through the HIYH connection to the UAE, VillageReach (Pain et al., 2024) secured a booth at the Dubai Health Forum to exhibit their project to a global audience. The representatives of two innovations (Medic Mobile and Fyodor Biotehnologies) spoke on a webinar in the series about COVID-19 hosted by the United African Congress on July 9, 2020, on the topic of "Innovations in Healthcare in Africa." Additionally, I described innovations in many of my presentations described in a later section in this chapter. HIYH was also featured in a blog post for the World Economic Forum (WEF) and in innumerable conference and events, described later in this chapter.

Another benefit to being featured by the HIYH platform is access to global networks, which in turn leads to further opportunities, such as mentorship, technical advice, and unlocking access to capital.

Examples of innovations

Several innovations in the HIYH platform are presented in this volume in separate chapters, which have technology-enabled services especially important in the ongoing COVID-19 response and going forward. These include Living Goods, reach52, Healthy Entrepreneurs, and VillageReach. For example, VillageReach uses drones in their "Drones for Health" solution as shown in Fig. 5, which saves time where it is more geographically challenging for cars, motorcycles or boats to reach, and also improves cost-effectiveness. It is critical, however, that the project includes community health workers who deliver in-person health services using the information supplied by the drones. This person-to-person contact is essential, leading to a main point throughout this volume that hybrid models where technological tools are supplemented by in-person contact are most effective (Kuriansky & Kakkattil, 2024).

Presentations

HIYH has been presented at numerous international conferences. These include important events sponsored by the UAE to showcase the "SDGs in Action" campaign and included reports about GC SDG3 (Health) and the HIYH platform.

One such presentation was an in-person side event about the "SDGs in Action" campaign held in a conference room at the United Nations during the 2019 United Nations General Assembly (UNGA) in New York City. The HIYH cofounders Kuriansky and Khalid made a presentation along with selected members of the other Global Councils, moderated by a representative of the UAE Mission to the UN, Ms. Hana Al-Hashimi.

The next year, a presentation featuring the "SDGs in Action" campaign included GC 3 among three panels of Global Council members. The event was held on-line—since the COVID-19 pandemic restricted travel—during the World Governments Summit, on December 9, 2020, as a

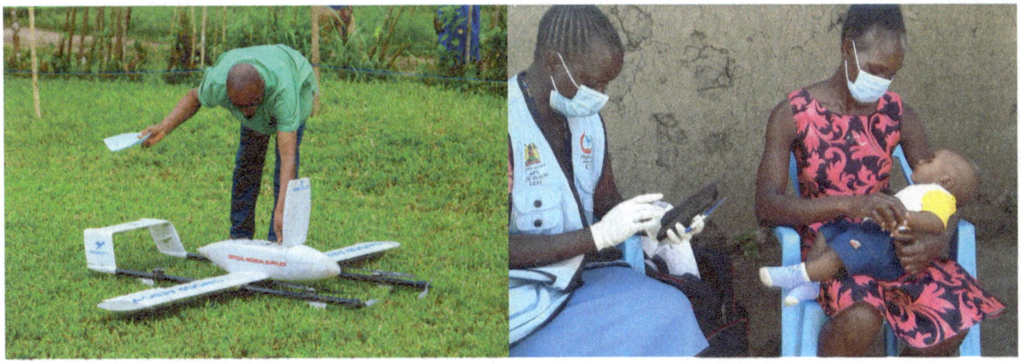

Fig. 5
Examples of innovations on the HIYH platform.

Fig. 6
Panel about "Realizing the Potential for All" as part of the 2020 WGS "SDGs in Action" with Dr. Kuriansky speaking about the Global Council on Health and the HIYH initiative, with chairs of other Global Councils. Pictured (top row, left to right: H.E. Mona Al Marri, Vice President, UAE Gender Balance Council (SDG5); Dr. Mohammed Mohieldin, special envoy on Financing the UN 2030 Agenda (SDG 1); Dr. Judy Kuriansky, "SDGs in Action" Global Council on Health (GC3); (bottom row, left to right): H.E. Tariq AlGurg, CEO of Dubai Cares (SDG4); and moderator Elizabeth M. Cousins, President and CEO of UN Foundation.

high-level annual meeting hosted by the government of the United Arab Emirates. In my role as GC3 cochair, I was on the panel entitled "Realizing Human Potential for All," as shown in Fig. 6.

Another presentation hosted by the UAE about the "SDGs in Action" campaign, where HIYH was acknowledged, was also held during the World Governments Summit, this time in person in Dubai, on March 29, 2022, in a session entitled, "Scaling New Heights for the Future of Humanity." The GC3 second term cochair, Dr. Mohamed Pate, was on a panel in this session entitled, "Global Councils on SDGs: Change Agents Weaving Endless Possibilities" along with Marie-Aimée Boury, cochair of the Finance Advisory Board (FinAB), as shown in Fig. 7. In his presentation, Dr. Pate noted that health is at the center of all the SDGs and that innovations play a key role in accelerating the achievement of the global goals. He reported about the focus of GC3 on the HIYH innovations, bringing them to scale, and exploring collaboration with the FinAB. He also mentioned the research project funded by Fondation Botnar which is examining structural constraints, including policy, regulations, and financial, which inhibit such upscaling,

At the next World Governments Summit, held in Dubai in February 2023, I represented Global Council 3, reporting about our progress, on a panel on "Global Councils on SDGs – Education,

Fig. 7

SDG3 term 2 co-chair Dr. Mohammad Pate (pictured second from left) on a panel at the World Governments Summit about the "SDGs in Action" campaign, with chair of the Finance Advisory Board Marie Aimée Boury (pictured third from left).

Health, and Technology," as shown in Fig. 8. My comments included addressing sustaining selected innovations on the "Health in Your Hands" platform, as well as the publication of this volume with many GC3 members contributing chapters about resilient health systems (Devadas et al., 2024a, 2024b; Dhatt et al., 2024; Hughsam et al., 2024; Pate & Jakab, 2024). Other deliverables are: progress in the working group examining structural constraints to scaling solutions mentioned above, as well as increased collaboration with the Finance Advisory Group, the Noor Dubai Foundation, and the WHO-led Alliance for Transformational Action on Climate and Health, looking ahead to the next COP meeting on climate change (Fig. 8) which was held in Dubai.

Other international fora provided an opportunity to talk about HIYH. For example, HIYH was highlighted in a presentation at the United Nations General Assembly event on 27 September 2019, entitled, "Achieving SDG 3 And Universal Health Coverage: Innovative Projects, Policies And Financing Models To Reach Those Left Furthest Behind." In this comprehensive program with governments and civil society speakers, HIYH cofounder Dr. Khalid gave an overview of the "Health In Your Hands" platform and introduced Dr. Ruth Ngechu, deputy country director of Living Goods, Kenya, who described their model of a public-private partnership operation in Kenya, Uganda, Burkina Faso and Sierra Leone that supports digitally empowered community health workers to serve underserved remote communities.

Fig. 8

Panelists at the 2023 World Governments Summit on "SDGs in Action – Education, Health and Technology" (pictured, left to right): Dr. Manal Taryam, CEO of Noor Dubai Foundation; Dr. Judy Kuriansky, "SDGs in Action" Global Council SDG 3 (Health); moderator Samer Costantini, Director of Communications MEA, Schenider Electric; Dr. Sonia Ben Jaafar, CEO of the Abdulla Al Ghurair Foundation; and Clint Brown, Director of Product Engineering at Esri.

In her presentation at that same event, on "SDGs in Action and the Global Councils inter-linkages," Dr. Radheya AlHashmi, Director of Policy Impact Department, Strategy and Innovation Sector, United Arab Emirates, Ministry of Cabinet Affairs and the Future, Prime Minister's Office, mentioned the importance of the HIYH project within the UAE campaign to advance the SDGs.

Additionally, I mentioned several of the innovations during my presentation for the UHC2030 Private Sector Constituency and UHC2030 Civil Society Engagement Mechanism (CSEM) on April 7, 2021, World Health Day, on "Partnering for UHC" (the flyer for which is shown in Fig. 9). My presentation showcased innovations on the HIYH platform: mothers2mothers, Living Goods, doctHERs, iKure, and InStrat Global Health Strategies.

Selected HIYH innovations were also featured in a webinar in the United African Congress series I co-organized throughout the pandemic, about "COVID-19 in Africa". In that webinar, in July 2020, representatives described how their projects pivoted during the pandemic to help prevent the spread but also serve their populations. The section of the webinar featuring presentations by the innovation representatives was presided over by youth Fellow, Ms. Layla AlNeyadi, whose participation in HIYH is described in the next section.

Fig. 9
Flyer for event on World Health Day, with Dr. Judy Kuriansky as a panelist for the UHC2030 Private Sector Constituency and the UHC2030 Civil Society Engagement Mechanism.

Involvement of youth

Consistent with the emphasis on youth participation by bodies at the United Nations and by the UAE, various students served as interns over the course of the development and presentations of the HIYH initiative that led to exceptional and rare opportunities. For example, Emirati native Layla Al Neyadi, who had been a student of mine in my class about "Psychology and the United Nations" at Columbia University Teachers College and whom I brought with me to Dubai to the WGS one year. As part of her Fellowship working on the HIYH innovation portfolio, she selected, and then introduced, several innovations for the webinar series described above, in the United African Congress series about "COVID-19 in Africa" in July 2020, to talk about how their projects were helping prevent spread of the virus. Also, she spoke about her experiences for a video about the initiative (at the link below). Further, she impressively was among the high-level speakers for the 2020 World Governments Summit. Reflecting about her experience with HIYH, Layla said, "Working on the project taught me the value of cross-sector collaboration and partnerships in fostering effective health systems. I also gained valuable insight into ways health innovations navigate complex systems to deliver healthcare to communities in diverse cultural settings. I now have an expanded perspective—where SDG3 about health and wellbeing is at the center of achieving the sustainable development goals but also in a holistic framework. This view is important for my future work back home in the UAE, and for my goal to impact the health of all people."

Progress and partnership

The HIYH team continued to maintain the HIYH platform and keep the information about the innovations updated. In 2021, the innovations were contacted to uncover their progress and also to identify how they had pivoted to address needs exposed and escalated during the COVID-19 pandemic crisis. This was done with several purposes in mind. One purpose was to identify those that would be included in this volume on "*Resilient Health Leveraging Technology and Social Innovations to Transform Healthcare for COVID-19 Recovery and Beyond*." Also, the HIYH team shortlisted 10 innovations from the portfolio with the most promise for scaling, and selected ones were presented by their founders to the second term members of the Global Council 3 over several virtual team meetings in 2022. Also, these were intended for consideration by the Finance Advisory Board of the "SDGs in Action" initiative to identify possible funding sources for scaling.

In a major part of the workstream of the GC 3 (besides this "Resilient Health" volume), several of the innovations in the HIYH portfolio were selected as case studies in a report funded by Fondation Botnar and carried out by the Springfield Centre, as part of the Global Council 3 Working Group on Structural Constraints to Scaling Innovations (SCSI). The report identified constraints related to politics, policies, laws and regulations, procurement procedures, financing or other factors which prevent scaling innovations, such as those in the HIYH platform, in alignment with the World Health Organization (WHO) Global Action Plan (GAP) and its accelerators. The innovations were Village Reach, Last MIle Health and Healthy Entrepreneurs. The innovations in the HIYH portfolio would also inform a major report, the Lancet Commission on the Future of Health and Economic Resilience in Africa (FHERA).

In a major step of the way forward, the HIYH co-founders Khalid and Kuriansky formed a strategic partnership with Pradeep Kakkattil, founder of the Health Innovation and Investment Exchange (HIEx), who is also co-editor with Kuriansky of this "Resilient Health" volume. HIYH is now cohosted on the HIEx website.

Fig. 10
HIYH founders Judy Kuriansky and Shariha Khalid advance partnership with HIEx founder Pradeep Kakkattil at the World Governments Summit in Dubai, February 2024.

The HiEx is an aligned platform with similar features to HIYH (Kakkattil, 2024), as it brings together policymakers, innovators and investors to leverage health innovations and investment to foster public-private partnerships that can help scale proven technologies and innovations to maximize their reach and health impact, especially for the most vulnerable communities. The HIYH/HIEx partnership was announced on November 30, 2021 at an event held in Geneva, Switzerland entitled, "New Horizons for investing in innovations for global health and resilient health systems" hosted by HIEx with Cross Border Impact Ventures (a women-led investment management firm offering opportunities in women's, adolescents and children's health) and the Geneva Innovation Movement. The event brought together key leaders from public and private sectors around the connecting theme of developing new and innovative models for scaling access to healthcare and building resilient health systems (as in the event title). The partnership was further consolidated and explored during meetings between the HIYH and HIEx founders during the World Governments Summit in Dubai in February 2024 (Fig. 10). An important step is that the HIYH portfolio is now being cohosted on the HIEx website. This partnership has several positive deliverables, in leveraging and maximizing the synergistic pipeline of innovations, opportunities for projects, and expertise across sectors and of the two initiatives. These multiple benefits presage bright outcomes for the future of healthcare.

Promotional video of HIYH: https://youtu.be/ph5ALDtE_EQ

Acknowledgments

Much appreciation is extended to Dr. Shariha Khalid, cofounder of HIYH and partner in the project over these many years, and to Mr. Ralph Mbonisi Mpofu for his excellent management of the HIYH portfolio, activities, and reports over years, and for their contributions to various sections of this chapter. Thank you to Joel Zinsou who edited the video about *Health in Your Hands* that so perfectly portrays sample innovations. Also, much gratitude to Russell Daisey who offered exceptionally wise support from the very beginning of the project and at all stages.

References

Coetzee, C., Kuriansky, J., & AlHammadi, R. (2024). The nexus model and the Young Leaders Program: Innovative models for the future of the SDGs in Action Campaign and the Global Councils on Sustainable Development Goals. In I. J. Kuriansky, & P. Kakkattil (Eds.), *Resilient health: Leveraging technology and social innovations to transform healthcare for COVID-19 recovery and beyond* (pp. 629–646). Elsevier.

Devadas, D., McGahan, A., & Theriault, A. (2024a). Financing instruments for innovators to scale health innovations in low- and middle-income countries. In J. Kuriansky, & P. Kakkattil (Eds.), *Resilient health: Leveraging technology and social innovations to transform healthcare for COVID-19 recovery and beyond* (pp. 663–677). Elsevier.

Devadas, D., Theriault, A., & McGahan, A. (2024b). Models to scale healthcare technologies for low- and middle-income countries. In J. Kuriansky, & P. Kakkttil (Eds.), *Resilient health: Leveraging technology and social innovations to transform healthcare for COVID-19 recovery and beyond* (pp. 329–337). Elsevier.

Dhatt, R., Preethi, J., & Thompson, R. (2024). Unlocking potential: Women's leadership and innovation for health. In *Resilient health: Leveraging technology and social innovations to transform healthcare for COVID-19 recovery and beyond* (pp. 703–712). Elsevier.

Diaz, D., & Kuriansky, J. (2024). Reimagining healthcare: Technological solutions to close the treatment gap and achieve equity in the wake of COVID-19. In J. Kuriansky, & P. Kakkattil (Eds.), *Resilient health: Leveraging technology and social innovations to transform healthcare for COVID-19 recovery and beyond* (pp. 941–954). Elsevier.

Hughsam, M., Claeson, M., Sinha, M., Kumar, S., Foote, M, & Zeitz. (2024). citiesRISE: A transdisciplinary, youth-driven approach to scale social and technological innovations in mental health and well-being locally and globally. In J. Kuriansky, & P. Kakkattil (Eds.), *Resilient health: Leveraging technology and social innovations to transform healthcare for COVID-19 recovery and beyond* (pp. 797–808). Elsevier.

Kakkattil, P. (2024). Breaking the silos: How the Health Innovation and Investment Exchange (HIEx) helps bridge the health innovation ecosystem. In J. Kuriansky, & P. Kakkattil (Eds.), *Resilient health: Leveraging technology and social innovations to transform healthcare for COVID-19 recovery and beyond* (pp. 979–987). Elsevier.

Khalid, S., Burgess, R. C., Kasonde, M., Erogbogbo, T., Sinha, S. Q., & Gatumia, B. (2024). The role of the private sector in achieving universal health coverage: Experiences from the UHC2030 private sector constituency. In J. Kuriansky, & P. Kakkattil (Eds.), *Resilient health: Leveraging technology and social innovations to transform healthcare for COVID-19 recovery and beyond* (pp. 541–560). Elsevier.

Kuriansky, J., & Kakkattil, P. (2024). Introduction to the volume on resilient health systems leveraging technology and social innovation to transform healthcare for pandemic recovery and beyond. In J. Kuriansky, & P. Kakkattil (Eds.), *Resilient health: Leveraging technology and social innovations to transform healthcare for COVID-19 recovery and beyond* (pp. 3–17). Elsevier.

Pain, A., Crouch, J., & Tillman, K. (2024). VillageReach: The COVID 411 initiative. In J. Kuriansky, & P. Kakkattil (Eds.), *Resilient health: Leveraging technology and social innovations to transform healthcare for COVID-19 recovery and beyond* (pp. 1005–1010). Elsevier.

Pate, M., & Jakab, Z. (2024). The state of the world in health and health systems. In J. Kuriansky, & P. Kakkattil (Eds.), *Resilient health: Leveraging technology and social innovations to transform healthcare for COVID-19 recovery and beyond* (pp. 19–36). Elsevier.

United Nations (n.d.) Transforming our world: The 2030 agenda for sustainable development. https://sdgs.un.org/2030agenda.

WHO (n.d.) Universal health coverage. https://www.who.int/health-topics/universal-health-coverage#tab=tab_1.

CHAPTER 84

VillageReach: The COVID 411 initiative

Amanda Pain, Jennifer Crouch, and Katlen Tillman
VillageReach, Seattle, WA, United States

Background

VillageReach's mission is to transform healthcare delivery to reach everyone. We work with governments and partners in Africa and have offices in the Democratic Republic of Congo (DRC), Mozambique, and Malawi. As part of our COVID-19 response, we launched the COVID 411 initiative in partnership with Focusing Philanthropy, a private foundation, to disseminate consistent, accurate information to limit the spread of COVID-19. Specifically, our goals were to:

(1) Provide accurate information about COVID-19 to help individuals build awareness and protect themselves from the coronavirus;
(2) Empower health workers with information about prevention and treatment; and
(3) Provide the government with data visibility into community concerns and questions related to the pandemic.

Table 1 shows our activities in the three countries, the DRC, Mozambique, and Malawi, the technology innovations implemented, the reach, and the results.

With the support of governments and partners in each of the three countries VillageReach designed COVID 411 using multiple channels and technologies with content developed in the local language(s) for each country (primarily French in the DRC, Portuguese in Mozambique, and Chichewa in Malawi) to achieve these goals. To reach communities, we used interactive voice response (IVR) messages, WhatsApp with chatbots, voice messages in local languages, and mobile health hotlines with trained hotline operators. To reach health workers, we used IVR messages, and for data visibility we developed data dashboards.

Between April 2020 and June 30, 2021, COVID 411 reached over 15,000 health workers remotely and answered over one million community questions related to COVID-19.

Resilient Health. https://doi.org/10.1016/B978-0-443-18529-8.00084-6

Table 1 Activities and outcomes of VillageReach tech-based innovations in response to COVID-19 in three African countries.

	DRC	Mozambique	Malawi
Activities	• Codeveloped COVID-19 content with MoH • Secured Facebook approval for WhatsApp in the DRC • Codeveloped dashboard	• Codeveloped COVID-19 content with MoH and FDC • Negotiated telecom agreement • Hardware and technology upgrades • Developed SOPs and conducted quality assurance • Trained hotline operators	• Codeveloped COVID-19 content with MoH • Supported surge staffing from 20 to 75 operators • Codesigned call back service for COVID-19 symptoms • Codeveloped dashboard
Result	• HealthAlert WhatsApp tool rolled out for COVID-19	• Health Center by Phone updated to respond to COVID-19 and upgraded with enhanced technology	• Health Center by Phone updated to respond to COVID-19
Audience	• Community—WhatsApp users	• Community—mobile phone users	• Community—mobile phone users
Geography	• 15 provinces	• Nationwide	• Nationwide

Problem/challenge being solved through this innovation

COVID 411 was rapidly implemented early in the pandemic to address fears, rumors, and myths related to COVID-19. By disseminating accurate information, we wanted to limit the spread of COVID-19. VillageReach works in hard-to-reach communities, where access to accurate health information is limited, and finding ways to connect these communities to the health system quickly is vital. This issue was acknowledged by the World Health Organization's Africa Infodemic Response Alliance (AIRA), which was launched in December 2020 (WHO, 2021).

Communities and health workers need consistent access to accurate and trustworthy health information to make informed health decisions and offer informed health services. This access to information is especially critical during a pandemic and COVID 411 was a vehicle to offer information on the symptoms, spread, prevention, and treatment of COVID-19 (including vaccines once these became available) to help people protect themselves and their communities.

How and where this innovation is being implemented. What technology is being used, at what level is it being implemented, who is being targeted, and who and how the service is being delivered

To meet the COVID 411 objectives we used existing and new technologies. For health workers and community health workers in Malawi, Mozambique, and the DRC, we developed remote learning modules on COVID-19 symptoms and risk mitigation using IVR messages sent via mobile phones.

To get accurate information to communities, get feedback from communities, and reduce the burden on health systems, we leveraged our Health Center by Phone (HCBP) digital solution. In Malawi, Chipatala cha pa Foni (CCPF) (VillageReach, 2019) is an existing nationwide health hotline, as shown in Fig. 1. Early in the pandemic, the president of Malawi designated CCPF as the primary source for COVID-19 information. To address the increased call volume during the pandemic we worked with the Malawi Ministry of Health (MoH) to develop IVR messages for the community in the national local languages and assisted the government in hiring additional hotline workers.

In Mozambique, we leveraged and adapted our HCBP solution to disseminate COVID-19 information through AlôVida, the national health hotline. AlôVida's original technology platform could not handle many calls and did not include IVR messages; therefore, we added these features during the pandemic to help increase reach.

In the DRC, we launched HealthAlert, a WhatsApp COVID-19 chatbot to complement an existing UNICEF emergency COVID-19 hotline. The chatbot had COVID-19-related content in French, Swahili, Lingala, Kikongo, and Tshiluba.

Fig. 1
CCPF Hotline worker accessing health information for a caller. *Photo Credit: Paul Joseph Brown.*

Impact of the work, with highlighted successes

By June 2021, COVID 411 answered over one million calls from community members and trained over 15,000 health workers. The initiative introduced a remote learning digital solution in all three countries, enhanced digital solutions in Malawi and Mozambique, and introduced the HCBP solution in the DRC. We also developed a database of health worker phone numbers for future remote learning opportunities and co-created data dashboards with governments to inform their COVID-19 response.

In Malawi, nearly 2100 health workers listened to at least one learning module, and the MoH opened a second hotline location to deal with the high community call volumes. We also used an incentive system in Malawi, offering health workers certification and airtime credits for completing all 12 learning modules which led to higher completion rates.

In Mozambique, over 6000 health workers listened to at least one learning module and in April and May of 2021 learning modules were pushed out to an additional 8000 teachers and community leaders. Through an agreement with Moz Telecom in April 2021, AlôVida is now functioning 24 h/day with the ability to take higher call volumes and offering concurrent IVR messages. In the DRC, over 4000 health workers listened to at least one learning module.

Challenges being faced (e.g., in policy, finance, technology, COVID-19, disconnects between sectors)

Reaching health workers was challenging because none of the three countries had existing databases of health workers or their phone numbers, and collecting accurate contact details was an on-going challenge.

Health workers reported difficulty following audio-only learning modules and only 68% of health workers completed one learning module. However, completion may have been higher as health workers reported listening to learning modules in a group, on one mobile phone. Other completion challenges faced by health workers included: being unavailable when the call came in, cost to call back if they missed the call, presentation speed being too fast, and lack of incentives.

In our community awareness campaigns, CCPF in Malawi experienced a rapid increase in call volumes, leaving calls unanswered until we could increase staff from 25 to 70 people. In Mozambique, improvements to the AlôVida platform were delayed because of technology challenges and stalled agreements with mobile network operators. In the DRC, setting up HealthAlert was delayed while the MoH sought permission from Facebook, which owns WhatsApp.

Strategic partners (government/public and private sector)

VillageReach believes in radical collaboration with government and partners to sustain impact at scale. We utilized our strong relationships with governments in Malawi, Mozambique, and the DRC to make this initiative a success, and each government was actively involved throughout the design and implementation of COVID 411. We worked with other partners[a] to create databases of health worker's phone numbers, and with Viamo, a social enterprise technology company which uses 3-2-1 messaging push technology,[b] to design course materials, evaluation tools, and data dashboards. We also worked with Praekelt.org, using their Turn platform[c] for HealthAlert, which was being used by millions of people to receive COVID-19 information.

Focusing Philanthropy, a funder that matched all individual donations during our COVID 411 fundraising campaign, was also a critical partner in this initiative.

Current business plan (funding sources and sustainability plans)

VillageReach develops solutions with government partners, and once proven impactful, transitions ownership and/or management to government or other partners to drive sustained impact. Our scaling strategy allows us to leverage new philanthropic funding to improve solutions for replication in new geographies.

In 2020 VillageReach transitioned CCPF to Malawi MoH ownership. We are currently seeking philanthropic funding to transform CCPF into a national telemedicine hub that will provide information to the community, train health workers, and enable referrals. VillageReach is establishing a community of practice to amplify integrated and sustaianable telelhealth implementations, in which Malawi will be seen as an exemplar country.

AlôVida is owned by the Mozambique government and operated by Fundação para o Desenvolvimento da Comunidade (Foundation for Community Development (FDC) in Maputo, Mozambique) with technical assistance from VillageReach. VillageReach will support WhatsApp messaging chatbot integration with AlôVida, with funding from Gavi INFUSE, which will help make WhatsApp an integral feature of HCBP implementations.

COVID 411 was funded through a coalition of foundations and individual donors whose gifts were matched by Focusing Philanthropy. HCBP has also been generously supported by donors

[a] John Hopkins University, n.d. FDC, GIZ, ONSE (USAID Maternal and Child Health Project in Malawi).

[b] The 3-2-1 technology requires no literacy and can reach any phone. The platform can incorporate questions to gauge learning.

[c] Turn is human operator-driven product that leverages the WhatsApp Business API (Application Protocol Interface) to engage users and drive impact at scale.

including Deutsche Gesellschaft für Internationale Zusammenarbeit (German Development Agency), Johnson & Johnson, Mulago Foundation, Preston-Werner Ventures, Dovetail Foundation, Gavi, and the United States Agency for International Development (USAID).

How COVID-19 has affected the work and response/pivot

We revised our COVID 411 goals in November 2020 largely due to coordination challenges with governments and other partners during a pandemic. Each country's COVID-19 response evolved as information about the coronavirus, its transmission, and vaccines evolved. This meant continuously adapting health worker learning modules and COVID-19 information being disseminated to communities.

Overall, despite the challenges, the immediate impact of COVID 411 is notable, and the potential for long-term impact is great with the introduction of new technologies and digital solutions each country can adapt these technologies to respond to a future health crisis.

Additional information

https://www.villagereach.org/our-impact/covid19/
https://www.youtube.com/watch?v=IFghMRWB5Qo

References

VillageReach. (2019). https://www.villagereach.org/wp-content/uploads/2019/06/Chipatala-cha-pa-Foni_Overview.pdf.

WHO, AFRO (2021). https://www.afro.who.int/aira.

CHAPTER 85

Realizing the transformative potential of Community Health Workers with digital innovation | A living goods case study

Mohini Bhavsar[a,b] **and Kanishka Katara**[a]
[a]*Living Goods, Nairobi, Kenya,* [b]*Grand Challenges Canada, Toronto, ON, Canada*

About the organization

Living Goods is an NGO that accompanies governments in achieving their national digital health visions for community health programs in Kenya, Uganda, and Burkina Faso. Since 2007, Living Goods (LG) has been working to close the gap between the potential that community health workers (CHWs) have to save and improve lives and the challenges they face regularly, including insufficient training and supervision, access to medicines, reliance on paper-based tools, and lack of compensation. Fig. 1 shows a Living Goods CHW visiting a village mother, in key person-to-person interaction that is part of the program.

Our organization is focused on an improved model for delivering care at the last mile, which we call DESC—Digitally Enabled, Equipped, Supervised, and Compensated. Based on programmatic evidence and field validation, we firmly believe that all of these DESC components are equally critical from the perspective of generating impact at the ground and while digital technology has inherent transformative potential it can only be actualized when it is catalyzed through all elements of the DESC framework. We consider digital and data as a cross-cutting and critical enabler in our community health work—a pillar which is central to our mission and program interventions. The technology we design and use supports what is first defined by a strong programmatic vision for care delivery. The range of activities led by the CHWs supported by LG are shown in Fig. 2, including use of disruptive technology tools at the point of delivering essential healthcare to hard-to-reach communities. CHWs act as advocates, educators, and caretakers, including for basic care and follow-up, as well as referrals in reproductive, maternal, newborn and child health.

Resilient Health. https://doi.org/10.1016/B978-0-443-18529-8.00085-8

Fig. 1
A Living Goods-supported community health worker in Uganda visiting a mother with her child.

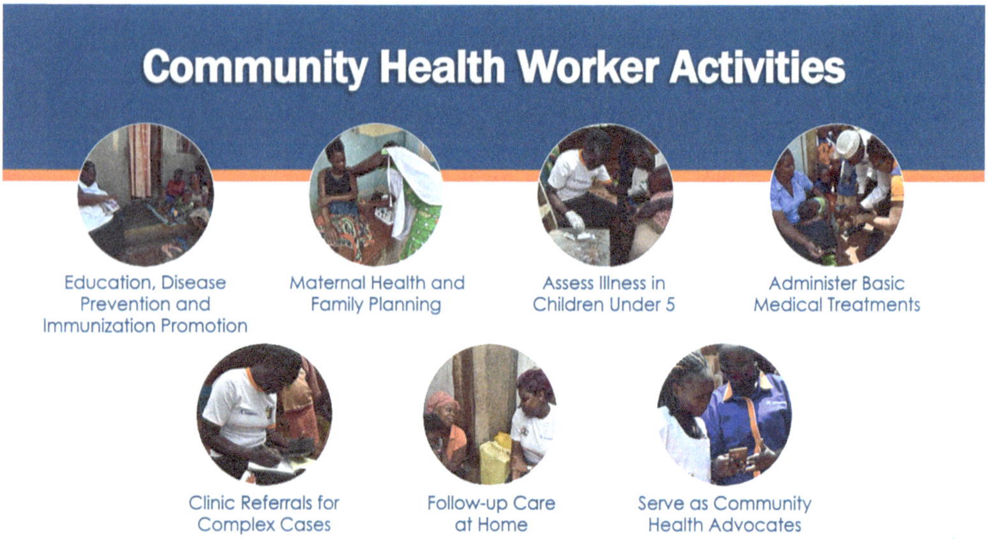

Fig. 2
Activities of the Living Goods-supported Community Health Workers.

In 2022 alone, more than 12,040 LG-supported CHWs treated and referred more than 3.56 million sick children and assisted 230,294 women to have safe pregnancies. In 2014, an external Randomized Controlled Trial (RCT) of our efforts in Uganda showed that our approaches to community health reduced child mortality by 27% (Nyqvist et al., 2019). LG's bold new strategic vision for community health focuses on the next frontier of impact: enabling civil society and government partners to improve health outcomes nationally through digitally enabled community health programs (Living Goods, 2021).

Background

In the quest for high-impact interventions to save lives, many decision-makers, implementers, and donors increasingly look to digital technologies and tools. Over the past decade, we have been searching for and developing such tools to transform the way frontline health workers deliver healthcare to reach more people with quality and affordable services. Despite the best intentions, our undertakings within the digital health community have led to a plethora of pilots that often end up in a 'virtual graveyard' after a short time in use.

When reflecting on why health systems around the world have not experienced a comprehensive digital transformation, clues can be found in the environment within which they function. The ways communities seek and receive healthcare and the ways providers deliver care have been largely static over the past decade. If we are to see true digital transformation of health systems, we need a cultural transformation to happen first.

Shifts in culture are notoriously hard. Something big needs to happen—something that cannot be ignored—for behaviors, patterns, and systems to change. For digital health, the COVID-19 pandemic was just that, acting as a trigger that disrupted the cultural inertia of health systems, health workers, and people. The pandemic continues to be a catalyst forcing health systems to change core cultural levers, including around policy and partnerships.

Since March 2020, everyone connected with the health sector—including CHWs—was forced to learn and adapt rapidly to changing circumstances. Data became a high-value resource to inform key decisions needed to flatten the curve of the pandemic. We also saw healthcare policies around the world evolve to approve vaccines in record time and make virtual care and telemedicine more accessible. Ministries of Health explored transformative partnerships by inviting ministries across sectors, private sector partners, and nonprofits to support and jointly design new services and products with them. Policy-making is culture-making, and these new ways of working and making decisions are ushering in change.

Policy change, establishing strategic partnerships, and building the foundational systems to move from descriptive to predictive data are examples of change that were slow to initiate and develop. COVID-19 created the environmental pressures that forced health systems to disrupt our traditional ways of working in these core areas, all of which are critical for digital transformation. Now the digital health industry is at another inflection point in our journey, but with different parameters to work with and a new emerging enabling environment available to us.

The enabling environment for digital transformation in community health is much better today than before. Today, many health workers have their own phones or know someone who does. Generally, CHWs have more digital literacy than previously, through the use of products like social media, though of course some are at the beginning of their digital journey. Access to electricity and connectivity is improving, through ICT investments, private sector involvement, and multisectoral approaches to creating fully digital economies in low- and middle-income settings.

Even though none of these essential enabling factors are universal, the landscape and the tools we have at our disposal are rapidly accelerating postpandemic. More than 40 countries in Africa have a published national digital health strategy (*Global strategy on digital health* 2020-2025, 2021). Our legacy practices of investing in digital health through siloed disease areas are shifting toward integrated care models powered by architecture that allows different digital systems to speak to each other. Donors are committing to consolidating their investments to sustain digital health innovations that can be achieved together rather than through one-off or conflicting projects they fund (*Digital health investment principles*, 2022). Governments are standing up technical working groups at the national level to prioritize their digital health needs and directing internal resources toward the long, arduous path of achieving their vision for change.

Problem/challenge being solved through this innovation

Given a more flexible enabling environment, the pandemic showed how quickly programs can pivot healthcare delivery and how rapidly existing digital solutions can be adapted to meet new programmatic objectives to maintain delivery of routine health services remotely, with limited to no in-person visits. At Living Goods, we made several strategic adjustments beginning in March 2020 to meet this new challenge, once the first cases of COVID-19 were reported in Kenya and Uganda. Following our DESC approach, our digital job aid (D) was accompanied by thoughtful changes in the program, by supplying a steady-state flow of essential medicines and training (E), remote and in-person models for supportive supervision (S), and changes to CHW compensation structures (C).

How and where this innovation is being implemented

In 2020 our first change was to adjust our digital job aids used by CHWs to support care being delivered remotely. Within four months of the pandemic, we launched our low-touch or no-touch digital workflows, which meant that CHWs could carry out many household check-ins virtually through phone calls instead of household visits. We also provided other support to CHWs, by equipping them with PPE and stocking their medical kits with free essential medicines to be given to families, and modified our CHW compensation structure to reflect the change in activities and other economic losses CHWs faced. These programmatic and technology modifications allowed more than 3000 CHWs in Kenya and Uganda to safely provide health services over the phone instead, such as screening for COVID-19, conducting assessments for sick children, or providing postnatal care to women. In-app messaging triggers and event tasks were enabled to flag cases which require follow-up by CHWs, who can reach out to clients via phone calls. This was especially critical during PPE shortages and for high-risk CHWs.

In parallel, in 2020, we tested two-way messaging for information campaigns and COVID-19 symptom assessments to the community members for a smaller subset of communities in Kenya and Uganda in a program we called Client-Initiated Health Assessments (CIHA). SMS text messages were a simple, cost-effective way of quickly reaching clients. The system supported caregivers to walk through a COVID-19 case screening algorithm via text that assesses their exposure to COVID-19 risk factors. The output from this engagement would trigger actionable tasks in the application of CHWs, program supervisors, and disease surveillance officers for further investigation. Our SMS solution delivered lifesaving prevention information and COVID-19 self-assessments to 13,886 unique care givers and community members and elicited a 92% response engagement in users interested in carrying out COVID-19 symptom screening, and 100% response to get more information on COVID-19 prevention knowledge. CIHA was completed by approximately 70% of users reached, identifying about 10% of users with potential symptoms of COVID-19, and linking them to follow-up testing and quarantine.

Impact of the work, with highlighted successes

In response to the pandemic, we deployed nearly six innovations in 2020 in a matter of a short eight months and discovered a window into a new frontier of community health programming. With an average of about one new innovation every six weeks, the breadth of Living Goods' learnings on potential new modalities to provide healthcare at the last mile, when faced with significant new environmental pressures, also expanded. The success from these rapid pivots has enabled CHWs to continue to deliver essential health services throughout the pandemic, and the number of monthly treatments and referrals which CHWs provided to children under five years of age spiked 84% in Kenya and 90% in Uganda from 2019 to 2020. These results have hints for the culture and systems change it takes at both program and organizational levels to build resilient, adaptable digital innovations that last.

In 2021 as the pandemic continued through new variants of concern, CHWs continued to use our low-touch, no-touch digital protocols and accompanying programmatic innovations. Our preliminary data for 2021 suggests that on average about 4-6% of all consultations in Kenya were carried out over the phone by LG-supported CHWs, initiated by a phone call. When disaggregating these figures by geographic regions, we observed that in places with wide geographic distances between households and health facilities, CHWs carried out twice as many virtual consultations than the average at LG-supported sites. Similarly, CHW supervision, was nearly entirely virtual, with high success, by increasing the frequency and efficiency of supervision to CHWs.

Our quest for adapting to the post-COVID-19 world led us to expanding our program in 2021, to include COVID-19 vaccine mobilization as part of our service delivery priorities for CHWs. As a result, for the first time, we grew our database of target community members supported. Where we had previously only targeted pregnant women, women of reproductive age and children under age 5, we expanded it to now include the elderly, adults, and adolescents for COVID-19 vaccines. We believe our understanding of under-5 immunization adoption, sensitization, and adherence strategies may translate into success strategies for COVID-19 vaccinations and other immunization programs for example.

Challenges being faced

While our innovative adaptations during the pandemic may hold insights for new sustained frontiers for virtual care delivered by CHWs and community health programs, the proportion of virtual consultations is still low compared to the traditional in-person model of care. The potential sustained behavior for CHWs adopting virtual consultations poses an interesting inquiry. Many services are easier to provide in person, such as malaria assessments with malaria rapid diagnostic tests and measuring mid-upper arm circumference (MUAC) for children. There are known challenges for virtual service delivery offering like telemedicine, such as limited consistent connectivity issues in remote areas and lack of access by some households to even simple phones and limited phone charging facilities. That said, remote capabilities provide critical flexibility for CHWs, and households, especially in times of crisis to enable self-triage or save time and travel costs for activities such as routine follow-ups or referrals to health facilities which are especially well suited for virtual care.

Strategic partners (government/public and private sector)

We are exploring this question further in the next two years with community health, technology, and government partners considering how COVID-19 disruptions around the world catalyzed a rapid growth of virtual care and telemedicine services as an effective delivery channel for health services and for which there is sustained acceptance by consumers and openness by regulators (Lee et al., 2021). Despite the promise, however, there are big questions around the effectiveness of virtual care at the last mile for the most underserved populations. Additional challenges include that these populations include those who have no or limited access to digital technology or adequate internet, cannot pay for the cost of services, or are not aware of or know how to access telehealth.

Our developing hypothesis is that hybrid models inclusive of virtual care can best respond to the needs, and increase access to care, for the underserved with the inclusion of CHWs. Most telemedicine providers in the private and public sector exclude CHWs as part of their service offering and modalities of care, thus missing the opportunity to connect to the last mile underserved pockets of the market. The research is clear that chronically underserved populations are more likely to accept a new modality of care if introduced by a trusted community member, like a CHW. Over the course of the next five years, Living Goods will explore our hypothesis and test designs for a CHW-in-the-loop virtual care approach as a first principle. While we aim to take inspiration from today's traditional telemedicine models—which primarily focus on connecting consumers directly to higher skilled physicians—we aim to develop a more inclusive model. Our approach will include health providers with more basic skills, such as CHWs, in the escalation pathway, so that the unique needs of rural and underserved communities can be met by a trusted community member.

Looking forward

Looking forward, over the next five years, Living Goods is working in this new reality to partner with governments in supporting 24,000 digitally enabled community health workers and supervisors to provide quality healthcare to 13 million people across three countries. This care addresses communities' basic health needs and CHWs will register every pregnant woman, conduct pre- and postnatal visits, treat sick children, provide family planning counseling, and track immunizations.

Digital transformation in healthcare is an ambitious goal and a long journey, which requires significant investments from governments, donors, and the private sector. The innovative mindset through open policy-making, transformative partnerships, and improving data-driven decision-making and better predictions catalyzed during the pandemic offers a window to enable digital transformation, and sustain and support new impactful innovations developed during the pandemic to take root with a greater probability to succeed.

It is our hope that we are at the beginning of a new era, a new cultural transformation postpandemic, that will bring digital enablement to community health programs for sustained impact and change as we manage COVID-19, help people access vaccines, and prepare for recovery. This is just the beginning of our innovation along new frontiers for CHWs to achieve their full potential as the last mile actors in the health system ecosystem.

References

(2022). *Digital health investment principles*. http://digitalinvestmentprinciples.org. (Accessed 2 June 2022).

Global strategy on digital health 2020-2025. (2021). Geneva: World Health Organization. Licence: CC BY-NC-SA 3.0 IGO.

Lee, S. G., Blood, A., Gordon, W., & Scirica, B. (2021). Disruptive and sustaining innovation in telemedicine: A strategic roadmap. *NEJM Catalyst*. https://catalyst.nejm.org/doi/full/10.1056/CAT.21.0311. (Accessed 2 June 2022).

Living Goods. (2021). *Living Goods launches new 5 year strategy*. https://livinggoods.org/media/living-goods-launches-new-5-year-strategy. (Accessed 2 June 2022).

Nyqvist, B., Martina, A. G., Svensson, J., & Yanagizawa-Drott, D. (2019). Reducing child mortality in the last mile: Experimental evidence on community health promoters in Uganda. *American Economic Journal: Applied Economics*, *11*(3), 155–192.

CHAPTER 86

Innovation in health delivery using technology: Addressing healthcare disparity in India

Sujay Santra and Tirumala Santra Mandal
iKure Techsoft Pvt. Ltd, Bengaluru, India

We at iKure realized that supplying more doctors overnight, or building hospitals to address the pandemic, would not solve the overwhelming need for care, especially for rural populations. Instead, technology was considered a catalyst, combined with Community Health Workers (CHWs) as an enabler, to drive efficiency at a population scale. Thus iKure used a simple cloud-hosted technology platform called Wireless Health Incident Monitoring System (WHIMS) that can be used by village women with basic education, who constituted a trained army of health workers. WHIMS is loaded on a tablet that collects patient's vitals by these CHWs. Supported with medic bags, the health workers carry various noninvasive, portable point-of-care diagnostic tools to measure vitals, diagnose various parameters like blood pressure, pulse, hemoglobin, blood glucose, and many more, which helps identify high-risk patients (Fig. 1).

The diagnostics are collected by the CHWs and shared in real time with trained doctors through a technology application who suggest treatments and prescribe medication through the WHIMS platform, who then relay this information to the patients, together with the necessary medicines; prescriptions; and counseling on nutrition, health, and hygiene.

The benefits of technology in this system of providing healthcare are significant. Using technology, especially in such an emergency as this COVID-19 pandemic, is an opportunity to address the healthcare gap. However, there is no substitute for a little dose of human caring touch, and a human being, such as the healthcare workers, who offers reassurance and comfort to reduce fears, with a smile and empathy. Also, this incredible catalyst of digital transformation in healthcare today also needs the right data to transform life "at the last mile" where patients are in remote settings. The COVID-19 pandemic is a wake-up call to look beyond existing capabilities of every village and every community. If we thought that the COVID-19 pandemic has compelled us to lose touch with each other, iKure's approach using

Resilient Health. https://doi.org/10.1016/B978-0-443-18529-8.00086-X

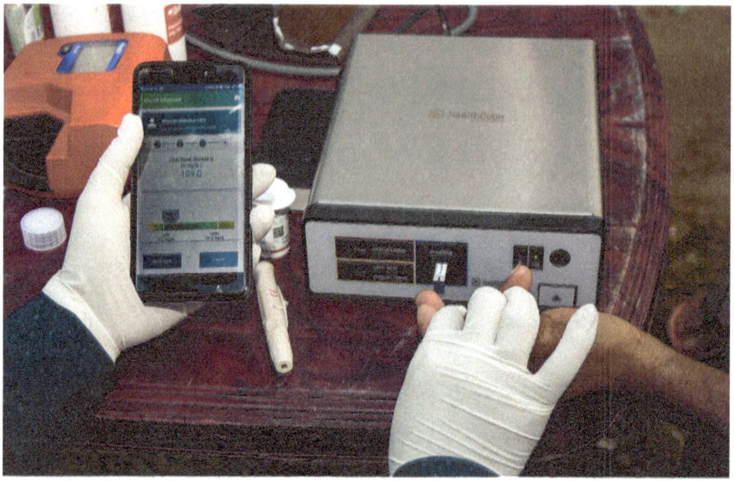

Fig. 1
iKure's Wireless Health Incident Monitoring System (WHIMS) with an intuitive graphical user interface that rural health workers with high school education can use.

technology and health workers for rural healthcare offers optimism, ensuring care while retaining the human touch even more than before.

These two stories, shared with our team members, represent the common problems faced by Indian people in accessing decent healthcare.

As Rahim Khan, 31 years old, was gathering his fishing net, the iKure team member noticed he seemed depressed. He relayed that his 2-year-old son at home was having breathing problems and needed medical care. But going to the distant hospital would cost Rahim more than a month of his livelihood, especially since the family was in dire financial straits since they were still picking up pieces after the devastating super cyclonic storm Amphan that happened in Sundarbans in Eastern India in 2020, which had left his meager acre of land saline and unusable, forcing him to change from farming to fishing in order to sustain his family.

Another man, Nioyi Chang, from Momching in the district of Nagaland, called his friend to take his ailing 7-month pregnant wife to the city clinic since she was suffering with severe vaginal bleeding. He lamented to his friend, "How will I save my wife?" Going to the city hospital would incur travel and money for food, lodging, and medical care, including testing, which all together would be huge out-of-pocket expenses for them which they could not afford. But, they would have to make it happen to save the lives of his wife and their yet-to-be-born baby.

The problems for these two cases demonstrate how the Indian healthcare scenario represents a spectrum of problems. With a contrasting landscape and diversity of demographics, India has hilly terrains, river deltas, and rough plains, distributed across 28 states, 8 territories, 700 tribes, and 100 languages, making access to healthcare difficult. With COVID-19, this spectrum widened devastatingly, posing uncertainty about how the postpandemic would be especially for deprived persons.

For example, Sundarbans in the Indo-Gangetic plains is the largest delta in the world that houses 102 islands, but only about half (54) are inhabited with the rest home to Royal Bengal tigers. Extreme geographic remoteness and poverty limit their access to healthcare.

The plight of the North-Eastern State in Nagaland is no different, where healthcare utilization is marked with substantial neglect. Nioyi Chang narrated his experience of taking his wife to the city clinic during COVID-19 as horrifying and torturous.

The availability of India's public healthcare is not adequately distributed, being not only fewer in overall number but operating with limited resources. As a result, during the COVID-19 situation, people with mild symptoms avoided seeing a doctor and visited only when faced with an emergency.

Public health utilization and health outcomes are predetermined by many factors. These include socioeconomic status, gender, class, and caste. For example, a women's reproductive age determines whether she could access a doctor or not, with older women at a disadvantage. Also, given long distances from major hospitals for remote persons, the chances of seeing a certified doctor are slim. Caste is another major predeterminer of health outcomes; for example, lower level of dietary intake, antenatal care, and child's immunization are handpicked and prioritized based on cultural norms, prejudice, and superstitions.

Access and utilization of healthcare varies in size and context across different regions under normal conditions, but changes happen drastically when a crisis like the pandemic hits.

At iKure, we realized that providing more medical doctors overnight or building hospitals to address the issue would not solve the problem of needs. The healthcare system in India is deeply entangled and highly personalized. Our Primary Health Centers (PHCs) remain ineffective and less utilized, combined with a shortage of health workforce which is inadequate to meet the huge demand and disparity of the large population.

Technology as an enabler to drive primary healthcare

The journey to address these inadequacies in India's rural healthcare needs led us at iKure to incorporate innovative technology approaches as a catalyst, combined with Community Health Workers (CHWs) as an enabler, to drive efficiency at a population scale.

iKure uses a simple cloud-hosted technology platform called Wireless Health Incident Monitoring System (WHIMS) that can be used by village women with basic education who are

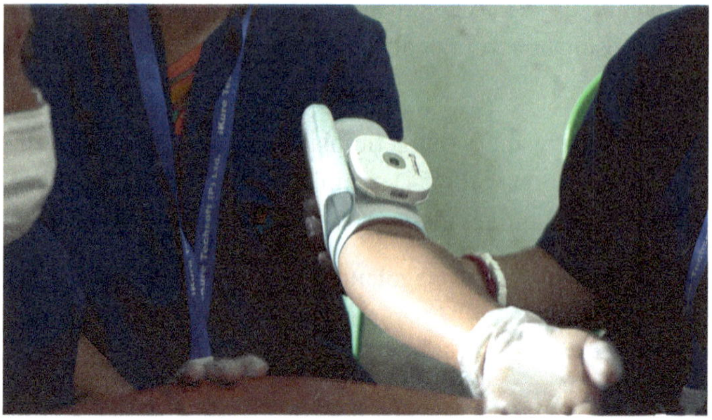

Fig. 2
iKure patient undergoing vital checkup using point-of-care device.

trained as health workers. Provided with medic bags, the health workers carry various noninvasive, portable point-of-care diagnostics tools when visiting people in the community including a WHIMS which is loaded on a tablet that collects patient's vitals. Thus they are able to measure vitals; diagnose various parameters like blood pressure, pulse, hemoglobin, blood glucose, and many more; and identify high-risk patients. Patients are diagnosed with portable point-of-care devices, as shown in Fig. 2.

The diagnostics collected by them are shared in real time with trained doctors, through the technology application, who then suggest treatments and prescribe medication through the WHIMS platform. The information is transferred to the health workers who share it with the patients, including medicines and prescriptions, and offer appropriate counseling on nutrition, health, and hygiene.

WHIMS operates through Hub and Spoke clinics set up. In order to ensure that quality healthcare is within everyone's reach, the Hub and spoke model is an innovative architecture that emphasizes the optimal utilization of scarce healthcare resources in low-income communities.

For iKure, the centralized distribution model consolidates information, equipment, and supplies at a central Hub and propagates the same to the peripheral centers called Spokes. The Spokes are manned by doctors and a health workforce who, under the guidance of the operational team, deliver essential services at hubs, spokes, and camps. Fig. 3 shows an iKure beneficiary undergoing diabetic retinopathy screening in a camp.

Such services include a physical doctor's consultation, maternal and childcare, advanced eye care, and techno-enabled pathology services, ensuring that these are available within the community where they are needed and at a time when it is essential, by skilled and empowered health workers.

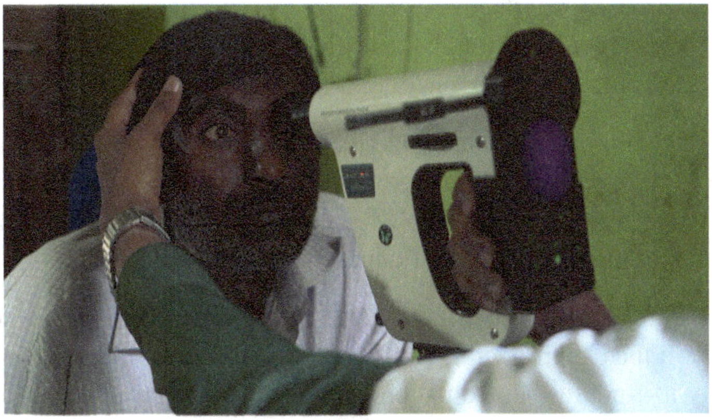

Fig. 3
An iKure beneficiary undergoing Diabetic Retinopathy screening.

Embracing telemedicine with a human touch

For healthcare providers like iKure, identifying the first step forward has been to recognize the problem and its underlying impact on the beneficiaries meant to be served. With the turbulent times of COVID-19, health services were directly and indirectly impacted. Noncommunicable disease screening, regular testing, and posthospitalized care radically decreased. Care for elderly people and pregnant women continued to suffer the most.

The public health system soon responded with large-scale telemedicine implementation to help mitigate the crisis of need. However, when the country was already suffering a stark shortage of infrastructure and health workforce, the platform has been slow. Also, telemedicine services of specialized care and disrupted technological connectivity were only accessible for privileged persons, leaving people at the lower class of society deprived.

Multilingualism has been also a hurdle, as each state in India onboarded its own doctors due to linguistic problems and thus the user's experience across different states varied widely.

For all the benefits of technology in healthcare, there is no substitute for a little dose of human touch. While telemedicine platform offered unparalleled capabilities in healthcare, we at iKure felt services also required a touch of compassion.

We knew that technology offers a valuable opportunity to address the healthcare gap, but it must also act as a friend who can take away people's humanly fear, with a smile and empathy. This meant that we had to shift to a hybrid telemedicine platform for our beneficiaries, providing in-person experiences along with the digital platform, to provide greater value to patient's health outcomes.

This works through a set of aided networks, such as community health workers who visit the patient, collect vitals, upload these data on the telemedicine platform, and initiate a consultation call with a medical professional doctor. The doctor reviews the vitals and communicates with the patient in his or her language through the health worker. Prescribed medicines are provided to the patient by the clinical staff physically present at their doorstep.

The CHWs working with us in this innovative approach understood patients' worries and addressed patients' concern with a human touch and smile, looking into the patient's eye and offering needed reassurance saying, "Don't worry, you will do well soon." Fig. 4 shows community health workers undergoing orientation with technology app.

In our experience at iKure, this mode of telemedicine picked up slowly. One issue is that patients felt far from the expert doctor. But steadily, assuring the communities that *clinicians* have moved *closer to*—not further from—patients, the next step was to introduce telemedicine in the existing Primary Healthcare Centers (PHCs) on an equal footing.

The hospitals were already overburdened, with no signs of abating the surge of the COVID-19 wave. The shift from hospitals to primary healthcare facilities was long awaiting.

The country of India boasts a strong network of *25,000* primary health centers and *5300* community health centers, spread across all regions and states. Given this potential, it was necessary to institute a high-performing infrastructure, to manage the epidemic and provide continuity of services.

To solve this problem, iKure's telemedicine got integrated into the government-run PHCs through collaborations, thus enabling last-mile connectivity and essential services in peri-urban and rural regions.

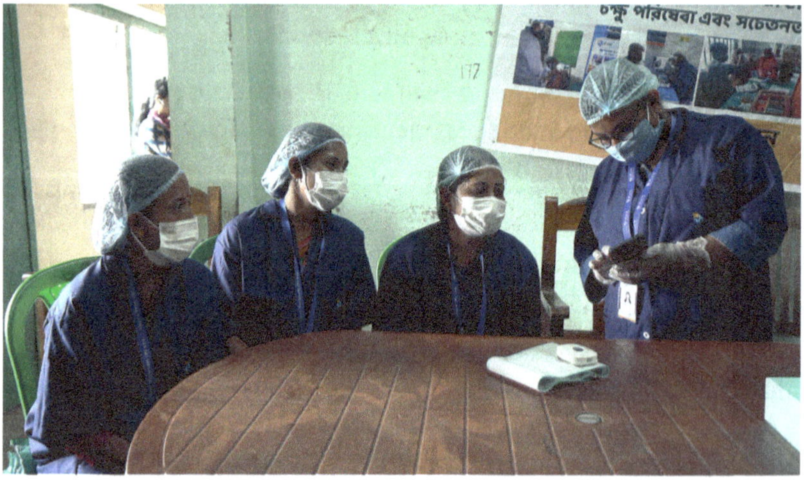

Fig. 4
iKure's community health workers undergoing technology training.

For example, the Naga Telehealth platform was able to serve the entire state of Nagaland during COVID-19, covering *192* Sub Health Centers and *19* Primary Health Centers. The technology platform was later adopted by the Jharkhand Government and has been running successfully across six subcenters in its second phase across the state.

Lessons learned: A collaborative ecosystem is the key to meet the pandemic crisis

The pandemic educated us about the need to break down the silos of healthcare system. Until now, what has been perceived as a system existing *within hospitals and clinics*, digital technology quickly prompted the system to operate *within home.*

Telemedicine forced the government and private players to act speedily to remove the walls between hospitals and the patient's doorstep. This was precisely iKure's aim, breaking silos and seeking integration of different players to make healthcare accessible to all.

The integrated approach set an example for others to follow by reinventing the wheel. iKure invited other actors to share their ability to respond collectively by facilitating remote monitoring services, early screening tests, and vaccination drive within the community as a collective measure.

Data as the lifeblood of healthcare system

Today, the impressive catalyst of digital transformations in healthcare needs health data to boost its effectiveness and reach, and to manage diseases for the future. With the emphasis on generating optimum health outcomes and the focus until now largely on building resources, services, and infrastructure, what got overlooked is the need for granular data. Data must be collected that comprises a comprehensive overview of a person's health.

Gaining access to the right data becomes more challenging if data now is collected through diagnostic devices, sensors, and wearables which follow a trodden path that is disconnected, unstructured, and fragmented, making it even more difficult to make effective healthcare decisions.

In another challenge, when data is machine ready, sharing data between actors in the healthcare ecosystem requires trust and ownership, governed by regulatory norms, privacy rules, and policies. Therefore, in the context of technology, building trust and collecting the right data will be the key to respond to the pandemic as recovery and variants continue.

iKure's association with Open Data Institute (ODI)—Microsoft through Peer Learning Network helped us reap the benefits of data for the public good. iKure was selected to work with different coactors through different lenses to solve significant problems around data.

This led us to work toward real-time knowledge generation using data and inclusion of patient's consent in the data access and utilization to make care *personalized.*

iKure's Living Lab in the town of Baruipur in Sundarbans will focus on exploring data capability through research partners, pharmaceuticals, and device partners. iKure's Living Lab is an effort to bring accessible and affordable healthcare services with technology innovations, training, and support of different stakeholders together to build a platform to solve complex healthcare challenges, specializing in health and innovations. The objective will be to facilitate a common platform for prototyping, testing, validation of health tech devices, research innovations, training and empowerment of grassroots community members as Village Level Entrepreneurs and iKure Community Health Activists, and nurturing health and wellness in the communities of extreme poverty. The lab will leverage data collected through everyday clinical practices as a source to create minimally biased *community-specific knowledge* and implement it in clinical practice in a more agile way.

Recognitions

iKure was presented at a United Nations virtual session in 2021 on "'Building Back Better' after COVID-19 in 2021 through Addressing Health Equity, Mental Health, and Well-being: Contributions of Psychological Science and Practice to Leave No One Behind" specifically on our topic of "A multistakeholder hybrid technology and in-person holistic health and mental health program for communities facing the COVID-19 surge in India with adaptations to other settings" that was coorganized by Dr. Judy Kuriansky and colleagues and sponsored by 16 Member States of the United Nations, WHO, and several NGOs (the International Association of Applied Psychology, American Psychological Association, and United African Congress among others) (reported in Chapter 10 of this volume, Kuriansky et al., 2024). Also, iKure was selected as the Top 50 Last Mile Responders for COVID-19 Response Alliance by the World Economic Forum in 2020. Further, iKure was selected as an innovation included on the "Health In Your Hands" (HIYH) platform, and given an opportunity to describe our approach at various of their forums and webinars during the pandemic, as well as in the HIYH video shown as part of their presentation at the World Governments Summit (which is presented in another chapter in this volume).

Our plans for the future

Going forward, iKure is aiming to build an open data platform where it can share data seamlessly and make sense of it all with the help of Artificial Intelligence. This capability will move the pointers toward continuum of care, benefitting patients and the community as a whole, as well as the medical community of professionals and community health workers.

As we realize the pandemic has overstretched already limited resources countrywide, the rural counterpart is battering the most. This requires a long-haul effort but is also a wake-up call to look beyond existing capabilities of every community and every village to confront any such crisis head on.

While we at iKure garnered immense support to spread our service reach across 10 states and 6400 villages in India, and touched base with Vietnam and MENA countries, we aim to serve 250 million persons by the year 2025, and create 20,000 frontline health workers. Our focus for the future is on building an integrated healthcare system using technology available to everyone far or near, poor or rich, but, the pandemic has forced us to realize that we also must not lose the human touch with each other. Ultimately, iKure is opening avenues to retain this human touch and use technology more than ever before. Using this integrated approach, we believe fundamentally that the way to touch life and to help people in health crises is to touch hearts through insuring good health and happiness.

Reference

Kuriansky, J., et al. (2024). Health equity, mental health, and partnerships to build back after COVID-19: Multistakeholder perspectives at a United Nations event. In J. Kuriansky & P. Kakkattil (Eds.). *Resilient health: Leveraging technology and social innovations to transform healthcare for COVID-19 recovery and beyond* (pp. 115–146). Elsevier.

CHAPTER 87

Adapting and scaling up mental health and psychosocial support response through innovative interventions to transform mental health care in Nepal for COVID-19 recovery and beyond

Kamal Gautam and Pitambar Koirala
Transcultural Psychosocial Organization Nepal (TPO Nepal), Kathmandu, Nepal

Background of COVID-19 and Nepal

A novel coronavirus named severe acute respiratory syndrome coronavirus 2 (SARS-CoV-2 or COVID-19) was first identified in Wuhan City of China in December 2019 (Zhu et al., 2020). The virus caused lower respiratory tract infection and started spreading progressively across the globe taking the form of a pandemic. COVID-19 was then declared as a Public Health Emergency of International Concern by the World Health Organization (WHO). The virus spread throughout the world, including the country of Nepal, causing severe economic and psychological trauma.

This chapter describes the mental health and psychosocial support (MHPSS) response of Transcultural Psychosocial Organization Nepal (TPO Nepal) to respond to the needs of the community at this drastic time and to prepare prevention measures. The activities include technological services as well as in-person offerings, all of which were extensive and comprehensive, and which represent coordination and collaboration with international agencies as well as with the Government of Nepal (GoN).

Nepal, a country situated in South Asia between India and China, is an Official Development Assistance (ODA) listed "least developed country" with one of the world's largest emigrant populations per capita. The Constitution of Nepal assures that health is a fundamental human right. After the promulgation of a new constitution in 2015, Nepal underwent a radical transformation from a unitary administrative system to three tiers of decentralized

Resilient Health. https://doi.org/10.1016/B978-0-443-18529-8.00087-1

governments: a federal government, seven provincial governments, and 753 local governments (Government of Nepal, 2015). Consistent with this, the healthcare structure also devolved into three tiers (federal, provincial, and local health systems) with more autonomy to local governments. At the local or community level (municipality or ward level), the primary contacts for health services are urban health clinics, health posts, and primary health care centers (Thapa et al., 2019).

Nepal's health system fundamentally suffers from constrained health care resources, including physical infrastructure and skilled human resources (Adhikari & Mishra, 2016; Ghimire et al., 2020). COVID-19 further unraveled the inadequacy and fragility of the health care delivery system across the country. Also, the COVID-19 pandemic and lockdown measures taken by nations worldwide to curb the pandemic brought a rise in mental health conditions. COVID-19-related stress, the uncertainty of times, social distancing measures, increase in hospitalizations and deaths due to COVID-19, loss of jobs, and financial security are among the many factors that led to a rise in mental health problems. Some groups of populations are reportedly more vulnerable to developing new mental health conditions, or worsening of preexisting mental illnesses, including those infected with COVID-19, the elderly, those with immuno-compromised status or with preexisting health issues, and health care workers (Tasnim et al., n.d.).

A systematic review and meta-analysis showed that the prevalence of anxiety and sleep disorders during the COVID-19 pandemic was in the range of 23.2%–47% and 2.3%–76.6%, respectively (Deng et al., 2021; Tasnim et al., n.d.). Stress, depression, irritability, insomnia, fear, confusion, anger, frustration, and stigma associated with quarantine were confirmed outcomes in a review of psychological sequelae in samples of quarantined people and health care providers in the United States (US). A few of these consequences persisted after the quarantine was lifted. The extended duration of isolation, inadequate resources, poor access to medical treatment, and ensuing financial losses were all found to be major stressors impacting mental well-being (Brooks et al., 2020). Another systematic review reported increased rates of symptoms of anxiety (6.33%–50.9%), depression (14.6%–48.3%), posttraumatic stress disorder (7%–53.8%), and stress (8.1%–81.9%) in the general population of countries such as Nepal, China, Turkey, Denmark, Spain, Italy, the United States, and Iran (Xiong et al., 2020).

Nepal reported the first case of COVID-19 infection in a 31-year-old Nepali student who returned from Wuhan University, China, in January 2020 (Bastola et al., 2020). The Government of Nepal responded to tackle the COVID-19 pandemic through the implementation of two key strategic approaches: public health prevention and hospital-based interventions (Ministry of Health & Population, 2020). A study identified several gaps in Nepal's health service preparedness against COVID-19 (Shrestha et al., n.d.) that emerged from systemic lack of coordination between the governance system and a gross scarcity of resources (human and nonhuman) to tackle the pandemic at all levels. Specifically, the lack of human resources, inadequate logistic chain management, and laboratory facilities for testing

COVID-19 appeared to have jeopardized the health system preparedness and escalated the pandemic in Nepal. Also, the suicide rate reportedly increased by 20% in the first month of the lockdown imposed by the government during the COVID-19 pandemic, compared to prelockdown months with suicidal tendencies linked to economic instability and crisis, social isolation, stigma, poverty, and unemployment (Poudel & Subedi, 2020). It can be undoubtedly concluded that individual health and public health, as well as emotional and social well-being, were all threatened by the COVID-19 pandemic. A study conducted in Nepal at the beginning of the widespread pandemic recommended that the Government of Nepal must initiate appropriate steps through learning from other nations and utilizing strategies to respond to the mental health epidemic that might emerge in the days to come (Gautam et al., 2020).

Transcultural Psychosocial Organization Nepal (TPO Nepal)

TPO Nepal (www.tponepal.org) is one of the leading mental health organizations in Nepal. It envisions conflict-resolved, resilient communities where local populations have adequate access to multidimensional MHPSS systems. Our mission is to promote psychosocial well-being and mental health of children and families in conflict-affected and other vulnerable communities, through development of sustainable, culturally appropriate, community-based psychosocial support systems. Our organization works on four major thematic areas: capacity building of MHPSS service providers, service delivery to beneficiaries, research, and advocacy with relevant stakeholders to mainstream MHPSS into the government system. During the COVID-19 pandemic, TPO Nepal contributed to initiate and scale up MHPSS responses in Nepal. A number of activities focusing on the four thematic areas of the organization were implemented. Innovative approaches were undertaken on capacity building, MHPSS service delivery, and advocacy.

Key activities and achievements of TPO Nepal in the thematic areas

Capacity building

Following the outbreak of COVID-19 in Nepal, TPO Nepal created a MHPSS team comprising MHPSS program managers, service providers, and a Monitoring and Evaluation (M & E) Officer. The MHPSS service providers included community-based psychosocial workers (CPSWs), community psychosocial counselors (PSCs), clinical psychologists, and a psychiatrist (the first author). With increasing demand for MHPSS services, the CPSWs and PSCs were trained on the provision of psychological first aid (PFA) (World Health Organization, 2013; https://interagencystandingcommittee.org/system/files/2020-05/Basic%20Psychosocial%20Skills-%20A%20Guide%20for%20COVID-19%20Responders.pdf), stress management and self-care (Doing what matters in times of stress: an illustrated guide, 2020), organizing group sessions, responding to helpline calls received, linking and making referrals to other available services and doing follow-ups. These techniques were based on our prior experience of operationalizing the helpline and the Standard Operating Procedure (SOP) for operationalizing helpline of TPO Nepal. These MHPSS service providers were mobilized after adequate capacity building to ensure quality of MHPSS services delivered to the beneficiaries and also that no harm was instituted. Routine periodic supervision and mentoring of all the cadres of MHPSS service providers was ascertained. Additionally, immediately after the imposition of lockdown, the MHPSS service providers and staff members of TPO Nepal were oriented periodically on COVID-19 with recommended precautionary measures at the workplace and seeking help. An Interim COVID-19 Management Guideline was developed and endorsed by the Management Committee of TPO Nepal. This was followed by orientation of all MHPSS staff members on this guideline.

In May, TPO Nepal invited an American psychologist and professor at Columbia University Teachers College, who was very familiar with the culture, to give a training to the staff on the topic of "Building Resilience During COVID-19: A Training for TPO Psychosocial Workers" which covered best practices for coping with crisis (to metaphorically achieve "psychological vaccination" as well as culturally relevant practices). The feedback was very positive, especially given the cultural sensitivity and the interactive activities, which can be applied personally as well as in practice with clients.

Advocacy and policy reforms

COVID-19 Mental Health and Psychosocial Support (MHPSS) preparedness and response framework

An increasing need and demand of MHPSS was seen following the onset of COVID-19 pandemic and imposition of lockdowns in Nepal. Under the leadership of the Ministry of Health & Population (MoHP) of Nepal, and with the technical support from the World Health Organization (WHO) Country Office Nepal, the MHPSS Subcluster of the Health Cluster developed a COVID-19 Preparedness and Response Framework in April 2020 to facilitate implementation of coordinated MHPSS response during COVID-19. This framework was developed with the following objectives:

- To enable stakeholders to plan and organize a set of minimum multisectoral responses to protect and improve people's mental health and psychosocial well-being during COVID 19.
- To identify a set of priority actions in order to provide focused attention by all MHPSS partner organizations.
- To increase efficiency in resource utilization and help in mobilizing additional resources.
- To coordinate the MHPSS response and enhance information sharing among humanitarian actors.

The framework consisted of sections including the background of COVID-19 and MHPSS, objectives of the framework, an action framework with strategic actions and activities and responsible/coordinating agencies, coordination, quality assurance, implementation, and funding. The following were the key strategic actions included in the framework:

i. Develop and disseminate mental health-related messages to the general public and to high-risk populations such as the elderly, people with multiple comorbidities, frontline responders, children, migrant workers, survivors of gender-based violence (GBV), and persons with disabilities including psychosocial disabilities.
ii. Reduce stigma and social discrimination associated with COVID-19.
iii. Address mental health needs of children and adolescents and prepare children, parents, and teachers to go back to school after the lockdown.
iv. Reduce the negative impacts of social isolation and quarantine.
v. Integrate MHPSS responses in health responses at all levels.
vi. Address mental health needs of health and social service providers.
vi. Strengthen primary health care services to provide essential mental health services in the aftermath of COVID-19.
viii. Ensure continuity of essential mental health services.
ix. Strengthen the capacity of local government for MHPSS response.
x. Apply a human rights-based approach in delivering mental health and psychosocial support services.
xi. Foster effective coordination, collaboration, and linkage with other organizations working in the COVID-19 response.
xii. Ensure that the interventions provided by all the partners are of acceptable standards.

TPO Nepal participated in the technical committee to draft and provide inputs in the finalization of the framework. This framework subsequently helped the MHPSS stakeholders to initiate and implement coordinated actions across the country to mitigate the psychosocial impacts of COVID-19. The MHPSS activities implemented, and services delivered by various government and nongovernment organizations including hospitals and universities, were reported to the Mental Health subcluster over biweekly meetings.

Psychological First Aid (PFA), stress management, and grief counseling training to frontline health workers

TPO Nepal provided technical assistance to the National Health Training Centre (NHTC), Ministry of Health & Population, in the development and endorsement of a training curriculum on PFA, stress management, and grief counseling. The training package was based on publications from WHO (Doing what matters in times of stress: an illustrated guide,

2020; World Health Organization, 2013) and the Inter-Agency Standing Committee (IASC) (https://interagencystandingcommittee.org/system/files/2020-05/Basic%20Psychosocial%20Skills-%20A%20Guide%20for%20COVID-19%20Responders.pdf). This package was subsequently rolled out in several government hospitals across all the provinces of country. TPO Nepal also provided support to the Psychiatrists' Association of Nepal (PAN) in building the capacity of facilitators to roll out the training to medical doctors. This technical assistance of TPO Nepal subsequently contributed to managing stress and burnout among healthcare providers while enhancing their capacities to offer support to patients needing care at their facilities.

Information, education, and communication through adaptation of global standards and mobilization of experts

Interim briefing note addressing mental health and psychosocial aspects of COVID-19 outbreak

This briefing note was developed by the Inter-Agency Standing Committee (IASC) Reference Group on Mental Health and Psychosocial Support in Emergency Settings (https://interagencystandingcommittee.org/iasc-reference-group-on-mental-health-and-psychosocial-support-in-emergency-settings). The IASC Reference Group, established in 2007, is a unique collaboration of United Nations (UN) agencies, national and international nongovernmental organizations, and international agencies providing mental health and psychosocial support in emergency settings. It produces guidelines and tools for interventions and coordination of activities. This briefing note, last updated in 2020, summarizes key mental health and psychosocial support (MHPSS) considerations in relation to the 2019 novel coronavirus (COVID-19) outbreak (Inter-agency Standing Committee, 2020). This briefing note, translated into Nepali by TPO Nepal, provided key guidance to MHPSS stakeholders in Nepal about planning and implementing MHPSS responses in Nepal.

Remote Psychological First Aid during a COVID-19 outbreak

Psychological First Aid (PFA) is a method of helping people in distress to make them feel calm and supported and to cope better with their challenges. It involves paying attention to reactions, active listening and, if needed, practical assistance, such as problem-solving, help to access basic needs, or referral to further available services for assistance. PFA is often used in humanitarian setting and serves as a first response to those affected. This guideline, developed by the International Federation of Red Cross and Red Crescent Societies (IFRC) in 2020 (International Federation of Red Cross and Red Crescent Societies. Psychosocial Centre (English), 2020), provides insight to set up and operationalize different kinds of remote services for the public during the time of a COVID-19 response. TPO Nepal translated and adapted the remote PFA guide in Nepal, which was widely used by MHPSS stakeholders to set up and operationalize remote methods of MHPSS service delivery to people affected by the COVID-19 outbreak in Nepal. Since the risk of COVID-19 contagion and government-imposed restrictions limited in-person PFA delivery, TPO Nepal adapted the virtual/remote platform for service delivery.

Basic psychosocial skills: A guide for COVID-19 responders

This guide, developed by the Inter-Agency Standing Committee's Mental Health and Psychosocial Support Reference Group (https://interagencystandingcommittee.org/system/files/2020-05/Basic%20Psychosocial%20Skills-%20A%20Guide%20for%20COVID-19%20Responders.pdf), is intended for use by health and social workers, emergency responders, and people who are providing support to vulnerable family members or members of their community. The guide offers practical advice for taking care of one's own mental well-being (e.g., ability to deal with stress, function productively), communicating with empathy (e.g., understanding others' feelings), and helping people who are suffering from psychological distress (e.g., to manage anxiety, sadness, worries, sleep disturbance). This guide was translated for use in Nepal by TPO Nepal. It has helped the responders to use basic psychosocial skills during MHPSS service delivery to the beneficiaries in community.

Press briefing and media interview

The first author of this chapter appeared in a press briefing organized by the Ministry of Health & Population in 2020, which was a live program broadcasted through the National Television viz. Nepal Television (NTV) of the Government of Nepal. The briefing focused on suicide and self-harm whereby information was disseminated to the large public audience on suicide statistics during COVID-19, causes, warning signs, preventive efforts, and seeking help. Similarly, a number of experts, psychosocial counselors, and program managers from TPO Nepal appeared in several media interviews about MHPSS, to disseminate information on the psychosocial impacts of COVID-19 and how to seek MHPSS care.

Podcast series

TPO Nepal was a technical partner in "Stress counselling through podcast in response to COVID-19" organized by the Youth Thinkers' Society, supported by UN WOMEN Nepal (https://msha.ke/ytscares). The podcasts consisted of interviews with MHPSS experts from TPO Nepal. This initial program led to the development of a series consisting of 20 different podcasts which focused on the MHPSS impacts of COVID-19 for specific populations, such as women and children, people in quarantine and isolation, media personnel, people with lived experience, persons with disabilities, the LGBTQI population, and the elderly. The podcasts were structured to provide information on the MHPSS impacts of COVID-19; demystify myths about COVID-19; identify steps to seek help; provide tips on self-care; and demonstrate stress management techniques, for example, deep breathing exercises, relaxation exercises, and grounding. The podcast series is available on YouTube (https://msha.ke/ytscares). Informal interactions with youth during consultation for preparing the episodes of the podcast series revealed that the series was beneficial to a large population, especially adolescents and youth.

Participation in multistakeholder webinars

TPO Nepal participated in several multistakeholder webinars over the course of the COVID-19 period, in collaboration with the Permanent Mission of Nepal to the United Nations, representing government; the International Association of Applied Psychology (IAAP) as a representative of civil society, also an advocacy body at the United Nations; media, through Cyber Headline which produced and broadcasted the events; youth, who founded NGOs in Nepal working on behalf of youth rights; and academia. The webinars were all organized by Ms. Sadikchhya (Meera) Khanal, Nepali graduate student at Columbia University Teachers College, with her professor, Dr. Judy Kuriansky (mentioned before having conducted the resilience training for TPO Nepal), described in detail in another chapter in this volume (Khanal & Kuriansky, 2024). The first webinar was held on June 13, 2020, entitled, "*Coping during COVID-19: Honoring Nepal and Advice about Mental Health and Wellbeing*," in which the first author of this chapter, Dr. Kamal Gautam, Executive Director of TPO Nepal, presented on tele-counseling services available in Nepal through quarantine centers and TPO Helpline. The collaboration was continued among the partners from the earlier phases over another event, for the Commission on the Status of Women, NGO, New York, with a focus particularly on women's issues, consistent with the theme of that conference. Further, TPO Nepal presented at a high-level global online event entitled, "*Building Back Better*" *after COVID-19* during the important United Nations High-Level Political Forum, in the section about "Voices from the Field" among selected similarly highly accomplished presenters reporting about the practical implementations of their work in psychosocial and physical health coping with the pandemic. Dr. Gautam presented about an important collaboration of TPO Nepal with the Ministry of Health and Population of Nepal that had just taken place in June, whereby TPO Nepal contributed to the development and roll out of training on stress and grief management for 500 government healthcare workers. This work is presented in another chapter in this volume (Khanal & Kuriansky, 2024) about this model of a multistakeholder project.

Innovations in MHPSS service delivery

Digital platform for MHPSS service delivery to health care providers

Within a few months following the onset of COVID-19 pandemic, the health care system of Nepal was overwhelmed (https://mewellnepal.org.np). Health workers in Nepal had been facing increased workload specifically due to contact tracing, providing services at quarantine and isolation centers, operating out- and in-patient departments, and inadequacy of staff.

Taking this into consideration, TPO Nepal initiated an innovative approach to deliver mental health and psychosocial support to health care providers, in the form of a web-based platform developed in collaboration with the MoHP and WHO Nepal. This platform provided health workers with services for screening mental health, psycho-educational materials, practical tips for stress management and coping, and taking a virtual appointment with specialists (a psychiatrist and clinical psychologist) whenever deemed necessary.

Remote MHPSS service delivery

TPO Nepal had been operationalizing a toll-free helpline at the phone number 16600102005 for 5 years prior to the pandemic. This helpline offers psychosocial support to individuals dialing in to the helpline and then links them with a specialist for consultation whenever required. Following the outbreak of COVID-19 and imposition of lockdown in Nepal, TPO Nepal extended its helpline hours to operate between 8 am and 6 pm every day. Similarly, TPO Nepal provided MHPSS services via other virtual platforms such as Viber, WhatsApp, and Zoom. These services were delivered by trained community psychosocial counselors, psychologists, and a psychiatrist. The sessions were individual as well as group based. Individual sessions were delivered on a one-to-one basis by the service providers of TPO Nepal. Group sessions were delivered to a group of 16–24 people at a time via Zoom. Group sessions were predominantly focused on stress management and self-care, managing burn out, care for caregivers, group psycho-education, and PFA.

MHPSS services at the quarantine and isolation centers

Thousands of Nepalese migrant workers returning from abroad were screened and kept at quarantine centers following arrival to Nepal during the COVID-19 outbreak. Similarly, those who tested positive for COVID-19 during contact tracing by health care providers were kept at isolation centers until they tested negative. People at those isolation and quarantine centers were living under poor conditions and inadequate supplies, and suffering with fear of contagion, dying, and stigmatization as well as worries about their family members and symptoms of psychological distress, including insomnia, anxiety, sadness, and sometimes suicidal ideation, thoughts, and attempts. Under this pretext, TPO Nepal mobilized trained community psychosocial counselors to those centers who organized group psycho-education sessions, where they offered PFA and tips on stress management and how to seek MHPSS support when needed. In addition, recommended precautionary measures were ascertained during in-person MHPSS service delivery at those centers. During this process, TPO Nepal collaborated with the respective government ministries, line agencies, and local authorities. In addition, trained psychosocial counselors and psychologists of TPO Nepal offered phone-based psychosocial support and follow-up services to individuals who tested positive and had been staying in isolation.

Research and evidence generation

COVID-19 social media survey

One month after the government-imposed lockdown following the COVID-19 outbreak in Nepal, TPO Nepal conducted an online survey to study the psychological impacts of COVID-19 among the Nepalese population. Data were collected through social media. A total of 2082 respondents completed the survey. Among them, only those respondents who were currently residing in Nepal ($n = 2014$) were included in the analysis. We excluded those who were not residing in Nepal at that time as they had different contexts of COVID-19, lockdown, restrictions, access to health care, etc. The results suggested that 50% of the respondents had at least one symptom of psychological distress, and 32% suffered from two or more symptoms of psychological distress such as restlessness, fearfulness, anxiety and worry, and sadness in the past two weeks preceding the survey date (Gautam et al., 2020).

Assessment of suicide and self-harm during COVID-19 pandemic in Nepal

TPO Nepal conducted a mixed-method study between July and September 2020 to assess the status of suicide and self-harm during COVID-19. Using secondary data from the Nepal Police, the study analyzed suicide trends over the previous four Nepalese fiscal years including the last 4 months post-COVID-19. The results showed that suicide rates had been increasing every year more so than ever following the COVID-19 pandemic in Nepal. Additionally, qualitative interviews were conducted with MHPSS service providers and stakeholders. The qualitative interviews revealed an increase in the prevalence of mental illness and associated risk factors for mental illness, suicide, and self-harm following COVID-19.

Conclusion and recommendations for the way forward

The COVID-19 pandemic not only posed a threat to physical health but also affected mental health and well-being of people. Additionally, the pandemic itself was novel and its various aspects were poorly understood, including preventing transmission, effectiveness of lockdowns, and managing widespread impacts. The health care delivery system of Nepal was already overwhelmed prepandemic. Under these adverse circumstances, services were vastly expanded and various alternative remote methods of service delivery were developed and adopted. TPO Nepal translated and adapted various globally available MHPSS standards and guides and then adopted a virtual platform of MHPSS service delivery. Tele-mental health, group psycho-education, PFA sessions, and services to people in quarantine and isolation were key activities delivered through digital platforms using the telephone, Viber, WhatsApp, and Zoom to minimize, but complement, in-person contacts. TPO Nepal provided technical assistance to the MoHP and collaborated with various government and nongovernment agencies (including UN agencies) to formulate the COVID-19 MHPSS Preparedness and Response Framework, develop a national-level training curriculum for frontline workers, and implement and scale up the necessary and urgent MHPSS response in Nepal.

References

Adhikari, B., & Mishra, S. R. (2016). Urgent need for reform in Nepal's medical education. *Lancet, 388*(10061), 2739–2740.

Bastola, A., Sah, R., Morales, A. J. R., Lal, B. K., Jha, R., Ojha, H. C., & Pandey, B. D. (2020). The first 2019 novel coronavirus case in Nepal. *The Lancet, 20*(3), 279–280. https://doi.org/10.1016/S1473-3099(20)30067-0.

Brooks, S. K., Webster, R. K., Smith, L. E., et al. (2020). The psychological impact of quarantine and how to reduce it: Rapid review of the evidence. *Lancet, 395*, 912–920.

Deng, J., Zhou, F., Hou, W., Silver, Z., Wong, C. Y., Chang, O., Huang, E., & Zuo, Q. K. (2021). The prevalence of depression, anxiety, and sleep disturbances in COVID-19 patients: a meta-analysis. *Annals of New York Academy of Sciences, 1486*(1), 90–111. https://doi.org/10.1111/nyas.14506. Epub 2020 Oct 2. PMID: 33009668; PMCID: PMC7675607.

Doing what matters in times of stress: an illustrated guide. (2020). *Doing what matters in times of stress: an illustrated guide*. Geneva: World Health Organization. Licence: CC BY-NC-SA 3.0 IGO.

Gautam, K., Adhikari, R. P., Gupta, A. S., et al. (2020). Self-reported psychological distress during the COVID-19 outbreak in Nepal: Findings from an online survey. *BMC Psychology, 8*, 127. https://doi.org/10.1186/s40359-020-00497-z.

Ghimire, U., Shrestha, N., Adhikari, B., et al. (2020). Health system's readiness to provide cardiovascular, diabetes and chronic respiratory disease related services in Nepal: analysis using 2015 health facility survey. *BMC Public Health, 20*(1), 1163.

Inter-agency Standing Committee. (2020). *Interim briefing note addressing mental health and psychosocial aspects of COVID-19 outbreak*. Geneva: Inter-agency Standing Committee.

International Federation of Red Cross and Red Crescent Societies. Psychosocial Centre (English). (2020). *Remote psychological first aid during a Covid-19 outbreak*. https://pscentre.org/?resource=remote-psychological-first-aid-during-the-covid-19-outbreak-interim-guidance-march-2020.

Khanal, M., & Kuriansky, J. (2024). Promoting resilience in the Nepali community in response to the COVID-19 pandemic: A multistage multistakeholder partnership addressing the United Nations Agenda. In J. Kuriansky, & P. Kakkattil (Eds.), *Resilient health: Leveraging technology and social innovations to transform healthcare for COVID-19 recovery and beyond* (pp. 601–627). Elsevier.

Poudel, K., & Subedi, P. (2020). Impact of COVID-19 pandemic on socioeconomic and mental health aspects in Nepal. *International Journal of Social Psychiatry*. 002076402094224. doi:10.1177/0020764020942247.

Shrestha N., Mishra S.R., Ghimire S., et al. Health system preparedness for COVID-19 and its impacts on frontline health-care workers in Nepal: A qualitative study among frontline health-care workers and policy-makers. Disaster Medicine and Public Health Preparedness doi:https://doi.org/10.1017/dmp.2021.204.

S. Tasnim, M. Rahman, P. Pawar, X. Chi, Q. Yu, L. Zou, A. Sultana, E. Lisako, J. McKyer, P. Ma, M.M. Hossain. Epidemiology of sleep disorders during COVID-19 pandemic: A systematic scoping review. medRxiv. https://doi.org/10.1101/2020.10.08.20209148.

Thapa R., Bam K., Tiwari P., Sinha T.K., Dahal S. Implementing federalism in the health system of Nepal: opportunities and challenges. International Journal of Health Policy and Management. 2019;8:195–8. https://doi.org/10.15171/IJHPM.2018.121.18.

World Health Organization. (2013). War Trauma Foundation and World Vision. In *Psychological first aid: Facilitator's manual for orienting field workers*. Geneva: WHO.

Xiong, J., Lipsitz, O., Nasri, F., Lui, L. M. W., Gill, H., Phan, L., Chen-Li, D., Iacobucci, M., Ho, R., Majeed, A., & McIntyre, R. S. (2020). Impact of COVID-19 pandemic on mental health in the general population: A systematic review. *Journal of Affective Disorders*. https://doi.org/10.1016/j.jad.2020.08.001.

Zhu, N., Zhang, D., Wang, W., Li, X., Yang, B., Song, J., & Tan, W. (2020). A novel coronavirus from patients with pneumonia in China 2019. *The New England Journal of Medicine, 382*, 723–733. https://doi.org/10.1056/NEJMoa2001017.

CHAPTER 88

The "Doctor for every citizen" Smart Telehealth Initiative: Increasing resilience and services in the UAE Dubai Health Authority during and beyond the COVID-19 pandemic

Ola Aldafrawy[a], Manal Mohamed Omran Taryam[b], Alia Mohammad Rafi Aldallal[a], Suad Hashim Ahmed[a], and Ashraf Swidan[a]

[a]Specialized Program Services Department (SPSD), Primary Health Care Sector (PHCS), Dubai Health, Dubai, United Arab Emirates, [b]Dubai Health Authority, Dubai, United Arab Emirates

Introduction

Telemedicine refers to the delivery and management of remote clinical medical services that is becoming increasingly popular as a way to deliver healthcare in an efficient, effective, and cost-effective way. This approach became even more necessary during the COVID-19 pandemic, when lockdowns and physical distancing became necessary in order to stop the spread of the virus. The United Arab Emirates has long been at the forefront of technological innovations in all fields, including health. Accordingly, this chapter describes the "Doctor for Every Citizen" telemedicine service that waslaunched in 2020 during the COVID-19 pandemic by the Dubai Health Authority, using the latest digital technology offering video or voice call options to deliver accessibility and convenience for medical staff and citizens while maintaining safety, dependability, and high quality care.

Background

The "Doctor for Every Citizen" initiative is outlined in the 5th article of the Fifty-Year Charter, launched by His Highness Sheikh Mohammed bin Rashid Al Maktoum, vice president and prime minister of the UAE and ruler of Dubai. The Fifty-Year Charter consists of nine Articles

Resilient Health. https://doi.org/10.1016/B978-0-443-18529-8.00088-3

that represent the vision of His Highness toward the future of Dubai and the ideal lifestyle of its citizens.

Importance: This initiative\service offers all Dubai citizens the best level of medical services anytime and anywhere, to sustain the welfare and happiness of all Dubai citizens. This initiative empowers the Dubai Health Authority (DHA) to become a leader in the region, using the latest digital technology for better patient care, enhancing efficiency in the health sector and better supporting medical professionals as well as providing patients with easy, convenient, and smart accessibility to care. Coincidentally, the service is environmentally responsible since eliminating the need for a building precludes the challenges to the climate of depending on physical health structures.

Problem/challenge

In the spirit of the UAE's commitment to modernizing society through technology, the Dubai Health Authority is offering better health services for people living in Dubai, which need to be delivered easily, with more accessibility and attention to low waiting time, reducing patient travel to medical facilities, and chances of contamination and infection, especially for high-risk individuals and "people of determination."

How and where this innovation is being implemented, technology being used, the target, and who and how the service is being delivered

Our technology used is OMNIChannel which provides a seamless experience whether online, by phone, or in person. This technology system with its mobile healthcare applications provides patients with options to approach a doctor and solve their problems faster than the traditional way, thereby becoming a highly economical method for patients to better access care. On the client side, we use the DHA "Smart Mobile Application," a service which opens a channel between the physician and the client, integrated with electronic medical records.

Impact/successes

Our "Doctor for Every Citizen" service offers many solutions for the healthcare crisis, proven to have significant positive outcome, as one of the new services in the country, now considered one of the foundations of our healthcare delivery (Swidan et al., 2022).

The service has seen exponential increase in demand since the beginning of the COVID-19 pandemic by almost 400%, reflecting the adoption of "smart services" into healthcare. Smart services are defined as utilizing intelligent technology into our traditional healthcare industry such as usage of DHA mobile applications to dynamically access patient information, directly connect people with the health services, and establish accessibility to care, while actively responding to medical ecosystem needs in an intelligent comprehensive manner. Patients have found easier accessibility to care, which has changed their mindset, so they are more willing to book their next medical appointment, and thus take care of their health. Patients can easily scroll through the list of physicians and get to know their doctor's background and interests, to be able to choose, and match, according to their needs.

The service runs 24/7, seven days a week, with flexibility to book, reschedule, or cancel appointments.

The service is also economically viable, since we are able to offer more care to more patients using telemedicine. Thus family medicine consultations—as one of the staples of our regular health services—are available without the need for us to construct a new building or space. In this way, the service can be considered as if it were a new standalone health center, except that it is present virtually. In addition, offering tele-services has the positive environmental impact of reducing our carbon footprint. The tele-service is extended to all overseas nationals, since the service operates 24 hours, making access possible in any time zone.

Challenges

Due to the increased demand during COVID-19 pandemic, we had to increase the number of work stations licensed for telemedicine, i.e., computer desktops with specified high quality camera and headphones. Also, telecom policy had to be developed and policies for telemedicine initiated. Manpower and capacity limitations were overcome by increasing training for the service.

Strategic partners

Our pharmacy partner is Dawaee Service, which offers free delivery of prescriptions and a fully remote service.

Business plan

Our business plan is that we:

(1) are online with "Dubai Paperless Strategy 2021" which focuses on being totally paperless, which was achieved 100%.

(2) added insurance companies coverage for easier accessibility to care for different sectors of the population.
(3) reduced client waiting time and travel to the clinic.
(4) added sick leave for patients.
(5) granted and ordered vaccinations.
(6) instituted a sustainability plan—which is the goal of the government for all Dubai innovations—including reducing our carbon footprint, since there is a reduction of in-person visits. Specifically, we calculated estimated customers' reduced travel of 692,420 km by eliminating the need to travel to and from the health center, and an estimated reduction of CO_2 emissions, as well as 90,014-kg reduction of disposable medical supplies and personal protective equipment.
(7) expanded the service to include specialist services in various medical fields, like "Smart Geriatric Health Services," "Smart Occupational Health Services," "Smart Multidisciplinary Diabetes Care Team," "Smoking Cessation Clinic," "Smart Post-natal Services," "Smart Osteoporosis Health Services," and "Smart Rehabilitation Services" and a "Smart Lifestyle Clinic."

How COVID-19 has affected our work and how we responded/pivoted

Efforts since the COVID-19 pandemic and expansion of the service included:

(1) Increase the number of stations from six to 16.
(2) Train 100+ physicians.
(3) Increase the total number of physicians providing the service from a total of 25 to 30, with rotations around the clock.
(4) Reduce consultation time from 20 to 15 minutes, to accommodate more clients.
(5) Provide free-of-charge accessibility to care for assessment of COVID-19 patients who were living at home, in quarantine, to reduce their anxiety.
(6) Conduct almost 40K+ COVID-19 consultations, with accessibility to care not limited to Dubai; in fact, other Emirates residents were utilizing the service to access care.
(7) Notify the Command-and-Control Center (for COVID-19 patients) with lists of close contacts for confirmed positive cases, to further aid in contact tracing and triage of risky and vulnerable cases for further assessment.
(8) Operate the "Doctor for Every Citizen" service in accordance with the national governmental campaign that took place during the lockdown, urging people to #StayHome, while continuing to provide direct access to vital and routine healthcare services.
(9) Connect with Dawaee pharmacy service for delivery of medication to patients in all Emirates.

(10) Conduct overseas telemedicine consultations.
(11) Increase utilization rate of appointment slots, which reached 97%.
(12) Train DFEC staff about skills related to mental health (e.g., psychiatry and psychology) in collaboration with Rashid Hospital, to provide holistic medical care consistent with the needs of citizens, and policy directives of the United Nations 2030 Agenda for Sustainable Development and of Universal Health Coverage and the standards and definition of total healthcare by the World Health Organization.
(13) Train DFEC staff in sign language for better communication with people of determination.

Conclusion and future plans

The approach in response to the medical emergency of the COVID-19 pandemic was shown to be a highly effective way to pivot to technological methodologies for healthcare delivery, consistent with the Dubai government's emphasis on technological innovation. Given the considerable success of the initiative, plans are to continue to monitor the efficiency of the service, expand its reach, and increase the offerings to coincide with citizens' needs for an effective model of a resilient health system.

Additional information

DHA website: www.dha.gov.ae
Social media accounts: DHA_Dubai

Reference

Swidan, A., Alnoon, N. A., Makki, I., Zidan, M., Alhammadi, H., Rahmani, N., Alameeri, A., Al Falasi, A., Fakhroo, A., Al Mulla, J., & Al Marzooqi, L. (2022). Effect of COVID-19 pandemic on patient utilization of the telemedicine services in Dubai. *Dubai Medical Journal*. https://doi.org/10.1159/000522346. https://www.karger.com/Article/Pdf/522346.

CHAPTER 89

Innovation: Digital healthcare model for noncommunicable diseases: The case of PharmAccess' NCD Care program, Ghana

Alex Israel Attachey and Maxwell Antwi
PharmAccess Foundation, Accra, Ghana

Background

In April 2020, PharmAccess Foundation with the Ministry of Health, Ghana; University of Ghana Medical Centre; and Luscii, a technology company, partnered together to launch 'CovidConnect' to complement Ghana's COVID-19 response. This digital service, built around an app, allowed for self-monitoring and symptom screening to identify which individuals required hospital care, while providing people with credible self-help information and care management, all in near real time.

Due to the app's versatility, it was adapted for noncommunicable diseases (NCDs) which are growing in prevalence in Ghana. By January 2021, this was piloted for remote monitoring of patients with hypertension and diabetes. Now called 'NCD Care,' the digital service complements Ghana's NCD strategy in line with the national health policy.

NCD Care has self-management tools and built-in protocols that allow diabetic and hypertensive patients to enter their daily blood pressure or blood sugar measurements into their mobile app on their phone, which is monitored and managed remotely by their doctors.

NCD Care allows:

- Patients to be delegated to appropriate care based on risk levels.
- Self-monitoring by patients, reducing frequency of clinic visits.
- Health information, lifestyle advice, and medication reminder messages sent to patients to improve adherence and self-management.
- Patients to review their data and become motivated to achieve their targets.

Resilient Health. https://doi.org/10.1016/B978-0-443-18529-8.00089-5

Problem/challenge

According to WHO, noncommunicable diseases (NCDs) kill 41 million people annually, with 77% in low- and middle-income countries. In addition to the epidemiological burden on health systems, NCDs threaten national economies, jeopardize global health security, and hamper development targets.

In Ghana, NCDs (such as cancer, cardiovascular disease, chronic respiratory diseases, and diabetes) cause 45% of all deaths. It has become impossible to sustainably include and retain the growing group of NCD patients in care using the current service and pricing model. There is also a need to help individuals modify behavior through lifestyle management, medicine adherence, and education.

The COVID-19 pandemic, with its associated isolation mandates, accelerated the acceptance of digital health technologies. Leveraging digital innovations, like NCD Care, helps address the rising prevalence of 'lifestyle associated' NCDs across Ghana and even across the entire continent of Africa, which have been brought on by social and economic changes. This chapter provides first evidence for a digital NCD case in Ghana.

How and where this innovation is being implemented, technology being used, target, who and how the service is being delivered

NCD Care is a digital application with built-in protocols for management and control of patients with diabetes and hypertension. The service is monitored remotely by a clinical team through a web interface, defined automated messages, and follow-up calls where required. The users' interface is a mobile app, in which patients enter their daily measurements and access responses from their doctors. The application also provides educational materials to users.

The app was developed by Luscii, a Dutch social enterprise, and implemented with PharmAccess and partners in Ghana, Nigeria, and Kenya as a means to identify and manage COVID-19 patients.

In Ghana, the application has been adapted for NCDs and is now used at the University of Ghana Medical Centre (UGMC), the 37 Military Hospital, and Greater Accra Regional Hospital (GARH), all public facilities and targets patients diagnosed with diabetes and hypertension.

Impact

NCD Care, the first of such initiative in Ghana, is a digital service used to reduce the burden on healthcare staff by allowing patients to digitally input their daily NCD measurements for remote monitoring and improved care quality. It also provides educational information and instigates behavioral changes.

A pilot is being carried out in three leading hospitals in Ghana, representing the country's social stratification. A considerable number of patients (N=500), including adults and adolescents, are enrolled. Initial outcomes show that on average, 55% of patients using "NCD Care" measure their health indicators at least once per week, while over 80% measured at least twice per month, and their systolic blood pressure values seem to stabilize over time. Overall, over 50% of patients using NCD Care have improved clinical outcomes or controlled.

In a qualitative survey conducted, physicians and nurses in the program emphasized that the digitally supported remote monitoring helps them work more efficiently, allowing them to monitor more patients with their same resources. This has far-reaching impact, as it reduces the burden on an already strained health staff and system.

Challenges

Ghana's healthcare system is predominantly geared toward identifying and treating communicable diseases, and the chronic nature of NCDs requires a new approach. The rise in NCDs compounds existing problems around healthcare delivery and financing in Ghana. Some of these challenges include limited access to quality NCD care and medication, low awareness on disease management, and avoidance of complications both among patients and at the primary healthcare level. In rural areas, tests and medication are often not available in facilities, resulting in patients incurring travel costs and long travel times. Many clinics are overburdened, without enough doctors to treat the growing number of NCD patients. And, there is the problem of medicine stockouts and access to quality medicines.

Sustainably financing NCD care is another major challenge, as potential payers and donors are often reluctant to cover NCD care and prevention in insurance packages because of the chronic nature of the conditions and unknown long-term costs. The innovation works around these problems.

Strategic partners

PharmAccess works with Ghana's National Health Insurance Authority (NHIA), which has 60% population coverage, supporting data analytics, digitizing claims (with a CLAIM-it app), and enrollment (10.8 to 16 million people from 2018 to 2021) by integrating the National Identification System.

We partner with the Healthcare Facilities Regulatory Agency (HEFRA) and the Christian Health Association of Ghana (CHAG). CHAG, with 340 facilities, provides a third of Ghana's healthcare services. Collectively, we partner with the FDA Ghana on a digital medicines' procurement platform, Med4All.

For CovidConnect and NCD Care, we partnered with various government, private sector, and academic partners, including the Ministry of Health, Luscii, the University of Ghana Medical Centre, the 37 Military Hospital, and the Greater Accra Regional Hospital.

Business plan

The digital service model that PharmAccess has developed creates transparency in NCD care costs and leverages digital tools to improve treatment adherence by providing technical assistance and capacity building. By increasing efficiency and lowering the cost of NCD care, the model supports the healthcare systems to respond to the increasing NCD burden.

PharmAccess has vast experience in developing and testing digital care models and financing approaches. Our program, "NCD Care," contributes to improved clinical care outcomes, promotes treatment adherence, and reduces costs. Patients are also empowered to self-manage with remote support and clinic support when needed.

This model contributes to Universal Health Coverage, by providing insight into costs of NCD care and piloting new models to improve cost-effectiveness, as well as improving adherence and preventing complications associated with NCDs like diabetes and hypertension. The innovation also contributes to the overall NCD knowledge base through research involving local and international research institutions.

Overall we are working with diverse partners to scale sustainably.

How COVID-19 sparked digital self-management of NCDs

During COVID-19, health services got overburdened, creating the challenge of early detection of suspected cases of COVID-19 and management of low-risk patients to free up capacity. COVID-19 deaths were mainly due to comorbidities; thus, the digital platform, CovidConnect, was adjusted to identify COVID-19 comorbidities as well.

A digital self-management tool with diabetes, hypertension protocols, and educational messages was added to our initial COVID-19 response, to reduce the burden of healthcare staff and improve the quality of care. The objective was to secure enhanced self-protection for people living with comorbidities and adherence to treatment of people living with COVID-19-related comorbidities, which became the useful new program of "NCD Care."

Additional information

https://ncdcare.net.gh/

CHAPTER 90

Redesigning healthcare to reach the 52% without access: reach52, a health-tech social enterprise

Logan Ansell
reach52, Singapore, Singapore

Background

reach52 is a healthtech social enterprise enabling rapid impact in rural low-resource contexts. We build networks of community health agents in disconnected communities, powered by an award-winning, data-driven tech platform, creating an ecosystem of businesses, governments, and social organizations to deliver targeted, scalable health campaigns across low- and middle-income countries (LMICs).

reach52's pioneering revenue model provides impact-on-demand for our B2B clients in the pharmaceutical, consumer health, and Fast-Moving Consumer Goods (FMCG) sectors. These clients register for longer-term subscriptions, billed only for the specific quantified outputs we achieve. These outputs support corporate social responsibility targets, along with commercial aims, through direct engagement with hard-to-reach populations in frontier and emerging markets. To achieve our impact, we recruit, train, and manage networks of digitally equipped health agents across select areas of Cambodia, India, Indonesia, Kenya, South Africa and the Philippines, as shown in Fig. 1. We offer reach52 agents a range of on-demand monetized tasks through the reach52 mobile application focusing on community health promotion, population screening, and health worker upskilling. Agents are then compensated based on the number of tasks completed and the tasks' associated dollar value. These Agents are supplemented by a growing network of partnered retailers, who support last-mile access to medicines and health commodities. By connecting local populations to a range of affordable healthcare products and services, we improve prevention, knowledge of risk factors, and build local health system capacity.

Resilient Health. https://doi.org/10.1016/B978-0-443-18529-8.00090-1

Fig. 1
A reach52 agent in Cambodia collects health data from a resident, using an offline-first app designed to provide access to a full range of health services in areas with low connectivity.

According to Edward Booty, CEO of reach52, "The pandemic has shone the light on how essential health system capacity building is, and how public and private partnerships are essential to enable this. At reach52, we want to continue to do our part to drive data-driven health campaigns at scale; building digital-first models that empower both government health providers and our private and non-profit partners."

Problem/challenge

reach52 addresses issues of health equity by increasing access and quality of services for disconnected rural populations. We are driven by the fact that 52% of the world lacks access to essential health services (a statistic from which we derive our name). There is a severe shortage of skilled healthcare professionals in most rural communities, where many residents live over an hour away from healthcare providers, adding another barrier to the affordability and accessibility of necessary healthcare. Many LMICs have seen an explosion in digitally supported health services, but these models are typically designed to support affluent urban populations. We see very little innovation in the way that health is being delivered for rural residents in the "missing middle," those who are typically too poor to afford top-tier care yet not poor enough to qualify for various social development programs. Further, these populations are typically underserved by existing government health systems, which often lack resources for prevention and treatment of chronic conditions. This is especially concerning as 80% of all deaths related to noncommunicable diseases (NCDs) are now occurring in LMICs.

How and where this innovation is being implemented, technology being used, target, who and how the service is being delivered

Our campaigns are powered by reach52 access, an offline-first Android app designed to support the work of our agents in the field, while providing powerful data insights for program administrators and B2B clients. As shown in Fig. 2, this data includes demographics and questions for a health profile. The use of our tool promotes a data-informed precision public health approach. Indicators can be tracked through the platform, simplifying monitoring and evaluation tasks and better ensuring that we are accurately measuring programs' outputs and impact. Under our programmatic approach: (1) reach52 agents "onboard" residents to our digital platform by collecting individual health/demographic data to create a health profile; (2) reach52 analyzes the aggregated data, working with stakeholders to develop corresponding interventions to respond to local health needs; and (3) in cooperation with local health systems, reach52 agents implement these community primary care interventions, across a range of disease areas. Services are targeted at individuals earning between US$3 and $8/day in rural areas of LMICs.

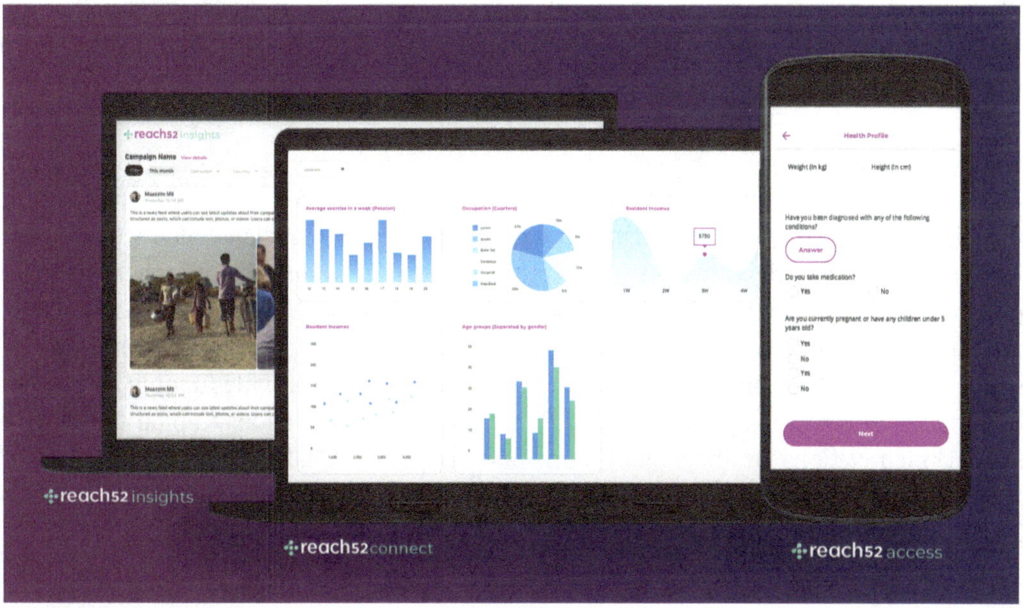

Fig. 2

Individual-level data are collected by agents using reach52's Android app, "reach52 access." We access the data and run campaigns through "reach52 connect," a web-based campaigns platform and partners track outcomes and impact achieved via "reach52 insights", our partner dashboard.

Impact

We now have 1.4 million beneficiaries across our six markets, implementing new and sustainable programs with leading global firms. reach52 has trained 14,000 agents to scale delivery of health products and services in noncommunicable diseases, infectious diseases, maternal and child health in rural remote areas via our reach52 access app. An external evaluation demonstrated that reach52's model generates US$2.10 of social value in our target communities for every US$1 invested. We operate programs across a range of disease areas. For example, these healthcare workers have now screened over 60,000 beneficiaries in diabetes and hypertension in a span of a year by equipping them with necessary skills and tools, with many enrolling themselves in our new community-based model of care programs that significantly improved their health. We continue to innovate to scale our reach by enabling health chatbots in Facebook and WhatsApp, with alert systems to promote health awareness and disease prevention reaching 100,000 users—all aimed to achieve the targets in universal healthcare.

Challenges

From a technology prospective, we are hampered by the lack of reliable, 4G internet connectivity in most of the communities where we operate. In response, we build our tech for fully offline functionality, but this creates limitations on the type of digital services we can offer. reach52 also runs up against regulatory barriers, as many of the countries where we operate do not have up-to-date legislation addressing services like remote consultation and ePharmacies.

Strategic partners

reach52 is a partner-driven organization, working in tandem with stakeholders to generate direct impact. Our work is always in cooperation with publicly funded local government health systems, typically through subnational partnership agreements. We are particularly proud of our partnerships with local community health workers, who support our implementation. We've worked with a range of multilateral organizations, who provide funding and technical assistance to reach52, including UNICEF Innovations, United Nations Capital Development Fund, United Nations Development Programme, and Grand Challenges Canada. We also partner closely with the private sector, including Johnson & Johnson, Unilever, Bayer AG, Sanofi, and Medtronic Labs.

Business plan

reach52 generates income through the sale of impact and insights subscriptions, supporting companies to achieve both CSR targets and commercial ambitions in emerging/frontier markets through our digitally equipped community agent network and our insights platform.

Most of our current clients are in the pharma and comsumer goods sectors, but we are forecasting significant growth through expanding our offerings to FMCG firms as well. Project outputs are developed at the outset with clients, who are billed on a monthly basis for quantified progress toward targets.

Effect of COVID-19, response/pivot

Associated lockdowns with the COVID-19 pandemic impacted our ability to operate our community network-based model. We temporarily discontinued the large-scale health promotion rallies and events which were a common feature of our work. It had also become challenging to travel door-to-door during certain periods. In response, we accelerated our use of SMS and tele-engagement, launching services suited for low-connectivity regions. We saw that health system partners had less capacity to collaborate on activities that weren't directly related to COVID-19 prevention and treatment. In some cases, this had meant delaying programs. However, the pandemic has created considerable interest in digital health, including updated regulatory environments, creating new opportunities for innovation and engagement.

Conclusion and next steps

reach52 continues to grow, with firm plans to further develop our end-to-end health access platform to enable partners and businesses to easily understand, engage and supply to hard-to-reach communities; and work with local stores and pharmacies to stock and offer affordable quality healthcare products. However, we recognize that our current impact pales in comparison to the enormity of the challenge. To this end, reach52 welcomes partnership with a full range of stakeholders in health. We firmly believe that achieving the SDGs by 2030 requires a coordinated, cooperative approach that includes and values social innovation and entrepreneurialism.

Additional information

Websites

www.reach52.com
https://www.straitstimes.com/life/taking-healthcare-to-the-poor

Videos

CNA feature
https://vimeo.com/727308071 Humans of reach52

CHAPTER 91

The new data value web paradigm for healthcare infrastructure in LMICs: The case of Nexleaf

Nithya Ramanathan and Erin Ross
Nexleaf Analytics, Los Angeles, CA, United States

Introduction

Nexleaf is a technology company that provides the data infrastructure required to monitor the hundreds of thousands of refrigerators required to deliver vaccines. Developing the vaccine is one hurdle; making sure those vaccines don't spoil is another. We build and deploy simple sensors that connect vaccine refrigerators to user-friendly data dashboards. Our technology is the "check engine" light for a vaccine fridge, so that health workers across the entire public health system in a country know when fridges fail and have the resources to fix them. At this point, our tech monitors vaccines for approximately 1 in 11 babies worldwide. However, the sensors aren't enough. Countries need to be in the driver's seat when it comes to vaccine distribution: this means they need the data for their entire vaccine supply chain, as well as the means to interpret that data, and to communicate that data, to make a case for where investments are needed in their vaccine supply chain.

At Nexleaf Analytics, we build data-gathering and analytics solutions to serve human health goals in low- and middle-income countries (LMICs). Our innovative technology suites achieve uptake and scale because we put our focus on serving health workers and administrators in the country with objective, automated data. By designing solutions that inform and assist Ministry of Health personnel at every level, we work together with health workers, countries, and multinationals to reimagine lifesaving equipment management and vaccine and drug delivery. Our mission took on even greater significance during the COVID-19 pandemic, and we are hopeful that new investments from this crisis can lay the groundwork for a much-needed overhaul in strengthening both health systems and data for health in LMICs.

Resilient Health. https://doi.org/10.1016/B978-0-443-18529-8.00091-3

Case study: ColdTrace for vaccine cold chain management

Over a decade ago, Nexleaf's small team of engineers and public health professionals began working with partners to bring low-cost Remote Temperature Monitoring (RTM) technology to vaccine refrigerators in rural health facilities in Asia and Africa. Nexleaf's RTM solution, called ColdTrace, gathers and uploads data automatically. The system alerts nurses in clinics if the temperature of the fridge gets too hot or too cold, endangering the potency of the vaccine doses stored inside. The data has been shown to prompt action from nurses charged with keeping the vaccines safe[a]. What's more, that same data is automatically fed to a dashboard that gives personnel at higher administrative levels an instantaneous view of the performance of their entire vaccine cold chain.

Once the data started flowing, we were amazed to see what it revealed and the potential it unlocked. For instance, ColdTrace uploads not only temperature data but also power availability data for every vaccine fridge on which it's installed. This means ColdTrace—a system designed for monitoring temperatures in vaccine fridges—gave Ministries of Health some of their first glimpses into the actual power conditions at health facilities around the country. ColdTrace reveals which facilities don't have enough hours of grid power, and thus require solar systems or backup power. It also offers information about which high-power facilities could sustain other lifesaving equipment such as oxygen concentrators.

The ability to leverage automated data to uncover and solve problems in unexpected ways is the key reason this type of data is uniquely powerful in the hands of country decision-makers. Ministries of Health started using ColdTrace data on their own, in innovative ways. Kenya used the data to determine where it was safest to store vaccines for an emergency polio vaccine campaign. When COVID-19 vaccines started coming into the country, Kenya used ColdTrace data—and emergent ColdTrace tools, like our Trek device for transport monitoring—to understand where to store valuable vaccines safely and to ensure shipments maintained safe temperatures during transport. It is now more clear than ever that countries need data tools to help them determine specific dimensions of their capacity for emergent health systems response—such as how many vaccine doses they can safely store, transport, and distribute—and we believe automated data is key to establishing this capability.

As we worked shoulder to shoulder with countries and vaccine system stakeholders to explore the potential for ColdTrace data, we began to see why and how automated, remote, real-time, time-series data was different from data solutions that had come before. And we began to form a more holistic understanding of the nature of successful, scalable, effective data interventions, and how health systems, health delivery, and even funding mechanisms for health are poised to evolve.

[a] Yavari, S., Prosser, W., Khanlawia, B, Ramanathan, N., Coelho, A., Bechtel, R., Dipuve A., Graham, E., Mvundura, M. Poster Presented at the 14th TechNet Conference, May 2015, Thailand. https://nexleaf.org/wp-content/uploads/2021/05/Nexleaf-RTM-RCT-Report.pdf

Data value web: Bringing simultaneous value to country stakeholders at every level

We see that the shift enabled by and required for data-driven, responsive health systems in LMICs is a transition from a value chain to a data value web. Currently, products are distributed across the supply chain almost blindly. Top-down mechanisms are used to prevent the administration of damaged vaccines, such as temperature-sensitive stickers that change color with too much heat exposure. But decision-makers don't have visibility into *where* failures along the supply chain could damage vaccines. They use proxies such as the age of a vaccine fridge to trigger equipment replacement, even though newer fridges fail frequently and older fridges can perform solidly for years. Countries have been searching for and paying for technology that will feed information back up, but these technologies often do so in ways that rely on tedious and error-prone manual data entry from already overworked personnel.

Our solution, outlined as follows, is to operationalize automated datastreams that sensors and the equipment itself can provide—or be paired with—in order to create a data value web that provides a holistic view into the entire system.

(1) Generating Data from the Bottom Up
First, we get the data flowing, deploying sensors that upload data automatically directly from the last mile. Next, we work with stakeholders at every level—from nurses and doctors to biomedical engineers to maintenance technicians to procurement experts—to build out use cases and tools. Whether it's via automated text message alerts, summary emails, or customized dashboard views, the data must be in the hands of the users so that the data is useful to them in doing their unique job. We iterate with users and focus on bringing utility to those health workers at the lowest level, closest to the equipment and the patients. Sustaining solutions overwhelmingly requires buy-in at every level, thus we provide value across the health system, so that each person is using the data to do their job better and thus is incentivized to maintain the flow of data.

(2) Providing a Shared Source of Truth
We operationalize objective data because it is unequivocal and incontrovertible: it provides an accepted and credible common source of truth. In the current paradigm, nurses in clinics have to call and describe the problem if they think something is wrong with a vaccine fridge, and often struggle to convince supervisors and technicians that malfunctions are occurring. With sensor data, that back-and-forth is eliminated since the fridge itself tells both the nurse and the manager how it's performing, and they can perform their role-specific remedies to address the situation. Also, sensor data flows continuously and uploads in real time, providing instantaneous, simultaneous data to all users about the state of the system moment to moment.
In other words, all data is not equal. A data value web relies on data that is perceived as objective (therefore unequivocal), instantaneous (so everyone has access to the same

blueprint game plan at the same time), and presents answers to questions that people actually care about and can act upon.

(3) Listening to Users
Critically, we engage users, rather than procurers and payers, in the complex multinational health funding system. RFPs for health information systems built from the top down don't get sustained by countries because they don't create enough value for users and don't improve the day-to-day management capabilities of health workers in the system. By focusing on ensuring users get value from (rather than just "buy into") the system, we design systems that are useful to personnel in country such that the high-quality data also can add value for the entire ecosystem (manufacturers, donors, etc.).

(4) Adding Value to the Global Health Ecosystem
Multinationals and funding agencies can indeed get real value from this approach, especially when it comes to monitoring and evaluation. It's much easier to measure the successes and failures of an intervention—and to locate the problems and fix what's not working—when objective high-quality data is available. Exploiting automated datastreams is also a smart way to reduce the cost of postmarket monitoring and other program evaluations. As long as multinationals understand that **collaborating** with governments on resolving barriers to success is necessary to achieving their goals, data can reveal the pain points and suggest solutions that can make the difference between success and failure.

(5) Building a Culture of Data
Most importantly, a functional data value web relies upon the use and proliferation of data, which we call a **culture of data**, rather than upon a business model of data capture. When those closest to any problem are armed with the information they need to inform their actions—and to make a case for the resources required to do their jobs better—the system becomes healthier and more resilient. It is thus absolutely crucial to ensure the data can flow through any pathway that has value to the country.

Conclusion: Fund objective, user-focused systems, and decolonize data

In private sector supply chains, the shift from data value chains to data value webs is already well underway. In contrast, health delivery in LMICs suffers from misalignments in the market, where the paying customers and users are often not the same entities. We at Nexleaf believe this means the shift to data webs in these health systems must be done more deliberately.

Nexleaf is currently expanding its work beyond the vaccine system, and we recently piloted power availability and power quality monitoring sensors in health facilities in Kenya. As the pilot wound down, the personnel working with us specifically asked us to leave the monitors in place after the pilot period. This is more evidence that automated data is useful, and the personal investment of people responsible for the day-to-day functioning of the health facility is what will sustain the solution as it sends data back up to the central Ministry of Health decision-makers. However, this type of evidence will take years to translate into meaningful epidemiological impacts. Funders must be willing to pay for the automated data interventions that in-country personnel want, use, need, and find useful. And, most importantly, we must all insist that the data belongs to the country.

The fact is, modern equipment can gather and send automated data. We expect WHO will finally begin requiring embedded data loggers in vaccine fridges in 2023, and objective data already underpins the storage and delivery of goods across the U.S. The proliferation of data-ready equipment and systems is likely the key to changing the global paradigm. However, too many equipment vendors have added selling the data the equipment collects back to the country as part of their business model. This stance is not only a huge impediment to creating a true data value web, but it is also an extractive model that seeks to drain countries of one of the most important resources of the 21st century: data itself.

Data is power, and LMICs must retain ownership of and authority over their data so that they can continue to build health systems they control, own, and sustain.

Nexleaf partners with country leaders to provide real-time visibility into their vaccine supply chains and works with global stakeholders to *listen* to countries and what their data is telling them. This visibility allows planners to identify which facilities can safely store vaccines immediately and where to prioritize resources, staff, and equipment to fill in any gaps.

What we need to be investing in are the health systems in LMICs, and in the people on the ground, who are closest to the problems and can build resilient solutions.

All of this stems from a belief in human potential, and in the ability of people on the ground to utilize data to build their *own* solutions.

CHAPTER 92

Big Data innovation: CLODS bridging the gap of access to quality healthcare in developing countries

Marjorie Abalos
CLODS Inc., New York, NY, United States

Introduction

Health systems throughout the globe have an urgent need to push for more rapid digital transformation to improve healthcare accessibility and effectiveness since data is the true lifeblood for the delivery of healthcare today and into the future. This is no longer an option for healthcare providers in developing countries; it is a public health imperative and national security priority. Immediate action must be taken by global government policymakers, organizations, business communities, and visionary individuals to find creative responses to accomplish this and ensure that impacts of current and future pandemics do not escalate and become a catalyst facilitating social, economic, human rights, terrorism, and global conflict crises. Such outcomes affecting our entire global population are today a clear and present danger and need to be recognized as potential "mass extinction" threats to be prevented, prepared for, and mitigated.

This chapter describes an innovative data solution being implemented in developing countries, and my personal journey to find and create such digital solutions to meet these needs. As a biomedical engineer by training and a social entrepreneur with a lifelong history of being engaged in social justice and public service projects, it is my life mission to bridge the gap of access to quality healthcare in developing countries and to create a global movement to accomplish this.

Background

As the COVID-19 pandemic lingers, outbreaks of Ebola and the Marburg virus recur, and threats of once-eradicated viruses such as monkeypox (World Health Organization, 2022a) and polio (Haseltine, n.d.) have reentered the global landscape, it is more apparent than ever that

Resilient Health. https://doi.org/10.1016/B978-0-443-18529-8.00092-5

health systems across the globe need to establish more resilient and integrated health ecosystems. This can be done most efficiently by leveraging digital health to be better equipped and to ensure global health security. As director-general of the World Health Organization Dr. Tedros Adhanom Ghebreyesus has stated, all countries need to "invest in more resilient health systems that can sustain essential health services during crises, including stronger health information systems" (World Health Organization, 2022b).

Throughout much of the world, specifically in developing countries that need the benefits of digital health technology the most, implementations of digital health systems similar to what exist in developed countries are still in their infancy.

My mission

As a social entrepreneur with a mission in life to "bridge the gap of access to quality healthcare in developing countries," I am the founder and CEO of CLODS Inc., an American information technology company that provides and maintains the digital infrastructure and health information networks for organization, government, and military health system operators in developing countries. With our numerous deployments in West Africa and project plans underway in Asia and the Americas, CLODS—an acronym for "Closed Loop Data System"—is well on its way to become a leading digital health solution for developing countries.

Over the years, during my extensive travels around the world to more than 50 countries, I had the opportunity and privilege to share my vision and enthusiasm for my mission in life with many leading global impactors from both the private and public sectors who have made true and measurable differences in the world affecting multitudes of people in their organization, government, or multinational corporate roles.

I also attentively listened, and heard their stories, insights, suggestions, and mentoring which fostered my pursuit to transform my personal mission in life into an impact generating "global movement" improving the delivery of healthcare by providing the digital infrastructure for implementations of digital health systems. This movement is growing organically and by my enlisting an ever-growing, multinational group of mission-aligned individuals and stakeholders including corporate, organization, and government entities and most importantly, their dedicated team members and personnel. Finding a sustainable solution to enable better healthcare delivery is no longer the sole responsibility of the public sector. There is collective responsibility across both the public sector and private sector to redefine traditional healthcare business models to find new pathways to sustainability (PricewaterhouseCoopers LLP, 2017).

I have spent the past five years in stealth mode piloting the CLODS System in the demographically and geographically selected West African country of Ghana. CLODS now has more than 60 healthcare facilities that have adopted the CLODS System or are under various phases of implementation under long-term Healthcare Management System Modernization

Programs with health system operators to transform their healthcare facilities to more streamlined and efficient operations resulting in greatly improved delivery of healthcare services and outcomes for their patients.

CLODS System implementations in Ghana have been proven operationally successful and sustainable. Since 2018, when our pilot Peki Government Hospital went live, implementations of the CLODS System have enabled healthcare facilities to transform and leapfrog their legacy "paper-based" operating system to a "paperless" integrated system that significantly reduces patient wait time; improves quality of patient care; reduces operating costs, expenses, and fraud; increases efficiency of service delivery and billing management; and provides the required data needed by governments for analytics to improve their population's health. To ensure long-term sustainability of a CLODS System implementation, a 10-year maintenance period for the entire system accompanies all implementations.

CLODS is scaling its CLODS System implementations, targeted for Nigeria and other African countries. CLODS also has traction underway in Indonesia and India, and projects planned for Mexico, Colombia, Brazil, and Tribal lands in the Americas.

Representative hospitals for each of the three major health system providers in the pilot country of Ghana where the CLODS System has been implemented and is being utilized are shown in Figs. 1–3. Fig. 1 shows a hospital operated by Ghana Armed Forces Medical Services. Fig. 2 shows a hospital operated by the Catholic Health Service Trust. Fig. 3 shows a hospital operated by the Ghana Health Service. In the aggregate, these health system operators oversee more than 2500 healthcare facilities.

Unique features of the CLODS system

The CLODS System digital infrastructure is unique and disruptive since it is an affordable turnkey, multicomponent comprehensive digital health solution that can be rapidly implemented simultaneously at multiple healthcare facilities. CLODS System implementations provide the deployment of software, hardware, connectivity, training personnel, and support services. Also, to ensure the system's sustainability long term, CLODS provides the ongoing maintenance of the entire system for an initial renewable period of 10 years, including software updates, scheduled replacement of required hardware components, connectivity, training, and support services.

Additionally, once the CLODS System digital infrastructure is implemented at a group of healthcare facilities for a health system operator, a nationwide, fiber-based CLODS Health Information Network is created for them, whereby they can securely share patient records and have the Big Data centralized at a CLODS data center. Thus four benefits of the CLODS System are that it (1) mirrors and renders "paperless" the entire workflow of the healthcare facility; (2) improves the delivery of healthcare at the ground level; (3) digitally generates in

Fig. 1
CLODS CEO Marjorie Abalos (center) surrounded by hospital administrators, staff, and personnel at the Ghana Armed Forces Medical Services hospital having "GONE LIVE" with the CLODS System on October 31, 2020, in Ho, Ghana.

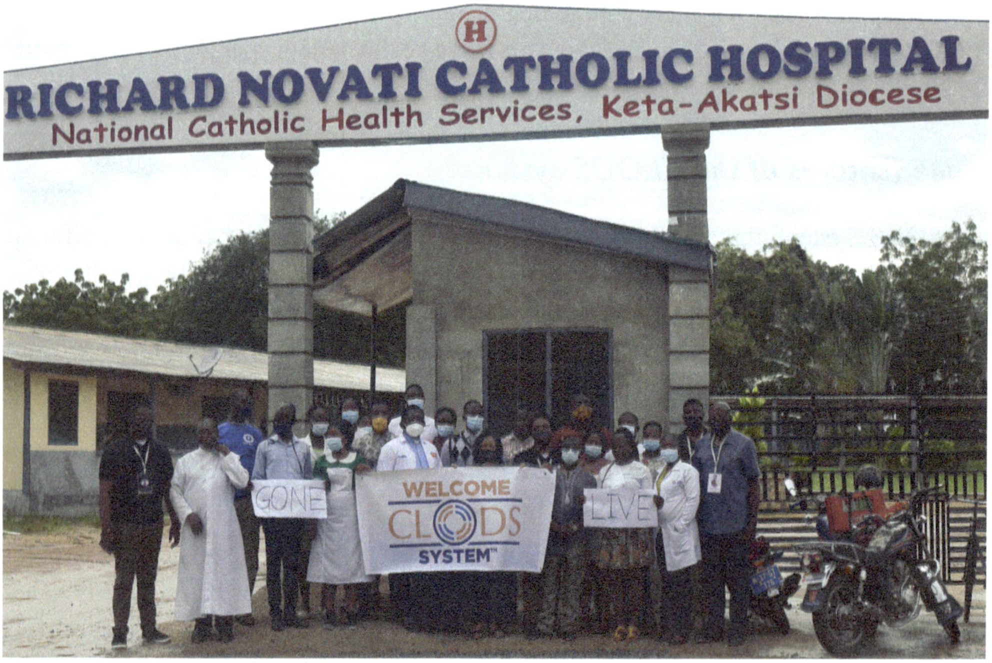

Fig. 2
CLODS CEO Marjorie Abalos (center) surrounded by hospital administrators, staff, and personnel at Richard Novati Catholic Hospital having "GONE LIVE" with the CLODS System on October 2, 2020, in Sogakope, Ghana.

Fig. 3
CLODS CEO Marjorie Abalos (center) surrounded by hospital administrators, staff, and personnel at Peki Government Hospital having "GONE LIVE" with the CLODS System on December 8, 2018, in Peki, Ghana.

Fig. 4
CLODS System takes healthcare facilities from "paper-based" to fully "paperless."

real time all the Big Data required for use locally and nationally to effectively operate healthcare facilities and health systems; and (4) provides the necessary real-time data flows for health system operators to allocate resources and conduct disease surveillance and analytics necessary to improve population health, advance national security, and to obtain domestic and global fundings for their health systems. Fig. 4 shows how the digital CLODS System is replacing the traditional "paper-based" system used at most healthcare facilities in developing countries.

A personal note about my journey

My journey generating measurable, large-scale social impact began more than 15 years ago when I had the privilege to address the 60th Anniversary Event at the United Nations in New York honoring its formation as the representative of my then youthful generation charged with making the United Nations more effective and more accountable to the peoples it exists to serve. Subsequently, I helped coordinate one of then-United Nations Secretary-General Kofi Annan's farewell addresses in December 2006 at the Harry S. Truman Presidential Library in Independence, Missouri (Fig. 5).

During my UN speech, I noted my belief that every human being has the fundamental right to the "highest attainable standard of health" (as stated in the WHO constitution) without distinction of race, religion, political belief, economic or social condition, and I publicly pledged to dedicate the next 30 years of my life to this purpose (Out4Good Ltd, n.d.). A decade later, in my talk released by the TED organization on Choosing a Mission in Life (La Sierra University, 2017), I shared this long and challenging journey and confirmed my ongoing dedication and commitment to my own personal mission to "bridge the gap of access to quality healthcare in developing countries."

Fig. 5

(left to right): UN secretary-general Kofi Annan and wife Nane Lagergren, CLODS CEO Marjorie Abalos, U.S. Congressman Dennis Moore, and Truman Presidential Library director Michael Devine.

To fulfill my mission, based upon my extensive travel experiences in more than 50 countries and a 20-year vocational background in global software-system deployment, I identified an urgently required global humanitarian need for expanding access to digital technologies in underserved communities that would enable better access, more affordable, and better quality healthcare.

After conducting extensive research, in 2013 I selected the West African country of Ghana as demographically, geographically, and geopolitically appropriate as the pilot country for my first digital health solution deployments. Some of the criteria used were based upon Ghana's main language being English, its central location in West Africa, its political stability and long-standing relationships with the United States and the United Kingdom, its skilled workforce, its population size, and the fact that there are only three major health system operators in the country.

At that time, in my positive but initially naive approach to this, I thought that accomplishing this initiative should not be too difficult, and that everyone would want to join my mission. I thought I could attract large global software providers who had not yet entered the African market to join me to serve the continent's exploding population of 1.4 billion people (Worldometer. Africa Population, n.d.). I thought this would be an easy win-win-win: the software company captures a material share of this market at its infancy, the healthcare providers and their patients benefit from the software deployment, and I accomplish my mission.

After spending more than a year doing groundwork, I learned that the major global software companies at the time were deluged with lucrative backorders in their respective developed countries and were not interested in joining me to address this awaiting underserved market. I also soon realized that even if the software was given to us to use, where is the hardware? And, if we were given the hardware, where is the connectivity? And even if these items were provided, where is the required capital and necessary human resources to help implement, train, operate, and maintain the system? Facing all these challenges, I realized that the only way to accomplish my mission was to take it upon myself to design a system that addresses all of these issues. Hence, the CLODS System was born.

My mission is now in process. CLODS System implementations have been adopted and are being utilized at numerous healthcare facilities affecting thousands of patients every day. Many more healthcare facilities are under various phases of deployment and CLODS now has implementations awaiting to begin on several continents.

To quote one of the greatest orators of all time, Martin Luther King Jr., "Well, I don't know what will happen now. We've got some difficult days ahead. But it doesn't matter with me now. Because I've been to the mountaintop… I just want to do God's will. And He's allowed me to go up to the mountain. And I've looked over. And I've seen the promised land. I may not get there with you. But I want you to know tonight, that we, as a people, will get to the promised land" (American Federation of State, County, and Municipal Employees, 2022). I draw a humbling analogy from these prophetic words, knowing that my global movement of mission-aligned stakeholders and individuals that are bridging the gap of access to quality healthcare in developing countries are fulfilling my mission in life, by affecting now—and well into the future—millions of patients, their families, and community members throughout the world by improving the delivery of their vital healthcare.

Marjorie Abalos is a global health visionary and social entrepreneur who is the founder and CEO of CLODS Inc. having created the CLODS System, a digital health solution specially built for developing countries that aggregates critically needed population health data in real time and generates Big Data affecting policies in healthcare delivery. She is also the board chair and cofounder of Out4Good Ltd., a nonprofit, nongovernmental organization and global research institute that mentors, pairs, and helps improve policy and decision-making processes. Previously, Ms. Abalos held senior management positions at software and renewable energy firms for Global 500 companies and startups. She began her career working in web development capacities for Internet startups in the telecom, web hosting, media, and advertising sectors after earning her bachelor's degree in biomedical/biochemical engineering from the University of Southern California where she worked on various groundbreaking projects utilizing cutting-edge biomedical imaging techniques. She has facilitated projects in many countries to develop process improvement plans that empowered millions of disenfranchised, underserved, and marginalized people to transform into the Information and Digital Age. She was also instrumental in deploying the first ATM networks in Vietnam and in addressing public and industrial environmental remediation issues in China. Ms. Abalos has been engaged in social justice and public service projects such as Amnesty International and the American Cancer Society's Relay for Life, as well as volunteering at hospitals, convalescent centers, and schools.

References

American Federation of State, County, and Municipal Employees. (2022). *"I've Been to the Mountaintop" by Dr. Martin Luther King, Jr.*. https://www.afscme.org/about/history/mlk/mountaintop. (Accessed 8 August 2022).

Haseltine, W. A. (2022). There may be a new polio epidemic on its way—If so, what we can do: Part III. *Forbes*. https://www.forbes.com/sites/williamhaseltine/2022/06/28/there-may-be-a-new-polio-epidemic-on-its-wayif-so-what-we-can-do-part-iii/?sh=39c054951fb1. (Accessed 8 August 2022).

La Sierra University. (2017). *Choosing a Mission in Life [Video]. YouTube*. https://www.youtube.com/watch?v=1LItDBzGZxE. (Accessed 8 August 2022).

Out4Good Ltd n.d. Marjorie Abalos United Nations Address & USA Gift Presentation [Video]. YouTube. https://www.youtube.com/watch?v=3qk0TPoJvm4. Published December 21, 2017. Accessed August 8, 2022.

PricewaterhouseCoopers LLP. (2017). *The Digital Healthcare Leap*. Available at https://www.pwc.com/gx/en/issues/high-growth-markets/assets/the-digital-healthcare-leap.pdf. (Accessed 8 August 2022).

World Health Organization. (2022a). *Monkeypox: what you need to know*. https://www.who.int/multi-media/details/monkeypox–what-you-need-to-know?gclid=EAIaIQobChMInbuTtrev-QIVWOd3Ch21YgOsEAAYASABEgL9GvD_BwE. (Accessed 8 August 2022).

World Health Organization. (2022b). *14.9 million excess deaths associated with the COVID-19 pandemic in 2020 and 2021*. https://www.who.int/news/item/05-05-2022-14.9-million-excess-deaths-were-associated-with-the-covid-19-pandemic-in-2020-and-2021. (Accessed 8 August 2022).

Worldometer. Africa Population. https://www.worldometers.info/world-population/africa-population/#:~:text=The%20current%20population%20of%20Africa,the%20latest%20United%20Nations%20estimates. Accessed August 8, 2022b.

CHAPTER 93

Innovation: Exploring Dimagi's Digital Platform CommCare: A widely adopted digital solution used by frontline workers worldwide

Gillian Javetski
Dimagi, Inc., Cambridge, MA, United States

Background

A leading social enterprise, Dimagi is committed to building and scaling sustainable, high-impact digital solutions that amplify the efforts of frontline workers and programs globally.

Dimagi's most significant contribution to date is the development of its flagship platform, CommCare. The most widely-deployed and evidence-based platform for enabling Frontline Workers, CommCare empowers organizations to build their own digital solutions to better deliver services, manage clients, and collect data. Built from the ground up to support the complexities and idiosyncrasies of frontline work, CommCare can meet nearly every frontline use case in any setting (including offline) at any scale, from pilot to nationwide programs. More than one million Frontline Workers have used CommCare applications to deliver critical frontline services across numerous sectors, including health, agriculture, social services, and more.

Problem/challenge

At least 50% of people around the world lack access to essential health services according to the World Bank and the World Health Organization (WHO). One contributing factor to this problem is that there are not enough specialists, especially in low-resource settings, to deliver the care needed to the global population. By 2030 the World Health Organization (WHO) estimates a projected shortfall of 18 million health workers. While expanding frontline worker programs alone is not sufficient to achieve Universal Health Care or to improve all general

Resilient Health. https://doi.org/10.1016/B978-0-443-18529-8.00093-7

well-being, they are a critical piece of the puzzle. One small but important piece of this complex web of actors is those organizations that develop technology to support frontline services. Dimagi believes that the best way the organization can contribute to achieving their vision is by building digital solutions that amplify frontline workers.

How and where this innovation is being implemented, technology being used, target, who and how the service is being delivered

In 2021 CommCare was used in more than 130 countries. Some of the largest CommCare programs are located in sub-Saharan Africa, Southeast Asia, and the United States.

Impact

CommCare is one of the most researched mobile data collection and service delivery platforms and supports an extensive evidence base of peer-reviewed studies, including several randomized controlled trials. Together, these studies have demonstrated CommCare's positive impact on strengthening frontline healthcare systems, frontline worker capabilities, and client outcomes. Read more about the CommCare Evidence Base here: https://www.dimagi.com/toolkits/commcare-evidence-base/.

Challenges

Due to the nature of global development, like many social impact oragnizations, Dimagi often faces challenges related to wanting to pursue long-term impact on short-term requirements. Public sector programs have significant budget uncertainty and can't count on the same consistency that is common in the more general, private sector market, including user satisfaction leading to long-term financial growth and sustainability. In order to address this, teams at Dimagi intentionally carve out space in projects to ensure that they are making a sustainable impact, guided by the company's High-Impact Growth framework. https://dimagi.com/blog/dimagis-5-year-strategy/

Partners

Dimagi works with a wide variety of partners ranging from governments and international agencies through small grassroots organizations and student groups. As a social enterprise, the organization is proud to support digital programs that are being self-driven at the community level all through national scale programs.

Business plan

Since its creation in 2002, Dimagi has historically focused almost exclusively on supporting frontline programs in low- and middle-income countries (LMICs). The majority of these programs are funded through the International Development market. Dimagi's business model is a blend of Product and Services Revenue streams that strongly leverages its position as a market leader in digital health. Dimagi's business model enables the organization to continue to grow its global reach through a low-cost Software-as-a-Service (SaaS) hosted offering, while keeping lower income markets as its primary focus. In 2020, Dimagi also entered new markets in high-income countries, including working at scale with several state governments in the United States.

Effect of COVID-19, response/pivot

During the height of the COVID-19 pandemic in 2020–2021, CommCare had been used as a COVID-19 platform by more than 40,000 users in over 30 countries. To support health programs with their response efforts, Dimagi made a set of "template applications" freely available for organizations to modify, including for contact tracing, vaccine distribution, lab supply tracking. Several of these were adapted by various national and state governments around the world. A report by Johns Hopkins University (https://publichealth.jhu.edu/2020/johns-hopkins-researchers-publish-assessment-of-digital-solutions-for-covid-19-response-in-low-and) determined that CommCare is one of the two most useful platforms for COVID-19 response.

Additional information

Website: www.dimagi.com

CHAPTER 94

Healthy Entrepreneurs: Transforming access to healthcare through an innovative model of community health entrepreneurship

Joost van Engen
Healthy Entrepreneurs, Utrecht, Netherlands

Background

Improving access to basic healthcare through Community Health Workers (CHWs) can prevent 3 million deaths a year. People are healthier, live longer, are more capable of working and contributing to the economy, and children develop better. In a shorter term, deploying community health workers decreases the burden on health centers and hospitals, reduces costs and identifies, as well as diagnoses and treats illnesses earlier. While CHWs have contributed significantly to improving health outcomes in well-funded and organized study settings, there are severe performance, retention, and sustainability problems. Most of these problems are related to the absence of a sustainable income, an inadequate supply chain of medicines and health products, as well as insufficient training and supervision, often depending on temporary donor funding. The sustainable community health entrepreneurship model by Healthy Entrepreneurs (HE) amplifies the effectiveness of the CHW system while increasing the reach (and supply) to the most isolated communities at the bottom of the pyramid. The HE model allows CHWs to generate an income, provides regular training for CHWs and can save millions of lives each year with proven interventions such as providing antibiotics for pneumonia, oral rehydration therapy with zinc for diarrhea, and malaria treatment. Through this model, such services by CHWs are continuously available in the most remote and hardest to reach areas of Sub-Saharan Africa (SSA).

Problem being solved

Access to basic healthcare services in rural areas of SSA countries is poor. Many families have to travel more than an hour to visit the nearest health facility often on treacherous roads or via expensive transport. Health facilities experience poor staff availability and stock-outs of

Resilient Health. https://doi.org/10.1016/B978-0-443-18529-8.00094-9

products. CHWs serve as a solution to overcome this access barrier. However, they experience severe performance, retention, and sustainability problems. The HE solution provides both a business opportunity for CHWs and improved healthcare in hard-to-reach, rural settings. CHWs are selected and trained as Community Health Entrepreneurs (CHEs) and onboarded into the network and onto the mobile IT platform. CHEs provide easy and affordable access to health information and products, allowing rural families to prevent and/or treat health threats at an early stage. CHEs have access to a medical expert at a distance through a toll-free line so that more primary health services are available at community level (e.g., hypertension and diabetes screening and treatment).

Fig. 1 shows how the telehealth system works. An entrepreneur will conduct health education and awareness sessions in her/his community and finds patients. With support of diagnostic tools (blood pressure machine, glucose meter, weighing scale, MUAC tape, etc.) the CHE collects the measurements and registers these in an application. A call is placed to the call center where the CHE shares the diagnostic results with medical experts (clinical officers, nurses). The medical experts collect diagnostic information in the system, and consult the patients for any additional information needed. After collecting all information, the medical expert will make a diagnosis whereby a referral to the nearest health facility can be made, or in case treatment is

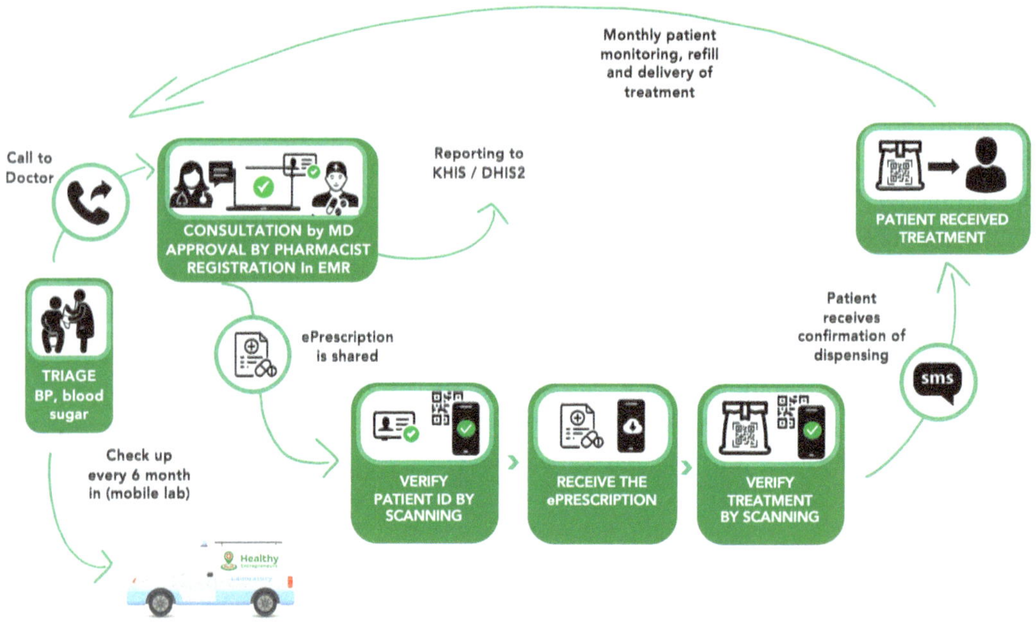

Fig. 1
The telehealth system of Healthy Entrepreneurs initiative.

possible a prescription will be shared to the entrepreneurs. After receiving the prescription, the CHE can distribute the right medication to the patient and plan a follow-up meeting.

How and where the innovation is being implemented, technology used, who and how the service is being delivered

HE trains and equips CHEs with business and health training, a smartphone, and a basket of products on credit. The CHEs can directly start their business. HE provides a reliable supply chain of products on a biweekly basis. The IT platform allows for qualitative health education, screening, and product ordering. HE is currently active in Uganda, Kenya, Tanzania, Burundi and Burkina Faso with a network of nearly 20,000 CHEs. HE targets CHWs who want to become, and show the potential to become, a successful CHE. HE improves their situation by increasing their income through the sales of the health products, by improving their motivation to execute their work, and by supporting them with supplies from the warehouse and additional training to provide reliable health service delivery to the rural communities in SSA countries (Fig. 2).

HE's indirect beneficiaries are the families living in rural, remote areas of SSA countries. Every CHE has a catchment area of 250 families/1250 people. Fig. 3 shows that the work area of Healthy Entrepreneurs, shown by female figures in green dresses, is off the main road (shown by the red line), into the rural villages.

Fig. 2
A CHE receives a new order from the warehouse.

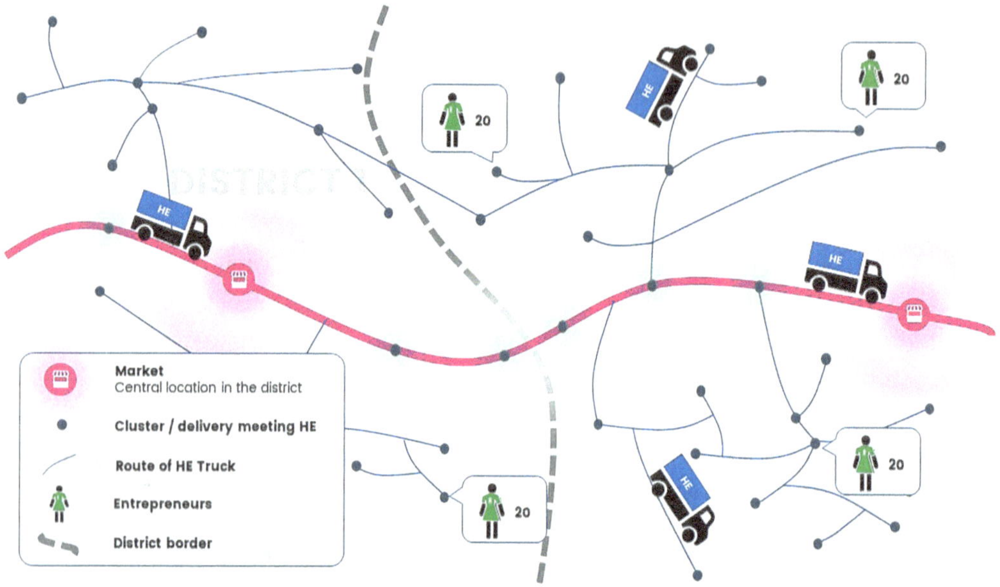

Fig. 3
Work area of the Community Health Entrepreneurs in the rural villages.

A CHE provides a number of services to her patients in the community. As shown in Fig. 4, she gives medication, tests vital signs, and provides counseling and education.

Impact and highlight of successes

In its 10 years of existence, HE's impact covers not only SDG 3 on health and well-being of the United Nations 2030 Sustainable Development Agenda, but also SDG 1 by reducing health expenditure costs for rural families, SDG 2 in providing access to fortified food and nutrition and information, and SDG 8 in providing jobs for CHWs who are mainly unpaid volunteers. Over the past years, the Erasmus University in the Netherlands and Makerere University in Uganda have conducted research on the impact of the HE model. The research showed significantly increased motivation and increased self-esteem among CHEs in comparison to regular CHWs. In the communities targeted by the CHEs, results showed more knowledge on family planning, maternal and child health, and HIV. Additionally, twice as many referrals were made in comparison to a nonintervention area. HE currently has a network of nearly 20,000 CHEs, reaching 25 million people in the African countries of Uganda, Kenya, Tanzania, Burundi and Burkina Faso. In total, HE had sold 35 million impact products and provided a substantial increase in income for CHEs and 55% cost savings in health expenditure for rural families by reducing transport costs to health services.

Fig. 4
CHEs in the community giving medication to her patient (upper left), testing her patient's blood pressure (upper right), counseling mother on maternal and child care (lower left), and providing health counseling and education to her community members (lower right).

Challenges

The main challenge HE faces is the traditional focus on providing free services at the community level. Free services in its current form are not sustainable and mostly dependent on external donor funding. This creates a situation in which services and products are irregularly provided to rural families, leaving them dependent. The HE model empowers local communities through sustained, reliable, and affordable healthcare provision through the HE business model, improving their health and well-being. International awareness on the role of social enterprises in health is still lacking. Especially national governments and stakeholders are not yet fully aware of the constructive role of social enterprises and how they could positively change health service provision and help them solve their challenges. Showcasing social enterprises on the national and international level is needed to embrace the full capacity of social entrepreneurship to deliver effective and affordable health services.

Strategic partners

HE works closely with national and local government and Ministries of Health in the implementing countries. MoUs with government for proper implementation and alignment are signed. HE also partners with NGOs, private sector, and social enterprises to improve the service delivery of CHEs. Examples of partners are AMREF, CARE, Philips Foundation, Bayer Foundation. Endorsements of UNICEF, WHO, and UNAIDS support HE in further awareness and acceptance of the model and scaling to new countries. The initiative "*Health in your Hands*" (described in another chapter in this book) supports HE in providing a platform to showcase its model, introduction to international stakeholders, and publications.

Current business plan?

The HE business model is unique in establishing a sustainable solution after the first investment. The small margins sold in large volumes cover HE operational expenses and provide the CHEs with a sustainable income. Healthy Entrepreneurs continues to supply the entrepreneurs with their biweekly orders and refreshes their knowledge during regular CHE cluster meetings. The revolving nature of the financing allows for continued impact after initial investments are made. HE uses a hybrid model of financing, which includes grants, equity, and debt. Grants are provided by the Dutch Ministry of Foreign Affairs, Dutch Postcode Lottery, CHAI, and Grand Challenges Canada (GCC), among others. Debt is provided by the Dutch Good Growth Fund. HE has launched a financing round of €15 million in equity, debt, and grants, allowing to grow the network to 92,500 CHEs by 2027 in 12 countries. To contribute to the United Nations 2030 Agenda for Sustainable Development and achieve the global goals (SDGs) by 2030, HE aims to have 250,000 CHEs active, with 150,000 CHEs in its own network and 100,000 CHEs deployed by third parties.

Effect of COVID-19 and response

The COVID-19 pandemic has had both a positive and a negative effect on HE's work, in that it restricted the onboarding of new CHEs, while at the same, spearheaded the business as rural families continued to buy from the CHEs. Due to the imposed government lockdown restrictions and subsequent cautions about in-person contact, movement of health personnel was limited. Yet, this provided the opportunity for HE to expand its telehealth services by supporting CHEs and CHW nationwide with a medical expert online at a distance. Advantages of this approach were that questions from people in rural communities concerning COVID-19 and other health issues could be easily received and tackled, with the result that HE saw a huge uptake in the toll-free line and medical consultations.

Additional information

www.healthyentrepreneurs.nl
https://www.youtube.com/channel/UCJb8RCu9TW3uzJNS9aCDZGg
https://academic.oup.com/heapol/article/34/9/676/5570869?login=false

The Amazon Hope Medical Programme: A mobile healthcare ship to "Leave No One Behind"

Paula Aljovin

Mental Health Team, Vine Trust Amazon Hope Medical Programme, Loreto, Peru; Department of Clinical Psychology, Teachers College, Columbia University, New York, NY, United States; International Association of Applied Psychology, Indianapolis, IN, United States

Introduction

Medical or hospital ships represent an innovative solution to increase capacity and provide healthcare to people living in remote inaccessible areas or to those who have been victims of natural and/or humanitarian crises. These ships are fully equipped with state-of-the-art healthcare supplies and equipment and trained full-service personnel. For example, since the 2004 Asian Tsunami, the U.S. Navy (USNS) has deployed hospital ships in support of humanitarian assistance missions. Every 2 years, the US Department of Defense (DoD) deploys the USNS *Mercy* (T-AH 19) and USNS *Comfort* (T-AH 20) hospital ships to perform humanitarian assistance missions (Licina, 2013) in the Pacific region and Latin America (Hameed et al., 2016; Scouten et al., 2017). Both ships incorporate interagency partners, partner nations, host nations, and nongovernmental organization (NGO) personnel in these missions. In early 2020, hospital ship USNS *Comfort* (T-AH-20) and USNS *Mercy* (T-AH 19) were deployed to New York City and Los Angeles, respectively, intended to accommodate patient overflow at overwhelmed local hospitals during the COVID-19 pandemic despite the mission failure largely attributed to complex protocols and bureaucratic hurdles (Schwirtz, 2020).

Like these USNS counterparts, the Chinese government launched its first hospital ship, in 2007, called the *Peace Ark*, and deployed it to African and Central American and Caribbean nations (Mizokami, 2014). Equipped similar to land-based level 3 hospitals, possessing a bed capacity of 300 (Ren et al., 2015), and designed to serve as a medical facility for mass casualties from natural disasters or acts of war (Boren et al., 2003), the *Peace Ark* has since succeeded in several humanitarian assistance missions, providing medical care for countries in Africa, Latin America, Asia, and the South Pacific islands (Ren et al., 2015).

Resilient Health. https://doi.org/10.1016/B978-0-443-18529-8.00095-0

Moreover *Mercy Ships* is an international charity that provides free healthcare in low-resourced settings and operates the largest nongovernmental hospital ships in the world, providing humanitarian aid like free healthcare, community development projects, community health education, mental health programs, agriculture projects, and palliative care for terminally ill patients. It was founded in 1938 by American missionaries Don and Deyon Stephens. *Mercy Ships* has operated in more than 57 developing nations and 18 developed nations around the world, with the largest current fleet, the *Africa Mercy*, servicing 15 West African nations (Harris, 2013).

The following chapter describes one such example in Peru, where *The Vine Trust* (VT), a Scotland-based international development charity, has deployed three medical ships since 2001. These ships aim to provide itinerant, free healthcare services to remote indigenous communities across the Peruvian Amazon basin. In order to reach remote communities with quality health and mental health services, we are implementing technological advances to the extent possible, while recognizing serious barriers to this critical step and examining how to overcome these.

Background

The Amazon rainforest comprises over 60% of the Peruvian territory, yet rural Amazonian communities have been historically excluded from the country's socioeconomic development (Ministerio de Salud [MINSA], 2012). They are often geographically isolated and live in extreme poverty, without access to clean drinking water, electricity, sanitation, healthcare, quality education, and communication services (Instituto Nacional de Estadística e Informática [INEI], 2018). Factors like poverty, social exclusion, and inequality representing powerful determinants of physical and mental health have been exacerbated during the COVID-19 pandemic.

The Amazonian region of Loreto, shown in Fig. 1, represents 28.7% of the Peruvian territory (MINSA, 2015), and 33.1% of its population lives below the poverty line (Ministerio de Desarrollo e Inclusión Social [MIDIS], 2021). Multiple rivers and tributaries crisscross the region, meaning that thousands of communities are only accessible by boat and located hours away from each other and from any health service. Peruvians below the poverty line have guaranteed health insurance, but access to care is limited due to geographic, financial, cultural, and health system barriers (Limaye et al., 2018). The region of Loreto has one of the lowest physician–patient ratios (4.4 per 10,000) and some of the worst health indicators in the country. For instance, in rural areas of Loreto, 60.4% of children below the age of 5 are anemic (INEI, 2020).

The region of Loreto is also becoming increasingly exposed to climate disasters, which bring about economic, sanitary, and social consequences. In past years, frequent and extreme

Fig. 1
Outline of the Region of Loreto, Peru. From "Great Rainforest or the Greatest Rainforest?," by A.K. Kistler, 2012, February 2. Center for International Environmental Law (CIEL). https://www.ciel.org/great-rainforest-or-the-greatest-rainforest/

hydrological events have affected the area. Heavy rains and floods have become more intense, affecting crops, keeping children from going to school, and even displacing entire communities due to the collapse of the river banks. Flooding has also led to the spread of water- and vector-borne diseases, including diarrhea, dengue, parasitic infections, and malaria. Floods are taking this water into the fields, while acid rain simultaneously damages the crops. During droughts, fish are becoming more scarce, smaller and thinner, affecting communities' diet and daily income. Water shortages force people to drink stagnant, murky water, and lead to rice and palm crop failure (Hofmeijer et al., 2013).

In terms of mental health, one of the few studies conducted by public institutions reports that Loreto's lifetime prevalence of depression and PTSD is 18% and 10.4%, respectively. Further, the percentage of women who have experienced abuse at some point in their lives is 33.1% (Instituto Nacional de Salud Mental Honorio Delgado Hideyo Noguchi [INSM], 2009).

The Vine Trust: *Connecting people to change lives*

Two decades ago, The VT, a Scotland-based international development charity, began its work in Peru and established the *The Amazon Hope Medical Programme*. Learning about the high mortality rates and suffering that members of remote Amazonian communities endure, they resolved to respond to this need by offering regular medical attention and health education (Vine Trust, 2022). Due to the geographic isolation of these villages, the VT realized they had to go out and reach the most remote communities in the Peruvian Amazon. Over the past 20 years, they have sailed three medical ships from the United Kingdom, across the Atlantic, through Brazil and into Peru. Two of these ships are still operating today: *The Amazon Hope 2* and the *Forth Hope*. Fig. 2 shows the *Forth Hope* hours before being deployed from Edinburgh towards Peru.

The use of a medical ship provides a unique and innovative approach to providing an itinerant healthcare service to isolated, rural communities. Mobile clinics are defined as "customized vehicles that travel to the heart of communities[...] which overcome barriers of time, money, and trust, and provide community-tailored care to vulnerable populations" (Hill et al., 2014, p. 261; Limaye et al., 2018). All of the VT ships are fully equipped, with medical and mental health consultation areas, a dental clinic, minor surgery room, pharmacy, laboratory, and all instruments and equipment necessary to provide quality service in primary healthcare, as shown in Fig. 3. Such a ship provides a hygienic and easily accessible platform for patients to receive free treatment, while allowing health professionals to work and be accommodated in a secure and modern environment. Regular expeditions involve a team of doctors, psychologists, dentists, an obstetrician, pharmacists, and nurses. The VT also receives international volunteers to create bilateral learning opportunities that promote sustainable development for vulnerable children and families.

Every month, the two ships leave a small port near the city of Iquitos, gateway to the most isolated communities of the region. Each ship embarks on its own 21-day expedition along a specific river basin. Every day, the ships stop at one or two communities and treat patients onboard. The healthcare staff works nonstop throughout the three weeks of navigation, and eventually returns to Iquitos, where they rest and replenish for 10 days before embarking on another expedition along a different river basin.

Fig. 2
Medical ship *The Forth Hope*. The Vine Trust, Ediburgh, Scotland. *Paula Aljovin.*

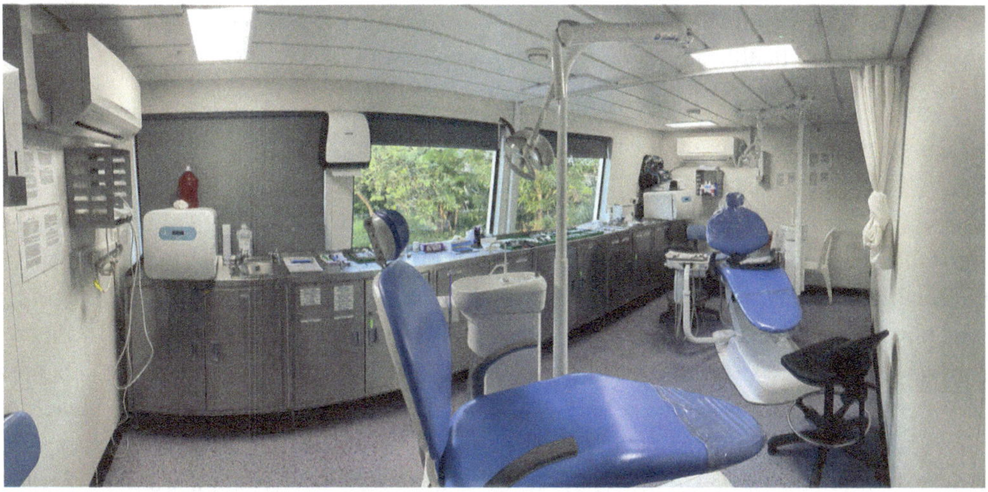

Fig. 3
Odontology consulting room at the *The Forth Hope*. Loreto, Peru. *Paula Aljovin.*

Most of the riverine communities we visit are considered native or indigenous and belong to various ethnic groups, including the *Kukama Kukamiria, Bora, Kichwa, Shipibo Konibo, Achuar,* and *Yagua*. With the peoples' lives intrinsically linked to the river, these communities fish, harvest, and trade in the iconic setting of the tropical Amazonian rainforest. While indigenous communities usually distrust the introduction of Western healthcare services, with which they are unfamiliar and perceive as foreign to their traditional culture, over the years, we have managed to build enough trust for them to use our medical services, which has unveiled an extreme demand for healthcare. Nonetheless, we never dismiss their own views of health and traditional medicine, but rather complement local treatments with Western medicine.

Throughout the last two decades of work, *The Amazon Hope Medical Programme* has provided over 2 million medical and dental consultations. Prior to this model of intervention, medical ships had never been implemented in the Peruvian Amazon basin. In 2013 the Peruvian Marines (Marina de Guerra del Perú) began implementing their own hospital ships, known as Itinerant Platform for Sustainable Social Action (Plataforma Itinerante de Acción Social Sostenible, PIAS), which serve several Amazonian regions (Oliveira, 2022) and complement our work.

In early 2020, just before COVID-19 hit Peru, my colleague Ms. Luna Dannon and I—who had both recently earned our undergraduate degrees in Clinical Psychology from Pontificia Universidad Catolica del Peru—joined one of the medical ships expeditions as volunteers. Given our training, we were asked by *The Amazon Hope Medical Programme* to evaluate the main mental health needs among members of 14 communities along the Ucayali River. We agreed to the challenge, despite the fact that we did not know what we would encounter, nor did we have any adapted tests to use in such faraway areas of the Amazon basin. We therefore designed an exploratory, qualitative study aiming to understand and assess people's main worries, needs, perceptions of well-being, and idioms of distress, with the ultimate goal of designing a mental health intervention to be included in the VT's services.

Our study (Dannon & Aljovin, 2020) found that mental health needs occur at distinct yet highly interdependent levels, namely the Macrosystem, Exosystem, Microsystem, and Individual system (Bronfenbrenner, 1975). This systems approach is consistent with the WHO's and Centers for Disease Control and Prevention's (CDC) use of the ecological framework to understand complex public health issues. Within the Macrosystem, our study identified traditional gender roles that endorse gender-based violence, and shared beliefs supporting the use of physical punishment toward children. Exosystem needs included threats to the environment, absence of primary health and mental health services, lack of sex education, extreme poverty, corruption, and absence of phone/internet reception. Within the Microsystem, we found partner violence, child abuse, and household overcrowding, along with strengths, like extended family support systems. At the level of the Individual System, we identified manifestations of depression, anxiety, and PTSD; COVID-19-related anxiety; history of loss, violence, and trauma during childhood; school dropout due to pregnancy or child labor; and alcohol abuse.

We also realized that the situation of the Amazonian indigenous people touched upon several of the United Nations Sustainable Development Goals (SDGs), including SDG-6 *Clean Water and Sanitation*, SDG-3 *Good Health and Well-being*, SDG-13 *Climate Action*, and SDG-5 *Gender Equality*. Barriers to development included policy failure, lack of financial resources, poor intersectorial relations, lack of access to the internet—which limits digital solutions—and systemic corruption. These are issues I would explore deeper when I took the course about the United Nations Agenda 2030 for Sustainable Development and Universal Health Coverage, "Psychology and the United Nations" given by Professor Dr. Judy Kuriansky, at Columbia University Teachers College, while I pursued my master's degree in Clinical Psychology to further my skills for my role on the ship.

As we left the Amazon when the first volunteer tour of duty ended and returned to Iquitos, the first cases of COVID-19 began to reach our country. A few days later, lockdowns were announced by the government, so we moved to the capital city of Lima, and the VT's medical ships were unfortunately no longer allowed to sail. Nonetheless, we began developing a mental health strategy to integrate into the medical program as soon as the lockdowns ceased, consistent with the current trend for integrating mental health into primary healthcare. Due to the exploratory study we had conducted prior to the pandemic, we knew that mental health needs were urgent and we worried about the possibility of COVID-19 further magnifying these. Our fears were justified, in that the medical and mental health treatment gap enlarged, with needs escalating and services diminishing.

Impact of the COVID-19 pandemic on the region of Loreto

The COVID-19 pandemic heightened all the preexisting issues that urban and rural Loreto were already facing. As of April 2022, Peru has had the world's highest number of COVID-19 deaths per capita (Statista, 2022; The New York Times, 2022; Worldometer, 2022). One of the most affected regions at the national level was Loreto, where the virus claimed the lives of the highest number of doctors and nurses in the country and collapsed the regional health system in less than a month. An overwhelming majority (80%) of health professionals were infected with COVID-19, leading to closure of numerous health centers in the midst of the pandemic due to lack of personnel (La Fondazione Albero Della Vita Onlus [FADV], 2020). Making matters worse, the COVID-19 pandemic struck Loreto when the region was in the midst of a strong dengue outbreak, which further contributed to the collapse of hospitals (La Fondazione Albero Della Vita Onlus [FADV], 2020). Iquitos, capital of Loreto and the world's largest city that cannot be reached by road, became the country's most devastated city during the pandemic, with 71% of its population got infected (Álvarez-Antonio et al., 2021; Nugent, 2020).

Over two-thirds (70%) of Loreto's remote, rural communities were also greatly affected by the pandemic (Radio Programas del Perú—RPP, 2020). Despite this need, the few health professionals who offered local services to the riverine communities were transferred to health centers in larger towns and cities, leaving villages to their own devices. As government-imposed lockdowns began, all river transportation was halted. Our medical ships could no longer sail and, as expected, the government response to our nation's most isolated villages was little to none. Despite the fact that riverine communities closed their borders and no one could enter or leave, the virus found its way in and swept from one house to the next. The handful of communities that had been blessed with their own health centers saw their nurses leave and medications quickly run out. In some villages, old people left their homes and went into the rainforest, where they built secluded shelters and relied on hunting, harvesting, and fishing for food (Fraser, 2021).

New alliances, new hope: USAID, the Amazon Hope Medical Programme *and CEDRO*

In early 2021, the Amazon Hope Medical Programme joined forces with CEDRO, a Peruvian NGO with over 35 years of work promoting healthy lifestyles and community development in Peru. CEDRO, with the technical and financial support of the U.S. Agency for International Development (USAID), was implementing the *Alliance for the Amazon against COVID-19* project, with the purpose of helping mitigate the health, social, and economic impacts of the pandemic in underserved Peruvian populations. With a long history of investing in Peru's healthcare systems and long-term development, the US government, mainly through USAID, allocated more than $75.5 million in newresources and provided two million doses of vaccines to assist Peru in itsemergency response to COVID-19 (USAID, 2022). Through these agencies together, once lockdowns were lifted, we managed to get the ships back on the Amazonian rivers and our health services resumed. We worked together to add the mental health project to the already-existing medical program and began running it in April 2021.

Throughout the last years (April 2021 through April 2022), the program has provided 95,078 primary healthcare consultations, including medical and dental consultations, family planning, lab tests, treatment for diabetes, hypertension, anemia, and antiparasitic care. We have helped take COVID-19 vaccinations to communities that the government was unable to reach and have provided over 3300 shots. 2164 have received antenatal care and 3846 children have undergone well-child checks (Ramírez et al., 2022).

The mental health project has three components: (1) research, (2) psychological services, and (3) community-level interventions.

Fig. 4
Clinical psychologist conducting an interview in a riverine community. Puinahua River, Loreto, Peru.
Joshua Gleiser. Photo taken with participants' consent.

Our work began with (1) research, which involved conducting hundreds of interviews used to map the main mental health needs within the communities. Every day, we visited a village, asked permission from the authorities, and interviewed members of the community, as shown in Fig. 4. Interviews were audio recorded, transcribed, and subsequently analyzed qualitatively, allowing us to build diagnoses of local psychosocial needs and strengths. We have mapped the mental health needs and strengths of the communities along four river basins: Bajo Ucayali, Bajo Amazonas, Puinahua, and Bajo Marañón. Each map includes data from over 25 communities. These findings allow us to cater our interventions to the specific needs of each region.

In terms of psychological services (2), the mental health team offers in-person psychosocial support and counseling sessions, either in our consulting room or at the patients' home. These sessions provide support, validate emotions, identify personal and community strengths, and offer people a safe space to share their stories, which are usually filled with violence and traumatic events, such as sexual abuse, child maltreatment, and early deaths of children, siblings, and others due to lack of healthcare. As of April 2022, we have provided over 1800 individual counseling sessions (Rodríguez, 2022).

Some patients required more constant monitoring due to high-risk issues, especially since their communities lacked mental health professionals in their health centers or schools, and professors were not well equipped or trained to deal with these kinds of cases. mental health professionals from the ship used text messaging services to keep in contact with such patients when the ship was sailing. Nonetheless, a frequent barrier was the lack of phone reception or its unpredictability, since even villages with reception used to lose it during storms and rains. In those cases, mental health professionals have tried to establish contact with local health professionals or school teachers who do outreach, visiting the community intermittently, providing them with advice on how to deal with specific cases. Many patients had been recently exposed to the sudden loss of a family member, either due to COVID-19, accidents, or violence (e.g., due to the Bretaña conflict between indigenous community members and police forces). Whenever there was a chance of a patient being able to access cell phone reception, they were immediately connected to volunteer psychotherapists in the city who would offer them free remote counseling sessions. These processes helped support the patients throughout the weeks that the ship was sailing in other areas.

Psychological services also include a series of group activities aimed to deal with mental health-related topics that had been identified in our needs assessments. Work with children includes art workshops, recognition and expression of emotions, and extracurricular academic support, especially important during the pandemic, since children missed in-person classes for over 2 years and had no way of connecting to remote lessons. Group activities with adolescents include workshops on sexuality and gender issues, one's life project and aspirations, recognition and expression of emotions, and increasing self-esteem. Adults are mostly offered workshops on child-rearing practices and partner relationships.

Community-level interventions (3) aim to establish a tight working alliance with community authorities, professors, and local healthcare professionals. Through discussion groups and training sessions, the ships' psychologists (we are a team of 6) talk about mental health, psychological first aid, gender-based violence, and reproductive/sexual health with community representatives, who are then charged with conveying main takeaways and applying these skills with members of their communities. Since the chosen topics have been identified through interviews with communities' members, participants are well aware of the pervasiveness of these issues within their villages. Addressing them helps authorities find concrete alternatives to some of the problems, for example, information on what to do in case of intimate partner violence or child abuse.

The promise of technology for the future

Since early 2022, USAID has helped finance the implementation of an internet satellite on one of our ships to allow connectivity to WiFi during expeditions. This helps our services in innumerable ways, including access to training and supervision. Currently, the health team onboard is receiving remote group sessions of emotional support from volunteer psychologists belonging to the *Psicologos Contigo* initiative in the capital city of Lima. This has allowed the medical and mental health team to have a safe space to process their experiences living on the ship and working with the communities.

Moreover, we will soon begin to hold virtual group sessions and workshops with community representatives to deal with component (3) of the mental health project. Technology will help us break geographical barriers and allow for mental health professionals who are not onboard to provide some of the workshops remotely.

Further, as suggested by the United Nations High-Level Panel on Digital Cooperation (2019), we are in urgent need of implementing and improving digital tools in order to achieve the SDGs and "leave no one behind." The digital transformation must be grounded in common human values, such as inclusiveness, respect, human rights, international law, human centeredness, transparency, and sustainability. With this in mind, we are currently in the process of designing software that will let us record patients' data digitally. As the Pan American Health Organization (Pan American Health Organization [PAHO], 2021) has recommended, the digital health transformation must pay close attention to local needs and contexts, which further highlights the importance of the mental health project's research component (1) as a way of assessing local needs, idioms of distress, and regional strengths.

At this point in time, due to current widespread lack of internet connectivity, it is difficult for our digital health records to sync with national databases. This is aggravated by the existing fragmentation within the national health system—divided into two public health networks and the private health sector—and by the fact that Peru is the country with the highest cost for internet use in Latin America (Gestion, 2019; Picodi, 2019).

We recognize that the implementation of teleconsultation and telemonitoring health services is a crucial way to help close the health treatment gap and to provide accessible, sustainable services to our remote populations. In fact, when the COVID-19 pandemic hit our country, the Peruvian government was forced to rapidly implement a telemedicine strategy (García-Villasante et al., 2020), yet, the insufficient internet system became a tremendous technological barrier for a nationwide telemedicine service (Carhuavilca Bonett et al., 2020), leaving thousands of rural communities behind. In the case of Amazonian rural communities, the lack of connectivity, electricity, and financial resources rendered them incapable of accessing the services.

An inclusive digital health transformation will require local and regional authorities to partner with the private sector and international agencies to design a viable way of providing connectivity to Peru's remotest areas. Hence, we will continue strengthening our working alliances with key stakeholders and communicating the need for multilateral cooperation. It is only through the joint effort of multiple actors that the services of our medical ship will be accelerated and that universal health coverage for our population has a chance to be achieved.

References

Álvarez-Antonio, C., Meza-Sánchez, G., Calampa, C., Casanova, W., Carey, C., Alava, F., Rodríguez-Ferrucci, H., & Quispe, A. M. (2021). Seroprevalence of anti-SARS-CoV-2 antibodies in Iquitos, Peru in July and August, 2020: A population-based study. *The Lancet Global Health*, *9*(7), e925–e931.

Boren, D., Forbus, R., Bibeau, P., et al. (2003). Managing critical care casualties on Navy's hospital ships. *Critical Care Nursing Clinics of North America*, *15*, 183–191.

Bronfenbrenner, U. (1975). Reality and research in the ecology of human development. *Proceedings of the American Philosophical Society*, *119*(6), 439–469.

Carhuavilca Bonett, D., Sánchez Aguilar, A., García Zanabria, J., Montoya Sánchez, L., & Cuetto Maza, M. (2020). *Producción Nacional*. Lima, Peru: Instituto Nacional de Estadística e Informática (INEI).

Dannon, L., & Aljovin, P. (2020). *Necesidades y recursos de salud mental en comunidades nativas del río Ucayali en Loreto, Perú: un estudio cualitativo exploratorio*. [Manuscript in preparation]. The Vine Trust, Programa Medico Esperanza Amazonica del Peru.

Fraser, B. (2021). *Peru's indigenous peoples struggle to survive Covid-19*. Dialogo Chino https://dialogochino.net/en/uncategorised/41388-perus-indigenous-peoples-struggle-to-survive-covid-19/.

García-Villasante, E., Baca-Carrasco, V., Gutierrez-Ortiz, C., & Pinedo-Torres, I. (2020). Diabetes care during COVID 19: Experience in telemedicine from a developing country. *Diabetes & Metabolic Syndrome*, *14*(5), 1519.

Gestion. (2019). *Perú Tiene el Precio Más Alto de América Latina para Acceder a Internet*. https://gestion.pe/economia/peru-entre-los-paises-de-america-latina-que-mas-paga-por-acceso-a-internet-noticia/.

Hameed, J. M., McCaffrey, R. L., McCoy, A., Brannock, T., Martin, G. J., Scouten, W. T., & Riddle, M. S. (2016). Incidence, etiology and risk factors for travelers' diarrhea during a hospital ship-based military humanitarian mission: Continuing promise 2011. *PLoS One*, *11*(5), e0154830. https://doi.org/10.1371/journal.pone.01548.

Harris, R. D. (2013). Radiology on the Africa Mercy, the largest private floating hospital ship in the world. *American Journal of Roentgenology*, *200*(2), W124–W129.

Hill, C. F., Powers, B. W., Jain, S. H., Bennet, J., Vavasis, A., & Oriol, N. E. (2014). Mobile health clinics in the era of reform. *The American Journal of Managed Care*, *20*(3), 261–264.

Hofmeijer, I., Ford, J. D., Berrang-ford, L., Zavaleta, C., Carcamo, C., Llanos, E., Carhuaz, C., Edge, V., Lwasa, S., & Namanya, D. (2013). Community vulnerability to the health effects of climate change among indigenous populations in the Peruvian Amazon: A case study from Panaillo and Nuevo Progreso. *Mitigation and Adaptation Strategies for Global Change*, *18*(7), 957–978. https://doi.org/10.1007/s11027-012-9402-6.

INEI. (2020). *Loreto: Encuesta Demográfica y de Salud Familiar 2019. [ENDES]*. Recuperado de: Lima, Peru: INEI. https://proyectos.inei.gob.pe/endes/2019/departamentales/Endes16/pdf/Loreto.pdf.

Instituto Nacional de Estadística e Informática [INEI]. (2018). *III Censo de Comunidades Nativas*. Lima, Peru: INEI. https://www.inei.gob.pe/media/MenuRecursivo/publicaciones_digitales/Est/Lib1598/TOMO_01.pdf.

Instituto Nacional de Salud Mental Honorio Delgado Hideyo Noguchi [INSM]. (2009). *Estudio epidemiológico de salud mental en la selva rural*. Lima, Peru: Informa general. https://www.insm.gob.pe/investigacion/archivos/estudios/2009-ASM-EESM-SR.pdf.

Kistler, A. K. (2012). *Outline of the region of Loreto, Peru. [Map]. Center for International Environmental Law (CIEL)*. https://www.ciel.org/great-rainforest-or-the-greatest-rainforest/.

La Fondazione Albero Della Vita Onlus [FADV] (2020) Impacto del COVID-19 en los niños, niñas, adolescentes y sus familias en la amazonia del Perú–Provincia de Maynas—Región Loreto. http://fadvamerica.org/wp-content/uploads/2021/02/Estudio-impacto-Covid19-NNA-Loreto-S.pdf.

Licina, D. (2013). Hospital ships adrift? Part 1: A systematic literature review characterizing US Navy Hospital Ship Humanitarian and Disaster Response, 2004-2012. *Prehospital and Disaster Medicine*, *28*(03), 230–238. https://doi.org/10.1017/s1049023x13000149.

Limaye, N. P., Blas, M. M., Alva, I. E., Carcamo, C. P., & García, P. J. (2018). The Amazon hope: A qualitative and quantitative assessment of a mobile clinic ship in the Peruvian Amazon. *PLoS One*, *13*(6), e0196988.

Ministerio de Desarrollo e Inclusión Social [MIDIS]. (2021). *Reporte Regional de Indicadores Sociales del Departamento de Loreto.* https://sdv.midis.gob.pe/redinforma/Upload/regional/Loreto.pdf.

Ministerio de Salud [MINSA]. (2012). *Comunidades indígenas: caracterización de su población situación de salud y factores determinantes de la salud.* Lima, Peru http://bvs.minsa.gob.pe/local/MINSA/2734.pdf.

MINSA. (2015). *Análisis de Situación de Salud de Loreto.* Lima, Peru http://www.dge.gob.pe/portal/Asis/indreg/asis_loreto.pdf.

Mizokami, K. (2014). *Peace Ark: onboard China's Hospital Ship.* U.S. Naval Institute (USNI) News https://news.usni.org/2014/07/23/peace-ark-onboard-chinas-hospital-ship.

Nugent, C. (2020). *Peru locked down hard and early. Why is its coronavirus outbreak so bad? Time.* https://time.com/5844768/peru-coronavirus/.

Oliveira, J. (2022). *La Plataforma Itinerante de Acción Social Sostenible (PIAS) Río Yavari de la Marina de Guerra del Perú.* Va de Barcos https://vadebarcos.net/2022/02/15/la-plataforma-itinerante-de-accion-social-sostenible-pias-rio-yavari-de-la-marina-de-guerra-del-peru/.

Pan American Health Organization [PAHO]. (2021). *From the evolution of information systems for health to the digital transformation of the health sector IS4H conference report.* https://iris.paho.org/bitstream/handle/10665.2/53364/PAHOEIHIS210006_eng.pdf?sequence=5&isAllowed=y.

Picodi. (2019). *El Precio de Internet en el Mundo.* https://www.picodi.com/pe/ahorro-magazine/el-precio-de-internet-en-el-mundo.

Radio Programas del Perú—RPP. (2020). *Lizardo Cauper: "Más del 70% de pueblos indígenas tienen síntomas de coronavirus".* https://rpp.pe/peru/actualidad/coronavirus-en-peru-lizardo-cauper-mas-del-70-de-pueblos-indigenas-tienen-sintomas-de-coronavirus-noticia-1280017.

Ramírez, R., Paima, M., & Pila, E. (2022). *2021-2022 Estadísticas Totales Atenciones en PMEAP* (Unpublished Raw Data). The Vine Trust, Programa Médico Esperanza Amazonica del Perú.

Ren, A.-J., Sun, T., Wang, H.-W., Ge, H.-Q., & Ye, H.-J. (2015). Radiology on board the PLA(N) Peace Ark hospital ship during a humanitarian medical relief mission to the Philippines. *Disaster Medicine and Public Health Preparedness*, *10*(01), 16–19. https://doi.org/10.1017/dmp.2015.93.

Rodríguez, R. (2022). *Seguimiento Trabajo Salud Mental, Eje 2: Atención, 2021 y 2022. [Unpublished Raw Data] The Vine Trust, Programa Médico Esperanza Amazónica del Perú.*

Schwirtz, M. (2020). The 1,000-bed comfort was supposed to aid New York. It has 20 patients. *The New York Times.* April 2 https://www.nytimes.com/2020/04/02/nyregion/ny-coronavirus-usns-comfort.html.

Scouten, W. T., Mehalick, M. L., Yoder, E., McCoy, A., Brannock, T., & Riddle, M. S. (2017). The epidemiology of operation stress during continuing promise 2011: A humanitarian response and disaster relief mission aboard a US Navy Hospital Ship. *Prehospital and Disaster Medicine*, *32*(04), 393–402. https://doi.org/10.1017/s1049023x17000218.

Statista. (2022). *Coronavirus (COVID-19) deaths worldwide per one million population as of April 26, 2022, by country.* https://www.statista.com/statistics/1104709/coronavirus-deaths-worldwide-per-million-inhabitants/.

The New York Times. (2022). *Coronavirus World Map: tracking the global outbreak.* https://www.nytimes.com/interactive/2021/world/covid-cases.html.

United Nations High-Level Panel on Digital Cooperation. (2019). *The age of digital interdependence: A report of the UN secretary-general's high-panel level on digital cooperation.* New York: UN.

USAID. (2022). *U.S. Government Assistance to Peru: COVID-19 response.* https://www.usaid.gov/peru/fact-sheets/us-government-assistance-peru-covid-19-response.

Vine Trust. (2022). *Medical services.* https://www.vinetrust.org/about/medical-services.

Worldometer. (2022). *Reported cases and deaths by country or territory.* https://www.worldometers.info/coronavirus/.

Further reading

Alvarez-Risco, A., Del-Aguila-Arcentales, S., & Yáñez, J. A. (2021). Telemedicine in Peru as a result of the COVID-19 pandemic: Perspective from a country with limited internet access. *The American Journal of Tropical Medicine and Hygiene*, *105*(1), 6–11. https://doi.org/10.4269/ajtmh.21-0255.

United Nations [UN]. (2020). *Report of the secretary general: roadmap for digital cooperation*. https://www.un.org/en/content/digital-cooperation-roadmap/assets/pdf/Roadmap_for_Digital_Cooperation_EN.pdf.

World Maps. (2021). *Political map of Peru [Map]. World Maps*. https://www.worldatlas.com/maps/peru.

CHAPTER 96

Innovative technology: The Enhatch intelligent surgery ecosystem

Ashley Porto
Enhatch, Hoboken, NJ, United States; Department of Clinical Psychology, Columbia University Teachers College, New York, NY, United States; International Association of Applied Psychology, Indianapolis, IN, United States

Background

The Intelligent Surgery Ecosystem harnesses the power of artificial intelligence to eliminate inefficiencies throughout every stage of the surgical workflow. This allows medical device companies to benefit from advanced technologies without the need for a significant investment, enabling them to remain relevant in the new era of digital ecosystems (Enhatch, 2022). The Intelligent Surgery Ecosystem is an open ecosystem of surgical technologies powered by artificial intelligence to personalize and streamline the entire clinical workflow, leading to faster and more effective outcomes for patients. Features include five steps:

- Plan: Convert 2D images such as, X-rays or CT scans, into patient-specific models
- Print: Utilize patient-matched instrumentation to reduce the number of surgical tools by 75%
- Deliver: Real-time visibility of case management
- Implant: Enable technologies to drive safer and smarter surgeries
- Learn: Utilize data to make more educated decisions based on insights

Problem/challenge

The Intelligent Surgery Ecosystem addresses inefficiencies in the operating room through its comprehensive technology platform. This end-to-end solution delivers the next generation of cutting-edge smart surgical devices and enterprise workflow solutions, enabling more efficient surgeries and improved outcomes. By leveraging advanced analytics, surgery time can be reduced by 50% and surgeons have the ability to become more accurate and precise by assessing analytics overtime. This will improve the quality of care for patients.

Resilient Health. https://doi.org/10.1016/B978-0-443-18529-8.00096-2

How and where this innovation is being implemented, technology used, target, who and how the service is being delivered

This innovation is designed to improve instrument delivery within surgery rooms as well as back offices to ensure that surgeries are properly planned for. This technology is targeted towards a variety of key stakeholders, including surgeons, product managers, directors of operations and logistics, as well as marketing executives. With its broad applicability and powerful capabilities, this innovation has the potential to transform the surgical landscape.

Impact/successes

Our work has had a profound impact on the lives of surgeons, providing them with powerful tools to plan for better outcomes of their patients through the use of our advanced technology and analytics. In addition, our inventory management technology provides high visibility of instrument cases and trays. With Enhatch's patient-specific instrumentation solution, medical device sales representatives can reduce the number of trays needed for surgery, minimizing disruptions in the operating room and improving overall surgical experience for both surgeons and patients. Enhatch's commitment to innovation is transforming the way surgeries are performed and helping to drive better outcomes for patients around the world.

Challenges

At Enhatch, we face intense competition in the medical device sector. While we pride ourselves on providing world-class technology, we recognize that companies are constantly seeking to develop their own in-house solutions. One of our greatest challenges is to stay ahead of the curve, ensuring that our technology remains cutting-edge and competitive in an ever-evolving marketplace. With cost and efficiency becoming increasingly important factors, we must continually innovate and improve our products to stay ahead of the competition. Although we face these challenges, we are committed to delivering the best possible technology solutions to our clients and driving better outcomes for patients.

Strategic partners

At Enhatch, we believe that collaboration is key to driving innovation and improving patient outcomes. Through partnerships with leading medical device companies, we have been able to develop more efficient processes that enable our partners to focus on their core competencies and create better outcomes for patients in the long run.

One of our current private sector partners is Numeta Capital, which helped us raise our Series A round funding. In 2022, we entered a partnership with US 3D printer manufacturer 3D Systems with the goal of making design and delivery of patient-specific models more efficient.

Business plan

Our company has been fortunate to secure funding from some of the most innovative and forward-thinking organizations in the industry including: Numeta Capital and US 3D printer manufacturer 3D Systems.

How COVID-19 has affected the work, and how we responded/pivoted

COVID-19 had an unexpected impact, forcing us quickly to adapt and find new ways to operate. However, this challenge has also presented an opportunity for our team to come together and discover new ways of working effectively, even while working remotely. As a result, we have been able to provide even better service to our customers and prospects, while also honing in on our preoperative planning capabilities to meet the evolving needs of the medical device industry. Despite the challenges presented by the pandemic, we remained committed to driving innovation and forging new partnerships. In fact, we were proud to announce our newest partnership with 3D Systems, a world leader for additive manufacturing. This partnership allows us to leverage their expertise and technology to further enhance the quality and effectiveness of our products, ultimately helping us to deliver even better outcomes for patients.

Additional information

www.enhatch.com

Reference

Enhatch. (2022). Intelligent surgery ecosystem. Retrieved June 10, 2022, from https://www.enhatch.com/intelligent-surgery.

CHAPTER 97

Reducing infections and needlestick injuries: The case of RETRAGO

Mauro Pantaleo and Omar ElFata
P&P Patents and Technologies, Rome, Italy

Premise

According to the World Health Organization (WHO), injections are one of the most common healthcare procedures. Injection practices worldwide, and especially in low- and middle-income countries (LMICs), include multiple, avoidable unsafe practices that ultimately lead to the large-scale transmission of bloodborne viruses among patients, healthcare providers, and the community at large. Reuse and unsafe injection administration practices also increase the risk of bacterial infections and abscesses at the injection site, which can cause long-term damage. Unsafe practices include the reuse of injection equipment, Accidental Needlestick Injuries (NSIs) in Healthcare Workers (HCWs), overuse of injections, and unsafe sharps waste management.

Each year there are millions of deaths and health expenditures lost due to infections arising from syringe reuse and accidental Needlestick Injuries (NSIs). For example, in Italy every day, ~300 HCWs sustain an injury involving a contaminated needle or sharp medical device (needlestick and sharps injuries, NSIs), totaling >100,000 accidents per year; yet, likely more, since only an estimated 45% are officially reported (PHASE Study Group, 2012). Similar figures are available for the United States where, before legislation mandated widespread implementation of needlestick prevention devices incorporating a safety mechanism to cover the tip after use (safety-engineered devices, SEDs), 385,000 NSIs were estimated to occur annually, 60% of which were unreported (Panlilio et al., 2004).

The spread of bloodborne viruses including human immunodeficiency virus (HIV), hepatitis B virus (HBV), and hepatitis C virus (HCV) occurs when injection equipment is reused for administering injections or accessing shared drugs. According to a 2000 literature study on the usage of injections in healthcare settings around the world, the proportion of injections given with unsterilized reused equipment ranged from 1.2% to 75% (Hutin et al., 2003). In 2000 WHO conducted a study on the global burden of illness caused by contaminated injections

Resilient Health. https://doi.org/10.1016/B978-0-443-18529-8.00097-4

delivered in healthcare settings, finding that 40% of the 16 billion injections were given with reused injection equipment, resulting in 21 million new HBV infections, 2 million new HCV cases, and roughly 260,000 new HIV cases.

Accidental Needlestick Injuries (NSIs) in healthcare workers occur while administering an injection or after the injection, as well as before, during, or after disposal (Jahan, 2005). For example, recapping contaminated needles is associated with NSIs (Motaarefi et al., 2016). According to the World Health Organization (WHO), 3 million accidental NSIs occurred in 2003, resulting in 37% of all new HBV cases in HCWs, 39% of new HCV cases, and 5.5% of new HIV cases (Prüss-Ustün et al., 2005).

It is estimated by the WHO that 16 billion injections are administered worldwide in developing and transitional countries (Safety of Injections, n.d.). The breakdown of use of these injections is approximately 5% for immunization, 5% contraceptive and other reasons, and the far majority (90%) are administered for medical reasons (Hutin et al., 2003).

Developing and transition countries are organized by the UN into 12 regional groups including 57 nations: "rapidly growing East Asian economies (or the 'Tigers'), Southeast Asia, China, South Asia, semiindustrialized 'Latin America' (including South Africa and Turkey with economic structures similar to their counterparts in the Western Hemisphere), the Andean countries, Central America and the Caribbean, Central and Eastern Europe, Russia and Ukraine representing the former USSR, 'representative' and 'other' countries in sub-Saharan Africa, and the Middle East" (Rada & Taylor, 2006).

Needlestick safety and prevention legislation

Injecting drug use is found in 179 of 206 nations throughout the world, according to a 2017 systematic study published in the Lancet Global Health, but less than half of these countries have implemented needlestick programs (Larney et al., 2017). There has been significant progress over recent years.

November 2020 marks the twentieth anniversary of Needlestick Safety and Prevention Act (NSPA) which was signed into law in the United States on November 6, 2000. The Occupational Safety and Health Administration (OSHA) endorsed the use of safe needles or needleless devices for the collection and/or withdrawal of body fluids and for the administration of fluids and medications.

In Europe, the Council Directive 2010/32/EU, "Prevention from sharp injuries in the hospital and healthcare sector," fully in force since 2013, protects HCWs from NSIs and their consequences, setting up integrated policies regarding risk assessment, risk prevention, training, education, and monitoring.

The WHO has been actively working to find a proactive and coordinated solution for the spread of bloodborne diseases, such as HIV or Hepatitis C, through unsafe medical devices. In 2015 the WHO launched a new policy on injection safety with a goal of the exclusive conversion to safety syringes by 2020 (World Health Organization, 2016). Unfortunately the COVID-19 pandemic forced this deadline to be postponed but has increased awareness on the issue of NSIs and syringe reuse.

Different types of safety syringes—Why RETRAGO is a game changer?

Historically the market shifted from the glass syringe—which was problematic to maintain sterility, painful, and difficult to mass produce—to the plastic syringe, which is hygienic, much less painful, and highly scalable.

There are many different variations of safety syringe designs. Official performance requirements and definitions have been added and developed over time. Early designs of safety syringes started with the introduction of the Auto Disable (AD) syringe for immunization (1999), which incorporates a safety mechanism to lock the syringe plunger after use. Later models introduced Reuse Prevention (RUP) features in 2006 and progressed to Sharps Injury Prevention (SIP) features in 2012. The International Organization for Standardization (ISO) has outlined the design requirements for safety syringes in the following ISO standards:

> ISO 7886-4:2018:
> Sterile hypodermic syringes for single use—Part 4: Syringes with reuse prevention feature
> ISO 23908:2011:
> Sharps injury protection—Requirements and test methods—Sharps protection features for single-use hypodermic needles, introducers for catheters, and needles used for blood sampling

Still, even by conforming to these international standards, there is varying performance of different safety syringes in practice.

However, these SIP and RUP safety features are not exclusive. Some safety syringes incorporate only SIP and some only RUP—and many have failed attempting to feasibly incorporate both into the same syringe.

In addition to the requirements laid out in the ISO standards, safety syringes should be designed by considering four de facto features in order to be safe and feasible to produce/sell.

The first feature is the *safety design and mechanism* itself. There are two main features in this category: RUP (Reuse Prevention) and SIP (Sharps Injury Prevention). The safety syringe must protect from both the needlestick injuries as well as prevent the reuse of the syringe itself. Most safety syringes have one of these features and rarely both.

The second feature is related to the *trigger or activation of the safety mechanism.* Most safety syringe designs are manually activated (called "active devices"), which means that after using the syringe, the nurse must manually activate the safety mechanism, e.g., manually recap a needle shield. In contrast, "passive devices" are automatically operated throughout the use of the device, which we have learned as an industry is far more effective since they reduce errors and any possibility to bypass the safety mechanism. Passive (fully automatic) safety syringes are associated with the lowest NSI rates in comparison to active (manually activated) designs (Tosini et al., 2010).

The third feature is related to *training*. It is all well and good if you have a syringe that is very safe, but if you need to retrain the nurse to use it, the safety syringe must work like a conventional syringe. If, for example, a car manufacturer were to release a car guaranteed to keep passengers safe and prevent any injuries, but required retraining drivers, it would be very difficult—if not impossible—to sell.

The last feature is the *scalability of the design.* Billions of syringes are produced every year, when you are incorporating a safety mechanism, you are adding to the design, which needs to be scalable and producible at those high production levels. This has always been the challenge with safety syringes, namely, balancing between the safety features and how fast you can mass produce them.

We addressed these features with *RETRAGO—an innovative reinvention of the syringe*, implementing safety mechanisms promoting infection prevention and control. After a single injection, the RETRAGO syringe automatically retracts the exposed needle and prevents the reuse of the syringe. RETRAGO is the first automatically retractable syringe that features Auto Disable (AD), Reuse Prevention (RUP), and Sharp Injury Protection (SIP) safety mechanisms in one device. The combination of these safety features not only protects patients, but also healthcare workers, waste management workers, and the community from the risk of Accidental Needlestick Injuries (NSI) and syringe reuse, potentially saving millions of lives.

RETRAGO was awarded by the UNAIDS Health Innovation Exchange (HIEx) as a top Global Health Innovation with the potential to deliver impact for countries reaching the Sustainable Development Goals (SDGs) and was launched during the 72nd World Health Assembly 2019 from 21st to 23rd May 2019 at the UN Palais des Nations (Geneva, Switzerland). This recognition, as well as the continued support of the HIEx team, has dramatically reduced the time to scale up our innovation.

The costs

The economic burden of syringe reuse and needlestick injuries has been difficult to estimate since studies vary widely in terms of setting, considered costs, and epidemiologic data.

The calculation of needlestick/sharps injury costs described here is viewed from the perspective of direct and indirect costs incurred by the healthcare organization to manage an exposed healthcare worker.

According to the Centers for Disease Control and Prevention (CDC):

1. *Direct costs:* Costs generally borne by a healthcare organization when an NSI occurs: baseline and follow-up laboratory testing, PEP, and other treatment eventually provided, including PEP side effect management.
2. *Indirect costs:* These costs include time and wages diverted to receiving or providing exposure-related care; lost productivity associated with reporting and receiving initial and follow-up treatment for the exposure; healthcare provider time to evaluate and treat an employee; and healthcare provider time to evaluate and test the source, including obtaining informed consent for testing if applicable.

Certain indirect intangible costs are not part of this definition, such as any pain and suffering or societal impact resulting from an exposure or seroconversion.

A systematic review of economic analyses related to NSIs following the CDC operative definition was published in peer-reviewed journal *Infection Control & Hospital Epidemiology* assessing a total of 14 studies—8 of which are data-driven and 4 modeling studies. The review calculates an aggregate (direct + indirect) cost of Int$747 per needlestick injury (Goldstein et al., 2005).

However, it is important to acknowledge that the data is only as good as the reporting—and unfortunately many injuries go unreported in a healthcare organization.

A real-world example is useful to understand the real impact of unsafe injections, taking Egypt as an example. Egypt has one of the highest infection rates of Hepatitis C virus and is categorized as endemic (Elgharably et al., 2016). It is a very dangerous virus because if you get infected and you do not get screened, you may not feel symptoms for years after infection. During that time, if you do not know you have the virus, one can imagine how many people you are potentially spreading it to.

The "100 Million Seha" campaign, which kicked off at a national scale in October 2018 under the auspices of President Abdel Fattah El-Sisi, aims at detecting and eliminating Hepatitis C in Egypt by 2023. The campaign has started providing free tests to all foreign nationals in Egypt including refugees and asylum seekers as of 17 March 2019, with the first refugee receiving free Hepatitis C medication a week later. Screening will include testing for HCV antibodies, and assessment for hypertension, diabetes, and obesity. To achieve this goal, all healthcare facilities in Egypt will participate, including 3,800 rural health units, all Ministry of Health district and central hospitals, health insurance organization facilities, university hospitals, and military health facilities. In all, more than 5,000 test sites will participate in the screening program. A screening and treatment program of this magnitude has not been attempted before and will be a large step toward disease elimination in Egypt.

Impact of the pandemic

By considering the COVID-19 vaccination campaign, the additional demand for syringes will continue to increase dramatically in the next years. Assuming that by 2022 the manufacturing capacity to reach herd immunity (80% of the global population) will be reached, additional 12.7, 12.8, 13.0, 13.1 billion doses of vaccines will be distributed in 2022, 2023, 2024, 2025, respectively (Banfi, 2021).

In addition to syringes for each dose, the vaccination campaign will require additional syringes also for diluting the vaccine. For example, the Pfizer vaccine comes in one vial which can make 5 vaccine shots. Each injection is 0.3 ml, so you need 0.5 ml or 1 ml syringe (per person). For each vial, one additional diluent syringe (3 ml or 5 ml) is needed.

Syringes with a special feature referred to as "zero dead space" are even more difficult to source. With each injection using a typical syringe, you have a small waste of the drug/vaccine in the "dead space" between the syringe tip and needle. That volume is wasted and does not actually flow into your body during an injection. Zero dead space syringes can extract even that last bit of drug, effectively eliminating any waste. This is very important because if you consider a Pfizer vaccine vial, for example, you can extract six doses instead of five when you use a zero dead space syringe compared to a normal syringe. That's almost 20% more, meaning vaccinating more people; every drop counts!

The United Nations International Children's Emergency Fund (UNICEF) has stockpiled more than 3 billion AD syringes to be distributed to Low- and Middle-Income Countries (LMICs) for the vaccination campaign. However, the UNICEF also estimates that there will be an additional shortage of up to 2.2 billion AD syringes, and that "Low- and middle-income countries—where this type of syringe is critical for safety—will bear the brunt of this shortage." The UNICEF has also called for the end to what it calls "syringe nationalism" (UNICEF, 2021) by countries who block exportation of syringes in order to prioritize local and future demand. The introduction of vaccine booster shots and global supply chain disruptions will further increase syringe and other essential medical device shortages.

The way forward

The world has benefited from the invention of the hypodermic syringe in ways we cannot fathom. For the first time, a medical device allowed for the administration of medicine directly into the bloodstream—forever changing healthcare. Syringes are not different to any product going through its development cycle. Just as automobiles introduced the seatbelt decades after their invention, syringes today being redesigned with safety functions. Infections by dangerous bloodborne diseases such as HIV and HCV can be reduced dramatically with the introduction of national sharps programs, legislation, and the exclusive use of safety syringes. The Egyptian "100 Million Seha" campaign was a great success in the screening and treatment of those diagnosed for free—the missing piece of this initiative is the safety syringe. Treatment using pharmaceutical drugs does not prevent new cases. Drugs are a reactive measure after an individual has contracted a disease from an unsafe injection. The exclusive use of effective safety syringes is a proactive measure that can reduce new infections, while treating existing cases in parallel. It is for this reason that Egypt was one of the first countries in Africa to decide to convert the market from standard syringes to safety syringes starting June 2020—a step in the right direction. Stakeholders in the entire value chain need to be actively involved, including healthcare providers, syringe manufacturers, policymakers, innovators, and investors in order to ultimately maximize health impact, especially for the most vulnerable communities.

References

Banfi, S. (2021). *Perspectives for the safety syringe market worldwide*. Milano: CSIL.

Elgharably, A., Gomaa, A. I., Crossey, M. M., Norsworthy, P. J., Waked, I., & Taylor-Robinson, S. D. (2016). Hepatitis C in Egypt—Past, present, and future. *International Journal of General Medicine*, *10*, 1–6. https://doi.org/10.2147/IJGM.S119301.

Goldstein, S. T., Zhou, F., Hadler, S. C., Bell, B. P., Mast, E. E., & Margolis, H. S. (2005). A mathematical model to estimate global hepatitis B disease burden and vaccination impact. *International Journal of Epidemiology*, *34*(6), 1329–1339. https://doi.org/10.1093/ije/dyi206.

Hutin, Y., Hauri, A., & Armstrong, G. (2003). Use of injections in healthcare settings worldwide, 2000: Literature review and regional estimates. *BMJ*, 327.

Jahan, S. (2005). Epidemiology of needlestick injuries among health care workers in a secondary care hospital in Saudi Arabia. *Annals of Saudi Medicine*, *25*(3), 233–238. https://doi.org/10.5144/0256-4947.2005.233.

Larney, S., Peacock, A., Leung, J., Colledge, S., Hickman, M., Vickerman, P., et al. (2017). Global, regional, and country-level coverage of interventions to prevent and manage HIV and hepatitis C among people who inject drugs: a systematic review. *The Lancet Global Health*, *5*(12), e1208-20.

Motaarefi, H., Mahmoudi, H., Mohammadi, E., & Hasanpour-Dehkordi, A. (2016). Factors associated with needlestick injuries in health care occupations: A systematic review. *Journal of Clinical and Diagnostic Research*, *10*(8), IE01–IE04. https://doi.org/10.7860/JCDR/2016/17973.8221.

Panlilio, A. L., Orelien, J. G., Srivastava, P. U., et al. (2004). EPINet Data Sharing Network. Estimate of the annual number of percutaneous injuries among hospital-based healthcare workers in the United States, 1997–1998. *Infection Control and Hospital Epidemiology*, *25*, 556–562.

PHASE Study Group. (2012). *Prevention of occupational exposure to biohazard resulting from accidental percutaneous injury (puncture, wound, cut) in the hospital and healthcare sector*. Technical overview and recommendations for the transposition and implementation of European Council Directive 2010/32/EU in Italy.

Prüss-Ustün, A., Rapiti, E., & Hutin, Y. (2005). Estimation of the global burden of disease attributable to contaminated sharps injuries among health-care workers. *American Journal of Industrial Medicine*, *48*, 482–490.

Rada, C., L. Taylor. 2006, Developing and transition economies in the late 20th century: diverging growth rates, economic structures, and sources of demand. UN/DESA, New York, NY.

Safety of Injections, n.d. Global facts & figures. https://cdn.who.int/media/docs/default-source/integrated-health-services-(ihs)/injection-safety/is_facts-figures.pdf?sfvrsn=ca369593_5.

Tosini, W., Ciotti, C., Goyer, F., Lolom, I., L'Hériteau, F., Abiteboul, D., … Bouvet, E. (2010). Needlestick injury rates according to different types of safety-engineered devices: Results of a French multicenter study. *Infection Control and Hospital Epidemiology*, *31*(4), 402–407. https://doi.org/10.1086/651301.

UNICEF. (2021). *Urgent action needed now to ensure sufficient COVID vaccine syringe supply to meet 2022 vaccination targets*. www.unicef.org. (Accessed 26 May 2022). https://www.unicef.org/press-releases/urgent-action-needed-now-ensure-sufficient-covid-vaccine-syringe-supply-meet-2022.

World Health Organization. (2016). *WHO guideline on the use of safety-engineered syringes for intramuscular, intradermal and subcutaneous injections in health care settings*. World Health Organization. https://apps.who.int/iris/handle/10665/250144.

CHAPTER 98

Addressing global healthcare challenges through proactive preventive care: The case of Proactiff

Kameshwar Eranki[†]
Proactiff Digital Healthcare Services Inc., Newark, CA, United States

The COVID-19 pandemic irrevocably altered the healthcare landscape and compounded the issues and challenges of how companies do business amid hurdles, in moving forward. Improving patient access, patient engagement, and patient outcomes are key healthcare challenges. Other challenges that need attention—that also offer opportunities—are numerous. They include, but are not limited to, disparate, siloed, heterogeneous systems; overload of paperwork; dealing with Electronic Health Record (EHR) dysfunction; erosion of trust; regulations; financial constraints; better and effective utilization of scarce skilled, trained resources; managing costs; providing flexibility; and establishing evidence-based intervention care.

Introduction

Proactiff Digital Healthcare Services Inc. is a Silicon Valley-based disruptive digital healthcare company that is transforming healthcare from reactive sick care to proactive preventive care. As its name implies, Proactiff is working on a paradigm shift from an archaic *reactive* healthcare model to a digital *proactive* integrated healthcare model to enrich and expand the healthcare industry as well as to deliver easy, accurate, efficient services to satisfy the patient economically.

Proactiff is also working on enhancing care delivery and improving patient outcomes for conditions that need constant surveillance, and enhancing hospital services through several technologies, for example IoT devices, cloud technologies, AI (Artificial Intelligence), and ML (machine learning) as well as remote patient monitoring and real-time care for consumers/patients with disruptive technologies and unique solutions.

[†] Deceased.

Resilient Health. https://doi.org/10.1016/B978-0-443-18529-8.00098-6

Proactiff empowers companies to take dynamic, digital, and decisive steps toward a new normal and also emphasizes the need of resilience in healthcare. To do this, for example, the Proactiff resilience program screens for, and shows how to build, coping skills and how to enhance the skills of relationships, mindfulness, social connectedness, psychosocial well-being, happiness, and compassion.

Why Proactiff?

The Proactiff digital healthcare platform is the *Largest Health Data Aggregator in the World*, which empowers you to take control of your health. Proactiff allows the upload of data in near real time of 30 IoT/PoC devices onto its platform. Examples of such devices are Google Fitbit, Apple Watch, Omron Blood Pressure Monitors, KardiaMobile EKG/ECG, Dexcom Continuous Blood Glucose Monitoring, Digital Body Fat Calipers, Digital Goniometer, and more.

Proactiff's platform is fully HIPAA compliant. The data privacy and security goes beyond the Personal Health Information (PHI), wherein the end-to-end encrypted platform encrypts and protects all the 4700 data points (as of now, for any individual). The data protection controls and standards are applicable to data in transit and data at rest.

The Proactiff's AI platform addresses and analyzes all forms of body composition analysis, anthropometric assessments, and diagnostic tests conducted in hospitals, private medical labs, and mobile labs, which then get sent to healthcare facilities and to private homes for patient convenience.

The platform helps individuals and organizations alike to collaborate with healthcare providers, physicians, nutritionists, exercise physiologists, and care providers, to keep consumers healthy, and to prevent, treat, and reverse disease progression of lifestyle diseases.

Strategic differentiators

The Proactiff digital healthcare platform has several strategic differentiators. These include:

- **Integrated care delivery model**—provides a common platform for the transition from a conventional or mainstream allopathic model, where medical professionals treat diseases, to an Integrative Care model which encompasses approaches like Ayurvedic, homeopathic, Tibetan medicine, yoga, meditation, and tai-chi, to drive evidence-based outcomes.
- **Perpetual digital health records**—provide control of their data to patients/consumers whereby users can securely access their data from anywhere and anytime.
- **Digital therapeutics**—drives evidence-based therapeutic interventions to prevent, manage, or treat a medical disorder or disease.

- **AI-enabled platform**—provides *contextual AI* that opens the door for personalized medicine by analyzing data points based on gender, age, weight, height, race, ethnicity, and related factors. As a result, the scope of AI-based clinical decision-making is obvious. This helps doctors, health professionals make quick and personalized decisions, as color-coded insights showcase "no risk," "low risk," and "severe risk" conditions as well as sending digital alert notifications. The conversational AI components (like chats) quickly deliver to consumers in a question-response way and improve ease of use.
- **Prebuilt cognitive behavioral health assessments**—Cognitive behavioral health is an integral part of the Proactiff digital healthcare platform and encompasses well-being, happiness, assessment of cognitive behavioral disorders, psychosocial well-being, mindset, relations, Quality of Life, mood, feelings, symptoms, and conditions (like anxiety, anger, stress, depression, PTSD, dissociative disorders, autism, alcoholism, substance use, postpartum screening, child mental health, ADHD, social connectedness and more).
- **Integrated telehealth**—Zoom integration with Proactiff ensures integrated care delivery through virtual care models, which are integrated into clinical care pathways.

Highlights—The Proactiff digital healthcare platform

- is the largest health data aggregator in the world.
- provides an AI platform which addresses 310+ diseases, covers every stage of treatment, and gets to the "root cause" through its AI Algorithmic analysis.
- allows patients to own and control their health records.
- can service all healthcare systems globally.
- is fully HIPAA compliant.
- includes reimbursement by Health Insurance (in the United States).
- allows the highly unique data aggregating feature of uploading data in near real time of thirty IoT/PoC devices onto its platform, with such devices as Apple Watch, Google Fitbit, and Dexcom Blood Glucose Monitoring.
- addresses and analyzes all forms of diagnostic tests conducted in hospitals, FDA or CLIA approved medical labs, as well as mobile labs which go to healthcare facilities and homes for patient convenience.
- automatically sends evidence-based digital alerts to the patient and physician if the algorithms determine concerning out of range medical lab results.
- can service all healthcare systems globally.
- services any form of healthcare platform including hospital systems, private practice healthcare clinics, nursing homes, and can be therefore used in all countries for all healthcare platforms.

Proactiff's platform includes reimbursement by health insurance for U.S.-based healthcare providers

The Proactiff platform is premapped with preapproved codes for health insurance for maximum reimbursement in the United States. Specifically, Proactiff implements the CMS guidelines (Centers for Medicare and Medicaid Services) overseeing payment and supports the CPT codes (Current Procedural Terminology) and ICD-10 codes (both coding for medical services and procedures to streamline reporting, increase accuracy and efficiency).

Globally, Proactiff can do the same for all other countries, following their provided guidelines.

Besides covering nondevice care services, Proactiff has prebuilt integrations with over 30 IoT/PoC devices as of today. These include integrated devices for blood pressure, heart rate, breath rate, SPO2, lipid profile, complete blood count, electrolyte profile, and more. This device data also uploads onto the Proactiff EHR platform, as the data is generated.

Among the current 30 IoT/PoC devices that the Proactiff platform incorporates are three proprietary devices developed by Proactiff, including Proactiff's unique "No blood draw Non-invasive" Blood Glucose Monitoring device. You simply put your finger into Proactiff's Glucose Blood Monitoring device, hold it for seconds, and have your Blood Glucose levels interpreted and uploaded in real time onto the Proactiff platform.

Aside from Proactiff's 30 IoT/PoC devices services described before, the nondevice care services on the Proactiff platform include:

a) Diagnostic tests (i.e., blood albumin tests, lipid tests, electrolyte tests, complete blood count, thyroid tests, etc.).
b) Diagnostic tests involving digital scans and digital videos (i.e., CT scans, MRIs, and Ultrasound tests).
c) Sophisticated and advanced medical testing (i.e., for the Gut, Mitochondrial, and Genome sequencing).

Proactiff's AI platform automatically sends evidence-based digital alerts to the patient and physician if the algorithms determine concerning out of range medical lab results. The AI platform determines if a person's latest test results are way beyond their acceptable range, and for abnormal conditions, flags them with the color code red, indicating a *severe risk* which drives digital health interventions. On the patient's Proactiff platform, the platform automatically sends a digital alert to the patient, the patient's physician, and the patient's family member/caregiver to notify them of the exceedingly out of range result(s) so that immediate action can be taken by the patient and physician.

Additionally, Proactiff's AI platform covers all stages of a person's healthcare, starting at primordial preventive care, to primary care, secondary care, tertiary care, and quaternary care (overmedicalization and polypharmacy).

Stopping the lifestyle diseases in its track and reversing the disease progression: The case of atherosclerosis

A problem in current medicine is that it is merely treats the symptom and not the root cause. In contrast, Proactiff gets to the root cause. Let's take the example of atherosclerosis. Traditional medicine treats atherosclerosis by giving statins (drugs), aspirin, and advice about some lifestyle changes. But in giving statins (essentially prescription medicine) there is always the fear of side effects of the medicine, coupled with immunity buildup to the medicine after years of treatment which renders the medicine less effective over the years.

In contrast, through Proactiff, atherosclerosis patients take a blood test to determine whether homocysteine levels in the blood are elevated. This, in turn, shows the direct cause of the plaque buildup in the coronary arteries.

The Proactiff treatment protocol goes even deeper than the homocysteine test to identify the "root cause(s)." Proactiff's AI algorithmic analysis determines the reasons why the homocysteine levels are elevated, causing the atherosclerotic plaque buildup. For example, the rise in the plaque buildup may be directly attributed to deficiency of nutrients, e.g., folic acid and Vitamin B6 and Vitamin B12. These nutrient deficiencies are the "ROOT CAUSE" of the atherosclerotic plaque buildup.

Upon identifying the root cause, the folic acid, Vitamin B6, and Vitamin B12 nutrient deficiencies could be brought back to normal through recommending nutritional supplements, naturopathy, allopathic medicine, and other nonprescription medicine treatments, under the supervision of lifestyle medicine doctors who would prescribe these approaches. Thus prescription medicine such as aspirin and/or statins in this case could possibly be averted in favor of this root cause treatment.

Once the folic acid, Vitamin B6, and Vitamin B12 levels are normalized through the treatment protocol before, this reduces the homocysteine levels so they no longer are elevated. The reduced homocysteine levels could directly make the plaque stop building up in the arteries. With such lifestyle changes, reversal of a lifestyle disease is now a possibility.

Community healthcare success story

To demonstrate the success of this approach, the following describes the case of delivery of healthcare diagnostic services and basic mental health services in three villages in Amethi, Uttar Pradesh, India.

Issues and challenges

The challenge for people in Amethi is that there is no primary care or diagnostics medical center. The nearest such center is about 70 km away at the district (county) hospital. Even simple diagnostic testing like checking blood pressure, heart rate or ECG, obesity, skin disorders is not available. Also, not available are basic mental health services, like checking for mood disorders, stress, anxiety, depression, and fatigue.

Solution

The digital health diagnostics services and basic health and mental checkups of Proactiff are available for all people in the age group of 17–84 years old. For example, a 84-year-old patient, Mr. Tripathi, was assessed for his health vital signs, like for other patients, e.g., blood pressure, heart rate, breath rate, SPO2, ECG, mood, fatigue, energy, and skin allergies. In another example, in Kakwa, Tikree, a 77-year-old patient, Mr. Shri Yadav ji, shown in Fig. 1, was analyzed for basic health status including his vital signs wherein blood pressure, heart rate, SPO2, ECG, body temperature were checked, as well as his mood, anxiety, stress, depression, depression, and fatigue. This is an important milestone given that there are no diagnostic centers in this rural area and people rely on the county (district) hospital. Similarly, a patient, Smt. Malathi Devi, was assessed for vital signs, basic mental health, and shoulder pain.

In another unique dimension of the Proactiff program, village center meetings (called choupal), as shown in Fig. 2, are conducted for people to congregate in the village center, where the community health is assessed.

The local support teams ensure the convergence of people to participate in the health checkups and ensure successful implementation of primordial primary care by the Proactiff team.

Those patients identified with potential health issues are escalated into secondary and tertiary care through referrals to other doctors, clinics, and district (county) hospital.

Proactiff health resilience program

Setting the context—Understanding the need of resilience in healthcare

Are you or your organization suffering with diminished patient care and underperforming financially due to burnout, chronic stress, miscommunication, and low morale or experiencing trauma, and a range of emotions such as feelings of grief and sadness at an individual level?

If this is true and you are wading through the emotions and effects of stress and pain and rebuilding lives, then you are called "resilient people." But what does personal resilience mean?

Fig. 1
Vital Signs Monitoring of 71-year-old man in community.

Fig. 2
Men gathering at Choupal Meeting and Health camp, Tikree, Amethi.

What about resilience in healthcare?

Individuals and organizations such as hospitals and government municipalities are ill prepared to address and overcome adversity, burnout, stress, fatigue, and improve communication. COVID-19 pandemic demonstrated this and there is a need for both individuals and organizations for coping skills, compassion, mindfulness, and empathy. And with most workers still reeling with the aftereffects, imparting coping skills is obvious. So is dire need for resilience in healthcare.

What's included in Proactiff health resiliency program? And how it works

Assessment screenings for anxiety, stress, emotional, psychosocial well-being, mood swings, feelings, mindset, positive state of mind, happiness, resilience, hope, inspiration, work–life balance, purpose of life, mindfulness, self-compassion, compassion to others are used to assess and based on the identified gaps therein, impart coping skills and managing stress.

The Proactiff resilience solution program focuses on individuals, healthcare systems, doctors as well as healthcare workers, municipality workers, police, and emergency response teams. Proactiff offers 30+ assessments in the resilience program. After the assessment screenings, the AI-enabled platform performs automatic evaluation, scoring, risk categorization, and dashboard visualizations. The interpretation of the results is in simple, easy to understand color codes of green (no risk, low risk), yellow (borderline), and red (serious risk). Based on the risk level, evidence-based intervention care delivers optimal health outcomes.

The Proactiff resilience program needs minimal staffing and is geared to support the impacted individuals and their families during extremely vulnerable times of pain, fear, and loss. The platform helps build higher levels of resilience.

Assessing resilience and how to build resilience in your healthcare

Proactiff resilience program offers several assessment screenings and tools. These include:

- WHO-5 Well-being Index (WHO5)
- Quality of Life Scale (QOLS)
- Self-monitoring Scale
- Big Five Questionnaire-10 Assessment (BFQ-10)
- Positive and Negative Affect Scale (PANAS)
- Scale of Positive and Negative Experience (SPANE)
- PERMA Profiler
- The Ryff Scales of Psychological Well-Being (RSPW)
- Mindset Assessment Scale
- Positive State of Mind Scale

- Subjective Happiness Sale
- Happiness Measure Scale
- The Conner-Davidson Resilience Scale (CD-RISC)
- Flourishing Scale
- Gratitude Questionnaire Scale
- The Adult Hope Scale
- Inspirational Scale
- Cantril Self-anchoring Striving Scale
- Affect Balance Scale
- The Grit Scale
- Ten Item Personality Measure
- Oxford Happiness Questionnaire
- Emotional intelligence Questionnaire
- The Holmes–Rahe Stress Scale
- Psychosocial Well-being Scale
- Emotional Empathy Scale
- Purpose in Life Test
- Mood and Anxiety Symptoms Questionnaire
- Mindfulness Assessment
- Self-Compassion Scale
- Compassionate Love Scale
- Psychological General Well-being Index (PGWI)
- Staying Healthy Assessment
- Alcohol Use Disorder Identification Test
- Screening Brief Intervention and Referral to Treatment (SBIRT)
- Brief COPE Inventory (BCOPE)
- Trauma Screen (CPSS) or
- Trauma Screening Questionnaire (TSQ)
- Epworth Sleep Scale (ESS)
- Pittsburgh Sleep Quality Index (PSQI)
- Rosenberg Self-Esteem Scale (RSES)
- General Health Questionnaire-12 (GHQ-12)
- SCOFF Questionnaire
- Malnutrition Screening Tool

The proactive difference—Blazing new trails of resilience healthcare

The combination of Connor–Davidson Resilience scale (CD-RISC) and other assessments measures the stress coping ability. In addition, the program screens:

- Your ability to adapt when change occurs
- Assess on close and secure relationship
- Ability to deal with whatever comes my way
- Paint humorous side of things when faced with problems
- How you bounce back after illness, injury, or other hardships
- Your beliefs
- How you make best effort, no matter what
- How you achieve goals even when there are obstacles
- How you don't give up even in hopeless situations
- How you focus and think clearly, even under pressure
- How you are not easily discouraged by failure
- How you make difficult decisions
- How to handle unpleasant or painful feelings like sadness, fear, and anger.
- Having a strong sense of purpose in life
- Feeling you are in control

Resilience training

Proactiff offers resilience training to doctors, healthcare professionals, patients, children, and parents. These steps are as follows:

Step 1: Take the assessments/screenings

Step 2: Build:

- Cognitive and emotional skills and flexibility
- Self-awareness and psychological mindedness
- Communication and relational capacity
- Self-efficacy and professional fulfillment
- Engaged supportive community

Step 3: Gain insights using dashboard visualizations

Step 4: Track and regulate stress

Step 5: Adopt resilience training for improving coping skills

Proactiff also tailors programs for organizations and implement a certification based on the "train the trainer model" for broader and consistent implementation of its Resilience training.

Snapshot of the resilience training program for healthcare professionals

The resilience training involves three sessions that refer to different skills. They are as follows:

Session 1—**Mindfulness**—focuses on understanding what resilience is and the technique of mindfulness to enhance resilience in times of increased stress.

Session 2—**Coping skills**—describes how our thoughts about the world and other people impact our feelings, and ways of coping with uncertainty.

Session 3—**Compassion**—introduces techniques for showing greater kindness toward self during difficult experiences, which can enhance resilience.

Conclusion

As people and governments have seen that COVID-19 pandemic irrevocably altered the healthcare landscape and compounded the issues and challenges of how companies do business, amid hurdles, in moving forward.

As improving patient access, patient safety, patient engagement, patient education, and patient outcomes are key healthcare challenges, Proactiff's digital healthcare platform, with the integrated Primordial preventive care, primary care, secondary care, tertiary care, and quaternary care, is delivering 24/7 care across urban, rural, and community healthcare settings and transforming the landscape. The social determinants of health (SDOH) barriers are overcome in this integrated care delivery model. In addition, patient engagement with self-care helps in achieving cost-effectiveness, providing reliable and consistent systems to the people.

CHAPTER 99

Advances in technology for hearing loss

Cherisse Mark, Patrick F. Connolly, and Daniel Schaudel
MED-EL Medical Electronics, Innsbruck, Austria

Background: Prevalence and health effects of hearing loss

Hearing loss (HL) affects more than 1.5 billion people worldwide, of whom 430 million have disabling hearing loss (World Health Organization (WHO), 2022). By 2050 these numbers are expected to rise to 2.5 billion and 700 million, respectively. HL is graded as mild, moderate, severe, or profound, based on measured hearing thresholds. Unaddressed HL can have a major impact on verbal communication development in children and on speech perception in children and adults (Graydon et al., 2019). This can have secondary effects upon education and employment (Graydon et al., 2019), as well as lead to social isolation (Shukla et al., 2020), stigma (Wallhagen, 2010), and decreased physical and mental quality of life (Lopez et al., 2011). Additionally, HL is associated with more rapid age-related cognitive decline (Loughrey et al., 2018) and is a major risk factor for dementia, while also being modifiable if treated (Livingston et al., 2020).

Technology and hearing loss

Multiple interventions currently exist or are in development to treat HL at all levels. Broadly, HL can be treated through compensation of defective afferent auditory transmission using hearing devices or through replacement of defective structures using gene- or cell-based therapies. The most effective current means of treatment are hearing devices. The earliest of these were amplification hearing aids, electronic versions of which have existed since the 19th century (Valentinuzzi, 2020) and which have, as expected, improved technologically over the years.

Cochlear implants for hearing loss

For more severe hearing loss, amplification may not provide sufficient hearing perception. For this reason, the cochlear implant (CI) was developed, which employs a series of electrodes to electrically stimulate the auditory nerve within the inner ear. The first attempts at electrical

Resilient Health. https://doi.org/10.1016/B978-0-443-18529-8.00099-8

stimulation of the auditory nerve began in the late 1950s, and single-channel CIs were first implanted in the 1960s by their inventor, American physician William House (O'Donoghue, 2013). The modern multichannel devices were developed in the late 1970s by two independent teams lead by Austrian electrical engineers Ingeborg and Erwin Hochmair and Australian professor of otolaryngology Graeme Clark (O'Donoghue, 2013). In 2013 Ingeborg Hochmair, Graeme Clark, and Blake Wilson were honored with the Lasker–DeBakey Clinical Medical Research Award in 2013 for the development of the modern cochlear implant.

CIs are composed of two separate devices: the implant and the audio processor. The implant consists of a stimulator and an array of electrodes, placed within the cochlea via a surgical procedure, which deliver electrical pulses to the auditory nerve transmitted to the brain through afferent nerve transmission, producing sound perception. Individuals may have an implant on one side only (unilateral) or on both sides (bilateral). The audio processor, an externally worn device, contains one or more microphones to pick up speech and environmental sounds, as well as electronics to convert these sounds into electrical pulses, transmitted across the skin to the implant. Given advances in modern technology, audio processors can also stream sounds directly from other devices using Bluetooth connectivity.

Other advances in technology have improved usability. For example, due to individual differences in HL and sensitivity to electrical stimulation, a CI must be tuned to each user in order to ensure maximum perception, clarity, and comfort, as well as to evaluate the course of their rehabilitation. This procedure, known as *programming* or *fitting*, is done by connecting the audio processor to a computer via a programming interface device, from which the audiologist can manipulate parameters such as the intensity of stimulation at each frequency channel. Readings can also be taken from the device to monitor its status, and physiological measurements from within the cochlea can also be made. Programming then accomplishes several necessary fittings.

The COVID-19 pandemic made it necessary to identify new opportunities for aftercare of CI users. The procedure of fitting a patient allows almost perfectly the use of a network connection and a hosted web service including an audiovisual link. Improvements in hosted web service technologies as well as continuously growing bandwidth globally have done its part.

Remote programming of cochlear implants

Another advance due to the COVID-19 pandemic is the increased use of remote programming (RP) of CIs, typically performed in an audiology clinic, but possible via a computer interface over a network connection with the user and audiologist in different locations. Advantages to this approach include reducing the need for travel, allowing a single audiologist to serve

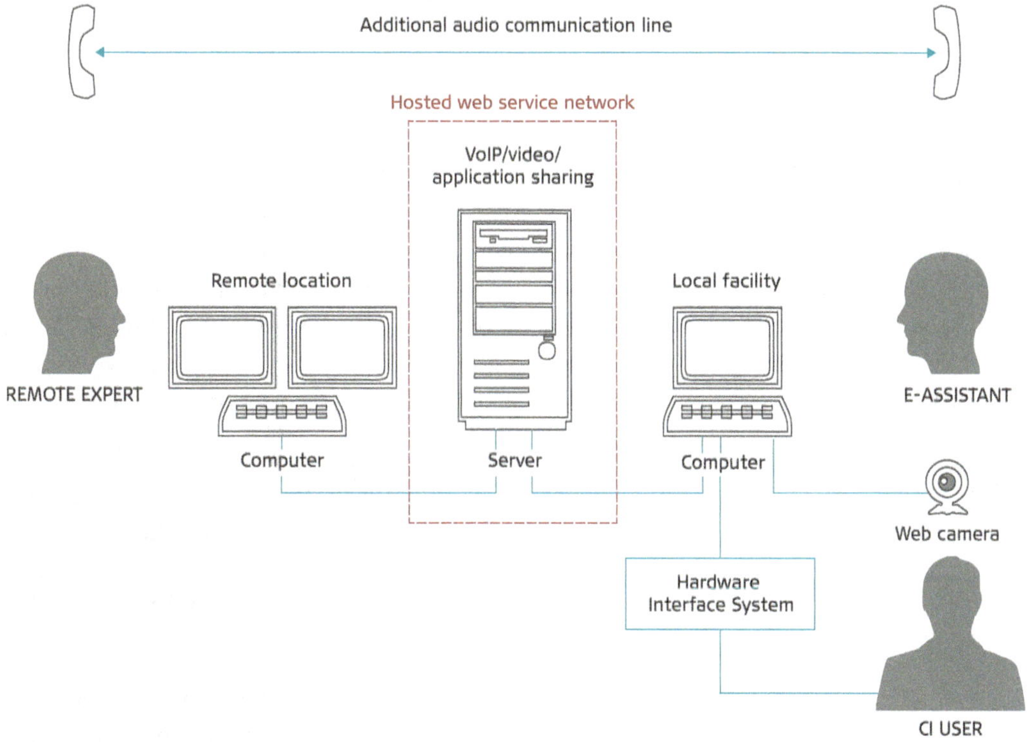

Fig. 1
Schematic of the remote programming protocol.

multiple patients from the same facility, reducing the burden for users who need frequent adjustments, and allowing audiologists to address users in underserved areas with more limited healthcare resources. The RP setup is outlined in Fig. 1.

RP is typically accomplished in the following way. The user travels to a local clinic for a follow-up fitting, where a locally present E-assistant manages the session. The E-assistant connects that audio processer to a computer via a programming interface device and establishes a remote network connection with the remote audiologist (also called remote expert). Then control over the programming software is assigned to the remote audiologist who can control the programming software on the local computer. The remote audiologist and user can also communicate directly using an audiovisual link via webcam and can also communicate via text.

The first pilot experiments in RP were conducted in 2009, and RP was approved in Europe by the Technical Inspection Association (TÜV) in 2009 and in the USA by the FDA in 2017 (Ramos et al., 2009; ASHA, 2018). It is safe and effective for both adult and pediatric users. Results of studies show that objective measurements and subjective benefit among users are similar between RP and face-to-face sessions, and audiologists find it to be convenient and useful (Eikelboom et al., 2014; Kuzovkov et al., 2014; Luryi et al., 2020; Ramos et al., 2009; Schepers et al., 2019; Slager et al., 2019). Regulatory changes in several jurisdictions have facilitated the use of RP and other e-health applications.

RP during the COVID-19 pandemic

During the COVID-19 pandemic, there were many severe disruptions to audiology services, such as suspensions of surgical procedures, and delays in rehabilitation and programming sessions, which led to negative outcomes both for existing and prospective CI users (Mostafa et al., 2021; Saxena et al., 2021; Telmesani et al., 2022; Vos et al., 2021). Likewise, the necessary use of face masks can impede lip reading between the user and audiologist.

In light of the rapidly evolving and vast societal problems during the COVID-19 pandemic, multiple regulatory changes were implemented in several jurisdictions to facilitate e-health applications. Among audiologists, the usage and perceived importance of e-health increased rapidly, to the point where 61% of audiologists reported using e-health applications (increasing from 42% prepandemic), and 74% expected to use such applications in the future (Eikelboom et al., 2022).

The pandemic also accelerated diffusion into clinical practice of RP and other e-health audiology. We and others conducted studies on the use of RP during this period and found that RP has highly favorable reviews among both users and audiologists. Measures of hearing outcomes are comparable between RP and conventional programming, as is the session duration. Connection-associated delays and disruptions were minimal. Most audiologists we surveyed rated the technique highly favorably and would recommend it to others.

Challenges

Potential challenges with implementing the RP approach include:

(1) The requirement for a relatively high-bandwidth internet connection
(2) Stable internet connection, both to facilitate application sharing and audiovisual communication
(3) Availability of electricity
(4) Connection latency, which can introduce a minimal delay in programming response times
(5) As with all digital services, a risk of malware, service disruption, and the security of transmitted information remain a potential concern
(6) Lack of direct face-to-face communication may be less desirable for some users and audiologists.

Future directions

Several potential future applications of RP infrastructure are possible. RP can be combined with other aspects of typical audiology follow-up in the same session, further reducing the need for frequent visits to an audiology center. Satellite-based internet connectivity may allow RP to be conducted in even more remote locations, allowing us to serve more underprivileged locales, and potentially freeing CI users from the need to remain within proximity of a major audiology facility.

Remote testing of the device during surgical implantation is also feasible. This would eliminate the need for an audiologist or technician to be physically present in the operating theatre, as has been successfully implemented (Yanov et al., 2018). The use of direct readouts of neural activity to assess auditory function (*objective measures*) has become common practice in recent years (Di Berardino et al., 2022; Somers et al., 2021) and can also be obtained during an RP session. These techniques are especially useful with individuals who cannot report their subjective sound perception, such as young children or those with communication disorders.

Finally, since CI users often participate in clinical and academic research, both in the development of device software improvements and in basic psychoacoustics research, RP infrastructure may allow users to participate in such research from convenient locations, when possible.

Conclusion and the way forward

In summary, despite the fact that the COVID-19 pandemic caused major disruptions in audiology services worldwide, we found that RP—which has been an established clinical practice for several years with multiple studies showing that it is safe and as effective as conventional fitting—served as a useful tool during the COVID-19 pandemic and helped sustain the continuity of audiological care for CI users throughout lockdowns. RP is not just a tool to facilitate continued patient care during societal disruptions; it has major advantages that incentivize its integration into standard clinical practice. It is anticipated that RP will remain useful during comparable future societal disruptions, and we encourage its continued adoption among audiology practitioners.

Over the past 40+ years, CIs have become the most widely used neural prosthesis, and the range of hearing loss indications could be increased dramatically over the coming years. Future topics include connected care and remote aftercare triggered by the pandemic as well as automated fitting utilizing physiological biosignals recorded in the auditory system (objective measures). This feature makes the fitting procedure independent from the patient's subjective feedback, which is particularly important for small children and patients who cannot give verbal feedback. These can reduce the workload of audiologists and increase the quality of fitting outcomes. The final goal is to continue expanding and improving access to audiology services to those over a billion persons worldwide who need them.

References

ASHA. (2018). FDA approves remote programming of cochlear implants. *The ASHA Leader*, *23*(2), 14. https://doi.org/10.1044/leader.nib1.23022018.14.

Di Berardino, F., Cavicchiolo, S., Del Carmen, F. M., Kontides, A., Lauss, K., & Zanetti, D. (2022). Maximum acceptable level for the determination of ECAP and ESRT in a paediatric population. *Cochlear Implants International*, 1–11. https://doi.org/10.1080/14670100.2022.2054097.

Eikelboom, R. H., Bennett, R. J., Manchaiah, V., et al. (2022). International survey of audiologists during the COVID-19 pandemic: use of and attitudes to telehealth. *International Journal of Audiology*, *61*(4), 283–292. https://doi.org/10.1080/14992027.2021.1957160.

Eikelboom, R. H., Jayakody, D. M., Swanepoel, D. W., Chang, S., & Atlas, M. D. (2014). Validation of remote mapping of cochlear implants. *Journal of Telemedicine and Telecare*, *20*(4), 171–177. https://doi.org/10.1177/1357633X14529234.

Graydon, K., Waterworth, C., Miller, H., & Gunasekera, H. (2019). Global burden of hearing impairment and ear disease. *The Journal of Laryngology and Otology*, *133*(1), 18–25. https://doi.org/10.1017/S0022215118001275.

Kuzovkov, V., Yanov, Y., Levin, S., et al. (2014). Remote programming of MED-EL cochlear implants: users' and professionals' evaluation of the remote programming experience. *Acta Oto-Laryngologica*, *134*(7), 709–716. https://doi.org/10.3109/00016489.2014.892212.

Livingston, G., Huntley, J., Sommerlad, A., et al. (2020). Dementia prevention, intervention, and care: 2020 report of the Lancet Commission. *Lancet*, *396*(10248), 413–446. https://doi.org/10.1016/S0140-6736(20)30367-6.

Lopez, D., McCaul, K. A., Hankey, G. J., et al. (2011). Falls, injuries from falls, health related quality of life and mortality in older adults with vision and hearing impairment—Is there a gender difference? *Maturitas*, *69*(4), 359–364. https://doi.org/10.1016/j.maturitas.2011.05.006.

Loughrey, D. G., Kelly, M. E., Kelley, G. A., Brennan, S., & Lawlor, B. A. (2018). Association of age-related hearing loss with cognitive function, cognitive impairment, and dementia: A systematic review and meta-analysis. *JAMA Otolaryngology–Head & Neck Surgery*, *144*(2), 115–126. https://doi.org/10.1001/jamaoto.2017.2513.

Luryi, A. L., Tower, J. I., Preston, J., Burkland, A., Trueheart, C. E., & Hildrew, D. M. (2020). Cochlear implant mapping through telemedicine—A feasibility study. *Otology & Neurotology*, *41*(3), e330–e333. https://doi.org/10.1097/MAO.0000000000002551.

Mostafa, A., ElRouby, I., & Alsoda, M. F. (2021). Challenges and solutions for cochlear implant rehabilitation during coronavirus disease 2019 pandemic. *Journal of Medicine in Scientific Research*, *4*(4), 271. https://doi.org/10.4103/jmisr.jmisr_44_21.

O'Donoghue, G. (2013). Cochlear implants—Science, serendipity, and success. *The New England Journal of Medicine*, *369*(13), 1190–1193. https://doi.org/10.1056/NEJMp1310111.

Ramos, A., Rodriguez, C., Martinez-Beneyto, P., et al. (2009). Use of telemedicine in the remote programming of cochlear implants. *Acta Oto-Laryngologica*, *129*(5), 533–540. https://doi.org/10.1080/00016480802294369.

Saxena, U., Chacko, G., Kumar, S. B. R., Nagabathula, V., & Damarla, V. (2021). Effect of COVID-19 on cochlear implant users: Parental perspectives. *Clinical Archives of Communication Disorders*, *6*(1), 12–17. Korean Academy of Speech-Language Pathologies 10.21849/cacd.2021.00353.

Schepers, K., Steinhoff, H. J., Ebenhoch, H., et al. (2019). Remote programming of cochlear implants in users of all ages. *Acta Oto-Laryngologica*, *139*(3), 251–257. https://doi.org/10.1080/00016489.2018.1554264.

Shukla, A., Harper, M., Pedersen, E., et al. (2020). Hearing loss, loneliness, and social isolation: A systematic review. *Otolaryngology and Head and Neck Surgery*, *162*(5), 622–633. https://doi.org/10.1177/0194599820910377.

Slager, H. K., Jensen, J., Kozlowski, K., et al. (2019). Remote programming of cochlear implants. *Otology & Neurotology*, *40*(3), e260–e266. https://doi.org/10.1097/MAO.0000000000002119.

Somers, B., Long, C. J., & Francart, T. (2021). EEG-based diagnostics of the auditory system using cochlear implant electrodes as sensors. *Scientific Reports*, *11*(1), 5383. Published 2021 Mar 8 https://doi.org/10.1038/s41598-021-84829-y.

Telmesani, L. M., Said, N. M., Mahrous, M. M., & Alrusayyis, D. F. (2022). The difficulties encountered by pediatric cochlear implant patients and their parents during the COVID-19 pandemic. *Audiology & Neuro-Otology*, *27*(1), 48–55. https://doi.org/10.1159/000517148.

Valentinuzzi, M. E. (2020). Hearing aid history: From ear trumpets to digital technology. *IEEE Pulse*, *11*(5), 33–36. https://doi.org/10.1109/MPULS.2020.3023833.

Vos, T. G., Brown, K. D., Buss, E., et al. (2021). Influence of postponed follow-up after cochlear implant activation during the COVID-19 pandemic on aided sound field detection and speech recognition. *Audiology & Neuro-Otology*, 1–8. https://doi.org/10.1159/000519908.

Wallhagen, M. I. (2010). The stigma of hearing loss. *The Gerontologist*, *50*(1), 66–75. https://doi.org/10.1093/geront/gnp107.

World Health Organization (WHO). (2022). *Fact sheet on deafness and hearing loss*. https://www.who.int/news-room/fact-sheets/detail/deafness-and-hearing-loss.

Yanov, Y., Kuzovkov, V., Sugarova, S., Levin, S., Lilenko, A., & Kliachko, D. (2018). Successful application and timing of a remote network for intraoperative objective measurements during cochlear implantation surgery. *International Journal of Audiology*, *57*(9), 688–694. https://doi.org/10.1080/14992027.2018.1479783.

CHAPTER 100

Building a state-of-the-art Healthcare-Medical City Center: Case study in Ethiopia

Mohammed Nurhussein

State University of New York Downstate Medical Center, Brooklyn, NY, United States; Ethio-American Doctors Group, Washigton, DC, United States; United Afrian Congress, New York, NY, United States

State-of-the-art hospital systems in Africa are direly needed. A group of Ethiopian doctors in the diaspora decided to do something about that for their country. Hence, the Ethiopian American Doctors Group was formed. Our goal is a fully equipped high tech facility, both brick and mortar and virtual outreach.

We've broken ground, and would be well on our way, if it weren't for the egregious and tragic war that ravages our land. We are encouraged by reports of an agreement to end the hostilities. Peace cannot come sooner for the country, for us, and for our health project.

Our plan is unique for Africa; therefore, our story is worth noting by the health community in the continent and the world.

Background: Breaking through barriers to breakthrough new health systems and hospitals in Africa

Emigration of skilled nationals (known as the "brain drain") has been bedeviling many low-income countries, especially in Africa. The African Union estimates losses in 2018 of 70,000 skilled workers, most of them in the health field who left for Europe, Australia, Canada, and the United States for better opportunities in their professional growth and advancement.

Thus a handful of Ethiopian doctors in the United States met in Washington, D.C., in 2012 to discuss ways to give back to their country that had invested so much in their education. Many of us have individually, or through charitable organizations, attempted to do that with negligible impact, sadly not making a dent in the health needs of the country. We decided to explore new and innovative ways to make a difference in our field of healthcare.

Resilient Health. https://doi.org/10.1016/B978-0-443-18529-8.00100-1

We pledged at this meeting to turn the brain drain into a "brain gain" through knowledge and technology transfer.

Thus the Ethiopian American Doctors Group (EADG) was formed to study the issue in depth and offer solutions. We agreed on an ambitious project: build a state-of-the-art hospital in Addis Ababa, the capital city of Ethiopia.

Patients seeking advanced medical care such as cardiovascular, neurosurgical, and cancer treatment were often driven to seek it abroad. They spent their meager resources, often mortgaging their homes and other assets, to fly far away to places like Thailand, India, and South Africa. The few who could afford it sought treatment in the USA and Europe.

Ethiopians spend over $80 million seeking healthcare in Thailand alone. This was a big drain on the country's limited foreign currency resource not to mention bankrupting the individual and family seeking the medical care.

The EADG hospital project

Our hospital will be a $100 million project. We agreed at the outset that it was to be sustainable and therefore needed to be a for-profit undertaking.

Our initial group of a dozen soon mushroomed to more than 350 healthcare professionals of Ethiopian origin living in the USA, Canada, Europe, and 14 other countries. Our requirement for membership was belief in its stated mission and willingness to invest, in money and services, to make it happen.

We raised close to $10 million in a short period of time (Fig. 1).

Our vision as stated is "to develop a high quality medical care through education and research for the people of Ethiopia, Africa and beyond."

Our project received enthusiastic support from the Ethiopian government under three successive prime ministers, recognizing its impact not only on healthcare and medical education in Ethiopia but also a vital investment in its economic development.

We intend to make sure the hospital construction and subsequent management meets world-class standards for healthcare delivery.

Our mission is to "build economically sustainable center of excellence hospital that would deliver internationally accredited standard of care and become a catalyst for change in how health care is delivered locally and in Africa."

We were able to acquire a lease from the government for a 150,000 sq. meters land in a rapidly developing section of Addis Ababa. The facility will consist of a 300-bed hospital that will have

Fig. 1
Architectural rendering of the hospital, a partial view. *From EADG website*

eight operating rooms, an 80-room primary and multispecialty ambulatory care services, a diagnostic center, and a cancer center of excellence. The remaining land will be developed to generate income for the hospital.

Laying the corner stone

After a yearlong intensive strategic planning, design, and review of the various phases in the development of the facility, the cornerstone for the project was finally laid by Prime Minister Hailemariam Desalegn accompanied by his Minister of Health and other officials of the Ethiopian government, on January 2017. Construction of the foundation was scheduled to begin within 3 months after that.

Leaving nothing to chance, we acquired the services of reputable US financial and architectural/engineering firms as well hospital technology, management, and staffing experts with whom we have been consulting all along since the early planning stages.

Once construction of the foundation began, we expanded the skeleton staff we had in Addis Ababa, to include a full-time manager, a full-time engineer to oversee the construction work, and an Ethiopian architectural/engineering consultant working with our American architectural firm. Our engineer who will supervise every step of the way is a highly regarded educator and practicing engineer with decades of experience in the US. He expects to make the hospital structure a green one by making it LEED (Leadership in Energy and Environmental Design) certified. There are few buildings even in the United States that have LEED Certification to qualify as a green building. We also have retained the services of a financial services firm located in Addis Ababa to liaise with Ethiopia's leading bank—Commercial Bank of Ethiopia which had shown a keen interest in our project right from its inception and agreed to be a partner in financing the project. The Trade and Development Bank—TDB is also a partner.

After all the bureaucratic hurdles were overcome, we were finally looking forward to get the construction going with the winning bid for the foundation work. The construction work of the foundation was completed in September 2020. The delay was partly caused by a variety of unexpected bureaucratic speed bumps. We were optimistic that we would be having a grand opening by December 2020 and eagerly looking forward to a gala event. But, our hopes were dashed. Political turmoil in the country culminating in a protracted civil war compounded by the global Covid 19 pandemic stalled the progress.

To date, we wait, to get back to building our dream.

Why for-profit?

Some of our members and other doctors before us who have attempted to give back either individually or as a group have been frustrated as they tried to establish a not-for-profit model hospital based on charitable contributions from benefactors. A few such hospitals have gone under for lack of funds to keep them going. Some could not finish their projects also for lack of funds.

We figured sustainability is key to such efforts and one way to make it sustainable is that it be a for-profit enterprise. A survey of our membership indicated that, for the overwhelming majority, personal profit was not the driving force behind their investment but rather the desire to bring to Ethiopia world-class healthcare services. One may rightly ask if expectation of profit is realistic in a poor developing country. We did our homework.

We identified three potential target populations

(1) *An emerging middle class*: This economic group has grown as a result of the rapid economic development in the country over the last three decades which shows that there is a local demand for international standard of healthcare. This is evidenced by the large number of Ethiopians seeking care in places, as I mentioned, like Thailand, India, Turkey, and South Africa, with the more affluent going to Europe and the US. (Interestingly,

Americans have been going to Thailand and India for procedures like bypass surgery and hip replacements where it is a third of the cost.)

(2) *Foreigners and expatriates*: Ethiopia is home to the African Union and the United Nations Economic Commission for Africa where there is a large contingent of international workforce. Ethiopia also maintains diplomatic relations with a vast majority of United Nations members represented by diplomats and their staff who would be reassured of a high standard of medical care in Addis from our hospital, rather than flying out of the country for their care. In fact, Addis Ababa ranks right behind Geneva and New York as home to an international community of workers. There is also a large number of nongovernmental international workforce in Ethiopia.

(3) *Tourists*: Ethiopia is increasingly becoming a tourist destination. Thus there has been an explosive growth in the hospitality services sector to accommodate the growing number of tourists arriving to explore the country's rich natural, historical, and cultural resources. According to World Bank Data, 184,000 tourists visited Ethiopia in 2004. That number jumped to close to a million tourists by 2017. We felt confident that visitors who spend two or three weeks in our country will feel reassured that their health needs will be met while visiting. In fact, our national airline, Ethiopian Airlines—which is a major international carrier—is a partner in our project.

Ethio-American doctors fund

"A decent provision for the poor is the true test of civilization," said Paul Farmer, American medical anthropologist and physician and humanitarian.

I like the quote as it pertains to our project in that it reflects the charitable energy we bring to our medical project.

Our Ethio-American Doctors Group has established a charitable arm whose funds will go toward financing the care of the indigent since our guiding principle is to treat anyone who walks our doors regardless of ability to pay. In addition to seeking help from donors, 10% of the profits that may accrue from the practice will go to the fund.

Technology and innovations

The hospital and the ambulatory care center as well as the cancer center and the diagnostic center will feature the "latest and the best" available in medical technology. The medical technology company ELSMED is our partner and will be responsible for the medical equipment needed. We are also partnering with Phillips. Our longtime partner, Global Health Services Network, will help with hospital operation and management.

Patient records will be held in the highest confidentiality and maintained in our electronic medical records (EMR).

It is our hope and expectation that the EADG Hospital and property will be a valuable asset in the community and beyond, a magnet for patients; medical and other health professionals; students, teachers, researchers, and entrepreneurs in the healthcare field; biotech and hi-tech industries; and investors.

CHAPTER 101

Wired for success: Using Whole Brain Thinking styles for innovation and solving global problems

Judy Kuriansky[a,b,c,d,e] and Ann Herrmann-Nehdi[f]

[a]Department of Clinical Psychology, Teachers College, Columbia University, New York, NY, United States, [b]Global Council SDG3 (Health), "SDGs in Action" Initiative, Dubai, United Arab Emirates, [c]International Association of Applied Psychology, Indianapolis, IN, United States, [d]United Nations Group of Friends of Mental Health and Well-being, New York, NY, United States, [e]World Council for Psychotherapy, Vienna, Austria, [f]Herrmann International, Forest City, NC, United States

Introduction

Innovation in healthcare requires breaking out of the box, thinking with new perspective, being creative, and taking risks.

How else could one come up with the idea of implanting chips in the brain to help paraplegic people walk, program an inanimate robot to perform delicate heart surgery, or create a bot that calms self-destructive thoughts.

Such innovations require the human brain to function at a high level, making new synaptic connections. Innovators reporting their creative solutions to dire healthcare needs, collated in the chapters of this volume, have been able to do just that process (Kuriansky & Kakkattil, 2024). This chapter explores the thinking style preferences that underlie this creativity. It is based on a system of styles that both authors have been using for decades to explain creativity, both in business and in personal relationships, and that has been intensively researched and applied worldwide by Herrmann International using data from their validated HBDI assessment (Herrmann & Herrmann-Nehdi, 2015). The goals of the schematic are to help individuals, managers, teams, and leaders be more productive and innovative, build skills, drive growth, and improve impact by making better decisions, effectively managing change, being more agile, and leveraging their untapped thinking potential.

Resilient Health. https://doi.org/10.1016/B978-0-443-18529-8.00101-3

The thinking categorization, based on brain research, is purposefully called "preference" because styles are not necessarily fixed. Also, while based on a classification reduced to a combination of only four styles seems oversimplified, the system actually works dramatically well even in the context of the complex way in which the brain works.

The four types are an elaboration of the much well known two-type division of right and left brain, whereby the left brain has traditionally stood for "logical" and the right brain by contrast, has been referred to as "creative". However, the latter is a misnomer, because Herrmann's research shows that all styles are creative. Also, the left/right dichotomy is overly simplistic. While the left-brain/right brain dichotomy has become so popularized and criticized, the he elaborated system presented here has solid supportive research and proven-useful applications in diverse settings with varied populations worldwide (Herrmann & Herrmann-Nehdi, 2015).

Ultimately, for real innovation to work—whether in a health setting as is the purview of this volume or in any other context—an idea needs to germinate and be implemented using the skills of all four styles described in the Whole Brain Thinking Model. This can manifest either within one person, or across a group of people in a team, where each person brings something from each style to the table.

This chapter presents Herrmann's four thinking style preferences, with some examples of innovations presented in chapters in this volume; outlines how the system can be applied for innovation, policy-making, and hypothetical solutions to achieve the global goals; and describes research that relates the styles to coping behaviors for resilience-building in the face of the COVID-19 pandemic.

The Whole Brain system

To be effective as an innovator and to manifest a vision into reality, it is important to understand your "thinking style preference" and that of all the stakeholders involved in the project, e.g., the people with whom you interact or who are on your team, the partners with whom you are developing or implementing an innovation, your users, and your funders. Certain styles mix well—as the saying goes, "birds of a feather flock together." It is also true that "opposites attract" but, can also clash.

The Whole Brain Thinking Model was originally designed and developed by Ned Herrmann who worked for GE for 25 years and subsequently applied his system to trainings for small businesses and multinational companies (e.g., Shell, IBM, DuPont, Hilton, Pfizer) for management development, product development and team-building. After years of research and computer testing, the process went online with an extensive questionnaire and research reports.

The four "styles" which emerged reflect relative dominance in four metaphorical quadrants. Although based on the brain, these are not located in one spot, as anyone familiar with neurology and the brain knows that complex connections occur for any one function.

Importance

Thinking style preference—or dominance—determines how you think, act, relate, and make choices and decisions. It is preferable to not be "stuck" in one style, nor to judge oneself, but to be flexible to tap into the qualities and abilities of each style when needed and appropriate. Of course, having all styles at your disposal is optimal, and Herrmann's research shows that each of us has some degree of preference for each style—even if one or more is more or less preferred.

The four styles

The four styles comprising the Whole Brain Thinking Model are shown in Fig. 1 and described in detail as follows. In the figure, they are labeled as A, B, C, and D to refer, respectively, to Analyzer focused thinkers, Organizer focused thinkers, Relational focused thinkers, and Visionary focused thinkers. Colors are assigned to refer to each (respectively, blue, green, red, and yellow) as well as adjectives that elaborate the style.

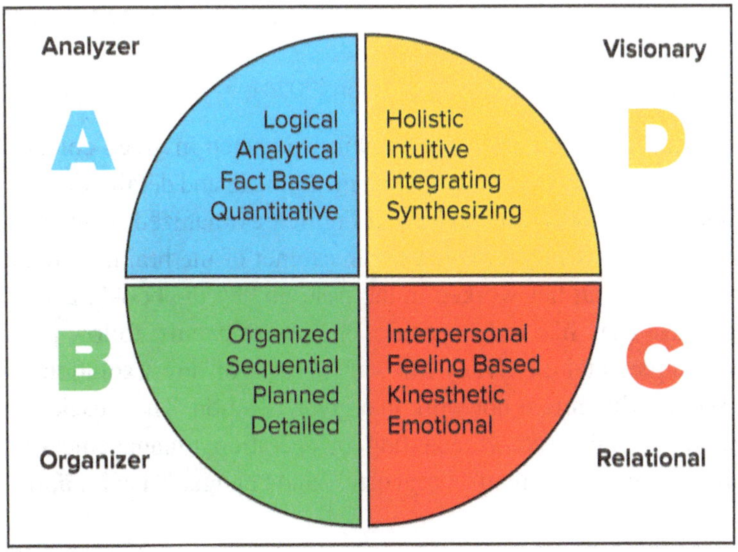

Fig. 1
The Whole Brain Thinking Model with four quadrants. *Source: The four-color, four-quadrant graphic and Whole Brain are registered Trademarks of Herrmann Global, LLC. 2015.*

Analyzer focused thinkers (A: Upper left thinking, depicted in blue color): People preferring this style like being logical, analytical, and fact-based, as shown in Fig. 1 in the upper left quadrant. They are rational and intellectual, and are good at these five steps of the problem-solving process: evaluating the problem, thinking it through carefully, analyzing options, making decisions, and coming up with solutions. In the COVID-19 context, they would require careful assessment of the needs on the ground and proceed to make decisions about what is required to improve or build a health system that is more resilient. They often have background training as scientists, lawyers, investment banking or engineering. They often prefer to "cut to the chase" rather than process emotions like Relationals would, and thus, can be seen as being not sensitive enough to other people's emotions or needs. Key words that appeal to them are "analyze" and "think." Their main focus is on problem-solving and making wise analyses and judgements.

Examples in chapters in this volume on resilient health address theoretical and structural issues like:

- social innovation for resilient health systems (Kpokiri, del Pilar-Labarda, & Tucker, 2024)
- inequalities of human rights in digital health (Imalingat & Mjwana, 2024)
- creating a "Connected Healthcare Ecosystem" through digitalization for an effective supply chain (Ramachandran, 2024)
- health equity to "build back better" after the pandemic (Kuriansky et al., 2024)
- principles of the applications of AR (augmented reality), VR (virtual reality), Mixed Reality, and the 5 A's applied to surgery (Shibata-Germanos, Ferenchick, Varalo, Semra, & Hachach-Haram, 2024)
- ethics in health and mental healthcare (Luxton, 2024).

Organizer focused thinkers (B: Lower left thinking, depicted in green color). People preferring this style are practical, well organized, methodical, and detail oriented. Adjectives to describe them, as shown in lower left quadrant of Fig. 1, are organized, sequential, planned, and detailed. They can be humorously called "the file cabinet of the brain." Structural about an innovation, they rely on what has worked in the past; go "by the book"; and value security, formulas, and formats. They like to be specific, emphasize details, follow procedures, and avoid vagueness or unpredictability. They will want to focus on how a solution can get executed and are good at seeing what might not work in a given solution. Their background training is often in the fields of administration, accounting, or operational management or data processing. Their key words are "plan", "organize", and "budget." Their approach is: What is the plan?

Examples of chapters from this volume listed reflect systems that emphasize the importance of data collection, management, and centralization; however, it is important to acknowledge that these innovations also clearly reflect considerable Analytic and Visionary thinking. In fact, all innovations emphasize the value of data in health systems to be effective, but for these, data is a central feature:

- the "Closed Loop Data System" (CLODS) (Abalos, 2024)
- The New Data Value Web Paradigm for Health Care Infrastructure in LMICs: The Case of Nexleaf (Ramanathan & Ross, 2024).

Relational focused thinkers (C: Lower right thinking, depicted in red color): As shown in the lower right quadrant of Fig. 1, adjectives that describe these people are emotional, kinesthetic, feeling-based, and interpersonal. They value expression, relationships, and communication and act on feelings rather than focusing on facts that Analyzers prefer. Humanistic, they care about how the innovation affects and helps people. They want to be "in the field" and "on the ground" reaching out to communities, especially those "left furthest behind." They often have a background as teachers, social workers, counselors, psychologists, and humanitarians. Relationals may be so concerned about people that they become overly sensitive to others' needs and fearful of offending anyone, but, they are good team players, anxious to form partnerships, and good at team-building since they see connections between people and work at making relationship work, making whatever compromises are needed. Their key questions are: Who will be impacted? and What's good for the people?

Since Relationals care about relationships, relevant innovations in this volume are those that focus on reaching out to the community and on community health workers. Also, since Relationals want to make partnerships, chapters relevant in this volume relate to partnership models, consistent with SDG 17 about partnerships called for in the UN Agenda 2030 for Sustainable Development. They know that the best approach involves multistakeholders and multilateralism (as called for consistently at the United Nations), because no one entity can solve big problems alone—much less design a resilient health system after the pervasive devastation of the pandemic.

Many innovations in this volume are hybrid models, which emphasize that technology and remote connection is not enough, and must be supplemented with reaching out to people in the community. Examples include:

- Realizing the transformative potential of Community Health Workers with digital innovation | A living goods case study (Bhavsar & Katara, 2024)
- Healthy Entrepreneurs: Transforming access to healthcare through an innovative model of community health entrepreneurship (van Engen, 2024)
- Innovation: Exploring Dimagi's Digital Platform CommCare: A widely adopted digital solution used by frontline workers worldwide (Javetski, 2024)
- citiesRISE: A transdisciplinary, youth-driven approach to scale social and technological innovations in mental health and well-being locally and globally (Hughsam et al., 2024)

- Citizen science: Technology tools to empower consumers for more resilient and sustainable healthcare systems during COVID-19 and beyond (Alhuwail & Alhouti, 2024)
- Empowering refugees through technology (Eluere, 2024)
- Platforms—like "Health in Your Hands" and "Health Innovation and Information Exchange," both web-based platforms that aggregate innovations, to make valuable connections (Kakkattil et al., 2024; Kuriansky, 2024a).

Visionary focused thinkers (D: Upper right thinking, depicted in yellow color): people with this preference have traditionally been the ones called "creative"; accordingly, in this dynamic Whole Brain Thinking model, they do like to "break out of the box." Visionaries like to "see" the big picture and even project into the future. They enjoy visual schematics namely, "picturing" solutions in contrast to data or numerical outlines preferred by Analytics or Organizers. As shown in the upper right quadrant of Fig. 1, their descriptive adjectives are holistic, intuitive, integrating, and synthesizing. Often free-spirited, they thrive on taking risks. They are often business innovators, strategic planners, artists and entrepreneurs. Like entrepreneurs, they enjoy coming up new ideas without necessarily knowing the details necessary to be carried out, which Organizers would manage. Like many CEOs, they project future growth. They might seem to others to be impulsive, yet, they move an idea, team, or organization forward, unafraid to try something new, to explore "What's possible?", and engage big ideas, like "How can we save the world?"

Examples from this book exemplifying Visionary focused thinking are:

- overviews of "The State of Health Today" and "The State of Mental Health Today" (Kopel & Fox, 2024; Pate & Jakab, 2024), as these require seeing "the big picture"
- future-thinking imagination and innovation in healthcare and all other areas of society in the United Arab Emirates, known as a hub for Innovation in systems (Al Rand et al., 2024) and individuals (Kuriansky, 2024b)
- technology innovations applied throughout society, e.g., by the government of Estonia, considered the "European silicon valley" (Annus & Merrimas, 2024)
- devising a road map for the future of technology in healthcare (Wallerstein & Venkatasubramanian, 2024)
- game-changing innovations and cutting-edge technologies (Greenshaw et al., 2024; Siddhanta & Kuriansky, 2024)
- solar energy as a way to connect when electricity is not available as can happen in remote communities (Osho & Sareen, 2024)
- Intelligent Virtual Agents, AI-enhanced virtual humans (Luxton & Watson, 2024)
- applying new concepts with which few people may be familiar, like "multi-omics" (Garbe & Sharma, 2024).

Overlap

All innovations in various chapters in this volume overlap these styles, even though they may emphasize one or the other. For example:

- The Permanent Mission of Mexico to the United Nations launched a mental health strategy during its tenure as an elected member to the UN Security Council, to highlight that access to mental health and psychosocial support services (MHPSS) must be at the forefront of sustainable peace and development (de la Fuente, Mendoza, & Guzik, 2024), and approach which clearly focuses on the well-being of people that Relationals care about but also on solutions to global issues characteristic of Visionaries.
- Activities of TPO Nepal (Gautam & Koirala, 2024) include research (analytic focused thinking), clinical services (relational focused thinking), research (organizational focused thinking), and advocacy with government about policy and forming strategic partnerships (visionary focused thinking).
- Models require activities of many thinking styles, including studies characteristic of Analyzers, innovative conceptualizations of Visionaries and connectivity of Relationals. An example is the Nexus model which brings together projects and partners focused on the Sustainable Development Goals of the UN Agenda that intersect with one another, creating synergistic ways they can enhance each other's ambitions (Coetzee et al., 2024).

Adapting

While a thinking style can be strongly preferred or even "dominant," any person is able to learn the skills involved in another style, if motivated. This is exemplified through the story of a student's development, evident in the multistakeholder project she codeveloped for resilience-building in Nepal that evolved from her semester project in the first author's class at Columbia University Teachers College, where she started out caring about the suffering of the people in her native country and ended up engaging varied stakeholders (including a high level government official she never suspected she could approach and would be highly receptive as well as several youth-based NGOs with innovative projects) as well as designing an assessment about the impact of a training for local counselors and analyzing the data, requiring skills she had to learn and which were not her thinking style preference at all (Khanal & Kuriansky, 2024). It is also interesting to note that a person can prefer one style at work and another in personal relationships, or when interacting with different people (e.g., reacting or adjusting to their style).

Gender issues

Stereotypes have persisted whereby males are thought to be more traditionally and frequently Analyzers, and women more Relational, but this dichotomy is changing. Consistent with the more traditional view, Analyzer focused thinkers tend to be more "command and control" style leaders focused primarily on results compared to Relational focused thinkers who prefer leading in a more personal and "transformational" style, engaging in cooperation, communication and inspiring others. Actually, either gender can be either style and adapt when needed.

To get to innovation

The most effective innovations use the "whole brain," tapping into all the thinking skills of each style, filling in any gaps either on one's own or in a team, thus compromising Whole Brain Thinking. The highest efficiency for innovation would engage all thinking styles, from all four styles: being analytic, organized, relational, or visionary, when necessary or appropriate. In problem-solving, a person would tap into skills from the four areas: theorizing, structuring, sharing, and exploring. Herrmann's research shows we all have access to all four styles and can stretch as needed. Also, an effective team would comprise individuals who contribute all four quadrant's qualities, to yield "the whole brain" solutions or products. A highly effective leader knows how to put such a team together recognizing that different thinking will complement their own preferences.

Each style has qualities others might find appealing or distasteful. For example, Organizational focused thinkers may have a plan but also be resistant to change, which would irk Visionaries. And, Analyzers who want to know the facts ("What happened?") can get frustrated with Relationals who want to share emotions ("How do I feel about what happened"?).

Similar thinking styles often gravitate to each other, when in reality they might really benefit from engaging with people or partners with opposite thinking styles, in order to get an innovation up and running. For example, Visionary focused thinkers very often irk Organizational focused thinkers (and vice versa) for what the latter perceive as coming up with impractical, risky "wild" ideas, but would gain from accepting and engaging this input to advance an initiative and new directions. By the same token, Visionary focused thinkers pushing for new ideas and modalities without first considering whether it works can become impatient and annoyed with Organizational focused thinkers who stall their imaginative process by making them stop to think about what may not be possible. In these combinations, someone like Organizers needs to worry about the risks, keep the books, and question realities when Visionaries can't be bothered and don't want anyone who "rains on their parade" and resists brainstorming.

In another potential and common conflict, Analyzer focused thinkers can appear offensively self-righteous and power hungry to Relational focused thinkers who prefer equality and equity, but would be wise to welcome the skills of effective strategizing that Analyzers bring to the table. Equally, Analyzers can find Relationals insufferable with their incessant concern for emotion and processing feelings when Analyzers are restrictive with feelings and want to just "get on with it."

All these combinations which can cause conflict can instead work well together if their respective style is understood and their respective value appreciated. For example, Visionaries may not be good at, or enjoy, detail work, and thus would be wise to embrace an Organizational focused thinker who can take notes, write an agenda, chart outlines, and track progress. Health innovators need an engineer who can figure out how to make an idea work, as in designing complex surgery instruments or constructing a drone.

Once you know your own and other's preferences you can react in complementary ways to avoid problems and enhance productivity.

A team can work together nicely, where a Visionary focused thinker comes up with an unusual idea, an Analyzer focused thinker figures out if it can work, an Organizational focused thinker documents the process and thinks about execution, while a Relational focused thinker considers the impact on the people being served. All have to participate in writing up the report describing the concepts and providing the proof.

Resilience study during COVID-19

Research by Herrmann revealed a relationship between Whole Brain Thinking style and personal coping with the COVID-19 pandemic for resilience-building consistent with the dominant characteristics of each style. As would be expected, Analyzer focused thinkers tended to cope by concerted efforts to comprehend and evaluate facts about the virus and information from experts, researchers, and scientists in order to make decisions about how to respond and be safe. They closely tracked the data and monitored and analyzed the financial and economic impacts. Those who prefer a more Organizational focused thinking style were more likely to do something to cope and exert control, like reorganizing their home or filing system, took extra precautions to prevent the risk of infection, and made plans, e.g., school- or work-related ("I went over what to do if someone got sick," "I made a daily schedule to keep track of our work and the kids school"). Consistent with their preferred focus on people, Relational focused thinkers prioritized self-care (e.g. doing meditation, yoga, exercise, walking, tele-therapy) and spent more quality time (when possible despite lockdowns) with family and friends (e.g., "I reached out to those I haven't talked to in a while"; "I strengthened my connection with friends by having virtual calls, chats and doing things"; "I spent more time enjoying my friends"). And, Visionary focused thinkers saw opportunities to use the time and explore new options like a new career, education plan, or hobby (writing a book, learning a musical instrument, cooking).

The system in action in a useful teaching experience

The first author's graduate course on "Psychology and the United Nations" at Columbia Univerity Teachers College gives the graduate students an exercise to come up with a solution to a global problem relevant to the Sustainable Development Goals of the United Nations agenda. They divide into teams of four or more, where each student is assigned to one thinking style to represent based on their scores on a short questionnaire. The team agrees on a problem they want to address, for example "eradicate hunger" (SDG 2); "end domestic violence" (a target in SDG 5); "provide health to a remote community in Africa" (SDG 3); or "stop global warming" (SDG 13), and in discussion together, they come up with a hypothetical solution to that problem, where each one contributes to the solution from the perspective of that thinking style. Then, each team presents to the class, describing their chosen problem, their solution and what each person contributed from their chosen thinking style. For example, the person representing the Organizational style would come up with a proposed budget, and the Relational representative would describe how their solution helps the target beneficiaries.

Policy-making

Whole Brain Thinking is distinctly involved in policy. Analytic focused thinking as well and Visionary thinking is important in the stage of strategizing. Organizational thinking evaluates past policies to build on those foundations; participates actively in the drafting and wordsmithing of the document, down to excruciating details like where to insert commas and semicolons; lists what as to be done to implement the policy and do follow-up and review, measuring the achievement of the goals and targets (as happens during the High-Level Political Forum where countries volunteer to present their progress)—in essence, determining compliance (a word used in other industries). Visionary focused thinking plays a part in every step of the policy process, especially ensuring the policy stands the test of time for the future aspirations of the country. Finally, Relational thinking brings in critical social perspectives and a viewpoint of how different constituents will respond to the given policy.

Thinking styles related to the UN Agenda

Being so familiar with the UN 2030 Agenda for Sustainable Development – in the preparatory stages leading up to the negotiations during 2014–2015 and throughout the Open Working Group (OWG) and the Intergovernmental Negotiations (IGN) until the passage of the Agenda by the governments in 2015, and then throughout the ensuing years of its implementation, advocating for the inclusion of mental health and well-being, the first author has conceived the agenda terminology as it relates to this thinking style scheme, as presented in Table 1.

Table 1 The thinking styles correlated to key words used by the United Nations and in the 2030 Agenda.

	• **Analyzer (A)** = "goals" "framework" "impact" "outcome" "justice" "sustainable development" "financing for development"
	• **Organizer (B)** = "agenda" "indicators" "disaggregated data" "systematic" "best practices" "data collection"
	• **Relational (C)** = "person-centered" "empowerment" "equality" "no one left behind", "the most vulnerable" "well-being"
	• **Visionary (D)** = "aspirational" "the future we want" "transforming our world" "for people and the planet" "now and into the future"

As presented in Table 1, the Analyzer focused thinking style relates to words like goals, framework, outcomes, and financing for development (FfD). Organizational-focused thinking is reflected in the push for disaggregated data and for best practices which are reassuring since they show a proven track record. The entirety of the agenda is "person centered" with the goal to "leave no one behind" which reflects Relational thinking style. And overall, the agenda itself is identified as "aspirational" consistent with the Visionary focused style, as it aims, reflected in the subtitle of the agenda, to "transform our world."

Conclusion

Truly effective innovations require varied skills and styles which can be classified into four thinking styles: Analyzer, Organizer, Relational, and Visionary focused thinkers, which comprise the Whole Brain Thinking model. This system has been extensively researched and shown to be useful when applied in diverse businesses and team-building. The technique is further useful as a hypothetical model to explore solutions to the global goals in the UN agenda and is consistent with the multilateral approach to effective solutions to global problems and achievement of the Sustainable Development Goals in the UN Agenda. Pilot research has also shown a relationship between the four thinking styles and coping behaviors in response to the COVID-19 pandemic. Teams of people who prefer one or the other of these approaches need to come together, appreciate, and build on each other's strengths, to develop and implement game-changing and cutting-edge technologies that advance resilient health systems required in these times when the world is still recovering from the impacts of the COVID-19 pandemic and needs to create new ecosystems.

References

Abalos, M. (2024). Big Data Innovation: CLODS bridging the gap of access to quality healthcare in developing countries. In J. Kuriansky, & P. Kakkattil (Eds.), *Resilient health: Leveraging technology and social innovations to transform healthcare for COVID-19 recovery and beyond* (pp. 1063–1071). Elsevier.

Al Rand, H. A. R., Madi, H., Alhemeiri, S., & AlObeidli, A. A. R. (2024). Healthcare innovation and leadership: Case study of the United Arab Emirates. In J. Kuriansky, & P. Kakkattil (Eds.), *Resilient health: Leveraging technology and social innovations to transform healthcare for COVID-19 recovery and beyond* (pp. 415–431). Elsevier.

Alhuwail, D., & Alhouti, A. (2024). Citizen science: Technology tools to empower consumers for more resilient and sustainable healthcare systems during COVID-19 and beyond. In J. Kuriansky, & P. Kakkattil (Eds.), *Resilient health: Leveraging technology and social innovations to transform healthcare for COVID-19 recovery and beyond* (pp. 593–599). Elsevier.

Annus, T., & Merrimas, K. (2024). The role of government in leading the pandemic response: The Estonian experience. In J. Kuriansky, & P. Kakkattil (Eds.), *Resilient health: Leveraging technology and social innovations to transform healthcare for COVID-19 recovery and beyond* (pp. 433–443). Elsevier.

Bhavsar, M., & Katara, K. (2024). Realizing the transformative potential of Community Health Workers with digital innovation | A living goods case study. In J. Kuriansky, & P. Kakkattil (Eds.), *Resilient health: Leveraging technology and social innovations to transform healthcare for COVID-19 recovery and beyond* (pp. 1011–1018). Elsevier.

Coetzee, C., Kuriansky, J., & Al Hammadi, R. (2024). The nexus model and the Young Leaders Program: Innovative models for the future of the SDGs in Action Campaign and the Global Councils on Sustainable Development Goals. In J. Kuriansky, & P. Kakkattil (Eds.), *Resilient health: Leveraging technology and social innovations to transform healthcare for COVID-19 recovery and beyond* (pp. 629–646). Elsevier.

de la Fuente, J. R., Mendoza, S. P., & Guzik, A. (2024). An unconventional security threat: Mental Health at the United Nations Security Council and Mexico's Response. In J. Kuriansky, & P. Kakkattil (Eds.), *Resilient health: Leveraging technology and social innovations to transform healthcare for COVID-19 recovery and beyond* (pp. 475–480). Elsevier.

Eluere, P. (2024). Empowering refugees through technology. In J. Kuriansky, & P. Kakkattil (Eds.), *Resilient health: Leveraging technology and social innovations to transform healthcare for COVID-19 recovery and beyond* (pp. 353–359). Elsevier.

Garbe, C., & Sharma, R. (2024). Multiomics technologies to increase resilience, adaptability, and affordability of integrated traditional and modern healthcare. In J. Kuriansky, & P. Kakkattil (Eds.), *Resilient health: Leveraging technology and social innovations to transform healthcare for COVID-19 recovery and beyond* (pp. 155–164). Elsevier.

Gautam, K, & Koirala, P. (2024). Adapting and scaling up mental health and psychosocial support response through innovative interventions to transform mental health care in Nepal for COVID-19 recovery and beyond. In J. Kuriansky & P. Kakkattil (Eds.), *Resilient health: Leveraging technology and social innovations to transform healthcare for COVID-19 recovery and beyond* (pp. 1029–1039). Elsevier.

Greenshaw, A., Noble, J. M., Sui, J., Cao, B., Li, T., Touros, J., & Grenier, R. (2024). Evolution in the health ecosystem: Cutting-edge technologies and important issues for improved health and mental health in the ongoing COVID-19 recovery. In J. Kuriansky, & P. Kakkattil (Eds.), *Resilient health: Leveraging technology and social innovations to transform healthcare for COVID-19 recovery and beyond* (pp. 361–372). Elsevier.

Herrmann, N., & Herrmann-Nehdi, A. (2015). *The whole brain business book, second edition: Unlocking the power of Whole Brain Thinking in organizations, teams, and individuals*. McGraw-Hill Education.

Hughsam, M., Claeson, M., Zeitz, L., Kumar, D., Foote, N., & Sinha. (2024). citiesRISE: A transdisciplinary, youth-driven approach to scale social and technological innovations in mental health and well-being locally and globally. In J. Kuriansky, & P. Kakkattil (Eds.), *Resilient health: Leveraging technology and social innovations to transform healthcare for COVID-19 recovery and beyond* (pp. 797–808). Elsevier.

Imalingat, T., & Mjwana, N. (2024). Digital health and human rights in an unequal world. In J. Kuriansky, & P. Kakkattil (Eds.), *Resilient health: Leveraging technology and social innovations to transform healthcare for COVID-19 recovery and beyond* (pp. 71–87). Elsevier.

Javetski, G. (2024). Innovation: Exploring Dimagi's Digital Platform CommCare: A widely adopted digital solution used by frontline workers worldwide. In J. Kuriansky, & P. Kakkattil (Eds.), *Resilient health: Leveraging technology and social innovations to transform healthcare for COVID-19 recovery and beyond* (pp. 1073–1075). Elsevier.

Kakkattil, P., Tiwari, A., & Jacob, B. (2024). Breaking the silos: How the Health Innovation and Investment Exchange (HIEx) helps bridge the health innovation ecosystem. In J. Kuriansky, & P. Kakkattil (Eds.), *Resilient health: Leveraging technology and social innovations to transform healthcare for COVID-19 recovery and beyond* (pp. 979–987). Elsevier.

Khanal, M., & Kuriansky, J. (2024). Promoting resilience in the Nepali community in response to the COVID-19 pandemic: A multistage multistakeholder partnership addressing the United Nations Agenda. In J. Kuriansky, & P. Kakkattil (Eds.), *Resilient health: Leveraging technology and social innovations to transform healthcare for COVID-19 recovery and beyond* (pp. 601–627). Elsevier.

Kopel, R., & Fox, C. (2024). Overview of the state of mental health in the COVID-19 recovery and beyond. In J. Kuriansky, & P. Kakkattil (Eds.), *Resilient health: Leveraging technology and social innovations to transform healthcare for COVID-19 recovery and beyond* (pp. 53–61). Elsevier.

Kpokiri, E., del Pilar-Labarda, M., & Tucker, J. (2024). Social innovation for resilient health systems. In J. Kuriansky, & P. Kakkattil (Eds.), *Resilient health: Leveraging technology and social innovations to transform healthcare for COVID-19 recovery and beyond* (pp. 63–69). Elsevier.

Kuriansky, J. (2024a). "Health in Your Hands": A platform for innovative healthcare solutions. In J. Kuriansky, & P. Kakkattil (Eds.), *Resilient health: Leveraging technology and social innovations to transform healthcare for COVID-19 recovery and beyond* (pp. 989–1004). Elsevier.

Kuriansky, J. (2024b). Leadership skills of His Highness Sheikh Mohammed bin Rashid Al Maktoum, Ruler of Dubai. In J. Kuriansky, & P. Kakkattil (Eds.), *Resilient health: Leveraging technology and social innovations to transform healthcare for COVID-19 recovery and beyond* (pp. 481–498). Elsevier.

Kuriansky, J., Altungy, P., Twose, G., Liébana, S., Maurice, S. R., Velo, C., … Guimarães, M. A. (2024). Health equity, mental health, and partnerships to build back after COVID-19: Multistakeholder perspectives at a United Nations event. In J. Kuriansky, & P. Kakkattil (Eds.), *Resilient health: Leveraging technology and social innovations to transform healthcare for COVID-19 recovery and beyond* (pp. 115–146). Elsevier.

Kuriansky, J., & Kakkattil, P. (Eds.). (2024). *Resilient health: Leveraging technology and social innovations to transform healthcare for COVID-19 recovery and beyond.* Elsevier.

Luxton, D. (2024). Technology ethics in healthcare delivery and public health: Current and emerging issues from the COVID-19 era. In J. Kuriansky, & P. Kakkattil (Eds.), *Resilient health: Leveraging technology and social innovations to transform healthcare for COVID-19 recovery and beyond* (pp. 193–203). Elsevier.

Luxton, D., & Watson. (2024). Intelligent virtual agents for current and emerging public health crises. In J. Kuriansky, & P. Kakkattil (Eds.), *Resilient health: Leveraging technology and social innovations to transform healthcare for COVID-19 recovery and beyond* (pp. 281–290). Elsevier.

Osho, Z., & Sareen, J. (2024). Accessing solar energy to deliver healthcare for all. In J. Kuriansky, & P. Kakkattil (Eds.), *Resilient health: Leveraging technology and social innovations to transform healthcare for COVID-19 recovery and beyond* (pp. 301–309). Elsevier.

Pate, M., & Jakab, S. (2024). The state of the world in health and health systems. In J. Kuriansky, & P. Kakkattil (Eds.), *Resilient health: Leveraging technology and social innovations to transform healthcare for COVID-19 recovery and beyond* (pp. 19–36). Elsevier.

Ramachandran, G. (2024). "Creating a digital spine for connected healthcare ecosystem": Digitalization for an effective supply chain. In J. Kuriansky, & P. Kakkattil (Eds.), *Resilient health: Leveraging technology and social innovations to transform healthcare for COVID-19 recovery and beyond* (pp. 291–300). Elsevier.

Ramanathan, N., & Ross, E. (2024). The new data value web paradigm for healthcare infrastructure in LMICs: The case of Nexleaf: The case of Nexleaf. In J. Kuriansky, & P. Kakkattil (Eds.), *Resilient health: Leveraging technology and social innovations to transform healthcare for COVID-19 recovery and beyond* (pp. 1057–1061). Elsevier.

Shibata-Germanos, S., Ferenchick, E., Varalo, J., Semra, K. E. A., & Hachach-Haram, N. (2024). Leveraging technology for health. In J. Kuriansky, & P. Kakkattil (Eds.), *Resilient health: Leveraging technology and social innovations to transform healthcare for COVID-19 recovery and beyond* (pp. 265–279). Elsevier.

Siddhanta, M., & Kuriansky, J. (2024). Game-changing digital mental health interventions: Cases in low- and middle-income countries and at-risk populations. In J. Kuriansky, & P. Kakkattil (Eds.), *Resilient health: Leveraging technology and social innovations to transform healthcare for COVID-19 recovery and beyond* (pp. 339–352). Elsevier.

van Engen, J. (2024). Healthy Entrepreneurs: Transforming access to healthcare through an innovative model of community health entrepreneurship. In J. Kuriansky, & P. Kakkattil (Eds.), *Resilient health: Leveraging technology and social innovations to transform healthcare for COVID-19 recovery and beyond* (pp. 1077–1083). Elsevier.

Wallerstein, D. A. M., & Venkatasubramanian, A. (2024). Technology innovation for global health: Past, present and future. In J. Kuriansky, & P. Kakkattil (Eds.), *Resilient health: Leveraging technology and social innovations to transform healthcare for COVID-19 recovery and beyond* (pp. 207–217). Elsevier.

Further reading

Devadas, D., Thériault, A., & McGahan, A. (2024). Models to scale healthcare technologies for low- and middle-income countries. In J. Kuriansky, & P. Kakkattil (Eds.), *Resilient health: Leveraging technology and social innovations to transform healthcare for COVID-19 recovery and beyond* (pp. 329–337). Elsevier.

CHAPTER 102

Teaching in-person or online: A graduate course on "Psychology and the United Nations" adapted during the COVID-19 pandemic

Judy Kuriansky

Department of Clinical Psychology, Teachers College, Columbia University, New York, NY, United States; International Association of Applied Psychology, Indianapolis, IN, United States; Global Council SDG3 (Health), "SDGs in Action" Initiative, Dubai, United Arab Emirates; Office of Ambassador Sidique Wai, Embassy of the Republic of Sierra Leone to the United States, Washington, DC, United States; United Nations Group of Friends of Mental Health and Well-being, New York, NY, United States

Introduction

The impact of the COVID-19 pandemic on education for both educators and students continues to be an issue of grave concern. Studies of the impact of online coursework have shown that diminished and more distant contact of students with both teachers and peers led to loneliness and declines in motivation, dedication to studies, student performance, and course completion. The United Nations Policy Brief: Education during COVID-19 and Beyond (United Nations, 2020) noted that closures of schools and other learning spaces have impacted 94% of the world's student population, and specifically up to 99% in low and lower-middle income countries. Two years later, at the UN Education Summit, the situation was still grim, with the pandemic having "harmed the learning of more than 90% of the world's children" (United Nations, 2022).

A few reports from the time the COVID-19 pandemic hit widespread proportions suggested some adaptive measures and positive outcomes when embracing online learning (Sohail, 2022). Four factors appear to affect the impact of online classes on the satisfaction and performance of students during the pandemic period of COVID-19: quality of the instructor, course design, feedback, and student expectations (Gopal et al., 2021).

Resilient Health. https://doi.org/10.1016/B978-0-443-18529-8.00102-5

Online courses can be referred to by many terms—online learning, open learning, web-based learning, computer-mediated learning, blended learning, m-learning—all involving use of a computer or other network-connected device to offer instruction from anywhere, anytime, in any rhythm, with varied technological means (Cojocariu et al., 2014). This approach can be synchronous ("live" meaning in real time) or asynchronous (whereby pretaped lectures are available to students to view at any time).

Most reports about the impact of online pedagogy have been about younger students, K-12, but growing literature addresses higher education, revealing both positive and negative outcomes (Cellini, 2021). This report adds to that at body of evidence, describing my experience teaching a 3-credit course on "Psychology and the United Nations" to masters-level graduate students at Columbia University Teachers College (TC), with comparisons and pros and cons of the online and in-person face-to-face experience which occurred in four time periods: before and during the COVID-19 pandemic, during an interim "hybrid" period (with most students in-person but some given school permission for synchronous online presence), and then when the school resumed required in-person classes (except for courses approved as fully online). Of note, the word "hybrid" refers commonly to a combination of the two approaches: in-person and web-based (with attendance simultaneously available online). Some issues discussed in this chapter can apply to any level of in-person or online teaching.

Background

Being in New York City, this course offers students an opportunity to attend events in-person at the United Nations headquarters and other related venues. However, during the COVID-19 pandemic, with mandated public building closures and imposed isolations, students were unable to enter these premises. Yet, this limitation actually opened up valuable opportunities to register for and attend meetings of the UN and related organizations held online which were previously not accessible for in-person attendance.

In the context of the United Nations and its global goals, which this class addresses, it is important to note that academic entities can be accredited to have consultative status with the United Nations Economic and Social Council, and association with the Department of Global Communications, allowing access to events, activities, and officials. A few universities have this ECOSOC status and accredited representatives (e.g., Temple, Fordham), as do a number of psychology-related NGOs. While an increasing number of psychology departments have courses on global health and mental health in general, few like mine described here focus directly on mental health and the United Nations, much less have high-level UN officials as guests as regularly, or offer as much access for students to UN events and delegates as mine does. Of those few taught by UN NGO ECOSOC-accredited representatives to UN headquarters in New York City, one is taught by a social worker, another is cotaught by two psychologists who both instruct online.

My course on "Psychology and the United Nations" is held weekly, as a 2-hour seminar on Tuesday evening, in the elegant Trustees room, with its wood paneling, chandeliers, and photos of distinguished past University luminaires. This environment befits the stateliness of the international topics being covered as well as receiving distinguished high-level guests, compared to other rooms in the college which are large brightly lit spaces with white walls. Most of those latter classrooms, however, have been outfitted with upgraded technology, whose features I sacrifice to be in the chosen space (Fig. 1).

By March 8, 2020, when the COVID-19 pandemic was reaching a peak spread, the TC president, Thomas Bailey, sent out a school-wide email alert about an alarming "widespread community transmission of COVID-19 in New York City" leading to the executive decision that as of March 10, and continuing after Spring Break (that next week), all teaching and learning will transition virtually, following the directive of Columbia president Lee Bollinger and other universities in the area, like NYU. One of my online classes is shown in Fig. 2, whereby I usually ended the class asking students to either wave to each other or make the "love" sign, in an effort to foster connection disrupted by mass separation not only through the online format, but by the lockdowns resulting in loneliness and other psychological distress caused by COVID-19.

Fig. 1
Fall 2022 class in the Trustees room (Professor Kuriansky on the left in the first row).

Fig. 2
Class on "Psychology and the United Nations" held online during the COVID-19 pandemic.

The transition from being in-person to online left little time to prepare for teaching virtually for those of us who had traditionally taught in-person. By June, the college sent a notice that online instruction would continue remotely into the Fall and Spring of 2021, with rare exceptions (for those needing to do fieldwork). By Spring 2022, the first 2 weeks of classes in January were scheduled online, but by the third week, in-person classes resumed, with mandates of mask-wearing and daily health screening. By the Fall 2022 semester, all classes were mandated to be held in-person, but proof of vaccination and booster was required for entry in the building. Approval for a medical or religious exemption required a weekly PCR COVID-19 test.

Teaching this course is directly in my purview, for many factors, including my 20 years as an advocate at the United Nations; involvement in successful advocacy to insure the inclusion of mental health and well-being in the UN Agenda (thanks to partnership with the Ambassador of Palau to the UN) and other international documents; friendship with high-level delegates; coorganizing innumerable UN-related events; and having done many interventions in countries worldwide (Kuriansky, 2016; Kuriansky & Zinsou, 2019).

Class overview

The class is described in the course catalogue as follows:

> An introduction to the role of psychologists at the United Nations and on the global stage, as a unique opportunity to explore and contribute to international psychological science, research and practice, and to learn about policies and programs "on the ground" worldwide relevant to mental health and well-being. Includes presentations, high-level guest speakers, and invitations to relevant events, with emphasis on an interdisciplinary perspective and students' pursuing their career goals and desire to "make a difference in the world."

The past few years have become ideal for this class topic, as mental health and well-being have become a priority on the international stage, even though those of us like me have been advocating about it for years at the UN. While momentum was already growing, the tragedy of the COVID-19 pandemic catapulted mental health into the limelight, given worldwide shocking statistics of the prevalence of anxieties, depression, suicide, and domestic violence. Even the Secretary-General of the UN António Guterres publicly mentioned attention to mental health several times as a "silver lining" of the dark cloud of the pandemic, a psychological concept about which I did some pilot research.

Modules in the class relate mental health and well-being within the nexus of the global issues of the United Nations 2030 Agenda for Sustainable Development (Coetzee et al., 2024), intersecting mental health and well-being (MHWB) with all the other Sustainable Development Goals (SDGs) like women's issues, education, climate change, and peace, as shown in Fig. 3, a schematic I designed showing mental health at the heart of the SDGs, which was used by me and the Ambassador of Palau to the UN Dr. Caleb Otto, when advocating to insure mental health and well-being was included in the UN 2030 Agenda (Kuriansky, 2016). Similar advocacy helped insure that MHWB was included related to the Political Agreement about Universal Health Coverage, the Global Compact on Migration (Kuriansky & Khurshid, 2023; Sandis & Bingham, 2023) and other international instruments.

Composition of the class

The class is like a mini-UN, in that students come from countries in regions all over the world, including North America (Canada, the United States); the Arab region (United Arab Emirates, Kuwait, Lebanon); North, South, West, and Eastern Europe (France, Germany, Portugal, Ireland, Hungary, Israel, Ukraine); Central and South America (Mexico, Brazil, Argentina, Peru, Colombia); East, Central and South Asia (Singapore, China, Japan, India, Pakistan, Nepal, Bhutan, Malaysia, Thailand, Turkey, and Iran); transcontinental Russia; CARICOM Caribbean countries (Jamaica, Haiti, Grenada); and Africa (Ghana, Ethiopia, Morocco, Uganda, Nigeria, Cameroon, Egypt, South Africa, the Democratic Republic of the Congo).

Fig. 3
Graphic showing mental health and well-being (target 3.4 of the UN Agenda) at the heart of the SDGs.

The course is offered as an elective to master's degree students through the Department of Clinical Psychology and Education, but is open to students in the Counseling program and any TC department (Art, Music, Industrial/Organizational Psychology, Transcultural Education, International Comparative Education, African studies) and even to the wider university (School of International and Public Affairs, Law School). The psychology students follow concentrations in trauma, global mental health, or SMBI (the Spirituality, Mind, Body Institute).

The topic of technology

A class module is always devoted to covering the topic of technology, its importance and examples of innovations—which became important during the COVID-19 time and in the ongoing recovery period—to:

(1) raise awareness about the use of technology to close the treatment gap—between the needs and lack of health and mental health services, to "reach those furthest behind" and "at the last mile" in remote communities in resource-poor developing countries with fragile health systems—which is the goal of the UN Agenda;
(2) build limited local capacity and resources, and integrate essential mental health components into a holistic healthcare system;
(3) leverage new sources of data and new approaches to collect and disseminate data; and new partnerships between civil society, the private sector, and academia;
(4) bridge the digital divide;
(5) provide a necessary link between health and mental health and technological innovation in important international campaigns. For example, the Global Council SDG 3 (Health) of the UAE's "SDGs in Action" initiative (for which I served as the immediate past co-chair), which is intended to advance achievement of the SDGs, adopted this volume within its work plan. Also, the *Health in Your Hands* initiative (which I cofounded) was designed as a platform to feature technology innovations which importantly are most effective when integrated with community outreach (featured in other chapters in this volume, Coetzee et al., 2024; Kuriansky, 2024); and
(6) encourage studies and projects related to COVID-19 issues and recovery. Examples of these include webinars on resilience and technology, as in the "COVID-19 and Africa" webinar series which I coproduced with my colleagues at the United African Congress, and for which I served as an expert commentator. A detailed report about that entire series was done by student Alexandria Rodriquez (Rodriquez, 2022).

The course module on technology covers technology at the United Nations and the importance of innovations, which are extensively presented throughout this volume by many projects around the world, including initiatives of the United Arab Emirates which particularly promotes innovation using technology (Al Rand et al., 2024; Coetzee et al., 2024; Kuriansky, 2024) and the impressive eMental Health International Collaborative covered in a chapter in this volume (Thapliyal, 2024) which includes specific involvement for universities in which some of my students have participated.

Technology and the UN

Technology is mentioned 41 times in the UN 2030 Agenda, for example, calling for its use in education (SDG 4) and in transfer from more developed to least developed countries to minimize inequities. Also, target 5.b in Goal 5 (SDG 5 about gender equality) calls upon States to "enhance use of technology to promote empowerment of women."

Additionally, targets in SDG 17 (about multistakeholder partnerships) include: 17.6, to "enhance North-South, South-South and triangular regional and international cooperation in technologies"; 17.7, to "promote the development, transfer, dissemination and diffusion of environmentally sound technologies to developing countries"; and target 17.8, to "fully operationalize the technology bank and science, technology and innovation capacity-building mechanism for least developed countries by 2017."

Recognizing the importance of technology in achieving the Global Agenda, the UN holds an annual Science, Technology and Innovation Forum. One year, my NGO (the International Association of Applied Psychology) coorganized an event for this conference on the topic of "Artificial Intelligence & Technology Tools for Mental Health, Well-being and Resilience" with examples being used for populations in Africa and Lebanon and for migrant workers in California as well as for responding to disasters. My students helped organize the event, and also two Chinese students presented their innovative project for posting translations of mental health information and journal articles in the WeChat format used in China (Kuriansky & Otto, 2017).

Overview of the course

It is my intention to introduce students to the diversified tripartite model of psychology with three domains: scientist, practitioner, and advocate. While the scientist-practitioner model is traditional, few students are exposed to being an advocate. The three roles now define my own career for the past 20 years, although I never expected to be an advocate, much less at the UN, until the then-executive of the International Association of Applied Psychology asked me out of the blue, "How would you like to be a representative at the UN?" While I had never visited the UN by that point, I followed my philosophy which I teach my students: "Say 'yes' and then figure it out." The role of being an advocate and drafting policy is now particularly appealing to students interested in social justice (Mallinckrodt et al., 2014).

My goals for the course are for students to:

- Understand the relevance of mental health and well-being, as well as clinical and counseling psychology, to the UN global agenda and other agreements.
- Appreciate the role of psychology and psychologists in policymaking as well as implementing programs on the international stage.
- Understand how personal relationships, the focus of clinical and counseling psychology, are a microcosm of a macrocosm of relationships between peoples and nations, toward personal, interpersonal, and world peace.
- Understand how the United Nations and related agencies and bodies work.
- Make valuable connections with global leaders in government and civil society.
- Learn how policy about mental health is accomplished by governments.
- Develop skills of advocacy, negotiation, and diplomacy.
- Learn about programs "on the ground": how they get developed, implemented, and evaluated as evidence-based.

- Explore personal interests and career opportunities, in fields of clinical work, education, international psychology, policy/advocacy and program implementation, humanitarian and development.
- Have valuable experiences, through writing short reports and a more intensive class project, to advance career objectives.
- Be creative, "present in the moment," and explore possibilities.
- Contribute to the class discussion with peers and expert guest speakers.
- Make the most of the experiences, the opportunities, and connections I offer.
- "Make a difference" in the world.

The course material focuses on policy and programs about mental health and well-being, as well as advances and what more needs to be done. Class presentations are didactic and also experiential. Students learn about my own and others' trainings and interventions around the world during natural disasters and epidemics (HIV, SARS, Ebola, and COVID-19). Much of my work, especially for children, relates to specific SDGs and uses healing expressive arts, which the students relate to, including simple, fun activities; movement; and indigenous original music, which I cowrite with my music partner, Russell Diasey. For example, anthems include "Rebati" (meaning "rebuilding") for Haiti, related to for SDG 13; "Every Woman, Every Child" related to the UN campaign of the same title and "Honor Our Girls" for the project of the First Lady of Sierra Leone Fatima Maada Bio against early child marriage, rape, and abuse of girls (Kuriansky, 2021) related to SDG 5; "Your Faith, My Faith" written for our World Interfaith Harmony week events; "Happy People Happy Planet" for the UN International Day of Happiness; and "Stand Up for Peace," related to SDG 16 and many UN peace celebrations like the International Day of Peace.

Three overriding messages are about hope, love, and togetherness, which align with themes for COVID-19 recovery. Songs about these themes ("Hope is Alive" and "One Love") are in the epigraph in this volume.

Guests

An appealing aspect of my class is visits by high-level guests, most of whom are good friends and colleagues, who give a short lecture and then engage in interactive dialogue with the students. These guests include Ambassadors/Permanent Representatives or Deputy Permanent Representatives from missions at the UN from all regions in the world (Portugal, Spain, the Netherlands, Hungary, Bangladesh, South Africa, Sierra Leone, Mexico, Colombia, Thailand, Denmark, Ireland, Bahrain, Iraq, Pakistan, Myanmar); the President of the General Assembly; representatives of UN agencies (World Health Organization, UNFPA) and UN Commissions (on the Status of Women); drafters of important UN international documents (Universal Health Coverage and the Global Compact for Migration). Guests are also leaders of civil society, the psychological community, and the public and private sector. To prepare for their visit, students

Fig. 4
Class with (front row, center, in white) Professor Kuriansky and to her left, the First Lady of Suriname and Father Wismick Jean-Charles. *Source: Gene Sampano.*

are given homework assignments to research the guest's background, country, and country's priorities and role at the UN.

In stark contrast to highly structured courses, flexibility in the syllabus is essential, for example, to accommodate high-level guests' schedules and last-minute availability to come in-person to class. This was the case for a class which was planned for a specific topic but had to be shifted last minute since two guests—on different topics—became available in-person. One of those guests was Father Wismick Jean-Charles, shown in Fig. 4, who was on his way from his home country of Haiti returning to Rome. Since I had been collaborating with him for years on disaster recovery in his country from the devastating earthquake of 2010 and later floods, the topic of that class would now cover humanitarian first response and the formation of the first Center for Spirituality and Mental Health (CESSA) in the region (Kuriansky, 2010, 2017), planned for a module later in the semester and with more preparation. In any case, his spirited presentation was greatly appreciated and highly relevant to many students' career goals (especially Haitian student Melissa Pierre who was fascinated to join CESSA).

On that same day, the First Lady of Suriname, Mrs. Mellisa Santokhi-Seenacherry, also shown Fig. 4, was in New York for the UN General Assembly and could fortuitously come to class in-person. This meant adjusting to two in-person guests back-to-back, on different topics. Yet, students adjusted well, even without much prepreparation and were fascinated with the First Lady's generous and lively account of her life as a professional female lawyer thrust into such a high position of leadership in government and becoming a female role model in a developing country.

It is often easier to make last-minute adjustments online than when guests need a staff or security accompaniment to come to the building.

Importantly, I offer students the opportunity to invite high-level guests, especially from the missions of countries of their birth or interest, which they never imagined they could accomplish, or that persons of such high level would respond to. For example, a student from South Africa, Avukile Zoya, persisted in reaching out, finally successfully, to confirm the visit of the Ambassador of South Africa, H.E. Mathu Joyini, for class on May 3, 2022, as shown in Fig. 5. The discussion valuably addressed women's rights, about which most students are especially interested; and highly relevant, Ambassador Joyini was elected to the prominent role of Chair (from the African States Group) of the 2023 Board of the Commission of the Status of Women.

Fig. 5
(Left to right) Student from South Africa Avukile Zoya, Ambassador of South Africa to the United Nations Mathu Joyini, and Professor Judy Kuriansky in class, May 3, 2022.

When guests come to class in-person, compared to online, the level of interaction and connection with the students is much more powerful than when online. This was viscerally apparent when Houry Guedelekian—chair of the NGO Committee on the Status of Women (NGO CSW/NY) and major advocate for Sustainable Development Goal 5 about the empowerment of women and girls and their rights—came to class in-person one semester compared to being online in another semester, sharing about having been subjected to early marriage, which she fervently advocates against (Fig. 6).

In the online format, when guests finish their appearance, they are basically disconnected. In contrast, when in-person, they often linger for photos and more interaction with the students "to know what young people think." Students also appreciate this extra time, since they have researched about the guest and submitted questions as homework (through a Google form) to be prepared for the Q and A session, but since that time is limited where not every student can pose their question, the extended interaction is invaluable for one-on-one discussion.

Since guests are usually engaged for a longer period of time when in-person than when online, many guests say they appreciate not only hearing the voice of youth, but being relaxed from their daily reserved diplomatic role. Two speakers distinctively said that speaking to the class is the highlight of their week. This may explain why in-person guests more often than those online say they would love to return. Figs. 7–9 show, respectively, speakers popular with the students who address topics particularly relevant to their interests, for example: on migration, by Fernando de la Mora from the Mission of Mexico who codrafted the Global Compact on

Fig. 6
Chair of the NGO Committee on the Status of Women (NGO CSW/NY) Houry Guedelekian addressing the class about rights of women and girls.

Fig. 7
Fernando de la Mora, delegate from the Mission of Mexico to the UN, addressing the class.

Fig. 8
Ambassador Mark Zellenrath from the Mission of the Kingdom of the Netherlands to the UN. addressing the class.

Fig. 9
Class with WHO Office at the UN with Director Werner Obermeyer (second row, center) with Technical Officer Ms. Eva Kiegele (to his left in white jacket) and myself, Professor Kuriansky (to his right in blue jacket).

Migration (de la Mora, 2023); on mental health in emergencies, by H.E. Ambassador Mark Zellenrath from the Mission of the Netherlands to the UN that hosted the "Mind the Mind" conference; and of course, about the World Health Organization, by the director of the WHO UN Office Mr. Werner Obermeyer and Technical Officer Ms. Eva Kiegele, who have a chapter in this volume (Obermeyer & Kiegele, 2024).

Having guests in-person has also led to valuable opportunities for students planning their future to go back home to advance mental health in their country. For example, the Permanent Representative of Pakistan to the UN, H.E. Ambassador Akram Munir, stayed after class to speak with three of my Pakistani students about internships and options in the health field (Fig. 10).

Many students express how amazed they are at how "approachable" such high-level guests are, and surprised at how they are able to connect to such distinguished guests' lives and careers. This was very much the case when H.E. Ambassador Sidique Wai, shown in Fig. 11, relayed his journey from being a civil society activist and community organizer for 25 years (including as liaison to the New York City Police Department) to being tapped by the President of his native country of Sierra Leone to become Chief of Mission and Ambassador to the United States in Washington, DC. The interest and excitement in the room to engage with a diplomat with such

Fig. 10

(Left to right) H.E. Ambassador Akram Munir, Permanent Representative of the Mission of Pakistan to the United Nations, Press Counselor Dr. Mariam Sheikh, and three Pakistani students from my class: Fatima Rizwan, Ayesha Liaqat, and Mariam Khurshid.

an important position and fascinating life story was palpable. His chapter about technology to advance health in Sierra Leone is in this volume (Wai & Kuriansky, 2024).

Guests often say they prefer (when schedule allows) the solo time of an externded in-person interaction. In fact, the online format is not as conducive to spirited interactive exchange in person between guests and students. Also, students online are more hesitant to speak up and interject in the conversation. Further, the in-person format allows - and requires - more time for a visit (at least one of the two hours of class), given the effort to come) compared to online appearances that can more easily be time-limited. Panels have been organized online, but rarely in-person; one successful in-person panel paired the Deputy Permanent Representative (DPR) of Bahrain to the UN, Hatem Abdhulhameed Hatem (a favorite repeat guest) with Felipe Queipo from the Department of Global Communications, for a module on Interfaith Harmony. a topic well related to both. This worked well since DPR Hatem describes Bahrain as distinctively a "multicultural, multifaith society" (as declared by King Hamad bin Isa Al-Khalifa) and as Felipe is the focal point for the "Focus on Faith" briefings linking faith traditions, religious organizations, and the UN's mission. The pairing, shown in Fig. 12, is also a powerful example of bringing government and civil society together, which is consistent with the UN emphasis on multistakeholder partnerships and which the students found inspiring.

Fig. 11
At class, H.E. Sidique Wai, Ambassador of the Republic of Sierra Leone to the United States (center, next to Dr. Kuriansky) with staff Mohamed Cozia (far left in blue jacket).

Some guests must be online when unavailable to come to campus, even if they are located in New York. This happened with the Permanent Representative of Thailand to the UN, Ambassador Suriya Chindawongse, as shown in Fig. 13, who, like in the situation mentioned before for South Africa, was invited by a student from the country, in this case, Thai Columbia Law School student in the class, Thorroong Phatiwet. Though being in-person would have been preferable, his participation was important given Thailand's support of mental health and their cofacilitation of the Universal Health Coverage Political Declaration.

Some desirable repeat guests who were once in New York but have returned to their country have to be online. For example, Her Excellency Katalin Bogyay, 15th Ambassador of the Mission of Hungary to the United Nations moved home to Hungary after completing her post at the UN, but is continuing major diplomatic roles, as founder of the Circle of Women Ambassadors and President of the International Association of UN Permanent Representatives, which she founded. As a musician, author, former television presenter, and Ambassador to UNESCO, she mesmerized the class about skills that overlap diplomacy and psychology, describing "cultural diplomacy" and "soft power" that require excellent listening, rapport, and communication skills, which students are learning in their clinical studies. A fan of psychology, Ambassador Bogyay has also been featured speaker at Psychology Day at the UN and is writing a book with me about psychology and diplomacy (Fig. 14).

Fig. 12
Deputy Permanent Representative of the Mission of Bahrain to the United Nations Hatem Abdulhameed Hatem (top row, center), and the UN Department of Global Communications Officer Felipe Queipo (top row, third from left) with some members of the class (with Professor Kuriansky in between both guests, in an abaya).

Another favorite online guest speaker located out of town (in the State of Washington) is psychologist Dr. David Luxton, an expert on tele-mental health, ethics, and artificial intelligence in behavioral and mental health, who is also the author of two chapters in this volume: one about ethics, and another about Intelligent Virtual Agents where facial expressions and nonverbal cues are scanned to develop reactions to coach in-need populations like veterans (Luxton, 2024; Luxton & Watson, 2024). Also on the topic of technology innovation which interests the students, the coeditor of this volume who is located in Geneva has presented to the class about his Health Innovation and Information Exchange platform which profiles innovations and is presented in this volume (Kakkattil et al., 2024).

Student projects

Students are required to do a semester report (usually on an event they have attended) and a semester project, which is a "deep dive" into a topic. They are encouraged to relate these to their professional growth and career goals. As such, projects span not only typical literature reviews but innovative initiatives in diverse topic areas.

Fig. 13
Permanent Representative of the Mission of Thailand to the United Nations, Ambassador Suriya Chindawongse, addressing the class on Zoom.

During the COVID-19 online period, such projects took on a decidedly more virtual nature. For example, several students explored their existing interests and work in technology and social innovation, some of which turned out to be appropriate for inclusion in this volume. These included innovations on a floating ship in Peru providing health services (Aljovin, 2024); with severely mentally ill patients (Fitapelli, 2024); and for surgical procedures (Porto, 2024a). Other students' projects included in this volume covered conferences on digital health and mental health presenting innovations in underreported countries and regions, specifically Central and South America and the Middle East (Diaz & Kuriansky, 2024; Irani et al., 2024; Lissinna, 2024; Otzoy-García et al., 2024). Also, given great interest in the workplace in COVID-19 recovery times, two students explored the web-based Mental Health program for UN staff with which I am familiar (Kuriansky et al., 2024). Other students pursued topics relating technology to their specific interest, for example, autism (Bikkani, 2024), sports (Barry & Kuriansky, 2024), families (Yoon, 2024), Artificial Intelligence and Augmented Reality (Porto, 2024b), tele-nursing (Knox, 2024), Pakistani female doctors (Rizwan, 2024), and creative online financing models given a student's desire to expand her own NGO (Bulanchuk & Kuriansky, 2024).

Fig. 14
15th Permanent Representative of Hungary to the United Nations, Ambassador Katalin Bogyay, speaking to the class online from her country.

Producing webinars during COVID-19

For their projects, many students during the COVID-19 period used tech tools and methods. For example, students produced webinar-based events, shown in Figs. 15–19 respectively, including:

- a webinar on "Our Post-Pandemic Future: The Importance of Cultivating Resilience and Identifying the Key Factors that impact Global & Personal Perceptions of Happiness" by two native Russian students, Elina Taboul and Svetlana Kiseleva, with guest speakers including myself and another TC professor of counseling who specializes in happiness, Dan Tomasulo, as well as the former Ambassador of Iraq to the UN, H.E. Hamid Al-Bayati, who founded the "Happiness" campaign at the UN, and a student leading a mediation
- a five-part webinar series on "Promoting resilience in the Nepali community in response to the COVID-19 pandemic: A multistage multistakeholder partnership addressing the UN Agenda," organized by Sadikchhya (Meera) Khanal, offering support for her native country, Nepal (Khanal & Kuriansky, 2024)

Fig. 15
Webinar on cultivating resilience and happiness during the pandemic organized by two students, with guest speakers Ambassador Emeritus of Iraq Hamid al Bayati (bottom row, right) and Professor Kuriansky (top row, center).

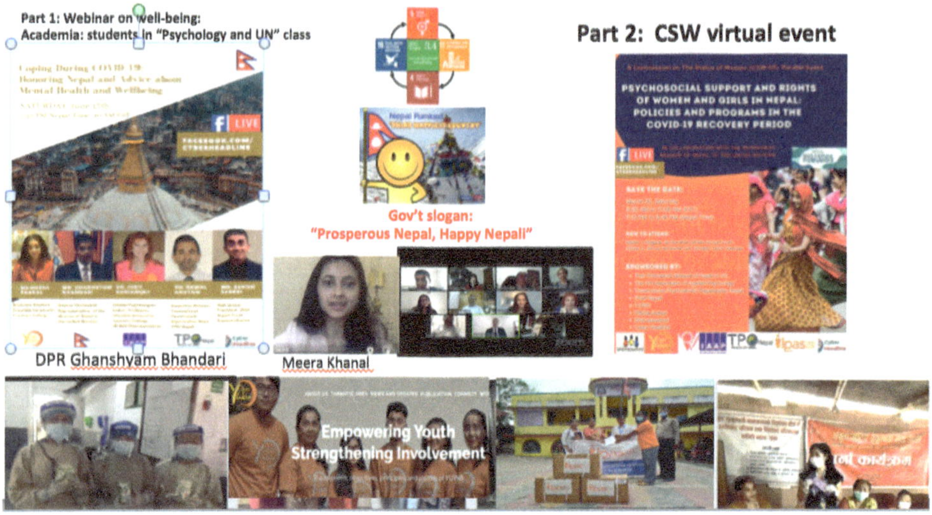

Fig. 16
Flyers and photos of participating NGOs on webinar to promote resilience for the Nepali community, organized and moderated by a student (middle row, center).

Fig. 17
Panel on resilience in the face of COVID-19 for China organized by three students from China.

Fig. 18
Training during COVID-19 for Chinese peer supporters led by Professor Kuriansky, organized by a student.

Fig. 19

One-day conference during the pandemic on change for people and the planet, organized by a student with Professor Kuriansky as one of the speakers.

- an online panel of experts about resilience in the face of COVID-19 for the Chinese population, in Chinese (with my presentation translated) organized by Yang Jiao, Xiya Li, and Jiawen Long
- a training led by me for youth volunteers who belong to a Chinese peer support program, organized by a student in my class
- a one-day online "Conference on Change for People and the Planet" promoting permaculture (SDG 2 about ending hunger) and mental health, with lectures and performances (Parks et al., 2022)

Given many students' interests in women's issues, several students produced webinars specifically for the 2021 and 2022 United Nations Commission on the Status of Women, NGO/NY Forum, held every March, and attended by thousands of registrants from around the world. These events were accessible to a considerably larger number of people when forced to be presented only in the online format due to the pandemic building closures and travel restrictions. Flyers for some of these events, shown in Fig. 20, were on the topics of:

- *Mitigating Transgenerational Trauma of Latina Migrant Women to Achieve Empowerment* by Pearce (2021)
- *Afghan Girl Refugees Coming to America: Approaches to Adjustment and Empowerment* by Smith (2023)

Fig. 20
Flyers for events organized by students for the UN Commission on the Status of Women NGO/NY Forum.

- *Mental health of refugee men* by Sharma (2021)
- *Mental Health, Gender Dynamic and Climate Change* by Liaqat (2022)
- *Psychosocial Support and Rights of Women and Girls in Nepal: Policies and Programs in the COVID-19 Recovery Period* by Khanal and Kuriansky (2024)
- *Women's Leadership Role and Psychosocial Perspective in Middle East Peacebuilding* by Yocheved Rabhan (Rabhan, 2021; Kuriansky et al., n.d.)

Other students used the tech medium of video, shown in Figs. 21–23 respectively, for example:

- producing a video summarizing a webinar in a series about "COVID-19 in Africa" organized by the United African Congress (UAC) of which I am a trustee and participated as an expert commentator, by Kidest Fikremarian

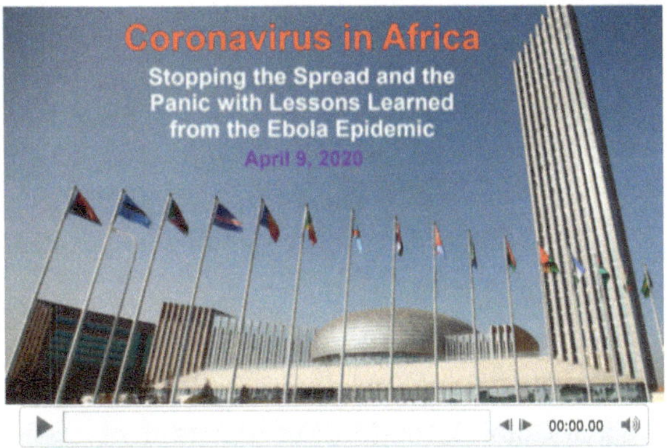

Fig. 21
Video produced by a student about the "COVID-19 in Africa" webinar series of the United African Congress.

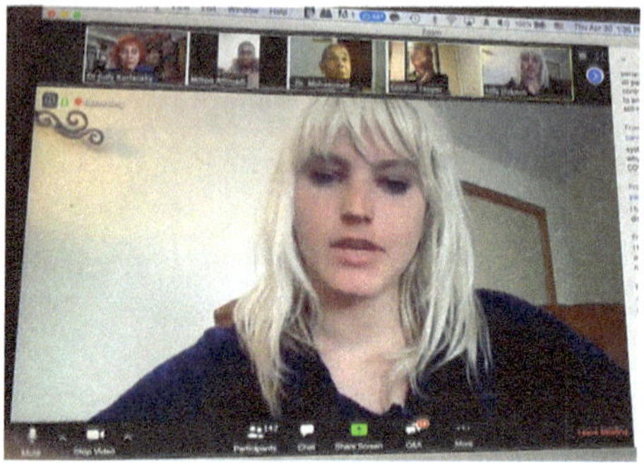

Student Lilly Dukich moderated the Q and A

Fig. 22
Student moderating the Q and A session of "COVID-19 in Africa" webinar in the United African Congress series during the pandemic.

Fig. 23
Video about Colorism produced by a student.

- participating in the UAC series mentioned before, moderating the Q and A session, and selecting and introducing innovations featured on that webinar series, by Lily Kukich
- producing a video about Colorism by Morgan Morrison

Some students had a rare opportunity to present their work related to technology at the United Nations. For example, before the COVID-19 closures, two Chinese students presented their

Fig. 24
Two of Dr. Kuriansky's Chinese students, Yanmeng Zou and Junting Yu, present their WeChat innovation at a UN Science, Technology and Innovation Forum event.

innovative project at an event, referred to above and shown in Fig. 24, that I organized for the UN Science, Technology and Innovation Forum on "Artificial Intelligence & Technology Tools for Mental Health, Well-being, and Resilience." The WeChat program they devised, which they named "Sleepless Society," presents American research and advice about psychology and counseling translated into Chinese to make that information accessible to Chinese youth who would not otherwise know about these issues.

International e-Collaboration

A unique online opportunity for the class evolved out of my cooperation with Anil Thapliyal, founder of the e-Mental Health International Collaborative mentioned earlier in this chapter and presented in another chapter in this volume (Thapliyal, 2024). The eMHIC mission—"Connect, Collaborate, Inspire, Make a Difference"—aligns with my own goals for my course. Under its Free Student Membership initiative, students (with an eligible school ID) get to learn all about eMental Health with free membership in the eMHIC community where they have access to all the usually paid eMHIC member benefits, including the knowledge bank, discussion forums, and webinar archive as well as chances to work on eMHIC projects, such as position statements, policy, and strategy, planning the International Conference, and reporting on these as part of their special semester project required for my course.

This program launched under their University Students-Focused Future Digital Mental Health Capability Development Initiative, in a partnership with the University of Otago in Dunedin, New Zealand, with clinical senior lecturer of Digital Health, Dr. Chris Paton, for students pursuing a Postgraduate Diploma in Digital Health. Paton, a UK-trained medical doctor specializing in clinical informatics and digital health, is head of the Global Health Informatics research group at the University of Oxford and is working on research projects based in Kenya and Vietnam. Students were required to browse the eMHIC Knowledge Bank, find an article of interest, and start a discussion thread about it in the forums for their peers and the eMHIC online community. The program was launched on July 14, 2022, yielding 17 topics and 223 posts. eMHIC awarded status badges based on the number of posts to encourage participation. Anil and his team presented a stimulating module for my class on e-mental health innovation and technologies.

International in-person opportunities for students

Besides opportunities described in the sections above in this chapter, students have unique and exceptional in-person opportunities from my course. These included one Emirati student Layla Al Neyadi who I took with me to the World Governments Summit in Dubai, where she met with leaders in the ministry in which she hoped to work when she returned home after her graduate studies and who expressed interest in her. I took another student, Julia Maney, with me to Dubai to the Expo 2020 Dubai and the Young Leaders Program where I was presenting during the World Governments Summit. I brought another student, Emma Wookey, with me to a conference in Vietnam where she presented a poster about her project, and a teaching assistant, Alexandra Margevich came with me to several conferences (in Mexico and Japan) held by the United Nations Office for Disaster Risk Reduction, as well as on missions to Haiti after natural disasters, about which she co-produced videos (Kuriansky et al., 2017a, 2017b). Such experiences are valuable for Ph.D. applications, and all these students were accepted into Ph.D. programs.

Also, students have joined NGO Committees associated with the UN, e.g., on aging, family, mental health, and migration. Others participated in a major field study about COVID-19 prevention in the Democratic Republic of the Congo. And, for a student whose dream was to be a Prime Minister of her native country of Jamaica, I introduced her to the Ambassador of Jamaica to the UN who invited her to events.

Technology training for teachers

Over time, TC had been upgrading its technological services, offering teachers courses about how to use programs and personal instruction. During the COVID-19 crisis, the Digital Future Institute (TCIT) ramped up its services, to teach professors like myself about teaching online and helping us migrate our syllabus to the CANVAS platform as opposed to paper syllabi (which I did). Trainings during the summer of 2020 included a media expert for training about computer and camera equipment packages, and room prep (lighting, positioning) for optimal webinar presentation. Of note, post-COVID TCIT trainings addressed the developing popularity of technologies (AI and ChatGPT) including methods, adaptations, and pros and cons of use.

The CANVAS platform, though decisively more elaborate with visually appealing elements consistent with my experience in media and high-level production, remains somewhat challenging. Making changes and edits can be difficult (even though a help line is available), especially since some require knowing html programming. Keeping an old-style simple outline on documents in my files or on google often seems preferable, allowing quicker cutting and pasting of material. Some students even find a very high-tech format complicated to navigate.

My experience with tech in class often caused stress. Since my classroom was not as technologically equipped and upgraded as were others in the college, various set-ups had to be tried. The media department tried an "OWL" mechanism to project the class for online guests to see the students, and then switched to an external camera. Sound was also challenging, e.g., turning off my computer sound source when using external microphones, to avoid feedback. Teaching online was relatively simple when projecting only a guest, as for Thailand Ambassador Chindawongse, but was more complicated when simultaneously showing another view (of the class or the PowerPoint). Fig. 25 shows the screen projection from my computer of an online guest, Ambassador Emeritus of Iraq to the UN, H.E. Hamid Al-Bayati, with two other scenes: one of my PowerPoint slides and the class captured by the "Owl" camera.

Fig. 25

Scenes on screen projection when guest is online, showing: (upper left) the speaker, Ambassador Emeritus of Iraq Hamid al-Bayati; the class (lower right); and a powerpoint slide (right).

Pros and cons online and in-person

Being that my pedagogy style involves keenly in tune with "energy," sensitive to the "temperature in the room," and being flexible, I prefer individual and group interchanges in-person compared to online. Teaching online is a much more linear experience, constraining one into an original plan, agenda, or outline, instead of allowing on-the-spot and spontaneous shifts in the topic or the order of PowerPoint slides. Yet, both scenarios have advantages and disadvantages, explored in the sections below. Interestingly, I have had experience with online teaching two decades ago, not forced as by the pandemic conditions, but necessary as I was in the USA and the students were all enrolled in a program in China. While research showed the effectiveness of the web-based format, at least occasional in-person contact and tutorials for supervision, support and connection were found valuable and preferred by both myself as the instructor and by the students (Ng et al., 2009).

Advantages of online classes

Despite the relatively impersonal nature of being online compared to in-person, since each student is in their own encapsulated environment, there are some advantages. Despite the separation caused by being in individual Zoom compartments, Zoom allows seeing faces up close and personal, unmasked, on the same plane and in the same dimension, compared to being in a seminar room where students can be scattered, block each other from view, or sit in the back of the room where they are not as clearly visible or audible as those up front. Also, the Zoom platform displays everyone's name below their face, so students can be identified, allowing me to easily learn their names and call on them by name. In-person, I have resorted at times to using nametags or name plates, so I and their peers learn their name.

The focus on faces also helps assure that students are "present" in class. To facilitate this, I request students to have their cameras "on." However, in one egregious case, a student logged in to class, thus appearing to be present but kept her camera off, telling me that her new baby on her lap would be distracting. Eventually, I discovered that she was never present during class at all and just kept her camera on and video off to appear present.

My "TV" format online

Given my experience as a television feature reporter and also previously host of half-hour shows and a guest on innumerable media shows, I think of my class as a television "show"—a TV term I often slipped into using when referring to the course. The "show," whether in-person or online, takes on different formats and tempos. When teaching in-person, guests are usually "booked" for the hour, especially since they made the trip to the college. In contrast, online segments can much more easily be shorter and lend themselves to a "magqazine"-type format, a TV term referring to a show where topics change in sequential "blocks" (a TV term for time periods).

The 2-hour-long class online became like a 2-hour TV "show," with an "open" (like on TV) to capture attention, consisting of a preview, possibly a video; breaking news relevant to the class; UN news (update about health; humanitarian crisis, as in Ethiopia; GA resolution or Security Council meeting); reference to some timely cultural date, event, or holiday (Nowruz, Hispanic week, Easter, Yom Kippur); followed by my PowerPoint presentation on the class topic, and then an introduction and presentation by a guest with student Q and A. I then sometimes ended the class with a "kicker," the TV term for a closing segment that is usually positive and uplifting, for example, an experiential bonding activity.

Following this TV news/magazine-show format for the online classes—being especially topical, reactive to news, and containing so multiple elements—required an enormous amount of preparatory work, sometimes throughout the entire week—much more than class in-person with far fewer elements and a slower pacing.

In some online classes, I "booked" up to four or five sequential guests, each presenting for 10 minute blocks of time. These short time periods were easier to control online than if they were happening in-person. Online guests were requested to log on at the beginning of the class (even if their appearance was later in the class), to check the connection and to "e-meet" each other. In these cases, they would be patient or truly enjoy meeting one another. This was the case, shown in Fig. 26, when the topic was about the Oceans related to mental health, in anticipation of the upcoming UN Oceans meeting hosted by Portugal and Kenya. In the opening segment, I had my student, who was from Portugal, describe the important connection of mental health to the oceans – that caring about oneself is preliminary to caring about the ocean, environment, and planet. This was a message which the Ambassador of Portugal, H.E. Francisco Duarte Lopes, the featured guest speaker, actually appreciated hearing about.

Opening slide of class and photo with delegates from UN Mission of Portugal

Fig. 26

Online class and the introductory powerpoint slide on the topic of climate change and ecopsychology featuring the Oceans and guests from the Mission of Portugal to the UN.

The Ambassador also enjoyed e-meeting another guest, the president of the Order of Portuguese Psychologists, Dr. Francisco Miranda Rodrigues, who had hosted a major international meeting of psychologists about climate change. Both guests enjoyed that I started the class with a video greeting from that meeting from Portugal's President, Marcelo Rebelo de Sousa, about the value of psychologists to society's health and well-being.

Advantages of in-person over online teaching

Interpersonal contact is so important in teaching, much less for student's mental health. In-person, my students often stay after class, sharing and bonding, sometimes for hours. Typically, I also stay for hours (until the guards insist that the school is closing) reviewing with my TAs and holding "office hours," answering students' questions, and going over papers and projects which in my view are more efficiently edited while in person than by just looking at edits in a google drive document.

Teaching in-person has a different, less pressured rhythm than being online. The in-person format is more conducive to communication that is slower, more flexible, more reactive to what is going on with the entire class, and more interactive, adding observations about body language which is not obvious online.

My teaching style is a hybrid of didactic lectures (a format which can be done as easily online) but including many active experiential elements requiring demonstration. Students welcome, and enjoy, these, as most students really want to be "in the field" and "on the ground," providing useful intervention. Activities, and some of these elements, had to be sacrificed when teaching online. Indeed, expressive arts is an important element in education and certainly in trauma recovery (Serlin et al., 2019) and my interventions around the world involve a toolbox of activities I have created and adapted, many of which depend on in-person connection (Art Saves Lives Foundation, 2019; Kuriansky, 2008, 2019; Kuriansky et al., 2016). Of note, these are aligned with the new definition of STEM (Science, Technology, Engineering and Math) education, which has become STEAM (adding the "A" for arts). While I was able to adapt some of these to online format (Khanal & Kuriansky, 2024), others were not possible; for example, two valuable activities for community cohesion and connection. In one, participants in a circle hold edges of a pen between them and then move, thereby being challenged to keep the pens from falling or, if they do, to recover and resume the circle, as a metaphor for how a community must help each other in a crisis. In another favorite, "The Web," as shown in Fig. 27, students toss balls of multicolored yarn to create a web and share the meaning of the interconnected threads symbolized by the varied yarn colors and the experience (e.g., "We are all together," "We are united despite our differences").

Fig. 27
An experiential activity about disaster recovery, connection and resilience that is essential to do in-person.

Another perfect example of an exceptionally valuable activity for the course that could only happen in-person is the "Fika," as shown in Fig. 28. Fika is the Swedish ritual of a "coffee or tea break" with refreshment at work or with friends and family, intended for socializing. Suggested by Niklas Nyblom who grew up in Sweden, the students brought a food from their culture and shared about their heritage—which turned out to be an exceptional learning experience about multiculturalism—consistent with the course theme—and also amazing bonding among the group.

Overall, my spontaneous style suits in-person teaching, consistent with vast experiences on TV talk shows, for example, being given Phil Donahue and Geraldo Rivera's own microphone to go into the audience and spontaneously ask people questions; walking around the classroom rather than lecturing from a seat and behind a computer; and creating activities for students that involve movement and interaction.

It is interesting to note that in contrast to clear views of students' faces when online, Fig. 27 shows that most students in-person still wearing masks, following the college recommendations at that time. This made it difficult to recognize students and also inhibited aspects of the

Fig. 28

Students sharing about their national food and culture during the Swedish Fika "coffee break" ritual. (Bottom row, right to left): Niklas Nyblom, describing the Swedish tradition and buns he cooked; Caitlyn Kamenitsu from Japanese background; Emily McGill with Lithuanian heritage; and Josefina Streeter from Peru.

interpersonal connection that has been well known as a negative outcome of the pandemic period but necessary for positive mental health and well-being.

These limitations led to discussions during class of the pros and cons of e-therapy, a model becoming increasingly popular during the pandemic times to allow patients access to therapy. More advanced models, and commercialized apps with programmed responses as discussed in other chapters in this volume, threaten to further depersonalize therapeutic contact. I made this point during the UN STI Forum noted above, in a demonstration, shown in Fig. 29, telling Alexa that I was (hypothetically) depressed. Her answer was first sympathetic, "I am sorry to hear that," and then offered advice to seek help if symptoms persist; however, I made the point that such responses only scratched the surface of addressing such a serious clinical problem.

Fig. 29
Dr. Kuriansky demonstrating an interaction with Alexa at an event about the use of technology to address depression during a United Nations event for the Science, Technology and Innovation Forum.

An exceptional example of the value of in-person teaching post-COVID

During 2023, the Mission of Mexico to the United Nations issued a "Call to Action" about mental health and wellbeing, and subsequently proposed a United Nations General Assembly Resolution on Mental Health and Psychosocial Support that led to historic success in its adoption on June 26, 2023 (de la Fuente et al., 2024). This campaign was exceptionally significant to me since Ambassador Caleb Otto of the Mission of Palau to the UN and I had drafted such a Resolution 10 years earlier (that got delayed), consistent with our advocacy at the negotiations among the UN delegates about inclusion of mental health and wellbeing in the UN Agenda 2030 for Sustainable Development. Participation of a team of my students described in this chapter to garner UN Missions support for the Mexico campaign (acknowledged as being very helpful) was notably an outcome of the lively in-person discussion when the Ambassador of the Mexico Mission, H.E. Dr. Juan Ramón de la Fuente, was a guest speaker, shown in Fig. 30. The successful outcome was certainly due to the class being in person rather than online, given the lively discussion both during and after class. In response to the Ambassador's presentation, a student from South Korea, June Chang, reported her unsuccessful efforts to secure her country's Ambassador as a class guest to address mental health. In response, Ambassador de la Fuente challenged the students to engage their home country to support the Resolution. The active interaction, spilling over into after-class discussion with the Ambassador sparked my idea to initiate a student campaign to engage all the UN missions to support the Resolution, as well as issues of youth and mental health in all agreements. In the resulting campaign, the student team sent several waves of emails to all 193 Missions which led

Fig. 30
Ambassador Juan Ramón de la Fuente of the Mission of Mexico to the United Nations with Dr. Judy Kuriansky, speaking to her class about mental health.

impressively to some in-person meetings and mission support. This invaluable TC student experience in advocacy then led to a similar second phase—including a team of youth from the International Association of Applied Psychology—to support other UN agreements and political declarations.

The hybrid model

In the Spring of 2021, the college decided to return to in-person teaching (after two initial weeks online) but to allow some students with permission for special medical or religious circumstances, to be online. The transition to this "hybrid model" presented challenges. Too many elements needed to be working well at the same time: a screen to show my PowerPoint; the other screen in the room for any remote guest; and sound working well for the students in the room, the guest, and the students logging in online. All this was coupled with using an OWL—a 360-degree video conference camera to capture the video and sound for everyone and especially for any online participants to "see" me and the class. An extra TV set was brought in to include the students, given online privileges, in the room, and a Zoom started so they could log on. This technical arrangement is shown in Fig. 31, when the ambassador of the Mission of South Africa to the UN, Her Excellency Mathu Joyini, was a guest in-person in the class. This plan was good in

Fig. 31

The "hybrid" model with in-person students and others given online permission to attend virtually and projected by Zoom on an external TV screen, in class with Professor Kuriansky (left in blue jacket) and H.E. Ambassador Mathu Joyini from the Mission of South Africa to the United Nations.

principle but did not always run smoothly, especially soundwise, nor did it allow recording the class. While the media department was accommodating to my needs, e.g., trying out two microphones (passing one around when students spoke), glitches and gaps prevailed; and media assistants had to be called, or stay in the class for periods of time, to be sure the equipment worked smoothly.

Breakout rooms. Students mention that they like breakout groups as a chance to get to know their peers, especially after the long COVID-19-imposed isolation. This feature is possible online, assigning students to a group "room," but is still preferable in-person. Also, teachers can only enter individual online rooms, and thus do not get an overall view of all the groups simultaneously as is possible in-person.

Attending UN events

An exceptionally valuable aspect of this class being held in New York City, the UN headquarters, is the opportunity for students to attend meetings in-person in that auspicious building. No amount of virtual reality, using headsets with surround sound and 360-degree visuals to simulate being present, can replace actually being there, with the sights and sensations of that international body. When lockdowns were lifted, students had unique opportunities to get "special event tickets" to attend events in the UN building, which as students told me, "This is the highlight of my life" and "It's been my dream to be here."

However, short of the elegance and cache of such live attendance, as I have told students, the shift that the COVID-19 pandemic forced to being online offered more opportunities for students to attend high-level meetings held by UN Missions with those delegates and other dignitaries and leaders conducted online. More such meetings were opened for registration to a broad audience who would never have such access if the meeting were held in-person in the UN building, when they are either closed to outsiders or limited in attendance given the room size.

Evaluation

The college sends students a voluntary course evaluation at the end of the semester for every class. Despite the challenges in technology noted before, and that use of technology was rated less highly than other (very highly rated) aspects, scores about the use of technology was still within average range. In the open-ended questions, mention of technology was made in only three instances over a year. On an anecdotal level, students have said that they prefer in-person classes, despite the fact that being online was convenient, for example, eliminated travel time and expense to get to the university.

Conclusions

Teaching online versus in-person at the higher education graduate level has advantages and disadvantages. Major challenges online are to maintain the quality of instruction and dynamism of class interaction. In my personal and professional estimation, my course on "Psychology and the United Nations" is more effective when in-person and face-to-face, although holding classes online once in a while over the semester can be valuable, when appropriate, to take advantage of the positive aspects of online teaching and accessibility of guest speakers. Moreover, the nature of the class interactive material, options for field visits (to the UN or other agencies), and my own instruction and presentation style impact the planning and flow of the course which is more compatible to the in-person versus web-based format. Regardless of the format, it is my goal to nurture the students as budding professionals to pursue their dreams, to continue advocacy promoting mental health and well-being in all policies and programs in the

UN, and to take over my role in advocacy for the next iteration of the UN Agenda for the ensuing period after 2030. After all, these youth are our hope for the future. Being fluent in and adaptable to both the online and in-person formats is essential, especially with an unpredictable future and rapidly advancing technologies affecting education.

References

Al Rand, H. A. R., Madi, H., Alhemeiri, S., & AlObeidli, A. A. R. (2024). Healthcare innovation and leadership: Case study of the United Arab Emirates. In J. Kuriansky, & P. Kakkattil (Eds.), *Resilient health: Leveraging technology and social innovations to transform healthcare for COVID-19 recovery and beyond* (pp. 415–431). Elsevier.

Aljovin, P. (2024). The Amazon Hope Medical Programme: A mobile healthcare ship to "Leave No One Behind". In J. Kuriansky, & P. Kakkattil (Eds.), *Resilient health: Leveraging technology and social innovations to transform healthcare for COVID-19 recovery and beyond* (pp. 1085–1098). Elsevier.

Art Saves Lives Foundation. (2019). *Mission report: Resilience, empowerment, and connection building training.* Report of training/workshops post-disaster in Sint Maarten. Retrieved from https://drive.google.com/open?id=17Gf6uOVZMD3N-Ak4igihhTRewTC5BoE1.

Barry, E., & Kuriansky, J. (2024). Scoring for Resilient Health: Innovations using technology in physical activity and sport. In J. Kuriansky, & P. Kakkattil (Eds.), *Resilient health: Leveraging technology and social innovations to transform healthcare for COVID-19 recovery and beyond* (pp. 373–385). Elsevier.

Bikkani, S. (2024). Autism spectrum disorder and technological interventions. In J. Kuriansky, & P. Kakkattil (Eds.), *Resilient health: Leveraging technology and social innovations to transform healthcare for COVID-19 recovery and beyond* (pp. 847–854). Elsevier.

Bulanchuk, N., & Kuriansky, J. (2024). Financing an NGO addressing mental health and poverty during and after a pandemic: My personal experience. In J. Kuriansky, & P. Kakkattil (Eds.), *Resilient health: Leveraging technology and social innovations to transform healthcare for COVID-19 recovery and beyond* (pp. 691–700). Elsevier.

Cellini, S. R. (2021). How does virtual learning impact students in higher education?. Brookings. https://www.brookings.edu/blog/brown-center-chalkboard/2021/08/13/how-does-virtual-learning-impact-students-inhigher-education/.

Coetzee, C., Kuriansky, J., & AlHammadi, R. (2024). The nexus model and the Young Leaders Program: Innovative models for the future of the SDGs in Action Campaign and the Global Councils on Sustainable Development Goals. In J. Kuriansky, & P. Kakkattil (Eds.), *Resilient health: Leveraging technology and social innovations to transform healthcare for COVID-19 recovery and beyond* (pp. 629–646). Elsevier.

Cojocariu, V.-M., Lazar, I., Nedeff, V., & Lazar, G. (2014). SWOT analysis of e-learning educational services from the perspective of their beneficiaries. *Procedia-Social and Behavioral Sciences, 116*, 1999–2003.

de la Fuente, J. R., Mendoza, S. P., & Guzik A. (2024). An unconventional security threat: Mental Health at the United Nations Security Council and Mexico's Response. In J. Kuriansky, & P. Kakkattil (Eds.), *Resilient health: Leveraging technology and social innovations to transform healthcare for COVID-19 recovery and beyond* (pp. 475–480). Elsevier.

de la Mora, F. (2023). Diplomatic psychology at ther United Nations: The Global Compact for Safe, Orderly and Regular Migration. In G. J. Rich, J. Kuriansky, U. Gielen, & D. Kaplin (Eds.), *Psychosocial experiences and adjustment of migrants: Coming to the USA* (pp. 339–363). Elsevier.

Diaz, D., & Kuriansky, J. (2024). Reimagining healthcare: Technological solutions to close the treatment gap and achieve equity in the wake of COVID-19. In J. Kuriansky, & P. Kakkattil (Eds.), *Resilient health: Leveraging technology and social innovations to transform healthcare for COVID-19 recovery and beyond* (pp. 941–954). Elsevier.

Fitapelli, B. (2024). Innovative technologies for mental healthcare: An application for severely mentally ill persons. In J. Kuriansky, & P. Kakkattil (Eds.), *Resilient health: Leveraging technology and social innovations to transform healthcare for COVID-19 recovery and beyond* (pp. 873–879). Elsevier.

Gopal, R., Singh, V., & Aggarwal, A. (2021). Impact of online classes on the satisfaction and performance of students during the pandemic period of COVID 19. *Education and Information Technologies*, *26*, 6923–6947. https://doi.org/10.1007/s10639-021-10523-1.

Irani, K., Siddhanta, M., Medhizadeh, T., & Kuriansky, J. (2024). Raising the profile of health technology in the Middle East and North Africa (MENA) region. In J. Kuriansky, & P. Kakkattil (Eds.), *Resilient health: Leveraging technology and social innovations to transform healthcare for COVID-19 recovery and beyond* (pp. 911–928). Elsevier.

Kakkattil, P., Tiwari, A., & Jacob, B. (2024). Breaking the silos: How the Health Innovation and Investment Exchange (HIEx) helps bridge the health innovation ecosystem. In J. Kuriansky, & P. Kakkattil (Eds.), *Resilient health: Leveraging technology and social innovations to transform healthcare for COVID-19 recovery and beyond* (pp. 979–987). Elsevier.

Khanal, S., & Kuriansky, J. (2024). Promoting resilience in the Nepali community in response to the COVID-19 pandemic: A multistage multistakeholder partnership addressing the United Nations Agenda. In J. Kuriansky, & P. Kakkattil (Eds.), *Resilient health: Leveraging technology and social innovations to transform healthcare for COVID-19 recovery and beyond* (pp. 601–627). Elsevier.

Knox, J. (2024). The crucial role of nurses in telehealth. In J. Kuriansky, & P. Kakkattil (Eds.), *Resilient health: Leveraging technology and social innovations to transform healthcare for COVID-19 recovery and beyond* (pp. 771–778). Elsevier.

Kuriansky, J. (2008). A clinical toolbox for cross-cultural counseling and training. In U. P. Gielen, J. G. Draguns, & J. M. Fish (Eds.), *Principles of multicultural counseling and therapy* (pp. 295–330). New York: Taylor and Francis/Routledge.

Kuriansky, J. (2010). Haiti pre- and post earthquake: Tracing professional and personal commitment past, present and future. International Psychology Bulletin, 14(2), 29–37. Spring http://internationalpsychology.files.wordpress.com/2013/01/ipb_spring_2010_4_27_10.pdf.

Kuriansky, J. (2016). Report of the IAAP United Nations team: Success in the New Global Agenda Advocacy. *Bulletin of the International Association of Applied Psychology*, *28*, 80–88.

Kuriansky, J. (2017). Reconstruire la raison et l'esprit en Haiti: Contributions du Centre de Spiritualité et de la Santé Mentale. In W. Jean-Charles (Ed.), *Santé mentale et psychothérapie en Haiti: Approches integratives et traitments multimodaux* (pp. 119–146). Coconut Creek, FL: EducaVision, Inc.

Kuriansky, J. (2019). Expressive arts for helping children heal in crisis and disaster. In I. A. Serlin, S. Krippner, & K. Rockefeller (Eds.), *Integrated health care for the traumatized: A whole person approach*. Lathan, NY: Rowman & Littlefield.

Kuriansky, J. (2021). *Record USA fundraiser for First Lady's "Hands Off Our Girls" initiative*. Black Star News. https://www.blackstarnews.com/global-politics/africa/%E2%80%9Chands-off-our-girls%E2%80%9D-usa-event-success-for-first-lady-of.

Kuriansky, J. (2024). "Health in Your Hands": A platform for innovative healthcare solutions. In J. Kuriansky, & P. Kakkattil (Eds.), *Resilient health: Leveraging technology and social innovations to transform healthcare for COVID-19 recovery and beyond* (pp. 989–1004). Elsevier.

Kuriansky, J., Jean-Charles, W., Margevich, A., & the CeSSA student team. (2017a). *Haiti post-2016 Hurricane: Community training workshop for children [Video file]*. https://www.youtube.com/watch?v=W14x9_nwVfQ.

Kuriansky, J., Khanal, S., Pearce, J., & Rabhan, Y. (n.d.). Young women must be at the forefront. Youth-led NGO CSW/NY forum: Events about migrant Latinex women, women in Middle East peace, and rights of women and girls in Nepal: Psychosocial issues in COVID-19 times. The NGO Reporter. http://globalngoexec.org/youth-led-ngo-csw-ny-forum-events-about-migrant-latinx-women-women-in-middle-east-peace-and-rights-of-women-and-girls-in-nepal-psychosocial-issues-in-covid-19-times/.

Kuriansky, J., & Khurshid, M. (2023). Psychosocial support for migration and refugees: Similarities and differences in international documents. In G. Rich, J. Kuriansky, U. P. Gielen, & D. Kaplin (Eds.), *Psychosocial experiences and adjustment of migrants: Coming to the USA* (pp. 391–418). Elsevier.

Kuriansky, J., Margevich, A, Jean-Charles, W., & Daisey, R. (2017b). Resilience and recovery in natural disasters and epidemics: Comparisons, challenges, and lessons learned from train-the-trainer projects. In G. Rich, & S. Sirikantraporn (Eds.), *Human strengths and resilience: Developmental, cross-cultural, and international perspectives* Lexington Books.

Kuriansky, J., Mehdizadeh, T., & Bikkani, S. (2024). Mental Health Matters at the United Nations: A web and in-person healthy workplace initiative and response during COVID-19 and beyond. In J. Kuriansky, & P. Kakkattil (Eds.), *Resilient health: Leveraging technology and social innovations to transform healthcare for COVID-19 recovery and beyond* (pp. 517–524). Elsevier.

Kuriansky, J., & Otto, J. (2017). *Artificial intelligence and technology tools for mental health, well-being and resilience.* Report on a Side Event held at the United Nations 16 May 2017 in the Margins of the UN Science, Technology & Innovation Forum. http://bit.ly/2mghjOH.

Kuriansky, J., Polizer, Y., & Zinsou, J. (2016). Children and Ebola: A model resilience and empowerment training and workshop. In J. Kuriansky (Ed.), *The psychosocial aspects of a deadly epidemic: What Ebola has taught us about holistic healing* (pp. 175–214). Santa Barbara, California: ABC-CLIO/Praeger.

Kuriansky, J., & Zinsou, J. (2019). Mental health and well-being at the United Nations: Activities, achievements and contributions of the IAAP New York Team. *Applied Psychology Around the World*, *1*(3), 16–28. https://iaapsy.org/site/assets/files/1082/apaw_vol1i3.pdf.

Liaqat, A. (2022) Mental health, women's empowerment and climate change.

Lissinna, H. (2024). Digital transformation of health featured during Digital Health Week 2021: Profiles of partnerships and health innovations using technology in countries around the world. In J. Kuriansky, & P. Kakkattil (Eds.), *Resilient health: Leveraging technology and social innovations to transform healthcare for COVID-19 recovery and beyond* (pp. 883–895). Elsevier.

Luxton, D. (2024). Technology ethics in healthcare delivery and public health: Current and emerging issues from the COVID-19 era. In J. Kuriansky, & P. Kakkattil (Eds.), *Resilient health: Leveraging technology and social innovations to transform healthcare for COVID-19 recovery and beyond* (pp. 193–203). Elsevier.

Luxton, D., & Watson. (2024). Intelligent virtual agents for current and emerging public health crises. In J. Kuriansky, & P. Kakkattil (Eds.), *Resilient health: Leveraging technology and social innovations to transform healthcare for COVID-19 recovery and beyond* (pp. 281–290). Elsevier.

Mallinckrodt, B., Miles, J. R., & Levy, J. J. (2014). The scientist-practitioner-advocate model: Addressing contemporary training needs for social justice advocacy. *Training and Education in Professional Psychology*, *8*(4), 303–311. https://doi.org/10.1037/tep0000045.

Ng, E., Kuriansky, J., Yuen, W. Y., So, Y. L., Ng, T. M., Chen, C. N., & Cheung, B. (2009). Logging on to learn: A web-based sexuality education training program. In E. Schroeder, & J. Kuriansky (Eds.), *4. Sexuality education: Past, present and future* Praeger.

Obermeyer, W., & Kiegele, E. (2024). The fundamental role of the United Nations General Assembly to foster political commitment in support of global health with the assistance of the World Health Organization. In J. Kuriansky, & P. Kakkattil (Eds.), *Resilient health: Leveraging technology and social innovations to transform healthcare for COVID-19 recovery and beyond* (pp. 499–515). Elsevier.

Otzoy-García, D., Iñigo Petralanda, M. I., & Aljovín. (2024). Raising the profile of digital health in Latin America: The RECAINSA digital health forum and network. In J. Kuriansky, & P. Kakkattil (Eds.), *Resilient health: Leveraging technology and social innovations to transform healthcare for COVID-19 recovery and beyond* (pp. 897–909). Elsevier.

Parks, S., Kuriansky, J., & Koch, P. (2022). Confronting the crisis of COVID-19: Evolution of a model for preservation of people and the planet. *International Perspectives in Psychology Special Issue: Psychology and the COVID-19 Pandemic – A Global Perspective*, *11*(3). https://doi.org/10.1027/2157-3891/a000053.

Pearce, J. (2021). Mitigating transgenerational trauma of latina migrant women to achieve empowerment. In *Webinar presented at the Commissionon the Status of Women CSW NGO/NY Forum, March 19* Online.

Porto, A. (2024a). Innovative technology: The Enhatch intelligent surgery ecosystem. In J. Kuriansky, & P. Kakkattil (Eds.), *Resilient health: Leveraging technology and social innovations to transform healthcare for COVID-19 recovery and beyond* (pp. 1099–1101). Elsevier.

Porto, A. (2024b). Artificial intelligence and augmented reality in COVID-19 recovery and beyond: The expansion of emerging technology in the medical industry. In J. Kuriansky, & P. Kakkattil (Eds.), *Resilient health: Leveraging technology and social innovations to transform healthcare for COVID-19 recovery and beyond* (pp. 255–264). Elsevier.

Rabhan, Y. (2021). Women's leadership role and psychosocial perspective in Middle East Peace building. In *Webinar presented at the Commission on the Status of Women NGO/NY Forum, March 21, 2021.*

Rizwan, F. (2024). Gender disparities faced by female doctors in Pakistan: Training but turning into trophy wives. In J. Kuriansky, & P. Kakkattil (Eds.), *Resilient health: Leveraging technology and social innovations to transform healthcare for COVID-19 recovery and beyond* (pp. 781–792). Elsevier.

Rodriquez, A. (2022). *The United African Congress webinar conference series on COVID-19 in Africa*. https://docs.google.com/document/d/135jXg8nViht0vR8aCEIVNbyxaT7Wy3lw/edit.

Sandis, E., & Bingham, J. K. (2023). Doing what matters: The NGO committee on migration. In G. Rich, J. Kuriansky, U. P. Gielen, & D. Kaplin (Eds.), *Psychosocial experiences and adjustment of migrants: Coming to the USA* Elsevier.

Serlin, I. A., Krippner, S., & Rockefeller, K. (2019). *Integrated health care for the traumatized: A whole person approach*. NY: Lathan.

Sharma, S. (2021). Mental health of refugee men. In *Webinar presented at the Commission on the Status of Women CSW NGO/NY Forum, March 21* Online.

Smith, A. (2023). Afghan girl reefugees comingto America. In G. J. Rich, J. Kuriansky, U. Gielen, & D. Kaplin (Eds.), *Psychosocial experiences and adjstment of migrants: Coming to the USA* (pp. 315–325). Elsevier.

Sohail, M. (2022). *Online learning: What next for higher education after COVID-19?*. World Economic Forum. https://www.weforum.org/agenda/2022/06/online-learning-higher-education-covid-19/.

Thapliyal, A. (2024). eMental Health International Collaborative: The global peak-body for digital mental health implementation. In J. Kuriansky, & P. Kakkattil (Eds.), *Resilient health: Leveraging technology and social innovations to transform healthcare for COVID-19 recovery and beyond* (pp. 929–940). Elsevier.

United Nations. (2020). *Policy brief: Education during COVID-19 and beyond*. https://unsdg.un.org/resources/policy-brief-education-during-covid-19-and-beyond.

United Nations. (2022). *Transforming education Summit*. https://www.un.org/en/transforming-education-summit/tes-summit-closing-press-release.

Wai, S., & Kuriansky, J. (2024). Health innovation and advances in the Republic of Sierra Leone. In J. Kuriansky, & P. Kakkattil (Eds.), *Resilient health: Leveraging technology and social innovations to transform healthcare for COVID-19 recovery and beyond* (pp. 445–462). Elsevier.

Yoon, D. M. (2024). Families and new technologies: Research, programs, and launch of event series celebrating the International Day of the family. In J. Kuriansky, & P. Kakkattil (Eds.), *Resilient health: Leveraging technology and social innovations to transform healthcare for COVID-19 recovery and beyond* (pp. 829–837). Elsevier.

PART 8

Resources, lessons learned, and the way forward

Technology and innovations are pointing the way to the future of resilient healthcare systems, in the recovery period of the COVID-19 pandemic. This period will, of course, be ongoing. The High-level meeting of the United Nations General Assembly in 2023 referred to this process as “Pandemic Prevention, Preparedness and Response (PPPR).”

Innovative approaches will evolve constantly, requiring stakeholders to keep apprised of developments, as well as connected to others for essential collaboration action. To aid in negotiating this ever-emerging landscape, this Part offers a chapter on resources collated into major categories: news/publications, conferences/expos, platforms, networks, virtual communities, incubators/accelerators, contests and awards, and academic study programs. Major reports with digital plans and agendas have been released by international organizations that are valuable resources; thus, their web access is listed. All these resources are gold mines for learning and for exploring ideas, support, collaborators, and partners.

Underlying all advancement is the requirement for innovative integrated investment—explained in a chapter by a team from the World Bank Group. This chapter builds on the chapters in Part IV about financing for health, but is appropriately in the final Part of this volume since it is essential for the way forward.

The Part, and volume, ends with an overview by the editors of this collection of contributions that puts the chapters together in a productive whole, pointing toward the brave new world of technology and innovation to insure resilient health systems for all peoples and countries worldwide.

What notables say

> *Never doubt that a small group of thoughtful, committed citizens can change the world; indeed, it's the only thing that ever has.*
>
> ***—Margaret Mead, American cultural anthropologist***

If you want to lift yourself up, lift up someone else.

—Booker T. Washington, author and advisor to Presidents

Individual commitment to a group effort—that is what makes a team work, a company work, a society work, a civilization work.

—Vince Lombardi, American football coach and executive in the National Football League

Individually, we are one drop. Together, we are an ocean.

—Ryunosuke Satoro, Japanese author

Our love is so big. It has room for everybody.

—Cari Tuna, president of the Open Philanthropy Project and Good Ventures

Forget about the fast lane. If you really want to fly, harness your power to your passion. Honor your calling. Everybody has one. Trust your heart, and success will come to you.

—Oprah Winfrey, American television personality and philanthropist

There is no passion to be found playing small—in settling for a life that is less than the one you are capable of living.

—Nelson Mandela, antiapartheid activist and first president of South Africa

Human beings, their ideas, innovations, dreams, and connections are the capital of the future. Because where great minds go today, great things will happen tomorrow.

—HH Sheikh Mohammed bin Rashid al Maktoum, vice president, prime minister, and minister of defence of the United Arab Emirates (UAE) as well as the ruler of Dubai and deputy chairman of the Executive Council of Dubai

CHAPTER 103

Investing in resilient health systems: A framework for prioritizing integrated investments in preparedness, response, and service delivery

Sulzhan Bali, Zara Shubber, Alethea Cook, and Feng Zhao
World Bank, Washington, DC, United States

Introduction

In May 2023, over three years after the start of the recent COVID-19 pandemic, the World Health Organization (WHO) declared an end to the pandemic's status as a global health emergency. While the emergency phase of COVID-19 has ended, it casts a long shadow in its wake. The COVID-19 pandemic (2020 to the present) has caused devastating impacts on health, which—alongside severe economic and social disruptions—have revealed structural weaknesses in the health systems of low- and high-income countries alike and a lack of global preparedness. The impacts are staggering: the global economy is projected to lose over US$22 trillion in cumulative global output by 2025 and the WHO has recorded over 6.95 million lives lost so far, although the true death toll may be as much as two to four times higher (Gopinath, 2021; Mathieu et al., 2023). And yet, the devastation wrought by COVID-19 may be surpassed by even graver health threats looming on the horizon.

The potential for infectious diseases to emerge and spread is growing exponentially as humans expand their footprint on the planet, encroaching on and altering natural habitats while accelerating population mobility, commerce, and climate change. Since 1980, the pace of emerging infectious diseases has accelerated at an annual rate of 6.7%, with the number of outbreaks rising to several hundred annually since 2000 (Berthe et al., 2022). Moreover, the growing frequency of infectious disease outbreaks is increasingly converging with long-term trends of population aging, rising chronic disease burdens, and climate change, to raise the risk

Resilient Health. https://doi.org/10.1016/B978-0-443-18529-8.00105-0

of syndemics—events in which two or more diseases adversely interact with each other and with political and economic conditions of poverty and inequality. The recent COVID-19 pandemic offers a bitter lesson in how these dynamics interact to cause tremendous human suffering, and underscores what is at risk should countries fail to adequately prepare for the next major outbreak.

The only way to prevent, prepare for, and manage these threats is by building resilient health systems to withstand shocks, prevent disruption in essential health services, and improve health outcomes between crises. Yet for decades, both individual countries and the international community have underinvested in preparedness and delayed health financing for emergency response—reacting only after an outbreak has struck. As evidenced by the far-reaching impacts of the recent COVID-19 pandemic, the world remains grossly unprepared to manage public health emergencies with pandemic potential.

The costs of significantly upgrading global preparedness and strengthening health system resilience, while substantial, are within reach for every country, even those with low incomes. As countries work to build resilient health systems in the shadow of COVID-19, this chapter describes key features of resilience and presents a three-tiered framework to help countries of all income levels prioritize integrated investments in preparedness, response, and service delivery based on their resilience impact, as set out in the World Bank (2022) report, *Change Cannot Wait: Building Resilient Health Systems in the Shadow of COVID-19.*

What is a resilient health system?

Despite lessons learned from prior health emergencies—including avian influenza, Ebola, and Zika—investments in preparedness have been not only insufficient, but also disconnected from foundational investments in health systems and universal health coverage (UHC). The destruction wrought by the recent COVID-19 pandemic highlights the urgent need to move toward a model of health system resilience that goes beyond emergency response to include both crisis preparedness and strong health service delivery. This chapter defines resilient health systems as integrated systems that are aware of threats, agile in response to evolving needs, absorptive of shocks, adaptive to minimize disruptions, and able to transform after a crisis, based on lessons learned (World Bank, 2022). Integral to this definition of resilience is the concept of integration or the system's ability to leverage relationships in the health sector by connecting preparedness with health service delivery. Box 1 summarizes seven key characteristics of a resilient health system: aware, agile, absorptive, adaptive, transformative, integrated, and multisectoral.

To achieve these qualities and advance resilience in the health sector, governments can adjust two main types of levers: enablers and core capacities. Enablers are cross-cutting factors that influence all aspects of health system performance, including governance; partnerships; health

BOX 1 Characteristics of a resilient health system.

Aware	Recognizes population health needs and risk drivers, detects threats, and maps strengths and weaknesses (e.g., through investments in multisectoral disease surveillance)
Agile	Responds to evidence and changing needs and uncertainty (e.g., prepositioned resources, procurement, and essential supplies; incorporation of risks in planning, including preparedness plans)
Absorptive	Manages crises and maintains core functions despite a need for redistribution of resources (e.g., emergency operation centers, capacity for contact tracing, and upskilling health workers)
Adaptive	Minimizes disruptions and maintains essential individual and population-based health services despite change in resources (e.g., reallocation of resources, use of telemedicine, altered standards of care, and rational use of personal protective equipment)
Transformative	Innovates and reorganizes structures and operations based on lessons learned during a crisis to reduce risk and improve function (e.g., the enactment of reforms based on lessons learned, simulations, and evaluations)
Integrated	Integrates essential public health functions with individual and clinical service delivery through investments; integrates health security and disease-specific planning with health systems strengthening initiatives
Multisectoral	Coordinates and draws value from partnerships to implement resilience-sensitive and specific interventions with allied health (e.g., environmental and animal health) and beyond the health sector (e.g., education, finance, transport, media, communications, and the private sector)

financing; human resources; and innovation, such as the deployment of digital health tools or new service delivery arrangements. The core capacities essential for resilience include health intelligence; health service delivery, in particular primary health care (PHC); community engagement and risk communication; and the health supply chain. Fig. 1 shows how these levers relate to each other and interact to enhance the seven key characteristics of resilient health systems. It further illustrates how investments in resilience promote progress not only toward countries' development of International Health Regulations (IHR) capacities, but also toward their advancement of the Sustainable Development Goals, including UHC. Ultimately, these investments can create a resilience dividend—economic benefits through averted crises and the promotion of human capital. This occurs through positive impacts on health and productivity; reduced mortality and morbidity from averting crises; and greater equity, trust, and confidence in health systems.

As shown in Fig. 1, enabling resilience and reaping this dividend requires addressing the broader determinants of health—including the economic, political, and sociocultural

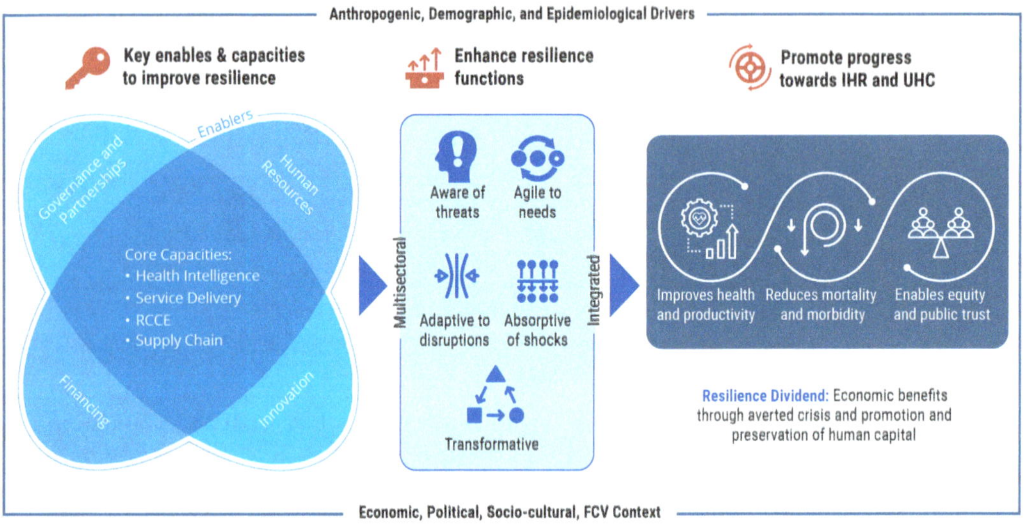

Fig. 1
Framework for health system resilience. RCCE, Risk communication and community engagement; IHR, International Health Regulations; UHC, universal health coverage.

environment—and the anthropogenic, demographic, and epidemiological drivers of health emergencies. As countries learned from the recent COVID-19 pandemic, health outcomes during a crisis are influenced by the interplay of broader dynamics such as socioeconomic and gender inequalities, aging populations, rising chronic comorbidities, and climatic and environmental change. Taken together, these trends constitute a powerful wakeup call to policymakers and political leaders to invest in building resilience and strengthening health equity now.

With the levers identified, action is required on the part of governments to operationalize health system resilience. Box 2 summarizes the key actions governments can take across these

BOX 2 Actions for governments to build more resilient health systems.

Governance—Governance has a profound impact on the trajectory of a crisis but is often neglected in preparedness. Key actions include:

- **i.** Review and strengthen decision-making processes based on evidence.
- **ii.** Invest in national public health institutions and update legal and regulatory frameworks.
- **iii.** Update preparedness plans and strategies to incorporate risk drivers and leverage learning from prior experiences.
- **iv.** Ensure leadership and command structure by enabling clarity of roles, capacity building, and establishing emergency operation centers.

BOX 2—Cont'd

Partnerships—Collaborative planning, strong regional networks, cross-sectoral partnerships, and community engagement are vital for resilience. Key actions include:

- **i.** Leverage regional institutions and partnerships for coordination, capacity building, and harmonization of policies.
- **ii.** Institutionalize whole-of-government approaches by setting up multisectoral and One Health platforms for planning and coordination of cross-sectoral activities.
- **iii.** Develop a fit-for-purpose private-sector engagement strategy and establish agreements to facilitate the private sector's role during crises.

Financing resilient health systems—Building resilient health systems requires strong, sustainable health financing that encompasses prevention, preparedness, and emergency response. Key actions include:

- **i.** Prioritize, ring-fence, and track investments in pandemic prevention, preparedness, and response (PPPR).
- **ii.** Leverage catalytic, complementary investments in resilience (e.g., joint planning tools).
- **iii.** Reduce financial barriers to access crisis interventions and essential health services.
- **iv.** Develop diverse and agile crisis-ready financing—create contingency funds and use innovative financing for disaster-risk layering.

Human resources—The recent COVID-19 pandemic has underscored how human resources, including health care workers, are at the heart of health systems but also revealed significant challenges, including staff shortages and maldistribution of trained professionals. Key actions include:

- **i.** Map health workforce needs and develop an evidence-based human resources for health strategy.
- **ii.** Diversify and repurpose the health workforce.
- **iii.** Increase professional capacity by cascading training to lesser-trained health workers.
- **iv.** Develop and sustain cadres of community health workers (CHWs) and train them for frontline PPPR roles.
- **v.** Build multidisciplinary competencies for PPPR through preservice and in-service trainings.
- **vi.** Protect frontline staff by providing access to infection, prevention, and control training; personal protective equipment; and information, education, and communication.

Innovation—Addressing health shocks requires innovation—such as technological advancements or new service delivery arrangements—to enable awareness, agility, absorptiveness, and adaptiveness. Key actions include:

- **i.** Update preparedness plans and national health strategies (e.g., digital health strategies, telehealth implementation guidelines) to include mechanisms for research and regulatory review processes during emergencies.
- **ii.** Spur people-centered innovation by engaging communities and supporting intermediary innovation platforms.
- **iii.** Invest in an agile regulatory system to accommodate and fast-track new medical technologies.
- **iv.** Build research, regulatory, and monitoring and evaluation capacity.
- **v.** Provide adequate support and funding for such innovations.

Continued

BOX 2—Cont'd

Health intelligence—Effective infectious disease surveillance and other core intelligence capacities are cornerstones of resilience. Health intelligence can help countries leverage early warning, detection, and data systems to identify and contain health threats. Key actions include:

i. Strengthen early warning surveillance and epidemic intelligence functions.
ii. Strengthen interconnected laboratory capacity and regional networks.
iii. Strengthen and integrate information systems, including for monitoring of service provision and disruptions at facilities.
iv. Leverage digital tools (e.g., electronic surveillance systems, mHealth contact tracing apps) and new technologies (e.g., genomic surveillance, wastewater surveillance) to enhance timely detection and notification.

Health service delivery—Health emergencies have far-reaching impacts on persons presenting with health and mental health conditions and exacerbate gaps and vulnerabilities in service delivery in many settings. Key actions include:

i. Assess readiness of health facilities for shocks and invest in crisis-ready facilities.
ii. Develop plans for continuity of essential services during crises and leverage alternate health service delivery sites and adapted models of care during crises, including through maintenance and broadening of successful telemedicine platforms.
iii. Strengthen patient referral systems and develop coordinated networks of health facilities.
iv. Build partnerships between public health agencies, health facilities, the private sector, and humanitarian agencies to scale service delivery in crises.
v. Invest in community-centered, comprehensive PHC integrated with mental health care and with integrated public health functions.

Risk communication and community engagement—Risk communication and community engagement are crucial for building individuals' trust in government messaging, interventions, and recommendations to prevent and manage threats. Key actions include:

i. Strengthen risk communication procedures, build capacity to communicate risks, and enable two-way communication.
ii. Empower communities by involving them in decision-making.
iii. Strengthen rumor monitoring, address misinformation, and build community trust in public health interventions.
iv. Engage community leaders and traditional healers in relevant settings to fortify government messaging.

Crisis-ready supply chains—Supply chains are vital but often undervalued or underaddressed components of resilient health systems. The pandemic highlighted both the critical importance and fragility of health supply chains. Key actions include:

i. Develop forecasting capacities, emergency logistics and supply chain management plans, rational-use guidance, prepositioned contingency procurement plans, and acquisition flexibilities for health emergencies.
ii. Leverage regional cooperation and domestic production for key medical goods.
iii. Tackle human-resource and infrastructure gaps to upgrade supply chains.

enablers and core capacities, which include governance, partnerships, financing resilient health systems, human resources, innovation, health intelligence, health service delivery, risk communication and community engagement (RCCE), and crisis-ready supply chains.

Technological innovation can play a transformative role in strengthening each of these enablers and core capacities. The use of digital health has the potential to improve health service delivery, for instance, through telemedicine or electronic health records; to strengthen risk communication and build community engagement using digital tools to counter misinformation; or to boost health intelligence with digital technologies such as mobile applications, location data, and electronic tags to facilitate planning, surveillance, contact tracing, quarantine, and clinical management. Such innovation can catalyze health system transformation while enabling resilience functions during a health emergency, but countries must develop the enabling environment needed to adopt, implement, and scale technological innovation. Countries can strengthen planning mechanisms to spur technological innovation by adopting national digital health strategies and guidance and by enacting legislation on the use of national electronic health records and privacy, with consideration of means to address the digital divide, inequality, and workforce and organizational barriers to innovation.

Where should countries invest first to build resilience?

A broad range of investments is needed to leverage these enablers and to strengthen health systems' core capacities for resilience. However, many countries now face shrinking public health budgets after health spending soared as part of their response to the recent COVID-19 pandemic, highlighting the need to set priorities for high-yield resilience spending. Now is the time for countries to strategically and sustainably invest in resilient health systems—before the next public health emergency strikes. By starting now, countries can make the task easier and less costly, optimize results, and sustain gains made during their COVID-19 response. To support countries in their investment decisions, this chapter presents a three-tiered framework designed to aid countries in prioritizing financing in areas that will maximize resilience in the event of a health shock, as shown in Fig. 2.

The aims and examples of the three tiers are as follows.

As shown in Fig. 2, Tier 1 is the foundation of precrisis pandemic defense and the cornerstone of health security. As the most important tier, investments for prevention and community preparedness yield the biggest "bang for the buck" in countries' resilience outcomes. Its focus is on upstream, preventive action to thwart potential outbreaks and its elements include well-functioning PHC and strong community-based surveillance and early warning systems. Tier 1 also includes emergency planning and preparedness, adequate health workforce capacity, strong public health institutions, supportive policies and regulations, robust data and information systems, and well-functioning procurement and delivery systems. Sustained

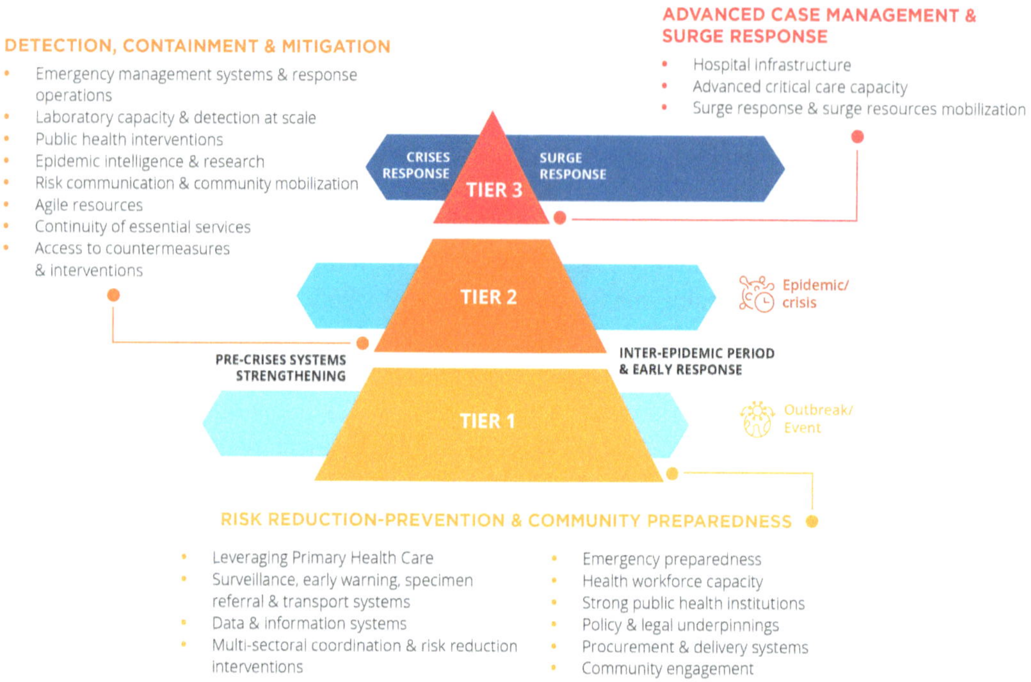

Fig. 2

Three-tiered investment framework to build resilient health systems. Outbreak: At least two epidemiologically linked cases or one case in an area where the disease does not exist. Epidemic: The occurrence and rapid increase of cases of disease in excess of what would normally be expected in a defined community, geographical area or season. While often used interchangeably with the term outbreak, an epidemic is considered to be the rapid increase and concurrent consolidation of multiple outbreaks over a wide geographical area. Precrises systems strengthening: Health systems strengthening before a crisis occurs. Interepidemic period and early response: Time between epidemics and during the early phases of an epidemic.

community engagement in health prevention and preparedness is critical to this tier, which also relies on effective multisectoral coordination. Thus this tier makes crucial relays between national policymaking and the frontlines of daily health action in communities. Examples of Tier 1 investments include reinforcing national public health institutes, bolstering legal and regulatory frameworks, mainstreaming an integrated One Health approach, ramping up community-based disease surveillance, and training PHC providers in health intelligence.

Tier 2: Detection, containment, and mitigation capabilities. With Tier 1 in place, countries also need health systems that can detect, contain, and mitigate outbreaks before they spread widely and reach epidemic or crisis proportions. Tier 2, shown in Fig. 2 as the middle tier, operates primarily during the early phase of an outbreak with a focus on implementing emergency management and response; identifying and protecting at-risk populations; scaling up laboratory capacity, testing, and detection; isolating suspected cases; conducting epidemic intelligence, surveillance, and contact tracing; ensuring continuity of essential health services; and implementing nonpharmaceutical interventions and risk communication in communities. Key to Tier 2 is ensuring agile resources are available to support outbreak response and help contain an outbreak early on. Examples of Tier 2 investments include strengthening disease surveillance systems, creating or upgrading national emergency operations centers, reinforcing testing capacities, and training CHWs to conduct contact tracing.

Tier 3: Advanced case management and surge response. This third layer of defense, shown at the top of Fig. 2, supports crisis response and surge response to a full-blown epidemic or health crisis. This tier includes surge response interventions and secondary and tertiary hospital interventions to ensure the infrastructure and capacity needed to provide advanced, critical care for complicated cases. Tier 3 relies on surge financing to quickly meet the extraordinary costs of a full-force epidemic or pandemic, making this the most expensive and least cost-effective tier.

While investment across all three tiers is important for health system resilience, the first two tiers are foundational. When faced with limited resources, the most impactful, cost-effective public investments a country can make are those that strengthen upstream, preventive interventions in the health system. This includes enhancing disease prevention, preparedness, early detection, and containment and mitigation capabilities with a focus on strong health service delivery. Strategic investments in these precrisis tiers will not only enhance the resilience of health systems to withstand future shocks, but also improve health outcomes between crises by preventing and managing health threats.

Applying lessons learned from COVID-19 and other health emergencies

Amid major economic contraction, heightened uncertainty, and shrinking health budgets following the recent COVID-19 emergency response, governments may ask how they can afford these investments. However, even low- and lower-middle-income countries have learned and applied the important lesson from past health crises that investment in preparedness is cost effective compared to the costliness of an uncontrolled outbreak, epidemic, or pandemic. The Government of Vietnam, for instance, passed a resolution specifying that at least 30% of the country's total health budget be allocated to preventive care and pandemic preparedness, following multiple outbreaks of severe acute respiratory syndrome, avian influenza, and Zika in recent decades (Osewe, 2021). This commitment guarantees stable financial resources for ongoing prevention and preparedness, while contingencies are also in place to ensure that reserve budget funds (set at 2%–4% of the annual health budget) are allocated and available for emergency response. Following Vietnam's example, countries at all income levels can seize the COVID-19 crisis as an opportunity to target investments at filling chronic financing gaps to improve resilience and health outcomes.

The recent COVID-19 pandemic has further shown that countries cannot afford to continue underinvesting in preparedness and health service delivery. Most countries' response to this COVID-19 pandemic relied heavily on the costly third tier of the investment framework. While a focus on surge response and advanced case management is reasonable during a crisis, ultimately, countries cannot achieve resilience without precrisis preparedness, which requires upstream investments in Tiers 1 and 2. As evidenced by the severe health and economic impacts of this COVID-19 pandemic, reactive strategies cannot achieve good health outcomes and are far more expensive—requiring emergency instruments and outsized resources—than a preemptive, cost-effective approach designed to thwart potential health crises before they arise.

As countries transition out of the emergency phase of this recent COVID-19 pandemic into long-term management, now is the time to strategically invest in building resilient health systems and apply the lessons learned from this pandemic as well as from other recent health emergencies. By starting now, countries of all income levels can break the painful cycle of panic and neglect and prepare for the next health emergency that could be around the corner. As COVID-19 has underscored, countries cannot afford to maintain a business-as-usual approach to emerging infectious diseases. To protect our health, communities, and economies from future catastrophes, countries and international donors must act now—before it is too late.

References

Berthe, F., Bali, S., & Batmanian, G. (2022). *Putting pandemics behind us: Investing in one health to reduce risks of emerging infectious diseases*. World Bank Group. https://openknowledge.worldbank.org/server/api/core/bitstreams/956a58be-ddd8-572f-8aac-df5ab453d7b2/content.

Gopinath, G. (2021, January 26). *A race between vaccines and the virus as recoveries diverge*. IMF Blog. https://www.imf.org/en/Blogs/Articles/2021/01/26/blog-weo-a-race-between-vaccines-and-the-virus-as-recoveries-diverge.

Mathieu, E., Ritchie, H., Rodés-Guirao, L., Appel, C., Giattino, C., Hasell, J., Macdonald, B., Dattani, S., Beltekian, D., Ortiz-Ospina, E., & Roser, M. (2023). *Excess mortality during the coronavirus pandemic (COVID-19)*. Our World in Data. https://ourworldindata.org/excess-mortality-covid.

Osewe, P. L. (2021). *Pandemic preparedness and response strategies: COVID-19 lessons from the Republic of Korea, Thailand, and Viet Nam*. Asian Development Bank. https://www.adb.org/sites/default/files/publication/743746/pandemic-preparedness-covid-19-lessons.pdf.

World Bank. (2022). *Change cannot wait: Building resilient health systems in the shadow of COVID-19*. World Bank. https://documents1.worldbank.org/curated/en/099352211032215942/pdf/IDU05f176a190f57604e3008700032d7a878f84f.pdf.

CHAPTER 104

Resources for resilient health systems

Kiyan Irani[a,b] and Judy Kuriansky[b,c,d,e,f,g,h]

[a]Department of Clinical and Counseling Psychology, Teachers College, Columbia University, New York, NY, United States, [b]International Association of Applied Psychology, Indianapolis, IN, United States, [c]Global Council SDG3 (Health), "SDGs in Action" Initiative, Dubai, United Arab Emirates, [d]World Council for Psychotherapy, Vienna, Austria, [e]United Nations Group of Friends of Mental Health and Well-being, New York, NY, United States, [f]Office of Ambassador Sidique Wai, Embassy of the Republic of Sierra Leone to the United States, Washington, DC, United States, [g]Mission of the Republic of Sierra Leone to the United Nations, New York, NY, United States, [h]Department of Clinical Psychology, Teachers College, Columbia University, New York, NY, United States

Introduction

Since the beginning of the 21st century, the world has witnessed an unprecedented technological revolution. Colloquially known as the "tech boom," this period is characterized by the rapid progression of digital innovations that result in significant industrial and societal changes. Within this realm, the healthcare industry has quickly become a dominant topic of discussion during the COVID-19 pandemic.

In a matter of months during the year 2020, the COVID-19 pandemic ravaged national healthcare systems, forcing all stakeholders to pivot from theoretical simulations and already-in-use models to sustainability trials and implementing innovations with real-world performance. While some approaches and nations found their healthcare techniques and infrastructure capable of withstanding the sudden influx of health and mental health demands triggered by the pandemic, others encountered overwhelming challenges and even collapsed under the pressure.

In the face of the pandemic, all stakeholders, including government, civil society, and private sectors, were involved in the response including mobilizing technological assistance. Some approaches were more successful than others. National healthcare systems least able to protect their citizens from the spread of COVID-19 and its multiple ensuing variants included those who suffered high rates of infection, hospitalizations, and deaths; took on more debt to sustain their healthcare systems than could reasonably be paid back in the foreseeable future; or required the most foreign aid during the pandemic. The most vulnerable members of the global family included those that were the least developed and already had fragile health systems with

Resilient Health. https://doi.org/10.1016/B978-0-443-18529-8.00103-7

weak infrastructure and low resources, and therefore had the least capacity to respond to the pervasive economic and social healthcare burden of the pandemic.

In reaction to the crisis, policymakers and healthcare professionals recognized a strong need to reinforce the cornerstones of the current digital health industry, with a particular emphasis on digital healthcare and technology, as well as to advance new approaches. Many product developers and service delivery ecosystems and networks mobilized as quickly as possible to keep up with the burgeoning health-tech evolution. This resource guide is intended to be used by those wishing to keep up-to-date with digital health innovation and technology, as well as by those aiming to introduce associated projects, initiatives and innovations, systems, or products to a broader audience.

The following resources in this chapter are categorized by ten major resource types: (1) news/publications/journals; (2) key policy reports and global agreements; (3) conferences/expos; (4) incubators/accelerators; (5) platforms; (6) networks; (7) virtual communities; (8) contests; (9) awards; and (10) academic study programs. Several resources cover health and eHealth as well as mental health and e-Mental health. It is important to note that this list is not intended to provide an exhaustive account of resources; a catalog of that magnitude would only mimic the arduous features of accessible online search engines. Furthermore, as technological applications and projects continue to emerge, so does the need for a more expansive resource guide. Thus, the current list is offered as a starting point to help navigate the current "infodemic" unleashed by the pandemic, as a means to unlock valuable and credible global health innovation and technology resources to date of this publication, as deemed by the editors of this volume and based on their extensive (albeit not exhaustive) contacts in, and knowledge of, the field.

Hundreds of potential resources for global health innovation and technology systems were scrutinized for their relevance, accessibility, usefulness, and credibility. Each resource in this chapter includes as much as was available for an overview of relevant information, including websites, participation benefits, applicable regions, event dates, application deadlines, subscription/partnership costs, and sponsors. Given the hybrid nature of this field, some categories overlap both by structure and offerings. Therefore, it is recommended that users of this catalog leverage multiple resource avenues to enhance synergistic gains.

Note that specifics of these resources (e.g., dates, locations) are subject to change; therefore, checking sources, including the website noted, is essential.

(1) News/publications/journals

Online health news platforms serve as a vital means for disseminating industry research and information about innovations and emerging trends in all sectors. Numerous sites, publications, and newsletters relevant to digital healthcare are available. Subscribing to these can provide helpful insight into industry-specific development and growth. Despite an overwhelming amount of information flooding inboxes daily, the notices from these services can be valuable in keeping up-to-date with advances in the health space and provide developers and partners with a tool for targeted exposure. Some sites are managed by staff, others accept submissions (be alert to any charges) and still others are offered as a value-add to members.

Mobi Health News

- **What it is**: A global digital health news source headquartered in the USA
- **What you get**: MobiHealthNews has proclaimed itself the most popular source for digital health news from scientists and researchers. As stated by Credible Mind—a third-party health tech information app—the agency is well respected in the scientific community. Typical users of the site include provider groups, pharmaceutical companies, tech and venture capital firms, and health-tech start-ups.
- **Subscription**: Free—requires a sign up to receive weekly email updates
- **News frequency**: Updated daily with new articles
- **Applicable regions**: All regions—with a particular emphasis on EMEA (Europe, the Middle East and Africa), Asia, and Australia
- **Website**: www.MobiHealthNews.com

Health Tech Digital

- **What it is**: An online resource for global health tech news, headquartered in the U.K.
- **What you get**: Access to concise and timely articles explicitly written for audiences interested in digital health and technology. Articles can be submitted by any author regardless of nationality but must pass an information vetting process by the editorial board.
- **Subscription**: Free; anyone can sign up to receive semiweekly email updates
- **News frequency**: Updated daily with new articles
- **Applicable regions**: All regions
- **Website**: www.HealthTechDigital.com

Digital Health Buzz

- **What it is**: A primarily online news platform that aims to circulate up-to-date stories, organizations, products, and events in the Digital Health, Healthcare IT, and Healthcare start-up space.
- **What you get**: Access to succinct and timely articles written for audiences interested in digital health and technology. Articles can be submitted by any author regardless of nationality. Compared to other popular industry news sources, DigitalHealthBuzz has a less strenuous vetting process for published material, including articles submitted by the general public through the agency's Twitter account.
- **Subscription**: Free; anyone can sign up to receive semiweekly email updates
- **News frequency**: Updated semiweekly
- **Applicable regions**: All regions—with a particular emphasis on news from the United States
- **Website**: www.DigitalHealthBuzz.com

The Lancet Digital Health

- **What it is**: A leading scientific journal focused on innovative and timely research on topics related to digital technology in clinical medicine, public health, and global health and mental health. The journal serves the digital health, clinical, and wider health communities by promoting high-quality science and supporting the ethical use of technologies and data in practice.
- **What you get**: The journal's "Alert" feature allows users to get real-time notifications of the most up-to-date peer-reviewed articles written for audiences interested in digital health and technology. Articles can be submitted for publication by any author regardless of nationality but must go through a thorough vetting and editing process. The journal's open-access content crosses subject boundaries, creating connections between health professionals, researchers, governments, and the private sector. The Lancet also conducts Commissions on various related health/mental health issues. Also, a free access COVID-19 resource center brings together research, reviews, commentary, news, and analysis from across the various *Lancet* journals, to assist health workers, policymakers, and researchers, which users can sign up for to receive email notifications (see: https://www.thelancet.com/coronavirus?dgcid=kr_pop-up_tlcoronavirus20)
- **Subscription**: Free; anyone can sign up to receive ad hoc publication email updates
- **News frequency**: Ongoing
- **Applicable regions**: All regions
- **Website**: www.TheLancet.com/journals/landig/home

JMIR Publications

- **What it is**: A set of open-access digital health resource journals advancing digital health and mental health, open science and new technologies, that disseminates information and ideas about innovations, ideas, protocols, and research to a wide audience, including experts, leaders, and researchers and also patients/consumers and other knowledge users.
- **What you get**: Access to the latest research in the field of digital health, including innovations in healthcare technologies, patient and caregiver education, public health, participatory medicine, gaming, biomedical engineering, and medical informatics, as well as reports, reviews, and an archive of over 10,000 papers on research and experiences related to the field.
- **Subscription**: Free
- **News frequency**: Ongoing
- **Applicable regions**: All regions
- **Website**: https://www.jmirpublications.com/

eHealth reporter Latin America

- **What it is**: A news portal exclusively dedicated to Digital Health in Latin America and also in Spain, in three language versions (Spanish, Portuguese, and English) covering new trends in the industry and sharing success stories and corporate news. It also intends to stimulate innovation through the support of regional Digital Health StartUps. It is published in partnership with the Healthcare Information and Management Systems Society (HIMSS) and serves as the official media partner for all Latin America events of the HIMSS.
- **What you get**: A weekly newsletter with the latest stories is sent to the entire community and a quarterly magazine is available for professionals to read about top issues in business, technology, and health.
- **Subscription**: Sign up on their website
- **News frequency**: Ongoing
- **Applicable regions**: All regions
- **Website**: http://ehealthreporter.com/en/home-english/?lang=en

Morning eHealth

- **What it is**: A newsletter that examines the latest new in eHealth politics and policy, from journalists at POLITICO, self-described as a "*global authority on the intersection of politics, policy, and power*." The customizable policy newsfeed and intelligence platform is intended for those who have a political, professional, or financial stake in politics and policy, meant to give access to exclusive nonpartisan reporting, analysis, and tools.
- **What you get**: A newsletter delivered daily for Politico Pro subscribers but weekly for others.
- **Subscription**: Free. The Pro platform involves a fee.
- **News frequency**: Ongoing
- **Applicable regions**: All regions
- **Website**: https://www.politico.com/morningehealth/

Frontiers in Digital Health

- **What it is**: A multidisciplinary open-access journal that publishes rigorously peer-reviewed research in digital health, positioning itself at the forefront of disseminating and communicating the latest scientific knowledge and impactful discoveries. The focus is on how to transform healthcare with innovative digital tools.
- **What you get**: Knowledge to arrive at "Medicine 4.0" in the age of Digital Health. Four specialty services include: Health Informatics, Health Technology Innovation, Personalized Medicine, and Connected Health. This multidisciplinary open-access journal publishes "rigorously peer-reviewed research" and positions itself (as do other

services) at the forefront of disseminating and communicating the latest scientific knowledge in the field of digital health.

- **Subscription**: Free. All articles are open access, able to be read and used.
- **Applicable regions**: All regions
- **Website**: https://www.frontiersin.org/journals/digital-health?utm_source=ad&utm_medium=ggl-src&utm_campaign=sub_ggl_fdgth&gclid=CjwKCAjw9NeXBhAMEiwAbaY4lsQ4ziZzWBpqFnosBcfcRXChX-2jEPJqmajDMl_KHMLCIV47iabAUxoCECQQAvD_BwE

World Health Organization (Varied) Reports on Resilient Health

- **What they are**: extensive reports on resilient health including toolkits and indicators.
- **Subscription**: Free. All articles online open access.
- **Websites**: https://www.who.int/teams/primary-health-care/health-systems-resilience
https://www.who.int/publications/i/item/9789240048751
https://www.who.int/publications/i/item/9789240088986
https://www.who.int/publications/i/item/WHO-UHL-PHC-SP-2021.01
https://www.frontiersin.org/research-topics/38177/health-systems-recovery-in-the-context-of-covid-19-and-protracted-conflict/magazine
https://www.who.int/publications/i/item/9789240088306

(2) Key policy reports and global agreements

- **The UN Agenda 2030 for Sustainable Development**: The Sustainable Development Goals are the blueprint to achieve a better and more sustainable future for all. The framework of 17 goals and 169 targets addresses the global challenges the world faces, including poverty, hunger, inequality (especially gender inequality), education, decent work, climate change, environmental protection, peace and justice, and of course, health and well-being. The SDGs were adopted in 2015 by the United Nations General Assembly and are intended to be achieved by 2030. Published by the United Nations, 25 September 2015, the Agenda is available at: https://sdgs.un.org/2030agenda. Also see: https://sustainabledevelopment.un.org/index.html
- **Universal Health Coverage (UHC).** UHC means that all people have access to the health and mental health services they need as well as when and where they need them, without financial hardship. It includes the full range of essential health services, from health promotion to prevention, treatment, rehabilitation, and palliative care. The Political Declaration for UHC was adopted in 2019. The World Health Organization (WHO) is supporting countries to develop their health systems to move toward and sustain UHC and to monitor progress. Available at: https://www.who.int/health-topics/universal-health-coverage#tab=tab_1
- **The Global Strategy on Digital Health** aims to strengthen health systems through the application of digital health technologies. It sets out a vision, strategic objectives, a

framework for action, and implementation principles to advance digital health. Published by the World Health Organization, 2021, it is available at: https://www.who.int/docs/default-source/documents/gs4dhdaa2a9f352b0445bafbc79ca799dce4d.pdf

- **The Digital Health Atlas** is a WHO global technology registry platform aiming to strengthen the value and impact of digital health investments, improve coordination, and facilitate institutionalization and scale. Available at: https://digitalhealthatlas.org/en/-/
- **The High-level Panel on Digital Cooperation** was convened by the UN Secretary-General to provide recommendations on how the international community could work together to optimize the use of digital technologies and mitigate the risks. In June 2019, the Panel published their report "The Age of Digital Interdependence" with a series of recommendations to improve digital cooperation. Available at: https://documents-dds-ny.un.org/doc/UNDOC/GEN/N20/102/51/PDF/N2010251.pdf?OpenElement
- **UN Innovation Toolkit**: Developed by the United Nations, for the UN, this digital platform includes 21 tools, step-by-step directions, worksheets, case studies, references, as well as a 27-question assessment which provides a diagnosis on every user's strengths and areas for improvement. Available at: https://www.uninnovation.network/un-innovation-toolkit
- **UN Innovation Network**: The UN Innovation Network (UNIN) is an informal, collaborative community of UN innovators interested in sharing their expertise and experience with others to promote and advance innovation within the UN System. The work of the UNIN is driven by the interests and needs of its members. For innovation practitioners, the network hosts regular knowledge sharing sessions to review and discuss the application of new innovation trends and how they can contribute to achieving the SDGs. It also helps entities share tools, resources, and best practices for innovating in the UN. Available at: https://www.uninnovation.network
- **Grand Challenges Canada**: Grand Challenges Canada supports "Ideas with Big Impact." Funded by the Government of Canada and other partners, Grand Challenges Canada funds innovators in low- and middle-income countries and Canada. Available at: https://www.grandchallenges.ca
- **Coronavirus Resilience Innovation Map**: The Coronavirus Resilience Innovation Map is a global directory with hundreds of innovations and solutions that help people cope and adapt to life amid the COVID-19 pandemic and connect with innovators. Available at: https://coronavirus.startupblink.com

(3) Conferences and expos

As investors and nonprofits continue to fund digital healthcare start-ups, conferences and expos provide a hub for innovators to learn from, and to network with, leaders in the field. Dozens of conferences are organized each year by various organizations, curated to advance digital healthcare technology. These events are attended by industry professionals, sponsors, academics, and others who understand the value of staying up-to-date with pioneering scholarship and emerging innovations, projects and products. Events also provide a platform for start-ups to demonstrate their projects' value-add and products' effectiveness to consumers and professionals within their community. Expos are also held as regional or global events dedicated to finding solutions to fundamental challenges facing healthcare and digital technology, purposefully held in various regions to reach specific but also diverse and large audiences. Attendees are also given a rare opportunity to assess their own and competitors' products before they get released to the general public.

HIMSS Global Health Conference and Exhibition

- **What it is**: Promoted as "the world's largest health information and technology event" with a mission "to reimagine health and wellness for everyone, everywhere," professionals in the digital global health ecosystem have traveled from over 60 countries annually to attend.
- **What you get**: Presenters and attendees benefit from "unparalleled global exposure and media coverage" and the opportunity to learn from and network with prominent leaders in the field.
- **Who can attend**: Anyone. Pricing per ticket ranges from US$325 for digital access to US$1395 for a combination of in-person and digital access.
- **When**: April (be sure to check)
- **Applicable regions**: All regions, with the conference held in the USA
- **Sponsors**: Healthcare Information and Management Systems Society, a member association committed to transforming the health-tech ecosystem
- **Website**: www.himss.org/global-conference

European Health Tech Innovation Week

- **What it is**: Called "Europe's most valuable health-tech event" by the *Financial Times*, this is a 5-day pan-European hybrid exhibit and conference showcasing leaders within healthcare and technology. As Europe's largest digital health showcase, there are 100k+ annual attendees, 450+ speakers, and 200+ exhibits. Though presenters are exclusively selected from European-based companies, individuals and businesses worldwide are welcome to attend.
- **What you get**: Presenters benefit from targeted exposure and the ability to present to industry leaders from all across the world, with a strong concentration of European-based digital health pioneers. Attendees are able to enhance their technical expertise, network with other leaders in their field, and attend presentations by industry executives.
- **Who can attend**: Anyone. Pricing for tickets range from €49 to €245 depending on the desired level of access at the event.
- **When**: May (be sure to check)
- **Applicable regions**: All regions with a particular emphasis on EMEA (anagram for Europe, the Middle East, and Africa)

- **Sponsors**: Giant Health, an international community focused on health technology
- **Website**: www.giant.health/european-health-tech-and-innovation-week

Digital Health Forum

- **What it is**: Organized as a global one-day showcase focused on three tenets of healthcare: equity, access, and sustainability. The conference is held virtually and as an homage to the global initiative of socialized healthcare.
- **What you get**: As one of the largest digital health conferences held each year, presenters benefit from targeted exposure and the ability to present to industry leaders worldwide. Attendees can glean valuable knowledge and advice from digital health and technology innovators.
- **Who can attend**: Anyone; admission is free and open to the public
- **When**: June (be sure to check)
- **Applicable regions**: All regions are welcome. Conference was held virtually during the COVID-19 pandemic, with attendance rising due to the virtual platform. Future events may be hybrid.
- **Sponsors**: G4A—a global accelerator dedicated to technology
- **Website**: www.g4a.health/events/digital-health-forum-2022

MENA Telehealth Conference

- **What it is**: Organized as a global telehealth conference highlighting implementations and applications of technological advances in the Middle East and North Africa (MENA) region. Leaders in the field present on wide-ranging topics that can generally be categorized into three major themes: regulation reform, business modeling, and improving healthcare outcomes.
- **What you get**: Presenters benefit from targeted exposure to dedicated healthcare and digital health professionals dedicated to improving health outcomes in the MENA region. Attendees are able to enhance their technical expertise, network with other leaders in their field, and attend presentations by industry executives. A chapter about this conference is in this volume (Otzoy-García, D., Iñigo Petralanda, M. I., Aljovín, Raising the profile of digital health in Latin America: The RECAINSA digital health forum and network (2024). In J. Kuriansky & P. Kakkattil. "Resilient Health: Leveraging Technology and Social Innovations to Transform Healthcare for COVID-19 Recovery and Beyond." Elsevier).
- **Who can attend**: Anyone; the event is held in person or virtually
- **When**: Early March (be sure to check)
- **Applicable regions**: targeted to MENA region but also available worldwide
- **Sponsors**: International Society for Telemedicine and eHealth
- **Website**: www.MenaTelehealth.com

Digital Health Week

- **What it is**: A week of presentations promoting advancements in digital health and the critical role in providing universal health coverage. This global expo is a conglomerate of thematic events to allow diverse stakeholders (including from civil society organizations, governments, private companies, and health institutions) the opportunity to come together to exchange ideas and experiences on digital health challenges and opportunities, and to discuss what needs to happen to ensure digital transformation of health to achieve universal health coverage.
- **What you get**: The week includes presentations on an exceptionally wide variety of topics in the digital health field. Participants can leverage their programs and content to enhance exposure of their events and tell their digital health stories to global audiences, build partnerships among a diverse group of stakeholders, and "bridge the technical-political divide" in digital health. A chapter about sessions in the conference, showing the diversity of interesting topics, is in this volume (Lissinna, H. Digital transformation of health featured during Digital Health Week 2021: Profiles of partnerships and health innovations using technology in countries around the world (2024). In J. Kuriansky & P. Kakkattil. "Resilient Health: Leveraging Technology and Social Innovations to Transform Healthcare for COVID-19 Recovery and Beyond." Elsevier).
- **Who can attend**: Anyone—related events are worldwide
- **When**: Dates can change each year, but the week event is generally held in the Fall.
- **Applicable regions**: All regions
- **Sponsors**: Transform Health
- **Website**: www.digitalhealthweek.co

Digital Health Africa Summit

- **What it is**: A two-day conference about healthcare services with digital health technology meant to address the fact that "digital health deployment in Africa is constrained by challenges such as poor coordination of mushrooming pilot projects, weak health systems, lack of awareness and knowledge about digital health, poor infrastructure such as unstable power supply, poor internet connectivity and lack of interoperability of the numerous digital health systems." The conference brought together thought leaders in healthcare communities desiring change and creating a conceptual framework needed for successful and sustainable deployment of digital health in Africa. Discussions centered on technologies such as the Internet of Things, wearables, and sensors, opening possibilities for monitoring individuals' health in remote areas.
- **What you get**: Awareness about health technologies in Africa. Videos of the agenda and presentations in the 2021 conference can be downloaded on their website by filling in a form. Event speakers included Chief Executive Officers of private health companies, a university Director of Health Services in Nigeria, and a Director of eHealth Hospital Services in a provincial department of health in South Africa.

- **Who could attend**: Free registration for anyone, but targeted at health start-ups and entrepreneurs, hospital executives, governments, medical and allied health professionals, medical students, healthcare institutions, and universities.
- **When**: was held in July 2021
- **Applicable regions**: All regions
- **Host/organizer**: BII World
- **Website**: https://www.biihealthtech.com/digihealthafrica/

Middle East HealthTech Innovation Summit

- **What it is**: A two-day conference with a preconference investor forum and awards ceremony. Presentations were on topics including "Adoption of digital health at scale during COVID" focusing on the UK; "Tackling diabetes with telehealth"; "Benefits of clinical data," presented by Malaffi, the UAE regions' Health Information Exchange of patient health information for public and private healthcare providers; "Women's health and Femtech opportunities"; "Digital transformation of healthcare in Abu Dhabi"; "HealthTech Investments and Attraction of Capital—A regulatory perspective" from the UAE Department of Health; standardizing record-keeping; and "Cold Plasma: the technology of the future." Awards were given in categories as: Best Health Informatic Platform (given to Children's Cancer Hospital in Egypt), Best Digital Patient Experience (to Decision Labs), Best Hospital Technology (King Faisal Specialist Hospital and Research Center in Saudi Arabia).
- **What you get**: Awareness about health technologies in the Middle East region. All 7 ½ hours of Day 1, 5½ hours of Day 2, the 6-hour preconference, and the 47-minute Award Presentation are available on the website. Also, pdfs of some presentations in their previous (2021) conference can be accessed and downloaded at: https://drive.google.com/drive/folders/1R_-ddZA0DEpZOASuqDUUv0QcqQXkTGXR
- **Who can attend**: Free registration for anyone, but targeted at health start-ups and entrepreneurs, hospital executives, governments, medical and allied health professionals, medical students, healthcare institutions, and universities.
- **When**: was held in September 2021 (check 2022 and beyond on website)
- **Applicable regions**: All regions
- **Host/organizer**: BII World
- **Website**: https://www.biihealthtech.com/healthtechinnovationme/

The United Nations Multistakeholder Forum on Science, Technology and Innovation (STI Forum)

- **What it is**: The STI Forum is organized as an annual meeting with an exclusive focus on science, technology, and innovation, held in person at the United Nations New York City headquarters or online. The themes align with issues at the United Nations in the global agenda related to the implementation of the Sustainable Development Goals and with the themes of the annual UN High-Level Political Forum on Sustainable Development. For example, the STI Forum 2022 theme was "Science, technology and innovation for building back better from the coronavirus disease (COVID-19) while advancing the full implementation of the 2030 Agenda for Sustainable Development." The meeting consists of roundtables, special events, and thematic sessions on topics like disruptive technologies, artificial intelligence, solar energy, and big data, some of which are related to health (but not all). The STI Forum is an important component of the Technology Facilitation Mechanism (TFM) that provides a mapping and gateway for information on science, technology, and innovation initiatives, mechanisms, and programs within and beyond the United Nations system.

- **What you get**: Information about best practices and lessons learned about science, technology, and innovation initiatives and policies by governments and other stakeholders in the field, in varied presentation and panels. The *Health in Your Hands* platform, described in a chapter in this volume, is an example of such an innovation, and was presented at this STI Forum in 2017, at an event sponsored by the United Arab Emirates (UAE), a member state of the UN. That project is part of the "SDGs in Action" campaign of the UAE. See: Kuriansky, J. "Health in Your Hands": A platform for innovative healthcare solutions (2024). In: J. Kuriansky, P. Kakkattil (Eds.). "Resilient Health: Leveraging Technology and Social Innovations to Transform Healthcare for COVID-19 Recovery and Beyond." Elsevier.
- **Who can attend**: Registration is required and open to only Member States of the United Nations, observer States, intergovernmental organizations and entities with observer Status in the General Assembly or the Economic and Social Council, United Nations system, major groups and other stakeholders, and media.
- **When**: Annually, typically in the first week of May (has been held in previous years in June)
- **Applicable regions**: All regions
- **Sponsors**: Organized by the United Nations interagency task force on STI for the Sustainable Development Goals (IATT), convened by the United Nations Division for Economic and Social Affairs (UN-DESA) and the United Nations Conference on Trade and Development (UNCTAD) and a 10-Member Group of high-level representatives appointed by the UN Secretary General.
- **Website**: https://sdgs.un.org/tfm/sti-forum

Dubai Health Forum

- **What it is**: While not exclusively about digital heath, Dubai in the United Arab Emirates is such an impressive hub for innovation that technological advances are covered in the Dubai Health Forum. The annual two-day event launched in 2017 brings together respected decision-makers, health practitioners, industry experts, scientists, doctors, health specialists, and innovators representing the wide spectrum of healthcare and showcases game-changing advances in health and well-being. The event aims to promote stronger relations between health institutions, enhance research, improve professional practices, expand the use of smart technologies and solutions, and create platforms to support the global, regional, and local industry in achieving these aims. The event is aligned with the fact that the global health community is currently focused on creating an integrated system capable of facing challenges, dealing with emergency situations, and enhancing people's well-being.

- **What you get**: Typically, the event draws more than 2000 participants, representing more than 30 countries with more than 30 sessions and about 60 speakers, offering a range of training workshops, lectures, and discussions. In 2022, about 10 main themes covered the pandemic, the "new normal," the path to recovery, vaccination, effective postpandemic care, sustainable development of the healthcare industry, the role of the media in the COVID-19 pandemic, digital health, genomics, space medicine, big data and artificial intelligence, value-based care, and health economics.
- **Who can attend**: Anyone
- **When**: Annually, about March, in Dubai
- **Applicable regions**: All regions
- **Sponsors**: Government of Dubai, Dubai Health Authority, under the patronage of His Highness Sheikh Mohammed bin Rashid Al Maktoum, Vice President, Prime Minister, and Ruler of Dubai, and H.H. Sheikh Hamdan bin Mohammed bin Rashid Al Maktoum, Crown Prince of Dubai and Chairman of the Dubai Executive Council.
- **Website**: https://www.dxbhealthforum.com/

(4) Incubators and accelerators

At their core, incubators operate as programs designed to help evolve start-ups during their initial growth period. Institutes that present these provide start-ups with mentorship, education, and shared physical workspaces to facilitate business development while minimizing operating costs. Incubators and accelerators often recruit companies that fit within specified industry parameters. Some incubators and accelerators mentioned in the following section specialize in enhancing venture capital opportunities for start-ups within the healthcare field. Incubators typically offer attractive funding and networking opportunities in exchange for equity. Most incubators require a formal application and interview to join and usually demand a fixed commitment period.

Cedars-Sinai Accelerator

- **What it is**: A 13-week incubator program that converts digital health ideas into products that can create lasting change. A key emphasis is to utilize direct connections between start-ups and healthcare implementers, such as physicians, nurses, and hospital administrators.

- **What you get**: The sponsor provides guaranteed funding of US$100***k*** to companies selected to participate in this incubator. A past seed-funding data reported the typical midpoint financing for chosen companies to be between US$500k and US$5mm, with some companies attaining over US$10mm throughout the duration of the program.
- **Who can apply**: Start-ups within the digital health field.
- **When**: Applications are accepted on a rolling basis. The company onboards new members when space becomes available (either due to companies graduating from the program or new funding becoming available).
- **Applicable regions**: All regions. One must attend the 13-week program in person at Cedars-Sinai Medical Center, a nonprofit academic healthcare organization based in Los Angeles, California, USA. International companies are encouraged to apply.
- **Sponsors**: Cedars-Sinai, a nonprofit medical center
- **Website**: www.CSaccelerator.com

Global Innovation through Science and Technology (GIST)

- **What it is**: Founded by the U.S. Department of State, GIST aims to connect domestic and international businesses with U.S. investors. Ultimately, this incubator strives to build the capacity and resilience of individual start-ups and the entrepreneurial ecosystems that support them.
- **What you get**: GIST aims to ensure that the U.S. has access to emerging technologies and ideas. The incubator also supports the economic prosperity and stability of the originating country by providing critical know-how, networking opportunities, and financial resources to each business. This incubator is distinctive in that it allows foreign companies the opportunity to apply for U.S. research and innovation grants.
- **Who can apply**: Any start-up
- **When**: Applications are accepted on a rolling basis
- **Applicable regions**: All regions, but the incubator primarily operates virtually from the U.S.
- **Sponsor**: The United States Department of State
- **Website**: www.GistNetwork.org/gist-business-incubation

Global Good Fund Fellowship

- **What it is**: A 12-month incubator program supporting the leadership development of social and health enterprise leaders across the globe.

- **What you get**: This incubator is unique in that each accepted participant is paired directly with an industry-specific coach and a C-suite business executive (who holds a senior leadership position in the company, e.g., CEO, CFO, COO) who serves as the primary mentor. The program provides leadership assessment resources, access to an impressive network of peers, and direct meetings with financial capital investors.
- **Who can apply**: Applications are open to any start-up. However, ideally, fellows are (1) under 40 years old, (2) currently holding leadership positions at their respective social enterprise, and (3) employed at or own their business for at least three years.
- **When**: Applications are accepted on a rolling basis.
- **Applicable regions**: All regions. Physical workspaces and headquarters are located in Massachusetts, USA; however, international companies have virtual attendance options.
- **Sponsors**: Global Good Fund, a global accelerator program that focuses on growing companies with social-focused missions
- **Website**: www.GlobalGoodFund.org/fellowship/fellowship-program

MedTech Innovator

- **What it is**: A global accelerator for medical device, digital health, and diagnostic companies, with the mission to improve the lives of patients by promoting corporate development and expansion within the healthcare system.
- **What you get**: Participants benefit from a remote-friendly environment, having operated as a virtual program since 2015. As a nonprofit, the accelerator does not require entry fees or equity for participation. The organization has an attractive track record; in 2021 the accelerator's alumni solidified US$1.5B (1.5 billion US dollars) in funding across 99 companies.
- **Who can apply**: Applications are open to preseed start-ups in the medical device, diagnostic, or digital health fields. Priority is given to companies that have not raised Series B funding or established a significant strategic partnership. Later-stage companies are still eligible for the showcase track of the program, a global competition for innovations in the same aforementioned fields.
- **Applicable regions**: US, Asia Pacific
- **Sponsors**: Asia Pacific Medical Technology Association (Asia region), Johnson & Johnson, RCT Ventures
- **Website**: www.MedTechInnovator.org

MEDX Xelerator

- **What it is**: A global medical device and digital health incubator with a targeted interest in interventional oncology, neurovascular applications, medical robotics, and digital solutions for remote therapy, with the mission to streamline the workflow of healthcare professionals and enhance the lives of patients suffering from unmet medical needs.

- **What you get**: The accelerator aims to leverage its strategic partnerships in the scientific and medical communities to progress start-ups through the initial growth phase and into successful medical ventures. Participants benefit from a guaranteed investment from MEDX of at least US$800 K for a two-year commitment and up to US$1.1MM after three years.
- **Who can apply**: Applications are open to any start-up in the medical device or digital health field. Priority is given to companies offering medical innovations in the vascular space, interventional cardiology, and other minimally invasive medical procedures.
- **When**: Applications are accepted on a rolling basis. From the time of initial contact, the application process takes about 6 to 8 months before official onboarding.
- **Applicable regions**: All regions
- **Sponsors**: MEDX Ventures Group, together with three partners: Boston Scientific, Intellectual Ventures' Invention Science Fund, and Sheba Medical Center
- **Website**: www.MedXelerator.com
- **Where to apply**: Partnership inquiries should be emailed to info@MedXelerator.com

Rockstart

- **What it is**: Europe's first full-year accelerator program, Rockstart, provides access to the healthcare ecosystem, capital, and coaching for entrepreneurs who are looking to revolutionize the health industry. Offices are in Amsterdam, Copenhagen, and Bogotá. The first event took place in Copenhagen in September 2022, followed by a three-day summit in Bogotá, Colombia in May 2023.
- **What you get**: Participants of the accelerator benefit from a hands-on approach with dedicated mentorship aimed at skill-building and corporate scaling. In collaboration with the Institute for Advanced Metropolitan Solutions (AMS Institute) and the City of Amsterdam, Rockstart launched "Amsterdam Circular", a new investor readiness program, designed to strengthen Amsterdam's early stage circular economy start-ups and their funding efforts. Amsterdam Circular is a two-month, cost-free program held in a hybrid format with in-person and online events, consisting of four to eight hours a week in workshops, training sessions, and other deep dives into building a lasting business. In this first-of-its-kind program, expert mentors offer consultation advice and personalized work with the start-ups themselves to achieve corporate development.
- **Who can apply**: Any start-up in the health space that has yet to receive Series B funding
- **When**: Applications are accepted on a rolling basis
- **Applicable regions**: All regions with a particular emphasis on European-based start-ups
- **Sponsors**: Rockstart
- **Website**: https://rockstart.com/
- **Where to apply**: partnership inquiries should be submitted through the online contact form found at www.RockStart.com/contact

Henry Ford Health Systems Global Technology Development Program

- **What it is**: This is a "world-class Innovations Program to translate leading edge discoveries into clinical use through best practices in areas as product development, technology commercialization, venture funding, and corporate engagement." Under the umbrella of the Henry Ford Institute, the Program aims to build synergistic healthcare technology business opportunities between the Henry Ford Health and growth-stage healthcare companies in the United States, and healthcare innovation regions around the globe. Henry Ford Health receives consideration (in the form of either equity, royalties, and/or fees) in return for its intellectual property contributions to companies that participate in the Program.
- **What you get**: Market introduction and investment support, as well as visibility on a platform to accelerate development and commercialization of viable healthcare technologies for U.S. and global markets and participation in the revitalization of Southeast Michigan's economy through healthcare technology business opportunities based upon adapting promising innovations.
- **Who can apply**: Privately held high potential growth-stage firms in the healthcare sector that are designing or developing a product or service aimed at U.S. market entry in all classes of healthcare products, including biotech, pharma, medical devices, and imaging; and global healthcare companies, including domestic companies with innovative products and ideas.
- **When**: On a rolling basis, through their application form.
- **Applicable regions**: All regions
- **Sponsor**: Henry Ford Health
- **Where to apply**: https://www.henryford.com/innovations/corporate/global-technology-development

(5) Platforms

Digital health platforms are initiative-focused projects that address urgent healthcare needs worldwide. A particular goal for many digital health platforms is to profile innovations and projects that provide much-needed healthcare to communities with little to no access, including those in rural or impoverished areas with fragile health infrastructure, as well as to facilitate connecting these market-based solutions to capital necessary to create sustainable health solutions.

Health Innovation Exchange (HIEx)

- **What it is**: Health Innovation Exchange (https://hiex.ch) was established as a UN-hosted partnership platform bringing together policymakers, innovators, and investors to leverage health innovations to save and improve lives around the world. Since 2019, HIEx has tackled the challenge of fragmented innovation ecosystems. HIEx fosters public-private partnerships that can help scale proven technologies and innovations to

maximize reach and health impact, especially for the most vulnerable communities. HIEx has an extensive partnership network globally, including governments, private sector companies, investors, academia, as well as innovation hubs, incubators, and accelerators.
- **Who can apply**: Any organization that is interested in contributing to the SDGs with an emphasis on SDG3 (Good health and well-being) in the United Nations Agenda 2030 for Sustainable Development
- **Partnership cost**: Free
- **What you get**: Healthcare policymakers/implementers can access support including in identifying innovations that are relevant to address their health challenges and connect with other health leaders across countries. Innovators and growth-stage health companies can receive support in connecting with markets, facilitated access to financing (grants and investments as appropriate), advisory services, and acceleration support. Investors can access a pipeline of proven health innovations and technologies that are relevant for scaling access to healthcare through a sustainable business model as well as access to a community of like-minded coinvestors committed to SDGs.
- **Applicable regions**: All regions—emphasis on the global South
- **Where to connect**: https://hiex.ch

For partnership inquiries, email: team@hiex.ch

Health In Your Hands (HIYH)

- **What it is**: HIYH is a platform that aggregates health innovations consistent with the goals of the United Nations Agenda 2030 for Sustainable Development to reach those "left furthest behind" and achieve Goal 3 of good health and well-being for all, as well as the goals of the Political Declaration of Universal Health Coverage, guided by the principles of the World Health Organization. The project is under the aegis of the Global Council on Health (SDG 3) of the United Arab Emirates's "SDGs in Action" campaign.
- **Who can apply**: The managers of the platform usually source innovations and invite them to apply, to assess whether they meet the criteria (impact, scalability, innovation, assisting in universal health coverage, research, use of technology, capacity building, communications, financial sustainability and profitability, viable management team, contribution to the health ecosystem), but it will consider organizations that reach out which contribute to the health ecosystem.
- **Partnership cost**: Free
- **What you get**: Innovations can leverage exposure and potential partnership opportunities promoted on a respected platform associated with the Global Council on Health of the UAE's SDGs in Action campaign.
- **Applicable regions**: All regions

- **Where to connect**: See chapter in this volume: Kuriansky J. "Health in Your Hands": A platform for innovative healthcare solutions (2024). In J. Kuriansky & P. Kakkattil. "Resilient Health: Leveraging Technology and Social Innovations to Transform Healthcare for COVID-19 Recovery and Beyond." Elsevier.

World Health Organization Integrated Data Platform (WIDP)

- **What it is**: WIDP is a global platform used to collect, analyze, and disseminate health system data. The platform intends to catalyze partnerships between WHO programs, connect governments to health systems, identify health-specific development opportunities, and work to ensure that investments are channeled to support universal health coverage.
- **What you get**: Members are granted access to World Health Organization's business coalition, healthcare and data collection staff, and proprietary health systems.
- **Who can apply**: Any organization, including sovereign entities that contribute to health ecosystems
- **Partnership cost**: Free (temporarily). Countries that do not have the capability to collect and analyze health information can request to provisionally use WIDP to collect quality data to inform decision-making. The data collected is proprietary to the country soliciting the services.
- **Applicable regions**: All regions
- **Where to connect**: www.who.int/tools/who-integrated-data-platform—partnership inquiries should be emailed to integrated-data@who.int

USAID's Innovation, Technology, and Research (ITR) Hub

- **What it is**: This Hub of USAID leverages "the promise of innovation, technology and research to bring about positive change and solve some of the world's most pressing challenges." They have four main aims, as stated on their website: to source, test, and scale proven solutions that impact millions of people; to open development to people everywhere with good ideas; to use data and evidence in decision-making; and to harness scientific and technological advances. They help innovators reach more users and secure needed funding to achieve their most possible impact. The activities of the ITR Hub supports USAID's goals to "reduce poverty, strengthen democratic governance, and help people solve their own development challenges," and efforts to "keep the Agency on the forefront of science and technology… integrate and apply innovative solutions to USAID's projects and programs around the world, and continually find ways to reach development objectives more effectively and sustainably."

- **What you get**: New partnerships, and connecting with others involved in innovative strategy and implementation. Also, they provide funding to pilot, test, and scale creative solutions that demonstrate evidence of impact, cost-effectiveness, and the potential to scale, improving millions of lives at a fraction of the usual cost, in the Development Innovation Ventures (DIV), an open innovation grant program. They also design challenges, prizes, and hackathons for development that inspire and incentivize creative problem-solvers around the world to test and scale solutions to a broad range of development challenges.
- **Who can apply**: diverse potential partners, including USAID missions, entrepreneurs, small businesses, researchers, universities, technology companies, government, civil society, faith-based organizations, and development professionals.
- **Where to connect**: https://www.usaid.gov/innovation-technology-research

Startup Health

- **What it is**: A global platform that aims to elevate companies dedicated to improving the health and well-being of all people worldwide. The platform utilizes a five-pronged approach to help companies accelerate and expand market share and progress out of the initial growth phase. The platform's stated approach includes: invest in every stage of growth, unite member entrepreneurs and their companies into a connected community, leverage media networks broadcast stories about community progress, build global partner networks to support social initiatives, and grow health initiative funds.
- **What you get**: Members benefit from a combination of community-driven entrepreneurial coaching and a dedicated media-spotlighting platform designed to increase an entrepreneur's probability of success. Their unique model for investing is designed to support entrepreneurs at any growth stage (including companies that have already solidified Series B funding). They advise that the best way to leverage the power of the platform is to actively participate, support, and collaborate with fellow members and take advantage of the resources provided.
- **Who can apply**: Any start-up dedicated to improving the health and well-being of any community
- **Partnership cost**: Free; however, there is a vetting process for potential partners and investments
- **Applicable regions**: All regions
- **Where to connect**: www.StartupHealth.com. Partnership inquiries should be submitted through the online contact form found at www.StartupHealth.com/getting-started

(6) Networks

Networks bring together individuals operating in complementary topic areas, in this case in the area of health and mental health, both broadly and also focused on technology innovation. Networks benefit from the group or association's convenings and opportunities to learn about what's going on in the field, and to find partners for advocacy, programs, or funding.

eMental Health International Collaborative (eMHIC)

- **What it is**: A global group of professionals and subject matter experts focused on disseminating knowledge and enhancing collaboration to tackle issues related to eMental health and addiction. The collaborative aims to support and improve mental health and addiction outcomes for all populations.
- **What you get**: Members are invited to online group forums and webinars chaired by subject matter experts in the mental health field; granted access to a vast industry-specific article database; and allowed exclusive marketing rights to the company's membership badge. Members may also submit a "Request to Collaborate" form for current projects that could benefit from corporate partnerships; if chosen, the company will showcase these projects in its bimonthly newsletter and LinkedIn (social media) account.
- **Who can apply**: Any organization that contributes to mental health ecosystems
- **Partnership cost**: Membership fees range from (USD) $9 for students to $49 for groups of up to 10 people
- **Applicable regions**: All regions
- **Website**: eMHICglobal.com/join. See the chapter specifically about eMHIC in this volume for more information: Thapliyal A. eMental Health International Collaborative: The global peak-body for digital mental health implementation (2024). In J. Kuriansky & P. Kakkattil. "Resilient Health: Leveraging Technology and Social Innovations to Transform Healthcare for COVID-19 Recovery and Beyond." Elsevier.
- **Where to connect**: partnership inquiries should be submitted through the online contact form found at www.emhicGlobal.com/contact-information

RECAINSA: The Central American Health Informatics Network

- **What it is**: RECAINSA is a network where professionals from the public, private, and academic sectors specifically from Central and Latin America can share experiences and lessons learned related to digital transformation in health, and learn about innovations, best practices, and successful experiences on topics related to digital health in the region. The organization actively promotes the strengthening of eHealth governance, training of leaders, and exchange of experiences, good practices and the dissemination of knowledge in digital health in the countries of Central America, that can also apply to other regions, but is focused on this region which has been heretofore under-recognized. RECAINSA holds a Digital Health Forum to showcase health and transformative digital technology research, products, and systems to revolutionize digital health in Latin America and to promote the region's digital health evolution. The intention is also to ensure adequate and inclusive Universal Health Coverage, in sync with the United Nations global agenda to "leave no one behind."

- **What you get**: Networking with innovative technology leaders in Central America and news about digital health and events, mainly relevant to the region but also worldwide. RECAINSA also offers connections to digital health stakeholders, policymakers, and programmers focused on Latin America, to foster partnerships worldwide as well as access to technology advances, blogs, and notifications about events.
- **Who can apply**: Anyone
- **Applicable regions**: Focused on Latin America, but open to all regions
- **Website**: www.recainsa.org. See the chapter in this volume for more information: Otzoy-García, D., Iñigo Petralanda, M. I., Aljovín, P. Raising the profile of digital health in Latin America: The RECAINSA digital health forum and network (2024). In J. Kuriansky & P. Kakkattil. "Resilient Health: Leveraging Technology and Social Innovations to Transform Healthcare for COVID-19 Recovery and Beyond." Elsevier.

The Mental Health Innovation Network (MHIN)

- **What it is**: MHIN is a website that represents a global community of mental health innovators, including researchers, practitioners, policymakers, service user advocates, and donors from around the world, who post their innovation as a way to share innovative resources and ideas to promote mental health and improve the lives of people with mental, neurological, and substance use disorders. MHIN was founded by United for Global Mental Health, headquartered in London, dedicated to raise awareness, reduce stigma, and increase support for mental health worldwide with trusted partners. MHIN aims to facilitate the development and upscale of effective mental health interventions and best practices. MHIN's core team is based at the Centre for Global Mental Health at the London School of Hygiene & Tropical Medicine and the World Health Organization's Department for Mental Health and Substance Abuse.
- **What you get**: Sharing innovations learning about what colleagues are doing in the field of mental health to share and replicate projects and resources, and connections with

innovators in mental health worldwide to build partnerships and leverage resources to scale programs. Members can submit and have access to resources posted by other members, for example, screening tools, training manuals, and survey instruments that might be useful to others in the field. Debate and discussion is facilitated by a comment function, available on all postings of innovations, resources, and many other content types across the website. Members can receive a newsletter, write a blog, contribute a podcast, or host a webinar or 'Ask the Expert' session.

- **Who can apply**: Anyone can sign up for free to become a member and post an organizational profile. While the posting of innovations is meant to be inclusive for everyone to learn from each other's experiences, those who want to submit their specific innovation use an online form to the MHIN team who apply some selection criteria based on relevance to the site audience and quality of evidence, and will respond with next steps.
- **Applicable regions**: Open to all regions
- **Website**: https://www.mhinnovation.net/. Also see: https://unitedgmh.org/

The Global Digital Health Network

- **What it is**: Established in 2009 by global health organizations for global health organizations, the Network is a 3900+ person-strong networking forum with members from 117 countries to share information, engage with the broader community, and provide leadership in digital health for global public health. Events and communications provide a space to engage with and learn from each other in the health field as well as with digital implementers in nonhealth sectors, to strengthen digital technologies. The network "envisions a world where technical innovation supports improved health and contributes to better quality, accessibility and sustainability of health services and health outcomes, particularly for underserved populations in low resource environments."
- **What you get**: The network hosts open meetings with presentations by/for members on digital health topics as a global forum to share experiences, accomplishments, resources and lessons learned; facilitate joint activities and partnerships among members locally, regionally, and globally, across sectors; build digital health skills and capacity; engage decision-makers and stakeholders working in international health and development at all levels to mobilize planning and resources for digital health and advocate for digital health; and leverage members' strengths and support to organize digital health efforts. The network holds monthly meeting and hosts an annual Global Digital Health Forum in December, connecting government stakeholders, digital health developers, researchers, donors, implementers, and field experts from across the globe. Recordings and presentations can be viewed and downloaded from the website, for example, a monthly meeting where experts from PATH and Cooper/Smith shared learnings collected and synthesized from five countries—Burkina Faso, Ethiopia, Malawi, South Africa, and

Tanzania—using a model for digital transformation for data use built upon the World Health Organization (WHO) and International Telecommunication Union (ITU) eHealth Strategy Building Blocks. Member can share their digital health transformation story on their website through a guest blog they will adapt it to their readership.

- **Who can apply**: Any organization involved in the healthcare industry
- **Applicable regions**: All are welcome.
- **Website**: https://www.globaldigitalhealthnetwork.org/

Healthcare Information and Management Systems Society (HIMSS)

- **What it is**: HIMSS is a nonprofit member-based society for 60+ years with operations in North America, Europe, the United Kingdom, the Middle East, and Asia-Pacific, "committed to reforming the global health ecosystem through the power of information and technology." It has a job center and publishes books on healthcare management, informatics, artificial intelligence, patient experience, and information management and systems.
- **What you get**: Different access is based on membership level and fee. The society offers connection to the global health ecosystem, events, newsletters, expertise in health innovation, public policy, workforce development, research, and digital health transformation to advise leaders, stakeholders, and influencers across the global health ecosystem on best practices. Members have access to a membership directory that can be used for individual, personal, and confidential reference purposes only.
- **Who can apply**: A fee-based membership in various categories of affiliations or partnerships (corporate organizational, nonprofits) with different perks and access, e.g., regular (US$199), chapter (US$39), student (US$30), or digital (US$39). Members include more than 110,000 individuals, 480 provider organizations, 470 nonprofit partners, and 650 health services organizations, including healthcare providers, payers, governments, start-ups, life sciences, and other health services stakeholders.
- **Applicable regions**: All are welcome. Headquartered in Chicago, Illinois, USA, with a policy center in Washington, DC, and global offices in Germany and Singapore.
- **Website**: https://www.himss.org/

Accenture

- **What it is**: This global professional services company has multifaceted capabilities in digital health, cloud systems, and healthcare security. The company provides strategy consulting, technology services, and operations services for national healthcare organizations and health start-ups. The organization also hosts a global health challenge—the Accenture Healthnet Innovation Challenge—that highlights innovative health start-ups in the aforementioned fields that serve North America.

- **What you get**: With annual revenue of >$50bn in 2021, the company leverages its leading market presence to invest in cutting-edge collaborative tools. The company's digital health platform specializes in establishing innovative instruments to enhance the automation of repetitive, albeit essential, clinical tasks. This directive is vital to decreasing healthcare bottlenecks and allowing clinicians to spend more time providing quality treatment to improve patient outcomes.
- **Who can apply**: Any organization involved in the healthcare industry
- **Partnership cost**: Given the company's vast healthcare capabilities, partnership cost is highly dependent on the collaboration's structure—ranging from pro bono services provided to nonprofits to collaborative business ventures valued at several million dollars.
- **When**: Applications for the Accenture HealthTech Innovation Challenge are due in June while the event itself is held in September (be sure to check dates)
- **Applicable regions**: All regions
- **Website**: www.accenture.com/us-en
- **Where to connect**: Partnership inquiries should be emailed to investor.relations@accenture.com

Health Hub Vienna

- **What it is**: Health Hub Vienna is a digital health platform that joins pharmaceutical companies, medical device manufacturers, private and public insurance companies, and healthcare start-ups to develop innovative solutions to the world's most pressing health system issues.
- **What you get**: The platform is managed by a prominent high-tech incubator with a 15-year track record of successfully developing and connecting start-ups in the health sciences. The platform has specialized in developing technological-based health innovations focused on improving patient outcomes.
- **Who can apply**: Any organization involved in the healthcare industry
- **Partnership cost**: Dependent on resources invested in facilitating a successful collaborative merge—typically 4% of the business agreement price
- **Applicable regions**: All regions
- **Website**: www.HealthHubVienna.at
- **Where to connect**: Partnership inquiries should be submitted through the online contact form found at www.HealthHubVienna.at/contact

Social Innovation in Health Initiative (SIHI)

- **What it is**: SIHI is a global network of passionate individuals, organizations, and institutions advocating for social innovation in health and advancing research in social innovation. The collaborative is actively engaged in implementing partnerships that are

engaged in conducting country research and building the capacity for social innovation within their countries.
- **Who can apply**: Any organization with benefits to the overarching goal of social innovations in health.
- **Partnership cost**: Free
- **What you get**: The organization is engaged with the World Health Organization and other international institutions; therefore, members are offered access to a global network of individuals tied to established health-centered entities.
- **Applicable regions**: All regions
- **Website**: www.SocialInnovationInHealth.org
- **Where to connect**: Current partnership contacts can be found online at www.SocialInnovationInHealth.org/about/get-in-touch

eMedEvents Network

- **What it is**: This global digital networking hub for healthcare professionals offers medical practitioners the opportunity to connect with contemporaries in their fields while searching up-to-date educational activities in one convenient spot.
- **Who can apply**: Any individual working or interested in the healthcare industry
- **Partnership cost**: Free
- **What you get**: The organization maintains a strong network of approximately 2.8 million healthcare professionals from over 150 countries worldwide. Community users are able to search for professionals by entering a location, conference title, or specific name and connect with these individuals directly through the website's various discussion boards or privately through a private online messaging system.
- **Applicable regions**: All regions
- **Where to connect**: www.eMedEvents.com/healthcare-speakers/network-with-peers-and-organizers

(7) Virtual communities

Virtual digital health communities allow professionals to exchange information and have valuable discussions on industry-specific current events. These forums allow users a flexible space to solicit timely feedback on proposals; collaborate with other leaders in the field to brainstorm solutions to complicated topics; and circumvent geographical, cultural, and social barriers. In the rapidly progressing digital health and technology field, it is particularly beneficial for start-ups to leverage the free, tailored commentary provided in online forum threads. Furthermore, participating in chat groups is intended to provide users a feeling of inclusion and camaraderie with other professionals within their field.

Reddit (HealthIT)

- **What it is**: Reddit is a vast collection of online forums composed of thousands of subcommunities, known as "subreddits." Each subreddit is dedicated to a specific topic or industry, such as digital health and technology. Registered members submit content to the site in the form of posts, which are voted up or down by other community members. Popularity within a subcommunity is directly proportional to the number of netted up-votes compared to down-votes.
- **What you get**: Since year-end 2021, Reddit has more than 430 million monthly active users, offers over 100 thousand subcommunities, and is ranked as the most popular online forum worldwide. The platform offers numerous subcommunities relevant to digital health and technology content, including "HealthIT," which currently has 19.4k active members.
- **Membership fees**: Free
- **Applicable regions**: All regions
- **Where to connect**: www.reddit.com/r/healthIT

Quora (Healthcare information technology)

- **What it is**: Quora is a global virtual community that solicits free answers from industry professionals. The platform's voting system utilizes a unique public vetting system to elevate well-liked answers to the top of subcommittee pages, providing users the advantage of avoiding the hassle of sorting through lengthy answer lists to find the most popular choices. Members of this forum can create publicly accessed user profiles that include credibility metrics such as level of education and attended universities, listed areas of expertise, and current job position.
- **What you get**: According to CNN Business, since year-end 2021, Quora has had more than 300 million monthly active users and is ranked as the most popular online forum in the United States, with 35% of the nation's adult population registered with active accounts. The platform offers numerous subcommunities relevant to digital health topics, including "Healthcare Information Technology," which currently has 55k followers.
- **Membership fees**: Free
- **Applicable regions**: All regions
- **Where to connect**: www.quora.com/topic/healthcare-information-technology

Lemmy

- **What it is**: Lemmy is a self-hosted virtual discussion platform that is not controlled by any hosting company; this means that there is no advertising, tracking, or covert algorithms collecting marketing data on its users. Similar to other popular virtual forums, content is organized into categorized subcommunities.

- **What you get**: Although Lemmy is a relatively new platform with only 2 years of operation, according to the *Financial Times,* Lemmy is a rapidly growing virtual forum that attracts many technologically inclined users worldwide. The organization's website invites anyone interested in participating in intellectual debates regarding the latest tech innovations to join. The community also allows members to pose questions on virtual bulletin boards, share news articles, and download documents from subcommittee postings.
- **Membership fees**: Free
- **Applicable regions**: All regions
- **Where to connect**: www.join-lemmy.org

(8) Contests

Recently, there has been an upsurge in the number of digital healthcare and technology competitions worldwide. Rather than seeking solutions to systemic health issues in-house, companies and organizations are increasingly creating contests to generate outside submissions for innovative ideas. Even market leaders utilize contests as a passive way to bring new development opportunities to their attention. Theoretical (system enhancement) and technical (product development) competitions provide a platform for individuals and companies alike to showcase their novel inventions and ideas.

Participating in a competition can provide considerable inherent value to a start-up, including soliciting valuable feedback from industry professionals, enhancing product visibility, and solidifying the product's narrative and brand. The primary benefit to any competition is to place within the "accolade zone"—the placement needed to achieve some kind of prize from the competition. However, any professional exposure can be valuable. In respected digital health competitions, top-performing teams are granted rewards that are often invaluable to kick-start product development. Such prizes include cash awards, media publicity, and renowned distinction from direct competitors, all of which can have value.

Given the overlap between awards and organizations, besides the following listed, see also MedTech Innovator and Accenture under the above topic of *Incubators and Accelerators* and *Collaboratives*, respectively, for additional related contests.

American Hospital Association Innovation Challenge

- **What it is**: This global pitch competition is focused on finding sustainable and scalable solutions that employ technology to combat social determinants of health. Pitch proposals range from mobile apps to new data collection tools. For example, a recent winner was

NewYork-Presbyterian—an American-based academic healthcare delivery system organization devoted to providing high-quality and compassionate care to patients worldwide—for their idea to leverage digital health tools to screen for and address social determinants of health (SDOH)-related risks for mothers and babies during the postpartum period.

- **What you get**: The top three proposals are awarded at these monetary levels: US$100k, US$25k, and US$15k
- **Who can apply**: Anyone
- **When**: Yearly challenge topics were traditionally announced in January, with proposals due by mid-February and final-round pitches presented in July. Due to the COVID-19 pandemic, the competition was postponed in 2020.
- **Applicable regions**: All regions, competition held in the USA
- **Sponsors**: First American Healthcare Finance, an American healthcare banking institution
- **Where to apply**: www.aha.org

PITCH

- **What it is**: As its name suggests, this is a global pitch competition and is meant to highlight the world's most promising start-ups. While the competition is not solely limited to digital healthcare start-ups, previous victors have won with digital health proposals. For example, GlucoActive, a Polish start-up that created a noninvasive wearable device for measuring blood glucose concentration, won first place in Collision's PITCH competition in 2019. The company used spectrophotometric methods to diagnose and monitor diabetes treatment plans in adolescents and adults.
- **What you get**: Winners are bestowed with a global distinction from Collision (a global annual technology conference) and international media coverage, and are showcased in front of a panel of respected venture capitalists.
- **Who can apply**: Any start-up
- **When**: Proposals are usually due by the beginning of February, with the event held in late June (be sure to check dates).
- **Applicable regions**: All regions, with the competition held in the USA
- **Sponsor**: Collision, an American-based annual tech conference
- **Where to apply**: www.CollisionConf.com/startups/pitch

Hello Tomorrow Global Challenge

- **What it is**: This entrepreneurial science and technology competition features 12 industry tracks (e.g., food, energy, AI), including Digital Health and Global Health contest categories. For example, RxAll, a Nigerian start-up tackling the global problem of counterfeit drugs, received the Global Challenge Digital Health Prize from BNP Paribas, a global partner of Hello Tomorrow, for their innovative digital platform that enables anyone to test the quality of pharmaceuticals using their proprietary nanoscanner technology and mobile app. Also, X-Therma Inc., a USA-based company, was awarded the "Global Health" Prize from G4A Bayer for making novel and highly advanced chemical technologies to solve the long-standing problem of cold storage and transportation of biological cells, tissues, and organs.

- **What you get**: Winners get the chance to pitch in front of hundreds of industry experts, interview with international media, and gain access to exclusive industry summits. Teams who place within the accolade zone win the following cash prizes: €100k (1st place), €30k (2nd place), and €20k (3rd place).
- **Who can apply**: The worldwide start-up competition is for early stage science and deeptech entrepreneurs. In 2019, 4500 applications came from 119 countries, indicating that the development of new technologies is more widespread than ever before.
- **When**: Proposals are due in late September, with the event held in March
- **Applicable regions**: All regions, competition held in France
- **Sponsors**: Hello Tomorrow Inc.—a global think-tank dedicated to technology
- **Where to apply**: www.hello-tomorrow.org/start-ups

(9) Awards

This section overlaps with contests mentioned before, but are mainly listed here as their title uses the word "award." Some awards are given without intense competitive submissions; others require a submission. Healthcare technology awards are given to those who have demonstrated superior proficiency and impact on the field. While many awards in this sector are designed to bestow recognition and prestige, some awards provide additional prizes such as direct financial outlays, grants, and partnership opportunities. In addition, a few awards also provide top performers consultation privileges with industry leaders.

HealthTech Innovation Awards

- **What it is**: This prominent global digital health tech award recognizes the efforts of various organizations, associations, and service providers in the field of healthcare services, and is meant to honor "pioneers and visionaries who are going the extra mile to transform the healthcare industry and making this world a better place to live in."
- **What you get**: Winners are recognized by a broad professional audience and are allowed exclusive rights to the award logo to be used on marketing material. Furthermore, all winners are listed in perpetuity on the company's website.

- **Who can apply**: Self-or-other nominations by hospitals, technologies, healthcare start-ups (e.g., platforms, wearable devices, marketplaces, apps).
- **When**: Applications are accepted until May. Winners are announced in November
- **Applicable regions**: All regions
- **Sponsors**: Health Tech Digital—a renowned digital health news agency
- **Where to apply**: www.BiiHealthTech.com/healthtech-innovation-awards/

MedTech Breakthrough Awards

- **What it is**: Considered one of the health-tech industry's foremost technology award programs, this recognition celebrates and champions innovation-focused entrepreneurship. MedTech Breakthrough Awards places distinction on health technology companies that demonstrate the greatest potential to change healthcare systems and delivery. Relevant digital health award categories include Medical Data, Internet Healthcare, and Mobile Communications and Tele-Health.
- **What you get**: National recognition in one of the world's leading digital health geo-hubs. Winners are also showcased to potential seed-funders and venture capitalists.
- **Who can apply**: Any individual, company, or organization involved in producing digital health and medical technology products and services
- **When**: Applications open in mid-November and close in early February each year
- **Applicable regions**: USA
- **Sponsor**: Tech Breakthrough—a leading market intelligence platform focused on global technology innovation and leadership
- **Where to apply**: www.MedTechBreakthrough.com/nominate

Health Tech Digital Awards

- **What it is**: Industry pioneers are recognized as champions in several digital healthcare categories, including Best Healthcare Technology Solution of the Year, Best Healthcare Information System, and, as of March 2021, Best COVID-19 Telecommunications Solution.
- **What you get**: Winners receive a customized trophy and are allowed marketing rights to the digital champion badge. A list of all winners is circulated in the corporate newsletter and posted on the news agency's popular social media channels.
- **Who can apply**: Winners are chosen by Health Tech Advisory Board judges based on overall industry impact and originality. The cost to submit an award entry is £150.00+ VAT. Categories are extensive, including solutions for training, medical imaging, mhealth applications, information systems; best use of AI robotics and blockchain technology;

men's, women's, aging, and mental health technology solutions; and COVID-19 health solutions.

- **When**: Award entry deadline is typically annually in May (it was extended to July in 2022)
- **Applicable regions**: All regions, with a particular emphasis on Europe, Middle-East, and Africa
- **Sponsors**: Health Tech Digital, a renowned digital health news agency
- **Where to apply**: www.HealthTechDigital.com

The Cartier Women's Initiative Science & Technology Pioneer Award

- **What it is**: Launched in 2021, this new award recognizes innovative disruptive solutions using unique, protected, or hard-to-reproduce technological or scientific advances that require significant research and development (R&D), a long commercialization cycle, large funding, technology risk, and complexity. The award is part of an entrepreneurship program driving change by empowering women entrepreneurs from any sector (or country) that has social or environmental impact, which includes but is not exclusive to health.
- **What you get**: Winners receive financial capital support of $30k, $60k, or $100k for first, second, or third place, respectively. Winners are granted one-on-one training in financial analysis, impact measurement, leadership communication, and can take advantage of executive coaching by the Coaching Fellowship, a nonprofit organization dedicated to developing women leaders in social and environmental change, Besides prize money, award recipients also receive a scholarship to a six-day INSEAD Social Entrepreneurship Executive Education Program and benefit from social capital support in the form of international exposure and media visibility and access to a community of 500+ experts and peers.
- **Who can apply**: The application is open to any women-owned and women-run business from any country and sector within science and technology, including healthcare and digital health. Other inclusion criteria include that the applicant must (1) exemplify English proficiency (both verbal and written), (2) commit to participating in the fellowship program from January to December in the year following the application, (3) operate as a for-profit entity, (4) be in the initial phase of its development or growth, (5) demonstrate the potential to have a strong positive impact in its industry, (6) have a woman as the chief applicant who holds an executive role in the company (i.e., CEO, COO, Managing Director).
- **When**: There were no open calls in 2022
- **Applicable regions**: All regions

- **Sponsors**: The Cartier Women's Initiative, an annual international entrepreneurship program geared toward empowering female impact entrepreneurs, with Cartier of course being the French luxury jewelry and accessories company. The company partnered with Expo 2020 Dubai to present the Women's Pavilion which showcased the contributions of women and their impact on society.
- **Where to apply**: www.cwi-portal.com/submit

The Global Health Impact Awards

- **What it is**: This is an award hosted by the Washington Global Health Alliance (WGHA), a fee-based membership organization headquartered in the state of Washington in the USA that offers a typical web platform, events, and newsletter, whereby the award celebrates people and organizations working to improve health equity and strengthen health systems around the world. It is included in this chapter as a typical example of how organizations and collectives give out awards at their fundraisers. The award was won in 2020 by VillageReach, an innovation profiled in this volume.
- **What you get**: Recognition at an award ceremony, as a Rising Star, Luminary Leader or Organizational Impact award, and an opportunity to publicize your innovation or project being recognized by a health-related organization.
- **Who can apply**: Any organization nominee that represents a diversity of race, culture, ethnicity, nationality, sexual orientation, gender identity and expression, discipline, and lived experience.
- **Applicable regions**: All regions
- **Sponsor**: Hosted by the Washington Global Health Alliance, with sponsors including the Bill and Melinda Gates Foundation.
- **Where to apply**: https://www.wghalliance.org/event/global-health-impact-awards/

(10) Academic study programs

An increasing number of academic institutions of higher learning and other levels, as well as schools of specialized training, are including courses or full degree studies in technology and health. These can be stand-alone or included in broader curricula. As digital health increasingly adds value in all aspects of prevention, diagnosis, and therapy, and research, and is a cross-cutting issue in many fields including pharmacology, biology, medicine, psychology, sociology, the involvement of academia is key to help study, inform, catalyze, and evolve the field in a critical and evidence-based way. Most of these programs emphasize innovation. Recognized programs in established educational institutions include, but are not limited to, the following. *In the U.S.*: The Johns Hopkins Center for Health Tech & Innovation in Baltimore, Maryland; The Center for Digital Health at the

Icahn Institute in the Icahn School of Medicine at Mount Sinai in New York and Center for Digital Health Innovation at UCSF (University of California at San Francisco). *In the UK*: The University College London Institute of Digital Health. *In the Netherlands*: Radboud University Medical Center REshape Center. An innumerable number of programs are available through a search on the internet to select a program based on factors such as location, cost, and targeted subject matter, for example at: https://digital.health/academic-digital-health-programs.

Conclusion and recommendations

This catalog enables interested stakeholders to identify and connect to relevant and trusted resources in the healthcare and technology field. Utilizing this chapter offers a way to avoid the plethora of potentially misrepresented material targeted to online users. Hundreds of resources for global health innovation and technology systems were scrutinized by the authors of this chapter for their relevance, accessibility, usefulness, and credibility. Some are well-known to the authors and coeditors of this volume. However, given the influx nature of this field, new resources will undoubtedly become available after this volume's publication. Also, the editors do not claim that this list is comprehensive of all resources possible. Therefore, it is recommended that users of this list utilize the information and referrals presented here as a guide for further exploration and inquiry and springboard their continued searches from professional entities affiliated to the listed resources. Due to the overlap in the presented categories' content, offerings, and structure, it is also recommended that users leverage multiple resource categories to derive the greatest value.

To ensure this list remains comprehensive and up-to-date in future volumes, we welcome readers to email any suggested additions or amendments to: healthinnovationbook@gmail.com

CHAPTER 105

Putting the pieces of the puzzle together: Health technology and resilient health systems now and for the future

Pradeep Kakkattil[a,b] **and Judy Kuriansky**[b,c,d,e,f,g,h,i]

[a]*Health Innovation and Investment Exchange (HIEx), Geneva, Switzerland,* [b]*Global Council SDG3 (Health), "SDGs in Action" Initiative, Dubai, United Arab Emirates,* [c]*Department of Clinical Psychology, Teachers College, Columbia University, New York, NY, United States,* [d]*International Association of Applied Psychology, Indianapolis, IN, United States,* [e]*Office of Ambassador Sidique Wai, Embassy of the Republic of Sierra Leone to the United States, Washington, DC, United States,* [f]*United Nations Group of Friends of Mental Health and Well-being, New York, NY, United States,* [g]*Mission of the Republic of Sierra Leone to the United Nations, New York, NY, United States,* [h]*Global Leadership Council, eMental Health International Collaborative (eMHIC), Auckland, New Zealand,* [i]*International Women in Radio and TV, Quezon City, Philippines*

When we conceptualized this volume about leveraging innovations and technology for resilient health systems, the world was a very different place. The second wave of the COVID-19 pandemic was at its peak and world economies and markets had shut down. Hundreds of people were dying every day. No country on the planet was untouched by the virus and the numbers of people getting infected with COVID-19 were spiraling out of control. Panic and helplessness were pervasive.

It didn't help that social media was filled with conflicting information, as was the political debate on mainstream television. The COVID-19 vaccine was not available. Political leaders were trying their best but had no clue about the right thing to do in responding to a rapidly growing pandemic. While lockdowns were promoted as the magic bullet that would "flatten the curve in two weeks," there was no perceptible stop to the spread of the virus, and more people—especially migrants, the poor, physically compromised and marginalized populations, were paying a heavy price with their lives.

The virus dramatically exposed the inadequacy and fragility of our health systems worldwide—in developed and developing countries alike—but also the fissures in society. For the first time, the world saw how fragile and ill-equipped our health infrastructure is—even in the wealthiest countries—and also how divided our politics were, and how deadly the combination of a virus and divisive politics could be. The world was not, and is still not, prepared for a pandemic of any sort.

Resilient Health. https://doi.org/10.1016/B978-0-443-18529-8.00104-9

Looking back, while we still see the challenges wrought by the pandemic, we also see significant progress. Hopelessness and fear has moved to hope and resilience, as the COVID-19 crisis brought the best of human ingenuity and resourcefulness to the fore. The crisis has clearly demonstrated that when pushed to the edge, scientists, healthcare workers, and ordinary citizens working with each other and with governments are capable of innovation and confronting any challenge. Over the last few years, we editors of this volume have had the privilege of working closely with hundreds of forward-thinkers and innovators—including experts in public health, frontline care providers, political leaders, academic leaders, community workers, investors, and people from all fields, disciplines, and walks of life who have taken risks, challenged the status quo, and developed solutions that have not only saved lives but have left an indelible mark in building a more resilient health ecosystem that is absolutely fundamental for facing future pandemics and securing the future of mankind. The COVID-19 pandemic has no doubt been a major agent of transformation in the health sector—and beyond—and that has stimulated brilliant minds to stretch their skills and caused some "ordinary" people to do extraordinary things—all of which continue to inspire us all.

However, such extended creative effort cannot be limited to times of crisis alone. What we saw during the COVID-19 crisis is not new—a daily struggle is normal for half the world every day. One out of two people globally don't have access to essential healthcare. The pandemic escalated this disastrous situation; the limited progress made over the last decades in scaling access to healthcare has been lost over the last years of the pandemic as several countries saw their fragile health infrastructure and precious capacities being lost during COVID-19. In fact, World Health Organization (WHO) estimates predict that by 2030, at least 5 billion people will not have access to essential healthcare—unless we do something radical today.

At the time this volume was submitted, the deadliest part of the disease has abated, but variations appeared, keeping us aware that the threat is not over. Fears prevail of another infection to come; perhaps, for example, a repeat of the COVID-19 crisis with the emerging monkeypox. Also, the world is in a general global crisis. Governments are in unrest and wars persist around the world, not the least of which in Ukraine and the Middle East. Economies are still not restored. Education is at least two years behind for students everywhere. Health systems are still inadequate, which need to be stepped up to recover and to be prepared for the future. Fortunately, the governments of the world are paying attention, by holding high-level meetings at the 2023 General Assembly to address issues of both Universal Health Coverage, and Pandemic Prevention, Preparedness and Response (PPPR). Now, more than ever, we need to be aware of, and build on, health solutions in the field of technology and social innovation.

While the sense of urgency was clear during the peak of the COVID-19 crisis, today, as the numbers of infections and deaths, notwithstanding ongoing variations of the virus, have declined, the slogan uttered by many politicians and other stakeholders throughout the times of crisis—namely, to "build back better"—is in danger of potentially ending up being exactly that, a slogan.

For that reason, we are pleased to offer the collection of perspectives and solutions in this volume to all the many stakeholders who can benefit and join in collective action to indeed "build back better, or better yet, build *forward* better." The brilliant contributions in this volume cover so many aspects of health in so many countries, giving us hope. There is so much innovation going on for potential benefit to all populations—which is why our original plan for 36 chapters ballooned into 105. And, there could have been many more contributions, as every day, interesting and valuable new developments emerge.

So, what does technology and innovation offer in building a resilient health system? From the chapters in this book and our conversations with experts across sectors, it is patently clear that the world today has the tools necessary to rapidly expand access to healthcare and to build a resilient system that is prepared for the next pandemic—which experts say can be imminent. This chapter presents important lessons learned and pathways to the future.

Infrastructure

A fundamental challenge that health systems struggle with is poor and outdated infrastructure. How can systems expand access on a scale and speed that is necessary to deliver universal health coverage? Infrastructure—including needed facilities community health centers, clinics, hospitals, labs, equipment, staff, and technologies —is one of the biggest barriers to scale. The traditional brick-and-mortar model of healthcare delivery is not designed to scale at the speed that is necessary to reach the 2030 Global Goals set by the ambitious United Nations 2030 Agenda for Sustainable Development and the Political Declaration for Universal Health Coverage. Moreover, much of available infrastructure is often concentrated in urban centers with vast swathes of rural areas almost completely devoid of any service.

This is where the rapid expansion of mobile telephony and data services globally gives hope—with 93% of the world having access to a mobile broadband network, including nearly 80% in sub-Saharan Africa (International Telecom Union, 2020). It is possible to conceive of building a digital infrastructure for health that leverages mobile infrastructure, and can rapidly expand access to healthcare even in the most remote parts of the world. While cost of data is still prohibitive, particularly in Africa, competition—from low-orbit satellites to the Africa 2 cable that should be in operation shortly—these costs will rapidly fall. This opens up the possibility of leveraging digital health—from use of telemedicine to digital diagnosis and improved real-time data—that with limited investments can truly transform healthcare access. There are several chapters in this volume that show the dramatic potential of digital health. For example, the chapter describing the innovation of Portal Telemedicina in Brazil clearly demonstrates what telemedicine can deliver when adding a layer of artificial intelligence, that has resulted in a rapid reduction of mortality rates by as much as 40%! Similarly, innovations like Living Goods reach out to community health workers in-person, with technology tools in hand, that allows collecting real-time community-led data that enables timely action. These examples prove the immense potential of innovation when deployed right. Indeed, the COVID-19 pandemic triggered rapid adoption of several of these approaches, which fortunately will continue to grow even in the "post"-COVID-19 era.

While distributed healthcare will be a major trend in the coming years, both in developed and developing economies, one major challenge that the developing world faces in scaling digital health is energy security for health infrastructure. According to the WHO, between 50% and 70% of health facilities in low- and middle-income countries do not have stable electricity supply. This not only means inability to leverage digital health but also limits the basics: lack of electricity means no running water, no refrigeration, and no lights at night. Communities have had to adapt; surely, some of us have seen images of babies being delivered using light from mobile phones. One of the biggest challenges faced in delivery of vaccines (including for COVID-19) is lack of a robust cold chain including refrigeration for storing vaccines and essential medicines.

Solar panels offer hope. These have the potential to not only address the challenges outlined in the chapter in this volume on solarizing healthcare, but also will help replace diesel generators currently in use, thereby significantly contributing to reducing pollution (Osho & Jageet, 2024). Powering up health facilities also has huge potential to improve real-time information on emerging outbreaks, that could help halt epidemics before they escalate out of control. Such a solution can also foster better collaboration across healthcare providers and better monitoring of health supplies and commodities.

A key lesson learned after looking at the various challenges in scaling infrastructure for health for all is the need for a multisectoral approach to building resilient systems for health. This means "a whole of society" approach, with governments,the public and private sectors, civil society, academia, media, and all stakeholders working together. The value of the private sector as a "partner" in the process is evident in several chapters of this volume. Concomitantly, the ongoing lack of trust between the public and private sector has severely hampered the potential of collaboration and will need to be leveraged better to reach sustainable development goals. Also, donors must step up to the plate. As chapters in this volume show, new financing models are evolving; money must flow, not shrink (as it has been), and must be seen, as in the chapter about mental health, as "an investment not an expenditure" (Kline et al., 2024). Donors also need to overcome their tendency to fund pilots that don't scale, and be more strategic in their investments.

Human resources

While significant progress has been made in addressing the inadequate number of health workers per population, their distribution continues to be a major challenge. Africa, which accounts for nearly a quarter of the global disease burden, accounts for less than 4% of the global health workforce—with a mere 2.9 medical doctors per 10,000 Africans compared to 36.6 in Europe (The 2022 update, Global Health Workforce Statistics, 2022). Moreover, the COVID-19 pandemic has decimated the health workforce—with an estimated 80,000–180,000 dying of COVID-19 infection between January 2020 and May 2021 (World Health Organization, 2021). Globally, we see a growing gap in health workforce, as healthcare workers in several countries indicate that they would like to leave the health sector because of the risks, fatigue, and burnout post-COVID-19.

How can technology and innovation address this widening gap? While there is need for a significant ramp up in investments to build human resources for health, several approaches described in this volume provide insights into how available resources can be amplified and how technologies could help maximize potential. For example, Proximie is an outstanding example of how surgical capacities can be leveraged globally while also rapidly strengthening local capacities. Similarly, we have already seen the huge potential that artificial intelligence brings to diagnostics and improved service delivery. One of the highlights of our conversations during the preparation of this volume was with a former Minister of Health who had managed to introduce AI for the diagnosis of tuberculosis, which had significantly reduced the workload on the radiologists. Similarly, Living Goods shows how to deploy health workers into the community to meet people "where they are" and provide these workers with digital job tools which improve their efficiency and productive patient outreach.

The COVID-19 pandemic also highlighted the role of traditional medicine and capacities in the traditional health sector. In the early days of the pandemic, when modern medicine provided little relief for the virus, a proportion of the population reverted back to traditional therapies and self-care. Importantly, the principle of traditional medicine which addresses body, mind, and spirit, resonated during the COVID-19 lockdowns. The chapters in this volume on Traditional Medicine describe the changing nature of traditional medicine as it adopts new technologies and innovation. While researching this volume, we have been impressed with the rapid growth and adoption of traditional medicine especially in India, with an independent Ministry of Ayush (traditional medicine) established that has truly led to a revival of traditional medicine in India. This is an important lesson for other countries that could find inspiration from the Ayush Ministry's efforts to standardize traditional practices with scientific rigor.

The efforts to date in building capacities have been suboptimal. Unless the global community invests in addressing the WHO-estimated shortfall of 18 million health workers by 2030, mostly in low- and lower-middle income countries, the world will not be secure. On the flip side, the pandemic emergency became an opportunity for countries to prioritize investments in this area; healthcare will continue to be one of the fastest growing sectors for employment, particularly for young women.

Lack of investments in building local manufacturing capacities, particularly in Africa, is a major challenge—much of it due to lack of trained manpower. As of 2019, 70%–90% of the drugs consumed in sub-Saharan Africa were imported. Moreover, Africa produces less than

0.1% of the world's vaccines while consuming nearly 25% of the global vaccine production (McKinsey, 2019). There is need for a quantum leap in investments and policies that enable the African continent to be less dependent on imports for even their basic drugs and diagnostics. The COVID-19 crisis has brought attention of the African Union leadership to the dependency and the risks involved. While there is commitment—including the establishment of the African Medicines Agency—this attention must be sustained in the post-COVID-19 era. In fact, Africa has an opportunity to enter the digital health innovation space by providing the right incentives for the nascent innovation ecosystems emerging on the continent.

The disconnect

One of the questions that we editors of this volume asked as we were developing this volume was: "With all the technologies availiable that can be transformative, why don't countries leverage them to deliver access to healthcare?" Ministers of health, and health leaders from low- and middle-income countries that we spoke to across regions consistently came back with one answer to this question: Whose problem are these innovations solving?

This reveals the classic clashing parallel worlds—where innovators were developing solutions to problems that they perceived existed and were disappointed that the public sector was not taking on their solutions, while the public sector felt that innovators were not understanding their challenges and their problems were not being addressed. While it's true that many of the technologies being developed are applicable for the North American and European market—driven by financial incentives—it's refreshing to see several innovators from the Global South coming up with solutions that are cutting edge, often low cost, robust, and solving a real problem within their context. We are delighted to showcase these in chapters in this volume.

Over the course of our work over years in this space, and then forming even deeper relationships when developing this volume, we are seeing some exceptionally bright start-ups and health innovators in developing countries around the world—for example, in India, Indonesia, the Middle East, Brazil, Kenya, Ghana, and Nigeria—all of whom have global ambitions and are giving innovators in California's Silicon Valley (the traditional major hub of technology innovation) some real competition. While some of these seek to solve local problems on scale, with aims to reach the more lucrative developed world markets, it is disappointing to note that some innovators in the Global North do not have a dual market strategy. This is a major concern—as developments in technologies and innovations will widen inequities between the Global North and South unless we build in incentives and invest in capacities in the Global South. For example, the COVID-19 crisis uncovered stark inequities in vaccine access and distribution. The blueprint for this is laid out in the UN Agenda 2030 which calls for technology transfer to close the disparity gap. But, inequities are likely to continue, and the gulf will certainly widen, as new technologies—including new techniques, therapies, and genomics—will emerge that radically change longevity and quality of life. One need only know the difference that has existed over three decades between the country with the highest life expectancy, namely, Hong Kong with 85 years, and the country with the lowest, namely, Chad at 54 years. Health, while not the only contributor to life expectancy, is a large factor. This gap is unacceptable.

Another lesson that the siloed world of health innovations needs to learn is the importance of listening. Healthcare providers have clear pain points that they would like innovators to address. While innovators are committed to solving the challenges, they can be disconnected from the reality on the ground experienced by the health providers and the people. Most multilaterals and health actors have failed to play a bridging role within the health innovation ecosystem. This multilateralism and working together are much needed at this time and need to be a priority at local and national levels.

Fragmentation and scale

Another major lesson learned is that for innovations to have impact, they need to be implemented on scale. Countries have in common that the health landscape has hundreds of innovations, with many of them based on apps funded by different donors—and in several cases, by different arms of the same donor—that have not scaled. Even where some demonstrate promise, financing has not been forthcoming, as donors are often comfortable investing in pilots but risk-averse on scaling. Scaling is an art and a science, a fine balance that the chapter in this volume on scaling models captures so well.

The fragmentation of the innovation landscape is a serious challenge in most of the developing world where the public sector is often the single largest market for healthcare. While health innovation strategies are being outlined in some of these countries, most have little or no funding, often remaining as plans that never see the day of light. One development that we are heartened to see is the development of WHO's digital health strategy—a summary of which is in this volume. Also, the African Union is finalizing a continental digital health strategy which is comprehensive and makes a clear attempt to consolidate gains and establish common frameworks for adoption of digital health on the continent.

International organizations and normative agencies have failed to help countries develop a pragmatic approach to adoption of innovation in health systems. It is disappointing that even today, there is lack of information on maturity of specific health innovations in different settings that are gold standard "best practices" that have proven to be effective, valid, and reliable. Even for primary health, where innovation and technologies have been tested in various settings, there is not a simple toolbox of robust and proven innovations that countries could choose from and adapt for their local contexts. Every country has been left to reinvent the wheel with little support or guidance, which leads to further fragmentation. The chapter in this volume on the Health Innovation Exchange (HIEx) platform outlines how the HIEx attempts to solve this problem, as a platform to aggregate these solutions and bridge partners for health impact (Kakkattil et al., 2024).

Also, while technologies itself are important, even more critical is the culture of innovation within health institutions. Ministries of health, and actors in the health sector overall, who are often risk-averse, have a culture that consistently avoids challenges and is comfortable with the status quo. International attention must focus on investing in building an innovation culture within existing rigid structures in any entity and within government ministries of health, to break out of this mold.

The role of communities and trust

The COVID-19 narrative was initially dominated by the development of a vaccine—targeted for within 12 months of the outbreak, and then, with millions being vaccinated to prevent further spread. That goal alone was a spectacular testament to human ingenuity in times of crisis to develop the vaccine speedily and then pool collective efforts to get it to people. However, when we look at vaccine hesitancy that also emerged across continents (people refusing to take the vaccine), it is clear that any innovation that does not engage citizenry in its design and scaling will remain suboptimal in its adoption.

COVID-19 is the first global pandemic that has seen the power of social media—its ability to share information in real time and rapidly with billions of people. But information about the virus kept changing and not all was accurate. An *infodemic* emerged, with myths about how COVID-19 spread, which was not helped by the fact that scientists were still figuring things out and often had contradictory messages. While fact-checking provided some relief, the damage usually was already done. The infodemic had another counter-productive effect, with some public debate critical in shaping popular opinion going underground, while still remaining influential to some people. "Cancel culture" spread faster than the virus, leading to further distrust and polarization. An important lesson learned is that policies shaping moderation of public debates, particularly in social spaces, must urgently be developed so that the public square for healthy scientific debate is maintained and does not become controlled by private sector entities, or for that matter lead to censorship by governments, irrespective of their intentions.

The COVID-19 era will be remembered as a time when people lost trust in public institutions, including governments and multilateral entities. A chapter in this volume about the government of Estonia—sometimes referred to as the "digital republic"— provides significant lessons on managing the narrative and trust during a crisis, and engaging citizens as equal stakeholders rather than as adversaries who deserve to be locked down (Annus & Merrimas, 2024).

Another major lesson learned is about the involvement of communities. Communities are central to adoption of technologies and innovation in all sectors, and in health in particular. The chapter in this volume by CARE International provides valuable insights about how communities can be engaged across the development processes (Deshmukh et al., 2024).

Innovations for various vulnerable communities presented in this volume have much to offer with unique techniques, partnerships, and insights for scaling. As one innovator told us, "Without communities owning our solution, innovations are like plants without roots—it may rise quickly but the moment the sun shines and it gets hot, they wither away." We fully concur—innovations need to engage communities from the problem identification to cocreating solutions, adaptations, and implementation for innovations to be sustainable.

Financing for mental health needs to be seen as an investment rather than as an expenditure. This is supported by studies showing that every US$1 spent on evidence-based care for depression and anxiety returns US$5 (World Health Organization, 2020).

Mental health

Recently, we spoke with a highly respected health minister from Africa about his priorities for his second term in office. He talked about the priority of his country to improve its supply chain, given that a third of the drugs and health commodities in most countries is lost due to supply chain inefficiencies (between 50% and 70% of vaccines are wasted due to supply chain issues according to a WHO report). After a short pause, clearly struggling about whether to speak this truth, he said, "If I had a choice, I would put mental health at the center of our health response."

The COVID-19 pandemic has surfaced the problem of mental health that until recently was brushed under the carpet. Countries and economies are struggling under the burden of mental health, with one country showing nearly 40% of adults suffering from depression and anxiety. He is not alone—several leaders agree on the need to address mental health as a priority.

Recognition of the importance of mental health at the United Nations is evident in the chapter in this volume by the Ambassador of the Mission of Mexico to the United Nations, H.E. Juan Ramón de la Fuente, and his staff (de La Fuente et al., 2024). Impressively, Mexico led the successful, and historic, campaign to get a resolution about mental health adopted by the Security Council in December 2022, and also a resolution about Mental Health and Psychosocial Support adopted by the United Nations General Assembly in June 2023.The second author of this chapter, Kuriansky, a psychology professor and mental health advocate for decades at the United Nations, with her students at Teachers College, Columbia University, assisted in this campaign to gain support of many countries who after the pandemic finally recognized the critical importance of addressing widespread public suffering and the importance of a holistic approach to resilient health care (combining physical and mental health). Fortunately there is growing recognition for the need to address mental health, evident in the updated WHO 2022 World Mental Health Report presented in a chapter in this volume, as well as in other conferences, networks and resources in chapters throughout this volume. This represents a "silver lining" in the dark cloud of the pandemic.

Technology is a major solution, but, we believe, it is a double-edged sword. On the downside, technology can contribute to a sense of isolation and disconnect for millions; yet at the same time, it has the potential for diagnosis and treatments to exponentially support mental health, with emerging and game-changing approaches, for all groups, including marginalized migrants and refugees, women, youth and aging, indigenous peoples, and persons with disabilities and disorders. A key lesson that emerges from the chapters in this volume, and from our experience, is that new technologies can be integrated into tried and tested approaches for positive mental health for all. In the presently ongoing debate about AI, we maintain that humans will never be replaced by machines for mental health but machines can help humans deliver better care. We have seen some interesting work on large language models in mental health that provides promise.

An evolving lesson is that the private sector is becoming more engaged in mental health services, as evidenced in the chapter in this volume about the workplace (Marquez, 2024). Importantly, attention is being paid to the needs of employees and their families. Also, greater investments in mental health are being made in related innovations, and also being considered by governments. Digital technologies are already impacting the delivery of positive mental health and well-being, and this trend needs to increase.

Financing innovation

Funding is essential. Chapters in this volume emphasize the lesson that building resilient systems for health requires significant investment. The World Bank estimates that the world's 54 poorest countries—home to 1.5 billion people—will face an annual US$176 billion gap in 2030 in funding for essential health services (World Bank, 2019). COVID-19 has clearly shown the human and economic cost of weak health systems (Bali et al., 2024).

This calls for innovations in financing healthcare. Several chapters address this in this volume and provide a valuable overview into what is happening now, and the potential role of the private investment world in scaling access to healthcare. While healthcare is seen as one of the most attractive areas for private investment, and has grown significantly year on year, more equitable distribution has to occur, since much of this is focused on Southeast Asia, China, and India, with Africa receiving a tiny fraction of this investment.

Health financing is clearly an area in need of innovation, including the role that traditional donors and development financing institutions should prioritize. Questions to address include: How can incentives be built to attract private investment to Africa where the need is greatest? In speaking to investors, we are convinced that there is need for further education and demystifying the perceived risks associated with investing in Africa. Traditional funders and local governments can also do better in enabling policies, building incentives, and leveraging instruments to derisk investments by private investors. This requires donors, development financing institutions, multilateral financing mechanisms, and investors to work together with local governments and the private sector much more closely, and to align interests.

Lessons about financing in these chapters show the need for greater investment in building innovation ecosystems in countries and incentivizing start-ups in the health sector. Some countries have done this well, like India's Biotechnology Industry Research Assistance Council (BIRAC) which over the last decade has enabled over 2000 start-ups in healthcare, with several of them scaling across the country and beyond. As the chapter about Amref in this book shows, Africa clearly can benefit from similar innovation enablers. A strong message from this volume is that donors should defintiely prioritize investing in the health space.

Advance at the United Nations

Notably, the importance of technology is being reflected at the United Nations in an historic 2024 document, the Global Digital Compact (GDC). The GDC presents an ambitious roadmap agreed to by the Member States of the United Nations to insure "an inclusive, open, safe, and secure digital future [to be achieved] through international cooperation and governance that closes digital divides and advances an equitable and inclusive digital world." It outlines commitments of governments about issues like assurances of safety, education, inclusivity, financing, and partnerships with other stakeholders—including civil society, the private sector, technical and academic communities as well as international and regional organizations—to design, launch and support initiatives that leverage technology to advance solutions for the UN Sustainable Development Goals.

Conclusion

In conclusion, over the last several years and after connecting with brilliant minds in the field of health and innovation represented in this volume, we see opportunities and cautions. We have the science, knowledge, innovation, technology, and tools necessary to build a resilient health system, even in severely constrained settings. We have the ability to deliver healthcare on a scale and speed that will save lives and improve the quality of lives. And, this can be achieved with reasonable investment—a fraction of what COVID-19 cost the world.

That is where we also see some cautions. While COVID-19 brought emergency attention to the state of the world's health and security threats, as the world recovers, we must not return to status quo. The world has never been as divided politically as it is today. Economies across the world are struggling, and the lack of trust across and within nations adds to the difficulties in making hard choices essential to build the health systems that can resist and reverse any new pandemics on the horizon. The brilliant minds, dedication, and ideas are here. We need to work together—across sectors and countries—to share resources and escalate progress for a better and healthy life for all.

As is clear from the impressive content in chapters in this volume, technology offers dazzling options for advancement in health, but is not a panacea. True value is unlocked in combination with person-to-person contact that comes from community engagement.

The choices we make today will determine how the world prepares and copes with coming challenges. It's no longer about *can we* build resilient health systems. It's really about *will we* and *when*? The answer is now.

References

Annus, T., & Merrimas, K. (2024). The role of government in leading the pandemic response: The Estonian experience. In J. Kuriansky, & P. Kakkattil (Eds.), *Resilient health: Leveraging technology and social innovations to transform healthcare for COVID-19 recovery and beyond* (pp. 433–443). Elsevier.

Bali, S., Shubber, Z., Cook, A., & Zhao, F. (2024). Investing in resilient health systems: A framework for prioritizing integrated investments in preparedness, response, and service delivery. In J. Kuriansky, & P. Kakkattil (Eds.), *Resilient health: Leveraging technology and social innovations to transform healthcare for COVID-19 recovery and beyond* (pp. 1193–1202). Elsevier.

de la Fuente, J. R., Mendoza, S. P., & Guzik, A. (2024). An unconventional security threat: Mental health at the United Nations Security Council and Mexico's Response. In J. Kuriansky, & P. Kakkattil (Eds.), *Resilient health: Leveraging technology and social innovations to transform healthcare for COVID-19 recovery and beyond* (pp. 475–480). Elsevier.

Deshmukh, M., Philbrick, W. C., & Machael, P. N. (2024). Emerging role of NGOs in social innovations: A paradigm shift. In J. Kuriansky, & P. Kakkattil (Eds.), *Resilient health: Leveraging technology and social innovations to transform healthcare for COVID-19 recovery and beyond* (pp. 561–571). Elsevier.

International Telecom Union. (2020). *Measuring digital development Facts and figures 2020*. Geneva: ITU.

Kakkattil, P., Tiwari, A., & Jacob, B. (2024). Breaking the silos: How the Health Innovation and Investment Exchange (HIEx) helps bridge the health innovation ecosystem. In J. Kuriansky, & P. Kakkattil (Eds.), *Resilient health: Leveraging technology and social innovations to transform healthcare for COVID-19 recovery and beyond* (pp. 979–987). Elsevier.

Kline, S., Sale, J., & Saxena, S. (2024). Financing for mental health: A smart investment, not an expense. In J. Kuriansky, & P. Kakkattil (Eds.), *Resilient health: Leveraging technology and social innovations to transform healthcare for COVID-19 recovery and beyond* (pp. 679–690). Elsevier.

Marquez, P. (2024). The imperative for mental health in the workplace post-COVID-19. In J. Kuriansky, & P. Kakkattil (Eds.), *Resilient health: Leveraging technology and social innovations to transform healthcare for COVID-19 recovery and beyond* (pp. 755–769). Elsevier.

McKinsey. (2019). Should sub-Saharan Africa make its own drugs? Retrieved from https://www.mckinsey.com/industries/public-and-social-sector/our-insights/should-sub-saharan-africa-make-its-own-drugs.

Osho, Z., & Sareen, J. (2024). Accessing solar energy to deliver healthcare for all. In J. Kuriansky, & P. Kakkattil (Eds.), *Resilient health: Leveraging technology and social innovations to transform healthcare for COVID-19 recovery and beyond* (pp. 301–309). Elsevier.

The 2022 update. (2022). *Global health workforce statistics*. Geneva: World Health Organization.

World Bank. (2019). *High-performance health financing for universal health coverage (Vol. 2): Driving sustainable, inclusive growth in the 21st century*. World: World Bank.

World Health Organization. (2020). *COVID-19 disrupting mental health services in most countries. WHO Survey*. https://www.who.int/news/item/05-10-2020-covid-19-disrupting-mental-health-services-in-most-countries-who-survey.

World Health Organization. (2021). *The impact of COVID-19 on health and care workers: a closer look at deaths*. Geneva: World Health Organization.

Index

Note: Page numbers followed by *f* indicate figures, *t* indicate tables, and *b* indicate boxes.

D

E

F

G

I

O

P

S

U

V